MEDICAL MICROBIOLOGY

SEVENTEENTH EDITION

A GUIDE TO MICROBIAL INFECTIONS:
PATHOGENESIS, IMMUNITY, LABORATORY DIAGNOSIS AND CONTROL

EDITED BY

David Greenwood BSc PhD DSc FRCPath
Emeritus Professor of Antimicrobial Science, University of Nottingham Medical School,
Nottingham, UK

Richard Slack MA MB BChir FRCPath FRIPH DRCOG
Senior Lecturer, Division of Microbiology, University of Nottingham Medical School,
Nottingham, UK; Honorary Consultant in Communicable Disease Control and Regional
Microbiologist, Health Protection Agency, East Midlands, UK

John Peutherer BSc MB ChB MD FRCPath FRCPE
Formerly Senior Lecturer, Department of Medical Microbiology, University of Edinburgh Medical
School; Honorary Consultant, The Royal Infirmary of Edinburgh NHS Trust, Edinburgh, UK

Mike Barer BSc MBBS MSc PhD FRCPath
Professor of Clinical Microbiology, Department of Infection, Immunity and Inflammation,
University of Leicester Medical School, Leicester, UK

CHURCHILL
LIVINGSTONE

ELSEVIER

EDINBURGH   LONDON   NEW YORK   OXFORD   PHILADELPHIA   ST LOUIS   SYDNEY   TORONTO   2007

CHURCHILL LIVINGSTONE
ELSEVIER

An imprint of Elsevier Limited

© 2007, Elsevier Limited. All rights reserved.

First edition 1925
Second edition 1928
Third edition 1931
Fourth edition 1934
Fifth edition 1938
Sixth edition 1942
Seventh edition 1945
Eighth edition 1948
Ninth edition 1953
Tenth edition 1960
Eleventh edition 1965
Twelfth edition (Vol. 1) 1973
Twelfth edition (Vol. 2) 1975
Thirteenth edition (Vol. 1) 1978
Thirteenth edition (Vol. 2) 1980
Fourteenth edition 1992
Fifteenth edition 1997
Sixteenth edition 2002
This edition 2007

ISBN: 978-0-443-10209-7
  Reprinted 2008 (twice), 2011
International Student Edition ISBN: 978-0-443-10210-3
  Reprinted 2009, 2010, 2011

British Library Cataloguing in Publication Data
A catalogue record for this book is available from the British Library

Library of Congress Cataloging in Publication Data
A catalog record for this book is available from the Library of Congress

ELSEVIER
your source for books, journals and multimedia in the health sciences
www.elsevierhealth.com

Working together to grow libraries in developing countries
www.elsevier.com | www.bookaid.org | www.sabre.org
ELSEVIER   BOOK AID International   Sabre Foundation

The publisher's policy is to use **paper manufactured from sustainable forests**

Printed in China

MEDICAL MICROBIOLOGY

SEVENTEENTH EDITION

Commissioning Editor: Timothy Horne
Development Editor: Hannah Kenner
Project Manager: Emma Riley
Design Direction: Erik Bigland
Illustrator: Antbits
Illustrations Manager: Bruce Hogarth

It is now more than 80 years since the first appearance of this textbook's illustrious forerunner – Mackie and McCartney's *Introduction to Practical Bacteriology as Applied to Medicine and Public Health*. When that classic text first appeared in 1925, the requirements of medical students, then thought to include a need to be fully conversant with laboratory methods, could be encompassed in a small-format handbook of less than 300 pages. At that time, virology scarcely existed, immunology was in its infancy, parasitology was regarded as a subject of study necessary only for prospective colonial doctors, effective treatment of microbial disease was almost non-existent and molecular biology was unknown. Medical students are, thankfully, no longer expected to be familiar with what goes on in microbiology laboratories, except insofar as they need to be able to use laboratory services and interpret the results emanating from them in an intelligent fashion, but this has been replaced by an unmanageable corpus of clinical knowledge that encompasses not only the burgeoning subjects of bacteriology, virology, mycology and para-sitology, but also the related disciplines of immunology, antimicrobial chemotherapy and epidemiology. Such has been the explosion in knowledge that one can only sympathize with today's student doctors in their struggle to master the basic facets of medical microbiology that will impinge daily on their professional lives.

Of course, all the information one could ever need is now available at the click of a mouse via the internet – that vast, unregulated agglomeration of the good the bad and the ugly – yet textbooks remain stubbornly popular with students, at least for core subjects. This should not be surprising, as a good textbook offers a uniquely user-friendly source of knowledge, assembled by experts familiar with the needs of students and presented in an accessible format, clearly written and logically arranged. Such, at any rate, have been the defining features of previous manifestations of this textbook, a tradition that we believe this new edition fully upholds.

As before, the 17th edition of *Medical Microbiology* strives to bridge the gap between texts that deal with microbiology in a traditional organism-based way and the more modern approaches that take microbial diseases as the starting point or attempt to view the subject from an immunological or epidemiological perspective. Thus, a thorough overview of microbial biology is followed by a consideration of the principles of the body's immunological response to various types of micro-organism, before moving on to consider bacteria, viruses, fungi and parasites individually. The content of these 'systematic' chapters is heavily biased towards understanding the associated diseases, their pathogenesis, clinical features, epidemiology and control. A final section integrates what has gone before in terms of the day-to-day practicalities of the diagnosis of infection, its treatment and avoidance.

In our experience, students of medicine and allied health-care sciences are among the most motivated of all students: it is rare to find one who does not harbour a genuine desire to become a safe and knowledgeable doctor or health-care professional. For our part we sincerely hope that we have provided a text that will help them to fulfil these ambitions in the context of the intrinsically fascinating subject of medical microbiology. For the first time, we have included 'key point' boxes in each chapter to highlight issues that are of particular relevance to the topic. These are intended to provide students with signposts to a wider understanding of key issues and are certainly not meant to represent 'all the student needs to remember' on a particular theme. To use them in such a way would impoverish the student's understanding of the subject and negate the whole purpose of the text. Make no mistake: a thorough knowledge of infection in all its guises is as necessary to the practice of medicine today as it ever was.

Finally, we must once again earnestly thank our inter-national team of contributors – some veterans of several previous editions, others new to the task and bringing welcome new insights – who, for little reward, find time in their very busy lives to share their expertise with our readership. We are also, as ever, grateful to Timothy Horne and his expert team in the Edinburgh office of Elsevier, who have seen the book through to publication with their usual courtesy and skill.

D.G.
R.C.B.S.
J.F.P.
M.R.B
Nottingham, Edinburgh, Leicester
October 2006

In addition to the specific suggestions provided at the end of individual chapters, the following internet sites, most of which offer diverse links to sources of further information, are recommended as the starting point for searches of information on microbial topics.

American Society for Microbiology:
www.asm.org

Centers for Disease Control and Prevention:
www.cdc.gov

CDC National Center for Infectious Diseases:
www.cdc.gov/ncidod/index.htm

Medscape:
www.medscape.com/InfectiousDiseases

Oregon Health Sciences University:
www.ohsuhealth.com

Health Protection Agency (formerly Public Health Laboratory Service):
www.hpa.org.uk

PubMed (National Library of Medicine):
www.ncbi.nlm.nih.gov/PubMed/

Society for General Microbiology:
www.sgm.ac.uk

University of Leicester:
http://www.le.ac.uk/bs/medbiobrochure/welcome.htm

World Health Organization:
www.who.int/en/

The Wellcome Trust Tropical Medical Resource produces excellent CD-ROMs under the general title *Topics in International Health* (www.wellcome.ac.uk/doc_WTx 022177.html). Currently available titles include:

Acute Respiratory Infection
Dengue
Diarrhoeal Diseases
HIV/AIDS (revised edition)
Leishmaniasis
Leprosy
Malaria (3rd edn)
Nutrition
Schistosomiasis
Sexually Transmitted Infections (2nd edn)
Sickle Cell Disease
Trachoma (2nd edn)
Tuberculosis

D. A. A. Ala'Aldeen
Molecular Bacteriology and Immunology Group
Division of Microbiology and Infectious Diseases
University Hospital
Queen's Medical Centre
Nottingham NG7 2UH
UK

R. P. Allaker
Oral Microbiology
Institute of Dentistry
Queen Mary, University of London
4 Newark Street
London E1 2AT
UK

H. S. Atkins
Department of Biomedical Sciences
Defence Science and Technology Laboratory
Porton Down
Salisbury SP4 0JQ
UK

M. R. Barer
Department of Infection, Immunity and
Inflammation
University of Leicester
Medical Sciences Building
University Road
Leicester LE1 9HN
UK

A. D. T. Barrett
Department of Pathology
University of Texas Medical Branch
301 University Boulevard
Galveston
Texas 77555-0609
USA

S. M. Burns
Department of Laboratory Medicine
Royal Infirmary of Edinburgh
51 Little France Crescent
Edinburgh EH16 4SA
UK

H. Chart
Laboratory of Enteric Pathogens
Health Protection Agency
61 Colindale Avenue
London NW9 5EQ
UK

A. Cockayne
Centre for Biomolecular Sciences
University of Nottingham
University Park
Nottingham NG7 2RD
UK

T. J. Coleman
Martins Croft
Eaton Bishop
Hereford HR2 9QD
UK

M. J. Corbel
National Institute for Biological Standards and
Control
Blanche Lane
South Mimms
Potters Bar
Hertfordshire EN6 3QG
UK

H. A. Cubie
Specialist Virology Centre
Royal Infirmary of Edinburgh
51 Little France Crescent
Edinburgh EH16 4SA
UK

W. D. Cubitt
Department of Microbiology
Great Ormond Street Hospital
Great Ormond Street
London WC1N 3JH
UK

N. A. Cunliffe
Department of Medical Microbiology and Genitourinary
Medicine
University of Liverpool
Duncan Building
Daulby Street
Liverpool L69 3GA
UK

J. M. Darville
Department of Medical Microbiology
Southmead Hospital
Westbury-on-Trym
Bristol BS10 5NB
UK

G. F. S. Edwards
Scottish Legionella Reference Laboratory
Stobhill NHS Trust
133 Balornock Road
Glasgow G21 3UW
UK

D. Goldberg
Health Protection Agency Scotland
Clifton House
Clifton Place
Sauchiehall Street
Glasgow G3 7LN
UK

J. R. W. Govan
Cystic Fibrosis Group
Centre for Infectious Diseases
University of Edinburgh Medical School
The Chancellor's Building
49 Little France Crescent
Edinburgh EH16 4SB
UK

J. M. Grange
UCL Centre for Infectious Disease and International
Health
Windeyer Institute
46 Cleveland Street
London W1T 4JF
UK

D. Greenwood
11 Perry Road
Sherwood
Nottingham NG5 3AD
UK

C. A. Hart
Department of Medical Microbiology and Genitourinary
Medicine
University of Liverpool
Duncan Building
Daulby Street
Liverpool L69 3GA
UK

J. Hood
Department of Clinical Microbiology
Glasgow Royal Infirmary
Glasgow G4 0SF
UK

H. Humphreys
Department of Clinical Microbiology
Royal College of Surgeons in Ireland
Education and Research Centre
Beaumont Hospital
Smurfit Building
Dublin 9
Ireland

J. W. Ironside
Neuropathology Laboratory
Western General Hospital
The Bryan Matthews Building
Crewe Road
Edinburgh EH4 2XU
UK

A. S. Johnson
Department of Microbiology and Immunology
Hereford Hospitals NHS Trust
County Hospital
Hereford HR1 2ER
UK

J. M. Ketley
Department of Genetics
University of Leicester
Adrian Building
University Road
Leicester LE1 7RH
UK

M. Kilian
Institute of Medical Microbiology & Immunology
University of Aarhus
The Bartholin Building
DK-8000 Aarhus C
Denmark

C. R. Madeley
Burnfoot
Stocksfield
Northumberland NE43 7TN
UK

R. C. Matthews
Department of Medical Microbiology
Manchester Royal Infirmary
Clinical Sciences Building
Manchester M13 9WL
UK

J. McLauchlin
Food Safety Microbiology Laboratory
Centre for Infections
Health Protection Agency
61 Colindale Avenue
London NW9 5EQ
UK

P. J. Molyneaux
Department of Medical Microbiology
Aberdeen Royal Infirmary
Foresterhill
Aberdeen AB25 2ZN
UK

P. Morgan-Capner
Crow Trees
Melling
Carnforth
Lancashire LA6 2QZ
UK

O. Nakagomi
Department of Molecular Microbiology and
Immunology
Nagasaki University
Nagasaki 852-8523
Japan

M. Norval
University of Edinburgh Medical School
Teviot Place
Edinburgh EH8 9AG
UK

M. M. Ogilvie
62/3 Blacket Place
Edinburgh EH9 1RJ
UK

J. S. M. Peiris
Department of Microbiology
University of Hong Kong
Queen Mary Hospital
University Pathology Building
Pokfulam Rd
Hong Kong
China

T. H. Pennington
Department of Medical Microbiology
University of Aberdeen
Medical School Building
Foresterhill
Aberdeen AB25 2ZD
UK

J. F. Peutherer
44 Bridge Road
Edinburgh EH13 0LQ
UK

T. L. Pitt
Centre for Infections
Health Protection Agency
61 Colindale Avenue
London NW9 5EQ
UK

N. W. Preston
36 Queen's Drive
Stockport SK4 3JW
UK

D. Reid
29 Arkleston Road
Paisley PA1 3TE
UK

P. Riegel
Laboratory of Bacteriology
University Hospitals of Strasbourg
3 Rue Koeberle
F-67000 Strasbourg
France

T. V. Riley
The University of Western Australia and
Division of Microbiology & Infectious Diseases
Path West Laboratory Medicine
Queen Elizabeth II Medical Centre
Nedlands 6009
Western Australia

P. Simmonds
Laboratory for Molecular and Clinical Virology
Centre for Infectious Diseases
Royal (Dick) Veterinary School
Summerhall
Edinburgh EH9 1QH
UK

M. P. E. Slack
Haemophilus influenzae Reference Unit
Respiratory and Systemic Infection Laboratory
Health Protection Agency
Centre for Infections
61 Colindale Avenue
London NW9 5EQ
UK

R. C. B. Slack
Health Protection Agency East Midlands North
Mill 3
Pleasley Vale Business Park
Mansfield
Nottinghamshire NG19 8RL
UK

J. Stewart
School of Biomedical Sciences
University of Edinburgh
Medical School
Teviot Place
Edinburgh EH8 9AG
UK

S. Sutherland
Houndwood House
Eyemouth
Scottish Borders TD14 5TW
UK

D. Taylor-Robinson
6 Vache Mews
Vache Lane
Chalfont St Giles
Buckinghamshire HP8 4UT
UK

R. W. Titball
Defence Science and Technology Laboratory
Porton Down
Salisbury SP4 0JQ
UK

K. J. Towner
Department of Clinical Microbiology
University Hospital
Queen's Medical Centre
Nottingham NG7 2UH
UK

D. H. Walker
Department of Pathology
University of Texas Medical Branch at Galveston
Galveston
Texas 77555-0609
USA

M. E. Ward
Mailpoint 814
Southampton General Hospital
Tremona Road
Southampton SO16 6YD
UK

D. W. Warnock
Division of Bacterial and Mycotic Diseases
National Center for Infectious Diseases
Centers for Disease Control and Prevention
1600 Clifton Road NE, Mailstop G-11
Atlanta
Georgia 30333
USA

S. C. Weaver
Department of Pathology
University of Texas Medical Branch at Galveston
301 University Boulevard
Galveston
Texas 77555-0609
USA

X.-J. Yu
Department of Pathology
University of Texas Medical Branch at Galveston
301 University Boulevard
Galveston
Texas 77555-0609
USA

CONTENTS

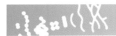

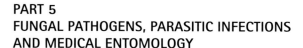

PART 1
MICROBIAL BIOLOGY

Microbiology and medicine

D. Greenwood and M. R. Barer

Read this paragraph, then close your eyes and think through the following: inside your intestines, in your mouth and on your skin, there reside more than 100 000 000 000 000 microbial cells – 100-fold more than the number of cells that make up the human body. We are not conscious of these colleagues any more than we are conscious of passing them round every time we shake hands, speak or touch a surface. Inoculation with just one microbe of the wrong type in the wrong way may kill you, yet we tolerate and indeed thrive on constant appropriate exposure to this unseen world.

Because microbes are generally hidden from our senses, an appreciation of microbiology and infection demands imagination. Our forebears who established the discipline lavished imagination on the problems they studied. Sadly it is lack of imagination that that now underpins serious problems such as hospital-acquired infection and antibiotic resistance.

Hard-won advances in microbiology have transformed the diagnosis, prevention and cure of infection and have made key contributions to improved human health and a doubling in life expectancy. The conquest of epidemic and fatal infections has sometimes seemed so conclusive that infections may be dismissed as of minor concern to modern doctors in wealthy countries. However, infection is far from defeated. In resource-poor countries, an estimated 10 million young children die each year from the effects of infectious diarrhoea, measles, malaria, tetanus, diphtheria and whooping cough alone. Many other classical scourges, such as tuberculosis, cholera, typhoid and leprosy, continue to take their toll. Although we have the potential to prevent nearly all of these deaths, political and social issues constantly hinder progress, and more effective and economic means of delivery provide a constant challenge.

Even in wealthy nations, infection is still extremely common: at least a quarter of all illnesses for which patients consult their doctors in the UK are infective and around one in ten patients acquire infection while in hospital, sometimes with multiresistant organisms. Global communications and changes in production systems, particularly those affecting food, can have a profound effect on the spread of infectious disease. The emergence of human immunodeficiency virus (HIV), new-variant Creutzfeldt–Jakob disease (CJD), severe acute respiratory syndrome (SARS) and avian influenza illustrate the need for continued vigilance.

The relative freedom of wealthy societies from fatal infections has been won through great struggles, which are all too easily forgotten. As generations grow up without the experience of losing friends and relatives through infection, so the balance of perceived risk and benefit looks different. So now, in addition to the old threats, which are ever present, we constantly face pressure to drop or modify measures such as public immunization. A historical understanding of infection is as important in maintaining and improving the present status as is knowledge of contemporary progress.

AN OUTLINE HISTORY OF MICROBIOLOGY AND INFECTION

Micro-organisms and infection

Infection and microbiology followed different strands of development for centuries (Fig. 1.1). We tend to map this story against the recorded efforts of prominent individuals, though many others doubtless contributed.

Ideas of infection and epidemics were recorded by Hippocrates, but it was nearly 2000 years before Girolamo Fracastoro (1483–1553) proposed in his classic tome 'De Contagione' that 'seeds of contagion' (as opposed to spirits in the ether) might be responsible. Quite separately, the early microscopists began to make observations on objects too small to be seen by the naked eye. Foremost among these was the Dutchman Antonie van Leeuwenhoek (1632–1723). With his remarkable home-made and hand-held microscope, he found many micro-organisms in materials such as water, mud, saliva and the intestinal contents of healthy subjects, and recognized them as living creatures ('animalcules') because they swam

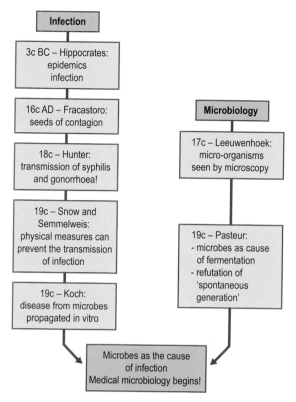

Fig. 1.1 Timelines for the history of infection and microbiology.

This established that *physical measures could prevent the transmission of infection*, a point further illustrated by Ignaz Semmelweis (1818–1865) in Vienna and others who showed that fatal streptococcal infections (puerperal fever) affecting mothers following childbirth could be substantially reduced if those attending the birth applied simple hygienic measures.

Later two towering figures, Louis Pasteur (1822–1895) and Robert Koch (1843–1910), played central roles in establishing the microbial causation of infectious disease. The brilliant French chemist Louis Pasteur crushed two prevailing dogmas: that the fermentation responsible for alcohol formation was a purely chemical process (by demonstrating that the presence of living micro-organisms was essential), and that life could be spontaneously generated (by showing that nutrient solutions remained sterile if microbes were excluded). Refutation of spontaneous generation established unequivocally the need for a *chain of transmission*. Pasteur made many other seminal contributions, including the identification of several causal agents of disease and the recognition that microbes could be rendered less capable of causing disease (less *virulent*) or *attenuated* by artificial subculture. He used the principle of attenuation to develop a successful vaccine against anthrax for use in animals. It was the influence of Pasteur's work that inspired the British surgeon Joseph Lister (1827–1912) to establish *antisepsis*, aimed at destroying the micro-organisms responsible for infection during surgery.

The other great founding father of medical microbiology, Robert Koch, came to microbiology through medicine. Working originally as a country doctor in East Prussia, he established the techniques required to isolate and propagate pure cultures of specific bacteria. His numerous contributions include establishment of the bacterial causes of anthrax, tuberculosis and cholera. He also formulated more precisely proposals first put forward by one of his mentors, Jacob Henle (1809–1885), describing how specific microbes might be recognized as the cause of specific diseases. These principles, often referred to as *Koch's postulates*, are used to substantiate claims that a particular organism causes a specific ailment. They require that:

- The organism is demonstrable in every case of the disease.
- It can be isolated and propagated in pure culture in vitro.
- Inoculation of the pure culture by a suitable route into a suitable host should reproduce the disease.
- The organism can be re-isolated from the new host.

For various reasons, universal application of the postulates is impossible and greater subtleties in establishing causal

about actively. That he saw bacteria as well as the larger microbes is known from his measurements of their size ('one-sixth the diameter of a red blood corpuscle').

Before the discipline of microbiology was formally established in the second half of the nineteenth century, three key aspects of infection were brought into stark relief by publicly acknowledged demonstrations:

1. John Hunter (1728–1793) inoculated secretions from sores around a prostitute's genitals into a penis (his own according to some sources) and demonstrated a *physical reality to the transmission of infection*, in this case syphilis and gonorrhoea (the prostitute had both, leading to a mistaken belief that the distinctive symptoms were manifestations of the same disease).

2. Edward Jenner (1749–1823) adapted the long-established oriental practice of variolation (inoculation with material from a mild case of smallpox) for the prevention of smallpox, by showing that cowpox was as effective and safer. The procedure, termed *vaccination* (Latin vacca = cow) established the concept of *immunization* in Europe.

3. John Snow (1813–1858) showed that, by preventing access to a water source epidemiologically linked to a cholera outbreak, further infections could be terminated.

relationships in infection are now recognized. Austin Bradford Hill (1897–1991) developed a sophisticated algorithm to recognize a biological gradient of association; most recently an approach to determining the role of specific molecules in pathogenesis has been enshrined in 'molecular Koch's postulates'.

Organisms for which Koch's postulates and later modifications have been fulfilled are clearly capable of inducing disease and are designated as *pathogens* to distinguish them from the vast majority of *non-pathogenic* micro-organisms. It should be emphasized that fulfilment of the postulates and the diagnostic process in which a given patient's illness is attributed to a known pathogen are profoundly different processes. In the former, many experiments are done to provide robust scientific evidence, whereas in the latter circumstantial evidence is obtained, which, in the light of experience, identifies a particular micro-organism as the most likely cause of the illness.

In the century following Pasteur and Koch's work, the list of specific human pathogens has extended to include several hundred organisms. Early on, fungal and protozoan pathogens were recognized, as were macroscopic agents including parasitic worms and insects. Technological breakthroughs, including tissue culture and electron microscopy, were required to enable recognition of viruses. In the early days, viral pathogens were termed filterable agents, because they passed through filters designed to retain bacteria. In many cases pathogens of insects, animals or even plants were described before their medical equivalents were recognized.

Many further advances in technology through the twentieth century provided more precise understanding of the nature and function of microbes. The revolution in molecular biology that followed the elucidation of the structure of DNA by James Watson, Francis Crick, Maurice Wilkins and Rosalind Franklin in 1953 ultimately enabled a leap forward in analytical capability. For three decades this did not radically change the understanding of microbes and infection. However, almost exactly a century after Pasteur and Koch initiated what has been called the 'golden era of bacteriology', three interconnected breakthroughs once again altered the perspective:

1. The recognition, principally by the American molecular biologist Carl Woese, that ribosomal ribonucleic acid (rRNA), which has essentially the same core structure in all cells, carries unique signatures indicating its evolutionary relationships. It transpires that all cellular forms of life can be classified according to the DNA sequence encoding their rRNA (rDNA). Determination of this sequence provides a means of identifying all microbes and has led to the discovery of a previously unsuspected third 'domain' of life, the *Archea* (see Ch. 2).

2. Technological advances made possible by molecular genetics. The molecular basis for the pathogenesis of infection now enables recognition of the specific roles of individual genes and their products in both the pathogen and the host. This offers the promise of new approaches to treatment and prevention of infection. The discovery of mobile genetic elements that convey genes from one organism to another (see Ch. 6) confronted our biological sense of what makes up an individual. Mobile bacterial genes encoding antibiotic resistance present a major problem to the practice of medicine.

3. The development of ultra-sensitive means of detecting specific DNA or RNA sequences and the development, by Kary Mullis, of the polymerase chain reaction (PCR) in 1986. The analytical capacity of nucleic acid amplification techniques offers the prospect that it may be possible to diagnose microbial disease routinely by these methods. However, there are many challenges to be met before this shift away from detection of micro-organisms by isolation in laboratory cultures can be accepted.

Our capacity to exploit these breakthroughs has been enhanced enormously by the development of high-throughput DNA sequencing by the double Nobel Laureate Fred Sanger in Cambridge.

Hygiene, treatment and prevention of infection

The work of Snow, Semmelweis, Lister and others led to an appreciation of the benefits of hygiene in the prevention of infection. Nursing practices rooted in almost obsessive cleanliness became the norm and *aseptic* practice (avoidance of contact between sterile body tissues and materials contaminated with live micro-organisms) was introduced to supplement the use of antisepsis. Before the advent of antibiotics, hygiene was a matter of life and death; institutes of hygiene were established around the world. When treatment of infection later became reliable and routine, hygiene standards were often allowed to drop, leading to present problems, notably with hospital-acquired infection.

The discovery of phagocytic cells and humoral immunity (antibodies) as natural defence mechanisms at the end of the nineteenth century led to a re-assessment of the response to infection. One outcome was the use of antibodies produced in one host for the protection of another (*serum therapy*). This produced some spectacular successes, notably in the life-saving use of antitoxin in diphtheria and tetanus. Unfortunately these foreign proteins often caused hypersensitivity reactions (*serum sickness*) and few diseases responded reliably to serum therapy. Nevertheless, the capacity of the immune system to achieve *selective toxicity* and observations by the brilliant German doctor Paul Ehrlich (1854–1915) that dyes used to stain infected tissues selectively labelled parasites in preference to host tissues contributed to the notion that systemic chemotherapy might be achievable.

In 1909 Ehrlich and his colleagues introduced the arsenical drug Salvarsan for the treatment of syphilis, but it fell short of his ideal of a *magic bullet* that would destroy the parasite without harming the host. A more important breakthrough than Ehrlich's came in 1935 with the publication of a paper by Gerhard Domagk (1895–1964) of the German dyestuffs consortium, IG Farbenindustrie. Domagk described the remarkable activity against strepto-cocci of a dye derivative, prontosil, which turned out to owe its activity to a sulphonamide substituent previously unsuspected of antibacterial activity. Earlier, in 1928, Alexander Fleming had accidentally discovered the antibacterial properties of a fungal mould *Penicillium nota-tum*, but he was unable to purify or exploit the therapeutic potential of his discovery. This was left to a team of scientists at Oxford led by the Australian experimental pathologist Howard Florey (1898–1968), heralding the start of the antibiotic era – the most important therapeutic development of the twentieth century.

Meanwhile, in America, the Ukrainian-born soil micro-biologist Selman Waksman (1888–1973) undertook a systematic search for antibiotic substances produced by soil micro-organisms that achieved its greatest success in 1943 with the discovery of streptomycin by one of his PhD students, Albert Schatz (1920–2005). The hunt for antibiotics intensified after the Second World War, yield-ing chloramphenicol, tetracyclines and many other natural, synthetic and semi-synthetic antibacterial compounds. Progress in the development of antiviral, antifungal and antiparasitic compounds has been much slower, despite the fact that two effective antiprotozoal drugs, quinine (cinchona bark) and emetine (ipecacuanha root), and some natural anthelminthic agents have been recognized for centuries. Therapeutic choice in non-bacterial infection consequently remains severely limited, although the human immunodeficiency virus (HIV) pandemic has stimulated much work in the antiviral field that has been rewarded with significant success. Meanwhile, the explosion of knowledge in immunology has renewed hopes that it may be possible to manipulate immunological processes triggered by infection to the benefit of the host.

SOURCES AND SPREAD OF INFECTION

To grasp adequately the ways in which the microbial world intersects with human lives it is necessary to understand different microbial lifestyles and the degree to which they depend on human beings. Thus there are some pathogens for which an association with man is essential in order for them to propagate, whereas

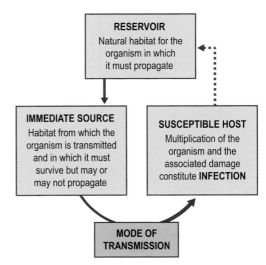

Fig. 1.2 Reservoir, immediate source and mode of transmission in infection.

for others human association is of little significance compared with their propagation in other species or environments. Microbes that depend on human beings are *obligate parasites*. A few actually need to cause disease to propagate themselves; these are termed *obligate pathogens*. In most cases, disease is accidental, or even detrimental, to the microbe's long-term survival. Viruses that cause disease in man are obligate parasites, although they often cause inapparent, subclinical or *asymptomatic* infection. Many viruses rely on infecting a particular host species. Smallpox was eradicated, not only because of the availability of an effective vaccine, but also because man was the only host. Some bacteria, fungi, protozoa and helminths are also species-specific. Among bacteria, the agent of tuberculosis, which is harboured by one-third of humanity, has an absolute requirement to cause disease for its natural transmission to continue.

Since Pasteur established the need for a chain of trans-mission in infection, it has been possible to fit the sources and spread of infection into a relatively simple framework. All infection recently transmitted has an *immediate source* and reaches the newly infected individual via one or more specific *mode(s) of transmission*. Behind these events, the organism, which of course does not care how we choose to classify it or its activities, lives and propagates in its natural habitat(s). These may or may not be the same as the immediate source but, in considering the control of infection, the natural habitat of the causal organism constitutes the *reservoir of infection*. These points are illustrated in Figure 1.2. Elimination of the organism from the reservoir will lead to eradication of the infection, whereas elimination from the immediate

Table 1.1   Examples of reservoirs, sources and modes of transmission

Infective disease	Agent of infection	Reservoir	Immediate source	Mode of transmission
Sore throat	*Streptococcus pyogenes*[a] (bacterium)	Human upper respiratory tract	Human upper respiratory tract	Exogenous: airborne droplets
Oral thrush	*Candida albicans* (fungus)	Most human mucosal surfaces	Normal microbiota of oral mucosa	Endogenous: Overgrowth in antibiotic-treated or immunocompromised patient
Tetanus	*Clostridium tetani* (bacterium)	Soil or animal intestine	Any environment contaminated with soil or animal faeces	Exogenous: penetrating injury
Syphilis	*Treponema pallidum* (bacterium)	Infected human beings	Patients with genital ulcers or secondary syphilis	Exogenous: sexual contact
Yellow fever[b]	Yellow fever virus (virus)	Monkeys	Usually infected human beings Occasionally monkeys	Exogenous: mosquito-borne
AIDS	Human immunodeficiency virus (virus)	Infected human beings	Usually human blood	Exogenous: mainly blood-borne and by sexual contact
Toxoplasmosis[b]	*Toxoplasma gondii* (protozoon)	Cats	Undercooked meat or contact with areas contaminated by cat faeces	Exogenous: ingestion

[a]One of many causes of sore throat; [b]example of a zoonosis. AIDS, acquired immune deficiency syndrome.

source, if this is distinct from the reservoir, provides one means by which control of infection can be achieved.

The mode of transmission can involve other infected individuals in the case of *contagious* infections: food in the case of foodborne infections; water in waterborne infection; aerosol generation by an infected individual; or contamination of an inanimate object (*fomites*) such as medical equipment or bed linen. The possible sources and modes of transmission of infection are enormous, and new variants are continually being recognized. Engagement of all health-care workers in recognizing and controlling these hazards is a vital part of medical practice. Fortunately, most infections are transmitted by well recognized pathways (Table 1.1), and these must be clearly understood and learnt.

In the public mind, most infection is seen as contagious, but a large proportion of infections result from *endogenous infection* with a bacterium or fungus that is normally resident in the patient concerned. These resident organisms constitute the *normal flora* or *microbiota* of the host. These abundant fellow travellers generally cause infection when they get into the wrong place, often as a result of traumatic wounds (including surgery) or other types of impairment of the host's ability to prevent the spread of organisms to sites where they may cause mischief. Disturbance of the normal flora by antibiotics may also allow unaffected *opportunist* pathogens

from the endogenous flora or the environment to cause infection.

In the case of endogenous infection, the reservoir and source of infection are the same and transmission is unnecessary. When infection comes from an external source it is termed *exogenous infection* and the reservoir reflects the natural habitat of the organism. Where other animals constitute that habitat the infection is termed a *zoonosis*. In many countries bacteria, protozoa, helminths and viruses are commonly transmitted by insects or other arthropods and are classified as *vector-borne* diseases.

Study of the ecology and transmission of disease, including infectious disease, is the province of the important public health discipline of *epidemiology*. Important tools include surveillance of the *prevalence* (total cases in a defined population at a particular time) and *incidence* (number of new cases occurring during a defined period) of disease. Knowledge of the ways in which microorganisms spread and cause disease in communities has produced vital insights that can be used to inform effective control programmes in hospitals and the wider community. Monitoring of the prevalence and incidence of infection on an institutional, local, national or global basis can similarly help in the formulation of policies that reduce the impact of specific infections (monitoring of influenza virus variants to forestall global pandemics is a good example) or of drug-resistant micro-organisms such

KEY POINTS

- Microbes are too small to be seen directly and special methods are needed to investigate them. In daily life and in clinical practice we are forced to use our imagination to understand how our behaviour influences and is influenced by them.
- Infections and microbes were considered as separate phenomena until the late nineteenth century when Pasteur reconciled previous observations on the physical requirements for the transmission of infection with the nature of microbes and established the necessity of a chain of transmission in infection.
- Some infections can be prevented by interrupting transmission and/or by immunization.
- The role of specific microbes in specific infective conditions may be established by propagating the microbe in pure laboratory culture and subsequently reproducing the disease in a suitable model.

- Molecular biology has opened up new ways of identifying microbes and establishing causality in infection.
- Transmission of infection is related to the *reservoir*, *immediate source* and *mode of transmission* of the causal agent.
- Approximately 10^{14} bacterial, fungal and protozoan cells live on and in healthy human bodies. Most are harmless or even beneficial. Those that cause disease in otherwise healthy individuals are termed *pathogens*. The *normal flora* or *microbiota* constitutes the reservoir and immediate source for *endogenous* infection. Infections in which the source of the causal organism is external are termed *endogenous* infections.
- Many infections can now be treated with antimicrobial agents that possess *selective toxicity*. None the less, infection remains the most common cause of morbidity and premature death in the world.

as those causing malaria, tuberculosis or staphylococcal infections. The World Health Organization and other national or international surveillance agencies carry out much of this important work and deserve full support. For, make no mistake, despite antibiotics, immunization and – for the fortunate – improved living conditions and effective health services, infection will remain the commonest cause of sickness and premature death for the foreseeable future.

RECOMMENDED READING

Brock T D (ed.) 1999 *Milestones in Microbiology*. American Society for Microbiology, Washington, DC

Bulloch W 1938 *The History of Bacteriology*. Oxford University Press, Oxford

Collard P 1976 *The Development of Microbiology*. Cambridge University Press, Cambridge

Cox F E G (ed.) 1996 *Illustrated History of Tropical Diseases*. Wellcome Trust, London

Foster W D 1970 *A History of Medical Bacteriology and Immunology*. Cox and Wyman, London

Grove D I 1990 *A History of Human Helminthology*. CAB International, Wallingford, Oxford

Mann J 1999 *The Elusive Magic Bullet: The Search for the Perfect Drug*. Oxford University Press, Oxford

Waterson A P, Wilkinson L 1978 *An Introduction to the History of Virology*. Cambridge University Press, Cambridge

Zinsser H 1935 *Rats, Lice and History*. Routledge, London

Internet sites

United Nations Children's Fund. *The State of the World's Children 2006*. http://www.unicef.org/sowc06/

World Health Organization. Infectious diseases. http://www.who.int/topics/infectious_diseases/en/

2 Morphology and nature of micro-organisms

M. R. Barer

Micro-organisms are beyond doubt the most successful forms of life; they have been here longest, they are the most numerous and their distribution defines the limits of the biosphere, encompassing environments previously thought incapable of sustaining life. Here, we are concerned with the tiny fraction of micro-organisms that form associations with human beings and these encompass the cellular entities, *bacteria, archaea, fungi* and *protozoa*, and the subcellular entities, viruses, viroids and prions. Whether the last three can be considered organisms or even living entities is a matter for debate. None the less, their transmissible nature, the immune responses they provoke and our inability to detect them with the naked eye place them firmly within the province of microbiology. The first two of these criteria also require us to consider some multicellular macroscopically visible organisms (members of the *helminths*) as agents of infection (see Ch. 62). Most of this chapter is concerned with bacteria. The subcellular entities are introduced briefly and the remaining medically significant groups are considered in more specialized chapters.

Medical microbiology has been founded on recognizing micro-organisms that are associated with human disease. This recognition has relied predominantly on two techniques:

- microscopy
- propagation in laboratory cultures.

Over the past 30 years, it has become possible to detect, describe and differentiate micro-organisms by biochemical and genetic methods, and this has had two profound effects on microbiology. First, due largely to the work of Carl Woese (see Ch. 1), it is now possible to make a reasonable assessment of the evolutionary relationships between micro-organisms, a task that previously could be achieved for macro-organisms only by examining fossil records (micro-organisms have not left interpretable fossils). Woese's approach led to the recognition of a whole 'new' *domain* (a group ranking above kingdom level) of cellular organisms, the *Archaea*, and a fundamental review of microbial classification (Fig. 2.1). Second,

the accumulation of molecular data describing microbes (including many complete genome nucleotide sequences) is underpinning the development of molecular detection methods. Thus morphology and cultural characteristics are regarded by many as secondary characteristics in the context of classification and identification, while molecular detection methods are steadily encroaching on microscopy and culture in clinical laboratories.

We are currently in a transitional, information gathering, phase. It seems likely that molecular descriptions and detection methods will come to dominate our view of the microbial world. However, at present it must be emphasized that the morphological classification provides a basic structure that is understood by clinicians and will remain as a basis for communication, at least in the medium term. Moreover, light and fluorescence microscopy remain competitively cheap and rapid compared with molecular methods as means of providing information relevant to the clinical management of infection. Finally, the discipline of medical microbiology requires the understanding of the basic structural properties and physiology of micro-organisms to underpin our approach to infections.

PROKARYOTIC AND EUKARYOTIC CELLS

Micro-organisms are microscopic in size and are usually unicellular. The diameter of the smallest body that can be resolved and seen clearly with the naked eye is about 100 μm. All medically relevant bacteria are smaller than this and a microscope is therefore necessary to see individual cells. When propagated on solid media, bacteria (and fungi) form macroscopically visible structures comprising at least 10^8 cells, which are known as *colonies*.

Woese's insights provided for the first time a coherent view of the evolutionary pathways behind the diversity of all living organisms. In particular, a satisfactory explanation is offered for the existence and diversity of *prokaryotic* and *eukaryotic* cells, and all living forms are

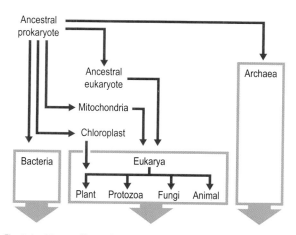

Fig. 2.1 Diagram illustrating proposed evolutionary pathways from a putative common ancestral prokaryote to the present. The Archaea were not recognized as a separate lineage until Woese's work.

seen to fall within three *domains* of life: the *Bacteria*, the *Archaea* and the *Eukarya*. Although the first two of these are prokaryotes (organisms without a membrane-bound nucleus), the *Archaea* share many characteristics with the *Eukarya*. This division is of practical significance, as the earlier the point of divergence, the greater the difference in metabolic properties between the present-day representatives of the two lineages. These differences can be exploited by directing treatments at processes unique to the target organism. Some key differences between the three domains of life are summarized in Table 2.1.

ANATOMY OF THE BACTERIAL CELL

The principal structures of a typical bacterial cell are shown in Figure 2.2. The *protoplast*, that is, the whole body of living material (*protoplasm*), is bounded peripherally by a very thin, elastic and semi-permeable cytoplasmic (or plasma) membrane (a conventional phospholipid bilayer). Outside, and closely covering this, lies the rigid, supporting *cell wall*, which is porous and relatively permeable. Cell division occurs by the development, from the periphery inwards, of a transverse cytoplasmic membrane and a transverse cell wall known as a *septum* or *cross-wall*.

The pattern of cell division and the structures associated with the cell wall and cytoplasmic membrane (collectively the cell envelope) combine to produce the cell morphology and characteristic patterns of cell arrangement. The recognition of these features by oil-immersion light microscopy remains of great practical value in making presumptive identifications of bacteria associated with human infections. Bacterial cells may have two basic shapes, spherical (*coccus*) or rod shaped (*bacillus*); the rod-shaped bacteria show variants that are comma shaped (*vibrio*), spiral (*spirillum* and *spirochaete*) or filamentous (Fig. 2.3).

The *cytoplasm*, or main part of the protoplasm, is a predominantly aqueous environment packed with ribosomes and numerous other protein and nucleotide–protein complexes. The cytoplasmic contents are not normally visible by light microscopy, which can only resolve objects more than 0.2 μm in diameter. Although bacterial cytoplasm has traditionally been viewed as devoid of

	Prokaryotes		Eukaryotes
Domains	Bacteria	Archaea	Eukarya
Major groups (examples only)	Gram positives, Proteobacteria	*Methanococcus, Thermococcus*	Fungi, entamoebae, ciliates, flagellates
Cell diameter	≈1 μm	≈1 μm	≈10 μm
Membrane-bound organelles	–	–	+ (e.g. mitochondria, nucleus, Golgi, etc.)
Chromosomes	Single, closed circular	Single, closed circular	Multiple, linear
Introns	Rare	Rare	Common
Transcription/translation	Coupled	Coupled	Compartmentalized
mRNA	Very labile	Very labile	Stable and labile
Ribosomes	70S	70S	80S
Protein synthesis inhibited by:			
Chloramphenicol	+	–	–
Diphtheria toxin	–	+	+
Peptidoglycan cell wall	+	–	–

Table 2.1   General characteristics of cellular micro-organisms in the three domains of life

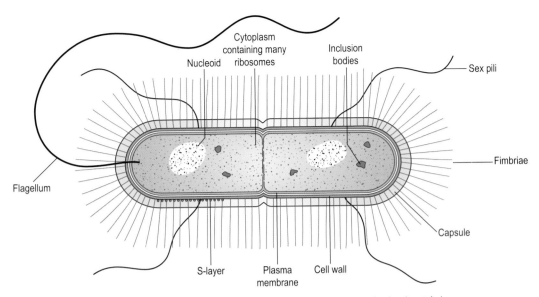

Fig. 2.2 Diagram illustrating the key features of bacterial cells. The S-layer is a variably demonstrated ordered protein layer.

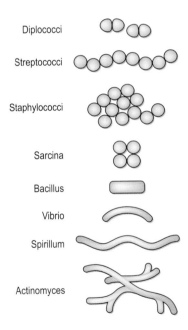

Fig. 2.3 The shapes and characteristic groupings of various bacterial cells.

structure, it is now clear that, like eukarya, bacteria have an extensive cytoskeletal network comprising actin- and tubulin-like (FtsZ) filaments. The importance of these is emerging in determining cell shape, division and spore formation (see below). The development of antimicrobials targeting these functions is eagerly awaited.

Some larger structures such as spores or *inclusion bodies* of storage products such as volutin (polyphosphate), lipid (e.g. poly-β-hydroxyalkanoate), glycogen or starch occur in some species under specific growth conditions. Specialized labelling techniques (generally requiring fluorescence imaging) enable visualization of the nuclear material or *nucleoid* and other structures (e.g. the form- ing cell division annulus). Figure 2.4 is an electron micrograph of a thin section of a dividing bacterial cell. Outside the cell wall there may be a protective gelatinous covering layer called a *capsule* or, when it is too thin to be resolved with the light microscope, a *microcapsule*. Soluble large-molecular material may be dispersed by the bacterium into the environment as *loose slime*. Some bacteria bear, protruding outwards from the cell wall, one or more kinds of protein-based filamentous appendages called *flagella*, which are organs of locomotion, and hair-like structures termed *fimbriae* or *pili*, which, via specific receptor–ligand interactions at their tip, mediate adhesion.

Bacterial nucleoid

The genetic information of a bacterial cell is mostly contained in a single, long molecule of double-stranded DNA, which can be extracted in the form of a closed circular thread about 1 mm long. The cell solves the problem of packaging this enormous macromolecule by condensing and looping it into a *supercoiled* state. As well

Fig. 2.4 Thin section of a dividing bacillus showing cell wall, cytoplasmic membrane, ribosomes, a developing cross-wall and a mesosome, a membranous structure now generally considered to be an artifact or reflecting some form of cell damage. (By courtesy of Dr P. J. Highton and the editors of *Journal of Ultrastructure Research.*) ×50 000.

cases) and the simultaneous requirement for multiple rounds of chromosome replication in one cell provide a mind-boggling challenge for the segregation machinery. (Imagine unravelling two 1-mm double-stranded threads inside a sphere 1 μm across – scale it up to metre threads and millimetre spheres if you wish.)

The bacterial nucleoid lies within the cytoplasm. This means that as DNA-dependent RNA polymerase makes RNA, ribosomes may attach and initiate protein synthesis on the still attached (nascent) messenger RNA. Synthesis of mRNA and protein (transcription and translation) are therefore seen to be directly coupled in bacteria. In contrast, complete transcripts in eukaryotic cells have to be spliced (to remove the non-coding introns) and capped with polyadenine (this rarely occurs with bacterial mRNA) before the post-transcriptionally modified message is translocated to the cytoplasm.

Ribosomes

Bacterial ribosomes are slightly smaller (10–20 nm) than those of eukaryotic cells and they have a sedimentation coefficient of 70S, being composed of a 30S and a 50S subunit (cf. 40S and 60S in the 80S eukaryotic counterparts). They may be seen with the electron microscope, and number tens of thousands in growing cells. Multiple ribosomes attach to single mRNA molecules to form *polysomes*. It was the nucleotide sequencing of DNA encoding small-subunit ribosomal RNA (rDNA) that led Woese to postulate the evolutionary pathways shown in Figure 2.1. Essentially all cellular organisms can now be classified at least down to genus level by their small-subunit rRNA (SSrRNA) nucleotide sequences. Subsequently it was recognized that, as growing cells contain so many ribosomes, it should be possible to detect unique identifying (or determinative) SSrRNA sequences by complementary in-situ hybridization with fluorescently labelled oligonucleotide probes. Indeed, it is now possible to apply this approach to natural samples and this has enabled recognition of bacteria that have never been grown in laboratory culture. Finally it should be noted that two key organelles in eukaryotes, mitochondria and chloroplasts, both contain 70S ribosomes. When the SSrRNA-encoding genes on the circular chromosomes of these organelles were sequenced, their original free-living bacterial origin was clearly indicated and the proposed endosymbiotic route by which they became organelles confirmed.

Cytoplasmic membrane

The bacterial protoplast is limited externally by a thin, elastic cytoplasmic membrane which is 5–10 nm thick and consists mainly of phospholipids and proteins. Its structure

as the chromosome, the bacterium may contain one or more additional fragments of *episomal* (extrachromosomal) DNA, known as *plasmids*. Bacteria are essentially haploid organisms with only one allele of each gene per cell, although there may be multiple copies of chromosomes and plasmids. Unlike the mitotic or meiotic divisions of eukaryotic cells, chromosomal segregation in bacteria at the time of cell division (or fission) does not involve structures that can be resolved by light microscopy. None the less, the speed at which replication can occur (cell divisions more frequently than one every 15 min in some

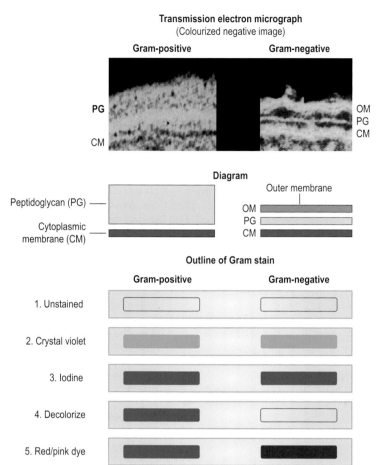

Transmission electron micrograph
(Colourized negative image)

Gram-positive   Gram-negative

PG   OM
       PG
CM   CM

Fig. 2.5 Electron micrograph and diagrams illustrating the basis for the Gram reaction in bacteria.

Diagram

Outer membrane

Peptidoglycan (PG) —   OM
                        PG
Cytoplasmic   CM
membrane (CM)

Outline of Gram stain

Gram-positive   Gram-negative

1. Unstained

2. Crystal violet

3. Iodine

4. Decolorize

5. Red/pink dye

can be resolved in some ultra-thin sections examined by electron microscopy. *Membranes* generally appear in suitably stained electron microscope preparations as two dark lines about 2.5 nm wide, separated by a lighter area of similar width. Integral, transmembrane, and peripheral or anchored proteins occur in abundance and perform similar functions to those described in eukaryotes (e.g. transport and signal transduction). A key feature differentiating prokaryotic cytoplasmic membranes from those of eukaryotes is their multifunctional nature. Thus, while in eukaryotic cells the endoplasmic reticulum and Golgi apparatus are involved in protein secretion, packaging and processing, and the mitochondrial inner membrane is the site of electron transport and oxidative phosphorylation, all of these functions must be performed by one membrane in prokaryotes. It is hardly surprising that prokaryotic cell membranes are relatively protein rich, allowing relatively little space for phospholipids.

Cell wall

The cell wall (Figs 2.5 & 2.6) encases the protoplast and lies immediately external to the cytoplasmic membrane. It is 10–25 nm thick, strong and relatively rigid, though with some elasticity, and openly porous, being freely permeable to solute molecules smaller than 10 kDa in mass and 1 nm in diameter. It is strong but elastic, and supports the weak cytoplasmic membrane against the high internal osmotic pressure of the protoplasm (25 and 5 atm. in Gram-positive and Gram-negative cells respectively) and maintains the characteristic shape of the bacterium in its coccal, bacillary, filamentous or spiral form.

Except under defined osmotic conditions, protoplast survival is dependent on the integrity of the cell wall. If the wall is weakened or ruptured, the protoplasm may swell from osmotic inflow of water and burst the weak cytoplasmic membrane. This process of lethal disintegration and dissolution is termed *lysis*.

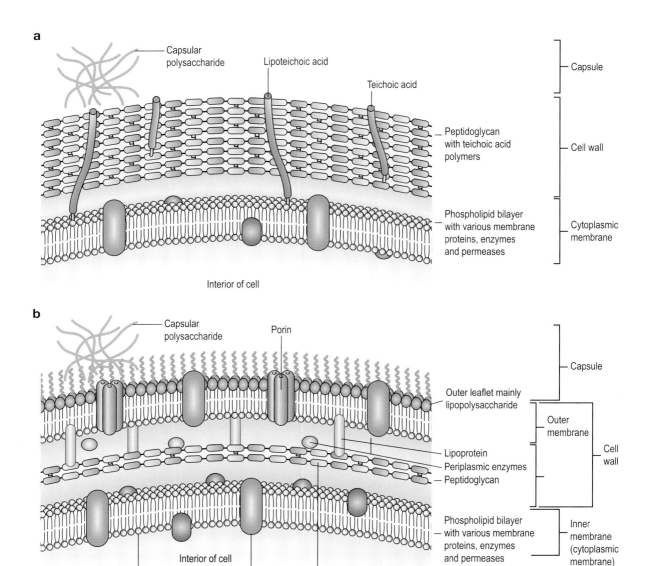

Fig. 2.6 The envelope of **a** the Gram-positive cell wall and **b** the Gram-negative cell wall.

The cell wall plays an important part in *cell division*. A transverse partition of cell wall material grows inwards, like a closing iris diaphragm, from the lateral wall at the equator of the cell and forms a complete *septum* (or crosswall) separating two daughter cells. This process can now be followed in live cells by fluorescence microscopy by imaging the cell division protein FtsZ.

The chemical composition of the cell wall differs considerably between different bacterial species, but in all species the main strengthening component is *peptidoglycan* (syn. *mucopeptide* or *murein*). Peptidoglycan is composed of *N*-acetylglucosamine and *N*-acetylmuramic acid molecules linked alternately in a chain (Fig. 2.7). This heteropolymer forms a

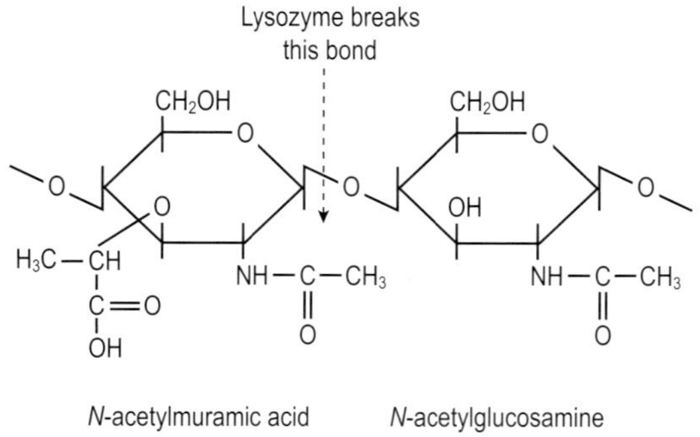

Lysozyme breaks
this bond

N-acetylmuramic acid   N-acetylglucosamine

Fig. 2.7 The basic building block of bacterial cell wall peptidoglycan. N-acetylmuramic acid is derived from N-acetylglucosamine by the addition of a lactic acid unit. Each N-acetylmuramic acid molecule is substituted with a pentapeptide; an N-acetylglucosamine molecule is joined to the muramylpentapeptide within the cell membrane and the unit is transferred to growth points in the existing peptidoglycan, where adjacent strands are cross-linked. (See also Fig. 2.8.)

single molecular continuous sac around the protoplast (described as the murein sacculus). The thickness of the peptidoglycan layer turns out to be of great practical importance in differentiating medically significant bacteria. In the late nineteenth century a Danish physician, Christian Gram, immortalized himself by devising a staining procedure that we now know distinguishes bacteria with a thick (Gram-positive) and a thin (Gram-negative) murein sacculus (see Fig. 2.5). The traditional classification of bacteria is fundamentally rooted in this dichotomy, which has, fortunately, largely been supported by rRNA-based classification.

The rapid (<5 min) Gram-stain procedure remains a cornerstone of day-to-day practice in detecting and identifying bacteria in clinical laboratories. *Gram's stain* distinguishes bacteria as 'Gram positive' or 'Gram negative', according to whether or not they resist decoloration with acetone, alcohol or aniline oil after staining with a triphenyl methane dye, such as crystal violet, and subsequent treatment with iodine. The Gram-positive bacteria resist decoloration and remain stained a dark purple colour. The Gram-negative bacteria are decolorized, and are then counterstained light pink by the subsequent application of safranin, neutral red or dilute carbol fuchsin. In routine diagnostic work a Gram-stained smear is often the only preparation examined microscopically, as it shows clearly the general morphology of the bacteria as well as revealing their Gram reaction. It should be noted that characteristically Gram-positive species may sometimes appear Gram negative under certain conditions of growth, especially in ageing cultures on nutrient agar or after exposure to antibiotics.

The N-acetylmuramic acid units of peptidoglycan each carry a short peptide, usually consisting of L-alanine, D-glutamic acid, either *meso*-diaminopimelic acid (in Gram-negative bacteria) or L-lysine (in Gram-positive bacteria) and D-alanyl-D-alanine. The wall is given its strength by cross-links that form between adjacent strands. These may be formed directly between the *meso*-diaminopimelic acid or L-lysine of one strand and the penultimate D-alanine of the next, or (the usual form in Gram-positive organisms) through an interpeptide bridge composed of up to five amino acids; in either case, the terminal D-alanine is lost in the cross-linking reaction (Fig. 2.8). Several antibiotics interfere with the construction of the cell wall peptidoglycan (see Ch. 5).

The bacterial cell wall also contains other components whose nature and amount vary with the species. Many Gram-positive bacteria contain relatively large amounts of *teichoic acid* (a polymer of ribitol or glycerol phosphate complexed with sugar residues) interspersed with the peptidoglycan; some of this material (*lipoteichoic acid*) is linked to lipids buried in the cell membrane.

Electron microscopy reveals that Gram-negative bacteria possess a second *outer membrane* external to the peptidoglycan layer. This is essentially another unit membrane in which the outer leaflet is composed of a molecule referred to as lipopolysaccharide (LPS). Like the cytoplasmic membrane, this membrane contains many associated proteins whose functions include selective permeability (porins) and attachment (adhesins). The outer membrane confers several important properties on Gram-negative bacteria:

- It protects the peptidoglycan from the effects of lysozyme (a natural body defence substance that cleaves the link between N-acetylglucosamine and N-acetylmuramic acid (see Fig. 2.7).
- It impedes the ingress of many antibiotics.

Components of the LPS, in particular the core structure, lipid A, form endotoxin, which, when released into the bloodstream, may give rise to endotoxic shock (see Ch. 13).

In addition to the basic Gram stain-related properties outlined above, a third type of cell envelope is characteristic of mycobacteria, a group that includes the causal agents of tuberculosis and leprosy. Mycobacteria are taxonomically Gram-positive bacteria, though they can rarely be demonstrated as such. The peptidoglycan layer is covalently linked on its outer aspect to arabinogalactan, which is itself substituted with unique lipids known as *mycolic acids*. These β-hydroxy fatty acids consist of 60 to 90 carbon residues and, together with non-covalently linked free lipids, form an extremely hydrophobic external layer. This layer has some properties in common with the Gram-negative outer membrane (indeed, porins

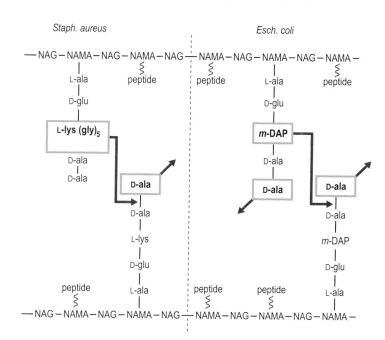

Staph. aureus

— NAG — NAMA — NAG — NAMA — NAG — | — NAMA — NAG — NAMA — NAG — NAMA —

Esch. coli

Fig. 2.8 Schematic representation of the peptidoglycan of a representative Gram-positive organism (*Staphylococcus aureus*) and a representative Gram-negative organism (*Escherichia coli*). Note that in the Gram-positive bacterium cross-linking occurs through a peptide bridge (pentaglycine in *Staph. aureus*), whereas direct cross-linking occurs in *Esch. coli*. In both cases the terminal D-alanine is lost. Not all peptides are engaged in cross-linking in *Esch. coli*, and carboxypeptidases remove redundant D-alanine residues. NAG, N-acetylglucosamine; NAMA, N-acetylmuramic acid; m-DAP, meso-diaminopimelic acid.

have recently been detected therein). The whole envelope structure confers the property of *acid-fast staining* by methods such as the Ziehl–Neelsen (ZN) and phenol–auramine procedures.

The ZN method is of great value in the detection of the tubercle bacillus and other mycobacteria. The mycolic acids referred to above provide a barrier to simple aqueous stains, but when permeability is altered by heating or phenol (or both), concentrated solutions of basic fuchsin, and the fluorescent dyes auramine and rhodamine can produce well-stained cells that subsequently resist decolorization by strong acid in alcohol. Any decoloured non-acid-fast organisms are counterstained in a contrasting colour with methylene blue or malachite green. Modifications of the ZN method are also useful for the demonstration of bacterial endospores and organisms such as *Nocardia* spp. and cysts of some protozoa, notably *Cryptosporidium* spp.

The cell envelope is a highly dynamic structure in growing bacterial cultures. Its components are subject to rapid turnover (synthesis, assembly, disassembly and degradation) and there is busy molecular traffic in and out of the cytoplasm. Although the diagrammatic and photographic representations here give it a somewhat monolithic and immutable character, the envelope and other surface structures can change very rapidly (within minutes) in response to environmental signals. The cell surface receives and transmits many signals from the surrounding environment, including those involving other bacteria, particularly those belonging to the same strain.

This latter phenomenon is known as *quorum sensing* and appears to be important in regulating gene expression in groups of bacteria.

The structures involved in the molecular traffic through the cell envelope are the subject of intense investigation. In particular, the outward secretion of proteins attracts much current interest. At least seven distinct processes have been identified; all involve impressive macromolecular complexes anchored in the cytoplasmic membrane. Of particular interest here are the *type III secretion systems* in Gram-negative bacteria. The fully assembled multiprotein complex spans the cytoplasmic membrane, the periplasmic space, the murein sacculus and the outer membrane, and in some cases projects from the cell surface into an adjacent host (human) cell. These impressive delivery systems are capable of injecting *effector molecules* into the host cell and thereby subvert the latter's function to the advantage of the microbe.

Extracellular polysaccharides: capsules, microcapsules and loose slime

Many bacteria have been demonstrated to possess a more or less continuous but relatively amorphous layer external to the Gram-negative and Gram-positive envelopes described above. Although these are detected quite readily in some bacteria grown under laboratory conditions, they are somewhat ephemeral in others. These structures appear to be important in mediating contact with potentially hostile

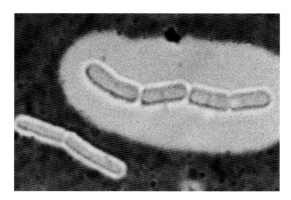

Fig. 2.9 *Bacillus megaterium*. A chain of bacilli with a large capsule, and a pair with a very small capsule. Wet film with India ink, ×3500.

environments and may be subject to strict environmental control.

When this layer is fully hydrated and resolvable by light microscopy, it is called a *capsule* (Fig. 2.9). When it is narrower, and detectable only by indirect, serological means or by electron microscopy, it may be termed a *microcapsule*. The capsular gel consists largely of water and has only a small content (e.g. 2%) of solids. In most species, the solid material is a complex polysaccharide, although in some species its main constituent is polypeptide.

Loose slime, or *free slime*, is an amorphous, viscid, colloidal material that is secreted extracellularly by some bacteria. In bacteria that also possess a demonstrable capsule, the slime is generally similar in chemical composition and antigenic character to the capsular substance. When slime-forming bacteria are grown on a solid culture medium, the slime remains around the bacteria as a matrix in which they are embedded, and its presence confers on the growths a watery and sticky 'mucoid' character. The slime is freely soluble in water and, when the bacteria are grown or suspended in a liquid medium, it passes away from them and disperses through the medium.

All of these features appear to have some role in interactions with the external environment. In some cases capsules have been shown to protect against phagocytosis, the lytic action of complement and bacteriophage invasion. In at least three instances antibodies directed against capsular antigens have been shown to protect against infection and, indeed, capsular preparations are used in several vaccines. Capsules also appear to have a role in protecting cells against desiccation. The production of extracellular polysaccharides in general provides a matrix within which *biofilm* formation can take place.

S-layers

A rather more structured (paracrystalline) protein layer has been demonstrated in some bacteria. This S-layer can be shown by electron microscopy and appears to share at least some functional properties with capsules.

Flagella and motility

Motile bacteria possess filamentous appendages known as *flagella*, which act as organs of locomotion. The flagellum is a long, thin filament, twisted spirally in an open, regular wave-form. It is about 0.02 µm thick and is usually several times the length of the bacterial cell. It originates in the bacterial protoplasm and the structure projects through the cell envelope. According to the species, there may be one, or up to 20, flagella per cell. In elongated bacteria the arrangement of the flagella may be *peritrichous*, or *lateral*, when they originate from the sides of the cell, or *polar*, when they originate from one or both ends. Where several occur on a cell, they may function coiled together as a single 'tail'. The external portion of a flagellum is essentially a polymer of a single protein, *flagellin*, whereas the basal region inserted into the cytoplasmic membrane comprises multiple subunits that anchor and power the organ. In a remarkably elegant manner, the flagellar motor is powered directly (as opposed to indirectly via adenosine triphosphate) by the proton gradient created across the cytoplasmic membrane by electron transport. In *Escherichia coli*, alternation between the anticlockwise and clockwise motion of the flagella effects, respectively, linear or tumbling motility. The intervals between these two patterns are modulated by chemical signals in the environment and the end result is that the bacterium shows *chemotactic behaviour* (movement towards or away from certain stimuli).

Flagella are invisible in ordinary light microscope preparations, but may be shown by the use of special staining methods, and in special circumstances by dark-ground illumination. Because of the difficulties of these methods, the presence of flagella is commonly inferred from the observation of motility. They can be demonstrated easily and clearly with the electron microscope, usually appearing as simple fibrils without internal differentiation (Fig. 2.10). In some preparations the flagellum appears as a hollow tube formed of helically twisted fibrils, and the flagella of some bacteria (e.g. vibrios) have an outer sheath. In the spirochaetes the flagellum is located in the periplasm and hence is referred to as an endo-flagellum, and this presumably underpins their characteristic spiral motion.

Motility is clearly important to many bacteria and probably serves mainly to place the cell in environments favourable to growth and free from noxious influences. In

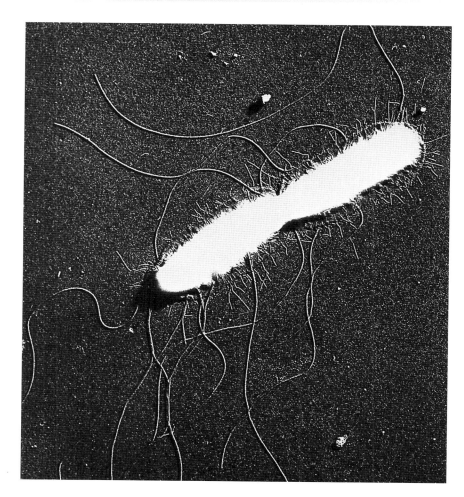

Fig. 2.10 *Salmonella enterica* serotype Typhi. Dividing bacillus from log-phase culture bears about 15 long wavy flagella and more than 100 short fimbriae. Note dense (white) shrunken protoplast surrounded by an empty fold of cell wall. Whole bacillus dried and shadow-cast. Electron micrograph, ×16 000. (From Duguid J P, Wilkinson J F 1961 Environmentally induced changes in bacterial morphology. *Symposia of the Society for General Microbiology* 11: 69–99.)

some cases possession of flagella is thought to contribute to the pathogenesis of disease.

Fimbriae and pili

Many bacteria possess filamentous appendages called *fimbriae* or *pili*. These terms are often used interchangeably, although the latter was originally reserved for structures involved in genetic exchange between bacteria (sex pili; see below). Fimbriae are far more numerous than flagella (e.g. 100–500, being borne peritrichously by each cell), and are much shorter and only about half as thick (e.g. varying from 0.1 to 1.5 μm in length and having a uniform width between 4 and 8 nm). They do not have the smoothly curved spiral form of flagella and are mostly more or less straight. They cannot be seen with the light microscope but are clearly seen with the electron microscope in preparations that have been metal-

shadowed or negatively stained with phosphotungstic acid (see Fig. 2.10).

Multiple types (e.g. types 1 and 2, P type, etc.) of fimbriae have been recognized according to their dimensions, antigenic and phenotypic properties. They have been studied most extensively in *Esch. coli*. In the medical and veterinary contexts, fimbriae are recognized to be important in mediating adhesion between the bacterium and host cells (classically this was recognized in the phenomenon of haemagglutination, a property of type 1, mannose-sensitive pili). In contrast, *sex pili* are structurally similar to other fimbriae but are longer and confer the ability to attach specifically to other bacteria that lack these appendages. Sex pili initiate the process of conjugation (see Ch. 6); they also act as receptor sites for certain bacteriophages described as being 'donor specific'.

It should be noted that fimbriae are not the only means by which bacteria can be involved in specific adhesion events. *Non-fimbrial adhesins* (generally proteins or glyco-proteins) are also important in this regard. Receptor-specific interactions are very important in infective disease, as they are thought to determine much of the tissue tropism of the pathological process.

Importance of microbial surface structures in infection

The structures described have significance for the function of bacteria and their identification by clinical microbiologists, but the surface structures of all micro-organisms (not just bacteria) are of critical importance in the process of infection. They are vital in initiating the contact that occurs during the encounter and establishment of infection (see Ch. 13). Moreover, in addition to substances secreted by micro-organisms, surface structures are exposed to the actions of the innate and adaptive immune systems (see Ch. 8) and, among pathogens, their composition, variability and function reflect these selection pressures.

THE BACTERIAL 'LIFE CYCLE'

Multicellular organisms have long been recognized to pass through many different stages. These may include many immature forms (cf. the larval stages of helminth parasitic worms) or dormant forms (e.g. plant seeds). Even protozoa and fungi may show multiple developmental stages. In contrast, bacteria have been viewed as growing (*vegetative*), stationary or dead. With the exception of spore-forming genera (see below), because they do not undergo morphological differentiation, bacterial cells have been considered essentially uniform in their properties. As indicated above, it is now recognized that all bacteria adapt extensively and rapidly to their environment. This adaptation takes place at both phenotypic (gene expression) and genotypic (genetic complement and arrangement) levels. It seems clear that there are many more physiological states in which bacteria can exist than previously acknowledged and that these states may influence the capacity of the immune system and antimicrobial agents to eliminate them. In particular, the possibility that non-sporulating bacteria are capable of dormancy (a reversible state of metabolic shutdown) has attracted much interest. Spore formation, however, is the key paradigm of differentiation and dormancy in bacteria.

The different forms taken on by organisms at different stages in the life cycle are of course important in their recognition. The range of basic cellular forms of bacteria was mentioned above. It is difficult to generalize, but important to mention that bacterial cell morphology does alter with the physiological state. Characteristically, the cells of bacteria that are growing rapidly are larger than their non- or slowly growing counterparts. Although this does not alter basic coccal morphology, it may make bacilli appear more intermediate (cocco-bacillary) or spherical (coccoid) in shape. Bacilli exposed to certain noxious influences (notably some antibiotics) may produce extended forms that are sometimes described as filamentous. Bacteria that are characterized as essentially filamentous produce a mat of intertwining filaments known as a mycelium (one characteristic form of fungal growth). This form of growth is also associated with fragmentation in which coccal forms may be released, and this results in highly *pleomorphic* cultures. All of these alternative growth forms are undoubtedly under the influence of environmental signals, although their identities have yet to be determined in most cases.

Bacterial spores

Some bacteria, notably those of the genera *Bacillus* and *Clostridium*, develop a highly resistant resting phase or *endospore*, whereby the organism can survive in a dormant state through a long period of starvation or other adverse environmental conditions (resuscitation of spores several thousand years old has been claimed). The process does not involve multiplication: in *sporulation*, each vegetative cell forms only one spore, and in subsequent *germination* each spore gives rise to a single vegetative cell. Geneticists have viewed sporulation as a paradigm of a simple differentiation process, and the key molecular processes required in *Bacillus subtilis* are now understood in great detail. In the face of sporulation stimuli, classically starvation or transition from growth to stationary phase, a programme of sequential expression of specific genes is triggered. The end result is a morphologically distinct structure, the endospore, within the *mother cell*.

In unstained preparations the spore is recognized within the parent cell by its greater refractility. It is larger than lipid inclusion granules and is often ovoid, in contrast to the spherical shape of the lipid granules. Mature ungerminated spores are 'phase bright' when viewed by phase-contrast microscopy; immature or germinated spores are 'phase dark'. When mature, the spore resists coloration by simple stains, appearing as a clear space within the stained cell protoplasm. Spores are slightly acid-fast and may be stained differentially by a modification of the ZN method. The appearance of the mature spores varies according to the species, being spherical, ovoid or elongated, occupying a terminal, subterminal or central position, and being narrower than the cell, or broader and bulging it. Spores of some

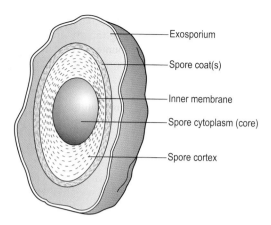

— Exosporium

— Spore coat(s)

— Inner membrane

— Spore cytoplasm (core)

— Spore cortex

Fig. 2.11 Cross-section of a bacterial spore. The core is surrounded by the inner spore membrane. The cortex, a laminated structure, is protected by a more resistant layer or multiple layers forming the spore coat. In some cases, a loose outer covering (exosporium) can be defined.

species have an additional, apparently loose, covering known as the *exosporium* (Fig. 2.11).

Spores are much more resistant than the vegetative forms to exposure to disinfectants, drying and heating. Thus, application of moist heat at 100–120°C or more for a period of 10–20 min may be needed to kill spores, whereas heating at 60°C suffices to kill vegetative cells. In the dry state, or in moist conditions unfavourable to growth, spores may remain viable for many years. The marked resistance of spores has been attributed to several factors in which they differ from vegetative cells: the impermeability of their cortex and outer coat, their high content of calcium and dipicolinic acid, their low content of water, and their very low metabolic and enzymic activity.

Reactivation of the spore is termed *germination* and it should be noted that this is not just a reversal of the process by which the spore was formed. Germination of the spore occurs in response to specific stimuli that are generally related to external conditions favourable to growth. It is irreversible and involves rapid degradative changes. The spore successively loses its heat resistance and its dipicolinic acid; it loses calcium, it becomes permeable to dyes and its refractivity changes. Spores that have survived exposure to severe adverse influences such as heat are much more exacting than normal spores in their requirements for germination. For this reason, specially enriched culture media are used when testing the sterility of materials, such as surgical catgut, that have been exposed to disinfecting procedures. In the process of germination, the spore swells, its cortex disintegrates, its coat is broken open and a single vegetative cell emerges.

The initiation of germination (*activation*) is incompletely understood. It is clear that the state of dormancy of spores may be altered by various treatments, such as transient exposure to heat at 80°C, so that germination can then proceed more rapidly in the individual cells or more completely in a spore population. Activation is distinct from germination and is reversible if germination does not proceed.

After germination, cell growth leading up to the formation of the first vegetative cell and before the first cell division is referred to as *outgrowth*. The conditions required for successful outgrowth may differ markedly from those that allow germination.

Conidia (exospores)

Some of the mycelial bacteria (*Actinomycetales*) and many filamentous fungi form *conidia*, resting spores of a kind different from endospores. The conidia are borne *externally* by abstriction from the ends of the parent cells (conidiophores), and are disseminated by the air or other means to fresh habitats. They are not especially resistant to heat and disinfectants.

Pleomorphism and involution

During growth, bacteria of a single strain may show considerable variation in size and shape, or form a proportion of cells that are swollen, spherical, elongated or pear shaped. This pleomorphism occurs most readily in certain species (e.g. *Streptobacillus moniliformis* and *Yersinia pestis*) in ageing cultures on artificial medium and especially in the presence of antagonistic substances such as penicillin, glycine, lithium chloride, sodium chloride in high concentrations, and organic acids at low pH. The abnormal cells are generally regarded as degenerate or *involution* forms; some are non-viable, whereas others may grow and revert to the normal form when transferred to a suitable environment. In many cases the abnormal shape seems to be the result of defective cell wall synthesis; the growing protoplasm expands the weakened wall to produce a grotesquely swollen cell, comparable to a spheroplast (see below), that later usually bursts and lyses.

Spheroplasts, protoplasts and L–forms

If bacteria have their cell walls removed or weakened while they are held in a solution of sufficient osmolarity to prevent them taking up water by osmosis, they may escape being lysed and, instead, may become converted into viable spherical bodies. If all the cell wall material has been removed from them, the spheres are *free protoplasts*. If they remain enclosed by an intact, but weakened, residual cell wall, they are called *spheroplasts*. Protoplasts and spheroplasts are osmotically sensitive; they vary in size with the osmotic pressure of the suspending medium and, if the medium is much diluted, they swell up and perish by lysis. In contrast to these laboratory generated forms, *L-forms* of bacteria may arise spontaneously and are also cell wall deficient. There is much controversy about the contribution of L-forms to infection. They are difficult to demonstrate as they do not stain with Gram or acid-fast methods and may not propagate in vitro. In common with another controversial area in bacteriology, the 'nanobacteria', L-forms may pass through standard bacteria-stopping filters.

THE NATURE AND COMPOSITION OF VIRUSES

Structure

The basic infectious particle of a virus is known as the *virion*. In the simplest viruses this consists of nucleic acid and a surrounding coat of protein called the *capsid*. Some viruses are enclosed within an *envelope*, derived from host cell membranes but modified by the inclusion of viral glycoproteins. The capsid is composed of distinct morphological units or *capsomeres*, which are assembled from viral proteins. Depending on the arrangement of these proteins, the capsomeres may be spherical, cylindrical or ring-like in appearance. The *nucleocapsid* is the combination of nucleic acid and capsid. The arrangement of the capsomeres around the nucleic acid determines the *symmetry* of the virion. When the capsomeres are applied directly to the helical nucleic acid, a coil-like structure with the appearance of a hollow tube is formed. Viruses with this arrangement are said to have *helical symmetry*. Most helical viruses enclose the nucleocapsid within an envelope and thus do not have a rigid appearance. The other major type is shown by the viruses with *icosahedral symmetry* (Fig. 2.12), in which the capsomeres are arranged as if lying on the faces of an icosahedron with 20 equilateral triangular faces and 12 corners or apices (Fig. 2.13). Capsomeres on the faces and edges of this figure are called hexons, as they always link with six adjacent capsomeres; those positioned at the apices are the pentons, as they always join to five capsomeres. Viruses with icosahedral symmetry have a rigid structure and, under the electron microscope, have a characteristic hexagonal outline with triangular faces. However, if the diameter of the virion is less than about 50 nm the particle will appear spherical. Many viruses

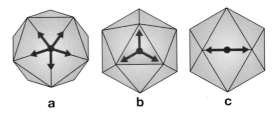

Fig. 2.12 An icosahedron viewed along its **a** five-fold, **b** three-fold and **c** two-fold axes of symmetry.

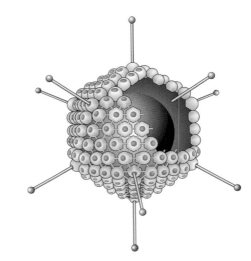

Fig. 2.13 Icosahedron of an adenovirus. The core of DNA is represented by a circular mass. Some of the pentamers at the 12 vertices have been indicated with protruding fibres and terminal knobs. The remaining 240 hexamer capsids are, for the most part, shown as compressed into hollow spheres linked to one another by divalent bonds. The hexagonal shape of a few of the capsomeres is seen in the centre of the diagram.

with icosahedral symmetry are enclosed by an outer envelope. The poxviruses are large and complex, and do not show either type of symmetry; they are referred to as complex. The size of virions varies considerably, from 25 to 300 nm, in different families (Fig. 2.14).

Viral nucleic acid

The commonest types of nucleic acid in viruses of human beings are single-stranded RNA and double-stranded DNA. However, both double-stranded RNA and single-stranded DNA occur in the reoviruses and parvoviruses, respectively. The genomes of RNA viruses may be present as a single strand as in paramyxoviruses, or as two copies as in the retroviruses, or exist as a specific

Fig. 2.14   Morphology of viruses.

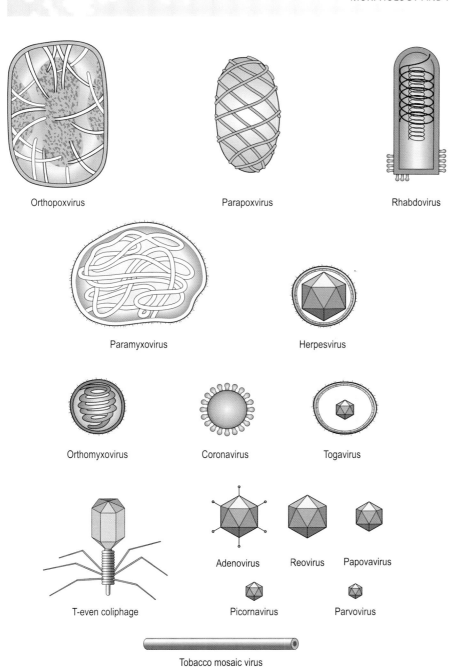

Orthopoxvirus

Parapoxvirus

Rhabdovirus

Paramyxovirus

Herpesvirus

Orthomyxovirus

Coronavirus

Togavirus

T-even coliphage

Adenovirus   Reovirus   Papovavirus

Picornavirus   Parvovirus

Tobacco mosaic virus

number of fragments as in the orthomyxoviruses and reoviruses. Circular molecules of DNA are present in the virions of papovaviruses and hepadnaviruses. The amount of nucleic acid in virions is constant for a particular virus but shows considerable variation; thus the genome can vary from 5 kilobase pairs in parvoviruses to 375 kilobase pairs in the largest poxviruses.

Virion enzymes

Several viruses carry essential enzymes in the virion. As discussed in Chapter 7, an RNA-dependent RNA polymerase or transcriptase is an essential component of the virion in several virus families, including the negative-strand RNA viruses. Among DNA viruses only the poxviruses carry a DNA-dependent RNA polymerase.

The hepadnaviruses have a virion polymerase complex that has some similarity to the reverse transcriptase complex found in the retroviruses.

Viral proteins

Analysis of the proteins produced in a cell during viral infection shows that some are essential components of the virion; these are the structural proteins and include capsid proteins and enzymes as well as basic core proteins that may be necessary to package the nucleic acid within the capsid. Other proteins such as enzymes are needed for the production of viral components but are not part of the virion; these are the non-structural proteins. The essential steps of virus attachment and penetration of the host cell are known to depend on regions of the outer capsid, such as the apical fibres and knobs of adenoviruses, or on parts of the envelope glycoproteins of viruses, such as influenza A and B and the human immunodeficiency virus.

Viroids, defective viruses and prions

Our concept of the organismal nature of infectious agents is stretched to the limit by these infective entities. Viroids are essentially circular RNA molecules that have been associated with several plant diseases; they do not encode proteins or possess a capsid. The hepatitis delta agent has some features in common with viroids and defective viruses (viruses that need the help of another virus for the formation of infectious particles). In the case of the delta agent these features result in an infective agent that can be transmitted in parallel with hepatitis B.

Prions are proteinaceous infective agents that are responsible for the transmissible spongiform encephalopathies (see Ch. 59). An increase in the number of prion proteins in a new host seems to result from the capacity of the introduced protein to induce abnormal conformational changes in a closely related host protein, rather than by replication. Accumulation of the protein in the induced conformation produces the characteristic pathology of the disease.

KEY POINTS

- Agents of infection include cellular organisms belonging to two of the three recently defined *domains* of life, the *Bacteria* and the *Eukarya*. The latter include fungi and protozoa. The subcellular entities *viruses*, *viroids* and *prions* also cause infection but depend on host cells and tissues for propagation.
- Bacterial and eukaryotic micro-organisms can be detected by light microscopy, whereas electron microscopy is required for viruses. Adult stages of multicellular eukaryotic agents of infection or infestation, such as *helminths* (worms) and insects, are generally visible to the naked eye.
- Most pathogenic bacteria can be recognized as either *Gram-positive* or *Gram-negative* after staining. These properties reflect, respectively, the relatively thick *peptidoglycan* layer and the thin peptidoglycan plus outer membrane cell wall structures possessed by cells belonging to these two groups.

- The mycobacteria, which include the global pathogens causing tuberculosis and leprosy, have a different staining property, described as *acid fast*.
- In addition to the cell wall, key bacterial structures with biological and medical significance include the *nucleoid, inclusion granules* and *spores* within the cell, and *flagella, fimbriae* or *pili*, and *capsules* on the cell surface.
- Bacterial *endospores* are highly resistant bacterial cells that result from a differentiation process in some Gram-positive bacteria.
- Viruses are obligate intracellular parasites that use the host cell's machinery to replicate. They contain a nucleic acid core comprising DNA or RNA (not both) in single- or double-stranded form.
- The core is surrounded by a protein *capsid* comprising multiple *capsomeres*; an envelope derived from the host cell membrane surrounds the capsid of some viruses.

RECOMMENDED READING

Armitage J P 1999 Bacterial tactic responses. *Advances in Microbial Physiology* 41: 231–291

Blair D F 1995 How bacteria sense and swim. *Annual Review of Microbiology* 49: 489–522

Frey D, Oldfield R J, Bridger R C 1979 *A Colour Atlas of Pathogenic Fungi.* Wolfe Medical, London

Madeley C R, Field A M 1988 *Virus Morphology*, 2nd edn. Churchill Livingstone, Edinburgh

Moat A G, Foster J W 2002 *Microbial Physiology*, 4th edn. Wiley-Liss, New York

Neidhardt F C, Ingraham J L, Schaechter M 1990 *Physiology of the Bacterial Cell: A Molecular Approach.* Sinauer, Sunderland, MA

Nikaido H, Vaara M 1985 Molecular basis of bacterial outer membrane permeability. *Microbiological Reviews* 49: 1–32

Olds R J 1975 *A Colour Atlas of Microbiology*. Wolfe Medical, London

Internet sites

CELLS alive! http://www.cellsalive.com/

Tree of Life Web Project. http://tolweb.org/tree/

University of California, Berkeley. http://www.ucmp.berkeley.edu/bacteria/bacteriamm.html

3 Classification, identification and typing of micro-organisms

T. L. Pitt

Micro-organisms may be classified in the following large biological groups:

1. Algae
2. Protozoa
3. Slime moulds
4. Fungi
5. Bacteria
6. Archaea
7. Viruses.

The algae (excluding the blue–green algae), the protozoa, slime moulds and fungi include the larger and more highly developed micro-organisms; their cells have the same general type of structure and organization, described as *eukaryotic*, as that found in higher plants and animals. The bacteria, including organisms of the mycoplasma, rickettsia and chlamydia groups, together with the related blue–green algae, comprise the smaller micro-organisms, with a simpler form of cellular organization described as *prokaryotic*. The archaea are a distinct phylogenetic group of prokaryotes that bear only a remote ancestral relationship to other organisms (see Ch. 2). As the algae, slime moulds and archaea are not thought to contain species of medical or veterinary importance, they will not be considered further. Blue–green algae do not cause infection, but certain species produce potent peptide toxins that may affect persons or animals ingesting polluted water.

The viruses are the smallest of the infective agents; they have a relatively simple structure that is not comparable with that of a cell, and their mode of reproduction is fundamentally different from that of cellular organisms. Even simpler are *viroids*, protein-free fragments of single-stranded circular RNA that cause disease in plants. Another class of infectious particles are *prions*, the causative agents of fatal neurodegenerative disorders in animals and man. These are postulated to be naturally occurring host cell membrane glycoproteins that undergo conformative changes to an infectious isoform (see Ch. 59).

TAXONOMY

Taxonomy consists of three components: *classification, nomenclature* and *identification*. Classification allows the orderly grouping of micro-organisms, whereas nomenclature concerns the naming of these organisms and requires agreement so that the same name is used unambiguously by everyone. Changes in nomenclature may give rise to confusion and are subject to internationally agreed rules. In clinical practice, microbiologists are generally concerned with identification – the correct naming of isolates according to agreed systems of classification. These components, together with taxonomy, make up the overarching discipline of *systematics*, which is concerned with evolution, genetics and speciation of organisms, and is commonly referred to as *phylogenetics*.

Protozoa, fungi and helminths are classified and named according to the standard rules of classification and nomenclature that have been developed following the pioneering work of the eighteenth century Swedish botanist Linnaeus (Carl von Linné). Large subdivisions (class, order, family, etc.) are finally classified into individual *species* designated by a Latin binomial, the first term of which is the *genus*, e.g. *Plasmodium* (genus) *falciparum* (species). Occasionally it is useful to recognize a biological variant with particular properties: thus, *Trypanosoma* (genus) *brucei* (species) *gambiense* (variant) differs from the variant *T. brucei brucei* in being pathogenic for man.

Bacteria are similarly classified, but the infinite variety of microbial life and the natural capacity of bacteria for variation and adaptation make rigid classification difficult. Identification is performed by the use of keys that allow the organization of bacterial traits based on growth or activity in a biochemical test system. Some tests are definitive of a genus or species, for example the universal production of catalase enzyme and cytochrome c, respectively, by *Staphylococcus* spp. and *Pseudomonas aeruginosa*. Other characters may be unique to individual species and serve to differentiate them from organisms with closely similar biochemical activity profiles. Some bacteria do not grow in the laboratory (leprosy bacillus, treponemes), and identification by genetic methods may be necessary. The taxonomic ranks used in the classification of bacteria are (example in parentheses):

- Kingdom (Prokaryotae)
- Division (Gracilicutes)
- Class (Betaproteobacteria)
- Order (Burkholderiales)
- Family (Burkholderiaceae)
- Genus (*Burkholderia*)
- Species (*Burkholderia cepacia*).

Some genera, such as *Acinetobacter*, have been subdivided into a number of genomic species by DNA homology analysis. Some are named and others are referred to only by a number. Many of the genomic species cannot be differentiated with accuracy by phenotypic tests. Another subgenus grouping in current usage recognizes species complexes, which are differentiated into genomovars by polyphasic taxonomic methods. A good example of this is the *B. cepacia* complex of organisms, which includes a very diverse group of organisms ranging from strict plant to human pathogens.

At present no standard classification of bacteria is universally accepted and applied, although *Bergey's Manual of Determinative Bacteriology* is widely used as an authoritative source. Bacterial nomenclature is governed by an international code prepared by the *International Committee on Systematic Bacteriology* and published as *Approved Lists of Bacterial Names* in the *International Journal of Systematic and Evolutionary Microbiology*; most new species are also first described in this journal, and a species is considered to be validly published only if it appears on a validation list in this journal.

The *International Committee on Taxonomy of Viruses* (ICTV) classifies viruses and publishes its reports in the journal *Archives of Virology*. Latin names are used wherever possible for the ranks family, subfamily and genus, but at present there are no formal categories higher than family, and binomial nomenclature is not used for species. Viruses do not lend themselves easily to classification according to Linnaean principles, and vernacular names still have wide usage among medical virologists. Readers are referred to the standard work on virus taxonomy *Classification and Nomenclature of Viruses* and the ICTV database website.

METHODS OF CLASSIFICATION

Adansonian or numerical classification

In most systems of bacterial classification, the major groups are distinguished by fundamental characters such as cell shape, Gram-stain reaction and spore formation; genera and species are usually distinguished by properties such as fermentation reactions, nutritional requirements and pathogenicity. The relative 'importance' of different characters in defining major and minor groupings is often purely arbitrary. The uncertainties of arbitrary choices are avoided in the Adansonian system of taxonomy. This system determines the degrees of relationship between strains by a statistical coefficient that takes account of the widest range of characters, all of which are considered of equal weight. It is clear, of course, that some characters, for instance cell shape or Gram-stain reaction, represent a much wider and permanent genetic commitment than other characters which, being dependent on only one or a few genes, may be unstable. For this reason, the Adansonian method is most useful for the classification of strains within a larger grouping that shares major characters.

By scoring a large number of phenotypic characters it is possible to estimate a *similarity coefficient* when shared positive characters are considered, or a *matching coefficient* when both negative and positive shared characters (matches) are taken into account. This numerical taxonomy is best performed on a computer that can calculate degrees of similarity for a group of different organisms; these data are displayed as a *similarity matrix* or a dendrogram tree (Fig. 3.1) (see also Owen 2004 in recommended reading list).

DNA composition

The hydrogen bonding between guanine and cytosine (G–C) base pairs in DNA is stronger than that between adenine and thymine (A–T). Thus, the melting or denaturation temperature of DNA (at which the two strands separate) is determined primarily by the G + C content. At the melting temperature, the separation of the strands brings about a marked change in the light absorption characteristics at a wavelength of 260 nm, and this is readily detected by spectrophotometry. There is a very wide range in the G + C component of bacterial DNA, varying from about 25 to 80 mol% in different genera. However, for any one species, the G + C content is relatively fixed, or falls within a very narrow range, and this provides a basis for classification.

DNA homology

Another approach to classification is to arrange individual organisms into groups on the basis of the *homology* of their DNA base sequences. This exploits the fact that double strands re-form (anneal) from separated strands during controlled cooling of a heated preparation of DNA. This process can be readily demonstrated with suitably heated homologous DNA extracted from a

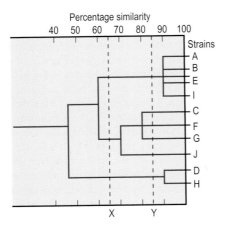

Fig. 3.1 Hierarchical taxonomic tree (*dendrogram*) prepared from similarity matrix data. The dashed lines X and Y indicate levels of similarity at which separation into genera and species might be possible.

single species, but it can also occur with DNA from two related species, so that hybrid pairs of DNA strands are produced. These hybrid pairings occur with high frequency between complementary regions of DNA, and the degree of hybridization can be assessed if labelled DNA preparations are used. Binding studies with messenger RNA (mRNA) can also give information to complement these observations, which provide genetic evidence of relatedness among bacteria. Organisms with different G + C ratios are unlikely to show significant DNA homology. However, organisms with the same, or close, G + C ratios do not necessarily show homology. A novel real-time polymerase chain reaction (PCR) (see pp. 31 and 77–78) has been described for estimation of G + C content.

Ribosomal RNA sequencing

The structure of ribosomal RNA (rRNA) appears to have been highly conserved during the course of evolution, and close similarities in nucleotide sequences reflect phylogenetic relationships. Advances in technology have made nucleotide sequencing relatively simple, and the rDNA sequences (and other genes) of most medically important bacterial species are available from a number of internet sites. Far fewer full-length gene sequences are known for 23S rRNA than for 16S rRNA and 16S–23S internal transcribed sequences. In practice the DNA of the test organism is extracted and amplified by PCR using universal primers. The DNA sequence of the product is determined and the sequence compared against databases to find the closest fit. It is generally accepted that sequence similarity between 0.5% and 1.0% (with the type species) is required for the identification of an unknown organism, and less than 97% similarity is a common cut-off point

for the differentiation of species. However, different species of *Mycobacterium* may exhibit more than 97% similarity in rDNA sequence. Commercial systems are available for bacterial species identification (MicroSeq; Applied Biosystems, Foster City, California, USA).

Nucleotide sequence variation in ribosomal genes is sometimes insufficient to discriminate between closely related species. Other candidate genes have been explored but the *recA* gene, which encodes a protein essential for repair and recombination of DNA, appears to be one of the best suited for phylogenetic analysis in that it defines evolutionary trees consistent with those observed for rRNA genes. Additional housekeeping genes used for phylogenetic studies include *rpoB* (RNA polymerase), *groEL* (heat shock protein) and *gyrB* (DNA gyrase), among others.

CLASSIFICATION IN CLINICAL PRACTICE

The identification of micro-organisms in routine practice requires a pragmatic approach to taxonomy. Table 3.1 outlines a simple, but practical, classification scheme in which organisms are grouped according to a few shared characteristics. Within these groups, organisms may be further identified, sometimes to species level, by a few supplementary tests. Protozoa, helminths and fungi can often be definitively identified on morphological criteria alone (see appropriate chapters).

Protozoa

These are non-photosynthetic unicellular organisms with protoplasm clearly differentiated into nucleus and cytoplasm. They are relatively large, with transverse diameters mainly in the range of 2–100 μm. Their surface membranes vary in complexity and rigidity from a thin, flexible membrane in amoebae, which allows major changes in cell shape and the protrusion of pseudopodia for the purposes of locomotion and ingestion, to a relatively stiff pellicle in ciliate protozoa, preserving a characteristic cell shape. Most free-living and some parasitic species capture, ingest and digest internally solid particles of food material; many protozoa, for instance, feed on bacteria. Protozoa, therefore, are generally regarded as the lowest forms of animal life, although certain flagellate protozoa are closely related in their morphology and mode of development to photosynthetic flagellate algae in the plant kingdom. Protozoa reproduce asexually by binary fission or multiple fission (*schizogony*), and some also by a sexual mechanism (*sporogony*). The most important groups of medical protozoa are the *sporozoa* (malaria parasites, etc.), amoebae and flagellates (see Ch. 61).

Table 3.1 Simple classification of some cellular micro-organisms of medical importance

Common group name	Normal genus names
EUKARYOTES	
Protozoa	
Sporozoa	*Plasmodium, Isospora, Toxoplasma, Cryptosporidium*
Flagellates	*Giardia, Trichomonas, Trypanosoma, Leishmania*
Amoebae	*Entamoeba, Naegleria, Acanthamoeba*
Other	*Babesia, Balantidium*
Fungi	
Mould-like	*Epidermophyton, Trichophyton, Microsporum, Aspergillus*
Yeast-like	*Candida*
Dimorphic	*Histoplasma, Blastomyces, Coccidioides*
True yeast	*Cryptococcus*
PROKARYOTES	
Bacteria	
Filamentous bacteria	*Actinomyces, Nocardia, Streptomyces, Mycobacterium*
'True bacteria'	
Gram-positive bacilli	Aerobes: *Corynebacterium, Listeria, Bacillus*
	Anaerobes: *Clostridium, Lactobacillus, Eubacterium*
Gram-positive cocci	*Staphylococcus, Streptococcus, Enterococcus*
Gram-negative cocci	Aerobes: *Neisseria*
	Anaerobes: *Veillonella*
Gram-negative bacilli	Aerobes:
	Enterobacteria – *Escherichia, Klebsiella, Proteus, Salmonella, Shigella, Yersinia*
	Pseudomonads – *Pseudomonas, Burkholderia, Stenotrophomonas*
	Parvobacteria – *Haemophilus, Bordetella, Brucella, Pasteurella*
	Anaerobes: *Bacteroides, Fusobacterium*
Gram-negative vibrios and spirilla	*Vibrio, Spirillum, Campylobacter, Helicobacter*
Spirochaetes	*Borrelia, Treponema, Brachyspira, Leptospira*
Mycoplasmas	*Mycoplasma, Ureaplasma*
Rickettsiae and chlamydiae	*Rickettsia, Coxiella, Chlamydia*

Fungi

These are non-photosynthetic organisms that possess relatively rigid cell walls. They may be saprophytic or parasitic, and take in soluble nutrients by diffusion through their cell surfaces.

Moulds grow as branching filaments (*hyphae*), usually between 2 and 10 μm in width, which interlace to form a meshwork (*mycelium*). The hyphae are coenocytic (i.e. have a continuous multinucleate protoplasm), being either non-septate or septate with a central pore in each cross-wall. Moulds reproduce by the formation of various kinds of sexual and asexual spores that develop from the vegetative (feeding) mycelium, or from an aerial mycelium that effects their air-borne dissemination (see Ch. 60).

Yeasts are ovoid or spherical cells that reproduce asexually by budding and also, in many cases, sexually, with the formation of sexual spores. They do not form a mycelium, although the intermediate *yeast-like fungi* form a pseudo-mycelium consisting of chains of elongated cells. The *dimorphic fungi* produce a vegetative mycelium in artificial culture, but are yeast-like in infected lesions. The higher fungi of the class *Basidiomycetes* (mushrooms), which produce large fruiting structures for aerial dissemination of spores, are not infectious for human beings or animals, although some species are poisonous.

Bacteria

The main groups of bacteria are distinguished by microscopic observation of their morphology and staining reactions. The Gram-staining procedure, which reflects fundamental differences in cell wall structure, separates most bacteria into two great divisions: *Gram-positive bacteria* and *Gram-negative bacteria* (see Ch. 2).

Details of structure provide a basis for a separate division into:

1. *Filamentous bacteria (Actinomycetes)*, most of which are capable of true branching and which may produce a type of mycelium.
2. *'True' bacteria*, which multiply by simple binary fission.
3. *Spirochaetes*, which divide by transverse binary fission.
4. *Mycoplasmas*, which lack a rigid cell wall.
5. *Rickettsiae* and *chlamydiae*, which are strict intracellular parasites.

Filamentous bacteria

These are sometimes referred to as 'higher bacteria'. A few are of medical interest as pathogens, and some produce antibiotics.

- *Actinomyces*. Gram positive, non-acid-fast, tend to fragment into short coccal and bacillary forms and not to form conidia; anaerobic (e.g. *Actinomyces israelii*).
- *Nocardia*. Similar to *Actinomyces*, but aerobic and mostly acid-fast (e.g. *Nocardia asteroides*).
- *Streptomyces*. Vegetative mycelium does not fragment into short forms; conidia form in chains from aerial hyphae (e.g. *Streptomyces griseus*).

- *Mycobacterium*. Acid-fast; Gram positive, but does not readily stain by the Gram method; usually bacillary, rarely branching; aerobic (e.g. *Mycobacterium tuberculosis*).

True bacteria

Most medically important bacteria fall into this group. They are classified on the basis of their shape:

- *cocci* – spherical, or nearly spherical, cells
- *bacilli* – relatively straight, rod-shaped (cylindrical) cells
- *vibrios and spirilla* – curved or twisted rod-shaped cells.

Cocci. The main groups of cocci are distinguished by their predominant mode of cell grouping and their reaction to Gram's stain. The different cocci are relatively uniform in size (usually about 1 μm in diameter). Some species are capsulate and a very few are motile.

1. *Streptococcus*. Gram-positive cells, mainly adherent in chains due to successive cell divisions occurring in the same axis (e.g. *Streptococcus pyogenes*); sometimes predominantly diplococcal (e.g. *Streptococcus pneumoniae*).

2. *Staphylococcus* and *Micrococcus*. Gram-positive cells, mainly adherent in irregular clusters due to successive divisions occurring irregularly in different planes (e.g. *Staph. aureus*).

3. *Sarcina*. Gram-positive cells, mainly adherent in cubical arrays of eight, or multiples thereof, due to division occurring successively in three planes at right angles (e.g. *Sarcina lutea*).

4. *Neisseria*. Gram negative cells, mainly adherent in pairs and slightly elongated at right angles to axis of pairs (e.g. *Neisseria meningitidis*).

5. *Veillonella*. Gram negative; generally very small cocci arranged mainly in clusters and pairs; anaerobic (e.g. *Veillonella parvula*).

Bacilli. The primary subdivision of the rod-shaped bacteria is made according to their staining reaction by the Gram method and the presence or absence of endospores.

1. *Gram-positive spore-forming bacilli*. Apart from some rare saprophytic varieties, the only bacteria to form endospores are those of the genera *Bacillus*, *Paenibacillus* (aerobic) and *Clostridium* (anaerobic). They are Gram positive, but liable to become Gram negative in ageing cultures. The size, shape and position of the spore may assist recognition of the species, for example the bulging, spherical, terminal spore ('drumstick' form) of *Clostridium tetani*.

2. *Gram-positive non-sporing bacilli*. These include several genera. *Corynebacterium* is distinguished by a tendency to slight curving, a club-shaped or ovoid swelling of the bacilli, and their arrangement in parallel or angular clusters caused by the snapping mode of cell division. *Erysipelothrix* and *Lactobacillus* are distinguished by a tendency to grow in chains and filaments, and *Listeria* by flagella that confer motility.

3. *Gram-negative bacilli*. This large grouping includes numerous genera such as the pseudomonads and the family Enterobacteriaceae ('coliform bacilli') as well as small, often pleomorphic, bacilli represented by *Haemophilus*, *Brucella*, etc. ('parvobacteria'), and anaerobes such as *Bacteroides* and *Prevotella*.

4. *Vibrios and spirilla*. Vibrios and the related campylobacters are recognized as short, non-flexuous, comma-shaped bacilli (e.g. *Vibrio cholerae*), and spirilla as non-flexuous spiral filaments (e.g. *Anaerobiospirillum*, '*Spirillum minus*'). They are Gram negative and mostly motile, having polar flagella and showing very active 'darting' motility.

Spirochaetes

These organisms differ from the 'true' bacteria in being slender flexuous spiral filaments that, unlike the spirilla, are motile without possession of flagella. The staining reaction, when demonstrable, is Gram negative. The different varieties are recognized by their size, shape, waveform and refractility, observed in the natural state in unstained wet films by dark-ground microscopy. Genera of medical importance include *Borrelia*, *Treponema* and *Leptospira*.

Mycoplasmas

These are prokaryotes that differ from 'true' bacteria in their smaller size and lack of a rigid cell wall, which leads to extreme pleomorphism and sensitivity to external osmotic pressure. The viable elements range from 0.15 to over 1 μm in diameter, the smallest being capable of passing through filters that retain conventional bacteria. Mycoplasmas can be cultivated on cell-free nutrient media, and are the smallest and simplest organisms capable of autonomous growth.

Rickettsiae and chlamydiae

The rickettsiae are rod-shaped, spherical or pleomorphic Gram-negative organisms. They are generally smaller than 'true' bacteria, but are still resolvable in the light microscope. Most are strict parasites that can grow only in the living tissues of a suitable animal host, usually intracellularly (e.g. *Rickettsia prowazekii*). Chlamydiae are similar to rickettsiae, but have a more complex intracellular cycle (e.g. *Chlamydia trachomatis*).

Viruses

Viruses usually consist of little more than a strand of DNA or RNA enclosed in a simple protein shell known as a *capsid*. Sometimes the complete nucleocapsid may be enclosed in a lipoprotein envelope derived largely from the host cell. Viruses are capable of growing only within the living cells of an appropriate animal, plant or bacterial host; none can grow in an inanimate nutrient medium. The viruses that infect and parasitize bacteria are termed *bacteriophages* or *phages*. A simple classification of the viruses that are involved in human disease is shown in Table 3.2.

IDENTIFICATION OF MICRO-ORGANISMS

Precise or *definitive* identification of bacteria is time consuming and contentious, and best carried out in specialized reference centres. For most clinical purposes, clear, rapid guidance on the likely cause of an infection is required and, consequently, microbiologists usually rely on a few simple procedures, notably microscopy and culture, backed up, when necessary, by a few supplementary tests to achieve a *presumptive* identification. Microscopy is the most rapid test of all, but culture inevitably takes at least 24 hours, sometimes longer. More rapid tests are constantly being sought, and some progress has been achieved with antigen detection methods and specific *gene probes* (see below).

Most specimens for bacteriological examination, whether from human beings, animals or the environment, contain mixtures of bacteria, and it is essential to obtain *pure cultures* of individual isolates before embarking on identification. Non-cultural methods, such as antigen detection or gene probes, do not have this disadvantage; however, they do have the potential limitation of being highly specific so that the investigator must know beforehand what it is necessary to look for.

Microscopy

Morphology and staining reactions of individual organisms generally serve as preliminary criteria to place an unknown species in its appropriate biological group. A Gram-stain smear suffices to show the Gram reaction, size, shape and grouping of the bacteria, and the arrangement of any endospores. An unstained wet film may be examined with dark-ground illumination in the microscope to observe the morphology of delicate spirochaetes; an unstained wet film, or 'hanging-drop', preparation is examined with ordinary bright-field illumination for observation of motility. Capsules surrounding bacterial cells are demonstrated by 'negative staining' with India

Table 3.2 Principal types of virus causing human disease

Type of virus	Examples
RNA viruses	
Orthomyxoviruses	Influenza A, B and C viruses
Paramyxoviruses	Parainfluenza viruses, mumps virus, measles virus, respiratory syncytial virus, Hendra virus, Nipah virus, human metapneumovirus
Rhabdoviruses	Rabies virus
Arenaviruses	Lassa virus
Filoviruses	Marburg and Ebola viruses
Togaviruses	Many arboviruses, rubella virus
Flaviviruses	Yellow fever virus, Dengue virus, Japanese encephalitis virus, West Nile virus, hepatitis C virus
Bunyaviruses	Hantaan virus
Coronaviruses	Human coronavirus, 229E (group 1), OC43 (group 2), NL63 (group 1-like novel corona viruses), SARS (zoonosis)
Caliciviruses	Norwalk-like viruses, hepatitis E virus
Picornaviruses	Enteroviruses: poliovirus (3 types), echovirus (31 types), Coxsackie A virus (24 types), Coxsackie B virus (6 types), enterovirus types 68–71, hepatitis A virus (type 72), rhinovirus, many serotypes
Retroviruses	Human immunodeficiency virus types 1 and 2, human T-lymphotropic virus types I and II
Reoviruses	Rotaviruses
DNA viruses	
Poxviruses	Variola, vaccinia, molluscum contagiosum virus, Orf virus
Herpesviruses	Herpes simplex virus types 1 and 2, varicella-zoster virus, cytomegalovirus, Epstein–Barr virus, human herpesvirus types 6, 7 and 8
Adenoviruses	Many serotypes
Papovaviruses	JC virus, BK virus, human papillomavirus
Hepadnaviruses	Hepatitis B virus
Parvoviruses	B19 virus, new human parvovirus

SARS, severe acute respiratory syndrome.

ink; the capsules remain unstained against the background of ink particles. To identify mycobacteria, or other acid-fast organisms, a preparation is stained by the Ziehl–Neelsen method or one of its modifications (see p. 15). The microscopic characters of certain organisms in pathological specimens may be sufficient for presumptive identification, for example tubercle bacilli in sputum, or *T. pallidum* in exudate from a chancre. However, many bacteria share similar morphological features, and further tests must be applied to differentiate them.

Cultural characteristics

The appearance of colonial growth on the surface of a solid medium, such as nutrient agar, is often very characteristic. Attention is paid to the diameter of the colonies, their outline, their elevation, their translucency (clear, translucent or opaque) and colour. Changes brought about in the medium (e.g. haemolysis in a blood agar medium) may also be significant. The range of conditions that support growth is characteristic of particular organisms. The ability or inability of the organism to grow in the presence (aerobe) or absence (anaerobe) of oxygen, in a reduced oxygen atmosphere (micro-aerophile) or in the presence of carbon dioxide, or on media containing selective inhibitory factors (e.g. bile salt, specific antimicrobial agents, or low or high pH) may also be of diagnostic significance (see Table 4.1).

Biochemical reactions

Species that cannot be distinguished by morphology and cultural characters may exhibit metabolic differences that can be exploited. It is usual to test the ability of the organism to produce acidic and gaseous end-products when presented with individual carbohydrates (glucose, lactose, sucrose, mannitol, etc.) as the sole carbon source. Other tests determine whether the bacterium produces particular end-products (e.g. indole or hydrogen sulphide) when grown in suitable culture media, and whether it possesses certain enzyme activities, such as oxidase, catalase, urease, gelatinase or lecithinase. Traditionally, such tests have been performed selectively and individually according to the recommendations of standard guides, such as the invaluable *Cowan and Steel's Manual for the Identification of Medical Bacteria*. However, today most diagnostic laboratories use commercially prepared microgalleries of identification tests which, though expensive, combine simplicity and accuracy. Test kits are now available for a number of different groups of organisms, including enterobacteria, staphylococci, streptococci and anaerobes. Other kits facilitate the testing of carbon source utilization, the assimilation of specific substrates and the enzymes produced by an organism.

On occasion, more elaborate procedures may be used for the analysis of metabolic products or whole-cell fatty acids. Indeed, a fully automated, fatty acid-based identification system, which combines high-resolution gas chromatography and pattern recognition software, is widely used to identify a variety of aerobic and anaerobic bacterial species. New profiles are added to a computerized database, thus increasing the sensitivity of the system. Mass spectrometric methods show promise for rapid identification, particularly matrix-assisted laser desorption ionization time-of-flight (MALDI-TOF) mass spectrometry. This offers the analysis of whole bacterial cultures for unique mass spectra from charged macromolecules by rapid, high-throughput testing with a rapidly growing database.

INDIRECT IDENTIFICATION METHODS

Gene probes

These are cloned fragments of DNA that recognize complementary sequences within micro-organisms (see p. 77). Binding is detected by tagging the DNA with a radioactive label, or with a reagent that can be developed to give a colour reaction, such as biotin. By selecting DNA fragments specific for features characteristic of individual organisms, gene probes can be tailored to the rapid identification of individual species in clinical material. The disadvantages of this approach are that the organism itself may be dead and a viable isolate is not made available for subsequent tests of susceptibility to antimicrobial agents, toxin production or epidemiological investigation.

Culture and preliminary identification of bacteria in the laboratory is time consuming and relatively labour intensive. Bacteria may also be uncultivable, slow growing, or fastidious in nutrient requirements. Nucleic acid techniques for the detection and identification of bacteria have evolved against this background; today numerous commercially available systems have been developed and many are in use in diagnostic laboratories. The technologies fall into three basic groups:

1. target amplification by PCR, transcription-mediated amplification, nucleic acid sequence-based amplification, etc.
2. probe amplification using ligase chain reaction or Q-beta replicase
3. signal amplification, as in branched DNA assay.

It is possible to detect the presence of an increasing number of species either by PCR of universal or specific gene targets or by hybridization with specific probes. As with conventional methods, nucleic acid technology has its limitations, the most frequent being contamination of a sample by post-amplification products. Other factors include operator skill, primer design, stringency of assay, presence of inhibitory compounds in the specimen and the ubiquity of the organism sought. The latter is fundamental to the interpretation of results as many bacterial pathogens occur naturally as commensals in certain body sites. The new technologies do, however, offer a considerable advantage over phenotypic methods in terms of sensitivity, and many optimized PCR systems claim to be able to detect as few as two to ten bacteria per millilitre of specimen, which is far below the threshold of conventional culture.

Nucleic acid assays for antimicrobial resistance genes are also in use and development. A recently described innovation may have the potential to detect, identify the species, subtype the organism and identify resistance genes on microscope slide smears. The technique uses peptide nucleic acid (PNA, pseudopeptides) with DNA binding capacity. The PNA molecules have a polyamide backbone instead of the sugar phosphate of DNA and RNA, and nucleotide bases attached to this backbone are able to hybridize by specific base pairing with complementary DNA or RNA sequences. PCR amplification of target genes on conventionally stained microscope smears is also possible, and this accelerates the prospect of very rapid and sensitive test methods limited only by the specificity of the primer for the target.

In the past 5 years, many prokaryotic genomic sequences have been completed. This has been paralleled by the development of high-density oligonucleotide arrays which consist of many thousands of different probes. These arrays are constructed by in-situ oligonucleotide synthesis on a glass support by photolithography or other methods. Hybridization of the prelabelled target nucleic acids to the bound probe is detected directly with fluorescein or radioactive ligands, or indirectly using enzyme conjugates. Areas of application for high-density arrays include DNA sequencing, strain genotyping, identifying gene functions, location of resistance genes, changes in mRNA expression and phylogenetic relatedness. Arrays with selected gene targets have recently been developed in an Eppendorf tube format. The chip is embedded in the bottom of the tube and carries optimized sets of oligonucleotide probes specific for certain organisms or for antimicrobial resistance genes or virulence factors. In this way chips can be customized for individual bacteria or groups of bacteria. All stages of the assay – sample preparation from agar-grown colonies, PCR amplification, hybridization, conjugation with reporter molecule and detection by automated image recording – are carried out in a single tube within 6–8 hours. An increasingly used development is that of real-time PCR, which combines sample amplification with a means for detection of the specific product by fluorescence so that both steps take place conveniently in a single reaction tube. The system has significant advantages over conventional PCR in terms of rapidity, simplicity and number of manual procedures; contamination is effectively eliminated by the tube being closed after amplification. The DNA product may be detected with a fluorescent dye, or increased specificity obtained with hybridization with fluorescence-labelled sequence-specific oligonucleotide probes. Quantification of the target DNA is also possible with this system, allowing estimation of viral or bacterial numbers in a specimen (see p. 78).

Recently fluorescence in-situ hybridization (FISH) has been used to detect bacteria directly in clinical specimens. This technique utilizes probes specific for target organisms in the sample without the need for culture and allows quantification of the cell count and morphology. Sensitivity and specificity vary according to the probe used, but if the probe is optimized it can rival conventional culture methods.

Antibody reactions

Species and types of micro-organism can often be identified by specific serological reactions. These depend on the fact that the serum of an animal immunized against a micro-organism contains antibodies specific for the homologous species or type that react in a characteristic manner (e.g. agglutination or precipitation) with the particular micro-organism. Such simple in-vitro tests have been used for many years in microbiology, notably in the formal identification of presumptive isolates of pathogens (e.g. salmonellae) from clinical material. The specificity and range of antibody tests have been greatly improved by the availability of highly specific *monoclonal antibodies*. These are produced by the *hybridoma* technique in which individual antibody-producing spleen cells are fused with 'immortal' tumour cells in vitro. The progeny of these hybrid cells produce only the type of antibody appropriate to the spleen cell precursor (see p. 123).

Latex agglutination

By adsorbing specific antibody to inert latex particles, a visible agglutination reaction can be induced in the presence of homologous antigen. This principle can be applied in reverse to detect serum antibodies. Latex-based kits are widely used for serological grouping of organisms and detection of toxins produced by bacteria during growth.

Enzyme-linked immunosorbent assay

In enzyme-linked immunosorbent assay (ELISA), a specific antibody is attached to the surface of a plastic well and material containing the test antigen is added. After washing, the presence of the antigen is detected by addition of more of the specific antibody, this time labelled with an enzyme that can initiate a colour reaction when provided with the appropriate substrate. The intensity of the colour change is related to the amount of antigen bound. The ELISA method may also be used in the reverse manner for the quantitative detection of antibodies, by adsorbing purified antigen to the well before adding test serum; in this case the enzyme-linked system used to detect the antigen–antibody reaction is a labelled anti-human globulin. In immunoglobulin (Ig) M antibody capture ELISA (MAC-ELISA), widely used in virological diagnosis for the detection of IgM, anti-human μ-chain antibody (usually raised in goats) is bound to the well. The test serum is added, and any IgM binds to the capture reagent; after washing, purified

Table 3.3 Panmictic versus clonal populations

	Reproduction	Recombination	Allele arrangement	Mutation	Selective pressures
Panmictic	Sexual*	Frequent	Segregated	Normal	Natural selection
Clonal	Asexual	Rare	Non-random association	Normal	Environmental

*Refers to recombination between genetic elements from different organisms in bacteria.

antigen (e.g. rubella antigen) is added, and this can be detected with an appropriate labelled antibody.

Haemagglutination and haemadsorption

Certain viruses, notably the influenza viruses, have the property of attaching to specific receptors on the surface of appropriate red blood cells. In this manner the virus particles act as bridges linking the red cells in visible clumps. In tissue culture, such haemagglutinins may appear on the surface of cells infected with a virus. If red cells are added to the tissue culture, they adhere to the surface of infected cells, a phenomenon known as *haemadsorption*. Red blood cells can also be coated with specific antibody so that they agglutinate in the presence of the homologous virus particle in a manner similar to that described for latex agglutination above.

Fluorescence microscopy and immunofluorescence

When certain dyes are exposed to ultraviolet light, they absorb energy and emit visible light; that is, they fluoresce. Tissues or organisms stained with such a dye and examined with ultraviolet light in a specially adapted microscope are seen as fluorescent objects; for example, auramine can be used in this way to stain *Mycobacterium tuberculosis*. Antibody molecules can be labelled by conjugation with a fluorochrome dye such as fluorescein isothiocyanate (which fluoresces green) or rhodamine (orange–red). When fluorescent antibody is allowed to react with homologous antigen exposed at a cell surface, this *direct immunofluorescence* procedure affords a highly sensitive method for the identification of the particular antigen. For this procedure it is necessary to have a specific antibody conjugate for each antigen; however, unconjugated antibody can be used and the reaction then detected by the addition of an antiglobulin conjugate, which will react with any antibody from the species in which the antibody was raised.

Immuno–polymerase chain reaction

This technique arose out of the fusion of antibody technology with PCR methods with the aim of enhancing the capability of antigen detection systems. In immuno-PCR, a linker molecule with bispecific binding affinity for DNA and antibodies is used to attach a DNA molecule (marker) to an antigen–antibody complex. This produces a specific antigen–antibody–DNA conjugate. The attached DNA marker can be amplified by PCR with appropriate primers, and the presence of amplification products shows that the marker DNA is attached to antigen–antibody complexes, indicating the presence of antigen. The enhanced sensitivity of immuno-PCR achieved over ELISA is reported to be in excess of 10^5, theoretically allowing as few as 580 antigen molecules (9.6×10^{-22} moles) to be detected.

TYPING OF BACTERIA

Different bacterial species often exhibit different population structures. Some species are characterized by highly diverse populations at one extreme and closely similar members at the other. The frequency of recombination of chromosomal genes (see Ch. 6) is considered the major determinant of a population structure of a given species, and this frequency ranges from absent to low to very high. Highly recombining populations are termed *panmictic*, in contrast to *clonal* populations where recombination is infrequent (Table 3.3). Species such as *Neisseria gonorrhoeae* and *Haemophilus influenzae* are naturally transformable, that is, they are able to take up DNA (foreign and native) from their environment, and their populations are characterized by a high frequency of recombination, segregation of alleles and relatively low mutation. In clonal populations such as *Salmonella enterica*, recombination is rare and there is non-random association of alleles in a background of limited genetic exchange. Mutations occur as a result of natural and selective pressures, but these are not sufficient to disrupt the clonal lineage and daughter cells continue to resemble the ancestral parent. Bacterial clones are therefore not identical to their parents but display a number of characteristics in common with their ancestors. Many species are characterized by considerable genetic diversity but with clonal expansion of a subpopulation. Some of these clones may be transient, although others may persist and spread nationally and globally.

By *typing* we identify a recognizable subdivision of a species that serves as a reference marker against which other isolates of the same species can be compared. A population of bacteria presumed to descend from a single bacterium, as found in a natural habitat, in primary cultures from the habitat, and in subcultures from the primary cultures, is called a *strain*. Each primary culture from a natural source is called an *isolate*. The distinction between strains and isolates may be important; for example, cultures of typhoid bacilli isolated from ten different patients should be regarded simply as ten different *isolates* unless epidemiological or other evidence indicates that the patients have been infected from a common source with the same *strain*. The ability to discriminate between similar strains may be of great epidemiological value in tracing sources or modes of spread of infection in a community or hospital ward, and various typing methods have been devised. Strains may be distinguishable only in minor characters and it is usually simpler to establish differences between isolates from a common source than unequivocally to prove their identity. Demonstration of an identical response by a single reproducible typing method is not proof that two strains are the same. However, the confidence with which similarity can be inferred is greatly increased if more than one typing method is used.

Typing may inform different levels of epidemiological investigation, ranging from micro-epidemiology (local investigation), macro-epidemiology (regional, national, international) to population structure analysis (evolution of strains and global patterns of spread). The data derived may assist in the control of infection by excluding sources, identifying carriers and establishing the prevalence of individual strains. Common reasons for microbial typing are to identify common or point sources, discriminate between mixed strain infections, distinguish re-infection from relapse, and occasionally to identify a type and disease association (e.g. *Escherichia coli* O157 and haemolytic uraemic syndrome, skin and throat types of group A *Str. pyogenes*, etc).

Typing methods should be reproducible both in the laboratory and clinically. The former is easily established by repeated tests on a sample of experimental strains, but should also be established in vivo by examining multiple pairs of isolates from single sources to determine the stability of the strain characteristics probed by the typing method used. A typing method should also discriminate adequately and clearly between different populations and be comprehensive, that is, assign most populations to a type. The typing data should be in a format that is easily assimilated into databases and should be able to be incorporated into the national picture to inform other workers in the field. Very few, if any, single typing methods will meet these criteria and hence there is a need to utilize different methods, preferably directed at unlinked targets and always in the context of an epidemiological investigation.

Biotyping

Biochemical test reactions that are not universally positive or negative within a species may define *biotypes* of the species, and these may be efficient strain markers. In practice biotyping is often less discriminatory than other strain typing methods and may be unstable because of loss of the property. Differences among strains may also be detected by variations in sensitivity to fixed concentrations of chemicals such as heavy metals, a process known as *resistotyping*. The nutritional requirements of the isolate (amino acids) for growth may also be used to define the *auxotype* of an isolate.

Serotyping

Many surface structures of bacteria (lipopolysaccharide and outer membrane, flagella, capsule, etc.) are antigenic, and antibodies raised against them can be used to group isolates into defined *serotypes*. Some species are characterized by numerous antigenic types and serotyping for these species is highly discriminatory, whereas for others conservation of antigen epitopes renders serotyping of little value for epidemiological purposes. Members of the species *Salmonella enterica* are defined by their somatic and flagellar serotypes (see Ch. 24). Capsular antigens may be associated with pathogenicity of the organism, and many vaccines protect the individual against infection by stimulating antibodies to capsular antigen epitopes. Agglutination of bacterial suspensions with rabbit antibodies is the most commonly used method for typing, but other techniques such as precipitation in agar gels, ELISA and capsular swelling (Fig. 3.2) may be used.

Phage typing

Bacteria often show differential susceptibility to lysis by certain bacteriophages. The phage adsorbs to a specific receptor on the bacterial surface and injects its DNA into the host. Phage DNA may become stably integrated into the bacterial chromosome and this state is referred to as *lysogeny*; phages capable of this are called *temperate phages*. In lysogeny, a small proportion of host cells express the phage genes, and some cell lysis and liberation of phage progeny occurs. Alternatively, the phage DNA may enter a replicative cycle, leading to the death of the host and the production of new phage particles. These *lytic* or *virulent* wild phages lyse the bacterium at the end of the replicative cycle and release a large number of daughter phage particles that infect neighbouring cells.

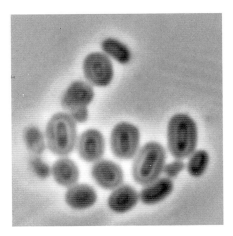

Fig. 3.2 Capsular swelling reaction of *Klebsiella pneumoniae*. Antibody adsorbed to capsule alters the refractive index, allowing visualization of the capsule around the cell within.

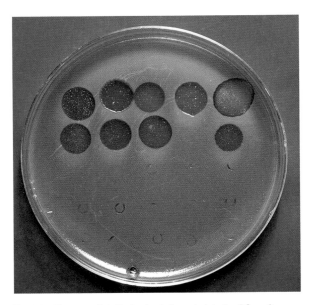

Fig. 3.3 Phage-mediated lysis of red pigmented strain of *Serratia marcescens*.

This process leads to visible inhibition of the growth of the host cells (Fig. 3.3). The *phage type* of the culture is identified according to the pattern of susceptibility to a set of lytic and/or temperate phages. Lytic phages may be readily recovered from sewage, waste and river water, and temperate phages may be released from a lysogenic strain by induction with ultraviolet radiation or chemical mutagens.

The critical factors governing the interpretation of phage typing results are discrimination and reproducibility. If the system is both highly discriminatory and reproducible, any differences in lysis patterns of isolates will indicate that they represent different strains. Schemes that utilize adapted phages (a single phage propagated in different strains) that are specific for a particular receptor site such as the Vi polysaccharide of *Salmonella enterica* serotype Typhi are relatively reproducible, and minor differences are significant and reproducible. On the other hand, a phage set comprising random unrelated phages may adhere to a number of different receptor sites and have different biological properties, and so is unlikely to be highly reproducible but may be adequately discriminating. This lack of stability results in the definition of broad phage groups rather than defined types, and is the case for *Staph. aureus* where at least two strong lytic reactions have to be present in the patterns of isolates before they can be termed distinct strains.

Bacteriocin typing

Bacteriocins are naturally occurring antibacterial substances, elaborated by most bacterial species, that are active mainly against strains of the same genus as the producer strain. Bacteriocin typing may define the spectrum of bacteriocins produced by field strains, or the sensitivity of these strains to bacteriocins of a standard panel of strains. Patterns of production or susceptibility to bacteriocins allow the division of species into *bacteriocin types*.

Protein typing

Bacteria manufacture thousands of proteins that can be visualized by electrophoresis in acrylamide gels in the presence of a strong detergent. The proteins separate according to molecular size and, after staining with a dye, the pattern of bands from each isolate can be compared. This system has been used successfully to type many bacterial and fungal species, but lacks reproducibility. Investigation of microbial populations by gel electrophoresis of metabolic enzymes, which can then be detected by specific substrates, has also in the past been applied widely for clonal analysis within species.

Restriction endonuclease typing

Restriction endonucleases are a family of enzymes that each cut DNA at a specific sequence recognition site, which may be rare or frequent in the DNA of the species being examined. The frequency with which an enzyme cuts in a particular species is dependent on the oligonucleotide sequence, the frequency of the restriction site, and the percentage G + C content of the species. For example, the recognition site of enzyme *Sma*I is 5'-CCC↓GGG-3' (↓ site of cleavage) and this cuts infrequently in the

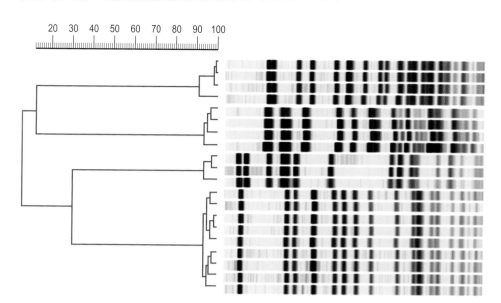

Fig 3.4 Pulsed-field gel electrophoresis profiles of *Xba*I digests of DNA of *Ps. aeruginosa* isolates. Dendrogram shows relative percentage similarity of profiles calculated with Pearson's coefficient.

AT-rich genome of *Staph. aureus*, whereas enzyme *Xba*I (5'-T↓CTAGA-3') is a rare cutter in most Gram-negative species with a high GC content. Both plasmid and chromosomal DNA can be analysed by this means. Frequent-cutting endonucleases generate numerous small fragments that can be resolved by conventional electrophoresis in agarose gel and detected by staining with a dye. The resolution of conventional agarose gel electrophoresis does not exceed 20 kb and optimal separation in standard length gels is achieved between 1 and 15 kb.

The large DNA fragments produced by infrequent-cutting enzymes need to be separated in special electrical fields with a pulsed current (*pulsed-field gel electrophoresis*; PFGE). In this technique, bacteria are encased in an agarose plug (to minimize shearing of DNA) and the cells are digested with proteinase K enzyme before the DNA is digested with the enzyme. By introducing a pulse or change in the direction of the electric field, fragments as large as 10 Mb can be separated. The time taken by fragments to reorient to the alternate electric field is proportional to their molecular size and where they migrate in the electric field. The most widely used apparatus is the contour-clamped homogeneous electric field (CHEF), which has 24 electrodes arranged in a hexagonal array. Run times are often of the order of 30–40 hours, but shorter, more rapid, protocols have been described. A number of factors influence the quality of results, including DNA quality and concentration, agarose concentration, voltage and pulse times, and buffer strength and temperature.

Interpretation of PFGE profiles can be problematic. For some species the criteria of Tenover (see recommended reading list) can be applied to establish the significance of differences in banding profiles of strains. As a rule of thumb, isolates from an incident under investigation that show no difference in profiles can be considered indistinguishable, those with one to three band differences as closely related, four to six bands as possibly related, and seven or more band differences as indicating distinct strains. However, this rule should be applied with a degree of caution as some species (e.g. *Enterococcus faecium*) can exhibit significant variation (six to ten band differences) apparently within members of the same clone. A number of computer-assisted analysis packages are available that calculate coefficients of similarity between strains and represent these as dendrograms (Fig. 3.4). Two commonly employed coefficients, the Jaccard and Dice, use the number of concordant bands in profiles and the total number of possible band positions to calculate the percentage similarity between the isolates. The Pearson coefficient gives the advantage that specific band positions do not have to be defined. A cut-off point of 85% similarity is often used but, as for the band difference rule, this should be set by experiment with related and unrelated strain sets.

Gene probe typing

DNA probes (see above) for strain typing consist of cloned specific, random or universal sequences that can detect restriction site heterogeneity in the target DNA. The detection of variation in rDNA gene loci is the basis of *ribotyping*, and this method has been universally applied to the typing of various species. Other commonly used probes are *insertion sequences* (lengths of DNA involved in transposition; see Ch. 6) that may define clonal structures of populations.

Polymerase chain reaction typing

PCR is a technique that allows specific sequences of DNA to be amplified. Multiple copies of regions of the genome defined by specific oligonucleotide primers are made by repeated cycles of amplification under controlled conditions. Such methods can be used to study DNA from any source. Several variations on the PCR theme have been described, and use of these techniques continues to expand and develop. PCR-mediated DNA fingerprinting makes use of the variable regions in DNA molecules. These may be variable numbers of tandem repeat regions or areas with restriction endonuclease recognition sequences. To perform PCR typing, it is necessary to know the sequences of the bordering regions so that specific oligonucleotide primers can be synthesized. Primers may be specific for a known sequence or be random. Random primers are extensively used in the techniques of *random amplification of polymorphic DNA* (RAPD) and *arbitrarily primed PCR* (AP-PCR). Both of these approaches have problems with reproducibility as a result of false priming, faint versus sharp bands and variation in electrophoretic migration of products. Repetitive sequence-based PCR (rep-PCR) indexes variation in multiple interspersed repetitive sequences in intergenic regions dispersed throughout the genome. An automated, standardized rep-PCR system has proven useful for strain typing of a number of species and is reported to give similar discrimination to PFGE (Bacterial Barcodes, Houston, Texas, USA). Amplified fragment length polymorphism is a DNA sequence-based technique that combines restriction endonuclease digestion with PCR. Incorporation of a fluorescent label and the use of a capillary DNA sequencer allows optimal standardization of reproducibility and resolution of single base-pair differences between genomes.

Multilocus sequence typing

This technique indexes allelic variation in several housekeeping genes by nucleotide sequencing rather than indirectly from the electrophoretic mobilities of their gene products, as was the case with its parent technique, multilocus enzyme electrophoresis. Housekeeping genes are not subject to selective forces as are variable genes and they diversify slowly. Multiple genes (usually seven) are employed to overcome the effects of recombination in a single locus, which might distort the interpretation of the relationship of the strains being compared. Multilocus sequence typing (MLST) can rightly be referred to as definitive genotyping as sequence data are unambiguous and databases of allelic profiles of isolates of individual species are accessible via the internet. The level of discri-

mination of MLST depends on the degree of diversity within the population to generate alleles at each locus, but some highly uniform species such as *M. tuberculosis* are not amenable to analysis by the technique. Recently, increased discrimination has been sought in virulence-associated genes necessary for survival and spread of the organism on the basis that these genes are exposed to frequent environmental changes and thus provide a higher degree of sequence variation. Intergenic regions of selected genes are amplified by PCR and a 500-bp internal fragment sequenced to identify allelic polymorphisms.

A variant of MLST termed multilocus restriction typing introduces restriction digestion of amplified housekeeping genes and removes the need for sequencing. The restriction fragment length polymorphisms (RFLPs) can be sorted into type patterns and reveal population structures similar to those with MLST.

Variable number tandem repeat analysis

Variable number tandem repeats (VNTRs) are short nucleotide sequences (20–100 bp) that vary in copy number in

---

KEY POINTS

- Taxonomy is the classification, nomenclature and identification of microbes (algae, protozoa, slime moulds, fungi, bacteria, archaea and viruses). The naming of organisms by genus and species is governed by an international code.
- Bacteria can be separated into two major divisions by their reaction to Gram's stain, and exhibit a range of shapes and sizes from spherical (cocci) through rod shaped (bacilli) to filaments and spiral shapes.
- In clinical practice, bacteria are classified by macroscopic and microscopic morphology, their requirement for oxygen, and activity in phenotypic and biochemical tests.
- Various diagnostic test systems are used to detect specific bacteria in clinical systems, including specific gene probes, reaction with antibodies in ELISA formats, immunofluorescence and, increasingly, PCR-based technology.
- Different bacterial species often exhibit different population structures, highly diverse (panmictic) or relatively uniform (clonal) depending mainly on the frequency of gene recombination (from external sources).
- Typing of bacterial isolates is necessary for epidemiological investigations in outbreaks and for surveillance, and a variety of phenotypic and genetic methods has evolved for the identification of strains.

bacterial genomes. They are thought to arise through DNA strand slippage during replication and are of unknown function. Separate VNTR loci are identified from published sequences and are often located in intergenic regions and annotated open reading frames. Primers are designed to amplify five to eight loci and the products

sequenced to generate a digital profile. VNTR typing is rapid and reproducible, and relatively simple to perform. Improved discrimination may be achieved by identification of more loci but there is debate about their stability over time.

RECOMMENDED READING

Barrow G I, Feltham R K A (eds) 1993 *Cowan and Steel's Manual for the Identification of Medical Bacteria*, 3rd edn. Cambridge University Press, Cambridge

Garrity G M (editor-in-chief) 2005 *Bergey's Manual of Systematic Bacteriology*, 2nd edn. Springer, New York

Kaufmann M E 1998. Pulsed-field gel electrophoresis. In: *Methods in Molecular Medicine*, Vol. 15: *Molecular Bacteriology: Protocols and Clinical Applications*. Woodford N, Johnson A P (eds), pp. 33–50. Humana Press, Totowa, NJ

Murray P R, Baron E J, Jorgensen J H, Pfaller M A, Yolken R H (eds) 2003 *Manual of Clinical Microbiology*, 8th edn. ASM Press, Washington, DC

Owen R J 2004 Bacterial taxonomics: finding the wood through the phylogenetic trees. *Methods in Molecular Biology* 266: 353–384

Spratt B G, Feil E J, Smith N H 2002 Population genetics of bacterial pathogens. In: *Molecular Medical Microbiology*, Sussman M (ed.), pp. 445–484. Academic Press, San Diego

Tenover F C, Arbeit R D, Goering R V et al 1995 Interpreting chromosomal DNA restriction patterns produced by pulsed-field gel electrophoresis: criteria for strain typing. *Journal of Clinical Microbiology* 33: 2233–2239

Van Regenmortel M H V, Fauquet C M, Bishop D H L (eds) 2000 *Virus Taxonomy. Classification and Nomenclature of Viruses*. Academic Press, San Diego

Woese C R 2000 Interpreting the universal phylogenetic tree. *Proceedings of the National Academy of Sciences of the USA* 97: 8392–8396

Internet sites

Genotyping database at Oxford University. http://www.mlst.net

National Center for Biotechnology information for rRNA sequence analysis. http://www.ncbi.nlm.nih.gov/Genbank

Ribosomal differentiation of medical microorganisms for rRNA sequence analysis. http://www.ridom-rdna.de

Sequence retrieval system. http://srs.embl-heidelberg.de:8000/srs5

Universal virus database of the International Committee on Taxonomy of Viruses. http://www.ncbi.nlm.nih.gov/ICTVdb

Woese's work at the National Academy of Sciences. http://www.pnas.org

4 Bacterial growth, physiology and death

M. R. Barer

Most of what we know about bacteria derives from their growth. Their ability to propagate may be seen as a supreme achievement that enables them to attain enormous populations at rates that are breathtaking from a human perspective. These properties underpin their capacity for change by mutation and the rapidity with which some infections develop.

Bacterial growth involves both an increase in the size of organisms and an increase in their number. Whatever the balance between these two processes, the net effect is an increase in the total mass (*biomass*) of the culture. Medical microbiologists have traditionally concentrated on the number of individuals in growth studies. Whether this emphasis on cell number is appropriate remains uncertain; none the less, it will be adopted here, as the number of individual bacteria involved is important in the course and outcome of infections and in the measurement of the effects of antibiotics.

Students of medicine may be surprised and even dismayed to hear that organisms as small as bacteria have a physiology. However, the complement of enzymes and the biochemical and biophysical processes occurring in a prokaryotic cell at any one time represent the product of genetic and biochemical control mechanisms that are every bit as sophisticated and tightly regulated as those in eukaryotic cells. Moreover, the recognition and definition of the mechanisms by which bacteria sense and adapt to nutritional and noxious stimuli in their environments have provided insights that are likely to translate into medically significant advances in the foreseeable future.

In some sense asexual organisms such as bacteria appear to be immortal, but bacterial death or loss of viability occurs in many natural settings. This has practical consequences, as only viable bacteria can initiate infections and most microscopic, molecular and immunological detection methods do not differentiate between live and dead organisms. Of course, we often need to assess the lethal effects of antibiotics and processes aimed at *sterilization*, *disinfection* and *antisepsis*. The practical approach to assessing the effects of antibiotics is introduced in Chapter 5, but the principles of sterilization and disinfection are introduced here.

Although this chapter discusses growth and physiology only from a bacteriological perspective, some of the principles are also applicable to fungi, particularly yeasts. The central difference between their growth is that cell division is generally achieved in bacteria by *binary fission* to produce identical offspring that cannot be distinguished as parents and progeny, whereas fungi divide by budding in the case of yeast growth and hyphal septation in the mould form. In contrast, the principles of sterilization and disinfection refer to all infective agents. Their application is considered further in Chapter 68.

BACTERIAL GROWTH

When placed in a suitable nutritious environment and maintained under appropriate physical and chemical conditions, a bacterial cell begins to grow; when it has manufactured approximately twice the amount of component materials that it started with, it divides. The range of specific components that define 'suitable' and 'appropriate' for all known bacteria (and *Archaea*) is so broad that it actually defines the global biosphere (those environments that can sustain life), and includes temperatures and pressures present at the opening of hydrothermal vents on the ocean floor to the outer reaches of the atmosphere. Although these conditions do not regularly occur in man, they serve to illustrate that no part of the body or medical device with which it may come in contact is too difficult for bacteria to colonize and that bacteria may lurk in surprising environmental niches. Conversely, the conditions required for some organisms to grow are so precise that, so far, we have not been able to reproduce them in artificial laboratory media. This applies to some well known organisms such as the agents of leprosy and syphilis, but also to many other potential pathogens about which we are beginning to learn through molecular methods that do not depend on growth. In fact, it is estimated that we have not yet isolated more than 1% of all the bacterial species that exist, and it is almost certain that there are many medically important organisms among the 'as yet uncultivated' micro-organisms.

As the central technique in bacteriology, growth in the laboratory has been used to serve many different purposes. From the clinical perspective, growth is used for detection and identification, and for the assessment of antibiotic effects, whereas scientific and industrial objectives are often served by growth in bulk to obtain sufficient biomass for detailed biochemical analysis and to produce the desirable products of the brewing and biotechnology industries.

Types of growth

In the laboratory, bacterial growth can be seen in three main forms:

1. By the development of *colonies*, the macroscopic product of 20 to 30 cell divisions of a single cell.
2. By the transformation of a clear broth medium to a turbid suspension of 10^7–10^9 cells per millilitre.
3. In *biofilm* formation, in which growth is spread thinly (300–400 μm thick) over an inert surface and nutrition obtained from a bathing fluid.

In natural systems only biofilms, such as those that develop on the surfaces of intravascular cannulae, appear to function in a manner comparable to biofilms produced in the laboratory, whereas colonies, the other form of *sessile* growth, rarely reach macroscopic dimensions. Turbid liquid systems caused by *planktonic* growth of a single organism are also a rarity in nature. Single organism infections affecting normally sterile sites in the body are one exception to this, whereas most natural microbial communities are complex assemblies of micro-organisms competing, and in many cases co-operating, to exploit the local resources. However, in spite of these unrepresentative features, pure growth of single organisms in *monocultures* to produce macroscopic colonies or high cell densities in broth offer great practical advantages and remain central techniques.

Growth phases in broth culture

Bacterial growth in broth has been studied in great detail and has provided a framework within which the growth state or growth phase of any given pure culture of a single organism can be placed; these phases are summarized in the idealized *growth curve* shown in Figure 4.1. When growth is initiated by inoculation into appropriate broth conditions, the number of cells present appears to remain constant for the *lag phase*, during which cells are thought to be preparing for growth. Increase in cell number then becomes detectable, and its rate accelerates rapidly until it is established at the maximum achievable rate for the available conditions. This is known as the *exponential phase*, because the number of

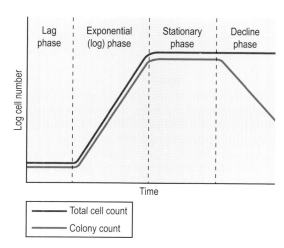

Fig. 4.1 Phases of growth in a broth culture.

cells is increasing exponentially with time. To accommodate the astronomic changes in number, the growth curve is normally displayed on a logarithmic scale, which shows a linear increase in log cell number with time (hence the older term, *log phase*). This log-linear relationship is sufficiently constant for a given bacterial strain under one set of conditions that it can be defined mathematically, and is often quoted as the *doubling time* for that organism. Doubling times have been measured at anything between 13 min for *Vibrio cholerae* and 24 h for *Mycobacterium tuberculosis*. On this basis it is not surprising that cholera is a disease that can kill within 12 h, whereas tuberculosis takes months to develop. A further consequence is that, when specimens are submitted to diagnostic laboratories for the detection of these organisms by culture, a result is usually available for *V. cholerae* the next day, whereas several weeks are required for conventional culture of *M. tuberculosis*.

It is often difficult to grasp fully the scale of exponential microbial growth; the message may be strengthened by considering that the progeny of a lecture theatre containing 150 students would exceed the global population of humanity (6×10^9) within 8.5 h if they were able to breed like *Escherichia coli*!

Exponential growth cannot be sustained indefinitely in a closed (*batch*) system with limited available nutrients. Eventually growth slows down, and the total bacterial cell number reaches a maximum and stabilizes. This is known as the *stationary* or *post-exponential phase*. At this stage it becomes important to know what method has been used to determine the growth curve. If a direct method that assesses the total number of cells present is used then the count remains constant. Such methods include counting cells in a volumetric chamber observed by microscopy, electronic particle counters and measurement of turbidity. If, however, the growth potential of the individual cells present in the culture is assessed by taking regular samples, making ten-fold dilutions of these and inoculating them on to agar, the number of *colony-forming units* (cfu) per unit volume can be determined at each sample time. Although such cfu counts closely parallel the results obtained by direct counting methods in the exponential and early stationary phases, a divergence begins to emerge towards the end of the latter; the total cell number remains constant whereas the colony count declines. This marks the beginning of the final, *decline phase*, in the sequence of growth states that can be observed in broth. The discrepancy between the total and cfu counts is conventionally held to represent the death of cells because of nutrient exhaustion and accumulation of detrimental metabolic end-products. However, there is some doubt concerning this interpretation (see below).

The study of bacterial growth in broth provides a valuable point of reference to which practical, experimental and routine diagnostic procedures are often related. For example, the length of the lag phase and rates of exponential growth in different circumstances are used to make predictions and contribute to safety standards for storage in the food industry. An important feature to emerge is that cultures inoculated with cells prepared at different stages in the growth curve yield different results. The exponential phase is the most reproducible and readily identified, and is therefore used most frequently. It can be extended in an open system known as *continuous culture* using a *chemostat* in which cells of a growing culture are harvested continuously and nutrients replenished continuously. Chemostat studies have provided very detailed information on the chemistry of microbial growth and the way in which different organisms convert specific substrates into biomass. The extraordinary efficiency of this process has made natural and genetically manipulated microbes a powerful resource for the biotechnology industry.

In contrast to growth in broth, far less is known about the state of the bacteria in a mature macroscopic colony on an agar plate. Such a colony presents a wide range of environments, from an abundance of oxygen and nutrients at the edge to almost no oxygen or nutrients available to cells in the centre. It is likely that all phases of growth are represented in colonies, depending on the location of a particular cell and the age of the culture. Although in practice colonies can be used reliably to inoculate routine tests of antimicrobial susceptibility in clinical laboratories, they cannot be considered a defined starting point for experimental work because they comprise such a heterogeneous population of cells. In fact, colonies are complex and dynamic communities in which cells at different locations can show startlingly different phenotypes. In spite of its complexity, the capacity for and quality of colonial growth of specific organisms on specialized media is central to the laboratory description of medically important bacteria.

MEDIA FOR BACTERIAL GROWTH

The media used in a medical diagnostic bacteriology laboratory have their origins, for the most part, back in the 'golden age of bacteriology' in the late nineteenth and early twentieth centuries. A vast amount of experience and knowledge has accrued from their use and, apart from better standardization and quality control in their production, little has changed in their basic design. The objectives of early medium design were to grow pathogenic bacteria, separate them from other organisms present in samples and, ultimately, differentiate their phenotypic properties so that they could be identified. A critical development was the introduction of solidifying agents, most particularly the largely indigestible poly-saccharide extract of seaweed known as agar. Alternative solidifying agents include gelatine and egg albumen. Before the development of solid media, pure cultures could be achieved only by dilution of inocula so that only one growing cell or clump of cells was present at the initiation of growth, a very laborious and unreliable procedure. In contrast, solid media in Petri dishes provided a growth substrate on to which mixed cultures could be inoculated and, provided the population density could be made low enough to allow development of well separated colonies, the different organisms present could be differentiated and subsequently separated into pure cultures.

Media used for isolation and identification of pathogens

The central features of media in medical bacteriology are:

1. a source of protein or protein hydrolysate, often derived from casein or an infusion of brain, heart or liver obtained from the nearest butcher
2. control of pH in the final product (after sterilization)
3. a defined salt content.

Early media often included blood or serum in an attempt to reproduce nutritional features present in the human body. Growth of some pathogens was found to be dependent on such supplements, and it was recognized that these relatively *fastidious* or *nutritionally exacting* organisms were dependent on *growth factors*. The identity of many of the growth factors is now known (e.g. haemin and several coenzymes), but blood often remains their most convenient source.

Selective and indicator media

Tremendous ingenuity has gone into designing growth media that provide information relevant to patient management as early as possible. There are two main approaches, both of which depend on adding supplements to the basal medium. *Selective media* contain substances such as bile salts or antibiotics that inhibit the growth of some organisms but have little or no effect on the organisms for whose isolation they were designed. They are essential for samples containing a normal microbial flora such as faeces. The inclusion of components or specific reagents that show whether the bacteria possess a particular biochemical property characterizes an *indicator medium*. Such media are critical to the rapid presumptive identification of isolates. Combinations of selective and indicator supplements in agar media have led to formulations with some remarkably elegant differential properties that effectively colour-code the colonies according to their biochemical properties and restrict growth to a desired range of organisms. Broth indicator media tend to be much simpler, as they generally require a pure inoculum of a single organism and reveal only one property per formulation. Broth media with selective properties are usually referred to as *enrichment media* as they change the balance of organisms inoculated in favour of the desired range of organisms, thereby enriching them.

Media for laboratory studies

Most of the objectives of a clinical diagnostic laboratory can be fulfilled with the range of media outlined above. However, the composition of these media is not defined, and this poses problems for some investigations, including the detailed analysis of antibiotic action. Wherever possible, such investigations are based on a *defined* or *synthetic medium* where every chemical component is carefully regulated. In genetic experiments use is often made of a *minimal medium* in which every component is required for the growth of the organism under investigation, so that if one component is removed growth cannot occur. Minimal media also prevent the growth of mutants that have additional nutritional requirements to those of the parent strain. For some organisms, particularly those that can grow outside the human body, minimal media may comprise as little as an ammonium salt to provide nitrogen, a carbon source, which in some cases can be as simple as methane or carbon monoxide, trace amounts of iron and other essential elements, and pH adjustment to within an appropriate range. Defined and minimal media generally have to be developed for small groups of closely related organisms and should not be used for other organisms.

Relatively well defined media are preferred, even for routine antibiotic tests, because quantitative aspects of bacterial biochemistry, growth and susceptibility to noxious stimuli can be influenced substantially by minor changes in medium composition. The use of fully defined media has underpinned almost all of what we know about bacterial physiology. Rather curiously, however, it is well recognized that defined media are often suboptimal for the recovery of bacteria from environments in which they have been stressed. This may reflect the support provided to injured bacteria by complex media. Defined media can really be optimized only for bacteria in a single physiological state, whereas complex media have greater potential to cope with the diversity of states present in natural samples.

BACTERIAL PHYSIOLOGY

The complement of processes that enable an organism to occupy and thrive in a particular environment places certain requirements on its physiology. Traditional descriptions of bacterial groups emphasize features that place a microbe in particular ecological niches. Thus we have *acidophiles* for organisms such as *Lactobacillus* spp. that grow at lower pH levels than most other organisms, and *halophiles* for organisms that grow at high salt concentrations. The environments that can be colonized by a pathogen are, of course, critical in determining its reservoirs and potential modes of transmission. More recently it has been recognized that individual bacteria are not restricted to a single physiological state. Rather, they respond to environmental stimuli and undergo *adaptive responses* that confer improved capacity for survival in adverse conditions. All of these properties sustain the *viability* of the organism. However, it has become apparent that our ability to measure viability by conventional means may be inadequate.

The specific means by which a particular organism obtains energy and raw materials to sustain its growth (its nutritional type) and the physical conditions it requires reflect its fundamental physiological characteristics. Placing an organism into the groups defined by these characteristics is an important step in its conventional classification.

Nutritional types

Traditionally, all living organisms have been divided into two nutritional groups: *heterotrophs* and *autotrophs*. The former depend on the latter to produce organic molecules by fixing carbon dioxide, predominantly by photosynthesis. Bacterial metabolism is now recognized to be so diverse that it cannot be encompassed by these two terms. Three basic features are used in the present terminology: the *energy source*, the *hydrogen donors* and the *carbon source*.

Energy for adenosine triphosphate (ATP) synthesis may be obtained from light in a *phototrophic* organism and from chemical oxidations in the case of a *chemotrophic* organism. The hydrogen donor type characterizes an organism as an *organotroph* if it requires organic sources of hydrogen and as a *lithotroph* if it can use inorganic sources (e.g. ammonia or hydrogen sulphide). Finally, the terms autotroph and heterotroph are reserved for the carbon source; the former can fix carbon dioxide directly whereas the latter require an organic source. In general, only the energy and hydrogen donor designations are referred to routinely by combining the two terms. Hence we refer to *chemo-organotrophs* (the vast majority of currently recognized medically important organisms) and *chemolithotrophs* (e.g. some *Pseudomonas* spp.). Surprisingly, there are even some *photolithotrophs* with medical significance; the cyanobacteria are now known to produce many toxins that can affect man.

Physical conditions required for growth

All living organisms use oxidation to transfer energy to compounds that participate in their internal biochemical and biophysical processes. Oxidation of a molecule is equivalent to the removal of hydrogen, and requires another molecule to receive electrons in the process. In *aerobic* respiration the final electron recipient in the oxidation process is molecular oxygen (i.e. O_2), whereas under *anaerobic* conditions (in the absence of oxygen) most medically important organisms use an organic molecule as the final electron recipient, and the oxidative process is referred to as *fermentation*. There are also some forms of anaerobic respiration that use inorganic electron acceptors such as nitrates. Respiration in this context is generally used to denote involvement of a membrane-associated electron transport chain in the oxidation. In the early period of development of life on Earth there was no oxygen in the atmosphere; thus, at this time, all bacteria were *anaerobes*. Subsequently, following the development of photo-autotrophic organisms, atmospheric oxygen became abundant, and organisms capable of using oxygen evolved.

Although aerobic metabolism is a more efficient means of obtaining energy than anaerobiosis, it is not without its cost. Some oxidation–reduction (redox) reactions occurring in the presence of oxygen commonly result in the formation of the reactive superoxide (O_2^-) and hydroxyl (OH^-) radicals as well as hydrogen peroxide (H_2O_2), all of which are highly toxic. To cope with this, aerobic organisms or *aerobes* have developed two enzymes that detoxify these molecules. *Superoxide dismutase* converts superoxide radicals to hydrogen peroxide ($2 O_2^- + 2H^+ \rightarrow H_2O_2 + O_2$), whereas *catalase* converts hydrogen peroxide to water and oxygen in the

reaction $2 H_2O_2 \rightarrow H_2O + O_2$. Possession or lack of these enzymes has the important consequence of defining the atmosphere necessary for growth and survival of different organisms. Moreover, when produced in large amounts, the enzymes also provide protection for pathogenic organisms against the reactive oxygen intermediates deliberately produced as a defence mechanism by phagocytic cells.

Growth atmosphere

These oxygen-related features underpin the major practical grouping of bacteria according to their atmospheric requirements (Table 4.1). Thus, *strict* or *obligate aerobes* require oxygen, usually at ambient levels ($\approx20\%$), and *strict* or *obligate anaerobes* require the complete absence of oxygen. Many organisms exhibit intermediate properties: *facultative anaerobes* generally grow better in oxygen but are still able to grow well in its absence; *micro-aerophilic* organisms require a reduced oxygen level ($\approx5\%$); *aerotolerant anaerobes* have a fermentative pattern of metabolism but can tolerate the presence of oxygen because they possess superoxide dismutase. Many medically important organisms are facultative anaerobes. There is a mixture of aerobic and anaerobic micro-environments in the human body, and the capacity to replicate in both is clearly advantageous. For obvious reasons, strict anaerobes are particularly associated with infection of tissues where the blood supply has been interrupted.

Among the various physical requirements for the growth of different bacterial groups, atmosphere assumes particular importance because, in practice, agar cultures from most clinical specimens are set up aerobically and anaerobically. Thus, when growth is first inspected after overnight incubation, the isolates can readily be differentiated into strict aerobes, anaerobes and facultative anaerobes according to the conditions under which they have grown. Various atmospheric conditions can also be obtained in broth media. If the medium is unstirred, strict aerobes tend to grow on the surface, micro-aerophiles just under the surface and anaerobes in the body of the medium away from the surface. Growth of anaerobes is often improved by the addition of a reducing agent such as cysteine or thioglycollate to mop up any free oxygen.

Growth temperature

The other significant physical condition for bacterial growth from the medical perspective is temperature (see Table 4.1). Pathogens that actually replicate on or in the human body must be able to grow within the temperature range of 20–40°C, and are generally referred to as *mesophiles*. Organisms that can grow outside this range

Table 4.1 Key descriptive terms used to categorize bacteria according to their growth requirements

Descriptive term	Property	Example
Growth atmosphere		
Strict (obligate) aerobe	Requires atmospheric oxygen for growth	*Pseudomonas aeruginosa*
Strict (obligate) anaerobe	Will not tolerate oxygen	*Bacteroides fragilis*
Facultative anaerobe	Grows best aerobically, but can grow anaerobically	*Staphylococcus* spp., *Esch. coli*, etc.
Aerotolerant anaerobe	Anaerobic, but tolerates exposure to oxygen	*Clostridium perfringens*
Micro-aerophilic organism	Requires or prefers reduced oxygen levels	*Campylobacter* spp., *Helicobacter* spp.
Capnophilic organism	Requires or prefers increased carbon dioxide levels	*Neisseria* spp.
Growth temperature		
Psychrophile	Grows best at low temperature (e.g. <10°C)	*Flavobacterium* spp.
Thermophile	Grows best at high temperature (e.g. >60°C)	*Bacillus stearothermophilus*[a]
Mesophile	Grows best between 20 and 40°C	Most bacterial pathogens

[a]Not a pathogen; its spores are very heat resistant and are used for testing the efficiency of heat sterilization.

are either *psychrophiles* (cold loving) or *thermophiles* (heat loving). The former may be capable of growth in food or pharmaceuticals stored at normal refrigeration temperatures (0–8°C), whereas the latter can be a source of proteins with remarkable thermotolerant properties, such as *taq* polymerase, the key enzyme used in the polymerase chain reaction. Organisms such as the leprosy bacillus that prefer lower growth temperatures are often associated with skin and superficial infections, whereas organisms that grow in the colon (often a few degrees warmer than normal body temperature) can grow well up to 44°C.

Extremophiles

Some bacteria require ostensibly bizarre physical conditions for growth. For example, barophiles isolated from the ocean floor may require enormous pressures before they can replicate. Such organisms are often referred to as *extremophiles*. The properties of these organisms serve to remind us that microbes have the potential to occupy any environmental niche where energy and nutrition are available. It should be noted that most extremophiles actually turn out to belong to the Archaea (see Ch. 2).

Bacterial metabolism

Although some bacteria are able to obtain their resources for growth in ways that seem alien to us, the core of their metabolism is essentially very similar to that of mammalian cells. The basic details of glycolysis, the tricarboxylic acid cycle, oxidative phosphorylation, ATP biosynthesis and amino acid metabolism are constant. Variations in the pathways that feed into and flow from these core processes are readily detected by what are

loosely termed *biochemical tests* in medical laboratories. These detect traits such as the ability to use individual carbohydrate sources to produce acid and the possession of specific enzymes.

The common nature of central catabolic and anabolic pathways in bacteria and higher organisms reflects the economy of biology and evolution. Processes that work well cannot be outcompeted and tend to be preserved in the genetic stock. Thus, many of the specific enzymes involved in bacterial metabolism show remarkable levels of conservation in their amino acid sequences across very substantial distances in evolutionary terms. DNA sequencing has enabled the identification of *molecular families* of proteins with a common evolutionary origin. In addition to the metabolic enzymes, it has been recognized that many transport proteins responsible for importing and exporting specific substrates into and out of the bacterial cytoplasm are closely related in their structure and mode of function to those present in mammalian cells. Of course, because bacteria generally have only one cell compartment in which to operate, the location of these proteins is often different; for example, as they have no mitochondria, the cytoplasmic membrane contains the components of the electron transport chain, and the proton gradient across the inner mitochondrial membrane is generated across the cytoplasmic membrane instead. This feature actually means that bacteria can perform some energy-requiring processes at the cell surface, notably flagellar rotation (motility), by directly exploiting the proton gradient rather than consuming ATP.

Aside from its role in identification and intrinsic biological interest, bacterial metabolism has real consequences for humans. In direct terms, the resident microflora have consequences in human health and disease. For example, the bacteria in dental plaque produce acid when presented with certain carbohydrate sources, and this acid is responsible for tooth decay; on the positive side, bacteria in the intestines deconjugate bile salts and thereby contribute to the enterohepatic circulation. It seems likely that the importance of such bioconversions will be recognized increasingly in the future. In particular, the role of bacteria in recovering nitrogen excreted into the colon in marginal human nutritional states and metabolic activity leading to the formation of carcinogens or other biologically active molecules are both areas where there is much room for further work.

Human beings are also indirectly affected by microbial metabolism. At one level, the chemistry of our environment has been shaped extensively by microbes; the original development of oxygen in our atmosphere, the availability of elemental sulphur and the flow of nitrogen are all critically dependent on microbial metabolism. Exploitation of microbial metabolism in industry has, of course, given us ethanol, and many of the other alcohols and acids that result from fermentation have commercial value. Finally, bacteria have been used to combat the deleterious effects of environmental pollution in the process referred to as *bioremediation*.

Adaptive responses in bacteria

The extent to which bacteria respond to environmental stimuli was originally recognized by monitoring gross phenotypic, biochemical and behavioural changes. Much of the genetic basis for how bacteria change their phenotypes was established in the 1960s and 1970s following on from the paradigm established for β-galactosidase regulation in *Esch. coli* by Jacob and Monod. The scale and rapidity (major changes can be seen in seconds) of bacterial responses became apparent through the 1980s and 1990s as the use of global analytical approaches that attempt to characterize the instantaneous expression of every gene the organism carries became established. At the translational level, the use of two-dimensional gel electrophoresis has underpinned the so-called *proteomic* approach. This technique reveals and separates most of the several hundred proteins that are being synthesized by a pure culture at a particular time. The catalogue of different proteins detected represents those proteins that the organism requires to function in the circumstances from which the sample was drawn. Assays of this type have shown that different sets of proteins are made in the exponential and stationary phases of the growth cycle and, indeed, in response to almost any environmental change. This finding underpins the recognition of just how different the phenotype of a single organism can be in different physiological states and reinforces the need to define the inoculum used in laboratory experiments. More recently the development of DNA arrays has enabled

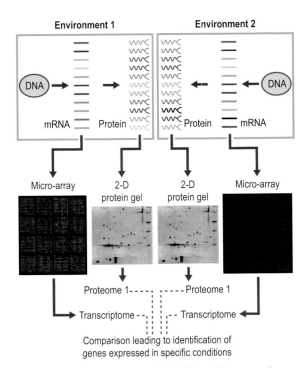

Fig. 4.2 Global strategies for identifying differentially expressed genes. Analysis at both mRNA and protein levels is preferred as there may be differences between the two. Individual spots on the two-dimensional (2-D) gels may be identified as specific gene products by mass spectrometry. Transcriptome analysis requires a representation of all the genes concerned in a DNA array and therefore needs prior knowledge (ideally a complete sequence) of the genome of the organism to be tested. These two global approaches provide a broad picture of the physiology of the organism under study.

global analysis of responses at the transcriptional level by detecting messenger RNA (mRNA) molecules relating to every gene in the organism in a single analysis. The complement of RNA species present in an organism at a given time is referred to as the *transcriptome*. The basic features of the comparative protein and mRNA analyses are outlined in Figure 4.2.

The comprehensive analyses achieved by transcriptome and proteome analyses followed on from the recognition that global genome analyses and comparisons or *genomics* (see Ch. 6) have the potential to explain many – some would say most – biological and medical phenomena. The complexities linking genotype to phenotype remain overwhelming in most instances; none the less, we have now entered an era where global analyses of mRNA, proteome and metabolic function (recently termed 'metabolomics') are being addressed enthusiastically in an integrative computational approach pooling data from different analyses in what has been termed the 'systems biology' approach.

The effects of specific sublethal but noxious stimuli on gene expression are the subject of intense current study. Each different stimulus leads to an adaptive *stress response*, which is to some extent specific to the stimulus applied. Heat shock (the effects of raising temperature to 45°C and above for a few minutes) has been studied most extensively. The newly synthesized proteins elicited in this response are referred to as *heat shock proteins*. When the amino acid sequences of the principal heat shock proteins were determined, they were found to belong to a molecular family now recognized in all prokaryotic and eukaryotic cells. Apart from their role in improving the ability of bacteria to survive heat shock, these proteins, by virtue of their similarity to analogous host cell antigens, seem to be involved in initiating autoimmune damage and immune dysfunction. A very important feature of the stress response in bacteria is that many of the stimuli used are prominent aspects of the stresses applied by the human immune system to an invading pathogen. Thus, acid stress is provided by the stomach and the hostile environment of phagolysosomes includes both oxidative and pH stress.

The information built up from studying stress responses has made it possible to identify sets of proteins that are made in response to several different stresses and those that appear exclusive to one stress. Together with other approaches, this has allowed recognition of *global regulatory systems* or *networks* within bacteria that are responsible for differential gene expression under different circumstances. The hierarchy of specific control mechanisms involved has spawned two important new terms, *stimulon* and *regulon*. A stimulon denotes all the genes whose expression is increased or decreased by a specific external stimulus, whereas a regulon refers to all the genes under the influence of a specific regulatory protein. A regulon may affect several operons (see Ch. 6), and there may be many regulons in one stimulon.

Regulatory networks have been identified in almost every area of bacterial physiology. Thus, in addition to the stimuli cited above, osmotic stress, cold shock, nutrient limitation (separate responses for carbon, nitrogen and phosphate), anaerobic and many other stimulons are recognized. These control systems are responsible for making sure the organism synthesizes only those proteins appropriate to its current circumstances. A particularly important medical example of this is the regulation of proteins concerned with an organism's progress in an infection (virulence factors). Equally important from the scientific perspective is the recognition that chemicals secreted by an organism can themselves act as regulatory stimuli to individuals of the same species in a way analogous to the pheromones released by insects.

Although it is still important to recognize that different organisms are particularly adapted to special environmental niches with descriptive terms such as mesophile, acidophile and halophile, the discovery of adaptive responses in bacteria has pushed us into an uncertain period where much of what has been established about the tolerance of micro-organisms to noxious stresses will have to be re-examined. Furthermore, as the extent to which bacteria modulate their phenotype according to their circumstances is now clear, the need for caution in concluding that any property detected in the laboratory is significant in a natural infection is unavoidably obvious.

Bacterial defence against noxious chemicals

The features outlined above all contribute to the well-being of bacteria. In the natural world micro-organisms encounter many chemicals that could cause their destruction and, in their 3.5 billion years on earth they have evolved numerous protective mechanisms. In clinical practice these are recognized as *biochemical mechanisms of antibiotic resistance*, and four basic categories are recognized:

1. *Preventing access*: achieved by low cell envelope permeability or efflux pumps affecting the chemical concerned.
2. *Destruction*: achieved by enzymes that modify or degrade the chemical.
3. *Lack of target*: many chemicals that damage bacteria work through specific targets. The target may be absent or be altered by mutation (see Ch. 6).
4. *Bypass of target*: in some cases an alternate or modified pathway can be used.

These mechanisms may be *intrinsic* to the organism concerned or they may be *acquired* through mutation or gene transfer (see Ch. 6).

Bacterial viability

A central feature of the general and adaptive physiology of bacteria is the capacity to preserve the viability of a particular organism. There is, however, a persistent problem – how do we define viability in practical terms? Traditionally, the operational definition of the capacity of a cell to form a colony on an appropriate agar medium (the colony or cfu count) has been almost universally accepted. It is also often expressed as the proportion of cells within a population that are capable of forming colonies. This has been used extensively in recognizing the *cidal* (lethal) and *static* (growth inhibitory) activities of antibiotics. In the former case, cfu counts decline, and in the latter they remain constant. However, it must be emphasized that viability is not a clearly measurable property. At the individual level it expresses the expectation that, in a suitable environment, a particular cell has the capacity to grow and undergo binary fission and that its progeny will have the same potential. The key assumption is that colony counts provide an accurate measure of viability.

The central problem can be stated as follows: it is self-evident that if a bacterial cell produces a colony it must have been viable, but to what extent is it true that a cell that fails to do this is non-viable or dead? Immediately contradictions to this proposal can be identified. The bacterial pathogens, such as *Mycobacterium leprae* and *Treponema pallidum*, which cannot be induced to form colonies on available agar media, are clearly viable. Similarly, all the 'as yet uncultivated organisms' (possibly as many as 99% of all bacterial species) are clearly able to propagate themselves. They simply have not been sporting enough to do it on our laboratory media. A further exception is the phenomenon of bacterial recovery from injury (e.g. cold or osmotic shock) in which colony counts can be shown to rise in the absence of cell division.

It is possible that some organisms that are readily cultivable may be able to switch to a physiological state in which they cannot be induced to form colonies. The current popular terminology for cells in this putative state is *viable but non-culturable* (VBNC).

Epidemiological and laboratory evidence provide some support for the existence of a VBNC state. In particular, the occurrence of several infectious diseases acquired from environmental sources, notably cholera, is at variance with our ability to recover the causal organisms from the implicated source. Environmental studies have demonstrated cells with immunological properties compatible with those of the cholera vibrio while failing to recover the organism in culture, and laboratory studies have indicated that the organism can persist in a non-culturable form. There is also evidence that non-culturable forms may revert to their 'normal' culturable state.

A major attraction of the VBNC hypothesis is that it may resolve a number of important mysteries in medical microbiology. In general, these are situations in which we know the organism must be present but are unable to culture it. This is particularly so with diseases such as tuberculosis that have latent phases.

The most significant problem for the VBNC hypothesis is that it has not been defined in physiological, biochemical or genetic terms. On the face of it, one might expect transition to the VBNC state to result from an adaptive response such as those described in the previous section. Alternatively, transition might be the result of a programme of gene expression such as that observed in spore formation or starvation. From this standpoint the stationary and decline phases in the growth cycle outlined above (when spore formation is induced in sporulating bacteria) may represent the initiation of and transition to a non-culturable phase rather than loss of viability (the traditional view). In spite of their popularity, these ideas

must presently be viewed as interesting speculations for which there is circumstantial but no conclusive evidence.

Measurement of viability has been of great practical value in medical microbiology. Colony counts performed to investigate the action of antibiotics and other disruptive influences such as heat, and those performed at different stages during experimental infections in animal models of human infections, have provided a wealth of valuable information. Moreover, there is no reason to doubt that this approach will continue to be extremely useful.

None the less, it is necessary to maintain a clear view of the limitations of bacterial culture as a measure of the organisms present in a sample and of their viability. Studies often use the term 'viability' when in fact growth on agar or in broth was measured, and confusion would be prevented if the terms 'culturability' and 'colony counts' were used instead.

We are now entering an era in which many diagnostic and investigational techniques may be replaced by molecular detection procedures. The fact that signals based on such techniques may come from culturable, dead and potentially VBNC cells should be recognized. Unravelling these three possibilities presents ample challenge for medical and non-medical microbiologists alike.

Bacterial death

Notwithstanding the problems outlined in the previous section, the ability to recognize and quantify bacterial death is of great practical significance in the practice of medicine. At present, except in highly defined circumstances, the cfu count remains the cornerstone for such measurements. In natural systems where no actively noxious environmental conditions pertain, if bacterial growth ceases, as in the stationary phase described above, after a variable period of time depending on the conditions and the organism concerned, then cfu counts begin to decline. In some cases this may lead to complete loss of viability, whereas in others a stable but lower cfu count is established. For example, *Esch. coli* appears to survive indefinitely in buffered salt solutions, the constant lysis of dying cells apparently providing for a balancing level of cell replication. Even after adaptation to starvation or other conditions leading to stasis, the rate at which viability is lost seems to follow a well defined pattern. Cells in stasis are clearly getting older and this provides a bacterial correlate of senescence. Although the study of bacterial cell senescence is relatively new, it is emerging that cumulative oxidative damage to cell proteins and other key macromolecules is one critical determinant of survival. This observation fits very well with the observation that one can often recover higher cfu counts of stressed facultative organisms on media containing catalase or other reagents that provide protection against reactive oxygen intermediates or following incubation under micro-aerophilic conditions.

It is not widely appreciated that many (probably all) bacteria carry genes encoding for programmed cell death. While several mechanisms are involved, these are distinct from the process of apoptosis that occurs in eukaryotic cells. These systems were first identified as toxin–antitoxin pairs functioning to maintain particular genes in the bacterial cell. However, it now seems likely that their occurrence cannot be explained solely on this basis. It should be noted that the activation of latent prophages (see Ch. 6) constitutes another endogenous mechanism by which bacteria can initiate their own demise.

In addition to killing bacteria with antibiotics, medical practice is frequently concerned with decontaminating locations and materials that have been in contact with infectious patients. Moreover, the safe practice of surgery, parenteral administration of therapy, and the preparation of media and sampling materials for bacteriological studies all require the reduction or complete elimination of bacteria from key locations and devices. Although the methods applied to remove live bacteria may be checked with tests of biological efficacy, the relatively predictable rate of decline achieved with specific methods and target organisms enables safe practice. Because the cfu count is relatively convenient, it has been used in the establishment of most methodologies. However, removal or destruction of all infective agents is necessary to achieve sterility, and tests directed to all of these are required to some extent in establishing safe practice.

STERILIZATION AND DISINFECTION

Key definitions

Sterilization. The inactivation of all self-propagating biological entities (e.g. bacteria, viruses, prions) associated with the materials or areas under consideration.

Disinfection. The reduction of pathogenic organisms to a level at which they no longer constitute a risk.

Antisepsis. Term used to describe disinfection applied to living tissue such as a wound.

Methods used in sterilization and disinfection

In practice, all processes of sterilization have a finite probability of failure. By convention, an article may be regarded as sterile if it can be demonstrated that there is a probability of less than one in a million of there being viable micro-organisms on it. As will be seen below, the level of microbial killing achieved by applying a particular method is dependent on the intensity with which the method is applied and its duration. Five main approaches are used.

Heat. The only method of sterilization that is both reliable and widely applicable is heating under carefully controlled conditions at temperatures above 100°C to ensure that bacterial spores are killed. There is some concern that even this temperature is insufficient to destroy prions. Shorter applications of lower temperatures, such as in pasteurization can effectively remove specific infection hazards.

Ionizing radiation. Both β (electrons) irradiation and γ (photons) irradiation are employed industrially for the sterilization of single-use disposable items such as needles and syringes, latex catheters and surgical gloves, and in the food industry to reduce spoilage and remove pathogens. Ultraviolet irradiation can be used to cut down the level of contamination, but is generally too mild to achieve sterility.

Filtration. Filters are used to remove bacteria and all larger micro-organisms from liquids that are liable to be spoiled by heating, for instance blood serum and antibiotic solutions in which contamination with filter-passing viruses is improbable or unimportant. Industrial scale filtration is used widely to reduce bacterial load and remove cysts of protozoa that are not killed by chlorination in the production of drinking water.

Gaseous chemical agents. Ethylene oxide is used mainly by industry for the sterilization of plastics and other thermolabile materials that cannot withstand heating. Formaldehyde in combination with subatmospheric steam is used more commonly in hospitals for reprocessing thermolabile equipment. Both processes carry toxic and other hazards for the user and the patient. Formaldehyde vapour on its own is used widely to decontaminate rooms and laboratory equipment.

Liquid chemical agents. Use of liquids such as glutaraldehyde is generally the least effective and most unreliable method. Such methods should be regarded as 'high-grade disinfection' only, to be applied when no other sterilization method is available, for example for heat-labile fibreoptic instruments such as flexible endoscopes. Various chemicals with antimicrobial properties are used as disinfectants. They are all liable to be inactivated by excessive dilution and contact with organic materials such as dirt or blood, or a variety of other materials. Nevertheless, they may provide a convenient method for environmental disinfection and other specific applications.

Choice of method

The choice of method of sterilization or disinfection depends on:

- the nature of the item to be treated
- the likely microbial contamination
- the risk of transmitting infection to patients or staff in contact with the item.

Choice is based on an assessment of risk according to different categories of patient (e.g. immunocompromised), the equipment involved and its application. The selection of sterilization, disinfection or simple cleaning processes for individual items of equipment and the environment should be agreed as part of the infection control policy of a hospital (see Ch. 68). The preferred option wherever possible, for both sterilization and disinfection, is heat rather than chemicals. This relates not only to the antimicrobial efficacy but to safety considerations, which are more difficult to control in some chemical processes. Wherever chemicals are to be used for disinfection and sterilization, the safety of persons involved directly or indirectly in the procedure must be considered. It should be remembered that all sterilizing and disinfecting agents have some action on human cells. No method should be assumed to be safe unless appropriate precautions are taken.

Measurement of microbial death

Every method used must be validated to demonstrate the required degree of microbial kill. With heat sterilization and irradiation, a biological test may not be required if the physical conditions are sufficiently well defined and controlled.

When micro-organisms are subjected to a lethal process, the number of viable cells decreases exponentially in relationship to the extent of exposure. If the logarithm of the number of survivors is plotted against the lethal dose received (e.g. time of heating at a particular temperature), the resulting curve is described as the *survivor curve*. This is independent of the size of the original population and is approximately linear. The linear survivor curve is an idealized concept and, in practice, minor variations, such as an initial shoulder or final tail, occur (Fig. 4.3).

D value

The *D value* or *decimal reduction value* is the dose required to inactivate 90% of the initial population. From Figure 4.3, it can be seen that the time (dose) required to reduce the population from 10^6 to 10^5 is the same as the time (dose) required to reduce the population from 10^5 to 10^4, that is, the D value remains constant over the full range of the survivor curve. Extending the treatment beyond the point at which there is one surviving cell does not give rise to fractions of a surviving cell but rather to a statement of the probability of finding one survivor. Thus, by extrapolation from the experimental data, it is possible to determine the lethal dose required to give a probability of less than 10^{-6}, which is required to meet the pharmacopoeial definition of 'sterile'. Note that in preparations intended for mass use, if the probability of a single live organism in a batch from which 10 million

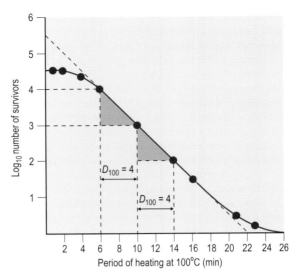

Fig. 4.3 Rate of inactivation of an inoculum of bacterial spores showing the decimal reduction time (*D* value) at 100°C and the non-linear 'shoulder and tail' effects.

doses are to be administered is 10^{-6}, then it is likely that around ten people will receive doses containing live organisms! Another consequence recognizable from Figure 4.3 is that the greater the number of microbes in the material to be sterilized, the longer the required exposure time. Thus, where efficient decontamination is the target, thorough cleansing can reduce the microbial load by several orders of magnitude and dramatically reduce both the time required and the level of certainty that sterility or adequate disinfection has been achieved.

Resistance to sterilization and disinfection

Many common factors affect the ability of micro-organisms to withstand the lethal effects of sterilization or disinfection processes. Factors specific to individual processes are considered in the description of those processes. In general, vegetative bacteria and viruses are more susceptible, and bacterial spores the most resistant, to sterilizing and disinfecting agents. However, within different species and strains of species there may be wide variation in intrinsic resistance. For example, within the Enterobacteriaceae, D values at 60°C range from a few minutes (*Esch. coli*) to 1 h (*Salmonella enterica* serotype Senftenberg). The typical D value for *Staphylococcus aureus* at 70°C is less than 1 min, compared with 3 min for *Staph. epidermidis*. However, an unusual strain of *Staph. aureus* has been isolated with a D value of 14 min at 70°C. Such variations may be attributed to morphological or physiological changes such as alterations in cell proteins or specific targets in the cell envelope affecting permeability.

Inactivation data obtained for one micro-organism should not be extrapolated to another; thus it should not be assumed that bactericidal disinfectants are also potent against viruses. The inactivation data for scrapie, bovine spongiform encephalopathy and Creutzfeldt–Jakob disease (CJD) suggest that prions are highly resistant agents, requiring six times the normal heat sterilization cycle (134°C for 18 min). This has led to requirements for the mandatory use of disposable instruments that are in direct contact with brain or other nervous tissue (including the retina) or tonsils where the risk of exposure to the prion causing CJD is high.

Owing to the adaptive processes described above, the conditions under which the micro-organisms were grown or maintained before exposure to the lethal process have a marked effect on their resistance. Organisms grown under nutrient-limiting conditions are typically more resistant than those grown under nutrient-rich conditions. Resistance usually increases through the late logarithmic phase of growth of vegetative cells and declines erratically during the stationary phase. Finally bacterial endospores, formed principally by *Bacillus* and *Clostridium* species, are relatively resistant to most processes. Similarly, fungal spores are more resistant than the vegetative mycelium, although they are not usually as resistant as bacterial spores. Bacterial spores were used to define the sterilization processes in current use, and preparations of bacterial spores (biological indicators) are used to monitor the efficacy of ethylene oxide sterilization, in which physical monitoring is inadequate. In general, disinfection processes have little or no activity against bacterial spores.

The micro-environment of the organism during exposure to the lethal process has a profound effect on its resistance. Thus, micro-organisms occluded in salt have greatly enhanced resistance to ethylene oxide; the presence of blood or other organic material will reduce the effectiveness of hypochlorite solution.

Sterilization by moist heat

Moist heat is much more effective than dry heat because hydrated proteins can be denatured with less energy than dehydrated semi-crystalline proteins. Further, where steam is used, its condensation delivers the latent heat of vaporization to the surface concerned. It is therefore necessary that all parts of the load to be sterilized are in direct contact with the water molecules in steam. Sterilization requires, in most cases, exposure to moist heat at 121°C for 15 min.

Moist heat sterilization requires temperatures above that of boiling water. Such conditions are attained under controlled conditions by raising the pressure of steam in a pressure vessel (*autoclave*). At sea level, boiling water at atmospheric pressure (1 bar) produces steam at 98–100°C, whereas raising the pressure to 2.4 bar increases the temperature to 125°C, and at 3.0 bar to 134°C. Conversely, at subatmospheric pressures, including those at higher altitude, water boils at lower temperatures.

Steam is non-toxic and non-corrosive, but for effective sterilization it must be *saturated*, which means that it holds all the water it can in the form of a transparent vapour. It must also be *dry*, which means that it does not contain water droplets. When dry saturated steam meets a cooler surface it condenses into a small volume of water and liberates the latent heat of vaporization. The energy available from this latent heat is considerable; for example, 6 litres of steam at a temperature of 134°C (and a corresponding pressure of 3 bar absolute) will condense into 10 ml of water and liberate 2162 J of heat energy. By comparison, less than 100 J of heat energy is released to an article by the sensible heat from air at 134°C.

Steam at a higher temperature than the corresponding pressure would allow is referred to as *super-heated steam*, and behaves in a similar manner to hot air. Conversely, steam that contains suspended droplets of water at the same temperature is referred to as *wet steam* and is less efficient. The presence of air in steam affects the sterilizing efficiency by changing the pressure–temperature relationship.

As can be seen from the foregoing, sterilization by moist heat requires delivery of steam at exactly the right temperature and pressure, and for the right time. This places considerable demands on the engineering and maintenance of autoclaves, and in critical situations such as provision of sterile materials for clinical practice their performance must be monitored continually and precisely. Physical measurements of temperature, pressure and time with thermometers and pressure gauges are recorded for every load, and periodic detailed tests are undertaken with temperature-sensitive probes (thermocouples) inserted into standard test packs. Biological indicators comprising dried spore suspensions of a reference heat-resistant bacterium, *Bacillus stearothermophilus*, are no longer considered appropriate for routine testing, although spore indicators are essential for low-temperature gaseous processes in which the physical measurements are not reliable.

Sterilization by dry heat

Dry heat is believed to kill micro-organisms by causing a destructive oxidation of essential cell constituents. Killing of the most resistant spores by dry heat requires a temperature of 160°C for 2 h. This high temperature causes slight charring of paper, cotton and other organic materials.

Incineration is an efficient method for the sterilization and disposal of contaminated materials at a high temperature. It has a particular application for pathological waste materials, surgical dressings, sharp needles and other clinical waste. Red heat is achieved by holding inoculating wires, loops and points of forceps in the flame of a Bunsen burner until they are red hot.

Hot air sterilizers are used to process materials that can withstand high temperatures for the length of time needed for sterilization by dry heat, but that are likely to be affected by contact with steam. Examples include oils, powders, carbon steel microsurgical instruments and empty laboratory glassware. The overall cycle of heating up and cooling may take several hours.

Disinfection by chemicals

Chemicals used in the environment or on the skin (*disinfectants* or *antiseptics*) cannot be relied on to kill or inhibit all pathogenic micro-organisms. The distinction between disinfectants and antiseptics is not clear-cut; an antiseptic can be regarded as a special kind of disinfectant that is sufficiently free from injurious effects to be applied to the surface of the body, though not suitable for systemic administration. Some would restrict the term *antiseptic* to preparations applied to open wounds or abraded tissue, and prefer the term *skin disinfection* for the removal of organisms from the hands and intact skin surfaces.

The efficacy of a particular method of chemical disinfection is heavily dependent on the concentration and stability of the agent; the number, type and accessibility of micro-organisms; the temperature and pH; and the presence of organic (especially protein) or other interfering substances.

In general, the rate of inactivation of a susceptible microbial population in the presence of an antimicrobial chemical is dependent on the relative concentration of the two reactants, the micro-organism and the chemical. The optimum concentration required to produce a standardized microbial effect in practice is described as the *in-use* concentration. Care must always be taken in preparing an accurate in-use dilution of concentrated product. Accidental or arbitrary overdilution may result in failure of disinfection.

The velocity of the reaction depends on the number and type of organisms present. In general, Gram-positive bacteria are more sensitive to disinfectants than Gram-negative bacteria; mycobacteria and fungal spores are relatively resistant, and bacterial spores are highly resistant. Enveloped or lipophilic viruses are relatively sensitive, whereas hydrophilic viruses such as poliovirus and other enteroviruses are less susceptible. Although difficult to test in vitro, there is evidence that hepatitis B virus is more resistant than other viruses (including human immunodeficiency virus) and most vegetative bacteria to the action of chemical disinfectants and heat.

Glutaraldehyde is highly active against bacteria, viruses and spores. Other disinfectants, such as hexachlorophane, have a relatively narrow range of activity, predominantly against Gram-positive cocci. Some disinfectants are more active or stable at a particular pH value; although glutaraldehyde is more stable under acidic conditions, use at a higher pH (8.0) improves the antimicrobial effect.

KEY POINTS

- Bacterial growth and multiplication is of practical value in the detection and identification of pathogens, and is generally a necessary component of infection.
- Bacteria divide asexually through a process of *binary fission*, passing through *lag*, *exponential* and *stationary* phases of *planktonic* growth in broth cultures. Bacterial growth can also be recognized in *sessile* form as *colonies* or *biofilms*. A given bacterial strain may have profoundly different physiological properties in each of these growth states.
- Recovery of pure bacterial cultures was greatly enhanced by the development of solidified agar media. Different medium designs enable *selection*, *enrichment*, *identification* or *defined* growth conditions.
- Different bacteria have evolved to grow and survive in widely differing habitats and these define their potential reservoirs and sources of infection. The growth atmospheres required by different bacteria are an important defining characteristic, and *obligate aerobes*, *obligate anaerobes*, *micro-aerophilic* and *facultative* organisms are recognized.
- Bacterial viability is generally recognized and quantified by detecting growth of single cells into colonies in colony-forming unit (cfu) counts. Discrepancies between cfu counts and the number of cells seen by microscopy have led to recognition that many cells in natural samples do not form colonies.
- Bacteria may die through senescence in stationary cultures, through genetically programmed or prophage-induced cell death, or as result of external noxious influences such as antibiotics or the deliberate processes of *sterilization* and *disinfection*.
- Sterilization involves the destruction of all propagating biological entities, whereas disinfection involves a reduction in microbial load to an acceptable level. Both processes can be achieved by application of *moist and dry heat, ionizing radiation, filtration, gaseous chemical agents* and *liquid chemical agents*.

Disinfectants may be inactivated by hard tap water, cork, plastics, blood, urine, soaps and detergents, or another disinfectant. Information should be sought from the manufacturer or from reference authorities to confirm that the disinfectant will remain active in the circumstances of use.

Maintenance of effective disinfection in large health-care facilities is a major challenge requiring management skills and technical understanding in equal measure. The selection of appropriate disinfectants and maintaining standards in practice is supported by the development of local disinfection policies. These points are considered further in Chapter 68.

RECOMMENDED READING

Barer M R, Harwood C R 1999 Bacterial viability and culturability. *Advances in Microbial Physiology* 41: 94–138

Block S S (ed.) 2000 *Disinfection, Sterilization and Preservation*, 5th edn. Lea and Febiger, Philadelphia

Moat A G, Foster J W 2002 *Microbial Physiology*, 4th edn. Wiley-Liss, New York

Musser J M, Deleo F R 2005 Toward a genome-wide systems biology analysis of host–pathogen interactions in group A streptococcus. *American Journal of Pathology* 167: 1461–1472

Neidhardt F C, Ingraham J L, Schaechter M 1990 *Physiology of the Bacterial Cell: A Molecular Approach*. Sinauer, Sunderland, MA

Roszaak D B, Colwell R R 1987 Survival strategies of bacteria in the natural environment. *Microbiological Reviews* 51: 365–379

Russell A D, Hugo W B, Ayliffe G A J (eds) 1998 *Principles and Practice of Disinfection, Preservation and Sterilisation*, 3rd edn, Blackwell Scientific, Oxford

Internet site

LabWork. Bacterial growth curve. http://www-micro.msb.le.ac.uk/LabWork/bact/bact1.htm

5 Antimicrobial agents

D. Greenwood and M. M. Ogilvie

Antimicrobial agents are used not only to treat bacterial diseases, but also infections with viruses, fungi, protozoa and helminths. These drugs have transformed the management of infectious disease, but none is free from unwanted side effects, and microbial resistance is a constant threat. Consequently, they must be used with discretion and understanding of their individual properties. The treatment of individual infections is dealt with in the appropriate chapters. The general strategy of antimicrobial chemotherapy, which is crucial to the control of antimicrobial drug resistance, is covered in Chapter 66.

Antibiotics are naturally occurring microbial products; synthetic compounds such as sulphonamides, quinolones, nitrofurans and imidazoles should strictly be referred to as *chemotherapeutic agents*. However, as some antibiotics can be manufactured synthetically whereas others are the products of chemical manipulation of naturally occurring compounds (*semi-synthetic antibiotics*), the distinction is ill defined. Nowadays the term *antibiotic* is used loosely to describe agents (mainly, but not exclusively, antibacterial agents) employed to treat systemic infection. Antimicrobial substances that are too toxic to be used other than in topical therapy or for environmental decontamination are referred to as *antiseptics* or *disinfectants* (see Ch. 4).

ANTIBACTERIAL AGENTS

The principal types of antibacterial agent are listed in Table 5.1. Because there are so many, it is convenient to group them according to their site of action.

Inhibitors of bacterial cell wall synthesis

As most bacteria possess a rigid cell wall that is lacking in mammalian cells, this structure is a prime target for agents that exhibit *selective toxicity*, the ability to inhibit or destroy the microbe without harming the host. However, the bacterial cell wall can also prevent access of agents that would otherwise be effective. Thus, the complex outer envelope of Gram-negative bacteria is impermeable to large hydrophilic molecules, which may be prevented from reaching an otherwise susceptible target.

Inhibitors of bacterial cell wall synthesis act on the formation of the peptidoglycan layer (Fig. 5.1). Bacteria that lack peptidoglycan, such as mycoplasmas, are resistant to these agents.

β-Lactam agents

Penicillins, cephalosporins and other compounds that feature a β-lactam ring in their structure fall into this group (Fig. 5.2). All of these compounds bind to proteins situated at the cell wall–cell membrane interface. These *penicillin-binding proteins* are involved in cell wall construction, including the cross-linking of the peptidoglycan strands that gives the wall its strength. Opening of the β-lactam ring by hydrolytic enzymes, collectively called *β-lactamases*, abolishes antibacterial activity. Many such enzymes are found in bacteria. Those elaborated by Gram-negative enteric bacilli are particularly diverse in their activity and properties. Most prevalent are the so-called *TEM β-lactamases* (TEM-1, TEM-2, etc.), numerous forms of which have evolved under selective pressure of β-lactam antibiotic use. Gram-negative bacteria able to produce enzymes that inactivate many different β-lactam antibiotics – so-called *extended-spectrum β-lactamases* (ESBLs) – sometimes become endemic in hospitals, causing serious problems, especially in patients in high-dependency units.

Penicillins. Benzylpenicillin (penicillin G; often called simply 'penicillin') exhibits unrivalled activity against staphylococci, streptococci, neisseriae, spirochaetes and certain other organisms. However, resistance, normally due to the production of β-lactamase, has undermined its activity against staphylococci and, to a lesser extent, gonococci. Bacteria, including staphylococci and pneumococci, that exhibit reduced susceptibility to penicillin by a non-enzymic mechanism are also encountered. Benzylpenicillin revolutionized the treatment of infection caused by some of the most virulent bacterial pathogens, but it also suffers from several shortcomings:

Table 5.1 Principal types of antibacterial agent (other than agents used exclusively in mycobacterial infection)

Agent	Site of action	Usual activity[a] against					
		Staphylococci	Streptococci	Enterobacteria	*Pseudomonas aeruginosa*	*Mycobacterium tuberculosis*	Anaerobes
Penicillins	Cell wall	(+)	+	v	v	–	+[b]
Cephalosporins	Cell wall	+	+	+	v	–	+[b]
Other β-lactam agents	Cell wall	v	v	+	v	–	v
Glycopeptides	Cell wall	+	+	–	–	–	+[c]
Tetracyclines	Ribosome	(+)	(+)	(+)	–	–	(+)
Chloramphenicol	Ribosome	+	+	+	–	–	–
Aminoglycosides	Ribosome	+	–	+	v	v	–
Macrolides	Ribosome	+	+	–	–	–	+
Lincosamides	Ribosome	+	+	–	–	–	+
Fusidic acid	Ribosome	+	+	–	–	–	+
Oxazolidinones	Ribosome	+	+	–	–	+	+
Streptogramins	Ribosome	+	+[d]	–	–	–	–
Rifamycins	RNA synthesis	+	+	+	–	+	+
Sulphonamides	Folate metabolism	(+)	(+)	(+)	–	–	–
Diaminopyrimidines	Folate metabolism	+	+	(+)	–	–	–
Quinolones	DNA synthesis	v	v	+	v	v	–
Nitrofurans	DNA synthesis	–	–	+	–	–	+
Nitroimidazoles	DNA synthesis	–	–	–	–	–	+

v, variable activity among different agents of the group.
[a]Usual spectrum of intrinsic activity; parentheses indicate that resistance is common.
[b]Poor activity against anaerobes of the *Bacteroides fragilis* group.
[c]Poor activity against most Gram-negative anaerobes.
[d]Poor activity against *Enterococcus faecalis*.

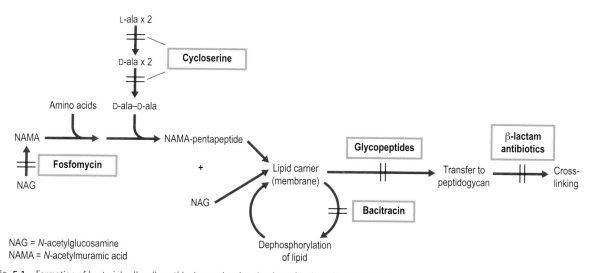

NAG = *N*-acetylglucosamine
NAMA = *N*-acetylmuramic acid

Fig. 5.1 Formation of bacterial cell wall peptidoglycan, showing the sites of action of inhibitors of the process.

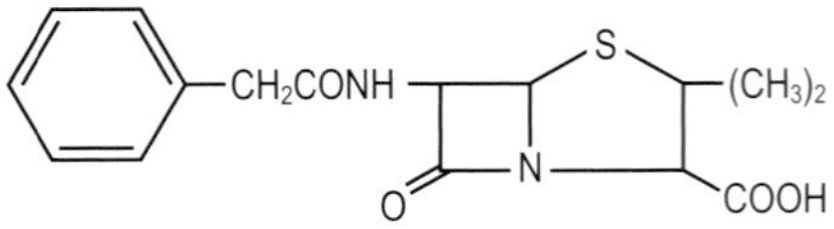

Benzylpenicillin
(a penam)

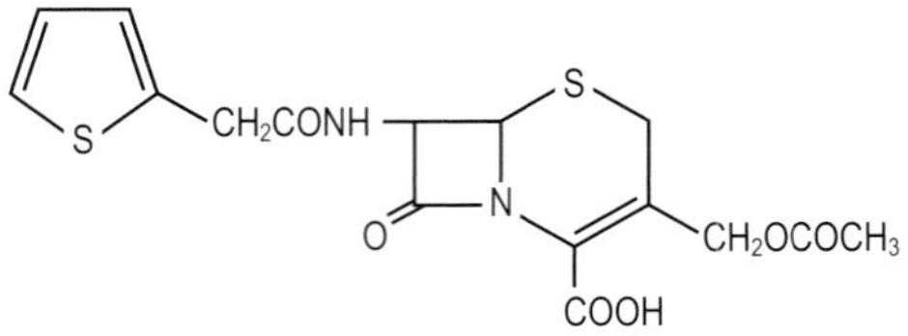

Cefalotin
(a cephem)

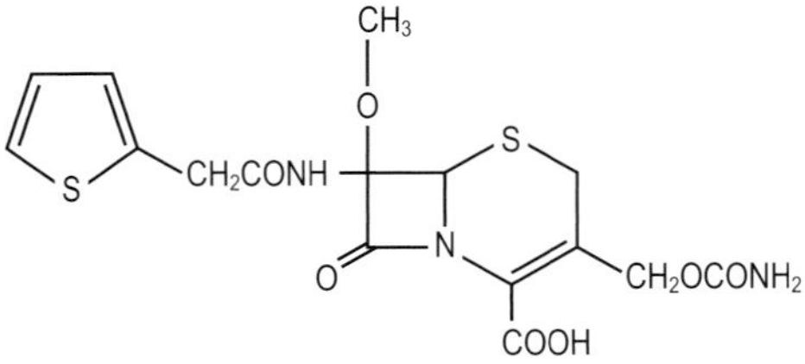

Cefoxitin
(a cephamycin)

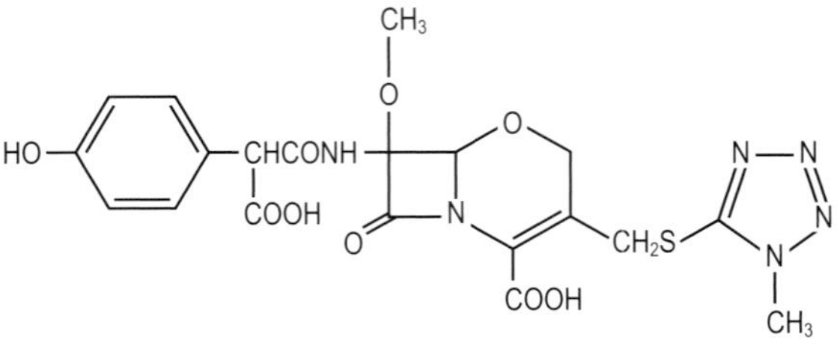

Latamoxef
(an oxa-cephem)

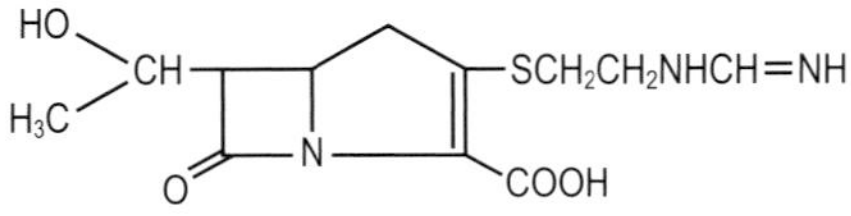

Imipenem
(a carbapenem)

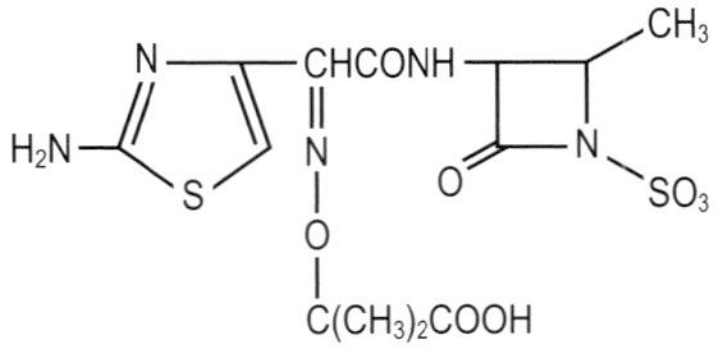

Aztreonam
(a monobactam)

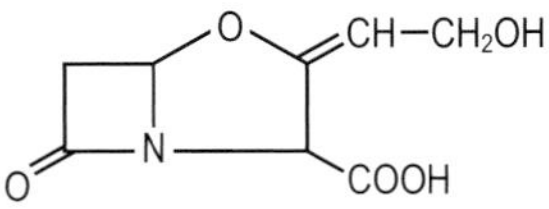

Clavulanic acid
(a clavam)

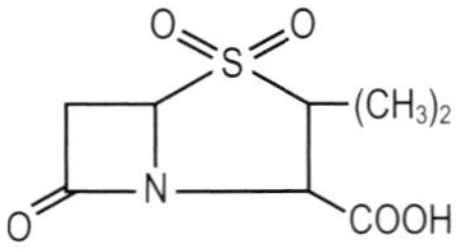

Sulbactam
(a sulphone)

Fig. 5.2 Examples of different types of molecular structure among β-lactam antibiotics.

- Breakdown by gastric acidity when given orally
- Very rapid excretion by the kidney
- Susceptibility to penicillinase (β-lactamase)
- Restricted spectrum of activity.

Further development of the penicillin family has been directed towards improving these properties. Crucial to this was the discovery that removal of the phenylacetic acid side-chain left intact the core structure, 6-aminopenicillanic acid, the starting point for the numerous semi-synthetic penicillins that have been produced (Fig. 5.3). Among the most important penicillins that followed the introduction of benzylpenicillin are:

- phenoxymethylpenicillin (penicillin V), which can be given orally
- procaine penicillin, a long-acting salt of benzylpenicillin
- flucloxacillin, a compound resistant to staphylococcal β-lactamase

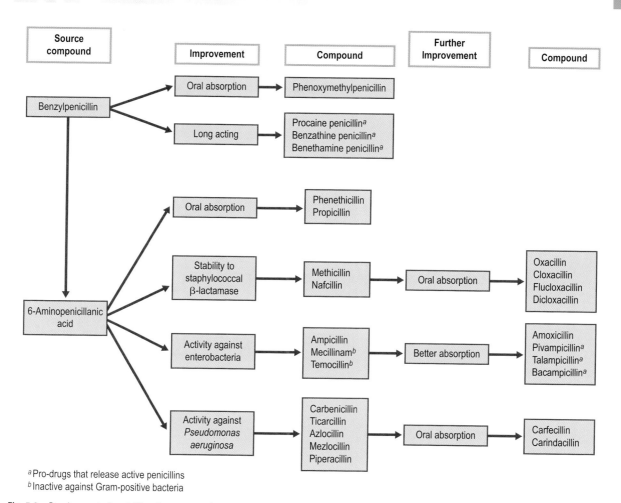

Fig. 5.3 Development of penicillins. 'Improvement' means a qualitative improvement, and is not intended to indicate that compounds listed together are necessarily equivalent.

- ampicillin and amoxicillin, which are active against some enterobacteria
- ticarcillin, azlocillin and piperacillin, which are active against *Pseudomonas aeruginosa*.

None of these compounds, with the exception of flucloxacillin (and related antistaphylococcal penicillins), exhibits stability to staphylococcal β-lactamase. *Methicillin-resistant Staphylococcus aureus* (MRSA; a term that has persisted although methicillin is now virtually obsolete) and other staphylococci that owe their resistance to alterations in the target penicillin-binding proteins are resistant to all penicillins and to all other β-lactam antibiotics.

Cephalosporins. Cephalosporins are close cousins of the penicillins, but the β-lactam ring is fused to a six-membered dihydrothiazine ring rather than the five-membered thiazolidine ring of penicillins. The additional carbon carries substitutions that may alter the pharmacological behaviour of the molecule, and sometimes its antibacterial activity. Some cephalosporins (e.g. cefalotin [formerly called cephalothin] and cefotaxime) carry an acetoxymethyl group on the extra carbon. This can be removed by hepatic enzymes to yield a less active derivative, but it is doubtful whether this has any therapeutic significance. Other cephalosporins (e.g. cefamandole, cefoperazone and the oxa-cephem, latamoxef) possess a methyltetrazole substituent. Use of compounds with this feature has been associated with hypoprothrombinaemia and bleeding in some patients.

Cephalosporins are generally stable to staphylococcal penicillinase (though they are differentially susceptible to hydrolysis by the various types of enterobacterial β-lactamase), but they lack activity against enterococci.

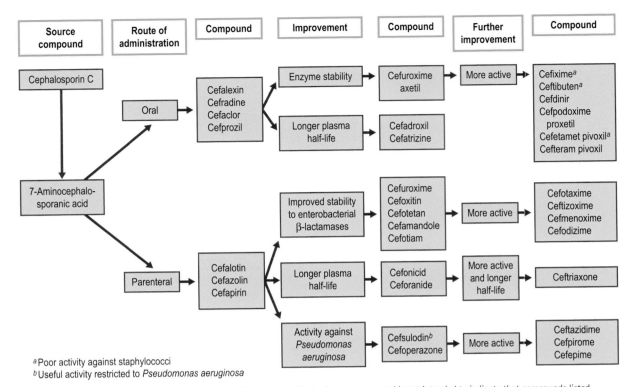

Fig. 5.4 Development of cephalosporins. 'Improvement' means a qualitative improvement, and is not intended to indicate that compounds listed together are necessarily equivalent.

[a] Poor activity against staphylococci
[b] Useful activity restricted to *Pseudomonas aeruginosa*

They exhibit a broader spectrum than most penicillins and are less prone to cause hypersensitivity reactions. The range of available derivatives is shown in Figure 5.4. Among the most important are:

- cefalexin and cefaclor, which can be given orally
- cefuroxime and cefoxitin, which are stable to many β-lactamases
- cefotaxime and ceftriaxone, which combine β-lactamase stability with high intrinsic activity
- ceftazidime and cefpirome, which additionally exhibit good activity against *Ps. aeruginosa*.

Other β-lactam agents. Various agents with diverse properties share the structural feature of a β-lactam ring with penicillins and cephalosporins (see Fig. 5.2):

- *Monobactams* (e.g. aztreonam) are monocyclic compounds with a spectrum that is restricted to aerobic Gram-negative bacteria.
- *Carbapenems* (e.g. imipenem and meropenem) have an unusually broad spectrum of activity, embracing most Gram-positive and Gram-negative aerobic and anaerobic bacteria. Imipenem is inactivated by a dehydropeptidase in the human kidney, and is co-administered with a dehydropeptidase inhibitor, cilastatin.
- *Oxa-cephems* (e.g. latamoxef) are broad-spectrum β-lactamase-stable compounds.
- The *clavam*, clavulanic acid, exhibits poor antibacterial activity, but has proved useful as a β-lactamase inhibitor when used in combination with β-lactamase-susceptible compounds (e.g. co-amoxiclav, the combination of amoxicillin and clavulanic acid).
- The *sulphones*, sulbactam and tazobactam, also act as β-lactamase inhibitors and are marketed in combination with ampicillin (or cefoperazone) and piperacillin, respectively.

Glycopeptides

Vancomycin and teicoplanin are large molecules that are unable to penetrate the outer membrane of Gram-negative bacteria, and the spectrum is consequently restricted to Gram-positive organisms. Their chief importance resides in their action against Gram-positive cocci with multiple resistance to other drugs. Enterococci and staphylococci, including MRSA, that exhibit resistance or reduced sensitivity to glycopeptides are being reported more frequently.

Table 5.2 Summary of the important differential properties of aminoglycoside antibiotics

Aminoglycoside	Activity against		Relative susceptibility to inactivation by bacterial enzymes	Relative degree of	
	Pseudomonas aeruginosa	Mycobacterium tuberculosis		Ototoxicity	Nephrotoxicity
Amikacin	+	+	±	++	+
Gentamicin	+	–	++	++	++
Kanamycin	–	+	++	++	++
Neomycin	–	±	++	+++	++
Netilmicin	+	–	+	+	+
Sisomicin	+	–	++	++	++
Streptomycin	–	+	++	+++	±
Tobramycin	+	–	++	++	++

Other inhibitors of bacterial cell wall synthesis

- Fosfomycin is an antibiotic with a simple phosphonic acid structure. It exhibits a fairly broad spectrum, notably against Gram-negative bacilli, and is used mainly for the treatment of urinary tract infection. Resistance arises readily in vitro.
- Bacitracin is active against Gram-positive bacteria, but is too toxic for systemic use. It is found in many topical preparations, and is also used in the laboratory in the presumptive identification of haemolytic streptococci of Lancefield group A (see p. 179).
- Cycloserine is an analogue of D-alanine used only as a second-line agent in infections with multiresistant strains of Mycobacterium tuberculosis.
- Isoniazid and some other compounds used to treat tuberculosis probably act by interfering with formation of the mycolic acids of the mycobacterial cell wall.

Inhibitors of bacterial protein synthesis

Bacterial ribosomes are sufficiently different from those of mammalian cells to allow selective inhibition of protein synthesis. Some of the agents that act at this level do, however, have an effect in eukaryotic cells, which have mitochondrial ribosomes similar to those of bacteria (see p. 11). Most are true antibiotics (or derivatives thereof) produced by Streptomyces species or other soil organisms.

Tetracyclines

These are broad-spectrum agents with important activity against chlamydiae, rickettsiae, mycoplasmas and, surprisingly, malaria parasites, as well as most conventional Gram-positive and Gram-negative bacteria. They prevent binding of amino-acyl transfer RNA (tRNA) to the ribosome and inhibit, but do not kill, susceptible bacteria. The various members of the tetracycline group are closely related and differ more in their pharmacological behaviour than in antibacterial activity. Doxycycline and minocycline are in most common use. Resistance has limited the value of tetracyclines against many Gram-positive and Gram-negative bacteria, but not against rickettsiae, chlamydiae and mycoplasmas to date. Tigecycline (a so-called glycylcycline) retains activity against many bacteria resistant to other tetracyclines.

Chloramphenicol

This compound and the related thiamphenicol also possess a very broad antibacterial spectrum. They act by blocking the growth of the peptide chain. Use of chloramphenicol has been limited to typhoid fever, meningitis and a few other clinical indications because of the occurrence of a rare but fatal side effect, aplastic anaemia. Thiamphenicol is said to lack this disadvantage, but is more likely to cause a reversible type of bone marrow toxicity.

Aminoglycosides

Streptomycin, the first antibiotic to be discovered by random screening of soil organisms, is predominantly active against enterobacteria and M. tuberculosis. Like other members of the aminoglycoside family it has no useful activity against streptococci, anaerobes or intracellular bacteria. The group also has in common a tendency to damage the eighth cranial nerve (ototoxicity) and the kidney (nephrotoxicity). The chief properties of aminoglycosides are shown in Table 5.2. They inhibit formation of the ribosomal initiation complex and also cause misreading of messenger RNA (mRNA). They are bactericidal compounds, and some, notably gentamicin and tobramycin, exhibit good activity against Ps. aeruginosa. Such compounds have been used widely, often in combination with β-lactam antibiotics, with which they interact synergistically, in the 'blind' treatment of sepsis in immunocompromised patients. Resistance may arise from ribosomal changes (streptomycin) or alterations in drug uptake. However, it is more often caused by bacterial enzymes that phosphorylate, acetylate or adenylate exposed amino or hydroxyl groups. Enzymic resistance consequently affects the various aminoglycosides differentially, depending on the possession of exposed groups that can be attacked by the enzyme involved. Amikacin is resistant to most of the common enzymes.

Macrolides

Macrolides are antibiotics in which a large macrocyclic lactone ring is substituted with some unusual sugars. They act by interfering with the translocation of mRNA on the bacterial ribosome. They are used mainly as antistaphylococcal and antistreptococcal agents, though some have wider applications. They have no useful activity against enteric Gram-negative bacilli. The original macrolide, erythromycin, is unstable in gastric acid and is usually administered orally as the stearate salt or as an esterified *pro-drug* (pharmacological preparations that improve absorption and deliver the active drug into the circulation). Salts suitable for intravenous administration are also available. Certain later macrolides, including clarithromycin, dirithromycin and roxithromycin, offer improved pharmacological properties.

The macrolactone ring of these compounds is composed of 14 atoms, but other macrolides have additional carbons conferring a 16-membered structure. These include oleandomycin, josamycin, midecamycin (the properties of which are similar to those of erythromycin), and spiramycin, which has some useful activity against the protozoan parasite, *Toxoplasma gondii*.

Some macrolides feature structural changes in the macrocyclic ring: in azithromycin, a compound distinguished by good tissue penetration and a long terminal half-life, the ring has been expanded by inclusion of a nitrogen atom to form an *azalide*; in telithromycin, a compound that retains activity against macrolide-resistant Gram-positive cocci, a keto function has been introduced to produce a *ketolide*.

Lincosamides

The original lincosamide antibiotic, lincomycin, has been superseded by a derivative, clindamycin, that is better absorbed after oral administration and is more active against the organisms within its spectrum. These include staphylococci, streptococci and most anaerobic bacteria, against which clindamycin exhibits outstanding activity. Enthusiasm for the use of clindamycin has been tempered by an association with the occasional development of severe diarrhoea, which sometimes progresses to a life-threatening pseudomembranous colitis (see *Clostridium difficile*, Ch. 22).

Lincosamides bind to the 50S ribosomal subunit at a site closely related to that at which macrolides act. Inducible resistance to macrolides caused by enzymic modification of the ribosomal binding site also renders the cells resistant to lincosamides (and streptogramins; see below), but only in the presence of macrolides, which alone are able to act as inducers.

Fusidic acid

The structure of fusidic acid is related to that of steroids, but the antibiotic is devoid of steroid-like activity. It blocks factor G, which is involved in peptide elongation. Fusidic acid has an unusual spectrum of activity that includes corynebacteria, nocardia and *M. tuberculosis*, but the antibiotic is usually regarded simply as an antistaphylococcal agent. It penetrates well into bone and has been used widely (generally in combination with a β-lactam antibiotic to prevent the selection of resistant variants) in the treatment of staphylococcal osteomyelitis.

Linezolid

Linezolid is classified chemically as an oxazolidinone. It is a narrow-spectrum anti-Gram-positive agent which acts by preventing the formation of the ribosomal initiation complex. It is used exclusively against MRSA and other Gram-positive cocci resistant to older agents. There is some preliminary evidence that it may be of use in drug-resistant tuberculosis.

Streptogramins

This is the collective name for a family of antibiotics that occur naturally as two synergistic components. They were formerly used mainly in animal husbandry, although one member of the group, pristinamycin, is available in some countries as an antistaphylococcal agent. Use was limited by poor solubility, but derivatives suitable for parenteral administration, quinupristin and dalfopristin, have been developed as a combination product. The combination exhibits bactericidal activity against most Gram-positive cocci, but has poor activity against *Enterococcus faecalis*.

Mupirocin

This is an antibiotic, produced by *Pseudomonas fluorescens*, that blocks incorporation of isoleucine into proteins. Its useful activity is restricted to staphylococci and streptococci; as it is inactivated when given systemically, it is used only in topical preparations.

Table 5.3   Types of quinolone antibacterial agent

Narrow-spectrum compounds[a]	Broad-spectrum compounds[b]	Compounds with further enhanced spectrum[c]
Acrosoxacin	Ciprofloxacin	Clinafloxacin
Cinoxacin	Enoxacin	Gatifloxacin
Flumequine	Levofloxacin	Gemifloxacin
Nalidixic acid	Lomefloxacin	Moxifloxacin
Oxolinic acid	Norfloxacin	Sparfloxacin
Pipemidic acid	Ofloxacin	Tosufloxacin
Piromidic acid	Pefloxacin	Trovafloxacin

[a]Spectrum restricted to enteric Gram-negative bacilli.
[b]Improved activity against *Pseudomonas aeruginosa* and Gram-positive cocci.
[c]Further improved activity against Gram-positive cocci and some anaerobes

Inhibitors of nucleic acid synthesis

A number of important antibacterial agents act directly or indirectly on DNA or RNA synthesis.

Sulphonamides and diaminopyrimidines

These agents affect DNA synthesis because of their role in folic acid metabolism. Folic acid is used in many one-carbon transfers in living cells, including the conversion of deoxyuridine to thymidine. During this process the active form of the vitamin, tetrahydrofolate, is oxidized to dihydrofolate, and this must be reduced before it can function in further reactions.

Sulphonamides are analogues of *para*-aminobenzoic acid, and prevent the condensation of this compound with dihydropteridine during the formation of folic acid. Diaminopyrimidines, which include the broad-spectrum antibacterial agent trimethoprim and the antimalarial compounds pyrimethamine and cycloguanil (the metabolic product of proguanil), prevent the reduction of dihydrofolate to tetrahydrofolate. Sulphonamides and diaminopyrimidines thus act at sequential stages of the same metabolic pathway and interact synergistically, although in bacterial infections trimethoprim is generally sufficiently effective, and less toxic, when used alone.

Sulphonamides are broad-spectrum antibacterial agents, but resistance is common and the group also suffers from problems of toxicity. The numerous sulphonamides exhibit similar antibacterial activity, but differ widely in their pharmacokinetic behaviour. They have largely been replaced by safer and more active agents, although the combination of sulfamethoxazole with trimethoprim (co-trimoxazole) is still used. Sulfadoxine or sulfadiazine combined with pyrimethamine are used in malaria and toxoplasmosis, respectively.

Quinolones

These drugs act on the α subunit of DNA gyrase. Their properties allow them to be categorized roughly into three groups (Table 5.3). Nalidixic acid and its early congeners are narrow-spectrum agents active only against Gram-negative bacteria. Their use is virtually restricted to urinary tract infection, although they have also been used in enteric infections and, in the case of acrosoxacin, in gonorrhoea. Later quinolones, such as ciprofloxacin and ofloxacin, which are 6-fluoro derivatives, display much enhanced activity and a broader spectrum, although activity against some Gram-positive cocci, notably *Streptococcus pneumoniae*, is unreliable. Continued development has produced compounds that lack the latter defect and, in some cases, exhibit further broadening of the spectrum and improved pharmacokinetic properties.

Quinolones are quite well absorbed when given orally and are widely distributed throughout the body. Extensive metabolization may occur, particularly with nalidixic acid and the older derivatives. Ciprofloxacin and other fluoroquinolones are used widely despite certain problems of toxicity, and resistance is becoming more prevalent.

Nitroimidazoles

Azole derivatives feature prominently among antifungal, antiprotozoal and anthelminthic agents. Those that exhibit antibacterial activity are 5-nitroimidazoles. At low redox (E_h) values they are reduced to a short-lived intermediate that causes DNA strand breakage. Because of the requirement for low E_h values, 5-nitroimidazoles are active only against anaerobic (and certain micro-aerophilic) bacteria and anaerobic protozoa. The representative of the group most commonly used clinically is metronidazole; similar derivatives include tinidazole, ornidazole and nimorazole.

Nitrofurans

The most familiar nitrofuran derivative is nitrofurantoin, an agent used exclusively in urinary tract infection. Other nitrofurans, including furazolidone, which is used in enteric infections, are marketed for a variety of purposes in some parts of the world. The mode of action of nitrofurans has not been elucidated, but it is probable that a reduced metabolite acts on DNA in a manner analogous to that of the nitroimidazoles.

Novobiocin

This compound acts on the β subunit of DNA gyrase (cf. quinolones). It was once used widely as a reserve antistaphylococcal agent, but is no longer favoured because of problems of resistance and toxicity.

Rifamycins

This group of antibiotics is characterized by excellent activity against mycobacteria, although other bacteria are also susceptible; staphylococci in particular are exquisitely sensitive. These compounds act by inhibiting transcription of RNA from DNA. Rifampicin, the best known member of the group, is used in tuberculosis and leprosy. Wider use has been discouraged on the grounds that it might inadvertently foster the emergence of resistance in mycobacteria. Rifapentine has similar properties, but exhibits a longer plasma half-life. Rifabutin (ansamycin) is used in infections caused by atypical mycobacteria of the avium-intracellulare group (see Ch. 19).

Miscellaneous antibacterial agents

Polymyxins

Polymyxin B and colistin (polymyxin E) act like cationic detergents to disrupt cell membranes. They exhibit potent antipseudomonal activity, but toxicity has limited their usefulness, except in topical preparations and bowel decontamination regimens. If systemic use is contemplated, a sulphomethylated derivative, colistin sulfomethate, is preferred.

Daptomycin

Daptomycin is a semi-synthetic lipopeptide antibiotic with activity against Gram-positive cocci. It has a minor role in the treatment of infections caused by multiresistant organisms.

Antimycobacterial agents

As well as streptomycin and rifampicin (see above), various agents are used exclusively for the treatment of mycobacterial infection. These include isoniazid, ethambutol and pyrazinamide, which are commonly found in antituberculosis regimens, and diaminodiphenylsulfone (dapsone) and clofazimine, which are used in leprosy. Cycloserine and p-aminosalicylic acid (PAS), which were formerly used in tuberculosis, have now been largely abandoned except for drug-resistant tuberculosis. Some fluoroquinolones and macrolides exhibit activity against certain mycobacteria, and may have a role in treatment.

ANTIFUNGAL AGENTS

Although fungi cause a wide variety of infections, relatively few agents are available for treatment, especially for the systemic therapy of serious mycoses. Superficial fungal infections of the skin and mucous membranes can often be treated with topical agents, including *polyenes*, such as nystatin, or *azole* derivatives, of which many (clotrimazole, miconazole, econazole, etc.) are marketed as vaginal pessaries and creams. For dermatophyte infections of the nails, oral therapy with griseofulvin or the allylamine derivative terbinafine is peculiarly suitable, as these agents are deposited in newly formed keratin.

Serious systemic disease caused by yeasts and other fungi is often treated with the polyene, amphotericin B, which is extremely toxic. Newer formulations of the drug, in which it is complexed with liposomes or lipids, are better tolerated. The pyrimidine analogue 5-fluorocytosine is active against many types of yeast, and is used in combination with amphotericin B in severe systemic yeast infections. Caspofungin is a member of the *echinocandin* class of agents that interfere with β-glucan synthesis in the fungal cell wall. It is active against various fungi, including *Candida*, *Aspergillus* and *Histoplasma* spp. (but not *Cryptococcus neoformans*), and is administered by intravenous infusion.

Azole derivatives exhibit the broadest spectrum of activity, embracing yeasts, filamentous fungi and dimorphic fungi, but few are suitable for systemic use. Those that are include the 2-nitroimidazole ketoconazole, and the triazoles itraconazole, fluconazole and voriconazole. Itraconazole and voriconazole exhibit useful activity against *Aspergillus fumigatus*. Fluconazole is well distributed after oral administration, and has been used successfully in systemic yeast infections, including cryptococcal meningitis.

The fungus *Pneumocystis jirovecii* (formerly *P. carinii*), long thought to be a protozoon, is not susceptible to conventional antifungal agents (see p. 618).

The spectrum of activity of the common antifungal compounds is shown in Table 5.4. Most act by interfering with the integrity of the fungal cell membrane, either by binding to membrane sterols (polyenes) or by preventing

Table 5.4   Summary of the spectrum of activity of antifungal agents

Agent	*Candida albicans*	*Cryptococcus neoformans*	Dermatophytes	*Aspergillus fumigatus*	Dimorphic fungi
Amphotericin B	+	+	−	+	+
Echinocandins	+	−	−	+	+[b]
Flucytosine	+	+	−	−	−
Griseofulvin	−	−	+	−	−
Imidazoles	+	+	+	−	+
Nystatin[a]	+	−	+	−	−
Terbinafine	−	−	+	+[b]	+[b]
Triazoles	+	+	+	(+)[c]	+

[a]For topical use only.
[b]Clinical efficacy not yet established.
[c]Itraconazole and voriconazole are active against *A. fumigatus*, but fluconazole is not.

Table 5.5   Antiviral agents in clinical use for infections other than human immunodeficiency virus

Compound	Mode of action	Indication
Aciclovir	Nucleoside analogue	Herpes simplex; varicella-zoster
Amantadine (and rimantadine)	Viral uncoating	Influenza A
Cidofovir	Nucleotide analogue	Cytomegalovirus retinitis
Famciclovir	Pro-drug of penciclovir	Herpes simplex; zoster
Fomivirsen	Antisense oligonucleotide	Cytomegalovirus retinitis
Foscarnet	Inhibition of DNA polymerase	Cytomegalovirus; aciclovir-resistant herpes simplex or varicella-zoster
Ganciclovir	Nucleoside analogue	Cytomegalovirus
Interferon-α	Immunomodulator	Chronic hepatitis B and C
Lamivudine	Nucleoside analogue	Chronic hepatitis B
Oseltamivir	Neuraminidase inhibitor	Influenza A and B
Penciclovir	Nucleoside analogue	Herpes simplex
Ribavirin	Nucleoside analogue	Respiratory syncytial virus; hepatitis C
Valaciclovir	Pro-drug of aciclovir	Herpes simplex; zoster
Valganciclovir	Pro-drug of ganciclovir	Cytomegalovirus
Zanamivir	Neuraminidase inhibitor	Influenza A and B

the synthesis of ergosterol (azoles and allylamines). Their use in individual fungal diseases is considered in Chapter 60.

ANTIVIRAL AGENTS

Compared with the number of agents available for the treatment of bacterial infection, there are relatively few antiviral agents. However, an increasing number are effective in the treatment and prophylaxis of a range of viral diseases. Those antiviral agents in clinical use are presented in Table 5.5 for agents used for diseases other than human immunodeficiency virus (HIV) infection, and in Table 5.6 for the equally long list of antiretroviral agents.

About half of all the antiviral agents presently available are nucleoside (or nucleotide) analogues, which are phosphorylated within cells to an active triphosphate and inhibit viral DNA synthesis. Some important antiviral compounds, representing a range of molecular structures, are illustrated in Figure 5.5.

Nucleoside analogues

Inhibitors of herpesvirus DNA polymerases

The most widely used antiviral agent, aciclovir (and the related penciclovir), is first phosphorylated to the monophosphate by a virus-encoded thymidine kinase produced in cells infected by the herpes simplex virus (HSV) or by varicella-zoster virus (VZV). Subsequent

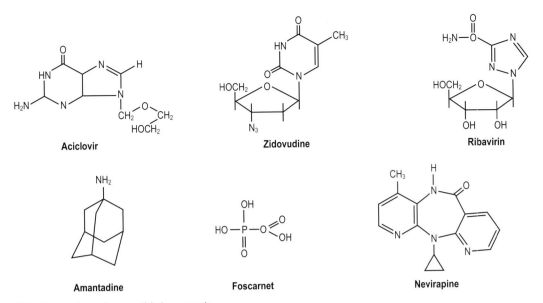

Fig. 5.5 Molecular structures of some antiviral compounds.

phosphorylation steps are completed by cellular kinases to form aciclovir triphosphate. This competes with the natural substrate for the viral DNA polymerase, becomes incorporated into the viral DNA chain and inhibits further DNA polymerase activity (Fig. 5.6). Aciclovir lacks a 3'-hydroxyl group on its acyclic side-chain, and therefore it cannot form a phosphodiester bond with the next nucleotide due to be added to the growing herpesvirus DNA chain, which is terminated prematurely. The lack of cellular toxicity of aciclovir is a consequence of two selective features:

1. Initial phosphorylation takes place only in virus-infected cells.
2. Aciclovir triphosphate inhibits viral (not cellular) DNA polymerase.

Aciclovir has an established record in the treatment of HSV and VZV disease, as prophylaxis against HSV reactivation after transplantation, and in the long-term suppression of recurrent genital herpes. Valaciclovir and famciclovir are oral pro-drug formulations of aciclovir and penciclovir, respectively. They provide improved systemic drug levels and require less frequent administration.

Ganciclovir, which exhibits preferential activity against another herpesvirus, cytomegalovirus (CMV), is activated by a CMV protein kinase in CMV-infected cells, but there is also considerable phosphorylation in uninfected cells, and ganciclovir is much more toxic than aciclovir. Ganciclovir also exhibits antiviral effects against human

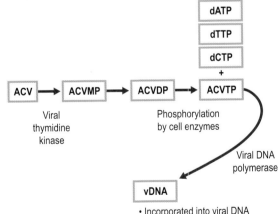

Fig. 5.6 Mode of activation and action of aciclovir (ACV). ACVMP, aciclovir monophosphate, ACVDP, aciclovir diphosphate; ACVTP, aciclovir triphosphate; dATP, deoxyadenosine triphosphate; dTTP, deoxythymidine triphosphate; dCTP, deoxycytidine triphosphate; vDNA, viral deoxyribonucleic acid.

herpesviruses HHV-6 and HHV-7, which do not respond to aciclovir. Valganciclovir is an oral pro-drug formulation providing a much improved systemic concentration of ganciclovir after oral administration.

Cidofovir is a nucleoside phosphonate and therefore does not require activation. It is converted in cells into the diphosphate, which inhibits CMV DNA polymerase. It shows activity against a range of DNA viruses, but its main clinical use is for CMV retinitis.

These antiviral agents inhibit only replicating herpesvirus and do not eliminate the latent virus. Reduced susceptibility is occasionally found in isolates of HSV, VZV or CMV from severely immunocompromised hosts.

Other inhibitors of viral RNA or DNA synthesis

Ribavirin is a nucleoside analogue that has no useful anti-herpes activity, but is used in a nebulized form in the treatment of respiratory syncytial virus infection and intravenously for Lassa fever. It is also used orally in combination with interferon for the treatment of chronic hepatitis C.

Lamivudine, one of the nucleoside analogues active against HIV reverse transcriptase (see below), also inhibits reverse transcriptase activity of hepatitis B virus (HBV), and reduces virus replication. Adefovir, a nucleotide phosphonate analogue, used as a pro-drug (adefovir dipivoxil), also exhibits activity against HBV, including lamivudine-resistant virus.

Non-nucleoside anti-herpes agents

Foscarnet, a pyrophosphate analogue, inhibits nucleic acid synthesis without requiring any activation, and is an important agent for treatment of herpesvirus infections that have become resistant to aciclovir or ganciclovir through mutations in the viral thymidine kinase (or protein kinase) gene. Fomivirsen is an antisense oligonucleotide that inhibits translation of mRNA into proteins. It has limited use as a local treatment for CMV retinitis.

Agents that block viral uncoating

Amantadine, a symmetrical amine compound, has long been known to inhibit influenza A virus replication, but its mode of action was discovered only much later when resistant mutants could be examined at the molecular level. Amantadine (and rimantadine, a similar drug) blocks an ion channel formed by the integral membrane protein (M2) of influenza A (but not influenza B) virus, preventing uncoating of the virus within cells.

Neuraminidase inhibitors

Zanamivir and oseltamivir are agents specifically designed to block the action of the influenza virus enzyme neuraminidase by occupying the catalytic site. They act on influenza A and B viruses. Neuraminidase is found on the surface of influenza virus, and these compounds act extracellularly, reducing the spread of influenza viruses locally. Zanamivir is taken by inhalation into the oropharynx; oseltamivir is the oral pro-drug form of a similar agent.

Interferons

Interferons, naturally occurring antiviral compounds produced by mammalian cells in response to viral infection, are now manufactured by genetic recombination. Interferon-α is used to reduce persistent carriage of hepatitis B and C viruses. The mode of action of interferon-α is complex, producing an antiviral state in cells to which it binds, and interfering with the production of virus in those cells (see Ch. 10). However, interferons also act through immune modulation, upregulating the expression of major histocompatibility complex (MHC) molecules on cell surfaces. This is a significant part of the action in clearing chronic hepatitis B.

Modification of interferon by addition of polyethylene glycol – *peginterferon* – results in sustained effects after just one dose per week. This agent appears to offer an effective treatment for chronic hepatitis C, a common form of chronic hepatitis, when combined with oral ribavirin.

Antiretroviral agents

The HIV pandemic has generated an enormous interest in potential antiviral agents. Five classes of drugs are now generally available (Table 5.6). All inhibit only replicating HIV and do not eliminate the integrated proviral DNA.

When used in the recommended combination regimens (see Ch. 55), these compounds can reduce HIV replication to levels undetectable in the circulation and postpone the appearance of acquired immune deficiency syndrome (AIDS) in infected patients.

Emergence of resistant mutants has been a problem with all classes of anti-HIV agent, and has stimulated developments in the field of antiviral susceptibility testing.

Nucleoside and nucleotide reverse transcriptase inhibitors

The earliest anti-HIV agent, zidovudine (azidothymidine), and several other compounds, including didanosine (dideoxyinosine), zalcitabine (dideoxycytidine), lamivudine and stavudine, are nucleoside analogues. New additions include abacavir and emtricitabine, and a nucleotide analogue,

Table 5.6 Antiretroviral agents in clinical use

Type of agent	Drug names
Nucleoside analogue reverse transcriptase inhibitor	Abacavir, didanosine, emtricitabine, lamivudine, stavudine, zalcitabine, zidovudine
Nucleotide analogue reverse transcriptase inhibitor	Tenofovir
Non-nucleoside reverse transcriptase inhibitor[a]	Delavirdine, efavirenz, nevirapine
Protease inhibitor	Atazanavir, amprenavir (and fosamprenavir), indinavir, lopinavir (formulated with ritonavir), nelfinavir, ritonavir, saquinavir, tipranavir
Fusion inhibitor	Enfuvirtide[a]

[a]Active against only HIV-1.

tenofovir. They are activated (phosphorylated) by cellular enzymes and inhibit the reverse transcriptase function of the viral polymerase of HIV-1 or HIV-2, with many terminating the chain of proviral DNA. Phosphorylation rates vary in different cell types, and between resting and replicating cells.

Unlike aciclovir, these nucleoside analogues are associated with some toxicity as there is less selectivity in their activation and action:

- Initial phosphorylation is by cellular kinases.
- Some inhibition of cellular (mitochondrial) DNA polymerase occurs.

Non-nucleoside reverse transcriptase inhibitors

These compounds, which include nevirapine, delavirdine and efavirenz, inhibit only HIV-1. They bind directly, without activation, away from the catalytic site of reverse transcriptase, but exert a structural change that inhibits its action. They are not incorporated into the DNA chain.

HIV protease inhibitors

These compounds act at a late stage in the viral cycle by interfering with the cleavage of essential polyprotein precursors. The numerous derivatives now available are listed in Table 5.6. The addition of protease inhibitors to reverse transcriptase inhibitors provides a 'highly active antiretroviral therapy' (HAART; see Ch. 55) and has led to immune reconstitution in many patients who had developed AIDS.

HIV fusion inhibitors

Only one example of this class of antiretroviral agents is presently available: enfuvirtide. This drug is a homologue of the short peptide sequence active in fusion of the HIV-1 envelope to the cell membrane after attachment, and it inhibits that final stage of the entry process. The drug has to be given by subcutaneous injection, and its use is reserved for patients who have limited options for treatment. It is not active against HIV-2.

Other inhibitors of the HIV entry process are under development, including drugs that block the co-receptors for attachment. Cellular as well as viral factors will influence the success of this approach.

ANTIPARASITIC AGENTS

The choice of agents for the treatment of protozoal and helminthic infections remains extremely limited. Part of the problem is that protozoa and helminths are very varied, reflecting diverse solutions to the problems of their specialized parasitic existence. Consequently, there are few 'broad-spectrum' antiprotozoal or anthelminthic agents, although some compounds exhibit a surprising range of activity. Thus the antiprotozoal nitroimidazoles, such as metronidazole, are active against *Entamoeba histolytica*, *Trichomonas vaginalis* and *Giardia lamblia* (see Ch. 61); benzimidazoles, such as mebendazole and albendazole, act against most intestinal nematodes; praziquantel not only exhibits good activity against all human schistosomes but also includes other trematodes and tapeworms in its spectrum (see Ch. 62). Most remarkably of all, ivermectin is active not only against many filarial worms and some intestinal roundworms but also against ectoparasites such as the scabies mite.

Antimicrobial agents commonly used against pathogenic protozoa and helminths are shown in Tables 5.7 and 5.8, respectively.

ANTIMICROBIAL SENSITIVITY TESTS

To establish the activity of an antimicrobial agent, microorganisms are tested for their ability to grow in the presence of suitable concentrations of the drug. Bacteria are the simplest to test and, in practice, routine sensitivity testing is usually reserved for the common, easily grown, bacterial pathogens. Tests of mycobacteria, fungi,

Table 5.7 Principal agents used against the major protozoan parasites of man

Species	Agent
Cryptosporidium parvum	Nitazoxanide
Entamoeba histolytica	Metronidazole Diloxanide furoate (Emetine)
Giardia lamblia	Metronidazole (Mepacrine)
Leishmania spp.	Sodium stibogluconate Amphotericin B Miltefosine
Plasmodium spp.	Chloroquine Quinine Pyrimethamine Proguanil Mefloquine Halofantrine Artemisinin and its derivatives
Toxoplasma gondii	Pyrimethamine + sulfadiazine
Trichomonas vaginalis	Metronidazole
Trypanosoma brucei ssp. rhodesiense and gambiense	Melarsoprol (Suramin) (Pentamidine) Eflornithine[a]
Trypanosoma cruzi	(Nifurtimox)

Compounds in parentheses are of limited value.
[a]Not active against T. brucei rhodesiense.

Table 5.8 Spectrum of activity of the principal anthelminthic agents

Agent	Active against
Benzimidazoles[a]	Intestinal nematodes
Diethylcarbamazine	Filarial worms
Ivermectin	Onchocerca volvulus; other filariae
Levamisole	Hookworms; Ascaris lumbricoides
Niclosamide	Tapeworms
Metrifonate	Schistosoma haematobium
Oxamniquine	Schistosoma mansoni
Piperazine	Ascaris lumbricoides; Enterobius vermicularis
Praziquantel	Schistosoma spp.; other trematodes; tapeworms
Pyrantel pamoate	A. lumbricoides; E. vermicularis; hookworms
Trivalent antimonials	Schistosoma spp.

[a]Include mebendazole, tiabendazole and albendazole.

viruses and some other organisms are available in certain laboratories and reference units. Genotypic assays for mutations associated with antiviral resistance are becoming more widely available in specialist virology centres. Similarly, genotypic assays that detect resistance mutations affecting antituberculosis therapy are available through reference centres.

Antibacterial agents

Potency of antibacterial agents is often expressed as the *minimum inhibitory concentration* (MIC): the lowest concentration of the agent that prevents the development of visible growth of the test organism during overnight incubation. Serial dilutions of the agent are prepared in a suitable broth or agar medium, and a standard inoculum of the test organism is added. If agar is used, many different isolates can be tested at the same time by spot inoculation of the plate. Broth dilution MIC titrations have the advantage that the *minimum bactericidal concentration* (MBC) can additionally be estimated by subculture of dilutions of the antibiotic above that in which inhibition has occurred overnight. The MBC is usually taken as the lowest concentration capable of reducing the original inoculum by a factor of 1000, for example from 10^5 colony-forming units (cfu)/ml to 10^2 cfu/ml or below. To establish the rate of killing, the number of viable organisms in broth cultures is measured at timed intervals after addition of appropriate concentrations of the antibiotic.

Antibiotic titrations are too laborious for the routine assessment of antibiotic activity in clinical practice, although a truncated form of the method, in which isolates are tested at agreed *break points* of sensitivity or resistance, is sometimes used. A common alternative method is the *disc diffusion test*. The culture to be examined is seeded confluently, or semi-confluently, over the surface of an agar plate, and paper discs individually impregnated with different antibiotics are spaced evenly over the inoculated plate. Antibiotic diffuses outwards from each disc into the surrounding agar and produces a diminishing gradient of concentration. On incubation, the bacteria grow on areas of the plate except those around the drugs to which they are sensitive. The width of the *zone of inhibition* is a rough measure of the degree of sensitivity to the drug. A highly standardized version of the disc diffusion method, the *Bauer–Kirby test*, is favoured in the USA and some other countries.

Characteristics of the growth medium (such as pH or the presence of antagonizing substances), the size of the bacterial inoculum and the conditions of the test may influence the results of sensitivity tests. In disc methods, the size of the inhibition zone is additionally affected by the diffusion characteristics of the antibiotic. It is therefore important that suitable control organisms are tested in parallel in all tests of antibiotic susceptibility. In *Stokes' method*, control is achieved by placing antibiotic discs at the interface between inocula of the test and control bacteria on the same plate. In this way the zone of inhibition obtained with the test organism can be compared directly with that of a known control, and variations in the cultural conditions or in disc content can be avoided.

Antiviral agents

Tests for antiviral susceptibility are still far from routine, being readily available only through reference or specialist laboratories. Experience with these assays is steadily increasing, and the clinical benefits of measuring antiviral susceptibility are gradually being established.

The detection of mutations associated with drug resistance is now a routine procedure in the management of HIV infection. Some reference laboratories have built up a great deal of experience with in-house sequencing of HIV genes and have generated large databases, which are shared on an international basis to allow for prediction of reduced susceptibility. Expert opinion on the significance of the mutations may usefully guide therapeutic decisions and is invaluable for selection of therapeutic combinations.

Phenotypic assays

Phenotypic assays measure the inhibitory effect of the antiviral agent on the clinical virus isolate. The plaque reduction assay for inhibition of HSV by aciclovir is one example. An effect is usually considered to be significant when virus replication or product formation is reduced by 50% compared to that found with no drug (50% inhibitory concentration: IC_{50}).

For HIV assays, inhibition of replication in cultures of peripheral blood mononuclear cells is preferred. There are also recombinant virus assays in which HIV genes of interest from the test isolate are put into a laboratory strain to provide a cheaper, faster assay.

Genotypic assays

These tests detect the presence of known resistance-associated mutations in viral genes (e.g. HIV reverse transcriptase or protease), from which reduced susceptibility is predicted. Some of these assays are available in commercial kit format, for example:

- a line probe assay by reverse hybridization of amplified viral fragments with a panel of oligonucleotides

- an automated sequencing programme detecting mutations in reverse transcriptase and protease genes.

Genotypic assays are more commonly used than phenotypic ones, being simpler, safer and less costly for routine use.

ASSAY OF ANTIMICROBIAL DRUGS

In most clinical circumstances in which antimicrobial agents are used it is not usually necessary to measure the concentrations achieved in blood or other body fluids (see Ch. 66). Such measurements are needed, however, during the development of a new agent to establish the pharmacokinetic behaviour of the drug.

Various techniques are available, including microbiological methods in which the antibiotic-containing material and known dilutions of the drug are titrated in parallel against a susceptible indicator organism. Analy-

---

KEY POINTS

- Most antimicrobial agents are active only against bacteria; smaller numbers of antifungal, antiviral, antiprotozoal and anthelminthic agents are available.
- Antimicrobial agents used in clinical medicine do not affect spores of bacteria or fungi, or latent viruses.
- The largest group of antibacterial agents are β-lactam compounds (penicillins, cephalosporins, etc.), most of which are semi-synthetic derivatives of naturally occurring antibiotics. Members of this group have widely different properties.
- Other widely used antibacterial agents include aminoglycosides, tetracyclines, macrolides, glycopeptides and quinolones.
- Most antifungal agents are suitable only for topical application. Agents used for systemic therapy of fungal infection include amphotericin B, certain azole derivatives, griseofulvin and terbinafine.
- The largest group of antiviral compounds are antiretroviral (anti-HIV) or anti-herpes agents, such as aciclovir, that act on nucleic acid synthesis.
- Combinations of three or more antiretroviral drugs can effectively keep HIV levels in the circulation down at the limit of detection.
- Most antiparasitic agents have specific activity against particular protozoa or helminths, although some have broader activity.
- Routine antibiotic sensitivity testing is useful in the case of common bacterial pathogens. The disc diffusion method is commonly used.

tical techniques such as high-pressure liquid chromatography may also be used.

In the few cases in which assays are needed in clinical practice, notably during treatment with aminoglycosides, immunochemical methods are often used with commercially available instrumentation.

With regard to antiviral drug assays, concerns over toxic levels of aciclovir or ganciclovir arise on occasion, and these can be measured. Therapeutic drug monitoring is also practised in relation to HIV therapy, mainly to test for compliance with the onerous regimens necessary in the management of this chronic condition.

RECOMMENDED READING

Finch R G, Greenwood D, Norrby S R, Whitley R J 2002 Antibiotic and Chemotherapy: Anti-infective Agents and their Use in Therapy, 8th edn. Churchill Livingstone, Edinburgh

Franklin T J, Snow G A 2005 *Biochemistry and Molecular Biology of Antimicrobial Drug Action*, 6th edn. Springer, New York

Greenwood D, Finch R G, Davey P G, Wilcox M H 2007 *Antimicrobial Chemotherapy*, 5th edn. Oxford University Press, Oxford

Kucers A, Crowe S M, Grayson M L, Hoy JF 1997 *The Use of Antibiotics: A Clinical Review of Antibacterial, Antifungal and Antiviral Drugs*, 5th edn. Hodder Arnold, London

Lorian V (ed.) 2005 *Antibiotics in Laboratory Medicine*, 5th edn. Lippincott, Williams & Wilkins, Baltimore

Russell A D, Chopra I 1996 *Understanding Antibacterial Action and Resistance*, 2nd edn. Ellis Horwood, Chichester

Scholar E M, Pratt W B 2000 *The Antimicrobial Drugs*, 2nd edn. Oxford University Press, New York

Internet sites

UK Department of Health, Standing Medical Advisory Committee on Antimicrobial Resistance. *The Path of Least Resistance*. http://www.advisorybodies.doh.gov.uk/smac1.htm

United States National Library of Medicine and National Institutes of Health. *Antibiotics*. http://www.nlm.nih.gov/medlineplus/antibiotics.html

World Health Organization. *Medicine Policy and Standards, Technical Cooperation for Essential Drugs and Traditional Medicine*. http://www.who.int/medicines/en

6 Bacterial genetics

K. J. Towner

GENETIC ORGANIZATION AND REGULATION OF THE BACTERIAL CELL

All properties of a bacterial cell, including those of medical importance such as virulence, pathogenicity and antibiotic resistance, are determined ultimately by the genetic information contained within the cell *genome*. This information is normally encoded by the specific sequence of nucleotide bases comprising the DNA of the cell. There are four common nucleotide bases in DNA: adenine, guanine, cytosine and thymine; it is the linear order in which these bases are arranged that determines the properties of the cell. With only a few exceptions, most of the genetic information required by the bacterial cell is arranged in the form of a single circular double-stranded chromosome. In *Escherichia coli* the chromosome is about 1300 μm long and occurs in an irregular coiled bundle lying free in the cytoplasm. The DNA is not associated with histone proteins as it is in eukaryotic cell chromosomes, although there are many DNA binding proteins involved in regulation of gene expression, including several proteins referred to as histone-like.

In addition to the single main chromosome, bacterial cells may also carry one or more small circular extra-chromosomal elements. These were originally called episomes, but are now termed *plasmids*. Plasmids replicate independently of the main chromosome in the cell. Although dispensable, they often carry supplementary genetic information coding for beneficial properties (e.g. resistance to antibiotics) that enable the host cell to survive under a particular set of environmental conditions.

A third source of genetic information in a bacterial cell can be provided by the presence of certain types of bacterial virus, *bacteriophages*. Bacteriophages consist essentially of just a protein coat enclosing the virus genome and, because they are unable to multiply in the absence of their bacterial host, generally they are lethal to their host cell. In some instances, they can enter a potentially long-term state of controlled replication, *lysogeny*, within the bacterial cell without causing lysis. In such a state the bacteriophage genome is referred to as a *prophage* and effectively becomes a temporary part of the total genetic information available to the cell and may consequently bestow additional properties on the cell.

Not all genetic material consists of double-stranded DNA. Some bacteriophages contain single-stranded molecules of either DNA or RNA, which can be either circular or linear in configuration. The genetic elements under consideration here, including bacteriophages, plasmids and members of the domain Bacteria, vary in size by more than three orders of magnitude, from about 3 to 5000 kilobases (kb) (Table 6.1).

Processes leading to protein synthesis

The character of a bacterial cell is determined essentially by the specific polypeptides that comprise its enzymes and other proteins. The DNA acts as a template for the *transcription* of RNA by RNA polymerase for subsequent protein production within the cell. In the transcription process the specific sequence of nucleotides in the DNA determines the corresponding sequence of nucleotides in the messenger RNA (mRNA). This, in turn, is then *translated* into the appropriate sequence of amino acids by ribosomes. Finally, the sequence of amino acids in the resulting polypeptide chain determines the configuration into which the polypeptide chain folds itself, which in many cases determines the enzymic properties of the completed protein. A segment of DNA that specifies the production of a particular polypeptide chain is called a *gene*, and the processes of transcription and translation leading to protein synthesis are collectively termed the *central dogma* of molecular biology. These processes are illustrated schematically in Figure 6.1.

Gene regulation

Most bacteria contain enough DNA to code for the production of between 1000 and 3000 different polypeptide chains – 1000 to 3000 different genes. However, during normal

Table 6.1 Examples of prokaryotic genetic elements

Genetic element	Type	Configuration	Size (kb)
Bacterial chromosome			
Escherichia coli	DNA	ds circular	3.8×10^3
Bacillus subtilis	DNA	ds circular	2.0×10^3
Plasmids			
R300B	DNA	ds circular	9.0
RK2	DNA	ds circular	60.0
Bacteriophages			
MS2	RNA	ss linear	3.6
Φ174	DNA	ss circular	5.4
T7	DNA	ds linear	40.4

ss, single-stranded; ds, double-stranded.

bacterial life, some polypeptides will be required only at particular stages, whereas others will be needed only when the cell is provided with a new or unusual growth substrate, or is confronted with a new challenge, such as an antibiotic. Thus, many antibiotic resistance mechanisms found in bacteria are *inducible* (see below). Protein production is an energy-intensive process, and therefore the expression of many genes is controlled actively within the cell to prevent wasteful energy consumption.

In bacteria the process of gene expression is regulated mainly at the transcriptional level, thereby conserving the energy supply and the transcription–translation apparatus. This is achieved by means of regulatory elements that either inhibit or enhance the rate of RNA chain initiation and termination for a particular gene. Numerous complex regulatory chain mechanisms are involved in co-ordinating the many biochemical reactions that proceed inside a cell, but related genes involved in a common regulatory system are often clustered on the bacterial chromosome. Such functional clusters are known as *operons*, of which the most well known example is the lactose operon of *Esch. coli* (Fig. 6.2).

For transcription to occur as the first stage in protein synthesis, RNA polymerase has to attach to DNA at a specific *promoter* region and transcribe the DNA in a fixed direction. This process can be switched off by the attachment of a *repressor* molecule to a specific region of the DNA, known as the *operator*. This lies between the promoter and the structural gene(s) being transcribed; the repressor then blocks the movement of the RNA polymerase molecule so that the genes downstream are not transcribed.

The repressor is often an allosteric molecule with two active sites. One recognizes the operator region so that the repressor can bind to it to prevent transcription. The other recognizes an *inducer* molecule. When the inducer is present, it binds to the repressor and alters simultaneously the binding specificity at the other site, so that the repressor no longer binds to the operator and transcription can resume.

There are many different variations on this basic regulatory system. For example, a repressor may be normally inactive, but activated by the end-product of a biosynthetic pathway; thus, only when the end-product is present in adequate concentration will the repressor combine with the operator and switch off transcription of the operon. Alternatively, regulation of certain operons involves proteins that bind to the DNA and assist RNA polymerase to initiate transcription. These are just a few examples, and other regulatory systems display both minor and major differences. Finally, it should be stressed that prokaryotic gene regulation frequently involves interwoven regulatory circuits that respond to a variety of different stimuli. As noted in Chapter 4, these networks of genes are organized into functional units known as *stimulons* when they respond to a particular stimulus and *regulons* when they are regulated by a single protein.

MUTATION

As bacteria reproduce by asexual binary fission, the genome is normally identical in all of the progeny. The DNA replication process is therefore very accurate, but occasional rare inaccuracies produce a slightly altered nucleotide sequence in one of the progeny cells. Such a mutation is heritable and will be passed on stably to subsequent generations. One of the fundamental requirements for evolution is that, although gene replication must normally be completely accurate to ensure stability, there must also be occasional variation to produce new or altered characters that could prove to be of selective value to the organism. Mutations may not produce any observable effect on the structure or function of the corresponding protein, but in a small proportion of cases an enzyme with altered specificity for substrates, inhibitors or regulatory molecules may be produced. This is the kind of mutation that is most likely to be of evolutionary value to an organism; indeed, many examples of acquired antibiotic resistance have been shown to be of this type (see Ch. 5). Other mutations may alter a gene so that a non-functional protein is formed; if this protein is essential to the cell then the mutation will be lethal.

As mutation may occur in any of the several thousand genes of the cell, and different mutations in the same gene may produce different effects in the cell, the number of possible mutations is very large. Particular mutations occur at fairly constant rates, normally between one per 10^4 and one per 10^{10} cell divisions. As a large bacterial colony contains at least 10^9 cells, even a 'pure' bacterial culture will contain many thousands of different mutations affecting many of the genes in the cell. Some of these mutations will be viable and could be selected by particular environmental conditions

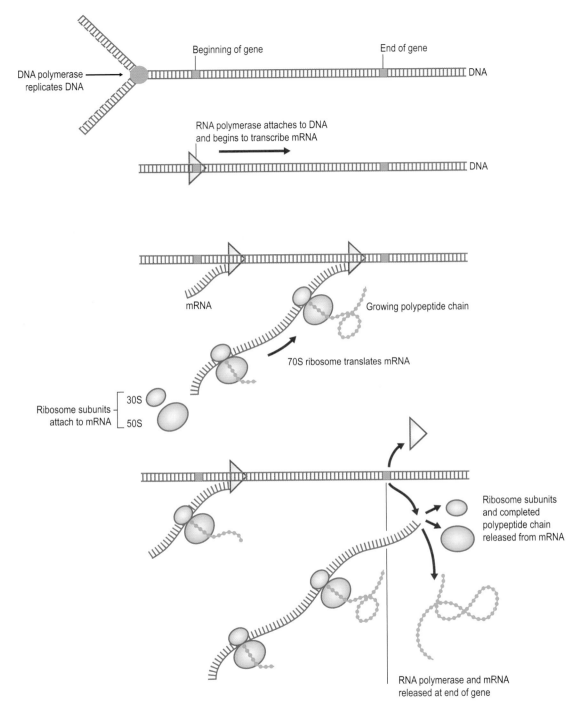

Fig. 6.1 The central dogma of molecular biology.

during subculture. For the same reason, in an infected patient, a variety of mutants will appear spontaneously in the population that grows from the few bacteria originally entering the body. Such mutations may enhance the ability of an organism to grow in the body, for example by conferring antibiotic resistance, enhanced virulence or altered surface antigens. In such a situation, cells with the mutation will rapidly outgrow cells without the mutation, so that selection of the mutant cells occurs and they soon become the predominant type.

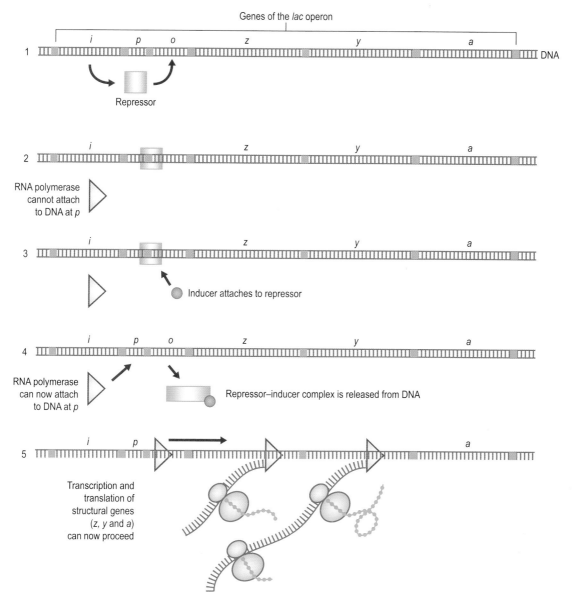

Fig. 6.2 The *lac* operon of *Escherichia coli*.
1. The *lac* repressor is produced from the *i* gene.
2. Binding of the repressor to the operator site (*o*) prevents transcription of the genes *z* (β-galactosidase), *y* (galactoside permease) and *a* (transacetylase).
3. The inducer (lactose, or a closely related derivative) can bind specifically to the repressor.
4. The repressor molecule is thereby altered at its operator-binding site and the repressor–inducer complex is released from the DNA.
5. RNA polymerase can now attach to the promoter site (*p*) and transcribe the structural genes of the *lac* operon. Note that, in bacteria, several genes may be transcribed into a single *polycistronic* mRNA molecule.

Phenotypic variation

The properties of a bacterial cell at a particular time are referred to as the *phenotype* of the cell. These properties are determined not only by its genome (*genotype*), but also by its environment. *Phenotypic variation* occurs

when the *expression* of genes is changed in response to the environment, for instance by the induction or repression of synthesis of particular enzymes. Such changes in gene expression underpin the differentiation process involved in sporulation (see p. 18), the differing phenotypes observed during different growth phases

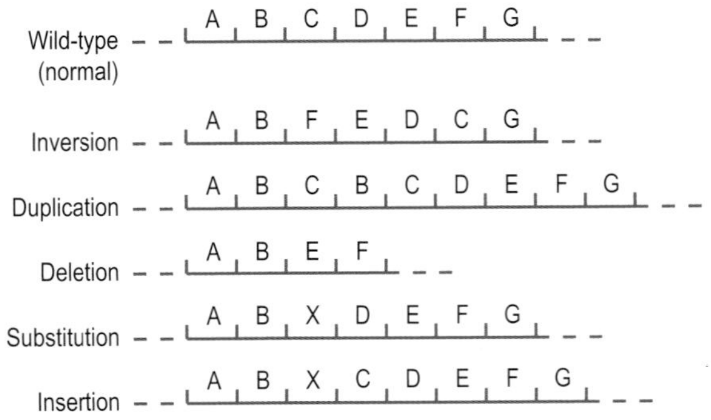

Fig. 6.3 Examples of types of mutation. The top sequence represents a portion of the wild-type chromosome from which the different mutational rearrangements shown below are derived.

and under different growth conditions (p. 39) and the physiological *adaptive* responses of bacteria (p. 44). The distinction between genotypic mutation and phenotypic variation is critical; the former is heritable and maintained through changes in environmental conditions, whereas the latter is reversible, being dependent on environmental conditions and altering when these change. Phenotypic variation is therefore not a form of mutation.

Types of mutation

Mutations can be divided conveniently into *multisite mutations*, involving extensive chromosomal rearrangements such as inversions, duplications and deletions, and *point mutations*, which are defined as only affecting one, or very few, nucleotides. The structure of DNA is such that point mutations can be divided into one of three basic types (Fig. 6.3):

1. substitution of one nucleotide for another
2. deletion of one or more nucleotides
3. insertion of one or more nucleotides.

Mutations occur spontaneously during replication of DNA, but most are corrected immediately by the editing apparatus of the cell. Occasional mutations, particularly those conferring a selective advantage to the cell, will be inherited stably by the progeny, but secondary mutations can occasionally restore the original nucleotide sequence. It is important to distinguish this relatively rare event of *back-mutation* from the separate process of *phase variation*, which is readily reversible and occurs with relatively high frequency in either direction, for example once per 10^3 cell divisions. The variation of certain Gram-negative bacteria between a fimbriate and a non-fimbriate phase, and the variation of flagellar antigens in *Salmonella enterica* serotypes, are examples of phase variation. These seem to involve special genetic regions that are specifically

inverted to yield alternative gene products and different phenotypes. In some cases it is known that promoters initiate RNA transcription in different directions to give a flip-flop type of action. The number of such switching systems is probably quite limited, but they have value to the organism in providing a mechanism to switch to a reversible alternative, as opposed to an irreversible change.

Until the invention of the polymerase chain reaction (PCR; see below and Ch. 3), the study of bacterial genetics was dependent largely upon the isolation and characterization of mutants in particular genes. Mutations may occur in any gene, but many individual mutations are lethal. Other mutations affect gene products that are essential only under particular cultural conditions. The detailed function and regulation of processes in the bacterial cell can often be analysed only by searching systematically for mutations that affect each separate step in a process. Thus, mutations can be found that produce increased resistance to almost any antimicrobial agent, and the study of such mutants is an essential step in understanding the modes of action of antibiotic agents and mechanisms of resistance to them.

GENE TRANSFER

A change in the genome of a bacterial cell may be caused either by a mutation in the DNA of the cell or result from the acquisition of additional DNA from an external source. DNA may be transferred between bacteria by three mechanisms:

1. transformation
2. conjugation
3. transduction.

Each of these mechanisms probably occurs at a low frequency in nature and may therefore have been of value in bacterial evolution. It should, however, be noted that the acquisition by bacteria of new properties following gene transfer is significant, as with mutation, only if the new genetic end-product is subject to favourable selection by the conditions under which the bacteria are growing.

Transformation

Most species of bacteria are unable to take up exogenous DNA from the environment; indeed, most bacteria produce nucleases that recognize and break down foreign DNA. However, bacteria in some genera, notably pneumococci, *Haemophilus influenzae* and certain *Bacillus* species, have been shown to be capable of taking up DNA either extracted artificially or released by lysis from cells of another strain. Cells are *competent* for transformation only under certain conditions of growth, usually in late log phase or, in *Bacillus* species, during sporulation. However, bacterial geneticists have also developed treatments by means of which organisms can be made artificially competent.

Once a piece of DNA has entered the cell by transformation, it has to become incorporated into the existing chromosome of the cell by a process of *recombination* in order to survive. This is a complex molecular process for which the transformed DNA must have been derived from a closely related strain, as pieces of DNA can normally recombine with the chromosome only when there is a high degree of nucleic acid similarity (*homology*).

Any gene may be transferred by transformation, as any fragment of a donor chromosome may be taken up by the recipient cells. However, a piece of DNA introduced into a cell by transformation will normally be relatively short, and will contain only a very small number of genes. For this reason, transformation is of limited use for studying the organization of genes in relation to one another (*genetic mapping*; see below).

Conjugation

Conjugation is a process by which one cell, the *donor* or male cell, makes contact with another, the *recipient* or female cell, and DNA is transferred directly from the donor into the recipient. Certain types of plasmids carry the genetic information necessary for conjugation to occur. Only cells that contain such a plasmid can act as donors; those lacking a corresponding plasmid act as recipients.

Transfer of DNA between cells by conjugation requires direct contact between donor and recipient cells. Plasmids capable of mediating conjugation carry genes coding for the production of a 1–2-µm long protein appendage, termed a *pilus*, on the surface of the donor cell. The tip of the pilus attaches to the surface of a recipient cell and holds the two cells together so that DNA can pass into the recipient cell. It is probable, but not absolutely certain, that transfer actually occurs through the pilus; alternatively, the pilus could act simply as a mechanism by which the donor and recipient cells are drawn together. Different types of pilus are specified by different types of plasmid and can therefore be used as an aid to plasmid classification.

In the vast majority of cases, the only DNA transferred during the conjugation process is the plasmid that mediates the process. It is thought that one strand of the circular DNA of the plasmid is nicked open at a specific site and the free end is passed into the recipient cell. The DNA is replicated during transfer so that each cell receives a copy. As donor ability is dependent upon having a copy of the plasmid, the recipient strain becomes converted into a donor, able to conjugate with further recipients and convert them in turn. In this way a plasmid may spread rapidly through a whole population of recipient cells; this process is sometimes described as infectious spread of a plasmid.

Mobilization of chromosomal genes by conjugation

Many different types of plasmid have the ability to transfer themselves. Some (but not all) plasmids also have the ability to mobilize the chromosomal genes of bacteria. The prototype plasmid of this type is the 'F factor' (fertility factor) of *Esch. coli*. The F factor is a plasmid that contains the basic genetic information for extrachromosomal existence and for self-transfer. Cells that contain the F plasmid free in the cytoplasm (F^+ cells) have no unusual characteristics apart from the ability to produce F pili and to transfer the F plasmid to F^- cells by conjugation. In a very small proportion of F+ cells, the F plasmid becomes inserted into the bacterial chromosome. Once inserted, the entire chromosome behaves like an enormous F plasmid, and hence chromosomal genes can be transferred in the normal manner to a recipient cell at a relatively high frequency. Such cultures are termed *high-frequency recombination* (Hfr) strains.

It is important to emphasize that the F plasmid system is confined to *Esch. coli* and other closely related enteric bacteria. However, many other plasmids are capable of mediating conjugation, and sometimes chromosome mobilization, not only in *Esch. coli* but also in other bacteria. For example, plasmid RP4 and its relatives have been used to mediate conjugation in a wide range of Gram-negative bacteria, and there have been reports of conjugation systems in Gram-positive bacteria, such as *Enterococcus faecalis*, and several *Streptomyces* species.

Transduction

The third known mechanism of gene transfer in bacteria involves the transfer of DNA between cells by bacteriophages. Most bacteriophages carry their genetic information (the phage genome) as a length of double-stranded DNA coiled up inside a protein coat. Other phages are known in which the phage genome consists of single-stranded DNA or RNA but, as far as is known, transducing phages all contain double-stranded DNA. Two major types of transduction are known to occur in bacteria: *generalized* and *specialized* transduction. In both types, bacterial genes are occasionally and accidentally incorporated into new phage particles. When such a phage particle subsequently infects a second bacterial cell, the DNA that enters the cell includes a short segment of chromosome from the original host. Bacterial genes have been *transduced* by the phage into the second cell. Genes can be transduced only between fairly closely related strains, as particular phages usually attack only a limited range of bacteria. As well as chromosomal genes, transducing bacteriophages may also pick up and transfer plasmid DNA. As an example, the penicillinase gene in staphylococci is usually located on a plasmid, and it may be transferred into other staphylococcal strains by transduction.

Lysogenic conversion

The presence of prophage DNA constitutes a genetic alteration to the host cell. Usually only the phage repressor gene is expressed, but in certain cases it can be demonstrated that other genes are also expressed by the host cell. For example, *Corynebacterium diphtheriae* produces diphtheria toxin only when it is lysogenized by β phage; the toxin is specified by one of the phage genes. This process is termed *lysogenic conversion*. It is probable that the production of many toxins by staphylococci, streptococci and clostridia is also dependent upon lysogenic conversion by specific bacteriophages. In such cases, lysogenic conversion not only gives the cell superinfection immunity, but also actively influences the virulence of the bacterium for humans.

PLASMIDS

Properties encoded by plasmids

As described above, plasmids are circular extrachromosomal genetic elements that may encode a variety of supplementary genetic information, including the information for self-transfer to other cells by conjugation. Not all plasmids can transfer themselves: the *non-conjugative class* of plasmids encodes neither donor pili nor transfer. They can, however, be *mobilized* by other conjugative plasmids present in the same donor cell. Apart from this optional transfer ability, all bacterial plasmids contain the basic genetic information necessary for self-replication and segregation into daughter cells at cell division. Plasmids seem to be ubiquitous in bacteria; many encode genetic information for such properties as resistance to antibiotics, bacteriocin production, resistance to toxic metal ions, production of toxins and other virulence factors, reduced sensitivity to mutagens, or the ability to degrade complex organic molecules.

Plasmid classification

Because of the vast range of plasmids, it is necessary to have a means of classification so that their distribution and epidemiology can be studied. Plasmids can be grouped initially according to the properties that they encode, but other physical and genetical methods are needed to study their spread and distribution.

As all plasmids are relatively small structures that are normally separate from the bacterial cell chromosome, it is possible to isolate them from the chromosome by physical techniques. Centrifugation and electrophoresis techniques allow the sizes of different plasmids to be compared directly. Plasmids of similar size that confer identical phenotypes on host cells can be compared by generating *restriction endonuclease fingerprints* from purified plasmid DNA. A restriction endonuclease is an enzyme that cuts the DNA molecule at, or near to, a specific nucleotide sequence to produce discrete DNA fragments that can be separated by gel electrophoresis. The pattern ('fingerprint') of fragments produced is dependent on the distribution of the specific DNA sequences recognized by the enzyme. Closely related plasmids will produce the same, or very similar, fingerprints, whereas unrelated plasmids will produce different fingerprints. Restriction endonucleases of different specificities may be needed to generate distinctive fingerprints.

An initial genetic test will distinguish groups of plasmids that are self-transmissible from those that are not. Linked with the question of transferability is the question of host range; for example, some groups can be transferred only between members of the enterobacteria, others can be transferred from the enteric bacteria to the *Pseudomonas* family, whereas others can be transferred between almost any Gram-negative bacteria. Similar host range relationships exist among plasmids of Gram-positive bacteria.

Once the host range of a plasmid has been determined, plasmids may be classified by *incompatibility testing*. This method relies on the fact that closely related plasmids are unable to coexist stably in the same bacterial cell. Plasmids that are sufficiently closely related to interfere with one another's replication in this manner are said to be *incompatible* and to belong to the same *incompatibility group*. In contrast, unrelated plasmids can coexist stably and are therefore said to belong to different incompatibility groups.

A final method for plasmid classification involves the use of specific virulent bacteriophages. All members of the same plasmid incompatibility group produce the same type of pilus for conjugation. For some incompatibility groups, specific virulent bacteriophages have been isolated that will adhere only to the type of pilus produced by that particular group of plasmids. Lysis by such a phage shows that a particular type of pilus is being produced, which in turn allows the identification of the group to which the plasmid contained in the cell belongs.

Plasmid epidemiology and distribution

Some plasmid groups have been identified in many different countries of the world, whereas others have so far been found only in a single bacterial species isolated from a solitary ecological niche. There seem to be two major ways in which plasmids spread:

1. by direct transfer from one bacterium to another in a particular micro-environment
2. by being carried in a particular host from one environment, such as a hospital, to another.

The epidemiological tracing of these pathways requires identification not just of the plasmids involved but also of their host bacterial strains.

GENETIC MAPPING

The location of genes with reference to one another and to their respective control regions by mapping techniques is an essential part of genetic analysis. Historically, this required the transfer of genetic material between different mutants, initially by conjugation to locate the approximate position of an unknown gene on the chromosome or a plasmid, followed by fine structure mapping with generalized transduction. By far the most extensive genetic map available is that of the *Esch. coli* strain K12 chromosome. It is now a relatively simple matter to map an 'unknown' gene on the chromosome of *Esch. coli* K12, but much more difficult for less intensively studied genera.

A more modern approach that can be used for all bacterial genera makes use of molecular techniques allowing the isolation, cloning and nucleotide sequencing of individual genes (see Ch. 3). Once a particular gene or sequence of DNA has been selected and analysed, it is possible to label it and use the gene as a *probe* in hybridization experiments. Alternatively, it is possible to synthesize a small stretch of nucleotides, termed an *oligonucleotide*, from within the overall gene sequence for use as a probe. *Hybridization* is a process in which two single strands of nucleic acid come together to form a stable double-stranded molecule. As long as the sequence of bases is complementary on each strand, the two strands will bind together by the formation of hydrogen bonds. Oligonucleotide probes are normally only 10 to 40 bases in length, and can be synthesized and chemically labelled by automated instruments designed specifically for this purpose. The labelled probe can then be used in hybridization experiments with relatively large fragments ('fingerprints') generated from the entire bacterial chromosome with rare-cutting restriction endonucleases. The probe will hybridize only to the chromosomal DNA fragment containing the original gene from which the probe was derived, and will therefore indicate the precise physical location of the original gene in relation to other known genes and control regions. Thus, it is now possible to elucidate the relationship between gene structure and function in the cell at the most fundamental molecular level.

GENETIC BASIS OF ANTIBIOTIC RESISTANCE

All of the properties of a micro-organism are determined ultimately by genes located either on the chromosome or on plasmids or on lysogenic bacteriophages. With regard to antibiotic resistance, it is important to distinguish between *intrinsic* and *acquired* resistance. Intrinsic resistance is dependent upon the natural insusceptibility of an organism. In contrast, acquired resistance involves changes in the DNA content of a cell, such that the cell acquires a phenotype (i.e. antibiotic resistance) that is not inherent in that particular species.

Intrinsic resistance

Organisms that are naturally insensitive to a particular drug will always exist. The most obvious determinant of bacterial response to an antibiotic is the presence or absence of the target for the action of the drug. Thus, polyene antibiotics such as amphotericin B kill fungi by binding tightly to the sterols in the fungal cell membrane and altering the permeability of the fungal cell. As bacterial membranes do not contain sterols they are intrinsically resistant to this class of antibiotics. Similarly, the presence of a permeability barrier provided by the cell envelopes of Gram-negative bacteria is important in determining sensitivity patterns to many antibiotics. Intrinsic resistance is usually predictable in a clinical situation and should not pose problems, provided an informed and judicious choice is made of appropriate antimicrobial therapy.

Acquired resistance

A problem of antimicrobial chemotherapy has been the appearance of resistance to particular drugs in a normally sensitive microbial population. An organism may lose its sensitivity to an antibiotic during a course of treatment. In some cases the loss of sensitivity may be slight, but often organisms become resistant to clinically achievable concentrations of a drug. Once resistance has appeared, the continuing presence of an antibiotic exerts a *selective pressure* in favour of the resistant organisms. Three main factors affect the frequency of acquired resistance:

1. the amount of antibiotic that is being used
2. the frequency with which bacteria can undergo spontaneous mutations to resistance
3. the prevalence of plasmids able to transfer resistance from one bacterium to another.

Chromosomal mutations

Random spontaneous mutations occur continuously at a low frequency in all bacterial populations, and some mutations may confer resistance to a particular antibiotic. The rate at which these mutations occur is not influenced by the antibiotic, but in the presence of the drug the resistant mutant can survive, grow and eventually become the predominant, or only, member of the population. The degree of resistance conferred by chromosomal mutation depends on the biological consequences of the mutation. With *single large-step mutations*, the drug target is altered by mutation so that it is totally unable to bind a drug, although it can still carry out its normal biological functions sufficiently well to permit the continued survival of the cell. This type of mutation occurs with streptomycin, but is otherwise not common clinically. More commonly, the target is altered so that it can no longer bind a drug as efficiently, although it still has some residual affinity. In such a case a higher concentration of antibiotic would be required to produce the same antimicrobial effect: the *minimum inhibitory concentration* (MIC) of the antibiotic for the organism would be increased. Once a slightly resistant organism has been produced, additional mutational events – each conferring an additional small degree of resistance – can eventually lead to the production of organisms that are highly resistant. This is called the *multistep pattern of resistance*. Spontaneous chromosomal mutation is of clinical importance in tuberculosis, in which mutants resistant to any single drug (e.g. streptomycin, rifampicin or isoniazid) are likely to be present in the patient before the start of treatment. If only one drug is given to the patient, the few resistant mutant bacteria will multiply and eventually cause a relapse of the disease. Combined therapy with several drugs to which the organism is sensitive is used in the treatment of tuberculosis, so that each drug kills the few mutants that are resistant to the other. The frequency with which double or triple mutations occur spontaneously in the same cell is so low as to be clinically insignificant.

There are many other examples of chromosomal mutations to antibiotic resistance that have assumed clinical importance. Bacterial enzymes called β-lactamases are commonly responsible for resistance to penicillins, cephalosporins and related antibiotics that contain a β-lactam ring (see Ch. 5). Mutations in the genes controlling the production of chromosomally encoded β-lactamases in Gram-negative bacteria can result in overproduction of these enzymes and consequent resistance to the cephalosporin antibiotics normally regarded as stable to β-lactamase.

Chromosomal mutations leading to antibiotic resistance are in many cases just as important clinically as the types of transferable resistance described in the next section.

Transferable antibiotic resistance

Of the three modes of gene transfer in bacteria, plasmid-mediated conjugation is of greatest significance in terms of drug resistance. Plasmids conferring resistance to one or more unrelated groups of antibiotics (*R plasmids*) can be transferred rapidly by conjugation throughout the population.

R plasmids were first demonstrated in Japan in 1959, when it was shown that resistance to several antibiotics could be transferred by conjugation between strains of *Shigella* and *Esch. coli*. Many surveys since then in all parts of the world have shown that R plasmids are common and widespread.

The way in which R plasmids are built up in vivo probably varies from case to case, but it is clear that simple transfer factors can pick up resistance genes and combine them with non-transmissible resistance plasmids to produce complex transmissible R plasmids that encode resistance to as many as eight or more different antimicrobial drugs. This process of plasmid evolution is accelerated considerably by genetic elements termed *transposons*. These are linear pieces of DNA, often including genes for antibiotic resistance, that can migrate between unrelated plasmids and/or the bacterial chromosome independently of the normal bacterial recombination processes. R plasmids can transfer themselves into a wide range of commensal and pathogenic bacteria. Once resistance to an antibiotic appears in any one of these species, the process of *transposition* assists the dissemination of the responsible gene between different R plasmids and subsequent distribution to other bacterial species.

A crucial question remains regarding the mechanism by which transposons acquire the resistance genes in the first place. The answer lies in the existence of a further class of genetic element, termed an *integron*. These elements form an essential 'building block' of many transposons and allow the rapid formation and expression of new combinations of antibiotic resistance genes in response to selection pressures. A detailed description of the properties of integrons lies outside the scope of this text (see Recommended Reading), but suffice it to say that these elements seem to provide the primary mechanism for initial antibiotic resistance gene capture and dissemination, certainly among Gram-negative bacteria.

As the prevalence of multiple-resistance R plasmids carrying transposons and integrons continues to increase, infections caused by a wide range of pathogens become more difficult to treat. In addition, R plasmids can also carry genes, for example for toxin production, that confer increased virulence on a bacterial cell. Thus, use of antibiotics may select for bacteria carrying plasmids that confer not only multiple drug resistance but also increased pathogenicity.

Control of antibiotic resistance

The major cause of the spread of genes conferring antibiotic resistance is the selection pressure brought about by the increased, and often indiscriminate, use of antibiotics in man and animals. Plasmid-encoded drug resistance is increased by the widespread use of antibiotics in animal husbandry, where antibiotics are used as animal feed supplements and whole animal populations may be treated, rather than an individual subject as occurs in medical practice. When R plasmids are present, the mass use of antibiotics fails to prevent the spread of resistance and selects R plasmids in the gut flora of the whole population of animals. Such R plasmids, evolved in farm animals, have the potential to spread to human commensal *Esch. coli*, followed by transfer to more important human pathogens.

It is important to minimize the use of antibiotics as much as possible and to reduce the chance of cross-infection. Rational use of antibiotics and sensible restriction of their availability in man and animals could prevent further spread of R plasmids and perhaps reduce their incidence. Some R plasmids are unstable and tend to lose resistance genes when the selection pressure is removed. R plasmids are also lost spontaneously from a small proportion of cells in a culture, as plasmid replication and segregation are not always precisely synchronous with chromosome replication and segregation. Cells that lose an R plasmid may have a slight metabolic advantage and may slowly outgrow drug-resistant organisms. Moreover, R plasmids that evolve in one species may be unstable in another or may transfer themselves to other organisms much less efficiently. Similarly, organisms that are adapted to the gut of a calf, pig or chicken may not establish readily in human beings. Such factors may help to contain the spread of R plasmids.

APPLICATIONS OF MOLECULAR GENETICS

Specific gene probes

Every properly classified species must, by definition, have somewhere on its chromosome a unique DNA sequence that distinguishes it from every other species. If this sequence can be identified, a specific labelled DNA probe (see above) can be used in a hybridization reaction to recognize pathogen-specific DNA released from clinical samples. Initial isolation of the infecting pathogen is not necessary and, consequently, DNA probes can be used to detect pathogens that cannot be cultured easily in vitro.

DNA probes have already been used successfully to identify a wide variety of pathogens, from simple viruses to pathogenic bacteria and parasites. Probes have also been developed that can recognize specific antibiotic resistance

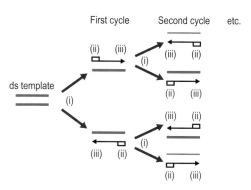

Fig. 6.4 Schematic outline of the polymerase chain reaction (PCR). Each cycle in the exponential reaction consists of three steps: (i) heat denaturation, typically at 94°C, to dissociate double-stranded (ds) DNA; (ii) annealing of primers (▭) at a temperature determined empirically for each individual PCR; (iii) elongation at an optimal temperature for thermostable polymerase activity, typically 72°C for *Taq* polymerase.

genes, so that antimicrobial susceptibility of an infecting organism can be determined directly without primary isolation and growth. Commercial kits incorporating DNA probes are now available to detect a range of bacteria and viruses.

Nucleic acid amplification technology

In the *polymerase chain reaction* (PCR) a thermostable DNA polymerase and two specific oligonucleotide primers are used to produce multiple copies of specific nucleic acid regions quickly and exponentially (Fig. 6.4). The specificity of the reaction is controlled by the oligonucleotide primers that direct replication of the intervening 'target' region. In an exponential reaction, the target sequence is amplified a million-fold or more within a few hours. Although PCR is the most widely used method, other amplification techniques for DNA and RNA molecules are available. Once an amplification reaction has occurred, a variety of methods is available to detect the amplified product, of which the simplest is to identify the product by size after electrophoresis and migration on an agarose gel. For many diagnostic applications, the simple visualization of an amplification product of characteristic size is a significant result, because it indicates the presence of the target DNA sequence in the original sample, but confirmation of sequence identity by specific hybridization tests is often required.

Amplification offers an exquisitely sensitive approach to the detection and identification of specific micro-organisms in a variety of sample types. Potentially, a characteristic DNA or RNA sequence from a single virus particle or bacterial cell can be amplified to detectable levels within a very short period of time. The method has received particular attention for detecting the presence of low numbers of bacteria or virus particles in clinical and environmental specimens. For example, although a diagnostic antibody response may take up to 8 weeks to develop in an individual infected with human immuno-deficiency virus (HIV), specific HIV sequences can be detected in a few hours by PCR, even if present at only 1 part per 100 000 human genome equivalents.

Real-time amplification

It is the detection process that discriminates real-time amplification from conventional PCR assays. The possibility of detecting the accumulation of amplification product in real time as the PCR progresses has been made possible by the labelling of primers, oligonucleotide probes (*oligoprobes*) or amplicons with molecules capable of fluorescing in the reaction tube. These fluorescent labels produce a change in signal at a specific wavelength following direct interaction with, or hybridization to, the amplicon. The signal produced is related to the amount of amplicon present at the end of each cycle and increases as the amount of specific amplicon increases (see Recommended Reading). Commercially available robotic nucleic acid extraction systems, combined with rapid thermal cyclers and instrumentation (e.g. LightCycler®, TaqMan®) capable of detecting and differentiating multiple amplicons, make real-time PCR an attractive and viable proposition for the routine diagnostic laboratory. Real-time PCR assays have been extremely useful for studying microbial agents of infectious disease. The greatest impact to date has been in the field of virology, where real-time assays have been used to detect rapidly a range of viruses in human specimens and to monitor quantitatively viral loads and response to antiviral therapy. Benefits to the patient can also be seen in bacteriology, where rapid detection of bacterial pathogens and/or antibiotic resistance genes can help to ensure the appropriate use of antibiotics, reduce the duration of hospital stay and minimize the potential for resistant strains of bacteria to emerge. Recent developments in real-time PCR have suggested a future in which rapid identification, quantification and typing of a range of microbial targets in single multiplex reactions will become commonplace.

Molecular typing of micro-organisms

Typing of micro-organisms is increasingly important for studying cross-infection and epidemiological relationships, particularly during outbreaks of nosocomial infection (see Ch. 68). Molecular fingerprinting methods are now the most commonly used techniques for assessing the relatedness of individual bacterial isolates in epidemiological studies. These techniques can be used to study any organism from which DNA can be prepared and offer the possibility of a unified approach to microbial typing that can be applied

immediately to a new epidemiological problem with no prior knowledge of the organisms being investigated.

A complete DNA sequence forms the ultimate reference standard for identifying micro-organisms and their subtypes. Increasing numbers of micro-organisms are now being sequenced and the knowledge gained is of immense value for research purposes. However, even with rapid automated sequencing techniques, it is highly unlikely that routine diagnostic laboratories will ever have the facilities or resources to sequence all their isolates of clinical or epidemiological interest routinely. One possibility might involve the sequencing of small relatively conserved regions of the genome that can provide diagnostic information, and such techniques for sequencing genes encoding 16S ribosomal RNA are sometimes used presumptively to identify 'unknown' or unculturable isolates derived from clinical specimens.

Recent developments in automation and sequencing chemistries have resulted in the development of a typing technique termed *multilocus sequence typing* (MLST). The technique relies on the analysis of sequence variation in a relatively small set of 'housekeeping genes' (usually about seven) that are present in all isolates of a particular bacterial species. The genes selected for analysis should be widely separated on the chromosome and should not be adjacent to genes that may be under selective pressure. Specific primers are designed that amplify c. 500-bp fragments of these genes, which are then sequenced to determine naturally occurring variation. The sequences can be compared with those already contained in worldwide databases in order to analyse both global and local epidemiology. MLST schemes and databases are already available for a range of major pathogens, including *Neisseria meningitidis*, *Streptococcus pneumoniae*, *Staphylococcus aureus* and *Haemophilus influenzae*. This is a powerful new approach for the characterization of micro-organisms, as it provides unambiguous data that are electronically portable between laboratories and can be used for global epidemiological studies.

Genomics

Entire genome sequences for pathogenic micro-organisms are increasingly becoming available. Combined with new methods for obtaining genome-wide mRNA expression data, a global view of changes in gene expression patterns in response to physiological alterations or manipulations of transcriptional regulators can now be obtained. An alternative approach for identification and typing is already available that involves the use of *DNA microarrays* (sometimes called '*DNA chips*'). These consist of a very large number of evenly spaced spots of DNA fixed to a microscope slide. Each spot is a unique DNA fragment transferred by a gridding robot from multi-well plates on to the slide. Such DNA chips may become commercially available and could then be hybridized in diagnostic laboratories with DNA extracts from 'unknown' isolates to yield distinctive patterns of hybridization on the slide. Such patterns would be readily amenable to computerized analysis and comparison with electronic databases. More advanced chips are 'nanolaboratories', in that nanotechnology can be used to perform PCR, chromatography, electrophoresis and other techniques on a single chip. Some chips allow multiplexed systems, so that several techniques can be used simultaneously with several samples. Microarrays have already been used to investigate genetic expression, polymorphisms and antibiotic resistance in pathogens such as *Mycobacterium tuberculosis*, *Esch. coli*, *Staph.*

aureus and *Helicobacter pylori*, not only in the context of epidemiological research, but also as a rapid diagnostic method for use with critically ill patients. Such techniques are not yet used routinely in diagnostic laboratories, but this is the beginning of a new molecular era for the diagnosis and treatment of infectious diseases, and these approaches offer the possibility of a very important shortcut to early diagnosis and treatment. The fundamental strategy of the current era of genomics is to move from studying single genes to the simultaneous study of a large number of interacting genes and their corresponding proteins. At the time of writing, this technology is in the early stages of development, but may ultimately revolutionize diagnostic microbiology in the twenty-first century.

KEY POINTS

- The properties of a bacterial cell are defined by the information encoded in its double-stranded DNA genetic complement and its interaction with the environment. The single circular chromosome that, in most cases, comprises the *genome*, carries most of this, but much additional information resides within extrachromosomal or *episomal* elements known as *plasmids*.

- Plasmids and several other mobile genetic elements, including *bacteriophages*, *transposons* and other *integron*-based entities, render the bacterial genetic complement highly susceptible to variation by the addition of new genes. Such elements may affect the medical significance of bacterial strains when the genes they encode affect antibiotic susceptibility or virulence.

- Mobile genetic elements are transferred between related bacteria through the processes of *transformation*, *conjugation* and *transduction*.

- Bacterial genomes are also susceptible to change through mutation, in which the primary nucleotide sequence of one or more genes or regulatory elements

is altered. Because bacterial replication produces very large cell numbers, the possibilities for generation and survival of advantageous mutations (commonly at frequencies of 10^{-7} to 10^{-10}), such as those conferring resistance to antibiotics, are of practical significance.

- These forms of *genotypic variation* should be distinguished from *phenotypic variation*, as the former are heritable whereas the latter are not, and reflect changes in gene expression and, in the case of *phase variation*, genetic rearrangements leading to altered expression patterns without changing the cell's genetic complement.

- Knowledge of the genetic complement of bacteria enables their specific detection, typing and recognition of key aspects of genotypic variation by the application of specific gene (hybridization) probes and by the use of targeted DNA amplification technologies such as the *polymerase chain reaction (PCR)*. The possibility of rapid detection (hours) of many different targets has been enhanced by the recent development of *real-time PCR*.

RECOMMENDED READING

Dijkshoorn L, Towner K J, Struelens M (eds) 2001 *New Approaches for the Generation and Analysis of Microbial Typing Data*. Elsevier, Amsterdam

Erlich H A 1989 *PCR Technology – Principles and Applications of DNA Amplification*. Stockton Press, New York

Goering RV 2000 The molecular epidemiology of nosocomial infection: past, present and future. *Reviews in Medical Microbiology* 11: 145–152

Hall R M, Collis C M 1998 Antibiotic resistance in Gram-negative bacteria: the role of gene cassettes and integrons. *Drug Resistance Updates* 1: 109–119

Innis M A, Gelfand D H, Sninsky J J, White T 1990 *PCR Protocols – A Guide to Methods and Applications*. Academic Press, San Diego

Mackay I M 2004 Real-time PCR in the microbiology laboratory. *Clinical Microbiology and Infection* 10: 190–212

Mazel D, Davies J 1999 Antibiotic resistance in microbes. *Cellular and Molecular Life Sciences* 56: 742–754

Rowe-Magnus D A, Mazel D 1999 Resistance gene capture. *Current Opinion in Microbiology* 2: 483–488

Spratt B G 1999 Multilocus sequence typing: molecular typing of bacteria pathogens in an era of rapid DNA sequencing and the internet. *Current Opinion in Microbiology* 2: 312–316

Summers D K 1996 *The Biology of Plasmids*. Blackwell, Oxford

Tenover F C 1988 Diagnostic deoxyribonucleic acid probes for infectious diseases. *Clinical Microbiology Reviews* 1: 82–101

7 Virus–cell interactions

M. Norval

Viruses are totally dependent on the cells they infect to provide the energy, metabolic intermediates and most (in some cases all) of the enzymes required for their replication. With advances in the techniques of molecular virology, crystallography and modelling, together with the classical methods of electron microscopy, titration and biochemical assay, it has become possible to study virus–cell interactions to a sophisticated level. The picture that has emerged, and is still emerging, is a fascinating one, as viruses are found to associate with, and affect, cells in a wide variety of ways. The range of possible interactions is indicated in Table 7.1. It is possible to divide these into three broad categories that show considerable overlap:

1. Viruses that infect and replicate within cells causing the cells to lyse when the progeny virions are released. This is called a *cytolytic cycle*; the infection is productive and the cell culture demonstrates *cytopathic effects*, which are often characteristic of the infecting virus. The host cells are termed *permissive*. In some instances viruses are produced from the infected cells but the cells are not killed by the process, that is, the infection is *productive* but *non-cytolytic*, and may become *persistent*.

2. Viruses that infect cells but do not complete the replication cycle. The infection is thus called *abortive* or *non-productive*. Abortive infections can be due to a mutation in the virus so that some essential function is lost, or to the production of defective interfering particles, or to the action of interferons. It may be possible to manipulate the conditions in vitro to obtain a *steady state* or *persistent* infection in which infected and uninfected cells coexist and there is some, generally limited, virus production.

3. Viruses that enter cells but are not produced by the infected cell. The virus is maintained within the cell in the form of DNA, which replicates in association with the host cell DNA. The host cell is termed *non-permissive* and the infection is *non-productive*. Occasionally this type of interaction results in *transformation*, where the cell exhibits many of the properties of a tumour cell. In other cases a *latent infection* ensues in which little or no viral gene expression is found, no viral replication occurs and the cell retains its normal properties.

THE CYTOLYTIC OR CYTOCIDAL GROWTH CYCLE

Although there are large differences in the details of the lytic growth cycle depending on the virus studied, and to some extent on the host cell, certain features are common, and a simplified description is given first. The quantitative aspects of virion production were determined initially using bacteriophages, but have now been ascertained for many animal viruses growing in vitro in cell culture. A one-step growth curve is obtained when samples are removed from an infected cell culture at intervals and assayed for the total content of infectious virus after artificial lysis of the cells (Fig. 7.1).

In the early part of the cycle, virus particles come into contact with the cells, and may then *attach* or *adsorb* to them. This marks the start of the *eclipse phase*. The virion then *enters* or *penetrates* into the host cell and is partially *uncoated* to reveal the viral genome. *Macromolecular synthesis* of viral components follows. This can often be divided into *early* and *late phases* separated by the replication of the viral nucleic acid. Early messenger RNA (mRNA) is first transcribed and translated

Table 7.1 The range of virus–cell interactions	
Type of infection/effect on cells	Comment
Cytolytic	Virus produced
Non-cytolytic (persistent)	Virus produced
Abortive	Virus not produced
Abortive (persistent)	Virus produced
Latency (persistent)	Viral nucleic acid present
Transformation (persistent)	Viral nucleic acid present

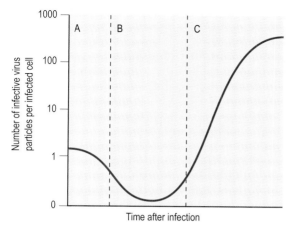

Fig. 7.1 Lytic growth cycle of a virus. Samples are removed from the infected culture at intervals and assayed for the total content of virus. Phase A, adsorption; phase B, eclipse; phase C, assembly and release.

into proteins. These are frequently non-structural proteins and enzymes required to undertake nucleic acid synthesis and the later stages of replication. Viral nucleic acid is then produced, followed by late mRNA transcription and translation. Most proteins synthesized at this stage are structural ones, and will make up part of the final virion. The eclipse phase ends with the *assembly* and *release* of newly formed virus particles. The cycle is shown in diagrammatic form in Figure 7.2, and can vary from as little as 8 h for some picornaviruses to more than 40 h for cytomegalovirus, a human herpesvirus.

Attachment (adsorption)

The initial interaction is by random collision and depends on the relative concentrations of virus particles and cells. Under in-vitro conditions, the ionic composition of the culture medium is an important factor as both viruses and cells are negatively charged at neutral pH and thus tend to repel one another. The presence of cations, such as Mg^{2+}, helps to promote close contact. Adsorption then takes place through *specific binding sites* on the virus and *receptors* on the plasma membrane of the cell. It is largely a temperature- and energy-independent process. Viruses vary widely in the range of cells to which they can adsorb, depending on the nature of the sites to which they attach and how widespread they are among cells of different types, tissues and species. The presence of the receptor determines whether the cell will be susceptible to the virus, but the cells must also be *permissive*, that is, for successful production of new virions, they need to contain the range of intracellular components required by the virus for its replication. The ability of a virus to enter and replicate in a particular cell type is called tissue or cell *tropism*.

Many cellular receptors are protein in nature but they can also be composed of carbohydrate or lipid. There is considerable interest at present in identifying receptors for particular viruses, as the attachment step is a potential target for antiviral therapy and could aid in the understanding of viral pathogenesis. Although specific receptors for selected viruses have been described, they have frequently been disputed, and it has become apparent in recent years that more than one type of receptor molecule may be required by the majority of viruses to complete the entry stage of the

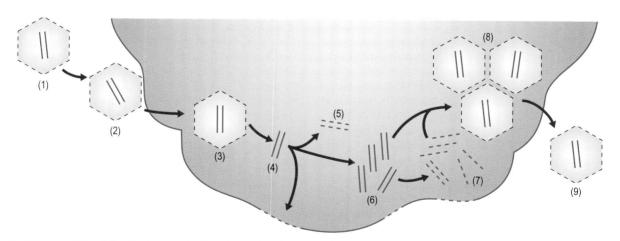

Fig. 7.2 A simplified viral replication cycle showing a hypothetical virus particle (1) attaching to the surface of a susceptible cell (2), entering into the cell (3), being uncoated (4), undergoing early transcription and translation (5), then replication of the viral nucleic acid (6), late transcription and translation (7), and, finally, assembly of new virus particles (8) and release from the cell (9).

Table 7.2   Examples of viruses that interact with multiple receptors on host cells

Virus	Binding receptor	Co-receptor/post-binding receptor
Human immunodeficiency virus 1	CD4	CCR5, CXCR4
Influenza virus A	Sialic acid	Unknown
Severe acute respiratory syndrome corona virus	Angiotensin-converting enzyme 2	Unknown
Herpes simplex virus type 1	Heparan sulphate	Herpesvirus entry-mediator A, nectin 1 and 2, various integrins
Epstein–Barr virus	CD21 (complement receptor)	HLA-II, various integrins
Respiratory syncytial virus	Heparan sulphate	ICAM-1
Foot and mouth disease virus	Sialic acid	Various integrins
Rotavirus	Sialic acid	Various integrins, heat shock protein 70
Poliovirus type 1	CD155 (immunoglobulin-like)	Unknown
Rabies virus	Nicotinic acetylcholine receptor	Neuronal cell adhesion molecule, p75 neurotrophin receptor

CD, cluster of differentiation; HLA, human leucocyte antigen; ICAM, intercellular adhesion molecule.

replication cycle. Indeed it is likely that there is a complex interaction between different functional domains of the virus and several receptor arrays. First there is the 'true' attachment step whereby the virus binds to the cell receptor, and then entry itself may involve a further set of receptors called *co-receptors* or *post-binding receptors*, acting either in succession or in parallel. These interactions frequently induce conformation changes in the surface proteins of the virus, exposing hidden domains that are required for the entry step (see next section). This more complicated view of the initial contact between the virus and the cell suggests that the binding receptor may not be the only determinant of tropism.

In Table 7.2, several examples of viruses that interact with multiple receptors on host cells are shown.

Entry (penetration)

Entry occurs immediately after attachment and, unlike adsorption, requires energy and does not occur at 0°C. The speed of this stage of the replication cycle varies between different viruses, some penetrating into cells in less than a second and others taking several minutes. In addition, the efficiency of the process varies from 50% of attached viruses entering successfully to less than 0.1%. Entry is complex and, despite much study, it is still not clear exactly what the steps are for the majority of viruses.

For viruses with envelopes, penetration is accomplished by membrane fusion catalysed by fusion proteins in the viral envelope. The fusions proteins that mediate entry have been divided into two categories. *Class I fusion proteins* include influenza haemagglutinin protein, human immunodeficiency virus (HIV) gp120 and paramyxovirus fusion protein. They are all cleaved into two pieces during synthesis, are found as trimeric spikes protruding from the surface of the viral particles and contain a fusion peptide characteristically composed of 20 hydrophobic amino acids. In some cases, for example HIV, fusion occurs at the cell surface at neutral pH with the activation energy being provided by receptor binding. In other cases, for example influenza virus, receptor-mediated endocytosis occurs. Receptors with adsorbed virus move together (patch) to pits coated with *clathrin* before moving into the cytosol to form small uncoated vesicles, which then fuse together as endosomes. A proton pump in the endosome lowers the pH to about 5. This acidic pH triggers the conformational change leading to the fusion of the viral and endosomal membranes, releasing the nucleocapsid into the cytosol. The endosomes combine with lysosomes, which eventually degrade any viral components contained within. This process is outlined in Figure 7.3. *Class II fusion proteins* are found in Flaviviruses (e.g. yellow fever virus, hepatitis C virus) and Togaviruses (e.g. rubella) They have different structural features from the class I fusion proteins, consisting of heterodimers. In the acid endosome following receptor-mediated endocytosis, reorganization of the protein takes place with formation of active homotrimers and insertion of a hydrophobic fusion loop into the target membrane.

Recent evidence indicates that viral entry via endocytosis can be independent of clathrin and dependent instead on *caveolae* or *lipid rafts*. These are areas of the membrane that are rich in cholesterol and sphingolipids. Penetration into the cytosol occurs through the endoplasmic reticulum. Papillomavirus is one example of a virus that enters through caveolae.

For viruses without envelopes, membrane penetration is poorly understood but, following exposure of hydrophobic sequences on attachment and endocytosis, they may enter the cytosol by lysing or creating a pore in the endosomal membrane.

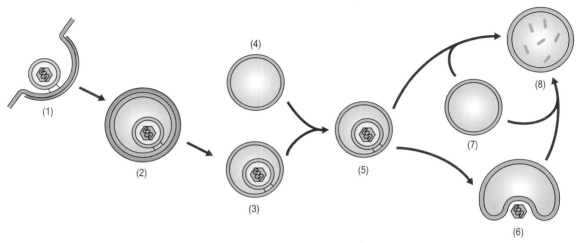

Fig. 7.3   Receptor-mediated endocytosis and entry of an enveloped virus. The virus attaches to specific receptors on the cell membrane (1) that patch at coated pits before being pinched off to form vesicles (2). These lose their coat (3) and fuse with other vesicles (4) to form endosomes (5). At the acid pH of endosomes, fusion of the viral envelope and the endosome membrane occurs, releasing the virus into the cytosol (6). (7) Fusion of the endosome with lysosomes leads to the final degradation of viral components and their return to the surface (8).

Uncoating

Uncoating can take place at several stages and sites in the cell and, generally, is not a well understood process. Some viruses undergo conformational changes on attachment that result in the opening of the capsid and release of selected viral proteins and viral nucleic acid into the cell. For example, crystallography studies of picornaviruses have revealed that the myristic acid groups at one end of the VP4 structural protein interact with the host membrane; this causes VP4 and the viral genome to exit from the capsid through a channel and to enter the cytosol. Enveloped viruses that enter by receptor-mediated endocytosis may be affected by the low pH in the endosome and the action of lysosomal enzymes. Uncoating can also take place in the cytosol or at the nuclear membrane. Reoviruses never fully uncoat, the viral genome remaining within a recognizable capsid structure. Poxviruses become uncoated in two stages. In the first, the outer layers and lateral bodies are removed within endosomal vesicles using host enzymes, and the core lies in the cytosol. Poxviruses carry their own DNA-dependent RNA polymerase, and this enzyme is used in the second stage to transcribe mRNA, which is translated into a special uncoating protein; this enables the final release of viral DNA from the core.

The final step in the complex uncoating process involves transport of the capsid (or the viral genome with, in some instances, viral enzymes and proteins) to the correct site in the cell to commence synthesis of the macromolecules that will comprise the new virions. Although details are not available for many viruses, it is clear that microtubules and microtubule-dependent motors are frequently involved in the transport. Some viruses stay in the cytosol for the remainder of their replication (e.g. poliovirus), but others proceed towards the nucleus where they are uncoated at the nuclear membrane before entry into the nucleus (e.g. herpesviruses) or enter the nucleus intact (e.g. papillomaviruses). Targeting to the nucleus depends on nuclear localization sequences found on the surface of the capsids. To gain access to the nucleus, the virus or its genome can either enter when the cell is undergoing mitosis (when the nuclear membrane is temporarily absent) or, more commonly, be delivered directly into the nucleoplasm though nuclear pore complexes.

Synthesis of viral components

The nucleic acid in viruses is either single or double stranded, circular or linear, in one piece or segmented. In addition, viruses vary enormously in their complexity, ranging from those with nucleic acid sufficient to code for only a few proteins, such as the papovaviruses, up to those coding for several hundred proteins, such as the poxviruses. Although every virus has a unique method of replicating and has a strict temporal control on the synthesis of components, each must present functional mRNA to the cell, so that new virally encoded polypeptides and nucleic acid can be synthesized using the normal cellular processes. Thus, only viruses that contain DNA and replicate in the nucleus can use solely cellular enzymes for transcription and translation. All other viruses must synthesize their mRNA by processes other than those

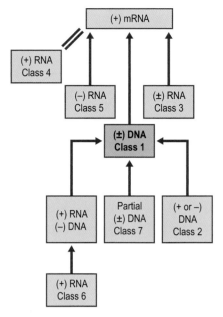

Fig. 7.4 Division of animal viruses into seven classes, based on mechanisms of transcription

found in uninfected cells. Seven different classes, six of which were first described by Baltimore in 1970, can be distinguished (Fig. 7.4). Conventionally in the scheme, nucleic acid of the same polarity or sense as mRNA is called 'positive' (+), and that of the opposite polarity or anti-sense is called 'negative' (−). Rather than be exhaustive, one or two illustrative examples from each class will now be described.

Class 1: Double-stranded DNA viruses

This comprises a very large group of viruses that contain double-stranded DNA in a linear form (e.g. herpesviruses, adenoviruses and poxviruses) or a circular form (e.g. papovaviruses). The poxviruses can be separated from the others, as their replication takes place entirely in the cytoplasm and they can code for all the factors required for their own transcription and genomic replication. In the remaining double-stranded DNA viruses, replication occurs in the nucleus and is dependent to some extent on host cell factors. Herpes simplex virus is used as an example (Fig. 7.5).

After uncoating at the nuclear pore, the viral nucleic acid enters the nucleus and, using the normal host cell mechanisms of transcription and translation, three groups of viral polypeptides are synthesized in a strict temporal fashion. They are called immediate early (α), early (β) and late (γ). A component in the virus particle (α-transcription initiation factor [α-TIF], a γ protein), acting as a transactivator, induces the transcription of the first set of mRNAs. A second component of the viral tegument called virion host shut-off protein (VHS) inhibits host cell macromolecular synthesis, and all the metabolic energy of the cell is turned towards the production of new virus particles. The genes coding for the α, β and γ proteins have been mapped on the genome and, whereas the β and γ genes tend to be scattered, the α genes are located together. Among the early gene products are thymidine kinase and a virus-specific DNA polymerase. Most of the late proteins are structural proteins that inhibit the synthesis of the α and β proteins. Between β and γ protein synthesis, new viral DNA begins to be made, probably by circularization using a rolling circle model.

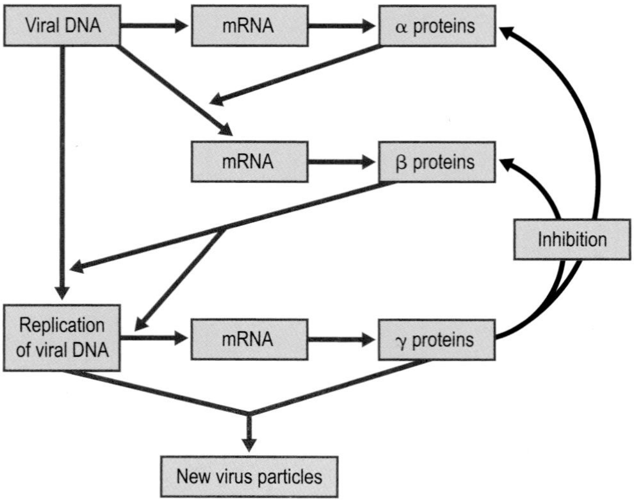

Fig. 7.5 Diagram of macromolecular synthesis during the replication of herpes simplex virus.

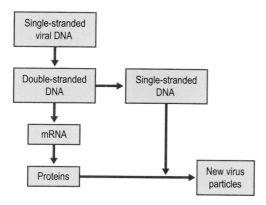

Fig. 7.6   Diagram of parvovirus replication.

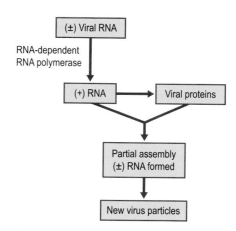

Fig. 7.7   Diagram of double-stranded RNA virus replication.

Class 2: Single-stranded DNA viruses

Parvoviruses comprise the sole family in this group. They are small, with DNA of about 5 kilobases. Some parvoviruses contain DNA of '–' polarity, and grow only in rapidly dividing cells; others contain either '+' or '–' DNA, and depend on co-infection with a helper virus for their replication. Parvoviruses use the cellular DNA polymerases to make the viral genome double stranded, called the replicative form. Priming is by the viral nucleic acid itself forming a loop at the 3′ terminus. This is followed by displacement of the parental DNA strand and synthesis of more DNA complementary to the template strand. Messenger RNAs are made using the appropriate DNA strand as the template, and are translated into viral proteins (Fig. 7.6).

Class 3: Double-stranded RNA viruses

This group includes the reoviruses and rotaviruses. All members have segmented genomes and each RNA segment codes for a single polypeptide. Replication of viral nucleic acid, transcription and translation occur solely in the cytoplasm without nuclear involvement at any stage. Each infectious virus carries its own RNA-dependent RNA polymerase, an enzyme unique to some RNA viruses and not found in uninfected cells. It enables the transcription of one strand (–) into mRNAs, subsequently translated into viral proteins. The transcription is thus asymmetric and conservative, that is, only mRNAs are formed and the parental duplex is not broken apart. Each mRNA is later encapsidated and copied once to form double-stranded molecules (Fig. 7.7). Several hours pass between the '+' and '–' strands of the new virus particles being synthesized. Thus the replication of the double-stranded DNA and RNA viruses is very different.

Class 4: '+' single-stranded RNA viruses

This class comprises a large group of viruses containing RNA of the same polarity as mRNA. Because they code

for all the proteins required during replication, the viral RNA extracted from the virions is infectious by itself. Poliovirus falls into this category, and is used as an example (Fig. 7.8). Macromolecular synthesis of viral components occurs entirely in the cytoplasm.

After entry of poliovirus into the cell, the viral RNA binds to ribosomes, acts as mRNA and is translated in its entirety into one large polypeptide. This is then proteolytically cleaved to give the products RNA polymerase and protease enzymes and new capsid proteins. Using the polymerase enzyme, '–'-strand RNA is synthesized with the genomic RNA as the template, and a temporary double-stranded RNA is formed, called the replicative intermediate. The replicative intermediate consists of complete '+' RNA and numerous partially completed '–' strands. When the '–' strands are ready, they can be used as templates to make more '+'-strand RNA. This is required as genomic RNA for assembly into new virus particles and for transcription into more viral proteins.

At the same time as viral replication, host cell protein synthesis and RNA synthesis are inhibited. Initiation of translation of cellular mRNA requires the participation of a cap-binding protein at the 5′ end. Poliovirus induces the cleavage of this protein, and thus halts the synthesis of cellular proteins. The RNA genome of poliovirus does not have such a cap although it has a small protein, called VPg, at the 5′ end. A special region near the 5′ end of the genome directs cap-independent initiation of protein synthesis.

Complex interactions between viral and cellular proteins are thought to determine how much viral RNA is used for new virus particles or is translated into protein. The capsid of poliovirus consists of 60 copies of each of four proteins, VP1, VP2, VP3 and VP4, forming the icosahedron. One of the first cleavages of the polyprotein produces VP1, which is then broken into VP0, VP3 and VP1. Finally, on assembly, VP0 is

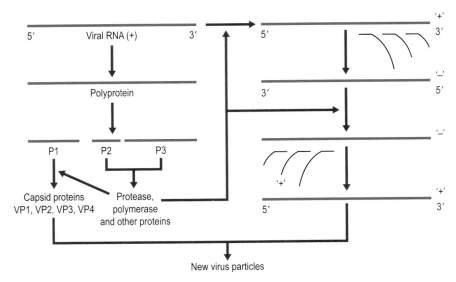

Fig. 7.8 Diagram of poliovirus replication.

cleaved into VP4 and VP2, a process catalysed by VP0 itself.

Class 5: '−' single-stranded RNA viruses

Viruses of this group have single-stranded RNA of '−' polarity and must carry their own RNA transcriptase complex to be infectious, as the normal cellular enzymes are unable to replicate RNA. Influenza virus is an example (Fig. 7.9). It contains eight segments of '−'-strand RNA, plus the RNA transcriptase complex within each virus particle.

After entry into the cell by receptor-mediated endocytosis, transcription to viral mRNA occurs in the nucleus. Influenza virus is the only '−'-strand RNA virus to replicate in the nucleus. To initiate transcription, a nucleotide sequence of about 10 to 13 bases, found at the 5' end of the cellular mRNAs and already capped, is used. This is cleaved from cellular mRNAs by an endonuclease activity of the viral RNA transcriptase complex. Thus, all the viral mRNAs have a 5'-terminal segment of the host cell mRNA.

Once the mRNAs have been generated, they are translated into polypeptides. Each genomic segment produces one mRNA, translated into one polypeptide except in two instances where, by RNA splicing of the original transcript, more than one mRNA is produced and therefore more than one protein. Unlike the transcription of mRNAs, the production of '+'-strand RNAs, required as intermediates to make the progeny '−'-strand RNAs, proceeds without the need for primers.

There is much trafficking of viral polypeptides in the cell; the haemagglutinin, neuraminidase and M_2 protein are inserted in the plasma membrane, and the M_1 pro-

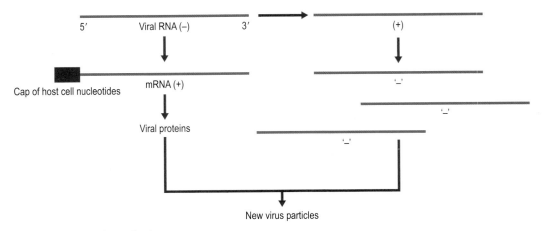

Fig. 7.9 Diagram of influenza virus replication.

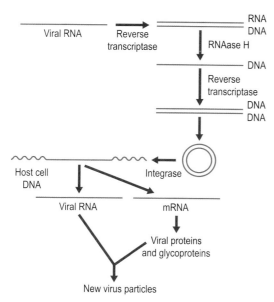

Fig. 7.10 Diagram of retrovirus replication.

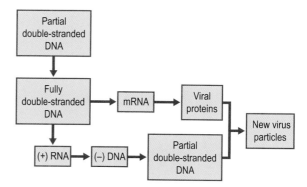

Fig. 7.11 Diagram of hepatitis B virus replication.

tein below this point on the membrane, whereas the nucleocapsid assembles around the viral RNAs in the nucleus.

Class 6: Retroviruses

Viruses of this group are unique as they contain single-stranded RNA (in the form of two identical subunits), yet they replicate via an integrated double-stranded DNA stage. Retroviruses are the only such family and the virus particles contain a reverse transcriptase complex, with RNA-dependent DNA polymerase activity, from which the name 'retrovirus' is derived. This enzyme is not found in uninfected cells.

After entry, synthesis of DNA complementary to the viral RNA occurs using the reverse transcriptase, originating at a primer binding site near the 5′ end of the viral genome. The primer is a specific transfer RNA (tRNA) and varies from one retrovirus to another (e.g. tRNA^{lys} in HIV). Transcription proceeds towards the 5′ end, and is probably continued by a jump across to the 3′ end of the same molecule. In addition to RNA-dependent DNA polymerase activity, the reverse transcriptase complex has ribonuclease (RNAase) H activity, that is, it is able to digest RNA from a DNA–RNA hybrid (Fig. 7.10). The resulting single-stranded DNA is then made double stranded, using the reverse transcriptase as enzyme and starting from a purine-rich sequence. Thus, a linear double-stranded DNA form is produced, first found in the cytoplasm.

The viral RNA has a short sequence of about 12 to 235 bases repeated at each end. During replication there is

generation of a longer repeat sequence, from 250 to 1000 nucleotides, at both ends of the DNA molecule. This is called the *long terminal repeat*; it contains the enhancer and promoter sequences controlling the expression of the viral genome as well as the sequence for the initiation of transcription. The linear double-stranded DNA is able to circularize, and is found in this form in the nucleus.

The next step is integration of the circular DNA into the host cell DNA. This is catalysed by an integrase carried by the virion. It is thought that the circular viral DNA is cleaved leaving staggered ends, and the cellular DNA similarly, to allow insertion of the viral DNA into the cellular DNA; the viral DNA is now called a *provirus*. The site of insertion is not thought to be specific. The provirus is co-linear with the original viral genome, and is always flanked by a 4–6-base pair direct repeat of the host DNA; this repeat is also found flanking transposons. The integrated state is a stable one and, as the DNA of the cell is replicated during cell growth, so the viral DNA is also replicated. Integration can result, on occasion, in cell transformation (see below).

The replication cycle is completed using the normal cellular RNA polymerase II to synthesize viral RNA and viral mRNAs, which are translated into polyproteins and processed into the final proteins found in the virus particle. A viral protease is responsible for many of these cleavages. The control of this stage is complex.

Class 7: Partial double-stranded DNA viruses

Hepadnaviruses are unique among the animal viruses in containing partial double-stranded DNA and replicating via an RNA intermediate, as shown in Figure 7.11. One example of this group is hepatitis B virus.

The first stage in the replication cycle is the production, in the nucleus, of fully double-stranded DNA, followed by the synthesis of single-stranded positive-sense RNA using the cellular DNA-dependent RNA polymerase. The RNA is transported into the cytoplasm and translated

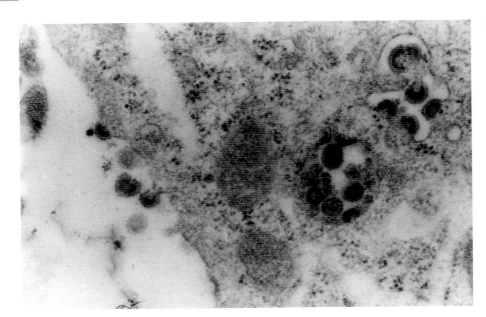

Fig. 7.12 Budding of retroviruses. ×70 000.

into the core protein, which encapsidates the RNA, together with newly synthesized viral RNA-dependent DNA polymerase (reverse transcriptase). Then, using this enzyme, a complementary negative strand of DNA is made, while the RNA is degraded. The DNA is next transcribed into positive-sense DNA, and is found as partial double-stranded DNA in the new virus particles.

Assembly and release

After synthesis of viral proteins and viral nucleic acid, there is a stage of assembly called *morphogenesis*, followed by *release* of virus particles, the productive phase of the infection. The release is either through *cell lysis*, or through *budding* without cell death in many instances. The former method is used by non-enveloped icosahedral viruses, and the latter by enveloped viruses. There is also some evidence for active release without cell lysis for some non-enveloped viruses. Generally the components that will constitute the new virions are produced in high quantities, and the assembly process is probably rather inefficient.

Viruses that are released by killing the cell depend on the cell disintegrating to let them out. For this type of virus, morphogenesis may occur spontaneously once the capsid proteins have been made, the specificity depending on the amino acid sequence of the proteins. Thus the structural proteins of the viruses can form capsomeres by themselves, which then aggregate to form the pro-capsid, a structure without nucleic acid. Often there is proteolytic cleavage of a capsid protein to form the final virus particle, as has been described above for poliovirus.

The precise nature of the interaction between the nucleic acid and the structural proteins that make up the capsid is not known, despite extensive study. It is possible that the viral nucleic acid is inserted into the procapsid through a pore, or it might cause a structural reorientation of the procapsid, thereby becoming internalized. Alternatively, the capsomeres may accumulate around a condensed core of nucleic acid as the nucleic acid is being synthesized.

The second method of assembly and release is by budding. This can take place through the plasma membrane, thus releasing the virions from the cell (e.g. orthomyxoviruses and retroviruses; Fig. 7.12), or through internal membranes, such as the inner nuclear membrane in the case of the herpesviruses (Fig. 7.13b), followed by fusion of the vesicles containing the viruses with the plasma membrane. Envelope glycoproteins specified by the virus are synthesized by essentially the same mechanism as cellular membrane glycoproteins. The viral proteins destined to become envelope proteins contain a sequence of 15 to 30 hydrophobic amino acids, known as the signal sequence. This sequence binds the growing polypeptide chain to a receptor on the cytoplasmic side of the rough endoplasmic reticulum and enables its passage through the membrane. Glycosylation occurs in the lumen of the rough endoplasmic reticulum, and the proteins are transported to the Golgi apparatus. There they are further glycosylated and acylated before transport to the plasma membrane, the direction probably determined by a sorting signal in the polypeptide sequence. In some cases, the signal directs the viral glycoprotein to one surface of the cell only; for example, orthomyxoviruses bud only from the outer (apical) surface of epithelial cells, whereas

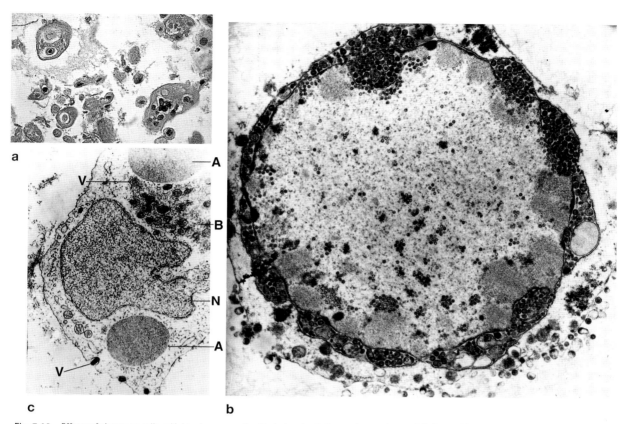

Fig. 7.13   Effects of viruses on cells. **a** Light microscopy of a skin lesion due to herpesvirus to show cell fusion and intranuclear inclusions (Cowdry type A). ×60. **b** Electron micrograph of a cell infected with herpes simplex virus. Assembly of capsids within nucleus – enveloped virus between layers of nuclear membrane. ×9000. **c** Type A (accumulation of viral protein) and type B (virus factory) inclusions (identified as A and B, respectively) in the cytoplasm of a poxvirus-infected cell. V, virus; N, the cell nucleus. ×700.

rhabdoviruses bud only from the inner (basal) surface. The viral glycoproteins are very important in terms of antigenicity as the hydrophilic domains protrude from the surface of the cell, with the N terminus being furthest away, and change its surface structure significantly. They remain anchored in the membrane via a hydrophobic domain near the carboxyl (C) terminus. After insertion into the membrane, the viral glycoproteins accumulate together to form oligomers; at the same time the host cell glycoproteins move away. At the C terminus of the viral glycoproteins there is frequently a short hydrophilic sequence that remains inside the cell and is assumed to interact with the internal components of the virus during assembly. Recent information indicates that lipid rafts function as microdomains for the accumulation of many viral glycoproteins, and may also initiate the actual budding sequence. It is not known how the nucleocapsids are directed to the assembly site. Once there, they are engulfed by the membrane; this process requires bending of the membrane, leading to its outward curvature. In the case of the class I fusion proteins described above,

a final cleavage of the glycoprotein is required to make the virus infectious. For example, the haemagglutinin H_0 of influenza virus is cleaved into two peptides, HA_1 and HA_2, linked by disulphide bridges. The bud is completed by the fusion of the two apposing membranes, and it finally separates from the plasma membrane of the host to become a new infectious virus. For retroviruses, budding occurs by highjacking a cellular pathway that normally creates vesicles that bud into late endosomal compartments called multivesicular bodies. These viruses then exit through the plasma membrane or via exosomes.

Several thousand virus particles can be produced per infected cell, although this number varies considerably with virus type and host cell type. The budding viruses tend to be released slowly over several hours, whereas the lytic ones are released together. Only a few of the newly formed virus particles are infectious, as indicated by a high ratio of particles to infectious virions. Presumably most do not have the correct complement of proteins, enzymes or viral nucleic acid, or have been assembled incorrectly.

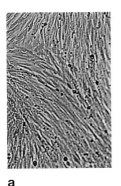 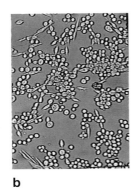 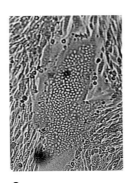

a　　　　　　　　b　　　　　　　　c

Fig. 7.14 Cytopathic effects: **a** uninfected fibroblast cells; **b** cell rounding due to herpes simplex virus; **c** syncytium formation or cell fusion due to respiratory syncytial virus. All unstained. ×65.

Evidence has accumulated to indicate that, for some viruses at least, perturbation of the normal cell metabolism during replication can stimulate cell death by *apoptosis*. Several virus-specific factors have been identified as inducers of apoptosis, such as by causing DNA strand breaks in the case of a parvovirus, by stabilization of the tumour suppressor gene product p53 in the case of Epstein–Barr virus, or by receptor signalling in the case of HIV. The advantage to the virus could be that the spread of the infection is enhanced as the entire cellular contents, including progeny viruses, are packaged into membrane-bound apoptotic bodies that are then taken up by adjacent cells. In contrast, other viral proteins have been revealed that block or delay apoptosis, presumably until sufficient progeny viruses have been produced within the cells. In this case, the factors target specific stages of the apoptotic pathway. For example, caspases are inhibited during poxvirus infections, the action of interferon is downregulated during influenza virus infections and p53 is destroyed during some human papillomavirus infections. Therefore the susceptibility of the host cell to apoptosis depends on the acute death pathways in the cell itself and the range of apoptotic modulators induced by the infecting viruses.

Microscopy of infected cells

It is possible to observe effects on the host cell microscopically. Firstly, there may be morphological changes called *inclusion bodies* in the infected cell, seen by altered staining characteristics. The inclusion bodies are nuclear or cytoplasmic and vary in their composition. They can consist of viral factories in which morphogenesis occurs, crystalline arrays of virus particles ready for release, overproduction of a particular viral protein or proteins, or some aberrant cellular structure, such as clumped chromatin. Virally encoded non-structural proteins are likely to be involved in forming the matrix of these structures and in recruiting viral components to them. Some examples of inclusion bodies are shown in Figure 7.13. Secondly, the cells may be killed by the viral infection. There are several

possible reasons for this, including factors produced by the virus that induce apoptosis (see above). It is likely that the accumulation of viral structural proteins is toxic for the cells in some cases. In addition, some viruses, such as herpes simplex virus and the poxviruses, inhibit host cell macromolecular synthesis from an early stage in the replication cycle, leading to structural and functional damage. Plasma membrane function and permeability change, and lysosomal membranes begin to break down, allowing leakage of the contents with degradative activity into the cytoplasm. There may also be marked effects on the cytoskeleton. These changes lead to a *cytopathic* effect, seen clearly in cell culture. It can take several forms, one of the commonest being *cell rounding* and subsequent detachment from the solid surface (Fig. 7.14b). Another is the formation of a *syncytium*, whereby the membranes of adjacent infected cells fuse and a giant cell is formed containing many nuclei (Fig. 7.14c). In some cases, the nuclei fuse to make hybrid cells, a property that used to be exploited in monoclonal antibody production.

NON-CYTOCIDAL PRODUCTIVE INFECTIONS

Some viruses are able to infect cells productively but the cells are not killed by the replication process. Viruses that are released by budding frequently come into this category. The cell type used for the infection is critical and, presumably, any inhibitory effect of the virus on the cellular metabolism does not take place. This type of interaction may lead to a *persistent infection* in which infected cells and viruses coexist over a long period of time. There will, however, be antigenic changes in the infected cells, often the insertion of viral glycoproteins in the plasma membrane. This can be exploited for the detection of a virus; for example, when influenza virus is cultured in monkey kidney cells, virus is produced from the cells but there is no immediate cytopathic effect. However, there is insertion of viral haemagglutinin into the plasma membrane during replication. Thus, when red blood cells of certain species are added to the infected cells, they adhere to the haemagglutinin and can be seen microscopically. This phenomenon is called *haemadsorption*.

ABORTIVE (NON–PRODUCTIVE) INFECTIONS

Some viruses are unable to infect cells because they cannot adsorb to them; the cells are therefore called *resistant*. In other cases viruses are able to infect cells but are not produced from them; such infection is called *non-productive* or *abortive*, and the cells are *non-permissive*. Often there is a block at one stage of the replication cycle owing to the absence of a cellular function essential for viral replication. This can also occur in a permissive cell if the viral genome itself is *defective* in some way, so that the replication cannot be completed. This can happen in two ways:

1. The virus is too small to code for all the proteins required for replication. An example is provided by the adeno-associated viruses belonging to the Parvoviridae family. They are unable to replicate on their own but depend on a second 'helper' virus infecting the same cell and providing the essential function that they lack. Adenoviruses act as the helper viruses and are thought to activate transcription of the parvovirus genome in the infected cell.

2. Abortive infections can arise as a result of viral mutation. In fact, only a few amino acid substitutions in selected proteins of the virus can change the nature of a lytic infection. In some cases the non-productive infection can be converted into a persistent one, in which new virus particles are synthesized; three examples are given below.

Temperature–sensitive mutants

Here the wild-type virus has mutated to produce a variant – a temperature-sensitive mutant – that lacks an essential gene function at the *non-permissive temperature*, normally above 38°C. This is thought to be due to the thermal instability of the secondary or tertiary structure of a particular protein. The temperature can be lowered, generally to below 35°C, to a level called permissive, where the defect is no longer functional and the infection becomes productive. However, temperature-sensitive mutants tend to be less cytopathic than the wild type, even at permissive temperatures, probably because they synthesize less mRNA and protein. Thus, within a cell culture, a persistent infection can result with a balance between infected and uninfected cells together with virus production. Temperature-sensitive mutants have been used to identify the gene responsible for a particular replication step and to map where functions lie on the viral genome. It is possible to place these mutants

into complementation groups where, at the non-permissive temperature, the defect in one temperature-sensitive mutant is compensated for by a second one with a defect in a different gene.

Defective interfering particles

It has been known for many years that, when cells are infected at high multiplicity, there will be among the progeny a number of virus particles with genomes shorter than normal, containing at least one deletion – so-called *defective interfering particles*. These particles cannot replicate themselves, although they are able to infect new cells. However, they can replicate in the presence of helper virus, often the parental virus, which compensates for the lack of a particular gene or gene cluster. The defective particles retain an origin of replication and the ability to form capsids. One of their important properties is that they *interfere* with the replication of normal parental viruses because, firstly, less time and energy are required to replicate the defective genome compared with the full-length genome and, secondly, the transcriptase complex has a greater affinity for the defective genome than the full-length one. Hence defective interfering particles, as their numbers increase, have a greater and greater effect on the replication of parental virus. It is possible to obtain an in-vitro cell culture in which infected and uninfected cells together with infectious virus and defective particles are in balance for prolonged period of time. Thus a steady state exists and the infection is persistent.

Abortive infections maintained by interferons

The final example of abortive infections arises from the action of *interferons* in infected cell cultures. Interferons are produced from virally infected cells and can protect other cells from attack by viruses. These molecules, of α or β type, inhibit various stages of the viral replication cycle, especially polypeptide synthesis (see Ch. 10 for details). In cell culture, persistent infections can be obtained when the antiviral effects of interferons protect sufficient cells from the cytolytic effects of viral replication to allow cells and viruses to coexist.

It should be noted that some viruses can mutate so that they replicate poorly, if at all, in some cell types in vitro and in vivo, or they replicate normally but are less virulent in vivo. These mutated strains are described as *attenuated* and are the basis of most live viral vaccines. For example, the three serotypes used in the Sabin poliovirus vaccine have attenuating mutations in the non-coding regions of the viral RNA, leading to a failure to replicate in the brain and less efficient replication than the wild-type strains at the primary site of infection in the gut. Attenuation can be achieved by culturing the viruses repeatedly in cells other than those of the normal host or by culturing at non-physiological temperatures.

LATENCY

Latency represents a type of persistence whereby the virus is present in the form of its genome only and there is limited expression of viral genes. The genome is found either integrated into the host cell chromosome or as a circular non-integrated episome. It is maintained throughout cell division when the host cell replicates. Latent infections are more common with DNA viruses than RNA viruses, perhaps because no mechanisms exist to maintain RNA for long periods of time intracellularly. One example of latency is provided by Epstein–Barr virus, which persists in B lymphocytes as episomal viral DNA with limited transcription of viral genes, probably around 11 protein products being expressed. For herpes simplex virus, latency occurs in neuronal cells where the latency-associated transcript is the only viral RNA abundantly expressed. The transcript can inhibit apoptosis and thus contribute to persistence of the virus. It is not clear at present whether any viral proteins are present during herpes simplex virus latency and whether the viral gene expression is substantially prevented by the immune response of the host. In some cases, specific stimuli trigger the reactivation of the virus from the latent state, and the infection becomes productive with the appearance of new virions.

TRANSFORMATION

In this type of virus–cell interaction, the virus infects the cell non-productively and is found in the form of viral DNA, either integrated in the host cell DNA or unintegrated, or in both states. The properties of the cells are changed dramatically, a process called *transformation*. Transformed cells have similar properties to tumour cells, and a detailed study of the mechanism of viral transformation has led to increased understanding of the molecular basis of cancer. Only members of some virus families are able to transform cells. These include herpesviruses, adenoviruses, hepadnaviruses, papovaviruses and poxviruses of the DNA viruses, and, of the RNA viruses, only retroviruses. The type of cell infected and the species are also important. It should be noted that transformation is a rare event: at most only 1 in 10^5 cells infected by a particular virus will become transformed.

Some of the main properties of transformed cells that distinguish them from normal cells are listed below:

- loss of contact inhibition of growth
- can grow to high saturation density
- less requirement for serum factors
- indefinite number of cell divisions
- expression of viral antigens
- absence of fibronectin
- fetal antigens often present
- changes in agglutinability by plant lectins
- induction of tumours in experimental animals.

One of the most striking changes is the loss of contact inhibition of growth, whereby cells that normally grow in an ordered fashion beside their neighbours and stop dividing when they touch one another now grow on top of each other and lose their orientation with respect to each other (Fig. 7.15). As a result they reach much higher densities. They have less requirement for serum factors in the medium and can be cultured in suspension without being attached to a solid surface (anchorage independent). Normal cells have a limited number of cell divisions, called the *Hayflick limit*, that they can undergo in vitro before apoptosis; for example, the Hayflick limit for cells taken from a human fetus is around 60 divisions. Transformed cells no longer have this limit and thus can grow and divide indefinitely. There are many changes in the surface properties of transformed cells. Often viral-specific antigens are found, particularly ones synthesized early in the replication cycle. Fibronectin, a surface glycoprotein thought to be important in keeping cells together in a tissue or organ, is no longer found. Commonly fetal antigens are expressed, and the agglutinability of cells by plant lectins changes, demonstrating alterations in the distribution of membrane glycoproteins. Finally, some transformed cells form tumours when injected into susceptible animals. Often these animals have to be immunocompromised in some way before tumours are produced, or the cells inserted into an immunologically protected site, such as the cheek pouch of the hamster. In addition, it is important to appreciate that by no means all transformed cells will form tumours. It is thought that there are degrees of transformation and that several stages have to be completed before the cells are fully transformed and equivalent to malignant tumour cells. Viral transformation may represent only the first step or a single step in such a pathway. There are no in-vitro markers that determine the degree of transformation; thus the potential ability of the transformed cell to produce tumours in experimental animals cannot be predicted, as yet.

All of the viruses that cause transformation in vitro have a similar interaction with the host cell (Fig. 7.16). The initial stages are exactly as described above for the productive infections. There is attachment, entry, uncoating and, in most but not all cases, selected viral

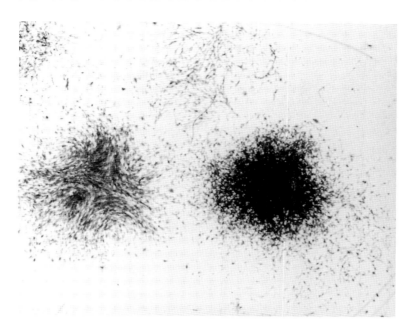

Fig. 7.15 Colonies of human embryo fibroblasts growing normally (left) and following viral transformation (right).

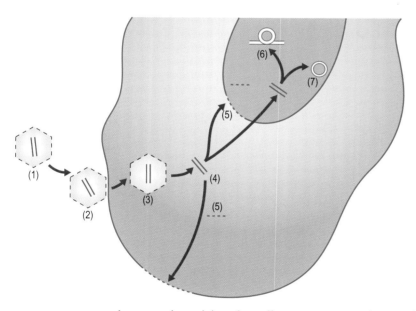

Fig. 7.16 Simplified diagram showing the events of viral transformation. The hypothetical virus (1) attaches (2), enters (3), is uncoated (4), and usually some early viral proteins are synthesized (5) followed by integration of the viral genome in the host cell DNA (provirus) (6) or formation of a circular non-integrated genomic DNA (episome) (7).

genes are expressed as proteins, giving the cell new antigenic properties. At this stage the viral nucleic acid becomes integrated in the host cell DNA, probably not at a specific site, or it circularizes and is maintained in a non-integrated episomal form in the nucleus. The association is a stable one, so that when the host cell DNA is replicated the viral nucleic acid is also replicated and the number of viral genome copies per cell remains constant over many cell generations. Thus, transformation is a heritable alteration. With some viruses, such as Epstein–Barr virus and the

papovaviruses, the whole viral genome is normally integrated, whereas with others, such as herpes simplex virus and adenoviruses, only part of the viral genome is integrated and the remainder is lost.

Recent work in this area has concentrated on the molecular events surrounding transformation and in analysing the functions of the viral proteins found in transformed cells. Two examples of transforming viruses, one RNA and the other DNA, are described briefly below to illustrate the approaches taken. Both are associated with human tumours.

The first is human T lymphotropic or T cell leukaemia virus type I (HTLV-I), a retrovirus, which is found in CD4$^+$ T cells of patients with adult T cell leukaemia. It is able to transform CD4$^+$ lymphocytes in vitro with integration of the DNA provirus. Genetic analysis has revealed that the viral genome can code for several non-structural proteins, including one of special interest called Tax, of molecular weight 40 kDa. This protein, which has no cellular homologue, is able to activate transcription in the long terminal repeat of the integrated virus. Tax has also been revealed to affect the transcription of a remarkable, and increasing, number of cellular genes at more distant sites that are involved in cell cycle progression and DNA repair. In short, it functions in a complex manner to promote cell proliferation, accumulate DNA damage and inhibit apoptosis. However, it is unlikely that Tax expression alone leads to the end-point of leukaemia and further, so far unexplained, molecular events occurring over a period of several years, are probably necessary.

The second example is human papillomavirus type 16 (HPV-16), found as integrated DNA in many cases of carcinoma of the cervix. In vitro this virus is able to transform most types of human epithelial cells, including keratinocytes. The viral proteins responsible for transformation are the products of two genes, *E6* and *E7*, which are found in cervical tumour cells: E6 protein interacts with p53, and E7 with retinoblastoma protein, thereby inactivating them. As both p53 and retinoblastoma protein act as cellular growth-suppressing proteins, loss of their functions is likely to lead to transformation. In addition, integration of the viral genome normally involves the disruption of the *E2* gene, the product of which is required to stop transcription of the E6 and E7 promoter, and therefore the continued expression of the E6 and E7 proteins results. Further properties of the E6 and E7 proteins include the inhibition of apoptosis, chromosome destabilization and, in vivo, various mechanisms to evade local immune responses.

KEY POINTS

- Viruses are completely dependent on the host cell for their replication.
- A part of the *capsid* (in the case of *non-enveloped viruses*) or *envelope* (in the case of *enveloped viruses*) binds to a specific receptor or receptors on the host cell to initiate entry of the virus into the cell.
- The interaction of viruses with cells can result in:
  (a) *production* of new virus particles with or without lysis of the host cells
  (b) *abortive infection*
  (c) *latency*
  (d) *transformation*.

- In the productive replication cycle, the sequential stages are *attachment* (adsorption), *entry* (penetration) into the cytosol, *uncoating*, synthesis of *viral macromolecules* (mRNA, proteins and genomes), *assembly* of new viral particles and their *release*.
- In latency, the virus persists as its genome, with limited expression of selected viral genes, sometimes as RNA only.
- In viral transformation, the virus persists as its genome, with expression of selected viral proteins that induce the host cell to behave like a tumour cell.

RECOMMENDED READING

Arrand J R, Harper D R (eds) 1998 *Viruses and Human Cancer*. BIOS Scientific, Oxford

Bangham C R, Kirkwood T B 1993 Defective interfering particles and virus evolution. *Trends in Microbiology* 1: 260–264

Cann A J 2001 *Principles of Molecular Virology*, 3rd edn. Academic Press, London

de la Torre J C, Oldstone M B A 1996 Anatomy of viral persistence: mechanisms of persistence and associated diseases. *Advances in Virus Research* 46: 311–343

Dimitrov D S 2004 Virus entry: molecular mechanisms and biomedical applications. *Nature Reviews Microbiology* 2: 109–122

Fields D N, Knipe D M, Howley P M (eds) 1996 *Virology*, 3rd edn. Lippincott-Raven, Philadelphia

Flint S J, Enquist L W, Krug R M, Racaniello V R, Skalka A M 2000 *Principles of Virology: Molecular Biology, Pathogenesis, and Control*. ASM Press, Washington

Harper D R 1998 *Molecular Virology*, 2nd edn. BIOS Scientific, Oxford

Munger K, Howley P M 2002 Human papillomavirus immortalization and transformation functions. *Virus Research* 89: 213–228

Nayak D P, Hui E K-W, Barman S 2004 Assembly and budding of influenza virus. *Virus Research* 106: 147–165

Roulston A, Marcellus R C, Branton P E 1999 Viruses and apoptosis. *Annual Review of Microbiology* 53: 577–628

Smith A E, Helenius A 2004 How viruses enter animal cells. *Science* 304: 237–242

Smyth M S, Martin J H 2002 Picornavirus uncoating. *Journal of Clinical Pathology: Molecular Pathology* 55: 214–219

Whittaker G R, Kann M, Helenius A 2000 Viral entry into the nucleus. *Annual Review of Cell and Developmental Biology* 16: 627–651

Internet site

Virus replication. http://www-micro.msb.le.ac.uk/3035/3035Replication.html

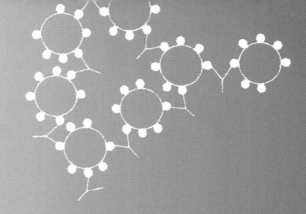

PART 2
INFECTION AND IMMUNITY

8 Immunological principles: antigens and antigen recognition

J. Stewart

An antigen is any substance capable of provoking the lymphoid tissues of an animal to respond by generating an immune reaction directed specifically at the inducing substance and not at other unrelated substances. The response is not to the entire molecule but to individual chemical groups within it that have a specific three-dimensional shape. The specificity of the response to these *antigenic determinants* or *epitopes* is an important characteristic of immune responses. The reaction of an animal to contact with antigen, called the *acquired immune response*, takes two forms: first, the *humoral* or *circulating antibody response* and, second, the *cell-mediated response*, and their characteristics are described in Chapter 9. Most of the information available on the specificity of the immune response comes from studies of the interaction of circulating antibody with antigen. An antibody directed against an epitope of a particular molecule will react only with this determinant or other very similar structures. Even minor chemical changes in the conformation of the epitope markedly reduce the ability of the original antibody to react with the altered material.

The term 'antigen', referring to substances that either act as stimulants of the immune response or react with antibody, is used rather loosely by immunologists. Use is made of the functional classification of antigens into:

- substances that are able to generate an immune response by themselves, which are termed immunogens
- molecules that are able to react with antibodies but are unable to stimulate their production directly.

The latter substances are often low molecular weight chemicals, termed *haptens*, that react with preformed antibodies but become immunogenic only when attached to large molecules, called *carriers*. The hapten forms an epitope on the carrier molecule that is recognized by the immune system and stimulates the production of antibody. In other words, the ability of a chemical grouping to interact with an antibody is not sufficient to stimulate an immune response. As we will see later, when discussing the sites on molecules recognized by cells of the immune system, all antigens can be considered to be composed of haptens on larger carrier structures. By convention, the term 'immune response' was used to refer to the acquired immune system and innate immunity was considered to be a rather non-specific, although relatively effective, defence against infection. In the past few years it has become evident that components of the innate immune system recognize a set of molecular signatures that have been termed *pathogen-associated molecular patterns* through pattern recognition receptors on the surface of cells (see Ch. 9).

GENERAL PROPERTIES OF ANTIGENS

A substance that acts as an antigen in one species of animal may not do so in another if it is represented in the tissues or fluids of the second species. This underlines the requirement that an antigen must be a foreign substance to elicit an immune response. For example, egg albumen, although an excellent antigen in rabbits, fails to induce an antibody response in fowl. The more foreign and evolutionarily distant a substance is from a particular species, the more likely it is to be a powerful antigen.

A widely recognized requirement for a substance to be antigenic in its own right, without having to be attached to a carrier molecule, is that it should have a molecular weight in excess of 5000 Da. It is, however, possible to induce an immune response to substances of lower molecular weight. For example, glucagon (molecular weight of 3800 Da) can stimulate antibody production, but only if special measures are taken such as the use of an *adjuvant* which gives an additional stimulus to the immune system. Very large proteins, such as the crustacean respiratory pigment haemocyanin, are very powerful antigens and are used widely in experimental immunology. Polysaccharides vary in antigenicity; for example, dextran with a molecular weight of 600 000 Da is a good antigen, whereas dextran with a molecular weight of 100 000 Da is not.

Some low molecular weight chemical substances appear to contradict the requirement that an antigen be large. Among these are picryl chloride, formaldehyde and drugs such as aspirin, penicillin and sulphonamides. These substances are highly antigenic, particularly when applied to the skin. The reason for this appears to be that such materials form complexes by means of covalent bonds with tissue proteins. The complex of such a substance, acting as a hapten, with a tissue protein acting as a carrier, forms a complete antigen. This phenomenon has important implications in the development of certain types of hypersensitivity (see Ch. 9).

ANTIGENIC DETERMINANTS

The immune system does not recognize an infectious agent or foreign molecule as a whole, but reacts to structurally distinct areas: antigenic determinants or epitopes. Thus, exposure to a micro-organism will generate an immune response to many different epitopes. The antiserum produced will contain different antibodies reactive with each determinant. This will ensure that an individual is protected from the micro-organism by producing a response to at least a few of the possible determinants. If the host reacted only to the organism as a whole, then failure to react to this one site would have dire consequences: it would not be able to eliminate the pathogen. Certain antibodies may react with an epitope composed of residues that can also be part of two other epitopes recognized by different antibodies (Fig. 8.1).

A response to antigen involves the specific interaction of components of the immune system, antibodies and lymphocytes, with epitopes on the antigen. The lymphocytes have receptors on their surface that function as the recognition units; on B lymphocytes surface-bound immunoglobulin is the receptor, and on T lymphocytes the recognition unit is known as the T cell receptor. The interaction between an antibody (or cell-bound receptor) and antigen is governed by the complementarity of the electron cloud surrounding the determinants. The overall configuration of the outer electrons, not the chemical nature of the constituent residues, determines the shape of the epitope and its complementary *paratope* (the part of the antibody or T cell receptor that interacts with the epitope). The better the fit between the epitope and the paratope, the stronger the non-covalent bonds formed and consequently the higher the affinity of the interaction.

Antigenic determinants have to be topographical, that is, composed of structures on the surface of molecules, and can be constructed in two ways. They may be contained within a single segment of primary sequence or assembled from residues far apart in the primary sequence

Fig. 8.1 Overlapping epitopes. Two epitopes (1 and 2) on an antigen induce the formation of three antibodies (A, B and C).

but brought together on the surface by the folding of the molecule into its native conformation. The former are known as *sequential* epitopes, and those formed from distant residues are *conformational* epitopes. The majority of antigenic structures recognized by antibodies depend on the tertiary configuration of the immunogen (conformational), whereas T cell epitopes are defined by the primary structure (sequential).

ANTIGENIC SPECIFICITY

Foreignness of a substance to an animal can depend on the presence of chemical groupings that are not normally found in the animal's body. Arsenic acid, for example, can be chemically introduced into a protein molecule and, as a hapten, acts as a determinant of antigenic specificity of the molecule. There are many examples in which antibodies are able to distinguish subtle chemical differences between molecules. Thus, antisera can distinguish between glucose and galactose, which differ only by the interchange of a hydrogen atom and a hydroxyl group on one carbon atom.

The ability of antibody (or T cell receptors) to form a high-affinity interaction with an antigen depends on intermolecular forces, which act strongly only when the two molecules come together in a very precise manner. The better the fit, the stronger the bond. An antibody molecule directed against a particularly shaped antigenic determinant might be able to react with another similar but not quite identical determinant, as shown in Figure 8.2. This type of cross-reaction does occur, but the strength of the bond between the two molecules is diminished in the case of the less well fitting determinant.

A common source of confusion concerning the specificity of antibodies arises when an antibody to a particular antigen is found to be capable of combining with an apparently

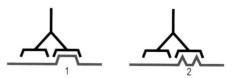

Fig. 8.2 Specificity and cross-reactions. Antibody produced in response to an antigen that contains epitope 1 will also combine with epitope 2.

Table 8.1 Physicochemical properties of human immunoglobulins

	Immunoglobulin isotype				
	IgA[a]	IgD	IgE	IgG	IgM[b]
Mean serum concentration (mg/dl)	300	5	0.005	1400	150
Mass (kDa)	160	184	188	160	970
Carbohydrate (%)	7–11	9–14	12	2–3	12
Half-life (days)	6	3	2	21	5
Heavy chain	α	δ	ε	γ	μ

The immunoglobulin serotype is determined by the type of heavy chain present. The different characteristics observed are also controlled by the heavy chain. Variation within a class gives rise to subclasses.
[a]IgA is also found as a dimer, and in secretions IgA is present in dimeric form associated with a protein known as secretory component.
[b]Data for IgM as a pentamer.

unrelated antigen. For example, glucose residues are present in many different types of molecule, and an antibody that binds to a glucose determinant in antigen X-glucose would be likely to react with the glucose group in antigen Y-glucose, provided the two determinants were equally accessible. The antibody directed against the glucose determinant is not a non-specific type of antibody but is simply reacting with an identical chemical determinant in another antigen molecule.

In laboratory practice, cross-reactivity is often found between antisera to certain bacterial antigens and antigens present on cells such as erythrocytes. Antigens shared in this way are known as *heterophile antigens*. The best known of the heterophile antigens is the Forssman antigen, which is present on the red cells of many species as well as in bacteria such as pneumococci and salmonellae. Another heterophile antigen is found in *Escherichia coli* and human red cells of blood group B individuals. These cross-reactivities are probably responsible for the generation of antibodies found in individuals of a certain blood group that bind to the red blood cells of individuals of a different blood group. These antibodies are known as *isohaemagglutinins* because they are able to bind the red blood cells and clump them together (i.e. cause agglutination).

IMMUNOGLOBULINS

Towards the end of the nineteenth century, von Behring and Kitasato in Berlin found that the serum of an appropriately immunized animal contained specific neutralizing substances, or antitoxins. This was the first demonstration of the activity of what are now known as *antibodies* or *immunoglobulins*. Antibodies are:

- glycoproteins
- present in the serum and body fluids

- induced when immunogenic molecules are introduced into the host's lymphoid system
- reactive with, and bind specifically to, the antigen that induced their formation.

The liquid collected from blood that has been allowed to clot is known as *serum*. It contains many different molecules but no cells or clotting factors. If serum is prepared from an animal that has been exposed to an antigen, it is known as an *antiserum* as it will contain antibodies reactive to the inducing antigen.

There are five distinct *classes* or *isotypes* of immunoglobulins: IgG, IgA, IgM, IgD and IgE. They differ from one another in terms of size, charge, carbohydrate content and, of course, amino acid composition (Table 8.1). Within certain classes there are subclasses that vary slightly in structure and function. These classes and subclasses can be separated from one another serologically, using antibody. If injected into the correct species they will induce the formation of antibodies that can be used to differentiate between the different isotypes.

Antibody structure

All antibody molecules have the same basic four-chain structure composed of two light chains and two heavy chains (Fig. 8.3). The light chains (molecular weight 25 000 Da) are one of two types designated κ and λ, and only one type is found in one antibody. The heavy chains vary in molecular weight from 50 000 to 70 000 Da, and it is these chains that determine the isotype. They are designated α, δ, ε, γ and μ for the respective classes of immunoglobulin (Table 8.1). The individual chains are held together by disulphide bridges and non-covalent interactions.

When individual light chains are studied, they are found to comprise two distinct areas or *domains* of approximately 110 amino acids. One end of the chain is identical in all members of the same isotype, and is termed the constant region of the

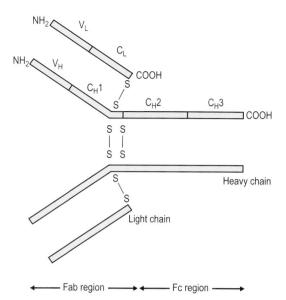

Fig. 8.3 Basic structure of an immunoglobulin molecule. See text for details.

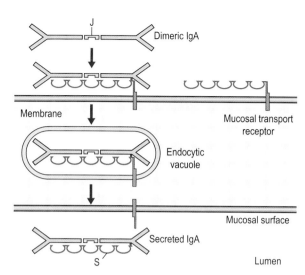

Fig. 8.4 Transport of secretory IgA. J, J chain; S, secretory component.

light chain, C_L. The other end shows considerable sequence variation, and is known as the variable region, V_L. The heavy chains are also split into domains of approximately the same size, the number varying between the five types of heavy chain. One of these domains will show considerable sequence variation (V_H), whereas the others (C_H) are similar for the same isotype. The tertiary structure generated by the combination of the V_L and V_H regions determines the shape of the antigen-combining site or paratope. As the two light and two heavy chains are identical, each antibody unit has two identical paratopes situated at the amino (N)-terminal end of the molecule that recognizes the antigen. The carboxyl (C)-terminal end of the antibody is the same for all members of the same class or subclass, and is involved in the biological activities of the molecule. The area of the heavy chains between the C_H1 and C_H2 domains contains a varying number of interchain disulphide bonds and is known as the *hinge region*. A number of enzymes cleave immunoglobulins at distinct points to generate different peptide fragments. Using these enzymes, antibodies can be divided into a Fab region ('fragment antigen binding') containing the paratope and an Fc region ('fragment crystallizable') that is similar for all antibodies of the same isotype.

Despite the differences between the various isotypes, as shown in Table 8.1, all antibody molecules are composed of the same basic unit structure, with the Fab portion containing the antigen-recognizing paratope and the Fc region carrying out the activities that protect the host (i.e. effector functions). The differences seen in the Fc region

between the various heavy chains are responsible for the different biological activities of the antibody isotypes.

IgG

This is the major immunoglobulin of serum, making up 75% of the total and having a molecular weight of 150 000 Da in man. Four subclasses are found in man – IgG1, IgG2, IgG3 and IgG4 – that differ in their relative concentrations, amino acid composition, number and position of interchain disulphide bonds, and biological function. IgG is the major antibody of the secondary response (see Ch. 9) and is found in both the serum and tissue fluids.

IgA

In man, most of the serum IgA occurs as a monomer, but in many other mammals it is found mostly as a dimer. The dimer is held together by a J chain, which is produced by the antibody-producing plasma cells. IgA is the predominant antibody class in seromucous secretions such as saliva, tears, colostrum at mucosal–epithelial surfaces in the respiratory, gastro-intestinal and genito-urinary secretions. This secretory IgA (sIgA) is always in the dimeric form and is composed of two basic four-chain units (two light chains and two heavy chains), a J chain and the secretory component. The secretory component is part of the molecule that transports the dimer produced by a submucosal plasma cell to the mucosal surface (Fig. 8.4). It facilitates passage through the epithelial cells and protects the secreted molecule from proteolytic digestion. There are two subclasses of IgA: IgA1 and IgA2.

IgM

IgM is a pentamer of the basic unit with μ heavy chains and a single J chain. Because of its large size, this isotype is confined mainly to the intravascular pool, and is the first antibody type to be produced during an immune response.

IgD

Many circulating B cells have IgD present on their surface, but IgD accounts for less than 1% of the circulating antibody. It is composed of the basic unit with δ heavy chains. The protein is very susceptible to proteolytic attack and therefore has a very short half-life in serum.

IgE

The IgE is present in extremely low levels in serum. However, it is found on the surface of mast cells and basophils, which possess a receptor specific for the Fc part of this molecule.

Antigen binding

The variability in amino acid sequence in the variable domains of light and heavy chains is not found over their entire length but is restricted to short segments. These segments show considerable variation, and are termed *hypervariable regions*. Hypervariable regions contain the residues that make direct contact with the antigen, and are referred to as *complementarity determining regions*. Although the remaining *framework* residues do not come into direct contact with the antigen, they are essential for the formation of the correct tertiary structure of the variable domain and maintenance of the integrity of the binding site. In both light and heavy chains there are three complementarity determining regions that, in combination, form the paratope.

The antigen and antibody are held together by various individually weak non-covalent interactions. However, the formation of a large number of hydrogen bonds and electrostatic, van der Waals' and hydrophobic interactions leads to a considerable binding energy. These attractive forces are active only over extremely short distances, and therefore the epitope and paratope must have complementary structures to enable them to combine. If the electron clouds overlap or residues of similar charge are brought together, repulsive forces will come into play. The balance of attraction against repulsion will dictate the strength of the interaction between an antibody and a particular antigen, that is, the affinity of the antibody for the antigen.

Antibody diversity

It is now known that an antigen selects from the available antibodies those that can combine with its epitopes. It therefore follows that an individual must have an extremely large number of different antibodies to cope with the vast array of different antigens present in the environment.

Immunoglobulin variability

The paratope is produced by the complementarity determining regions of the light and heavy chains generating a specific three-dimensional shape. Any light chain can join with any heavy chain to produce a different paratope. Thus, theoretically, with 10^4 different light chains and 10^4 different heavy chains, 10^8 different specificities could be generated.

The germline DNA is the structure of the gene as it is inherited. All cells in the body contain all the inherited genes, but different genes become active in different cells at different times. The functional immunoglobulin genes within B lymphocytes, the cells that differentiate to antibody-producing plasma cells, are formed by gene rearrangements and recombinations. These events give rise to the production of different variable domains in each B lymphocyte. Once a functional gene has been constructed, no other rearrangements are allowed to take place within this cell. This dictates that one particular cell will produce antibodies with an identical antigen-combining site, and is known as *allelic exclusion*. There is evidence that the gene segments for the variable region of immunoglobulins are particularly susceptible to mutations. This can lead to subtle changes in specificity and/or affinity that are important as an immune response develops (see Ch. 9).

When a B lymphocyte is first stimulated by antigen it produces IgM. As the immune response develops, the class of antibody changes. However, the immunoglobulin produced will have the same variable domain and therefore bind to the same antigen. All that is altered, or *switched*, is the heavy-chain constant region. Thus the progeny of a single B cell will produce different immunoglobulin isotypes as the response to a particular antigen develops, but each will have the same paratope.

Secreted and membrane immunoglobulins

At different stages in its development a B cell produces immunoglobulins that have to be inserted into the membrane or secreted. The membrane-bound immunoglobulin is used as the antigen receptor of the B cell, and a cell that binds antigen through this molecule will then secrete immunoglobulin of the same specificity. The only difference between the two types of antibody is to be found

Table 8.2 Biological properties of human immunoglobulins

Function	IgA	IgD	IgE	IgG1	IgG2	IgG3	IgG4	IgM
Neutralize	++	−	−	+	+	+	+	+++
Complement fixation	±[a]	−	−	++	+	+++	−	+++
Binding to phagocytes	±[b]	−	−	+++	±	+++	+	−
Binding to mast cells	−	−	+++	−	−	−	+	−
Enter tissues	−	−	−	+	+	+	+	−
Placental transfer	−	−	−	+	±	+	+	−
Protects mucosal surfaces	+	−	−	−	−	−	−	−

These activities are determined by the Fc portion of the molecules.
[a]IgA activates the alternative pathway.
[b]Receptors for the Fc portion of IgA have been found on neutrophils and alveolar macrophages.

at the C terminus, where the membrane form has an additional part, the transmembrane portion.

Antibody function

Knowledge gained from the structural studies discussed above has gone some way towards an understanding of the biological activities of the immunoglobulin molecule.

The primary function of an antibody is to bind the antigen that induced its formation. Apart from cases where this results in direct neutralization (e.g. inhibition of toxin activity or of microbial attachment), other effector functions must be generated. The binding of antigen is mediated by the Fab portion, and the Fc region controls the biological defence mechanisms. For every antibody the paratope is different, and different epitopes will therefore be recognized. However, for every antibody of the same isotype, the heavy-chain constant domains are the same, and they therefore all perform the same functions (Table 8.2).

Neutralization

Because antibodies are at least divalent, they can form a complex with multivalent antigens. Depending on the physical nature of the antigen these *immune complexes* exist in various forms (Fig. 8.5). If the antibody is directed against surface antigens of particulate material such as micro-organisms or erythrocytes, *agglutination* will occur. This results in a clump or aggregate that isolates the potential pathogen, stops its dissemination and stimulates its removal by other mechanisms. If the antigen is soluble, the size of the complex will determine its physical state. Small complexes remain soluble, whereas large complexes form *precipitates*.

As might be expected from knowledge of the structure of IgM, its ten combining sites make it a very efficient agglutinating antibody molecule. Rabbit IgM has been

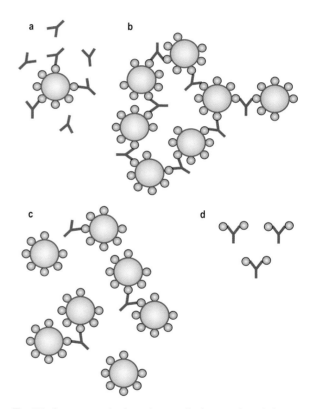

Fig. 8.5 Immune complex formation: a antibody excess; b equivalence; c antigen excess; d monovalent antigen.

shown to be more than 20 times as active as IgG in bringing about bacterial agglutination. Because of its size, IgM is confined largely to the bloodstream and probably plays an important role in protecting against blood invasion by micro-organisms. Certain sites on micro-organisms are critical to the establishment of an infection. Antibody bound to these sites interferes with attachment processes and can, therefore, stop infection by the microbe. The binding of an antibody to functionally important residues in toxins neutralizes their harmful effects.

Complement activation

Activation of the complement system is one of the most important antibody effector mechanisms. The complement cascade is a complex group of serum proteins that mediate inflammatory reactions and cell lysis, and is discussed more fully in Chapter 9. The Fc portion of certain isotypes (see Table 8.2), once antigen has been bound, will activate complement; this requires that C1q, a subunit of the first complement component, cross-links two antibody Fc portions. For this to happen the two regions must be in close proximity. It has been calculated that a single IgM molecule is 1000 times more efficient than IgG. This is because two IgG molecules must be close together for complement activation. A large number of IgG molecules would be required for this to occur if the epitopes were spread. Not all isotypes activate complement, presumably because they do not have the required amino acid sequence, and therefore tertiary structure, in the Fc portion. C1q binds to residues in the C_H3 domain of IgM and the C_H2 domain of IgG. Some isotypes, when interacting with antigen, can activate the alternative pathway of complement that does not use C1 but gives rise to the same biological activities.

Cell binding and opsonization

The Fc portion of certain immunoglobulin isotypes is able to interact with various cell types (see Table 8.2). Antibodies specific for particular antigens, such as bacteria, play a valuable role by binding to the surface and making the antigen more susceptible to phagocytosis and subsequent elimination. This process is known as *opsonization*, and again is mediated by the Fc portion of the antibody. A specific conformation on the Fc region of certain isotypes is recognized by Fc receptors on the surface of the phagocyte. The important residues are in the C_H2 domain near the hinge region. Individually, the interactions are not strong enough to signal the uptake of the antibody molecule; therefore, free immunoglobulin is not internalized. However, when an antigen is coated by many antibody molecules, summation of all the interactions stimulates phagocytosis or other effector mechanisms.

Certain phagocytic cells have receptors for activated complement components – *complement receptors*. If the binding of antibody to the antigen activates the complement cascade, various complement components are deposited on the antigen–antibody complex. Phagocytic cells that have receptors for these complement components then ingest the complexes.

The above-mentioned processes require that the antibody is first complexed with antigen. However, certain cell types can bind free antibody. Mast cells and basophils have Fc receptors that are specific for IgE. These cells perform a protective function but are also involved in the hypersensitivity reactions described in Chapter 9. In man, IgG has the ability to cross the placenta and reach the fetal circulation. This is a passive process involving specific Fc receptors. This route is limited to primates, whereas, in ruminants, immunoglobulin from colostrum is absorbed through the intestinal epithelium. Another Fc-mediated mechanism, already described, is the selective transport of IgA into mucosal secretions.

ANTIGEN RECOGNITION

The immune system has evolved to protect us from potentially harmful material but it must not respond to self molecules. Two separate recognition systems are present:

- humoral immunity
- cell-mediated immunity.

Antibody is the recognition molecule of humoral immunity. This glycoprotein is produced by plasma cells and circulates in the blood and other body fluids. Antibody is also present on the surface of B lymphocytes. The interaction of this surface immunoglobulin with its specific antigen is responsible for the differentiation of these cells into antibody-secreting plasma cells. Antibody molecules, whether free or on the surface of a B cell, recognize free native antigen.

This contrasts dramatically with the situation in cell-mediated immunity; the T lymphocyte antigen receptor binds only to fragments of antigen that are associated with products of the major histocompatibility complex (MHC). T cell recognition of antigen is said to be MHC restricted. These MHC products are present on the surface of cells; therefore T cells recognize only cell-associated antigens. This MHC-restricted recognition mechanism has evolved because of the functions carried out by T lymphocytes. Some T cells produce immunoregulatory molecules, lymphokines, some of which influence the activities of host cells and others directly kill infected or foreign cells. Therefore, it would be inefficient or dangerous to produce these effects in response to either free antigen or antigen sitting idly on some cell membrane. The joint recognition of MHC molecules and antigen ensures that the T cell makes contact with antigen on the surface of the appropriate target cell.

B cell receptor

Antibody is found free in body fluids and as a transmembrane protein on the surface of B lymphocytes (i.e.

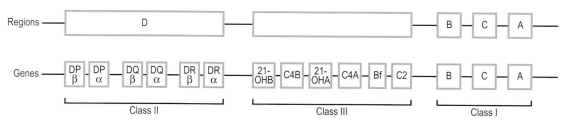

Fig. 8.6 Human MHC gene map.

surface immunoglobulin), where it acts as the B cell antigen receptor. The antibody present on the surface of the B cell is exactly the same molecule as is secreted when the cell develops into a plasma cell, except for the extreme C-terminal end as described above. It should be noted that the molecules present on the cell surface are present as monomers even though they are secreted in a polymeric form.

T cell receptor

The complex on T lymphocytes that is involved in antigen recognition is composed of a number of glycoprotein structures. Some of these molecules have been named systematically by CD (cluster of differentiation) nomenclature using antibodies. These generic names are used in preference to other symbols sometimes found in the literature.

The T cell antigen receptor is a heterodimer composed of an α and β or a γ and δ chain. Each chain contains a variable and constant domain, transmembrane portion and cytoplasmic tail. The variable domain folds to form a paratope that interacts with antigenic peptides associated with MHC molecules on the cell surface. The majority of T cells use the α–β heterodimer in antigen recognition. The role of cells that possess the γ–δ molecules is unknown, but they may be involved in the immune response to particular types of antigens at specific anatomical sites. The T cell receptor is the molecule that is responsible for the recognition of specific MHC–antigen complexes, and is different for every T cell. Genetic rearrangements of germline genes, similar to those seen in B cells, produce functional T cell receptors.

CD3 is present on all T cells and is non-covalently linked to the T cell receptor. The CD3 complex is thought to be involved in signal transduction, leading to cell activation, when a ligand binds to the T cell receptor. CD4 and CD8 are mutually exclusive molecules. They are present on T cells that are restricted in their recognition of antigen by MHC class II and class I molecules, respectively. Owing to their almost exclusive correlation with a specific MHC class, it is thought that these molecules bind to non-polymorphic determinants on the MHC molecules.

MAJOR HISTOCOMPATIBILITY COMPLEX

The MHC is the part of the genome that codes for molecules that are important in immune recognition, including interactions between lymphoid cells and other cell types. It is also involved in the rejection of allografts. The MHCs of a number of species have been studied, although most is known about those of the mouse and man.

The gene complex contains a large number of individual genes that can be grouped into three classes on the basis of the structure and function of their products. The molecules coded for by the genes are sometimes referred to as *MHC antigens* because they were first defined by serological analysis (i.e. using antibodies).

The MHC of man is known as *human leucocyte group A* (HLA), and in mice it is referred to as *histocompatibility 2* (H2).

Gene organization

The genes that code for the HLA molecules are found on the short arm of chromosome 6. They are arranged over a region of between 2000 and 4000 kilobases in size, containing sufficient DNA for more than 200 genes. The MHC genes are contained within regions known as A, B, C and D (Fig. 8.6).

MHC class I molecules consist of two non-covalently associated polypeptide chains. A single gene that codes for the larger chain is present in the A, B and C regions, whereas the smaller chain, known as β_2-microglobulin, is coded for elsewhere in the genome.

MHC class II molecules are composed of two chains, both of which are coded for within the D region. There are three class II molecules, DP, DQ and DR.

The class III genes are grouped together in a region between D and B. These genes code for a number of complement components and cytokines, but most have nothing to do with the immune system.

MHC antigen structure and distribution

The MHC class I molecule is a dimer composed of a glycosylated transmembrane protein, of molecular weight

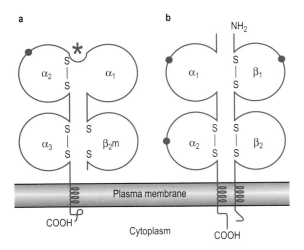

Fig. 8.7 Structure of MHC class I and class II molecules. Schematic representation of **a** class I and **b** class II molecules as found in the plasma membrane. β_2m, β_2-microglobulin; ●, carbohydrate moieties; *, antigen-binding cleft.

45 000 Da, coded for within the MHC, linked to a smaller protein, β_2-microglobulin (Fig. 8.7a). The globular protein formed by these two peptides is present on the surface of all human nucleated cells, except neurones. β_2-Microglobulin is required for the processing and expression of MHC-encoded molecules on the cell membrane. The MHC-encoded class I glycoprotein folds into three globular domains (α_1, α_2 and α_3) held in place by disulphide bonds and non-covalent interactions. These globular domains are found on the outer surface of the cell. There is a short cytoplasmic tail and a transmembrane portion. β_2-Microglobulin is non-covalently associated with the α_3 domain.

The MHC class II molecules consist of two poly-peptide chains (α and β) held together by non-covalent interactions (Fig. 8.7b). They have a much more limited cellular distribution, being limited to the surface of certain cells of the immune system. In man, they are normally found on dendritic cells, B lymphocytes, macrophages, monocytes and activated T lymphocytes. Each chain is composed of two extracellular domains, a transmembrane portion and a cytoplasmic tail.

These two types of molecule are folded into domains of a similar overall structure to immunoglobulin and, along with other molecules of the immune system involved in recognition processes, are thought to have evolved from a common ancestral molecule. A number of members of this *immunoglobulin supergene family* are depicted in Figure 8.8. MHC class II molecules, some interleukin receptors and Fc receptors are also included in the family.

The MHC antigens of each class have a similar basic structure. However, fine structural differences can be detected in the α_1 and α_2 domains of class I molecules and in α_1 and β_1 domains of class II molecules. These domains form a cleft on the outermost part of the molecules in

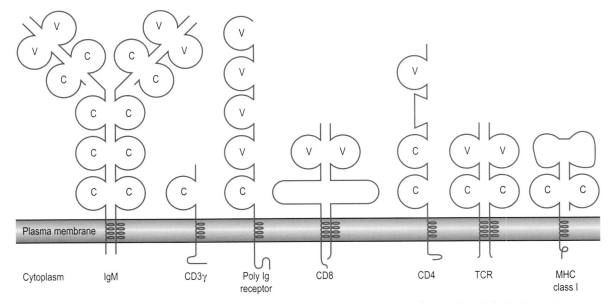

Fig. 8.8 The immunoglobulin supergene family. A number of molecules involved in the immune system display striking similarities in overall structure. Regions similar to immunoglobulin domains are shown as circles; those related to variable and constant domains are designated V and C, respectively.

which antigen fragments are found. The variations found are due to differences in the amino acid sequence and can be detected serologically. The variable residues give rise to different three-dimensional shapes on the MHC molecules. This, in turn, influences the selection of which antigen fragments can bind to a particular MHC molecule.

There are, therefore, many different forms of these molecules that can be identified in a population – they are highly *polymorphic*. Thus, it is highly unlikely that two individuals will have exactly the same MHC antigens. The MHC molecules of a particular individual can be given a designation using tissue-typing reagents. So, each chromosome of an individual will contain the genes that code for an A, B, C, DP, DQ and DR molecule. As the MHC genes are co-dominant, the products of both alleles are expressed on the cell surface. All of the nucleated cells in the body therefore express multiple copies of two HLA-A, two HLA-B and two HLA-C molecules. On certain cell types there will also be HLA-DP, -DQ and -DR molecules that were inherited from both parents.

Function

The MHC class I and II molecules are essential for immune recognition by T lymphocytes which can bind to antigens only when associated with these molecules. The different classes of molecule are involved in antigen recognition by different T cell types or subsets:

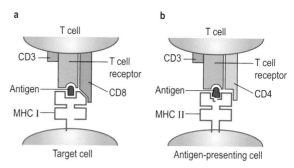

Fig. 8.9 Molecules involved in T cell recognition. **a** Antigen fragments that associate with class I molecules are recognized by T cells that have the CD8 molecule. **b** Antigen fragments that associate with MHC class II are recognized by T cells that have the CD4 molecule on their surface.

- T lymphocytes that have CD4 molecules on their surface recognize antigen in association with MHC class II molecules.
- T lymphocytes that have CD8 molecules are restricted by MHC class I molecules.

The T lymphocyte subsets perform different functions, but the division is not absolute. The one thing that they have in common is that they recognize, through their T cell receptor complex (CD3, CD4 or CD8, TCR), antigen fragments in association with MHC molecules (Fig. 8.9). In general terms:

KEY POINTS

- The immune system recognizes molecules known as antigens. Recognition provokes a response that may be immediate and relatively non-specific (*innate immunity*) or may develop over time and become progressively more specific (*acquired immunity*).
- Antigens generally comprise multiple *epitopes*, each eliciting separate but specific immune responses that, in acquired immunity, may be mediated by *antibodies* (*humoral immunity*) or by T lymphocytes, which are responsible for *cell-mediated immunity*.
- Antibodies are glycoproteins termed *immunoglobulins* based on a heterodimer structure of heavy (H) and light (L) chains. By recombination in the cells responsible for antibody synthesis (B lymphocytes), highly polymorphic (variable) regions of H and L chains are brought together such that binding specificity for an enormous range of epitopes is achieved.
- Five main heavy chain types provide the five different immunoglobulin classes: IgG, IgA, IgM, IgE and IgD

(in descending order of their relative abundance in human serum). The different classes serve different functions.
- Molecules that elicit innate immune responses display pathogen-associated molecular patterns that are recognized by pattern recognition receptors.
- In acquired immunity, antigens that elicit a humoral response are recognized by immunoglobulin molecules on B lymphocytes.
- Antigen recognition leading to cell-mediated immunity depends on T lymphocyte receptors. These are activated by epitopes resulting from intracellular processing of antigens. These epitopes are presented to T cells in combination with major histocompatibility complex (MHC) molecules on the surface of host cells.
- The two major effectors of cell-mediated immunity are T lymphocyte subsets termed CD4 and CD8 cells; these are, respectively, stimulated by epitopes presented in the context of MHC class II and class I molecules.

- CD4-positive (CD4$^+$) cells produce molecules, lymphokines, that stimulate and support the production of immune system cells.
- CD8$^+$ cells are involved in the destruction of virally infected cells (see Ch. 9) and the destruction of tissue grafts from MHC-incompatible donors.

CD4$^+$ cells produce molecules that stimulate the growth and differentiation of cells. These molecules are most effective over short distances, as they will be more concentrated. This happens when the two cells involved are actually joined together or in close proximity. The stimulation of CD4$^+$ T cells by antigen fragments on the surface of a responsive cell, or on a cell in the vicinity of a responsive cell, greatly increases the effectiveness of the messenger molecules produced by the T cell. In cell-mediated cytotoxicity the CD8$^+$ T cell has to bind to the infected cell so that the correct cell is killed. Therefore, the correct functioning of T lymphocytes requires direct contact with other cells. This interaction is mediated through T cell receptor recognition of antigen bound to MHC molecules on the host cell surface.

RECOMMENDED READING

Abbas A K, Lichtman A H 2003 *Cellular and Molecular Immunology*, 5th edn. Saunders, Philadelphia
Janeway C A 1993 How the immune system recognizes invaders. *Scientific American* Sept: 41–47

Internet site

Kimball's Biology Pages. http://biology-pages.info
http://www.cat.cc.md.us/courses/bio141/lecguide/unit3/index.html

9 Innate and acquired immunity

J. Stewart

The environment contains a vast number of potentially infectious organisms – viruses, bacteria, fungi, protozoa and worms. Any of these can cause damage if they multiply unchecked, and many could kill the host. However, the majority of infections in the normal individual are of limited duration and leave little permanent damage. This fortunate outcome is due largely to the *immune system*.

The immune system is split into two functional divisions. *Innate immunity* is the first line of defence against infectious agents, and most potential pathogens are checked before they establish an overt infection. If these defences are breached, the acquired immune system is called into play. *Acquired immunity* produces a specific response to each infectious agent, and the effector mechanisms generated normally eradicate the offending material. Furthermore, the adaptive immune system remembers the particular infectious agent and can prevent it causing disease later.

THE IMMUNE SYSTEM

The immune system consists of a number of organs and several different cell types. All cells of the immune system – tissue cells and white blood cells or *leucocytes* – develop from pluripotent stem cells in the bone marrow. These haemopoietic stem cells also give rise to the red blood cells or *erythrocytes*. The production of leucocytes is through two main pathways of differentiation (Fig. 9.1). The *lymphoid* lineage produces T lymphocytes and B lymphocytes. Natural killer (NK) cells, also known as large granular lymphocytes, probably also develop from lymphoid progenitors. The *myeloid* pathway gives rise to mononuclear phagocytes, monocytes and macrophages, and granulocytes, basophils, eosinophils and neutrophils, as well as platelets and mast cells. Platelets are involved in blood clotting and inflammation, whereas mast cells are similar to basophils but are found in tissues.

Lymphoid cells

Lymphocytes make up about 20% of the white blood cells present in the adult circulation. Mature lymphoid cells are long lived and may survive for many years as memory cells. These mononuclear cells are heterogeneous in size and morphology. The typical small lymphocytes comprise the T and B cell populations. The larger and less numerous cells, sometimes referred to as large granular lymphocytes, contain the population of NK cells. Cells within this population are able to kill certain tumour and virally infected cells (natural killing) and destroy cells coated with immunoglobulin (antibody-dependent cell-mediated cytotoxicity).

Morphologically it is quite difficult to distinguish between the different lymphoid cells and impossible to differentiate the subclasses of T cell. As these cells carry out different processes, they possess molecules on their surface unique to that functional requirement. These molecules, referred to as cell markers, can be used to distinguish between different cell types and also to identify cells at different stages of differentiation. The different cell surface molecules have been systematically named by the CD (cluster of differentiation) system; some of those expressed by different T cell populations are shown in Table 9.1. These CD markers are identified using specific monoclonal antibodies (see p. 123). The presence of these specific antibodies on the cell surface is then visualized using labelled antibodies that recognize the first antibody.

Myeloid cells

The second pathway of development gives rise to a variety of cell types of different morphology and function.

Mononuclear phagocytes

The common myeloid progenitor in the bone marrow gives rise to *monocytes*, which circulate in the blood and

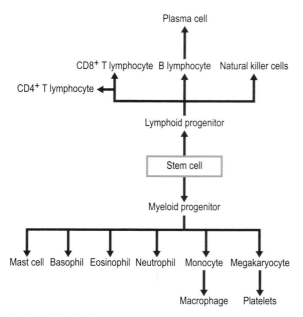

Fig. 9.1   Cells of the immune system.

Table 9.1	Major T lymphocyte markers	
Marker	Distribution	Proposed function
CD2	All T cells	Adherence to target cell
CD3	All T cells	Part of T cell antigen–receptor complex
CD4	Helper subset (T_H)	MHC class II-restricted recognition
CD7	All T cells	Unknown
CD8	Cytotoxic subset (T_c)	MHC class I-restricted recognition

CD, cluster of differentiation; MHC, major histocompatibility complex.

migrate into organs and tissues to become *macrophages*. The human blood monocyte is larger than a lymphocyte and usually has a kidney-shaped nucleus. This actively phagocytic cell has a ruffled membrane and many cytoplasmic granules. These *lysosomes* contain enzymes and molecules that are involved in the killing of micro-organisms. Mononuclear phagocytes adhere strongly to surfaces and have various cell membrane receptors to aid the binding and ingestion of foreign material. Their activities can be enhanced by molecules produced by T lymphocytes, called *lymphokines*. Macrophages and monocytes are capable of producing various complement components, prostaglandins, interferons and *monokines* such as interleukin (IL)-1 and tumour necrosis factor.

Lymphokines and monokines are collectively known as *cytokines*.

Polymorphonuclear leucocytes

These cells are sometimes referred to as *granulocytes* and are short-lived cells (2–3 days) compared to macrophages, which may survive for months or years. They are classified as *neutrophils*, *eosinophils* and *basophils* on the basis of their histochemical staining. The mature forms have a multilobed nucleus and many granules. Neutrophils constitute 60–70% of the leucocytes, but also migrate into tissues in response to injury or infection.

Neutrophils.   These are the most abundant circulating granulocyte. Their granules contain numerous micro-bicidal molecules and the cells enter the tissues when a chemotactic factor is produced, as the result of infection or injury.

Eosinophils.   Eosinophils are also phagocytic cells, although they appear to be less efficient than neutrophils. They are present in low numbers in a healthy individual (1–2% of leucocytes), but their numbers rise in certain allergic conditions. The granule contents can be released by the appropriate signal, and the cytotoxic molecules can then kill parasites that are too large to be phagocytosed.

Basophils.   These cells are found in extremely small numbers in the circulation (<0.2%) and have certain characteristics in common with tissue *mast cells*. Both cell types have receptors on their surface for the Fc portion of immunoglobulin (Ig) E, and cross-linking of this immunoglobulin by antigen leads to the release of various pharmacological mediators. These molecules stimulate an inflammatory response. There are two types of mast cell: one is found in connective tissue and the other is mucosa associated. Mast cells and basophils are both derived from bone marrow, but their developmental relationship is not clear.

Platelets

Platelets are also derived from myeloid progenitors. In addition to their role in clotting, they are involved in inflammation.

INNATE IMMUNITY

The healthy individual is protected from potentially harmful micro-organisms in the environment by a number of effective mechanisms, present from birth, that do not depend upon prior exposure to any particular micro-organism. The innate defence mechanisms are non-specific in the sense that they are effective against a wide range of potentially infectious agents. The characteristics

Table 9.2 Characteristics and determinants of innate and acquired immunity

Innate immunity	Acquired immunity
Non-specific	Specific
No change with repeat exposure	Memory
Mechanical barriers	
Bactericidal substances	
Natural flora	
Humoral	
Acute-phase proteins	Antibody
Interferons	
Lysozyme	
Complement	
Cell-mediated	
Natural killer cells	T lymphocytes
Phagocytes	

and constituents of innate and acquired immunity are shown in Table 9.2.

FEATURES OF INNATE IMMUNITY

The components of the innate immune system recognize structures that are unique to microbes. These include complex lipids and carbohydrates such as peptidoglycan of bacteria, lipopolysaccharides of Gram-negative bacteria, lipoteichoic acid in Gram-positive bacteria and mannose-containing oligosaccharides found in many microbial molecules. Other microbial specific molecules include double-stranded RNA found in replicating viruses and unmethylated CpG sequences in bacteria. Therefore, the innate immune system is able to recognize non-self structures and react appropriately but does not recognize self structure, so the potential of autoimmunity is avoided. The microbial products recognized by the innate immune system, known as pathogen-associated molecular patterns (PAMPs), are essential for survival of the micro-organisms and cannot easily be discarded or mutated. Different classes of micro-organism express different PAMPs that are recognized by different pattern recognition receptors (PRRs) on host cells and circulating molecules (Table 9.3). One group of PRRs that are still being characterized are the Toll-like receptors. Mammalian Toll-like receptors are expressed on different cell types that are components of the innate defences, including macrophages, dendritic cells, neutrophils, mucosal epithelial cells and endothelial cells. Recognition of microbial components by these receptors leads to a variety of outcomes, including cytokine release, inflammation and cell activation.

Innate defences act as the initial response to microbial challenge and can eliminate the micro-organism from the host. However, many microbes have evolved strategies to overcome innate defences, and in this situation the more potent and specialized acquired immune response is required to eliminate the pathogen. The innate immune system plays a critical role in the generation of an efficient and effective acquired immune response. Cytokines produced by the innate immune system signal that infectious agents are present and influence the type of acquired immune response that develops.

DETERMINANTS OF INNATE IMMUNITY

Species and strains

Marked differences exist in the susceptibility of different species to infective agents. The rat is strikingly resistant to diphtheria, whereas the guinea-pig and man are highly

Table 9.3 Examples of pathogen-associated molecular patterns (PAMPs) and pattern recognition receptors (PRRs) in innate immunity

PAMP	Source	PRR	Response
Sugars (mannose)	Microbial glycoproteins and glycolipids	Mannose receptors	Phagocytosis
		Mannose-binding protein	Complement activation
		Lectin-like receptors	Phagocytosis
N-formylmethionyl peptides	Bacterial protein synthesis	N-formylmethionyl peptides receptors	Chemotaxis and phagocyte activation
Phosphorylcholine	Microbial membranes	C-reactive protein	Complement activation
Lipoarabinomannan	Yeast cell wall	Toll-like receptor 2	
Lipoteichoic acid	Gram-positive bacterial cell wall	Toll-like receptor 2	Macrophage activation
Lipopolysaccharide	Gram-negative bacterial cell wall	Toll-like receptors 4 and 2	Cytokine production
Unmethylated CpG necleotides	Bacterial DNA	Toll-like receptor 9	
dsRNA	Replicating viruses	Toll-like receptor 3	Type 1 interferon production

dsRNA, double-stranded ribonucleic acid.

susceptible. The rabbit is particularly susceptible to myxomatosis, and human beings to syphilis, leprosy and meningococcal meningitis. Susceptibility to an infection does not always imply a lack of resistance to disease caused by the micro-organism. For example, although man is highly susceptible to the common cold, the infection is overcome within a few days. In some diseases, it may be difficult to initiate the infection, but once established the disease can progress rapidly, implying a lack of resistance. For example, rabies occurs in both human beings and dogs but is not readily established as the virus does not ordinarily penetrate healthy skin. Once infected, however, both species are unable to overcome the disease. Marked variations in resistance to infection have been noted between different strains of mice, and it is possible to breed, by selection, rabbits of low, intermediate and high resistance to experimental tuberculosis.

Individual differences and influence of age

The role of heredity in determining resistance to infection is well illustrated by studies on tuberculosis in twins. If one homozygous twin develops tuberculosis, the other twin has a three in one chance of developing the disease, compared with a one in three chance if the twins are heterozygous. Sometimes genetically controlled abnormalities are an advantage to the individual in resisting infection as, for example, in a hereditary abnormality of the red blood cells (sickling). These red blood cells cannot be parasitized by *Plasmodium falciparum*, thus conferring a degree of resistance to malaria in affected individuals.

Infectious diseases are often more severe in early childhood and in young animals; this higher suscep-tibility of the young appears to be associated with immaturity of the immunological mechanisms affecting the ability of the lymphoid system to deal with and react to foreign antigens. This is also the time when infectious agents are encountered for the first time (primary exposure), and a memory-acquired immune response cannot be called upon to aid elimination. In certain viral infections (e.g. polio and chickenpox), the clinical illness is more severe in adults than in children. This may be due to a more active immune response producing greater tissue damage. In the elderly, besides a general waning of the activities of the immune system, physical abnormalities (e.g. prostatic enlargement leading to stasis of urine) or long-term exposure to environmental factors (e.g. smoking) are common causes of increased susceptibility to infection.

Hormonal influences and sex

There is decreased resistance to infection in those with diseases such as diabetes mellitus, hypothyroidism and adrenal dysfunction. The reasons for this decrease have not yet been clarified but may be related to enzyme or hormone activities. It is known that glucocorticoids are anti-inflammatory agents, decreasing the ability of phagocytes to ingest material. They also have beneficial effects by interfering in some way with the toxic effects of bacterial products such as endotoxins.

There are no marked differences in susceptibility to infections between the sexes. Although the overall incidence and death rate from infectious disease are greater in males than in females, both infectious hepatitis and whooping cough have a higher morbidity and mortality in females.

Nutritional factors

The adverse effects of poor nutrition on susceptibility to certain infectious agents are not now seriously questioned. Experimental evidence in animals has shown repeatedly that inadequate diet may be correlated with increased susceptibility to a variety of bacterial diseases, associated with decreased phagocytic activity and leucopenia. In the case of viruses, which are intracellular parasites, mal-nutrition may have an effect on virus production, but the usual outcome is enhanced disease as a result of impaired immune responses, especially the cytotoxic responses.

MECHANISMS OF INNATE IMMUNITY

Mechanical barriers and surface secretions

The intact skin and mucous membranes of the body afford a high degree of protection against pathogens. In conditions where the skin is damaged, such as in patients with burns and after traumatic injury or surgery, infection can be a serious problem. The skin is a resistant barrier because of its outer horny layer consisting mainly of keratin, which is indigestible by most micro-organisms, and thus shields the living cells of the epidermis from micro-organisms and their toxins. The relatively dry condition of the skin and the high concentration of salt in drying sweat are inhibitory or lethal to many micro-organisms.

The sebaceous secretions and sweat of the skin contain bactericidal and fungicidal fatty acids, which constitute an effective protective mechanism against many potential pathogens. The protective ability of these secretions varies at different stages of life, and some fungal 'ring-worm' infections of children disappear at puberty with the marked increase of sebaceous secretions.

The sticky mucus covering the respiratory tract acts as a trapping mechanism for inhaled particles. The action of cilia sweeps the secretions, containing the foreign material, towards the oropharynx so that they are swallowed;

in the stomach the acidic secretions destroy most of the micro-organisms present. Nasal secretions and saliva contain mucopolysaccharides capable of blocking some viruses.

The washing action of tears and flushing of urine are effective in stopping invasion by micro-organisms. The commensal micro-organisms that make up the natural bacterial flora covering epithelial surfaces are protective in a number of ways:

- Their very presence uses up a niche that cannot be used by a pathogen.
- They compete for nutrients.
- They produce byproducts that can inhibit the growth of other organisms.

It is important not to disturb the relationship between the host and its indigenous flora.

Commensal organisms from the gut or bacteria normally present on the skin can cause problems if they gain access to an area that they do not normally populate. An example of this is urinary tract infection resulting from the introduction of *Escherichia coli*, a gut commensal, by means of a urinary catheter. Some commensal organisms possessing low virulence (see p. 170) that are provided with the circumstances by which to cause infections may be referred to as *opportunistic pathogens*. Infections with these opportunists are quite widespread, often appearing as a result of medical or surgical treatment that breaches the innate defences or reduces the host's ability to respond.

Humoral defence mechanisms

A number of microbicidal substances are present in the tissue and body fluids. Some of these molecules are produced constitutively (e.g. lysozyme), and others are produced in response to infection (e.g. acute-phase proteins and interferon). These molecules all show the characteristics of innate immunity: there is no recognition specific to the micro-organism beyond the distribution of the molecule(s) detected, and the response is not enhanced on re-exposure to the same antigen.

Lysozyme

This is a basic protein of low molecular weight found in relatively high concentrations in neutrophils as well as in most tissue fluids, except cerebrospinal fluid, sweat and urine. It functions as a mucolytic enzyme, splitting sugars off the structural peptidoglycan of the cell wall of many Gram-positive bacteria and thus causing their lysis. It seems likely that lysozyme may also play a role in the intracellular destruction of some Gram-negative bacteria. In many pathogenic bacteria the peptidoglycan of the cell wall appears to be protected from the access of lysozyme

by other wall components (e.g. lipopolysaccharide). The action of other enzymes from phagocytes or of complement may be needed to remove this protection and expose the peptidoglycan to the action of lysozyme.

Basic polypeptides

A variety of basic proteins, derived from tissues and blood cells, have some antibacterial properties. This group includes the basic proteins called spermine and spermidine, which can kill tubercle bacilli and some staphylococci. Other toxic compounds are the arginine- and lysine-containing proteins protamine and histone. The bactericidal activity of basic polypeptides probably depends on their ability to react non-specifically with acid polysaccharides at the bacterial cell surface.

Acute-phase proteins

The concentration of acute-phase proteins rises dramatically during an infection. Microbial products such as endotoxin can stimulate macrophages to release IL-1, which stimulates the liver to produce increased amounts of various acute-phase proteins, the concentrations of which can rise over 1000-fold. One of the best characterized acute-phase proteins is *C-reactive protein*, which binds to phosphorylcholine residues in the cell wall of certain micro-organisms. This complex is very effective at activating the classical complement pathway. Also included in this group of molecules are α_1-antitrypsin, α_2-macroglobulin, fibrinogen and serum amyloid A protein, all of which act to limit the spread of the infectious agent or stimulate the host response.

Interferon

The observation that cell cultures infected with one virus resist infection by a second virus (viral interference) led to the identification of the family of antiviral agents known as *interferons*. A number of molecules have been identified; α- and β-interferons (see Ch. 5) are part of innate immunity, and γ-interferon is produced by T cells as part of the acquired immune response (see Ch. 10).

Complement

The existence of a heat-labile serum component with the ability to lyse red blood cells and destroy Gram-negative bacteria has been known since the 1930s. The chemical complexity of the phenomenon was not appreciated by early workers, who ascribed the activity to a single component, called complement. Complement is in fact composed of a large number of different serum proteins present in low concentration in normal serum. These molecules are present in an inactive form but can be activated to form an enzyme cascade: the product of the first reaction is the catalyst of the next and so on.

Approximately 30 proteins are involved in the complement system, some of which are enzymes, some are control molecules and others are structural proteins with no enzymatic activity. A number of the molecules involved are split into two components (a and b fragments) by the product of the previous step. There are two pathways of complement activation, the alternative and classical, that lead to the same physiological consequences:

- opsonization
- cellular activation
- lysis.

The two pathways use different initiation processes. Component C3 forms the connection between the two pathways, and the binding of this molecule to a surface is the key process in complement activation.

Classical pathway

The classical pathway of activation leading to the cleavage of C3 is initiated by the binding of two or more of the globular domains of the C1q component of C1 to its ligand: immune complexes containing IgG or IgM and certain micro-organisms and their products. This causes a conformational change in the C1 complex that leads to the auto-activation of C1r. The enzyme C1r then converts C1s into an active serine esterase that acts on the thioester-containing molecule C4 to produce C4a and a reactive C4b (Fig. 9.2). C4a is released and some of the C4b becomes attached to a surface. C2 binds to the surface-bound C4b, becomes a substrate for the activated C1 complex, and is split into C2a and C2b. The C2b is released, leaving C4b2a – the classical pathway C3 convertase. This active enzyme then generates C3a and the unstable C3b from C3. A small amount of the C3b generated binds to the activating surface and acts as a focus for further complement activation. Activation of the classical pathway is regulated by C1 inhibitor and by a number of molecules that limit the production of the 'C3 convertase'.

The so-called *lectin pathway* is initiated by mannose-binding lectin (a secreted PRR) attaching to the surface of a micro-organism. This leads to the production of C4b2a and the generation of C3b on the activating surface.

Alternative pathway

Intrinsically, C3 undergoes a low level of hydrolysis of an internal thioester bond to generate C3b. This molecule complexes, in the presence of Mg^{2+} ions, with factor B, which is then acted on by factor D to produce C3bBb. This is a 'C3 convertase', which is capable of splitting more C3 to C3b, some of which will become membrane bound.

The initial binding of C3b generated by either the classical or the alternative pathway leads to an amplification loop that results in the binding of many more C3b molecules to the same surface. Factor B binds to the surface-bound C3b to form C3bB, the substrate for factor D – a serine esterase – which is present in very low concentrations in an already active form. The cleavage of factor B results in the formation of the C3 convertase, C3bBb, which dissociates rapidly unless it is stabilized by the binding of properdin (P), forming the complex C3bBbP. This convertase can cleave many more C3 molecules, some of which become surface bound. This amplification loop is a positive feedback system that will cycle until all the C3 is used up unless it is regulated carefully.

Regulation

The nature of the surface to which the C3b is bound regulates the outcome. Self cell membranes contain a number of regulatory molecules that promote the binding of factor H rather than factor B to C3b. This results in the inhibition of the activation process. On non-self structures the C3b is protected, as regulatory proteins are not present, and factor B has a higher affinity for C3b than factor H at these sites.

Thus the surface of many micro-organisms can stabilize the C3bBb by protecting it from factor H. In addition, another molecule, properdin, stabilizes the complex. The deposition of a few molecules of C3b on to these surfaces is followed by the formation of the relatively stable C3bBbP complex. This C3 convertase will lead to more C3b deposition. Immune complexes composed of certain immunoglobulins (e.g. IgA and IgE) also function as protected sites for C3b and activate complement by the alternative pathway. Poor activation surfaces are made more susceptible to deposition by the presence of antibody that generates C3b by the classical pathway.

Membrane attack complex

The next step after the formation of C3b is the cleavage of C5 (Fig. 9.3). The 'C5 convertases' are generated from C4b2a of the classical pathway and C3bBb of the alternative pathway by the addition of another C3b molecule. These membrane-bound trimolecular complexes selectively bind C5 and cleave it to give fluid-phase C5a and membrane-bound C5b. The formation of the rest of the membrane attack complex is non-enzymatic. C6 binds to C5b, and this joint complex is released from the C5 convertase. The formation of C5b67 generates a hydrophobic complex that inserts into the lipid bilayer in the vicinity of the

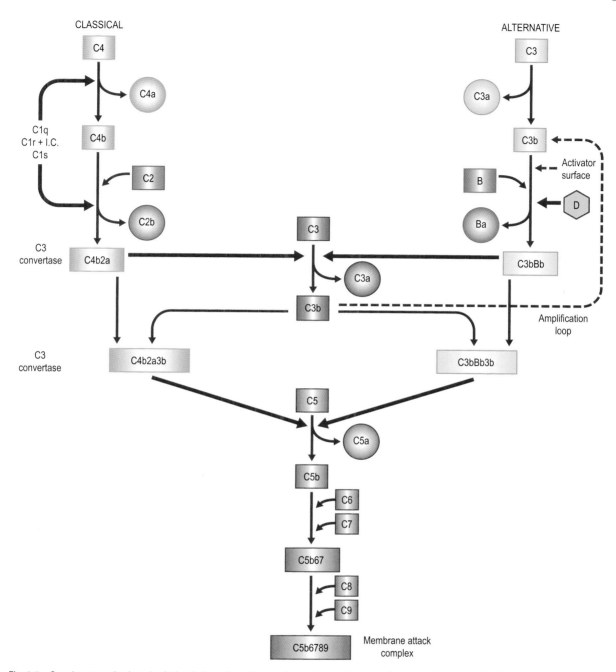

Fig. 9.2 Complement activation: classical and alternative pathways. Enzymatic reactions are indicated by thick arrows. I.C., immune complex.

initial activation site. Usually this is on the same cell surface as the initial trigger, but occasionally other cells may be involved. Therefore 'bystander' lysis can take place, giving rise to damage to surrounding tissue. There are a number of proteins present in body fluids to limit this potentially dangerous process by binding to fluid-phase C5b67. C8 and C9 bind to the membrane-inserted complex in sequence, resulting in the formation of a lytic polymeric complex containing up to 20 C9 monomers. A small amount of lysis can occur when C8 binds to C5b67, but it is the polymerized C9 that causes the most damage.

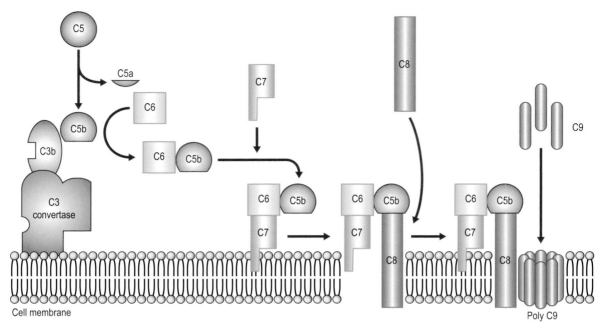

Fig. 9.3 Membrane attack complex.

Functions

The activation of complement by either pathway gives rise to C3b and the generation of a number of factors that can aid in the elimination of foreign material.

The complete insertion of the membrane attack complex into a cell will lead to membrane damage and lysis, probably by osmotic swelling. Some thin-walled pathogens, such as trypanosomes and malaria parasites, are killed by complement-mediated lysis. Some Gram-negative bacteria can be killed by complement in conjunction with lysozyme. However, complement-mediated lysis is of limited importance as a bactericidal mechanism compared with phagocyte destruction of bacteria. Inherited deficiencies of the terminal components are associated with infection by gonococci and meningococci, which can survive inside neutrophils and for which complement-mediated killing is important.

Phagocytic cells have receptors for certain complement components that facilitate the adherence of complement-coated particles. Therefore, complement is an *opsonin*, and in certain circumstances this attachment may lead to phagocytosis.

Two of the molecules released during the complement cascade, C3a and C5a, have potent biological activities. These molecules, known as *anaphylatoxins*, trigger mast cells and basophils to release mediators of inflammation (see below). They also stimulate neutrophils to produce reactive oxygen intermediates, whereas C5a on its own is a chemo-attractant and acts directly on vascular endothelium to cause vasodilatation and increased vascular permeability.

Cells

Phagocytes

Micro-organisms entering the tissue fluids or bloodstream are rapidly engulfed by *neutrophils* and *mononuclear phagocytes*. In the blood the latter are known as *monocytes*, whereas in the tissues they differentiate into *macrophages*. In connective tissue they are known as *histiocytes*, in kidney as *mesangial cells*, in liver as *Kupffer cells*, in bone as *osteoclasts*, in brain as *microglia*, and in the spleen, lymph node and thymus as the *sinus-lining macrophages*.

The essential features of these cells are that they:

- are actively phagocytic
- contain digestive enzymes to degrade ingested material
- are an important link between the innate and acquired immune mechanisms.

Part of their role in regard to acquired immunity is that they can process and present antigens, and produce molecules that stimulate lymphocyte differentiation into effector cells.

The role of the phagocyte in innate immunity is to engulf particles (phagocytosis) or soluble material (pinocytosis), and digest them intracellularly within specialized vacuoles.

The macrophages present in the walls of capillaries and vascular sinuses in spleen, liver, lungs and bone marrow serve an important role in clearing the bloodstream of foreign particulate material such as bacteria. So efficient is this process that the repeated finding, generally by sensitive broth culture, of a few bacteria or yeasts in the bloodstream usually indicates that there is a continuing release of micro-organisms from an active focus such as an abscess or the heart valve vegetations found in bacterial endocarditis.

The ability of macrophages to ingest and destroy micro-organisms can be impaired or enhanced by depression or stimulation of the phagocyte system. Some micro-organisms, such as mycobacteria and brucellae, can resist intracellular digestion by normal macrophages, though they may be digested by 'activated' ones.

Chemotaxis

For phagocytic cells to be effective, they must be attracted to the site of infection. Once they have passed through the capillary walls they move through the tissues in response to a concentration gradient of molecules produced at the site of damage. These chemotactic factors include:

- products of injured tissue
- factors from the blood (C5a)
- substances produced by neutrophils and mast cells (leukotrienes and histamine)
- bacterial products (formyl-methionine peptides).

Neutrophils respond first and move faster than monocytes.

Phagocytosis

Phagocytosis involves:

- recognition and binding
- ingestion
- digestion.

Phagocytosis may occur in the absence of antibody, especially on surfaces such as those of the lung alveoli and when inert particles are involved. Cell membranes carry a net negative charge that keeps them apart and stops autophagocytosis. The hydrophilic nature of certain bacterial cell wall components stops them passing through the hydrophobic membrane. To overcome these difficulties the phagocytes have receptors on their surface that mediate the attachment of particles coated with the correct ligand. Phagocytes have receptors for the Fc portion of certain immunoglobulin isotypes and for some components of the complement cascade. The presence of these molecules, or *opsonins*, on the particle surface markedly enhances the ingestion process and, in some cases, digestion. Whether mediated by specific

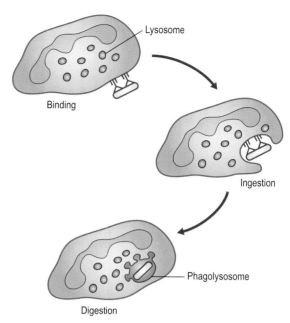

Fig. 9.4   Stages in phagocytosis.

receptors or not, the foreign particle is surrounded by the cell membrane, which then invaginates and produces an *endosome* or *phagosome* within the cell (Fig. 9.4).

The microbicidal machinery of the phagocyte is contained within organelles known as *lysosomes*. This compartmentalization of potentially toxic molecules is necessary to protect the cell from self-destruction and produce an environment where the molecules can function efficiently. The phagosome and lysosome fuse to form a *phagolysosome* in which the ingested material is killed and digested by various enzyme systems.

Ingestion is accompanied by enhanced glycolysis and an increase in the synthesis of proteins and membrane phospholipids in the phagocyte. After phagocytosis there is a respiratory burst consisting of a steep rise in oxygen consumption. This is accompanied by an increase in the activity of a number of enzymes and leads to the reduction of molecular oxygen to various highly reactive intermediates, such as the superoxide anion ($O_2^{\cdot-}$), hydrogen peroxide (H_2O_2), singlet oxygen ($O^{\cdot}$) and the hydroxyl radical ($OH^{\cdot}$). All of these chemical species have microbicidal activity and are termed oxygen-dependent killing mechanisms. The superoxide anion is a free radical produced by the one-electron reduction of molecular oxygen; it is very reactive and highly damaging to animal cells, as well as to micro-organisms. It is also the substrate for superoxide dismutase, which generates hydrogen peroxide for subsequent use in microbial killing. Myeloperoxidase uses hydrogen peroxide and halide ions, such as iodide or chloride, to produce at least two bactericidal systems. In one, halogenation (incorporation of iodine or chlorine) of the bacterial cell wall leads to death of the organism. In the second mechanism, myeloperoxidase and hydrogen peroxide damage the cell wall by converting amino acids into aldehydes that have antimicrobial activity.

Within phagocytes there are several oxygen-independent mechanisms that can destroy ingested material. Some of these enzymes can damage membranes. For example, *lysozyme* and *elastase* attack peptidoglycan of the bacterial cell wall, and then hydrolases are responsible for the complete digestion of the killed organism. The cationic proteins of lysosomes bind to and damage bacterial cell walls and enveloped viruses, such as herpes simplex virus. The iron-binding protein *lactoferrin* has antimicrobial properties. It complexes with iron, rendering it unavailable to bacteria that require iron for growth. The high acidity within phagolysosomes (pH 3.5–4.0) may have bactericidal effects, probably resulting from lactic acid production in glycolysis. In addition, many lysosomal enzymes, such as acid hydrolases, have acid pH optima. There are significant differences between macrophages and neutrophils in the killing of micro-organisms. Although macrophage lysosomes contain a variety of enzymes, including lysozyme, they lack cationic proteins and lactoferrin. Tissue macrophages do not have myeloperoxidase but probably use catalase to generate the hydrogen peroxide system. Normal macrophages are less efficient killers of certain pathogens, such as fungi, than neutrophils. The microbicidal activity of macrophages can, however, be greatly improved after contact with products of lymphocytes, known as lymphokines.

Once killed, most micro-organisms are digested and solubilized by lysosomal enzymes. The degradation products are then released to the exterior.

Natural killer (NK) cells

NK cells recognize changes on virus-infected cells and destroy them by an extracellular killing mechanism. After binding to the target cell, by an as yet undefined mechanism, the NK cell produces molecules that damage the membrane of the infected cell, leading to its destruction.

Natural killing is a function of several different cell types. This activity is performed by cells described as large granular lymphocytes and also by cells with T cell markers, macrophage markers and others that do not have the characteristics of any of the main cells of the immune system. Natural killing is present without previous exposure to the infectious agent and shows all the characteristics of an innate defence mechanism. NK cells have also been implicated in host defence against cancers. They are thought to recognize changes in the cell membranes of transformed cells in a mechanism similar to that used

to combat virus infection. Natural killing is enhanced by interferons that appear to stimulate the production of NK cells and also increase the rate at which they kill the target cells.

Eosinophils

Eosinophils are polymorphonuclear leucocytes with a characteristic bi-lobed nucleus and cytoplasmic granules. They are present in the blood of normal individuals at very low levels (<1%), but their numbers increase in patients with parasitic infections and allergies. They are not efficient phagocytic cells, although their granules contain molecules that are toxic to parasites. Large parasites such as helminths cannot be internalized by phagocytes and therefore must be killed extracellularly. Eosinophil granules contain an array of enzymes and toxic molecules active against parasitic worms. The release of these molecules must be controlled so that tissue damage is avoided. The eosinophils have specific receptors, including Fc and complement receptors, that bind the labelled target (i.e. antibody or complement-coated parasites). The granule contents are then released into the space between the cell and the parasite, thus targeting the toxic molecules onto the parasite membrane.

Temperature

The temperature preference of many micro-organisms is well known, and it is therefore apparent that temperature is an important factor in determining the innate immunity of an animal to some infectious agents. It seems likely that the pyrexia that follows so many different types of infection can function as a protective response against the infecting micro-organism. The febrile response in many cases is controlled by IL-1 produced by macrophages as part of the immune response.

Inflammation

A number of the above factors are responsible for the process of *acute inflammation*. This is the reaction of the body to injury, such as invasion by an infectious agent, exposure to a noxious chemical or physical trauma. The signs of inflammation are redness, heat, swelling, pain and loss of function. The molecular and cellular events that occur during an inflammatory reaction are:

- vasodilatation
- increased vascular permeability
- cellular infiltration.

These changes are brought about mainly by chemical mediators (Table 9.4), which are widely distributed in a sequestered or inactive form throughout the body and are

Table 9.4 Mediators of inflammation

Mediator	Main source	Function
Histamine[a]	Mast cells, basophils	Vasodilatation, increased vascular permeability, contraction of smooth muscle
Kinins (e.g. bradykinin)	Plasma	Vasodilatation, increased vascular permeability, contraction of smooth muscle, pain
Prostaglandins	Neutrophils, eosinophils, monocytes, platelets	Vasodilatation, increased vascular permeability, pain
Leukotrienes	Neutrophils, mast cells, basophils	Vasodilatation, increased vascular permeability, contraction of smooth muscle, induction of cell adherence and chemotaxis
Complement components (e.g. C3a, C5a)	Plasma	Cause mast cells to release mediator C5a as chemotactic factor
Plasmin	Plasma	Breaks down fibrin, kinin formation
Cytokines	Lymphocytes, macrophages	Chemotactic factors, colony-stimulating factors, macrophage activation

[a]In rodents, 5-hydroxytryptamine (serotonin) is present in mast cells and basophils.

released or activated locally at the site of inflammation. After release they tend to be inactivated rapidly, to ensure control of the inflammatory process.

There is increased blood supply to the affected area owing to the action of vasoactive amines, such as histamine and 5-hydroxytryptamine, and other mediators stored within mast cells. These molecules are released:

- as a consequence of the production of the anaphylatoxins (C3a and C5a) that trigger specific receptors on mast cells
- following interaction of antigen with IgE on the surface of mast cells
- by direct physical damage to the cells.

Other mediators, such as bradykinins and prostaglandins, are produced locally or released by platelets. The vasodilatation causes increased blood supply to the area, giving rise to redness and heat. The result is an increased supply of the molecules and cells that can combat the agent responsible for the initial trigger.

The same molecules, vasoactive amines, prostaglandins and kinins, increase vascular permeability, allowing plasma and plasma proteins to traverse the endothelial lining. The plasma proteins include immunoglobulins and molecules of the clotting and complement cascades. This leaking of fluid causes swelling (oedema), which in turn leads to increased tissue tension and pain. Some of the molecules themselves, for example prostaglandins and histamine, stimulate the pain responses directly. The inflammatory exudate has several important functions.

Bacteria often produce tissue-damaging toxins that are diluted by the exudate. The presence of clotting factors results in the deposition of fibrin, creating a physical obstruction to the spread of bacteria. The exudate is drained continuously by the lymphatic vessels, and antigens, such as bacteria and their toxins, are carried to the draining lymph node where immune responses can be generated.

The production of chemotactic factors, including C5a, histamine, leukotrienes and molecules specific for certain cell types, attracts phagocytic cells to the site. The increased vascular permeability allows easier access for neutrophils and monocytes, and the vasodilatation means that more cells are in the vicinity. The neutrophils arrive first and begin to destroy or remove the offending agent. Most are successful but a few die, releasing their tissue-damaging contents to increase the inflammatory process. Mononuclear phagocytes arrive on the scene to finish off the removal of the residual debris and stimulate tissue repair.

When the swelling is severe there may be loss of function to the affected area. If the offending agent is quickly removed, the tissue will soon be repaired. The inflammatory process continues until the conditions responsible for its initiation have been resolved. In most circumstances this occurs fairly rapidly, with an acute inflammatory reaction lasting for a matter of hours or days. If, however, the causative agent is not easily removed or is reintroduced continuously, chronic inflammation will ensue with the possibility of tissue destruction and complete loss of function.

ACQUIRED IMMUNITY

Micro-organisms that overcome or circumvent the innate non-specific defence mechanisms or are administered deliberately (i.e. active immunization) come up against the host's second line of defence: *acquired immunity*. To give expression to this acquired form of immunity it is necessary that the antigens of the invading micro-organism come into contact with cells of the immune system (macrophages and lymphocytes) and thereby initiate an immune response specific for the foreign material. The cells that respond are pre-committed, because of their surface receptors, to respond to a particular epitope on the antigen. This response takes two forms, *humoral* and *cell mediated*, which usually develop in parallel. The part played by each depends on a number of factors, including the nature of the antigen, the route of entry and the individual who is infected.

Humoral immunity depends on the appearance in the blood of antibodies produced by plasma cells.

The term 'cell-mediated immunity' was originally coined to describe localized reactions to organisms mediated by T lymphocytes and phagocytes rather than by antibody. It is now used to describe any response in which antibody plays a subordinate role. Cell-mediated immunity depends mainly on the development of T cells that are specifically responsive to the inducing agent, and is generally active against intracellular organisms.

Specific immunity may be acquired in two main ways:

1. induced by overt clinical infection or inapparent clinical infection
2. deliberate artificial immunization.

This is *active acquired immunity*, and contrasts with *passive acquired immunity*, which is the transfer of pre-formed antibodies to a non-immune individual by means of blood, serum components or lymphoid cells.

Actively acquired immunity is long lasting, although it may be circumvented by antigenic change in the infecting micro-organism. Passively acquired immunity provides only temporary protection. Passive immunity may be transferred to the fetus by the passage of maternal antibodies across the placenta.

TISSUES INVOLVED IN IMMUNE REACTIONS

For the generation of an immune response, antigen must interact with and activate a number of different cells. In addition, these cells must interact with one another. The cells involved in immune responses are organized into tissues and organs in order that these complex cellular interactions can occur most effectively. These structures are collectively referred to as the *lymphoid system*, which comprises lymphocytes, epithelial and stromal cells arranged into discrete capsulated organs or accumulations of diffuse lymphoid tissue. Lymphoid organs contain lymphocytes at various stages of development and are classified into primary and secondary lymphoid organs.

The primary lymphoid organs are the major sites of lymphopoiesis. Here, lymphoid progenitor cells develop into mature lymphocytes by a process of proliferation and differentiation. In mammals, T lymphocytes develop in the thymus, and B lymphocytes in the bone marrow and fetal liver. It is within the primary lymphoid organs that the lymphocytes acquire their repertoire of specific antigen receptors in order to cope with the antigenic challenges that the individual receives during its life. It is also within these tissues that self-reactive lymphocytes are eliminated to protect against autoimmune disease.

The secondary lymphoid organs create the environment in which lymphocytes can interact with one another and with antigen, and then disseminate the effector cells and molecules generated. Secondary lymphoid organs include lymph nodes, spleen and mucosa-associated lymphoid tissue (e.g. tonsils and Peyer's patches of the gut). These organs have a characteristic structure that relates to the function they carry out, with areas composed of mainly B cells or T cells.

DEVELOPMENT OF THE IMMUNE SYSTEM

In man, lymphoid tissue appears first in the thymus at about 8 weeks of gestation. Peyer's patches are distinguishable by the fifth month, and immunoglobulin-secreting cells appear in the spleen and lymph nodes at about 20 weeks. From this time onwards, IgM and IgD are synthesized by the fetus (Fig. 9.5). At birth the infant has a blood concentration of IgG comparable to that of the maternal circulation, having received IgG but not IgM via the placenta. The rate of synthesis of IgM in the infant increases rapidly within the first few days of life but does not reach adult levels until about a year. Serum IgG does not reach adult levels until after the second year, and IgA takes even longer. There is an actual drop in the level of IgG from birth due to the decay of maternal antibody, with the lowest levels of total IgG at around 3 months of age. This corresponds to an age of marked susceptibility to a number of infections. Cell-mediated immunity can be stimulated at birth, but these reactions may not be as powerful as in the adult.

LYMPHOCYTE TRAFFICKING

Lymphocytes differentiate and mature in the primary lymphoid organs and then enter the blood lymphocyte

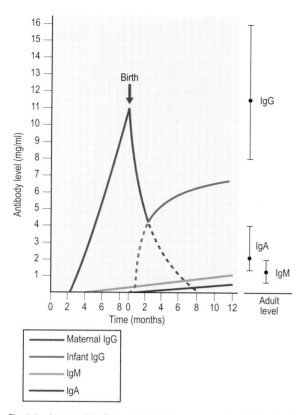

Fig. 9.5 Immunoglobulin levels in the fetus and neonate. Adult levels of the major isotypes are shown as normal ranges with mean serum levels.

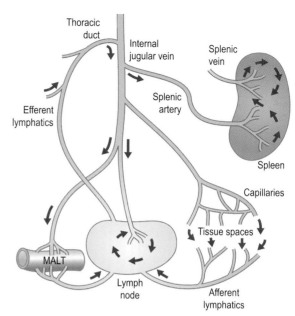

Fig. 9.6 Lymphocyte recirculation. MALT, mucosa-associated lymphoid tissue.

pool. B cells are produced in the bone marrow and mature there before proceeding via the circulation to the secondary lymphoid organs. T cell precursors leave the bone marrow and mature in the thymus before migrating to the secondary lymphoid organs. Once in the secondary lymphoid tissues, the lymphocytes do not remain there but move from one lymphoid organ to another through the blood and lymphatics (Fig. 9.6). One of the main advantages of this *lymphocyte recirculation* is that during the course of a natural infection the continual trafficking of lymphocytes enables many different lymphocytes to have access to the antigen.

Only a very small number of the lymphocytes will recognize a particular antigen. Pathogens can enter the body by many routes, but must be carried from the site of infection to the secondary lymphoid tissues where they are localized and concentrated on the dendritic processes of macrophages or on the surface of antigen-presenting cells. If the infection is in the tissues, antigen is carried in the lymphatics to the draining lymph node. Under normal conditions there is a continuous active flow of lymphocytes through lymph nodes, but when antigen and antigen-reactive cells enter there is a temporary shut-down of the exit. Thus,

antigen-specific cells are preferentially retained in the node draining the source of the antigen. This is partly responsible for the swollen glands (lymph nodes) that can sometimes be found during an infection. Microbes present on mucosal surfaces are taken up by specialized cells known as M cells, and are then delivered to the mucosa-associated lymphoid tissues such as the tonsils and Peyer's patches. Blood-borne antigens are trapped in the spleen. The passage of lymphocytes through an area where antigen has been localized facilitates the induction of an immune response. Lymphocytes with appropriate receptors bind to the antigen and become activated. Once activated, the lymphocytes mature into effector cells. In the case of B lymphocytes they become plasma cells and secrete antibody. T lymphocytes leave the secondary lymphoid tissue and return to the site of infection to destroy the infectious agent.

There is evidence for non-random migration of lymphocytes to particular lymphoid compartments. For example, lymphocytes that home to the gut are selectively transported across endothelial cells of venules in the intestine. It appears that lymphocytes have specific molecules on their surface that preferentially interact with endothelial cells in different anatomical sites. A lymphocyte that was initially stimulated by antigen in a Peyer's patch will migrate to the draining lymph node, respond, and memory cells will be produced. It is important that these memory cells migrate back to the area where the same pathogen might be encountered again. Therefore, they are found preferentially in the mucosa-associated lymphoid tissue.

CLONAL SELECTION

During their development in the primary lymphoid tissues both T and B lymphocytes acquire specific cell surface receptors that commit them to a single antigenic specificity. For T cells this receptor remains the same for its life, but the surface immunoglobulin on B cells can be modified as a result of somatic mutations. In the B cell this is mirrored in the modification of the antibody produced by the cell on exposure to its specific antigen. The lymphocytes are activated when they bind specific antigen and then proliferate, differentiate and mature into effector cells.

The lymphocytes reactive to any particular antigen are only a small proportion of the total pool. Therefore, antigen binds to the small number of cells that can recognize it and selects them to proliferate and mature so that sufficient cells are formed to mount an adequate immune response. A cell that responds to an antigenic trigger and proliferates will give rise to cells with a genetically identical make-up (i.e. *clones*). This phenomenon is therefore known as *clonal selection*.

Lymphocyte receptors, generated in the primary lymphoid tissues, are created in a random fashion, so there is no reason why some could not recognize 'self' molecules. An obviously important attribute of the immune system is that it is able to discriminate between 'self' and 'non-self'. During development, any lymphocyte with a receptor that binds strongly to self molecules is eliminated.

CELLULAR ACTIVATION

When an individual is exposed to foreign material, selected lymphocytes respond. B lymphocytes proliferate and differentiate into antibody-producing plasma cells and memory cells. T lymphocytes are stimulated to become effector cells that can directly eliminate the foreign material or produce molecules that help other cells to destroy the pathogen. The type (immunity or tolerance) and magnitude of the response, if generated, depends on a number of factors, including the nature, dose and route of entry of the antigen, and the individual's genetic make-up and previous exposure to the antigen.

The first stage in the production of effector cells and molecules is activation of the resting cells. This involves various cellular interactions with maturation of the response, leading to a co-ordinated, efficient production of effector T cells, immunoglobulin and memory cells.

Cross-linking of the B cell antigen receptor, surface immunoglobulin, is the initial trigger for activation.

When this happens, a number of biochemical changes are instigated. These changes probably act through protein kinases that cause the synthesis of RNA and ultimately immunoglobulin production. In a number of cases this is all that is required to stimulate antibody production. However, for the majority of antigens this initial cross-linking is not enough, and molecules produced by T cells are also required.

Thymus-independent antigens

A number of antigens will stimulate specific immunoglobulin production directly. These T-independent antigens are of two types: *mitogens* and certain large molecules.

Mitogens are substances that cause cells, particularly lymphocytes, to undergo cell division (i.e. proliferation). Certain glycoproteins, called lectins, have mitogenic activity. These molecules have specificity for sugars; they bind to the cell surface and activate all responsive cells. The response to the mitogens is therefore polyclonal, as lymphocytes of many different specificities are activated. However, at low concentrations these mitogens do not cause polyclonal activation but can lead to the stimulation of specific B cells. Lipopolysaccharide is an example of a B cell mitogen.

Some large molecules with regularly repeating epitopes, for instance polymers of D-amino acids and simple sugars such as pneumococcal polysaccharide and dextran, can interact directly with the B cell surface immunoglobulin. They may also be held on the surface of specialized macrophages in secondary lymphoid tissues, and the B cells interact with them there. The multiple repeats of the epitope interact with a large number of surface immunoglobulin molecules; the signal that is generated is sufficient to stimulate antibody production.

The immune response generated to these antigens tends to be similar on each exposure, that is, IgM is the main antibody and the response shows little memory. This suggests that class switch and memory production require additional factors (products of T lymphocytes).

Thymus-dependent antigens

Many antigens do not stimulate antibody production without the help of T lymphocytes. These antigens first bind to the B cell, which must then be exposed to T cell-derived lymphokines (helper factors) before antibody can be produced. For the second activation signal (i.e. help) to be targeted effectively at the B cell, the T and B cells must be in direct contact. For this to happen, the B and T cell epitopes must be linked physically. However, T cells only recognize antigen that has been processed and presented in association with products of the major

Table 9.5   Antigen-presenting cells in the lymph nodes

Area	Antigen-presenting cell	Antigen
Subcapsular marginal sinus	Marginal zone macrophage	T-independent antigens
Follicles and B cell areas	Follicular dendritic cells	Antigen–antibody complexes
Medulla	Classical macrophages	Most antigens
T cell areas	Interdigitating dendritic cells	Most antigens

histocompatibility complex (MHC), so it is impossible for native antigen to form a bridge between surface immunoglobulin and the T cell receptor. The B cell binds to its epitope on free antigen, but there is no site on this molecule to which the T cell can bind, because it requires antigen associated with MHC products. The answer to this problem can be seen when the requirements for antigen presentation in T cell recognition are considered.

Antigen processing and presentation

The development of an antibody response to a T-dependent antigen requires that the antigen becomes associated with MHC class II molecules (i.e. *processed*) and expressed on the cell surface (i.e. *presented*) in a form that helper T cells can recognize.

All cells express MHC class I molecules, but class II molecules are confined to cells of the immune system – the antigen-presenting cells. These cells present antigen to MHC class II-restricted T cells (the CD4-positive [CD4+] population) and therefore play a key role in the induction and development of immune responses. Within lymph nodes, different antigen-presenting cells are found in each of the main areas (Table 9.5).

There are a large number of antigen-presenting cells in the body, most of which constitutively express MHC class II molecules. Other cells, such as T lymphocytes and endothelium, can be induced to express MHC class II molecules by suitable stimuli such as lymphokines. The relative importance of each type depends on whether a primary or secondary response is being stimulated and on the location. The most studied antigen-presenting cells are the macrophages and dendritic cells. However, it is now apparent that in certain situations B cells may be important antigen-presenting cells. The relative importance of B cells becomes greatest during secondary responses, especially when the antigen concentration is low. Here the B cells can specifically engulf antigen via their surface immunoglobulin. In a primary response, specific B cells are at a low frequency and their receptors are of low affinity; in this situation macrophages and dendritic cells are probably most important.

The key feature of all antigen-presenting cells is that they can ingest antigen, degrade it and present it, in the context of MHC class II molecules, to T cells. The antigen is taken into the antigen-presenting cells and enters the endocytic pathway. Before it is destroyed completely, peptide fragments are taken to a structure called the *compartment for peptide loading*. MHC class II molecules are synthesized within the endoplasmic reticulum and are also transported to the compartment for peptide loading, where they associate with the processed antigen. The MHC class II molecule with the bound peptide is then transported to the cell surface.

T cell activation

The activation of resting CD4+ T cells requires two signals. The first is antigen in association with MHC class II molecules, and the second is the *co-stimulatory signal*. The generation of the first of these signals has just been discussed (i.e. antigen presentation). The second signal is delivered by the same antigen-presenting cell that gave the first signal. The co-stimulatory signal is mediated by the interaction of a molecule on the antigen-presenting cell engaging with its receptor on the T cell. The best characterized pairing is B7 on the antigen-presenting cell and CD28 on the T cell.

When both of these signals are generated, biochemical changes occur within the T cell, leading to RNA and protein synthesis. The responsive cells progress through the cell cycle from the G_0 to the G_1 phase. The cells start to express IL-2 receptors and produce IL-2, a T cell growth factor that causes the expansion of the responsive T cell population. IL-2 was originally thought to be the only T cell growth factor, but it is now known that IL-4 and IL-1 can support T cell growth, although they are not as potent. After about 2 days, IL-2 synthesis stops, whereas IL-2 receptors remain for up to a week if the cell is not reactivated. Therefore, there is a built-in limitation on T cell growth and clonal expansion. When stimulated, T cells secrete IL-2, which interacts with IL-2 receptors to mediate growth. This can be in an '*autocrine*' fashion if the same cell that released the IL-2 is stimulated. If the responding cell is in the vicinity of the producer, the stimulation is in a '*paracrine*' manner. IL-2 is not present at detectable levels in the blood; therefore no '*endocrine*' activity is involved (i.e. action at a distant site). The end-result is the production of a large number of activated CD4$^+$ T lymphocytes.

The other main type of T lymphocyte is the CD8$^+$ T cell. Antigen recognition by these cells is restricted by MHC class I molecules. Again, these cells require two signals to be activated: (1) antigen fragment in association with MHC class I and (2) the co-stimulatory signal. A cell that 'sees' both of these signals responds by clonal expansion and differentiation into a fully active effector T cell.

B cell activation

Mitogens and T-independent antigens have an inherent ability to drive B cells into division and differentiation. T-dependent responses rely on T cells and their products to control the antibody class, affinity and memory. The first cells to be activated are CD4$^+$ T cells that recognize the antigen in association with MHC class II molecules (see above). These cells respond to the signal of the antigen fragment–MHC complex, and produce a variety of lymphokines that act on B cells.

As far as B cell development is concerned, the antigen-stimulated cells develop under the influence of IL-4 (previously known as B cell stimulation factor), which is produced by closely adherent T cells. IL-5 and IL-6 then bring the cells to a state of full activation with terminal differentiation into an immunoglobulin-producing plasma cell. All this happens within a germinal centre of a lymph node secondary follicle that has evolved to facilitate the necessary cellular and molecular interactions.

Therefore, for both B and T cell activation, two stimuli are required:

1. The recognition of antigen makes sure that only those cells that will be effective against the foreign material are recruited.
2. The provision of the co-stimulatory signal has evolved to control the process and aid discrimination of 'self' and 'non-self'.

HUMORAL IMMUNITY

Synthesis of antibody

On exposure to antigen, antibody production follows a characteristic pattern (Fig. 9.7). There is a lag phase during which antibody cannot be detected. This is the time taken for the interactions described above to take place and for antibody to reach a level that can be measured. There is then an exponential rise in the antibody level or titre. This log phase is followed by a plateau with a constant level of antibody, when the amount produced equals the amount removed. The amount of antibody then declines,

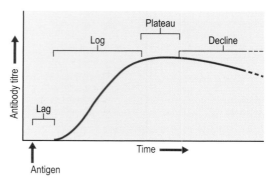

Fig. 9.7 Pattern of antibody production following antigen exposure.

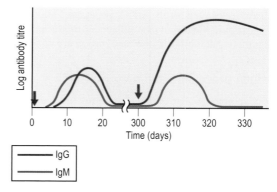

Fig. 9.8 Primary and secondary antibody response. The level of serum IgM and IgG detected with time after primary immunization (day 0) and challenge (day 300) with the same antigen.

owing to the clearing of antigen–antibody complexes and the natural catabolism of the immunoglobulin.

If the response is to a T-dependent antigen, the B cells can switch to the production of another isotype; for example, in a primary response IgM gives way to IgG production. This process is under the control of T cells, as the class of antibody produced depends on signals from the T cell. At some point, again under the control of T cells, a proportion of the antigen-reactive cells develop into memory cells. These cells react if the epitope is encountered again.

There are a number of differences in the reaction profile on second and subsequent exposures to an antigen compared with the primary response (Fig. 9.8). There is a shortened lag and an extended plateau and decline. The level and affinity of antibody produced are much increased, and antibody is mostly of the IgG isotype. Some IgM is generated, but it will follow the same pattern as in the primary response.

When first introduced, the antigen selects the cells that can react with it. However, before antibody is produced the B cell must differentiate into a plasma cell, involving the interactions already described. The B cells that are stimulated in the primary response synthesize IgM. With time, class switch will occur in some of the B cells, leading to the production of other isotypes. Somatic mutations occur, giving rise to *affinity maturation* through selection of cells bearing high-affinity receptors as the amount of antigen in the system falls. Memory cells are also produced. An equilibrium is reached whereby there is a balance between the amount of antibody synthesized and the amount used. Various mechanisms then come into play to turn off the response when it is no longer needed (see below). The simplest is the removal of the stimulant (antigen). Thus the production of antibody is stopped and there is a natural decline in antibody levels.

On subsequent exposure the responding cells (i.e. memory cells) are at a different level of activation and are present at an increased frequency. Therefore, there is a shorter lag before antibody can be detected; the main isotype is IgG. The level of antibody produced is ten or more times greater than during the primary response. The antibody is present for an extended period and has a higher affinity for antigen due to affinity maturation. As is seen in Figure 9.8, some IgM is also produced during a secondary response. This immunoglobulin is produced by the activation of B cells that were not present in the lymphocyte pool on the previous exposure but have developed since. The development of these cells follows the characteristics of a primary response, and they will give rise to a secondary response if the antigen is encountered again.

Monoclonal antibodies

When an antigen is introduced into the lymphoid system of a mouse, all the B cells that recognize epitopes on the antigen are stimulated to produce antibody. The serum of the immunized animal is known as a polyclonal antiserum, as it is the product of many clonally derived B cells. Even when highly purified antigen is used, the antiserum produced will contain a number of antibodies that react to the antigen and others that interact with antigens encountered naturally by the animal during this time. It is extremely difficult to purify the antibodies of interest from this complex mixture, but it is possible to fuse single plasma cells with a myeloma (a tumour) cell line to form a hybridoma that will grow in tissue culture. These cells will all be identical and therefore secrete the same antibody, a *monoclonal antibody*.

Human monoclonal antibodies are potentially of value in patient treatment. As starting material, peripheral blood or secondary lymphoid tissue such as tonsils have been used. It is impossible, for ethical reasons, to expose human subjects to most of the antigenic material that would be required to induce useful antibodies. Therefore, cells are only available from patients with certain diseases, such as tumours, infections and autoimmune diseases, or from individuals who have received immunizations.

All the molecules in a monoclonal antibody preparation have the same isotype, specificity and affinity, in contrast to the polyclonal antiserum produced by the inoculation of antigen into an experimental animal. In addition, the same polyclonal antiserum can never be reproduced, not even when using the same animal. However, monoclonal antibodies are defined reagents that can be produced indefinitely and on a large scale. They provide a standard material that can be used in studies ranging from the identification and enumeration of different cell types to blood typing and diagnosis of disease. They are also used increasingly in attempts to treat and prevent disease.

CELL-MEDIATED IMMUNITY

Specific cell-mediated responses are mediated by two different types of T lymphocyte. T cells that have the CD8 molecule on their surface recognize antigen fragments in association with MHC class I molecules on a target cell and cause cell lysis. MHC class II-restricted recognition is seen with T cells that have the CD4 marker. These cells secrete lymphokines when stimulated by the antigen–MHC class II complex. CD4+ T cells are involved in two main activities:

1. Cell-mediated reactions, as the lymphokines can aid in the elimination of foreign material by recruiting and activating other leucocytes and promoting an inflammatory response.
2. The generation and control of an immune response, as some of the lymphokines produced are growth and differentiation factors for T and B cells.

The other cell types, NK cells and phagocytes, that can participate in cell-mediated defence mechanisms have been described.

Cell–mediated cytotoxicity

Certain subpopulations of lymphoid and myeloid cells can destroy target cells to which they are closely bound. The stages and processes involved are similar for the different cell types, although the molecules that mediate the recognition of the target by the effector differ.

Cytotoxic T lymphocytes

Cytotoxic T cells (Tc cells) are small T lymphocytes derived from stem cells in the bone marrow. These cells mature in the thymus. Most cells that mediate MHC-restricted cytotoxicity are CD8+, and therefore recognize

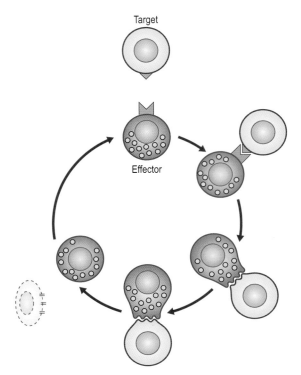

Target

Effector

Fig. 9.9 Mechanism of cell–mediated cytotoxicity. The effector cell has a receptor that is able to bind to a target cell that possesses the appropriate ligand (▲).

antigen in association with MHC class I antigens. Some are CD4+, and therefore MHC class II restricted.

MHC-unrestricted cytotoxic cells

A number of partially overlapping cell populations are able to carry out MHC-unrestricted killing. These include NK cells, lymphokine-activated killer (LAK) cells and killer (K) cells.

Most cells that have the capacity to perform natural killing have the morphology of large granular lymphocytes and a broad target range. The receptor on the NK cell and the structures that they recognize on the target have not been fully characterized. NK cells have been shown to produce a number of cytokines, including γ-interferon.

Several types of cell are able to destroy foreign material by antibody-dependent cell-mediated cytotoxicity. The cells that carry out this activity have a receptor for the Fc portion of immunoglobulin and are therefore able to bind to antibody-coated targets.

Lytic mechanism

Three distinct phases have been described in cell-mediated cytotoxicity (Fig. 9.9):

- binding to target
- rearrangement of cytoplasmic granules and release of their contents
- target cell death.

Once the effector–target conjugate has been formed, the cytoplasmic granules appear to become rearranged and concentrated at the side of the cell adjacent to the target. The granule contents are then released into the space between the two cells. There are at least three different types of molecule stored within the granules that can cause cell death. T cells and NK cells contain perforin, which is a monomeric protein related to the complement component C9. In the presence of Ca^{2+} ions the monomers bind to the target cell membrane and polymerize to form a transmembrane pore. This upsets the osmotic balance of the cell and leads to cell death. The granules also contain at least two serine esterases that may play a role in destroying the target cell. Several other toxic molecules are produced by cytotoxic cells, including tumour necrosis factor (TNF)-α, lymphotoxin (TNF-β), γ-interferon and NK cytotoxic factor. The process is unidirectional, with only the target cell being destroyed. The effector cell can then move on and eliminate another target cell.

Lymphokine production

The other arm of cell-mediated immunity is dependent on the production of lymphokines from antigen-activated T lymphocytes. These molecules, produced in an antigen-specific fashion, can act in an antigen-non-specific manner to recruit, activate and regulate effector cells with the potential to combat infectious agents.

The first documented reference to the production of lymphokines is credited to Robert Koch in 1880. Injection of purified antigen (tuberculin) into the skin of immune individuals produced a reaction that peaked within 24–72 h. The response was characterized by reddening and swelling, and accompanied by the accumulation of lymphocytes, monocytes and basophils. Because of the time course of the reaction, this response has become known as *delayed-type hypersensitivity* (DTH), and the cells responsible were called delayed-type hypersensitivity T lymphocytes (T_{DTH} or T_D cells). These cells are identical to the helper T (T_H) cell subset as far as antigen recognition is concerned. CD4+ T cells, usually still referred to as T_H cells, are therefore capable of mediating both helper activities and so-called delayed hypersensitivity reactions by producing lymphokines. Although the term 'delayed hypersensitivity' suggests a disease process, the production of lymphokines has a physiological function, and only in some situations do pathological consequences occur.

Cytokines are biologically active molecules released by specific cells that elicit a particular response from other

Table 9.6 Examples of some cytokines that are of importance in the immune system

Cytokine	Source	Target	Main effects
IL-1	Macrophages Endothelial cells Some epithelial cells	T lymphocytes Tissue cells	Fever Inflammation T cell activation Macrophage activation Stimulates acute-phase protein production
IL-2	T lymphocytes	T lymphocytes NK cells B lymphocytes	Proliferation
IL-4	T_H2 cells Mast cells	B lymphocytes T lymphocytes Mast cells	Stimulates proliferation, differentiation and class switch in B cells Differentiation and proliferation of T_H2 cells Mast cell growth
IL-8	Macrophages Endothelial cells	Neutrophils	Chemotaxis
IL-13	T_H2 cells	Macrophages	Inhibits macrophage activation and activities
TNF-α	Macrophages T lymphocytes	Macrophages Tissue cells	Fever Inflammation Macrophage activation Stimulates acute-phase protein production Kills certain tumour cells
Type I IFN (α and β)	Virus-infected cells	Tissue cells	Antiviral effect Induction of MHC class I Antiproliferative effects Activation of NK cells
IFN-γ	T lymphocytes (T_H1 and T_c) NK cells	Leucocytes and tissue cells	Macrophage activation Induction of MHC class I and II Antibody class switch Antiviral effect
GM-CSF	T lymphocytes Macrophages Endothelial cells Fibroblasts	Immature and committed progenitor cells in bone marrow	Stimulates division and differentiation Macrophage activation

Many of the molecules detailed above act synergistically to produce their biological effects.
IL, interleukin; TNF, tumour necrosis factor; IFN, interferon; GM-CSF, granulocyte–macrophage colony-stimulating factor; NK, natural killer; MHC, major histocompatibility complex.

cells on which they act. A number of these regulatory molecules produced by lymphocytes (lymphokines) and monocytes (monokines) are shown in Table 9.6. The responses caused by these substances are varied and interrelated. In general, cytokines control the growth, mobility and differentiation of lymphocytes, but they also exert a similar effect on other leucocytes and some non-immune cells.

The exact signals and mechanisms controlling the activation of T cells and the release of lymphokines are not known. The balance between the different lymphokines produced determines the response generated. CD4+ T cells can be divided into two main types depending on the profile of lymphokines they secrete. The T_H2 subset produces IL-4 and IL-5, which act on responsive B cells with antibody production as the main feature of the response. The T_H1 subset secretes mainly IL-2 and γ-interferon. The production of IL-2 stimulates T cell growth, whereas γ-interferon has multiple effects, including macrophage activation.

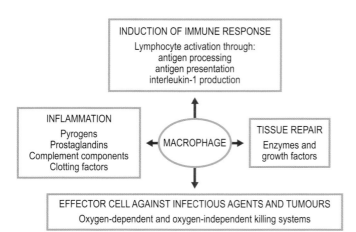

Fig. 9.10 The central role of macrophages.

Role of macrophages

Macrophages are able to carry out a remarkable array of different functions (Fig. 9.10). They play a key role in several aspects of cell-mediated immunity, being involved at the initiation of the response, as antigen-presenting cells, and as effector cells having microbicidal and tumoricidal activities. They also produce a number of cytokines (or more precisely monokines) that function as regulatory molecules. These monokines contribute to inflammation and fever, and affect the functioning of other cells. Macrophages can also produce various enzymes and factors that are involved in reorganization and repair following tissue damage. However, as they contain many important biological molecules, they can themselves cause damage if these enzymes and factors are released inappropriately.

Many of these activities are enhanced in macrophages that have been 'activated' by exposure to lymphokines, such as γ-interferon produced by T cells. Macrophage activation is a complex process that probably occurs in stages, with different effector functions being expressed at different stages. Macrophages from different sites in the body show different characteristics; they are heterogeneous and have different activation requirements.

γ-Interferon is a powerful macrophage-activating molecule that increases the uptake of antigens by an enhanced expression of Fc and complement receptors; the activities of intracellular enzymes involved in killing are also raised. As γ-interferon causes an increase in MHC class II expression, there will be an enhanced presentation of antigen to CD4+ T cells. This leads to the production of more lymphokines and more effective elimination of the offending material.

CD4+ T cells secrete the lymphokines that activate macrophages. Therefore, the presentation of antigen by an antigen-presenting cell leads to the production of lymphokines by T_H1 cells *specific* for the antigen involved. The lymphokines produced then activate any responsive macrophage in the vicinity of the responding cells. The activation process appears to depend on the presence of a number of lymphokines that act synergistically to induce activation. For example, pure IL-2, IL-4 or γ-interferon is unable to induce resistance to infection, but if γ-interferon is combined with any of the others then resistance is observed.

Macrophages and monocytes themselves are capable of producing a number of important cytokines. These monokines include:

- IL-1
- IL-6
- various colony-stimulating factors
- TNF-α.

TNF-α and IL-1, acting independently and together, have effects on many leucocytes and tissues. TNF-α is responsible for the tumoricidal activity of macrophages but is also implicated in the elimination of certain bacteria and parasites. It has a synergistic effect with γ-interferon on resistance to a number of viral infections.

GENERATION OF IMMUNE RESPONSES

As discussed above, the generation of humoral and cell-mediated responses requires the recognition of antigen, by the responding cell, as the first signal and a co-stimulation second signal. T_H cells, as has been emphasized, recognize antigen fragments only when in association with MHC class II molecules. The distribution of MHC class II molecules is limited, in normal situations, to certain cells of the immune system: the antigen-presenting cells. In certain circumstances non-lymphoid cells can present antigens if they are induced to express MHC class II molecules. To stimulate a T_H cell the antigen must be taken into the cell and re-expressed on the surface in association with MHC class II molecules. As the T cell antigen receptor recognizes antigen fragments bound to the MHC molecules, the antigen-presenting cell must also be able to process the antigen. MHC class I- and class II-restricted recognition by CD8+ and CD4+ T cells requires antigen processing. The pathways that lead to association of an antigen fragment with a particular restriction element are not fully understood.

Immune responses are generated in secondary lymphoid tissues, such as lymph nodes. As a number of cells and molecules must all interact, the architecture of the secondary lymphoid tissue has evolved for the efficient induction of an immune response. In a secondary immune response the cells involved are at a different stage of activation: they are memory cells, having already been exposed to antigen. Therefore, the growth factor signals may not be so critical, although antigen in association with MHC class II molecules is still required. In this situation B cells are important as antigen-presenting cells.

CD8+ T cells, as we have seen, recognize antigen fragments associated with MHC class I molecules. All cells have MHC class I molecules on their surface and are therefore expected to be capable of presenting antigen fragments to cytotoxic T cells, which are, for the most part, MHC class I restricted. The antigen fragments derived from endogenously synthesized molecules, for instance from a virus, are produced at a site distinct from the endocytic vesicles where exogenous antigens are processed.

At or around the site of protein synthesis, endogenously produced antigen fragments become associated with the newly produced MHC class I molecules. The MHC class II molecule picks up internalized antigen within the compartment for peptide loading (CPL), as it moves to the cell surface. In this compartment, antigen fragments cannot bind to the MHC class I molecules because these molecules have already associated with endogenously produced antigenic fragments within the endoplasmic reticulum and never go to the CPL. Thus the site where the antigen is processed determines whether it will associate with MHC class I or class II molecules. This separation of processing pathways explains why CD4+ and CD8+ T cells are involved in the destruction of exogenous and endogenous antigens, respectively.

If a particular antigen does not become associated with either MHC class I or class II molecules, no T-dependent immune response will be directed against that antigen. As MHC class II molecules are involved in the initiation of immune responses by presenting antigen fragments to T_H cells, they can control whether or not a response takes place. It has been clearly shown that the level of an immune response to a particular antigen is controlled by the MHC class II molecules. The genes that code for these molecules (MHC class II genes) have therefore been referred to as *immune response genes*.

It should be obvious that if an antigen cannot associate with the MHC class II molecules of an individual then no immune response will be generated. As the MHC molecules are polymorphic, the cells of some individuals will present, and therefore respond to, certain antigen fragments, whereas cells from other individuals will not. Fortunately, more than one antigenic fragment can be generated from each pathogen; otherwise individuals who did not respond to the particular sequence would be vulnerable to that micro-organism. In addition, individuals have at least six different MHC class II genes and therefore an increased chance that some fragments will bind to at least one of their MHC class II molecules. Variations in the levels and specificity of response occur in individuals who have different MHC class II molecules and have therefore produced different MHC–antigen complexes on their cells.

Immune response gene effects can also be controlled at the level of the T cell receptor. If an individual does not have a T cell with a receptor that recognizes a particular antigen–MHC complex, no response will be generated. The T cell receptor repertoire is generated in the thymus, where the genes of the immature T cells are rearranged to give rise to a functioning receptor. T cells that cross-react too strongly with self molecules are deleted, as are cells whose receptors do not interact with self MHC molecules. Therefore, T cells that interact weakly with MHC molecules are selected to mature and leave the thymus. When these cells later come across an antigen–MHC complex, the presence of antigen strengthens the weak T cell receptor–MHC interaction, leading to a stimulatory signal being transmitted to the T cell. If, for some reason, T cells that respond to a particular MHC–antigen configuration have been deleted, suppressed or not formed, no immune response will be generated to that antigen. There is what is known as a 'hole' in the T cell repertoire.

CONTROL OF IMMUNE RESPONSES

An antigen can induce two types of response: immunity or tolerance. Tolerance is the acquisition of non-reactivity towards a particular antigen. The generation of immunity or tolerance depends largely on the way in which the immune system first encounters the antigen. Once the immune system has been stimulated, the cells involved proliferate and produce a response that eliminates the offending agent. It is then important to dampen down the reacting cells; various feedback mechanisms operate to bring this about.

Role of antigen

The primary regulator of an immune response is the antigen itself. This makes sense, as it is important to initiate a response when antigen enters the host and once it has been eliminated it is wasteful, and in some cases dangerous, to continue to produce effector mechanisms.

Role of antibody

Many biological systems are controlled by the product inhibiting the reaction once a certain level has been reached. This type of negative feedback is seen with antibody, which may act by blocking the epitopes on the antigen so that it can no longer stimulate the cell through its receptor.

As antibody levels rise there is competition between free antibody and the B cell receptor. Consequently, only those B cells that have a receptor with a high affinity for antigen will be stimulated and therefore produce high-affinity antibody. For this reason antibody feedback is thought to be an important driving force in affinity maturation.

Regulatory T cells

T_H cells control the generation of effector cells by producing helper factors. However, the factors that stimulate the expansion of B and T cell numbers do not work indefinitely. Maturation factors are also produced that control terminal differentiation into effector cells. Under the influence of these latter lymphokines, the action of the proliferation factors is inhibited mainly by making the effector cell unresponsive to their effects.

Other T lymphocytes have been described that provide negative signals to the immune system. Suppressor T cells (T_S cells) limit the development of antibody-producing cells and effector T cells. The activity of suppressor cells can involve both the production of soluble factors and direct cell–cell interactions.

TOLERANCE

Two forms of tolerance can be identified: *natural* and *acquired* tolerance. The non-response to self molecules is due to natural tolerance. If this tolerance breaks down and the body responds to self molecules, an autoimmune disease will develop. Natural tolerance appears during fetal development when the immune system is being formed. In experimental animals the introduction of foreign material at the time of birth leads to tolerance.

Acquired tolerance arises when a potential immunogen induces a state of unresponsiveness to itself. This has consequences for host defences, as the presence of a tolerogenic epitope on a pathogen may compromise the ability of the body to resist infection.

An antigen can induce different effects on the two arms of the immune system. During an infection the host is exposed to a variety of antigenic determinants on a micro-organism. These epitopes are present at differing concentrations and possibly at different times during the infection. The epitopes can act as either immunogens or tolerogens. Therefore, it is possible that the antibody response to a particular antigen may be quite pronounced while the cell-mediated response may be lacking, or vice versa. Alternatively, both arms of the immune response may be stimulated or tolerized.

Generally, high doses of antigen tolerize B cells, whereas minute doses given repeatedly tolerize T cells. For acquired tolerance to be maintained, the tolerogen must persist or be administered repeatedly. This is probably necessary because of the continuous production of new T and B cells that must be made tolerant.

Several mechanisms play a role in the selective lack of response to specific antigens. As each lymphocyte has a receptor with a single specificity, the elimination of a specific cell will render the individual tolerant to the epitope it recognizes and leave the rest of the repertoire untouched. This mechanism relies on self molecules interacting with the receptor and causing their elimination. It is proposed that during lymphocyte development the cell goes through a phase in which contact with antigen leads to death or permanent inactivation. Immature B cells encountering antigen for the first time are particularly susceptible to tolerization in the presence of low doses of antigen. The requirement for two signals in the stimulation of B cells and the generation of effector T cells can give rise to tolerance. Both cell types require stimulation via the antigen receptor and 'help' from a specific T cell. If the helper factors are not produced, the responding cells will be functionally deleted. Therefore, the elimination of self-reactive T cells in the thymus during T cell maturation is an important step in maintaining a state of tolerance. Tolerance can also be induced by active suppression. Some T cells are capable of inducing unresponsiveness by acting directly on B cells or other T cells. These T_S cells can be antigen-specific and probably produce signals that actively suppress cells capable of responding to a particular antigen.

It was originally thought that unresponsiveness to self was controlled by the elimination of all self-reactive cells before they matured. This cannot be true, as self-reactive B cells are found in normal adult animals. It is thought that these B cells are controlled by a lack of T cell help, that is, T_H cells have been eliminated. It is likely

Table 9.7   The compromised host

Predisposing factor	Effect on immune system	Type of infection
Immunosuppression for transplant or cancer	Diminished cell-mediated and humoral immunity	Lung infections, bacteraemia, fungal infections, urinary tract infections
Viral immunosuppression (e.g. measles, human immunodeficiency virus, Epstein–Barr virus)	Impaired function of infected cells	Secondary bacterial infections, opportunistic pathogens
Tumour of immune cells	Replacement of cells of the immune system	Bacteraemia, pneumonia, urinary tract infections
Malnutrition	Lymphoid hypoplasia Decreased lymphocytes and phagocyte activity	Measles, tuberculosis, respiratory infections, gastro-intestinal infections
Breakdown of tissue barriers (e.g. surgery, burns, catheterization)	Breach innate defence mechanisms	Bacterial infections, opportunistic pathogens
Inhalation of particles due to employment or smoking	Damage to cilia, destruction of alveolar macrophages	Chronic respiratory infections, hypersensitivity reactions

that T_s cells play a subordinate role, acting as a back-up mechanism.

IMMUNODEFICIENCY

The immunologically competent cells of the lymphoid tissues derived from, renewed by and influenced by the activities of the thymus, bone marrow and other lymphoid tissues can be the subject of disease processes. The deficiency states seen are either due to defects in one of the components of the system itself, or secondary to some other disease process affecting the normal functioning of some part of the lymphoid tissues. Deficiency of one or more of the defence mechanisms can be inherited, developmental or acquired. The types of infections and diseases seen in patients with immunodeficiencies relate to the role the affected component plays in the normal situation. An individual whose immune system has been depressed in any of these ways is said to be *immunocompromised*. The compromised host is prone to infectious diseases that the normal individual would easily eradicate or not succumb to in the first place. Some examples of predisposing factors are given in Table 9.7.

Defective innate defence mechanisms

Defects in phagocyte function take two forms:

1. Where there is a quantitative deficiency of neutrophils that may be congenital (e.g. infantile agranulocytosis) or acquired as a result of replacement of bone marrow by tumour cells or the toxic effects of drugs or chemicals.

2. Where there is a qualitative deficiency in the functioning of neutrophils which, while ingesting bacteria normally, fail to digest them because of an enzymatic defect.

Characteristic of these diseases is a susceptibility to bacterial and fungal, but not viral or protozoan, infections. Among the enzyme deficiency disorders are *chronic granulomatous disease* and the *Chédiak–Higashi syndrome*.

The complement system can also suffer from certain defects in function leading to increased susceptibility to infection. The most severe abnormalities of host defences occur, as would be expected, when there is a defect in the functioning of C3. Severe deficiency or absence of C3 is associated with increased susceptibility to infection, particularly septicaemia, pneumonia, meningitis, otitis and pharyngitis.

Defective acquired immune defence mechanisms

Primary immunodeficiencies

Primary deficiencies in immunological function can arise through failure of any of the developmental processes from stem cell to functional end cell. A complete lack of all leucocytes is seen in *reticular dysgenesis* due to a defect in the development of bone marrow stem cells in the fetus. A baby born with this defect usually dies within the first year of life from recurrent, intractable infections. Defects in the development of the common lymphoid stem cell give rise to severe combined immunodeficiency. Both T and B lymphocytes fail to develop, but functional phagocytes are present.

There are several types of B cell defect that give rise to hypogammaglobulinaemias, that is, low levels of γ-globulins (antibodies) in the blood. Deficiency of immunoglobulin synthesis is almost complete in X-linked infantile hypogammaglobulinaemia (*Bruton's disease*). Male infants suffer from severe, chronic, bacterial infections once maternal antibody has disappeared. There is an absence, or deficiency, of all five classes of serum immunoglobulin. Therefore, the defect is thought to be caused by the absence of B cell precursors or their arrest at a pre-B cell stage. Cell-mediated immune mechanisms function normally and the patients seem to be able to handle viral infections relatively well.

Partial defects in immunoglobulin synthesis have been described affecting one or more of the immunoglobulin classes. In the *Wiskott–Aldrich syndrome*, which is inherited as an X-linked recessive character, there are low levels of IgM but IgA and IgE levels are raised. Patients are susceptible to pyogenic infections, along with recurrent bleeding and eczema. The bleeding is due to reduced platelet production (thrombocytopenia), and the allergy-related eczema is linked to the increased IgE levels. In patients with *dysgammaglobulinaemia* there is a deficiency in only one antibody class. Some patients have reduced levels of IgA whereas the other isotypes are normal. These patients have an increased incidence of infections in the upper and lower respiratory tracts, where IgA is normally protective.

Individuals with T cell defects tend to have more severe and persistent infections than those with antibody deficiencies. A lack of T lymphocytes is often associated with abnormal antibody levels, as T_H cells are involved in the generation and control of humoral immunity. Patients with T cell defects suffer from viral, intracellular bacterial, fungal and protozoan infections rather than acute bacterial infections. In the *DiGeorge syndrome* (congenital thymic aplasia) the person is born with little or no thymus. Individuals who survive develop recurrent and chronic infections, including pneumonia, diarrhoea and yeast infections, once passive maternal immunity has waned.

Secondary immunodeficiencies

Acquired deficiencies can occur secondarily to a number of disease states or after exposure to drugs and chemicals.

Deficiency of immunoglobulins can be brought about by excessive loss of protein through diseased kidneys or via the intestine in protein-losing enteropathy. Malnutrition and iron deficiency can lead to depressed immune responsiveness, particularly in cell-mediated immunity. Medical and surgical treatments such as irradiation, cytotoxic drugs and steroids often have undesirable effects on the immune system. Viral infections are often immunosuppressive. For example, measles, human immunodeficiency and other viruses infect cells of the immune system.

In contrast to the deficiency states just described, raised immunoglobulin levels are found in certain disorders of plasma cells due to malignant proliferation of a particular clone or group of plasma cells. In these conditions, such as chronic lymphocytic leukaemia and multiple myeloma, malignant clones each produce one particular type of antibody. There is usually a decreased synthesis of normal immunoglobulins and an associated deficiency in the immune response to acute bacterial infections. These B lymphoproliferative disorders contrast with the situation in *Hodgkin's disease*, a reticular cell neoplasm in which the patients show defective cell-mediated immunity and are susceptible to viruses and intracellular bacteria.

HYPERSENSITIVITY

Immunity was first recognized as a resistant state that followed infection. However, some forms of immune reaction, rather than providing exemption or safety, can produce severe and occasionally fatal results. These are known as *hypersensitivity reactions* and result from an excessive or inappropriate response to an antigenic stimulus. The mechanisms underlying these deleterious reactions are those that normally eradicate foreign material, but for various reasons the response leads to a disease state. When considering each of the four hypersensitivity states it is important to remember this fact and consider the underlying defence mechanism and how it has given rise to the observed immunopathology.

Various classifications of hypersensitivity reactions have been proposed; probably the most widely accepted is that of Coombs and Gell. This recognizes four types of hypersensitivity that are considered in turn.

Type I: anaphylactic

If a guinea-pig is injected with a small dose of an antigen such as egg albumin, no adverse effects are noted. If a second injection of the same antigen is given intravenously after an interval of about 2 weeks, a condition known as *anaphylactic shock* is likely to develop. The animal becomes restless, starts chewing and rubbing its nose, begins to wheeze, and may develop convulsions and die. The initial injection of antigen is termed the sensitizing dose, whereas the second injection causes anaphylactic shock. Such a reaction is seen in human beings after a bee-sting or injection of penicillin in sensitized individuals. Localized reactions are seen in patients with hay fever and asthma. In all of these situations the host responds to the first injection by producing IgE, and it is the level of IgE produced to a particular antigen that determines whether an anaphylactic reaction will occur on re-exposure to the same antigen. Asthma results from a similar response in the respiratory tract.

The biologically active molecules that are responsible for the manifestations of type I hypersensitivity are stored within mast cell and basophil granules or are synthesized after cell triggering. The signal for the release or production of these molecules is the cross-linking of surface-bound IgE by antigen. The release of these molecules, vasoactive amines and chemotactic factors, is responsible for the symptoms of type I hypersensitivity. IgE has been implicated in the control of parasitic worms; the importance of this is discussed in Chapter 11.

Type II: cytotoxic

Type II reactions are initiated by the binding of an antibody to an antigenic component on a cell surface. The antibody is directed against an epitope, which can be a self molecule or a drug or microbial product passively adsorbed on to a cell surface. The cell that is covered with antibody is then destroyed by the immune system. A variety of infectious diseases caused by salmonellae and mycobacteria are associated with haemolytic anaemia. There is evidence, particularly in studies of salmonella infection, that the haemolysis is due to an immune reaction against bacterial endotoxin that becomes coated on to the erythrocytes of the patient.

Type III: immune complex

As discussed above, when a soluble antigen combines with antibody the size and physical form of the immune complex formed depends on the relative proportions of the participating molecules and is affected by the class of antibody. Monocytes and macrophages are very efficient at binding and removing large complexes. These same cell types can also eliminate the smaller complexes made in antibody excess, but are relatively inefficient at removing those formed in antigen excess. Type III hypersensitivity reactions appear when there is a defect in the systems involving phagocytes and complement that remove immune complexes, or when the system is overloaded and the complexes are deposited in tissues. This latter situation occurs when antigens are never completely eliminated, as with persistent infection with an organism, autoimmunity and repeated contact with environmental factors.

The tissue damage that results from the deposition of immune complexes is caused by the activation of complement, platelets and phagocytes – in essence, an acute inflammatory response. In general, the degree and site of damage depend on the ratio of antigen to antibody.

At equivalence or slight excess of either component, the complexes precipitate at the site of antigen injection or production and a mild, local type III hypersensitivity reaction occurs (e.g. the *Arthus reaction*). In contrast, the complexes formed in large antigen excess become soluble and circulate, causing more serious systemic reactions (e.g. *serum sickness*), or eventually deposit in organs, such as skin, kidneys and joints. The type of disease and its time course depends on the immune status of the individual.

The local release of antigens from an infectious organism can cause a type III reaction. A number of parasitic worms, although undesirable, cause little or no damage. However, if the worm is killed it can become lodged in the lymphatics, and the inflammatory response initiated by antigen–antibody complexes causes a blockage of lymph flow. This leads to the condition of elephantiasis in which enormous swellings can occur. In some cases of tuberculosis, sarcoidosis, leprosy and streptococcal infections, vascular inflammatory lesions are seen mainly in the legs. These are variously referred to as *erythema nodosum*, *nodular vasculitis* and *erythema induratum*, and may be due to the deposition of immune complexes and the development of an Arthus reaction.

In systemic disease the clinical manifestations depend on where the immune complexes form or lodge – skin, joints, kidney and heart being particularly affected.

Drugs such as penicillin and sulphonamides can cause type III reactions. The most susceptible patients develop rashes (urticarial, morbilliform or scarlatiniform), pyrexia, arthralgia, lymphadenopathy and perhaps nephritis some 8–12 days after being given the drug. It is likely that similar events occur in many bacterial and viral infections (see Chs 12 and 10, respectively).

Type IV: cell-mediated or delayed

This form of hypersensitivity can be defined as a specifically provoked, slowly evolving (24–48 h), mixed cellular reaction involving lymphocytes and macrophages. The reaction is not brought about by circulating antibody but by sensitized lymphoid cells. This type of response is seen in a number of allergic reactions to bacteria, viruses and fungi, in contact dermatitis and in graft rejection. The classical example of this type of reaction is the tuberculin response that is seen following an intradermal injection of a purified protein derivative (see p. 210) from tubercle bacilli in immune individuals. An indurated inflammatory reaction in the skin appears about 24 h later and persists for a few weeks. In humans the injection site is infiltrated with large numbers of mononuclear cells, mainly lymphocytes, with about 10–20% macrophages. Most of these cells are in or around small blood vessels. The type IV hypersensitivity state arises when an inappropriate or exaggerated cell-mediated response occurs.

Cell-mediated hypersensitivity reactions are seen in a number of chronic infectious diseases caused by mycobacteria, protozoa and fungi. Because the host is unable to eliminate the micro-organism, the antigens persist and give rise to a chronic antigenic stimulus. Thus, continual release of lymphokines from sensitized T cells results in the accumulation of large numbers of activated macrophages that can become epithelioid cells. These cells can fuse together to form giant cells. Macrophages express antigen fragments on their surface in association with MHC class I and II molecules, and are therefore the targets of T_c cells and stimulate more lymphokine production. This whole process leads to tissue damage with the formation of a *chronic granuloma* and resultant cell death.

Penicillin sensitization is a common clinical complication following topical application of the antibiotic in ointments or creams. This and other substances that cause contact sensitivity are not themselves antigenic and become so only in combination with proteins in the skin. The Langerhans cells of the epidermis are efficient antigen-presenting cells favouring the development of a T cell response. These cells pick up the newly formed antigen in the skin and transport it to the draining lymph node where a T cell response is stimulated. Here, the specific T cells are stimulated to mature, and then return to the site of entry of the offending material and release their lymphokines. In a normal situation these would help to eliminate a pathogen, but in this case the continual or subsequent exposure to the foreign material leads to an inappropriate response. The reaction site is characterized by a mononuclear cell infiltrate peaking at 48 h. The clinical symptoms in these contact dermatitis lesions include redness, swelling, vesicles, scaling and exudation of fluid (i.e. eczema).

AUTOIMMUNITY

A fundamental characteristic of the immune system of an animal is that it does not, under normal circumstances, react against its own body constituents. Mechanisms, as we have seen, exist that allow the immune system to tolerate self and destroy non-self. Occasionally these mechanisms break down and autoantibodies are produced.

Genetic factors appear to play a role in the development of autoimmune diseases and there is a strong association between several autoimmune diseases and particular HLA (human leucocyte group A) specificities, suggesting that immune response gene effects may be involved.

There are a number of examples where potential autoantigenic determinants are present in exogenous material. These preparations may provide a new carrier, a T cell-stimulating determinant that provokes autoantibody formation. The encephalitis sometimes seen after rabies vaccination with the older vaccines is thought to result from a response directed against the brain that is stimulated by heterologous brain tissues present in the vaccine.

KEY POINTS

- The cells of the immune system are divided into lymphoid and myeloid lineages. The former include T lymphocytes and their subsets identified by CD markers, B lymphocytes and natural killer (NK) cells. The myeloid lineage includes the neutrophils, eosinophils and basophils as well as the monocyte/macrophage series and platelets.
- Innate immunity depends on physical, physiological and chemical barriers to infection, on the response to injury and on detection of pathogen-associated molecular patterns (PAMPs) by pattern recognition receptors (PRRs). Phagocytic cells and the enzyme cascade known as *complement* are key effectors responding to PAMPs and components of *acute inflammation*.
- Acquired immunity depends on specific recognition of antigens either directly by antibodies on the surface of B cells or through presentation of processed antigens in the context of MHC molecules by host cells to T cells. In contrast to innate immunity, on re-exposure the responses are faster, more vigorous and more specific.
- Acquired immune responses are driven by the availability of antigen. As they mature, only cells with high-affinity receptors for the antigen are stimulated to divide. The expanded clones of antigen-specific cells are said to have resulted from *clonal selection*.
- Lymphocytes are activated by antigen and the appropriate combination of *cytokines*, signalling molecules secreted by other lymphocytes and by macrophages.
- Humoral acquired immunity leads to antigen–antibody complexes that neutralize key aspects of microbial activity either directly or through the activation of complement, *opsonization* and directed cytotoxicity.
- Cell-mediated immunity generates cytotoxic T lymphocytes (CD8+), which directly kill cells containing intracellular pathogens, and helper T cells (CD4+), which secrete lymphokines that stimulate other effector aspects of immunity.
- Inherited and acquired defects in the immune system lead to immunodeficiencies that make individuals more susceptible to certain infections.
- Damage due to immune reactions may reflect attempts to eliminate micro-organisms or self antigen-directed (*autoimmune*) reactions.

Micro-organisms are a source of cross-reacting antigens, sharing antigenic determinants with tissue components. This may be an important way of inducing autoimmunity. The group A streptococcus, which is closely associated with rheumatic fever, shares an antigen with the human heart. Heart lesions are a common finding in rheumatic fever, and anti-heart antibody is found in just over 50% of patients with this condition. Nephritogenic strains of type 12 group A streptococci carry surface antigens similar to those found in human glomeruli, and infection with these organisms has been associated with the development of acute nephritis. Some of the immunopathology seen in Chagas' disease has been attributed to a cross-reaction between *Trypanosoma cruzi* and cardiac muscle.

Autoimmunity can be induced by bypassing T cells. Self-reactive cells can be stimulated by polyclonal activators that directly activate B cells. A number of micro-organisms or their products are potent polyclonal activators; however, the response that is generated tends to be IgM and to wane when the pathogen is eliminated. Bacterial endotoxin, the lipopolysaccharide of Gram-negative bacteria, provides a non-specific inductive signal to B cells, bypassing the need for T cell help. A variety of antibodies are present in infectious mononucleosis, including autoantibodies, as a result of the polyclonal activation of B cells by Epstein–Barr virus.

RECOMMENDED READING

Abbas A K, Janeway C A 2000 Immunology: improving on nature in the twenty-first century. *Cell* 100: 129–138
Abbas A K, Lichtman A H 2003 *Cellular and Molecular Immunology*, 5th edn. Saunders, Philadelphia
Janeway C A, Travers P, Walport M, Shlomchik 2005 *Immunobiology; the Immune System in Health and Disease*, 6th edn. Current Biology, London
Staros E B 2005 Innate immunity: new approaches to understanding its clinical significance. *American Journal of Clinical Pathology* 123: 305–312

Internet sites

CELLS alive! http://www.cellsalive.com/
Microbiology and Immunology On-line. University of South Carolina School of Medicine. http://pathmicro.med.sc.edu/book/immunol-sta.htm
Cytokines & Cells Online Pathfinder Encyclopaedia (COPE). http://www.copewithcytokines.de/

10 Immunity in viral infections

J. Stewart

The host response to an invading virus depends on the characteristics of the infectious agent and where it is encountered. In many cases viral infections are sub-clinical, that is, symptomless. A vast array of host defence mechanisms work in a concerted way to protect the individual from viruses and to eliminate them if an infection occurs. In several instances, virus-induced immune responses may have immunopathological consequences.

THE RESPONSE TO VIRAL INFECTIONS

Interferons

At the time of the discovery of interferon in 1957, the term was used to identify a factor produced by cells in response to viral infection that protected other cells of the same species from attack by a wide range of viruses. It is now clear that this activity is mediated by members of a family of regulatory proteins.

In man, as in a number of other species, there are three main interferons:

1. α-interferon (IFN-α), produced mainly by peripheral blood mononuclear cells
2. β-interferon (IFN-β), produced predominantly by fibroblasts
3. γ-interferon (IFN-γ), a lymphokine produced in response to a specific antigenic signal.

IFN-α and IFN-β are known as type 1 interferons and are considered part of the innate defences whereas IFN-γ is referred to as type 2 interferon.

There is only one gene for IFN-β and one for IFN-γ, but there are at least 18 different IFN-α genes coding for 14 functional proteins. All the IFN-α genes are closely related and clustered on chromosome 9, close to the IFN-β gene; the IFN-γ gene is on chromosome 12. The production of interferons is under strict inductional control. IFN-α and IFN-β are produced in response to the presence of viruses and certain intracellular bacteria. Double-stranded RNA may be the important inducer acting through Toll-like receptors (see p. 109) to signal the production of type 1 interferons. IFN-γ, which has an extensive role in the control of immune responses, is produced by antigen-activated T lymphocytes and natural killer (NK) cells (see pp. 123–125).

To exert their biological effects these molecules must interact with cell surface receptors. IFN-α and IFN-β share a common receptor, whereas IFN-γ binds to its own specific receptor. After binding to the cell surface receptors, interferons act by rapidly and transiently inducing or upregulating some cellular genes and downregulating others. The overall effect is to inhibit viral replication and activate host defence mechanisms.

The antiviral activity is mediated by the interferon released from a virus-infected cell binding to a neighbouring cell and inducing the synthesis of antiviral proteins (Fig. 10.1). Interferons are extremely potent in this function, acting at femtomolar (10^{-15} M) concentrations. They can inhibit many stages of the virus life cycle – attachment and uncoating, early viral transcription, viral translation, protein synthesis and budding. Many new proteins can be detected in cells exposed to interferon, but major roles have been proposed for two enzymes that inhibit protein synthesis: 2′,5′-oligoadenylate synthetase (2,5-A synthetase) and a protein kinase. The activity of both of these enzymes is dependent on double-stranded RNA (dsRNA) provided by viral intermediates in the cell. The protein kinase is responsible for the phosphorylation of histones and the protein synthesis initiation factor eIF2. This leads to the inhibition of protein synthesis within interferon-stimulated cells as a result of inhibition of ribosome assembly. The 2,5-A synthetase is strongly induced in human cells by all three types of interferon and forms 2′,5′-linked oligonucleotides of adenosine from adenosine triphosphate (ATP). These oligonucleotides activate a latent cellular endonuclease that degrades both messenger and ribosomal RNA, with a resultant inhibition of protein synthesis. The requirement for the presence of dsRNA for the full expression of these responses safeguards uninfected cells from the damaging effects of the enzymes. Apart from these well

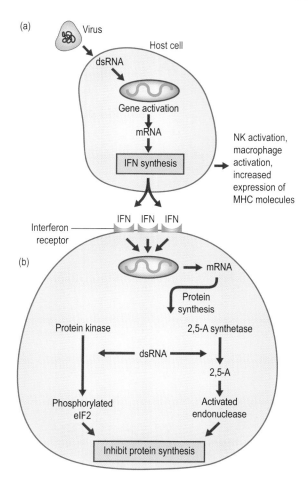

Fig. 10.1 Proposed mechanisms of **a** induction of synthesis of interferon (IFN)-α and IFN-β, and **b** production of resistance to virus infection. dsRNA, double-stranded ribonucleic acid; mRNA, messenger RNA; 2,5-A, 2′,5′-oligoadenylate; MHC, major histocompatibility complex; NK, natural killer cells; eIF, eukaryotic initiation factor.

malignant cell growth. A number of clinical trials have shown that IFN-α is active against some human cancers, especially those of haemopoietic origin.

Interferons are able to modify immune responses by:

- altering the expression of cell surface molecules
- altering the production and secretion of cellular proteins
- enhancing or inhibiting effector cell functions.

One of the main ways in which interferons control immune responses is by the induction or enhancement of major histocompatibility complex (MHC)-encoded molecules.

Class I MHC genes are upregulated by all types of interferon, as is the production of β_2-microglobulin. IFN-γ induces and increases the expression of MHC class II antigens. In addition, interferons can induce or enhance the expression of Fc receptors and receptors for a number of cytokines. These activities increase the efficiency of antigen recognition and lead to a more effective immune response.

Interferons have also been implicated in the control of B cell responses. When added in vitro or in vivo they can suppress or enhance primary or secondary antibody responses, depending on the dose and time of addition. The regulatory effects seem to be on the B cells themselves, on increased antigen presentation and through an effect on regulatory T cells.

A number of immune effector cells act by killing infected target cells. The cytotoxicity of macrophages, neutrophils, T cells and NK cells is enhanced by interferons. IFN-γ produced by T lymphocytes is capable of activating macrophages to kill intracellular bacteria. This lymphokine has all the activities of the molecule that used to be known as macrophage-activating factor (MAF). NK cells are able to destroy a range of syngeneic, allogeneic and xenogeneic cells in an MHC-unrestricted fashion (see below). All three types of interferon increase NK cell activity in vitro and in vivo, not only by recruiting pre-NK cells to become actively lytic but also by increasing the spectrum of cells lysed. The mechanisms by which interferons make cells cytotoxic are not clear, but it is of interest that interferons can stimulate the production of cytotoxins such as TNF.

TNF has also been reported to have several antiviral activities similar to those of IFN-γ, but working through a different pathway. However, if both TNF and IFN-γ are added together, a synergistic effect is seen. If TNF is added to cells after viral infection, it can lead to their destruction even though the cells are normally resistant to TNF. This effect is also synergistic with IFN-γ. Certain viruses have been shown to trigger the release of TNF from mononuclear cells, and it seems likely that this cytokine is an important host response to viral infections.

Certain cell-mediated reactions are also part of the innate defences against viral infection. The structures

characterized changes, many other changes occur in cells treated with interferons. Some viral proteins can inhibit the interferon response.

Some effects of interferon are virus-specific. The Mx protein induced in mice by IFN-α/β is specifically involved in the resistance to influenza virus infection. There is a related protein in human cells, and it is possible that some of the other interferon-induced proteins may confer resistance to specific virus types. Resistance of IFN-γ-treated cells to the parasite *Toxoplasma gondii* is associated with induction of the enzyme indolamine dioxygenase, which catabolizes the essential amino acid tryptophan. However, interferons have effects on host cell growth and differentiation. Interferons, particularly IFN-α and IFN-β, are potent inhibitors of normal and recognized by NK cells are not known, but changes in the level of expression of MHC class I molecules on the infected cells are important. A number of viruses, especially those causing latent infections, evade immune recognition by interfering with the MHC class I processing pathway. Cells infected with these viruses do not display MHC class I molecules on their surface and are therefore not recognized by cytotoxic T cells. However, the infected cells are recognized by NK cells. The formation of a close conjugate between the NK cell and the target induces the effector cell to produce molecules that lead to the death of the infected cell by apoptosis. Antibodies that recognize NK cells have been used to deplete these cells in mice. It was found that the treated mice were more susceptible to murine cytomegalovirus than were normal mice. Natural killing may form a first-line defence against viral attack, most importantly by herpesviruses, before the acquired immune response is generated. Natural killing is increased by interferons (both the number of effector cells and their killing potential) and, therefore, these two innate defence mechanisms appear to work together to protect the host from viral infection.

The interferon response is rapid and helps to protect the host until acquired responses develop. Interferons induce a febrile response and this may also be important in inhibiting viral growth in some infections with viruses that have a low ceiling temperature for growth.

Acquired immunity

The response to viral antigens is almost entirely T cell dependent. Immunodeficiencies involving T cells are always characterized by markedly enhanced susceptibility to viral infections. However, this tells us little about the effector mechanisms involved, as T cells are required for both antibody production and cytotoxic reactions.

The viral epitopes to which the immune system responds have been studied to give an insight into the mechanisms involved in the host response to these pathogens and also to aid in the development of better vaccines. The recognition of viral antigens is similar to that for all foreign material. B cells and immunoglobulin are able to combine with exposed epitopes, and processed viral fragments presented in the context of MHC molecules are recognized by T cells.

Antigen-specific B cells can act as antigen-presenting cells and thereby generate an immune response. B cells present antigen to T cells and in return are stimulated by growth and differentiation molecules. Intramolecular help may explain hapten–carrier effects. The uptake of an intact virion means that the B cell is able to present peptides derived from internal proteins to T cells (Fig. 10.2). Thus, a B cell that is specific for a surface antigen can receive

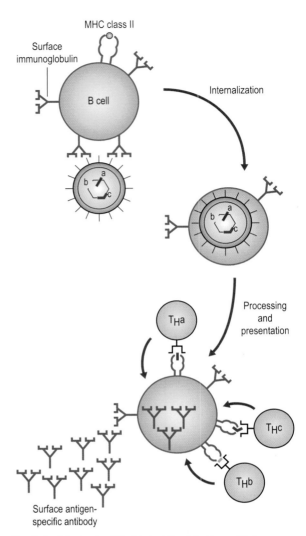

Fig. 10.2 Intrastructural T cell help. A B cell that is specific for a surface component of a virus binds the virus through its receptor, surface immunoglobulin. The virus contains three potential T cell epitopes (a, b and c) within its nucleocapsid. The entire virus can be internalized, and all viral polypeptides will be processed and fragments re-expressed in association with MHC class II molecules (presentation) on the B cell surface.

help from a T cell specific for another molecule as long as it is present within the same particle – intrastructural help.

In most cases, exogenous virus proteins (i.e. those derived from an extracellular virus and taken into a cell) are presented in the context of MHC class II molecules and stimulate T_H cells. Cells that are supporting viral replication express virus-derived peptides in association with MHC class I molecules (the endogenous pathway). The fact that some endogenously produced viral proteins are presented in the context of MHC class II molecules suggests that there can be some overlap between the MHC class I and class II pathways.

Humoral immunity

There are several ways in which antibody against viral components can protect the host. Antibodies cannot enter cells, and therefore are ineffective against latent viruses and those that spread directly from cell to cell. They do, however, bind to extracellular viral epitopes. These epitopes can be on intact virions or on the surface of infected cells. The binding of antibody to free virus can inhibit a number of processes essential to virus replication. Antibodies can block binding to the host cell membrane, and thus stop attachment and penetration. The immunoglobulins IgG and IgM have this important function in serum and body fluids, and IgA can neutralize viruses by a similar mechanism on mucosal surfaces. Antibody can also work at stages after penetration. Uncoating, with its release of viral nucleic acid into the cytoplasm, can be inhibited if the virion is covered by antibody.

Antibody can also cause aggregation of virus particles, thus limiting the spread of the infectious particles and forming a complex that is readily phagocytosed. Complement can aid in the neutralization process by opsonizing the virus or directly lysing enveloped viruses. In certain cases complement alone can inactivate viruses. Some retroviruses have a protein that can act as a receptor for C1q, and other viruses have been reported to activate the alternative pathway. In some infections, viral proteins remain on the surface of the cell after entry or become associated with the cell membrane during replication. Antibodies against these molecules can cause cell lysis by the classical pathway, but an intact alternative pathway is necessary to amplify the initial triggering by the antibody-dependent pathway. In certain situations antibody-mediated reactions are not always of benefit (see below). Antibodies are also capable of modulating or stripping viral antigens from the cell surface, allowing the infected cell to avoid destruction by other effector mechanisms.

Viral infections, particularly those caused by enteroviruses, are frequent and severe when humoral immunity is impaired, as in certain inherited immunodeficiency states. In Bruton-type deficiencies, poliomyelitis may develop after vaccination with the live virus vaccine; meningo-encephalitis, caused by echovirus and coxsackievirus, may also be seen. In many situations, viruses seem able to escape the humoral defence mechanisms. Some viruses become latent (e.g. herpesviruses) and are reactivated despite the presence of circulating antibody, as they can pass directly from cell to cell. Other escape mechanisms include antigenic variation in which the antigenic structure of the virus (e.g. influenza type A)

changes so that antibodies formed to the previous strain are no longer effective.

In viral infections the efficiency of antibody depends largely on whether the virus passes through the bloodstream outside host cells to reach its target organ. Poliovirus crosses the intestinal wall, enters the bloodstream to cause a cell-free viraemia, and passes to the spinal cord and brain where it replicates. Small amounts of antibody in the blood can neutralize the virus before it reaches its target cells in the nervous system.

In comparison, in viral diseases such as influenza and the common cold the viruses do not pass through the bloodstream. These infections have a short incubation period, their target organ being at the site of entry into the body, namely the respiratory mucous membranes. In this type of infection a high level of antibody in the blood is relatively ineffective in comparison with its effect on blood-borne viruses. In this case the antibody must be present in the mucous secretions at the time of infection. There are very low levels of IgG or IgM in secretions, but IgA has been shown to be responsible for most of the neutralizing activity present in nasal secretions against rhinoviruses and other respiratory tract viruses.

One consequence of this is that conventional immunization methods using killed virus or viral subunits, which produce high levels of circulating antibody, are unlikely to be effective against viruses that attack the mucous membranes. Some considerable effort is being directed at developing methods for stimulating local production of IgA in the mucous membranes themselves. Live virus vaccines are effective in this respect, and the intranasal administration of a live-attenuated influenza virus vaccine is an attempt to overcome this problem. The high degree of immunity provided by the oral polio vaccine is due in part to locally produced antibody in the gut neutralizing the virus before it attaches to cell receptors to cause infection. The presence of IgA against polio has been demonstrated in faeces, duodenal fluid and saliva. There is evidence to suggest that parenteral administration of a killed virus may give rise to a secretory IgA response if the individual has previously been exposed to the virus or has received an oral-attenuated vaccine. So there appears to be a link between the systemic immune system and local mucosal immunity.

Humoral immunity does play a major protective role in polio and a number of other viral infections, and is probably the predominant form of immunity responsible for protection from reinfection. Passively administered antibody can protect against several human infections, including measles, hepatitis A and B, and chickenpox, if given before or very soon after exposure. Immunity to many viral infections is lifelong. This may occur because antibodies are boosted by occasional re-exposure to the virus.

Cell-mediated immunity

The destruction of virus-infected cells is an important mechanism in the eradication of virus from the host. Antibody can neutralize free virions, but once these agents have entered the cell other strategies are employed. The destruction of an infected cell before progeny particles are released is an effective way of terminating a viral infection. For this process to occur the immune system must recognize the infected cell, and various types of effector cell have evolved to mediate these processes.

As viral proteins are synthesized within the cell some of these molecules are processed into small peptides. These endogenously produced antigen fragments become associated with MHC class I molecules, and this complex is then transported to the cell surface where it acts as the recognition unit for cytotoxic T (T_c) lymphocytes. Most T_c cells have a receptor that binds to fragments of the virus sitting in the cleft of an MHC class I molecule. This T cell also has the CD8 molecule on its surface. Some T_c cells are restricted in their recognition of antigen by MHC class II molecules and therefore have the CD4 molecule. Once these T_c cells have bound to the infected cell they release molecules that induce apoptosis.

Many viruses, such as poliovirus and papillomavirus, replicate and produce fully infectious particles inside the cell. These viruses are liberated from the infected cell as it disintegrates. However, other viruses do not wait for the cell to die but are released by a process of budding through the cell membrane. During their replication, virus-encoded molecules (viral antigens) are inserted into the host cell membrane, and the nucleocapsid becomes associated with these molecules. The virus particle finally acquires an envelope as it is released. Such viruses include herpesviruses, alphaviruses, flaviviruses, retroviruses, hepadnaviruses, orthomyxoviruses and paramyxoviruses. Viral antigens often appear on the cell surface very early in the replicative cycle, many hours before progeny virus is liberated. In cells infected with herpes simplex virus, as many as five different viral glycoproteins appear on the cell surface. In other virus infections, such as those caused by poxviruses and papovaviruses, the viral particles are not released by budding, but viral antigens appear on the cell surface. These molecules, including those that are not incorporated into the released virion, can therefore act as signals indicating the presence of virus within a cell. If antibody binds to these cell surface viral antigens, the infected cell can be destroyed by antibody-dependent cell-mediated cytotoxicity. The effector cells have an Fc receptor that recognizes the Fc portion of immunoglobulin bound to viral antigens present on the infected cell surface. This interaction brings the two cells close together, and toxic molecules are released on to the target cell membrane, causing cell death.

CD4+ and CD8+ T cells can produce various lymphokines when stimulated by antigen, including molecules that are active in the elimination of virus (e.g. IFN-γ and TNF), and others that generally increase the effectiveness of the immune system by attracting cells to the site of infection, stimulating the production of more cells and supporting their growth. Macrophages are activated, leading to enhanced microbicidal activities and the production of monokines.

Induction of an immune response

The precise nature of the acquired immune reactions that are generated in response to infection depends to a great extent on the site of infection, type of virus, previous exposure to the agent and the genetic make-up of the host. Both humoral and cell-mediated responses are produced to all infections. The importance of genetic factors is illustrated by the severe *X-linked recessive lymphoproliferative syndrome* in which fatal infectious mononucleosis results from the unrestricted replication of Epstein–Barr virus, as affected males have reduced numbers of normal lymphocytes.

The virus or viral components entering the peripheral tissues or being produced there are carried to the draining lymph node either free in the lymph or in cells such as Langerhans' cells. Once in the secondary lymphoid tissues the free virus is taken up by macrophages and processed; viral peptides associate with MHC class II molecules and are transported to the cell surface. Langerhans' cells that enter the lymph node become interdigitating cells and present viral peptides, associated with MHC class II molecules, on their surface. A T_H cell will bind to the peptide–MHC class II complex, expand clonally and produce helper factors. The proportions of the different lymphokines produced determines the type and level of the response generated. B cells will enter the node, and those that bind antigen are stimulated into antibody production. Other T cells will enter the lymph node. If these cells have a receptor that interacts with an antigen fragment–MHC class I complex, they will respond to the growth factors present in the node and proliferate and mature into effector T_c cells. After a time, the products of the immune response leave the node to circulate round the body and localize at the site of infection. As the response progresses the pathogen is eliminated, the tissue repaired and memory cells generated. Finally, when all of the antigen has been eliminated, the immune response is terminated. In some instances virus-induced immune responses may have immunopathogical consequences.

IMMUNOPATHOLOGY

Viruses have evolved a multitude of mechanisms for exploiting weaknesses in the host immune system and avoiding – sometimes actually subverting – immune mechanisms. Some viruses are so successful in avoiding host defences that they persist in the host indefinitely, sometimes in a latent form without producing disease.

One of the most important strategies developed by viruses is to infect cells of the immune system itself. The effect of this is often to disable the normal functioning of the cell type that has been infected. Many common human viruses, including rubella, mumps, measles and herpes viruses, infect cells of the immune system, as does human immunodeficiency virus (HIV). The consequences of viral infection of cells of the immune system have been categorized in two ways:

1. Infections that cause temporary immune deficiency to unrelated antigens and sometimes to the antigens of the infecting virus. It is known that infection with influenza, rubella, measles and cytomegalovirus predisposes to bacterial and other infections. This is sometimes associated with depressed immunoglobulin synthesis and interference with the antimicrobial functions of phagocytes.

2. Permanent depression of immunity to unrelated antigens and occasionally to antigens of the infecting virus. Acquired immune deficiency syndrome is an example of such a disease, where the patient becomes susceptible to otherwise harmless protozoa, bacteria, viruses and fungi.

Viruses have also developed other mechanisms to avoid the immune system. These include:

- antigenic variation
- release of antigens
- production of antigens at sites that are inaccessible to the immune system.

Viruses that cannot enter and replicate within phagocytic cells will be destroyed if they are engulfed by a neutrophil or macrophage. As neutrophils are short-lived cells they do not usually give rise to progeny virus. On the other hand, monocytes and macrophages are long-lived cells and can be responsible for disseminating a virus throughout the body. Viruses that do replicate within macrophages must escape from the phagosome very rapidly before it fuses with the lysosome. Reovirus infection of macrophages is actually helped by the lysosomal enzymes, which initiate 'uncoating' of the virus and therefore enhance viral replication.

A virus is relatively safe from immune destruction as long as it remains within the cell and allows only very low or no viral antigen expression on the infected cell membrane. This is what happens in latent infections where herpes simplex or varicella-zoster virus is present in the dorsal root ganglion.

Antibody can actually remove viral antigens from cell membranes, as cross-linking of antigens on the cell surface can lead to their internalization by capping. Here the antigens complexed with the antibody are drawn to one pole of the cell and internalized or shed into the surrounding tissue. Capping occurs on brain cells infected with measles virus in subacute sclerosing panencephalitis. Viruses that move from cell to cell without entering the extracellular fluid can also escape the action of antibodies, as can those passed from cell to cell by cell division.

A number of infections continually shed virus into external secretions, such as saliva, milk or urine. As long as the infected cell forms virus only on the luminal surface of the mucosa, cells of the immune system and antibody will be unable to destroy the infected cell. IgA present in the secretions may neutralize the virus, but this class of antibody does not activate complement efficiently, so the cell will not be lysed. A similar situation applies to epidermal infections with wart virus. The infected cell is keratinized and about to be released from the surface of the body before any virus or viral antigens are produced. The infected cell is therefore isolated from the host's immune cells.

During the course of an infection various antibodies are formed against different epitopes on a virus. These antibodies are of differing affinities and also stimulate different effector functions. Antibodies against some of the epitopes will neutralize the virus, but other antibodies will be against unimportant epitopes or be of an ineffective isotype that may fail to neutralize the virus, and may actually aid in its infectivity by allowing uptake of virus–antibody complexes via Fc receptors or cause tissue damage through immune complex disease. Soluble antigens liberated from infected cells may 'mop up' free antibody so that it can no longer interact and destroy extracellular virus. Whether the small particles present in the serum of patients and carriers with hepatitis B virus infections function in this way is not known, although patients in the prodromal phase may suffer from rash, myalgia and arthralgia. Polyarteritis nodosa and glomerulonephritis can also occur, all suggestive of immune complex formation.

Susceptibility to infection is generally greater in the very young and very old because of a weaker immune response. However, the immunopathology tends to be less severe. In the very young, infections can spread rapidly and prove fatal without the clinical and pathological changes seen in adults. Latent infections are kept under control by the immune system, and in older people the infections show an increased incidence of activation (e.g. zoster, or shingles). Immunological immaturity makes the neonate highly susceptible to many viral infections. Maternally derived antibody provides passive protection for 3–6 months, after which time the infant is at risk of infection; respiratory and alimentary tract infections are frequent.

Physical and physiological differences may also contribute to age-related disease susceptibility. Respiratory infections in old age are probably a bigger problem than in young adults because of weaker respiratory muscles and a poorer cough reflex. In the young the airways are narrower and more easily blocked by secretions and exudate. Infants, because of their low body-weight, show signs of distress from loss of fluid and electrolytes, so that fever, vomiting and diarrhoea tend to be very serious at this time of life. Often the reasons for the differences between infants and adults are not known. Respiratory syncytial virus causes severe illness in the early months of life with croup, bronchiolitis and bronchopneumonia, despite the presence of maternal IgG. In adults the virus usually causes a mild upper respiratory tract infection.

Certain viral infections produce a milder disease in children than in adults (e.g. varicella, mumps, poliomyelitis and Epstein–Barr virus infections). Varicella often causes pneumonia in adults, and mumps may involve the testes and ovaries after puberty, giving rise to orchitis and oophoritis. Epstein–Barr virus is excreted in saliva, and in developing countries most individuals are infected early in life, usually asymptomatically. In developed countries where childhood infection is less common, first infection may be delayed to adolescence or early adulthood, when salivary exposure occurs during kissing. In this age group, Epstein–Barr virus infection gives rise to glandular fever. It is not clear why these infections are more severe in adults, but it may be linked to the more powerful immune defences giving rise to immunopathological sequelae in adults.

There are also age-related differences in the incidence of infections. It is not surprising that most infections are commonest in childhood when the individual is exposed to the micro-organisms for the first time.

Antigenic variation

A micro-organism can avoid the acquired immune response by periodically changing the structure of molecules that are recognized by the host immune system. The immune system selects the variants by not being able to mount an immune response against them before they are shed. The micro-organism will only be able to change a component in a way that does not alter the functioning of the molecule. The molecules involved can be active enzymes, recognition molecules or structural proteins. HIV shows considerable variation in parts of the envelope glycoprotein within a given individual.

The significance of antigenic variation is well illustrated by influenza viruses. Here changes in the surface glycoproteins are linked to the occurrence of epidemics of infection (see Ch. 49).

With influenza virus the infection is localized to the respiratory tract where the principal protection against reinfection is secretory IgA. The virus-specific IgA still present at the mucosal surface a few years after infection can protect against the original infecting virus but may be insufficient to deal with an antigenic variant despite antigenic overlap. Thus, in effect, IgA levels become a selective pressure, which will allow infection by the mutant, and antigenic drift occurs. The very short incubation period (1–3 days) is more rapid than the secondary antibody response so that the IgA levels cannot be boosted in time to abort infection.

Antigenic variation is likely to be an important viral adaptation for overcoming host immunity in long-lived species such as humans where there is a need for multiple reinfection of the same individual if the virus is to survive and the virus is unable to become latent. In shorter-lived animals such as mice and rabbits a susceptible population appears quickly enough to maintain the infectious cycle.

Persistence of virus

Certain viruses give rise to a persistent infection, which is held in check as long as the immune system remains intact.

Chickenpox is a persistent infection characterized by latency in that there is apparent recovery from the original infection but the virus can reappear later in life when a localized eruption, shingles, results. Other herpesviruses, cytomegalovirus and Epstein–Barr virus also persist after infection. If the carrier's immune system remains intact, there will be no evidence of disease. However, cytomegalovirus causes many problems in immunosuppressed patients. The polyomavirus, JC, usually causes asymptomatic infections, but in the immunosuppressed it has been found in areas of destruction in the central nervous system; the disease is progressive multifocal leucoencephalopathy.

In other persistent infections, the immune system contributes to the pathology of the disease, often over a period of years. Thus, in subacute sclerosing panencephalitis, persistence of measles virus in neurones triggers their destruction by the host's immune system. Similarly, the chronic active hepatitis seen in those carriers of hepatitis B who continue to produce virions appears to be caused by a T_c cell response to viral antigens present in the hepatocyte membrane. In both examples, several years may elapse before symptoms appear.

VACCINES

Natural infection with a virus is an extremely effective means of giving lifelong immunity from the disease. In most cases, where there is one virus type, this means that second attacks are extremely rare. The memory of the immune system ensures that, for these infections, a secondary response can be generated before the virus has time to cause the disease. The level of immunity needed to protect an individual depends on the incubation period of the virus and its life cycle. For viruses with very short incubation periods, a high level of protective immunity must be present before exposure to the infective agent. In the case of a virus with a long incubation period (10–20 days), the immune system has time to generate a protective response.

It is also important to consider the type of immune response that will be protective against different viruses. If antibody gives protection then steps must be taken to ensure that the material to be used for immunization contains the correct epitopes. A denatured antigen will not generate antibodies that can combine with the native virus. Sometimes the chemical treatment of the antigen may destroy important components. Thus the original killed measles virus vaccine did not contain the fusion protein. As a result, vaccinees suffered enhanced disease when exposed to the virus as viral replication and spread could occur in the presence of antibody to the other viral proteins. If T cell immunity is important, the vaccine must be in a form that will give rise to peptides in the correct compartment of the cell to produce antigen fragments in association with MHC-encoded products. It will have to associate with the MHC class II molecules to generate help and with MHC class I molecules to stimulate effector cell formation. For antibody production, T cell help is also required. Therefore, for an antibody response, a killed

vaccine may be sufficient but, when T cell immunity is required, a live-attenuated vaccine is needed.

Vaccination has been responsible for the elimination of smallpox and for reducing the incidence of other viral diseases. It should be possible to control many viral diseases, but with some the problem is more difficult. New technologies and a better understanding of the immune system are helping with this task.

KEY POINTS

- *Interferons* provide a key innate defence to viral infection. Their overall effect is to inhibit viral replication.
- Antibodies act on extracellular virus to prevent establishment of infection. In view of their intracellular replication, viruses are particularly targeted by cell-mediated immunity.
- Many of the symptoms and signs of virus infection reflect immune responses to viral antigens (immunopathology) rather than direct damage due to viral replication.
- Many viral infections induce a temporary immune suppression rendering the host susceptible to bacterial infection, whereas others such as HIV produce a more permanent effect.
- There are many vaccines that protect against viral infections. Smallpox has been eradicated by vaccination and other eradications are feasible.

RECOMMENDED READING

Alcami A, Koszinowski U H 2000 Viral mechanisms of immune evasion. *Immunology Today* 21: 447–455

Biron C A 1999 Initial and innate responses to viral infections – pattern setting in immunity or disease. *Current Opinion in Microbiology* 2: 374–381

Mims C, Nash A, Stephen J 2001 *Mims' Pathogenesis of Infectious Disease*, 5th edn. Academic Press: London

Stark G R, Kerr I M, Williams B R G, Silverman R H, Schreiber R D 1998 How cells respond to interferons. *Annual Review of Biochemistry* 67: 227–264

Internet site

All the virology on the WWW. http://www.virology.net/garryfavwebindex.html

11 Parasitic infections: pathogenesis and immunity

J. Stewart

By convention the term 'parasitic diseases' refers to those caused by protozoa, worms (helminths) and arthropods (insects and arachnids). Such parasites affect many hundreds of millions of people in tropical parts of the world and are responsible for many severe and debilitating diseases (see Chs 61–63).

These infections are associated with a broad spectrum of effects. Some are due to the parasites themselves and others are a consequence of the host response to the invader. The nature and extent of the pathological effects are dependent upon the site and mode of infection, and also on the level of the parasite burden.

The development of protective immunity to such parasites is more complicated than that to bacteria and viruses because of the complicated life cycles of the parasites involved.

PATHOGENIC MECHANISMS

As with other infectious agents, the site occupied by a parasite is important. A host will survive with a large number of lung flukes, *Paragonimus* spp., in the lungs, but infection in the brain may cause far more serious effects. The severity of disease depends not only on the degree of infection but also on the physiological state of the host. A lowering of general health, due to malnutrition for example, predisposes to more serious consequences following infection by parasites.

Mechanical tissue damage

Physical obstruction of anatomical sites leading to loss of function can be a major component of the diseases caused by parasites. The intestinal lumen can be blocked by worms such as *Ascaris lumbricoides* or tapeworms, and filarial parasites (*Wuchereria bancrofti* and *Brugia malayi*) can obstruct the flow of lymph through lymphatics.

Intestinal infection with the tapeworm *Taenia solium* is usually of little consequence, but the eggs may develop into larvae (cysticerci) in humans, causing cysticercosis.

The cysticerci can be found in muscle, liver, eye or, most dangerously, the brain. Hydatid cysts, the larval stage of the dog tapeworm (*Echinococcus granulosus*) in man, may reach volumes of 1–2 litres, and such masses can cause severe damage to an infected organ.

Physiological effects

Large numbers of *Giardia* spp. covering the walls of the small intestine can lead to malabsorption, especially of fats. Competition by parasites for essential nutrients leads to host deprivation. Thus, depletion of vitamin B_{12} by the tapeworm *Diphyllobothrium latum* sometimes leads to pernicious anaemia. Other forms of anaemia result from blood loss, especially in hookworm infection, and from red blood cell destruction in malaria.

Some parasites produce metabolites that may have profound effects on the host. *Trypanosoma cruzi* secretes a neurotoxin that affects the autonomic nervous system. Malaria parasites are thought to produce a metabolite with vasoconstrictor activity.

Tissue damage

The presence of parasites can result in the release of proteolytic enzymes that damage host tissues. Ulceration of the intestinal wall occurs in amoebic dysentery, and trophozoites (the active, motile forms of a protozoan parasite) can penetrate deep into the wall of the intestine to reach the blood and hence the liver, lungs and brain where secondary amoebic abscesses may occur. The skin damage caused by skin-penetrating helminths, such as *Strongyloides stercoralis* and hookworms, can also permit entry of other infectious agents.

The host reaction to parasites and their products can evoke immunological reactions that may lead to secondary damage to host tissues. This is seen in schistosomiasis, where the host response to parasite eggs in tissues leads to the formation of a granuloma with subsequent tissue destruction through fibrosis. Other types of hypersensitivity reaction can be generated in various parasite infections (see below).

IMMUNE DEFENCE MECHANISMS

The large size of parasites means that they display more antigens than bacteria or viruses to the immune system. When the parasite has a complicated life cycle, some of these antigens may be specific to a particular stage of development. Parasites have evolved to be closely adapted to the host, and most parasitic infections are chronic and show a degree of host specificity. For example, the malaria parasites of man, birds and rodents are confined to their own particular species. An exception to this is *Trichinella spiralis*, which is able to infect many animal species.

In the natural host there is no single defence mechanism that acts in isolation against a particular parasite. In turn the parasite will have evolved a number of strategies to evade elimination. In general terms, cell-mediated immune mechanisms are more effective against intracellular protozoa, whereas antibody, with the aid of certain effector cells, is involved in the destruction of extracellular targets. Again, because of the life cycles of some parasites, either cell-mediated or humoral immunity may be of greater importance at different states of their development.

Innate defences

Several of the innate or natural defence mechanisms that are active against bacteria and viruses are also effective against parasitic infections. The physical barrier of the skin protects against many parasites but is ineffective against those transmitted by a blood-feeding insect. In addition, other parasites, such as schistosomes, have evolved active mechanisms for penetrating intact skin. Certain individuals are genetically less susceptible to certain parasites. Thus, individuals with the sickle cell trait have a genetic defect in their haemoglobin that causes a mechanical distortion in their erythrocytes. This somehow leads to the destruction of intracellular malaria parasites. The Duffy blood group antigen is the attachment site for one of the plasmodium parasites. Individuals who lack this determinant are therefore protected from malaria caused by *Plasmodium vivax*.

Several other non-specific host defence mechanisms are involved in the control of parasitic infections. These include direct cellular responses by monocytes, macrophages and granulocytes, and by natural killer cells. The byproducts of acquired immune reactions enhance the antiparasitic activity of these cells. For example, certain protozoan parasites infect macrophages. In particular, *Leishmania* spp. are obligate parasites of mononuclear phagocytes, and are completely dependent upon macrophages, where they survive in the phagolysosome. The parasite appears to be able to survive within non-stimulated resident macrophages, whereas they are destroyed in activated macrophages.

Complement, through activation by the alternative pathway, is active against a number of parasites, including adult worms and active larvae of *T. spiralis* and schistosomula of *Schistosoma mansoni*. The spleen is thought to be active in the elimination of intracellular parasites, as its filtering of infected erythrocytes is thought to remove intracellular plasmodium.

Macrophages

Macrophages play an important role in the elimination and control of parasitic protozoa and worms. They secrete monokines, such as interleukin (IL)-1, tumour necrosis factor and colony-stimulating factors that affect not only T cells and antibody production but also granulocytes. However, other monokines, for instance prostaglandins, are immunosuppressive. Macrophages are also phagocytic cells and function as such in the elimination of parasites. In the same way as in the eradication of other infectious agents, opsonins greatly enhance this process. Once internalized, the parasite is killed, by oxygen-dependent and -independent mechanisms, and digested. Many of the molecules produced by macrophages are cytotoxic, and when produced in close proximity to a parasite will kill it. Specific antibody, immunoglobulin (Ig) G and IgE, can mediate the attachment of the macrophage to the surface of parasites that are too large to internalize but are vulnerable to antibody-dependent cell-mediated cytotoxicity. Acting as antigen-presenting cells, they can aid elimination by helping in the initiation of an immune response.

In addition to lymphokines some products of parasites themselves, such as those produced by *Tryp. brucei* and the malaria parasite, can cause macrophage activation. These may be direct effects or result from the production of monokines such as tumour necrosis factor.

In some parasitic infections the immune system is unable to eradicate the offending organism. The body reacts by trying to isolate the parasite within a granuloma. In this situation there is chronic stimulation of those T cells specific for antigens on the parasite. The continual release of lymphokines leads to macrophage accumulation, release of fibrogenic factors, stimulation of granuloma formation and, ultimately, fibrosis. Granuloma formation around schistosome eggs can occur in the liver, and this response is thought to benefit the host by isolating host cells from the toxic substances produced by the eggs. It can, however, lead to pathological consequences if the damage to the liver leads to loss of liver function.

Granulocytes

Neutrophils and eosinophils are thought to play a role in the elimination of protozoa and worms. The smaller parasites can be phagocytosed and destroyed by both oxygen-dependent and -independent processes. The phagocytic capacity of neutrophils is superior to that of eosinophils. Both cell types possess receptors for the Fc portion of immunoglobulin and for various complement components, so the presence of opsonins increases phagocytosis. Extracellular destruction of large parasites can occur by antibody-dependent cell-mediated cytotoxicity.

Neutrophils are attracted to sites of inflammation and will clear the offending parasite. They have been reported to be more effective than eosinophils at eliminating several species of nematode, including *T. spiralis*, although the relative importance of the two cell types may depend on the class of antibody present.

Eosinophilia and high levels of IgE are characteristics of many parasitic worm infections. It has been suggested that eosinophils are especially active against helminths, and IgE-dependent degranulation of mast cells has evolved to attract these cells to the site where the parasite is localized. The eosinophilia is T cell dependent and the lymphokines that induce production of the cells also cause an increase in their activation state. These effector cells are attracted to the site by chemotactic factors produced by mast cells (see below). Once at the site they degranulate in response to perturbation of their cell membrane induced by antibodies and complement bound to the surface of the parasite. The toxic molecules are therefore released on to the surface of the target and cause its destruction.

Mast cells

The mediators stored and produced by mast cells play an important role in eliminating worm infections. Parasite antigens cause the release of mediators from mast cells; these molecules induce a local inflammatory response. Chemotactic factors are produced and attract eosinophils and neutrophils. Thus the IgE-dependent release of mast cell products helps in expulsion of the worm. The number of mucosal mast cells rises during a parasitic worm infestation due to a T cell-dependent process.

Platelets

Platelet activation results in the release of molecules that are toxic to various parasites, including schistosoma, *Toxoplasma gondii* and *Tryp. cruzi*. The release process does not require antibody, although IgE-dependent cytotoxicity is possible, but seems to involve acute-phase proteins. The cytotoxic potential of platelets is enhanced by various cytokines, including γ-interferon and tumour necrosis factor.

Acquired immunity

An individual with a parasitic infection mounts a specific response against the invading parasite. These immune reactions generate antibody and effector T cells directed against specific parasite antigens. Memory B and T cells are also produced. For a number of reasons, described below, much acquired immunity is ineffective in protecting the host against recurrent infection. However, in certain cases, such as amoebiasis and toxoplasmosis, immunity to reinfection is fairly complete. In schistosomiasis, the presence of surviving adult forms protects against further infections. However, this may be an effect of the parasite and not of the host.

Antibody

The specific immune response to parasites leads to the production of antibody. Infection by protozoan parasites is associated with the production of IgG and IgM. With helminths there is, in addition, the synthesis of substantial amounts of IgE. IgA is produced in response to intestinal protozoa, such as *Entamoeba histolytica* and *Giardia lamblia*.

In addition to these specific T-dependent responses, a non-specific hypergammaglobulinaemia is present in many parasitic infections. Much of this non-specific antibody is the result of polyclonal B cell activation by released parasite antigens acting as mitogens. This response is ineffective at counteracting the parasite and can enhance the pathogenicity by causing the production of autoantibodies, and may actually lead to a diminished specific response due to B cell exhaustion. It has also been reported that some parasite molecules are T cell mitogens. This could lead to the generation of autoreactive T cells or activation of suppressor responses.

There are a number of mechanisms by which specific antibody can provide protection against and control parasitic infections (Table 11.1). As with viral infections, antibody is effective only against extracellular parasites and where parasite antigens are displayed on the surface of infected cells. Antibody can neutralize parasites by combining with various surface molecules, blocking or interfering with their function. The binding of antibody to an attachment site stops the infection of a new host cell. The agglutination of blood parasites by IgM may occur, leading to the prevention of spread, as in the acute phase of infection with *Tryp. cruzi*. Toxins and enzymes produced by certain parasites add to their pathogenicity, and antibodies that inhibit these molecules protect the host from damage and also affect the infection process directly. Intracellular parasites have evolved a number of mechanisms that allow them to survive in this environment. Antibodies against the molecules that aid the parasite in these activities, for

Table 11.1 Humoral defence mechanisms against parasite infections

Mechanism	Effect	Parasite
Neutralization	Blocks attachment to host cell	Protozoa
	Acts to inhibit evasion mechanisms of intracellular organisms	Protozoa
	Binding to toxins or enzymes	Protozoa and worms
Physical interference	Obstructs orifices of parasite	Worms
	Agglutination	Protozoa
Opsonization	Increases clearance by phagocytes	Protozoa
Cytotoxicity	Complement-mediated lysis	Protozoa and worms
	Antibody-dependent cell-mediated cytotoxicity	Protozoa and worms

instance to escape from endosomes or inhibit lysosomal fusion, lead to removal of the intruder by the phagocyte. Parasitic worms are multicellular organisms with defined anatomical features that are responsible for functions such as feeding and reproduction. Antibodies that block particular orifices (e.g. oral and genital) interfere with critical physiological functions and may cause starvation or curtail reproduction.

Antibodies can bind to the surface of parasites and cause direct damage, or interact with complement leading to cell lysis. Antibody also acts as an opsonin and hence increases uptake by phagocytic cells. In this context, complement activation leads to enhanced ingestion due to complement receptors. Macrophage activation leads to the expression of increased Fc and complement receptors, so phagocytosis is enhanced in the presence of macrophage-activating factors. Phagocytes play an important role in the control of infections by *Plasmodium* spp. and *Tryp. brucei*.

Antibody-dependent cell-mediated cytotoxicity has been shown to play a part in infections caused by a number of parasites, including *Tryp. cruzi, T. spiralis, S. mansoni* and filarial worms. The effector cells – macrophages, monocytes, neutrophils, eosinophils and natural killer cells – bind to the antibody-coated parasites by their Fc and complement receptors. Close apposition of the effector cell and the target is necessary because the toxic molecules produced are non-specific and may damage host cells. A major basic protein from eosinophils damages the tegument of schistosomes and other worms, causing their death. It appears that different cell types and immunoglobulin isotypes are active against different developmental stages of parasites. Eosinophils are more effective at killing newborn larvae of *T. spiralis* than other cells, whereas macrophages are very effective against microfilariae.

T cells

The importance of T cells in countering protozoan infections has been shown using nude (athymic) or T cell-depleted mice, which have a reduced capacity to control trypanosomal and malarial infections. The transfer of spleen cells, especially T cells, from immune animals gives protection against most parasitic infections. The type of T cell that is effective depends on the parasite. CD4+ T cells transfer protection against *Leishmania major* and *L. tropica*, and may be necessary for the elimination of other parasites. The intracellular parasite of cattle, *Theileria parvum*, is destroyed by cytotoxic T (T_c) cells.

CD4+ T cells may act by providing help in antibody production, but they also secrete various lymphokines that interact with other effector cells. CD8+ cells may be cytotoxic in certain situations, but these cells also produce a variety of lymphokines. IL-2 production has been shown to be deficient during parasitic infections, such as malaria and trypanosomiasis. Administration of IL-2 to mice infected with *Tryp. cruzi* reduces parasitaemia and increases survival.

Colony-stimulating factors (e.g. IL-3 and granulocyte–monocyte colony-stimulating factor) are also produced by activated T cells. These molecules act on myeloid progenitors in the bone marrow, causing increased production of neutrophils, eosinophils and monocytes. They also increase the activity of these cells; the monocytosis and splenomegaly in malaria are caused by these T cell-derived molecules. The accumulation of macrophages in the liver as granulomata in schistosomiasis and the eosinophilia that is characteristic of worm infestations are also T cell-dependent phenomena.

In certain cases the production of lymphokines may have adverse effects. *Leishmania* infects macrophages, and the release of molecules that stimulate the production of more host cells may potentiate the infection.

γ-Interferon does not inhibit or kill parasites directly, although multiplication of the liver stages of the malaria parasite is inhibited by γ-interferon, possibly through interaction with its receptor on the surface of hepatocytes. γ-Interferon is a potent macrophage activation factor and is probably involved in the resistance and elimination of intracellular parasites, such as *Toxoplasma gondii* and *Leishmania* spp. Activated macrophages are more effective killers and can destroy intracellular parasites before they establish themselves within the cell.

Table 11.2 Parasite escape mechanisms

Escape mechansim	Organisms
Intracellular habitat	Malaria parasites, trypanosomes and *Leishmania* spp.
Encystment	*Toxoplasma gondii* and *Trypanosoma cruzi*
Resistance to microbicidal products of phagocytes	*Leishmania donovani*
Masking of antigens	Schistosomes
Variation of antigen	Trypanosomes and malaria parasites
Suppression of immune response	Most parasites (e.g. malaria parasites, *Trichinella spiralis* and *Schistosoma mansoni*)
Interference by antigens	Trypanosomes
Polyclonal activation	Trypanosomes
Sharing of antigens between parasite and host (molecular mimicry)	Schistosomes
Continuous turnover and release of surface antigens of parasite	Schistosomes

EVASION MECHANISMS

All animal pathogens, including parasitic protozoa and worms, have evolved effective mechanisms to avoid elimination by host defence systems (Table 11.2).

Seclusion

Many parasites inhabit cells or anatomical sites that are inaccessible to host defence mechanisms. Those that attempt to survive within cells avoid the effects of antibody but must possess mechanisms to avoid destruction if the cell involved is capable of destroying them. *Plasmodium* spp. inhabit erythrocytes, whereas toxoplasmas are less selective and infect non-phagocytic cells as well as phagocytes. A number of different ways of avoiding destruction in macrophages have evolved. *Leishmania donovani* amastigotes are able to survive and metabolize in the acidic environment (pH 4–5) found in phagolysosomes, and *Tox. gondii* is able to inhibit the fusion of lysosomes with the parasite-containing phagosome.

L. major has a similar escape mechanism by attaching to a phagocyte complement receptor (CR1) that does not trigger the respiratory burst. Activation of the complement system by protozoan parasites seems to be a common mechanism for achieving attachment to target cells. *Tryp. cruzi* trypomastigotes can infect T cells of both the CD4 and CD8 subsets, and may be similar to retroviruses in using receptor molecules on the T cell surface for penetration.

The effectiveness of macrophages in the elimination of *Tryp. cruzi* depends upon the stage of development of the parasite. Trypomastigotes are able to escape from the phagocytic vacuole and survive in the cytoplasm, whereas epimastigotes do not escape and are killed. Macrophages are also the preferred habitat of *Leishmania* spp., which multiply in the phagolysosome where they are resistant to digestion.

In an immune host these evasion mechanisms are less effective because of the presence of antibody and lymphokines. The ability to resist complement destruction also appears to be important. For example, *L. tropica* is easily killed by complement and causes only a localized self-healing lesion in the skin, whereas a disseminating, often fatal, disease is seen with *L. donovani*, which is ten times more resistant to complement killing. Large parasites such as helminths cannot infect individual cells; however, they can still achieve anatomical seclusion. *T. spiralis* larvae avoid the immune system by encysting in muscle; intestinal nematodes live in the lumen of the intestine.

Evasion

Parasites may avoid recognition by:

- Antigenic variation
- Acquiring host-derived molecules.

African trypanosomes have the capacity to express more than 100 different surface glycoproteins. By producing novel antigens throughout their lives, these parasites continuously evade the immune system. By the time the host has mounted a response against each new antigen, the parasite has changed again. Plasmodia pass through several discrete developmental stages, each with its own particular antigens. A similar situation is seen in certain helminths such as *T. spiralis*. As a result, each new stage of the life cycle is seen by the host as a 'new' infective challenge.

A number of parasites are known to adsorb host-derived molecules on to their surface. This is thought to mask their own antigens and enable them to evade immunological attack.

Parasitic protozoa and worms also use devices to avoid immune destruction. Certain parasites retain a surface coat, or glycocalyx, that blocks direct exposure of its surface antigens.

Immunosuppression

Parasites are not always able to evade detection and many have evolved mechanisms to suppress or divert immune reactions. Some parasites produce or generate molecules that act against cells of the immune system. Thus, the larvae of *T. spiralis* produce a molecule that is cytotoxic to lymphocytes, and schistosomes can cleave a peptide from IgG, thereby decreasing its effectiveness. During many parasitic infections a large amount of antigenic material is released into the body fluids, and this may inhibit the response to or divert the response away from the parasite. High antigen concentrations can lead to tolerance by clonal exhaustion or clonal deletion. The immune complexes formed can also inhibit antibody production by negative feedback via Fc receptors on plasma cells. Many of these released molecules are polyclonal activators of T and B cells. This leads to the production of non-specific antibody, impairment of B cell function and immunosuppression. It has also been proposed that many parasites can cause unresponsiveness by activating immune suppressor mechanisms.

In many cases the immunosuppression has been attributed to macrophage dysfunction associated with antigen overload or the presence of intracellular parasites. In addition to the non-specific immunosuppression there can be parasite-specific effects. Mice infected by *Leishmania* spp. show antigen-specific depression of lymphokine production. As this genus inhabits macrophages and is partly controlled by activation of these cells by lymphokines, the effect is a diminished response against the pathogen.

Schistosomes have a receptor for part of the antibody molecule. They also release several proteases that cleave antibody molecules and release products that prevent macrophage activation. A schistosome-derived inhibitory factor suppresses T cell activity and is believed to allow other parasites to survive the effects of T cells and may explain the inefficiency of cytotoxic T cells in damaging the parasite.

When tested in a lymphocyte proliferation assay, peripheral blood lymphocytes of patients infected with *Plasmodium falciparum* are unresponsive to antigen prepared from the parasite, and in nearly 40% of the patients this persists for more than 4 weeks. Patients infected with *P. falciparum* show a suppression of lymphocyte reactivity that is not related to the degree of parasitaemia or severity of the clinical illness. The depressed lymphocyte reactivity is associated with a loss of both CD4$^+$ and CD8$^+$ lymphocytes from the peripheral blood. Once the parasite has been cleared, the response returns to normal. An even more sophisticated strategy has been evolved by *Leishmania mexicana* and *L. donovani*, which use IL-2 to stimulate their own growth. Mammalian epidermal growth factor has also been shown to stimulate the growth of certain trypanosomes in vitro.

IMMUNOPATHOLOGY

The immune response to parasites is aimed at eliminating the organisms, but many of the host reactions have pathological effects.

The IgE produced in parasitic worm infections can have severe effects on the host if it stimulates excessive mast cell degranulation (type I hypersensitivity). Anaphylactic shock can occur if a cyst ruptures and releases vast amounts of antigenic material into the circulation of a sensitized individual. Asthma-like symptoms occur in *Toxocara canis* infections when larvae of worms migrate through the lungs.

The polyclonal B cell activation seen with many parasitic infections can give rise to autoantibodies. In trypanosomiasis and malaria antibodies against red blood cells, lymphocytes and deoxyribonucleic acid (DNA) have been detected. Host antigens incorporated into the parasite, as an immune evasion mechanism, may stimulate autoantibody production by giving rise to T cell help and overcoming tolerance. In Chagas' disease about 20% of individuals develop progressive cardiomyopathy and neuropathy of the digestive tract that is believed to be autoimmune in nature. These effects are thought to result from cross-reactivity between antibody or T cells responsive to *Tryp. cruzi* and nerve ganglia.

Immune complex-mediated disease occurs in malaria, trypanosomiasis, schistosomiasis and onchocerciasis. The deposition of immune complexes in the kidney is responsible for the nephrotic syndrome of quartan malaria.

Enlargement of the spleen and liver in malaria, trypanosomiasis and visceral leishmaniasis is associated with increased numbers of macrophages and lymphocytes in these organs. The liver, renal and cardiopulmonary pathology of schistosomiasis is related to cell-mediated responses to the worm eggs. Symptoms similar to those seen in endotoxaemia induced by Gram-negative bacteria are found in the acute stages of malaria.

The non-specific immunosuppression discussed above may explain why individuals with parasite infections are especially susceptible to bacterial and viral infections.

VACCINATION

No effective vaccine for humans has so far been developed against parasitic protozoa and worms, mainly because of the complex parasite life cycles and their sophisticated adaptive responses. As protection depends in many cases on both antibody and cell-mediated reactions, a vaccine must induce long-lived B and T cell immunity. In addition, because recognition by T cells is genetically restricted, the vaccine preparation must stimulate T cells from most haplotypes, preferably without suppressor epitopes. Because of the immunopathology seen in many parasite infections, antigens that induce a potentially damaging response must be avoided.

A much better understanding of the biological mechanisms underlying the natural history of parasitic diseases is required before it will be possible to control these globally important diseases.

KEY POINTS

- Parasites cause disease by diverse mechanisms including mechanical damage, physiological disturbance, tissue destruction and immunopathology.
- Some parasitic worm infestations are associated with raised levels of IgE and eosinophilia. These responses may provide some protection.
- Some parasites may evade immune responses by seclusion away from the immune system, by antigenic variation, by acquiring host-derived molecules or by immunosuppression.
- Despite major efforts, particularly those directed towards malaria, there are no available effective vaccines against parasitic infections.

RECOMMENDED READING

Mims C, Nash A, Stephen J 2001 *Mims' Pathogenesis of Infectious Disease*, 5th edn. Academic Press, London

Internet site

DPDx. Parasites & Health. http://www.dpd.cdc.gov/dpdx/HTML/Para_Health.htm

Immunity in bacterial infections

J. Stewart

Modern medical science has managed to subdue many of the classical infectious diseases, but has helped to create new ones that result from interference with normal host defence mechanisms, consequent upon medical and surgical procedures such as chemotherapy, catheterization, immunosuppression and irradiation. Infections that develop in this way are known as *iatrogenic* (physician-induced) diseases.

It is important to differentiate infection from disease. A host may be infected with a particular micro-organism and be unaware of its presence. If the microbe reproduces itself to such an extent that toxic products or sheer numbers of organisms begin to harm the host, a disease process has developed. Potentially pathogenic bacteria such as pneumococci, streptococci and salmonellae are found in the nose, throat or bowel; this is known as the carrier state and is a source of infection to other individuals. These bacteria may also cause disease in the carrier if they enter a vulnerable tissue.

HOST DEFENCES

Few organisms can penetrate intact skin, and the various other innate defence mechanisms are extremely efficient at keeping bacteria at bay. When bacteria do gain access to the tissues, the ability of the host to limit damage and eliminate the microbe depends on the generation of an effective immune response against microbial antigens. In most cases the host defences are directed against external components and secreted molecules. Bacteria are surrounded by a cytoplasmic membrane and a peptidoglycan cell wall. Associated with these basic structures there can be a variety of other components such as proteins, capsules, lipopolysaccharide or teichoic acids. There are also structures involved in motility or adherence to the cells of the host (see Ch. 2). These are some of the components to which the immune system directs its response. In general, peptidoglycan is attacked by lysosomal enzymes, and the outer lipid layer of Gram-negative bacteria by cationic proteins and complement. Specific antibodies can bind

to flagella or fimbriae, affecting their ability to function properly, and can inactivate various bacterial enzymes and toxins. Antibodies therefore interfere with many important bacterial processes but, ultimately, phagocytes are needed to destroy and remove the bacteria (Fig. 12.1). In some situations cell-mediated responses are required.

Inflammation

Having successfully avoided the innate immune mechanisms that protect the individual (mechanical barriers, antibacterial substances and phagocytosis, described in Chapter 9), a bacterium starts to proliferate in the tissues. The presence of bacteria-specific molecules is recognized by pattern recognition receptors, leading to cytokine release. These cytokines, along with tissue-damaging toxic bacterial products, trigger an inflammatory reaction. The resulting increase in vascular permeability leads to an exudation of serum proteins, including complement

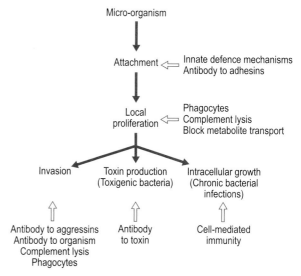

Fig. 12.1 Scheme showing the progress of infection and immunological defence mechanisms.

components, antibodies and clotting factors, as well as phagocytic cells. The phagocytes are attracted to the site of inflammation by chemotactic factors. Anaphylatoxins generated by complement activation further increase vascular permeability and encourage exudation of fluid and cells at the site of inflammation. Many of these mediators also cause vasodilatation, thereby increasing blood flow to the area.

Many types of micro-organism (e.g. staphylococci and streptococci) are dealt with effectively by the phagocytes. The intensity and duration of the inflammatory process that is stimulated depends on the degree of success with which the micro-organism initially establishes itself. This, in turn, depends on the extent of the injury, the amount of associated tissue damage and the number and type of micro-organisms introduced. A localized abscess may arise at the site of infection.

If bacteria are not eliminated at the site of entry and continue to proliferate, they pass via the tissue fluids and lymphatics to the draining lymph node, where a specific acquired immune response is generated. The antibody and effector cells generated will leave the node to return to the area of infection to eliminate the bacteria. Some capsule micro-organisms, such as pneumococci, are able to resist phagocytosis and are not dealt with effectively until large amounts of antibody have been made. These 'mop up' the released capsular polysaccharide, and phagocytosis occurs. Other micro-organisms produce exotoxins, and effective immunity to exotoxins requires the development of specific antibodies against the toxin (i.e. *antitoxin*).

The types of infection described above are usually referred to as *acute* infections, and contrast with the protracted or *chronic* infections usually induced by bacteria that have adapted to survive within the cells of the host. Included among these are tuberculosis and leprosy, brucella infections and listeriosis. In these infections, cell-mediated immunity plays a predominant part in the final elimination of the micro-organism.

Humoral immunity

The attachment of a micro-organism to an epithelial surface is a prerequisite for the development of an infectious process (see p. 156). A first line of attack by antibody could be to inhibit colonization by stopping attachment.

The immunoglobulin IgA can stop colonization of the mucosal surface if it interferes with the attachment molecules (*adhesins*) present on the bacterial surface. IgA does not activate complement very efficiently; therefore, an inflammatory reaction is not stimulated. Damage to the gut wall during an inflammatory reaction would allow the entry of many potential pathogens into vulnerable tissues.

Many micro-organisms owe their pathogenic abilities to the production of *exotoxins*. Among diseases dependent on this type of mechanism are diphtheria, cholera, tetanus and botulism. Antibodies acquired by either immunization or previous infection, or given passively as antiserum, are able to *neutralize* bacterial toxins.

Many bacterial exotoxins are enzymes, and protective antibody can prevent interaction of the enzyme with its substrate. The antibody can bind directly to the active site of the enzyme or to adjacent residues and inhibit by steric hindrance. Antibody may also act by stopping activation of a zymogen into an active enzyme, interfere with the interaction between the toxin and its target cell, or bind to a site on the molecule, causing a conformational change that destroys the enzymatic activity.

The direct binding of antibody to a bacterium can interfere with normal bacterial functioning in numerous ways. Antibody can kill bacteria on its own or in conjunction with host factors and cells. To survive and multiply, bacteria must ingest nutrients and ions mainly by specific transport systems. Antibodies that affect the activity of specific transport systems will deprive the bacteria of their energy supply and other essential chemicals. Some bacteria are invasive, moving into the tissues aided by enzymes that they produce. Invasion can also be inhibited by antibody that attaches to the flagella of the micro-organism in such a way as to affect its motility. Antibodies can agglutinate bacteria, and formation of the aggregate will impede spread of the organism.

In addition, the formation of an immune complex of bacteria and antibody will stimulate phagocytosis and complement activation.

When a particle is coated with antibody, a large number of Fc portions are exposed to the outside. This increases the chance that the particle will be held in contact with the phagocyte long enough to stimulate phagocytosis. The interaction of multiple ligands increases the overall affinity of the binding and, if antibody and complement components are present on the same particle, the binding is even stronger.

The bacteria are internalized and attacked by the oxygen-dependent and oxygen-independent killing mechanisms within the phagocyte. Phagocytes are also responsible for the removal and digestion of bacteria that have been killed extracellularly. Bacteria are susceptible to the lytic action of complement, which may be activated by bacterial components. The presence of antibody on the bacterial cell surface further stimulates the activation of complement. In certain circumstances, antibodies in conjunction with other bactericidal molecules lead to more efficient bacterial destruction. Gram-negative organisms are normally resistant to the action of lysozyme, probably because of the lipopolysaccharide component of the cell. The action of antibody and complement is thought to expose the underlying cell wall, which is then attacked by the lysozyme.

Cell-mediated immunity

Ultimately, all bacteria will be engulfed by a phagocyte, either to be killed or removed after extracellular killing. The host defence mechanisms of macrophages and monocytes can be enhanced by various activating stimuli, including microbial products such as the muramyl dipeptide found in many cell walls and trehalose dimycolate from *Mycobacterium tuberculosis*. The chemotactic formyl-methionyl peptides have been shown to increase the activities of various macrophage functions. The linkage of chemotaxis to activation has the advantage that the cell being attracted to the site of tissue injury will be better equipped to deal with the insult. Endotoxins present in the cell wall of Gram-negative bacteria and various carbohydrate polymers, such as β-glucans, are potent macrophage activators.

The immune system is also active in the production of macrophage-activating factors. In particular, lymphokines, produced by T lymphocytes, are often required to potentiate bacterial clearance both by attracting phagocytes to the site of infection and by activating them. The most important activator is γ-interferon, although tumour necrosis factor and colony-stimulating factors have also been implicated (see Ch. 9). All T lymphocytes produce some lymphokines when stimulated, and the balance of the different factors produced dictates the effect on surrounding cells. The overall effect of the lymphokines is to increase the effectiveness of host defence mechanisms, but if these molecules are produced in excess or to an inappropriate signal then a type IV hypersensitivity reaction can occur.

EVASION

Once a micro-organism becomes established in the tissues, having escaped the innate defence mechanisms, it can often make use of a number of evasion strategies that protect it from the immune reactions of the host. Pathogenic bacteria produce a rather ill-defined group of bacterial products called *aggressins* and *impedins*, possession of which is associated with virulence (see Ch. 13). If antibody to such substances is present, the pathogenicity of the micro-organism is likely to be reduced.

Intracellular bacteria

Bacteria that can survive and replicate within phagocytic cells are at an advantage as they are protected from host defence mechanisms. Some micro-organisms reside intracellularly only transiently, whereas others spend most, if not all, of their life inside cells. An intracellular lifestyle demands potent evasion mechanisms to sur-

vive this hostile environment. Some bacteria, such as *Mycobacterium leprae*, have become so accustomed to their intracellular environment that they can no longer live in the extracellular space.

Listeria monocytogenes, the causative agent of listeriosis, can survive and multiply in normal macrophages, but is killed within macrophages activated by lymphokines released from T lymphocytes. Listeriosis is most commonly seen in immunocompromised patients, pregnant women and neonates, in whom a lack of adequate T cell-derived macrophage-activating factors is probably the critical factor. *Salmonella* spp. and *Brucella* spp. can also survive intracellularly. Mycobacteria have a waxy cell wall that is very hydrophobic. This external surface is very resistant to lysosomal enzymes and persists for a long time even when the bacteria have been killed. In addition, these micro-organisms have evolved other strategies to evade destruction. *M. tuberculosis* secretes molecules that inhibit lysosome/phagosome fusion, and *M. leprae* can escape from the phagosome and grow in the cytoplasm. The cell wall of both of these bacteria contains lipoarabinomannan, which blocks the effects of γ-interferon on macrophages.

T lymphocytes represent the major host defence against intracellular pathogens. In many cases the bacteria themselves do not directly harm the host: the pathogenesis is caused by the immune response. After invasion of the host, intracellular bacteria are taken up by macrophages, evade intracellular killing and multiply. During intracellular replication some microbial molecules are processed and presented on the surface of the infected cell in association with major histocompatibility complex (MHC) gene products. The exact nature of the microbial molecules and the processing mechanism are not known. The processing appears to produce mainly antigen fragments in association with MHC class II molecules. This complex on the cell surface is recognized by specific CD4+ T cells, which are stimulated to release lymphokines. These molecules in turn activate the macrophage so that the intracellular bacteria are killed.

It has also been proposed that peptides derived from the bacteria can become associated with MHC class I molecules. This complex leads to destruction of the infected cell by CD8+ T cells. These cells also produce lymphokines that can aid in the elimination of the infection. The lymphokines produced attract blood monocytes to the site and activate them. If these newly recruited cells take up the released mycobacteria they are more likely to destroy them as they will be in an activated state. The accumulation of macrophages also causes the formation of a granuloma, which will prevent dissemination of the bacteria to other sites in the body. As conditions for the survival of the pathogens become less suitable, they stop replicating and die. Some intracellular bacteria infect cells that are unable to destroy them. In this situation a more aggressive immune response may be needed to remove the pathogen and this can cause pathogenic damage unless kept under control (see below).

IMMUNOPATHOLOGY

The immune response to an organism leads to some tissue damage through inflammation, lymph node swelling and cell infiltration. Sometimes the damage caused by the immune system is very severe, leading to serious disease and death. Rheumatic fever can follow group A streptococcal infections of the throat and is believed to be due to antibodies formed against a streptococcal cell wall component cross-reacting with cardiac muscle or heart valve. Myocarditis develops a few weeks after the throat infection and can be restimulated if the patient is reinfected with different streptococci.

Immune complex disease (type III hypersensitivity) is frequently associated with bacterial infection. Infective endocarditis caused by staphylococci and streptococci is associated with circulating complexes of antibody and bacterial antigen. Detection of these complexes can be helpful in diagnosis, but they may lead to joint and kidney lesions, vasculitis and skin rashes. Immune complexes may play a role in the pathogenesis of leprosy, typhoid fever and gonorrhoea.

Effects of endotoxin

Endotoxin interacts with cells and molecules of inflammation, immunity and haemostasis.

- Fever is induced by interleukin-1, produced by the liver in response to endotoxin, acting on the temperature-regulating hypothalamus.
- The action of lipopolysaccharide on platelets and activation of Hageman's factor causes disseminated intravascular coagulation with ensuing ischaemic tissue damage to various organs.
- Septic shock occurs during severe infections with Gram-negative organisms when bacteria or lipopolysaccharide enter the bloodstream.

Endotoxin acts on neutrophils, platelets and complement to produce, both directly and through mast cell degranulation, vasoactive amines that cause hypotension. The mortality rate is very high.

Recognition of endotoxin through Toll-like receptor 4 (see p. 109) causes macrophages to produce large quantities of potent cytokines such as interleukin-1, tumour necrosis factor and colony-stimulating factors. Endotoxin also causes polyclonal activation of B cells and can stimulate natural killer cells and other cell types to produce γ-interferon. A substantial part of the pathogenesis of endotoxin shock

is probably due to the production of these molecules by cells of the immune system. In small amounts, endotoxin may actually be beneficial to the host, but when present in excess the results can be disastrous.

Mycobacterial disease

Activated macrophages secrete a variety of biologically active molecules, including:

- proteases
- tumour necrosis factor
- reactive oxygen intermediates that are harmful to the surrounding tissue.

Tissue destruction is an inevitable side effect of this important mechanism of resistance. In the acute phase of a response this is likely to be tolerated; however, in the case of resistant organisms such as mycobacteria, the process may become chronic and the tissue destruction extensive. Mycobacterial components are still able to stimulate a response after the bacterium has been killed, because they persist for a long time, adding to the tissue damage.

There is evidence that lysis of infected cells may also occur. At first sight this may appear beneficial. Such a direct effect may be particularly relevant in the case of obligate intracellular pathogens such as *M. leprae*. Release into the hypoxic centre of a productive granuloma may also be fatal for *M. tuberculosis*, which is highly sensitive to low oxygen pressures. The same cytolytic event may also result in microbial discharge from the granuloma into surrounding capillaries or alveoli, and hence facilitate dissemination to other parts of the body or to other individuals. Lysis of infected cells causes tissue destruction, the severity of the effects depending on the importance of the tissue involved. *M. leprae* infects the Schwann cell, an irreplaceable component of the peripheral nervous system. Although the presence of *M. leprae* does not appear to affect the host cell to any extent, the presence of activated macrophages releasing toxic molecules or direct lysis by cytotoxic cells constitutes a major pathological mechanism in leprosy.

If the leprosy bacillus is released from a lysed nonphagocytic cell to be engulfed by an activated macrophage, the bacterium will be eliminated. Therefore, macrophage activation and target cell lysis can be beneficial as well as detrimental to the host.

The cell-mediated response that has the potential to eliminate these infections will give rise to type IV hypersensitivity reactions if the antigen is not removed efficiently. Chronic production of lymphokines causes granuloma formation, which with time can lead to fibrosis and loss of organ function. This type of response is particularly prevalent in patients with tuberculosis.

KEY POINTS

- Phagocytosis and complement activation are key innate defences against bacterial infections associated with acute inflammation.
- Antibodies are notably effective against bacterial cell surface and extracellular virulence factors such as toxins. Cell-mediated immunity is essential for defence against intracellular bacteria.
- Intracellular survival and growth provides one means of immune evasion for bacteria.
- Bacterial lipoplysaccharide and a number of other cell wall components can elicit uncontrolled innate immune responses that are potentially fatal to the host in the form of *septic shock*.

RECOMMENDED READING

Mims C, Nash A, Stephen J 2000 *Mims' Pathogenesis of Infectious Disease*, 5th edn. Academic Press, London

Patrick S, Larkin M J 1995 *Immunological and Molecular Aspects of Bacterial Virulence*. Wiley, Chichester

Internet site

Bayer HealthCare. Infections: prevention and treatment. http://infections.bayer.com/en/index.html

13 Bacterial pathogenicity

D. A. A. Ala'Aldeen

Pathogenicity, or the capacity to cause disease, is a relatively rare quality among microbes. It requires the attributes of *transmissibility* or communicability from one host or reservoir to a fresh host, *survival* in the new host, *infectivity* or the ability to breach the new host's defences, and *virulence*, a variable that is multifactorial and denotes the capacity of a pathogen to harm the host. Virulence (see Ch. 14) in the clinical sense is a manifestation of a complex parasite–host relationship in which the capacity of the organism to cause disease is considered in relation to the resistance of the host.

TYPES OF BACTERIAL PATHOGEN

Bacterial pathogens can be classified into two broad groups, *opportunists* and *primary pathogens*. Both groups have a broad spectrum of virulence capabilities, hence their overlap.

Opportunistic pathogens

These rarely cause disease in individuals with intact immunological and anatomical defences. Only when such defences are impaired or compromised, as a result of congenital or acquired disease or by the use of immunosuppressive therapy or surgical techniques, are these bacteria able to cause disease. Many opportunistic pathogens (e.g. coagulase-negative staphylococci) are part of the normal human flora, carried on the skin or mucosal surfaces where they cause no harm and may actually have a beneficial effect by preventing colonization by other potential pathogens. However, the introduction of these organisms into anatomical sites where they are not normally found, or removal of competing bacteria by the use of broad-spectrum antibiotics, may allow their localized multiplication and subsequent development of disease.

Primary pathogens

These bacteria are capable of establishing infection and causing disease in previously healthy individuals with intact immunological defences. However, they may more readily cause disease in individuals with impaired defences.

The above classification is applicable to the vast majority of pathogens. However, there are exceptions and variations within both categories of bacterial pathogens. Different strains of any bacterial species can vary in their genetic make-up and virulence. For example, the majority of *Neisseria meningitidis* strains are harmless commensals and are considered to be opportunistic bacteria; however, some hypervirulent clones of the organism can cause disease in the previously healthy individual. Conversely, people vary in their genetic make-up and susceptibility to invading bacteria, including meningococci.

VIRULENCE DETERMINANTS

Both opportunistic and primary pathogens possess *virulence determinants* or *aggressins* that facilitate pathogenesis. Possession of a single virulence determinant is rarely sufficient to allow the initiation of infection and production of pathology. Many bacteria possess several virulence determinants, all of which play some part at various stages of the disease process. In addition, not all strains of a particular bacterial species are equally pathogenic. For example, although six separate serotypes of encapsulated *Haemophilus influenzae* are recognized, serious infection is almost exclusively associated with isolates of serotype b. Moreover, even within serotype b isolates, 80% of serious infections are caused by six of more than 100 clonal types.

Different strains of a pathogenic species may cause distinct types of infection, each associated with possession of a particular complement of virulence determinants. Different strains of *Escherichia coli*, for example, cause several distinct gastro-intestinal diseases, urinary tract infections, septicaemia, meningitis and a range of other minor infections (see Ch. 26).

Expression and analysis of virulence determinants

Many pathogens produce an impressive armoury of virulence determinants in vitro. However, relatively early in the study of pathogenesis, it was appreciated that a knowledge of the behaviour of the pathogen in vivo is crucial to an understanding of virulence.

Animal models have been used to compare the virulence of naturally occurring variants differing in the expression of a particular determinant, and have provided much useful information. However, not all human clinical syndromes can be reproduced in animals, and extrapolation from animal studies to man can be misleading. Furthermore, the possibility that observed differences in virulence may be due to additional cryptic phenotypic or genotypic variations cannot always be excluded. Molecular techniques have been used to construct *isogenic* mutants of bacteria that differ only in the particular determinant of interest, and these constructs have allowed more detailed analysis of the role of such components in pathogenesis. More recently, comparative analysis of bacterial genome sequences have revealed the extent of genetic variation, as well as genetic mobility among bacterial strains.

Most studies of bacterial virulence determinants are by necessity performed in model systems in vitro. However, growth conditions in vitro differ significantly from those found in tissues, and as the expression of many virulence determinants is influenced by environmental factors, it is essential that such studies use cultural conditions that mimic as closely as possible those found in the host and that, where possible, confirmatory evidence is obtained that the phenomena observed actually occur during human infection.

Genetic studies have shown that expression of several different virulence determinants in a single bacterium is sometimes regulated in a co-ordinated fashion. Iron limitation, the situation encountered in host tissues, is one environmental stimulus that co-ordinately increases the production of many bacterial proteins, including virulence determinants such as haemolysin of *Esch. coli* and diphtheria toxin of *Corynebacterium diphtheriae*. In other bacteria, such as *Staphylococcus aureus* and *Pseudomonas aeruginosa*, some virulence determinants are expressed exclusively or maximally during the stationary phase of growth. Expression of these factors is associated with the production of inducer molecules or pheromones in the bacterial culture that accumulate as the bacteria grow until a threshold level is reached and gene expression is triggered – a process known as *quorum sensing*. The ability to regulate production of virulence determinants may save energy in situations in which expression is not required (e.g. in the environment) and quorum sensing may be important in establishing

a sufficiently large population of bacteria in tissue to guarantee survival of the infecting organism. It is also clear that most organisms express some proteins only when in direct contact with host cells.

Molecular studies have also allowed mechanisms of transmission of virulence determinants to be investigated (see Ch. 6). Virulence determinants encoded by genomic DNA sequences, plasmids, bacteriophages and transposons have been reported. It is interesting that, in nature, these genetic elements can move between related organisms, transfer virulence factors (e.g. toxins) horizontally, and transform the recipient bacteria to more adapted or more virulent pathogens. Apart from these genetic elements, there are other mechanisms by which some bacteria can exchange virulence genes. For example, neisseriae recognize and take up DNA fragments that contain specific sequences (uptake sequences) and incorporate them into their own genomes. In this way they can either vary the structure of an existing gene or, in the process, acquire a new set of genes. The genomes of several bacterial pathogens have been fully sequenced. The data reveal that several bacteria have acquired very large stretches of foreign DNA (often called pathogenicity islands) that contain virulence-related genes. This further demonstrates that the microbial population consists of a vibrant, kinetic and highly interactive community. In this community, bacteria evolve continuously and new pathogens, or old pathogens with newly acquired capabilities, may emerge as a result.

Establishment of infection

Potential pathogens may enter the body by various routes, including the respiratory, gastro-intestinal, urinary or genital tract. Alternatively, they may enter tissues directly through insect bites, or by accidental or surgical trauma to the skin. Many opportunistic and several primary pathogens are carried as part of the normal human flora, which acts as a ready source of infection in the compromised host. For many primary pathogens, however, transmission to a new host and establishment of infection are more complex processes. Transmission of respiratory pathogens, such as *Bordetella pertussis*, may require direct contact with infectious material, as the organism cannot survive for any length of time in the environment. Sexually transmitted pathogens such as *Neisseria gonorrhoeae* and *Treponema pallidum* have evolved further along this route, and require direct person-to-person mucosal contact for transmission. Man is the only natural host for these pathogens, which die rapidly in the environment. The source of infection may be individuals with clinical disease or subclinically infected *carriers*, in whom symptoms may be absent or relatively mild either because

Table 13.1 Examples of fimbriae produced by Gram-negative pathogens

Designation	Bacterium
Common (type 1)[a]	Enterobacteriaceae Uropathogenic *Escherichia coli*
CFA I, CFA II (CS1, CS2, CS3) E8775 (CS4, CS5, CS6)	Enterotoxigenic *Esch. coli* from humans
K88 K99 F41	Enterotoxigenic *Esch. coli* from animals
Pap-G, Prs-G	Uropathogenic *Esch. coli*
P fimbriae X-adhesins (S, M)	Pyelonephritogenic *Esch. coli*
N-methylphenylalanine fimbriae	*Pseudomonas, Neisseria, Moraxella, Bacteroides, Vibrio* species

See text for abbreviations and explanation
[a]Mannose-sensitive fimbriae.

the disease process is at an early stage or because of partial immunity to the pathogen.

In contrast, for many gastro-intestinal pathogens such as *Salmonella*, *Shigella* and *Campylobacter* species, the primary source is environmental, and infection follows the ingestion of contaminated food or water. Many of these organisms also infect other animals, often without harmful effect, and these act as a *reservoir of infection* and source of environmental contamination (see Ch. 1).

Colonization

For many pathogenic bacteria, the initial interaction with host tissues occurs at a mucosal surface, and colonization – the establishment of a stable population of bacteria in the host – normally requires adhesion to the mucosal cell surface. This allows the establishment of a focus of infection, which may remain localized or may subsequently spread to other tissues. Adhesion is necessary to avoid innate host defence mechanisms such as peristalsis in the gut and the flushing action of mucus, saliva and urine which removes non-adherent bacteria. For invasive bacteria, adhesion is an essential preliminary to penetration through tissues. Successful colonization also requires that bacteria are able to acquire essential nutrients, particularly iron, for growth.

Adhesion

Adhesion involves surface interactions between specific receptors on the mammalian cell membrane (usually carbohydrates) and *ligands* (usually proteins) on the bacterial surface. The presence or absence of specific *receptors* on mammalian cells contributes significantly to tissue specificity of infection. Non-specific surface properties of the bacterium, including surface charge and hydrophobicity, also contribute to the initial stages of the adhesion process.

Several different mechanisms of bacterial adherence have evolved, all utilizing specialized cell surface organelles or macromolecules that help to overcome the natural forces of repulsion that exist between the pathogen and its target cell.

Fimbrial adhesins

Electron microscopy of the surface of many Gram-negative and some Gram-positive bacteria reveals the presence of numerous thin, rigid, rod-like structures called *fimbriae*, or *pili*, that are easily distinguishable from the much thicker bacterial flagella (see p. 17). Fimbriae are involved in mediating attachment of some bacteria to mammalian cell surfaces. Different strains or species of bacteria may produce different types of fimbriae, which can be identified on the basis of antigenic composition, morphology and receptor specificity (Table 13.1). A broad division can be made between fimbriae in which adherence in vitro is inhibited by D-mannose (*mannose-sensitive fimbriae*) and those unaffected by this treatment (*mannose-resistant fimbriae*).

The antigenic composition of fimbriae can be complex. For instance, two fimbrial antigens called colonization factor antigen (CFA) I and II have been detected in enteropathogenic *Esch. coli* strains. CFA-II consists of three distinct fimbrial antigens designated as coli surface (CS) antigens 1, 2 and 3. Another *Esch. coli* strain, E8775, has been found to produce three other CS antigens, CS4, CS5 and CS6. Pyelonephritogenic *Esch. coli* isolates produce a group of adhesins called X-adhesins; two fimbrial types designated S and M on the basis of receptor specificity have been identified in this group.

The evolutionary significance of such heterogeneity may be that the ability of an individual bacterium to express several different types of fimbria allows different target receptors to be used at different anatomical sites of the infected host. In vitro, production of fimbriae is influenced by cultural conditions such as incubation temperature and medium composition, which may switch off the production of fimbriae or induce a phase change from one fimbrial type to another.

For some fimbriae the association with infection is clear. Thus the K88 fimbrial antigen is clearly associated with the ability of *Esch. coli* K88 to cause diarrhoea in pigs; pigs lacking the appropriate intestinal receptors are spared the enterotoxigenic effects of *Esch. coli* strains of this type. In many other instances the association between production of fimbriae and infection remains putative at present. Production of fimbriae is controlled by either chromosomal or plasmid genes.

The structure of one of these fimbrial types – *type 1* or *common fimbriae* – has been studied in detail. These consist of aggregates of a structural protein subunit called *fimbrillin* (or *pilin*) arranged in a regular helical array to produce a rigid rod-like structure 7 nm in diameter, with a central hole running along its length.

A highly conserved minor protein, located at both the tip and at intervals along the length of the fimbriae, mediates specific adhesion. Type 1 fimbriae bind specifically to D-mannose residues. Their role in vivo remains controversial; however, they may be involved in the pathogenesis of urinary tract infections.

Other Gram-negative bacteria, including those of the genera *Pseudomonas*, *Neisseria*, *Bacteroides* and *Vibrio*, produce fimbriae that share some homology, especially in the amino-terminal region of the fimbrillin subunits (the so-called *N-methylphenylalanine fimbriae*). These fimbriae have been shown to act as virulence determinants for *Ps. aeruginosa* and *N. gonorrhoeae*.

Non-fimbrial adhesins

Non-fimbrial adhesins include protein or polysaccharide structures that are surface exposed on the bacterial cell and/or secreted. Protein-based adhesins include the filamentous haemagglutinin of *B. pertussis*, a mannose-resistant haemagglutinin from *Salmonella enterica* serotype Typhimurium and a fibrillar haemagglutinin from *Helicobacter pylori*. Outer membrane proteins are involved in the adherence of many, if not most, bacteria.

Exopolysaccharides present on the surface of some Gram-positive bacteria are also involved in adhesion. For example, *Streptococcus mutans*, which is involved in the pathogenesis of dental caries, synthesizes a homo-polymer of glucose that anchors the bacterium to the tooth surface and contributes to the matrix of dental plaque. Actinomyces may adhere to other oral bacteria – a process called *co-aggregation*. Teichoic acid and surface proteins of coagulase-negative staphylococci mediate adherence of the bacterium to prosthetic devices

and catheters, contributing to increasing numbers of hospital-acquired infection. Continued growth following attachment to these biomaterials may result in formation of a *biofilm* (see Ch. 4), which may hinder successful antibiotic treatment by restricting access of drugs to the bacterium.

Flagella act as adhesins in *Vibrio cholerae* and *Campylobacter jejuni*. Bacterial motility is also thought to be important in chemotaxis of these organisms and *H. pylori* towards intestinal cells, and in penetration of these bacteria through the mucous layer during colonization.

Binding to fibronectin

Fibronectin is a complex multifunctional glycoprotein found in plasma and associated with mucosal cell surfaces, where it promotes numerous adhesion functions. Many pathogenic bacteria bind fibronectin at the bacterial surface, and for some organisms fibronectin has been shown to act as the cell surface receptor for bacterial adhesion. In *Streptococcus pyogenes*, lipoteichoic acid mediates attachment of the bacterium to the amino terminus of the fibronectin molecule. Attachment of *Staph. aureus* to cell surfaces also involves the amino terminus of fibronectin, but the bacterial ligand appears to be protein in this instance. *T. pallidum* also binds fibronectin. The significance of the interaction with fibronectin in the pathogenesis of syphilis and many other bacterial diseases needs further clarification. Binding of bacterial pathogens to a number of other connective tissue proteins including collagen, laminin and vitronectin has also been described.

Consequences of adhesion

In addition to preventing loss of the pathogen from the host, adhesion induces structural and functional changes in mucosal cells, and these may contribute to disease. For example, adherence of enteropathogenic *Esch. coli* (EPEC) to epithelial cells induces rearrangements of the cell cytoskeleton causing loss of microvilli and localized accumulation of actin, without subsequent invasion of the host cell. In contrast, adherence of *H. pylori* to gastric epithelial cells causes enhanced production of the pro-inflammatory chemokine interleukin (IL)-8, which contributes to gastric pathology. In both cases, these changes involve the induction of intracellular signalling pathways triggered after binding of the bacteria to specific receptors on the epithelial cell surface. Adhesion of bacteria to mammalian cells may also induce changes in bacterial gene expression.

Invasion

Once attached to a mucosal surface, some bacteria exert their pathogenic effects without penetrating the tissues of the host: toxins, other aggressins and induction of intracellular signalling pathways mediate tissue damage at local or distant sites. For a number of pathogenic bacteria, however, adherence to the mucosal surface represents but the first stage of the invasion of tissues. Examples of organisms that are able to invade and survive within host cells include mycobacteria and those of the genera *Salmonella, Shigella, Escherichia, Yersinia, Legionella, Listeria, Campylobacter* and *Neisseria*. Cell invasion confers the ability to avoid humoral host defence mechanisms and potentially provides a niche rich in nutrients and devoid of competition from other bacteria. However, the survival of bacteria in professional phagocytes, such as macrophages or polymorphonuclear leucocytes, depends on subverting intracellular killing mechanisms that would normally result in microbial destruction (see below). For some bacteria (*Neisseria meningitidis*), penetration through or between epithelial cells allows dissemination from the initial site of entry to other body sites.

Uptake into host cells

The initial phase of cellular invasion involves penetration of the mammalian cell membrane; many intracellular pathogens use normal phagocytic entry mechanisms to gain access.

Shigellae invade colonic mucosal cells but rarely penetrate deeper into the host tissues. Inside the cell, they are surrounded by a membrane-bound vesicle derived from the host cell. Soon after entry, this vesicle is lysed by the action of a plasmid-encoded haemolysin (haemolysis is just one way of detecting a membrane-damaging toxin), and the bacteria are released into the cell cytoplasm. *Listeria monocytogenes* produces a heat shock protein with a similar function, termed *listeriolysin*. Once free in the cell cytoplasm, shigellae multiply rapidly with subsequent inhibition of host cell protein synthesis. Several hours later the host cell dies, and bacteria spread to adjacent cells where the process of invasion is repeated.

In contrast to shigellae, most salmonellae proceed through the superficial layers of the gut and invade deeper tissues, in particular cells of the reticulo-endothelial system such as macrophages. Salmonellae also occupy a host-derived vesicle, but this does not lyse. Instead, several vesicles coalesce to form large intracellular vacuoles. These vacuoles traverse the cytoplasm to reach the opposite side of the cell and initiate spread to adjacent cells and deeper tissues. For both *Salmonella* and *Shigella* species, bacterial transcription and translation are required for invasion.

Role of cell receptors

The availability of specific receptors defines the types of host cell that are involved. As a result some pathogens can invade a wide range of cell types, whereas others have a more restricted invasive potential. Specific host receptors for some of the invasive pathogens have been identified. For example, *Legionella pneumophila* and *Mycobacterium tuberculosis* adhere to complement receptors on the surface of phagocytic cells. A specific receptor for *Yersinia pseudotuberculosis* belongs to a family of proteins termed *integrins* that form a network on the surface of host cells to which host proteins such as fibronectin can bind. Mimicry of the amino acid sequence (Arg-Gly-Asp) of fibronectin that mediates attachment to the integrins may represent a common mechanism of effecting intracellular entry.

The ability to utilize integrins may not be restricted to intracellular bacteria. The filamentous haemagglutinin of *B. pertussis* may use the fibronectin integrin to mediate attachment in the respiratory tract.

Survival and multiplication

To cause disease, most micro-organisms must survive on an epithelial surface, within a lumen or within host tissues and, at some stage multiply. Survival depends to a large extent on the organism's ability to avoid, evade or resist host defences. Multiplication depends on acquiring the nutrients needed and there are many barriers to success. The most extensively studied nutritional challenge to pathogens is iron acquisition.

Avoidance of host defence mechanisms

Colonization by bacterial pathogens, particularly of normally sterile areas of the body, results in the induction of specific and non-specific humoral and cell-mediated immune responses designed to eradicate the organism from the site of infection. Products of the organism may be chemotactic for phagocytic cells that are attracted to the site. Moreover, complement components may directly damage the bacterium and release peptides chemotactic for phagocytic cells. Other humoral antibacterial factors include lysozyme and the iron chelators transferrin and lactoferrin. Lysozyme is active primarily against Gram-positive bacteria but potentiates the activity of complement against Gram-negative organisms. Transferrin and lactoferrin chelate iron in body fluids, and reduce the amount of free iron to a level below that necessary for bacterial growth.

Pathogenic bacteria have evolved ways of avoiding or neutralizing these highly efficient clearance systems. As most of the interactions between the bacterium and the immune effectors involve the bacterial surface, resistance to these effects is related to the molecular architecture of the bacterial surface layers.

Capsules

Many bacterial pathogens need to avoid phagocytosis; production of an extracellular capsule is the most common mechanism by which this is achieved. Virtually all the pathogens associated with meningitis and pneumonia, including *H. influenzae, N. meningitidis, Esch. coli* and *Str. pneumoniae*, have capsules, and non-capsulate variants usually exhibit much reduced virulence. Most capsules are polysaccharides composed of sugar monomers that vary among different bacteria. Polysaccharide capsules reduce the efficiency of phagocytosis in a number of ways:

1. In the absence of specific antibody to the bacterium, the hydrophilic nature of the capsule may hinder uptake by phagocytes, a process that occurs more readily at hydrophobic surfaces. This may be overcome if the phagocyte is able to trap the bacterium against a surface, a process referred to as *surface phagocytosis.*
2. Capsules prevent efficient opsonization of the bacterium by complement or specific antibody, events that promote interaction with phagocytic cells. Capsules may either prevent complement deposition completely or cause complement to be deposited at a distance from the bacterial membrane where it is unable to damage the organism.
3. Capsules tend to be weakly immunogenic and may mask more immunogenic surface components and reduce interactions with both complement and antibody. In some cases, for example serogroup B *N. meningitidis* and serotype K1 *Esch. coli*, the capsular polysaccharide may mimic host polysaccharides moieties (e.g. brain sialic acid) and be seen as self antigen.

Streptococcal M protein

The M protein present on the surface of *Str. pyogenes* is not a capsule but functions in a similar manner to prevent complement deposition at the bacterial surface. The M protein binds both fibrinogen and fibrin, and deposition of this material on the streptococcal surface hinders the access of complement activated by the alternative pathway.

Resistance to killing by phagocytic cells

Some pathogens not only survive within macrophages and other phagocytes, but may actually multiply intracellularly. The normal sequence of events following phagocytosis involves fusion of the *phagosome* in which the bacterium is contained with *lysosomal granules* present in the cell cytoplasm. These granules contain enzymes and cationic peptides involved in oxygen-dependent and oxygen-independent bacterial killing mechanisms (see p. 115).

Different organisms use different strategies for survival (Table 13.2). *M. tuberculosis* is thought to resist intracellular killing by preventing phagosome–lysosome

Table 13.2 Some strategies adopted by bacteria to avoid intracellular killing

Species	Method
Mycobacterium tuberculosis	Prevents phagosome–lysosome fusion
Salmonella serotype Typhi	Fails to stimulate oxygen-dependent killing
Staphylococcus aureus	Produces catalase to negate effect of toxic oxygen radicals
Legionella pneumophila	Inhibits phagosome–lysosome acidification

fusion; other bacteria are able to resist the action of such lysosomal components after fusion. Some organisms stimulate a normal respiratory burst but are intrinsically resistant to the effects of the potentially toxic oxygen radicals produced. Production of catalase by *Staph. aureus* and *N. gonorrhoeae* is thought to protect these organisms from such toxic products. The smooth lipopolysaccharide of many bacterial pathogens is also thought to contribute to their resistance to the effects of bactericidal cationic peptides present in the phagolysosome.

Antigenic variation

Variation in surface antigen composition during the course of infection provides a mechanism of avoidance of specific immune responses directed at those antigens. This strategy is most highly developed in blood-borne parasitic protozoa, such as trypanosomes, but is also exhibited by bacteria. Pathogenic *Neisseria*, for example, are capable of changing surface antigens using three highly efficient mechanisms. These are mutation of individual amino acids, phase variation (switching genes on and off) and horizontal exchange of DNA material. *N. meningitidis* can avoid the killing effect of antibodies against its major porin (PorA) by mutating amino acids and/or acquiring parts or all of its *porA* gene from another meningococcal strain. The organism can switch off the expression of its capsule or its immunogenic proteins by shifts in the nucleotide sequence encoding them. The latter varies as a result of recombination or mutation during DNA replication.

Another mechanism of antigen variation in *Neisseria* is the genetic rearrangements demonstrated in the fimbriae. Usually only one complete fimbrillin gene is expressed, although there may be several incomplete 'silent' gene sequences present on the chromosome. Movement of the incomplete gene sequences to an expression locus results in synthesis of a protein that may differ antigenically from the original. Alternatively, variant fimbrillin gene DNA may be acquired from other strains of the same species by transformation to allow new genes to be constructed by recombination at the expression site.

The borreliae that cause relapsing fever use a similar strategy to generate antigenic variation in their outer surface proteins. Other bacteria show strain-specific antigenic variability, for example group A streptococci produce up to 75 antigenically distinct serotypes of M protein.

The capacity for variation in surface antigens allows for longer survival of an individual organism in a host and means that antibody produced in response to infection by one strain of a pathogen may not protect against subsequent challenge with a different strain of that bacterium. This makes the variable antigens elusive targets for protective antibodies, and the development of vaccines based on inhibition of attachment or generation of opsonic or bactericidal antibodies is particularly difficult for these organisms.

Immunoglobulin A proteases

Several species of pathogenic bacteria that cause disease on mucosal surfaces produce a protease that specifically cleaves immunoglobulin A (IgA), the principal antibody type produced at these sites. These proteases are specific for human IgA isotype I. Nearly all of the pathogens causing meningitis possess an IgA protease and a polysaccharide capsule enabling them to persist on the mucosal surface and resist phagocytosis during the invasive phase of the disease.

Serum resistance

To survive in the bloodstream, bacteria must be able to resist lysis as a result of deposition of complement on the bacterial surface. In the Enterobacteriaceae, resistance is primarily due to the composition of the lipopolysaccharide present in the bacterial outer membrane. *Smooth* colonial variants that possess polysaccharide 'O' side chains are more resistant than *rough* colonial variants that lack such side chains (see below). The side chains sterically hinder deposition of complement components on the bacterial surface. Conversely, however, some O-chain polysaccharides activate complement by an alternative pathway leading to lysis of the bacterial cell. In *N. meningitidis* group B and *Esch. coli* K1, sialic acid capsules prevent efficient complement activation and, in *N. gonorrhoeae*, complement binds but forms an aberrant configuration in the bacterial outer membrane so that it is unable to effect lysis.

Iron acquisition

The concentration of free iron in bodily secretions is below that required for bacterial growth because it is chelated by high-affinity mammalian iron-binding proteins such as transferrin and lactoferrin. To multiply in body fluids or on mucous membranes, bacteria must therefore obtain iron, and many bacterial pathogens have evolved efficient mechanisms for scavenging iron from mammalian iron-binding proteins. Bacteria such as *Esch. coli*, *Klebsiella pneumoniae* and some staphylococci produce extracellular iron chelators called *siderophores* for this purpose. Others including *N. meningitidis*, *Haemophilus parainfluenzae*, *H. influenzae* type b, *Staphylococcus epidermidis* and *Staph. aureus* have specific receptors for transferrin or lactoferrin, or both, on their surfaces, and are able to bind these proteins and their chelated iron directly from body fluids. Production of siderophores, their cell surface receptors, and receptors for transferrin, lactoferrin and other mammalian iron-binding proteins is iron-regulated and occurs mainly under conditions of iron restriction.

Two other mechanisms of iron acquisition from mammalian iron chelators have been described. Some *Bacteroides* species remove iron by proteolytic cleavage of the chelator. In *L. monocytogenes*, reduction of the Fe^{3+} ion to Fe^{2+} reduces the affinity for the chelator sufficiently for it to be removed by the bacterium.

Many bacteria express receptors for binding and/or internalizing other mammalian iron-containing molecules, such as haem, haemoglobin and haemoglobin–haemopexin complexes. These mammalian molecules are located intracellularly and can be released by bacterial haemolysins that lyse the red cells or are utilized directly by intracellular organisms.

Damage or dysfunction

In order for an infection to become apparent there must be sufficient host damage or dysfunction for the individual to become symptomatic. It is particularly important to appreciate that there are generally two components to this: the direct effect of the organism and the host response. In many cases, this largely immune-mediated damage (immunopathology) can predominate. This may reflect an excessive innate response as in septic shock (see below), or the adaptive response as in tuberculosis (see Ch. 18). The most obvious means by which bacteria cause host damage or dysfunction is by the production of toxins.

Toxins

In many bacterial infections part or all the characteristic pathology of the disease is caused by *toxins*. Toxins may exert their pathogenic effects directly on a target cell or may interact with cells of the immune system resulting in the release of immunological mediators (cytokines) that cause pathophysiological effects (see Chs 8, 9 and 12). Such effects may not always lead to the death of the

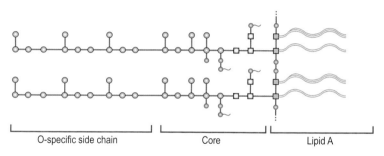

Fig. 13.1 Diagrammatic representation of the structure of bacterial lipopolysaccharide. (After Rietschel E T, Galanos C, Lüderlitz O 1975 Structure, endotoxicity and immunogenicity of the lipid A component of bacterial lipopolysaccharide. In: Schlessinger D (ed.) *Microbiology* – 1975, pp. 307–314. American Society for Microbiology, Washington, DC, 1975.)

O-specific side chain	Core	Lipid A

- ○ = various sugar residues
- □ = ketodeoxyoctonate (KDO)
- ▨ = glucosamine
- ○~ = phosphoethanolamine
- ～ = fatty acid residues

target cell but may selectively impair specific functions. Substances that have toxic physiological effects on target cells in vitro do not necessarily exert the same effects in vivo, but a number of toxins have been shown to be responsible for the typical clinical features of bacterial disease.

Two major types of toxin have been described: *endotoxin*, which is a component of the outer membrane of Gram-negative bacteria, and *exotoxins*, which are produced extracellularly by both Gram-negative and Gram-positive bacteria.

Endotoxin

Endotoxin, also called lipopolysaccharide or lipo-oligosaccharide, is a component of the outer membrane of Gram-negative bacteria and is released from the bacterial surface via outer membrane vesicles (blebs) following natural lysis of the bacterium or by disintegration of the organism in vitro. Lipopolysaccharide is anchored into the bacterial outer membrane through a unique molecule termed *lipid A* (Fig. 13.1). Covalently linked to lipid A is an eight-carbon sugar, ketodeoxyoctonate, in turn linked to the chain of sugar molecules (saccharides) that form the highly variable O antigen structures of Gram-negative bacteria. On bacteriological media, bacteria carrying lipopolysaccharide with O antigen form *smooth* colonies with hydrophilic surfaces; in contrast those carrying lipopolysaccharide without the O antigen form *rough* colonies with hydrophobic surfaces.

The term *endotoxin* was originally introduced to describe the component of Gram-negative bacteria responsible for the pathophysiology of *endotoxic* shock, a syndrome with a high mortality rate, particularly in immunocompromised or otherwise debilitated individuals. Endotoxin activates complement via the alternative pathway, but most of the biological activity of the molecule is attributable to lipid A. Both endotoxin and lipid A are potent activators of macrophages, resulting in the induction of a range of cytokines involved in the regulation of immune and inflammatory responses (see Ch. 9). Although endotoxin from Gram-negative organisms remains central to our understanding of *septic shock*, other structural and secreted components of bacteria that interact with pattern recognition receptors (see Ch. 9) can contribute to the pathogenesis of this clinical syndrome. Thus the multi-system effects, often involving complement, blood clotting factor and kinin activation together with extensive cytokine release, should not be seen as exclusive to Gram-negative infection.

Exotoxins

Exotoxins, in contrast to endotoxin, are diffusible proteins secreted into the external medium by the pathogen. Most pathogens secrete various protein molecules that facilitate adhesion to, or invasion of, the host. Many others cause damage to host cells. The damage may be physiological, for example cholera toxin promotes electrolyte (and fluid) excretion from enterocytes without killing the cells, or pathological, where the toxin (e.g. diphtheria toxin) inhibits protein synthesis and induces cell death. Exotoxins vary in their molecular structure, biological function, mechanism of secretion and immunological properties. The list of bacterial exotoxins is now endless and increasing; however, they are often classified by their mode of action on animal cells:

- Type I (membrane acting) toxins bind surface receptors and stimulate transmembrane signals, and include the *super-antigenic* toxins.
- Type II (membrane damaging) toxins directly affect membranes, forming pores or disrupting lipid bilayers.

Table 13.3 Some effects of bacterial exotoxins

Toxic effect	Examples
Lethal action	
Effect on neuromuscular junction	*Clostridium botulinum* toxin A
Effect on voluntary muscle	Tetanus toxin
Damage to heart, lungs, kidneys, etc.	Diphtheria toxin
Pyrogenic effect	
Increase in body temperature and polyclonal T cell activation	Super-antigenic exotoxins of *Staphylococcus aureus* and *Streptococcus pyogenes* (e.g. staphylococcal toxic shock syndrome toxin 1)
Action on gastro-intestinal tract	
Secretion of water and electrolytes	Cholera and *Escherichia coli* enterotoxins
Pseudomembranous colitis	*Clostridium difficile* toxins A and B
Bacillary dysentery	Shigella toxin
Vomiting	*Staph. aureus* enterotoxins A–E
Action on skin	
Necrosis	Clostridial toxins; staphylococcal α-toxin
Erythema	Diphtheria toxin; streptococcal erythrogenic toxin
Permeability of skin capillaries	Cholera enterotoxin; *Esch. coli* heat-labile toxin
Nikolsky sign[a]	*Staph. aureus* epidermolytic toxin
Cytolytic effects	
Lysis of blood cells	*Staph. aureus* α-, β- and δ-lysins; leucocidin Streptolysin O and S *Clostridium perfringens* α and θ toxins
Inhibition of metabolic activity	
Protein synthesis	Diphtheria toxin; shiga toxin

[a]Separation of epidermis from dermis.

- Type III (intracellular effector) toxins translocate an active enzymatic component into the cell and modify an intracellular target molecule.

Examples of exotoxins and their effects on target cells are shown in Table 13.3. Bacteria secrete proteins using various mechanisms, five of which (again named types I–V) are relatively well understood. Figure 13.2 shows types I and V; these are relatively well characterized. In type I, at least three proteins get together to form a channel through which large molecules (such as hae-molysin of *Esch. coli*) are exported. In type V, however, a single precursor protein that consists of three domains finds its way across the inner and outer membranes, and cleaves itself off the cell. These latter proteins are called *autotransporters*. A typical example is the IgA1 protease of *Neisseria* spp.

Enterotoxins cause symptoms of gastro-intestinal disease, including diarrhoea, dysentery and vomiting. In some cases the disease is caused by ingestion of preformed toxin in food, but in most cases colonization of the intestine is required before toxin is made.

Cholera toxin and heat-labile toxins of enterotoxigenic *Esch. coli* (ETEC) do not induce inflammatory changes in the intestinal mucosa, but perturb the processes that regulate ion and water exchange across the intestinal epithelium (see Chs 26 and 30). In contrast, the enterotoxins of *Clostridium difficile, C. perfringens* type A and *Bacillus cereus* cause structural damage to epithelial cells, resulting in inflammation. Other gastro-intestinal pathogens such as enteropathogenic *Esch. coli* (EPEC) mediate damage by ill defined mechanisms following close contact between the bacterium and the cell surface.

B. pertussis, the causative agent of whooping cough, produces various extracellular products, including a tracheal cytotoxin that inhibits the beating of cilia on tracheal epithelial cells, pertussis toxin, which exhibits several systemic effects, and an adenylate cyclase that interferes with phagocyte function.

Another group of toxins causes damage to subepithelial tissues following penetration and multiplication of the pathogen at the site of infection. Many of these toxins also inhibit or interfere with components of the host immune system. Membrane-damaging toxins such as staphylococcal

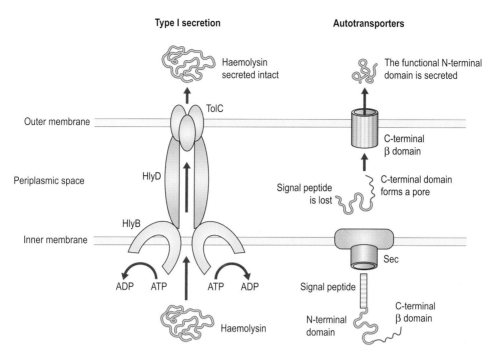

Type I secretion

Haemolysin
secreted intact

TolC

Outer membrane

HlyD

Periplasmic space

HlyB

Inner membrane

ADP   ATP   ATP   ADP

Haemolysin

Autotransporters

The functional N-terminal
domain is secreted

C-terminal
β domain

C-terminal domain
forms a pore

Signal peptide
is lost

Sec

Signal peptide

N-terminal
domain

C-terminal
β domain

Fig. 13.2   Diagrammatic representation of type I and type V (autotransporter) secretion of exotoxins across the bacterial cell membrane. Hly, haemolysin; Tol, special receptor; ADP, adenosine diphosphate; ATP, adenosine triphosphate.

α and β toxins, streptolysin O and streptolysin S, and *C. perfringens* α and θ toxins inhibit leucocyte chemotaxis at subcytolytic concentrations, but cause necrosis and tissue damage at higher concentrations.

Systemic effects of toxins

Some toxins cause damage to internal organs following absorption from the focus of infection. Included in this category are the toxins causing diphtheria, tetanus and botulism, and those associated with streptococcal scarlet fever and staphylococcal toxic shock syndrome. The diphtheria toxin, the gene for which is bacteriophage encoded, inhibits protein synthesis in mammalian cells. Tetanus toxin, in contrast, exerts its effect by preventing the release of inhibitory neurotransmitters whose function is to prevent overstimulation of motor neurones in the central nervous system, resulting in the convulsive muscle spasm characteristic of tetanus. Diphtheria and tetanus toxins represent the sole determinant of disease and are neutralized by specific antitoxin antibody. As a result, vaccination with diphtheria and tetanus toxoids (formalin-inactivated toxins) is highly effective (see Ch. 69).

Botulism results from the ingestion of preformed toxin produced by *Clostridium botulinum* in food contaminated with this bacterium, and is not a true infectious disease. The toxic activity is due to a family of serologically distinct polypeptide neurotoxins that prevent release of acetylcholine at neuromuscular junctions, resulting in the symptoms of flaccid paralysis. These toxins have been used clinically in treating squints and muscle spasm.

Other toxins cause disseminated multi-system organ damage. Such pathology is seen in staphylococcal *toxic shock syndrome* caused by certain strains of *Staph. aureus* that produce a toxin designated *toxic shock syndrome toxin 1* (TSST-1). This toxin belongs to a group of functionally related proteins collectively referred to as *superantigens*, which include the staphylococcal enterotoxins, staphylococcal exfoliative toxin and streptococcal pyrogenic exotoxin A. These molecules are potent T cell mitogens whose reactivity with lymphocytes induces cytokine release; they may initiate tissue damage by mechanisms similar to those postulated to account for Gram-negative endotoxic shock (see p. 161).

Other extracellular aggressins

Many bacteria secrete a range of enzymes that may be involved in the pathogenic processes.

Proteus spp. and some other bacteria that cause urinary tract infections produce *ureases* that break down urea in the urine, and the release of ammonia may contribute to the pathology. The urease produced by the gastric and duodenal pathogen *H. pylori* is similarly implicated in the virulence of the organism. *L. pneumophila* produces a metalloprotease thought to contribute to the characteristic pathology seen in legionella pneumonia.

Many other degradative enzymes, including *mucinases, phospholipases, collagenases* and *hyaluronidases*, are produced by pathogenic bacteria. Many non-pathogenic bacteria also produce such enzymes, and their role in pathogenesis requires further clarification.

An understanding of the basic mechanisms of pathogenesis is important for the design of new or improved vaccines and appropriate therapies. Such knowledge is also invaluable in the analysis of 'new' bacterial pathogens that are recognized from time to time. However, for some bacterial diseases, for example syphilis, such approaches have still not defined the mechanisms of pathogenesis or the virulence determinants involved, and new strategies employed by such successful pathogens may yet be discovered.

KEY POINTS

- Opportunist pathogens require a defect in host defence before they cause disease, whereas primary pathogens affect otherwise healthy individuals. The possession of virulence determinants generally differentiates pathogens from non-pathogens and, in turn, their number and potency separate opportunist from primary pathogens.
- Expression of virulence determinants is carefully regulated and may involve a form of chemical communication between bacteria known as *quorum sensing*.
- Adhesins are often involved the establishment of an infection. Bacterial adhesion may be mediated by fimbrial or non-fimbrial adhesins and generally involves interactions with host cell surface receptors or surface associated proteins such as fibronectin.
- Invasive pathogens gain entry and spread either by subverting host uptake mechanisms or by tissue disruption. Once established, pathogens must deploy a variety of mechanisms to avoid host defences.
- Multiplication within host tissues requires specific mechanisms to gain essential nutrients such as iron. Most pathogens make *siderophores* that compete with the host's high-affinity iron transport and storage systems.
- Host damage may be direct, via the release of toxins, or indirect, via the effects of the host's innate and adaptive immune responses (immunopathology).
- Endotoxins and exotoxins are recognized. Endotoxin is synonymous with the lipopolysaccharide or lipo-oligosaccharide of Gram-negative bacteria, and sufficient amounts elicit a cascade of responses leading to endotoxic shock.
- Exotoxins are proteins that cause damage or dysfunction by signalling at host cell membranes (type I), by damaging membranes (type II) or by entering target cells and directly altering function (type III).

RECOMMENDED READING

Alouf J E, Freer J H 1999 *Bacterial Protein Toxins*, 2nd edn. Academic Press, London

Mims C, Nash A, Stephen J 2000 *Mims' Pathogenesis of Infectious Disease*, 5th edn. Academic Press, London

Hacker J, Hessemann J (eds) 2000 *Molecular Infection Biology: Interaction Between Microorganisms and Cells*. John Wiley, Hoboken, NJ

Salyers A A, Whitt D D 2002 *Bacterial Pathogenesis: A Molecular Approach*, 2nd edn. ASM Press, Washington

Williams P, Ketley J, Salmond G 1998 *Bacterial Pathogenesis*, Vol. 28. Academic Press, London

14 The natural history of infection

M. R. Barer

The purpose of this chapter is to link the basic properties of micro-organisms with the patterns of infective disease experienced in public health and clinical practice, and with the tissue and organ pathology that can be observed. Although there are numerous exceptions to the general patterns described, the intention is to give an underlying structure that can be used to make sense of the many different types of organism and the diseases they cause. The term 'natural history' is used in two senses here: first, to denote an overall biological consideration of the life cycle of the infective agent and how this intersects with the human host and, second, to consider the process of infection from the point of the encounter between the agent and the susceptible host through to its outcome.

MEETINGS BETWEEN HUMAN BEINGS AND MICRO-ORGANISMS

The vast majority of micro-organisms do not form stable associations with human beings. Clearly pathogens must do so, at least temporarily. However, it is worth briefly considering how important this association is to the micro-organism concerned. In some cases, man constitutes the only environment in which micro-organisms can survive (i.e. they are obligate parasites of human beings). Thus, man is the reservoir and immediate source of the infections caused by this group (see Ch. 1). In other cases colonization or infection of human beings may be entirely incidental to the life cycle of the organism. These organisms may need to live in animals, as in the case of zoonoses, or in specific environmental reservoirs. Their life cycles in these habitats are critical to the epidemiology of the infections they cause. This biological perspective is difficult to avoid when considering the complex life cycles of parasitic protozoa and helminths, but such considerations are equally applicable to bacterial and viral pathogens.

Obligate pathogens

These organisms have to cause disease in human beings in order to continue to survive and propagate. This is true for most viruses that cause human disease and for which man is the only natural host. The major caveat to this is that the degree to which these agents cause symptomatic infections can vary over a very wide range. Thus asymptomatic infections with smallpox were virtually unknown, whereas they are very common with polio. This contributed substantially to the eradication of smallpox as it was relatively straightforward to identify where transmission was taking place. Among bacterial pathogens, *Mycobacterium tuberculosis* is a prime example of an agent that has to cause symptomatic disease in order to survive and propagate. In some cases pathogenicity reflects an early stage in the development of host–parasite relations, with pathogens evolving towards a more benign association with their host. Clearly, if a parasite kills all of its potential hosts then it has destroyed its own habitat. However, this does not always hold true. Some pathogens actually become more virulent as a means of increasing their potential to survive.

Accidental or incidental pathogens

This term applies to many bacterial pathogens. Causing disease confers no obvious biological advantage on the organism and indeed may be a dead end. There are two groups of bacterial pathogen for which this is probably the case. The first group have their natural habitat in man but cause disease in only a small minority; these include the major pathogens of bacterial pharyngitis (*Streptococcus pyogenes*), acute pneumonia (*Streptococcus pneumoniae*) and the principal agents of acute pyogenic meningitis (*Streptococcus pneumoniae, Neisseria meningitidis* and *Haemophilus influenzae* type b). The second group have a habitat (or reservoir) in nature but, if they encounter a susceptible host in a particular way, infection may ensue. For example, the agent of cholera, *Vibrio cholerae*, lives in brackish water and causes human disease only when ingested. *Clostridium tetani,* the agent of tetanus, probably propagates in animal gastrointestinal tracts and infects wounds contaminated by soil containing animal excreta.

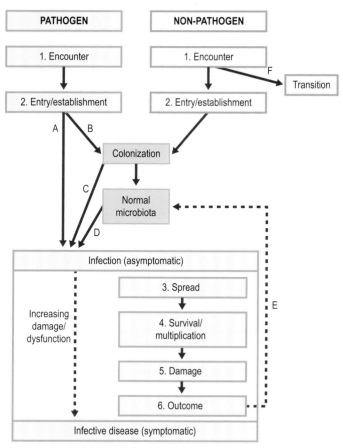

Fig. 14.1 Pathways to and stages of infection following encounter between a host and a micro-organism. The blue sector includes the possible outcomes for a pathogen, the yellow a non-pathogen, and the green either of these. An organism detected in a diagnostic laboratory might reflect one or more of the stages identified, including transition. The possibility of transition from yellow to blue zones reflects the lack of a rigid division between pathogen and non-pathogen, and one aspect of opportunism. Detection in a diagnostic laboratory is most likely following colonization or multiplication. Infection may become apparent as a result of many different pathways: A, directly, without any colonization phase; B, many pathogens colonize first then proceed to infection either C after a brief period of colonization or D after a sustained period in which they live as *commensals* as part of the normal microbiota (progression here is not inevitable). Note that infection becomes symptomatic only when the level of damage or dysfunction is sufficient. At the end of the symptomatic infection a small number of pathogens may enter the normal microbiota (E) and the convalescent host is then described as a *carrier* for the organism concerned. F, Some organisms may simply be passing through and form no stable association with the host.

Pathogens in the environment

Whatever the method of acquisition, the organism must survive long enough to encounter a susceptible human host if it is going to cause human disease. The dynamics of pathogen survival in various environments are relevant to the control of infection. The capacities of different pathogens to survive and propagate in food and water are of particular concern, as is survival in aerosols and through desiccation and many other common environmental stresses. These properties provide the biological basis for the transmission of infection and many opportunities for improved control of specific pathogens.

STAGES OF INFECTION

Most infections can be broken down into a core series of steps:

- encounter
- entry and establishment
- spread
- survival and multiplication

- damage
- outcome.

It should be noted that the virulence determinants described in Chapter 13 were related to all but the first and last of these stages. Most pathogens can cause infection only via a limited set of routes (see above). Thus *Vibrio cholerae* must be ingested; it cannot cause infection if rubbed on the skin. Human immunodeficiency virus (HIV) must gain access to circulating CD4$^+$ cells via a parenteral route, and so on. Some general points concerning the passage through alternative stages of infection are made in Figure 14.1. Note that, although the simple direct pathway (A) reflects the norm for an exogenous infection, there are intermediates between this and D, endogenous infection. A single organism may be capable of following multiple routes to infection. For example, *Staphylococcus aureus* may be introduced exogenously into a wound. Around 30% of individuals are colonized with this organism at any point in time but in only one-third of these (10%) does the organism appear to be a member of the normal microbiota. Both temporary and permanent relationships may provide for endogenous infections due to *Staph. aureus*.

Table 14.1   Normal human microbiota[a]

Location[b]	Composition[c]	Abundance[d]
Dry skin (face, forearm)	Gram-positive anaerobes (e.g. propionibacteria)	10^2
Moist skin (axilla, groin)	Staphylococci (esp. coagulase negative); corynebacteria; Gram-negatives rare but more frequent after prolonged hospital stay	10^{6-7}
Oropharynx	Anaerobes; streptococci; neisseria; candida	10^9
Small intestine	Anaerobes, lactobacilli, peptostreptocccus, porphyromonas	10^{5-7}
Large intestine	Anaerobes, clostridium, bacteroides; enterobacteria; enterococci, candida, protozoa	10^{9-11}
Vagina	Anaerobes, lactobacillus; streptococci, candida	10^{8-9}

[a]*The* term microbiota is preferred to 'flora' as the latter refers back to a period when bacteria were classified with plants.
[b]These are extremely broad. In practice each micro-niche in the body constitutes a different environment colonized with different organisms; for example, the microbiota associated with the lumen and the mucosa of the gut are different.
[c]A very rough introductory guide. Note the predominance of anaerobes.
[d]Per square centimetre of surface or gram of fluid.

Many opportunistic pathogens become part of the normal microbiota before they cause infection. They may be assisted in colonizing a new host by interventions such as repeated use of antibiotics. This appears to be the case with *Pseudomonas aeruginosa*, which is resistant to most routinely used antibacterial agents, and is probably also the case for methicillin-resistant *Staph. aureus* (MRSA) and vancomycin-resistant enterococci (VRE). Most members of the normal microbiota appear to have very little capacity to cause disease. Their number and composition varies in different parts of the body (Table 14.1) and it is clear that a 'healthy' normal microbiota provides some protection against invading pathogens. One of the major challenges in clinical bacteriology is to differentiate between organisms that are present as innocent bystanders in infection and those causing disease.

The survival of mankind to the present day reflects the fact that most untreated infections are not fatal. Indeed, many human genes, particularly those concerned with the immune response, clearly reflect the selection pressure provided by infection. Before the development of antibiotics and immunization at least 50% of deaths were attributable to infection (this is still the case in many resource-poor countries). Nevertheless, owing to our inherent and highly efficient defence mechanisms, many infections resolve without medical intervention. All doctors must have some skill in recognizing those infections for which an intervention is unnecessary. This is particularly important in the case of antibiotic use because of the dangers of encouraging resistance.

PATHOLOGICAL PATTERNS OF INFECTION

All of the foregoing reflects a set of proposed mechanisms that, by and large, fit and make sense of the available facts concerning the epidemiology and detailed pathogenesis of infection. In this section the link to clinical practice is developed by describing the pattern of pathology directly observable in various infections. Most infections can be placed into one of four patterns:

1. Toxin mediated (mainly bacterial).
2. Acute (including acute viral syndromes and acute pyogenic bacterial infections).
3. Subacute (many virus and several 'atypical' bacterial infections).
4. Chronic (chronic viral infections, chronic granulomatous bacterial, fungal and parasitic infections).

Simply by recognizing the basic characteristics of a suspected infection against these four possibilities, the possible range of causal agents can be narrowed down substantially. Their features are summarized and exemplified from the perspective of bacterial infections in Table 14.2.

Toxin–mediated bacterial infections

This was the first recognized pathogenic mechanism in bacterial infection and resulted in early successful therapeutic and preventive measures. When a single toxin is responsible for most of the features of an infection, the dysfunction or damage is often distant from the site of bacterial multiplication, the disease may be reproduced by administration of the pure toxin alone and it can be prevented with antibodies directed against the toxin. The clostridial diseases tetanus and botulism are toxin mediated. In the latter case, as in several other forms of food poisoning, ingestion of only the toxin is required, so many cases of botulism are not strictly infections. Once the pathogen has grown and produced toxin, the onset of disease can be very rapid.

Table 14.2 Patterns in the presentation and pathology of bacterial infection

Pattern	Examples
Toxin–mediated disease	
Pathology often distant from site of bacterial growth	Diphtheria, tetanus
Protective immunity may be mediated by anti-toxin antibodies alone	Staphylococcal food poisoning, cholera
Disease may be fully reproduced by administering the toxin alone	Pseudomembranous colitis
Acute pyogenic infection	
Generally rapid growing organisms	Streptococcal pharyngitis
Interaction with innate immune system and acute inflammation predominates	Staphylococcal abscess, bacterial meningitis, lobar pneumonia, acute cystitis
Where immune damage occurs, it is 'post-infective'	Post-streptococcal glomerulonephritis
Subacute infection	
No pattern to growth rate	Subacute bacterial endocarditis
Site of infection may be only partially accessible to the immune system	'Atypical' pneumonia
Immunopathology often in parallel with direct effects of organism	
Chronic (granulomatous)	
Bacterial growth rate often moderate or slow	Tuberculosis, brucellosis
Organisms often survive and grow intracellularly	
Immune damage occurs with infection – predominantly cell mediated	(Some fungal and parasitic infections have this pattern)

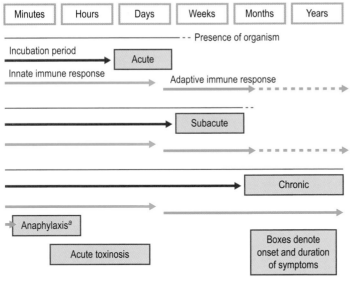

Fig. 14.2 Timescales for key events in infection.
[a]Anaphylaxis is included for comparison and to illustrate the timescale when the adaptive immune response is primed against the antigen(s) concerned.

It is often possible to abolish the biological activity of toxins without affecting their immunogenicity. Such *toxoids* were among the first effective immunizations against bacterial infection. Diphtheria and tetanus toxoid vaccines have controlled these infections in the UK. In life-threatening toxin-mediated disease, the administration of pre-formed antibodies can be life-saving. Antibiotics are not effective in treating established disease, but may prevent further toxin formation. In the special case of *Escherichia coli* O157 infections, however, some antibiotics actually stimulate further synthesis of toxin.

Acute pyogenic bacterial infections

Pyogenic means pus inducing. Pus is composed primarily of live and dead neutrophil polymorphs. The presence of pus generally reflects an acute inflammatory process and activation of the innate immune system. The inflammatory process may be localized, as in the formation of an abscess, or more disseminated through tissue planes. Anything more than a trivial acute pyogenic infection is usually accompanied by an increase in the blood neutrophil count. The acuteness of these infections is reflected in their rapid onset. Accordingly the bacteria that cause them generally grow rapidly, producing visible colonies within 24 h of inoculation. Medical intervention is most effective when given early in infection before the development of acquired immunity, which, when successful, terminates the illness. Serological evidence of acquired immunity cannot be used in the diagnosis of infection during its acute phase. Occasionally, immunopathology occurs after the causal organism is no longer detectable in the host; classic examples are post-streptococcal glomerulonephritis and rheumatic fever. Similarly, Guillain–Barré syndrome, a paralytic disease, sometimes follows acute campylobacter infection.

Many bacteria that cause acute pyogenic infections also produce toxins. Thus there may be both acute pyogenic and toxin-mediated components to the damage and dysfunction that develops. This is particularly true of staphylococcal and streptococcal infections. The complex mixture of the pathogenic processes attributable to different virulence determinants can make the most severe of these infections very difficult to treat.

Subacute bacterial infections

These have a more insidious onset than acute infections and are accompanied by less prominent signs of acute inflammation. Classically, bacterial endocarditis was described as subacute, although this is no longer considered a suitable catch-all term for this type of infection. Because such diseases have a more protracted course, the adaptive immune response often contributes to damage. Hence subacute forms of bacterial endocarditis are often accompanied by immune complex-mediated pathology, whereas *Mycoplasma pneumoniae* infection (a form of atypical pneumonia) may be accompanied by several different immunopathological reactions reflecting specific immune responses (see Ch. 41).

Chronic granulomatous bacterial infections

When bacterial infections persist over months or even years they tend to elicit a pathological entity known as a *granuloma*. Granulomas are a common form of localized cell-mediated immune response directed to antigens or other foreign bodies that appear to be refractory to elimination from tissues. An ordered accumulation of lymphocytes and macrophages occurs around a central focus in a manner that, to the experienced pathologist, can be more or less specific to the eliciting stimulus. Persisting bacterial infections, notably those due to mycobacteria (e.g. tuberculosis) and, to a lesser extent *Brucella* spp., produce chronic granulomatous infections. The agents concerned are generally slow growing and have the capacity to survive inside host cells, notably macrophages. Cell-mediated

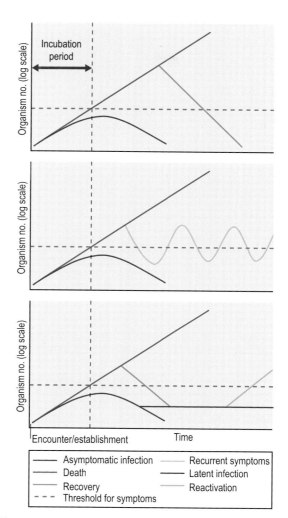

Fig. 14.3 A model for incubation periods, progression and latency related to organism number.

immunopathology (delayed-type hypersensitivity; see Ch. 9) is a prominent feature of these infections.

Timing of key events in infection

As different infections proceed at different rates, the timing of the symptoms, their relation to immune responses and the ability to detect the causal agent all vary. The *incubation period*, the time between the encounter with the pathogen and the onset of symptoms, is an important practical consideration in understanding and managing infection. This is characteristic for different pathogens and can be vital in determining whether an individual is still at risk of developing disease after exposure to a particular agent. Incubation periods for the four patterns of infection discussed above are illustrated in Figure 14.2, along with the

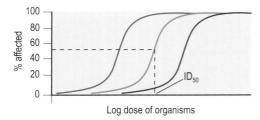

Fig. 14.4 Quantitative relationships between organism dose and outcome in experimental infection. The organism has been administered by a single route to a uniform host population. Note that the outcome (percentage affected – which could be percentage infected, percentage who died or many other endpoints) depends on the dose. This approach allows virulence to be compared between strains of a particular micro-organism. A more virulent organism shifts the curve to the left (blue curve) and a less virulent organism to the right (red curve). Lesser or greater host resistance would, respectively, have the same effect. The approach allows for the specific recognition of virulence determinants and the effects of immunization. The dose required to produce the specified endpoint in 50% of the target population is often reproducible and can be used for statistical comparisons. The 50% infected endpoint is known as the ID_{50}.

time-frames over which immune responses and presence of the pathogen are expected. A more dynamic view of individual infections is shown in Figure 14.3, in which the additional concepts of recurrent, latent and reactivated infections are illustrated. Figure 14.3 introduces the notion that the progression of an infection is related to the numbers of the pathogen. Although many other factors are involved, the concept is useful because it illustrates how, in some rapidly developing infections, the interval between the onset of symptoms and death may be short. The slope of increasing pathogen numbers clearly also reflects the balance between growth of the pathogen and the efforts of the immune system to resist. Accordingly, when the immune system is suppressed, progression may be exceptionally rapid and the response must be equally so.

VIRULENCE AND INFECTIVITY

By now it should be apparent that what makes a pathogen more or less *virulent* is, in most cases, extremely complex.

None the less, when infections can be studied in animal experimental systems, virulence can be seen as a quantifiable property. As the dose administered to a group of susceptible hosts is increased, the number acquiring infections also increases in a fairly well defined dose–response relationship (Fig. 14.4). Not only does recognition of this relationship help explain why, when a group of individuals is exposed to a pathogen only some get infected, it also provides a framework for understanding the effects of immunization and immune deficiency, as well as a systematic basis for identifying individual traits that contribute to virulence and host resistance.

KEY POINTS

- Symptomatic infection is a rare outcome when human beings and micro-organisms meet. None the less, some organisms, known as *obligate pathogens*, must cause disease to survive; this applies to most viral pathogens. Many bacterial pathogens appear to derive little benefit from causing infection.
- Micro-organisms that form short- or long-term associations with humans do so in a number of recognizable forms and stages, including: *entry/establishment, colonization, commensalism, spread, survival and multiplication, damage,* and *carriage.* Where these associations lead to bacterial disease, the different stages are often associated with the virulence determinants described in Chapter 13.
- Infections can generally be recognized in one of four categories: *toxin mediated, acute, subacute* and *chronic.*
- The capacity of a particular micro-organism to cause disease is known as its *virulence.* Where suitable experimental models are available, we can recognize virulence quantitatively by the ability of a low number of organisms to produce infection or death in the host population. Such measurements can be useful in identifying virulence determinants and in developing vaccines.

RECOMMENDED READING

Burnet F M 1970 *Natural History of Infectious Diseases.* Cambridge University Press, Cambridge

Ewald P W 1996 *Evolution of Infectious Disease.* Oxford University Press, USA

Mims C, Nash A, Stephen J 2000 *Mims' Pathogenesis of Infectious Disease,* 5th edn. Academic Press, London

Salyers A A, Whitt D D 2002 *Bacterial Pathogenesis: A Molecular Approach,* 2nd edn. ASM Press, Washington, DC

PART 3
BACTERIAL PATHOGENS AND ASSOCIATED DISEASES

15 Staphylococcus

Skin infections; osteomyelitis; food poisoning; foreign body infections; MRSA

H. Humphreys

Sir Alexander Ogston, a Scottish surgeon, first showed in 1880 that a number of human pyogenic diseases were associated with a cluster-forming micro-organism. He introduced the name 'staphylococcus' (Greek: *staphyle* = bunch of grapes; *kokkos* = grain or berry), now used as the genus name for a group of facultatively anaerobic, catalase-positive, Gram-positive cocci. Staphylococci are resistant to dry conditions and high salt concentrations, and are well suited to their ecological niche, which is the skin. They may also be found as part of the normal flora of other sites such as the upper respiratory tract, and are commonly present on animals.

The major pathogen within the genus, *Staphylococcus aureus*, causes a wide range of major and minor infections in man and animals (Table 15.1) and is characterized by its ability to clot blood plasma by action of the enzyme *coagulase*. There are at least 30 other species of staphylococci, all of which lack this enzyme. These coagulase-negative staphylococci are skin commensals that can cause opportunistic infections associated with prostheses or foreign bodies (usually due to *Staph. epidermidis*), and urinary tract infections (*Staph. saprophyticus*). The presence of methicillin-resistant *Staph. aureus* (MRSA) in many hospitals has become a major public health issue, with concern expressed by patients and members of the public about the clinical implications.

STAPHYLOCOCCUS AUREUS

Description

Staph. aureus is a Gram-positive coccus about 1 μm in diameter. The cocci are usually arranged in grape-like clusters (Fig. 15.1). The organisms are non-sporing, non-motile and usually non-capsulate. When grown on many types of agar for 24 h at 37°C, individual colonies are circular, 2–3 mm in diameter, with a smooth, shiny surface; colonies appear opaque and are often pigmented (golden-yellow). The main distinctive diagnostic features of *Staph. aureus* are:

- Production of an extracellular enzyme, *coagulase*, which converts plasma fibrinogen into fibrin, aided by an activator present in plasma.
- Production of thermostable nucleases that break down DNA.
- Production of a surface-associated protein known as *clumping factor* or *bound coagulase* that reacts with fibrinogen.

Various commercial systems are available that rapidly identify staphylococci. They are particularly useful for screening large numbers of strains.

Pathogenesis

Staph. aureus is present in the nose of 30% of healthy people and may be found on the skin. It causes infection most commonly at sites of lowered host resistance, such as damaged skin or mucous membranes.

Virulence factors

Staph. aureus strains possess a large number of cell-associated and extracellular factors, some of which contribute to the ability of the organism to overcome the body's defences and to invade, survive in and colonize the tissues (Table 15.2). Although the role of each factor is not fully understood individually, it is likely that they are

Table 15.1 Infections caused by *Staph. aureus*

Pyogenic infections	Toxin-mediated infections
Boils, carbuncles	Scalded skin syndrome
Wound infection	Pemphigus neonatorum
Abscesses	Toxic shock syndrome
Impetigo	Food poisoning
Mastitis	
Bacteraemia	
Osteomyelitis	
Pneumonia	
Endocarditis	

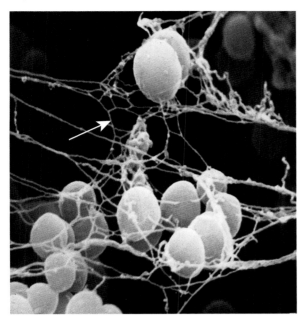

Fig. 15.1 Scanning electron micrograph of staphylococci in a biofilm enmeshed in exopolysaccharide (arrow). Original magnification ×15 000. (Courtesy of Dr R. Bayston, University of Nottingham.)

Table 15.2 Some virulence factors of *Staph. aureus*

Virulence factor	Activity
Cell wall polymers	
Peptidoglycan	Inhibits inflammatory response; endotoxin-like activity
Teichoic acid	Phage adsorption; reservoir of bound divalent cations
Cell surface proteins	
Protein A	Reacts with Fc region of IgG
Clumping factor	Binds to fibrinogen
Fibronectin-binding protein	Binds to fibronectin
Exoproteins	
α-Lysin	Impairment of membrane permeability; cytotoxic effects on phagocytic and tissue cells
β-Lysin	
γ-Lysin	
δ-Lysin	
Panton–Valentine leucocidin	Dermo-necrotic
Epidermolytic toxins	Cause blistering of skin
Toxic shock syndrome toxin	Induces multi-system effects; superantigen effects
Enterotoxins	Induce vomiting and diarrhoea; superantigen effects
Coagulase	Converts fibrinogen to fibrin in plasma
Staphylokinase	Degrades fibrin
Lipase	Degrades lipid
Deoxyribonuclease	Degrades DNA

responsible for the establishment of infection, enabling the organism to bind to connective tissue, opposing destruction by the bactericidal activities of humoral factors such as complement, and overcoming uptake and intracellular killing by phagocytes.

Animal experiments support the view that no single virulence factor is pre-eminent in overcoming host resistance and establishing a focus of staphylococcal infection. However, particular exotoxins are responsible for the symptoms of certain syndromes.

Staphylococcal toxins

Enterotoxins. Enterotoxins, types A–E, G, H, I and J, are commonly produced by up to 65% of strains of *Staph. aureus*, sometimes singly and sometimes in combination. These toxic proteins withstand exposure to 100°C for several minutes. When ingested as preformed toxins in contaminated food, microgram amounts of toxin can, within a few hours, induce the symptoms of staphylococcal food poisoning: nausea, vomiting and diarrhoea.

Toxic shock syndrome toxin (TSST-1). This was discovered in the early 1980s as a result of epidemiological and microbiological investigations in the USA of *toxic shock syndrome*, a multi-system disease caused by staphylococcal TSST-1 or enterotoxin, or both. A link was established with the use of highly absorbent tampons in menstruating women, although non-menstrual cases are now as common. The absence of circulating antibodies to TSST-1 is a factor in the pathogenesis of this syndrome.

TSST-1 and the enterotoxins are now recognized as *superantigens*, that is, they are potent activators of T lymphocytes resulting in the liberation of cytokines such as tumour necrosis factor, and they bind with high affinity to mononuclear cells. These characteristics partly explain the florid and multi-system nature of the clinical conditions associated with these toxins.

Epidermolytic toxins. Two kinds of epidermolytic toxin (types A and B) are commonly produced by strains, belonging mainly to phage group II (see below), that cause blistering diseases. These toxins induce intraepidermal blisters at the granular cell layer. Such blisters range in severity from the trivial to the distended blisters of *pemphigus neonatorum*. The most dramatic manifestation of epidermolytic toxin is the *scalded skin syndrome* in small children, where the toxin spreads systemically in individuals who lack neutralizing antitoxin. Extensive areas of skin are affected, which, after the development of a painful rash, slough off; the skin surface resembles scalding (Fig. 15.2).

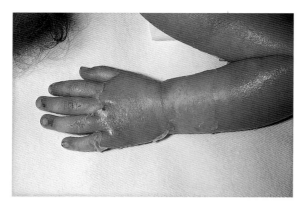

Fig. 15.2 Scalded skin syndrome (toxic epidermal necrolysis). (Courtesy of Dr L G Millard, Queen's Medical Centre, Nottingham.)

Epidemiology

Sources of infection

Infected lesions. Large numbers of staphylococci are disseminated in pus and dried exudate discharged from large infected wounds, burns and secondarily infected skin lesions, and in sputum coughed from the lung of patients with bronchopneumonia. Direct contact is the most important mode of spread, but airborne dissemination may also occur. Cross-infection is an important method of spread of staphylococcal disease, particularly in hospitals, and scrupulous hand washing is essential in preventing spread. Food handlers may similarly introduce enterotoxin-producing food poisoning strains into food.

Healthy carriers. *Staph. aureus* grows harmlessly on the moist skin of the nostrils in about 30% of healthy persons, and the perineum is also commonly colonized. Organisms are spread from these sites into the environment by the hands, handkerchiefs, clothing, and dust consisting of skin squames and cloth fibres. Some carriers, called *shedders*, disseminate exceptionally large numbers of staphylococci.

During the first day or two of life most babies become colonized in the nose and skin by staphylococci, and transmission from babies to nursing mothers, who then develop mastitis, is well described.

Animals. Animals may disseminate *Staph. aureus* and so cause human infection, for example milk from a dairy cow with mastitis, causing staphylococcal food poisoning.

Modes of infection

Acquisition of infection may be *exogenous* (from an external source) or *endogenous* (from a carriage site, or minor lesion elsewhere in the patient's own body). It is important to remember that the body surfaces of human beings and animals are the main reservoir. Although not spore forming, staphylococci may remain alive in a dormant state for several months when dried in pus, sputum, bed clothes or dust, or on inanimate surfaces such as floors. They are fairly readily killed by heat (e.g. moist heat at 65°C for 30 min), by exposure to light and by common disinfectants.

Methicillin-resistant Staph. aureus (MRSA)

MRSA is endemic in many hospitals throughout the world and particularly affects vulnerable patients, such as those who have undergone major surgery and patients in the intensive care unit. Although 50–60% of patients with MRSA are merely colonized (i.e. they carry the bacteria but do not have symptoms or an illness), serious infections such as those involving the bloodstream, respiratory tract and bones or joints do occur. These infections are then more difficult to treat than infections caused by methicillin-susceptible isolates, and MRSA can spread easily among patients in hospital. Methicillin resistance is mediated through the *mecA* gene, which encodes a unique penicillin-binding protein.

Community-acquired strains have been described that can cause soft tissue infections; these strains often produce the Panton–Valentine leucocidin. They can be distinguished from endemic hospital strains, from which it is believed that they have arisen.

The control and prevention of MRSA involves early and reliable detection in the laboratory through surveillance, patient isolation when admitted to hospital, good professional practice by all health-care workers (including compliance with hand hygiene guidelines), effective hospital hygiene programmes and the sensible use of antibiotics. Such measures have been very successful in Scandinavia and some other countries, but the prevalence of MRSA elsewhere, including the UK, Ireland and southern Europe, is much higher.

Laboratory diagnosis

One or more of the following specimens should be collected to confirm a diagnosis:

- *Pus* from abscesses, wounds, burns, etc. is much preferred to swabs.
- *Sputum* from patients with pneumonia (e.g. post-influenzal or ventilator-associated pneumonia); bronchoscopic specimens are increasingly used in critically ill patients.
- *Faeces* or *vomit* from patients with suspected food poisoning, or the remains of implicated foods.

- *Blood* from patients with suspected bloodstream infection (bacteraemia), such as septic shock, osteomyelitis or endocarditis.
- *Mid-stream urine* from patients with suspected cystitis or pyelonephritis.
- *Anterior nasal* and *perineal swabs* (moistened in saline or sterile water) from suspected carriers; nasal swabs should be rubbed in turn over the anterior walls of both nostrils.

The characteristic clusters of Gram-positive cocci can often be demonstrated by microscopy, and the organisms cultured readily on blood agar and most other media. The tube or slide coagulase test is performed to distinguish *Staph. aureus* from coagulase-negative species. Molecular methods such as the polymerase chain reaction (PCR) have been developed, but are still being evaluated to determine their role in routine laboratory practice.

Typing

Most staphylococcal infections are sporadic, but outbreaks occur, especially in hospitals. The identification of an outbreak strain, by determining whether all the isolates are of the same type, is an important aspect in the investigation of a source. Strains of *Staph. aureus* can be differentiated into different *phage types* by observation of their pattern of susceptibility to lysis by a standard set of *Staph. aureus* bacteriophages (viruses that infect bacteria). Virulent phages cause lysis of staphylococci and thus produce a clearing in the lawn of growth. Phage types are designated according to the phages able to cause this effect, and there is international agreement on the interpretation of results. Thus, a strain of type 3B/3C/55 is one that is lysed only by phages 3B, 3C and 55.

Many strains of MRSA are non-typable with standard and additional or experimental phages. Consequently, phage typing is being supplemented by genotypic methods such as PCR, pulsed-field gel electrophoresis, ribotyping and gene sequencing (see Ch. 3). Unfortunately, there are as yet no internationally agreed criteria for assessing the results in the same way as there is for phage typing.

Treatment

Sensitivity to antibiotics

Staph. aureus and other staphylococci are inherently sensitive to many antimicrobial agents (Table 15.3). Among the most active is benzylpenicillin, but about 90% of strains found in hospitals are now resistant. Resistance to penicillin depends on production of the enzyme penicillinase, a β-lactamase that opens the β-lactam ring. Penicillinase also

Table 15.3 Antibiotics and staphylococci

Active agents	Agents lacking useful activity
Penicillins[a]	Aztreonam
Cephalosporins	Polymyxins
Aminoglycosides[b]	Mecillinam
Tetracyclines	Nitroimidazoles
Macrolides	Quinolones[c]
Lincosamides	
Glycopeptides	
Fluoroquinolones[c]	
Rifampicin[b]	
Fusidic acid[b]	
Trimethoprim	
Chloramphenicol	
Carbapenems	
Oxazolidinones	

[a]Resistance common (see text).
[b]Usually used in combination, for example with flucloxacillin.
[c]For categorization of quinolones, see Table 5.3 and associated text.

inactivates most of the other penicillins, although a few, including methicillin (used for laboratory testing and not for therapy), oxacillin, cloxacillin and flucloxacillin, are stable to the enzyme. Cephalosporins and β-lactamase inhibitors are also stable to penicillinase (see Ch. 5).

MRSA strains are resistant to all β-lactam agents, and often to other agents such as the aminoglycosides and fluoroquinolones. Glycopeptides (vancomycin or teicoplanin) are the agents of choice in the treatment of systemic infection with MRSA, but these agents are expensive and may be toxic. Isolates of MRSA with reduced susceptibility or full resistance to glycopeptide antibiotics are uncommon, but have been detected sporadically. These isolates have either thickened cell walls (reduced susceptibility) or the *vanA* gene (fully resistant), and can be difficult to detect in the routine diagnostic laboratory.

Choice of antibiotic for therapy

Pending receipt of susceptibility test results, the treatment of severe infections suspected to be caused by *Staph. aureus* should be started with flucloxacillin unless MRSA is endemic locally, in which case a glycopeptide such as vancomycin is indicated. Erythromycin, clindamycin or vancomycin (or teicoplanin) is indicated if the patient is allergic to penicillin. Fusidic acid and rifampicin are not used alone in serious infections, because mutation to resistance arises readily. It is usually necessary to remove an infected source, such as an intravascular catheter, or drain an abscess as part of the treatment.

Infections caused by bacteria that exhibit reduced susceptibility to glycopeptides may be treated (if susceptible)

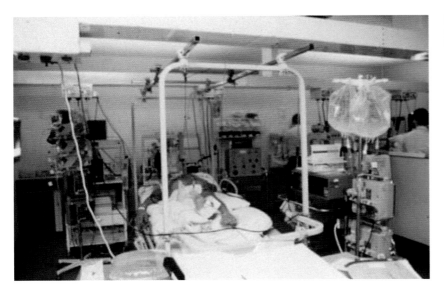

Fig. 15.3 Patients in the intensive care unit who require multi-organ support are at particular risk of MRSA.

with other antistaphylococcal agents. Newer agents such as quinupristin–dalfopristin, linezolid and daptomycin (see pp. 58 and 60) are promising, but experience is limited.

Life-threatening toxin-mediated disease, such as toxic shock syndrome, requires major medical support such as intravenous fluids to prevent multi-organ failure, often best provided in the intensive care unit (Fig. 15.3).

COAGULASE–NEGATIVE STAPHYLOCOCCI

Coagulase-negative staphylococci comprise a large group of related species commonly found on the surface of healthy persons, in whom they are rarely the cause of infection. More than 30 species are recognized, but few are commonly incriminated in human infection. *Staph. epidermidis* accounts for about 75% of all clinical isolates, probably reflecting its preponderance in the normal skin flora. Other species include *Staph. haemolyticus*, *Staph. hominis*, *Staph. capitis* and *Staph. saprophyticus* (a cause of urinary infection in young women). The emergence of coagulase-negative staphylococci as major pathogens reflects the increased use of implants such as cerebrospinal fluid shunts, intravascular lines and cannulae, cardiac valves, pacemakers, artificial joints, vascular grafts and urinary catheters, and the increasing numbers of severely debilitated patients in hospitals.

Description

Coagulase-negative staphylococci are morphologically similar to *Staph. aureus*, and the methods for isolation are the same. Colonies can be distinguished from *Staph.*

aureus by their failure to coagulate plasma and by their lack of clumping factor and deoxyribonuclease. Because *Staph. epidermidis* may contaminate clinical specimens, care has to be exercised in assessing its significance, especially from superficial sites. When isolated from sites such as blood or cerebrospinal fluid, further specimens should be obtained to confirm its clinical significance.

Coagulase-negative staphylococci are opportunistic pathogens that cause infection in debilitated or compromised patients such as premature neonates and oncology patients, often by colonizing biomedical devices such as prostheses, implants and intravascular lines. They cause particular problems in:

- cardiac surgery (prosthetic valve endocarditis)
- patients fitted with cerebrospinal fluid shunts (meningitis)
- continuous ambulatory peritoneal dialysis (peritonitis)
- immunocompromised patients (e.g. bloodstream infection)
- women during the reproductive years (urinary tract infection with *Staph. saprophyticus*).

Pathogenesis

Production of an exopolysaccharide, allowing adherence and subsequent formation of a multi-layered biofilm, appears to be essential for the pathogenesis of device-related *Staph. epidermidis* infection. A complex array of inter-related chemical messengers controls expression of polysaccharide and drives intercellular adhesion and biofilm formation. Attachment is enhanced by the presence of matrix proteins, such as fibronectin and fibrinogen. There is considerable interest in the development of implantable devices such as prosthetic heart valves or cerebrospinal shunts that are less prone to adherence by *Staph. epidermidis* and subsequent biofilm formation. This may be achieved by altering the structure of the polymer or by incorporating antibacterial agents on the surface of the device.

Treatment

Antibiotic treatment of coagulase-negative staphylococcal infections is complicated because susceptibility is gene-rally unpredictable. Strains resistant to penicillin, penicillinase-stable penicillins, gentamicin, erythromycin and chloramphenicol are common. If a strain is the cause of systemic infection, vancomycin or teicoplanin should be used. Rifampicin in combination with a glycopeptide is occasionally useful in treating central nervous system infections. Vancomycin resistance is rare, but one species, *Staph. haemolyticus*, is often resistant to teicoplanin. Uncomplicated urinary tract infection caused by *Staph. saprophyticus* usually responds to trimethoprim or one of the fluoroquinolones.

KEY POINTS

- Staphylococci are commonly found on the skin of healthy individuals. *Staph. aureus* is present in the nose of 30% of healthy people but can cause infections where there is lowered host resistance (e.g. damaged skin).
- Many virulence factors have been described for *Staph. aureus*, but for most a specific role has not been determined. Exceptions include the enterotoxins, toxic shock syndrome toxin and the epidermolytic toxins.
- Organisms are spread from colonized sites (e.g. skin) by hands, clothing, dust and desquamation from the skin.
- Methicillin-resistant *Staph. aureus* (MRSA) is increasingly prevalent in hospitals, causes the same range of infections as methicillin-susceptible isolates, and is of increasing concern to patients and the public.
- *Staph. aureus* is inherently susceptible to many antibiotics, but flucloxacillin and vancomycin or teicoplanin are the agents of choice to treat methicillin-susceptible and methicillin-resistant *Staph. aureus* infections, respectively.
- Coagulase-negative staphylococci (e.g. *Staph. epidermidis*) are major pathogens involving prosthetic implants such as intravascular lines or cardiac valves; the pathogenesis involves biofilm production.
- Device removal is usually required for successful treatment of infections caused by coagulase-negative staphylococci, as well as appropriate antibiotics, such as vancomycin or teicoplanin.

RECOMMENDED READING

Ala'Aldeen D, Hiramatsu K 2004 *Staphylococcus aureus: Molecular and Clinical Aspects*. Horwood Publishing, Chichester

Anstead G M, Owens A D 2004 Recent advances in the treatment of infections due to resistant *Staphylococcus aureus*. *Current Opinion in Infectious Diseases* 17: 549–555

Hiramatsu K 2001 Vancomycin-resistant *Staphylococcus aureus:* a new model of antibiotic resistance. *The Lancet Infectious Diseases* 1: 147–155

Kloos W E, Bannerman T L 1994 Update on clinical significance of coagulase-negative staphylococci. *Clinical Microbiology Reviews* 7: 117–140

Lowy F D 1998 *Staphylococcus aureus* infections. *New England Journal of Medicine* 339: 520–532

Mack D 1999 Molecular mechanisms of *Staphylococcus epidermidis* biofilm formation. *Journal of Hospital Infection* 43(Suppl): S113–S125

Working Party Report 2006 Revised guidelines for the control of meticillin-resistant *Staphylococcus aureus* infection in hospitals. *Journal of Hospital Infection* 63(Suppl 1): S1–S44.

Internet sites

UK Health Protection Agency. *Staphylococcus aureus*. http://www.hpa.org.uk/infections/topics_az/staphylo/menu.htm

European Antimicrobial Resistance Surveillance System (Various documents including annual reports on prevalence of antibiotic resistance) http://www.rivm.nl/earss/

16 Streptococcus and enterococcus

Pharyngitis; scarlet fever; skin and soft tissue infections; streptococcal toxic shock syndrome; pneumonia; meningitis; urinary tract infections; rheumatic fever; post–streptococcal glomerulonephritis

M. Kilian

Streptococci is the general term for a diverse collection of Gram-positive cocci that typically grow as chains or pairs (Greek: *streptos* = pliant or chain; *coccos* = a grain or berry) (Fig. 16.1). Virtually all of the streptococci that are important in human medicine and dentistry fall into the genera *Streptococcus* and *Enterococcus*. Commensal streptococci of the oral cavity are common causes of subacute bacterial endocarditis. Occasional opportunistic infections are associated with other genera of streptococci, such as *Peptostreptococcus* (see Ch. 36) and *Abiotrophia* ('nutritionally variant streptococci').

Streptococci are generally strong fermenters of carbohydrates, resulting in the production of lactic acid, a property used in the dairy industry. Most are facultative anaerobes, although peptostreptococci are obligate anaerobes. Streptococci do not produce spores and are nonmotile. They are catalase negative.

CLASSIFICATION

The genus *Streptococcus* includes important pathogens and commensals of mucosal membranes of the upper respiratory tract and, for some species, the intestines. The genus *Enterococcus*, which is also an intestinal commensal, is related to the other streptococci, but is classified separately.

The genus *Streptococcus* includes more than 50 species. With few exceptions, the individual species are exclusively associated, as either pathogens or commensals, with man or a particular animal. The genus consists of six clusters of species (Table 16.1), each of which is characterized by distinct pathogenic potential and other properties:

- The *pyogenic* group includes most species that are overt human and animal pathogens.

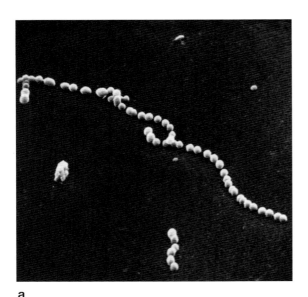

a

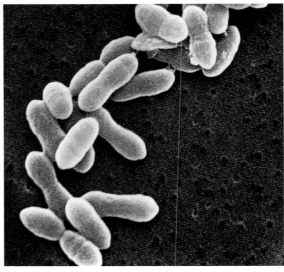

b

Fig. 16.1 Scanning electron micrograph of a *Str. pyogenes* showing typical chain formation (original magnification ×2000) and b *Str. pneumoniae* showing typical diplococcus formation (original magnification ×7000). (Courtesy of A P Shelton, University Hospital, Nottingham.)

Table 16.1 *Streptococcus* species of clinical importance			
Phylogenetic group	Species	Lancefield group	Type of haemolysis[a]
Pyogenic group	*Str. pyogenes*	A	β
	Str. agalactiae	B	β
	Str. equisimilis	C	β
Mitis group	*Str. pneumoniae*	O	α
	Str. mitis	O	α
	Str. oralis	Not identified	α
	Str. sanguinis	H	α
Anginosus group	*Str. anginosus*	G, F (and A)	α or β
	Str. intermedius		α
Salivarius group	*Str. salivarius*	K	None
Bovis group	*'Str. bovis'*	D	α or none
Mutans group	*Str. mutans*	Not designated	None
	Str. sobrinus	Not designated	None

[a]On horse blood agar.

- The *mitis* group includes commensals of the human oral cavity and pharynx, although one of the species, *Streptococcus pneumoniae*, is also an important human pathogen.
- The *anginosus* and *salivarius* groups are part of the commensal microflora of the oral cavity and pharynx.
- The *bovis* group belongs in the colon.
- The *mutans* group of streptococci colonizes exclusively the tooth surfaces of man and some animals; some species belonging to this cluster are involved in the development of dental caries.

Virtually all of the commensal species, including the enterococci, are opportunistic pathogens, primarily if they gain access to the bloodstream from the oral cavity or from the gut.

Haemolytic activity

Early attempts to distinguish between pathogenic and commensal streptococci recognized different types of reactions around colonies on blood agar plates. Colonies of streptococci belonging to the pyogenic group are generally surrounded by a clear zone, usually several millimetres in diameter, caused by lysis of red blood cells in the agar medium induced by bacterial haemolysins. This is called β-*haemolysis* and constitutes the principal marker for potentially pathogenic streptococci in cultures of throat swabs or other clinical samples. Accordingly, the pyogenic streptococci are also referred to as *haemolytic streptococci*.

In contrast, most commensal streptococci give rise to a green discoloration around colonies on blood agar. This phenomenon is termed α-*haemolysis*, although not caused

by haemolysis. The factor causing the green discoloration is not a haemolysin, but hydrogen peroxide, which oxidizes haemoglobin to the green methaemoglobin.

Collectively, commensal streptococci are often called '*viridans streptococci*', which refers to their α-haemolytic property (*viridis* = green). Not quite logically, this term also includes the few streptococci, such as those of the salivarius and mutans groups, that induce neither α- nor β-haemolysis. Moreover, in common usage, the term excludes *Str. pneumoniae*, although this species is also α-haemolytic.

Lancefield grouping

An important method of distinguishing between pyogenic streptococci is the serological classification pioneered by the American bacteriologist Rebecca Lancefield, who detected different versions of the major cell wall polysaccharide among the pyogenic streptococci.

This polysaccharide can be extracted from streptococci with hot hydrochloric acid and the different forms can be distinguished by precipitation with specific antibodies raised in rabbits. The polysaccharide is referred to as the group polysaccharide and identifies a number of different groups labelled by capital letters (Lancefield groups A, B, C, etc.).

Among the pyogenic streptococci the individual serological groups are, with few exceptions, identical to distinct species (Table 16.1). Subsequently, serological grouping has also been applied to viridans streptococci and to enterococci, and the number of serological groups has been extended to a total of 21 (A–H and K–W). However, this has limited practical significance because there is no direct correlation between individual serogroups of viridans streptococci and species.

STREPTOCOCCUS PYOGENES

This species, which consists of Lancefield group A streptococci, is among the most prevalent of human bacterial pathogens. It is associated exclusively with human infections. It causes a wide range of suppurative infections in the respiratory tract and skin, life-threatening soft tissue infections and certain types of toxin-associated reactions. Some of these infections may, in addition, result in severe non-suppurative sequelae due to adverse immunological reactions induced by the infecting streptococci.

The spectrum of infections caused by *Str. pyogenes* resembles that of *Staphylococcus aureus*, but the clinical characteristics associated with these two groups of pyogenic cocci are often distinct. Similarities and differences can be explained by the virulence factors expressed by the two species.

Pathogenesis

Virulence factors

Strains of *Str. pyogenes* express a large arsenal of virulence factors involved in adherence, evasion of host immunity and tissue damage (Fig. 16.2). Although some factors are expressed by all clinical isolates, others are variably present among *Str. pyogenes* strains. This variation is due to the horizontal transfer of virulence genes among strains, primarily by transduction (see p. 73), and probably explains the temporal variations in the prevalence of severe infections and sequelae. It furthermore explains differences in the virulence of individual strains and, to some extent, the different clinical pictures that may be associated with infections due to *Str. pyogenes*. Many of the virulence factors of *Str. pyogenes* are also expressed by some of the other species of pyogenic streptococci. In some species pathogenic for animals the corresponding virulence factors are expressed in a form specifically adapted to interact with a particular host.

Adhesion. Interaction with host fibronectin, a matrix protein on eukaryotic cells, is considered the principal mechanism by which *Str. pyogenes* binds to epithelial cells of the pharynx and skin. The structure that recognizes host fibronectin is located on the F protein,

which is one of the many proteins expressed on the surface of *Str. pyogenes* (Fig. 16.2). The interaction between the streptococcal F protein and host cell fibronectin also mediates internalization of the bacteria into host cells.

In addition to the F protein, surface-exposed lipoteichoic acid and M proteins appear to be involved in adherence to mucosal and skin epithelial cells.

M proteins. The ability of *Str. pyogenes* to resist phagocytosis by polymorphonuclear leucocytes is largely due to the cell surface-exposed M protein. The M protein is anchored in the cytoplasmic membrane, spans the entire cell wall, and protrudes from the cell surface as fibrils (Fig. 16.2). Acquired resistance to infection by *Str. pyogenes* is the result of antibodies to the M protein molecule in secretions and plasma. However, as a result of genetic polymorphism in the gene encoding the M protein, the most distal part of the protein shows extensive variability among strains. As a consequence, individuals may suffer from recurrent *Str. pyogenes* infections with strains expressing different versions of the M protein. More than 100 different types of M protein have been identified by serological means and gene sequencing.

Some strains produce two different M proteins with antiphagocytic activity, and some an additional structurally related M-like protein. All of these proteins can bind

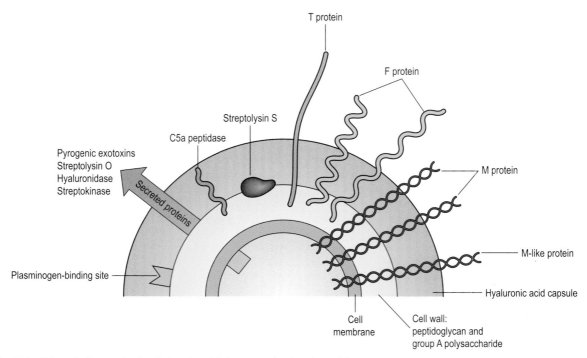

Fig. 16.2 Schematic diagram showing the location of virulence-associated products of *Str. pyogenes*.

various plasma proteins of the host, including fibrinogen, plasminogen, albumin, immunoglobulin (Ig) G, IgA, the proteinase inhibitor α_2-macroglobulin and some regulatory factors from the complement system (factor H and C4b-binding protein). As well as masking the bacterial surface with host proteins, some of these affinities are responsible for the ability of M proteins to resist phagocytosis. Thus, factor H can destabilize the important opsonin C3b when deposited on the bacterial surface. Likewise, the C4b-binding protein inhibits surface complement deposition by stimulating degradation of both C4b and C3b.

Str. pyogenes can release some of the M proteins with its own cysteine protease. If shed into the circulation the M protein forms complexes with fibrinogen, which by indirect pathways induces the release of inflammatory mediators from neutrophils. This process plays an important role in the leakage of plasma into the tissues and lungs, and the ensuing low blood pressure seen in some invasive infections with *Str. pyogenes*.

Capsule.　Some strains of *Str. pyogenes* form a capsule composed of hyaluronic acid. Such strains grow as mucoid colonies on blood agar and are highly virulent in animal models. Although capsule production is rare among isolates from uncomplicated pharyngitis, a significant proportion of isolates from severe infections have a capsule. Like other bacterial capsules, the capsule has an antiphagocytic effect. The relative significance of the M protein and the capsule as antiphagocytic factors differs among strains.

The capsule is identical to the hyaluronic acid of the connective tissue of the host and is not immunogenic. In this way the bacteria can disguise themselves with an immunological 'self' substance.

C5a peptidase.　The C5a peptidase, which is also found in human pathogenic strains of *Str. agalactiae*, is present on the surface of all strains of *Str. pyogenes*. It specifically cleaves, and thereby inactivates, human C5a, one of the principal chemo-attractants of phagocytic cells.

Streptolysins.　*Str. pyogenes* produces two distinct haemolysins, termed streptolysins O (oxygen labile) and S (serum soluble), both of which lyse erythrocytes, polymorphonuclear leucocytes and platelets by forming pores in their cell membrane.

Streptolysin O belongs to a family of haemolysins found in many pathogenic bacteria. Intravenous injection into experimental animals causes death within seconds, as the result of an acute toxic action on the heart. Streptolysin O may play a role in the pathogenesis of post-streptococcal rheumatic fever. Serum antibodies can be demonstrated after streptococcal infection, particularly after severe infections.

Streptolysin S is responsible for the β-haemolysis around colonies on blood agar plates. It can also induce the release of lysosomal contents with subsequent cell death after engulfment by phagocytes. In contrast to streptolysin O, it is not immunogenic.

Pyrogenic exotoxins.　Most strains of *Str. pyogenes* produce one or more toxins that are called *pyrogenic exotoxins* because of their ability to induce fever. Three, designated SPE-A, SPE-B and SPE-C, have been characterized extensively, but there are several others. Purified SPE-A causes death when injected into rabbits and is the most toxic of the three, but SPE-B also causes myocardial necrosis and death in experimental animals.

The genes for SPE-A and SPE-C are transmitted between strains by bacteriophage, and stable production depends on lysogenic conversion in a manner analogous to toxin production by *Corynebacterium diphtheriae* (see p. 74). Even among strains that possess the genes, the quantity of toxin secreted varies dramatically.

SPE-A and SPE-C are also called erythrogenic toxins, as they are responsible for the rash observed in patients with scarlatina. They are genetically related to *Staph. aureus* enterotoxins and, like these, have superantigen activity. By cross-linking MHC class II molecules on antigen-presenting cells and the Vβ domain of the antigen receptor on a subset of T lymphocytes, these toxins cause comprehensive (antigen independent) activation of helper T lymphocytes. The result is massive release of pro-inflammatory cytokines such as interleukin (IL)-1 and IL-2, tumour necrosis factor (TNF)-α and interferon-γ. These cytokines may cause a variety of clinical signs, including inflammation, hypotensive shock and organ failure seen in some patients with severe streptococcal disease. The different clinical outcome of infections with the same *Str. pyogenes* strain appears to be due to the fact that the toxins bind preferentially to certain major histocompatibility complex (MHC) class II tissue types.

Unlike SPE-A and SPE-C, all strains of *Str. pyogenes* produce SPE-B, which is a potent cysteine proteinase capable of cleaving many host proteins.

None of the pyrogenic exotoxins is associated unambiguously with any of the clinical syndromes caused by *Str. pyogenes*. However, most isolates from episodes of severe invasive disease and toxic shock-like syndrome produce SPE-A, and the protease SPE-B appears to be responsible for the extensive tissue destruction observed in many patients with severe invasive infections, including toxic shock-like syndrome.

Hyaluronidase.　*Str. pyogenes* and several other pyogenic streptococci use a secreted hyaluronidase to degrade hyaluronic acid, the ground substance of host connective tissue. This property may facilitate the spread of infection along fascial planes. During infections, particularly those involving the skin, serum antibody titres to hyaluronidase show a significant rise.

Streptokinase. Streptokinase, also known as fibrino-lysin, is another spreading factor. It is expressed by all strains of *Str. pyogenes* and co-operates with a surface-expressed plasminogen-binding site on the bacteria. Once host plasminogen is bound to the bacterial surface, it is activated to plasmin by streptokinase. Thus, in contrast to *Staph. aureus*, which aims at hiding behind a wall of coagulated plasma (fibrin), *Str. pyogenes* employs host plasmin to hinder the build-up of fibrin barriers. As a result, soft tissue infections due to *Str. pyogenes* are more diffuse, and often rapidly spreading, than the well localized abscesses that typify staphylococcal infections.

Deoxyribonucleases (DNAases). At least four distinct forms of DNAases are produced by *Str. pyogenes*. The enzymes hydrolyse nucleic acids and may play a role as spreading factors by liquefying viscous exudates.

Clinical features

Although a general decrease in the prevalence of serious infections with *Str. pyogenes* has occurred since the mid nineteenth century, there has been a resurgence in severe streptococcal infections and increased mortality due to streptococcal sepsis since the early 1980s.

The most common route of entry of *Str. pyogenes* is the upper respiratory tract, which is usually the primary site of infection and also serves as a focus for other types of infection. Spread from person to person is by respiratory droplets or by direct contact with infected wounds or sores on the skin. Not all individuals colonized by *Str. pyogenes* in the upper respiratory tract develop clinical signs of infection.

After an acute upper respiratory tract infection, the convalescent patient may carry the infecting streptococci for some weeks. Only a few healthy adults carry *Str. pyogenes* in the respiratory tract, but the carriage rate in young school children is just over 10%. It may be considerably higher before or during an epidemic.

Non-invasive streptococcal disease

The most common infections caused by *Str. pyogenes* are relatively mild and non-invasive infections of the upper respiratory tract (*pharyngitis*) and skin (*impetigo*). In the USA more than 10 million cases of non-invasive *Str. pyogenes* infection are estimated to occur annually.

Pharyngitis. This is the most common infection caused by *Str. pyogenes*. Clinical signs such as abrupt onset of sore throat, fever, malaise and headache generally develop 2–4 days after exposure to the pathogen. The posterior pharynx is usually diffusely reddened, with enlarged tonsils that may show patches of grey–white exudate on their surface and, sometimes, accumulations of pus in the crypts. The local inflammation results in swelling of cervical lymph nodes. Occasionally, tonsillar abscesses develop; this is a very painful condition and potentially dangerous as the pathogen may spread to neighbouring regions and to the bloodstream.

Despite the significant symptoms and clinical signs, differentiating streptococcal pharyngitis ('*strep throat*') from viral pharyngitis is impossible without microbiological or serological examination. Culture studies show that 20–30% of cases of pharyngitis are associated with *Str. pyogenes*.

Some cases of streptococcal sore throat are caused by species belonging to Lancefield groups C and G.

Scarlet fever. Pharyngitis caused by certain pyro-genic exotoxin-producing strains of *Str. pyogenes* may be associated with a diffuse erythematous rash of the skin and mucous membranes (Fig. 16.3). The condition is known as *scarlet fever* or *scarlatina*. The rash develops within 1–2 days after the first symptoms of pharyngitis and initially appears on the upper chest, then spreading to the extremities. After an initial phase with a yellowish-white coating, the tongue becomes red and denuded ('*strawberry tongue*').

Between 1860 and 1870 the mean annual death rate from scarlet fever in England and Wales was close to 2500 per million of the population. Since then a steady decline in incidence has been observed. By the end of the twentieth century the number of scarlet fever cases in England and Wales had fallen below 25 per million, and death is now extremely rare.

Skin infections. *Str. pyogenes* may cause several types of skin infection, sometimes in association with *Staph. aureus*. The superficial and localized skin infection, known as *impetigo* or *pyoderma*, occurs mainly in children (Fig. 16.4). It primarily affects exposed areas on the face, arms or legs. The skin becomes colonized after contact with an infected person and the bacteria enter the skin through small defects. Initially, clear vesicles develop, which within a few days become pus filled. Secondary spread is often seen as a result of scratching.

Potentially more severe is the acute skin infection *erysipelas* (*erythros* = red; *pella* = skin), which occurs in the superficial layers of the skin (cellulitis) and involves the lymphatics. The infection is characterized by diffuse redness of the skin, and patients experience local pain, enlargement of regional lymph nodes and fever. Untreated, the infection may spread to the bloodstream, and was often fatal before antibiotics became available.

Invasive soft tissue infections

Severe, sometimes life-threatening, *Str. pyogenes* infec-tions may occur when the bacteria get into normally sterile parts of the body. The most severe forms of invasive infections are *necrotizing fasciitis*, streptococcal *toxic*

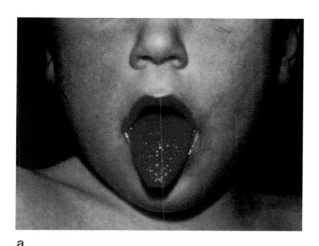

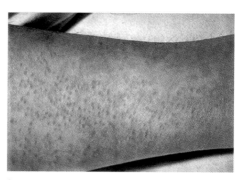

Fig. 16.3. The characteristic erythematous rash on **a** tongue and **b** skin associated with scarlet fever.

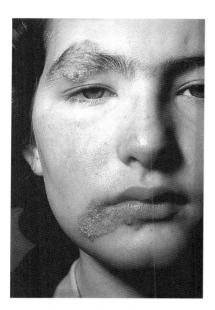

Fig. 16.4 Impetigo. (Courtesy of Dr L G Millard, Queen's Medical Centre, Nottingham.)

shock syndrome and *puerperal fever*, all of which are associated with bacteraemia.

Worldwide, rates of invasive infection increased from the mid 1980s to the early 1990s. In 2002, around 9000 cases of invasive *Str. pyogenes* infection occurred in the USA. Although these infections may occur in previously healthy individuals, patients with chronic illnesses such as cancer and diabetes, and those on kidney dialysis or receiving steroids, have a higher risk. Even with antibiotic treatment death occurs in 10–13% of all invasive cases:

45% of patients with toxic shock syndrome and 25% of patients with necrotizing fasciitis.

Necrotizing fasciitis. This infection progresses very rapidly, destroying fat and fascia. Although *Str. pyogenes* gains entry to these tissues through the skin after trauma, often of a minor nature, the skin itself may show only minimal signs of infection and may indeed be spared (Fig. 16.5). Systemic shock and general deterioration occur very quickly. The disease affects the fit young person with no obvious underlying pathology, as well as the immunocompromised.

The clinical diagnosis may be difficult because *Staph. aureus* and anaerobes such as *Clostridium perfringens* can produce a similar clinical picture. Streptococci can be isolated from the blood, blister fluid and cultures of the infected area.

Streptococcal toxic shock syndrome. Patients with invasive and bacteraemic *Str. pyogenes* infections, and in particular necrotizing fasciitis, may develop streptococcal toxic shock syndrome. The disease, which was first described in the late 1980s, is the result of the release of streptococcal toxins to the bloodstream. A striking feature of this acute fulminating disease is severe pain at the site of initial infection, usually the soft tissues. The additional clinical signs resemble those of staphylococcal toxic shock syndrome (see p. 173) and include fever, malaise, nausea, vomiting and diarrhoea, dizziness, confusion and a flat rash over large parts of the body. Without treatment, the disease progresses to shock and general organ failure.

Other suppurative infections. Historically, *Str. pyogenes* has been an important cause of puerperal sepsis. However, since the introduction of antibiotic therapy, this and other suppurative infections such as lymphangitis, pneumonia and meningitis are relatively rare.

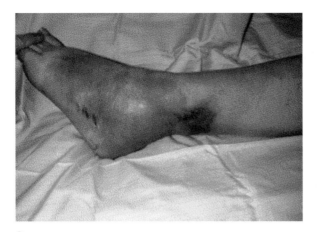

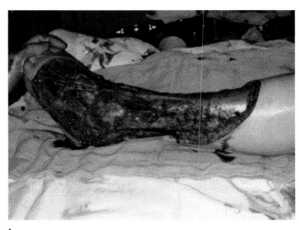

a   b

Fig. 16.5   Necrotizing fasciitis **a** before and **b** after surgical exploration and debridement. (Courtesy of Dr M Llewelyn, reproduced with permission from Elsevier.)

Bacteraemia. *Str. pyogenes* is the second most common (after *Str. agalactiae*) of the pyogenic streptococci isolated from blood cultures. Bacteraemia is seen regularly in patients with necrotizing fasciitis and toxic shock syndrome, but rarely as a complication to pharyngitis and local skin infections. Once in the blood, *Str. pyogenes* multiplies with incredible speed (doubling time 18 min), and the mortality rate approaches 40%. The potential complications include acute endocarditis leading to heart failure.

Non-suppurative sequelae

Two serious diseases may develop as sequelae to *Str. pyogenes* infections:

1. *Rheumatic fever*, a potential sequela to pharyngitis (including scarlet fever)
2. *Acute glomerulonephritis*, which is primarily, but not exclusively, associated with skin infections.

Both are caused by immune reactions induced by the streptococcal infection. The first clinical signs appear 1–5 weeks after the infection and at a time when the bacteria may have been eradicated by the immune system or as a result of antibiotic therapy.

Clinical correlations suggest that *psoriasis* may also be triggered by *Str. pyogenes* throat infection. Preliminary evidence supports the hypothesis that some streptococcal superantigens cause disruption of immunological tolerance of a $CD8^+$ T cell subset that recognizes cross-reactive epitopes on M proteins and skin keratin.

Rheumatic fever. This manifests as an inflammation of the joints (arthritis), heart (carditis), central nervous system (chorea), skin (erythema marginatum) and/or subcutaneous nodules. Polyarticular arthritis is the most common manifestation, whereas carditis is the most serious as it leads to permanent damage, particularly of the heart valves.

Rheumatic fever is a major cause of acquired heart disease in young people throughout the world. The incidence of rheumatic heart disease worldwide ranges from 0.5 to 11 per 1000 of the population. New cases are relatively rare in most of Europe, but increased incidences have been observed among the aboriginal populations of Australia and New Zealand, and in Hawaii and Sri Lanka. Outbreaks of rheumatic fever have also been seen in the USA.

The disease is autoimmune in nature and is believed to result from the production of autoreactive (and poly-specific) antibodies and T lymphocytes induced by cross-reactive components of the bacteria and host tissues. Therefore, repeated episodes of *Str. pyogenes* infection increase the severity of the disease. The major antigens involved are myosin, tropomyosin, laminin and keratin in the human tissues, and the group A antigen (a polymer of *N*-acetylglucosamine) in the *Str. pyogenes* cell wall in addition to epitopes on some variants of surface M proteins.

Acute post-streptococcal glomerulonephritis. The clinical manifestations include:

- coffee-coloured urine caused by haematuria
- oedema of the face and extremities
- circulatory congestion caused by renal impairment.

Unlike rheumatic disease, outbreaks of post-streptococcal acute glomerulonephritis have continued to decline in most parts of the world. Regions that still exhibit a high incidence of this disease include Africa, the Caribbean, South America, New Zealand and Kuwait.

Post-streptococcal glomerulonephritis is usually referred to as an immune complex-mediated disease. However, the exact pathogenesis is not clear. Several mechanisms have been proposed, including:

- immune complex deposition in the glomeruli
- reaction of antibodies cross-reactive with streptococcal and glomerular antigens
- alterations of glomerular tissues by streptococcal products such as streptokinase
- direct complement activation by streptococcal components that have a direct affinity for glomerular tissues.

Disease is associated with a limited number of M types of *Str. pyogenes*, and there is evidence that particular variants of streptokinase are crucial nephritogenic factors.

Unlike rheumatic fever, there is a general absence of individual recurrences, suggesting that antibodies to nephritogenic factors protect against disease rather than the opposite.

STREPTOCOCCUS AGALACTIAE

Str. agalactiae is equivalent to Lancefield group B streptococci. Its primary human habitat is the colon. It may be carried in the throat and, importantly, 10–40% of women intermittently carry *Str. agalactiae* in the vagina.

Previously, *Str. agalactiae* was recognized primarily as a cause of bovine mastitis (*agalactia*, want of milk). However, since 1960 it has become the leading cause of neonatal infections in industrialized countries and is also an important cause of morbidity among peripartum women and non-pregnant adults with chronic medical conditions. Among β-haemolytic streptococci, *Str. agalactiae* is the most frequent isolate from blood cultures.

Pathogenesis

Virulence factors

Str. agalactiae produces several virulence factors, including haemolysins, capsule polysaccharide, C5a peptidase (only human pathogenic strains), hyaluronidase (not all strains), and various surface proteins that bind human IgA and serve as adhesins.

Nine different types of the capsular polysaccharide have been identified (Ia, Ib and II–VIII). The serotype most frequently associated with neonatal infections is type III, whereas infections in adults are more evenly distributed over the different serotypes.

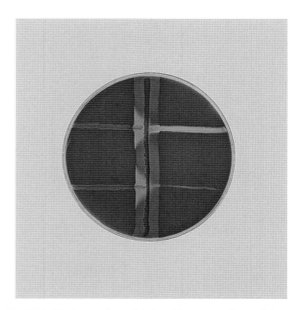

Fig. 16.6 Blood agar culture of strains of *Str. pyogenes* (group A) (upper right), *Str. equisimilis* (group C) (lower right) and *Str. agalactiae* (group B) (upper and lower left) surrounding a vertical streak of *Staph. aureus*. The two *Str. agalactiae* strains show a positive CAMP reaction.

Among the haemolysins produced by *Str. agalactiae*, one, known as the CAMP factor (so-called because it was originally described by Christie, Atkins and Munch-Petersen), plays an important role in the recognition of this species in the laboratory. The CAMP factor lyses sheep or bovine red blood cells pretreated with the β-toxin of *Staph. aureus* (Fig. 16.6). Purified CAMP factor protein is fatal to rabbits when injected intravenously.

Clinical features

Infection in the neonate

Two different entities are recognized:

1. *Early-onset disease*, most cases of which present at or within 12 h of birth.
2. *Late-onset disease*, presenting more than 7 days and up to 3 months after birth.

Early-onset disease. This results from ascending spread of *Str. agalactiae* from the vagina into the amniotic fluid, which is then aspirated by the infant and results in septicaemia in the infant or the mother, or both. Infants borne by mothers carrying *Str. agalactiae* may also become colonized during passage through the vagina, but early-onset disease is an uncommon outcome (about 1% of cases).

Depending on the site of initial contamination, neonates may be ill at birth or develop acute and fulminating illness a few hours, or a day or two, later. The clinical symptoms include lethargy, cyanosis and apnoea; when septicaemia progresses, shock ensues and death will occur if treatment is not instituted quickly. Meningitis and pulmonary infection may be associated.

As a result of improved recognition and prompt treatment of babies with symptoms, the fatality rate has been reduced to less than 10%. However, considerable morbidity persists among some survivors, especially those with meningitis.

Risk factors for neonatal colonization and infection are:

- premature rupture of membranes
- prolonged labour
- premature delivery
- low birth-weight
- intrapartum fever.

The immune status of the mother, and hence the level of maternal IgG antibodies in the infant, appears to be more important than the degree of colonization of the mother's genital tract by *Str. agalactiae*.

It is possible that *Str. agalactiae* itself may cause premature rupture of membranes as a result of secretion of proteases and activation of local inflammation.

Late-onset disease. Purulent meningitis is the most common manifestation, but septic arthritis, osteomyelitis, conjunctivitis, sinusitis, otitis media, endocarditis and peritonitis also occur. The incidence of invasive infection is higher among pre-term infants than among those born at term.

The pathogenesis is distinct from that of early-onset disease. There is usually no history of obstetric complications and the disease is unrelated to vaginal colonization in the mother. Many cases are acquired in hospital. Ward staff can be carriers of *Str. agalactiae*, and contamination of the baby may occur during nursing procedures, with subsequent baby-to-baby spread. Mastitis in the mother has also been described as a source of infection.

Infections in the adult

Ascending spread of *Str. agalactiae* leading to amniotic infection may result in abortion, chorio-amnionitis, post-partum sepsis (endometritis) and other infections (e.g. pneumonia) in the postpartum period in young, previously healthy women.

Str. agalactiae is also a frequent cause of infection in certain risk groups of non-pregnant adults. Disease may manifest as sepsis, pneumonia, soft tissue infections such as cellulitis and arthritis, and urinary tract infections complicated by bacteraemia. The risk factors in these patients are diabetes mellitus, liver cirrhosis, renal failure, stroke and cancer. Older age, independent of underlying medical conditions, increases the risk of invasive *Str. agalactiae* infection.

OTHER PYOGENIC STREPTOCOCCI

Group C and G streptococci

Several species of streptococci that carry the Lancefield group C or G antigens occasionally cause infections similar to those of *Str. pyogenes*. Most of these species have their primary habitat in horses, cattle and pigs, but some strains have become adapted to the human host and then produce some of the virulence factors primarily associated with *Str. pyogenes*. They can cause epidemic sore throat, especially in communities such as schools, nurseries and institutions, often associated with consumption of unpasteurized milk. Group C streptococci have been associated with acute glomerulonephritis but not with rheumatic fever. The most important of the group C streptococci in human medicine is *Str. equisimilis*.

Group R streptococci

Str. suis serotype 2 (group R streptococci) cause septicaemia and meningitis in pigs. They belong to a phylogenetic lineage separate from the other pyogenic streptococci. They occasionally infect people in contact with contaminated pork or infected pigs, and may cause septicaemia, meningitis and respiratory tract infections. Abattoir workers, butchers and, to a lesser extent, those involved in domestic food preparation are at risk.

STREPTOCOCCUS PNEUMONIAE

Str. pneumoniae, commonly called the pneumococcus, is a member of the oropharyngeal flora of 5–70% of the population, with the highest isolation rate in children during the winter months. In contrast to other streptococci, *Str. pneumoniae* generally occurs as characteristic diplococci (see Fig. 16.1). Although closely related genetically to the commensal *Str. mitis* and *Str. oralis*, *Str. pneumoniae* is an important pathogen. It primarily causes disease of the middle ear, paranasal sinuses, mastoids and lung parenchyma, but may spread to other sites, such as the joints, peritoneum, endocardium, biliary tract and, in particular, the meninges.

Str. pneumoniae is genetically very flexible because of frequent recombination between individual strains. Gene transfer is by transformation and may result in the expression of a different capsular serotype. Experimental transfer of capsule genes in pneumococci was the basis of the original demonstration that DNA contains the genetic information in cells.

Pathogenesis

Virulence factors

Capsule. The capsular polysaccharide is a crucial virulence factor. The capsule is antiphagocytic, inhibiting complement deposition and phagocytosis where type-specific opsonic antibody is absent. A total of 90 different capsular serotypes have been identified.

The serotypes are designated by numbers, and those that are structurally related are grouped together (1, 2, 3, 4, 5, 6A, 6B, etc.). The different serotypes differ in virulence. Thus, about 90% of cases of bacteraemic pneumococcal pneumonia and meningitis are caused by some 23 serotypes.

IgA1 protease. Like the two other principal causes of bacterial meningitis (*Neisseria meningitidis* and *Haemophilus influenzae*), pneumococci produce an extracellular protease that specifically cleaves human IgA1 in the hinge region. This protease enables these pathogens to evade the protective functions of the principal immunoglobulin isotype of the upper respiratory tract.

Pneumolysin. Pneumococci produce an intracellular membrane-damaging toxin known as pneumolysin, which is released by autolysis. Pneumolysin inhibits:

- neutrophil chemotaxis
- phagocytosis and the respiratory burst
- lymphocyte proliferation and immunoglobulin synthesis.

In experimental models it induces the features of lobar pneumonia and contributes to the mortality associated with this disease. Pneumolysin is immunogenic and might be suitable for a new pneumococcal vaccine.

Autolysin. When activated, the pneumococcal autolysin breaks the peptide cross-linking of the cell wall peptidoglycan, leading to lysis of the bacteria. Autolysis enables the release of pneumolysin and, in addition, large amounts of cell wall fragments. The massive inflammatory response to these peptidoglycan fragments is an important component of the pathogenesis of pneumococcal pneumonia and meningitis.

Clinical features

Predisposing factors

Most *Str. pneumoniae* infections are associated with various predisposing conditions. Although occasional clusterings of pneumococcal infections are recognized, person-to-person spread is uncommon.

Pneumonia results from aspiration of pneumococci contained in upper airway secretions into the lower respiratory tract; for example when the normal mechanisms of mucus entrapment and expulsion by an intact glottic reflex and mucociliary escalator are impaired. This situation may arise in:

- disturbed consciousness in association with general anaesthesia, convulsions, alcoholism, epilepsy or head trauma
- respiratory viral infections, such as influenza
- chronic bronchitis and other forms of chronic bronchial sepsis.

Other predisposing disease states in which pneumococcal pneumonia may be the terminal event include:

- valvular and ischaemic heart disease
- chronic renal failure
- diabetes mellitus
- bronchogenic and metastatic malignancy
- advancing age.

Immune deficiencies that predispose to pneumococcal infection include:

- hypogammaglobulinaemia
- asplenia or hyposplenism
- malignancies such as multiple myeloma.

In these conditions there is either a relative or absolute deficiency of opsonic antibody activity or an inability to induce a sufficient type-specific antibody response. *Tuftsin*, a naturally occurring tetrapeptide secreted by the spleen, also plays a role in combating pneumococcal sepsis; particularly at risk are those deficient in splenic activity:

- congenital asplenia
- traumatic removal
- functional impairment (e.g. homozygous sickle cell disease).

Human immunodeficiency virus (HIV) infection carries an increased risk of bacterial infections, including those caused by the pneumococcus, particularly in children.

Acute infections of the middle ear and paranasal sinuses occur in otherwise healthy children, but are usually preceded by a viral infection of the upper respiratory tract leading to local inflammation and swelling, and obstruction of the flow from these sites.

Pneumonia

Str. pneumoniae is the most frequent cause of pneumonia. The estimated annual incidence is 1–3 per 1000 of the population, with a 5% case fatality rate. Pneumococcal pneumonia follows aspiration with subsequent migration through the bronchial mucosa to involve the peribronchial lymphatics. The inflammatory reaction is focused primarily within the alveolus of a single lobule or lobe, although multilobar disease can also occur. Contiguous spread commonly results in inflammatory involvement of the pleura; this may progress to empyema.

Pericarditis is an uncommon but well recognized complication. Occasionally, lung necrosis and intrapulmonary abscess formation occur with the more virulent pneumococcal serotypes. Bacteraemia may complicate pneumococcal pneumonia in up to 15% of patients. This can result in metastatic involvement of the meninges, joints and, rarely, the endocardium.

The mortality rate from pneumococcal pneumonia in those admitted to hospital in the UK is approximately 15%. It is increased by age, underlying disease, bloodstream involvement, metastatic infection and certain types of pneumococci with large capsules (e.g. serotype 3).

Otitis media and sinusitis

Middle ear infections (otitis media) affect approximately half of all children between the ages of 6 months and 3 years; approximately one-third of cases are caused by *Str. pneumoniae*. Disease occurs after acquisition of a new strain to which there is no pre-existing immunity. The prevalence is highest among children attending primary school, where there is a constant exchange of pneumococcal strains.

Meningitis

Str. pneumoniae is among the leading causes of bacterial meningitis. It is assumed that invasion arises from the pharynx to the meninges via the bloodstream, as bacteraemia usually coexists. Meningitis may occasionally complicate pneumococcal infection at other sites, such as the lung.

The incidence of pneumococcal meningitis is bimodal and affects children less than 3 years of age and adults of 45 years and above. The fatality rates are 20% and 30%, respectively, considerably higher than those associated with other types of bacterial meningitis.

COMMENSAL STREPTOCOCCI

Viridans streptococci

The viridans streptococci, and in particular the species of the mitis and salivarius groups, are dominant members of the resident flora of the oral cavity and pharynx in all age groups. They play an important role by inhibiting the colonization of many pathogens, including pyogenic streptococci. This is achieved by two different mechanisms:

Table 16.2 Factors predisposing to infective endocarditis

Cardiac factors	Non-cardiac factors
Rheumatic heart disease	Dental manipulations[a]
Atherosclerotic heart disease	Endoscopy
Congenital heart disease	Intravenous drug abuse
Cardiac surgery	Intravenous cannulae and shunts
Prosthetic heart valves	Sepsis
Previous endocarditis	

[a]Procedures in which bleeding occurs.

1. production of bacteriocins (see p. 34)
2. production of hydrogen peroxide (also responsible for α-haemolysis).

Most strains secrete bacteriocins. Experimental implantation of strains of *Str. salivarius* with strong bacteriocin activity prevents colonization with *Str. pyogenes* in humans.

Mitis group. *Str. mitis*, *Str. oralis* and *Str. sanguinis*, among other viridans streptococci, colonize tooth surfaces as well as mucosal membranes. Because of their presence in the bacterial deposits (dental plaque) on tooth surfaces, these species may enter the bloodstream during dental procedures such as tooth extraction or vigorous tooth cleaning, particularly when the gingival tissue is inflamed.

In healthy individuals, bacteria of such low virulence are cleared from the circulation within 1 h. However, in patients with various predisposing conditions (Table 16.2), in particular heart valve damage due to post-streptococcal rheumatic fever, the circulating streptococci may settle in a niche protected from phagocytic cells. Local growth on the surface of heart valves eventually causes scarring and functional deficiency. As the disease progresses over several months, it is referred to as *subacute bacterial endocarditis*. It is usually accompanied by intermittent fever. Disruption of bacteria from the cardiac vegetations may cause embolic abscesses in various organs, including the brain. Until endocarditis due to skin staphylococci became more prevalent as a result of intravenous drug abuse, viridans streptococci were the most frequent causes of infective endocarditis.

Str. mitis and *Str. oralis* are increasingly recognized as causes of often fatal septicaemias in immunocompromised patients.

Mutans group. *Str. mutans* and *Str. sobrinus* exclusively colonize tooth enamel and do not occur until tooth eruption. Their proportions in the biofilm forming on tooth surfaces (dental plaque) are closely related to sugar consumption, and they are a major cause of dental caries because of their ability to produce large amounts of lactic acid even at pH values below 5.0. Like most other plaque streptococci they may cause subacute bacterial endocarditis.

Anginosus group. *Str. anginosus, Str. intermedius* and several other ill defined taxa are regular members of the commensal bacteria on tooth surfaces, in particular in the gingival crevices. They are often isolated from abscesses and other opportunistic purulent infections.

Bovis group. Some of the species of this group are present in the human gut. They occasionally cause bacteraemia and subacute endocarditis. These infections are often associated with colonic carcinoma, which jeopardizes the barrier function of the intestinal wall.

Enterococcus species

As indicated by the name, members of the genus *Enterococcus* have their natural habitat in the human intestines. The species most commonly associated with human disease are *E. faecalis* and *E. faecium*. The diseases with which they are associated are:

- urinary tract infection
- infective endocarditis
- biliary tract infections
- suppurative abdominal lesions
- peritonitis.

E. faecalis and *E. faecium* are important causes of wound and urinary tract infection in hospital patients and may cause sporadic outbreaks. Bacteraemia carries a poor prognosis as it often occurs in patients with major underlying pathology and in those who are immunocompromised.

LABORATORY DIAGNOSIS

Collection of specimens

The diagnosis of streptococcal infections is established by demonstrating the presence of the pathogen in throat or skin swabs, pus, blood cultures, cerebrospinal fluid (CSF), expectorates or urine according to the site of infection.

Particular problems are associated with the collection of expectorates for detection of *Str. pneumoniae*. Although it is the commonest cause of community-acquired pneumonia, sputum culture is positive in only about 20% of cases for several reasons.

- There may be difficulty in obtaining an expectorated specimen; postural drainage or inhaled aerosolized saline can encourage its production. Sputum may also be obtained by bronchoscopy in patients who are ventilated.
- Previous antibiotic treatment substantially reduces the chance of isolating the pneumococcus from the sputum.

The adequacy of the sputum sample should be confirmed by microscopy, which should show a predominance of pus cells rather than squamous epithelial cells from the mouth. The specimen is then homogenized with an agent such as *N*-acetylcysteine which permits semi-quantitative culture.

In pneumococcal meningitis the CSF is often macroscopically cloudy. The cell count is usually increased markedly and shows a predominance of polymorphonuclear leucocytes. Typical Gram-positive diplococci can commonly be demonstrated, sometimes in enormous numbers, by Gram-stain examination of a CSF deposit. The appearance is often typical, and a presumptive diagnosis can be made to allow appropriate therapy to be started before the identity of the organism is confirmed by culture.

Blood cultures are of value in patients with invasive streptococcal infections. This is also the case in patients with suspected pneumococcal pneumonia, particularly when this is severe, as up to 15% of patients are bacteraemic. Detection of bacteria by direct plating or microscopy of blood is not feasible owing to their low density.

Other body sites that may merit investigation according to the clinical presentation include joint and peritoneal fluids. Tympanocentesis provides the possibility of establishing the microbial cause of otitis media, but as most of these infections settle spontaneously, or with the assistance of a few days' antibiotic treatment, tympanocentesis is not usually necessary.

Cultivation and identification

Unlike staphylococci, streptococci lack the enzyme catalase, which releases oxygen from hydrogen peroxide. Catalase-negative Gram-positive cocci are therefore likely to be streptococci. The appearance of cocci in obvious chains (see Fig. 16.1) is another useful criterion, but the length of the chains varies with the species and conditions of growth. Optimal chain formation is seen in broth cultures. There may be marked variation in size and shape, particularly in older cultures or in direct smears from purulent exudates.

The primary cultivation medium for streptococci is blood agar, supplemented, whenever enterococci are suspected, with an agar medium (e.g. MacConkey's medium) selective for enterobacteria. Streptococci of aetiological significance usually predominate in the culture even when the sample is taken from a site with a resident microflora.

The pyogenic streptococci are detected initially by their β-haemolytic activity. The colonies are about 1 mm in diameter and, in contrast to those of staphylococci, lack pigment. Colonies of pneumococci are α-haemolytic, smooth, and may vary in size according to the amount of capsular polysaccharide produced; those of serotypes 3 and 37 are usually larger than the rest and have a watery or mucoid appearance. During prolonged incubation, autolysis of bacteria within the flat pneumococcal colonies results in a typical subsidence of the centre ('*draughtsman colonies*').

Species identification of pyogenic streptococci is based largely on serological detection of group antigens by immune precipitation or co-agglutination techniques. An additional test that is helpful in the presumptive identification of *Str. pyogenes* is the bacitracin sensitivity test. In contrast to most other streptococci, *Str. pyogenes* is uniformly sensitive and large inhibition zones are formed round bacitracin discs on blood agar. Likewise, *Str. agalactiae* can be identified presumptively by the CAMP reaction (see Fig. 16.6).

Pneumococci are distinguished from other α-haemolytic streptococci by their characteristic sensitivity to optochin (ethylhydrocupreine). Growth of pneumococci is inhibited around an optochin disc applied to an inoculated blood agar plate. With few exceptions, other α-haemolytic streptococci are not inhibited. In doubtful cases the identity of pneumococci is confirmed by demonstrating bile solubility, autolysis or reactivity to a poly-specific antiserum ('omni-serum') against capsular polysaccharides.

Other streptococci are identified by biochemical characteristics, such as the ability to ferment various carbohydrates and hydrolyse amino acids. However, some of the viridans streptococci are notoriously difficult to identify.

Enterococci are unique among the streptococci in their ability to grow on bile-containing media.

Antigen detection

Numerous commercial kits are available for the detection of *Str. pyogenes* directly in throat swabs without cultivation. These diagnostic kits use specific antibodies to detect the group A antigen in the material on the swab. They allow practitioners to test whether a throat infection is caused by *Str. pyogenes*, but the sensitivity and specificity of individual tests vary.

Antibody detection

Detection of antibodies against antigens of *Str. pyogenes* is an important means of establishing the diagnosis of post-streptococcal rheumatic fever and glomerulonephritis. In many cases the initiating infection in the throat or on the skin is no longer present.

Immune responses vary depending on whether the original focus is the throat or skin. Antibodies against streptolysin O (the ASO test) are used to document antecedent streptococcal infection in the throat of patients with clinical signs of rheumatic fever. A significant increase in antibody titre appears 3–4 weeks after initial exposure to the micro-organism. Detection of increased levels of serum antibodies to streptococcal hyaluronidase and DNAase B is also of diagnostic importance.

ASO estimation is unreliable in pyoderma-associated acute glomerulonephritis. A raised ASO titre is not observed in these patients, perhaps because lipids present in the skin inactivate the streptolysin O. Detection of antibodies against streptococcal DNAase B is recommended as a diagnostic tool in these patients.

Typing of streptococci

Strains of *Str. pyogenes* can be subdivided into serological types. The most comprehensive typing scheme is based on structural differences in the highly variable surface M protein. More than 100 different M types may be distinguished with type-specific antisera or by sequence differences in the gene (*emm*) encoding the M protein. An alternative typing scheme is based on the surface protein known as the T antigen.

Apart from serving epidemiological purposes, particular M types are associated with particular types of infection. Thus, certain M types are more commonly associated with skin infections than mucosal infections. Recent increases in the rate and severity of invasive *Str. pyogenes* infections (toxic shock syndrome and necrotizing fasciitis) have been associated primarily with serotypes M-1 and M-3. Rheumatic fever is often, but not exclusively, associated with M serotypes 1, 3, 5, 6 and 18.

Pneumococci are typed on the basis of the differences in capsular polysaccharides, of which 90 have been described. The addition of India ink to a suspension of pneumococci shows the presence of the capsule as a clear halo around the organisms. Mixing a suspension of pneumococci with type-specific antisera increases the visibility of the capsule in the microscope, and is the basis of the *quellung reaction* or *capsular swelling test*. Serotyping of pneumococci is carried out mainly in reference laboratories.

TREATMENT

Penicillin resistance has never been detected in *Str. pyogenes*. As a result, benzylpenicillin (penicillin G) or oral phenoxymethylpenicillin (penicillin V) are the drugs of choice for treatment of infections with *Str. pyogenes*. Antibiotic sensitivity tests are currently unnecessary when that species is identified as the infecting organism. In cases of hypersensitivity to penicillin, erythromycin is usually the second choice, but resistance occurs and is common in some countries.

Treatment for 3–5 days limits the effect of severe attacks of streptococcal infection and prevents suppurative complications such as otitis media, although the streptococci are eliminated from the infected area only if treatment is continued for 10 days. Treatment of uncomplicated throat infections with *Str. pyogenes* is not warranted in areas where post-streptococcal sequelae are rare.

Surgery is essential to remove damaged tissue in case of necrotizing fasciitis, as antibiotic penetration of the infected area is poor. Clindamycin is preferred to penicillin because it inhibits protein synthesis, including production of exotoxin.

Most strains of *Str. agalactiae* are susceptible to penicillins, macrolides and glycopeptides.

Although streptococci are intrinsically resistant to aminoglycosides, these agents interact synergically with penicillins and the combination is often used in the treatment of streptococcal and enterococcal endocarditis.

Pneumococci and viridans streptococci are often resistant to penicillins owing to mutations in the target penicillin-binding proteins. These mutations have accumulated in strains of *Str. mitis* and *Str. oralis*, and the altered genes have subsequently been transferred by genetic transformation to *Str. pneumoniae*. High-level penicillin resistance (minimum inhibitory concentration above 2 mg/l) was first recognized in 1977 in South Africa, where it was responsible for an epidemic of pneumococcal meningitis unresponsive to penicillin. The incidence of penicillin resistance is quite variable geographically and reflects the local level of antibiotic usage.

Most pneumococcal infections with strains exhibiting intermediate-level resistance to penicillin (minimum inhibitory concentration 0.1–1 mg/l) respond to high-dose therapy; an exception is meningitis, because of problems of penetration into the CSF. Penicillin resistance in pneumococci and other viridans streptococci is often linked to resistance to several other antibiotics. Resistance to erythromycin, tetracycline and chloramphenicol is not uncommon, and tolerance even to vancomycin has been reported.

The dose of penicillin necessary to treat susceptible pneumococcal infection is determined largely by pharmacological factors at the site of infection. For example, pneumococcal pneumonia responds to doses of penicillin as low as 0.3 g (0.5 mega-units) twice daily, whereas pneumococcal meningitis requires much higher doses. In patients unable to tolerate penicillin, erythromycin is the most widely used alternative agent for respiratory pneumococcal infections.

Unlike other streptococci, enterococci are intrinsically resistant to cephalosporins. Sensitivity to penicillins and other antibiotics varies widely, and clinical isolates must be tested for their susceptibility. Vancomycin resistance has been observed in enterococci and is a problem in high-dependency areas of some hospitals.

PREVENTION AND CONTROL

Hygienic measures

Skin infections with *Str. pyogenes* are usually associated with poor hygiene, and can to a large extent be prevented by standard hygienic measures. Late-onset neonatal infections with *Str. agalactiae* may also be prevented or significantly reduced by standard aseptic nursing procedures.

Likewise, hygiene is the most important preventive measure in relation to dental caries, which can be prevented largely by regular tooth-brushing with a fluoride-containing dentifrice.

Chemoprophylaxis

Prophylactic use of antibiotics is relevant in some streptococcal infections. As the primary attack of rheumatic fever usually occurs during childhood, long-term penicillin prophylaxis until adulthood is recommended to reduce the risk of further attacks and further heart injury. This is not the case in patients with acute glomerulonephritis, because of the lack of recurrences.

Two different approaches are used to prevent early-onset neonatal *Str. agalactiae* infections:

1. A risk-based strategy in which women of unknown colonization status receive intrapartum antibiotic prophylaxis in case of: threatened delivery at less than 37 weeks' gestation; premature rupture of the membranes; intrapartum fever; or previous delivery of a child who developed neonatal infection.

2. A screening-based approach in which all pregnant women at 35–37 weeks' gestation are screened for *Str. agalactiae* colonization in vaginal and rectal specimens. All identified carriers are offered intrapartum chemoprophylaxis.

Intravenous or intramuscular penicillin is the agent of choice because its antimicrobial spectrum, narrower than that of ampicillin, reduces the likelihood of resistance developing in other bacteria and selection of other potential pathogens. A cephalosporin is an appropriate alternative for patients with penicillin allergy as increasing proportions of *Str. agalactiae* are resistant to erythromycin and clindamycin.

Patients at risk of developing infective endocarditis (see Table 16.2) should be given prophylactic antibiotics in association with dental procedures that lead to bleeding. The current international recommendations are amoxicillin 1 h before dental treatment or, in case of penicillin allergy, clindamycin. If the patient has been on long-term penicillin prophylaxis, the oral streptococci are likely to have reduced susceptibility to penicillins, and clindamycin or vancomycin is recommended as the alternative. It is imperative that patients at risk maintain healthy periodontal conditions and that the amount of dental plaque is kept to a minimum.

Vaccines

Pyogenic streptococci

Attempts to develop a vaccine against *Str. pyogenes* infections have been hampered by two problems:

1. The considerable antigenic diversity of the M protein and other vaccine candidate antigens.
2. The potential immunological cross-reactivity of many of the antigens with host tissue components.

Several strategies are currently being tested, including oral vaccination.

A vaccine against neonatal *Str. agalactiae* infection based on protein-conjugated type III capsular polysaccharide is being tested for use primarily in women of reproductive age. However, additional serotypes are increasingly prevalent.

Pneumococci

Before the widespread availability of effective antimicrobial drugs the treatment of pneumococcal infections was based on the use of type-specific antiserum. This reduced the mortality rate associated with bacteraemic pneumococcal pneumonia, but not to the same extent that penicillin was subsequently shown to achieve. However, it indicated that type-specific antibody had a role in the control of pneumococcal disease and led to a variety of prototype vaccines. The vaccine that has been in use for many years contains a mixture of 23 polysaccharide serotypes chosen according to the prevalence of serotypes responsible for bacteraemic pneumococcal infection. It offers protection against 90% of isolates.

Like other vaccines based on pure polysaccharides, the immunogenicity of the multivalent vaccine is inadequate in those below 2 years of age and in those immunosuppressed as a result of malignancy, steroid therapy or other chronic disease. To overcome this problem, pneumococcal vaccines containing capsular polysaccharide coupled to a

carrier protein are now available. These vaccines increase the immunogenicity of the polysaccharide by rendering the response dependent on T lymphocyte help. The current conjugate vaccine includes only seven of the capsular polysaccharides, although vaccines with more comprehensive coverage are anticipated. Some countries contemplate inclusion of the new vaccine in the childhood vaccination programme.

Immunization is recommended for various groups at risk of pneumococcal disease, particularly those with congenital or surgical asplenia and those with hereditary haemoglobinopathies such as sickle cell disease, as pneumococcal infection can be fulminant in these patients. Vaccine efficacy is not complete and many clinicians also prescribe oral phenoxymethylpenicillin as long-term chemoprophylaxis in this high-risk group. In some countries, including the USA and the UK, the vaccine is recommended for those over 65 years of age, with or without previous ill health, although there have been difficulties in establishing scientifically the efficacy in this group.

KEY POINTS

- *Str. pyogenes* (group A streptococcus) is among the most prevalent of human bacterial pathogens.
- *Str. pyogenes* infections range from sore throat, scarlet fever and superficial skin infections to invasive soft tissue infections and septicaemia.
- *Str. pyogenes* produces several superantigenic extracellular toxins that are involved in the pathogenesis of the rash associated with scarlet fever and streptococcal toxic shock syndrome.
- Rheumatic fever and acute glomerulonephritis are potential immune-mediated sequelae of infections with *Str. pyogenes*.
- *Str. agalactiae* (group B streptococci) causes neonatal septicaemia and meningitis.
- *Str. pneumoniae* is a common cause of pneumonia, middle ear infections and meningitis.
- Commensal streptococci of the oral cavity are frequent causes of subacute bacterial endocarditis.

RECOMMENDED READING

Cunningham M W 2000 Pathogenesis of group A streptococcal infections. *Clinical Microbiology Reviews* 13: 470–511

Douglas C W I, Heath J, Hampton K K, Preston F E 1993 Identity of viridans streptococci isolated from cases of infective endocarditis. *Journal of Medical Microbiology* 39: 179–182

Kilian M 2005 *Streptococcus and Lactobacillus*. In Borriello P, Murray P R, Funke G (eds) *Topley and Wilson's Microbiology and Microbial Infections*, 9th edn, Vol. 2, pp. 833–881. Arnold, London

Llewelyn M, Cohen J 2002 Superantigens: microbial agents that corrupt immunity. *The Lancet Infectious Diseases* 2: 156–162

Marsh P D, Martin M W 1999 *Oral Microbiology*, 3rd edn. Wright, Oxford

Mitchell T J 2003 The pathogenesis of streptococcal infections: from tooth decay to meningitis. *Nature Reviews* 1: 219–230

Reinert R R, Reinert S, van der Linden M, Cil M Y, Al-Lahham A, Appelbaum P 2005 Antimicrobial susceptibility of *Streptococcus pneumoniae* in eight European countries from 2001 to 2003. *Antimicrobial Agents and Chemotherapy* 49: 2903–2913

Schrag S J, Zywicki S, Farley M M et al 2000 Group B streptococcal disease in the era of intrapartum antibiotic prophylaxis. *New England Journal of Medicine* 342: 15–20

Schuchat A 1999 Group B streptococcus. *Lancet* 353: 51–56

Stevens D L, Kaplan E L (eds) 2000 *Streptococcal Infections. Clinical Aspects, Microbiology, and Molecular Pathogenesis.* Oxford University Press, Oxford

Tomasz A 2000 *Streptococcus pneumoniae. Molecular Biology and Mechanisms of Disease.* Mary Ann Liebert, Larchmont, NY

Internet site

Centers for Disease Control and Prevention. Disease listing. http://www.cdc.gov/ncidod/dbmd/diseaseinfo/

Coryneform bacteria, listeria and erysipelothrix

Diphtheria; listeriosis; erysipeloid

J. McLauchlin and P. Riegel

CORYNEFORM BACTERIA

The term *coryneform* is used to describe aerobic, non-sporing and irregularly shaped Gram-positive rods. According to this broad definition, they include bacteria of the genus *Corynebacterium* with a typically club-shaped morphology (Greek κορσψνη = club), environmental bacteria showing coccoid forms such as *Rhodococcus*, *Gordonia* and *Brevibacterium* species, and preferentially anaerobic bacteria of the genera *Actinomyces* (see Ch. 20), *Actinobaculum*, *Arcanobacterium* and *Propionibacterium*, which exhibit some branched forms.

CORYNEBACTERIUM DIPHTHERIAE

The major disease caused by *C. diphtheriae* is *diphtheria*, an infection of the local tissue of the upper respiratory tract with the production of a toxin that causes systemic effects, notably in the heart and peripheral nerves. Diphtheria has virtually disappeared in developed countries following mass immunization, but is still endemic in many regions of the world. Skin infections are prevalent in some countries. Non-toxigenic strains have been associated with endocarditis, meningitis, cerebral abscess and osteoarthritis throughout the world.

Description

C. diphtheriae, like other members of the genus, are non-motile, non-spore-forming, straight or slightly curved rods with tapered ends. They are Gram-positive, but easily decolorized, particularly in older cultures. Cells often contain metachromatic granules (polymetaphosphate), which stain bluish-purple with methylene blue. Snapping division produces groups of cells in angular and palisade arrangements that create a 'Chinese character' effect. *C. diphtheriae* is aerobic and facultatively anaerobic, growing best on a blood- or serum-containing medium at 35–37°C with or without carbon dioxide enrichment. On agar medium containing tellurite, colonies of *C. diphtheriae* are characteristically black or grey after 24–48 h.

Biotypes of *C. diphtheriae* named *gravis*, *intermedius* or *mitis* are genomically similar variants exhibiting distinct biochemical features and cultural morphology. Bacilli of the gravis biotype are usually short, whereas those of biotype mitis are long and pleomorphic; biotype intermedius ranges from very long to short rods. In broth medium, *C. diphtheriae* biotype *gravis* forms a pellicle and a granular deposit, whereas *C. diphtheriae* biotype *mitis* produces a diffuse turbidity. The biotype *intermedius* forms no pellicle, but a fine granular deposit can be observed.

Pathogenesis

To cause disease *C. diphtheriae* must:

- invade, colonize and proliferate in local tissues
- be lysogenized by a specific β-phage, enabling it to produce toxin.

In the upper respiratory tract, diphtheria bacilli elicit an inflammatory exudate and cause necrosis of the cells of the faucial mucosa (Fig. 17.1). The diphtheria toxin possibly assists colonization of the throat or skin by killing epithelial cells or neutrophils.

The organisms do not penetrate deeply into the mucosal tissue and bacteraemia does not usually occur. The exotoxin is produced locally and spread by the bloodstream to distant organs, with a special affinity for heart muscle, the peripheral nervous system and the adrenal glands.

C. diphtheriae can colonize the throats of people who have been immunized against diphtheria or who have become immune as a result of natural exposure, but usually no pseudomembrane develops.

The diphtheria toxin is a heat-stable polypeptide, composed of two fragments: A (active) and B (binding). The toxin binds to a specific receptor on susceptible cells and enters by receptor-mediated endocytosis. The A subunit is cleaved and released from the B subunit as it inserts and passes through the membrane into the

Fig. 17.1 Diphtheritic membrane on throat. From Conlon, C., Snydman, D. Mosby's Color Atlas and Text of Infectious Diseases. Edinburgh: Mosby Elsevier, 2000. Courtesy of Nigel Day.

cytoplasm. Fragment A catalyses the transfer of adenosine disphosphate (ADP)-ribose from nicotinamide adenine dinucleotide (NAD) to the eukaryotic elongation factor 2, which inhibits the function of the latter in protein synthesis. Inhibition of protein synthesis is probably responsible for both the necrotic and neurotoxic effects of the toxin. Production of toxin by lysogenized *C. diphtheriae* is enhanced considerably when the bacteria are grown in low iron conditions. Other factors such as osmolarity, amino acid concentrations and pH have a role.

The *Schick test*, an intradermal injection of stabilized diphtheria toxin, was formerly used to determine individual susceptibility to the toxin. A localized erythema and, sometimes, severe local reactions reaching maximum size and intensity in 2–4 days, indicate that there is little or no neutralizing antitoxin in the tissue, and that the individual is susceptible to diphtheria. Absence of a reaction indicates immunity. Tissue culture neutralization tests, enzyme-linked immunosorbent assay (ELISA) and passive haemagglutination assay to measure serum antitoxin levels are now preferred. For epidemiological purposes the minimum protective level is considered to be 0.01 international units (IU) of diphtheria antitoxin per millilitre in a serum sample. A level of 0.1 IU/ml is desirable for individual protection.

Non-toxigenic strains of *C. diphtheriae* may cause pharyngitis and cutaneous abscesses. Systemic disease, including endocarditis, septic arthritis and osteomyelitis, has also been reported. The virulence factors of these strains remain unknown. Conversion of a non-toxigenic strain to a toxigenic strain by phage infection can occur in human populations.

Clinical features

The incubation period of diphtheria is 2–5 days, with a range of 1–10 days. At first, patients present with malaise, sore throat and moderate fever. A thick, adherent green *pseudomembrane* is present on one or both tonsils or adjacent pharynx. In nasopharyngeal infection, the pseudomembrane may involve nasal mucosa, the pharyngeal wall and the soft palate. In this form, oedema involving the cervical lymph glands may occur in the anterior tissues of the neck, a condition known as *bullneck diphtheria*.

Laryngeal involvement leads to obstruction of the larynx and lower airways. Organisms multiply within the membranes and toxaemia is prominent. The patient is gravely ill, with a weak pulse, restlessness and confusion. Intoxication takes the form of myocarditis and peripheral neuritis, and may be associated with thrombocytopenia. Visual disturbance, difficulty in swallowing, and paralysis of the arms and legs also occur but usually resolve spontaneously. Complete heart block may result from myocarditis. Death is most commonly due to congestive heart failure and cardiac arrhythmias.

Cutaneous diphtheria occurs mostly in tropical countries. The lesion is usually characterized by an ulcer covered by a necrotic pseudomembrane and may involve any area of the skin. Although the organism usually produces toxin, systemic toxic manifestations are uncommon.

Diagnosis

The diagnosis is made on clinical grounds, supported by a history of diphtheria among contacts, lack of prior immunization or travel in countries where diphtheria is endemic.

The role of the laboratory is to confirm the diagnosis by recovery of *C. diphtheriae* in culture followed by appropriate tests for detection of toxin production (Fig. 17.2). The clinician should inform the laboratory of the presumptive diagnosis of diphtheria because isolation of *C. diphtheriae* requires special media. Material for cultures should be obtained on a swab from the inflamed areas surrounding the pseudomembranes.

Direct microscopy of a smear is unreliable because *C. diphtheriae* is morphologically similar to other coryneforms. The recommended media include blood agar and a selective medium containing tellurite. Identification is based on carbohydrate fermentation reactions and enzymatic activities. Commercial kits such as the API Coryne strip provide a reliable identification.

Toxigenicity testing is essential. Production of diphtheria toxin is demonstrated by the agar immunoprecipitation test (*Elek test*; Fig. 17.3) or by the tissue culture cytotoxicity assay, which has replaced the virulence test in guinea-pigs. The toxin gene can be detected by the polymerase chain reaction (PCR). This test shows excellent correlation with guinea-pig virulence, although there is the rare possibility of a false-positive PCR assay if the strain harbouring the *tox* gene is unable to express it. The detection of the *tox* gene by PCR directly from clinical specimens is feasible. All biotypes are potentially toxigenic.

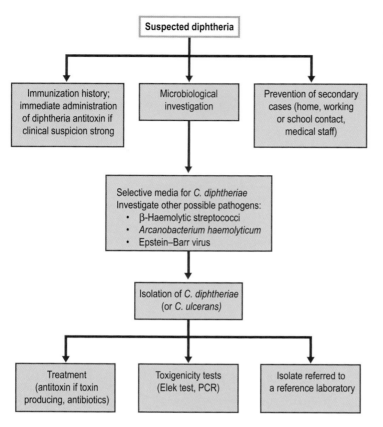

Fig. 17.2 Algorithm for the management of a suspected case of diphtheria. *Note*: Antitoxin treatment should not await laboratory confirmation, which may take several days. Photograph courtesy of Dr A. Efstratiou, Central Public Health Laboratory, London.

Measurement of antibodies to diphtheria toxin in serum collected before administration of antitoxin may support the diagnosis when cultures are negative. An algorithm for the management of suspect cases of diphtheria is shown in Figure 17.2.

Treatment

If diphtheria is strongly suspected on clinical grounds, treatment should not await laboratory confirmation, which may take several days (Fig. 17.2). Diphtheria antitoxin (hyperimmune horse serum) is given, as antibiotics have no effect on preformed toxin which rapidly diffuses from the local lesions and soon becomes irreversibly bound to tissue cells. Because antitoxin neutralizes only circulating toxin, it should be administered promptly.

Treatment with parenteral penicillin or oral erythromycin eradicates the organism and terminates toxin production. *C. diphtheriae* is universally sensitive to penicillins but some strains are resistant to erythromycin, tetracyclines and rifampicin. Erythromycin may be preferred to penicillin for elimination of the bacilli from the throat, particularly in treatment of persistent carriers. Some strains are tolerant to the bactericidal action of penicillins, and treatment of complicated infections should contain an association with an aminoglycoside.

Patients should be placed in strict isolation, nursed by staff whose immunization history is documented and have daily platelet counts and electrocardiography.

Epidemiology

Diphtheria has virtually disappeared in developed countries following mass immunization in the 1940s, but is still endemic in many regions of the world. About 50 000 cases of diphtheria occurred in the newly independent states of the former Soviet Union during 1990–1996, leading to infection in short-term visitors from western Europe. Other countries that have experienced outbreaks of diphtheria in recent years include China, Ecuador, Algeria, South-East Asia and the eastern Mediterranean. There were 124 cases of diphtheria, with eight deaths in the UK between 1970 and 1994, since when no deaths have been reported. In the USA, only 45 cases were reported during 1980–1995. In 2003, a total of 896 cases were reported from the World Health Organization European Region; 99% were from eastern Europe. Three cases were reported from the UK. In 2002, one case of diphtheria was reported in the USA but more toxigenic strains were referred to North American reference laboratories. Non-

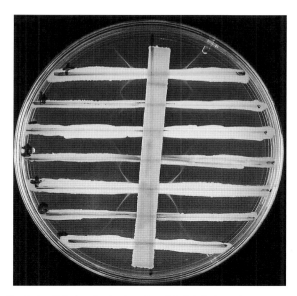

Fig. 17.3 Elek plate for the detection of *C. diphtheriae* toxin production. Cultures are streaked horizontally, then overlaid with an antitoxin-impregnated strip. Toxin and antitoxin diffuse into the culture during incubation, and precipitin lines develop where toxin and antitoxin are present in a critical ratio. Positive reactions in test cultures are indicated by precipitin lines that arc with those produced by positive controls. The *C. diphtheriae* cultures are (top to bottom): National Collection of Type Cultures (NCTC) strain 10648 (positive control); test culture (positive); NCTC 10356 (negative control); NCTC 3984 (weak positive control); NCTC 10648 (positive control); test culture (negative); NCTC 10356 (negative control). (Courtesy of Dr A Efstratiou, Health Protection Agency, London.)

toxigenic strains capable of causing mild disease continue to circulate throughout the world.

Infection is confined to man and usually involves contact with a diphtheria case or a carrier. The most important mode of spread is person-to-person transmission by aerosolized droplets when an infected person coughs, sneezes or talks, or by direct contact with skin lesions. Most clinical infections are probably contracted from carriers rather than symptomatic patients. Prolonged close contact with an infected person and intimate contact increases the likelihood of transmission.

Acquired immunity to diphtheria is due primarily to toxin-neutralizing antibody (antitoxin). Passive immunity in utero is acquired transplacentally and can last for 1 or 2 years after birth. Active immunity can probably be produced by a mild or subclinical infection in infants who retain some maternal immunity. Unimmunized children under 15 years old are most likely to contract diphtheria. The disease is also found among adults whose immunization was neglected. The mortality rate is highest among young children and in people aged over 40 years. Skin infections caused by *C. diphtheriae* may result in early development of natural immunity against the disease.

C. diphtheriae persists longer in skin lesions than in the tonsils or nose, and cutaneous diphtheria appears to be more contagious than respiratory diphtheria. Untreated people who are infected with the diphtheria bacillus can be contagious for up to 2 weeks, but seldom for more than 4 weeks. If treated with appropriate antibiotics, the contagious period can be limited to less than 4 days. *C. diphtheriae* can survive in the environment in dust and on dry vomits for several months, and transmission via vomits has been documented. Animal-to-man transmission and food-borne transmission by consumption of contaminated foods such as raw milk have been described, but are very rare.

Control

High population immunity achieved through mass immunization (at least 95% coverage in children and at least 90% coverage in adults) is the most effective measure to control epidemic diphtheria. Immunization with diphtheria toxoid was first introduced in 1923. Large-scale immunization programmes introduced in the 1940s reduced the incidence of diphtheria dramatically, although the disease was not eradicated completely. Immunization schedules are discussed in Chapter 69.

Prevention of secondary cases by the rapid investigation of close contacts is essential. These investigations should include ascertainment of the immunization histories of all home and school contacts. Primary courses of immunization or a booster are given if necessary.

OTHER MEDICALLY IMPORTANT CORYNEBACTERIA

The non-diphtheria corynebacteria ('*diphtheroids*') are diverse and comprise strictly aerobic bacteria usually isolated from the environment, as well as facultative or preferentially anaerobic bacteria, which are commensals of the skin and mucous membranes. The principal species involved and the main clinical syndromes associated with infection are shown in Table 17.1.

Corynebacterium ulcerans

C. ulcerans has been isolated from raw milk and can cause mastitis in cattle. In man, it is seen almost exclusively in cases of exudative pharyngitis, but occasional soft tissue infections occur. *C. ulcerans* can produce a toxin that is 95% identical to the diphtheria toxin, causing a diphtheria-like illness. It seems likely that many human infections are transmitted by a dog or cat. Therapy involves the administration of appropriate antibiotics, such as

Table 17.1 Habitat and disease associations of corynebacteria

Organism	Major habitat	Disease association
Corynebacterium diphtheriae	Throat, skin	Diphtheria (toxigenic strains), wound infections, bacteraemia, endocarditis
C. ulcerans	Human throat and skin Animals: raw milk	Man: diphtheria (toxigenic strains), pharyngitis and wound infection Cattle: mastitis
C. pseudotuberculosis	Sheep, horses, goats	Man: lymphadenitis Animals: abscesses and abortion
C. jeikeium	Skin	Bacteraemia, endocarditis; infection of foreign bodies and CSF shunts
C. urealyticum	Skin, urinary tract	Urinary tract infection, pyelonephritis, endocarditis
C. amycolatum	Man and animals	Man: bacteraemia, endocarditis, peritonitis and wound infection Cattle: mastitis
C. glucuronolyticum	Urinary tract of man and animals	Urogenital tract infection
C. minutissimum	Skin, urinary tract	Erythrasma, bacteraemia
C. striatum	Respiratory tract, skin	Respiratory tract infection, wound infection, bacteraemia
C. pseudodiphtheriticum	Respiratory tract	Respiratory tract infection, endocarditis
C. kroppenstedtii	Unknown	Breast abscess, granulomatous mastitis
Arcanobacterium haemolyticum	Throat	Pharyngitis, skin ulcers, endocarditis
Rhodococcus equi	Animals, soil	Pulmonary infection and soft tissue infection

penicillins or erythromycin, and of diphtheria antitoxin in the case of diphtheria-like disease.

Corynebacterium pseudotuberculosis

C. pseudotuberculosis is primarily an animal pathogen and rarely infects man. It causes caseous lymphadenitis in sheep and goats, and abscesses or ulcerative lymphangitis in horses. Human infections occur mainly in patients with animal contact. Infection usually presents as a subacute or chronic granulomatous lymphadenitis involving the axillary or cervical nodes, but pneumonias have been described. Some strains are lysogenized by bacteriophages of *C. diphtheriae* and thus produce diphtheria toxin, but no clinical cases of diphtheria-like disease have been attributed to *C. pseudotuberculosis* infection. Treatment requires prolonged antibiotic therapy with erythromycin, penicillins or tetracycline, and surgical drainage or excision.

Corynebacterium jeikeium

C. jeikeium (formerly CDC coryneform group JK) is part of the normal skin flora, particularly in inguinal, axillary and rectal areas. Colonization by antibiotic-resistant strains is unusual in healthy individuals, but is common in hospital patients, particularly those who are neutropenic or receiving antibiotics. Most infections are associated with skin damaged by wounds or invasive devices. Such infections include:

- prosthetic valve endocarditis
- bacteraemia associated with infected long-term intravenous cannulae

- peritonitis in patients on peritoneal dialysis
- septicaemia and local infection following insertion of an epicardial pacemaker
- central nervous system infection in patients with ventriculoperitoneal or atrial shunts for hydrocephalus.

Most infections occur in patients in hospital for prolonged periods and who have received broad-spectrum antimicrobial therapy. Spread is through environmental contamination, the hands of ward staff or autoinfection.

Treatment

Most isolates of *C. jeikeium* recovered from infections are highly resistant to penicillins and cephalosporins in vitro. Even with susceptible isolates, penicillin is incompletely bactericidal, but aminoglycoside-sensitive strains can be eradicated successfully with combined penicillin and aminoglycoside therapy. Systemic amoxicillin, gentamicin, rifampicin or ciprofloxacin can be used if the isolate is susceptible. Resistance to aminoglycosides and macrolides has been reported in more than 60% of isolates and resistance to fluoroquinolones is variable.

Glycopeptides are the drugs of choice for treating serious infections. *C. jeikeium* is sensitive to glycopeptides and these antibiotics are bactericidal. Combinations of vancomycin with gentamicin have been used to treat infective endocarditis. Peritonitis secondary to peritoneal dialysis and meningitis related to shunts can be treated with intraperitoneal or intrathecal vancomycin, respectively.

Corynebacterium urealyticum

C. urealyticum (formerly CDC coryneform group D-2) is a frequent skin colonizer, mainly in hospital patients. The groin, abdominal wall and axilla are most frequently colonized. This micro-organism is associated with urinary tract infections, particularly with alkaline-encrusted cystitis and pyelitis related to its strong urease production. Infection is a consequence of the use of broad-spectrum antibiotics for patients with underlying conditions that predispose to urinary tract infection. The organism may also cause pyelonephritis and is an infrequent cause of endocarditis, osteomyelitis or soft tissue infection.

Like *C. jeikeium*, *C. urealyticum* is usually highly resistant to most antimicrobial agents, except glycopeptides. Vancomycin, tetracyclines, erythromycin and norfloxacin have proven effective in treatment. Prolonged treatment with appropriate antibiotics, acidification of the urine, and removal of crusts is essential for proper management of encrusted cystitis.

Corynebacterium amycolatum

C. amycolatum is a human skin commensal similar to other corynebacteria, but lacks cell wall mycolic acids. Its biochemical characteristics are variable and it is often misidentified. Strains isolated from hospital patients may be multiresistant to antibiotics except glycopeptides. *C. amycolatum* has been reported as causing bacteraemia, endocarditis, peritonitis and wound infection.

Corynebacterium glucuronolyticum

C. glucuronolyticum (syn. *C. seminale*) is most commonly isolated from men with prostatitis and urethritis, but can be also isolated from the female genital tract. It is commonly isolated from semen specimens, especially in sexually experienced men. It exhibits strong β-glucuronidase activity and some strains produce urease. It is usually sensitive to antibiotics, although tetracyclines and macrolides are the most effective in vitro.

Corynebacterium minutissimum

C. minutissimum is believed to be the cause of erythrasma, a relatively common and localized infection of the stratum corneum that produces reddish-brown scaly patches in intertriginous sites. Lesions usually involve the groin, toeweb and axillae, and fluoresce coral red when examined by Wood's light. The organism can be cultured from skin scrapings, but the diagnosis is usually based on clinical aspects and the characteristic fluorescence. More serious infections have been described, including bacteraemia and breast abscess. Some infections attributed to *C. minutissimum* may have been caused by *C. amycolatum*.

C. minutissimum is sensitive to penicillins; susceptibility to erythromycin is variable.

Corynebacterium striatum

This species is part of the normal flora of the nose and skin. It is a rare cause of pulmonary infection, particularly in patients with chronic obstructive airway disease or those who are intubated. Transmission to mechanically ventilated patients in an intensive care unit has been documented. It has also been isolated from blood, catheter tips, wounds, leg ulcers, peritoneal fluid, urine, semen, vaginal exudate and placental tissues. *C. striatum* is sensitive to penicillins and glycopeptides; susceptibility to aminoglycosides, ciprofloxacin, erythromycin and rifampicin is variable. Many isolates are resistant to cephalosporins.

Corynebacterium pseudodiphtheriticum

C. pseudodiphtheriticum is a commensal of the human nasopharynx. It is occasionally associated with respiratory tract infections, including tracheobronchitis, necrotizing tracheitis, pneumonia and lung abscess. Most isolates come from patients with endotracheal tubes or chronic obstructive pulmonary disease. It has also been reported to cause endocarditis in patients with prosthetic valves or pre-existing valvular damage. *C. pseudodiphtheriticum* is usually susceptible to most antibiotics except erythromycin.

Corynebacterium kroppenstedtii

C. kroppenstedtii was first described in 1998, after isolation of a single strain from human sputum. Later, when an association was found between corynebacterial infection and granulomatous mastitis, most of the corynebacteria were identified as *C. kroppenstedtii*. Isolation of the species requires Tween-supplemented media and prolonged incubation. *C. kroppenstedtii* is sensitive to many antibiotics including penicillins. Treatment of granulomatous mastitis is usually based on steroids, but addition of antibiotics is appropriate.

Arcanobacterium haemolyticum

A. haemolyticum is phylogenetically related to *Actinomyces* spp. (see Ch. 20). It causes pharyngitis and chronic skin ulcers. Cases of cellulitis, osteomyelitis, brain abscesses and endocarditis have occasionally been described. The species produces at least two extracellular toxins, phospholipase D and a haemolysin.

Most patients are young adults who present with sore throat; some have membranous exudates and peri-tonsillar abscesses. The organism is rarely found in healthy individuals, but occurs in about 2% of symptomatic 15–25-year-olds with pharyngitis. Infection cannot be differentiated from streptococcal pharyngitis on clinical findings alone. *A. haemolyticum* is often isolated in association with streptococci of the *Streptococcus anginosus* group (see p. 189). A scarlatiniform rash occurs in half of the patients with pharyngitis, perhaps caused by a toxin genetically related to the erythrogenic toxin of *Str. pyogenes*. Erythromycin or other macrolides seem to be effective in treatment. *A. haemolyticum* is sensitive to penicillin, but treatment failure has been documented.

Rhodococcus equi

R. equi is a pathogen of horses, pigs and cattle. It is a rare cause of severe pulmonary infections in patients with the acquired immune deficiency syndrome, neoplastic diseases or renal transplants. Most infections develop insidiously, with fever and respiratory symptoms difficult to distinguish from mycobacterial infection. Infections are often recurrent and refractory to treatment, and may be associated with pleural effusion and bacteraemia. The diagnosis is usually established from bronchoscopy specimens, pleural fluid cultures or blood cultures.

R. equi is usually sensitive to tetracyclines, macrolides, rifampicin, imipenem and vancomycin, but resistance to penicillins has been reported. Treatment includes surgical drainage when feasible and prolonged therapy with an antibiotic combination such as erythromycin and rifampicin or imipenem and vancomycin, established by in-vitro tests.

Other coryneform bacteria

- *C. accolens* is usually recovered from respiratory specimens.
- *C. afermentans* ssp. *lipophilum* and CDC coryneform groups G and F-1 may be isolated from a variety of sources, including blood, wound, semen and urine.
- *C. argentoratense*, *C. propinquum*, *C. matruchotii* and *C. durum* have been isolated from the throat, but no pathogenic role has been demonstrated.
- *C. aurimucosum* (syn. *C. nigricans*) exhibits black-pigmented colonies. It has been isolated from genital specimens of women with complications of pregnancy.
- *C. bovis* is commonly isolated from bovine mastitis, but is rarely encountered in human infection.
- *C. macginleyi* strains have been isolated from the eye, often in association with infection.
- *C. xerosis* has been confused with *C. amycolatum* and is very rare.

- *Rothia dentocariosa* is commonly isolated from respiratory tract specimens and has been associated with endocarditis and brain abscess.
- *Turicella otitidis* and *C. auris* have been isolated from ears of healthy patients and those with ear infections.
- Several species of *Arthrobacter* and *Actinobaculum* have been recovered from patients with urinary tract infections.

LISTERIA

Organisms of the genus *Listeria* are non-sporing Gram-positive bacilli. The genus contains six species, but almost all cases of human listeriosis are caused by *L. monocytogenes*. The disease chiefly affects pregnant women, unborn or newly delivered infants, the immunosuppressed and elderly. It is transmitted predominantly by the consumption of contaminated food. *L. ivanovii* is associated with about 10% of infections in livestock animals. *L. ivanovii* and *L. seeligeri* have been associated with a very small number of human infections.

Listeria spp. grow well on a wide variety of non-selective laboratory media and some species, including *L. monocytogenes*, exhibit β-haemolysis on blood agar. The bacilli are non-motile at 37°C, but exhibit characteristic 'tumbling' motility when tested at 25°C.

LISTERIA MONOCYTOGENES

Description

L. monocytogenes is genetically similar to other *Listeria* species, but can be differentiated by biochemical tests. Thirteen serotypes (serovars) are recognized and can be further characterized by phenotypic (bacteriophage typing; multilocus enzyme electrophoresis) or genotypic (plasmid profiling; random amplified polymorphic DNA analysis; pulsed-field gel electrophoresis; amplified fragment length polymorphism typing; direct DNA sequencing) methods.

Most cases of human listeriosis are caused by serovars 4b, 1/2a and 1/2b. Large food-borne outbreaks have been caused predominantly by serovar 4b strains.

The properties of the organism favour food as an agent in transmission of listeriosis. It grows in a wide range of foods having relatively high water activities (A_w >0.95) and over a wide range of temperatures (0–45°C). Growth at refrigeration temperatures is relatively slow, with a maximum doubling time of about 1–2 days at 4°C. Multiplication in food is restricted to the pH range 5–9. *L. monocytogenes* is not sufficiently heat resistant to survive pasteurization.

Pathogenesis

L. monocytogenes is an intracellular parasite, and it is in this environment that the pathogen gains protection and evades some of the host's defences. However, the host has a number of strategies to deal with such parasites. Non-specific mechanisms of resistance are important as first lines of defence once the mucous membranes have been breached. Human neutrophils and non-activated macrophages can phagocytose and kill the bacteria. Protective immunity in man probably depends on T lymphocytes, with antibodies playing little or no role.

L. monocytogenes enters phagocytic and non-phagocytic cells and a listerial surface protein, *internalin* (reminiscent of the M protein of *Str. pyogenes*), is involved with the initial stages of invasion on all cell types. After internalization, *L. monocytogenes* becomes encapsulated in a membrane-bound compartment. In the phagocyte, most cells in the phagocytic vacuole are probably killed. However, those surviving in the phagocytic vacuole, and those in the membrane-bound compartment of non-professional phagocytes, mediate the dissolution of the vacuole membrane by means of a haemolysin (listeriolysin O), and in addition, possibly, the action of a phospholipase C.

In the host cell cytoplasm, where bacterial growth occurs, the organism becomes surrounded by polymerized host cell actin. The ability to polymerize actin preferentially on the older pole of the listeria cell with a surface protein (ActA) subverts the host cell's cytoskeleton and confers intracellular motility to the bacterium. The resulting 'comet tail'-like structure pushes the bacterium into an adjacent mammalian cell, where it again becomes encapsulated in a vacuole. A listerial lecithinase is involved with dissolution of these membranes; the haemolysin may also contribute in this process. Intracellular growth and movement in the newly invaded cell is then repeated. Similar sets of virulence genes are present in *L. ivanovii* and *L. seeligeri*.

Clinical aspects of infection

L. monocytogenes principally causes intra-uterine infection, meningitis and septicaemia. The incubation period varies widely between individuals from 1 to 90 days, with an average for intra-uterine infection of around 30 days.

Infection in pregnancy and the neonate

Listeriosis in pregnancy is classified by fetal gestation at onset, as this correlates best with the clinical features, microbiology and prognosis. Neonatal infection is divided into disease of early (<2 days old), intermediate (3–5 days old) and late (>5 days old) onset.

Maternal listeriosis occurs throughout gestation, but is rare before 20 weeks of pregnancy. The mother is usually previously well with a normal pregnancy. Pregnant women often have very mild symptoms (chills, fever, back pain, sore throat and headache, sometimes with conjunctivitis, diarrhoea or drowsiness), but may be asymptomatic until the delivery of an infected infant. Symptomatic women may have positive blood cultures.

Cultures from high vaginal swabs, stool and midstream urine samples, together with pre- or post-natal antibody tests, are of little help in diagnosis. With the onset of fever, fetal movements are reduced, and premature labour occurs within about 1 week. There may be a transient fever during labour, and the amniotic fluid is often discoloured or stained with meconium. Culture of the amniotic fluid, placenta or high vaginal swab after delivery invariably yields a heavy growth of *L. monocytogenes*. Fever resolves soon after birth, and the vagina is usually culture negative after about 1 month. Although the outcome of infection for the mother is invariably benign, the outcome for the infant is more variable. Abortion, stillbirth and early-onset neonatal disease are common, depending on the gestation at infection. However, maternal infection without infection of the offspring can occur and even progress to placental infection without ill effects for the fetus.

Repeated pregnancy-associated infections are exceedingly rare, and an association between listeria carriage and habitual abortion has not been substantiated.

Early neonatal listeriosis is predominantly a septicaemic illness, contracted in utero. In contrast, late neonatal infection is predominantly meningitic and may be associated with hospital cross-infection. The main characteristics of these two forms are summarized in Table 17.2. Early-onset disease represents a spectrum of mild to severe infection, which can be correlated with the microbiological findings. Those neonates who die from infection usually do so within a few days of birth and have pneumonia, hepatosplenomegaly, petechiae, abscesses in the liver or brain, peritonitis and enterocolitis.

Late-onset disease is the third commonest form of meningitis in neonates. The cerebrospinal fluid (CSF) protein content is almost always raised and the glucose level reduced. The total number of white cells is increased but the counts are variable; neutrophils usually predominate, but lymphocytes or monocytes may be the main cell type. In about 50% of Gram films, bacteria, which may resemble rods or cocci, are seen.

Adult and juvenile infection

Listeriosis in children older than 1 month is very rare, except in those with underlying disease. In adults and juveniles the main syndromes are central nervous system infection, septicaemia and endocarditis. Most cases occur in immunosuppressed patients receiving steroid or cytotoxic

Table 17.2 Characteristics of neonatal infection with *L. monocytogenes*

	Type of infection	
	Early	Late
Onset after delivery	<2 days	>5 days
Maternal factors[a]	Common	Rare
Source of infection	Intrauterine infection acquired haematogenously from mother	Hospital-acquired from early-onset case, post-natal environment or maternally acquired during delivery
Signs/symptoms	Disseminated infection Cardiopulmonary distress Central nervous system signs Vomiting and diarrhoea Hepatosplenomegaly Skin rash	Meningitis Irritability Poor appetite Fever
Laboratory findings	Leucocytosis or leucopenia Thrombocytopenia Mottling on chest radiography Increased fibrinogen	Leucocytosis; occasional radiographic changes CSF: total protein and white cell count raised; glucose level lowered
Sites of isolation	Blood, superficial sites and amniotic fluid; less commonly gastric aspirate, CSF and HVS	Commonly CSF; rarely blood
Mortality rate	30–60%	10–12%

[a]Obstetric problems; low birth-weight; maternal fever; abnormal amniotic fluid.
CSF, cerebrospinal fluid; HVS, high vaginal swab.

therapy or with malignant neoplasms. However, about one-third of patients with meningitis and around 10% with primary bacteraemia are immunocompetent.

Meningitis

The clinical presentation is the same in all groups, but progression is more rapid in immunocompromised subjects. A peripheral blood leucocytosis occurs, and the CSF white blood cell count is raised. The CSF glucose level is low and the protein level is raised; a very high protein concentration may be a poor prognostic indicator. Gram stains of the CSF are often negative, and the clinical features of infection are such that it is not possible to tell listerial meningitis from meningococcal or pneumococcal infection. However, *L. monocytogenes* is isolated from blood cultures in most cases.

In the rare cases of encephalitis, cerebritis or cerebral abscesses, the CSF may be normal, but the white blood cell count is often raised mildly and the protein level is slightly increased, with a low glucose concentration. The Gram film and culture are usually negative. Blood cultures are the main source of the organism in many of these patients.

Bacteraemia and endocarditis

Primary bacteraemia is more common in men than in women, and occurs most often in patients with haematological malignancy or a renal transplant. A few patients develop central nervous system infection, which has a poor prognosis. Infective endocarditis is twice as common in men as in women. The main predisposing factors are prosthetic valves or damaged natural valves, but some patients belong to other risk groups.

Gastro-enteritis

Several food-borne outbreaks of acute gastro-enteritis with fever have been described. The foods associated with these outbreaks have been diverse, but heavily contaminated by the bacterium. Symptoms develop in 1–2 days. Large numbers of *L. monocytogenes* are present in the stool, and a few patients develop serious systemic infection. The ability to cause gastro-enteritis may be specific to certain strains.

Other infections

Rarer manifestations of listeriosis include arthritis, hepatitis, endophthalmitis, cutaneous lesions and peritonitis in patients on continuous ambulatory peritoneal dialysis. Pneumonia occurs in renal transplant recipients and other groups of patients.

Epidemiology

Incidence

Most western countries report infection rates of 1–10 cases per million of the population per year. Pregnancy and neonatal disease account for 10–20% of cases. Among these, 15–25% of infections lead to abortion and stillbirth, and about 70% are neonatal infections. In about 5% of maternal infections bacteraemia occurs and the fetus is not affected.

The incidence of infection increases with age so that the mean age of adult infections is over 55 years. Men are more commonly infected than women over the age of 40 years and, as women are infected in the child-bearing years, the overall sex distribution is more or less equal. Immunosuppression is a major risk factor for both the epidemic and sporadic forms of listeriosis and probably accounts for the increasing incidence with age. Human immunodeficiency virus disease is a predisposing factor in some areas. The peak incidence of human disease occurs in July, August or September. Most patients live in urban areas and usually have no exposure to animals.

L. monocytogenes, like other *Listeria* species, has been isolated from numerous sites, including soil, sewage, water and decaying plant material, where it can survive for more than 2 years. Although the true home of listeria is probably in the environment, these organisms are also found in excreta of apparently healthy animals, including man. Up to 5% of healthy adults may have the organism in their faeces. Faecal carriage in man probably reflects consumption of contaminated foods and is likely to be transitory.

Numerous types of raw, processed, cooked and ready-to-eat foods contain *L. monocytogenes*, usually at levels below 10 organisms per gram. The unusual tolerance of the bacterium to sodium chloride and sodium nitrite, and the ability to multiply (albeit slowly) at refrigeration temperatures makes *L. monocytogenes* of particular concern as a post-processing contaminant in long-shelf-life refrigerated foods. Even when present at high levels in foods, spoilage or taints are not generally produced. The widespread distribution of *L. monocytogenes* and the ability to survive on dry and moist surfaces favour post-processing contamination of foods from both raw product and factory sites.

Transmission

Most cases are sporadic and in only a few is a route of infection identified. The consumption of contaminated foods is the principal route of transmission. Microbiological and epidemiological evidence supports an association with many food types (dairy, meat, vegetable, fish and shellfish) in both sporadic and epidemic listeriosis. Foods associated with transmission often show common features:

- the ability to support the multiplication of *L. monocytogenes* (relatively high water activity and near-neutral pH)
- relatively heavy contamination (>10^3 organisms per gram) with the implicated strain
- processed with an extended (refrigerated) shelf-life
- consumed without further cooking.

Outbreaks of human listeriosis involving more than 100 individuals have occurred, some lasting for several years. This is likely to represent a long-term colonization of a single site in the food manufacturing environment as well as the long incubation periods shown by some patients. Sites of contamination within food processing facilities involved in human infection have included wooden manufacturing equipment, wooden and metallic shelving, porous conveyor belts, cool-room condensates and floor drains. *L. monocytogenes* survives well in moist environments with organic material, and it is from such sites that contamination of food occurs during processing. Epidemiological typing is invaluable for the identification of common source food-borne outbreaks and for tracking the bacterium in the food chain.

Listeriosis transmitted by direct contact with the environment, infected animals or animal material is relatively rare. Papular or pustular cutaneous lesions have been described, usually on the upper arms or wrists, in farmers or veterinarians 1–4 days after attending bovine abortions. Infection is invariably mild and usually resolves without antimicrobial therapy, although serious systemic involvement has been described. Conjunctivitis in poultry workers has also been reported.

Hospital cross-infection between newborn infants occurs. Typically, an apparently healthy baby (rarely more than one) develops late-onset listeriosis 5–12 days after delivery in a hospital in which an infant with congenital listeriosis was born shortly before. The same strain of *L. monocytogenes* is isolated from both infants and the mother of the early-onset case, but not from the mother of the late-onset case. The cases are usually delivered or nursed in the same or adjacent rooms, and consequently staff and equipment are common to both. There is little evidence of cross-infection or person-to-person transmission outside the neonatal period.

Treatment

L. monocytogenes is susceptible to a wide range of antibiotics in vitro, including ampicillin, penicillin, vancomycin, tetracyclines, chloramphenicol, aminoglycosides and co-trimoxazole. There is little agreement about the best treatment, but many patients have been treated successfully with ampicillin or penicillin with or without an aminoglycoside.

Ampicillin and penicillin are probably equivalent agents for the treatment of meningitis. Co-trimoxazole is an excellent alternative. Aminoglycosides interact synergistically with penicillin or ampicillin and improve mortality rates in experimental animals, but no such evidence exists for human infection. Chloramphenicol, when used alone, is less effective than penicillin or ampicillin and may result in relapses. The combination of chloramphenicol with penicillin or ampicillin has resulted in increased mortality. Cephalosporins are ineffective.

The combination of ampicillin and an aminoglycoside is used most commonly in other forms of neonatal and adult listeriosis. Vancomycin is a useful alternative in bacteraemia.

No significant change in the antimicrobial susceptibility of *L. monocytogenes* has been recognized over the past 40 years, and resistance to any of the agents recommended for therapy is unlikely.

Prognosis

The mortality rate in late neonatal disease is about 10%. In contrast, the mortality rate in early disease is 30–60%, and about 20–40% of survivors develop sequelae such as lung disease, hydrocephalus or other neurological defects. Early use of appropriate antibiotics during pregnancy may improve neonatal survival.

The mortality rate in adult infection is about 20–50% in meningitis, 5–20% in primary bacteraemia and 50% in infective endocarditis. About 25–75% of patients surviving central nervous system infection suffer sequelae such as hemiplegia and other neurological defects.

ERYSIPELOTHRIX

Erysipelothrix is a genus of aerobic, non-sporing, non-motile, Gram-positive bacilli. The genus comprises at least three species: *E. rhusiopathiae*, *E. inopinata* and *E. tonsillarum*. *E. rhusiopathiae* causes economically important disease in domestic animals, notably pigs. Human infections from *E. rhusiopathiae* are rare, but present as a localized cutaneous infection (*erysipeloid*), which occasionally becomes diffuse and may lead to septicaemia and endocarditis. Infection is most often associated with close animal contact and usually occurs in such occupational groups as butchers, abattoir workers, veterinarians, farmers, fisherfolk and fish-handlers.

The organism is cultured most often from biopsies, aspirates or blood. The bacilli are short (1–2 µm), but may produce long filamentous forms resembling lactobacilli. Growth is improved by incubation in 5–10% carbon dioxide. Colonies on blood agar are α-haemolytic.

Penicillin and other β-lactam antibiotics are effective. Erythromycin and clindamycin offer suitable alternatives, but *E. rhusiopathiae* is resistant to vancomycin.

KEY POINTS

- The genus *Corynebacterium* includes *C. diphtheriae*, which causes the toxin-mediated pharyngeal infection diphtheria. The infection has largely been controlled by widespread use of a toxoid vaccine.
- Diphtheria toxin is encoded by a lysogenic bacteriophage and has cardiac and neurotoxic effects. Non-toxigenic strains are occasionally isolated from diverse infections.
- Other medically significant corynebacteria include *C. ulcerans*, which causes exudative pharyngitis, and *C. jeikeium*, which is part of the normal skin flora and is an opportunistic pathogen of hospital (particularly neutropenic) patients.
- Members of the genus *Listeria* are environmental bacilli and include *L. monocytogenes*, an important cause of disease in domestic animals, that also causes severe human infections, predominantly affecting the central nervous system, in the unborn or newly delivered infants, the immunosuppressed and elderly.
- Listeriosis is transmitted predominantly by the consumption of ready-to-eat foods; the agent is able to grow in a variety of foods at refrigeration temperatures.
- *Erysipelothrix rhusiopathiae* causes economically important disease in domestic animals, notably pigs. Occasional human infections occur, and present as a cellulitis.

RECOMMENDED READING

Begg N 1994 *Manual for the Management and Control of Diphtheria in the European Region.* World Health Organization, Copenhagen

Brooke C J, Riley T V 1999 *Erysipelothrix rhusiopathiae*: biology, epidemiology and clinical manifestations of an occupational pathogen. *Journal of Medical Microbiology* 48: 789–799

Cossart P, Sansonetti P J 2004 Bacterial invasion: the paradigms of enteroinvasive pathogens. *Science* 304: 242–248

Efstratiou A, George R C 1996 Microbiology and epidemiology of diphtheria. *Reviews in Medical Microbiology* 7: 31–42

Efstratiou A, Engler K H, Mazurova I K, Glushkevich T, Vuopio-Varkila J, Popovic T 2000 Current approaches to the laboratory diagnosis of diphtheria. *Journal of Infectious Diseases* 181: S138–S145

Farber J M, Peterkin P I 1991 *Listeria monocytogenes*, a food-borne pathogen. *Microbiology Reviews* 55: 476–511

Funke G, von Graevenitz A, Clarridge J E, Bernard K A 1997 Clinical microbiology of coryneform bacteria. *Clinical Microbiology Reviews* 10: 125–159

Hof H, Nichterlein T, Kretschmar M 1997 Management of listeriosis. *Clinical Microbiology Reviews* 10: 345–357

Low J C, Donachie W 1997 A review of *Listeria monocytogenes* and listeriosis. *Veterinary Journal* 153: 9–29

McLauchlin J 1997 The identification of *Listeria* species. *International Journal of Food Microbiology* 38: 77–81

Robson J M, McDougall R, van der Valk S, Waite E D, Sullivan J J 1998 *Erysipelothrix rhusiopathiae*: an uncommon but ever present zoonosis. *Pathology* 30: 391–394

Schlech W F 1997 *Listeria* gastroenteritis: old syndrome, new pathogen. *New England Journal of Medicine* 336: 130–132

18 Mycobacterium

Tuberculosis; leprosy

J. M. Grange

The name of the genus *Mycobacterium* (fungus-bacterium) is an allusion to the mould-like pellicles formed when members of this genus are grown in liquid media. This hydrophobic property is due to their possession of thick, complex, lipid-rich, waxy cell walls. A further important characteristic, also due to their waxy cell walls, is *acid-fastness*, or resistance to decolorization by a dilute mineral acid (or alcohol) after staining with hot carbol fuchsin or other arylmethane dyes.

There are more than 100 named species of mycobacteria (listed and regularly updated on the website *www.bacterio. net*), which are divisible into two major groups, the slow and rapid growers, although the growth rate of the latter is slow relative to that of most other bacteria. The leprosy bacillus *Mycobacterium leprae*, which has never convincingly been grown in vitro, and members of the slow growing *Mycobacterium tuberculosis* complex are obligate pathogens. The other species are environmental sapro-phytes, some of which can cause opportunist disease (see Ch. 19).

MYCOBACTERIUM TUBERCULOSIS COMPLEX

'M. tuberculosis complex' refers to a closely related group of variants of a single species. All cause *tuberculosis*, a chronic granulomatous disease affecting man and many other mammals. The complex includes:

- *M. tuberculosis*, which causes most human tuberculosis
- *M. caprae*, isolated from goats and the cause of a few cases of tuberculosis in veterinary surgeons
- *M. bovis*, the principal cause of tuberculosis in cattle and many other mammals
- *M. microti*, a rarely encountered pathogen of voles and other small mammals, but not of human beings
- *M. africanum*, which appears to be intermediate in form between the human and bovine types. It causes human tuberculosis and is found mainly in equatorial Africa. Type 1 is more common in west Africa and

has several features in common with *M. bovis*; type 2 is mainly of east African origin and more closely resembles *M. tuberculosis*.

DESCRIPTION

Members of the *M. tuberculosis* complex (*tubercle bacilli*) are non-motile, non-sporing, non-capsulate, straight or slightly curved rods about 3×0.3 μm in size. In sputum and other clinical specimens they may occur singly or in small clumps, and in liquid cultures they often grow as twisted rope-like colonies termed *serpentine cords* (Fig. 18.1).

Tubercle bacilli are able to grow on a wide range of enriched culture media, but Löwenstein–Jensen (LJ) medium is the most widely used in clinical practice. This is an egg–glycerol-based medium to which malachite green dye is added to inhibit the growth of some conta-minating bacteria and to provide a contrasting colour

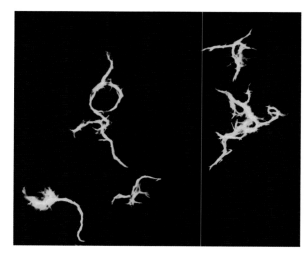

Fig. 18.1 Fluorescent stained microcolonies of *M. tuberculosis* showing 'serpentine cord' formation.

against which colonies of mycobacteria are easily seen. Strains of *M. bovis* grow poorly, or not at all, on standard LJ medium but grow much better on media containing sodium pyruvate in place of glycerol. Agar-based media or broths enriched with bovine serum albumin are also used, particularly in automated culture systems.

On subculture, human tubercle bacilli usually produce visible growth on LJ medium in 2–4 weeks, although on primary isolation from clinical material colonies may take up to 8 weeks to appear. Colonies are of an off-white (buff) colour and (except for the very rarely encountered 'Canetti' variant which has smooth colonies) usually have a dry breadcrumb-like appearance. Growth is characteristically heaped up and luxuriant or 'eugonic', in contrast to the small, flat 'dysgonic' colonies of *M. bovis*.

The optimal growth temperature of tubercle bacilli is 35–37°C, but they fail to grow at 25°C or 41°C. Most other mycobacteria grow at one or other, or both, of these temperatures.

All mycobacteria are obligate aerobes, but *M. bovis* grows better in conditions of reduced oxygen tension. Thus, when incorporated in soft agar media, *M. tuberculosis* grows on the surface whereas *M. bovis* grows as a band a few millimetres below the surface. This provides a useful differentiating test. Various other differential characteristics of human tubercle bacilli are shown in Table 18.1.

Tubercle bacilli survive in milk and other organic materials and on pasture land so long as they are not exposed to ultraviolet light, to which they are very sensitive. They are also heat sensitive and are destroyed by pasteurization. Mycobacteria are susceptible to alcohol, formaldehyde, glutaraldehyde and, to a lesser extent, hypochlorites and phenolic disinfectants. They are considerably more resistant than other bacteria to acids, alkalis and quaternary ammonium compounds.

PATHOGENESIS

The tubercle bacillus owes its virulence to its ability to survive within the macrophage rather than to the production of a toxic substance. The mechanism of virulence is poorly understood and is almost certainly multifactorial. The immune response to the bacillus is of the cell-mediated type, which, if mediated by type 1 T helper (T_H) cells, leads to protective immunity, although the presence of T_H2 cells facilitates tissue-destroying hypersensitivity reactions and progression of the disease process. The nature of the immune responses following infection changes with time, so that human tuberculosis is divisible into primary and post-primary forms with quite different pathological features.

Primary tuberculosis

The site of the initial infection is usually the lung, following the inhalation of bacilli. These bacilli are engulfed by alveolar macrophages in which they replicate to form the initial lesion or *Ghon focus*. Some bacilli are carried in phagocytic cells to the hilar lymph nodes where additional foci of infection develop. The Ghon focus together with the enlarged hilar lymph nodes form the *primary complex*. In addition, bacilli are seeded by further lymphatic and haematogenous dissemination in many organs and tissues, including other parts of the lung. When the bacilli enter the mouth, as when drinking milk containing *M. bovis*, the primary complexes involve the tonsil and cervical nodes (*scrofula*; Fig. 18.2) or the intestine, often the ileocaecal region, and the mesenteric lymph nodes. Likewise, the primary focus may be in the skin, with involvement of the regional lymph nodes. This form of tuberculosis, termed *prosector's wart*, was an occupational disease of anatomists and pathologists.

Within about 10 days of infection, antigens of *M. tuberculosis* are processed by antigen-presenting cells, activated by bacterial components termed adjuvants, and presented to antigen-specific T lymphocytes which undergo clonal proliferation. The activated T cells release cytokines, notably interferon-γ, which, together with calcitriol, activate macrophages (Fig. 18.3) and cause them to form a compact cluster, or granuloma, around the foci of infection (Fig. 18.4). These activated macrophages are

Table 18.1 Some differential characteristics of tubercle bacilli causing human disease

Species	Atmospheric preference	Nitratase	TCH	Pyrazinamide
M. tuberculosis	Aerobic	Positive	Resistant[a]	Sensitive
M. bovis	Micro-aerophilic	Negative	Sensitive	Resistant
M. bovis BCG	Aerobic	Negative	Sensitive	Resistant
M. africanum I	Micro-aerophilic	Negative	Sensitive	Sensitive
M. africanum II	Micro-aerophilic	Positive	Sensitive	Sensitive

[a]Strains from southern India may be sensitive.
TCH, thiophen-2-carboxylic acid hydrazide.

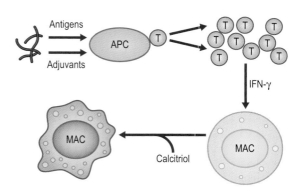

Fig. 18.3 Antigens of *M. tuberculosis* are, after priming by adjuvants, processed by the antigen-presenting cell (APC) and presented to T helper lymphocytes (T), which are activated and proliferate to form a clone. T cell-produced interferon (IFN)-γ and calcitriol activate the resting macrophages (MAC).

Fig. 18.2 Tuberculous cervical lymphadenitis (scrofula) with sinus formation in an Indonesian woman.

Fig. 18.4 The granuloma of primary tuberculosis.

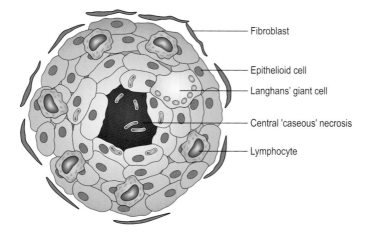

- Fibroblast
- Epithelioid cell
- Langhans' giant cell
- Central 'caseous' necrosis
- Lymphocyte

termed *epithelioid cells* because of their microscopical resemblance to epithelial cells. Some of them fuse to form multinucleate giant cells. The centre of the granuloma contains a mixture of necrotic tissue and dead macrophages, and, because of its cheese-like appearance and consistency, is referred to as *caseation*.

Activated human macrophages inhibit the replication of the tubercle bacilli, but there is no clear evidence that they can actually kill them. Being metabolically very active, the macrophages in the granuloma consume oxygen, and the resulting anoxia and acidosis in the centre of the lesion probably kills most of the bacilli.

Granuloma formation is usually sufficient to limit the primary infection: the lesions become quiescent and surrounding fibroblasts produce dense scar tissue, which may become calcified. Not all bacilli are destroyed; some remain in a poorly understood dormant form that, when reactivated, causes post-primary disease. Programmed cell death (apoptosis) of bacteria-laden cells by cytotoxic T cells and natural killer (NK) cells may also contribute to protective immunity.

In a minority of cases one of the infective foci progresses and gives rise to the serious manifestations of primary disease, including progressive primary lesions

Table 18.2   Stages of primary tuberculosis in childhood

Stage	Time (from onset)	Characteristics
1	3–8 weeks	Primary complex develops and tuberculin conversion occurs
2	2–6 months	Progressive healing of primary complex
		Possibility of pleural effusion
3	6–12 months	Possibility of miliary or meningeal tuberculosis
4	1–3 years	Possibility of bone or joint tuberculosis
5	3–5 years or more	Possibility of genito-urinary or chronic skin tuberculosis

Adapted from Miller F J W 1982 *Tuberculosis in Children*. Churchill Livingstone, Edinburgh.

(particularly in infants; Table 18.2), meningitis, pleurisy and disease of the kidneys, spine (*Pott's disease*) and other bones and joints. If a focus ruptures into a blood vessel, bacilli are disseminated throughout the body with the formation of numerous granulomata. This, from the millet seed-like appearance of the lesions, is known as *miliary tuberculosis*.

Tuberculin reactivity

About 6–8 weeks after the initial infection, the phenomenon of tuberculin conversion occurs. This altered reactivity was discovered by Robert Koch while attempting to develop a remedy for tuberculosis. When tuberculous guinea-pigs were injected intradermally with living tubercle bacilli, the skin around the injection site became necrotic within a day or two and was sloughed off, together with the bacilli. Koch then found that the same reaction occurred when he injected *old tuberculin* – a heat-concentrated filtrate of a broth in which tubercle bacilli had been grown. This reaction became known as the *Koch phenomenon*, and its characteristic feature is extensive tissue necrosis.

Although Koch's tuberculin proved unsuccessful as a therapeutic agent, it formed the basis of the widely used tuberculin test (see below).

Post-primary tuberculosis

In many individuals, the primary complex resolves and the only evidence of infection is a conversion to tuberculin reactivity. After an interval of months, years or decades, reactivation of dormant foci of tubercle bacilli or exogenous reinfection may lead to post-primary tuberculosis, which differs in several respects from primary disease (Table 18.3). Endogenous reactivation may occur spontaneously or after an intercurrent illness or other condition that lowers the host's immune responsiveness (see below). For unknown reasons, reactivation or reinfection tuberculosis tends to develop in the upper lobes of the lungs. The same process of granuloma

Table 18.3   Main differences between primary and post-primary tuberculosis in non-immunocompromised patients

Characteristic	Primary	Post-primary
Local lesion	Small	Large
Lymphatic involvement	Yes	Minimal
Cavity formation	Rare	Frequent
Tuberculin reactivity	Negative (initially)	Positive
Infectivity[a]	Uncommon	Usual
Site	Any part of lung	Apical region
Local spread	Uncommon	Frequent

[a]Pulmonary cases.

formation occurs, but the necrotic element of the reaction causes tissue destruction and the formation of large areas of caseation termed *tuberculomas*. Proteases liberated by activated macrophages soften and liquefy the caseous material, and an excess of tumour necrosis factor and other immunological mediators causes the wasting and fevers characteristic of the disease.

The interior of the tuberculoma is acidic and anoxic, and contains few viable tubercle bacilli. Eventually, however, the expanding lesion erodes through the wall of a bronchus, the liquefied contents are discharged and a well aerated cavity is formed. The atmosphere of the lung, with a high carbon dioxide level, is ideal for supporting the growth of the bacilli, and huge numbers of these are found in the cavity walls. For this reason, closure of the cavities by collapsing the lung, either by artificial pneumothorax or by excising large portions of the chest wall, was a standard treatment for tuberculosis in the pre-chemotherapy era.

Once the cavity is formed, large numbers of bacilli gain access to the sputum, and the patient becomes an open or infectious case. This is a good example of the transmissibility of a pathogen being dependent upon the host's immune response to infection. Surprisingly, about 20% of cases of open cavitating tuberculosis resolve without treatment.

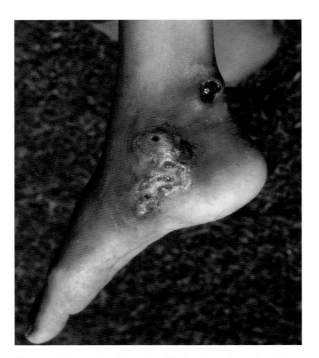

Fig. 18.5 Tuberculosis of the ankle with sinus formation and overlying involvement of the skin.

In post-primary tuberculosis, dissemination of bacilli to lymph nodes and other organs is unusual. Instead, spread of infection occurs through the bronchial tree so that secondary lesions develop in the lower lobes of the lung and, occasionally, in the trachea, larynx and mouth. Bacilli in swallowed sputum cause intestinal lesions. Secondary lesions may also develop in the bladder and epididymis in cases of renal tuberculosis. Post-primary cutaneous tuberculosis (*lupus vulgaris*) usually affects the face and neck. Untreated, it is a chronic condition leading to gross scarring and deformity. Most cases of lupus vulgaris were caused by *M. bovis* and this condition is now rarely seen in developed countries. Some cases of cutaneous tuberculosis are secondary to sinus formation between tuberculous lymph nodes and the skin (*scrofuloderma*) and other structures including bones and joints (Fig. 18.5).

Immunocompromised individuals

Reactivation tuberculosis is particularly likely to occur in immunocompromised individuals, including the elderly and transplant recipients; it often occurs early in the course of human immunodeficiency virus (HIV) infection. Tuberculosis acts synergistically with HIV to lower the patient's immunity and it is a defining condition

for the acquired immune deficiency syndrome (AIDS). As a result, even when tuberculosis is treated effectively in HIV-positive patients, the mortality rate from other AIDS-related conditions is high, with many dying within 2 years. Cavity formation is unusual in the more profoundly immunocompromised patients, emphasizing the importance of the immune response in this pathological process. Instead, diffuse infiltrates develop in any part of the lung. In contrast to post-primary disease in non-immunocompromised individuals, lymphatic and haematogenous dissemination are common. Sometimes there are numerous minute lesions teeming with tubercle bacilli throughout the body – a rapidly fatal condition termed *cryptic disseminated* tuberculosis. The interval between infection and development of disease is considerably shortened in immunocompromised persons.

The tuberculin test and other immunological tests

Although Robert Koch's attempts to use old tuberculin as a remedy for tuberculosis failed, an Austrian physician, Clemens von Pirquet, used the Koch phenomenon as an indication of bacterial 'allergy' resulting from previous infection. Individuals with active tuberculosis were usually tuberculin positive, but many of those with disseminated and rapidly progressive disease were negative. This led to the widespread but erroneous belief that tuberculin reactivity is an indicator of immunity to tuberculosis.

Old tuberculin caused non-specific reactions and has been replaced by *purified protein derivative* (PPD). This is given by:

- intracutaneous injection (Mantoux method)
- a spring-loaded gun which fires six prongs into the skin through a drop of PPD (Heaf method)
- single-test disposable devices with PPD dried on to prongs (tine tests).

The biological activity of tuberculin is compared with international standards and activity expressed in international units (IU). In the UK, solutions of PPD for Mantoux testing are supplied as dilutions of 1 in 10 000, 1 in 1000 and 1 in 100, which correspond to 1, 10 and 100 IU in the injected dose of 0.1 ml. The standard test dose is 10 IU, but those suspected of having tuberculosis, and who are therefore likely to react strongly, may first be tested with 1 IU. Undiluted PPD is supplied for use with the Heaf gun.

The tuberculin test is widely used as a diagnostic test, although its usefulness is limited by its failure to distinguish active disease from quiescent infections and past BCG (bacille Calmette–Guérin; see p. 214) vaccination. In addition, exposure to various mycobacteria in the environment may induce low levels of tuberculin reactivity. In the USA, where BCG is not used, more diagnostic reliance is placed on the tuberculin test. The test is used in epidemiological studies as outlined below.

Numerous attempts have been made to develop serological tests for the disease with limited success; a more promising approach is the detection of interferon-γ-producing peripheral blood T cells that respond to specific antigens of *M. tuberculosis*.

LABORATORY DIAGNOSIS

The definitive diagnosis of tuberculosis is based on detection of the causative organism in clinical specimens by microscopy, cultural techniques or the polymerase chain reaction (PCR) and its various derivatives.

Specimens

The most usual specimen for diagnosis of pulmonary tuberculosis is sputum but, if none is produced, bronchial washings, brushings or biopsies and early-morning gastric aspirates (to harvest any bacilli swallowed overnight) may be examined. Tissue biopsies are homogenized by grinding for microscopy and culture. Cerebrospinal fluid, pleural fluid, urine and other fluids are centrifuged and the deposits examined.

Microscopy

Use is made of the acid-fast property of mycobacteria to detect them in sputum and other clinical material. In the Ziehl–Neelsen (ZN) staining technique, heat-fixed smears of the specimens are flooded with a solution of carbol fuchsin (a mixture of basic fuchsin and phenol) and heated until steam rises. After washing with water, the slide is flooded with a dilute mineral acid (e.g. 3% hydrochloric acid) and, after further washing, a green or blue counterstain is applied. Red bacilli are seen against the contrasting background colour. In some methods, the acid is diluted in 95% ethanol rather than water. This gives a cleaner background but, contrary to a common belief, it does not enable tubercle bacilli to be distinguished from other mycobacteria. Fluorescence microscopy, based on the same principle of acid-fastness, is increasingly used and is much less tiring for the microscopist. Modifications of the various staining techniques are used to examine tissue sections.

Cultural methods

As sputum and certain other specimens frequently contain many bacteria and fungi that would rapidly overgrow any mycobacteria on the culture media, these must be destroyed. Decontamination methods make use of the relatively high resistance of mycobacteria to acids, alkalis and certain disinfectants. In the widely used *Petroff method*, sputum is mixed well with 4% sodium hydroxide for 15–30 min, neutralized with potassium dihydrogen orthophosphate and centrifuged. The deposit is used to inoculate LJ or similar media. Specimens such as cerebrospinal fluid and tissue biopsies, which are unlikely to be contaminated, are inoculated directly on to culture media. As an alternative to chemical decontamination, mixtures of antibiotics that kill fungi and all bacteria other than mycobacteria may be added to the culture media. These are used principally in the automated culture systems described below.

Inoculated media are incubated at 35–37°C and inspected weekly for at least 8 weeks. Cultures of material from skin lesions should also be incubated at 33°C. Any bacterial growth is stained by the ZN method and, if acid-fast, subcultured for further identification.

A more rapid bacteriological diagnosis is achievable by use of commercially available automated systems. Systems that detect colour changes in dyes induced by the release of carbon dioxide, or the unquenching of fluorescent dyes on the consumption of oxygen by metabolizing bacilli, have replaced the earlier radiometric method.

The first step in identification is to determine whether an isolate is a member of the *M. tuberculosis* complex. These organisms:

- grow slowly
- do not produce yellow pigment
- fail to grow at 25°C and 41°C
- do not grow on egg media containing *p*-nitrobenzoic acid (500 mg/l).

Strains differing in any of these properties belong to other species.

Nucleic acid technology

Nucleic acid probes for identification of the *M. tuberculosis* complex and, specifically, *M. tuberculosis*, and also for certain other species, are commercially available. They are not sufficiently sensitive to detect mycobacteria in clinical specimens and are used to identify mycobacteria cultivated by conventional techniques.

Amplification of specific nucleic acid sequences in specimens is achievable by PCR and related techniques, some of which are commercially available. Problems of low sensitivity, 'false-positive' reactions and cross-contamination have been largely overcome by the introduction of closed-system, isothermal techniques for amplification of species-specific 16S ribosomal ribonucleic acid (RNA).

Most members of the *M. tuberculosis* complex contain 1 to 20 copies of the insertion sequence IS6110, which

Table 18.4 In-vivo activity of antituberculosis drugs		
Sterilizing	Bactericidal	Bacteristatic
Rifampicin	Isoniazid	Ethionamide
Pyrazinamide	Streptomycin	Prothionamide
	Ethambutol[a]	Thiacetazone
	Quinolones	p-Aminosalicylic acid
	Macrolides	Cycloserine

[a]In early stages of therapy.

has been used to develop DNA fingerprinting (see p. 74) methods for epidemiological purposes. Alternatively, detection of spacer oligonucleotides, short DNA sequences found around the sites of the insertion sequences, is useful for typing isolates (spoligotyping). Use of these techniques has shown that *M. tuberculosis* is divisible into several genotypes or families with different geographical distributions. There is also some evidence that these genotypes differ in their virulence, with one of them, the Beijing genotype, appearing to be more virulent than many others.

Drug susceptibility testing

Several methods have been described.

- Assessment of growth inhibition on solid media containing various dilutions of the drug, in comparison with test strains. As the methods depend on observation of visible growth, results are not available until several weeks after isolation of the organism.
- Radiometric or non-radiometric automated systems, which provide results more quickly.
- Nucleic acid technology, which is even more rapid. Kits capable of detecting about 95% of mutations to rifampicin resistance caused by mutations in the *rpoB* gene are available commercially.

TREATMENT

The antituberculosis drugs are divisible into three groups (Table 18.4):

1. Bactericidal drugs that effectively sterilize tuberculous lesions.
2. Bactericidal drugs that kill tubercle bacilli only in certain situations. Streptomycin is effective in neutral or alkaline environments, such as the cavity wall; pyrazinamide is active only in acidic environments, such as within macrophages and in dense inflammatory tissue; and isoniazid kills bacilli only when they are replicating.

3. Bacteristatic drugs, which are of limited usefulness and are not included in standard drug regimens.

Mutation to drug resistance occurs at a rate of about one mutation every 10^8 cell divisions. Successful therapy requires the prevention of the emergence of drug-resistant strains by the simultaneous use of at least two drugs to which the organism is sensitive.

The earlier 2-year regimen of streptomycin, isoniazid and p-aminosalicylic acid has been replaced by much more acceptable orally administered regimens based on an initial intensive 2-month phase and a 4–6-month continuation phase, depending on the World Health Organization (WHO) treatment category (Table 18.5). Most patients are in category I when diagnosed and the most widely used regimen therefore consists of an intensive phase of rifampicin, isoniazid, pyrazinamide and ethambutol for 2 months, followed by the first two drugs to a total of 6 months. Ideally, the drugs are given daily, but to ensure compliance by supervising the administration of the drugs they may be given thrice weekly during the continuation phase or throughout.

The response to therapy of drug-susceptible tuberculosis is divisible into three phases:

1. During the first week or two, the large numbers of actively replicating bacilli in cavity walls are killed, principally by isoniazid, but also by rifampicin and ethambutol. The patient rapidly ceases to be infectious, and prolonged hospitalization with barrier nursing is now rarely necessary.
2. In the following few weeks the less active bacilli within macrophages, caseous material and dense, acidic, inflammatory lesions are killed by pyrazinamide and rifampicin.
3. In the continuation phase any remaining dormant bacilli are killed by rifampicin during their short bursts of metabolic activity. Any rifampicin-resistant mutants that arise and start to replicate are killed by isoniazid.

Short-course regimens are usually well tolerated by the patient. Isoniazid, rifampicin and pyrazinamide are all potentially hepatotoxic, and rifampicin may cause an influenza-like syndrome, which, paradoxically, is more likely to occur when the drug is given intermittently. Rifampicin antagonizes the action of several drugs and also oral contraceptives – an important point to be considered when treating women. There is also mutual antagonism between rifampicin and the antiretroviral drugs used for HIV infection (see p. 63 and Ch. 55). Isoniazid may cause mild psychiatric disturbances and peripheral neuropathy, particularly in alcoholics, but these are usually preventable by prescribing pyridoxine (vitamin B_6). Ethambutol is toxic to the eye and, although this is rare with standard dosages, care is required; the patient should be informed of this possibility.

Table 18.5 WHO treatment categories and recommended short-course drug regimens

Treatment category	Definition	Initial phase[a]	Continuation phase[a]
I	New sputum smear-positive; smear-negative with extensive lung involvement; severe non-pulmonary disease	Preferred: 2HRZE Optional: 2(HRZE)$_3$ or 2HRZE	Preferred: 4HR or 4(HR)$_3$ Optional: 4(HR)$_3$ or 6HE
II	Sputum smear-positive after relapse, treatment failure or interruption of treatment	Preferred: 2HRZES then 1HRZE Optional: 2(HRZES)$_3$ then 1(HRZE)$_3$	Preferred: 5HRE Optional: 5(HRE)$_3$
III	New smear-negative (other than category I) and less severe non-pulmonary disease	Preferred: 2HRZE Optional: 2(HRZE)$_3$ or 2HRZE	Preferred: 4HR or 4(HR)$_3$ Optional: 4(HR)$_3$ or 6HE
IV	Chronic cases	Specially designed standard or individualized regimens	

[a]The preceding numeral refers to the duration of therapy in months and the subscript numeral 3 indicates thrice-weekly doses.
H, isoniazid; R, rifampicin; Z, pyrazinamide; E, ethambutol; S, streptomycin.

The emergence of drug-resistant strains (acquired resistance) during adequately supervised short-course chemotherapy is uncommon; most relapses are due to drug-sensitive bacilli. Multidrug-resistant strains are defined as those resistant to isoniazid and rifampicin; they are sometimes also resistant to other drugs and pose serious problems. The regimens for multidrug-resistant tuberculosis are based on in-vitro drug susceptibility tests. Useful agents include the newer fluoroquinolones and macrolides and amoxicillin with a β-lactamase inhibitor such as clavulanic acid. With careful and caring supervision, the majority of patients can be cured, although HIV seropositivity is a poor prognostic factor.

EPIDEMIOLOGY

Human tuberculosis is transmitted principally by inhalation of bacilli in moist droplets coughed out by individuals with open pulmonary tuberculosis. Dried bacilli in dust appear to be much less of a health hazard. The sputum of patients with positive findings on microscopy contains at least 5000 tubercle bacilli per millilitre; these patients are considerably more infectious than those who have negative findings on microscopy. Most transmission of the disease occurs within households or other environments where individuals are close together for long periods.

For epidemiological purposes, the incidence of infection by the tubercle bacillus in a community is calculated from the number of tuberculin-positive individuals in different age groups, provided they have not received BCG vaccination. The annual infection rate gives an indirect measure of the number of open or infectious individuals in the community. Although subject to many variables, a 1% annual infection rate indicates that there are around 50 infectious cases in every 100 000 members of the community. The prevalence of tuberculosis in a region is not the same as the annual infection rate, because only a minority of infected individuals develop overt disease and they often become ill several years after the initial infection.

Determinations of the incidence of tuberculosis worldwide are notoriously difficult. According to current (2005) WHO estimates:

- one-third of the world's population has been infected.
- about 100 million individuals are infected annually and more than 8 million develop overt disease.
- each year 4 to 5 million become open or infectious and around 2 million die.

Tuberculosis is relatively uncommon in the industrially developed nations, with an annual infection rate of 0.3–0.1% or less, but the previous decline has halted and in several countries there has been an increase in notifications since around 1990. The disease is often concentrated in certain 'hot spots', particularly deprived inner city areas. In the developing countries the annual infection rates are usually between 2% and 5%.

Owing to the impact of the HIV/AIDS pandemic the incidence of tuberculosis is increasing in some countries and predictions of future trends are worrying.

- In 2003, 12% of all cases of tuberculosis were HIV related, rising to over 30% in sub-Saharan Africa with a country-to-country range of 20–70%.

- In 2003, there were an estimated 3 million AIDS-related deaths and tuberculosis was the cause of death in 30%.
- A person dually infected by the tubercle bacillus and HIV has an 8% annual risk of developing active tuberculosis. The risk of developing tuberculosis after infection or reinfection is much higher, approaching 100%, in patients with AIDS.
- In 2003 there were an estimated 40 million HIV-infected people worldwide, of whom one in three were also infected with the tubercle bacillus and around 1 million developed active tuberculosis.
- The interval between infection and overt disease is considerably shortened, resulting in explosive mini-epidemics, in institutions where HIV-positive persons congregate. Some such mini-epidemics have involved multidrug-resistant strains, notably in the USA.

Bovine tuberculosis is spread from animal to animal, and sometimes to human attendants, in moist cough spray. About 1% of infected cows develop lesions in the udder, and bacilli excreted in the milk can then infect people who drink it. Heat treatment, or pasteurization, prevents milk-borne disease. In most developed countries, the incidence of bovine tuberculosis has been reduced drastically by regular tuberculin testing of herds and the slaughter of reactors, but the total eradication has been prevented by infection of cattle from wild animals, notably badgers in the UK and possums in New Zealand. Human disease due to *M. bovis* is very rare in developed countries and is usually the result of reactivation of old lesions. Cattle have occasionally been infected by farmworkers with open tuberculosis due to *M. bovis*. The prevalence of the disease in cattle in the developing nations is poorly documented. Person-to-person spread of *M. bovis* leading to active disease is uncommon, although a few instances of such spread among HIV-positive persons have been reported.

CONTROL

Human tuberculosis is preventable:

- by the early detection and effective therapy of the open or infectious individuals in a community
- by lowering the chance of infection by reducing overcrowding
- to a limited extent, by vaccination.

Active case finding involves a deliberate search, often on a house-to-house basis or in workplaces, for suspects with a chronic cough of a month or more in duration. Merely waiting for patients with symptoms to seek medical attention is much less effective, even when supported by

Table 18.6 Variations in the protective efficacy of BCG vaccinations in nine major trials

Country or population	Age range of vaccinees (years)	Protection (%)
North American Indian	0–20	80
UK	14–15	78
Chicago, Illinois, USA	Neonates	75
Puerto Rico	1–18	31
South India (Bangalore)	All ages	30
Georgia and Alabama, USA	>5	14
Georgia, USA	6–17	0
Illinois, USA	Young adults	0
South India (Chingleput)	All ages	0[a]

[a]Some protection demonstrated in those vaccinated neonatally on 15-year follow-up.

education programmes. Regular chest examination by mass miniature radiography detects fewer than 15% of individuals with tuberculosis and its use is now restricted to certain high-risk situations.

The most important factors affecting the incidence of tuberculosis are socio-economic ones, particularly those leading to a reduction of overcrowding in homes and workplaces. In developing countries it is estimated that each patient with open tuberculosis infects about 20 contacts annually, whereas in Europe the corresponding figure is two or three.

Vaccination

BCG is a living attenuated vaccine derived from a strain of *M. bovis* by repeated subculture between the years 1908 and 1920. This species was selected rather than *M. tuberculosis* on the dubious assumption that it was of limited virulence in humans. The vaccine was originally given orally to neonates but is now given by intracutaneous injection.

The protective efficacy of BCG varies enormously from country to country. Early trials in the UK indicated that administration of BCG to schoolchildren affords 78% protection, but a major trial in south India involving individuals of all ages found no protection (Table 18.6). This difference appears to be the result of prior exposure of the population to environmental mycobacteria, which, in some regions, confer some protection, but in others induce inappropriate immune reactions that antagonize protection. For this reason, vaccination given neonatally, before environmental sensitization occurs, is now gaining popularity.

When BCG vaccination is introduced into a region it has an immediate impact on the incidence of the serious but non-infectious forms of childhood tuberculosis such as meningitis, but has little impact on the annual infection rate in the community as the smear-positive source cases arise mostly from among the older, unvaccinated, tuberculin-positive members of the community. Vaccination has not therefore proved to be an effective control measure.

Being a living vaccine, serious infections and even disseminated disease may occur in immunocompromised persons. Thus, BCG should never be given to persons known to be HIV positive. Many attempts are currently being made to develop alternative vaccines, particularly non-viable subunit ones.

Prophylactic chemotherapy

In true prophylactic chemotherapy, drugs are administered to uninfected individuals who are in unavoidable contact with a patient with open tuberculosis. The main example is a baby born to a mother with the disease. More usually, it refers to therapy, principally with isoniazid alone for 9–12 months, given to individuals who have been infected but show no signs of active disease. In the UK, such therapy may be given to unvaccinated children who have converted to tuberculin positivity after exposure to a source case. The use of prophylactic chemotherapy in older tuberculin reactors is controversial and virtually restricted to countries such as the USA where BCG vaccination is no longer given. Treatment may be shortened to 4 months by the addition of rifampicin, but an even shorter (2 months) regimen of rifampicin, isoniazid and pyrazinamide was associated with an unacceptably high risk of liver toxicity and has been abandoned. When infected with both *M. tuberculosis* and HIV, persons who are tuberculin positive are much more responsive to preventive therapy than those who are tuberculin negative, and the WHO thus recommends prophylactic therapy only for the former.

MYCOBACTERIUM LEPRAE

Leprosy is a particularly tragic affliction as the nature of the infection often causes severe disfigurement and deformity which, throughout history, have led to social ostracism or even total banishment of those suffering from the disease. The prevalence of leprosy, and changing trends in prevalence, are not easily determined. The number of registered patients on treatment declined from more than 10 million in 1982 to just over half a million in 2003 (a 17% drop from the 2002 figure), but this may be because of changes in case definition and the much shorter duration of therapy. The annual detection rate of new cases has not shown such a steep decline, although this may reflect a higher case detection rate. About 2–3 million patients have been cured bacteriologically, but they have residual deformities requiring lifelong care.

Once of worldwide distribution, the disease was reported in 122 mainly tropical countries in 1985, but by 2002 elimination (by definition a prevalence of less than 1 in 10 000 of the population) had been achieved in all except 14 countries. These remaining countries are in Africa, Asia and Latin America, with 90% of all registered persons living in India, Brazil, Nepal, Madagascar, Mozambique and Myanmar. The disease was long endemic in the British Isles; Robert the Bruce of Scotland was one of its victims. The last British patient to acquire the disease in this country died in the Shetland islands in 1798. In Norway the disease persisted into the twentieth century; Armauer Hansen first described the causative organism in that country in 1873.

Leprosy is often cited as a disease of great antiquity but, in fact, literary and skeletal evidence of this very characteristic disease go back no further than 500 BC. Biblical leprosy was almost certainly not the same as the disease that now bears this name.

DESCRIPTION

M. leprae has never convincingly been cultivated in vitro; this has been attributed to a loss or disruption of many of the genes required for metabolism. The genome of *M. leprae* is smaller than that of cultivable mycobacteria, and about half of its genes are defective counterparts of functional ones found in other mycobacteria. In the 1970s it was shown that armadillos infected with *M. leprae* often developed extensive disease, with up to 10^{10} bacilli in each gram of diseased tissue. This animal has therefore provided sufficient bacilli for research projects and for the production of a skin test reagent, leprosin-A. Limited replication, yielding 10^6 bacilli after 6–8 months, also occurs in the mouse footpad, and this has been used for testing the sensitivity of bacilli to antileprosy drugs.

Leprosy bacilli resemble tubercle bacilli in their general morphology, but they are not so strongly acid-fast. In clinical material from lepromatous patients, the bacilli are typically found within macrophages in dense clumps. A characteristic surface lipid, peptidoglycolipid-1 (PGL-1), has been extracted from *M. leprae*, and its unique carbohydrate antigenic determinant has been synthesized. Specific PCR primers for diagnostic purposes have been synthesized.

PATHOGENESIS

The principal target cell for the leprosy bacillus is the Schwann cell. The resulting nerve damage is responsible

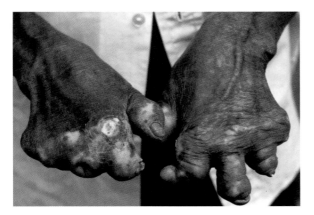

Fig. 18.6 Borderline tuberculoid leprosy. Trophic changes in the hands secondary to nerve damage, vasculitis and anaesthesia.

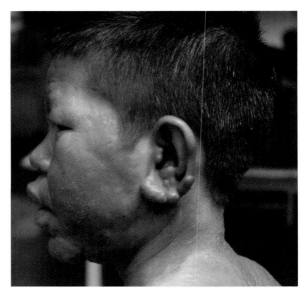

Fig. 18.7 Lepromatous leprosy. Nodular swelling of face, enlargement of ear lobes and loss of eyebrows.

for the main clinical features of leprosy: anaesthesia and muscle paralysis. Repeated injuries to, and infections of, the anaesthetic extremities leads to their gradual destruction (Fig. 18.6). Infiltration of the skin and cutaneous nerves by bacilli leads to the formation of visible lesions, often with pigmentary changes.

The first sign of leprosy is a non-specific or indeterminate skin lesion, which often heals spontaneously. If the disease progresses, its clinical manifestation is determined by the specific immune responsiveness of the patient to the bacillus, and there is a distinct immunological spectrum (Table 18.7).

- Hyper-reactive *tuberculoid* (TT) leprosy, with small numbers of localized skin lesions containing so few bacilli that they are not seen on microscopy and an inappropriately intense granulomatous response that often damages major nerve trunks.
- Anergic *lepromatous* (LL) leprosy, in which the skin lesions are numerous or confluent and contain huge numbers of bacilli, usually seen as clusters or globi within monocytes. Cooler parts of the body, such as the ear lobes, are particularly heavily infiltrated by bacilli (Fig. 18.7). There is no histological evidence of an immune response.
- Intermediate forms classified as borderline tuberculoid (BT), mid-borderline (BB) or borderline lepromatous (BL).

Destruction of the nasal bones may lead to collapse of the nose (Fig. 18.8). In addition, large numbers of leprosy bacilli are discharged in nasal secretions in multibacillary disease. The eye is frequently damaged by direct bacillary invasion, uveitis or corneal infection secondary to paralysis

Table 18.7 Characteristics of the five points on the spectrum of leprosy					
	TT	BT	BB	BL	LL
Bacilli seen in skin	−	±	+	++	+++
Bacilli in nasal secretions	−	−	−	+	+++
Granuloma formation	+++	++	+	−	−
Reaction to lepromin	+++	+	±	−	−
Antibodies to *M. leprae*	±	±	+	++	+++
Main phagocytic cell	Mature epithelioid	Immature epithelioid	Immature epithelioid	Macrophage	Macrophage
In-vitro correlates of CMI	+++	++	+	±	−
Type 1 reactions	−	+	+	+	−
Type 2 reactions	−	−	−	±	++
CMI, cell-mediated immunity; see text for other abbreviations.					

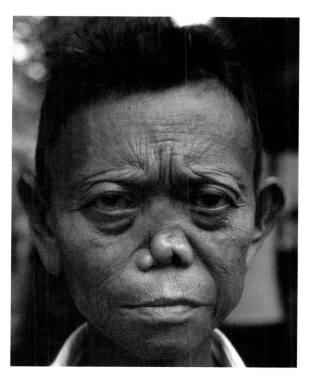

Fig. 18.8 Treated lepromatous leprosy. The nodularity of the skin has resolved on treatment but the absence of eyebrows and the nasal collapse remain.

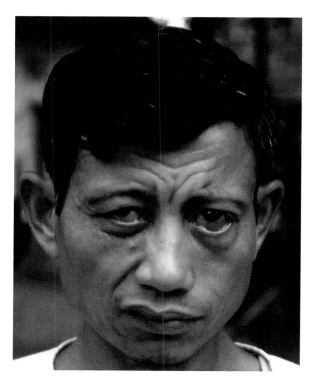

Fig. 18.9 Tuberculoid leprosy. The only feature on presentation was paralysis of the left facial nerve, resulting in loss of the left nasolabial fold and an inability to close the eye, predisposing to corneal damage.

of the eyelids (Fig. 18.9). Blindness is a common and tragic complication of untreated leprosy.

Additional tissue damage in leprosy is caused by immune reactions resulting from delayed hypersensitivity (Jopling type 1 reactions) or a vasculitis associated with the deposition of antigen–antibody complexes (Jopling type 2 reactions, *erythema nodosum leprosum*) (Table 18.8). The former, which occurs in borderline cases (BT, BB and BL), may rapidly cause severe and permanent nerve damage, and requires urgent treatment with anti-inflammatory agents and, sometimes, surgical decompression of a greatly swollen nerve.

LABORATORY DIAGNOSIS

The clinical diagnosis may be confirmed by histological examination of skin biopsies and by the detection of acid-fast bacilli in nasal discharges, scrapings from the nasal mucosa and *slit-skin smears*. The latter are prepared by making superficial incisions in the skin, scraping out some tissue fluid and cells, and making smears on glass slides. Smears are obtained from obvious lesions, the ear lobes and apparently unaffected skin. Secretions and skin

smears are stained by the ZN method, and the number of bacilli seen in each high-power field may be recorded as the *bacillary index*; however, for usual practical purposes, patients with clinically active leprosy but in whom no bacilli are seen on slit-skin smear examination are described as having *paucibacillary* disease, and those who are positive at any site are described as having *multibacillary* disease. This is an important distinction for the selection of treatment (see below).

It is widely assumed, but unproven, that leprosy bacilli that stain strongly and evenly are viable whereas those that stain weakly and irregularly are dead. The percentage of the former gives the *morphological index*, which declines during chemotherapy. An increase in the morphological index is a useful indication of non-compliance of the patient and the emergence of drug resistance. Where facilities exist, the PCR may be used to detect *M. leprae* in clinical specimens.

TREATMENT

Multidrug therapy based on dapsone, rifampicin and clofazimine is highly effective. The regimen is deter-

Table 18.8 Main characteristics of the reactions in leprosy

Characteristic	Type 1 (reversal reaction)	Type 2 (erythema nodosum leprosum)
Immunological basis	Cell mediated	Vasculitis with antigen–antibody complex deposition
Type of patient	BT, BB, BL	BL, LL
Systemic disturbance	No (or mild)	Yes
Haematological changes	No	Yes
Proteinuria	No	Frequently
Relation to therapy	Usually within first 6 months	Rare during first 6 months

See text for explanation of abbreviations.

Table 18.9 WHO recommendations for multidrug therapy

Type of leprosy	Drug	Dose (mg)	Frequency	Total duration (months)
Paucibacillary	Rifampicin	600	Monthly, supervised	6
	Dapsone	100	Daily, unsupervised	
Multibacillary	Rifampicin	600	Monthly, supervised	12–24
	Dapsone	100	Daily, unsupervised	
	Clofazimine plus	300	Monthly, supervised	
	clofazimine	50	Daily, unsupervised	
Multibacillary: alternative regimen	Rifampicin	600	Monthly, supervised	24
	Ofloxacin	400	Monthly, supervised	
	Minocycline	100	Monthly, supervised	

mined by whether the patient has paucibacillary or multibacillary disease (Table 18.9). Clofazimine causes skin discoloration, particularly in fair-skinned people. If this results in the patient refusing the drug, a combination of rifampicin, ofloxacin and minocycline, administered monthly for 24 months for multibacillary disease, may be used instead. A single dose of rifampicin–ofloxacin–minocycline has been advocated for the treatment of single-lesion paucibacillary leprosy.

The treatment of leprosy demands far more than the administration of antimicrobial agents. It is often necessary to:

- correct deformities
- prevent blindness and further damage to anaesthetic extremities
- treat reactions with anti-inflammatory drugs
- attend to the patient's social, psychological and spiritual welfare.

EPIDEMIOLOGY

Once thought to be restricted to humans, leprosy has been reported in chimpanzees and sooty mangabey monkeys in Africa, and in free-living armadillos in Louisiana, USA. It was also thought that leprosy was transmitted by skin-to-skin contact but it now appears more likely that the bacilli are disseminated from the nasal secretions of patients with lepromatous leprosy. In addition, the blood of patients with lepromatous leprosy contains enough bacilli to render transmission by blood-sucking insects a definite, though unproven, possibility.

As in the case of tuberculosis, transmission of bacilli from patients with multibacillary leprosy to their contacts occurs readily, but only a minority of those infected develop overt disease. The infectivity of patients with paucibacillary leprosy is much lower. Leprosy often commences during childhood or early adult life but, as the incubation period is usually 3–5 years, it is rare in children aged less than 5 years.

Epidemiological studies on the prevalence and transmission of leprosy have been aided by skin test reagents, of which there are two types:

1. *Lepromins*, which are prepared from boiled bacilli-rich lepromatous lesions.
2. *Leprosins*, which are ultrasonicates of tissue-free bacilli extracted from lesions (the suffixes -H and -A are used to denote human and armadillo origins, respectively).

These reagents elicit two types of reaction:

1. The *Fernandez reaction* is analogous to tuberculin reactivity and appears in sensitized subjects 48 h after skin testing. It is best seen with leprosins.

2. The *Mitsuda reaction* is a granulomatous swelling that appears about 3 weeks after testing with lepromin. This reaction is indicative of the host's ability to give a granulomatous response to antigens of *M. leprae*, and is positive at or near the TT pole.

Skin testing is of limited diagnostic value, but is useful in epidemiological studies to establish the extent of infection in contacts and in the community, and in classifying patients according to the immunological spectrum.

CONTROL

The most effective control measure in leprosy, as in tuberculosis, is the early detection and treatment of infectious cases. This requires that patients should attend for therapy as soon as signs of the disease appear. Unfortunately, owing to the stigma associated with the disease, many patients delay seeking treatment until they have infected many contacts and developed irreversible disfigurement and handicap. Women, in particular, are likely to conceal their disease.

No living attenuated vaccines have been prepared for *M. leprae*, but BCG vaccine seems to protect against leprosy in regions where it protects against tuberculosis, strongly suggesting that protection is induced by common mycobacterial antigens. A vaccine prepared from an environmental mycobacterium (strain Mw) has been shown to afford protection in India.

KEY POINTS

- The mycobacteria are characterized by thick lipid-rich cell walls and the 'acid-fast' staining property.
- Most of the 100 or so named mycobacterial species are environmental saprophytes, although some occasionally cause disease. However, *M. tuberculosis* and *M. leprae* are obligate pathogens.
- Tuberculosis is a chronic intracellular infection characterized by granuloma formation. Most individuals (about 90%) do not develop symptomatic disease and probably remain latently infected.
- The lung is the usual site of initial infection and disease, but non-pulmonary forms occur. *Post-primary* tuberculosis is characterized by gross tissue necrosis leading to pulmonary cavity formation and aerogenous infectivity.
- One-third of the world's population has been infected by the tubercle bacillus and about 100 million individuals are newly infected annually.
- Diagnosis is made by clinical and radiological examination, tuberculin testing and detection of the tubercle bacillus by acid-fast staining, culture and

PCR. Therapy is based on a regimen of several drugs, usually for a period of 6 months. BCG is an attenuated live vaccine; its protective efficacy varies greatly between geographical regions and its overall impact on tuberculosis control is limited.
- *M. leprae* has never been cultivated in vitro. In leprosy, the nerves and skin are the principal sites of disease, resulting in deformities and visible lesions.
- Transmission is probably by the aerogenous route rather than by touch.
- *Tuberculoid* leprosy is characterized by excessive granuloma formation in the presence of a very small (*paucibacillary*) bacterial load, whereas *lepromatous* leprosy is characterized by a huge (*multibacillary*) bacterial load and little or no immune reactivity.
- Diagnosis is by clinical examination, microscopic detection or PCR of bacilli in skin or nasal smears and biopsies. Multidrug treatment regimens are highly effective and BCG vaccination appears to offer some degree of protection.

RECOMMENDED READING

Davies P D O (ed.) 2003 *Clinical Tuberculosis*, 3rd edn. Arnold, London

Grange J M, Lethaby J I 2004 Leprosy of the past and today. *Seminars in Respiratory and Critical Care Medicine (Tuberculosis and Other Mycobacterial Infections)* 25: 271–282

Grange J M, Yates M D, de Kantor I N 1996 *Guidelines for Speciation within the* Mycobacterium tuberculosis *Complex*, 2nd edn. World Health Organization, Geneva

Madkour M M (ed.) 2003 *Tuberculosis*. Springer, Berlin

Sansarricq H 2004 *Multidrug Therapy Against Leprosy: Development and Implementation Over the Past 25 Years*. World Health Organization, Geneva

World Health Organization 2003 *Treatment of Tuberculosis. Guidelines for National Programmes*, 3rd edn. World Health Organization, Geneva

World Health Organization Leprosy Group 2003 Leprosy. In Cook G C, Zumla A (eds) *Manson's Tropical Diseases*, 21st edn, pp. 1065–1084. Saunders, Edinburgh

CD-ROMs

Wellcome Trust 1999 *Tuberculosis* (Topics in International Health). CAB International, Wallingford

Wellcome Trust 1999 *Leprosy* (Topics in International Health). CAB International, Wallingford

Internet sites

Tuberculosis

World Health Organization. Tuberculosis. http://www.who.int/topics/tuberculosis/en/ and http://www.who.int/tb/en/

TB Alert. http://www.tbalert.org

Montefiore Medical Center, New York. http://www.tuberculosis.net

New Jersey Medical School Global Tuberculosis Institute. http://www.umdnj.edu/globaltb/home.htm

Centers for Disease Control and Prevention. Division of Tuberculosis Elimination. http://www.cdc.gov/nchstp/tb/

International Union Against Tuberculosis and Lung Disease. http://www.iuatld.org

Leprosy

World Health Organization. Elimination of leprosy as a public health problem. http://www.who.int/lep

International Federation of Anti-Leprosy Associations. http://www.ilep.org.uk

International Leprosy Association. Global Project on the History of Leprosy. http://www.leprosyhistory.org

Leprosy Information Service of Netherlands Leprosy Relief (Infolep). http://infolep.antenna.nl

19 Environmental mycobacteria

Opportunist disease

J. M. Grange

In addition to the tubercle and leprosy bacilli there are more than 100 recognized species of mycobacteria (listed on the website *www.bacterio.net*) that normally exist as saprophytes of soil and water. Termed environmental (or non-tuberculous) mycobacteria, some species occasionally cause opportunist disease in animals and man. Although such opportunist pathogens were described towards the end of the nineteenth century, their classification remained in a state of chaos for over 50 years and they were often called *anonymous mycobacteria*. Order began to replace chaos in 1959 when a botanist, Ernest Runyon, described four groups of mycobacteria associated with human disease according to their production of yellow or orange pigment and their rate of growth. Thus, mycobacterial species were divided into *photochromogens*, which develop pigment in (or after exposure to) light, *scotochromogens*, which become pigmented in the dark, *non-chromogens* and *rapid growers*. The latter species produce visible growth on Löwenstein–Jensen medium within 1 week on subculture, although growth on primary culture of clinical material often takes considerably longer; they may be photochromogens, scotochromogens or non-chromogens.

DESCRIPTION

Photochromogens

This group contains the following species:
- *Mycobacterium kansasii*, which grows well at 37°C and is isolated principally from patients with pulmonary disease. On microscopy, it often appears elongated and distinctly beaded (Fig. 19.1).
- *M. simiae*, which, like *M. kansasii*, grows at 37°C and is occasionally involved in pulmonary disease.
- *M. marinum*, the cause of a warty skin infection known as *swimming pool granuloma* or *fish tank granuloma*. It grows poorly, if at all, at 37°C, and cultures from skin lesions should be incubated at 33°C. Microscopically, it resembles *M. kansasii*.

Scotochromogens

Most slowly growing scotochromogens isolated from sputum or urine are of no clinical significance. These include *M. gordonae* (formerly *M. aquae*), which is frequently found in water and is a common contaminant of clinical material. It is, however, a rare cause of pulmonary disease. Two species are more usually associated with disease:

1. *M. scrofulaceum* is associated principally with scrofula or cervical lymphadenitis, but also causes pulmonary disease.
2. *M. szulgai*, an uncommon cause of pulmonary disease and bursitis, is a scotochromogen when incubated at 37°C but a photochromogen at 25°C.

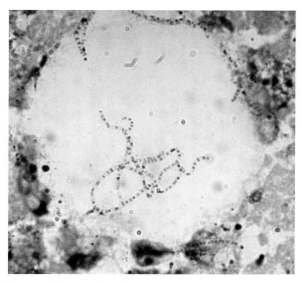

Fig. 19.1 *Mycobacterium kansasii* in sputum, showing elongated and beaded appearance. (Courtesy of Professor John Stanford.)

Non-chromogens

The most prevalent and important opportunistic pathogens are in a group known as the *M. avium* complex (MAC). This group consists of:

- *M. avium avium* (the avian tubercle bacillus), a pathogen of birds and also a cause of lymphadenitis in pigs as well as occasional disease in various other wild and domestic animals.
- *M. avium intracellulare*, principally a human pathogen.
- *M. avium paratuberculosis*, the cause of chronic hypertrophic enteritis or Johne's disease of cattle. There have been claims, which remain to be substantiated, that it is a cause of Crohn's disease in man. It produces little or no mycobactin, a lipid-soluble iron-binding compound essential for growth, and this substance must be added to media used for cultivation.
- *M. avium sylvaticum*, a cause of tuberculosis in wood pigeons. Like *M. paratuberculosis* it needs mycobactin for growth.
- *M. avium lepraemurium*, the cause of a leprosy-like disease of rats, mice and cats.

Clinical isolates are either *M. avium avium* or *M. avium intracellulare* and are the cause of lymphadenitis and pulmonary lesions. They are also a cause of disseminated disease in profoundly immunosuppressed patients, notably in those with the acquired immune deficiency syndrome (AIDS). Most clinical isolates are smooth and easy to emulsify; 28 serotypes have been delineated by agglutination by specific antisera, although untypable strains are not uncommon. For reasons that are not understood, isolates from patients with AIDS are of serotypes 1, 4 and 8, which, on DNA analysis, show homology with *M. avium* rather than *M. intracellulare*. By contrast, isolates from patients with other predisposing conditions are of a wider range of types.

Other non-chromogens include:

- *M. ulcerans*, a very slowly growing species that grows in vitro only at 31–34°C. Colonies are non-pigmented or a pale lemon-yellow colour. *M. ulcerans* is the cause of *Buruli ulcer*; unlike other mycobacterial pathogens it produces a toxin that causes tissue necrosis and is involved in the pathogenesis of the disease. *M. shinshuense*, a very rare cause of skin ulcers in Japan and China, is a variant of *M. ulcerans*.
- *M. xenopi*, originally isolated from a xenopus toad, is a thermophile that grows well at 45°C. It is principally responsible for pulmonary lesions in man. Most reported cases have been from London and south-east England and northern France. Two phylogenetically similar species, *M. celatum* and *M. branderi*, have been described.

- *M. malmoense* is a cause of pulmonary disease and lymphadenitis. It grows very slowly, often taking as long as 10 weeks to appear on primary culture, and is therefore likely to be missed if cultures are discarded earlier. For unknown reasons, disease due to this pathogen is increasing in several European countries. In some regions, notably northern England, it has become one of the most frequently isolated of the environmental mycobacteria from patients with pulmonary disease.
- *M. terrae* (the radish bacillus), *M. nonchromogenicum* and *M. triviale* are sometimes grouped as the *M. terrae* complex. They are very rare causes of pulmonary disease and, in the case of *M. terrae*, post-traumatic lesions.
- *M. haemophilum*, characterized by its growth requirement for haem or other sources of iron, is a rare cause of granulomatous or ulcerative skin lesions in xenograft recipients and other immunocompromised individuals, and of lymphadenitis in otherwise healthy children.
- *M. genevense*, a very slow growing organism occasionally isolated from patients with AIDS with disseminated mycobacterial disease, and from pet and zoo birds.

Rapid growers

Four rapidly growing non-chromogenic species are well recognized human pathogens: *M. chelonae*, *M. abscessus* (formerly regarded as a variant of *M. chelonae*), *M. fortuitum* and *M. peregrinum* (formerly a variant of *M. fortuitum*). They occasionally cause pulmonary or disseminated disease but are principally responsible for post-injection abscesses and wound infections, including corneal ulcers.

Most of the many other rapidly growing species are pigmented and, although disease due to them is exceedingly rare, they frequently contaminate clinical specimens. They are found on the genitalia and gain access to urine samples, although, contrary to a common belief, *M. smegmatis* is rarely found in this site. *M. flavescens*, which grows more slowly than other members of this group and is therefore sometimes classified as a slow grower, is an occasional cause of post-injection abscesses.

PATHOGENESIS

Compared with tubercle bacilli, environmental mycobacteria are of low virulence and, although human beings are in regular contact with these organisms and therefore frequently infected, overt disease is uncommon except in those who are profoundly immunosuppressed. Five main types of opportunist mycobacterial disease have been described in man (Table 19.1):

1. localized lymphadenitis
2. skin lesions following traumatic inoculation of bacteria
3. tuberculosis-like pulmonary lesions
4. tuberculosis-like non-pulmonary lesions
5. disseminated disease.

Lymphadenitis

This is caused by a number of different species that vary in relative frequency from region to region. The *M. avium* complex is the predominant cause worldwide. Some reports claim a high incidence of *M. scrofulaceum*, but these strains were probably misidentified members of the *M. avium* complex. In most cases a single node, usually tonsillar or pre-auricular, is involved, and most patients are children aged less than 5 years (Fig. 19.2). Unless contra-indicated by the risk of nerve damage, excision of the node, usually performed for diagnostic purposes, is almost always curative. This disease was very rare where children were vaccinated with BCG (bacille Calmette–Guérin) neonatally, but the incidence increased considerably in countries where such vaccination was terminated. Lymphadenitis occasionally occurs as part of a more disseminated infection, particularly in individuals with AIDS.

Skin lesions

Three main types have been described: post-injection (and post-traumatic) abscesses, swimming pool granuloma and Buruli ulcer.

Post-injection and post-traumatic abscesses

These are usually caused by the rapidly growing pathogens *M. abscessus*, *M. chelonae*, *M. fortuitum* and, less frequently, *M. peregrinum*. Abscesses occur sporadically, particularly in the tropics, or in small epidemics when these bacteria contaminate batches of injectable materials. Abscesses develop within a week or so, or for up to a year or more, after the injection. They are painful, may become quite large – up to 8 or 10 cm in diameter – and may persist for many months. Treatment is by drainage with curettage or total excision. Chemotherapy is not required unless there is local spread of disease or multiple abscesses, as may occur in insulin-injecting diabetics.

More serious lesions, also usually due to these rapidly growing pathogens, have followed surgery, particularly procedures involving insertion of prostheses such as heart valves, and corneal infections have followed ocular injuries or surgery (Fig. 19.3). Infections by *M. terrae* have occurred in farmers and others who have been injured while working with soil.

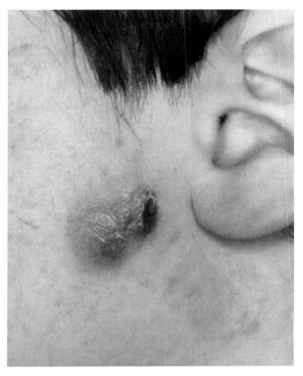

Fig. 19.2 Pre-auricular lymphadenitis due to an environmental mycobacterium in a child. (Courtesy of Professor Ali Zumla, University College London.)

Table 19.1 Principal types of opportunist mycobacterial disease in man, and the usual causative agents

Disease	Usual causative agent
Lymphadenopathy	M. avium complex
	M. scrofulaceum
Skin lesions	
Post-traumatic abscesses	M. chelonae
	M. abscessus
	M. fortuitum
	M. peregrinum
Swimming pool granuloma	M. marinum
Buruli ulcer	M. ulcerans
Pulmonary disease and solitary non-pulmonary lesions	M. avium complex
	M. kansasii
	M. xenopi
	M. malmoense
Disseminated disease	
AIDS related	M. avium complex
	M. genevense
Non-AIDS related	M. avium complex
	M. chelonae
	M. abscessus

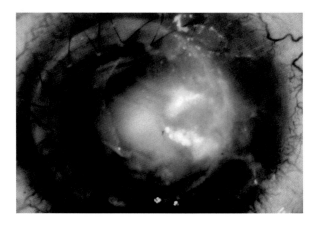

Fig. 19.3 Keratitis due to *M. chelonae* following corneal grafting. The cornea shows a characteristic 'cracked windscreen' appearance. (Reproduced by permission of Elsevier from Khooshabeh R, Grange J M, Yates M D, McCartney A C E, Casey T A 1994 A case report of *Mycobacterium chelonae* keratitis and a review of mycobacterial infections of the eye. *Tubercle and Lung Disease* 75: 377–382.)

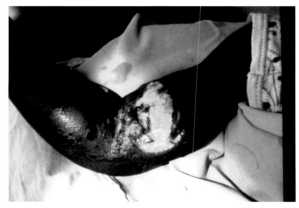

Fig. 19.4 Buruli ulcer. Undermined ulcer overlying the biceps and swelling of the surrounding tissues. (Courtesy of Dr Alan Knell, Wellcome Tropical Institute.)

Swimming pool granuloma

This is also known as *fish tank granuloma* and, occasionally, *fish fancier's finger*, and is caused by *M. marinum*. Those affected are mostly users of swimming pools, keepers of tropical fish and others involved in aquatic hobbies. The bacilli enter scratches and abrasions, and cause warty lesions similar to those seen in skin tuberculosis. The lesions, which usually occur on the knees and elbows of swimmers and on the hands of aquarium keepers, are usually localized, although secondary lesions sometimes appear along the line of the dermal lymphatics. This is termed *sporotrichoid spread* as a similar phenomenon occurs in the fungus infection sporotrichosis (see Ch. 60). The disease is usually self-limiting, although chemotherapy with minocycline, co-trimoxazole or rifampicin with ethambutol hastens its resolution.

Buruli ulcer

This disease, caused by *M. ulcerans*, was first described in Australia, although the name is derived from the Buruli district of Uganda where a large outbreak was investigated extensively in the 1960s. Buruli ulcer occurs in several tropical countries, notably Ghana, Nigeria, Congo Kinshasa, Côte d'Ivoire, Togo, Mexico, Malaysia and Papua New Guinea, and is limited to certain localities, characteristically low-lying areas subject to periodic flooding. *M. shinshuense*, a rare cause of similar lesions in Japan and China, is a variant of *M. ulcerans*. There is strong though indirect evidence that *M. ulcerans* is introduced from the environment into the human dermis

by minor injuries, particularly by spiky grasses. It is becoming a major public health problem in central and western equatorial Africa, where the number and severity of cases has increased; in some areas, its incidence is exceeding that of tuberculosis and leprosy. This increase does not appear to be related to infection with the human immunodeficiency virus (HIV) and the cause for the increase remains uncertain.

The first manifestation of the disease is a hard cutaneous nodule, which may be itchy. The nodule enlarges and develops central softening and fluctuation owing to necrosis of the subcutaneous adipose tissue caused by an exotoxin, which also has immunosuppressive properties. The overlying skin becomes anoxic and breaks down, the liquefied necrotic contents of the lesion are discharged, and one or more ulcers with deeply undermined edges are thereby formed (Fig. 19.4). At this stage the lesion is teeming with acid-fast bacilli and there is no histological immunological evidence of an active cell-mediated immune response. During this anergic stage the lesion may progress to an enormous size, sometimes involving an entire limb or a major part of the trunk.

For unknown reasons, the anergic phase may eventually give way to an immunoreactive phase when a granulomatous response develops in the lesion, the acid-fast bacilli disappear and immune reactivity to antigens of *M. ulcerans* is detectable. Healing then occurs but the patient is often left with considerable disfigurement and disability caused by extensive scarring and contractures.

The early, pre-ulcerative, lesions are easily treatable by excision and primary closure by suture. Ulcerated lesions require excision and skin grafting. In the anergic phase, the excision must be extensive enough to remove all disease to avoid recurrences. Treatment with various antibacterial agents has been attempted but with unconvincing results.

Pulmonary disease

This is seen most frequently in middle-aged or elderly men with lung damage caused by smoking or exposure to industrial dusts. It also occurs in individuals with congenital or acquired immune deficiencies, malignant disease and cystic fibrosis, but a substantial minority of cases occur in persons with no apparent underlying localized or generalized disorder. The disease may be caused by many species, although the most frequent are the *M. avium* complex and *M. kansasii*. Other causative organisms in the UK are *M. xenopi* and *M. malmoense*, with the former being more common in the south of the country and the latter in the north.

Localized non-pulmonary disease

Localized non-pulmonary lesions resembling those seen in tuberculosis are uncommon. Disease involving the meninges, bone (including the spine) and joints, kidney and male genitalia has been described. Mycobacterial peritonitis is a serious complication of peritoneal dialysis.

Disseminated disease

In the 1980s and early 1990s, up to a half of all persons dying from AIDS in the USA had disseminated mycobacterial disease, almost always due to the *M. avium* complex. This AIDS-related disease was also common in Europe, although it was uncommon in Africa. The introduction of highly active antiretroviral therapy (see Ch. 55) has led to a substantial decline in the incidence of this disease in developed nations.

The source of the causative organisms is uncertain. Some workers consider that disease is due to reactivation of dormant foci of infection acquired in childhood, whereas others argue that it results from recent infection from the environment; these explanations are not mutually exclusive. Acid-fast bacilli are readily isolated from bone marrow aspirates, intestinal biopsies, blood and faeces. A few cases of AIDS-related disseminated disease have been caused by other species, including *M. genavense*.

Disseminated disease occurs occasionally in individuals with other congenital or acquired causes of immunosuppression, including renal transplantation. Again, the *M. avium* complex is the usual cause, but disseminated disease due to other infections, notably *M. chelonae*, has occurred in recipients of renal transplants and in other immunocompromised patients (Fig. 19.5).

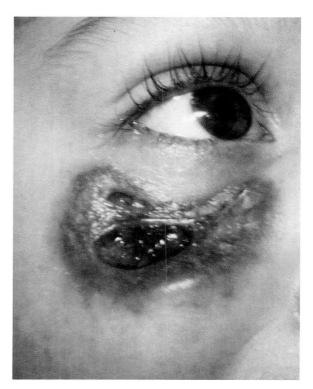

Fig. 19.5 Skin ulcer on the face due to disseminated *M. chelonae* in an immunocompromised child. (Courtesy of Dr Kurt Schopfer.)

LABORATORY DIAGNOSIS

Most environmental mycobacteria can be detected microscopically and cultured on media suitable for *M. tuberculosis* (see pp. 206–207). Great care must be taken to differentiate true disease from transient colonization or superinfection. In particular, there are no clinical or radiological characteristics that reliably differentiate pulmonary disease caused by opportunist mycobacteria from tuberculosis, and the diagnosis is therefore made by isolation and identification of the pathogen. As a general rule, a diagnosis of opportunist mycobacterial disease may be made when a heavy growth of the pathogen is repeatedly isolated from the sputum of a patient with compatible clinical and radiological features, and in whom other causes of these features have been carefully excluded. Fibre-optic bronchoscopy is a useful aid to diagnosis as lesions may be directly biopsied.

Identification is usually undertaken by specialist reference laboratories. There is no universally recognized identification scheme, although reliance is usually placed on cultural characteristics (rate and temperature of growth and pigmentation), various biochemical reactions and resistance to antimicrobial agents. Many reference centres now use nucleic acid-based methods to identify known species and to delineate new ones. The most discriminative methods are based on the detection of sequence differences in the gene coding for 16S ribosomal ribonucleic acid (RNA). DNA probes for the more commonly encountered species are available commercially, and species-specific polymerase chain reaction (PCR) primers have been described.

TREATMENT

Most slowly growing environmental mycobacteria are resistant to many antituberculosis drugs in vitro, although infections often respond to various combinations of these drugs. Pulmonary disease caused by the *M. avium* complex, *M. xenopi* and *M. malmoense* often responds to a regimen of rifampicin and ethambutol – provided that both drugs are given for 18–24 months. Treatment is, however, unsuccessful in a substantial minority of cases, and localized lesions are surgically excised when possible. Limited experience indicates that the newer macrolides and fluoroquinolones are efficacious, but adverse reactions are more common.

Pulmonary and non-pulmonary disease due to the rapidly growing species *M. abscessus*, *M. chelonae*, *M. fortuitum* and *M. peregrinum*, have been treated successfully by various combinations of erythromycin, newer macrolides, sulphonamides, trimethoprim, amikacin, gentamicin, imipenem, extended-spectrum cephalosporins and fluoroquinolones. In-vitro susceptibility tests are not always a reliable guide of clinical response and some infections, notably keratitis due to *M. chelonae*, frequently relapse and require surgical intervention, despite in-vitro drug susceptibility.

Drug regimens based on a macrolide (clarithromycin or azithromycin) and ethambutol are effective against AIDS-associated disease caused by the *M. avium* complex. There is, however, evidence that reduction of the viral load by use of antiretroviral agents contributes more to remission of such disease than antibacterial therapy.

EPIDEMIOLOGY

Mycobacteria are widely distributed in the environment and are particularly abundant in wet soil, marshland, streams, rivers and estuaries. Some species, such as *M. terrae*, are found in soil whereas others, including *M. marinum* and *M. gordonae*, prefer free water. Many species, including potential pathogens such as the *M. avium* complex, *M. kansasii* and *M. xenopi*, are able to colonize piped water supplies. Human beings are therefore regularly exposed to mycobacteria as a result of drinking, washing, showering and inhalation of natural aerosols. Such repeated subclinical infection may induce sensitization to tuberculin and other mycobacterial skin-testing reagents. There is also evidence that contact with environmental mycobacteria profoundly affects the subsequent ability of BCG vaccine to induce protective immunity, thereby explaining the great regional differences in the efficacy of this vaccine (see p. 214).

The number of cases of disease due to environmental mycobacteria relative to cases of tuberculosis increases in regions where the latter is uncommon and declining in incidence. In south-east England, about 5% of mycobacterial disease is due to environmental species; the proportion is much higher in some parts of the USA. In addition, the absolute incidence is increasing as a result of the growing number of immunocompromised individuals, notably patients with AIDS.

A few 'epidemics' of falsely diagnosed mycobacterial pulmonary disease and urinary tract infection have resulted from the collection of sputum and urine specimens in containers rinsed out with water from taps colonized by mycobacteria. Inadequate cleaning of endoscopes has also led to mycobacterial contamination of clinical specimens and diagnostic confusion. Likewise, false-positive sputum smear examinations for acid-fast bacilli have occurred when staining reagents were prepared from contaminated water.

CONTROL

The incidence and type of disease in any region are determined by the species and numbers of mycobacteria in the environment and the opportunities for human transmission. Unlike tuberculosis, person-to-person transmission of opportunist mycobacterial disease rarely, if ever, occurs. Thus the incidence of such disease is independent of that of tuberculosis and unaffected by public health measures designed to control the latter.

KEY POINTS

- Environmental or non-tuberculous mycobacteria (about 100 species) are divisible into slow and rapid growers. Human disease is often opportunistic (e.g. in AIDS and chronic pre-existing lung disease) and attributable to slow growers, notably members of the *M. avium* complex.
- Infection is from the environment, and transmission control methods applied to tuberculosis are ineffective.
- Presentations include localized lymphadenitis, post-inoculation (injection or trauma) skin lesions, tuberculosis-like pulmonary lesions, solitary non-pulmonary lesions and disseminated disease.
- Lymphadenitis occurs in otherwise healthy young children (<6 years of age) and in immunosuppressed persons.

- Skin diseases include swimming pool granuloma (*M. marinum*), Buruli ulcer (*M. ulcerans*) and post-inoculation abscesses caused by various rapid growers.
- Diagnosis is based on a range of simple tests or by molecular techniques including DNA probes or specific PCR primers. Repeated detection in the face of progressive disease assists in distinguishing pathogenic association from contamination or colonization.
- Therapy depends on the causative organism and is largely empirical as few clinical trials have been conducted and drug susceptibility tests are not very reliable.

RECOMMENDED READING

Asiedu K, Scherpbier R, Raviglione M 2000 *Buruli Ulcer.* Mycobacterium ulcerans *Infection.* World Health Organization Global Buruli Ulcer Initiative, Geneva

Banks J, Campbell I A 2003 Environmental mycobacteria. In Davies P D O (ed.) *Clinical Tuberculosis*, 3rd edn, pp. 439–448. Arnold, London

Falkinham J O 2003 The changing pattern of nontuberculous mycobacterial disease. *Canadian Journal of Infectious Diseases* 14: 281–286

Grange J M 1996 *Mycobacteria and Human Disease*, 2nd edn. Arnold, London

Park H, Jang H, Kim C, Chung B, Chang C L, Park S K, Song S 2000 Detection and identification of mycobacteria by amplification of the internal transcribed spacer regions with genus- and species-specific PCR primers. *Journal of Clinical Microbiology* 38: 4080–4085

Research Committee of the British Thoracic Society 2001 First randomised trial of treatments for pulmonary disease caused by *M. avium intracellulare*, *M. malmoense*, and *M. xenopi* in HIV negative patients: rifampicin, ethambutol and isoniazid versus rifampicin and ethambutol. *Thorax* 56: 167–172

Wansbrough-Jones M H 1999 Non-tuberculous or atypical mycobacteria. In James D G, Zumla A (eds) *The Granulomatous Disorders*, pp. 189–204. Cambridge University Press, Cambridge

Internet site

Chest Medicine On-line (Priory Lodge Education Ltd.). Environmental mycobacteria. http://www.priory.com/envmyco1.htm

20 Actinomyces, nocardia and tropheryma

Actinomycosis; nocardiasis; Whipple's disease

J. M. Grange

Gram-positive bacteria with branching filaments that sometimes develop into mycelia are included in the rather loosely defined order Actinomycetales. Although mostly soil saprophytes, five genera, *Actinomyces*, *Nocardia*, *Actinomadura*, *Propionibacterium* and *Bifidobacterium* (the latter two are considered briefly in Ch. 36), occasionally cause chronic granulomatous infections in animals and man. Members of the genera *Gordona*, *Oerskovia*, *Rothia* and *Tsukamurella* very rarely cause similar infections.

Tropheryma whippeli, the causative agent of Whipple's disease, has been shown to be an actinomycete on the basis of nucleic acid studies. Another genus, *Streptomyces*, is an extremely rare cause of disease, but is the source of several antibiotics. Repeated inhalation of thermophilic actinomycetes, notably *Faenia rectivirgula* and *Thermoactinomyces* species, causes extrinsic allergic alveolitis (farmer's lung, mushroom worker's lung, bagassosis) in those who are occupationally exposed to mouldy vegetable matter.

ACTINOMYCES

DESCRIPTION

Actinomyces are branching Gram-positive bacilli. They are facultative anaerobes, but often fail to grow aerobically on primary culture. They grow best under anaerobic or micro-aerophilic conditions with the addition of 5–10% carbon dioxide. Almost all species are commensals of the mouth and have a narrow temperature range of growth of around 35–37°C. They are responsible for the disease known as *actinomycosis*. Three-quarters of human cases are caused by *Actinomyces israelii*. Less common causes include *A. gerencseriae*, *A. naeslundii*, *A. odontolyticus*, *A. viscosus*, *A. meyeri*, *Arachnia propionica* and members of the genus *Bifidobacterium*.

Concomitant bacteria, notably a small Gram-negative rod, *Actinobacillus actinomycetemcomitans*, but also *Hae-*

mophilus species, fusiforms and anaerobic streptococci, are sometimes found in actinomycotic lesions, although their contribution to the pathogenesis of the disease, if any, is unknown. *Act. actinomycetemcomitans* is a rare cause of endocarditis.

PATHOGENESIS

Actinomycosis is a chronic disease characterized by multiple abscesses and granulomata, tissue destruction, extensive fibrosis and the formation of sinuses. Within diseased tissues the actinomycetes form large masses of mycelia embedded in an amorphous protein–polysaccharide matrix and surrounded by a zone of Gram-negative, weakly acid-fast, club-like structures. These clubs were once thought to consist, at least in part, of material derived from host tissue, but it now appears that they are formed entirely from the bacteria. The mycelial masses may be visible to the naked eye and, as they are often light yellow in colour, they are called *sulphur granules*. In older lesions the sulphur granules may be dark brown and very hard because of the deposition of calcium phosphate in the matrix. Various species of actinomyces may colonize diseased tissue, such as lung cancer, but sulphur granules are not seen.

The principal forms of human actinomycosis are:

- Cervicofacial infection, which accounts for more than half of reported cases; the jaw is often involved. The disease is endogenous in origin; dental caries is a predisposing factor, and infection may follow tooth extractions or other dental procedures. Men are affected more frequently than women, and in some regions the disease is more common in rural agricultural workers than in town dwellers, probably owing to lower standards of dental care in the former.
- Thoracic actinomycosis commences in the lung, probably as a result of aspiration of actinomyces from the mouth. Sinuses often appear on the chest wall, and the ribs and spine may be eroded. Primary

endobronchial actinomycosis is an uncommon complication of an inhaled foreign body.

- Abdominal cases commence in the appendix or, less frequently, in colonic diverticula.
- Pelvic actinomycosis occurs occasionally in women fitted with plastic intra-uterine contraceptive devices.

The lymphatics are not usually involved, but haematogenous spread to the liver, brain and other internal organs occurs occasionally. Involvement of bone is uncommon in human actinomycosis and is usually the result of direct extension of adjacent soft tissue lesions.

LABORATORY DIAGNOSIS

Specimens should be obtained directly from lesions by open biopsy, needle aspiration or, in the case of pulmonary lesions, by fibre-optic bronchoscopy. Examination of sputum is of no value as it frequently contains oral actinomycetes. Material from suspected cases is shaken with sterile water in a tube. Sulphur granules settle to the bottom and may be removed with a Pasteur pipette. Granules crushed between two glass slides are stained by the Gram and Ziehl–Neelsen (modified by using 1% sulphuric acid for decolorization) methods, which reveal the Gram-positive mycelia and the zone of radiating acid-fast clubs. Sulphur granules and mycelia in tissue sections are identifiable by use of fluorescein-conjugated specific antisera.

For culture, suitable media, such as blood or brain–heart infusion agar, glucose broth and enriched thioglycollate broth, are inoculated with washed and crushed granules. Contamination is reduced by the incorporation of metronidazole and nalidixic acid or cadmium sulphate in the media. Cultures are incubated aerobically and anaerobically for up to 14 days. After several days on agar medium, *A. israelii* may form so-called *spider colonies* that resemble molar teeth. The identity may be confirmed by biochemical tests, by staining with specific fluorescent antisera or by gas chromatography of metabolic products of carbohydrate fermentation.

TREATMENT

Actinomyces are sensitive to many antibiotics, but the penetration of drugs into the densely fibrotic diseased tissue is poor. Thus, large doses are required for prolonged periods, and recurrence of disease is not uncommon. Surgical debridement reduces scarring and deformity, hastens healing and lowers the incidence of recurrences. Penicillin-based regimens, such as injectable penicillin for 2–6 weeks, followed by oral penicillin V or amoxi-

Table 20.1 Differences between the genera *Actinomyces* and *Nocardia*

Actinomyces spp.	*Nocardia* spp.
Facultative anaerobes	Strict aerobes
Grow at 35–37°C	Wide temperature range of growth
Oral commensals	Environmental saprophytes
Non-acid-fast mycelia	Weakly acid-fast
Endogenous cause of disease	Exogenous cause of disease

cillin for 3–12 months depending on clinical response, are frequently used. Shorter regimens are also effective, including amoxicillin with clavulanic acid for 2 weeks for cervicofacial disease or 4 weeks for thoracic and abdominal disease. The clavulanic acid is required because lesions are often concomitantly infected with β-lactamase-producing bacteria. Alternative agents include tetracyclines, clindamycin, macrolides and imipenem. Additional drugs, including aminoglycosides and and metronidazole, may be required when concomitant organisms are present.

NOCARDIA

DESCRIPTION

The nocardiae are branched, strictly aerobic, Gram-positive bacteria that are closely related to the rapidly growing mycobacteria. Like the latter, but unlike actinomyces, they are environmental saprophytes with a broad temperature range of growth. The properties of nocardiae and actinomycetes are compared in Table 20.1. Most isolates are acid-fast when decolorized with 1% sulphuric acid.

Many species of nocardia are found in the environment, notably in soil, but human opportunist disease is almost always caused by *Nocardia asteroides*, so named because of its star-shaped colonies, *N. farcinica*, *N. brasiliensis*, *N. otitidis caviarum*, *N. nova* and *N. transvalensis*. Other species are encountered occasionally in profoundly immunosuppressed patients.

A related group of non-acid-fast species are assigned to the genus *Actinomadura*, which includes the species *Actinomadura madurae*, a common cause of Madura foot (see below).

PATHOGENESIS

Nocardiae, principally *N. asteroides*, are uncommon causes of opportunist pulmonary disease, which usually, but not always, occurs in immunocompromised individuals, including those receiving post-transplant immunosuppressive therapy or chemotherapy for cancer and those with acquired immune deficiency syndrome (AIDS). Corticosteroid therapy is a strong risk factor. As a result, the frequency and diversity of clinical manifestations of nocardial disease has increased over the past few decades. Pre-existing lung disease, notably alveolar proteinosis, also predisposes to nocardial disease. The infection is exogenous, resulting from inhalation of the bacilli. The clinical and radiological features are very variable and non-specific, and diagnosis is not easy. In most cases there are multiple confluent abscesses with little or no surrounding fibrous reaction, and local spread may result in pleural effusions, empyema and invasion of bones. In some cases the disease is chronic, whereas in others it spreads rapidly through the lungs. Secondary abscesses in the brain and, less frequently, in other organs occur in about one-third of patients with pulmonary nocardiasis. Acute dissemination with involvement of many organs occurs in profoundly immunosuppressed persons, notably those with AIDS.

Nocardiae also cause primary post-traumatic, post-operative or post-inoculation cutaneous infections (primary cuteneous nocardiasis). The most frequent cause is *N. brasiliensis* but some cases are caused by *N. asteroides* or other species. In the USA and the southern hemisphere, but rarely in Europe, cutaneous infections may result in fungating tumour-like masses termed *mycetomas*.

Madura foot is a chronic granulomatous infection of the bones and soft tissues of the foot resulting in mycetoma formation and gross deformity. It occurs in Sudan, north Africa and the west coast of India, principally among those who walk barefoot and are therefore prone to contamination of foot injuries by soil-derived organisms. A common causative organism is *Actinomadura madurae*, but Madura foot is also caused by other actinomycetes including *Streptomyces somaliensis* and by fungi (see Ch. 60).

LABORATORY DIAGNOSIS

A presumptive diagnosis of pulmonary nocardiasis may be made by a microscopical examination of sputum. In many cases the sputum contains numerous lymphocytes and macrophages, some of which contain pleomorphic Gram-positive and weakly acid-fast bacilli, and occasional extracellular branching filaments. Nocardiae are not so easily seen in tissue biopsies stained by the Gram or modified Ziehl–Neelsen methods, but may be seen in preparations stained by the Gram–Weigert or Gomori methenamine silver methods.

Nocardiae grow on blood agar, although growth is better on enriched media including Löwenstein–Jensen medium, brain–heart infusion agar and Sabouraud's dextrose agar

containing chloramphenicol as a selective agent. Growth is visible after incubation for between 2 days and 1 month; selective growth is favoured by incubation at 45°C. Colonies are cream, orange or pink coloured; their surfaces may develop a dry, chalky appearance, and they adhere firmly to the medium.

Identification of species is usually undertaken in reference laboratories. Method used include: high-pressure liquid chromatography; analysis of restriction fragment length polymorphisms in the 16S and 32S ribosomal RNA (rRNA) genes; or, in the case of common species such as *N. farcinica*, detection of species-specific DNA fragments by a polymerase chain reaction (PCR).

TREATMENT

The most widely used regimen is sulfamethoxazole with trimethoprim (co-trimoxazole) for 3 months or more, although this prolonged course often causes adverse drug reactions. In addition, some strains, especially of *N. farcinica*, are resistant to sulphonamides. An alternative regimen is high-dose imipenem with amikacin for 4–6 weeks. Limited experience indicates that minocycline, cephalosporins and amoxicillin–clavulanate combinations are also effective. Drug susceptibility testing is subject to several variables and no standardized methods have been proposed. Mycetomata due to nocardiae are much easier to treat than those due to fungi. Even long-standing cases with extensive mycetoma formation respond well to chemotherapy.

TROPHERYMA WHIPPELII

Whipple's disease is a rare multi-system disease with symptoms including diarrhoea and malabsorption, often with arthralgia, lymphadenopathy and fever. The intestine is principally affected but involvement of the lung, heart, skeletal muscle and central nervous system (CNS) occurs in a (probably underestimated) minority of patients. Most cases occur in middle-aged white males. The causative organism, *Tropheryma whippelii,* has a depleted genome and was originally cultivated in human embryonic lung cells, but sequencing of the genome has permitted the development of a medium for its cultivation in vitro. Diagnosis is usually made by histological examination of periodic acid–Schiff stained intestinal biopsies, electron microscopy, and PCR detection of a unique DNA sequence coding for 16S rRNA. The usual treatment consists of a 2-week course of streptomycin with benzylpenicillin (or, in the case of CNS involvement, ceftriaxone) followed by co-trimoxazole for 1 year. Relapses involving the CNS have responded to a 1-year course of chloramphenicol.

KEY POINTS

- Several species of *Actinomyces* cause chronic lesions characterized by multiple abscesses, granulomata, tissue destruction, extensive fibrosis and sinus formation.
- More than half of human cases occur in the cervicofacial region and often involve the jaw. Other cases occur in the thorax, abdomen and pelvis.
- Lesions may contain various concomitant organisms of doubtful significance to pathogenesis.
- Diagnosis is made by observation of the characteristic 'sulphur granules' in clinical material and by culture of the organism. Treatment is usually by prolonged penicillin-based regimens.

- Many species of *Nocardia* have been described but only a few, notably *N. asteroides*, cause human disease. Lung involvement in immunocompromised individuals predominates. Secondary abscesses, notably in the brain, occur in about one-third of patients.
- Post-inoculation cutaneous infections (primary cuteneous nocardiasis) may result in fungating tumour-like masses termed mycetomas.
- *Tropheryma whippelii*, the cause of Whipple's disease, was one of the first bacterial pathogens identified by 16S rRNA studies.
- Whipple's disease is a rare multi-system disease with principal involvement of the intestine, resulting in diarrhoea and malabsorption.

RECOMMENDED READING

Dutly F, Altwegg M 2001 Whipple's disease and '*Tropheryma whippelii*'. *Clinical Microbiology Reviews* 14: 561–583

Hay R J 2003 Nocardiasis. In Weatherall D J, Cox T M, Firth J D, Benz E J (eds) *Oxford Textbook of Medicine*, 4th edn, pp. 588–590. Oxford University Press, Oxford

Lederman E R, Crum N F 2004 A case series and focused review of nocardiosis. Clinical and microbiological aspects. *Medicine (Baltimore)* 83: 300–313

Marth T, Raoult D 2003 Whipple's disease. *Lancet* 361: 239–246

Schaal K P 2003 Actinomycosis. In Weatherall D J, Cox T M, Firth J D, Benz E J (eds) *Oxford Textbook of Medicine*, 4th edn, pp. 585–588. Oxford University Press, Oxford

Scott G 2003 Actinomycosis. In Cook G C, Zumla A I (eds) *Manson's Tropical Diseases*, 21st edn, pp. 1091–1093. Saunders, Edinburgh

Internet site

European Network on *Tropheryma whippelii* Infection. Whipple's Disease. http://www.whipplesdisease.info/

21 Bacillus

Anthrax; food poisoning

R. W. Titball and H. S. Atkins

The genus *Bacillus* originally included all rod-shaped bacteria, but now comprises only large, spore-forming, Gram-positive bacilli that form chains and usually grow aerobically or anaerobically. They are common environmental organisms, frequently isolated in laboratories as contaminants of media or specimens. *Bacillus anthracis*, the cause of *anthrax*, is the most important pathogen of the group. The organism holds a crucial place in the history of medical microbiology:

- Robert Koch's work on anthrax showed that a causative organism could be isolated from the blood of infected animals, artificially grown in pure culture and then used to reproduce the disease in animals. This led to the development of the present-day methods of isolation and identification of bacteria, and to the formulation of *Koch's postulates* (see Ch. 1).
- Louis Pasteur showed that animals could be actively immunized by infecting them with cultures of *B. anthracis* that had been attenuated by growth at 43°C.

Although rare in the industrialized nations, the cases of anthrax in the USA caused by the contamination of mail with *B. anthracis* spores is a reminder of the potential for this pathogen to be used as a biological weapon.

Bacillus cereus may contaminate food, especially rice, in large numbers and is commonly implicated in episodes of food poisoning. An atypical strain of *B. cereus* has been implicated as the causative agent of an anthrax-like disease in man. Other species of *Bacillus* are less often incriminated as human pathogens, usually in the immunocompromised.

BACILLUS ANTHRACIS

DESCRIPTION

B. anthracis is a large (4–8 × 1–1.5 µm) non-motile, sporing bacillus. The spores, which form readily when the bacterium is grown on certain artificial media, are oval, refractile and central in position. The temperature range for growth is 12–45°C (optimum 35°C); it grows on all ordinary media as typical colonies with a wavy margin and small projections, the so-called *medusa head* appearance. Table 21.1 lists some of the differences between *B. anthracis* and other important members of the genus *Bacillus*.

Genome sequences of several strains of *B. anthracis* have been determined. In addition to the chromosome, strains harbour two plasmids termed pX01 and pX02, encoding the toxin and capsule, respectively. Most of the chromosomal genes of *B. anthracis* are also found in *B. cereus*. A plasmid similar to pX01 has been found in some *B. cereus* strains (see below).

PATHOGENESIS

Anthrax is a *zoonosis* – a disease of animals transmissible secondarily to man. Human beings are relatively resistant to infection with *B. anthracis* and anthrax arises most commonly by inoculation through the skin of material from infected animals or their products. The disease is usually a consequence of the exposure of a susceptible host to spores of the bacillus. Spores are not found in host tissues, but appear on exposure of the vegetative cells to oxygen in the air. During naturally occurring disease this is the result of the leakage of infected blood and other body fluids from the corpse.

B. anthracis spores introduced into the body by abrasion, inhalation or ingestion are phagocytosed by macrophages and transported from the site of infection to regional lymph nodes, where the spores germinate and vegetative bacteria multiply. The bacilli then enter the bloodstream causing massive septicaemia, with up to 10^8 colony-forming units/ml of blood.

Virulence factors

The pathogenicity of *B. anthracis* depends primarily on two major virulence factors:

Table 21.1 Distinguishing properties of some important *Bacillus* species

Property	B. anthracis	B. cereus[a]	B. subtilis	B. stearothermophilus
Cell size	Large (6 × 1.5 mm)	Large	Small (3 × 0.6 mm)	Small
Motility	–	+	+	+
Capsule	+	–	–	–
Mouse pathogenicity	+++	+	–	–
Anaerobic growth	+	+	–	±
Optimal growth temperature (°C)	35	30	37	55

[a]Atypical strains may cause an anthrax-like disease.

1. the poly-D-glutamic acid capsule
2. the toxin complex comprising three proteins: the *protective antigen, oedema factor* and *lethal factor.*

The capsule appears to enhance the virulence of *B. anthracis* by inhibiting the phagocytosis of vegetative cells in the extracellular environment of the lymphatic system and bloodstream. It is mainly the action of the toxin that mediates damage to the host. The three components of the tripartite toxin combine to form two binary toxins, the *oedema toxin* and *lethal toxin,* formed by association of the protective antigen with the oedema factor and lethal factor, respectively. In each case, the protective antigen binds to host cells and facilitates the entry of the associated oedema or lethal factor (Fig. 21.1).

The oedema toxin is thought to be responsible for the characteristic localized swelling associated with cutaneous anthrax. Oedema factor is a calmodulin-dependent adenylate cyclase that catalyses the production of intracellular cyclic adenosine monophosphate (cAMP) from host adenosine triphosphate (ATP), inducing interleukin (IL)-6 and inhibiting tumour necrosis factor (TNF)-α in monocytes. In addition to disrupting cytokine responses, the oedema toxin may also increase host susceptibility to infection by impairing neutrophil function. However, it is the lethal toxin that is believed to play the major role in damage to the host and death. Lethal factor is a zinc metalloprotease that inactivates mitogen-activated protein kinase kinase, particularly in macrophages. The lethal toxin stimulates macrophages to produce IL-1β and TNF-α. During infection, IL-1β accumulates within macrophages and TNF-α is released. As the concentration of lethal toxin increases later in the infection process, macrophage lysis produces a sudden release of IL-1β, causing shock and death.

Thus, the pathogenesis of anthrax is related to the sensitivity of macrophages to:

- the antiphagocytic activity of the capsule
- the adenylate cyclase activity of the oedema toxin
- the metalloprotease activity of the lethal toxin.

Clinical features

Cutaneous anthrax

The primary lesion is often described as a *malignant pustule* because of its characteristic appearance (Fig. 21.2). Coagulation necrosis of the centre of the pustule results in the formation of a dark-coloured *eschar* that is later surrounded by a ring of vesicles containing serous fluid and an area of oedema and induration, which may become extensive. In patients with severe toxic signs and widespread oedema, the prognosis is poor.

Inhalation anthrax

This is a consequence of the inhalation of spores and is the most acute form of the disease, associated with a high mortality rate. The infectious human dose by the air-borne route is in the range 25 000 to 55 000 spores. Although sometimes incorrectly referred to as pneumonic anthrax, the disease does not develop as a bronchopneumonia.

Inhalational anthrax carries a high mortality rate owing to the intense inflammation, haemorrhage and septicaemia that result from the multiplication of organisms in bronchi and spread to the lungs, lymphatics and bloodstream. The production of toxins and the considerable bacterial load that rapidly occurs in the terminal septicaemic phase produce increased vascular permeability and hypotension similar to endotoxic shock.

In the past, cases of inhalation anthrax were reported in persons working in industries handling animal skins, hides and wool, giving rise to the name of *wool-sorter's disease*. In the UK inhalational anthrax from this source was eliminated by the close control of imported wool and fleeces, and by the disinfection of suspect materials. An accident at a former Soviet Union biological weapons factory in the 1980s resulted in the release of spores into the air. In the town of Sverdlosk, located downwind of the factory, 79 cases of human inhalation anthrax were recorded, some up to 6 weeks after exposure; even with antibiotic therapy there were 68 deaths.

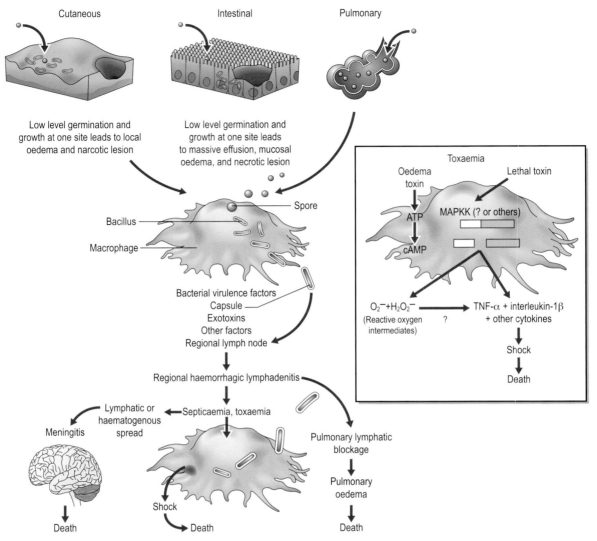

Fig. 21.1 Pathophysiology of anthrax. *Bacillus anthracis* spores reach a primary site in the subcutaneous layer, gastro-intestinal mucosa or alveolar spaces. For cutaneous and gastro-intestinal anthrax, low-level germination occurs at the primary site, leading to local oedema and necrosis. Endospores are phagocytosed by macrophages, and germinate. Macrophages containing bacilli detach and migrate to the regional lymph node. Vegetative anthrax bacilli grow in the lymph node, creating regional haemorrhagic lymphadenitis. Bacteria spread through the blood and lymph and multiply, causing severe septicaemia. High levels of exotoxins are produced that are responsible for overt symptoms and death. In a small number of cases, systemic anthrax can lead to meningeal involvement by means of lymphatic or haematogenous spread. In cases of pulmonary anthrax, peribronchial haemorrhagic lymphadenitis blocks pulmonary lymphatic drainage, leading to pulmonary oedema. Death results from septicaemia, toxaemia or pulmonary complications and can occur 1–7 days after exposure. The inset shows the effects of anthrax exotoxins on macrophages. See text for explanation. ATP, adenosine triphosphate; cAMP, cyclic adenosine monophosphate; IL, interleukin; MAPKK, mitogen-activated protein kinase kinase; TNF, tumour necrosis factor. (Reproduced with permission from Dixon et al 1999 © Massachusetts Medical Society, USA.)

Intestinal anthrax

The intestinal form of anthrax occurs among pastoralists who may be forced through poverty to eat infected animals that have been found dead. An individual may suffer after a day or so from haemorrhagic diarrhoea, and dies rapidly from septicaemia. Often these episodes occur as small outbreaks in a family or village. Because some individuals may suffer only cutaneous lesions, the recognition of the microbial cause of the outbreak is easily made clinically.

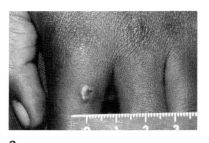
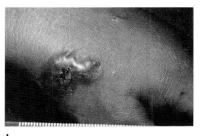
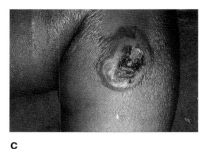

a   b   c

Fig. 21.2 Stages in the development and resolution of cutaneous anthrax lesions: **a** as first seen; **b** on day 2 or 3; **c** on day 6. (Images kindly provided by Dr Peter Turnbull.)

Naturally occurring infections of animals

All mammals are susceptible to anthrax although, in general, carnivores are relatively resistant to the disease. Herbivores are highly susceptible and omnivores show an intermediate level of resistance. Disease in wild herbivores or domesticated animals is usually septicaemic and follows the ingestion of spores along with coarse vegetation. The latter probably predisposes to trauma of the intestinal tract, allowing entry of the spores into the host. More rarely infection occurs after the inhalation of dust into the respiratory tract and, as in human disease, through skin abrasions leading to malignant pustules.

Although cases of anthrax are usually localized and sporadic, there may be large outbreaks of disease in wild or domesticated livestock. In the UK, the disease is occasionally found in livestock, chiefly cattle. During the period 1981–2001 there were 14 human cases of anthrax in the UK, all of which were the cutaneous form of the disease. Disease in pigs is atypical, appearing as a chronic disease with few fatalities. Sudden death in any herbivore should be treated with suspicion, and a veterinary officer summoned to examine the carcass without a post-mortem examination. A blood slide should be taken for Gram or methylene blue staining. Under the Anthrax Order the animal must remain on the farm and be incinerated on site if found to be positive. Deep burial in quicklime is an alternative method of disposal, but the spores remain viable for many years and may subsequently contaminate pasture and infect grazing animals. Very large numbers of bacilli are present in the terminal stages of disease in animals so that widespread dissemination may occur on death. These highly infectious animals serve as a source of anthrax both by direct spread to another beast and by contamination of the environment.

Animals models of disease

Various animals, including mice, guinea-pigs, rabbits and non-human primates have been used as models of human anthrax. Several strains of mice are highly susceptible to anthrax, but the response between infecting dose and morbidity is not always clear. To resolve this problem, the use of a partially attenuated strain of *B. anthracis* in mice deficient in complement C5 (e.g. A/J strain mice) has been proposed.

Guinea-pigs are extremely susceptible to anthrax, both by injection and by inhalation. After subcutaneous challenge the animal usually dies within 2–3 days, showing a marked inflammatory lesion at the site of inoculation and extensive gelatinous oedema in the subcutaneous tissues. Large numbers of the bacilli are present in the local lesion, and are also profusely present in the heart, blood and capillaries of the internal organs. They are especially numerous in the spleen (Fig. 21.3), which is enlarged and soft, giving rise to the description *splenic fever* in the ox and the German name for the organism – *Milzbrandbazillus* (spleen-destroying bacillus). It is questionable whether mice, guinea-pigs or rabbits are good models of human inhalational anthrax, as the physiology of the human respiratory tract is very different.

Studies in non-human primates have provided significant insight into the pathogenesis of inhalational anthrax, but their use is now considered to be ethically dubious. The estimation of the infectious dose in man is based largely on studies in non-human primates. Small particles of less than 5 μm containing spores are deposited on the alveolar walls and taken up by phagocytes. These phagocytes migrate to the draining lymph nodes, which become inflamed and enlarged. Surprisingly, ungerminated spores in phagocytes can remain in these lymph nodes for at least 100 days after exposure.

LABORATORY DIAGNOSIS

Clinical specimens

The fully developed malignant pustule may be difficult to swab and the central necrotic area gives a poor yield. Fluid aspirated from the surrounding vesicles, when present, is more likely to yield anthrax bacilli. Specimens should be taken before antibiotic therapy has been instituted.

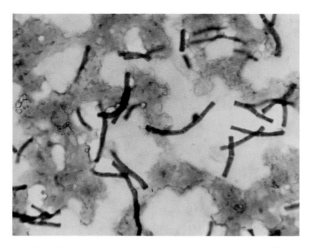

Fig. 21.3 Guinea-pig spleen imprint showing typical anthrax bacilli.

In laboratories unfamiliar with the disease, additional precautions for staff safety need to be organized. However, the relative ease of clinical diagnosis, given the characteristic appearance and occupational exposure, often makes laboratory confirmation superfluous.

Gram's stain may show typical large Gram-positive bacilli, and culture on blood agar yields the large, flat, greyish colonies with the characteristic 'medusa head' appearance. Staining of films from these colonies shows long chains of Gram-positive bacilli, some containing spores. Demonstration of non-motility, gelatin liquefaction, growth in straight chains and enhanced growth aerobically, as seen in the characteristic *inverted fir tree* appearance in a gelatin stab, will generally identify *B. anthracis* completely.

Serological diagnosis by enzyme-linked immuno-sorbent assay (ELISA) may be of value retrospectively, but is seldom used diagnostically. For the rapid identification of bacteria and diagnosis of disease, the polymerase chain reaction should be used. Because *B. anthracis* possesses a number of unique genes, the selection of suitable gene targets is not problematic. Toxin production can be demonstrated by immunological or gene probe methods in reference laboratories.

Environmental samples

It is occasionally necessary to isolate *B. anthracis* from potentially contaminated material such as animal hair, hides or soil. The heat treatment of aqueous extracts of these materials at 60°C for 1 h kills all except spore-forming bacteria and fungi. However, the isolation of *B. anthracis* may still be difficult because of the large number of spores of non-pathogenic *Bacillus* species found in the environment. For some soil types selective agars have been developed that allow preferential growth of *B. anthracis*.

TREATMENT

B. anthracis is susceptible in vitro to a wide range of antimicrobial agents and many have been used successfully for the treatment of anthrax in humans. Penicillin remains the drug of choice, as β-lactamase-producing strains of *B. anthracis* are rare. Most strains are also sensitive to macrolides, aminoglycosides, tetracyclines and chloramphenicol. Ciprofloxacin (or a similar fluoro-quinolone) is recommended as prophylaxis or early treatment for those considered at greatest risk of exposure following a large-scale release of anthrax spores in a deliberate attack.

The efficacy of antibiotics for the treatment of disease is dependent largely on the time at which therapy commences. Treatment with antibiotics is generally ineffective when septicaemic disease has developed and, in the later stages of disease, the general principles applied to the management of any patient in shock are more important than antimicrobial therapy. Because septicaemia develops rapidly in the case of intestinal or inhalational anthrax, these forms of the disease can be especially difficult to treat. It has been common practice to isolate patients with anthrax because of its highly infectious reputation. In fact, human-to-human spread does not occur. The use of specific antiserum is not recommended presently although, in the future, immunotherapy may play an important role in the treatment of septicaemic individuals.

EPIDEMIOLOGY AND CONTROL

The worldwide incidence of anthrax in man and animals is unknown, but many countries in Africa and Asia regularly report cases to the World Health Organization. When the disease is established in livestock and pasture becomes heavily contaminated with spores, an *enzootic* focus is created. Occasionally the disease may erupt in large numbers of domestic animals with associated human cases. In this situation anthrax is *epizootic*. Areas in which anthrax used to be enzootic, such as much of Europe, have been able to control the disease by controlling livestock and animal feeds (especially bone meal), and by strict regulations on the importation of animal hides.

The ability of the bacterium to persist in the soil is a consequence of the formation of spores that are able to survive for long periods in the environment. An indication of the robustness of the spores is their ability to survive exposure to chemical disinfectants and heat. For example, the spores will resist dry heat at 140°C for 1–3 h and boiling or steam at 100°C for 5–10 min. Autoclaving at 121°C (15 lb/in²) destroys them in 15 min.

Heavy contamination of soil exists in enzootic foci in many parts of the world, and spores may be recovered many years after the last known case. Artificial contamination of Gruinard island off the north-west coast of Scotland occurred in 1942–1943 as a result of tests of a biological warfare bomb containing live anthrax spores. Even by 1979 spores could still be detected in a 3-hectare area of the island. In the 1980s the area was decontaminated by burning the vegetation and spraying with 5% formaldehyde in sea water. By 1987, the ground was declared anthrax-free and, after reseeding, sheep were able to graze safely. In enzootic areas, disinfection of soil is not a practical control measure, and pastures known to be heavily contaminated should not be used for grazing animals. The organism multiplies when the soil pH is greater than 6 and when early rain has been followed by a long dry spell. Control in animals depends upon:

- early diagnosis
- isolation and incineration of infected animals
- use of vaccines.

In countries in which the disease is relatively rare in animals, contamination with imported materials is the commonest form of human infection. In general, the infectivity of *B. anthracis* for man is not of a high order, and when a case occurs in a factory spores are often widely distributed in large numbers in the dust and air. Anthrax is a recognized industrial hazard that, in the UK, is notifiable to Consultants in Communicable Disease Control and to the Health and Safety Executive. There is control on the importation of animal hides and hair, which are disinfected if considered infected. Workers at risk of exposure in the leather or wool industries, and veterinarians, may be offered immunization routinely and antibiotic prophylaxis if exposed to a known risk.

Immunization

Live-attenuated bacilli were first used by Louis Pasteur in May 1881, when he confounded professional scepticism in a famous public demonstration of the efficacy of a live vaccine at a farm at Pouilly-le-Fort in France. The 25 sheep that were vaccinated with heat-attenuated live bacilli and then inoculated with virulent anthrax material resisted infection, whereas 22 of 25 sheep acting as controls succumbed within 48 h. Thousands of sheep, cattle and horses were subsequently vaccinated, and the mortality rate among domesticated animals fell dramatically. Later, the Sterne strain of *B. anthracis*, which lacks the pX02 plasmid, was adopted for the immunization of animals, and this vaccine is still in use today.

Live vaccines are not considered to be sufficiently safe for human use and alternative vaccines based on the anthrax toxin have been developed. The currently licensed human vaccines are based on an alum precipitate of culture supernatant fluid. Predominantly they contain protective antigen, but there are also traces of lethal factor and oedema factor, which might account for the transient side-effects experienced by some vaccinees.

New recombinant protective antigen preparations free of toxin components may give better immunity and fewer adverse reactions; several types are in development.

BACILLUS CEREUS

DESCRIPTION

B. cereus is a large Gram-positive bacillus that resembles *B. anthracis*, except that it is motile and lacks the glutamic acid capsule. Like other members of the genus it is a saprophyte and frequents soil, water and vegetation. *B. cereus* closely resembles *B. anthracis* in culture, forming large, grey, irregular colonies described as *anthracoid*. Large inocula injected into laboratory animals may cause death but without the haemorrhagic appearance of anthrax, and blood smears do not show the characteristic pink capsule with McFadyean's stain.

PATHOGENESIS

B. cereus is most commonly associated with food poisoning, but the organism can also cause post-traumatic ophthalmitis, which requires rapid, aggressive management locally.

An atypical strain capable of causing a disease that resembled inhalation anthrax has been described. This strain appears to have acquired the toxin-encoding pX01 plasmid, and a plasmid encoding a polysaccharide capsule. The capsule presumably fulfils the same role as the poly-D-glutamic acid capsule on the surface of *B. anthracis*. Preliminary animal studies suggest that the strain is as virulent as *B. anthracis*. Its origins are presently unclear.

Food poisoning

Spores of *B. cereus* are particularly heat-resistant and most strains produce toxins. The organism is widespread in the environment and is found in most raw foods, especially cereals such as rice. Enormous numbers of organisms (up to 10^{10} organisms/g) may be found in contaminated food (commonly lightly cooked Chinese dishes), leading to two types of food poisoning:

1. Cases in which vomiting, occurring within 6 h of ingestion, is the main symptom. It is caused by preformed

toxin, which is a low molecular weight, heat- and acid-stable peptide that can withstand intestinal proteolytic enzymes.

2. A diarrhoeal form of food poisoning, occurring 8–24 h after ingestion, similar to enteritis caused by *Escherichia coli* or *Salmonella enterica* serotypes. This is caused by enterotoxins, which, like the *Clostridium perfringens* enterotoxin, are heat labile and formed in the intestine.

LABORATORY DIAGNOSIS

In the case of food poisoning, laboratory confirmation is easy if suspect food is available for testing. High numbers of *B. cereus*, often exceeding 10^8 organisms/g, are sufficient to make the diagnosis in the absence of other food-poisoning bacteria. Large facultatively anaerobic Gram-positive bacilli that produce anthracoid colonies on blood agar after overnight incubation at 37°C are almost certain to be *B. cereus*. Food reference laboratories are able to confirm identification and type if necessary.

Methods for the laboratory identification of *B. cereus* strains that cause an inhalation anthrax-like disease have yet to be devised. It is likely that genetic tests will be effective in identifying these atypical strains.

TREATMENT

Both the emetic and diarrhoeal syndromes associated with *B. cereus* are short lived and no specific treatment is needed. Most sufferers, even those with underlying conditions, seldom come to any harm. Acute symptoms last less than 24 h and recovery on a reduced diet and fluids is rapid. In comparison, the single case of inhalation anthrax-like disease caused by *B. cereus* was fatal. However, it is likely that antibiotic regimens for the treatment of anthrax would be equally effective for the treatment of disease caused by atypical strains of *B. cereus* strains.

CONTROL

Food poisoning caused by *B. cereus* is easily prevented by proper cooling and storage of food. Ideally, all dishes should be freshly prepared and eaten. Rice, in particular, should not be stored for long periods above 10°C.

OTHER BACILLUS SPECIES

Bacillus subtilis, *Bacillus pumilis* and *Bacillus licheniformis* have been implicated in causing food poisoning

similar to that due to *B. cereus*. They do not appear to form toxins, but some strains produce antibacterial peptides, such as the antibiotic bacitracin, which may facilitate growth in the intestinal tract. *Bacillus polymyxa* is the source of the antibiotic polymyxin.

B. cereus, *B. subtilis* and, rarely, other members of the genus may be found in wounds and tissues of immuno-compromised or burned patients. These opportunist pathogens are also common contaminants of specimens and laboratory media, so that the interpretation of significance is sometimes difficult. When found in numbers in normally sterile sites, such as blood or cerebrospinal fluid, these otherwise insignificant pathogens require specific treatment. Most strains produce abundant β-lactamase, which differs from the enzyme found in staphylococci.

STERILIZATION TEST BACILLI

Bacillus stearothermophilus was, until the discovery of archaebacteria in hot springs, the most heat-resistant organism known. Spores withstand 121°C for up to 12 min, and this has made the organism ideal for testing autoclaves that run on a time–temperature cycle designed to ensure the destruction of spores. Strips containing *B. stearothermophilus* are included with the material being autoclaved, and are subsequently examined by culture for surviving spores. The organism grows only at raised temperatures, typically between 50°C and 60°C; there is hardly any growth below 40°C. *Bacillus globigi*, a red-pigmented variant of *B. subtilis*, has been used to test ethylene oxide sterilizers, and *B. pumilis* has been used to test the efficacy of ionizing radiation.

KEY POINTS

- Bacteria belonging to the *Bacillus* genus are common environmental Gram-positive bacilli.
- *Bacillus anthracis* causes cutaneous, inhalational or ingestional anthrax.
- The pathogenicity of anthrax depends on the capsule and toxin.
- Antibiotic treatment of inhalational or ingestional anthrax is difficult.
- *B. anthracis* Sterne strain (toxin-negative) is used as an animal vaccine.
- A recombinant protein vaccine for anthrax is in development.
- *Bacillus cereus* commonly causes food poisoning.
- A *B. cereus* strain with anthrax toxin genes causes anthrax-like disease.

RECOMMENDED READING

Dixon T C, Meselson M, Guillemin J, Hanna P C 1999 Anthrax – a review. *New England Journal of Medicine* 341: 815–826

Drobniewski F A 1993 *Bacillus cereus* and related species. *Clinical Microbiology Reviews* 6: 324–338

Granum P E 1994 *Bacillus cereus* and its toxins. *Journal of Applied Bacteriology Symposium Supplement* 23: 61S–66S

Hoffmaster A R, Ravel J, Rasko D A et al 2004 Identification of anthrax toxin genes in *Bacillus cereus* associated with an illness resembling inhalation anthrax. *Proceedings of the National Academy of Sciences of the USA* 101: 8449–8454

Little S F, Ivins B E 1999 Molecular pathogenesis of *Bacillus anthracis* infection. *Microbes and Infection* 2:131–139

Nass M 1999 Anthrax vaccine. Model of a response to the biologic warfare threat. *Infectious Disease Clinics of North America* 13: 187–208, viii

Turnbull P C B, Hugh-Jones M E, Cosivi O 1999 World Health Organization activities on anthrax surveillance and control. *Journal of Applied Microbiology* 87: 318–320

Working Group on Civilian Biodefense 1999 Anthrax as a biological weapon. *Journal of the American Medical Association* 281: 1735–1745

22 Clostridium

Gas gangrene; tetanus; food poisoning; pseudomembranous colitis

T. V. Riley

The clostridia are Gram-positive spore-bearing anaerobic bacilli. Most species are saprophytes that normally occur in soil, water and decomposing plant and animal matter; they play an important part in natural processes of putrefaction. Some, such as *Clostridium perfringens* and *C. sporogenes*, are commensals of the animal and human gut. On the death of the host, these organisms and other members of the intestinal flora rapidly invade the blood and tissues, and initiate the decomposition of the corpse. The genus has undergone a major taxonomic revision, but this has had little impact on clostridia of medical significance. Pathogenic species include:

- *C. perfringens, C. septicum* and *C. novyi*, the causes of gas gangrene and other infections. *C. perfringens* is also associated with a form of food poisoning.
- *C. tetani*, the cause of tetanus.
- *C. botulinum*, the cause of botulism.
- *C. difficile*, the cause of pseudomembranous colitis and antibiotic-associated diarrhoea.

GENERAL DESCRIPTION

The clostridia are typically large, straight rods with slightly rounded ends. Pleomorphic forms, including filaments or elongated cells, club and spindle-shaped forms (*clostridium* is Latin for 'little spindle') are commonly seen in stained smears from cultures or wounds. They are Gram positive, but may appear to be Gram negative. All produce spores, which enable the organisms to survive in adverse conditions, for example in soil and dust and on skin.

Most species are obligate anaerobes: their spores do not germinate and growth does not normally proceed unless a suitably low redox potential (E_h) exists. A few species grow in the presence of trace amounts of air, and some actually grow slowly under normal atmospheric conditions.

Clostridia are biochemically active, frequently possessing both saccharolytic and proteolytic properties, although in varying degrees. Many species are highly toxigenic. The toxins produced by the organisms of tetanus and botulism attack nervous pathways and are referred to as neurotoxins. The organisms associated with gas gangrene attack soft tissues by producing toxins and aggressins, and are referred to as histotoxic. *C. difficile* and some strains of *C. perfringens* produce enterotoxins.

CLOSTRIDIUM PERFRINGENS

DESCRIPTION

C. perfringens is a relatively large Gram-positive bacillus (about $4–6 \times 1$ μm) with blunt ends. It is capsulate and non-motile. It grows quickly on laboratory media, particularly at high temperatures (approximately 42°C), when the doubling time can be as short as 8 min. It can be identified by the *Nagler reaction*, which exploits the action of its phospholipase on egg yolk medium; colonies are surrounded by zones of turbidity, and the effect is specifically inhibited if *C. perfringens* antiserum containing α-antitoxin is present on the medium. Typical food-poisoning strains produce heat-resistant spores that can survive boiling for several hours, whereas the spores of the type A strains that cause gas gangrene are inactivated within a few minutes by boiling.

GAS GANGRENE

C. perfringens is the commonest cause of gas gangrene, although various other species of clostridia, including *C. septicum, C. novyi* type A, *C. histolyticum* and *C. sordellii* are occasionally implicated, either alone or in combination. Gas gangrene is almost always a polymicrobial infection involving anaerobes and facultative organisms.

The disease is characterized by rapidly spreading oedema, myositis, necrosis of tissues, gas production and profound toxaemia occurring as a complication of wound infection. The diagnosis is made primarily on clinical grounds with laboratory confirmation.

The main source of the organisms is animal and human excreta, and spores of the causative clostridia are distributed widely. Infection usually results from contamination of a wound with soil, particularly from manured and cultivated land. However, it may be derived indirectly from dirty clothing, street dust, and even the air of an operating theatre if the ventilating system is poorly designed or improperly maintained. The skin often bears spores of *C. perfringens*, especially in areas of the body that may be contaminated with intestinal organisms.

Pathogenesis of gas gangrene

Impairment of the normal blood supply of tissue with a consequent reduction in oxygen tension may allow an anaerobic focus to develop. The patient's condition may deteriorate rapidly with the development of severe shock (Fig. 22.1).

Crushing of tissue and the severing of arteries in accidental injuries, rough handling of tissue and over-zealous clamping during surgery, or shock waves from gunshot injuries may compromise the microcirculation in an extensive area of tissue and prejudice tissue perfusion. The presence of devitalized or dead tissue, blood clot, extravasated fluid, foreign bodies and coincident pyogenic infection are all factors that promote the occurrence of gas gangrene in a wound. The spores of the clostridia and their vegetative bacilli cannot readily initiate infection in healthy tissues, presumably because the E_h is too high, and the organisms are unable to avoid destruction and clearance by phagocytosis. Predisposing host factors include debility, old age and diabetes.

When clostridial infection has been initiated in a focus of devitalized anaerobic tissue, the organisms multiply rapidly and produce a range of toxins and aggressins. These damage tissue by various necrotizing effects, and some have demonstrable lethal effects. They spread into adjacent viable tissue, particularly muscle, kill it, and render it anaerobic and vulnerable to further colonization, with the production of more toxins and aggressins.

- Hyaluronidase produced by *C. perfringens* breaks down intercellular cement substance and promotes the spread of the infection along tissue planes.
- Collagenase and other proteinases break down tissues and virtually liquefy muscles. The whole of a muscle group or segment of a limb may be affected.
- α-Toxin, a phospholipase C (lecithinase), is generally considered to be the main cause of the toxaemia associated with gas gangrene, although other clostridial species can produce similar manifestations.

In puerperal infections or in cases of septic abortion, the organisms may gain access from faeces-contaminated perineal skin or contaminated instruments to necrotic or

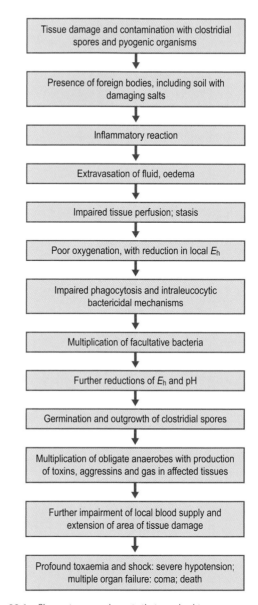

Fig. 22.1 Circumstances and events that may lead to gas gangrene.

devitalized tissues in the uterus or adnexa. Here they set up a dangerous and often fulminating pelvic infection, possibly with prompt invasion of the bloodstream. There may be intravascular haemolysis and anuria.

C. perfringens may also participate in peritoneal infections that occur as a result of extension of pathogens from the alimentary tract, as in cases of gangrenous appendicitis or intestinal obstruction or mesenteric thrombosis.

If a preparation of adrenaline (epinephrine) used for injection is contaminated with clostridial spores, the combination of an infective inoculum with the local ischaemia that follows the injection may be catastrophic. Gas gangrene may be a complication of surgical operations on the lower limb or hip of a patient whose blood supply is inadequate to maintain oxygenation in the post-operative period.

Clostridia may be associated with less severe forms of infection without the toxaemia and aggression of gas gangrene. Moreover, potentially pathogenic anaerobes may be cultivated from a wound that never shows any sign of gas gangrene, and sometimes the laboratory isolation may be attributed to the germination of a few contaminating spores when the specimen was being processed. Thus, the onus is on the clinician to relate a laboratory report to the patient's circumstances.

Clinical clues include *crepitus*, the sponge-cake consistency caused by small bubbles of gas in the adjacent tissues. In the early stages, the patient has an anxious, frightened appearance. Local pain is increased, and there is swelling of the affected tissues. Toxaemia and shock supervene, and the patient's conscious state drifts to drowsiness and into coma. Prompt diagnosis and intensive surgical and antimicrobial treatment greatly influence the patient's chance of survival and may avoid the loss of the affected limb, but all devitalized tissue must be excised (see below).

Laboratory diagnosis

If there are sloughs of necrotic tissue in the wound, small pieces should be transferred aseptically into a sterile screw-capped bottle and examined immediately by microscopy and culture. Specimens of exudate should be taken from the deeper areas of the wound where the infection seems to be most pronounced. Gram smears are prepared. If gas gangrene exists, typical Gram-positive bacilli may predominate, often with other bacteria present in a mixed infection. However, there is usually a pronounced lack of inflammatory cells. Initiation of treatment should not await a full laboratory report and early discussion with the bacteriologist is crucial. A direct smear of a wound exudate is often of great help in providing evidence of the relative numbers of different bacteria that may be participating in a mixed infection or may merely be present as contaminants, but the distinction is not invariably easy and joint discussions are important.

Treatment

Prompt and adequate surgical attention to the wound is of the utmost importance (Fig. 22.2).

- Sutures are removed, and necrotic and devitalized tissue is excised with careful debridement.
- Fascial compartments are incised to release tension.

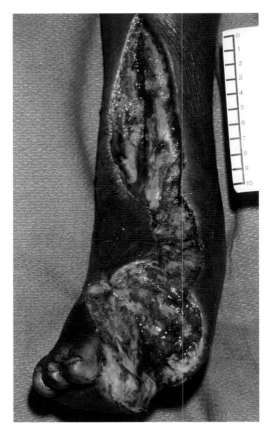

Fig. 22.2 Gas gangrene. Wound after debridement.

- Any foreign body is found and removed.
- The wound is not resutured but is left open after thorough cleansing, and loosely packed.

Antibiotic therapy is started immediately in very high doses. This must take account of the likely coexistence of coliform organisms, Gram-positive cocci and faecal anaerobes. Accordingly, penicillin, metronidazole and an aminoglycoside may be given in combination. Alternatively, clindamycin plus an aminoglycoside or a broad-spectrum antibiotic, such as meropenem or imipenem, may be considered. Much intensive supportive therapy is needed.

Enthusiastic claims have been made for the efficacy of hyperbaric oxygen therapy, but clinical trials have given conflicting results. Patients are placed in a special pressurized chamber where they breathe oxygen at 2–3 atm. pressure for periods of 1–2 h twice daily on several successive days. This may limit the amount of radical surgery needed.

A polyvalent antiserum containing *C. perfringens*, *C. septicum* and *C. novyi* antitoxins was used formerly, but has been replaced by intensive antimicrobial therapy.

Prophylaxis

Surgical wounds

C. perfringens is normally present in large numbers in human faeces, and its spores are found regularly on the skin, especially of the buttocks and thighs. As clostridial spores are very resistant to most disinfectants, they are likely to survive normal pre-operative skin preparation and persist in the area of the planned incision. The numbers can be reduced by more prolonged skin preparation with the sustained action of an antiseptic such as povidone–iodine for a day or two; this procedure has a place in orthopaedic surgery.

When inevitable skin contamination is combined with circumstances that predispose to devitalization of tissue and reduced oxygen tension, a patient may be vulnerable to the development of post-operative gas gangrene. These circumstances arise when an elderly patient or a patient with vascular insufficiency undergoes major surgery to the hip or lower limb. Perioperative antimicrobial prophylaxis with penicillin is justified in such cases.

Accidental wounds

The prevention of gas gangrene in accidentally sustained wounds must take account of the endogenous factors noted above and the exogenous sources of clostridial spores and vegetative forms in soil, on contaminated clothing, etc. In addition, there is an increased risk of anaerobic infection developing when foreign bodies such as soil, clothing, metal (nails, wire, bullets, shrapnel) and skin are driven into devitalized tissues. Prompt and adequate surgical attention is of paramount importance, but prophylactic administration of benzylpenicillin for patients presenting with serious contaminated wounds is also worthwhile. Prophylaxis may be omitted if the state of the wound and the patient's general condition are expertly and frequently monitored throughout the recovery period.

FOOD POISONING

Carrier rates for 'typical food poisoning strains' of *C. perfringens* range from about 2% to more than 30% in different groups that have been surveyed across the world. These bacteria also occur in animals; thus, meat is often contaminated with heat-resistant spores. When meat is cooked in bulk, heat penetration and subsequent cooling is slow unless special precautions are taken. Heat-resistant spores may survive and, during the cooling period, germinate in the anaerobic environment produced by the cooked meat and multiply. Anyone who eats this will consume the equivalent of a cooked meat broth culture of the organism. The organisms are protected from the gastric acid by the protein in the meal and pass in large numbers into the intestine.

Ingestion of large numbers of viable organisms is necessary for the production of the typical disease syndrome, which is mediated by an enterotoxin that is released when sporulation occurs in the gut. Typical symptoms are abdominal cramps beginning about 8–12 h after ingestion, followed by diarrhoea. Fever and vomiting are not usually encountered and symptoms generally subside within a day or two. No specific treatment is indicated. The carrier state persists for several weeks, but this should not be regarded as an indication for exclusion from any duties, as carriers may be quite numerous in various communities.

The vehicle of infection is usually a pre-cooked meat food that has been allowed to stand at a temperature conducive to the multiplication of *C. perfringens*. Although the heat resistance of spores of typical food-poisoning strains ensures their survival in cooked foods, and presumably accounts for the association of these strains with most of the reported outbreaks of *C. perfringens* food poisoning, similar trouble can be caused by classical heat-sensitive type A strains if they gain access to food during the cooling period under conditions suitable for their subsequent multiplication.

Laboratory diagnosis

This can be difficult as some people carry large numbers of *C. perfringens*. Diagnosis depends upon the isolation of similar strains of *C. perfringens* from the faeces of patients and from others at risk who have eaten the suspected food, and from the food itself. Numbers usually exceed 10^6 organisms/g faeces. The isolates can be sent to a reference laboratory for special typing to prove their relatedness.

Prevention

The occurrence of this type of food poisoning is an indictment of the catering practices concerned, as food has to be mishandled to allow the chain of events to take place. Nevertheless, *C. perfringens* is among the commonest causes of food poisoning.

COLITIS

A sporadic diarrhoeal syndrome, usually occurring in elderly patients during treatment with antibiotics, has been described. The circumstances differ substantially from those of *C. perfringens* food poisoning. An enterotoxin with a cytopathic effect can be detected in the patient's faeces.

ENTERITIS NECROTICANS (PIG–BEL)

A subgroup of *C. perfringens* type C that produces heat-resistant spores is the cause of a disease that affects New Guinea natives when they have pork feasts. The method of cooking the pork allows the clostridia to survive. When the contaminated meat is eaten along with a sweet potato vegetable that contains a proteinase inhibitor, a toxin (the β-toxin) is able to act on the small intestine to produce a necrotizing enteritis. A successful vaccination programme has reduced the incidence of pig-bel dramatically.

CLOSTRIDIUM SEPTICUM

DESCRIPTION

The bacilli are generally large, but pleomorphic forms are common. The organism is actively motile with numerous peritrichous flagella. It is one of the less exacting anaerobes and grows well at 37°C on ordinary media. Spores are readily formed and, as they develop, various shapes arise ranging from swollen Gram-positive 'citron bodies' to obviously sporing forms in which the oval spores may be central or subterminal and are clearly bulging.

PATHOGENESIS

C. septicum is one of the gas gangrene group of clostridia. It occurs harmlessly in the human intestine, but if the integrity of the gut epithelium is impaired, for instance by leukaemic infiltration, bacteraemia may occur. Cyclic or other forms of neutropenia are also associated with spontaneous, non-traumatic gas gangrene that begins with a bacteraemic phase. *Typhlitis*, a rapidly fatal terminal ileal infection and septicaemia in immunocompromised patients, is most commonly associated with *C. septicum*.

Intramuscular injection of cultures into laboratory animals produces a spreading inflammatory oedema, with slight gas formation in the tissues. Organisms invade the blood and the animal dies within a day or two. Smears from the liver show long filamentous forms and citron bodies. *C. septicum* produces several toxins (α, β, δ and ε). The α-toxin, which has lethal, haemolytic and necrotizing activity, appears to be the most important; it does not have phospholipase C activity and thus differs from the α-toxin of *C. perfringens*.

CLOSTRIDIUM NOVYI

C. novyi resembles *C. perfringens* in morphology, but is larger, more pleomorphic and more strictly anaerobic; it is readily killed when vegetative cells are exposed to air. The spores are oval, central or subterminal. The organism occurs widely in soil and is associated with disease in man and animals. There are four types – A, B, C and D – distinguished on the basis of permutations of the toxins and other soluble antigens they produce. Only type A strains are of medical interest as they cause some cases of gas gangrene in man. In 2000, an outbreak of *C. novyi* type A infections among injecting heroin users killed 13 people in the UK.

C. novyi gas gangrene is associated with profound toxaemia. Culture filtrates are highly toxic and possess at least four active substances (α, β, δ and ε toxins) that account for the various haemolytic, necrotizing, lethal, phospholipase and lipase activities of this organism.

CLOSTRIDIUM SPOROGENES

This clostridium is widely distributed in nature as a harmless saprophyte. Its spores may survive boiling for periods up to 6 h and the organism is often encountered in mixed cultures after preliminary heating to select heat-resistant pathogens. Although its presence in wound exudates may accelerate an established anaerobic infection by enhancing local conditions, it is not a pathogen in its own right and does not cause gas gangrene.

CLOSTRIDIUM TETANI

DESCRIPTION

The tetanus bacillus is a motile, straight, slender, Gram-positive rod. A fully developed terminal spore gives the organism the appearance of a drumstick with a large round end. Gram-negative forms are usually encountered in stained smears. It is an obligate anaerobe that grows well in cooked meat broth and produces a thin spreading film when grown on enriched blood agar. The spores may be highly resistant to adverse conditions, but the degree of resistance varies with the strain. Spores of some strains resist boiling in water for up to 3 h. They may resist dry heat at 160°C for 1 h, and 5% phenol for 2 weeks or more. Iodine (1%) in water is said to kill the spores within a few hours, but glutaraldehyde is one of the few chemical disinfectants that is assuredly sporicidal.

PATHOGENESIS

If washed spores are injected into an animal they fail to germinate and are removed by phagocytosis. The germination of spores and their outgrowth depend upon reduced oxygen tension in devitalized tissue and non-viable material in a wound, as discussed earlier in relation to gas gangrene (see pp. 241–242). When infection occurs, often assisted by the simultaneous growth of facultatively anaerobic organisms in a mixed inoculum, the tetanus bacillus remains strictly localized, but tetanus toxin is elaborated and diffuses, as described below.

Toxins

C. tetani produces an oxygen-labile haemolysin (tetanolysin), but the organism's neurotoxin (tetanospasmin) is the essential pathogenic product. Strains vary in their toxigenicity; some are highly toxigenic. Most strains produce demonstrable toxin after culture in broth for a few days.

The gene that encodes the neurotoxin is located on a 75-kilobase (kb) plasmid. The toxin is synthesized as a single polypeptide with a molecular weight of 150 000 Da, which undergoes post-translational cleavage into a heavy chain and a light chain linked by a disulphide bond. The estimated lethal dose for a mouse of pure tetanospasmin is 0.0001 µg. It is toxic to man and various animals when injected parenterally, but not by the oral route.

When tetanus occurs naturally, the tetanus bacilli stay at the site of the initial infection and are not generally invasive. Toxin diffuses to affect the relevant level of the spinal cord (local tetanus) and then to affect the entire system (generalized tetanus). These stages, including the intermediate one of 'ascending tetanus', are demonstrable in experimental animals, but the stages tend to merge in their clinical presentation in man.

The toxin is absorbed from the site of its production in an infective focus, but may be delivered via the blood to all nerves in the body. The heavy chain mediates attachment to gangliosides and the toxin is internalized. It is then moved from the peripheral to the central nervous system by retrograde axonal transport and trans-synaptic spread. The tendency for the first signs of human tetanus to be in the head and neck is attributed to the shorter length of the cranial nerves. In fact, descending involvement of the nervous system is seen as the tetanus toxin takes longer to traverse the longer motor nerves and also diffuses in the spinal cord.

Once the entire toxin molecule has been internalized into presynaptic cells, the light chain is released and affects the membrane of synaptic vesicles. This prevents the release of the neurotransmitter γ-aminobutyric acid.

Motor neurones are left under no inhibitory control and undergo sustained excitatory discharge, causing the characteristic motor spasms of tetanus. The toxin exerts its effects on the spinal cord, the brainstem, peripheral nerves, at neuromuscular junctions and directly on muscles.

Clinical features of tetanus

Cases of tetanus have been reported in which the infection was apparently associated with a superficial abrasion, a contaminated splinter or a minor thorn prick. Indeed, gardening enthusiasts form one of the recognized risk groups.

In some cases the site of infection is assumed to be in the external auditory meatus; thus, *otogenic tetanus* may be attributable to over-zealous cleansing of the meatus with a small stick. In other patients, the site of infection remains undiscovered; this condition is referred to as *cryptogenic tetanus*. Tetanus infection may also occur in or near the uterus in cases of septic abortion.

Tetanus neonatorum follows infection of the umbilical wound of newborn infants (see below). Cases of *post-operative tetanus* have been attributed to imperfectly sterilized catgut, dressings or glove powder, and sometimes to dust-borne infection of the wound at operation.

The onset of signs and symptoms is gradual, usually starting with some stiffness and perhaps pain in or near a recent wound. In some cases the initial complaint may be of stiffness of the jaw (*lockjaw*). Pain and stiffness in the neck and back may follow. The stiffness spreads to involve all muscle groups; facial spasms produce the 'sardonic grin', and in severe cases spasm of the back muscles produces the opisthotonos (extreme arching of the back) beloved of textbook writers. The period between injury and the first signs is usually about 10–14 days, but there is a considerable range. A severe case with a relatively poor prognosis shows rapid progression from the first signs to the development of generalized spasms. Sweating, tachycardia and arrhythmia, and swings in blood pressure, reflect sympathetic stimulation, which is not well understood but creates problems of management.

TREATMENT

The patient remains conscious and requires skilled sedation and constant nursing. If generalized spasms are worrying, the patient is paralysed and ventilated mechanically until the toxin that has been taken up has decayed; this may take some weeks.

The patient is given 10 000 units of human tetanus immunoglobulin (HTIG) in saline by slow intravenous infusion. Full wound exploration and debridement is arranged, and the wound is cleansed and left open with a loose pack. Penicillin or metronidazole is given for as long as considered necessary to ensure that bacterial growth and toxin production are stopped. The antitoxin and antibiotics are given immediately, and preferably before surgical excision, but delay must be avoided.

LABORATORY DIAGNOSIS

Gram smears of the wound exudate and any necrotic material may show the typical 'drumstick' bacilli, but this is not invariably so, and thus provides only presumptive evidence as other organisms that resemble *C. tetani* have terminal spores. Simple light microscopy is often unsuccessful; immunofluorescence microscopy with a specific stain is possible but not generally available.

Direct culture of unheated material on blood agar incubated anaerobically is often the best method of detecting *C. tetani*. There are various other tricks that exploit the organism's motility and fine spreading growth; sometimes these are vitiated by overgrowth with *Proteus* species. Material from the wound or from a mixed sporing subculture may be heated at various temperatures and for various times to exclude non-sporing bacteria; the heated specimens are then seeded on to solid media and incubated anaerobically. Tetanus may be produced in mice by subcutaneous injection of an anaerobic culture prepared from wound material; control mice are protected with tetanus antitoxin.

EPIDEMIOLOGY

Tetanus bacilli may be found in the human intestine, but infection seems to be derived primarily from animal faeces and soil. The organism is especially prevalent in manured soil and for this reason a wound through skin contaminated with soil or manure deserves special attention. However, tetanus spores occur very widely and are commonly present in gardens, sports fields and roads, in the dust, plaster and air of hospitals and houses, on clothing and on articles of common use.

Spores of *C. tetani* and other anaerobes may be embedded in surgical catgut and other dressings. However, the sterility of surgical catgut (prepared from the gut of cattle and sheep) is now rigorously controlled.

Tetanus ranks among the major fatal infections. During the 1980s there were between 800 000 and 1 million deaths annually from tetanus, of which 400 000 were due to neonatal tetanus. The incidence varies enormously from country to country, and is inversely related to socio-economic development and standards of living, preventive medicine and wound management. In developed countries, the reported incidence of adult and childhood tetanus is low. There is a direct relationship with fertile soil and a warm climate; thus, people living in the agricultural areas of developing tropical and subtropical countries are exposed to severe challenges associated with poor hygiene, lack of shoes, neglect of wounds and inadequate immunization.

In addition, some local customs promote the occurrence of tetanus:

- treatment of the umbilical cord stump with primitive applications that include animal dung
- tying of the umbilical cord itself with primitive ligatures
- ear-piercing and other operations performed with unsterile instruments.

Fatality rates may exceed 50% and neonatal tetanus carries a very high mortality rate. Case fatality rates can be greatly reduced to less than 10% by modern methods of treatment in specialist centres. Unfortunately, such expensive skilled help is available for only a small proportion of patients. Under-privileged people in countries with poorly developed or expensive medical services are at greatest risk, and are least likely to receive sophisticated assistance; 80% of deaths occur in Africa and South-East Asia. A maternal vaccination campaign mounted by the United Nations Children's Fund (UNICEF) during the past 10 years has reduced the incidence of neonatal tetanus by 50%.

PREVENTION AND CONTROL

Prompt and adequate wound toilet and proper surgical debridement of wounds are of paramount importance in the prevention of tetanus. There is an increased risk that tetanus spores may germinate in a wound if cleansing is delayed and if sepsis develops. Clean superficial wounds that receive prompt attention may not require specific protection against tetanus and it is unreasonable to insist that every small prick or abrasion requires protection with antitoxin or antibiotic.

Routine practice should take account of the local incidence of tetanus and the individual circumstances of the case. It is wise to recommend specific prophylaxis for a non-immunized patient with a deep wound, puncture or stab wound, ragged laceration, a wound with much bruising and any devitalized tissue, a wound that is already septic, or a bite wound or other type of wound that is likely to be heavily contaminated. Figure 22.3 shows an approach that reflects current thinking in the UK.

The need for passive immunization is avoided if the patient is known to be properly immunized against tetanus (see Ch. 69). A patient may be regarded as immune for 6 months after the first two injections, or for 5–10 years after a planned course of three injections (or a subsequent

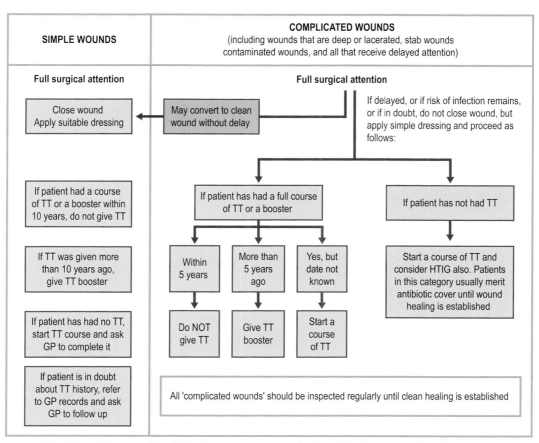

TT = Adsorbed tetanus toxoid   HTIG = human tetanus immunoglobulin   GP = General practitioner, family doctor

Fig. 22.3 Wound management guidelines, with special reference to the prevention of tetanus.

booster injection) of adsorbed tetanus toxoid. Tetanus antitoxin should not be given to immune patients, but their active immunity may be enhanced when necessary by giving a dose of tetanus toxoid at the time of injury if the circumstances justify it.

A patient is considered non-immune if there is no history of having had an injection of tetanus toxoid or if only one injection has been given. Take care: a patient may recall having had 'a tetanus shot', but this may have been a previous dose of antitoxin for passive protection (which is transient and cannot be boosted by toxoid). If more than 6 months have elapsed after a course of two injections, or more than 10 years after a full primary course of three injections of adsorbed toxoid (or a booster injection), the patient should be regarded as non-immune. A patient is non-immune if more than 2–3 weeks have elapsed since a previous injection of equine antitoxin, or more than 6–8 weeks in the case of homologous (human) antitoxin. Non-

immunity should be assumed if there is any doubt about the immunization history.

Passive immunization with antitoxin

HTIG (homologous antitoxin) is available for passive protection and now supersedes equine antitoxin (heterologous antitoxin), which was associated with occasional adverse reactions. However, equine antiserum should not be prematurely discarded in countries that do not yet have HTIG. The prophylactic dose of HTIG is 250–500 units by intramuscular injection.

Combined active–passive immunization

A non-immune patient receiving passive protection with HTIG after injury may be given the first dose of a course of active immunization with adsorbed toxoid at the same time, provided the injections are given from different syringes and into contralateral sites. The active immunization course must subsequently be completed.

Antibiotic protection

The prophylactic administration of antibiotics to all cases of open wounds is not recommended, although the use of penicillin or clindamycin is justified in some cases when there is a significant risk of infection. This precaution must not replace prompt and adequate surgical wound toilet.

CLOSTRIDIUM BOTULINUM

DESCRIPTION

C. botulinum is a strictly anaerobic Gram-positive bacillus. It is motile and has spores that are oval and subterminal. It is a widely distributed saprophyte found in soil, vegetables, fruits, leaves, silage, manure, the mud of lakes and sea mud. Its optimal growth temperature is about 35°C, but some strains have been shown to grow and produce toxin at temperatures as low as 1–5°C.

The widespread occurrence of *C. botulinum* in nature, its ability to produce a potent neurotoxin in food, and the resistance of its spores to inactivation combine to make it a formidable pathogen. Spores of some strains withstand boiling in water (100°C) for several hours. They are usually destroyed by moist heat at 120°C within 5 min. Spores of type E strains (see below) are usually much less heat resistant. Insufficient heating in the process of preserving foods is an important factor in the causation of botulism. The resistance of the spores to radiation is of special relevance to food processing.

PATHOGENESIS

Botulism is a severe, often fatal, form of food poisoning characterized by pronounced neurotoxic effects. Botulinum toxins are among the most poisonous natural substances known. Seven main types of *C. botulinum*, designated A–G, produce antigenically distinct toxins with pharmacologically identical actions. All types can cause human disease, but types A, B and E are most common. The importance of this point is that, if antitoxin is given to a patient in an emergency, only the type-specific antitoxin will be effective.

The disease has been caused by a wide range of foods, usually preserved hams, large sausages of the salami type, home-preserved meats and vegetables, canned products such as fish, liver paste, and even hazelnut purée and honey. Traditional dishes such as fish or seal's flipper fermented in a barrel buried in the ground cannot be recommended by a bacteriologist! Type E strains are particularly but not invariably associated with a marine source, whereas type A and type B strains are usually associated with soil.

Foods responsible for botulism may not exhibit signs of spoilage. The pre-formed toxin in the food is absorbed from the intestinal tract. Although it is protein, it is not inactivated by the intestinal proteolytic enzymes. After absorption into the bloodstream, botulinum toxin binds irreversibly to the presynaptic nerve endings of the peripheral nervous system and cranial nerves, where it inhibits acetylcholine release.

Botulinum toxin and bioterrorism

Botulinum toxin is categorized as a biothreat level A biological warfare agent. Introduction of toxin into a target population by contaminating food or water is unlikely to succeed because of logistical problems, but botulinum toxin can also cause disease by inhalation.

Clinical features

The period between ingestion of the toxin and the appearance of signs and symptoms is usually 1–2 days, but it may be much longer. There may be initial nausea and vomiting. The oculomotor muscles are affected, and the patient may have diplopia and drooping eyelids with a squint. There may be vertigo and blurred vision.

There is progressive descending motor loss with flaccid paralysis but with no loss of consciousness or sensation, although weakness and sleepiness are often described. The patient is thirsty, with a dry mouth and tongue. There are difficulties in speech and swallowing, with later problems of breathing and despair. There may be abdominal pain and restlessness. Death is due to respiratory or cardiac failure.

Wound botulism

Rare cases of wound infection with *C. botulinum* resulting in the characteristic signs and symptoms of botulism have been recorded.

Infant botulism

The 'floppy child syndrome' describes a young child, usually less than 6 months old, with flaccid paralysis that is ascribed to the growth of *C. botulinum* in the intestine at a stage in development when the colonization resistance of the gut is poor. There are various grades of the syndrome. Some cases have been attributed to the presence of *C. botulinum* spores in honey; when the honey was given as an encouragement to feed, the ingested spores were able to germinate and produce toxin in the infant gut.

LABORATORY DIAGNOSIS

The organism or its toxin may be detected in the suspected food, and toxin may be demonstrated in the patient's blood by toxin–antitoxin neutralization tests in mice. Samples of faeces or vomit may also yield such evidence. Take care: bear in mind that botulinum toxin is very dangerous – specialist help should be summoned and the laboratory alerted.

TREATMENT

The priorities are:

- to remove unabsorbed toxin from the stomach and intestinal tract
- to neutralize unfixed toxin by giving polyvalent antitoxin (with due precautions to avoid hypersensitivity reactions to the heterologous antiserum)
- to give relevant intensive care and support.

CONTROL

Great care must be taken in canning factories to ensure that adequate heating is achieved in all parts of the can contents. Home canning of foodstuffs should be avoided. The amateur preservation of meat and vegetables, especially beans, peas and root vegetables, is dangerous in inexperienced hands. Acid fruits may be bottled safely in the home with heating at 100°C, as a low pH is inhibitory to the growth of *C. botulinum*.

A prophylactic dose of polyvalent antitoxin should be given intramuscularly to all persons who have eaten food suspected of causing botulism. Injecting three doses of mixed toxoid at intervals of 2 months can produce active immunity, but the very low incidence of the disease under normal conditions does not justify this as a routine. Active immunization should be considered for laboratory staff who might have to handle the organism or specimens containing the organism or its toxin.

CLOSTRIDIUM DIFFICILE

DESCRIPTION

C. difficile is a motile Gram-positive rod with oval subterminal spores. It commonly occurs in the faeces of

Fig. 22.4 Pseudomembranous colitis.

neonates and babies until the age of weaning, but it is not generally found in adults.

The organism produces an enterotoxin (toxin A) and a cytotoxin (toxin B); some strains produce a third, binary, toxin. It is a proven cause of antibiotic-associated diarrhoea, occasionally leading to a life-threatening condition, *pseudomembranous colitis* (Fig. 22.4). There is almost always a history of previous antibiotic therapy. Clindamycin and lincomycin are associated with a particularly high risk. Extended-spectrum cephalosporins are also commonly incriminated and are much more important in terms of quantities used; however, there is virtually no antimicrobial drug (including the newer quinolones) that has escaped blame.

LABORATORY DIAGNOSIS

C. difficile can be isolated from the faeces by enrichment and selective culture procedures. Toxin B can be detected in the patient's faeces by testing extracts against cell monolayers of susceptible cells, or both toxins may be demonstrated by immunological methods, such as enzyme-linked immunosorbent assay (ELISA). Toxin A-negative strains of *C. difficile* occasionally cause disease and these are not detected by some currently available ELISA kits.

TREATMENT

It is essential to discontinue the antibiotic that is presumed to have precipitated the disease and to suppress the growth and toxin production of *C. difficile* by giving oral metronidazole or vancomycin. Pseudomembranous colitis may be fatal if it is not quickly recognized and treated.

EPIDEMIOLOGY

Evidence usually suggests that the organism is acquired from an exogenous source by a patient whose intestinal colonization resistance has been compromised by antibiotic exposure. Patients with antibiotic-associated diarrhoea tend to spend lengthy periods in hospital. Re-infection occurs in 20–50% of cases as it may take the gut 2–3 months to normalize after perturbation.

PREVENTION

Clinical awareness is the keynote. *C. difficile* is the most common cause of hospital-acquired diarrhoea. If a patient develops diarrhoea after at least 48 h in hospital, especially while taking antibiotics, the possibility of *C. difficile*-associated diarrhoea must be considered. If several cases occur in a hospital unit, cross-infection should be considered as the hospital environment can become extensively contaminated with *C. difficile* spores. The existing antibiotic and infection control policies of the unit should be reviewed.

KEY POINTS

- All clostridial infections are characterized by toxin production by the infecting species.
- The major toxins produced by the species are neurotoxins affecting nervous tissue, histotoxins affecting soft tissue and enterotoxins affecting the gut.
- *C. perfringens* is the major cause of gas gangrene.
- Tetanus, particularly neonatal tetanus, is still a major public health issue in developing countries.
- Botulism occurs predominantly as a severe form of food poisoning.
- *C. difficile* is the most common cause of diarrhoea in hospital patients in the developed world.

RECOMMENDED READING

Brazier J S 1995 The laboratory diagnosis of *Clostridium difficile*-associated disease. *Reviews in Medical Microbiology* 6: 236–245

Caya J G, Agni R, Miller J E 2004 *Clostridium botulinum* and the clinical laboratorian. *Archives of Pathology and Laboratory Medicine* 128: 653–662

Collee J G, van Heyningen S 1991 Systemic toxigenic diseases (tetanus, botulism). In Duerden B I, Drasar B S (eds) *Anaerobes in Human Disease*, pp. 372–394. Arnold, London

Farrar J J, Yen L M , Cook T et al 2000 Tetanus. *Journal of Neurology, Neurosurgery & Psychiatry* 69: 292–301

Hobbs B C, Roberts D 1993 *Food Poisoning and Food Hygiene*, 6th edn. Arnold, London

Rood J I, McClane B A, Songer J G, Titball R W 1997 *The Clostridia – Molecular Biology and Pathogenesis*, 1st edn. Academic Press, San Diego

Sakurai J 1995 Toxins of *Clostridium perfringens. Reviews in Medical Microbiology* 6: 175–185

Shapiro R L, Hatheway C, Swerdlow D L 1998 Botulism in the United States: a clinical and epidemiologic review. *Annals of Internal Medicine* 129: 221–228

Wilkins T D, Lyerly D M 2003 *Clostridium difficile* testing: after 20 years, still challenging. *Journal of Clinical Microbiology* 41: 531–534

23 Neisseria and moraxella

Meningitis; septicaemia; gonorrhoea; respiratory infections

D. A. A. Ala'Aldeen

The genus *Neisseria* contains two important human pathogens: *Neisseria meningitidis* (meningococcus) and *N. gonorrhoeae* (gonococcus). They are Gram-negative diplococci and obligate human pathogens, typically found inside polymorphonuclear pus cells of the inflammatory exudate. Although similar in terms of morphological and cultural characters, they are associated with entirely different diseases:

- *N. meningitidis* causes a range of diseases embraced by the term *invasive meningococcal disease*. Most common are purulent meningitis (variously called *epidemic cerebrospinal meningitis, cerebrospinal fever* or, because of the purpuric rash that is sometimes present, *spotted fever*) and an acute septicaemic illness with a petechial rash in the presence or absence of meningitis. About one-third of cases of meningococcal disease present as septicaemia; meningitis (with or without septicaemia) accounts for most others.
- *N. gonorrhoeae* causes the sexually transmitted disease *gonorrhoea*, which most commonly presents as a purulent infection of the mucous membrane of the urethra in men and the cervix uteri in women. In the newborn the gonococcus may give rises to a purulent conjunctivitis, and in young girls a vulvovaginitis. Disseminated gonococcal infection, which is recognized by a rash and evidence of blood spread, may also occur, more commonly in women.

Other members of the *Neisseria* genus are common commensals of the upper respiratory tract, which is also the reservoir of the meningococcus and, occasionally, the gonococcus.

N. lactamica and *N. polysacchareae* are frequently isolated from the nasopharynx. They are of low pathogenicity, but *N. lactamica* is occasionally isolated from blood or cerebrospinal fluid (CSF). Additional commensal species include *N. subflava* (of which there are several biovars), *N. sicca*, *N. mucosa* and *N. flavescens*. Two others, *N. elongata* and *N. weaveri*, are unusual in being rod shaped.

Moraxellae are non-fermentative organisms that may be coccoid or rod shaped. There is still uncertainty as to their taxonomic position. Some strains resemble *Acinetobacter* spp. and other glucose non-fermenters (see p. 298). The most important member of the group, *Moraxella catarrhalis* (formerly known as *Branhamella catarrhalis*), is a common commensal of the upper respiratory tract and an opportunist pathogen.

NEISSERIA

Description

N. meningitidis and *N. gonorrhoeae* are very similar in their morphological and cultural characters. They are Gram-negative, oval cocci occurring in pairs with the apposed surfaces flat or slightly concave (bean shaped), and with the axis of the pair parallel and not in line as in the pneumococcus. In pus from inflammatory exudates, such as CSF or urethral discharge, many diplococci are found in a small proportion of the polymorphonuclear cells. This is more marked with the gonococcus than with the meningococcus. Extracellular cocci also occur and there may be considerable variation in their size and intensity of staining.

Non-pathogenic neisseriae grow on ordinary nutrient media, but meningococci and gonococci do not grow on plain nutrient agar or at room temperature. However, the addition of heated blood (or ascitic fluid) to nutrient agar ensures a good growth of colonies from infected material, provided incubation is carried out, preferably at 35–36°C, in a moist atmosphere containing 5–10% carbon dioxide. Growth is rather slow (more so with the gonococcus) but, grey, glistening, slightly convex colonies of 0.5–1.0 mm in diameter appear in 8–24 h. Incubation should, however, be continued for a further 24 h, when the colonies will be much larger and the gonococcus, in particular, tends to have a slightly roughened surface and a crenated margin.

Both meningococci and gonococci express pili (fimbriae). These are multi-functional appendages associated with attachment of the organism to mucosal surfaces, twitching motility and acquisition of DNA.

Colonies of meningococci and gonococci react quickly in the test for cytochrome oxidase; non-pathogenic neisseriae react more slowly. Species identification depends on carbohydrate utilization reactions. Commercial kits that rapidly detect pre-formed enzymes are commonly used for identification.

Genome analysis

Genome sequence analysis of *N. meningitidis* and *N. gonorrhoea* shows that they are highly homologous at the protein coding level, although each species has several hundred unique genes. The organisms have evolved highly efficient ways of varying their genes and antigens. These include mutation, phase variation (switching genes on and off) and horizontal transfer of genes within the same strain or with other related species. Among the key phase-variable antigens is the capsular polysaccharide, which is often switched off. This may help avoid bactericidal anticapsular antibodies and prolong bacterial survival.

NEISSERIA MENINGITIDIS

Classification

Meningococci are divisible into 13 serogroups, based on antigenic variations in their capsular polysaccharides. Most common are serogroups A, B, C and W-135. Other serogroups that are rarely associated with disease include serogroups D, H, I, K, L, 29E, X, Y and Z. The serogroup is usually determined by a slide agglutination test with absorbed group-specific antisera.

Meningococcal serotypes and serosubtypes are determined by specific monoclonal antibodies raised against the antigenically hypervariable outer membrane proteins (porins) PorB and PorA. Strain identities are designated by their serogroup (e.g. B), serotype (e.g. 15) and serosubtype (e.g. P1.16). As a result of genetic variations in the antigens used for typing, non-typable strains are often isolated. In these cases, the corresponding porin genes are sequenced and classified. Immunotyping, based on antigenic variation of lipo-oligosaccharides, is sometimes carried out.

Multilocus enzyme electrophoresis (electrotyping), based on the presence and electrophoretic mobility of isoforms of cytosol enzymes, is also used to discriminate among strains. Use of this method has enabled the subdivision of genetically related strains of serogroup A to subgroups, and serogroups B and C to lineages and clonal complexes. It has been established that meningococcal

populations are genetically highly diverse and that only a few 'hypervirulent' strains are associated with disease worldwide. For example, clonal complexes ET-5 and ET-37 dominate the disease-causing serogroup B and C strains, respectively, in the UK and the rest of Europe. These dominant complexes are changing slowly with time, and an increase in the incidence of disease appears to coincide with the introduction of a new clonal complex to a particular community.

A more precise and less cumbersome genetic typing scheme is multilocus sequence typing in which relatively small DNA fragments from seven meningococcal 'housekeeping' genes are amplified by the polymerase chain reaction (PCR) and sequenced. Gene sequences from diverse meningococcal strains representing various clonal complexes are then aligned. Alleles of each gene are assigned numbers, and strains with particular combinations of allelic mixes of all the seven genes are then allocated a sequence type. Multilocus sequence typing has a greater discriminatory power than electrotyping; otherwise, the two methods serve the same purpose.

Pathogenesis

The natural habitat of the meningococcus is the human nasopharynx, and transmission is largely via close (intimate, kissing) contact. The carrier:case ratio varies in different outbreaks and with the strain and population surveyed. Bacterial, environmental and host factors are probably all equally important in disease.

Bacterial factors.

Some meningococcal strains (members of the hypervirulent lineages) are more able than others to cause invasive disease. The prevalence of meningococcal infection fluctuates over time, as does the relative proportion of serogroups (Fig. 23.1), serotypes, serosubtypes and clonal lineages.

Environmental factors

Environmental risk factors include overcrowding, seasonal variation and passive (and active) smoking. Meningococcal infections tend to occur during the winter months. Household contacts of a case are 500 to 800 times more likely to develop meningococcal infection than the general population.

Host factors

For largely unknown reasons, some individuals are more susceptible to infection than others and some do worse than others. The increase in immunity observed with increasing age is likely to be due to asymptomatic carriage, often for

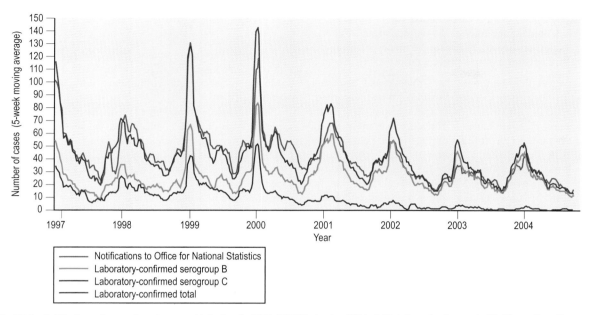

Fig. 23.1   Notifications of cases of meningococcal infection for 1997–2004 (England and Wales). (Data from the Communicable Disease Surveillance Centre, Health Protection Agency.)

many months, of virulent or avirulent meningococci or other neisseriae. About two-thirds of cases occur in the first 5 years of life. The peak prevalence of disease is in the first year and there is a smaller peak in adolescence (Fig. 23.2). The incidence is slightly higher in men than in women.

The absence of bactericidal antibody in the blood is believed to be the factor most closely related to susceptibility to clinical infection. Group-specific (anti-capsular) antibody is protective and this is the basis of the success of vaccination against meningococcal disease. Antibodies to immunogenic and surface-exposed outer membrane proteins also protect, but the range of antigens involved in this particular effect, and the role of non-specific defence mechanisms in preventing clinical disease, are ill understood. The complement system is important, as shown by the recurrent attacks of meningococcal infections in complement-deficient subjects.

Other host determinants such as lectin-binding proteins have been identified. Genetic differences in the ability to express certain cytokines may also influence the outcome of infection. Deficiency of either protein C, or its cofactor protein S, is associated with an increased risk of purpura fulminans. Patients with severe meningococcal disease have decreased endothelial expression of the two receptors, thrombomodulin and protein C receptor, which are essential for the activation of protein C.

Outbreaks of meningococcal meningitis require at least three factors:

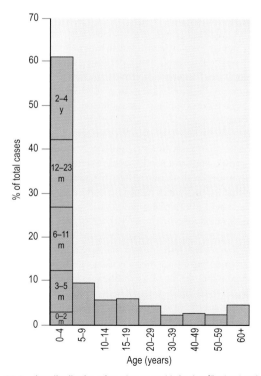

Fig. 23.2   Age distribution of meningococcal infection (England and Wales 1998–1999, Public Health Laboratory Service isolates).

1. a population of susceptible individuals who lack bactericidal antibodies to the current strains
2. a high transmission rate from person to person
3. a virulent strain of meningococcus.

The factors that determine the virulence and communicability of a meningococcus are still not fully understood.

The route of spread from the nasopharynx to the meninges is controversial. The organism may spread directly through the cribriform plate to the subarachnoid space by the perineural sheaths of the olfactory nerve. Much more probably, it passes through the nasopharyngeal mucosa to enter the bloodstream and finally reaches the CSF by crossing the cerebrovascular endothelial cells.

Pathophysiology

During meningococcal septicaemia there are signs and symptoms of circulatory failure, multi-organ dysfunction and coagulopathy. There is increased vascular permeability and vasodilatation that results in capillary leak syndrome with peripheral oedema. Loss of intravascular fluid and plasma proteins results in hypovolaemia and reduced venous return, and hence reduced cardiac output, hypotension and reduced perfusion of vital organs. Systemic hypoxia, acidosis, and gross electrolyte and metabolic impairment eventually culminate in multi-organ dysfunction. At the molecular level, the underlying pathophysiology of meningococcal sepsis is complex and involves numerous interactive cascades, including cytokine, chemokine, host cell receptors, and coagulation and complement components.

Meningococcal lipo-oligosaccharide (endotoxin), or more specifically its lipid A moiety, is thought to be primarily responsible for septic shock, extensive tissue damage and multi-organ dysfunction by stimulating the release of inflammatory mediators including tumour necrosis factor-α and a series of interleukins and other cytokines. In addition, meningococcal endotoxin seems to trigger a number of major intravascular cascade systems, including coagulation, fibrinolysis, complement and kallikrein–kinin. Although endotoxin is largely membrane associated, it is shed from live organisms in large quantities via outer membrane vesicles, which also contain outer membrane proteins. Increasing levels of circulating endotoxin are associated with increasing seriousness of disease and concentrations exceeding 700 ng/l are associated with fulminant septic shock, disseminated intravascular coagulation (DIC) and a high fatality rate. Endotoxin seems to trigger, directly or indirectly, the release of extremely high levels of monocyte tissue factor activity in severely septicaemic patients, suggesting that monocytes are responsible for the overwhelming DIC.

Clinical manifestations

Meningococci are able to cause a wide range of clinical syndromes varying in severity from a transient mild sore throat to meningitis or acute meningococcal septicaemia, which can cause death within hours of the appearance of symptoms.

Bacteraemia with or without sepsis, meningococcal septicaemia with or without meningitis, meningo-encephalitis, chronic meningococcaemia, pneumonia, septic arthritis, pericarditis, myocarditis, endocarditis, conjunctivitis, panophthalmitis, genito-urinary tract infection, pelvic infection, peritonitis and proctitis are among the diseases caused by meningococci. Meningitis and/or septicaemia are by far the most common presentations of disease. It is important to remember that the clinical picture can progress from one end of the spectrum to the other during the course of disease.

A significant number of the patients who recover end up with permanent neurological sequelae, including intellectual impairment, cranial nerve deficits and deafness (due to auditory nerve damage). The mortality rate from meningococcal disease varies between 5% and 70% depending on a number of factors including the severity of disease, the speed with which it develops, the organs involved, the age and immune status of the patient, the socio-economic status, the standard of health care, and the speed with which the disease is diagnosed and antibiotics administered.

Laboratory diagnosis

In any suspected meningococcal infection, blood culture must be undertaken; if meningitis is suspected, a lumbar puncture should be performed as soon as possible unless there are signs of raised intracranial pressure. As many patients receive penicillin before admission to hospital, lumbar puncture yields more positive cultures than blood. Where it is available it is preferable to use the PCR to detect DNA in blood or CSF.

Typically the CSF contains Gram-negative diplococci, which can be recognized by microscopical examination of the stained centrifuged deposit. At a later stage they may be sparse, and even apparently absent, in films stained by Gram's method.

The CSF is cultured on heated blood (chocolate) agar and on blood agar. In the absence of visible meningococci, glucose broth may be added to the remaining sediment of the centrifuged deposit to facilitate the isolation of very sparse organisms. Cultures are incubated overnight at 37°C in an atmosphere of 5–10% carbon dioxide. Sugar utilization tests or commercial kits are used to identify any Gram-negative diplococci. The oxidase test is performed on colonies on solid medium. Direct slide agglutination with specific antisera may be carried out on suspensions of colonies picked from solid medium.

Non-cultural methods are being used increasingly because early antibiotic treatment reduces the chance of successful culture. Meningococcal capsular polysaccharide may be demonstrated in CSF by counter-current electrophoresis, latex agglutination and PCR.

Cultured isolates should be sent, on slopes of chocolate agar, to a reference laboratory for serogrouping and serotyping, procedures that provide important epidemiological information.

Meningococci may be found in genital sites, and it is important to identify these organisms accurately and differentiate them from gonococci.

Treatment

Treatment should begin at the point of first contact as soon as meningococcal disease is suspected, even before the patient is transferred to hospital or investigated.

Intravenous penicillin, cefotaxime or ceftriaxone are the drugs of choice. Although these agents do not cross uninflamed meninges well, they readily pass into CSF when inflammation is present. Chloramphenicol is effective, but risks of blood dyscrasia have limited its use.

In the absence of organisms in the Gram-stained CSF deposit, it is wise to give therapy with cefotaxime or ceftriaxone, which cover *Haemophilus influenzae* and *Streptococcus pneumoniae*, the other two principal causes of meningitis in childhood after the neonatal period.

Almost all clinical isolates of meningococci in the UK are presently sensitive to benzylpenicillin. However, meningococci of reduced susceptibility to penicillin have been reported in some countries and are likely to become more common; it is therefore important that accurate sensitivity testing is carried out on all meningococcal isolates from clinical disease.

At the end of a course of therapy with penicillin it is important to give eradicative treatment with rifampicin or ciprofloxacin because penicillin does not eradicate meningococci from the nasopharynx and a patient returning home as a carrier may infect others. This probably does not apply to ceftriaxone or cefotaxime, which also eradicate throat carriage.

The mortality rate in septicaemic illness may range from 14% up to 50% in some outbreaks. The rate in meningitis is about 2–7%. Bad prognostic signs are:

- the presence of coma on admission to hospital
- a rapidly coalescing purpuric rash
- signs of shock.

Occasionally in the most fulminating forms of septicaemia there may not be time for the rash to develop before death. In such cases meningococci are isolated from the blood in life or post mortem, and haemorrhagic adrenals are seen at autopsy, these being characteristic of the *Waterhouse–Friderichsen syndrome*.

Epidemiology

Meningococcal infections occur worldwide and are notifiable in most countries. Around 5–10% of general populations are normally carriers of meningococci. In communities in which outbreaks of invasive meningococcal disease are occurring, the carriage rate of the epidemic strain may range from 20% to as high as 50% or more. Some studies have shown that a sharp increase in the carrier rate of group A or other pathogenic groups of meningococci precedes the occurrence of clinical cases.

Carriage rates are usually higher in crowded institutions, such as universities, prisons or army barracks, than in the community. There is a significant increase in carriage by university students in their first term when they live in shared accommodation on campus. Active and passive smoking appear to increase the risk of meningococcal carriage.

The incidence of meningococcal infection is increasing.

- Serogroup A is most prevalent in sub-Saharan Africa and the Middle East, with dramatic epidemics in many parts of the world.
- Serogroup B strains are dominant among pathogenic meningococci in most industrialized countries, followed by group C strains. Epidemics of group B have been reported from Norway, and of group C from Brazil, Africa and China.
- Serogroup W-135 is present in small numbers worldwide, but has been associated with outbreaks following the pilgrimage to Mecca (the Hajj).

Control

Chemoprophylaxis

Outbreaks of disease may be controlled by chemoprophylaxis alone, thus eradicating the organism from carriers, or combined with vaccination. Until the emergence of resistant strains, sulphonamides were effective, but rifampicin is now the preferred drug for children although it is effective in eradicating meningococcal carriage in only 80–90% of the population treated. After prophylaxis, rifampicin-resistant strains may be found in a small number of patients and may, on rare occasions, give rise to disease in contacts. Ciprofloxacin is used widely as a prophylactic for adolescents and adults as a single oral dose. All household and other intimate (e.g. mouth kissing) contacts of a case should be given chemoprophylaxis as a routine.

Vaccination

Resistance to meningococcal infection is closely associated with the possession of bactericidal antibodies that may be maternal in origin or actively produced in response to carriage. Vaccines containing the native, unconjugated, group-specific capsular polysaccharide of meningococci of groups A, C, Y and W135 are available and are good immunogens. However:

- the protection provided is group specific.
- it lasts at most 3 years.
- it does not prevent meningococcal carriage.
- the vaccines are poorly immunogenic in children aged under 2 years.

These shortcomings have limited the use of these vaccines largely for travel to parts of the world in which epidemics due to group A or C meningococci are in progress or are to be expected.

One approach to the management of epidemics of meningococcal infection due to group A or C meningococci has been to vaccinate all the population at risk and to give penicillin to any clinical cases. Any cases that arise before an immune response to the vaccine has commenced are also treated.

The development of protein-conjugated group C vaccines has changed the outlook for prevention. In the UK, vaccination has been introduced into the routine programme and clinical infection with group C has been significantly reduced, particularly in those aged under 18 years. This vaccine also reduces group C meningococcal carriage in the population. Conjugate quadrivalent groups A, C, Y and W-135 vaccines are under development.

Trials are in progress on several vaccines against group B meningococci, but it has been more difficult to find a suitable immunogenic epitope. The group B capsular polysaccharide is a poor immunogen; alternative protein-based vaccines have undergone clinical trials but with disappointing results.

NEISSERIA GONORRHOEAE

The name 'gonorrhoea' derives from the Greek words *gonos* (seed) and *rhoia* (flow), and described a condition in which semen flowed from the male organ without erection. It became apparent that gonorrhoea was associated with sexual promiscuity, one of the diseases celebrated by being named after the Roman goddess of love, Venus. Indeed, gonorrhoea is a classical venereal disease, being spread almost exclusively by sexual contact, having a short incubation period and being relatively easy to diagnose and treat.

Gonococci are as antigenically heterogeneous as meningococci. Classically, strains are characterized by *auxotyping*, which recognizes requirements for specific nutrients, such as arginine, proline, hypoxanthine and uracil. Panels of monoclonal antibodies that recognize specific proteins are also used to divide strains into various serovars. Epidemiological typing makes use of both methods as well as newer molecular techniques, such as restriction fragment length polymorphisms in genes encoding ribosomal ribonucleic acid (rRNA) (ribotyping) and the separation of large DNA fragments by pulsed-field gel electrophoresis.

Gonorrhoea is the second most commonly diagnosed bacterial sexually transmitted infection in the UK. The highest infection rates are seen in men aged 20–24 years and women aged 16–19 years.

Pathogenesis

N. gonorrhoeae is exclusively a human pathogen, although chimpanzees have been infected artificially. It is never found as a normal commensal, although a proportion of those infected, particularly women, may remain asymptomatic. These individuals may develop systemic or ascending infection at a later stage.

Gonorrhoeal infection is generally limited to superficial mucosal surfaces lined with columnar epithelium. The areas most frequently involved are the cervix, urethra, rectum, pharynx and conjunctiva. Squamous epithelium, which lines the adult vagina, is not susceptible to infection by the gonococcus. However, the prepubertal vaginal epithelium, which has not been keratinized under the influence of oestrogen, may be infected. Hence, gonorrhoea in young girls may present as vulvovaginitis.

The commonest clinical presentation is acute urethritis in the male a few days after unprotected vaginal or anal sexual intercourse. Dysuria and a purulent penile discharge make most sufferers seek treatment rapidly. A few men have relatively minor symptoms, which may disappear rapidly. Truly asymptomatic infection is rare in the active male. However, up to 5% may carry the organism without apparent distress, and this is more common with certain types of gonococci. Rectal and pharyngeal infection is less often symptomatic and may be discovered only after tracing contacts.

Endocervical infection is the most common form of uncomplicated gonorrhoea in women. Such infections are usually characterized by vaginal discharge and sometimes by dysuria. The cervical os may be erythematous and friable, with a purulent exudate. In women with vaginal infection, only half may have symptoms of discharge and dysuria. Most seek attention because of their partner's symptoms, or as part of contact tracing or screening of high-risk individuals. Local complications include abscesses in Bartholin's and Skene's glands. Rectal infections (proctitis) with *N. gonorrhoeae* occur in about

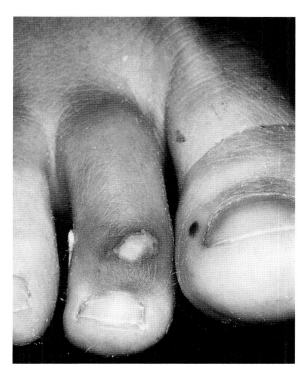

Fig. 23.3 Skin lesions in disseminated gonococcal infection.

one-third of women with cervical infection and are rarely symptomatic. In contrast, gonococcal proctitis in homosexual men is often symptomatic.

Asymptomatic carriage in women is common, especially in the endocervical canal. At menstruation or after instrumentation, particularly termination of pregnancy, gonococci ascend to the fallopian tubes to give rise to *acute salpingitis*, which may be followed by *pelvic inflammatory disease* and a high probability of sterility if treated inadequately. Peritoneal spread occurs occasionally and may produce a perihepatic inflammation (*Fitz-Hugh–Curtis syndrome*). Some auxotypes of *N. gonorrhoeae* spread more widely and give rise to disseminated gonococcal infection.

Disseminated infection is seen more commonly in women, who may present with painful joints, fever and a few septic skin lesions on their extremities (Fig. 23.3). The diagnosis of a venereal disease may not be obvious and isolation of gonococci from joint fluid, blood culture and skin aspirates requires particular care. The organisms are invariably present in the cervix, but in many cases antibiotics have been given before the diagnosis is considered. Rarely, disseminated gonococcal infection may present as endocarditis or meningitis.

Babies born to infected women may suffer *ophthalmia neonatorum*, in which the eyes are coated with gonococci as the baby passes down the birth canal. A severe purulent eye discharge with peri-orbital oedema occurs within a few days of birth. If untreated, ophthalmia leads rapidly to blindness. It may be prevented in areas of high prevalence by the instillation of 1% aqueous silver nitrate in the eyes of newborn babies. Alternatively, topical erythromycin can be used; this has the advantage of being active against chlamydia and less toxic.

Vulvovaginitis in prepubertal girls occurs either in conditions of poor hygiene or by sexual abuse; it should always be investigated carefully and the child put in touch with social services and other professionals capable of dealing with this difficult condition.

Laboratory diagnosis

Cultivation of *N. gonorrhoeae* from sites with scanty commensals, such as the male urethra, rarely presents any problems, and Gram staining of a smear is usually 95% sensitive. *N. gonorrhoeae* is intolerant of drying and temperature changes; it readily undergoes autolysis. It is a fastidious microbe, requiring humidity, 5–7% carbon dioxide and complex media for growth. Ideally, exudate is taken directly from the patient on to appropriate preheated, freshly prepared solid media and immediately placed in a carbon dioxide incubator. This is usually possible only in specialized clinics with sufficient patient numbers to justify the expense. Where there is likely to be any delay, transport media must be used to carry the material on swabs.

The combination of oxidase-positive colonies and Gram-negative diplococci provides a presumptive diagnosis. Fluorescent-antibody staining, co-agglutination, specific biochemical tests and DNA probes may be used for confirmation. DNA probes have also been used to detect gonococci in urethral and cervical specimens. PCR-based methods are available in some specialized laboratories.

Treatment

The susceptibility of isolates of *N. gonorrhoeae* to commonly used antibiotics varies so much that regular testing is essential.

Penicillin, especially in slow-release intramuscular forms such as procaine penicillin, remains the preferred therapy in many parts of the world. Small decreases in susceptibility can be overcome by increasing the size of the single dose. However, by the 1970s the dose of penicillin required to cure simple acute gonorrhoea in men in some parts of the world had reached an impossibly large injection.

Strains of *N. gonorrhoeae* that are completely resistant to penicillins are now common throughout the world, although the prevalence varies from country to country.

Table 23.1 Resistance to antimicrobial agents among isolates of *N. gonorrhoeae* in the UK

Antimicrobial agent	Percentage resistant	
	2004	2003
Penicillin	11.2	9.7
Ciprofloxacin	14.1	9.0
Cefixime or ceftriaxone	0	0
Azithromycin	1.8	0.9
Tetracycline	44.5	38.2
Spectinomycin	0.2	Rare
Multiresistant[a]	12.9	6.8

Data from the Health Protection Agency.
[a]Resistant to two or more of the listed agents.

These strains possess the gene coding for the TEM-type β-lactamase commonly found in *Escherichia coli*.

Ceftriaxone or cefixime are recommended as first-line therapy in the UK, but these drugs are expensive and may not be affordable in developing countries. Alternatives to cephalosporins and penicillin include fluoroquinolones (e.g. ciprofloxacin), azithromycin, tetracyclines, co-amoxiclav and spectinomycin. Use of these antibiotics is often limited in places where inappropriate use has led to a high prevalence of resistance. Resistance, or reduced susceptibility, to most of the commonly used antibiotics is increasing in the UK (Table. 23.1).

Single-dose therapy appears adequate for uncomplicated cases of acute genital gonorrhoea in men and women. There are obvious advantages to this approach in obtaining complete compliance and stopping the chain of infection. In disseminated gonococcal disease and any complicated infection, treatment for 7–10 days is necessary.

Epidemiology and control

Acute gonorrhoea is usually easily diagnosed and treated, and was well controlled in much of the world until the 1960s. The remarkable changes in travel, migration, sexual licence and availability of oral contraceptives rapidly reversed this process so that there was an increase in gonorrhoea and non-specific genital infection (caused mainly by chlamydiae) every year until scares about the acquired immune deficiency syndrome in the 1980s temporarily halted the rise. Barrier methods of contraception, condoms in particular, greatly reduce the rate of transmission.

The keys to control of gonorrhoea are:

- rapid diagnosis
- use of effective antibiotics
- tracing, examination and treatment of contacts.

Unfortunately, in many places, inappropriate self-medication has contributed to widespread antimicrobial resistance. Inability to treat contacts ensures the spread of the disease and re-infections.

There is no effective vaccine to prevent gonorrhoea. The organism does not possess a capsule and its immunogenic outer membrane proteins are antigenically variable. These, and lack of suitable animal models, have hampered vaccine development.

MORAXELLA

The genus *Moraxella* is a member of the family Neisseriaceae. The organisms are characterized as asaccharolytic, oxidase-positive, catalase-positive short rods, coccobacilli or, in the case of *M. catarrhalis*, diplococci. They are commensals of mucosal surfaces and occasionally give rise to opportunistic infections. *M. lacunata* is occasionally encountered as a cause of 'angular' blepharoconjunctivitis.

The related genus *Kingella* contains organisms that differ from the moraxellae in being catalase-negative, glucose-fermenting coccobacilli. In common with some other Gram-negative rods, such as *Cardiobacterium hominis*, *Eikenella corrodens* and *Actinobacillus actinomycetemcomitans*, *Kingella* species (usually *K. kingae*) are sometimes found in endocarditis. They have also been implicated in joint infections.

MORAXELLA CATARRHALIS

Pathogenesis

M. catarrhalis is a respiratory tract commensal and, as with other members of the upper respiratory tract flora such as the pneumococcus and *H. influenzae*, it can gain access to the lower respiratory tract in patients with chronic chest disease or compromised host defences. *M. catarrhalis* is commonly isolated from sputum, and a pathogenic role is suspected only when the sputum contains large numbers of pus cells and Gram-negative diplococci, and when culture yields a heavy growth of *M. catarrhalis* in the absence of other recognized respiratory pathogens.

As well as causing chest infection itself, *M. catarrhalis* may also protect other respiratory pathogens from the action of penicillin or ampicillin by producing β-lactamase.

M. catarrhalis has been incriminated in otitis media and sinusitis, and is occasionally isolated from blood of immunocompromised patients.

Laboratory diagnosis

Sputum is examined by Gram film. In true infections, large numbers of Gram-negative diplococci may be seen dispersed between the pus cells. Sputum is cultured on media suitable for the isolation of other potential respiratory pathogens (e.g. blood agar and chocolate agar) and incubated in 5% carbon dioxide overnight. In situations in which it is held to be pathogenic, *M. catarrhalis* is predominant in culture.

M. catarrhalis produces rough, circular, convex colonies that can be lifted off intact with a wire loop from agar culture medium. Colonies are oxidase positive. In common with other moraxellae, they do not ferment sugars and are easily differentiated from neisseriae. Growth on nutrient agar at 22°C has been suggested as

a differential characteristic, but clinically significant isolates of *M. catarrhalis* may not grow in these conditions. Tests for deoxyribonuclease and for butyrate esterase are positive. At least 50% of strains produce β-lactamase.

Treatment

M. catarrhalis is sensitive to amoxicillin, combined with clavulanic acid (co-amoxiclav) in the case of β-lactamase-producing strains, and also to cephalosporins, tetracyclines, macrolides and fluoroquinolones.

KEY POINTS

- *Neisseria meningitidis* (meningococcus) and *N. gonorrhoeae* (gonococcus) are obligate human parasites.
- *N. meningitidis* lives commensally in the nasopharynx, is transmitted via close kissing contact, and causes disease that varies in severity from mild sore throat to meningitis, septicaemia or septicaemic shock (circulatory failure, multi-organ dysfunction and coagulopathy). Lipo-oligosaccharide (endotoxin) is largely responsible for the deranged host response underpinning shock.
- Treatment is by intravenous administration of penicillin or ceftriaxone. Prophylactic antibiotics (e.g. rifampicin or ciprofloxacin) can be given to contacts to eradicate carriage and control outbreaks.
- Of the 13 serogroups, groups A, B, C and W-135 cause more than 90% of cases. Vaccines are available against all except group B, which is most prevalent in the developed world.

- *N. gonorrhoeae* causes the sexually transmitted disease gonorrhoea. Asymptomatic carriage in women is common, but the organism may give rise to acute salpingitis, which may be followed by pelvic inflammatory disease and a high probability of sterility if inadequately treated.
- In the UK, cephalosporins such as ceftriaxone are the drugs of choice. Use of antibiotics is limited in some countries by a high prevalence of resistance.
- Early diagnosis, effective treatment and contact tracing are key to preventing the spread of disease. There is no effective vaccine to prevent gonorrhoea.
- *M. catarrhalis* is an upper respiratory tract commensal that occasionally causes lower respiratory tract infection.

RECOMMENDED READING

Ala'Aldeen D A A, Turner D P J 2006 *Neisseria meningitidis*. In *Principles and Practice of Clinical Bacteriology*, 2nd edn, Gillespie S, Hawkey P (eds). John Wiley, Chichester
Bignell C J 2001 European guideline for the management of gonorrhoea. *International Journal of STD and AIDS* Suppl 3: 27–29
Bilukha O O, Rosenstein N 2005 Prevention and control of meningococcal disease. *Morbidity and Mortality Weekly Report* 54 (RR07): 1–21. Online. Available: http://www.cdc.gov/mmwr/preview/mmwrhtml/rr5407a1.htm
Barker R M, Shakespeare R M, Mortimore A J, Allen N A, Solomon C L, Stuart J M 1999 Practical guidelines for responding to an outbreak of meningococcal disease among university students based on experience in Southampton. *Communicable Disease and Public Health* 2: 168–173
Health Protection Agency 2005 *The Gonococcal Resistance to Antimicrobials Surveillance Programme Steering Group Year 2004 Report*. Health Protection Agency, London

Karalus R, Campagnari A 2000 *Moraxella catarrhalis*: a review of an important human mucosal pathogen *Microbes and Infection* 2: 547–559
Public Health Laboratory Service Meningococcus Forum 2002 Guidelines for public health management of meningococcal disease in the UK. *Communicable Disease and Public Health* 5: 187–204
Rosenstein N E, Perkins B A, Stephens D S, Popovic T, Hughes J M 2001 Meningococcal disease. *New England Journal of Medicine* 344: 1378–1388

Internet site

Health Protection Agency. Guidelines and Advice – Meningococcal. http://www.hpa.org.uk/infections/topics_az/meningo/guidelines.htm

24 Salmonella

Food poisoning; enteric fever

H. Chart

There are well over 2000 different antigenic types of salmonella. They were originally classified as separate species, but it is now generally accepted that they represent serotypes (serovars) of a single species, *Salmonella enterica*. Various subspecies are recognized, but most of the serotypes that infect mammals are found in a subspecies also designated *enterica*. For example, the full designation of the serotype formerly called *Salmonella enteritidis* is: *Salmonella enterica* subspecies *enterica* serotype Enteritidis. This cumbersome nomenclature is often abbreviated to *Salmonella* Enteritidis and this convention will be followed here in considering those salmonellae responsible for human infections.

Certain serotypes are a major cause of food-borne infection worldwide. Most infections are relatively benign and restricted to the intestinal tract, causing a short-lived diarrhoea, but some salmonellae cause life-threatening systemic disease. In England and Wales annual isolations of selected serotypes from man almost trebled, from 8939 to 26212 during the 1980s, largely due to the emergence of strains of *Salmonella* Enteritidis belonging to phage type 4 (PT 4) in chickens and poultry products. Since 1998 there has been a steady decline in the numbers of isolations of *S*. Enteritidis PT 4 (Fig. 24.1), largely due to vaccination of broiler flocks and restrictions of the importation of eggs from abroad.

In developing countries salmonellae are not as important a cause of community-acquired diarrhoea. However, infections with *S. enterica* serotypes Typhi and Paratyphi, which are encountered mainly as imported infections in developed countries, remain prevalent in other parts of the world.

DESCRIPTION

Salmonellae are typical members of the Enterobacteriaceae: facultatively anaerobic Gram-negative bacilli able to grow on a wide range of relatively simple media and distinguished from other members of the family by their biochemical characteristics and antigenic structure. Their normal habitat is the animal intestine.

Antigens

Typical strains of *S. enterica* express two sets of antigens, which are readily demonstrable by serotyping. Long-chain lipopolysaccharide (LPS) comprises heat-stable polysaccharide commonly known as the *somatic* or *O antigens*. These molecules are located in the outer membrane and are anchored into the cell wall by antigenically conserved lipid A and LPS-core regions. The long-chain LPS molecules exhibit considerable variation in sugar composition and degree of polysaccharide branching, and this structural heterogeneity is responsible for the large number of serotypes. Salmonellae are usually highly motile when growing in laboratory media (Fig. 24.2), and flagellar protein subunits contain the epitopes that

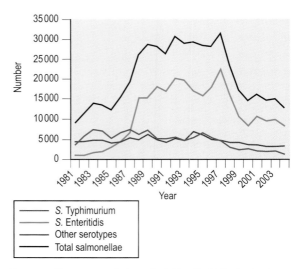

Fig. 24.1 Human cases of salmonellosis identified in England and Wales during 1981–2004 (Health Protection Agency data).

form the basis of the flagella-based serotyping scheme generally known as the *H antigens*. In most strains of *S. enterica* the flagella exhibit the property of diphasic variation, whereby one of two genetically distinct flagellar structures are expressed. When one flagellar structure is expressed it contains *phase 1* antigens, whereas when the other set is operative *phase 2* antigens are synthesized.

Certain serotypes of *S. enterica* express a surface polysaccharide, of which the Vi (virulence) antigen of *Salmonella* Typhi is the most important example. As the polysaccharide may encapsulate the entire bacterium, antibodies designed to recognize the LPS antigens may be prevented from binding; this can occasionally make detection of the O antigens difficult.

The various O antigens of salmonellae are numbered with Arabic numerals. The flagellar antigens of phase 1 are designated by lower-case letters, and those of phase 2 by a mixture of lower-case letters and Arabic numerals. The antigenic structure of any serotype of salmonella is thus expressed as an antigenic formula, which has three parts, describing the O antigens, the phase 1 H antigens and the phase 2 H antigens, in that order. The three parts are separated by colons, and the component antigens in each part by commas; for example, the distinctive antigenic formula of *Salmonella* Enteritidis is 1, 9, 12: g, m: 1, 7.

The original Kauffmann–White scheme, which elegantly catalogued salmonellae (but named them as individual species), placed them into some 30 groups on the basis of shared O antigens, and further subdivided the groups into clusters with H antigens in common. Some salmonellae, such as *Salmonella* Typhi (9, 12, [Vi]: d–), express only one flagellar phase. Some of the commoner serotypes are shown in Table 24.1.

Typing methods

Phage typing schemes developed by the Laboratory of Enteric Pathogens of the UK Health Protection Agency have proved extremely useful for discriminating within strains of *S. enterica* serotypes Typhimurium, Virchow, Enteritidis and Typhi. The numbers of phages comprising the various schemes are constantly being increased to improve strain discrimination.

Characterization of strains of *S. enterica* is also assisted by determining their distinctive patterns of resistance to a range of antibiotics. This has been of particular relevance to certain strains of *Salmonella* Typhimurium (notably definitive type (DT) 104) characterized by resistance to ampicillin, chloramphenicol, aminoglycosides and co-trimoxazole.

Pulsed-field gel electrophoresis analysis, of the entire *Salmonella* genome following digestion with selected enzymes, has facilitated strain discrimination in outbreak situations.

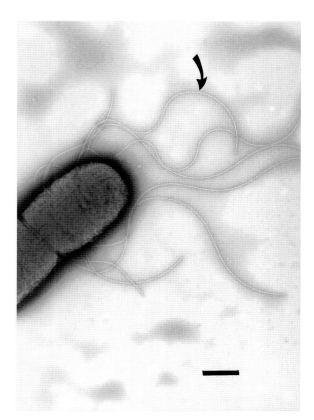

Fig. 24.2 Electron micrograph of *Salmonella* Enteritidis. Arrow indicates the flagella that carry the 'H' antigens. Bar = 0.3 μm.

Table 24.1 Antigenic structure of some representative salmonellae

Serotype	'O' antigens	'H' antigens	
		Phase 1	Phase 2
Typhi	9, 12, [Vi]	d	–
Paratyphi B	1, 4, 5, 12	b	1, 2
Typhimurium	1, 4, 5, 12	i	1, 2
Enteritidis	1, 9, 12	g, m	1, 7
Virchow	6, 7	r	1, 2
Kedougou	1, 13, 23	i	1, w
Hadar	6, 8	Z10	e, n, x
Heidelberg	1, 4, 5, 12	r	1, 2
Infantis	6, 7, 14	r	1, 5
Newport	6, 8, 20	e, h	1, 2
Panama	1, 9, 12	l, v	1, 5
Dublin	1, 9, 12	g, p	–

Host range and pathogenicity

Strains of *S. enterica* are widely distributed in nature. All vertebrates appear capable of harbouring these bacteria in their gut, and certain serotypes have also been isolated from a wide range of arthropods, including flies and cockroaches. Most animal infections seem to range from those without symptoms to those resulting in self-limiting gastro-enteritis of variable severity. Some strains, such as those belonging to serotype Typhimurium, show a wide host range and can be isolated from many different animal species. A small number of strains, the *host-adapted* serotypes, are much more restricted in the species they inhabit, and show a different spectrum of illness.

Among the host-adapted serotypes, Typhi and Paratyphi A, B and C are rarely, if ever, isolated from animals other than man. Paratyphi B, although essentially a human pathogen, is occasionally isolated from cattle, pigs, poultry, exotic reptiles and other animals, although cycles of transmission in these hosts have not been demonstrated. Human infection with these organisms is characterized by a long incubation period of 10–14 days, followed by a septicaemic illness, *enteric fever*, quite unlike the diarrhoea and vomiting that are characteristic of food poisoning.

Other salmonellae adapted to particular animal hosts include Cholerae-suis (pigs), Dublin (cattle), Gallinarum-pullorum (poultry), Abortus-equi (horses) and Abortus-ovis (sheep). These are all responsible for considerable morbidity, mortality and economic loss among domestic animals. All can cause human illness, but only strains of Cholerae-suis and Dublin do so regularly. Strains of *Salmonella* Typhimurium DT 104 isolated from human infection appear to have originated from bovine sources; however, whether these strains can be considered as host-adapted to farm animals remains to be established. The rest of the 2000 or so serotypes of salmonellae show no apparent host preference. The extent to which they cause human infection appears to result from their prevalence in domestic food animals at any particular time and on the opportunity for contamination of food in which further multiplication can take place. In developed countries most human infections are caused by a relatively small number of locally prevalent serotypes.

PATHOGENESIS

In common with many pathogenic enteric bacteria, strains of *S. enterica* require a range of pathogenic mechanisms to enable them to survive passage through the digestive tract, colonize a host and cause disease. The lack of a good animal model of salmonella pathogenicity has impeded research into how these bacteria cause the symptoms of disease, but elucidation of the entire gene sequence of strains of *Salmonella* Enteritidis and *Salmonella* Typhimurium (http://www.sanger.ac.uk/Projects/Salmonella/) should allow a better understanding of the pathogenic mechanisms carried by these organisms. In general terms, infection is initiated

by the ingestion of a sufficient number of organisms to survive the stomach acid and the effects of digestive bile prior to colonization of the gut mucosa, and to express the mechanisms resulting in overt disease.

Infective dose

For human infections, the number of bacteria that must be swallowed in order to cause infection is uncertain and varies with the serotype. The accepted dictum that large inocula of these bacteria are required for induction of human illness is based largely on volunteer studies. In most of these the median infective dose for most serotypes, including Typhi, has varied from 10^6 to 10^9 viable organisms. However, investigation of outbreaks suggests that in natural infection the infective dose might be below 10^3 viable organisms.

Many factors are thought to influence the infective dose. There appears to be considerable strain-to-strain variation in virulence even within a single serotype. Systematic variation in pathogenicity between serotypes is less easy to demonstrate outside the host-adapted strains. The vehicle of ingestion may also influence pathogenesis. Organisms ingested in water and other drinks may be carried through the stomach relatively rapidly, and evade the effect of gastric acid. Similarly, the administration of antacids, or the effects of gastric resection, reduces the infective dose. Bacteria within particles of food would also evade the action of stomach acids.

Host factors

Host factors are also likely to be important, particularly the immune status of the host. Host variables can, however, be confusing; for example, age-specific isolation rates for salmonellae, as for some other gut pathogens, are higher for children less than 1 year of age than for any other age group, but this reflects the fact that a higher proportion of infections are investigated in this age group.

Initiation of infection

Once salmonellae enter the lumen of the intestine they need to be able to tolerate the action of digestive bile and compete with the prevailing gut flora for adhesion sites on the gut mucosa. Certain serotypes, such as *Salmonella* Typhimurium, express type 1 fimbriae, which enable them to adhere to α-mannose-containing molecules on the microvilli of the ileal mucosa; however, surprisingly little is known about the range of fimbriae expressed. Strains of *Salmonella* Enteritidis are thought to express at least three different fimbrial structures (Fig. 24.3).

Salmonella serotypes such as Typhimurium and Enteritidis also express an adhesion mechanism that does not

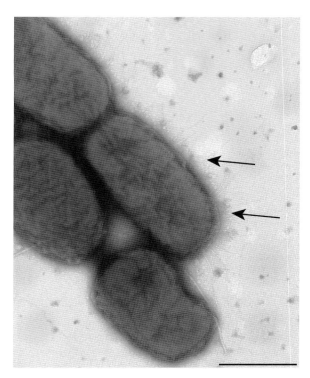

Fig. 24.3 Electron micrograph of *Salmonella* Enteritidis. Arrows indicate fimbriae. Bar = 0.5 µm. From Zuckerman AJ, Principles and practice of clinical virology 5E, 2004, John Wiley & Sons Ltd

involve fimbriae. Certain strains of enteric bacteria carry deoxyribonucleic acid (DNA) sequences that encode several pathogenic mechanisms, termed *pathogenicity islands*. In common with strains of Verocytotoxin-producing *Escherichia coli* belonging to serogroup O157 (see Ch. 26), strains of *Salmonella* Typhimurium and *Salmonella* Enteritidis have pathogenicity islands that encode an adhesion mechanism comprising both a bacterial adhesin and the adhesin receptor, which is translocated into the host intestine. This process enables these bacteria to insert their own binding site into the gut, unlike fimbriae, which require host-derived binding sites located in the intestinal wall.

Attachment to the host mucosa is followed by degeneration of the microvilli to form breaches in the cell membrane through which the salmonellae enter the intestinal epithelial cells. For certain strains, further multiplication in these cells and in macrophages of the Peyer's patches follows. Some bacteria penetrate into the submucosa and pass to the local mesenteric lymph nodes. All of the clinical manifestations of infection with salmonella, including diarrhoea, begin after ileal penetration where inflammation of the ileal mucosa results in the efflux of water and electrolytes resulting in diarrhoea. For strains of *Salmonella* Typhi, infection involves invasion of the bloodstream and various organs.

CLINICAL SYNDROMES

Although salmonellae can cause a wide spectrum of clinical illness there are four major syndromes, each with its own diagnostic and therapeutic problems:

- enteric fever
- gastro-enteritis
- bacteraemia with or without metastatic infection
- the asymptomatic carrier state.

Although uncommon, infection with *Salmonella* can result in sequelae, including reactive arthritis (*Reiter's syndrome*).

Enteric fever

Enteric fever is usually caused by strains of *Salmonella* Typhi or *Salmonella* Paratyphi A, B or C. The clinical features tend to be more severe with *Salmonella* Typhi (*typhoid fever*). After penetration of the ileal mucosa the organisms pass via the lymphatics to the mesenteric lymph nodes, whence after a period of multiplication they invade the bloodstream via the thoracic duct. The liver, gall bladder, spleen, kidney and bone marrow become infected during this primary bacteraemic phase in the first 7–10 days of the incubation period. After multiplication in these organs, bacilli pass into the blood, causing a second and heavier bacteraemia, the onset of which approximately coincides with that of fever and other signs of clinical illness. From the gall bladder, a further invasion of the intestine results. Peyer's patches and other gut lymphoid tissues become involved in an inflammatory reaction, and infiltration with mononuclear cells, followed by necrosis, sloughing and the formation of characteristic typhoid ulcers occurs.

Onset

The interval between ingestion of the organisms and the onset of illness varies with the size of the infecting dose. It can be as short as 3 days or as long as 50 days, but is usually about 2 weeks. The onset is usually insidious. Early symptoms are often vague: a dry cough and epistaxis associated with anorexia, a dull continuous headache, abdominal tenderness and discomfort are among the most common symptoms. Diarrhoea is uncommon and early in the illness many patients complain of constipation.

Progression

In the untreated case the temperature shows a step-ladder rise over the first week of the illness, remains high for 7–10 days and then falls during the third or fourth week.

Physical signs include a relative bradycardia at the height of the fever, hepatomegaly, splenomegaly and often a rash of *rose spots*. These are slightly raised, discrete, irregular, blanching, pink macules, 2–4 mm in diameter, most often found on the front of the chest. They appear in crops of up to a dozen at a time and fade after 3–4 days, leaving no scar. They are characteristic of, but not specific for, enteric fever.

Relapse

Apparent recovery can be followed by relapse in 5–10% of untreated cases. Relapse is usually shorter and of milder character than the initial illness, but can be severe and may be fatal. Severe intestinal haemorrhage and intestinal perforation are serious complications that can occur at any stage of the illness.

Morbidity and mortality

Classical typhoid fever is a serious infection that, when untreated, has a mortality rate approaching 20%. It is notoriously unpredictable in its presentation and course. Mild and asymptomatic infections are not uncommon. In endemic areas, and particularly where it coexists with schistosomiasis, chronic infection can present with fever of many months' duration, accompanied by chronic bacteraemia. Occasionally, diarrhoea may dominate the picture from the outset, particularly in paratyphoid infections, which sometimes present as typical gastro-enteritis no different from that caused by most *S. enterica* serotypes.

Gastro-enteritis and food poisoning

Acute gastro-enteritis is characterized by vomiting, abdominal pain, fever and diarrhoea. It can be caused by ingestion of a wide variety of bacteria or their products, several viruses, and a number of vegetable toxins and inorganic chemicals. The term *bacterial food poisoning* is conveniently restricted to cases and epidemics of acute gastro-enteritis that are caused by the ingestion of food contaminated by bacteria or their toxins. It is an important feature of bacterial food poisoning that the bacteria need the opportunity to multiply in the food to reach an infective concentration before being eaten. Infections such as hepatitis A or bacillary dysentery, in which food may be an incidental vector, are not usually considered to be examples of food poisoning. Strains of *S. enterica* commonly cause food poisoning worldwide.

Clinical features

The most common clinical manifestation of infection with non-invasive salmonella serotypes is diarrhoea, often accompanied by headache, malaise and nausea. The incubation period is usually 8–48 h, the onset abrupt, and the clinical course short and self-limiting. Symptoms vary from the passage of two or three loose stools, which may be disregarded by the sufferer, to a severe and prostrating illness with the frequent passage of watery, green, offensive stools, fever, shivering, abdominal pain and, in the most severe cases, dehydration leading to hypotension, cramps and renal failure. Vomiting is rarely a prominent feature of the illness.

Severe infections occur most often in the very young and the elderly, although mild subclinical infections also occur in these age groups. Infections with certain serotypes in those already ill or debilitated from other causes are likely to be more severe and life-threatening. In most cases the acute stage is over within 2–3 days, although it may be more prolonged. Persistent or high fever suggests bacteraemia, possibly with metastatic infection.

Bacteraemia and metastatic disease

Bacteraemia is a constant feature of enteric fever caused by strains of *Salmonella* Typhi and Paratyphi, and can occur as a rare complication of infection with other salmonellae. Transient bacteraemia occurs in up to 4% of cases of acute gastro-enteritis, but in most cases the organisms are cleared from the bloodstream without ill effect.

Occasionally, dissemination of the bacilli throughout the body results in the establishment of one or more localized foci of persisting infection, especially where pre-existing abnormality makes a tissue or organ vulnerable. Atherosclerotic plaques within large arteries, damaged heart valves, joint prostheses and other implants are all susceptible to metastatic infection.

Osteomyelitis is most often found in long bones, costochondral junctions and the spine. Multiple bony sites may be affected, and sickle cell anaemia is an important predisposing factor. Suppurative arthritis can occur either as an extension of contiguous osteomyelitis or as a primary infection.

Meningitis is a particularly serious complication of infection in neonates and very young children. Abscess formation can occur in almost any organ or tissue. Even in the absence of obvious tissue damage, the ability of salmonellae to enter and survive within macrophages and other cells, particularly in the liver and biliary tree, but also in bone marrow and the kidney, leads occasionally to persistent infection and the chronic carrier state.

The prolonged carrier state

Most people infected with salmonella continue to excrete the organism in their stools for days or weeks after complete clinical recovery, but eventual clearance of the bacteria from the body is usual. A few patients continue to excrete the salmonellae for prolonged periods. The term *chronic carrier* is reserved for those who excrete salmonellae for a year or more. Chronic carriage can follow symptomatic illness or may be the only manifestation of infection. It can occur with any serotype, but is a particularly important feature of enteric fever: up to 5% of convalescents from typhoid and a smaller number of those who have recovered from paratyphoid fever become chronic carriers, many for a lifetime. The bacilli are most commonly present in the gall bladder, less often in the urinary tract, and are shed in faeces and sometimes in urine. The long duration of the carrier state enables the enteric fever bacilli to survive in the community in non-epidemic times and to persist in small and relatively isolated communities.

Age and sex are important determinants of the frequency of carriage, at least of *Salmonella* Typhi. After enteric fever, less than 1% of patients under 20 years old become carriers, but this proportion rises to more than 10% in patients over 50 years of age. At all ages women become carriers twice as often as men.

The duration of excretion following infection with other salmonellae is less well documented, but more than 50% of patients stop excreting the organisms within 5 weeks of infection, and 90% of adults are culture negative at 9 weeks. The duration of excretion is significantly greater in children aged under 5 years, but virtually all permanent carriers are adults.

LABORATORY DIAGNOSIS

Selective media, such as desoxycholate–citrate agar or xylose–lysine desoxycholate agar, are used for the isolation of salmonella bacteria from faeces. Fluid enrichment media, such as tetrathionate or selenite broth, are also useful to detect small numbers of salmonellae in faeces, foods or environmental samples. Suspicious colonies from the culture plates are tested directly for the presence of *Salmonella* somatic (O) antigens by slide agglutination and subcultured to peptone water for the determination of flagellar (H) antigen structure and further biochemical analysis. A presumptive diagnosis of salmonellosis can often be made within 24 h of the receipt of a specimen, although confirmation may take another day, and formal identification of the serotype by a suitable reference laboratory takes several more days. A negative report must await the result of enrichment cultures – at least 48 h.

Enteric fever

Blood culture

The organisms may be recovered from the bloodstream at any stage of the illness, but are most commonly found during the first 7–10 days and during relapses. The organisms can also be recovered from the blood clot from a sample taken for serological tests. The clot is digested with streptokinase or minced, and incubated in broth.

Stool and urine culture

Specimens of faeces and urine should also be submitted for examination, although the isolation of salmonella from either of these specimens may indicate merely that the patient is a carrier. In typhoid fever, patients' stools may contain salmonella from the second week and urine cultures from the third week of the infection. In paratyphoid B infections the clinical course may be much shorter than in typhoid; diarrhoea may occur early and stool cultures are often positive in the first week of the illness.

Serological tests

Infections with both invasive and non-invasive serotypes may induce specific serum antibodies to *Salmonella* surface antigens, although serological tests have been applied extensively only in the diagnosis of infection with *Salmonella* Typhi. The Widal agglutination test, formerly used for the detection of specific O, H and Vi antigens, has been largely replaced by sensitive and specific methods such as enzyme-linked immunosorbent assay (ELISA) and immunoblotting. The results of serology should be considered in the light of antibodies to the somatic and flagellar antigens of *Salmonella* Typhi in the healthy population and knowledge of previous vaccination or a history of previous infection.

Cross-reacting antibodies from previous exposure to other salmonellae may confuse the results of serodiagnosis. The O antigens of *Salmonella* Typhi are expressed by other serotypes such as Enteritidis (see Table 24.1). Antibodies specific for group 'd' flagellar antigens can be used to differentiate infections caused by *Salmonella* Typhi from other serotypes that share O antigens.

Use of serology in the search for typhoid carriers, for example in the routine examination of food handlers and waterworks employees, is of doubtful value.

Food poisoning

The laboratory diagnosis of bacterial food poisoning depends on isolation of the causal organism from samples of faeces or suspected foodstuffs. The more common food-poisoning serotypes, such as Enteritidis or Typhimurium, may be characterized more fully by phage typing and antibiotic resistance typing (see above). Strains can be differentiated further by plasmid and pulsed-field gel electrophoresis typing so that the isolates from patients may be matched with those from the infected food and from a suspected animal source.

TREATMENT

Enteric fever

The introduction of chloramphenicol in 1948 transformed a life-threatening illness of several weeks' duration associated with a mortality rate of more than 20% into a short-lasting febrile illness with a mortality rate of less than 2%. Many patients can be treated adequately with oral chloramphenicol from the outset. Initial intravenous therapy with the drug may be necessary for the more severely ill patient, who may have anorexia, abdominal distension and, perhaps, vomiting. The intramuscular route gives inadequate blood levels. Treatment should be maintained for 14 days because relapse is more frequent with a shorter course.

The problem of bone marrow toxicity and the emergence of plasmid-mediated chloramphenicol resistance in many parts of the world prompted the search for alternative agents. Among these, amoxicillin and co-trimoxazole are as effective as chloramphenicol, and are used widely. Since 1989, however, simultaneous resistance to choramphenicol, ampicillin and trimethoprim has become increasingly common in strains of *Salmonella* Typhi in several endemic areas, and imported multiresistant typhoid is being encountered worldwide. Ciprofloxacin has emerged as the drug of choice for the treatment of adult typhoid, and is proving equally effective and free from side effects in children.

Gastro-enteritis

Management of salmonella gastro-enteritis includes replacement of fluids and electrolytes, and control of nausea, vomiting and pain. Drugs to control the hypermotility of the gut are contra-indicated; they may give symptomatic relief for a while, but it is easy to transform a trivial gastro-enteritis into a life-threatening bacteraemia by paralysing the bowel.

Antibiotics have no part to play in the management in most cases. Randomized, placebo-controlled, double-blind studies have failed to show any benefit from any antibiotic on the duration and severity of the diarrhoea or the duration of fever; some antibiotics seemed to prolong the carrier state. If a patient is clearly at increased risk of bacteraemia and generalized invasion, an antibiotic may protect against this serious complication. Such patients include infants under 3 months of age, patients with a malignancy, haemoglobinopathy or chronic gastro-intestinal disease such as ulcerative colitis (especially when treated with steroids), and patients who are immunosuppressed for other reasons. Treatment with an appropriate agent (see above) should continue until the gastro-enteritis has resolved completely.

Salmonella bacteraemia

Established salmonella bacteraemia requires aggressive antimicrobial treatment with ciprofloxacin, chloramphenicol, co-trimoxazole or high-dose ampicillin. Uncomplicated bacteraemia should be treated for 10–14 days. A careful search for focal metastatic disease should be undertaken, especially when relapse follows cessation of treatment. Surgical drainage of metastatic abscesses may be required, with surgical intervention if heart valves or large vessels are affected.

In salmonella meningitis in infancy, treatment with chloramphenicol or ampicillin may be unsuccessful in up to a third of cases caused by sensitive strains. As cefotaxime and ceftriaxone penetrate into the cerebrospinal fluid reasonably well and are highly active against most salmonellae, they offer an effective alternative to the more conventional agents.

Resistance to any of the drugs used to treat invasive infection may occur, so treatment should be supported by susceptibility testing whenever possible.

Chronic asymptomatic carriers

The chronic carrier state presents a particularly difficult therapeutic challenge. The principal site of carriage is the biliary tract, and concomitant biliary disease has significant implications for therapy. When the patient has chronic cholecystitis or gallstones, antibiotics alone are most unlikely to eradicate the infection. Cholecystectomy together with appropriate antibiotic treatment results in cure in about 90% of cases, but has significant risk, not least from metastatic infection from dissemination of the organisms during surgery.

In the absence of biliary disease, prolonged courses of ampicillin, amoxicillin, co-trimoxazole or ciprofloxacin may cure up to 80% of carriers. However, it is difficult to justify even moderately heroic efforts to cure a condition that has little if any ill effect on the individual and not much direct public health importance. The chronic human carrier is the principal reservoir of enteric fever salmonellae, but even in the developing world direct person-to-person spread by asymptomatic carriers is uncommon. 'Typhoid Mary', who is reputed to have caused many infections by her cooking, was an exception and must have had peculiar personal habits. Normal personal hygiene, adequate sanitation and a reliable supply of potable water are the real safeguards against enteric fever. Prolonged carriage of other *S. enterica* serotypes is of even less public health importance, and rarely if ever justifies exclusion from any employment, or intrusive efforts to eradicate the infection.

EPIDEMIOLOGY AND CONTROL

The typhoid and paratyphoid bacilli are essentially human parasites. Human beings are the reservoir host and most infections can be traced to a human source, or at least to a source of human sewage. All other salmonellae have animal hosts.

Enteric fever

Incidence

In developing countries *Salmonella* Typhi is common, but data are incomplete. It has been estimated that between 10 and 500 cases of typhoid per 100 000 population occur annually throughout the developing world. The almost universal introduction of chlorination of public water supplies, and improved techniques for the detection of cases and carriers, have markedly reduced the prevalence of enteric fever elsewhere.

A total of 203 cases were reported in England and Wales in 2003, an incidence similar to that reported from other parts of the developed world of about 0.2 cases per 100 000 population. Most contract their infection by travel to endemic areas or close family contact with excreters. Paratyphoid infections show a similar pattern.

Sanitation

The control of enteric fever is in theory straightforward. Cases occur from the ingestion of food or water contaminated with human sewage carrying typhoid or paratyphoid bacilli. Outbreaks occur when numbers of people are infected from a primary source. This may be a single meal or a briefly contaminated water supply, or may be a water supply or source of food contaminated and available for ingestion over a longer period. Secondary transmission from patients infected by the primary source is rare.

Enteric fever is a public health problem only where the wide public availability of wholesome drinking water and the provision of adequate means for the proper disposal of human excreta do not exist. In these circumstances the organisms can be spread widely from carriers and from convalescent and sick persons, possibly helped by cultural factors such as food habits, occupation and personal behaviour. In such communities the selection of control measures may be difficult, and provision of adequate sanitation and pure water may need to be supplemented by prophylactic immunization.

Vaccination

Heat-killed, phenol-preserved whole-cell vaccines containing a mixture of cultures of Typhi, Paratyphi A and Paratyphi B (TAB) have been used for many years in countries with a high endemic level of typhoid fever. Such preparations confer considerable, although not absolute, protection against typhoid for about 3 years. There has always been doubt about the value of the paratyphoid components, as well as concern about the unpleasant side effects associated with the large antigenic load of the triple vaccine, so that monovalent typhoid vaccines are now preferred. Capsular (Vi) polysaccharide vaccines have largely replaced whole-cell vaccines. Alternatively, an oral live-attenuated typhoid vaccine is used.

Travellers to endemic areas in which there are high carriage rates and poor standards of hygiene should be offered immunization, especially if they intend to visit rural areas or to live 'rough'. However, the risk to the air traveller with full board at a reputable hotel is so small that typhoid immunization may be unnecessary. The risk of infection in southern Europe does not justify recommending the vaccine to the many millions travelling there on holiday every year, but vaccination is recommended for travel to Eastern Europe, especially for backpacking holidays outside the main tourist areas.

Salmonella food poisoning

No other zoonosis is as complex in its epidemiology and control as salmonellosis. The biology of different *S. enterica* serotypes varies widely and epidemiological patterns differ greatly between geographical areas, depending on climate, population density, land use, farming practices, food-processing technologies and consumer habits.

Sources

The carcasses or products (cooked meats, eggs, milk) of naturally infected domestic animals are the commonest sources of food poisoning. Flesh may be infected when an ill, septicaemic animal is slaughtered, but in most cases the salmonellae important in human infection cause only mild or inapparent infection in their animal hosts, and abattoir and shop cross-contamination with intestinal contents from a carrier animal is a more important hazard. Poultry (particularly hens), ducks and turkeys are the most significant reservoirs of food poisoning salmonellae in the UK. Pigs share the honours with poultry in much of northern Europe, whereas beef cattle are important sources in the USA. Duck eggs have always been a problem, as they can be infected in the oviduct before the egg is shelled. Hen eggs acquire their shells higher in the oviduct and are less commonly infected in this way, although they can be contaminated on the outside if laid on soil contaminated by infected hen faeces.

Food contamination

Rats and mice are commonly infected with food-poisoning salmonellae and may contaminate human food with their faeces. Food poisoning may occasionally be caused by food contaminated by a human case or carrier. Thus, even if the foodstuff is initially free from salmonellae, the chance of contamination 'from the hoof to the home' is high, and the more sophisticated the manipulation of the food, the greater the chance of contamination. For example, one egg containing salmonellae, if eaten by an individual, probably will not give rise to an infection. On the other hand, if such an egg is pooled with others free from salmonellae, as in preparing a mayonnaise or a Hollandaise sauce for communal consumption, and if conditions of temperature and time allow multiplication of the salmonellae, there will be the potential for an outbreak of infection among those who eat the contaminated food.

A clean carcass can be contaminated at the abattoir by instruments or by hanging in contact with an infected carcass in the chilling hall or during transportation to the wholesale and retail butchers' premises. The aggregation of calves or pigs in holding pens greatly increases the occurrence of cross-infection before slaughter. Infection in pigs is most often due to the feeding of swill containing infected animal matter, a practice that is difficult to prevent.

There is a close correspondence between the types of salmonellae prevalent in animals, especially pigs and poultry, and the types causing human infection. Strains of *Salmonella* Typhimurium, which is a primary pathogen in a wide variety of animals, are common, and *Salmonella* Enteritidis has become prevalent in the UK in poultry flocks. Strains belonging to serotypes such as Heidelberg, Brandenburg, Panama and Virchow are rarely found in animals and yet may be responsible for widespread human infections. More needs to be learned about the epidemiology of such infections.

Food preparation

Infection of food with salmonellae is not in itself sufficient to cause food poisoning. It is necessary for the infected food to be moist and to be held long enough under conditions that will allow the bacteria to grow, for instance overnight in a warm kitchen or several days in a cool larder. If the food is then eaten without further cooking, infection may follow. Although cooking of liquid foods will render them safe, cooking of solid foods often fails to do so because of the relatively poor rate of heat penetration into the food. A cold or chilled joint of meat, a poultry carcass or a large meat pie may be heated in an oven until the surface is well cooked, while the central part is still insufficiently heated to destroy vegetative bacteria.

Outbreaks

Food-poisoning incidents occur most dramatically as explosive outbreaks among members of a community sharing communal meals, as in factories, hospitals or schools or at a celebratory feast, although sporadic incidents affecting a single family or a single person are much more common and much more difficult to track to a source. In the UK, salmonellae account for about 75% of the incidents of food poisoning in which a causal agent is identified. However, this may only reflect the fact that salmonellosis is relatively easy to confirm bacteriologically and is therefore more readily recognized than food poisoning caused by other agents, which often escapes diagnosis.

Surveillance

Efficient surveillance of diseases caused by *Salmonella* has been facilitated by accurate strain discrimination achieved with serotyping, phage typing, antimicrobial resistance typing and pulsed-field gel electrophoresis. A European network of laboratories, Enter-Net (http://www.hpa.org.uk/hpa/inter/enter-net_background.htm) has been established to enable rapid identification of outbreaks.

Prevention

The principles of the prevention of salmonella food poisoning are:

- raising of animals free from infection
- elimination of contamination by rodents at all levels of food production
- prevention of contamination by human handlers at the wholesale, retail and hotel levels.

The barriers of economic husbandry, out-of-date premises and, perhaps the most important of all, the need for continuing education of food handlers at all levels of production make implementation of these principles difficult. To reduce the incidence of food poisoning, whether due to salmonellae or to other bacteria, two basic precepts must be observed:

1. Raw foodstuffs of animal origin, which are always potentially contaminated, must never have direct or indirect contact with cooked foods.

2. Foodstuff thought to be contaminated should be treated or held under temperature conditions that prevent the organisms from growing.

Cooked foods should be served and eaten immediately after cooking and while still hot, or cooled rapidly and held at refrigerator temperature until eaten, so that at least the inoculum eaten by any individual is small. It is often found that the food incriminated in an outbreak had been cooked several hours or even a day or two before and then left at room temperature before being reheated immediately before serving. This procedure ensures that salmonellae that survived the initial cooking or gained access to the food from contaminated kitchen surfaces or implements had excellent opportunities to multiply in the interval before being warmed up for consumption.

As in most other endeavours to control the spread of infection, the human element is the weakest link in the chain, so that health education, particularly of food handlers, is a most important and continuing requirement.

KEY POINTS

- There are more than 2000 different antigenic types of *Salmonella*; those pathogenic to man are serotypes of *Salmonella enterica*.
- Most serotypes of *S. enterica* cause food-borne gastro-enteritis and have animal reservoirs.
- *S. enterica* serotypes Typhi and Paratyphi cause typhoid fever.
- Typhoid and other serious systemic salmonella infections are treated with amoxicillin, co-trimoxazole, ciprofloxacin or chloramphenicol.
- Antibiotics have no place in the management of salmonella gastro-enteritis unless invasive complications are suspected.
- Clean water, sanitation and hygienic handling of foodstuffs are the keys to prevention.

RECOMMENDED READING

Bell C, Kyriakides A 2002 *Salmonella: A Practical Approach to the Organism and its Control in Foods.* Blackwell Science, London
Chart H, Cheesbrough J S, Waghorn D J 2000 The serodiagnosis of infection with *Salmonella typhi. Journal of Clinical Pathology* 53: 851–853
House D, Bishop A, Parry C, Dougan G, Wain J 2001 Typhoid fever: pathogenesis and disease. *Current Opinion in Infectious Diseases* 14: 573–578
Threlfall E J 2002 Antimicrobial drug resistance in *Salmonella*: problems and perspectives in food- and water-borne infections. *FEMS Microbiology Reviews* 26: 141–148
Zhang S, Kingsley R A, Santos R L et al 2003 Molecular pathogenesis of *Salmonella enterica* serotype typhimurium-induced diarrhea. *Infection and Immunity* 71: 1–12

Internet sites

Health Protection Agency. Salmonella. http://www.hpa.org.uk/infections/topics_az/salmonella/menu.htm
Rijksinstituut voor Volksgezondheid en Milieu (on behalf of the European Commission). Community Reference Laboratory for Salmonella. http://www.rivm.nl/crlsalmonella/
Centers for Disease Control and Prevention. *Salmonella* Infection (Salmonellosis). http://www.cdc.gov/ncidod/diseases/submenus/sub_salmonella.htm

25 Shigella

Bacillary dysentery

H. Chart

Dysentery, the bloody flux of biblical times, is a clinical entity characterized by the frequent passage of blood-stained mucopurulent stools. Aetiologically it is divisible into two main categories, amoebic and bacillary. Both forms are endemic in most countries with a warm climate. Bacillary dysentery, caused by members of the genus *Shigella*, is also prevalent in many countries with temperate climates.

DESCRIPTION

The genus *Shigella* is subdivided on biochemical and serological grounds into four species:

- *Shigella dysenteriae* – serogroup A
- *Sh. flexneri* – serogroup B
- *Sh. boydii* – serogroup C
- *Sh. sonnei* – serogroup D.

Strains of *Shigella* spp. are typical members of the Enterobacteriaceae and are closely related to the genus *Escherichia*. Studies focusing on the ancestral lineage of these bacteria suggest that *Sh. boydii* is related only distantly to other members of the genus *Shigella*, which are more closely related to enteroinvasive serotypes of *Esch. coli* (see p. 280).

Microscopically, in stained preparations, shigellae are Gram-negative bacilli indistinguishable from other enterobacteria. They are non-motile and non-capsulate. Culturally they are similar to most other enterobacteria. The distinction between strains of *Shigella* spp. and *Esch. coli* depends on a limited number of diagnostic tests including motility, production of lysine decarboxylase and the utilization of citrate. *Sh. dysenteriae* type 1 produces Shiga toxin and this differentiates it from other members of the genus.

The antigenic structure is complex. There are 13 serotypes of *Sh. dysenteriae* of which type 1 is much the most important as a cause of severe bacillary dysentery. *Sh. boydii* can be subdivided into 18 specific serotypes. The six serotypes of *Sh. flexneri* can each be further subdivided

into subtypes. Strains of *Sh. sonnei* are serologically homogeneous, and a variety of other markers such as the ability to produce specific colicines, the carriage of drug resistance or other plasmids, or lysogeny by a panel of bacteriophages are used to discriminate between strains for epidemiological purposes.

CLINICAL FEATURES

The incubation period is usually between 2 and 3 days, but may be as long as 8 days. The onset of symptoms is usually sudden, and frequently the initial symptom is abdominal colic. This is followed by the onset of watery diarrhoea, and in all but the mildest cases this is accompanied by fever, headache, malaise and anorexia. Many episodes resolve at this point, but others progress to abdominal cramps, tenesmus and the frequent passage of small volumes of stool, predominantly consisting of bloody mucus. The symptoms typically last about 4 days, but may continue for 14 days or more. Infection may affect the nervous system resulting in seizures and encephalitis. Shigellosis is occasionally associated with the development of Reiter's syndrome (reactive arthritis).

The severity of the clinical illness is to some extent associated with the species involved. Infection with *Sh. dysenteriae* 1 is usually associated with a severe illness in which prostration is marked and, in young children, may be accompanied by febrile convulsions. Members of the *Sh. flexneri* and *Sh. boydii* groups may also cause severe illness. In contrast, dysentery associated with *Sh. sonnei* (*Sonnei dysentery*) in an otherwise healthy person may be confined to the passage of a few loose stools with vague abdominal discomfort, and the patient often continues at school or work.

Strains of *Sh. dysenteriae* 1 have been responsible for many cases of the haemolytic uraemic syndrome that accompanies outbreaks of dysentery in several countries. The condition, with its triad of haemolytic anaemia, thrombocytopenia and acute renal failure, can be caused by many pathogens (in particular, *Esch. coli* O157; see

p. 280). It is associated with complement activation and disseminated intravascular coagulation, and in some parts of the world is one of the commonest forms of acute renal failure in children.

Death from bacillary dysentery is uncommon in the developed world; it occurs mostly at the extremes of life or in individuals who are suffering from some other disease or debilitating condition.

PATHOGENESIS

Shigella spp. are pathogens of man and other primates, and the pathogenesis of infection with these bacteria and entero-invasive *Esch. coli* (EIEC; see p. 280) is very similar. The infective dose is small: bacillary dysentery may follow the ingestion of as few as ten viable bacteria. The site of infection is the M cells in the Peyer's patches of the large intestine. Strains of *Shigella* spp. are non-motile and it is not known how the bacteria reach and adhere to M cells.

Association with the intestinal mucosa initiates mucosal inflammation leading to apoptosis, which is thought to facilitate the invasion of the M cells, after which the bacteria are phagocytosed. The shigellae multiply within the epithelial cells and spread laterally into adjacent cells and deep into the lamina propria. The infected epithelial cells are killed, and the lamina propria and submucosa develop an inflammatory reaction with capillary thrombosis. Patches of necrotic epithelium are sloughed and ulcers form. The cellular response is mainly by polymorphonuclear leucocytes, which can be seen readily on microscopic examination of the stool, together with red cells and sloughed epithelium.

Dysentery bacilli rarely invade other tissues. Transient bacteraemia can occur but septicaemia with metastatic infection is rare.

Pathogenic mechanisms

In common with most bacteria, shigellae require a range of pathogenic mechanisms to cause disease. They are tolerant to the conditions of low pH encountered in the human stomach and the action of bile.

Pathogenic strains of shigellae, like entero-invasive *Esch. coli*, carry a plasmid of 100 to 140 MDa, which encodes the pathogenic mechanisms involved with eukaryotic cell invasion. Expression of the mechanisms encoded on the virulence plasmid are thermoregulated such that strains become invasive when growing at 37°C but not at 30°C, and regulation is by both plasmid and chromosomally located elements. Plasmid-encoded proteins are required for bacteria to break free from cellular endosomes and for the migration between epithelial cells.

Long-chain lipopolysaccharide plays a role in virulence by preventing the effects of serum complement. The lipid A component has been implicated in causing localized cytokine release, and the resultant inflammatory response and cellular disruption enable these bacteria to enter intestinal cells. *Sh. flexneri* and *Sh. sonnei* express an aerobactin-mediated high-affinity iron uptake system; however, as *Sh. dysenteriae* 1 does not express this siderophore, the role of aerobactin in the pathogenesis of the disease is unclear.

Shiga toxin

Sh. dysenteriae 1 produces a potent protein toxin (*Shiga toxin*) very similar to Verocytotoxin (VT)-1 expressed by strains of Verocytotoxigenic *Esch. coli* (VTEC; see p. 280); however, in contrast to VTEC, the genes encoding Shiga toxin are located on the chromosome. Expression of Shiga toxin has been shown to be iron regulated, with toxin production increasing under conditions of iron restriction.

Shiga toxin is a subunit toxin comprising an A portion and five B subunits. The A subunit possesses the biological activities of the toxin, and the B subunits mediate specific binding and receptor-mediated uptake. In common with the Verocytotoxins of *Esch. coli*, Shiga toxin binds to globotriosylceramide (Gb_3) molecules present on the surface of certain eukaryotic cells. During pathogenesis, release of the inflammatory mediators tumour necrosis factor and interleukin-1 increases the number of Gb_3 receptors on the surface of eukaryotic cells, enhancing the binding of toxin to these cells.

Like Verocytotoxin, Shiga toxin becomes internalized by host cells and remains active within endosomes, eventually reaching the Golgi apparatus. Within the host cell the A subunit divides to form portions A_1 and A_2; the A_1 portion of the toxin prevents protein synthesis and causes cell death. Haemolytic uraemic syndrome is thought to be caused by the action of Shiga toxin on kidney tissues; however, Shiga toxin has also been shown to have neurotoxic properties and the role of this toxin in the pathogenesis of bacillary dysentery remains to be elucidated fully.

LABORATORY DIAGNOSIS

A specimen of faeces is always preferable to a rectal swab. Rectal swabs do not allow adequate macroscopic and microscopic examination of the stool, and unless taken properly and bearing obvious faecal material may be no more than a swab of peri-anal skin. Moreover, because of drying of the swab, pathogenic species die quite rapidly, and may not survive transport to the laboratory.

The faeces are inoculated on desoxycholate citrate agar or MacConkey agar. Mucus, if present in the specimen, may be used as the inoculum. After overnight incubation, pale non-lactose-fermenting colonies are tested by standard biochemical and sugar utilization tests to differentiate them from other enterobacteria. Identity is confirmed by agglutination tests with species-specific antisera, and then with type-specific sera unless the strain is *Sh. sonnei*.

A multiplex polymerase chain reaction has been developed to detect genes encoding the invasion plasmid antigen and an aerobactin-mediated iron uptake system located on the virulence plasmid. The test aids the differentiation of species of *Shigella* from entero-invasive *Esch. coli*. Plasmid pattern analysis and colicine typing can also be used to characterize strains of *Sh. sonnei*.

Shiga toxin can be detected by Vero and HeLa cell tests (see p. 281) and immunoassays designed for the Verocytotoxin produced by certain *Esch. coli* strains.

Patients infected with *Sh. dysenteriae* 1 produce serum and salivary antibodies to the lipopolysaccharide antigens, but tests for the antibodies are not available routinely.

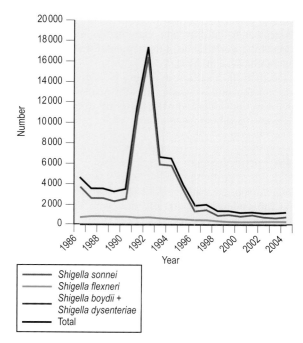

Fig. 25.1 Faecal isolates of *Sh. sonnei, Sh. flexneri, Sh. boydii* and *Sh. dysenteriae* in England and Wales 1986–2004. (Health Protection Agency data.)

TREATMENT

Most cases of shigella dysentery, especially those due to *Sh. sonnei*, are mild and do not require antibiotic therapy. Maintaining good nutrition is essential. Symptomatic treatment with the maintenance of hydration by use of oral rehydration salt solution (see Table 30.1, p. 312) is all that is required. As with salmonella infections, drugs that impair gut motility should be avoided.

Treatment with a suitable antibiotic is necessary in the very young, the aged or the debilitated, and in those with severe infections. Oral ampicillin, co-trimoxazole or ciprofloxacin is appropriate, provided the drug is shown to be active in vitro. There is no evidence that antibiotics reduce the period of excretion of the organisms, and they should not be used in the asymptomatic person, either prophylactically or in attempts to hasten clearance after recovery.

EPIDEMIOLOGY

During the twentieth century, infections due to *Sh. dysenteriae* 1 and *Sh. boydii* declined and strains of *Sh. sonnei* became dominant in the UK and other European countries. In England and Wales more than 16 000 cases of Sonnei dysentery were recorded in 1992. The incidence then declined steadily to an annual average of fewer than 1000 notified cases between 1998 and 2004 (Fig. 25.1). Infections caused by *Sh. flexneri* have also steadily

declined to around 250 cases/year, but are still more common than those caused by *Sh. boydii* (50–100 cases/year) or *Sh. dysenteriae* (40–50 cases/year), most of which are contracted abroad.

Globally, shigellosis remains a major problem particularly in tropical areas of the developing world, where shigellosis is endemic. It has been estimated that some 5 million cases require hospital treatment and about 600 000 die every year. Young children are particularly vulnerable.

Sources and spread

Bacillary dysentery is highly contagious and is usually spread by the faecal–oral route. The case or carrier, after contaminating his or her hands while cleansing at toilet, may contaminate the lavatory flush-handle, door knobs, washbasin taps, hand towels and other objects that, when handled by another individual, allow transfer of the dysentery bacteria to the recipient's hands and to the mouth. The carrier may also infect bedding and, in the case of young children, may contaminate toys. Dysentery bacilli are also liberated into the air in an aerosol when an infected loose stool is flushed from the toilet and, after settling on the surfaces of toilet seats, furniture and surroundings, may survive for some days in a moist atmosphere. Infection can also occur among those who indulge in sexual practices involving anal–oral contact. Contamination of foods, particularly those that are to be consumed raw, are a major source of infection.

An important feature of the epidemiology of bacillary dysentery in the UK and other countries with good environmental sanitation is that the main patient group involved is school-aged children, particularly primary school children. Neglect of toilet hygiene by children at school undoubtedly plays a part. The disease can be endemic among adults living in residential institutions where high standards of hygiene are difficult to maintain, and has in the past been a scourge in gaols and in armies in the field. The seasonal distribution of bacillary dysentery in the UK is bimodal, with the highest incidence in spring and a second peak in October and November. The incidence is at its lowest in summer, when school children are on holiday. As this is the period when flying insects are most abundant, this suggests that in the UK insects play little, if any, part in transmission. In certain settings, where disposal of human faeces is inadequate, flies may also serve as vectors for the spread of *Shigella* bacteria.

Occasional epidemics of bacillary dysentery have been traced to water supplies when chlorination of the supply has not been instituted or has been defective. Such waterborne epidemics are usually spectacular in the large numbers of people simultaneously infected and in the speed with which they can be terminated when the water supply is adequately treated. Epidemic infection may also follow the contamination of milk or ice cream.

CONTROL

The mild and often fleeting nature of the clinical illness associated with *Sh. sonnei* infection means that frequently the patient with bacillary dysentery remains ambulant and follows his or her daily labour and leisure pursuits, remaining in circulation as a disperser of the causal organism. The pressure on toilet facilities, particularly in schools, allows hand-to-mouth spread of the bacilli. The provision of washbasins in the same compartment as the toilet pedestal would allow some reduction in spread, especially if flushing mechanisms and washbasin taps could be operated by foot instead of by hand.

During diarrhoea, faecal soiling of the fingers can be heavy, and hand washing, although very important, will at best only reduce the numbers of bacteria present. The normal disinfecting effect of the skin fatty acids and competing skin organisms may take up to half an hour to destroy the rest. People with dysentery should stay off work and, as far as possible, out of circulation until the symptoms have subsided, especially when their work involves the preparation of food or direct contact with other people. Asymptomatic carriers are far less important in the spread of this disease and seldom, if ever, need to be excluded from any employment.

Control of an outbreak

Outbreaks of dysentery in schools and other institutions are notoriously difficult to control. In nursery school outbreaks, infection is usually widespread before the first cases have been notified, with considerable environmental contamination. Some children will be incubating the infection; others will have recovered from the diarrhoea but still be excreting the organisms. In this situation the acute case is far more important in the spread of infection than the symptomless excreter. There is little to be gained by trying to ascertain bacteriologically who is infected and who is not once the cause of the outbreak has been established, as there is little reason to exclude an asymptomatic carrier and it is pointless to seek confirmation of a clearly symptomatic case. The practical course is usually to keep children away from school.

Having determined the exclusion policy it is important to try to stop hand-to-hand spread among those who remain at school. Supervision of children using the lavatory, supervised hand washing before meals, frequent disinfection of toilets, including seats, lavatory chain and door handles, and the general use of paper towels all play a part.

Similar principles can be applied to outbreaks in residential institutions or a hospital ward. The most important single factor in all these situations is the need for adequate communication. Teachers, nurses, parents and all who are involved in trying to control the outbreak need to have explained to them exactly how the infection spreads, and the reasons for the measures taken or not taken. Finally, the temptation to use antibiotics prophylactically in an attempt to limit the spread must be resisted.

KEY POINTS

- *Shigella* species cause bacillary dysentery.
- The infective dose is very small.
- *Sh. dysenteriae* type 1 produces a toxin that resembles the Verocytotoxin of certain strains of *Esch. coli* and is responsible for the most serious forms of shigellosis.
- *Sh. flexneri*, *Sh. boydii* and *Sh. sonnei* cause enteric disease of varying severity.
- Sonnei dysentery is the most prevalent form of shigellosis in developed countries.
- Most cases of shigellosis do not require antibiotics. Treatment with ciprofloxacin is indicated in severe cases. Ampicillin, tetracyclines and trimethoprim are suitable alternatives, but resistance is common.

RECOMMENDED READING

Ashkenasi S 2004 Shigella infections in children: new insights. *Seminars in Pediatric Infectious Diseases* 15: 246–252

Bennish M L, Harris J R, Wojtyniak B J, Struelens M 1990 Death in shigellosis: incidence and risk factors in hospitalized patients. *Journal of Infectious Diseases* 161: 500–506

Dorman C J, Porter M E 1998 The *Shigella* virulence gene regulatory cascade: a paradigm of bacterial gene control mechanisms. *Molecular Microbiology* 29: 677–684

Hale T L, Formal S B 1987 Pathogenesis of shigella infections. *Pathology and Immunopathology Research* 6: 117–127

Kingcombe C I, Cerqueira-Compos M L, Farber J M 2005 Molecular strategies for the detection, identification and differentiation between enteroinvasive *Escherichia coli* and *Shigella* spp. *Journal of Food Protection* 68: 239–245

Lopez E L, Prado-Jimenez V, O'Ryan-Gallardo M, Contrini M M 2000 Shigella and Shiga toxin-producing *Escherichia coli* causing bloody diarrhea in Latin America. *Infectious Disease Clinics of North America* 14: 41–65

Nhieu G T, Sansonetti P J 1999 Mechanism of *Shigella* entry into epithelial cells. *Current Opinion in Microbiology* 2: 51–55

Niyogi S K 2005 Shigellosis. *Journal of Microbiology* 43: 133–143

Sansonetti P J 1998 Molecular and cellular mechanisms of invasion of the intestinal barrier by enteric pathogens: the paradigm of *Shigella*. *Folia Microbiologica* 43: 239–246

Sasakawa C 1995 Molecular basis of pathogenicity of *Shigella*. *Reviews in Medical Microbiology* 6: 257–266

Internet sites

Virtual Museum of Bacteria. Shigella. http://www.bacteriamuseum.org/species/shigella.shtml

Centers for Disease Control and Prevention. Shigellosis. http://www.cdc.gov/ncidod/dbmd/diseaseinfo/shigellosis_g.htm

Health Protection Agency. Shigella. http://www.hpa.org.uk/infections/topics_az/shigella/menu.htm

World Health Organization. Shigella. http://www.who.int/topics/shigella/en/

26 Escherichia

Urinary tract infection; travellers' diarrhoea; haemorrhagic colitis; haemolytic uraemic syndrome

H. Chart

Strains of *Escherichia coli* and related Gram-negative 'coliform' bacteria predominate among the aerobic commensal flora in the gut of human beings and animals. These bacteria are present wherever there is faecal contamination, a phenomenon that is exploited by public health microbiologists as an indicator of faecal pollution of water sources, drinking water and food. The species encompasses a variety of strains, which may be purely commensal or possess combinations of pathogenic mechanisms that enable them to cause disease in man and other animals.

DESCRIPTION

Strains of *Esch. coli* are usually motile and often fimbriate (Fig. 26.1). Some, especially those from extra-intestinal infections, may produce a polysaccharide capsule. They grow well on non-selective media and usually (with the exception of certain Verocytotoxin-producing strains) ferment lactose, producing large red colonies on MacConkey agar. They grow over a wide range of temperature (15–45°C); some strains are more heat resistant than other members of the Enterobacteriaceae and may survive 60°C for 15 min or 55°C for 60 min. Certain strains are haemolytic when grown on media containing suitable erythrocytes.

Esch. coli can be differentiated from other enteric Gram-negative bacteria by the ability to utilize certain sugars and by a range of other biochemical reactions. Many characteristic biochemical reactions, such as indole production and the formation of acid and gas from lactose and other carbohydrates, take place at 44°C as well as at 37°C.

The term *coliform* is commonly used to refer to any member of the Enterobacteriaceae, although many workers, especially those in fields related to public health, prefer to reserve it to describe genera or species of enterobacteria that normally ferment lactose, using the term *non-lactose fermenter* (NLF) to describe the other bacteria. Neither term has any taxonomic validity.

DNA–DNA recombination studies show that *Esch. coli* and *Shigella* spp. form a single genetic group, and it is therefore to be expected that intermediate strains will occur. Certain of these, which are non-motile and anaerogenic, and often ferment lactose late or not at all, have caused difficulties in classification. They were formerly included in the so-called 'Alkalescens-Dispar' group, but are now considered to be atypical forms of *Esch. coli*.

Antigenic structure

Serotyping is based on the distribution of lipopolysaccharide (LPS) or somatic (O) antigens, and flagellar (H) and capsular (K) antigens, as detected in agglutination assays with specific rabbit antibodies. More than 180 different O antigens have been described and new 'O' groups continue to emerge. Serotyping detects cross-reactions as a result of shared epitopes on the LPS expressed by strains of *Esch. coli*, and may occur with organisms belonging to the genera *Brucella, Citrobacter, Providencia, Salmonella, Shigella* and *Yersinia*.

More than 50 H antigens have been identified. Most are monophasic, but rare diphasic strains have been reported.

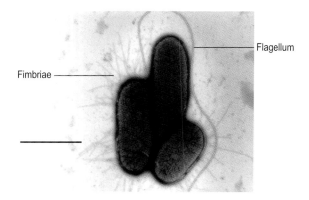

Fig. 26.1 Electron micrograph of *Esch. coli* showing flagellum and fimbriae (arrowed). Bar = 0.5 µm.

Table 26.1 K antigens of *Esch. coli*

Property	Group I	Group II
Molecular weight (Da)	>100 000	<50 000
Acidic component	Hexuronic acid, pyruvate	Glucuronic acid, phosphate, KDO, NeuNAc
Heat stability (100°C, pH 6)	All stable	Mostly labile
O groups	08, 09	Many
Chromosome site	*His*	*SerA*
Expressed at 17–20°C	Yes	No
Electrophoretic mobility	Low	High

KDO, ketodeoxyoctonate; NeuNAc, *N*-acetylneuraminic acid.

There are only a few significant cross-reactions between them and with the H antigens of other members of the Enterobacteriaceae. Because certain strains of *Esch. coli* cease to express flagella during growth in vitro, strains may need to be grown in semi-solid agar (Craigie tubes) to induce flagella expression.

The term 'K antigen' was first used collectively for surface or capsular antigens that prevent flagellar-specific antibodies from binding to the somatic antigens. In the past these antigens were divided into three classes (L, A and B) according to the effect of heat on the agglutinability, antigenicity and antibody-binding power of bacterial strains that express them. In modern usage, 'K antigen' refers to the acidic polysaccharide capsular antigens, and those of *Esch. coli* may be divided into two groups (groups I and II; Table 26.1) that largely correspond to the former A and L antigens.

Fimbrial antigens

Like many other members of the Enterobacteriaceae, *Esch. coli* may produce fimbriae, and strains may express both sex pili and more than one type of fimbrial structure (see p. 156). Within a given culture, there may exist individual cells with fimbriae and others with none, and there is reversible variation between the fimbriate and the non-fimbriate phase.

Type 1 fimbriae can mediate adhesion to a wide range of human and animal cells that contain the sugar mannose. Such adhesion might be involved in pathogenicity and there are some examples of this. Filamentous protein structures resembling fimbriae cause mannose-resistant haemagglutination, and there is good evidence to suggest that they play an important part in the pathogenesis of diarrhoeal disease and in urinary tract infection. They include the K88 antigen found in strains causing enteritis of pigs, the K99 antigen found in strains causing enteritis of calves and lambs, and the colonization factor antigens (CFAs) or coli surface (CS) antigens expressed by enterotoxigenic *Esch. coli* (ETEC) that cause human diarrhoeal disease.

Fimbriae that are of importance in urinary tract infection and cause mannose-resistant haemagglutination are distinguished according to their receptor specificities. These include the P fimbriae that bind specifically to receptors present on the P blood group antigens of human erythrocytes and uroepithelial cells.

PATHOGENESIS

Strains of *Esch. coli* possess a range of different pathogenic mechanisms. The polysaccharides of the O and K antigens protect the organism from the bactericidal effect of complement and phagocytes in the absence of specific antibodies. However, in the presence of antibody to K antigens alone, or to both O and K antigens, opsonization may occur.

Many strains express haemolysin(s), and in general strains of *Esch. coli* isolated from human extra-intestinal infections are more likely to be haemolytic than strains isolated from the faeces of healthy human beings. Haemolysin production is an important pathogenic mechanism for releasing essential ferric ions bound to haemoglobin, and the expression of certain haemolysins has been shown to be regulated by iron.

Strains of *Esch. coli* can express siderophores, such as enterobactin, which readily remove ferric ions from mammalian iron transport proteins such as transferrin and lactoferrin (see Ch. 13). Some strains also express the siderophore, aerobactin; this may be plasmid-mediated. The ability of strains of *Esch. coli* to acquire ferric ions is a recognized pathogenic mechanism. Expression of the aerobactin-mediated iron-uptake system is a common feature of strains isolated from patients with septicaemia, pyelonephritis and lower urinary tract infection. Strains of *Esch. coli* may also utilize siderophores produced by certain species of fungi (e.g. ferrichrome, coprogen, rhodoturulic acid) to acquire iron from environmental sources.

CLINICAL SYNDROMES

Urinary tract and septic infections

Esch. coli is the most common cause of acute, uncomplicated urinary tract infection outside hospitals, as well as causing hospital-associated urinary tract sepsis. These bacteria may also cause neonatal meningitis and septicaemia, sepsis in operation wounds and abscesses in a variety of organs.

As many as 80% of *Esch. coli* strains that cause neonatal meningitis and 40% of those isolated from infants with septicaemia but without meningitis express a K1 antigen. Strains possessing the K1 or the K5 antigen may be more virulent than those with other K antigens, as they share structural identity with host components.

Strains that cause urinary tract infection often originate from the gut of the patient, with infection occurring in an ascending manner. The ability of *Esch. coli* to infect the urinary tract may be associated with fimbriae that specifically mediate adherence to uroepithelial cells.

Epidemiology

Urinary tract infection occurs more frequently in women than in men because the shorter, wider, female urethra appears to be less effective in preventing access of the bacteria to the bladder. Sexual intercourse may be a predisposing factor. The high incidence in pregnant women can be attributed to impairment of urine flow due partly to hormonal changes and partly to pressure on the urinary tract. Other causes of urinary stagnation that may predispose to urinary tract infection include urethral obstruction, urinary stones, congenital malformations and neurological disorders, all of which occur in both sexes. In men, prostatic enlargement is the most common predisposing factor. Catheterization and cystoscopy may introduce bacteria into the bladder and therefore carry a risk of infection.

Most urinary tract infections are thought to be caused by organisms originating from the patient's own faecal flora. However, the prevalence of various serotypes of *Esch. coli* in urinary tract infections varies with geographical location, suggesting that *Esch. coli* causing such infections are specific pathogens for the urinary tract. Pathogenic strains, possibly transmitted in contaminated foods, are able to colonize the bowel, and in individuals with predisposing factors may cause a urinary tract infection. The prevalence of infections due to a particular strain may therefore increase for a time in a locality.

Laboratory diagnosis

Clinical specimens may be stained by Gram's method for microscopical examination, and are cultured on MacConkey agar or other suitable media. In the case of suspected urinary tract infection, culture is semi-quantitative; in acute *Esch. coli* infections the organism is generally present in pure culture at a count of 10^5 or more per millilitre of urine.

Treatment and control

In the absence of acquired resistance, *Esch. coli* is suceptible to many antibacterial agents, including ampicillin, cephalos-

Table 26.2 The major groups of diarrhoea-causing *Esch. coli*

Pathogenic group	Common serogroups
Enteropathogenic *Esch. coli* (EPEC)	O26, O55, O86, O111, O114, O125, O126, O127, O128, O142
Enterotoxigenic *Esch. coli* (ETEC)	O6, O8, O15, O25, O27, O63, O119, O125, O126, O127, O128, O142
Entero-invasive *Esch. coli* (EIEC)	O78, O115, O148, O153, O159, O167
Verocytotoxin-producing *Esch. coli* (VTEC)	O26, O28ac, O111, O112ac, O124, O136, O143, O144, O152, O157[a], O164
Entero-aggregative *Esch. coli* (EAggEC)	More than 50 'O' serogroups

[a]Other serogroups are far less common than O157 in human disease.

porins, tetracyclines, quinolones, aminoglycosides, trimethoprim and sulphonamides. Many strains, however, have acquired plasmids conferring resistance to one or more of these drugs, and antimicrobial therapy should be guided by laboratory tests of sensitivity if possible.

Uncomplicated cystitis usually responds to minimal treatment with oral agents such as trimethoprim or nitrofurantoin, but more serious infections require specific antimicrobial therapy based on laboratory results. In particular, bacterial meningitis is a medical emergency and vigorous early treatment with cefotaxime and gentamicin is required.

Urinary catheterization and cystoscopy require rigorous aseptic technique to minimize the introduction of bacteria into the bladder. Bladder irrigation and systemic treatment with antimicrobial agents has been used in catheter-associated infections, but such treatment is seldom more than palliative and encourages infections with resistant organisms.

Diarrhoea

Although *Esch. coli* is normally carried in the gut as a harmless commensal, it may cause gastro-intestinal disease ranging in severity from mild, self-limiting diarrhoea to haemorrhagic colitis and the associated, potentially life-threatening, *haemolytic uraemic syndrome*. Such strains fall into at least five groups, each associated with specific serotypes (Table 26.2) and with different pathogenic mechanisms:

1. *Enteropathogenic Esch. coli* (EPEC), which cause infantile enteritis, especially in tropical countries
2. *Enterotoxigenic Esch. coli* (ETEC), which are responsible for community-acquired diarrhoeal disease in areas of poor sanitation and are the commonest cause of travellers' diarrhoea

3. *Entero-invasive Esch. coli* (EIEC), which cause an illness resembling shigella dysentery in patients of all ages

4. *Verocytotoxin-producing Esch. coli* (VTEC), which cause symptoms ranging from mild, watery diarrhoea to haemorrhagic colitis and haemolytic uraemic syndrome

5. *Entero-aggregative Esch. coli* (EAggEC), which cause chronic diarrhoeal disease in certain developing countries.

Enteropathogenic Esch. coli (EPEC)

Pathogenesis. EPEC strains were originally identified epidemiologically as a cause of diarrhoeal disease in infants; certain strains belonging to characteristically EPEC serogroups, such as O26 and O111, were later shown to acquire the genes for expression of Verocytotoxin and are therefore classified as VTEC (see below).

Colonization of the upper part of the small intestine occurs in infantile enteritis associated with EPEC. Electron microscopy of intestinal biopsy specimens has shown that the bacteria become intimately associated with the mucosal surface and are partially surrounded by cup-like projections ('pedestals') of the enterocyte surface. In areas of EPEC attachment the brush border microvilli are lost. Adhesion to the gut wall and the subsequent mucosal damage has been termed an 'attaching and effacing' lesion. The genes responsible are located on a pathogenicity island located on the *Esch. coli* chromosome. One of the proteins involved, intimin, appears to be the key adhesin. EPEC use a novel mechanism of adhesion in which the receptor for the adhesin is synthesized by the bacteria and inserted in the host gut wall to provide a binding site for intimin.

Laboratory diagnosis. Stool specimens are plated on media such as MacConkey agar. Bacteria fermenting lactose are identified as *Esch. coli* and serotyped based on somatic and flagellar antigens. Strains belonging to EPEC-associated serogroups may be putative EPEC, but only detailed molecular methods for detecting EPEC-associated genes can provide an accurate identification.

Epidemiology. Since 1971 few epidemics of EPEC enteritis have been reported in the UK or the USA, although a satisfactory explanation for the decline has not been put forward. Strains responsible for sporadic cases that continue to occur in the UK, especially in the summer months, possess the same pathogenic mechanisms as those that caused earlier outbreaks.

EPEC enteritis is common in communities with poor hygiene where sporadic cases and frequent outbreaks occur in the general community as well as in institutions. The importance of EPEC as a cause of enteritis in adults is difficult to evaluate because few laboratories perform the relevant tests.

Table 26.3 Differential properties of heat-labile toxins (LT) of *Esch. coli*

	LT-I	LT-II
Cell changes		
CHO cells	+	+
Y1 cells	+	+
Vero cells	+	+
Molecular weight (kDa)		
A subunit	26	28
B subunit	11.5	11.8
Isoelectric point	8.5	6.8[a], 5.4[b]
Genetic location	Plasmid	Chromosome
Action	Binds to gangliosides	Activates cAMP

[a]LT-IIa; [b]LT-IIb.
CHO, Chinese hamster ovary; cAMP, cyclic adenosine monophosphate.

Enterotoxigenic Esch. coli (ETEC)

Pathogenesis. ETEC produce a heat-stable enterotoxin or a heat-labile enterotoxin, or both. In addition, they usually express fimbriae that are specific for the host animal species and that enable the organisms to adhere to the epithelium of the small intestine. Infection is usually of brief duration, often beginning with the rapid onset of loose stools and accompanied by variable symptoms, including nausea, vomiting and abdominal cramps.

Heat-labile enterotoxin (LT) is closely related to the toxin produced by strains of *Vibrio cholerae*. There are two main forms, termed LT-I and LT-II (Table 26.3). Different forms of LT-I associated with human, porcine and chicken infection have been described; similarly, two forms of LT-II (LT-IIa and LT-IIb) have been detected. Although these toxins have a degree of structural variation, they are all subunit protein toxins comprising one A subunit and five B subunits with molecular weights of 26 000–28 000 and 11 500–11 800 Da, respectively. The mechanism by which diarrhoea is caused is identical to that of cholera toxin (see p. 310).

In contrast to LTs, the heat-stable enterotoxin (STs) of *Esch. coli* (Table 26.4) have a low molecular weight which confers heat stability and poor antigenicity. There are two major classes, designated ST-I (ST$_a$) and ST-II (ST$_b$). Variants of ST-I have been associated with porcine and human infections. ST-I was originally detected by an infant mouse test in which secretion occurs in the intestine within 4 h of intragastric administration. It can now be detected by immunoassay. Alternatively, the genes can be detected by molecular methods. This toxin activates guanylate cyclase activity, resulting in an increase in the level of cyclic guanosine monophosphate (cGMP). The activity of ST-I is rapid, whereas LTs act

Table 26.4 Differential properties of heat-stable toxins (ST) of *Esch. coli*

	ST-I (ST$_a$)	ST-II (ST$_b$)
Molecular weight (kDa)	2	5
Infant mouse test	+	–
Methanol	Soluble	Insoluble
Pig intestinal loop	+	+
Rabbit ileal loop	+	–
Rat gut loop	+	–
Action	Activates cGMP	Unknown (cyclic nucleotides)

cGMP, cyclic guanosine monophosphate.

after a lag period. The mechanism of secretion caused by ST-I, via cGMP, is not fully understood but calcium appears to play a role. ST-I is plasmid encoded, and these plasmids may also encode the genes for LT, adhesive factors and antibiotic resistance.

ST-II is distinguished from ST-I by its biological activity and its insolubility in methanol. It stimulates fluid accumulation in ligated intestinal loops of piglets, but not in the infant mouse test. The mechanism of action is not known but it appears not to act via cyclic adenosine monophosphate (cAMP) or cGMP. Molecular methods have superseded animal models for detecting these organisms.

Enterotoxin alone is not sufficient to enable *Esch. coli* to cause diarrhoea. The organism must first bind to specific receptors on the mucosal surface of the epithelial cells of the small intestine This adhesion is usually mediated by fimbriae expressing *colonization factor antigens* (CFAs) or *coli surface* (CS) antigens. Plasmids

that simultaneously carry genes for both CFAs and enterotoxin production have been described.

The first colonization factor to be recognized in *Esch. coli* was a fimbrial antigen, K88, controlled by a transferable plasmid. Several more have subsequently been discovered in human strains of ETEC and, no doubt, others remain to be discovered. The properties of some of the more important colonization factors are shown in Table 26.5.

Laboratory diagnosis of ETEC

Detection of LT. Traditionally, tissue culture assays with monolayers of mouse adrenal cells (Y1), Chinese hamster ovary (CHO) or African green monkey kidney (Vero) cells were used for detecting LT (see Table 26.3). Toxin present in culture supernates has a cytotonic effect on these cells, producing characteristic changes in cell morphology. This is in contrast to the cytotoxic effect of Verocytotoxin (see below).

Immunological techniques for the detection of LT include enzyme-linked immunosorbent assay (ELISA) and a solid-phase radio-immunoassay (RIA). For the ELISA, plates are coated with ganglioside GM$_1$, which is used to 'capture' LT present in culture supernates. Bound toxin is then detected with toxin-specific rabbit antibodies. A precipitin test (the *Biken* test) performed directly on bacterial colonies growing on a special agar medium may be suitable for use in field laboratories. Rabbit antibodies specific for LT are incorporated into the agar culture medium. As the bacteria grow and secrete LT, the toxin binds to the anti-LT antibodies, forming a precipitin line. Commercial latex particle agglutination tests are also available for screening.

Table 26.5 Properties of some important human colonization factors

Colonization factor	Components	Mannose-resistant haemagglutinin			Fimbrial type	Associated toxin	Associated serogroups
		Human	Bovine	Guinea-pig			
CFA-I		+	+	–	Rod-like	ST or ST/LT	04, 07, 015, 020, 025, 063, 078, 090, 0104, 0110, 0126, 0128, 0136, 0153, 0159
CFA-II	CS1	(+)	+	–	Rod-like	ST/LT	06, 0139
	CS2	–	+	–	Rod-like	ST/LT	06
	CS3	–	+	–	Fibrillar	ST/LT	08, 078, 080, 085, 0115, 0128, 0139, 0168
CFA-III		–	–	–	Rod-like	LT	025
CFA-IV	CS4	+	+	–	Rod-like	ST/LT	025
	CS5	+	+	+	Helical	ST	06, 029, 092, 0114
	CS6	–	–	–	None	ST or LT	0115, 0167, 025, 027, 079, 089, 092, 0148, 0153, 0159, 0169

CS, coli surface antigen; ST, heat-stable toxin; LT heat-labile toxin; (+), weak reaction.

Detection of ST. Many enterotoxigenic strains of *Esch. coli* produce ST alone, and it is essential to include tests for ST production in any survey of enterotoxigenicity. Injection of both LTs and STs into ligated ileal loops of rabbits leads to the accumulation of fluid. The action of ST can be distinguished by its relative stability to heat and by the rapidity of its action.

So far it has proved impossible to devise tissue culture tests for ST, and until recently the most widely used method for detecting ST-I was the infant mouse test. The intestines are removed after the injection of culture supernates, and the ratio of gut weight to remaining body-weight is used as an objective measure of fluid accumulation. ELISA tests with monoclonal antibody specific for ST have largely replaced the infant mouse test.

Molecular methods. The polymerase chain reaction (PCR) and gene probes are currently used for the detection of STs and LTs of ETEC.

Epidemiology. In developing countries ETEC are a major cause of death in children under the age of 5 years. These strains also commonly cause diarrhoea in travellers visiting countries where ETEC are endemic.

The sources and modes of spread of ETEC infection in countries with a warm climate are not well understood, but it seems likely that water contaminated by human or animal sewage plays an important part in the spread of infection.

Entero-invasive Esch. coli (EIEC)

Pathogenesis. EIEC, like shigella (see Ch. 25), cause disease by invading intestinal epithelium. Infection is by ingestion; only a small number of bacteria need to be swallowed as they are relatively resistant to gastric acid and bile, and pass readily into the large intestine where they multiply in the gut lumen. The bacteria pass through the overlying mucous layer, attach to the intestinal epithelial cells and are carried into the cell by endocytosis into an endocytic vacuole, which then lyses. The ability to cause the vacuole to lyse is an important virulence attribute, as organisms unable to do this cannot spread to neighbouring cells. After lysis of the vacuole the bacteria multiply within the epithelial cell and kill it. Spread to neighbouring cells leads to tissue destruction and consequent inflammation, which is the underlying cause of the symptoms of bacillary dysentery.

Pathogenicity depends on both chromosomal and plasmid genes. A large plasmid carries genes for the expression of outer membrane proteins that are required for invasion as well as genes necessary for the insertion of these proteins into the cell membrane. Plasmid genes are also required for the ability to escape from the endocytic vacuole and to invade contiguous host cells. Chromosomal genes encoding pathogenic mechanisms include those required for the expression of long-chain LPS and those encoding an aerobactin-mediated iron-sequestering system.

Laboratory diagnosis of EIEC. The original Sereny test, in which the bacteria are tested for the ability to cause conjunctivitis in guinea-pigs, has been superseded by tissue culture methods that exploit the ability of the bacteria to invade monolayers of HEp-2 or HeLa cells. These tests are in turn being replaced by molecular methods to detect the genes encoding the invasion plasmid antigen (*ipaH*) and aerobactin expression (*iuc*).

Epidemiology. The epidemiology and ecology of EIEC have been poorly studied, but there appears to be no evidence of an animal or environmental reservoir. Surveys suggest that they cause about 5% of all diarrhoeas in areas of poor hygiene. In the UK and USA, outbreaks are occasionally described, especially in schools and hospitals for the mentally handicapped. Infections are usually food-borne but there is also evidence of cross-infection. The most common serogroup is O124.

Verocytotoxigenic Esch. coli (VTEC)

Strains of *Esch. coli* expressing a protein cytotoxic for Vero cells were discovered in 1977. Once epidemiologists were aware of VTEC, the importance of these bacteria in human disease became apparent and a link was established with two diseases of previously unknown aetiology: *haemorrhagic colitis* and *haemolytic uraemic syndrome*. Outbreaks were first recognized in the USA in 1982, and strains of VTEC belonging to serogroup O157 emerged as the major cause. Since then, outbreaks and sporadic cases have been reported in several other countries and VTEC belonging to many other serotypes have been described. Well publicized major episodes in the USA, Canada, Japan and Scotland have heightened general awareness of the importance of this disease.

Owing to the similarity in structure between Verocyto-toxin and Shiga toxin expressed by *Shigella dysenteriae* 1 (see p. 271), VTEC have also been termed Shiga toxin-producing *Esch. coli*. They are also sometimes referred to as enterohaemorrhagic *Esch. coli*.

Pathogenesis. Human VTEC infection can be associated with a range of clinical symptoms from mild, non-bloody diarrhoea to the severe manifestations of haemolytic uraemic syndrome; a wide spectrum of illness can occur even within a single outbreak.

Haemorrhagic colitis is a grossly bloody diarrhoea, usually in the absence of pyrexia. It is usually preceded by abdominal pain and watery diarrhoea. Haemolytic uraemic syndrome is characterized by acute renal failure, micro-angiopathic haemolytic anaemia and thrombocytopenia. It occurs in all age groups, but is more common in infants and young children, and is a major cause of renal failure

Table. 26.6 Differential properties of Verocytotoxins expressed by *Esch. coli*

	VT1	VT2	VT2v[a]
Synonym	SLT1	SLT2	SLT2v
Cytotoxicity			
Vero cells	+	+	+
HeLa cells	+	+	−
Molecular weight (kDa)			
A subunit	32	35	33
B subunit	7.7	10.7	7.5
Genes phage-encoded	+	+	−

[a]Human and porcine variants.

in childhood. The syndrome is usually associated with a prodromal bloody diarrhoea, but an 'atypical' form occurs without a diarrhoeal phase. It may be accompanied by thrombotic thrombocytopenic purpura in which the clinical features are further complicated by neurological involvement and fever.

VTEC have also been implicated as a cause of disease in animals, particularly calves and pigs.

Verocytotoxin (VT). The biological properties, physical characteristics and antigenicity of VT are very similar to those of Shiga toxin, produced by strains of *Sh. dysenteriae* type 1 (see p. 271), but the genes encoding VT in *Esch. coli* are carried on a lamda-like bacteriophage whereas those encoding Shiga toxin in *Sh. dysenteriae* are located on the chromosome. Serological tests have revealed two antigenically distinct forms, termed VT1 and VT2. Antibodies prepared to VT1 neutralize Shiga toxin, whereas antibodies specific for VT2 do not. Variant forms of VT2 have been described in strains of human and porcine origin. The genes controlling production of these variant toxins are not phage encoded, and the toxin receptor also differs from that used by VT1 and VT2 (Table 26.6).

Like Shiga toxin, VT1 and VT2 comprise A and B subunits. For both toxins the A subunit possesses the biological activities of the toxin and the B subunits mediate specific binding and receptor-mediated uptake of the toxin. VT1 and VT2 bind to globotriosylceramide (Gb$_3$) molecules present on the surface of certain eukaryotic cells. In contrast VT2 variant toxins bind to globotetraosylceramide (Gb$_4$). During infection with VTEC the inflammatory mediators, tumour necrosis factor and interleukin-1, in combination with LPS, increase the number of ceramide receptors on the surface of eukaryotic cells, enhancing the binding of VT to these cells. Infection also results in expression of VT molecules that bind to Gb$_3$ receptors located in the kidneys, leading to haemolytic uraemic syndrome. Molecules of Gb$_3$ have

been detected on mammalian erythrocytes that have P^k antigens. About 75% of the human population carries this antigen, and such people may have protection from developing haemolytic uraemic syndrome.

VT1 and VT2, like Shiga toxin, are cytotoxic for Vero and HeLa cells, although certain VT2 variant toxins do not bind to HeLa cells as this particular cell line does not express Gb$_3$ receptors.

VT causes a direct, dose-dependent, cytotoxic effect on human umbilical cord endothelial cells in culture; actively dividing cells are the most sensitive. Micro-angiopathy of the capillaries is a characteristic renal lesion in haemolytic uraemic syndrome, supporting the hypothesis that vascular endothelial cells are primary targets for VT.

Once bound to the eukaryotic cell surface, the holo-toxin becomes internalized by host cells and remains active within endosomes. The toxin eventually reaches the Golgi apparatus by mechanisms as yet unknown. At some point within the host cell, the A subunit becomes enzymically 'nicked' to form portions A$_1$ (28 kDa) and A$_2$ (4 kDa); the A$_1$ portion of the toxin prevents protein synthesis and results in cell death.

Strains of O157 VTEC express an 'attaching and effacing' phenotype and, in common with strains of EPEC, the genes involved are located on a pathogenicity island located on the *Esch. coli* chromosome (see p. 278). Although the mechanisms of bacterial adhesion expressed by O157 VTEC are similar to those of EPEC, antigenic variation in, for example, the respective intimin proteins, has been detected.

Laboratory diagnosis. The proportion of VTEC in the faecal flora may be low, often less than 1%, so that testing of individual colonies from culture plates may not always detect the presence of the pathogen. DNA probes for the genes encoding VT1 and VT2 have been developed, and by using these probes in colony hybridization tests several hundred colonies from each faecal specimen can be examined, giving a considerable increase in sensitivity. PCR tests with VT-specific primers have also been used to detect VTEC.

Although 95% of *Esch. coli* ferment sorbitol, O157 VTEC do so only slowly. This property is exploited by replacing the lactose present in MacConkey agar with sorbitol. Most strains of O157 VTEC produce colourless colonies after overnight incubation and these can be tested with an O157 LPS-specific antiserum in a simple agglutination assay. Identification of putative strains of *Esch. coli* O157 must be confirmed by routine bacteriological examination and serotyping. Toxigenicity is confirmed by gene probes, by PCR, by testing strains for a cytotoxic effect on Vero cells or by a VT-specific ELISA.

Infection with VTEC O157 results in high levels of serum antibodies to the O157 LPS antigens. These antibodies may be detected for several months after infection, providing valuable retrospective evidence of exposure or recovery from acute infection. Polyacrylamide gel electrophoresis and immunoblotting remain the most sensitive means of detecting antibodies to *Esch. coli* O157 LPS, although latex agglutination (available as a commercial kit) can be used as a rapid test. Patients may also produce salivary antibodies to the LPS antigens, but these are present for only a comparatively short period of time.

Epidemiology. Outbreaks of infection with VTEC have occurred in the community, in nursing homes for the elderly and in day care centres for young children. The most severe clinical manifestations are usually seen in the young and the elderly. More than 800 cases a year are reported in the UK each year.

Food is an important source. In several outbreaks of haemorrhagic colitis due to O157 VTEC products as diverse as hamburger meat, cooked meat products, unpasteurized apple juice, unpasteurized milk and radish sprouts have been implicated. VTEC have also been isolated from healthy heifers on farms associated with milk-borne incidents, and it is now generally accepted that cattle are a major reservoir.

Entero-aggregative Esch. coli (EAggEC)

EAggEC are characterized by their ability to adhere to particular laboratory-cultured cells, such as HEp-2, in an aggregative or 'stacked brick' pattern, a property usually encoded on a 60-MDa plasmid. Strains of EAggEC were first reported in 1987 as a cause of chronic diarrhoea in malnourished young children living in Chile, and were reported subsequently in many other countries. A study in England showed that strains of EAggEC were common in the faeces of apparently healthy members of the population, but EAggEC diarrhoea is comparatively rare in industrialized countries. Occasional outbreaks have occurred in Europe, including the UK, and travellers to endemic regions may become infected.

Pathogenesis. The mechanisms by which EAggEC cause diarrhoeal illness are poorly understood. Volunteer studies have failed to identify the infective dose, and the site of adhesion within the human host has not been determined. The characteristic pattern of adhesion to HEp-2 cells may be a putative pathogenic mechanism. Although strains of EAggEC may express fimbriae, adhesion to HEp-2 cells can also occur in the absence of these structures and for certain strains adhesion involves cell surface charge. Some, but not all, strains produce an ST-like toxin. Similarly, some isolates express haemolysins, an aerobactin-mediated high-affinity iron uptake system and a range of haemagglutinins; however, apart from the ability to adhere to HEp-2 cells in the stacked-brick formation, these strains are generally quite distinct.

Epidemiology. Strains of EAggEC belong to very diverse combinations of O and H type, and even within an outbreak of diarrhoeal disease strains with several different serotypes may be isolated. The diversity of serotypes and pathogenic mechanisms observed suggests that the genes encoding the aggregative phenotype may be accepted readily by strains of commensal and potentially pathogenic strains of *Esch. coli*, causing these bacteria to be considered as EAggEC.

Laboratory diagnosis. Although tissue culture tests are laborious, the pattern of adhesion to HEp-2 cells remains the key assay for detecting EAggEC. The test involves allowing strains of *Esch. coli* to adhere to cell monolayers in vitro and observing the pattern of adhesion by microscopy. Aggregative adhesion gene probes and PCRs have proved useful as comparatively rapid screening methods as a prelude to HEp-2 adhesion tests, but these tests remain specialized.

Treatment and control of *Esch. coli* enteritis

General measures

As with most diarrhoeal disease, the early administration of fluid and electrolytes is the most important single factor in preventing the death of the patient in severe infections. Despite the potentially serious consequences of VTEC infection, the use of antibiotics in this condition is controversial.

The most effective means of preventing infection is to avoid exposure to the infecting agent. Contaminated food and water are probably the most important vehicles of ETEC infection in developing countries. The provision of safe supplies of water together with education in hygienic practice in the handling and production of food, particularly that given to young children, are essential. Travellers to countries with poor hygiene, especially in the tropics, should select eating places with care and, if possible, should consume only hot food and drinks, or bottled water. Self-peeled fruits are probably safe, but salads should be avoided. Unheated milk should always be considered unsafe.

The spread of infantile enteritis in hospitals and nurseries is mainly from patient to patient, generally on the hands of attendants, or from contaminated infant feeds. It can be prevented only by very strict hygiene. Infected patients, and recently admitted patients suspected of being infected, must be isolated by barrier nursing techniques to prevent faecal spread. In some cases outbreaks can be terminated only by closing the ward or nursery and cleaning thoroughly before reopening.

VTEC infections are acquired most frequently from meat, unpasteurized milk and direct contact with animals. Food-borne infections should be avoided by normal food hygiene with particular attention to processing and handling cooked meat products separately from raw meat, and the thorough cooking of raw meats, especially if minced.

Vaccination

The most extensive studies of the use of vaccines have so far been in the veterinary field. A potential vaccine for human use has been prepared by cross-linking a synthetically produced ST with the non-toxic B subunit of LT, but trials were inconclusive. Tests in the rat showed that the vaccine protects against subsequent challenge with ST or LT or with organisms that produce them. In human volunteers, oral administration causes an increase in antitoxin levels in serum samples and jejunal aspirates.

Inhibition of enterotoxin activity.

Several substances such as activated charcoal, bismuth subsalicylate and non-steroidal anti-inflammatory drugs inhibit or reverse the secretory effects of enterotoxins in experimental animals and may be of value in the prevention or treatment of diarrhoea. Clinical trials have shown some benefit from such substances, but more work needs to be done to establish the optimal conditions for their use.

An experimental silica-based compound has been advocated for treating patients suspected of being infected with VTEC. The compound is designed to bind VT synthesized by bacteria present in the intestine, preventing the toxin from entering the patient's tissues. However, as symptoms follow the effects of toxin, treating a patient with toxin chelators may have only limited value.

Antimicrobial prophylaxis

A number of antimicrobial drugs reduce the incidence of diarrhoea in travellers to tropical areas. These include doxycycline, trimethoprim, norfloxacin and other fluoroquinolones. However, the widespread use of antibiotic prophylaxis has been criticized both on the grounds of drug toxicity and because of the possibility that the development and spread of drug resistance might be encouraged among a variety of enteropathogens. In addition, certain antimicrobial drugs have been shown to increase expression of VT by strains of VTEC, so that administration of antibiotics may be counterproductive.

OTHER ESCHERICHIA SPECIES

Esch. blattae was first described among bacteria isolated from the gut of the cockroach. It differs from *Esch. coli* both in oxidizing gluconate and fermenting malonate, as well as in failing to form indole, acidify mannitol or sorbitol, or produce β-galactosidase. It would probably be better placed in another genus. The species has not been reported from human clinical specimens. *Esch. fergusonii*, *Esch. hermanii* and *Esch. vulneris* have been recovered from various clinical specimens, especially faeces and wounds, but their clinical significance is usually unclear.

KEY POINTS

- *Escherichia coli* forms a consistent component of the normal intestinal microbiota.
- Different strains of *Esch. coli* carry a range of chromosomal and/or episomal genes encoding pathogenic mechanisms that enable them to cause a diverse range of infections.
- *Esch. coli* is the commonest cause of urinary tract infection.
- Enterotoxigenic *Esch. coli* causes a cholera-like illness.
- Verocytotoxin-producing *Esch. coli* belonging to serogroup O157 are a major cause of kidney failure in young children.
- Enteropathogenic *Esch. coli* cause infantile enteritis.
- Infection with entero-invasive *Esch. coli* resembles that caused by *Shigella dysenteriae* type 1.
- Entero-aggregative *Esch. coli* cause chronic diarrhoeal illness.
- Antibiotic treatment is appropriate in urinary infection and serious sepsis, but most enteric infections are managed conservatively.

RECOMMENDED READING

Chart H 1998 Toxigenic *Escherichia coli*. *Journal of Applied Microbiology* 84: 77S–86S

Chart H 2000 Clinical significance of Verocytotoxin-producing *Escherichia coli* O157. *World Journal of Microbiology and Biochemistry* 16: 719–724

Chart H, Jenkins C 1999 The serodiagnosis of infections with Verocytotoxin-producing *Escherichia coli*. *Journal of Applied Microbiology* 86: 731–740

Deisingh A K, Thompson M 2004 Strategies for the detection of *Escherichia coli* O157:H7. *Journal of Applied Microbiology* 96: 419–429

Huang D B, Okhuysen P C, Jiang Z-D, DuPont H L 2004 Enteroaggregative *Escherichia coli*: an emerging enteric pathogen. *American Journal of Gastroenterology* 99: 383–389

Kaper J B 1998 EPEC delivers the goods. *Trends in Microbiology* 6: 169–173

Kaper J B, Nataro J P, Mobley H L T 2004 Pathogenic *Escherichia coli*. *Nature Reviews* 2: 123–140

Sanchez J, Holmgren J 2005 Virulence factors, pathogenesis and vaccine protection in cholera and ETEC diarrhea. *Current Opinions in Immunology* 17: 1–11

Internet sites

Centers for Disease Control and Prevention. *Escherichia coli* Infection. http://www.cdc.gov/ncidod/diseases/submenus/sub_ecoli.htm

Health Protection Agency. *Escherichia coli*. http://www.hpa.org.uk/infections/topics_az/ecoli/menu.htm

27 Klebsiella, enterobacter, proteus and other enterobacteria

Pneumonia; urinary tract infection; opportunist infection

H. Chart

The genera described in this chapter conform to the general definition of the Enterobacteriaceae in that they are aerobic or facultatively anaerobic, ferment glucose and produce catalase but not oxidase. Together with organisms of the genera *Salmonella* (see Ch. 24), *Shigella* (see Ch. 25), *Escherichia* (see Ch. 26) and *Yersinia* (see Ch. 35), they are commonly referred to as enterobacteria. Species of clinical interest are listed alphabetically in Table 27.1, along with common synonyms.

KLEBSIELLA

CLASSIFICATION

In the past the name *Klebsiella aerogenes* was used for the non-motile, capsulate, gas-producing strains commonly found in human faeces and in water. These probably corresponded to strains described in the nineteenth century as *Bakterium lactis aerogenes*, referred to later as *Bacterium aerogenes*, and subsequently transferred to the genus *Klebsiella*. Unfortunately, the term '*Bact. aerogenes*' (later *Aerobacter aerogenes*) was also used by water bacteriologists to refer to motile organisms that are now classified as *Enterobacter* species. In an attempt to resolve the resultant confusion, some taxonomists adopted 'pneumoniae' as the species name for the non-motile aerogenes-like organisms, although it had earlier been used to designate certain biochemically atypical *Klebsiella* strains isolated from the respiratory tract of man and animals. The name *K. pneumoniae* has now been accepted formally, and appears in the Approved Lists of Bacterial Names, whereas *K. aerogenes* is omitted. The name *K. pneumoniae* is therefore used for the species as a whole, and the most frequently encountered, biochemically typical form of it is referred to as *K. pneumoniae* subspecies *aerogenes*. The atypical respiratory strains are included in the subspecies *ozaenae*, *pneumoniae* and *rhinoscleromatis*. A further species, *K. oxytoca*, is encountered occasionally in clinical specimens.

DESCRIPTION

Members of the genus *Klebsiella* are capsulate Gram-negative rods about 1–2 μm long. Strains can be differentiated by simple biochemical tests. Capsular material is produced in greater amounts on media rich in carbohydrate.

Table 27.1 Principal genera and species of Enterobacteriaceae of clinical interest

Genus	Species	Synonyms
Citrobacter	*C. amalonaticus*	*Levinea amalonatica*
	C. freundii	
	C. koseri	*C. diversus, L. amalonatica*
Edwardsiella	*E. tarda*	*E. anguillimortifera*
Enterobacter	*Ent. aerogenes*	*K. mobilis*
	Ent. cloacae	
	Ent. agglomerans	*Erwinia herbicola*
Escherichia[a]		
Hafnia	*H. alvei*	*Ent. alvei, Ent. hafniae*
Klebsiella	*K. oxytoca*	
	K. pneumoniae	
	ssp. *aerogenes*	*K. aerogenes*
	ssp. *ozaenae*	*K. ozaenae*
	ssp. *pneumoniae*	*K. pneumoniae*
	ssp. *rhinoscleromatis*	*K. rhinoscleromatis*
	K. ornithilytica	
Morganella	*M. morganii*	*Pr. morganii*
Proteus	*Pr. mirabilis*	
	Pr. vulgaris	
Providencia	*Prov. alcalifaciens*	
	Prov. rettgeri	*Pr. rettgeri*
	Prov. stuartii	
Salmonella[a]		
Serratia	*S. liquefaciens*	*Ent. liquefaciens*
	S. marcescens	
	S. odorifera	
Shigella[a]		
Yersinia[a]		

[a]See appropriate chapter.

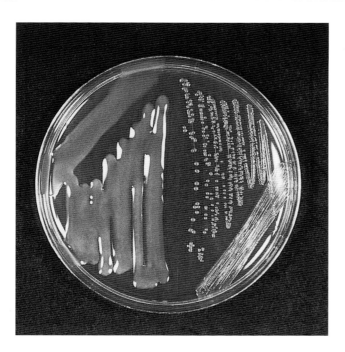

Fig. 27.1 Mucoid (left) and non-mucoid (right) variants of *Klebsiella* species on carbohydrate-rich medium. (Courtesy of George Sharp and Richard Edwards, Queen's Medical Centre, Nottingham.)

In these conditions the growth on agar is luxuriant, greyish white and extremely mucoid (Fig. 27.1). The polysaccharides of the different capsular types are all complex acid polysaccharides, and usually contain glucuronic acid and pyruvic acid. They resemble the K antigens of *Escherichia coli* (see p. 276).

Klebsiellae are non-motile but most strains express fimbriae. The organisms grow at temperatures of between 12°C and 43°C (optimum 37°C) and are killed by moist heat at 55°C for 30 min. They may survive drying for months and, when kept at room temperature, cultures remain viable for many weeks. They are facultative anaerobes, but growth under strictly anaerobic conditions is poor. There is no haemolysis of horse or sheep red cells.

Antigenic structure

About 80 capsular (K) antigens are presently recognized. Types K1, K2, K3, K5 and K21 are of particular significance in human disease, and the prevalence of these types limits the usefulness of capsular serotyping as an epidemiological tool.

Five different somatic or O antigens occur in various combinations with the capsular antigens. Four of the five *Klebsiella* O antigens are identical to or related to *Esch. coli* O antigens. It is therefore possible to divide *Klebsiella* strains into a small number of groups which may be further subdivided into capsular types, but this is of little practical value in the classification of the capsulate members of the genus.

There is some association between antigenic structure, biochemical activities and habitat. Members of capsular types 1–6 occur most frequently in the human respiratory tract. Considerable overlap occurs between *Klebsiella* antigens and those of unrelated organisms. Capsular type 2, for example, is immunologically similar to the type 2 pneumococcus.

Typing methods

Many *Klebsiella* strains produce bacteriocins, which appear to be distinct from colicins because they have no action on *Esch. coli*. They have a narrow range of activity on other klebsiellae and epidemiological analysis may be improved by the use of bacteriocins as an adjunct to capsular serotyping. Phage typing has also been used, and biotyping has been recommended instead of serotyping for use in less well equipped laboratories.

Molecular typing methods based on those described in Chapter 3 have also been developed for epidemiological analysis. Capsular antigens are usually detected by means of the capsular 'swelling' reaction, but various other techniques are used, including counter-current immuno-electrophoresis and enzyme-linked immunosorbent assay (ELISA).

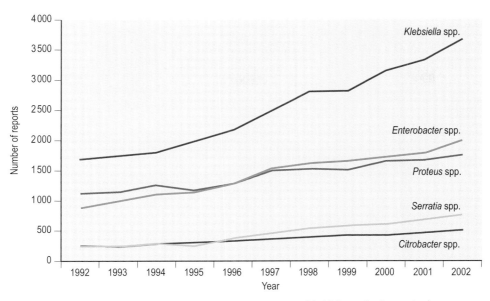

Fig. 27.2 Gram-negative bacteraemia. Laboratory reports, England and Wales 1992–2002 (Health Protection Agency data).

PATHOGENESIS

Klebsiella spp. are a fairly common cause of urinary tract infection and occasionally give rise to cases of severe bronchopneumonia, sometimes with chronic destructive lesions and multiple abscess formation in the lungs (*Friedländer's pneumonia*). Cases are sporadic and usually occur in members of the general population rather than in hospital patients. In many cases there is also bacteraemia, and the mortality rate is high. This condition is usually but not always associated with members of capsular types 1–5, which are often biochemically atypical.

However, the main importance of klebsiellae as human pathogens is in causing infections in hospital patients; the strains responsible are nearly always biochemically typical members of *K. pneumoniae* ssp. *aerogenes*, most of which belong to higher-numbered capsular types. Clinical sepsis develops in surgical wounds and in the urinary tract; a number of patients have bacteraemic infections (Fig. 27.2) and some die.

Colonization of the respiratory tract is very common in hospital patients receiving antibiotics, but its clinical significance is often difficult to assess. Some debilitated patients develop bronchopneumonia in which a *Klebsiella* sp. appears to be the primary infecting agent. Klebsiellae are naturally resistant to many antibiotics and readily acquire resistance to most others. The emergence of the species as an important cause of infection in hospitals is undoubtedly related to the use of antibiotics.

The *ozaenae* and *rhinoscleromatis* subspecies of *K. pneumoniae* take their names from the diseases with which they are associated. Rhinoscleroma is a chronic upper respiratory tract disease that occurs in many parts of the world, where it is associated with prolonged exposure to crowded and unhygienic conditions. The lesions occur in the nose, larynx, throat and, to a lesser extent, the trachea, and consist of granulomatous infiltrations of the submucosa. Ozaena (atrophic rhinitis) is an uncommon, chronic disease of the nasal mucosa, in which *K. pneumoniae* ssp. *ozaenae* is of disputed significance.

Pathogenic mechanisms

Long-chain lipopolysaccharide (LPS) protects strains from the action of serum complement, and complex polysaccharide capsules confer protection against phagocytosis. Adhesion to host tissues has been attributed to the expression of a range of fimbrial and non-fimbrial adhesins. In common with other members of the Enterobacteriaceae, klebsiellae express type 1 fimbriae that exhibit mannose-sensitive haemagglutination. Type 3 fimbriae also cause haemagglutination of erythrocytes pretreated with tannin. Additional fimbrial structures (type 6 and KPF-28) and non-fimbrial adhesins have been described. Adhesins CF29K and CS31A are plasmid-encoded adhesins that enable strains to adhere to cultured human intestinal cell lines.

In common with many enteric bacteria, klebsiellae express an enterobactin-mediated iron-sequestering system which uses ferric siderophore receptors antigenically related to those expressed by strains of *Esch. coli*. Rare strains of klebsiella may also express a plasmid-encoded aerobactin-mediated high-affinity iron-uptake system.

Heat-labile and heat-stable toxins have been described in strains isolated from burns patients; however, the role played by these toxins in the pathogenesis of disease remains to be determined.

TREATMENT

Clinical isolates of *Klebsiella* characteristically produce a β-lactamase that renders them resistant to ampicillin, amoxicillin and other broad-spectrum penicillins, but combinations of these drugs with β-lactamase inhibitors such as clavulanic acid are usually effective. Klebsiellae are normally susceptible to cephalosporins, especially β-lactamase-stable derivatives such as cefuroxime and cefotaxime, and to fluoroquinolones. Resistance to chloramphenicol and tetracyclines varies from strain to strain; they are often sensitive to gentamicin and other aminoglycosides, but transferable enzymic resistance to aminoglycosides is common in some hospitals. Multi-resistant strains of *Klebsiella* species sometimes cause serious hospital infection problems.

Klebsiella infection of the urine often responds to trimethoprim, nitrofurantoin, co-amoxiclav or oral cephalosporins. Pneumonia and other serious infections require vigorous treatment with an aminoglycoside or a cephalosporin such as cefotaxime.

Vaccines based on klebsiella LPS and capsular polysaccharides have been considered for the prevention of infections with these organisms. The long-chain LPS extends beyond the capsule, which might otherwise mask it. However, the toxic lipid-A component appears to be obstructing the use of LPS vaccines; similarly, capsular vaccines are being hampered by the range of different capsular antigen types expressed by strains of klebsiella.

ENTEROBACTER

DESCRIPTION

Enterobacter species have many features in common with those of the genus *Klebsiella*, but are readily distinguished by their motility, although non-motile variants occur occasionally. Several species are recognized; *Enterobacter aerogenes* and *Ent. cloacae* are the most important clinically. *Ent. agglomerans*, an anaerogenic, yellow-pigmented organism, formerly known as *Erwinia herbicola*, is also encountered occasionally. Strains of *Ent. amnigenus* and *E. asburiae* have also been isolated from human infections.

The colonies of *Enterobacter* strains may be slightly mucoid. In general, their fermentative activity is more limited than that of typical klebsiellae. *Ent. aerogenes* is usually able to express a lysine decarboxylase enzyme, but not arginine decarboxylase, whereas *Ent. cloacae* decarboxylates arginine, but not lysine.

Serotyping and phage typing schemes have been developed to characterize strains of *Enterobacter* spp. associated with hospital infections. Various molecular methods are also used.

PATHOGENESIS

The normal habitat of *Enterobacter* spp. is probably soil and water, but the organisms are occasionally found in human faeces and the respiratory tract. Infection of hospital patients occurs, but *Enterobacter* spp. are a much less important cause of hospital infection than *Klebsiella* spp. Most infections are of the urinary tract, although members of the genus are an important cause of bacteraemia in some hospitals (Fig. 27.2).

The pathogenic mechanisms expressed by strains of *Enterobacter* spp. are poorly understood. In common with certain strains of *Klebsiella*, they express type 1 and type 3 fimbriae. Most strains also express an aerobactin-mediated iron-uptake system, which is generally associated with extra-intestinal human bacterial pathogens. Strains may produce a haemolysin resembling the α-haemolysin produced by strains of *Esch. coli*. Very rarely strains hybridize with gene probes for Verocytotoxin 1 (see p. 281).

An outer membrane protein, termed OmpX, may be a pathogenic factor for strains of *Ent. cloacae*. This protein appears to reduce production of porins, leading to decreased sensitivity to β-lactam antibiotics, and might play a role in host cell invasion.

TREATMENT

Enterobacter strains produce a chromosomal β-lactamase with cephalosporinase activity and are nearly always highly resistant to penicillins and many cephalosporins. Many are also resistant to tetracyclines, chloramphenicol and streptomycin, although most are sensitive to other aminoglycosides, including gentamicin. Most strains are susceptible to fluoroquinolones, co-trimoxazole and carbapenems. *Enterobacter* strains differ from *Serratia* strains in being sensitive to the polymyxins.

HAFNIA

DESCRIPTION

Hafnia alvei was formerly placed in the genus *Enterobacter*, but DNA–DNA hybridization studies show that this organism deserves a separate generic status. Three DNA-related groups have been recognized and 16S ribosomal RNA (rRNA) gene sequencing has indicated intra-species grouping, but at present the genus contains only the one species.

Strains of *H. alvei* are motile, non-capsulate, Gram-negative rods that grow well on general laboratory media at 30–37°C. All *H. alvei* strains are lysed by a single phage, which does not affect other members of the Enterobacteriaceae. There is an antigenic scheme that includes many serotypes.

PATHOGENESIS

Strains are isolated from the faeces of man and other animals, and are also found in sewage, soil, water and dairy products. They are occasionally encountered as opportunist pathogens from blood urine or wound infections. Some strains carry the genes encoding the ability to cause 'attaching and effacing' lesions of intestinal cells, as described for strains of enteropathogenic *Esch. coli* (EPEC; see p. 278).

TREATMENT

Strains are usually sensitive to aminoglycosides, fluoroquinolones and carbapenems.

SERRATIA

DESCRIPTION

Although numerous *Serratia* species have been described, *S. marcescens* is the one most commonly encountered in clinical specimens, especially cases of bacteraemia (see Fig. 27.2). Several others, including *S. liquefaciens* (formerly known as *Ent. liquefaciens*) and *S. odorifera*, are sometimes isolated. They vary considerably in size. Even on the same type of medium, a single strain may at one time give rise to coccobacillary morphology and at another to rods indistinguishable from other enterobacteria. Capsules are not normally formed, but capsular material is formed on a well aerated medium poor in nitrogen and phosphate.

Serratia spp. are motile Gram-negative rods that grow well on laboratory media at 30–37°C and utilize most carbohydrates with the production of acid and gas. Certain strains of *S. marcescens* produce red pigmented colonies on agar. The pigment, prodigiosin, is formed only in the presence of oxygen and at a suitable temperature, which is not necessarily the same as that for optimal growth. Certain organisms unrelated to *S. marcescens*, including an actinomycete and certain Gram-negative rods isolated from sea water, also form prodigiosin.

Typing methods

Several typing methods have been developed, including bacteriocin and phage typing. Strains may also be differentiated by multilocus enzyme electrophoresis and molecular methods such as ribotyping and pulsed-field gel electrophoresis. Plasmid typing has been used, but only a limited proportion of strains carry plasmids.

PATHOGENESIS

S. marcescens is widely distributed in nature, but faecal carriage is uncommon in the general human population. Pigmented strains may cause concern by giving rise to red colours in food or by simulating the appearance of blood in the sputum or faeces. Pigmented and non-pigmented strains are found occasionally in the human respiratory tract and in faeces. Most infections occur in hospital patients; they include infections of the urinary and respiratory tracts, meningitis, wound infections, septicaemia and endocarditis. Some strains become established endemically in hospitals and may cause outbreaks of infection. *Serratia* spp. can multiply at ambient temperatures in fluids containing minimal nutrients, and outbreaks have followed the introduction of the organisms directly into the bloodstream in contaminated transfusion fluids. Only a small proportion of the strains responsible for infection are pigmented.

The other *Serratia* species also occur commonly in the natural environment, especially in water; *S. liquefaciens* and *S. odorifera* are occasionally found in clinical specimens.

Pathogenic mechanisms

Serratia spp. may express a range of fimbrial haemagglutinins, although the tissue specificity of these adhesins is unknown. Some strains express cell surface components causing these bacteria to be highly hydrophobic, and this may be involved in adhesion to eukaryotic cell surfaces. An iron-regulated haemolysin has been described, but a role for this toxin in the pathogenesis of disease has not been demonstrated. *Serratia* spp. also express an enterobactin-mediated high-affinity iron-uptake system and some may acquire ferric ions through an aerobactin-mediated iron-uptake system. Extracellular enzymes may be responsible for host tissue damage. Toxins resembling *Esch. coli* Verocytotoxin and heat-labile toxin have been described.

TREATMENT

Serratia strains are commonly resistant to cephalosporins. Resistance to ampicillin and gentamicin is variable, but many strains destroy these antibiotics enzymically. An aminoglycoside, such as gentamicin, is usually the most reliable first-line choice. Fluoroquinolones or carbapenems may be useful in recalcitrant cases.

PROTEUS AND RELATED GENERA

CLASSIFICATION

The history of the genera *Proteus*, *Providencia* and *Morganella* is inextricably linked, and they are best considered together. The organism first isolated by Morgan was formerly included in the genus *Proteus*, but genetic evidence and enzyme studies show that a separate genus is justified with a single species, *Morganella morganii*.

There has been similar debate over the taxonomy of the remaining species in this group, with successive proposals to combine and separate the genera *Proteus* and *Providencia*. The organisms once known as biotypes A and B of *Proteus inconstans* are now regarded as separate species of the genus *Providencia*: *Providencia alcalifaciens* and *Prov. stuartii*. As a result of genetic studies, *Proteus rettgeri* has also been transferred to the genus *Providencia* as *Prov. rettgeri*, even though it resembles *Proteus* spp. rather than other organisms of the genus *Providencia* in producing urease. This leaves only *Pr. vulgaris* and *Pr. mirabilis* in the genus *Proteus*.

DESCRIPTION

There is considerable morphological variation, but in agar-grown cultures the microscopical appearance is much like that of other coliform bacteria.

All grow well on laboratory nutrient media. A notable property of *Pr. vulgaris* and *Pr. mirabilis* strains is the ability to swarm on solid media: the bacterial growth spreads progressively from the edge of the colony and eventually covers the whole surface of the medium. This swarming characteristically takes place in a discontinuous manner, with each period of outward progress followed by a stationary period (Fig. 27.3). A number of methods

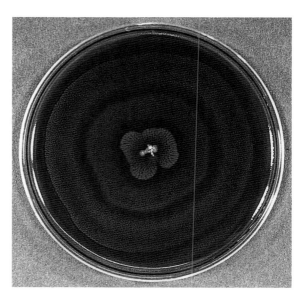

Fig. 27.3 Swarming growth of *Pr. mirabilis* inoculated centrally on to a blood agar plate and incubated overnight at 37°C. (Courtesy of George Sharp and Richard Edwards, Queen's Medical Centre, Nottingham.)

have been devised to inhibit swarming, mainly to avoid interference with the isolation of clinically more important organisms.

The various species are differentiated by standard biochemical tests. *M. morganii* and *Proteus* and *Providencia* species have the almost unique ability to deaminate amino acids oxidatively; this is shown by growing the organism in a medium containing phenylalanine, from which phenylpyruvic acid is formed (the PPA test).

Typing methods

Phage-typing, bacteriocin-typing and serotyping schemes have been developed for *Proteus* and *Providencia* species. Swarming *Proteus* strains exhibit the Dienes phenomenon (the mutual inhibition of swarming), and this forms the basis for a precise method of differentiation among such strains. Test organisms are inoculated on to the surface of an agar plate, and those that show no line of demarcation in areas where the swarming growths meet are regarded as identical.

PATHOGENESIS

Strains of *Pr. mirabilis* are a prominent cause of urinary tract infection in children and of bacteraemia (see Fig. 27.2). Indole-producing strains of *Proteus* and *Providencia* are usually isolated from hospital patients, especially in elderly men following surgery or instrumentation. Septicaemia generally occurs only in patients with serious underlying conditions or as a complication of urinary tract surgery, but outbreaks of septicaemia, often with meningitis, may occur among the newborn in hospitals. A variety of other infections, usually of surgical wounds or bedsores, occur in hospitals and are usually considered to originate from the gut flora.

M. morganii is uncommon in human disease but occasionally causes infections in hospital patients.

Pathogenic mechanisms

These bacteria are characteristically highly motile and chemotaxis may play a part in pathogenesis. Strains of *Proteus* spp. may also express calcium-dependent and calcium-independent haemolysins in addition to a range of proteases such as an IgAase.

Proteus spp. and other urease-producing organisms create alkaline conditions in the urine and may provoke the formation of calculi (stones) in the urinary tract.

TREATMENT

Most strains of *Pr. mirabilis* do not produce β-lactamase; they are consequently moderately sensitive to benzylpenicillin and fully sensitive to ampicillin, and most other β-lactam antibiotics. *Pr. vulgaris* strains are usually resistant to penicillins and many cephalosporins, although they may be sensitive to β-lactamase-stable derivatives such as cefotaxime. All strains are resistant to polymyxins and tetracyclines. *Proteus* and *Providencia* strains are inherently sensitive to aminoglycosides, but resistance, which may be due to enzymic or non-enzymic mechanisms, is now common.

Pr. mirabilis urinary tract infections usually respond to ampicillin or trimethoprim but nitrofurantoin is not effective. Treatment of infection associated with renal stones is often unsuccessful. Serious infection with other *Proteus*, *Providencia* or *Morganella* strains can often be treated with an aminoglycoside or a cephalosporin such as cefotaxime. However, susceptibility is unpredictable and treatment should be guided by laboratory findings.

CITROBACTER

DESCRIPTION

The genus *Citrobacter* was first proposed for a group of lactose-negative or late lactose-fermenting coliform bacteria that share certain somatic antigens with sal-monellae and were also known as the Ballerup-Bethesda group. These organisms are now known as *Citrobacter freundii*. Other species included in the genus are *C. koseri* (formerly known as *C. diversus*) and *C. amalonaticus*. They grow well on ordinary media and are unpigmented. Mucoid forms sometimes occur.

Serotyping schemes have been developed and antigenic diversity in expression of somatic (LPS) antigens is recognized. Antigen–antibody cross-reactions between strains of *Citrobacter* spp. and strains of *Salmonella* and *Esch. coli* have been described. For example, certain strains of *C. freundii* share LPS epitopes with *Esch. coli* expressing O157 antigens.

PATHOGENESIS

Citrobacter spp. are often found in human faeces and may be isolated from a variety of clinical specimens. They do not often give rise to serious infections but may cause bacteraemia (see Fig. 27.2).

C. koseri occasionally causes neonatal meningitis; in this condition there is a high mortality rate and the formation of cerebral abscesses is common.

Pathogenic mechanisms expressed by strains of *Citrobacter* spp. are poorly understood. Strains of *C. koseri* express type 1 (mannose-sensitive) fimbriae and occasional strains produce a form of *Esch. coli* Verocytotoxin type 2.

TREATMENT

C. freundii is usually sensitive to aminoglycosides, fluoroquinolones and chloramphenicol; sensitivity to ampicillin, tetracycline and cephalosporins varies. Resistance to aminoglycosides occurs frequently among *C. koseri* strains. Choice of treatment should be based on laboratory tests of susceptibility.

EDWARDSIELLA

Edwardsiella tarda, *E. hoshinae* and *E. ictaluri* belong to a group enterobacteria with distinctive properties, originally described as the 'Asakusa group'. Human infections were first reported in 1969. Members of the genus are small, motile, Gram-negative rods. They are facultative anaerobes and grow optimally at 37°C. Strains are biochemically inactive compared with other members of the Enterobacteriaceae but will utilize glucose. They grow well on ordinary media but produce only small colonies 0.5–1 mm in diameter after 24 h.

Edwardsiella spp. are principally associated with freshwater environments and can be isolated from healthy amphibia, reptiles and fish. *E. tarda* is most frequently associated with human disease. Two biotypes of *E. tarda* exist in nature with most human infections caused by biotype 2. Wound infection is most common, but meningitis and septicaemia have been reported. This bacterium is rarely found in the faeces of healthy people, but a higher isolation rate has been found in patients with diarrhoea. Some strains produce a heat-stable toxin causing fluid accumulation in the infant mouse test developed for detecting *Esch. coli* heat-stable toxin I.

OTHER GENERA

Various other Gram-negative bacilli, more or less related to those described in this chapter, surface occasionally in clinical specimens, usually from seriously ill, immunologically vulnerable patients. Identification is best left to specialist reference laboratories. Where a pathogenic role is suspected, priority should be given to the often unpredictable antimicrobial susceptibility pattern of the isolate so that appropriate treatment can be started quickly.

KEY POINTS

- *Klebsiella* spp. usually cause urinary tract infections but may cause bronchopneumonia and septicaemia.
- *Klebsiella* spp. are a major cause of hospital-acquired infections.
- *Enterobacter* spp. have many features in common with *Klebsiella* spp.
- *Hafnia alvei* is closely related to *Enterobacter* spp. and is an opportunist pathogen.
- *Serratia* spp. are opportunistic pathogens causing respiratory and wound infections, meningitis and septicaemia.
- Strains of *Proteus*, *Providentia* and *Morganella* are closely related. They are regularly isolated from urinary tract infections.
- Antibiotic susceptibility of enterobacteria is unpredicatable and treatment should be guided by laboratory results.

RECOMMENDED READING

Lund B M, Sussman M, Jones D, Stringer M R (eds) 1988 *Enterobacteriaceae in the Environment and as Pathogens.* Society of Applied Bacteriology Symposium Series, No. 17. Blackwell Scientific, Oxford (published as a supplement to *Journal of Applied Bacteriology* 65)
O'Hara C M, Brenner F W, Miller J M 2000 Classification, identification and clinical significance of *Proteus*, *Providencia*, and *Morganella*. *Clinical Microbiology Reviews* 13: 534–546
Podschun R, Ullmann U 1998 *Klebsiella* spp. as nosocomial pathogens: epidemiology, taxonomy, typing methods, and pathogenicity factors. *Clinical Microbiology Reviews* 1998 11: 589–603
Williams P, Tomas J M 1990 The pathogenicity of *Klebsiella pneumoniae*. *Reviews in Medical Microbiology* 1: 196–204

Internet sites

Health Protection Agency. Bacteraemia. http://www.hpa.org.uk/infections/topics_az/bacteraemia/menu.htm
Virtual museum of bacteria. *Klebsiella pneumoniae*. http://www.bacteriamuseum.org/species/Kpneumoniae.shtml
MicrobeWiki. Enterobacter. http://microbewiki.kenyon.edu/index.php/Enterobacter

28 Pseudomonads and non–fermenters

Opportunist infection; cystic fibrosis; melioidosis

J. R. W. Govan

The term *pseudomonads* describes a large group of aerobic, non-fermentative, Gram-negative bacilli belonging to more than 100 species that were originally contained within the genus *Pseudomonas*. Most are saprophytes found widely in soil, water and other moist environments. The group includes species pathogenic for man, animals, plants and insects.

Historically, the genus *Pseudomonas* has been used as a microbial repository for an extensive range of new species, often with diverse phenotypic characteristics. Recent comprehensive analyses of the group have led to revised classifications; many species have been allocated to new genera including *Burkholderia, Comamonas, Steno-trophomonas, Ralstonia, Pandoraea* and *Brevundimonas* (Table 28.1).

Pseudomonas aeruginosa is the species most commonly associated with human disease but *Burkholderia* (formerly *Pseudomonas*) *pseudomallei* is an important pathogen in tropical countries. Members of the *Burkholderia cepacia* complex, in particular *Burkholderia cenocepacia* and *Burkholderia multivorans*, are important pathogens in immunocompromised patients, particularly in individuals with cystic fibrosis or chronic granulomatous disease. *Steno-trophomonas maltophilia* also infects immunocompromised patients, with a mortality rate reaching 60% in patients with haematological malignancies. Several other pseudo-monads are occasionally isolated from clinical specimens as opportunistic pathogens, including *Ps. putida, Ps. fluo-rescens, Ps. stutzeri* and various 'glucose non-fermenters'.

There are several reasons for the pre-eminence of *Ps. aeruginosa* as an opportunistic human pathogen:

- its adaptability
- its innate resistance to many antibiotics and disinfectants
- its armoury of putative virulence factors
- an increasing supply of patients compromised by age, underlying disease or immunosuppressive therapy.

The *Ps. aeruginosa* genome is unusually large (6.3 Mb); *Burkholderia pseudomallei* (7.2 Mb) and *B. cenocepacia* (7.7 Mb) have even bigger, multi-replicon genomes, equalling those of protozoa (Fig. 28.1). Genomic and proteomic studies of these genomes could provide valuable insight into the versatility, virulence and resistance of these important pathogens.

PSEUDOMONAS AERUGINOSA

DESCRIPTION

Ps. aeruginosa is a non-sporing, non-capsulate, Gram-negative bacillus; it is usually motile by virtue of one or two polar flagella. It is a strict aerobe but can grow anaerobically if nitrate is available as a terminal electron acceptor. The organism grows readily on a wide variety of culture media over a wide temperature range and emits a sweet grape-like odour that is easily recognized. Most strains produce diffusible pigments; typically,

Table 28.1 The principal genera and species of pseudomonads of clinical interest	
Genus	Species
Pseudomonas	*Ps. aeruginosa*
	Ps. fluorescens
	Ps. putida
	Ps. stutzeri
	Ps. mendocina
Burkholderia	*B. pseudomallei*
	B. mallei
	B. cepacia complex[a]
	B. gladioli
Comamonas	*C. acidovorans*
	C. testosteroni
Stenotrophomonas	*Sten. maltophilia*
Ralstonia	*R. pickettii*
Brevundimonas	*Brev. diminuta*
Pandoraea	*P. apista*

[a]See text pp. 297–298.

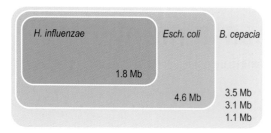

Fig. 28.1 Genome size of *Burkholderia cepacia* compared with that of *Haemophilus influenzae* and *Escherichia coli*.

the colony and surrounding medium is greenish-blue due to production of a soluble blue phenazine pigment, *pyocyanin*, and the yellow–green fluorescent pigment *fluorescein* or *pyoverdin*, which acts as a major siderophore; additional pigments include *pyorubrin* (red) and *melanin* (brown). Some 10–15% of *Ps. aeruginosa* strains produce pigment only when grown on pigment-enhancing media. Individual colonies vary from dwarf to large mucoid types; most commonly they are relatively large and flat with an irregular surface, a translucent edge and an oblong shape with the long axis parallel to the line of inoculum.

Ps. aeruginosa differs from members of the Entero-bacteriaceae by deriving energy from carbohydrates by an oxidative rather than a fermentative metabolism. It appears inactive in carbohydrate fermentation tests, and only glucose is utilized. However, all strains give a rapid positive oxidase reaction (within 30 s) and this is a useful preliminary test for non-pigmented strains.

Typing for epidemiological purposes, or to investigate clonal relationships between isolates, has traditionally relied on phenotypic markers such as serospecificity of the lipopolysaccharide (LPS), susceptibility to phages and the profile of bacteriocin production. Bacteriocin typing is simple to perform and allows characterization of most isolates, including the mucoid and LPS rough phenotypes. For non-mucoid isolates with smooth LPS, simple LPS-based serotyping methods are available. The problem of instability in the phenotypic markers used in serotyping or bacteriocin typing has been circumvented by the development of restriction endonuclease typing with pulsed-field gel electrophoresis – currently the 'gold standard' – and multilocus sequence typing (see Ch. 3).

PATHOGENESIS

Ps. aeruginosa can infect almost any external site or organ. Most community infections are mild and super-ficial, but in hospital patients infections are more common, more severe and more varied.

Community infections

Infections such as otitis externa and varicose ulcers are often chronic, but not disabling. In contrast, corneal infections resulting from contaminated contact lenses or other sources are very painful and can be rapidly destructive. Recreational and occupational conditions associated with pseudomonas infections include *jacuzzi* or *whirlpool rash* (an acute self-limiting folliculitis) and industrial eye injuries, which may lead to panophthalmitis.

Hospital infections

Infection in hospital patients is usually localized, as in catheter-related urinary tract infection, infected ulcers, bedsores, burns and eye infections. In patients compromised by age or immunosuppressing diseases such as leukaemia and acquired immune deficiency syndrome (AIDS), and in those treated with immunosuppressive drugs or corti-costeroids, pseudomonas infections frequently become generalized, and the organism may be cultured from the blood or from many organs of the body post mortem. *Ps. aeruginosa* is not a major cause of Gram-negative septicaemia or necrotizing pneumonia but is associated with a high mortality rate when these infections occur in neutropenic patients. Septicaemic infections are characterized by the black necrotic skin lesions known as *ecthyma gangrenosum*.

Cystic fibrosis

The lungs of individuals with cystic fibrosis are particularly susceptible to infections caused by *Ps. aeruginosa* and members of the *Burkholderia cepacia* complex. In these patients, asymptomatic pulmonary colonization with typical non-mucoid forms of *Ps. aeruginosa* is followed by the emergence of mucoid variants producing a gelatinous polysaccharide (Fig. 28.2). Debilitating episodes of pulmonary exacerbation due to mucoid *Ps. aeruginosa*, growing as bacterial biofilms, are the main cause of morbidity and mortality in cystic fibrosis.

Virulence factors

Most strains produce two exotoxins, exotoxin A and exo-enzyme S, and a variety of cytotoxic substances including proteases, phospholipases, rhamnolipids and the blue pigment pyocyanin. An alginate-like exopolysaccharide composed of D-mannuronic acid and L-glucuronic acid is responsible for the mucoid phenotype (Fig. 28.3). The

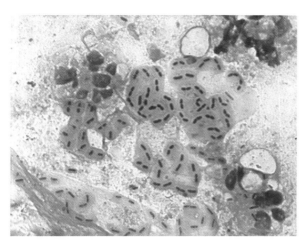

Fig. 28.2 Gram-stained sputum from patient harbouring mucoid *Ps. aeruginosa*. Note spawn-like microcolony and adjacent phagocytes. (Original magnification ×400.)

Fig. 28.3 Alcohol extraction of alginate from mucoid *Ps. aeruginosa*.

individual importance of these putative virulence factors depends upon the site and nature of infection.

- Proteases play a key role in corneal ulceration.
- Exotoxin and proteases (in particular elastase) are important in burn infections and septicaemia.
- Alginate and quorum sensing molecules are associated with the formation and architecture of biofilms in chronic pulmonary colonization.
- Pyochelin and fluorescein (pyoverdin) act as important bacterial siderophores.

Production of fluorescein in vivo allows *Ps. aeruginosa* to compete with mammalian iron-binding proteins such as transferrin. In association with pyocyanin, it gives rise to the characteristic blue–green pus of pseudomonas-infected wounds and the old species name *pyocyanea* (Latin = blue pus), as well as the present *aeruginosa* (Latin = verdigris; blue–green). The intracellular action of exotoxin A is similar to that of diphtheria toxin fragment A (see Ch. 17, p. 194).

LABORATORY DIAGNOSIS

Ps. aeruginosa grows on most common culture media. However, for specific investigations, material should be cultured on a selective medium that enhances the production of the characteristic blue–green pigment, pyocyanin. For enrichment culture from water or soil, a selective medium containing acetamide as the sole carbon and nitrogen source may be used. Resultant colonies are usually easily identified. About 10% of isolates do not produce detectable pigment even on suitable media; in such cases, a rapid positive oxidase test provides presumptive evidence of the identity. Formal identification can be made with an appropriate multitest system such as API 20NE. Tests for serum antibodies against pseudomonas antigens have no place in diagnosis except in patients with cystic fibrosis or chronic obstructive pulmonary disease, where increasing antibody levels correlate directly with immune-mediated lung damage.

TREATMENT

Ps. aeruginosa is normally resistant to many anti-microbial agents. Among β-lactam compounds, pipera-cillin, ticarcillin, ceftazidime and carbapenems are usually active, as are aminoglycosides such as gentamicin and tobramycin. Aminoglycosides are often used in combination with a β-lactam antibiotic. This provides the potential for antibacterial synergy and reduced antibiotic resistance, but there is little evidence of clinical superiority. Monotherapy with broad-spectrum β-lactam agents such as ceftazidime and imipenem is used, but resistance to these agents occurs.

Fluoroquinolones such as ciprofloxacin exhibit good activity against *Ps. aeruginosa* and penetrate well into most tissues, but resistance may develop. Unlike most antipseudomonal agents, ciprofloxacin can be administered orally. Polymyxins are usually active, but they are toxic and generally reserved for topical application. Aerosolized delivery of colistin (polymyxin E) or tobramycin into the lungs is clinically effective in the management of chronic *Ps. aeruginosa* lung infection in individuals with cystic fibrosis.

Passive vaccination may be useful for the treatment of septicaemia and burn infections due to *Ps. aeruginosa*, which are associated with a high mortality rate. Active vaccination to prevent pulmonary colonization would be desirable for patients with cystic fibrosis and several conjugate vaccines are in clinical trial.

EPIDEMIOLOGY

Ps. aeruginosa can be isolated from a wide variety of environmental sources. Its ability to persist and multiply in an extraordinary variety of moist environments and equipment (e.g. humidifiers) in hospital wards, bathrooms and kitchens is of particular importance in cross-infection control. Consumption of salad vegetables contaminated with pseudomonas is a potential risk for immunocompromised patients in intensive care units.

The organism is resistant to, and may multiply in, many disinfectants and antiseptics commonly used in hospitals. It can be a troublesome contaminant in pharmaceutical preparations and may cause ophthalmitis following the faulty chemical 'sterilization' of contact lenses.

Most hospital-acquired infections with *Ps. aeruginosa* originate from exogenous sources but some patients suffer endogenous infection, particularly of the urinary tract. Healthy carriers usually harbour strains in the gastro-intestinal tract, but in the open community the carriage rate seldom exceeds 10%. In contrast, acquisition of *Ps. aeruginosa* in a hospital is rapid, and up to 30% of patients may excrete the organisms within 2 days of admission.

Burned patients are especially at risk; the presence of the organism in ward air, dust and in eschar shed from the burns suggests that infection can be air-borne. However, contact spread is probably more important than the airborne route. Transmission may occur directly via the hands of medical staff, or indirectly via contaminated apparatus. Severely burned patients and those with chest injuries who require artificial ventilation are very susceptible to *Ps. aeruginosa*; pulmonary infection frequently precedes septicaemia, which is often fatal. Intensive care equipment

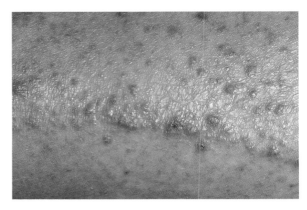

Fig. 28.4 Jacuzzi or whirlpool rash caused by *Pseudomonas aeruginosa*.

can be difficult to clean, and there is little doubt that infection can be spread by this source.

Severe eye infections may result from contaminated contact lenses, industrial eye injuries or the introduction of contaminated medicament during ophthalmic procedures. Similarly, in oil exploration, diving operations are occasionally aborted because of outbreaks of troublesome ear infection.

The warm, moist and aerated conditions under which *Ps. aeruginosa* thrives are ideally met in poorly maintained whirlpools or jacuzzis. Ear infections and an irritating folliculitis (jacuzzi rash; Fig. 28.4) may be acquired from such sources.

Epidemics of gastro-intestinal infection can occur in newborn and young infants in maternity units and paediatric wards, and may result from contaminated milk feeds. Other important *Ps. aeruginosa* infections in infants include septicaemia and meningitis.

CONTROL

Prevention is easier than cure; once *Ps. aeruginosa* has gained access to the hospital environment, or has established infection, it is notoriously difficult to eradicate. Three guidelines to control infection are offered in the knowledge that it is not always easy to put them into practice:

1. Patients at a high risk of acquiring infection with *Ps. aeruginosa* (e.g. a patient being evaluated for organ transplantation) should not be admitted to a ward where cases of pseudomonas infection are present.

2. Antimicrobial and other therapeutic substances and solutions must be free from *Ps. aeruginosa*. Particular danger exists when multidose ointments, creams or eye drops are used to treat several individuals over a period. Initially, the preparation may be sterile but contamination can easily occur between uses, and *Ps. aeruginosa* can

multiply readily at a range of temperatures in many medicaments.

3. In hospital units, episodes of cross-infection due to a single strain may occur as sporadic infections in individual patients over a period of months or years. For this reason, if facilities are available, it is advantageous to 'fingerprint' all clinically relevant isolates by a suitable typing system to identify epidemic strains.

BURKHOLDERIA PSEUDOMALLEI

B. pseudomallei is an environmental saprophyte that causes melioidosis, a devastating tropical infection of man and animals that is endemic in South-East Asia and northern Australia, and whose incidence has increased over the past two decades. A related organism, *B. mallei*, causes *glanders*, a potentially fatal infectious disease of horses, mules and donkeys. Laboratory-acquired infection with either organism is a serious risk and both species are included in hazard group 3.

B. pseudomallei is easily cultured and produces characteristic wrinkled colonies after several days of growth on nutrient agar; fresh cultures emit a characteristic pungent odour of putrefaction. The organism is oxidase positive and motile but, unlike some other pseudomonads, does not produce diffusible pigments.

MELIOIDOSIS

B. pseudomallei is found in soil and surface water in rice paddies and monsoon drains; isolation rates are highest during the rainy season and in still rather than flowing water. Human infection is acquired mainly percutaneously through skin abrasions or by inhalation of contaminated particles. The clinical manifestations of melioidosis are protean and range from a subclinical infection, diagnosed by the presence of specific antibodies, to a chronic pulmonary infection that may resemble tuberculosis and lead eventually to a fulminating septicaemia with a mortality rate of 80–90%.

In north-eastern Thailand, *B. pseudomallei* is responsible for 20% of all community-acquired septicaemia. Virtually every organ can be affected, and hence melioidosis has been called the 'great imitator' of other infectious diseases. Melioidosis commonly presents as pyrexia and, in endemic areas, serological testing for *B. pseudomallei* is important in the evaluation of pyrexia of unknown origin. The organism can survive intracellularly within the reticulo-endothelial system, and this may account for latency and the emergence of symptoms many years after exposure. Suppurative parotitis is a characteristic presentation of melioidosis in children.

Early diagnosis and appropriate antibiotic therapy are key to successful management. *B. pseudomallei* can be difficult to identify in a clinical laboratory with little experience of the organism. The organism may be observed, usually in very small numbers, as small, bipolar-stained Gram-negative bacilli in exudates, and may be cultured from sputum, urine, pus or blood on selective media. At present the most robust identification scheme combines initial screening (Gram's stain and positive oxidase test) with monoclonal antibody agglutination and gas–liquid chromatography of bacterial fatty acid esters. Polymerase chain reaction (PCR) methods are also available and may be incorporated in this schema.

Treatment

Intravenous ceftazidime, followed by a combination of co-trimoxazole and doxycycline, is emerging as the treatment of choice for severe melioidosis. Imipenem has sometimes been effective when treatment with ceftazidime has proved unsuccessful. Prolonged treatment is necessary to avoid relapse. The ability of *B. pseudomallei* to survive and multiply in phagocytes may explain the difficulty of treating melioidosis with antibiotics that are effective against the organism in vitro and the frequent relapses seen when the duration of treatment is insufficient.

BURKHOLDERIA CEPACIA COMPLEX

B. cepacia causes soft rot of onions (Latin, *cepia* = onion). However, these highly adaptable bacteria have also emerged as important human pathogens causing life-threatening respiratory infection in immunocompromised patients, in particular those with chronic granulomatous disease or cystic fibrosis. Among the cystic fibrosis population and their carers, anxiety over *B. cepacia* is based on the innate multiresistance of the organism to antibiotics, its transmissibility by social contact, and the risk of *cepacia syndrome*, an acute, fatal necrotizing pneumonia, sometimes accompanied by bacteraemia, which occurs in approximately 20% of infected patients.

Taxonomic studies show that isolates presumptively identified as 'B. cepacia' belong to at least nine phenotypically similar but genetically distinct species. These genomovars are referred to collectively as the *B. cepacia* complex and presently comprise: *B. cepacia*, *B. multivorans*, *B. cenocepacia*, *B. stabilis*, *B. vietnamiensis*, *B. dolosa*, *B. ambifaria*, *B. anthina* and *B. pyrrocinia*. All *B. cepacia* complex species, which include plant and animal pathogens, can be isolated from human and environmental sources, but 90% of human infections are caused by *B. multivorans* and *B. cenocepacia*.

Accurate identification of the *B. cepacia* complex is vital for optimal management of cystic fibrosis lung disease, in particular for the implementation of infection control procedures to reduce patient-to-patient spread, and in the surgical and clinical management of patients selected for lung transplantation. Culture and laboratory identification from clinical specimens and from environmental sites is challenging, as most commercially available multitest kits used in diagnostic laboratories are presently unreliable. Isolates of *Ps. aeruginosa*, *B. gladioli*, *Achromobacter xylosoxidans*, *Stenotrophomonas maltophilia*, *Acinetobacter baumanni* and various other species have been misidentified as *B. cepacia* complex. Use of selective media is essential for primary culture and presumptive phenotypic identification by API 20NE should be confirmed by various 16S ribosomal DNA (rDNA)- and *recA*-based PCR assays.

B. cepacia complex isolates produce several putative virulence determinants, including proteases, catalases, haemolysin, exopolysaccharide, cable pili and other adhesins, type III and IV secretion, intracellular growth and innate antibiotic resistance. In keeping with the pronounced inflammation seen in infection, LPS from these bacteria has an unusual lipid A structure. In-vitro studies have shown that pro-inflammatory cytokines such as tumour necrosis factor are stimulated tenfold more by *B. cepacia* complex LPS than by *Ps. aeruginosa* LPS.

Organisms of the *B. cepacia* complex are probably the most nutritionally adaptable of all pseudomonads and provide a striking example of soil saprophytes and phytopathogens that have emerged as an important threat to susceptible human hosts. Ironically, these organisms have also attracted much interest in agriculture as biopesticides and plant growth promoters; they are also used in bioremediation to break down recalcitrant herbicides and pollutants in contaminated soils. Some isolates display an unusual form of antibiotic resistance by an ability to use penicillin as a sole carbon and energy source!

In vitro, some *B. cepacia* complex isolates are susceptible to antipseudomonal agents and to trimethoprim, tetracycline and chloramphenicol. The carbapenem, meropenem, is typically the most active agent. However, human infections are usually intractable to therapy unless combinations of several antibiotics are used. Strict infection control measures, including segregation, are necessary to limit the spread of highly transmissible strains.

GLUCOSE NON-FERMENTERS

The heterogeneous group of aerobic Gram-negative bacilli commonly referred to as *glucose non-fermenters* is taxonomically distinct from the carbohydrate-fermenting Enterobacteriaceae and the oxidative pseudomonads. Their clinical relevance is based on their role as opportunistic pathogens in hospital-acquired infections and their intrinsic resistance to many antimicrobial agents. They grow easily on common culture media, but unequivocal identification may be difficult as most species are relatively inert in the biochemical tests used in identification of Gram-negative bacteria. Susceptibility to antibiotics is very variable and treatment should be based on the results of laboratory tests.

- *Acinetobacter* species are saprophytes found in soil, water and sewage, and occasionally as commensals of moist areas of human skin. The organisms survive well in the hospital environment, and now account for around 10% of nosocomial infections in intensive care units in Europe. Serious infections, including meningitis, osteomyelitis, wound infections (including war wounds), pneumonia and septicaemia, are most commonly associated with *Acinetobacter baumannii*. Patients in intensive care units are at particular risk. Isolates are often inherently resistant to many antimicrobial agents, including β-lactam agents (other than carbapenems), aminoglycosides and quinolones.

KEY POINTS

- Pseudomonads are aerobic, saprophytic and innately resistant bacteria causing opportunist infections in man, plants and insects. *Pseudomonas* and *Burkholderia* are regularly isolated from human infections.
- *Pseudomonas aeruginosa* is frequently isolated and causes infections that may be trivial or life-threatening.
- Mucoid forms of *Ps. aeruginosa* are a major cause of chronic debilitating and life-threatening respiratory infections in individuals with cystic fibrosis.
- *Burkholderia pseudomallei* is the causative agent of melioidosis, a systemic infection of man and animals endemic in South-East Asia and northern Australia.
- The *Burkholderia cepacia* complex causes life-threatening pulmonary infections in individuals with cystic fibrosis or chronic granulomatous disease.
- The term 'glucose non-fermenters' includes *Acinetobacter* spp. and describes a group of saprophytes that are distinct from the oxidative pseudomonads. They cause hospital-acquired opportunist infections, particularly in intensive care units.

- *Alcaligenes* and *Achromobacter* species are saprophytes found in moist environments, including those in hospital wards, and are associated with a range of hospital-acquired opportunistic infections, including septicaemia and ear discharges.
- *Eikenella corrodens* is a commensal of mucosal surfaces. It may cause a range of infections, in particular endocarditis, meningitis, pneumonia, and infections of wounds and various soft tissues.

- *Chryseobacterium* (formerly *Flavobacterium*) *meningosepticum* is a saprophyte whose natural habitat is soil and moist environments, including nebulizers; it may cause opportunistic nosocomial infections, particularly in infants. As the name suggests, this species is associated with meningitis, and has been responsible for high mortality in epidemic outbreaks.

RECOMMENDED READING

Bergogne-Bérézin E, Towner K J 1996 *Acinetobacter* spp. as nosocomial pathogens: microbiological, clinical and epidemiological features. *Clinical Microbiology Reviews* 9: 148–165

Dance D A B 2000 Melioidosis as an emerging global problem. *Acta Tropica* 74: 115–119

Govan J R W 2006 *Burkholderia cepacia* complex and *Stenotrophomonas maltophilia*. In Bush A, Alton E, Davies J, Griesenbach U, Jaffe A (eds) *Cystic Fibrosis in the 21st Century*. Progress in Respiratory Research, Vol. 34, pp. 145–152. Karger, Basel

Govan J R W, Hughes J, Vandamme P 1996 *Burkholderia cepacia*: medical taxonomic and ecological issues. *Journal of Medical Microbiology* 45: 395–407

Holden M T G, Titball R W, Peacock S J et al 2004 Genomic plasticity of the causative agent of melioidosis, *Burkholderia pseudomallei*. *Proceedings of the National Academy of Sciences of the USA* 101: 1420–1425

LiPuma J J 1998 *Burkholderia cepacia*: management issues and new insights. *Chest Medicine* 19: 473–486

Mahenthiralingam E, Urban T A, Goldberg J B 2005 The multifarious, multireplicon *Burkholderia cepacia* complex. *Nature Reviews* 3: 1–14

Internet sites

Pseudomonas Genome Project. http://www.pseudomonas.com

International *Burkholderia cepacia* Working Group. http://www.go.to/cepacia

29 Campylobacter and helicobacter

Enteritis; gastritis; peptic ulcer

J. M. Ketley

Campylobacter and *Helicobacter* are phylogenetically related, spirally shaped, flagellate bacteria. They are specially adapted to colonizing mucous membranes and are able to penetrate mucus with particular facility. In most industrialized countries *Campylobacter jejuni* is the most frequently identified cause of acute infective diarrhoea, and as a result it causes much morbidity and economic loss. The discovery of *Helicobacter pylori* in 1983 has been hailed as the most significant advance in gastroduodenal pathology of the twentieth century. Infection of the stomach with this bacterium is essentially the cause of 'idiopathic' peptic ulceration and a notable risk factor for the development of gastric cancer.

CAMPYLOBACTER

Campylobacters were first isolated in 1906 from aborting sheep in the UK. Originally thought to be vibrios, they were later placed in their own genus with *C. fetus* as the type species. The discovery that *C. jejuni* and *C. coli* commonly cause acute enteritis in man was not made until the late 1970s.

Several other species, such as *C. upsaliensis*, *C. lari* and the closely related bacterium *Arcobacter butzleri*, are occasionally associated with diarrhoea, mainly in children in developing countries. *C. fetus* is a major cause of abortion in sheep and cattle worldwide. It is a rare cause of human fetal infection and abortion, and occasionally causes bacteraemia in patients with immune deficiency. Several other species of campylobacter, notably *C. concisus* and *C. rectus*, are associated with periodontal disease.

CAMPYLOBACTER JEJUNI AND C. COLI

Description

C. jejuni and *C. coli* are small, spiral, Gram-negative rods with a single flagellum at one or both poles (Fig. 29.1); this endows the bacteria with exceptionally rapid darting motility. They are unusually sensitive to oxygen and superoxides, yet oxygen is essential for growth, so micro-aerophilic conditions must be provided for their cultivation. They are often called 'thermophilic campylobacters' because they grow best at 37–42°C. They undergo coccal transformation under adverse conditions. Campylobacters are inactive in many conventional biochemical tests, including metabolism of sugars, but they are strongly oxidase positive. Under laboratory conditions they are easily destroyed by heat and other physical and chemical agents.

Campylobacters readily take up naked DNA from their surroundings and are consequently genetically diverse. *C. jejuni* and *C. coli* are closely related and phenotypically homogeneous. Differentiation was difficult until polymerase chain reaction (PCR) tests for speciation were developed.

A capsular polysaccharide forms the major antigen for a serotyping system. Bacteriophages able to infect campylobacters are also used in strain typing. Genetic

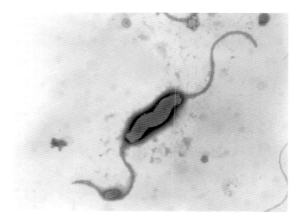

Fig. 29.1 Electron micrograph of *C. jejuni* showing single unsheathed bipolar flagella (original magnification ×11 500). (Photomicrograph by Dr A. L. Curry and D. M. Jones, Manchester Public Health Laboratory.)

'fingerprinting' methods have been developed and are replacing older methods for epidemiological studies.

Pathogenesis

C. jejuni accounts for 90–95% of human campylobacter infections in most parts of the world. Infection is acquired by ingestion of as few as 500 to 800 organisms. The jejunum and ileum are the first sites to become colonized, and the infection extends distally to affect the terminal ileum and, usually, the colon and rectum. The organisms can invade host cells and consequently penetrate past the epithelial cell layer. In well developed infections, mesenteric lymph nodes are enlarged, fleshy and inflamed, and there may be transient bacteraemia. Histological examination of the mucosa shows an acute neutrophil polymorphonuclear leucocyte response, oedema and sometimes superficial ulceration. These changes are indistinguishable from those seen in salmonella, shigella or yersinia infections.

Understanding of the mechanisms by which campylobacters cause enteric disease is sparse. Colonization of the intestine requires factors such as chemotactic motility, iron-uptake systems and several potential adhesins. Diarrhoea is likely to result from disruption of the intestinal mucosa due to cell invasion by campylobacters and the production of toxin(s) (Fig. 29.2). Both *C. jejuni* and *C. coli* produce at least one toxin, known as cytolethal distending toxin. The toxin blocks the cell cycle of host cells, but its precise role is not yet clear.

The complete genome sequence of two *C. jejuni* strains has been determined and has revealed the existence of many genes responsible for the production of various glycans, as well as the presence of short hypervariable sequences: so-called 'homopolymeric tracts', which often affect the expression of the genes (many of them responsible for glycan biosynthesis) in which they reside. The amount of the genome invested in glycan biosynthesis indicates an important role for glycans in campylobacters. The genes responsible for glycosylation of the lipo-oligosaccharide, capsule and flagellum are highly variable, suggesting a role in avoidance of host immune responses or adaptation to other environmental changes. In contrast, genes involved in general protein glycosylation are highly conserved and correspond to those found in eukaryotes. Disruption of glycosylation pathways in *C. jejuni* affects host cell invasion and intestinal colonization. Extensive glycosylation may reflect molecular mimicry of host epitopes as part of a strategy to avoid host immune responses.

Immune response

Specific humoral antibodies appear within 10 days of onset, peak in 2–4 weeks, and then decline rapidly. Most of the antibody is in the form of immunoglobulin (Ig) G, but healthy persons exposed to repeated infection show a progressive increase in IgA, which provides substantial immunity. Mild clinical symptoms, such as mild watery

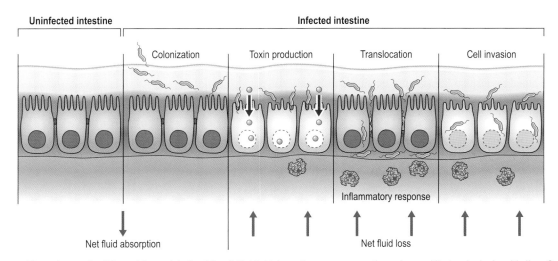

Fig. 29.2 The pathogenesis of *Campylobacter jejuni* and *C. coli*. Net fluid absorption occurs across the undamaged ileal and colonic epithelium. On infection of the host intestine, campylobacters use chemotactic motility to reach the mucous layer and colonize the epithelial surface. Campylobacters translocate directly across the cell layer, invade host cells and produce a cytotoxin. Cell damage disrupts the integrity of the mucosal cell layer and stimulates an inflammatory response. Disruption of the absorptive function of the epithelium and, possibly, stimulation of secretion leads to diarrhoeal disease.

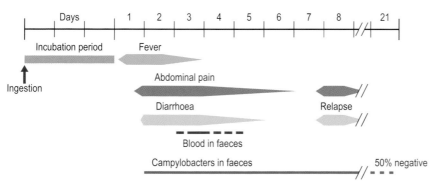

Fig. 29.3 Typical course of untreated campylobacter enteritis of average severity. There is considerable variation among individual patients.

diarrhoea or even asymptomatic colonization, may reflect increased levels of immunity.

Clinical features

The typical features of campylobacter enteritis are shown in Figure 29.3. The average incubation period is 3 days, with a range of 1–7 days. The illness may start with abdominal pain and diarrhoea, or there may be an influenza-like prodrome of fever and generalized aching, sometimes with rigors and sweating. Abdominal pain and diarrhoea are the main symptoms. Nausea is common, but vomiting is less pronounced. Severe watery diarrhoea may lead to prostration. Leucocytes are almost always present in the faeces, and frank blood may be apparent. Symptoms usually resolve within a few days, but excretion of bacteria may continue for several weeks. Prolonged carriage occurs only in patients with immunodeficiency. Campylobacter enteritis cannot be distinguished clinically from salmonella or shigella infection, but abdominal pain tends to be more severe in campylobacter infection. Indeed, a common reason for patients with campylobacter enteritis to be admitted to hospital is suspected acute appendicitis. Occasionally, illness starts with symptoms of colitis without preceding ileitis, which can make it difficult to distinguish from acute ulcerative colitis. In less developed regions *Campylobacter* infection often presents as milder diarrhoea and asymptomatic colonization is common. Clinical differences are mostly due to host immune status, although variation in virulence between strains has also been described.

Complications

Two conditions may arise 1–2 weeks after the onset of illness:

1. Reactive (aseptic) arthritis, which affects 1–2% of patients. It typically affects the ankles, knees and wrists, and, although it may be incapacitating, it is ultimately self-limiting.
2. Guillain–Barré syndrome, a form of peripheral polyneuropathy. This is much less frequent, but may cause serious and potentially fatal paralysis lasting for several months. The problem is thought to be the production of antibodies to certain epitopes present in the lipo-oligosaccharide of some *Campylobacter* strains that cross-react with the myelin in nerve sheaths, causing demyelination.

Laboratory diagnosis

Faecal specimens should be refrigerated pending delivery to the laboratory. Specimens sent by post should be placed in an appropriate transport medium.

Microscopy

The motility and morphology of campylobacters are sufficiently characteristic for a rapid presumptive diagnosis to be made by direct microscopy of fresh faeces, in either wet preparations or stained smears. This is occasionally useful, but is not done routinely.

Culture

Isolation of campylobacters from faeces requires some form of selective culture to inhibit competing faecal flora. Charcoal-based blood-free agar containing cefoperazone or other selective antimicrobial agents is widely used. Culture on non-selective media can be achieved by inoculating cellulose acetate membranes (pore size 0.45–0.65 μm) laid on the surface of the agar; campylobacters are small enough to swim through the membrane, which is removed before incubation. This method has the advantage of detecting fastidious campylobacter species

Fig. 29.4 Sources and transmission of *C. jejuni* and *C. coli*. Top boxes: animal reservoirs and sources of infection (the sheep has just given birth to a dead campylobacter-infected lamb). Left-hand box: transmission by direct occupational contact (farmer, butcher, poultry processor). Right-hand box: transmission by direct domestic contact (puppy or kitten with campylobacter diarrhoea; intra-familial spread, mainly from children). Central box: indirect transmission through consumption of untreated water, raw milk, raw or undercooked meat and poultry, food cross-contaminated from raw meats and poultry; possible transmission from flies. (From document VPH/CDD/FOS/84.1 by permission of the World Health Organization, which retains the copyright.)

that are sensitive to the antimicrobial agents used in selective media.

Plates are incubated in closed jars or specialized incubators with the oxygen tension lowered to 5–15% and carbon dioxide raised to 1–10%. Incubation at 42–43°C gives added selectivity against other faecal flora and more rapid growth of *C. jejuni* and *C. coli*, but may exclude other campylobacter species. Plates are incubated for 48 h. Colonies are typically flat with a tendency to spread on moist agar.

Serology

Serology can be useful in patients presenting with aseptic arthritis or the Guillain–Barré syndrome after a bout of diarrhoea that was not investigated. Group-specific complement fixation tests and enzyme-linked immunosorbent assay (ELISA) detect recent infection with *C. jejuni* or *C. coli*.

Epidemiology

Incidence

Campylobacter enteritis is the commonest form of acute infective diarrhoea in most developed countries. In the

UK, laboratory reports indicate an annual incidence of about 1 per 1000 population, but the true figure is probably much higher. The disease occurs in people of all ages and is often associated with travel. There is a high morbidity rate in young adults, and in some studies an unexplained excess of male patients in this group (male:female ratio 1.7:1).

In developing countries, infection is hyperendemic, and children are repeatedly exposed to infection from an early age. By the time they are 2–3 years old, children have developed substantial immunity that lasts into adulthood. The maintenance of immunity probably depends on continuous exposure to infection.

Sources and transmission

C. jejuni and *C. coli* are found in a wide variety of animal hosts, particularly birds, an adaptation that is reflected in their high optimum growth temperature. Figure 29.4 summarizes the principal sources and routes of transmission to human beings. There is a constant shedding of the bacteria from wild birds and other animals into the surface water of lakes, rivers and streams, in which campylobacters can survive for many weeks at low temperatures. Farm animals are often infected from such sources and flies have also been implicated in spread. Cattle are commonly infected, and raw milk often becomes contaminated. The distribution of raw or inadequately pasteurized milk and untreated water has caused major outbreaks of campylobacter enteritis, some affecting several thousand people.

At least 60% of chickens sold in shops are contaminated with campylobacters, and broiler chickens are thought to account for about 50% of human infections in industrialized countries. Red meats are less often contaminated.

Properly cooked poultry and meats do not pose a risk; it is what happens to them before they are cooked that can create problems. Cross-contamination from these raw products to other foods, such as bread and salads, probably accounts for most infections. Unlike salmonellae, campylobacters do not multiply in food, so explosive food-poisoning outbreaks are rare. Most infections are sporadic or confined to one household.

Direct contact with infected animals or their products can also give rise to infection. Spread of infection between individuals is of relatively minor importance.

Treatment

Campylobacter enteritis is usually self-limiting and patients seldom require more than fluid and electrolyte replacement. Antimicrobial treatment should be reserved for patients with severe or complicated infections.

Erythromycin is effective if given early in the disease. Ciprofloxacin and other fluoroquinolones are also effective, but resistance rates are rising. Many strains are sensitive to metronidazole and other nitroimidazoles, an unusual property among bacteria that are not strict anaerobes.

Control

The wide distribution of campylobacters in nature precludes any possibility of reducing the reservoir of infection. Efforts must be directed to interrupting transmission. The purification of water and the heat treatment of milk are obvious and basic measures. The control of infection in broiler chickens merits high priority, but the means to achieve this are beset with difficulty. Terminal γ-irradiation of carcasses eliminates campylobacters and other pathogens, but public acceptability is a problem. Public education on basic hygiene in the handling of raw meats, especially poultry, needs to be strongly promoted.

HELICOBACTER

Remarkably, *H. pylori*, which colonizes roughly half of the world's population, remained undiscovered until

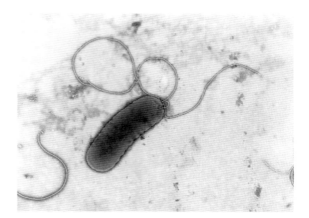

Fig. 29.5 *H. pylori* showing multiple sheathed unipolar flagella (original magnification ×11 500). (Photomicrograph by Dr A. L. Curry and D. M. Jones, Manchester Public Health Laboratory.)

1983 when Warren and Marshall in western Australia overturned the dogma that bacteria could not colonize the stomach. The discovery revolutionized the treatment of duodenal and gastric ulcers and earned them the 2005 Nobel Prize for Medicine.

Nearly 20 species of *Helicobacter* are now recognized. One group, the gastric helicobacters, colonize the stomachs of animals: the monkey, cat, dog, ferret and cheetah each harbour their own species. The enterohepatic group colonize the intestines and liver of a wide range of animals. *H. cinaedi* and *H. fennelliae* are associated with proctitis in homosexual men.

Less common helicobacters, originally named '*H. heilmannii*', are found in the human stomach and are associated with gastritis, ulcers and malignancy. Molecular studies suggest transmission from an animal source. The bacteria are more tightly spiralled than *H. pylori*, with up to 12 sheathed polar flagella. They belong to two groups: one closely related to a pathogenic gastric helicobacter of pigs, and the other to canine and feline helicobacters.

HELICOBACTER PYLORI

Description

H. pylori is a Gram-negative, spirally shaped bacterium, 0.5–0.9 μm wide by 2–4 μm long. Like campylobacters, it is strictly micro-aerophilic and requires carbon dioxide for growth, but it has a tuft of sheathed unipolar flagella, unlike the unsheathed flagella of campylobacters (Fig. 29.5; cf. Fig. 29.1). It is biochemically inactive in most conventional tests, but produces an exceptionally

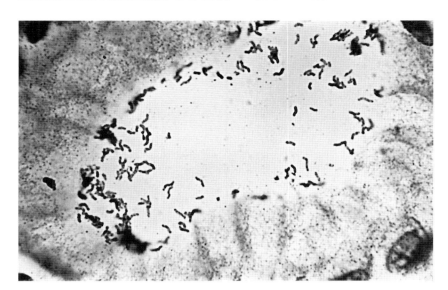

Fig. 29.6 Section of gastric mucosa showing colonization with *H. pylori* (Warthin–Starry silver stain; original magnification ×1400). (Photomicrograph by Mr G. H. Green, Worcester Royal Infirmary.)

powerful urease, almost 100 times more active than that of *Proteus vulgaris*, which is vital to its survival in the stomach. *H. pylori* undergoes coccal transformation even more rapidly than *C. jejuni* when exposed to adverse conditions and it is even more fragile, a point of importance when referring samples to a laboratory.

Antigens and strain typing

Although various antigens are expressed by *H. pylori*, serotyping is of limited practical value. However, its genetic diversity can be exploited by molecular typing based on DNA analysis. Like campylobacters, *H. pylori* exhibits considerable genetic diversity arising from a high mutation rate and frequent recombination events. Such a high level of variation means that a host is colonized by a population of closely related bacteria analogous to the 'quasi-species' observed with certain viruses.

Pathogenesis

Site of infection

H. pylori is a highly adapted organism that lives only on gastric mucosa. Colonization ceases abruptly where gastric mucosa ends, for example in areas of intestinal metaplasia in the stomach. Conversely, areas of gastric metaplasia elsewhere in the gut, notably the duodenum, may become colonized with *H. pylori*, thus setting the scene for ulceration.

The gastric antrum is the most favoured site, but other parts of the stomach may be colonized, especially in patients taking an acid-lowering drug such as an H_2 antagonist or proton pump inhibitor. The bacteria are present in the mucus overlying the mucosa. Although gastric acid is potentially destructive to *H. pylori*, protection is provided by its powerful urease, which acts on the urea passing through the gastric mucosa to generate ammonia, and this may neutralize acid around the bacteria. Colonization often extends into gastric glands, but the bacteria do not invade the mucosa (Fig. 29.6). Where they are numerous, the underlying mucosa usually shows a superficial gastritis of the type known as chronic active or type B gastritis (not to be confused with type A atrophic auto-immune gastritis of pernicious anaemia). This is characterized by infiltration with chronic inflammatory cells and polymorphonuclear leucocytes. Mucus-secreting foveolar cells often show damage where the bacteria are numerous.

The course of infection

After an incubation period of a few days, patients suffer a mild attack of acute achlorhydric gastritis with symptoms of abdominal pain, nausea, flatulence and bad breath. Colonization results in the formation of mucosa-associated lymphoid tissue (MALT) and infiltration of polymorphonuclear leucocytes, together producing the active gastritis. Symptoms last for about 2 weeks, but hypochlorhydria may persist for up to a year. Despite a substantial humoral antibody response, infection and chronic active gastritis persist, but after several decades there may be a progression to atrophic gastritis. Conditions are then inhospitable for *H. pylori*, which disappears or is much reduced in number.

Associated disease

The outcome of infection by *H. pylori* reflects an inter-action between strain virulence, host genotype and environmental factors. Despite the presence of chronic active gastritis, most infections are symptomless, and endoscopic appearances of the stomach are normal. However, in some infected persons the chronic active gastritis forms a launch-pad for more serious clinical outcomes, such as gastric and duodenal ulcers, non-ulcer dyspepsia and gastric malignancies.

Peptic ulceration. *H. pylori* is actively involved in the pathogenesis of peptic ulceration unrelated to non-steroidal anti-inflammatory agents or the Zollinger–Ellison syndrome. Infection with *H. pylori* is virtually a prerequisite for ulceration, and elimination of infection allows healing of ulcers without recurrence. Any recurrence of ulceration is almost always associated with recrudescence of infection.

The topographical pattern of gastritis is a predictor of clinical outcome. In antral predominant gastritis, hyperacidity induced by *H. pylori* promotes duodenal gastric metaplasia and this leads to colonization of the duodenum, inflammation and finally ulceration. With corpus-predominant gastritis or pangastritis, host interactions lead to the suppression of acid production. Consequently duodenal ulceration is not evident, although hypoacidity can lead to epithelial changes and gastric gland atrophy that increase the risk of gastric ulceration.

Non-ulcer dyspepsia. Some cases of non-ulcer dyspepsia are probably caused by *H. pylori* as eradication has a small but significant effect on dyspepsia and prevents the development of peptic ulcers in some patients. Although there is currently no way of identifying such patients, a 'test and treat' strategy is justified on economic grounds.

Gastric cancer. Atrophic gastritis resulting from longstanding infection with *H. pylori* is associated with an increased risk of developing gastric cancer. In addition, gastric MALT lymphoma is strongly associated with *H. pylori* infection, and in most cases complete regression has been observed after elimination of the bacteria.

Other disease. *H. pylori* infection has statistically been associated with several conditions outside the digestive tract. Among them are coronary heart disease, iron deficiency anaemia and cot death. Although these are of great potential importance, the links remain unproven because of possible confounding factors.

Virulence factors

Some strains of *H. pylori* produce a vacuolating toxin (VacA) and a cytotoxin (CagA). As well as providing acid protection, urease may also promote colonization in other ways. In addition, host factors strongly influence the outcome of infection.

The *cagA* gene is a marker for strains that confer an increased risk of both peptic ulceration and gastric malignancy, although other factors play a role as strains lacking the toxin can still cause gastritis. The gene forms part of a pathogenicity island, which also encodes a secretion system capable of injecting bacterial macromolecules, including CagA, into host cells. The injected CagA protein is phosphorylated by a host kinase and subsequently interacts with various signal transduction pathways to affect epithelial cell morphology and behaviour. An anti-apoptotic effect may aid bacterial persistence on the gastric epithelium.

The vacuolating toxin has been associated with pore formation in host cell membranes, the loosening of the tight junctions between epithelial cells (thus affecting mucosal barrier permeability), and specific immune suppression. The *vacA* gene is highly conserved among *H. pylori* strains, but exhibits a high degree of polymorphism. Each allele encodes different VacA types associated with an increased risk of certain patterns of gastro-duodenal disease.

Laboratory diagnosis

Definitive tests for *H. pylori* infection depend on finding the organism in specimens of gastric mucosa obtained by biopsy. In practice, non-invasive tests are used for initial screening.

Non-invasive tests

Serology. Serological tests, mostly based on ELISA or latex agglutination, detect antibodies to *H. pylori* or its products and are used to screen patients with dyspepsia. They are less useful for screening children and are unreliable for excluding infection in elderly patients, or as a test for cure in patients who have received treatment (owing to variable persistence of antibody). Rapid bed-side tests of whole blood are available but their accuracy is poor.

Urea breath test. This test detects bacterial urease activity in the stomach by measuring the output of carbon dioxide resulting from the splitting of urea into carbon dioxide and ammonia. A capsule of urea labelled with carbon-14 or -13 is fed to the patient, and the emission of the isotope in carbon dioxide subsequently exhaled in the breath is measured. Patients infected with *H. pylori* give high readings. The test has excellent sensitivity and specificity, but carbon-14 is radio-active, albeit weakly, so it is not used in children. Carbon-13 is not radio-active, but a mass spectrometer is needed for its assay, which means that samples must be sent to a specialist laboratory, with attendant costs. Availability of an analytical spectrometer that is less expensive and easier to use may alter this situation.

Faecal antigen test. Stool antigen tests that detect *H. pylori* antigens in faeces are available. Tests based on the use of monoclonal antibodies are more accurate than polyclonal antibody tests and have the potential to supplant serology for routine screening.

Polymerase chain reaction (PCR). DNA probes for the direct detection of *H. pylori* in gastric juice, faeces, dental plaque and water supplies have been developed. Some can also detect genes expressing antibiotic resistance and presence of the *cagA* pathogenicity island. Newer versions can detect *H. pylori* within a few hours. Present methods are unsuitable for general use because clinical samples may contain compounds that inhibit the reaction.

Invasive tests

Collection of specimens. Ideally, patients for endo-scopy should not have received antibiotics or proton pump inhibitors for 1 month before the test. Mucosal biopsy specimens are taken from the gastric antrum within 5 cm of the pylorus, and preferably also from the body of the stomach. For maximum sensitivity, duplicate specimens are taken: one for histopathology (placed in fixative); the other for culture (placed in the neck of a sterile bottle made humid by adding a tiny amount of normal saline). Specimens for culture must be processed as soon as possible, certainly on the same day, or placed in transport medium.

Biopsy urease test. This is a simple and cheap test that can be performed at the bedside. A biopsy specimen is placed into a small quantity of urea solution with an indicator that detects alkalinity resulting from the formation of ammonia. Most infected patients (70%) give a positive result within 2 h; 90% after 24 h. Newer tests have faster reaction times and a test with monoclonal antibody promises higher sensitivity and specificity.

Histopathology and microscopy. Histopathology pro-vides a permanent record of the nature and grading of a patient's gastritis as well as detecting *H. pylori*. Organisms can be seen in sections stained with haematoxylin and eosin, but more specific stains make the task easier (see Fig. 29.6). The bacteria can also be seen in smears of biopsy material stained with Gram's stain. Fluorescein-based molecular probes under development are potentially able to detect *H. pylori* and its virulence factors.

Culture. Culture is no more sensitive than skilled micro-scopy of histological sections, but has several advantages: isolates can be tested for antimicrobial resistance and typed for epidemiological studies; information about the presence of virulence factors can inform clinical outcome.

Selective agars and incubation conditions similar to those used for campylobacters are used for primary isolation (see pp. 302–303). Sensitivity is increased if a non-selective medium is used in parallel. High humidity is essential. Plates are left undisturbed for 3 days and incubated for a week before being discarded as negative. *H. pylori* forms discrete domed colonies, unlike the effuse colonies of *C. jejuni* and *C. coli*.

Epidemiology

Man appears to be the sole reservoir and source of *H. pylori*. How infection is transmitted is unknown, but it is presumed to be by the oral–oral or, possibly, faecal–oral route. Volunteer studies indicate that the adult infectious dose is relatively high, but infections resulting from lower doses may resolve quickly whereas higher doses lead to persistent infection.

Infection rates are strongly related to poor living conditions and overcrowding during childhood. There is a steady rise in seropositivity with increasing age (about 50% infected by the age of 60 years in industrialized countries), but this may simply mean that older people spent their childhood under poorer conditions than younger people, rather than a steady acquisition during life. In developed nations progressively fewer children are becoming colonized, but most children in developing countries are infected by the time they reach puberty. High rates of infection correlate broadly with high rates of gastric cancer.

Inmates of psychiatric units and orphanages and profes-sional staff carrying out endoscopy examinations show higher than average rates. Nosocomial infection from inadequately disinfected endoscopes has also occurred.

Treatment

The presence of *H. pylori* infection does not necessarily mean that attempts should be made to eliminate it. Two unequivocal indications for treatment are:

- peptic ulcer disease
- gastric MALT lymphoma.

Possible indications are:

- patients with non-ulcer dyspepsia refractory to conventional treatment
- patients with a family history of gastric carcinoma.

H. pylori is sensitive to most β-lactam antibiotics, macro-lides, tetracyclines and nitroimidazoles, but resistant to trimethoprim. It is also sensitive to bismuth subcitrate or subsalicylate and partially sensitive to the acid-lowering proton pump inhibitors omeprazole and lansoprazole.

To eradicate *H. pylori* infection, at least two anti-microbial agents must be given in combination with an acid-lowering agent (triple therapy). A popular regimen is a 1-week course of the macrolide clarithromycin, plus amoxicillin (or metronidazole) and omeprazole (or lansoprazole); the advantage, if any, of longer courses varies between studies. These regimens eliminate *H. pylori* in about 90% of patients. Similar regimens are used with children but eradication rates are generally lower than those in adults.

A recrudescence of infection demands a repeat biopsy with culture and sensitivity testing of the infecting strain. Treatment failure may result from poor compliance or antibiotic resistance. High rates of resistance to metronidazole or clarithromycin, or both, are evident among strains in some regions. Newer fluoroquinolones have shown promise in *H. pylori* eradication and changes in treatment regimen may improve compliance and eradication rates.

Control

Social deprivation is the dominant factor governing the prevalence of *H. pylori* infection. In western society, social advancement has brought about a reduction of infection, and peptic ulcer disease is on the decline. However, this is far from the case in developing countries, where a cheap and effective vaccine would be valuable, particularly for the prevention of gastric cancer.

KEY POINTS

- *Campylobacter jejuni*, a food-borne pathogen generally associated with faecal contamination of food or water, is the most frequently recognized cause of bacterial enteritis.
- This flagellate, spiral and toxigenic micro-aerobe is capable of invading host cells.
- The generally self-limiting clinical presentation includes acute abdominal pain followed by diarrhoea with blood and leucocytes, and antibiotic treatment is required only in severe cases.
- *Helicobacter pylori* is a flagellated spiral micro-aerobe causing peptic ulcer and gastritis. Infection is a risk factor for gastric cancer.
- It produces a cell-damaging toxin and a system that alters host cell signal transduction pathways.
- The transmission route is unclear. Colonization and disease rates are falling in industrialized countries.
- Treatment is by eradication of *H. pylori* using a combination of antibiotics and proton pump inhibitors.

RECOMMENDED READING

Achtman M, Suerbaum S (eds) 2001 *Helicobacter pylori: Molecular and Cellular Biology.* Horizon Bioscience, Norwich
Blaser M J, Atherton J C 2004 *Helicobacter pylori:* biology and disease. *Journal of Clinical Investigation* 113: 321–333
Danesh J, Pounder R E 2000 Eradication of *Helicobacter pylori* and non-ulcer dyspepsia. *Lancet* 355: 766–767
Everest P H, Ketley J M 2002 *Campylobacter.* In Sussman M (ed.) *Molecular Medical Microbiology*, pp. 1311–1330. Academic Press, London
Ketley J M, Konkel M E (eds) 2005 *Campylobacter: Molecular and Cellular Biology.* Horizon Bioscience, Norwich
Martin A C, Penn C W 2002 *Helicobacter pylori.* In Sussman M (ed.) *Molecular Medical Microbiology*, pp. 1331–1356. Academic Press, London
Nachamkin I, Blaser M J (eds) 2000 *Campylobacter*, 2nd edn. ASM Press, Washington
Skirrow M B, Blaser M J 2002 *Campylobacter jejuni.* In Blaser M J, Smith P D, Ravdin J I et al (eds) *Infections of the Gastrointestinal Tract*, 2nd edn, pp. 719–741. Lippincott, Williams & Wilkins, Philadelphia

van Vliet A H M, Ketley J M 2001 Pathogenesis of enteric *Campylobacter* infection. *Journal of Applied Microbiology* 90: 45S–56S

Internet sites

Health Protection Agency. Campylobacter. http://www.hpa.org.uk/infections/topics_az/campy/menu.htm
Health Protection Agency. *Helicobacter pylori*. http://www.hpa.org.uk/infections/topics_az/helicobacter/menu.htm
Centers for Disease Control and Prevention. Campylobacter Infections. http://www.cdc.gov/ncidod/dbmd/diseaseinfo/campylobacter_g.htm
Centers for Disease Control and Prevention. *Helicobacter pylori* and Peptic Ulcer Disease. http://www.cdc.gov/ulcer/index.htm
Centers for Disease Control and Prevention. *Helicobacter pylori* infections. http://www.cdc.gov/ncidod/dbmd/diseaseinfo/hpylori_g.htm
World Health Organization. Campylobacter. http://www.who.int/mediacentre/factsheets/fs255/en/
The Helicobacter Foundation. http://www.helico.com

Vibrio, mobiluncus, gardnerella and spirillum

Cholera; vaginosis; rat bite fever

H. Chart

VIBRIO

The *Vibrio* genus includes more than 30 species that are commonly found in aquatic environments. Some cause disease in human beings as well as in marine vertebrates and invertebrates. The most important pathogens of man are *Vibrio cholerae, V. parahaemolyticus* and *V. vulnificus*, but various other species are occasionally implicated as opportunist pathogens. Historically, vibrios have been associated almost exclusively with epidemic and pandemic cholera caused by a particular antigenic form of *V. cholerae*.

DESCRIPTION

Vibrios are short Gram-negative rods, which are often curved and actively motile by a single polar flagellum (Fig. 30.1). Nearly all produce the enzyme oxidase and give a positive indole reaction. The genus can be divided into non-halophilic vibrios, including *V. cholerae* and other species that are able to grow in media without added salt, and halophilic species such as *V. parahaemolyticus* and *V. vulnificus* that require salt for growth. Vibrios grow readily on ordinary media provided that their requirements for electrolytes are met, and grow best when abundant oxygen is present. Most grow at 30°C but some of the halophilic species grow poorly at 37°C, whereas *V. cholerae, V. parahaemolyticus* and *V. alginolyticus* grow at 42°C.

Vibrios have a low tolerance to acid and prefer alkaline conditions (growth range pH 6.8–10.2, optimum pH 7.4–9.6).

VIBRIO CHOLERAE

Description

More than 130 different O serogroups have been described; the classical cause of epidemic cholera possesses the O1 antigen, and is known as *V. cholerae* O1. Strains of other serogroups are collectively known as 'non-O1 *V. cholerae*' and correspond to strains formerly known as *non-agglutinable vibrios* or *non-cholera vibrios*. Some of these strains can cause diarrhoea in man. All strains of *V. cholerae* share the same flagellar (H) antigen.

Strains of V. *cholerae* O1 may be further subdivided on the basis of their O antigens into the subtypes *Inaba* and *Ogawa*. Some strains possess determinants of both of these subtypes and are known as subtype *Hikojima*.

There are two biotypes of *V. cholerae* O1: the *classical* and *El Tor* biotypes. The El Tor variant is distinguished from the classical biotype by the ability to express a haemolysin and by resistance to polymyxin B. The two biotypes can also be recognized by their differential susceptibility to specific phages.

V. cholerae O139, which has emerged as a new epidemic strain (see below), may have evolved from *V. cholerae* O1, but with a modified lipopolysaccharide structure.

Pathogenesis

Clinical manifestations

Cholera is characterized by the sudden onset of effortless vomiting and profuse watery diarrhoea. Although vomiting is a common feature, the rapid dehydration and hypovolaemic shock, which may cause death in 12–24 h, are related mainly to the profuse, watery, colourless stools with flecks of mucus and a distinctive fishy odour – *rice water stools* – which contain little protein and are very different from the mucopurulent blood-stained stools of bacillary dysentery. Anuria develops, muscle cramps occur, and the patient quickly becomes weak and lethargic with loss of skin turgor, low blood pressure and an absent or thready pulse. There are, however, all grades of severity, and milder cases cannot be distinguished clinically from other secretory diarrhoeas. Symptomless infections are common.

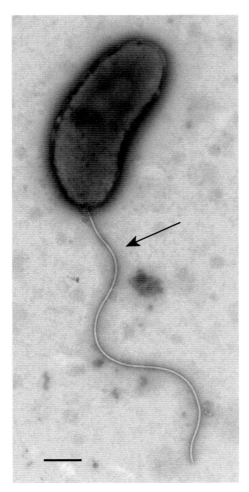

Fig. 30.1 Electron micrograph of *Vibrio cholerae* showing typical short rod morphology with a single polar flagellum (bar = 2 µm).

Pathogenic mechanisms

V. cholerae requires two major pathogenic mechanisms to cause disease:

1. the ability to produce cholera toxin
2. expression of toxin-co-regulated pili.

The sequence of events leading to cholera is confined to the gut. The cholera vibrios are ingested in drink or food and, in natural infections, the dose must often be small. After passing the acid barrier of the stomach the organisms begin to multiply in the alkaline environment of the small intestine, where they migrate towards epithelial cells, facilitated by active motility and the production of mucinase and other proteolytic enzymes. Once the organism has penetrated the mucous layer it adheres to

the enterocyte surface. In addition to toxin, co-regulated pili, haemagglutinins and lipopolysaccharide have also been implicated in adhesion.

Once adherent, the bacteria produce a potent enterotoxin known as *cholera toxin*. The toxin consists of five B subunits (molecular weight 11 600 Da) and a single A subunit (molecular weight 27 200 Da), and has structural, functional and antigenic similarity to heat-labile toxin expressed by strains of enterotoxigenic *Escherichia coli* (ETEC; see p. 278). The A subunit is made up of two peptides (A_1 and A_2) linked by a single disulphide bridge. The B subunit binds to sugar residues of ganglioside GM_1 on the cells lining the villi and crypts of the small intestine. It is thought that insertion of the B subunit into the host cell membrane forms a hydrophilic transmembrane channel through which the toxic A subunit can pass into the cytoplasm. Reduction of the disulphide bond releases the A_1 portion of the molecule, which has enterotoxic activity. The cholera enterotoxin causes the transfer of adenosine diphosphoribose (ADP ribose) from nicotinamide adenine dinucleotide (NAD) to a regulatory protein, which is part of the adenylate cyclase enzyme responsible for the generation of intracellular cyclic adenosine monophosphate (cAMP). The result is irreversible activation of adenylate cyclase and overproduction of cAMP. This in turn causes inhibition of uptake of Na^+ and Cl^- ions by cells lining the villi, together with hypersecretion of Cl^- and HCO_3^- ions. This blocks the uptake of water, which normally accompanies Na^+ and Cl^- absorption, and there is a passive net outflow of water across mucosal cells, leading to serious loss of water and electrolytes.

V. cholerae may express one or more of four haemolysins: thermo-stable direct haemolysin, El Tor haemolysin, thermo-labile haemolysin and thermo-stable haemolysin. The latter three are found in pathogenic strains of *V. cholerae*, and all are thought to contribute to virulence.

Some strains of *V. cholerae* O1 that cause diarrhoea produce a toxin that differs in antigenic nature, receptor site, mode of action and genetic homology from cholera toxin.

Non-O1 V. cholerae

Non-O1 strains of *V. cholerae* cause mild, sometimes bloody, diarrhoea, often accompanied by abdominal cramps. Symptoms may occasionally be severe, in which case the disease resembles cholera. Wound infections may occur in patients exposed to aquatic environments, and bacteraemia and meningitis have been reported.

Non-O1 *V. cholerae* strains may elaborate a wide range of virulence factors, including enterotoxins, cytotoxins, haemolysins and colonizing factors. A few strains produce cholera toxin.

Fig. 30.2   Areas reporting indigenous cholera to the World Health Organization 1959–1994. (Redrawn from data in: Public Health Laboratory Service 1991 *Communicable Disease Report Review* 1: R48–R50 and WHO 1995 *Weekly Epidemiological Record* 70: 201–208.)

Legend:
- 1959 – 1962
- 1963 – 1971
- 1972 – 1981
- 1982 –1994

Laboratory diagnosis

Stool specimens are inoculated into alkaline peptone water, in which vibrios grow rapidly and accumulate on the surface. After incubation for 3–6 h a loopful from the surface is inoculated on to a suitable solid medium such as thiosulphate–citrate–bile salts–sucrose (TCBS) agar. On this medium *V. cholerae* forms yellow sucrose-fermenting colonies, which are tested for the enzyme oxidase and for agglutination with rabbit antibodies specific for the O1 lipopolysaccharide antigens before biochemical confirmation. A fluorogenic test based on the production of lysyl aminopeptidase by *Vibrio* spp. has been developed.

Cholera toxin is detected by the same tissue culture assays and immunological techniques described for *Esch. coli* heat-labile toxin (see p. 279). Polymerase chain reaction (PCR) assays have been developed for each of the four haemolysins.

Epidemiology

V. cholerae O1

A series of six pandemics of cholera, originating in the Bengal basin, ravaged the world in the nineteenth and twentieth centuries. Subsequently, cholera was contained within the endemic foci and surrounding areas of India and Bangladesh until 1961, when a seventh pandemic due to the El Tor biotype of *V. cholerae* O1, originally isolated from pilgrims at the quarantine station known as El Tor, began to supplant the classical biotype in India. By 1973 the El Tor biotype had entirely displaced the classical biotype in Bangladesh and spread to Indonesia, the Far East and Africa. In 1991 it reached South America, where the first epidemic in that subcontinent of the twentieth century occurred in Peru. By December 1993 more than 820 000 cases of cholera, with almost 7000 deaths, had occurred, and the epidemic had involved all Latin American countries except Uruguay (Fig. 30.2). Since 1993, the number of cases in the western hemisphere has continued to fall. The vast majority of cases of cholera now occur in Africa and Asia.

During the 1980s the classical biotype once again replaced the El Tor biotype as the epidemic strain in Bangladesh, but it does not seem to have spread to other countries.

Infection is generally spread by contaminated water or foods such as uncooked seafood or vegetables. The source of the contamination is usually the faeces of carriers or patients with cholera, but contamination can probably sometimes occur from natural aquatic reservoirs. Cholera is characteristically an infection of crowded communities with poor standards of hygiene and shared communal water supplies such as tanks, ponds, canals or rivers used for bathing, washing and household use. Outbreaks occur either as explosive epidemics, usually in non-endemic areas, or as protracted epidemic waves in endemic areas. The seasonal incidence is fairly consistent in different endemic regions, although the climatic conditions during epidemic waves may be distinctive for each region. For example, in Bangladesh the cholera season (November to February) follows the monsoon rains and ends with the onset of the hot dry months. In Calcutta the main epidemic wave (May to July) rises to its peak in the hot dry season and ends with the onset of the monsoon, but extends inland to neighbouring states during the rainy season.

Spread of infection is facilitated by the high ratio of symptomless carriers to clinical cases, which varies from 10:1 to 100:1 depending on living conditions and biotype. Symptomless carriers occur much more frequently in El Tor than in classical infections.

V. cholerae O139

In 1992 cases of cholera indistinguishable from disease caused by *V. cholerae* O1 were reported in Madras, India. By mid-January 1993 similar isolates were found in neighbouring Bangladesh, and these rapidly spread north, following the course of the major rivers and raising fears of a new pandemic. In spring 2002, these fears were realized when an estimated 30 000 cases occurred in Dhaka, Bangladesh.

The new strain did not agglutinate with antisera to any of the O serogroups and was assigned to a new serogroup, O139 Bengal. However, it closely resembles *V. cholerae* O1 El Tor biochemically and physiologically, and may eventually attain the status of a subtype (*Bengal*) of that organism alongside the subtypes *Inaba* and *Ogawa*. Strains of *V. cholerae* O139 have been shown to be associated with free-living aquatic amoebae and other members of the zooplanktonic flora, which may act as a reservoir for the organism.

Other non-O1 V. cholerae serogroups

Strains belonging to serogroups other than O1 occur widely in aquatic environments, and many infections are associated with exposure to saline environments or consumption of seafood. They appear to survive and multiply better than *V. cholerae* O1 in a wide range of foods, and it is likely that food-borne outbreaks occur as with other more common enteric pathogens.

Treatment

In cholera absolute priority must be given to the life-saving replacement of fluid and electrolytes. Oral rehydration therapy is often sufficient, but severe cases may require intravenous rehydration. The World Health Organization (WHO) promotes the use of oral rehydration therapy in all cases of severe diarrhoea, including that due to *V. cholerae*. The recommended formula is shown in Table 30.1.

Tetracyclines, chloramphenicol and co-trimoxazole reduce the period of excretion of *V. cholerae* in the stools of patients with cholera. Tetracycline is often given to

Table 30.1 Formulation of oral rehydration solution recommended by the World Health Organization

Constituent	Amount (g)[a]
Sodium chloride	2.6
Potassium chloride	1.5
Trisodium citrate	2.9
Glucose (anhydrous)	13.5

[a]To be dissolved in 1 litre of clean drinking water.

reduce environmental contamination and the risk of cross-infection. Tetracycline-resistant strains of *V. cholerae* O1 El Tor began to appear in the 1970s and were soon followed by the appearance of strains resistant to a wide range of antimicrobial agents, including penicillins, streptomycin, chloramphenicol, sulphonamides and trimethoprim.

Control

Public health measures used in the control of any disease spread by faecal contamination are of value in the control of cholera. Most important are the provision of safe drinking water supplies and the proper disposal of human faeces.

Although cholera can be life threatening it is easily prevented and treated. Cholera among travellers is rare and few infections are reported from industrialized countries.

Vaccines

The immune response to cholera is directed against the bacterium rather than the toxin. It is specific for a given serotype. Infection with the classical biotype is followed by almost complete immunity for several years, but infection with the El Tor biotype confers little or no immunity. It is not surprising, therefore, that the development of effective vaccines has proved difficult.

Traditional whole-cell vaccines are not very effective and are no longer recommended for travellers. Oral vaccines that combine purified B subunit and killed whole cells are now widely available and appear to be safe and protective.

VIBRIO PARAHAEMOLYTICUS

Description

V. parahaemolyticus is a halophilic vibrio and does not grow in the absence of sodium chloride. This property is conveniently demonstrated by inoculating the organism on to cystine–lactose–electrolyte-deficient (CLED) agar, which supports the growth of non-halophilic vibrios but not halophilic species. Strains of *V. parahaemolyticus* from clinical specimens generally form green, non-sucrose-fermenting colonies on TCBS agar, but sucrose-fermenting strains are found in estuarine and coastal waters.

Strains associated with gastro-enteritis usually cause haemolysis of human red cells in Wagatsuma's agar, a special medium containing mannitol. This haemolysis is known as the *Kanagawa phenomenon*. *V. parahaemolyticus* produces thermo-labile haemolysin and thermo-stable haemolysin, but their role in pathogenicity or the Kanagawa phenomenon is not known.

Pathogenesis

V. parahaemolyticus can cause explosive diarrhoea, but symptoms usually abate after 3 days. Other symptoms include abdominal pain, nausea and vomiting, and there may be blood in the stools. A few extra-intestinal infections have been reported, particularly from wounds.

Kanagawa-positive strains produce a heat-stable cytotoxin and cause diarrhoea in volunteers. In contrast, volunteers have ingested 10^{10} cells of Kanagawa-negative strains without ill effects. Kanagawa-positive strains also adhere to human intestinal cells, whereas Kanagawa-negative strains do not. The heat-labile enterotoxin causes morphological changes in tissue culture cells resembling those caused by cholera toxin and the heat-labile enterotoxin of *Esch. coli*.

Most strains from seafood and the environment are Kanagawa negative, although positive colonies can usually be found if a sufficient number are tested. It is likely that a few Kanagawa-positive strains multiply selectively in the human intestine as infection develops and predominate in the stools of patients with diarrhoea.

Laboratory diagnosis

The faeces of patients with a history of recent consumption of seafood may be examined by the methods used for *V. cholerae*. In the examination of seafood and sea or estuarine waters for halophilic species, including *V. parahaemolyticus*, enrichment culture in alkaline peptone water containing 1% sodium chloride is used.

Epidemiology

V. parahaemolyticus is a common cause of diarrhoea in Japan and South-East Asia. It also causes illness associated with seafood in many other countries, including the USA and the UK. The organisms are common in fish and shellfish and in the waters from which they are harvested. Infections occur more frequently in the warmer months when the organisms are most prevalent in the aquatic environment. There is a particular risk associated with the consumption of raw seafood prepared and eaten in Japanese-style restaurants.

Extra-intestinal infections are always associated with exposure to the aquatic environment or handling of contaminated seafood.

Treatment and control

V. parahaemolyticus is usually susceptible to antimicrobial agents used to treat enteric infections, but patients with diarrhoea generally require only fluid replacement therapy. Infection can be avoided by normal food hygiene procedures and by refrigeration of seafood to reduce the possibility of bacterial multiplication. For wound infections and septicaemia, the most effective antimicrobial agents include tetracyclines, ciprofloxacin, ceftazidime and gentamicin.

VIBRIO VULNIFICUS

This halophilic species differs from other vibrios by utilizing lactose. There are three biotypes of *V. vulnificus* based on physiological, biochemical and serological properties. Biotype 1 is the predominant human pathogen.

Pathogenesis

There are three distinct clinical syndromes:

1. Rapid onset of fulminating septicaemia followed by the appearance of cutaneous lesions. More than 50% of those with primary septicaemia die. The condition is invariably associated with the consumption of raw shellfish. It is thought that the organisms enter the bloodstream by way of the portal vein or the intestinal lymph system. Elderly males with liver function defects due to alcohol abuse and people with iron-overload conditions are particularly susceptible, but any deficiency in the immune system may be a contributing factor.

2. A rapidly progressing cellulitis following contamination of a wound sustained during exposure to salt water. Infections of this kind occur in otherwise healthy persons as well as in the debilitated, and are characterized by wound oedema, erythema and necrosis, which progresses to septicaemia only occasionally. The infection can be rapidly fatal.

3. Acute diarrhoea following the consumption of shellfish. This is less common; victims generally have mildly debilitating underlying conditions. Death is rare.

Pathogenic mechanisms

A capsular polysaccharide and long-chain lipopolysaccharide enable pathogenic strains to resist phagocytosis and the killing effects of human serum complement. Strains of *V. vulnificus* have been reported to produce a siderophore (vulnibactin), which would indicate the presence of a high-affinity iron-uptake systems, but many strains are unable to obtain ferric ions bound to human transferrin. Patients with iron overload disorders such as haemochromatosis are highly susceptible; high levels of free serum iron are thought to facilitate the rapid septicaemia observed in patients infected with this organism.

Several toxins may contribute to tissue damage. They include a metalloprotease, a collagenase, a mucinase and a cytotoxin. A vascular permeability factor has also been described. Strains of *V. vulnificus* express El Tor haemolysin and thermo-labile haemolysin, but a role in pathogenesis has not been demonstrated.

Epidemiology

Infections occur most frequently in areas where the water temperature remains high throughout the year, such as the mid-Atlantic and Gulf coast states of the USA. They are much more common during the warmer months of the year when *V. vulnificus* is most abundant. Wound infections are associated with injuries sustained in the aquatic environment, whereas septicaemic infections are associated with the consumption of raw shellfish.

Treatment

V. vulnificus wound infections and primary septicemia require early antimicrobial treatment to reduce morbidity and mortality from the illness and to prevent complications. The most effective antimicrobial agents include tetracyclines, fluoroquinolones such as ciprofloxacin, ceftazidime and gentamicin. Because of the high case-fatality rates, it is particularly important for clinicians to suspect *V. vulnificus* wound or bloodstream infections in persons with shellfish or warm seawater exposure and a history of chronic liver disease or conditions of iron overload.

VIBRIO ALGINOLYTICUS

Description

Vibrio alginolyticus is a halophilic organism formerly regarded as biotype 2 of *V. parahaemolyticus*. It fails to grow on CLED agar but grows in the presence of 10% sodium chloride. It forms large, yellow (sucrose-fermenting) colonies on TCBS. There is pronounced swarming on non-selective solid media.

Pathogenesis

V. alginolyticus causes wound and ear infections. Clinical features include mild cellulitis and a seropurulent exudate. The pathogenic mechanisms are not fully understood although genetic homology between *V. alginolyticus*, *V. cholerae* and *V. parahaemolyticus* has been shown.

Epidemiology

This organism is widely distributed in sea water and seafood and is probably the most common vibrio found in these sources in the UK. It occurs in large numbers throughout the year. Infections are invariably associated with exposure to sea water. Strains appear to be sensitive to ciprofloxacin.

OTHER VIBRIOS

- *Photobacterium* (formerly *Vibrio*) *damsela* is a halophilic marine vibrio found in tropical and semi-tropical aquatic environments. It is associated with severe infections of wounds acquired in warm coastal areas.
- *Vibrio fluvialis* (previously known as group F vibrios, one biotype of which is now regarded as a separate species, *V. furnissii*) is easily confused with *Aeromonas hydrophila*, but can be differentiated by growth on media containing 6% sodium chloride. Patients experience diarrhoea, abdominal pain, fever and dehydration. Low numbers of *V. fluvialis* can be isolated from fish and shellfish, and from warm sea water. It seems likely that infection is from contaminated seafood. Strains of *V. fluvialis* express El Tor haemolysin, and a vaculolating toxin acting on HeLa cells, but the role of these in pathogenesis has not been demonstrated.
- *Vibrio hollisae* is associated with bacteraemia and diarrhoea, especially in warm coastal areas, such as the Gulf of Mexico. Infections are strongly associated with the consumption of raw seafood. The organism is difficult to isolate because it grows poorly on TCBS agar. Strains of *V. hollisae* exhibit gene sequences homologous with those encoding the thermo-stable haemolysin of *V. parahaemolyticus*.
- *Vibrio mimicus* is a non-halophilic vibrio named for its similarity to *V. cholerae* and occurrence in similar environments. Most isolates are from the stools of patients who develop gastro-enteritis after consumption of raw oysters, although a few cases of otitis media have also been reported. In vitro, strains of *V. mimicus* express thermo-stable direct

haemolysin, El Tor haemolysin and thermo-labile haemolysin.

- *Vibrio furnissii* was first isolated in 1983, and has been most commonly isolated from stool samples. It has occasionally been implicated in gastroenteritis.

Other aquatic organisms that are probably related to vibrios include *Aeromonas* spp. and *Plesiomonas shigelloides*. *Aeromonas* spp., notably *A. hydrophila*, have been implicated in diarrhoea and occasionally cause more serious infection in compromised individuals. *A. salmonicida* is an economically important pathogen of fish. *P. shigelloides* is an organism of uncertain taxonomic status that sometimes causes water-borne outbreaks of diarrhoea in warm countries.

MOBILUNCUS

The name *Mobiluncus* was first proposed for a group of curved, motile, Gram-variable, anaerobic bacteria isolated from the vagina of women with bacterial vaginosis. Its taxonomic position is uncertain. Studies of 16S RNA suggest that the genus is most closely related to *Actinomyces*.

DESCRIPTION

There are two species: *M. curtisii* and *M. mulieris*. The former is short (mean length 1.5 µm) and Gram-variable, whereas the latter is long (mean length 3.0 µm) and Gram negative. Both have multiple flagella originating from the concave aspect of the cells. Cell wall studies have revealed no outer membrane, but it has been suggested that both species are in fact Gram negative. Two subspecies have been proposed for *M. curtisii*, ssp. *curtisii* and *holmesii*.

PATHOGENESIS

Bacterial vaginosis

Mobiluncus spp. are isolated from 97% of women with bacterial vaginosis (non-specific vaginitis) and is rarely found in the vagina of healthy women. Bacterial vaginosis appears to be a polymicrobial infection, with certain organisms playing a key role, especially when they overgrow the lactobacilli of the normal flora, raising the vaginal pH above 4.5. The condition is characterized by the presence of a thin, homogeneous vaginal discharge with a characteristic 'rotten fish' smell. This becomes more pronounced on alkalization, and can be evoked by placing a drop of potassium hydroxide solution on the

fresh exudate on a slide or the speculum used for the vaginal examination. The characteristic smell is ascribed to amines produced by one or more of the bacterial species that form the complex microbial flora of the vagina.

Mobiluncus are frequently found in association with *Gardnerella vaginalis* (see below) and with other organisms that may also be of aetiological importance. It appears that both the combination of species and their relative numbers are of importance in the development of the syndrome. The organisms may be isolated from the urethra of male consorts of infected women but do not persist in men once condom use is implemented.

Mobiluncus spp. are occasionally isolated from extragenital sites, especially from breast abscesses.

Pathogenic mechanisms

The mechanisms that allow *Mobiluncus* spp. to cause disease are poorly understood. They express pili and are able to obtain iron from lactoferrin, but the role of these in the pathogenesis of disease is speculative. Strains of *Mobiluncus* spp. express a relatively thermo-stable toxin that is cytotoxic for Vero (African green monkey) cells, but a role in pathogenesis has not been demonstrated. It has been possible to infect primates experimentally and animal studies may help to elucidate the processes involved in the pathogenesis of vaginitis.

LABORATORY DIAGNOSIS

Microscopy of fresh unstained vaginal exudates reveals epithelial cells covered with adherent bacteria (*clue cells*). *Mobiluncus* spp. can be grown in Columbia blood broth and peptone–starch–dextrose broth containing 10% horse serum. The organisms are essentially anaerobic but will grow slowly in 5% oxygen in nitrogen. They do not produce oxidase or catalase but are saccharolytic with the production of succinic and acetic acids.

Molecular methods have been developed to provide evidence of infection; a multiplex PCR comprising primers for both *Mobiluncus* spp. and *G. vaginalis* has been applied to cases of bacterial vaginosis.

TREATMENT

Mobiluncus spp. are susceptible to many antimicrobial agents, including ampicillin, benzylpenicillin, clindamycin, erythromycin and nitroimidazoles. Treatment of bacterial vaginosis aims to restore the normal vaginal flora by eliminating *Mobiluncus* and other organisms that may be involved. Oral or intravaginal metronidazole and clindamycin have been successfully used. Although *M. mulieris* is more susceptible than *M. curtisii* to metronidazole, treatment with this drug appears to eliminate all *Mobiluncus* species in patients with vaginosis.

GARDNERELLA

Gardnerella vaginalis (formerly *Corynebacterium vaginale* or *Haemophilus vaginalis*) is commonly isolated together with *Mobiluncus* spp. in bacterial vaginosis. It has been implicated in cases of cervical cancer and infections of the urinary tract, but because the organism is frequently present in the vagina of asymptomatic patients its role in disease is equivocal.

DESCRIPTION

G. vaginalis is a non-motile, non-sporing, micro-aerophilic coccobaccilus. It is Gram variable, but because the cell wall contains lipopolysaccharide it appears to be Gram negative.

LABORATORY DIAGNOSIS

G. vaginalis grows on various media, such as Columbia agar containing colistin and nalidixic acid as selective agents. Plates are incubated at 35–37°C in 5–10% carbon dioxide for 2–3 days. On media containing human erythrocytes, the organism produces zones of β-haemolysis. *G. vaginalis* does not produce oxidase or catalase, but ferments starch and hydrolyses hippurate, and these properties provide a means of presumptive identification.

PATHOGENESIS

Whether *G. vaginalis* can cause disease in isolation is unclear. The organisms may simply flourish in the vaginal environment provided by other bacteria. The ability to lyse human red cells offers a mechanism for acquiring metabolic iron and may aid multiplication. Similarly, *G. vaginalis* can acquire ferric ions from human lactoferrin, the main iron-carrying protein of mucosal surfaces. Strains produce various mucinases such as sialidase and proline dipeptidase, which are thought to damage the vaginal mucosa as part of the pathogenic process. Patients infected with *G. vaginalis* produce antibodies of the IgA class to the haemolysin, but whether these can be used for serodiagnosis is not known.

TREATMENT

For the treatment of bacterial vaginosis, see *Mobiluncus* (above).

SPIRILLUM MINUS

The organism commonly known as *Spirillum minus* is of uncertain taxonomic position. Along with *Streptobacillus moniliformis* (see p. 341) it is one of the causes of rat bite fever in humans.

DESCRIPTION

S. minus is a spiral Gram-negative organism about 2–5 μm in length and 0.2 μm in diameter. Longer forms of up to 10 μm may be observed. The regular short coils have a wavelength of 0.8–1.0 μm. The organisms are very actively motile, showing darting movements like those of a vibrio. The movement is due to polar flagella, which vary in number from one to seven at each pole. The organisms can be demonstrated in fresh specimens by dark-ground illumination or by staining with Giemsa, Wright's or Leishman's stains. Although there have been many unconfirmed claims, the organism has not been cultivated on artificial media and many of its properties are unknown. The organism can be cultured in vivo by intraperitoneal injection into guinea-pigs or mice.

LABORATORY DIAGNOSIS

In rat bite fever, *S. minus* may be demonstrated in the local lesion, in the regional lymph glands or in the blood by direct microscopy or by animal inoculation.

PATHOGENESIS

The clinical syndrome of rat bite fever begins with an acute onset of fever and chills 1–4 weeks after the animal bite, although infection without fever may occur. The bite usually heals before the onset of symptoms but it often re-ulcerates. Local lymphadenopathy and lymphangitis develop with the onset of fever and systemic disease. A generalized rash with large brown to purple macules is usually observed, but some patients present with urticarial lesions. A roseolar rash may spread from the area of the original bite. Fever usually declines within 1 week before returning again after a few days; the fever may then recur in an episodic fashion for months or even years.

Endocarditis, meningitis, hepatitis, nephritis and myocarditis are rare complications. In most untreated cases, symptoms resolve within 2 months, after six to eight episodes of fever, although up to 6.5% of untreated cases may be fatal.

EPIDEMIOLOGY

S. minus occurs naturally in wild rats and other rodents, causing bacteraemia. Human rat bite fever occurs mainly in Africa, Japan (where it is known as *sodoku*) and the Far East. There have been a few reports from Europe and the USA. The disease is prevalent in laboratory workers who handle rats, and in children who live in rat-infested homes.

TREATMENT

Infections respond to treatment with penicillin, erythromycin and tetracyclines. In the rare case of endocarditis the addition of an aminoglycoside may be of value.

KEY POINTS

- *Vibrio cholerae* belonging to serogroup O1 is the causative agent of epidemic cholera.
- *V. cholerae* O139 is emerging as a cause of epidemic cholera.
- Cholera toxin is the key pathogenic mechanism, causing extensive loss of water and electrolytes in the form of rice-water stools; death from cholera can be prevented with rehydration therapy.
- *V. parahaemolyticus* is a major cause of diarrhoea in Japan and South-East Asia; infection is associated with the consumption of seafood.
- Infection with *V. vulnificus* may result in rapid-onset and fatal septicaemia, particularly in people with conditions of iron overload, and is associated with the consumption of seafood.
- *Mobiluncus* spp. and *Gardnerella vaginalis* are a major cause of bacterial vaginosis; changes in the normal vaginal flora appear to permit these organisms to cause disease.
- *Spirillum minus* causes rat bite fever.

RECOMMENDED READING

Chiang S-R, Chuang Y-C 2003 *Vibrio vulnificus*: clinical manifestations, pathogenesis and antimicrobial therapy. *Journal of Microbiology, Immunology, and Infection* 36: 81–88

Forsum U, Holst E, Larsson P G, Vasquez A, Jakobsson T, Mattsby-Baltzer I 2005 Bacterial vaginosis – a microbiological and immunological enigma. *Acta Pathologica, Microbiologica et Immunologica Scandinavica* 113: 81–90

Jenkins S G 1988 Rat-bite fever. *Clinical Microbiology Newsletter* 10: 57–59

Powell J L 1999 Vibrio species. *Clinics in Laboratory Medicine* 19: 537–552

Sack D A, Sack R B, Nair G B, Siddique A K 2004 Cholera. *Lancet* 363: 223–233

Sanchez J, Holmgren J 2005 Virulence factors, pathogenesis and vaccine protection in cholera and ETEC diarrhea. *Current Opinion in Immunology* 17: 1–11

Spiegal C A 2002 Bacterial vaginosis. *Reviews in Medical Microbiology* 13: 43–51

Stewart-Tull D E S 2001 Vaba, Haiza, Kholera or Cholera: in any language still a disease of seven pandemics. *Journal of Applied Microbiology* 91: 580–591

Strom M S, Paranjpye R N 2000 Epidemiology and pathogenesis of *Vibrio vulnificus*. *Microbes and Infection* 2: 177–188

Toranzo A E, Santos Y, Barja J L 1997 Immunization with bacterial antigens: *Vibrio* infections. *Developments in Biological Standardization* 90: 93–105

Internet sites

Centers for Disease Control and Prevention. *Vibrio vulnificus*. http://www.cdc.gov/ncidod/dbmd/diseaseinfo/vibriovulnificus_g.htm

Centers for Disease Control and Prevention. *Vibrio parahaemolyticus*. http://www.cdc.gov/ncidod/dbmd/diseaseinfo/vibrioparahaemolyticus_g.htm

Health Protection Agency. Cholera. http://www.hpa.org.uk/infections/topics_az/cholera/menu.htm

Todar's Online Textbook of Bacteriology. *Vibrio cholerae* and Asiatic Cholera. http://textbookofbacteriology.net/cholera.html

31 Haemophilus

Respiratory infections; meningitis; chancroid

M. P. E. Slack

Haemophilus influenzae is associated with a variety of invasive infections such as meningitis, epiglottitis, pneumonia and septic arthritis, and localized disease of the respiratory tract including bronchitis and otitis media. *H. influenzae* biogroup aegyptius (formerly *H. aegyptius*) is a cause of epidemic conjunctivitis and Brazilian purpuric fever. Other haemophili of medical importance include *H. ducreyi*, the causative organism of chancroid, and *H. parainfluenzae*, *H. aphrophilus* and *H. paraphrophilus*, three organisms that are occasionally encountered in patients with infective endocarditis and various other miscellaneous conditions. The generic name (= 'blood-loving') relates to the inability of these organisms to grow on culture media unless whole blood or certain of its constituents are present.

During the influenza pandemic of 1889–1892, Pfeiffer noted the constant presence of large numbers of small bacilli in the sputum of patients affected with the disease and suggested that the bacillus was the causative agent. Although the true aetiological agent of influenza was shown in 1933 to be a virus, it remains possible that secondary infection with *H. influenzae* contributed to the high mortality rate seen in the 1889–92 and 1918–19 pandemics.

DESCRIPTION

Haemophili are small, pleomorphic, Gram-negative rods or coccobacilli with occasional longer, filamentous forms (Fig. 31.1). Some strains of *H. influenzae* produce a polysaccharide capsule, which is demonstrable by capsule stains and a *Quellung* reaction (swelling of the capsule) with type-specific antisera. There are six capsular types, designated a–f, which can be identified by a polymerase chain reaction (PCR) method. The most important is type b, a polymer of ribosyl ribitol phosphate.

All species of *Haemophilus* are catalase and oxidase positive; they reduce nitrate to nitrite and ferment glucose. Patterns of acid production from other carbohydrates are used to differentiate the species. *H. influenzae* can be divided into eight biotypes on the basis of indole production, urease activity and ornithine decarboxylase reactions. Biotypes I–III are the most common, and most invasive (type b) organisms are biotype I.

Growth requirements

Growth depends on a requirement for two factors, termed X and V:

- X factor (haemin) is required for the synthesis of cytochrome *c* and other iron-containing respiratory enzymes. Unlike most bacteria, haemin-dependent haemophili cannot synthesize protoporphyrin from δ-aminolaevulinic acid.
- V factor is nicotinamide adenine dinucleotide (NAD), NAD phosphate or certain unidentified precursor compounds. It is essential for oxidation–reduction processes in cell metabolism.

Different species of *Haemophilus* require either or both of these factors, and growth factor dependence forms an important step in identification.

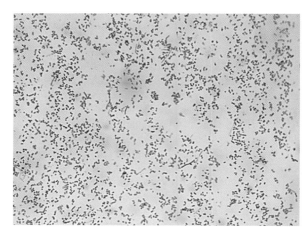

Fig. 31.1 Gram stain of culture of *Haemophilus* influenzae: Gram-negative pleomorphic coccobacilli.

Table 31.1 Clinical spectrum of *Haemophilus influenzae* infections

Type of infection	Age group	Strains
Invasive	90% children <4 years[a]	90% *H. influenzae* type b[a] 10% non-capsulate strains 1% types e and f
Neonatal and maternal	Neonates; pregnant and parturient women	>90% non-capsulate strains
Non-invasive respiratory	Children and adults	>90% non-capsulate strains

[a]Percentages observed before the introduction of Hib conjugate vaccine; since the introduction of routine vaccination, the epidemiology has changed (see text).

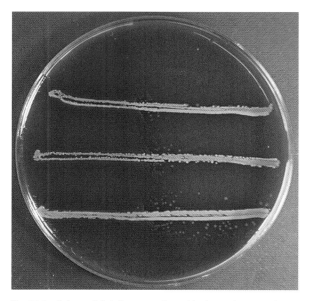

Fig. 31.2 Culture of *H. influenzae* on horse blood agar demonstrating 'satellitism'. Growth is stimulated in the vicinity of a streak of *Staphylococcus aureus*, which synthesizes V factor (see text).

Ordinary blood agar contains X and V factors, but growth of *H. influenzae*, which requires both factors, is poor. Growth is enhanced if the medium is supplemented with NAD. Streaking an organism that excretes this substance (e.g. *Staphylococcus aureus*) across the surface of the agar stimulates growth in its vicinity (satellitism; Fig. 31.2). Heating blood agar for a few minutes at 70–80°C until it turns brown (*chocolate agar*) much improves the growth of *H. influenzae*. This process removes serum NADase, which limits the amount of V factor, and also liberates extra X and V factors from the red cells into the medium. X factor is heat stable, but heating of media at 120°C for several minutes destroys V factor.

Anaerobic growth considerably reduces the haemin requirement of X-dependent species. *H. aphrophilus* has a requirement for carbon dioxide, but this character may be lost on subculture. *H. influenzae* does not require a carbon dioxide-enriched atmosphere, but often grows better in such conditions.

PATHOGENESIS

H. influenzae is associated with two types of infection which are quite distinct in their epidemiological profiles: invasive infections and non-invasive infections (Table 31.1).

Invasive infections

Most invasive infections are caused by *H. influenzae* type b (Hib). Meningitis is the most common manifestation, but *H. influenzae* also causes epiglottitis, septic arthritis, osteomyelitis, pneumonia and cellulitis. In some cases the patient develops a bacteraemia without a clearly defined focus of infection.

These infections are unusual in the first 2 months of life, but are otherwise seen mainly in early childhood. Most cases occur in children under 2 years of age, but acute epiglottitis has a peak incidence between 2 and 4 years of age. The incidence of invasive disease in children has been reduced dramatically in countries where a conjugate Hib vaccine has been introduced, and consequently the epidemiology is changing (see below).

The polysaccharide capsule is the major virulence factor for Hib. When the organism invades the bloodstream, the capsule enables the organisms to evade phagocytosis and complement-mediated lysis in the non-immune host. The rarity of infections in the first 2 months of life correlates with the presence of maternal capsular antibodies, and the occurrence of infection in early infancy with the absence of such antibodies. As the prevalence and mean level of capsular antibodies in the population rise, *H. influenzae* type b infections become less common (Fig. 31.3).

What determines whether acquisition of type b organisms in a susceptible host will lead to asymptomatic carriage and the stimulation of protective antibodies, or to the induction of invasive disease, is unclear. However, animal experiments suggest that when invasion occurs the organism penetrates the submucosa of the nasopharynx and establishes systemic infection through the bloodstream.

The type b capsular polysaccharide facilitates all phases of the invasion process. Other virulence factors that may be involved include:

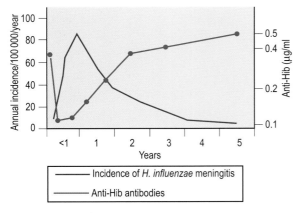

Fig. 31.3 Incidence of *H. influenzae* meningitis during the first 5 years of life and the corresponding mean level of anti-*H. influenzae* type b (Hib) capsular polysaccharide antibodies. (From Peltola H, Käyhty H, Sivonen A, Mäkelä H 1977 *Haemophilus influenzae* type b capsular polysaccharide vaccine in children: a double blind field study of 100 000 vaccinees 3 months to 5 years of age in Finland. *Pediatrics* 60: 730–737. © 1977 by the American Academy of Pediatrics.)

- fimbriae, which assist attachment to epithelial cells
- immunoglobulin (Ig) A proteases, which are also involved in colonization
- outer membrane proteins and lipopolysaccharide, which may contribute to invasion at several stages.

Initiation of invasive infection may be potentiated by intercurrent viral infection. Host genetic factors and immunosuppression may also play a role. It is unclear whether it is exposure to *H. influenzae* type b, or some other organism (e.g. *Escherichia coli* K100) possessing cross-reacting antigens, that usually stimulates natural protective antibody production.

Non-capsulate *H. influenzae* occasionally cause invasive disease. Since the introduction of routine infant immunization with conjugate Hib vaccine, invasive disease caused by non-capsulate strains has become more common than that caused by Hib in the UK. Meningitis and septicaemia due to non-capsulate *H. influenzae* is sometimes seen in the neonate. Infections with non-capsulate strains may also occur in children and adults. Pneumonia and bacteraemia are the commonest manifestations, often in patients with an underlying disease, notably chronic lung disease or malignancy. The highest rates occur in patients aged over 60 years, and the case fatality rate is high.

Invasive infections due to *H. influenzae* of serotypes other than b (principally types e and f) are rare. The spectrum of disease is similar to that seen with type b.

Non-invasive disease

H. influenzae produces a variety of local infections, which are often associated with some underlying physiological or anatomical abnormality. Most are caused by non-capsulate strains. The commonest are:

- otitis media
- sinusitis
- acute exacerbations of chronic obstructive airway disease
- conjunctivitis.

Acute sinusitis and otitis media are usually initiated by viral infections, which predispose to secondary infection with potentially pathogenic components of the resident microbial flora. The mechanisms may involve:

- obstruction to the outflow of respiratory secretions
- decreased clearance of micro-organisms via the normal mucociliary mechanism
- depression of local immunity.

Acute exacerbations of chronic obstructive airway disease are similarly initiated by acute viral infections. Respiratory viruses compromise an already impaired mucociliary clearance mechanism in patients with chronic lung disease and allow bacterial colonization of the lower respiratory tract. In this situation *H. influenzae* can establish purulent infection and further damages pulmonary function by a direct toxic effect on cilia.

LABORATORY DIAGNOSIS

Gram-stained smears of cerebrospinal fluid, pus, sputum or aspirates from joints, middle ears or sinuses can provide a rapid presumptive identification. Haemophili tend to stain poorly, and dilute carbol fuchsin is a better counterstain than neutral red or safranin.

The viability of *H. influenzae* in clinical specimens declines with time, particularly at 4°C. Specimens should therefore be transported to the laboratory and cultured without delay. Chocolate agar is a good, general purpose, culture medium and can be used without further supplementation for specimens obtained from sites that would normally be expected to be sterile. Plates should be incubated in an aerobic atmosphere enriched with 5–10% carbon dioxide.

Specimens of expectorated sputum inevitably become contaminated by upper respiratory flora, commonly including *H. influenzae*, and the finding of the organism in such specimens cannot be automatically taken to imply involvement in disease. Support for the significance of *H. influenzae* is provided if, in a purulent sample, the organism is present as the predominant isolate, or

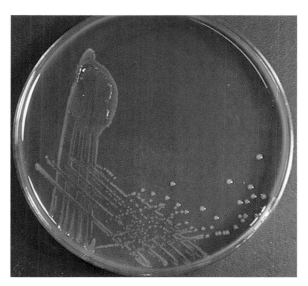

Fig. 31.4 Culture of *H. influenzae* on chocolate agar.

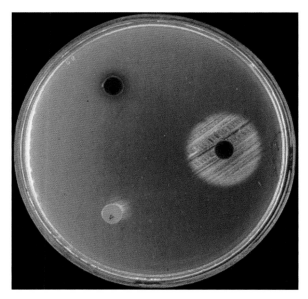

Fig. 31.5 Determination of the growth factor requirement of *H. influenzae*. Growth around the disc containing both X and V factors (right-hand disc), but not round discs of the individual factors (left-hand discs), indicates that the organism is *H. influenzae* (see text).

in a viable count of over 10^6 colony-forming units per millilitre. Addition of bacitracin (10 international units (IU)/ml) facilitates the selective isolation of *H. influenzae* from mixed cultures of respiratory organisms. Obtaining bronchial secretions by broncho-alveolar lavage reduces the problem of contamination with commensal organisms.

The temptation to obtain throat swabs in patients with suspected acute epiglottitis should be resisted, as attempts to obtain the sample may precipitate complete airway obstruction. Blood cultures are usually positive in this condition.

Blood culture is indicated for patients with suspected invasive disease. Visual examination of the bottles cannot be relied upon to indicate positive growth. Bottles must be routinely subcultured on to solid medium unless an automated detection system is used.

Identification

H. influenzae grows poorly on blood agar. On chocolate agar the colonies are smooth, grey or colourless (Fig. 31.4), with a characteristic seminal odour. Confirmation of the identity depends on demonstrating a requirement for one or both of the growth factors, X and V:

- *H. influenzae* requires both X and V factors.
- *H. parainfluenzae* requires V factor alone.
- *H. aphrophilus* and *H. ducreyi* require X factor alone.

The culture is plated on nutrient agar that is deficient in both X and V factor, and paper discs containing X factor, V factor and X + V factor are placed on the surface of the agar. After

overnight incubation, growth is observed around the discs supplying the necessary growth factors (Fig. 31.5).

PCR techniques are used to identify *Haemophilus* species in clinical specimens and as confirmatory tests on isolates. The capsular type of *H. influenzae* isolates is determined by slide agglutination with type-specific antisera, or a PCR-based method.

Antigen detection

The detection of type b polysaccharide antigen in body fluids or pus is useful, particularly in patients who received antibiotics before specimens were obtained. A rapid latex agglutination test with rabbit antibody to type b antigen is used most commonly.

In the absence of confirmatory cultures, the results should be regarded with caution as some serotypes of *Streptococcus pneumoniae* and *Esch. coli* may share similar antigens.

Antibiotic sensitivity tests

Accurate determination of the antibiotic susceptibility of *H. influenzae* requires careful standardization of the methodology. Disc tests are less reliable for detecting enzyme-mediated ampicillin (β-lactamase) and chloramphenicol (chloramphenicol acetyltransferase) resistance than microbiological or biochemical techniques that demonstrate antibiotic inactivation.

TREATMENT

H. influenzae is usually susceptible to ampicillin (or amoxicillin), chloramphenicol and tetracyclines. Among cephalosporins, compounds such as cefuroxime, cefotaxime and ceftriaxone are highly active. Other antibiotics active against *H. influenzae* include co-amoxiclav, ciprofloxacin, azithromycin and clarithromycin.

Ceftriaxone (or a related cephalosporin such as cefotaxime) is the antibiotic of first choice for the treatment of meningitis and acute epiglottitis. It is bactericidal for *H. influenzae*, achieves good concentrations in the meninges and cerebral tissues, and is highly effective.

Enzymic resistance to ampicillin is now encountered in up to 20% of type b strains in the UK. For this reason ampicillin should not be used as a single agent in meningitis when *H. influenzae* is a possibility and the results of sensitivity tests are not available. Resistance to chloramphenicol may also be encountered, but in most parts of the world this remains uncommon.

Antibiotic therapy is only one component of the clinical management of patients with haemophilus meningitis and full supportive care is required to achieve the most favourable outcome. Skilled medical and nursing care is also vital in the management of acute epiglottitis, where maintenance of a patent airway is crucial.

For the treatment of less serious respiratory infections, such as otitis media, sinusitis and acute exacerbations of chronic bronchitis, oral antibiotics such as amoxicillin, co-amoxiclav and clarithromycin are all effective. β-Lactamase-mediated amoxicillin resistance is seen in about 20% of invasive non-capsulate isolates in the UK.

EPIDEMIOLOGY OF INVASIVE DISEASE

Non-capsulate *H. influenzae* are present in the nasopharynx or throat of 25–80% of healthy people; capsulate strains (about half of which are capsular type b) are present in 5–10%. *H. influenzae* type b is an important cause of serious systemic infection in children throughout the world. Meningitis is more common in winter months, in families of low socio-economic status and in household contacts of a case. The disease is usually seen in the youngest member of a family and uncommonly in children who have no siblings. Household contacts of patients with invasive disease have an increased risk of acquiring infection if they are less than 5 years of age. The risk for children under 2 years of age is 600–800-fold higher than the age-adjusted risk for the general population.

Outbreaks of infection have been described in close communities, such as nursery schools. Contact with a case in a day care centre or nursery has also been associated with increased attack rates in children under 2 years of age, although the calculated risk is lower than that seen in household contacts.

Very high incidence rates have been reported in certain populations of Australian Aborigines, American Indians and Inuits. By contrast, very low rates have been reported in Hong Kong Chinese. It is possible that socio-economic considerations are important in determining such racial differences, but host genetic factors may also play a role. Immunosuppression, whether iatrogenic or associated with malignancies (especially Hodgkin's disease), asplenia or agammaglobulinaemia, also predispose to invasive disease.

The mortality rate associated with *H. influenzae* meningitis is around 5%. Neurological sequelae, including intellectual impairment, seizures and profound or severe hearing loss, may be present in 10–20% of survivors.

Haemophilus influenzae type b disease in the UK

Before the introduction of conjugate Hib vaccine into the infant immunization schedule in 1992, approximately 1500 cases of invasive Hib disease, including 900 cases of meningitis, occurred in the UK every year, with 60 deaths. Immunization rates have remained high (about 93%), and between 1992 and 1999 *H. influenzae* type b disease in children less than 5 years old fell by 95% (Fig. 31.6). In 1998 only 21 cases of invasive Hib disease were reported in England and Wales. From 1999 there was a small but gradual increase in the number of cases, most notably in fully immunized children born in 2000 and 2001, but also in older children and adults. In 2003 children over 6 months and under 4 years of age were offered a booster dose of Hib vaccine. The campaign had a marked effect on the incidence of invasive Hib disease, most dramatically in 1–4-year-olds, but also in older children and adults.

CONTROL

Active immunization

Early haemophilus vaccines consisting of purified type b capsular polysaccharide were poorly immunogenic in children less than 2 years old and in patients with immune deficiency. Conjugate vaccines in which the polysaccharide is covalently coupled to proteins such as tetanus toxoid, a non-toxic variant of diphtheria toxin, *Neisseria meningitidis* outer membrane protein or diphtheria toxoid produce a lasting anamnestic response, which is not age-related and may be effective in high-risk patients who respond poorly to polysaccharide vaccine alone. In the UK, *H. influenzae* type b vaccine is offered routinely to infants at 2, 3 and 4 months of age as part of a pentavalent diphtheria, tetanus, pertussis, polio, Hib vaccine (see Ch. 69). A Hib booster dose (in combination with meningococcus C vaccine), given at 12 months of age, was introduced in 2006. Immunization of

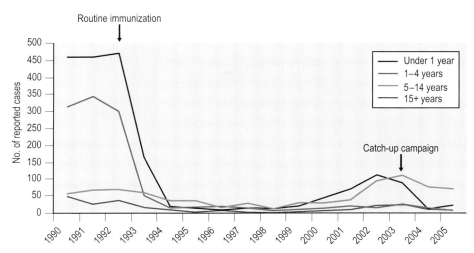

Fig. 31.6   Incidence of invasive *H. influenzae* type b infections in England and Wales by age group, October 1990 to June 2005. (Data from Health Protection Agency Centre for Infections.)

infants significantly reduces pharyngeal carriage of Hib, but has no effect on the carriage of other capsular types or non-capsulate strains.

Conjugate Hib vaccine is recommended for children and adults with splenic dysfunction, because they are at increased risk of invasive Hib infection.

Prophylaxis

Rifampicin (20 mg/kg daily, up to a maximum of 600 mg daily) given orally once daily for 4 days eradicates carriage of *H. influenzae*. The drug has been used to prevent secondary infection in household and nursery contacts, but conclusive evidence of efficacy is not available.

Unvaccinated siblings of the index case who are less than 10 years old should be immunized with conjugate vaccine. If there are household contacts who are at risk (children under 4 years of age or any individual with immunosuppression), they and the index case should be given chemoprophylaxis.

When two or more cases of Hib disease occur in a nursery or playgroup within 120 days, chemoprophylaxis should be offered to all room contacts – carers and children. Any unvaccinated children under 10 years of age should be offered Hib vaccine.

The widespread use of conjugate vaccine may soon render chemoprophylaxis unnecessary.

HAEMOPHILI OTHER THAN *H. INFLUENZAE*

H. influenzae biogroup aegyptius

This organism, formerly known as the *Koch–Weeks bacillus* and *H. aegyptius*, is now classified as a subgroup of *H. influenzae*. It is indistinguishable from *H. influenzae* biotype III in routine tests, but can be identified by a PCR method. It causes a purulent conjunctivitis and *Brazilian purpuric fever*, a clinical syndrome first recognized in Brazil in 1984, in which conjunctivitis proceeds to an overwhelming septicaemia resembling fulminating meningococcal infection. Ampicillin in combination with chloramphenicol has been successful when treatment has been started sufficiently early.

H. ducreyi

Haemophilus ducreyi is responsible for a sexually transmitted infection, *chancroid*, which is most prevalent in tropical regions, particularly Africa and South-East Asia. Patients present with painful penile ulcers (*soft sore* or *soft chancre*) and inguinal lymphadenitis. Typical small Gram-negative bacilli can be seen in material from the ulcers or in pus from lymph node aspirates. It is likely that the lesions of chancroid have facilitated the transmission of human immunodeficiency virus (HIV) in some tropical countries.

Haemophilus ducreyi is an extremely fastidious organism requiring specialized culture media. A multiplex PCR has been developed for the simultaneous amplification of DNA targets from *H. ducreyi*, *Treponema pallidum* and herpes simplex virus types 1 and 2.

Chancroid is best treated with azithromycin, ceftriaxone, ciprofloxacin or erythromycin. Azithromycin and ceftriaxone have the advantage of single-dose therapy. Resistance to sulphonamides, trimethoprim and tetracyclines has reduced the usefulness of these agents. Strains with intermediate resistance to ciprofloxacin or erythromycin have been reported. Treatment failures are much more likely in patients with concurrent HIV infection.

An unrelated Gram-negative rod, *Calymmatobacterium granulomatis*, causes a somewhat similar sexually transmitted disease, *granuloma inguinale* or *donovanosis*, in parts of the tropics. Intracellular organisms, known as *Donovan bodies* (not to be confused with the Leishman–Donovan bodies of leishmaniasis), can be demonstrated in the stained smears from the lesions. Tetracyclines are usually used in treatment.

Other haemophili

Various other species of *Haemophilus*, notably *H. parainfluenzae*, *H. aphrophilus* and *H. paraphrophilus*, are occasionally implicated in human disease, notably infective endocarditis, but also dental infections, lung abscess and brain abscess. Endocarditis is usually treated successfully with a combination of ampicillin and gentamicin.

KEY POINTS

- Most strains of *Haemophilus influenzae* are non-capsulate but some strains possess a polysaccharide capsule (types a–f).
- *H. influenzae* type b (Hib) is a major human pathogen that causes invasive infections, including meningitis and epiglottitis.
- Non-capsulate strains cause approximately 10% of invasive infections and 90% of non-invasive respiratory infections, including otitis media and acute exacerbations of chronic obstructive airway disease.
- Some 20% of *H. influenzae* strains are ampicillin resistant (β-lactamase mediated); ceftriaxone is now the treatment of choice for invasive disease.
- Conjugate Hib vaccine is routinely offered to infants at 2, 3, 4 and 12 months in the UK.
- *H. ducreyi* causes chancroid, sexually transmitted genital ulcers that are common in Africa and South-East Asia but rare in the UK. Single-dose therapy with azithromycin or ceftriaxone is effective.
- Chancroid lesions may facilitate the transmission of HIV.

RECOMMENDED READING

Al-Tawfiq J A, Spinola S M 2002 *Haemophilus ducreyi*: clinical diagnosis and pathogenesis. *Current Opinion in Infectious Disease* 15: 43–47

Heath P T, McVernon J 2002 The UK Hib vaccine experience. *Archives of Diseases in Childhood* 86: 396–399

Jacobs M R 2003 Worldwide trends in antimicrobial resistance among common respiratory tract pathogens in children. *Pediatric Infectious Disease Journal* 22(Suppl 8): S109–S119

Kelly D F, Moxon E R, Pollard A J 2004 *Haemophilus influenzae* type b conjugate vaccines. *Immunology* 113: 163–174

Moxon E R 1992 Molecular basis of invasive *Haemophilus influenzae* type b disease. *Journal of Infectious Diseases* 165(Suppl 1): S77–S81

Peltola H 2000 Worldwide *Haemophilus influenzae* type b disease at the beginning of the 21st century: global analysis of the disease burden 25 years after the use of the polysaccharide vaccine and a decade after the advent of conjugates. *Clinical Microbiology Reviews* 13: 302–317

Swaminathan B, Mayer L W, Bibb W F et al for the Brazilian Purpuric Fever Study Group 1989 Microbiology of Brazilian purpuric fever and diagnostic tests. *Journal of Clinical Microbiology* 17: 605–608

Trotter C L, Ramsay M E, Slack M P E 2003 Rising incidence of *Haemophilus influenzae* type b disease in England and Wales indicates a need for a second catch-up vaccination campaign. *Communicable Disease and Public Health* 6: 55–58

Internet sites

Clinical Effectiveness Group (Association for Genitourinary Medicine and the Medical Society for the Study of Venereal Diseases). 2001 National Guideline for the Management of Chancroid. http://www.bashh.org/guidelines/2002/chancroid_0901b.pdf

UK Department of Health. Important changes to the Childhood Immunisation Programme. http://www.dh.gov.uk/assetRoot/04/13/71/75/04137175.pdf

UK Department of Health. *Haemophilus influenzae* type b (Hib) Immunization. http://www.dh.gov.uk/assetRoot/04/13/79/06/04137906.pdf

Todar's Online Textbook of Bacteriology. *Haemophilus influenzae*. http://textbookofbacteriology.net/haemophilus.html

32 Bordetella

Whooping cough

N. W. Preston and R. C. Matthews

The genus *Bordetella* constitutes one of the groups of very thin ovoid or rod-shaped Gram-negative bacilli, often described as parvobacteria. The genus contains two notable human pathogens, *Bordetella pertussis* and *B. parapertussis*, which cause one of the most frequent and serious bacterial respiratory infections of childhood in communities not protected effectively by vaccination.

DESCRIPTION

Bordetellae used to be classified in the genus *Haemophilus*. However, growth is not dependent on either of the nutritional factors X and V (see p. 318), and *B. parapertussis* and *B. bronchiseptica* do not require blood for their growth. The three species resemble one another in being small Gram-negative bacilli, in causing infection of the respiratory tract, and in sharing some surface antigens.

Bordetella pertussis

This is the most fastidious of the bordetellae. It produces toxic products that must be absorbed by a culture medium containing charcoal, or starch, or a high concentration of blood as in the original Bordet–Gengou medium. Because agglutination forms an important part of identification, a smooth growth is essential, and this is best provided by charcoal–blood agar. *B. pertussis* is a strict aerobe, with an optimal growth temperature of 35–36°C. Even under these conditions it usually takes 3 days for colonies to be visible to the naked eye.

Typical colonies are shiny, greyish white, convex and with a butyrous consistency. By slide agglutination, they react strongly with homologous (pertussis) antiserum, and weakly or not at all with parapertussis antiserum, depending on the specificity of the reagent. Subculture reveals no growth on nutrient agar.

The organism produces three major agglutinogens (1, 2 and 3), which can be detected by the use of absorbed single-agglutinin sera. Factor 1 is common to all strains; the three serotypes pathogenic to man (type 1,2, type 1,3 and type 1,2,3) also possess factor 2 or factor 3, or both factors, and these type-specific agglutinogens have a role in immunity to infection.

Bordetella parapertussis

This organism is readily distinguished from *B. pertussis* by its ability to grow on nutrient agar, with the production of a brown diffusible pigment after 2 days (Table 32.1). It also grows more rapidly than *B. pertussis* on charcoal–blood agar, and is agglutinated more strongly by parapertussis than by pertussis antiserum.

B. parapertussis usually causes less severe illness than *B. pertussis*, and is uncommon in most countries, although occasionally it has been responsible for outbreaks of whooping cough.

Bordetella bronchiseptica

Colonies of this species are visible on nutrient agar after overnight incubation; it differs from the other species by also being motile and by producing an obvious alkaline reaction in the Hugh and Leifson medium that is used to differentiate oxidative from fermentative action on sugars. It is therefore placed by some taxonomists in the genus *Alcaligenes*; however, it is readily distinguished from the intestinal commensal *Alcaligenes faecalis* by its rapid hydrolysis of urea.

Although rarely encountered in human infection, *B. bronchiseptica* is a common respiratory pathogen of animals, especially laboratory stocks of rodents. Because it shares antigens with other bordetellae, animals must be checked for freedom from bordetella antibody before they are used in the preparation of specific antisera. For this reason, sheep or donkeys have sometimes been used in preference to rodents.

Table 32.1 Differential properties of *B. pertussis* and *B. parapertussis*

Property	B. pertussis	B. parapertussis
Duration of incubation to yield visible colonies	3 days	2 days
Growth on nutrient agar	None	Good
Pigment diffusing in medium	None	Brown
Slide agglutination with:		
Pertussis antiserum	Strong	Weak
Parapertussis antiserum	Weak	Strong

PATHOGENESIS

Whooping cough is a non-invasive infection of the respiratory mucosa, with man as the only natural host. In a typical case, an incubation period of 1–2 weeks is followed by a 'catarrhal' phase with a simple cough but no distinctive features. Within about a week, this leads into the 'paroxysmal' phase, with increasing severity and frequency of paroxysmal cough, which may last for many weeks and be followed by an equally prolonged 'convalescent' phase.

In the initial preventable stage of the infection there is colonization of the ciliated epithelium of the bronchi and trachea with vast numbers of bacteria whose agglutinogens play a vital and type-specific role in attachment. *B. pertussis* produces a tracheal cytotoxin that paralyses the cilia and leads to paroxysms of coughing as an alternative means of removing the increased mucus. Another bacterial product, called pertussis toxin, is responsible for the characteristic lymphocytosis in uncomplicated whooping cough. A subsequent increase in the number of neutrophils, together with fever, suggests broncho-pneumonia or other secondary infection, perhaps with pyogenic cocci. Blockage of airways may cause areas of lung collapse, and anoxia may lead to convulsions, although with modern intensive care the disease is rarely fatal.

Clinical features

Per-tussis (severe cough) has been recognized as a clinical entity for several centuries. Typically, the child suffers many bouts of paroxysmal coughing each day; during these, with no pause for air intake, the tongue is protruded fully, fluids stream from the eyes, nose and mouth, and the face becomes cyanotic; when death seems imminent, a final cough clears the secretions and, with a massive inspiratory effort, air is sucked through the narrowed glottis, producing a long high-pitched whoop – hence the term *whooping cough*. Such attacks often terminate with vomiting. Between them the patient does not usually appear ill.

If a characteristic attack is witnessed, a diagnosis of pertussis is usually made on clinical grounds alone. However, the illness is often mild and atypical, especially in:

- older children and adults
- younger children who have been incompletely immunized
- very young infants partially protected by maternal antibody.

In these cases, the laboratory has a vital role in diagnosis, because similar coughing may be caused by a variety of viruses and such illness is generally mild and of short duration. Here, the term *pseudo-whooping cough* has been applied aptly. False diagnosis may create a popular impression of pertussis as a trivial disease; it is recommended that, in the absence of positive bacterial culture, whooping cough should not be diagnosed for paroxysmal coughing lasting less than 3 weeks. With genuine pertussis, the illness is likely to persist for months rather than weeks. Furthermore, because pertussis vaccine cannot be expected to protect against viral infection, estimates of vaccine efficacy require an accurate diagnosis. Thus, a study in the UK found an efficacy of 93% against pertussis confirmed by bacterial culture, compared with only 82% for cases diagnosed solely on clinical criteria.

In developing countries, whooping cough is still a major cause of death; however, in developed countries, concern is focused on a very prolonged and frightening illness with possible respiratory and neurological sequelae, on the anxiety and exhaustion of parents, and on the heavy use of hospital and community medical resources.

Experimental infection in animals

Some, though not all, of the features of human disease have been produced in animals. Thus, marmosets and rabbits develop catarrh during prolonged colonization of the respiratory tract; they produce a similar range of agglutinin responses to vaccination, and this immunity shows evidence of serotype specificity.

However, the mouse, which has long been used in the evaluation of pertussis vaccine potency, does not show these features. It can be infected and even killed by degraded organisms of serotype 1, which have lost the type-specific agglutinogens (2 and 3) that are necessary for human infection. It also reveals additional properties of pertussis toxin that are not seen in humans: histamine sensitization and islets activation (including hyperinsulinaemia and hypoglycaemia). Furthermore, pertussis toxin and other components of the organism – filamentous haemagglutinin and pertactin – are virulence factors in the mouse, but their role in human infection is uncertain. It is therefore necessary to interpret experimental evidence from mice or other small rodents with caution.

LABORATORY DIAGNOSIS

Bacterial culture

Because atypical clinical cases occur frequently, laboratory confirmation of the diagnosis is often essential. Bacterial culture has the highest specificity of the tests available. In the absence of really effective therapy, accuracy in the diagnosis is more important than speed. Bacterial culture has the additional advantage that the isolate can be serotyped and genotyped, and thus provide valuable epidemiological information.

So rarely has a positive culture been obtained from a healthy person, other than one incubating the disease, that a false-positive result can be discounted. Moreover, with good technique of swabbing and culture, the organism can be recovered up to 3 months from the onset of illness when coughing persists. This casts doubt on the widespread belief that the bacterium is eliminated in a few weeks, and has implications for the transmission of infection.

Although the disease is mainly in the lower respiratory tract, the organism can be recovered readily from the nasopharynx. 'Cough plates' and postnasal swabs are unsatisfactory because of overgrowth by commensal bacteria. A pernasal swab acquires fewer commensals, and these can be suppressed by penicillin (0.25 units/ml = 0.15 mg/l) or cefalexin (30 mg/l) in the charcoal–blood agar plate; higher concentrations may suppress bordetellae. Pernasal swabs on flexible wire are available commercially; the tip is directed downwards and towards the midline, passing gently along the floor of the nose for about 5 cm (depending on the patient's age) until stopped by the posterior wall of the nasopharynx. Practice in swabbing is necessary – initially with a co-operative adult! If old enough, patients should be warned to expect a tickling sensation but no pain; a child's head should be held steady. Ideally, a segment of the culture plate should be inoculated immediately after withdrawal of the swab. The use of transport medium reduces the isolation rate. A single swab may yield a negative culture, but isolation rates of up to 80% may be achieved by taking specimens on several successive days.

In the laboratory, the inoculum is spread to give separate colonies, and the plate is incubated for at least 7 days before being discarded as negative. Because of the prolonged incubation, the medium should have a depth of 6–7 mm (40 ml in a 9-cm dish) and a bowl of water may be placed in the incubator to reduce drying of the culture. Cefalexin tends to give 'rough' growth, which may have to be subcultured on cefalexin-free medium for reliable serological identification.

Detection of bacterial antigens or DNA

Bordetella antigens may be detected in serum and urine in tests with specific antiserum. Alternatively, bacteria in nasopharyngeal secretions are labelled with fluorescein-conjugated antiserum and examined by ultraviolet microscopy. This method has the theoretical advantage, compared with culture, of detecting dead bordetellae. However, false-negative results are likely unless the patient's own antibody is removed from the bacteria by enzyme before application of the fluorescent reagent. Moreover, false-positive results may occur because of serological cross-reactions with organisms such as staphylococci, yeasts, haemophili and moraxellae, some of which resemble bordetellae microscopically. Bordetella antiserum should be absorbed with these organisms, but appropriate reagents are not readily available. Reports on the high specificity of this test lack conviction in the absence of reliable evidence on the true diagnosis.

There have been numerous studies of the polymerase chain reaction (PCR) in the detection of bordetella DNA in nasopharyngeal specimens, by the use of various primers. However, the method is relatively expensive and technically demanding compared with culture; as yet, there is a lack of consensus on its diagnostic reliability, with the need to detect both *B. pertussis* and *B. parapertussis*, and then to distinguish between them. Moreover, to be of epidemiological value, these methods would need modification to enable them to identify the serotype of the infecting strain of *B. pertussis*.

Furthermore, a test with very high sensitivity may merely detect the transient presence of a small number of bacteria attempting to colonize the mucosa before being eliminated from an immune host; such scanty bacteria would constitute a minimal risk of transmission to contacts.

Detection of bordetella antibody

Sera and nasopharyngeal secretions can usefully be examined for antibody. However, a negative result does not exclude pertussis because the serological response is often slow and weak, especially in very young children. More importantly, a positive result needs careful interpretation because antigens are shared with other organisms (see above). Even with insensitive tests, such as agglutination, pertussis antibodies are readily detected in the sera of healthy persons. More sensitive techniques, such as enzyme-linked immunosorbent assay (ELISA), are liable to increase the number of false-positive results and thereby give spurious respectability to a diagnosis that should rightly be 'pseudo-whooping cough'. The need at present in serological diagnosis is not for greater sensitivity but for greater specificity. Even then, the detected antibody may be an 'anamnestic' response to previous pertussis infection or vaccination, provoked non-specifically by a current, antigenically unrelated, illness.

Nevertheless, bacterial agglutination may be a useful guide in the serodiagnosis of pertussis, provided that sera are absorbed with type 1 organisms and titrated for the more specific agglutinins 2 and 3, and that paired sera are taken about 3 weeks apart to detect a greater than four-fold rise in titre. This requires a serum sample of at least 0.5 ml on each occasion.

Differential blood count

Although lymphocytosis is a characteristic response to pertussis infection, many cases of true pertussis do not develop a significant increase in circulating lymphocytes; conversely, there are so many other causes of lymphocytosis that a positive result lacks diagnostic specificity.

TREATMENT

Antimicrobial drugs

Most antibiotics have little or no clinical effect when the infection is well established, even though the organism may be sensitive in vitro.

The drug of choice is erythromycin (or one of the newer macrolides such as clarithromycin), which may reduce the severity of the illness if given before the paroxysmal stage. If given for at least 14 days, erythromycin sometimes eliminates the organism and so reduces the exposure of contacts. However, positive cultures are frequently obtained after short periods of erythromycin therapy.

Erythromycin may also be given to protect non-vaccinated infants, although it seems unrealistic to expect this treatment to be maintained throughout the several months that the older sibling (or adult) may remain infectious.

Appropriate antibiotics should, of course, be administered to patients who show signs of secondary bacterial infection.

Other measures

Cough suppressants and corticosteroids may control the paroxysms, but may be harmful by encouraging retention of secretions. Cyanosis and anoxia can be reduced by avoiding sudden noises, excitement and excessive medical examination, which tend to precipitate paroxysms. Mucus and vomit should be removed to prevent their inhalation.

Treatment with pertussis immunoglobulin has been tried, but with limited success, probably because such materials have never been checked for the presence of all three agglutinins.

Because of the dearth of effective therapy for whooping cough, the widespread use of pertussis vaccine is of supreme importance (see below).

EPIDEMIOLOGY

Source and transmission of infection

Most new cases arise from patients (usually children, occasionally adults) with typical symptoms, presumably because the paroxysmal cough provides an efficient means of droplet dissemination. Atypical cases have only a minor role in transmission; long-term asymptomatic carriage is unknown. The degree of contact is important: 80–90% of non-immune siblings exposed in the household become infected, compared with less than 50% of non-immune child contacts at school. Antibiotic therapy may reduce transmission, but is not completely effective.

Incidence and mortality

Pertussis infection occurs worldwide, affects all ages, and is a major cause of death in malnourished populations. In developed countries the mortality rate has gradually declined with a combination of improved socio-economic conditions, availability of intensive care in hospitals, and antibiotic therapy to combat secondary infection. However, the latter constitute an unnecessary use of medical resources for a disease that is eminently preventable by vaccination.

The disease is most severe and the morbidity rate highest in the first 2 years of life; most fatal cases are in infants less than 1 year old. Even very young babies are not immune: maternal antibody does pass to the fetus, but it rarely contains all three agglutinins and protection is incomplete.

Although one attack usually confers long-lasting immunity, infection with a different serotype of the organism can occur subsequently.

The disease occurs in epidemic waves at about 4-year intervals – the time needed to build up a new susceptible population after the 'herd' immunity produced by an epidemic. Figure 32.1 illustrates the pattern for England and Wales, where whooping cough has been a notifiable disease since 1940. The maintenance of a 4-year cycle presumably results from the interaction of various factors, such as the degree of artificial immunity produced by high vaccination rates, and the levels of natural immunity that follow either large epidemics or a high background incidence of endemic pertussis in inter-epidemic intervals.

Figure 32.1 also illustrates how variations in the rates of uptake of pertussis vaccine have affected the incidence

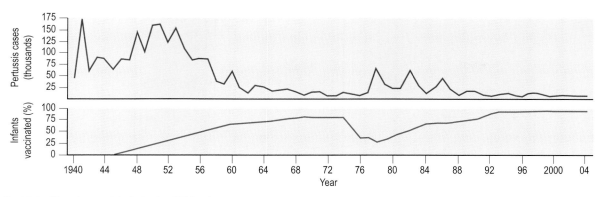

Fig. 32.1 Whooping cough in England and Wales since 1940.

Table 32.2 Occurrence of serotypes of *B. pertussis* in different communities

	Serotypes of *B. pertussis* that cause whooping cough		
	1,2 ←———→[a] 1,2,3 ←———→[a]		(1)[b],3
Prevalence in non-vaccinated communities	High		Low
Prevalence in communities vaccinated with type 1,2	Low		High[c]
Prevalence in communities vaccinated with type 1,3 or type 1	High[c]		Low
Relative incidence in communities vaccinated with type 1,2,3 (before complete elimination)	Lower		Higher

[a]The arrows indicate spontaneous reversible variation of serotype.
[b]The parentheses (1) indicate a weak factor 1 component, so that type 1,3 does not protect against type 1,2 infection, and vice versa.
[c]Infection occurs even in vaccinated children.

of whooping cough more than the steady improvement in the general health of the population that continued throughout the period. After the gradual introduction of pertussis vaccination during the 1950s, there was a steady reduction in the size of epidemics until the 1970s. Unfounded fear of brain damage caused a loss of faith in the vaccine, and three large epidemics occurred before the slow restoration of confidence in the vaccine began to take effect.

Prevalence of serotypes

The three serotypes of *B. pertussis* pathogenic for man are liable to spontaneous reversible variation, as shown by the arrows in Table 32.2. Fimbriae are readily demonstrable on strains possessing agglutinogen 2, aiding colonization of the respiratory mucosa; serotypes 1,2 and 1,2,3 predominate among the strains isolated from patients in unvaccinated communities.

In the 1960s, many countries used type 1,2 vaccine, and these two serotypes were suppressed. However, type 1,3 organisms, which have a weak factor 1 component, became predominant in these countries and caused infection even in vaccinated children before the vaccine was modified in the late 1960s by the addition of agglutinogen 3. Similarly, countries that used a vaccine deficient in agglutinogen 2, or in both agglutinogens 2 and 3, saw a predominance of type 1,2 infection, even in vaccinated children.

To be effective, it seems, a vaccine must contain all three agglutinogens, as recommended by the World Health Organization. However, because the agglutinin 3 response is usually the weakest (when type 1,2,3 vaccine is used), type 1,3 organisms are the last to be eliminated from a community with an effective vaccination programme.

Genotypes

It is not possible to trace the spread of infection by serotyping isolates because there are only three serotypes and they undergo spontaneous variation. In contrast, more than 40 genotypes have been demons-

Table 32.3 Plain and adsorbed pertussis vaccine

Vaccine	Immune response	Adverse reactions
Plain	Weaker	More
Adsorbed (on to adjuvant)	Stronger	Fewer

trated in macro-restriction profiles by pulsed-field gel electrophoresis of DNA digests. Isolates within a household have been shown to belong to the same genotype, but the technique is probably too expensive and time consuming for routine use.

CONTROL

Treatment and quarantine

Antibiotics and immunoglobulins currently available are not very effective for the treatment of patients or the protection of contacts. Because patients typically disseminate the organism for many weeks or months, and because children are infectious even before the most characteristic symptoms develop, control of the disease by quarantine is unrealistic.

Vaccination

Vaccination is safe and more than 90% effective, and is strongly recommended. The vaccines in general use are suspensions of whole bacterial cells, killed by heat or chemicals, and are administered by deep intramuscular injection. Adsorption of the bacteria on to an adjuvant, such as aluminium hydroxide, enhances the immune response (particularly important with factor 3) and also causes fewer adverse reactions (Table 32.3).

Three conditions are essential for good protection:

1. Presence of all three agglutinogens in the vaccine
2. Use of adsorbed vaccine (i.e. with adjuvant)
3. A minimum of three doses, at monthly intervals.

Since there is little passive protection from the mother and effective active immunity cannot be achieved until after the third injection of vaccine, the first dose is given as soon as a good response can be obtained. In many countries, therefore, the first injection is recommended at 3 months of age. In the UK and some other countries, concern for the vulnerability of the young infant prompted a start at 2 months, although the immune response is somewhat weaker and there have been doubts about the effectiveness of this early start. An upsurge of invasive infection with *Haemophilus influenzae* type b followed

in the Netherlands when the policy of giving the first dose at 3 months was changed to 2 months. For these and other aspects of vaccination, see also Chapter 69.

Safety of pertussis vaccine

Minor adverse reactions occur in about one-half of vaccinated children, and can be considered as part of the normal immune response. Parents should therefore be warned to expect possible erythema and local swelling, slight feverishness and crying. Of much more concern are possible neurological sequelae, but the National Childhood Encephalopathy Study in the UK and several other studies have shown that pertussis vaccination seems merely to trigger the manifestation of neurological disorders that would occur in any case, and there is no firm evidence that the vaccine causes serious long-term adverse side effects.

Contra-indications to pertussis vaccination

Severe adverse reaction to a previous dose has been considered the only firm contra-indication. A current feverish illness (not merely snuffles) is cause for postponement until the child is well. Parental concern over a possible neurological contra-indication is due reason for consultation with a paediatrician. Allergy is not a contra-indication, neither is age – children who missed vaccination in infancy may receive the normal three-dose course. Vaccination is also sometimes advised for adults, such as nurses and doctors in appropriate hospitals.

Acellular pertussis vaccine

Although doubts about the efficacy and safety of whole-cell pertussis vaccine have largely passed, the urge to identify the essential protective components persists. Trials in different countries (including one with type 1,2 prevalence and another with type 1,3 prevalence), and with vaccines containing various components, may reveal a correlation between the protection of children and the response to individual pertussis antigens and type-specific agglutinogens.

A large-scale trial with good diagnostic criteria in Sweden showed that antibodies to pertussis toxin and filamentous haemagglutinin do not confer protection in the child, although these antigens have been incorporated in nearly all acellular vaccines because they provide mouse protection (see above). This trial did indicate a correlation between agglutinin titres and protection of children, and it showed that whole-cell vaccine had a higher efficacy than any of the acellular vaccines used. Moreover, the whole-cell vaccine has an additional adjuvant effect on other antigens given simultaneously;

in the UK an increase in *Haemophilus influenzae* type b infections followed the replacement of whole-cell pertussis vaccine with an acellular product owing to a reduced antibody response.

An eventually satisfactory acellular vaccine may be too expensive for routine use, especially for developing countries where the need is greatest.

Eradication

Vaccination aims at a herd immunity, which breaks the cycle of transmission because the organism dies before finding a new susceptible host (see Ch. 69).

In several countries, good whole-cell vaccine is available. Eradication is possible, but even if high levels of vaccination of infants are maintained it will be some years before adequate herd immunity is achieved within the child population.

KEY POINTS

- Whooping cough is caused by *Bordetella pertussis*, and occasionally by *B. parapertussis*.
- Paroxysmal cough, typically with vomit and whoop, lasts for several weeks or even months.
- Pertussis is a human disease (most severe in children) and has no animal or environmental reservoir.
- Clinical diagnosis is unreliable in mild cases.
- Bacterial culture is the optimal method of laboratory confirmation.
- Therapy for established cases is unsatisfactory.
- Immunity to *B. pertussis* infection is serotype specific.
- Vaccine containing the three serotype antigens is safe, and is effective if the vaccination schedule is optimal.
- Eradication of whooping cough by childhood vaccination is an achievable goal.

RECOMMENDED READING

Cherry J D, Brunell P A, Golden G S, Karzon D T 1988 Report of the Task Force on Pertussis and Pertussis Immunization – 1988. *Pediatrics* 81: 939–984

Editorial 1992 Pertussis: adults, infants, and herds. *Lancet* 339: 526–527

Jenkinson D 1995 Natural course of 500 consecutive cases of whooping cough: a general practice population study. *British Medical Journal* 310: 299–302

Khattak M N, Matthews R C 1993 Genetic relatedness of *Bordetella* species as determined by macrorestriction digests resolved by pulsed-field gel electrophoresis. *International Journal of Systematic Bacteriology* 43: 659–664

Preston N W 1993 Eradication by vaccination: the memorial to smallpox could be surrounded by others. *Progress in Drug Research* 41: 151–189

Preston N W 2000 Pertussis (whooping-cough): the road to eradication is well sign-posted but erratically trodden. *Infectious Diseases Review* 2: 5–11

Preston N W 2003 Why the rise in *Haemophilus influenzae* type b infections? *Lancet* 362: 330–331

Preston N W, Matthews R C 1995 Immunological and bacteriological distinction between parapertussis and pertussis. *Lancet* 345: 463–464

Preston N W, Matthews R C 1998 Acellular pertussis vaccines: progress but déjà vu. *Lancet* 351: 1811–1812

Salmaso S and the Eurosurveillance editorial team 2004 Pertussis vaccine schedules across Europe. *Eurosurveillance* 9: 70–71

Thomas M G, Lambert H P 1987 From whom do children catch pertussis? *British Medical Journal* 295: 751–752

Wardlaw A C, Parton R (eds) 1988 *Pathogenesis and Immunity in Pertussis*. Wiley, Chichester

33 Legionella

Legionnaires' disease; Pontiac fever

J. Hood and G. F. S. Edwards

The Legionellaceae are Gram-negative rods whose natural habitat is water. There are more than 50 genetically defined species, of which much the most important is *Legionella pneumophila*. This species can be subdivided on the basis of deoxyribonucleic acid (DNA) relationships into three subspecies:

- *L. pneumophila* ssp. *pneumophila* and *L. pneumophila* ssp. *fraseri*, which have been described in human disease
- *L. pneumophila* ssp. *pascullei*, which has so far been isolated only from the environment.

Eighteen *Legionella* species have been associated with human disease (Table 33.1), but most infections are caused by just one of the many serogroups of *L. pneumophila*: serogroup 1. Other serogroups and species, such as *L. micdadei*, *L. bozemanii* and *L. longbeachae*, account for a few cases; other legionellae are rare causes of infection. Human beings do not normally carry legionellae. Infection is usually acquired accidentally and the disease is not transmissible from person to person.

Legionellae give rise to two main clinical syndromes:

1. *Legionnaires' disease*, a pneumonia that may progress rapidly unless treated with appropriate antibiotics. In previously healthy subjects the mortality rate is about 10%, but in those with nosocomial infection the rate may be much higher.

2. *Pontiac fever*, a brief febrile influenza-like illness that may be slow to resolve fully, but does not cause death.

Legionellae have rarely been associated with other infections such as prosthetic valve endocarditis or wound infection, but these are usually nosocomial infections.

DESCRIPTION

In biological material (e.g. sputum or lung) or in water deposits legionellae are short rods or coccobacilli, but in culture they become longer and are sometimes filamentous.

Although they are weakly Gram-negative, they stain poorly with Gram's stain, but may be stained by a silver impregnation method. Specific fluorescent antibody stains are used diagnostically. The organisms may be numerous, particularly in patients who are infected in hospital or who are immunosuppressed, but are often present only in very small numbers in the scanty sputum that is characteristic of legionnaires' disease. They have not been demonstrated in patients with Pontiac fever.

Legionellae are exacting in their growth requirements and grow best on buffered charcoal yeast-extract agar (BCYE), which contains iron plus cysteine as an essential growth factor. Some legionellae grow better in the presence of 2.5–5% carbon dioxide at 35–36°C. Colonies usually appear after incubation for 48 h to 5 days, but species other than *L. pneumophila* may take up to 10 days. Colonies have a 'cut glass' appearance on examination

Table 33.1 *Legionella* species associated with human disease

Species	Number of serogroups	Autofluorescence under ultraviolet light
L. anisa	1	+
L. birminghamensis	1	–
L. bozemanii	2	–
L. cincinnatiensis	1	–
L. dumoffii	1	+
L. feeleii	2	–
L. gormanii	1	+
L. hackeliae	2	–
L. jordanis	1	–
L. lansingensis	1	–
L. longbeachae	2	–
L. maceachernii	1	–
L. micdadei	1	–
L. parisiensis	1	+
L. pneumophila	16	–
L. sainthelensi	2	–
L. tucsonensis	1	+
L. wadsworthii	1	–

+, Blue–white fluorescence; –, no autofluorescence.

under the plate microscope. Colonies of some *Legionella* species show blue–white (or red – not yet seen in species isolated from man) autofluorescence on illumination with long-wave ultraviolet light (see Table 33.1).

Species and serogroups within species have specific heat-stable lipopolysaccharide antigens, and further identification depends on serological examination, usually by the fluorescent antibody test or by slide agglutination with specific antisera. Species may be differentiated further on the basis of serology into serotypes within serogroups, but this applies, at the present time, almost entirely to *L. pneumophila* serogroup 1. Subtyping is usually by the use of monoclonal antibodies in an immunofluorescence test. Similarities or differences between strains of the same serogroup may be revealed by genetic studies, and this may be useful in the investigation of possible outbreaks.

PATHOGENESIS

Legionnaires' disease

Infection is almost always due to *L. pneumophila* serogroup 1. The illness is characterized by:

- an incubation period of 2–10 days
- high fever
- respiratory distress
- confusion, hallucinations and, occasionally, focal neurological signs.

Once infection is established the patient develops pneumonic consolidation with an outpouring of proteinaceous fibrinous exudate, containing macrophages and polymorphs, into the alveoli. Despite the outpouring of cells, patients usually produce little sputum. Infection may extend to involve two or more lobes of the lung and renal impairment leading to renal failure may occur. The severity of the disease may range from a rapidly progressing fatal pneumonia to a relatively mild pneumonic illness. The mechanism of the distant toxic effects of the infection on the nervous system is unknown.

Patients who are debilitated, for example by immunosuppression or surgery, are more prone to infections, and these are usually more severe than those encountered as sporadic community cases. Smoking is a predisposing factor. The deterioration in body defences associated with ageing is also important: the disease is most common in those over the age of 40 years, with a peak in the 60–70-year age group. The only generally accepted mode of infection is inhalation of an aerosol of fine water droplets containing the organism, but some believe that aspiration of water containing legionellae can also lead to infection. There is presently no evidence that ingestion plays a part in pathogenesis, although the demonstration

Table 33.2 Diagnostic tests for legionella infection

Nature of test	Test	Appropriate specimen
Detection of whole organism	Culture FAT Gene probes	Sputum or other pathological material
Detection of soluble antigen	ELISA	Urine
Detection of antibody	FAT ELISA	Serum

FAT, fluorescent antibody test; ELISA, enzyme-linked immunosorbent assay.

of *L. pneumophila* in bowel contents raises this as a possibility.

Animal models provide some insight into the train of events in the lungs. Guinea-pigs infected by inhalation of an aerosol containing legionellae develop a lobular pneumonia that rapidly becomes confluent. Instillation of a protease produced by *L. pneumophila* into the lungs of guinea-pigs produces a pneumonia that appears to be the same as that caused by inhalation of intact bacteria.

At the cellular level, legionellae are engulfed by monocytes and can survive therein as intracellular parasites. The duration of intracellular parasitization is unknown but the persistent excretion of group-specific legionella antigen in the urine of some patients, as opposed to its more transient appearance in most cases of legionella pneumonia, suggests that in some patients intracellular parasitism may be prolonged. There is, however, no evidence of a chronic carrier state or chronic infection.

Pontiac fever

The pathogenesis of this non-pneumonic, non-fatal form of legionella infection is not understood. Living legionellae have been isolated from sources of infection associated with outbreaks of Pontiac fever, and legionella antigen has been demonstrated in the urine of some cases, indicating that appreciable quantities of legionellae have been involved in the infection.

LABORATORY DIAGNOSIS

The tests used are listed in Table 33.2. As in any case of pneumonia, respiratory secretions (sputum, bronchial aspirate or washings), as well as pleural fluid, lung biopsy or autopsy material, should be examined by microscopy and culture. Gram-stained films are of little value except to demonstrate the presence of other pathogens and organisms that may interfere with the isolation of legionellae. Legionellae have occasionally been isolated from blood culture, but this is not a rewarding routine procedure.

Immunofluorescent staining with monoclonal or poly-clonal antisera is a specific diagnostic method but legionellae are usually hard to find in the scanty sputum produced by patients. Specific cultures are made on BCYE medium with and without antibiotics added to suppress other respiratory tract flora. Potentially contaminated material such as sputum or post-mortem material may also be heated at 50°C for 30 min in order to diminish growth by less heat-stable respiratory tract organisms that may inhibit growth of legionellae in culture. In heavy infections, legionellae may appear on BCYE media, but not on standard media, after incubation for 48 h at 36°C in air, preferably enriched with 2.5% carbon dioxide. Some of the less common legionellae may take longer to grow and cultures should not be discarded until after 10–14 days of incubation. Colonies having a 'cut glass' appearance by plate microscopy (as well as those fluorescing blue–white under ultraviolet light) are Gram stained, and single colonies are subcultured on to blood agar or cysteine-deficient medium to show that they will *not* grow on these media. Cultures are identified by use of specific antisera in an immunofluorescence test.

Antigen tests

The examination of urine for legionella antigen by enzyme-linked immunosorbent assay (ELISA) is a rapid and specific method of identifying *L. pneumophila* as the likely cause of a pneumonia. Most legionella infections are now diagnosed by urine antigen tests, but failure to detect urinary antigen does not exclude infection with legionellae other than *L. pneumophila* serogroup 1.

Serology

Although antibodies take at least 8 days to develop after the onset of infection, some patients may not reach hospital until this period has elapsed, so it is worthwhile examining serum for antibodies to *L. pneumophila* on admission to hospital. Further sera should be taken at intervals to show the development of antibodies or a rise in antibody titre. Antibodies usually develop after 8–10 days of illness and then increase in titre, but some patients may not produce antibody for some weeks or, rarely, for several months.

- A four-fold or greater rise in antibody titre in a typical clinical case indicates infection with legionella.
- A single titre of 256 or more is presumptive of infection.

In some cases proven by culture, lower titres may be found, especially when death occurs early in the illness. Antibody may persist for months or years and can be a source of confusion, as may cross-reacting antibody produced by some patients with *Campylobacter* infection. At present the only fully validated antibody test is that for infection by *L. pneumophila* serogroup 1, although patients proven by bacterial culture to be infected by other legionellae produce antibodies to the infecting strain.

TREATMENT

High-dose intravenous erythromycin is the standard therapy in legionella pneumonia. Azithromycin exhibits better activity than erythromycin in vitro and penetrates well into cells and lung tissue. It may emerge as the drug of choice if its clinical efficacy is confirmed.

In severe cases a macrolide may be supported by rifampicin and, possibly, by the addition of ciprofloxacin, although evidence for the value of this agent is not well documented.

Susceptibility testing of clinical isolates is rarely justi-fied. It is technically demanding and, as person-to-person transmission of infection does not occur, antibiotic use is unlikely to contribute to the development of resistance in other patients.

EPIDEMIOLOGY

In 1976 an outbreak of 182 cases of pneumonia, mainly affecting members of the American Legion, occurred at a convention in Philadelphia. This form of pneumonia became known as *legionnaires' disease*, and the bacterium associated with it as *L. pneumophila*. Since that time, legionella pneumonia has been recognized as the only acute *bacterial* pneumonia that may occur in outbreak form. This is due to the dissemination of the bacteria in aerosols, which may travel as much as 1–2 km from the source.

Infected aerosols are usually generated from warm water sources, typically:

- the ponds in cooling towers of refrigeration plants in air-conditioning systems
- domestic hot water systems in hotels and hospitals
- warm water in nebulizers and oxygen line humidifiers
- whirlpool spa baths and showers.

Legionellae are engulfed by, and survive within, free-living amoebae, and the bacteria may be protected from drying and disinfectants when present in amoebic cysts. Community outbreaks may occur on a fairly large scale and the source of infection in such outbreaks is invariably a cooling tower in which the bacteria are harboured in the water of the pond associated with the apparatus. Smaller outbreaks have been associated with domestic water supplies in hospitals, spas and hotels, and also with whirlpool spas. So-called sporadic cases may, on careful epidemiological examination, prove to be associated with cooling towers. About a third of patients with legionnaires' disease in the UK acquire their infection abroad, usually from domestic water supplies in hotels, and apparently sporadic cases account for many others.

Legionnaires' disease accounts for a small but significant number of patients with pneumonia admitted to hospital. The disease is more prevalent in late summer and autumn. This may be due to an increase in bacterial numbers in warm water both in natural sources and in cooling towers.

The route and source of infection of Pontiac fever are the same as in legionnaires' disease. Pontiac fever may affect all age groups, including children. The attack rate is high, with almost all of those exposed to the infection source being affected, whereas in legionnaires' disease the attack rate is low.

CONTROL

There is no vaccine. Nevertheless, unlike most forms of bacterial pneumonia, the disease may be prevented by the eradication of *Legionella* species in the various kinds of water source that may give rise to aerosol production. It is therefore important that any outbreak, or even one case occurring in hospital, is investigated to try to identify possible sources of an infectious aerosol so that it can be eradicated. Information about cases must be notified to epidemiological centres so that any association between cases may be established; once the source has been identified, legionellae can be eradicated from water in several ways:

- heat
- disinfection with chlorine or other biocides, including chlorine dioxide
- copper–silver ionization.

Water systems in hotels and hospitals should be managed so that hot water is heated to above 60°C before distribution and does not lie stagnant and cooling in the pipes. As legionellae do not multiply in cold water, cold water supplies should be kept below 20°C; water should not be allowed to stagnate, as it may warm up and allow the multiplication of legionellae. It may be necessary in some infected water systems to dose continuously with a suitable biocide (see above) to maintain suppression of growth. Cooling towers should be disinfected with chlorine or other biocides in a way that ensures that the growth of legionellae and possible supporting organisms, such as algae or amoebae, is suppressed.

KEY POINTS

- *Legionella* species are water-borne bacilli and include *L. pneumophila,* which is responsible for a form of pneumonia known as legionnaires' disease, and a less serious influenza-like illness called Pontiac fever.
- Many serogroups of *L. pneumophila* are recognized, but human infection is almost always caused by serogroup 1.
- Legionnaires' disease is diagnosed by demonstrating the organism in sputum or soluble antigen in urine.
- High-dose macrolide, e.g. erythromycin, with or without rifampicin, is used for treatment.
- Suppression of the organism in air-conditioning systems and water supplies in public buildings (common immediate sources) is central to control of the disease.

RECOMMENDED READING

Amsden G W 2005 Treatment of legionnaires' disease. *Drugs* 65: 605–614
Bartlett C L R, Macrae A D, Macfarlane J T 1986 *Legionella Infections.* Edward Arnold, London
Fields B S, Benson R F, Besser R E 2002 Legionella and legionnaires' disease: 25 years of investigation. *Clinical Microbiology Reviews* 15: 506–526

Harrison T G, Taylor A G (eds) 1988 *A Laboratory Manual of Legionella.* Wiley, Chichester
Stout J E, Yu V L 1997 Legionellosis. *New England Journal of Medicine* 337: 682–687
Winn W C 1988 Legionnaires' disease: historical perspective. *Clinical Microbiology Reviews* 1: 60–81

34 Brucella, bartonella and streptobacillus

Brucellosis; Oroya fever; trench fever; cat scratch disease; bacillary angiomatosis; rat bite fever

M. J. Corbel

BRUCELLA

The genus *Brucella* comprises a group of Gram-negative coccobacilli that can infect a wide range of mammals ranging from rodents to killer whales. They are of particular zoonotic and economic importance as a cause of highly transmissible disease in cattle, sheep, goats and pigs. Infection in pregnant animals often leads to abortion, and involvement of the mammary glands may cause the organisms to be excreted in milk for months or even years. Human infections arise through direct contact with infected animals, including handling of infected carcasses; indirectly from a contaminated environment; or through consumption of infected dairy produce or meat.

Brucellosis is a typical zoonosis, and person-to-person infection does not play a significant role in transmission. Infection may remain latent or subclinical, or it may give rise to symptoms of varying intensity and duration. Brucellosis can present as an acute or subacute pyrexial illness that may persist for months or develop into a focal infection that can involve almost any organ system. The characteristic intermittent waves of increased temperature that gave the name *undulant fever* to the human disease are now usually seen only in long-standing untreated cases.

DESCRIPTION

Classification

The *Brucella* genus comprises a group of closely related bacteria that probably represent variants of a single species. For convenience these have been classified into nomen species that differ from one another in their preferred animal host, genetic arrangement, phage sensitivity pattern, and oxidation of certain amino acids and carbohydrates. The main human pathogens are *Brucella abortus, B. melitensis, B. suis* and *B. canis*. The first three may be further subdivided into biovars associated with various animal hosts. *B. abortus* has a preference for cattle and other Bovidae, *B. melitensis* for sheep and goats. The first three biovars of *B. suis* preferentially infect pigs, whereas the fourth and fifth biovars have reindeer or caribou and rodents, respectively, as natural hosts. The biovars differ in their sensitivity to dyes, in production of hydrogen sulphide and in agglutination by sera monospecific for A and M epitopes. Molecular typing methods may also be used to differentiate subtypes down to the level of individual strains.

Strains designated as *B. cetaceae* and *B. pinnipediae,* isolated from seals, dolphins, porpoises and killer whales, appear to be pathogenic for man.

Morphology

Brucellae are Gram-negative coccobacilli or short bacilli, 0.5–0.7 µm wide by 0.6–1.5 µm long. They occur singly, in groups or short chains. They are non-motile, non-capsulate and non-sporing.

Culture characteristics

Brucella spp. are aerobic. However, *B. ovis* and many strains of *B. abortus*, when first cultured, are unable to grow without the addition of 5–10% carbon dioxide. All strains grow best at 37°C in a medium enriched with animal serum and glucose.

On clear solid medium, smooth, transparent and glistening ('honey droplet') colonies appear after several days. However, the organisms can mutate, especially in liquid media, forming 'rough' colonies on subculture. There is a corresponding loss in virulence and an antigenic change, so that they are no longer readily agglutinated by homologous antisera prepared against normal smooth strains. Identification can be made by the polymerase chain reaction (PCR) with appropriate primers or by a combination of biochemical, cultural, phage typing and serological tests. Rapid gallery tests may misidentify *Brucella* spp. and this has resulted in laboratory-acquired infections; they are not recommended.

Sensitivity and survival

Brucellae may be killed at a temperature of 60°C for 10 min, but dense suspensions, such as laboratory cultures, can require more drastic heat treatment to ensure their inactivation. Infected milk is rendered safe by efficient pasteurization. Brucellae are very sensitive to direct sunlight and moderately sensitive to acid, so that they tend to die out in sour milk and in cheese that has undergone lactic acid fermentation. The organisms can survive in soil, manure and dust for weeks or months, and remain viable in dead fetal material for even longer. They have been isolated from butter, cheese and ice-cream prepared from infected milk. They may survive in carcass meat, pork and ham for several weeks under refrigeration. Pickling and smoking reduce survival. They are susceptible to common disinfectants if used at appropriate concentration and temperature. They are sensitive in vitro to a wide range of antibiotics, only a few of which are effective therapeutically.

Antigenic structure

In all smooth strains the dominant surface antigen is a lipopolysaccharide (LPS) O chain, which, depending on the three-dimensional structure of the polysaccharide portion, forms A, M or C epitopes. These are common to all smooth species, but the distribution of A and M depends on biovar. Rough strains do not produce the O chain but have a common R epitope. The LPS has endotoxin activity and elicits limited antibody-mediated protection. More complete immunity is dependent on cell-mediated and cytotoxic responses elicited by ribosomal and other proteins.

PATHOGENESIS

The incubation period is usually about 10–30 days, although infection may persist for several months before causing any symptoms. *B. melitensis* and *B. suis* tend to cause more severe disease than *B. abortus* or *B. canis*. Infection by any species may give rise to a variety of symptoms and, without the fluctuating temperature to act as a guide, diagnosis may be difficult.

Brucellae can enter the body through skin abrasions, through mucosal surfaces of the alimentary or respiratory tracts, and sometimes through the conjunctivae, to reach the bloodstream by way of regional lymphatics. The organisms are facultative intracellular parasites and subsequently localize in various parts of the reticulo-endothelial system with the formation of abscesses or granulomatous lesions, resulting in complications that may involve any part of the body. Brucellae surviving within cells may cause relapses of acute disease, or a chronic syndrome may develop that is associated with continued illness and vague symptoms of malaise, low-grade fever, lassitude, insomnia, irritability and joint pain. Such 'chronic brucellosis' may follow an acute attack or develop insidiously over several years without previous acute manifestations. It rarely responds to antibiotic therapy and is probably a post-infectious response similar to the 'myalgic encephalomyelitis syndrome'.

The signs and symptoms of brucellosis are not specific. Pointers to the diagnosis are a history of occupational exposure or recent travel to endemic areas with consumption of milk products.

LABORATORY DIAGNOSIS

Brucellosis is confirmed in man by isolating the organisms from blood or other tissue samples and by serological and other tests. In animals, culture may be attempted from abortion material, placenta, milk, semen or samples of lymphoid tissue, mammary gland, uterus or testis collected post mortem.

Brucellae are easily transmitted by aerosols, ingestion and percutaneous inoculation. Samples suspected to contain brucellae must be treated as high risk. Cultures must be handled under containment conditions appropriate to class 3 pathogens.

Blood culture

When brucellosis is suspected, blood culture should be attempted repeatedly, not only during the febrile phase. Because the organisms may be scanty, at least 10 ml of blood should be withdrawn on each occasion, 5 ml being added to each of two blood culture bottles containing glucose–serum broth. One of these bottles should be incubated in an atmosphere containing 10% carbon dioxide. Preliminary lysis and centrifugation of the blood improves the isolation rate. Other materials such as bone marrow, solid tissue samples or exudates are also suitable for culture.

Subculture should be made on to serum–dextrose agar every few days; alternatively, a two-phase Castaneda culture system, in which the broth is periodically allowed to flow over agar contained within the blood culture bottle, may be used. Blood cultures should be retained for 6–8 weeks before being discarded as negative. Automated blood culture systems may also be used.

B. melitensis and *B. suis* are more frequently isolated from blood than are *B. abortus* or *B. canis*.

Serological tests

In the absence of positive cultures, the diagnosis of brucellosis usually depends on serological tests, the

Table 34.1 Results of serological tests used in the diagnosis of brucellosis

Type of brucellosis	Agglutination test	Mercaptoethanol test	Complement fixation test	ELISA
Acute	+	±	+	+
Chronic	(±)	+	+	+
Past infection	(–)	–	–	(–)

(–), weak or negative. ELISA, enzyme-linked immunosorbent assay.

results of which tend to vary with the stage of the infection (Table 34.1).

In some rural communities the sera from a proportion of the normal population are reactive in low dilutions in serological tests because of previous subclinical infection.

Standard agglutination test

This test usually becomes positive 7–10 days after onset of symptoms. During the acute stage of the disease, levels of agglutinins associated with both immunoglobulin (Ig) M and IgG continue to rise. As high-titre sera may not cause agglutination in low dilution (the *prozone* effect), a range of serum dilutions from 1 in 10 to greater than 1 in 1000 should be made.

As the disease progresses from the acute to the chronic phase and the organisms become localized intracellularly in various parts of the body, the IgM antibodies decrease; the agglutination titre falls and may become undetectable even while the patient is still ill. The absence of agglutination therefore does not rule out the possibility of infection. Persisting IgG and IgA antibodies that are no longer capable of agglutinating may be detected by complement fixation, antiglobulin or enxyme-linked immunosorbent assay (ELISA) tests. In latent or chronic infection, the complement fixation test is likely to be positive, whereas in cases of past infection it is negative.

Mercaptoethanol test

Low-titre agglutinins due to residual IgM may persist for many months or even years after the infection has cleared. The mercaptoethanol test is carried out simultaneously and in the same manner as the standard agglutination test except that the saline diluent contains 0.05 M 2-mercaptoethanol. The agglutinating ability of IgM, and sometimes IgA, is destroyed by 2-mercaptoethanol, and therefore agglutination in this test is indicative of the continuing presence of IgG and the likelihood of persisting infection.

Enzyme-linked immunosorbent assay

The ELISA for IgG and IgA antibodies shows a good correlation with active disease, especially in long-standing infection. It has largely replaced the anti-human globulin (Coombs') test, formerly used for detecting non-agglutinating (IgG) antibodies.

The sera of persons who have been immunized against cholera and of those who have antibodies to *Francisella tularensis* may give false-positive reactions in the agglutination test against brucellae. More extensive false-positive cross-reactions are produced by infection with *Yersinia enterocolitica* O9, and to a lesser extent *Salmonella* O30 and *Escherichia coli* O157. Western blotting against whole cell protein extracts may be useful for differentiation.

Other diagnostic tests

PCR methods can detect *Brucella* specifically and also give an indication of species and biovar. Promising results have been obtained in clinical studies.

The Rose Bengal plate test, a rapid slide agglutination test with a buffered stained antigen, is widely used as a screening test in farm animals, but also gives good results in human brucellosis. It is not affected by prozones or immunoglobulin switching. Positive results should be confirmed by a quantitative method.

The brucellin skin test, similar to the tuberculin test (see p. 210), does not differentiate active from past or subclinical infection and is no longer recommended.

TREATMENT

Brucella infections respond to a combination of streptomycin or gentamicin and tetracycline, or to rifampicin and doxycycline. Tetracycline alone is often adequate in mild cases. Fluoroquinolones may be used in combination with rifampicin or tetracyclines but are not recommended for monotherapy. Treatment should be continued for at least 6 weeks. Co-trimoxazole and rifampicin can be used in children. In patients with endocarditis and neurobrucellosis, a combination of a tetracycline, aminoglycoside and rifampicin is recommended.

Serum antibody titres usually decline sharply after effective treatment. The chronic post-infectious form without localizing lesions responds poorly to treatment.

EPIDEMIOLOGY

B. abortus has been eradicated from cattle in most developed countries, although there has been a resurgence in some parts of Europe. It was formerly common in dairy farmers, veterinarians and abattoir workers, but is now rare. Nearly all human cases in the UK are now acquired abroad; most are caused by *B. melitensis*, which is still prevalent in mediterranean countries, the Middle East, central and southern Asia, and parts of Africa and South America.

Human brucellosis due to *B. suis* is largely an occupational disease arising from contact with infected pigs or pig meat. It was once common in the USA, chiefly among those who handled raw meat shortly after slaughter. It occurs in feral pigs in Australia and the USA, and is a hazard to hunters. It is widespread in domesticated pigs in various African, Asian and South American countries; biovar 4 is found only in the Arctic regions of North America and Russia.

Brucellae have potential as agents of biological warfare or bioterrorism and this possibility should be borne in mind in the event of unexplained outbreaks.

CONTROL

The live-attenuated *B. abortus* strain S19 vaccine has been used to protect cattle from abortion and so reduce spread of the disease. It can interfere with subsequent diagnostic serology and has been widely replaced by the rough strain *B. abortus* RB51, which may give comparable protection, but does not induce interfering antibodies and is less hazardous to man.

The live-attenuated smooth strain *B. melitensis* Rev I is used to protect sheep and goats from *B. melitensis* infection. Vaccination of pigs is not widely practised, although the attenuated *B. suis* strain 2 has been used in China.

Human vaccination is not recommended, as effective and non-reactogenic vaccines are not currently available.

Pasteurization eliminates the risk of brucellosis from the consumption of infected milk or milk products. However, there remains the possibility of infection due to contact with infected animals or their tissues. Veterinary surgeons, farmers and laboratory workers are particularly at risk.

Eradication depends on elimination of the infection from domestic animals by a policy of compulsory testing of the animals and slaughtering of positive reactors.

BARTONELLA

The genus *Bartonella*, which is genetically related to *Brucella*, comprises at least 20 species of very small Gram-negative bacilli, most of which have been implicated in various febrile and localizing diseases in man.

- *Bartonella bacilliformis* is the cause of Oroya fever or Carrion's disease and verruga peruana. It is spread by sandflies.
- *Bart. quintana* is the cause of *louse-borne trench fever*.
- *Bart. henselae* and *Bart. clarridgeiae* are the most common causes of *cat scratch disease* and can be transmitted by fleas and possibly ticks.

Other *Bartonella* species and subspecies have been identified as pathogens of dogs and other mammals. Some, including *Bart. vinsoni*, *Bart. vinsoni* ssp. *berkhoffii*, *Bart. vinsoni* ssp. *apruensis* and *Bart. elisabethae*, occasionally cause fever, bacteraemia and endocarditis in man. *Bart. grahami* has been implicated in ocular disease.

It is becoming apparent that the various species can cause a similar range of syndromes involving many organ systems. As the genus extends, it is likely that other species will be implicated in human disease.

BARTONELLA BACILLIFORMIS

Bart. bacilliformis is responsible for outbreaks of a severe and often fatal disease of man in the mountainous regions of Peru, Colombia and Ecuador. The name Oroya fever was given after an epidemic of the disease in 1870 during the building of a railway between Lima and Oroya, when 7000 labourers died within a few weeks. The infection is spread by sandflies, usually *Lutzomyia verrucarum* and *L. peruensis*.

After recovery from Oroya fever the patient may develop a skin eruption known as verruga peruana. Individuals may remain bacteraemic and act as reservoirs of infection long after recovery from the illness, or after asymptomatic infection, which probably occurs in more than 50% of those exposed. *Bart. bacilliformis* is pathogenic only to man.

Description

Bart. bacilliformis is a small, strictly aerobic, Gram-negative coccobacillus. The organisms occur singly, in pairs, chains or clumps. In older cultures they tend to be extremely pleomorphic. They are motile through a cluster of about ten flagella situated at one end of the cell. The organism grows best at 25–28°C and at pH 7.8.

Pathogenicity

After an incubation period of about 20 days, Oroya fever presents as a high fever followed by progressively severe anaemia due to blood cell destruction. There may be enlargement of the spleen and liver, and haemorrhages into the lymph nodes.

The case fatality rate in untreated cases may be over 40%, although the overall fatality rate for all forms of the infection is probably only 0.1%. Verruga peruana may occur without the initial attack of Oroya fever or it may develop several weeks after recovery. It consists of a pleomorphic skin eruption of round, elevated, hard nodules that may become secondarily infected, producing ulcers and haemorrhagic lesions. The rash usually appears mainly on the legs, arms and face, although all parts of the body may be affected. The condition may persist for as long as a year, but is rarely fatal.

Laboratory diagnosis

In both Oroya fever and verruga peruana, bartonellosis is confirmed by demonstrating the organisms in smears of blood or tissue aspirates stained by Giemsa or immunofluorescent stain. They are seen packing the cytoplasm of the cells and adhering to the cell surfaces.

Bartonella spp. are dangerous pathogens and should be handled under class 3 containment conditions. *Bart. bacilliformis* is readily cultured in semi-solid nutrient agar supplemented with rabbit serum and haemoglobin, similar to that used for the culture of leptospires. Visible growth may take up to 10 days. PCR or serology is used for identification.

Blood culture should be carried out at all stages of infection. It may be difficult to isolate the organisms from the blood when the verruga stage has developed, and culture from the skin lesions is rarely satisfactory.

PCR tests offer a rapid and reliable means of diagnosis, but are usually available only from reference laboratories. Antibodies to *Bart. bacilliformis* can be detected by various serological tests. They are common in inhabitants of endemic areas and are not necessarily diagnostic of active disease.

Treatment

Chloramphenicol can drastically reduce the mortality rate in Oroya fever and the frequently associated salmonella infections. Penicillin, streptomycin, tetracyclines, fluoroquino-

lones and newer macrolides such as clarithromycin may also be effective in uncomplicated cases. Blood transfusion may be necessary in severe cases of anaemia.

Control

No vaccine is available. Insecticides are used to eliminate the sandfly vector in likely breeding sites inside and outside houses and surrounding areas. As the insects bite only at night, individuals may protect themselves by withdrawing from affected areas at nightfall.

BARTONELLA QUINTANA

This organism was formerly classified among the rickettsiae as *Rochalimaea quintana*. However, unlike the rickettsiae, these organisms can grow in cell-free media and they tend to be epicellular rather than strictly intracellular parasites of man. Unlike *Bart. bacilliformis* the organism does not possess flagella, although it may exhibit a twitching movement owing to fimbriae.

Bart. quintana was first identified as the cause of the febrile illness known as *trench fever* among the troops in the First World War. It is transmitted by the body louse, *Pediculus humanus*, under unhygienic living conditions and is not uncommon among homeless people in some countries. Trench fever is a bacteraemic condition typically associated with periodic febrile episodes lasting for about 5 days. *Bart. quintana* has also been implicated in cases of angiomatosis and endocarditis. The organism may be isolated from the blood of patients by culture on blood agar. *Bart. vinsoni* and its subspecies are similar and can cause an identical syndrome.

BARTONELLA HENSELAE

Bart. henselae has been isolated from the blood and lymph nodes of patients suffering from *cat scratch disease*, a severe condition of regional lymphadenopathy and fever resulting from the scratch or bite of an infected cat. Cat fleas and ticks may be responsible for transmission. *Bart. clarridgeiae*, which can be differentiated from *Bart. henselae* by its flagella, can cause an identical syndrome.

An organism known as *Afipia felis* has also been implicated in a small proportion of cases of cat scratch disease. It is morphologically similar to *Bart. henselae*, but differs biochemically, genetically and in being culturally less fastidious.

Bart. henselae and, less frequently, *Bart. quintana* and other species have been identified in the blood and tissues of individuals suffering from two severe clinical syndromes associated with human immunodeficiency virus (HIV) or other immunosuppressant conditions:

1. *bacillary angiomatosis*, which produces proliferative vascular lesions in the skin, regional lymph nodes and various internal organs
2. *bacillary peliosis*, which affects the liver and spleen.

Diagnosis

Bart. henselae may be cultured from the pus or lymph node samples of patients with cat scratch disease, and from blood, lymphoid tissue, liver and spleen of patients with bacillary angiomatosis or peliosis. In tissue sections the organisms are best demonstrated by silver stain or an immunospecific stain. ELISA, with various protein antigens, is the most useful serological test.

Most *Bartonella* species can be detected and differentiated by PCR methods.

Treatment

Tetracyclines, aminoglycosides, chloramphenicol and clarithromycin are all effective against bartonella infections. Treatment may need to be prolonged for at least 6 weeks.

STREPTOBACILLUS MONILIFORMIS

Streptobacillus moniliformis is one of the causes of *rat bite fever* in man, the other being *Spirillum minus* (see p. 316). It is a common commensal of the nasopharynx of rodents, and sometimes causes epizootic disease in mice and rats, resulting in otitis media, multiple arthritis and swelling of the feet and legs. Laboratory workers who handle rodents are most at risk. Rarely, outbreaks of infection occur as a result of the ingestion of milk or other food contaminated by rats.

DESCRIPTION

S. moniliformis is Gram-negative, non-motile, non-capsulate and highly pleomorphic. The organisms appear as short bacilli, 1–3 µm in length, forming chains interspersed with long filaments that may show oval or spherical lateral swellings.

It is a facultative anaerobe that benefits from added carbon dioxide and a moist atmosphere. It grows best at 37°C and pH 7.6. Culture media must contain blood, serum or ascitic fluid. Loeffler's serum medium is satisfactory. After incubation for 2 days, discrete, granular, greyish yellow colonies 1–5 mm in diameter are visible on the surface, and minute 'fried egg' colonies appear in the depth of the medium. The latter are L-phase variants that have little or no virulence for laboratory animals. They develop spontaneously and are thought to have a defective mechanism for cell wall formation.

Sensitivity

S. moniliformis is killed in 30 min by a temperature of 55°C. In culture it survives for only a few days, although in serum broth at 37°C it may remain viable for as long as 1 week. With the exception of the L-forms, *S. moniliformis* is susceptible to penicillin, and both forms are sensitive to streptomycin and tetracycline.

PATHOGENICITY

In man the organism usually enters the body through wounds caused by rodent bites. It multiplies and invades the lymphatics and bloodstream, causing a feverish illness with severe toxic symptoms and sometimes complications such as arthritis, endocarditis and pneumonia.

Infection acquired by ingestion of contaminated water, milk or food is known as *Haverhill fever*, a condition characterized by fever, sore throat, rash, polyarthritis and erythema. The duration of the illness varies from a few days to several weeks. Endocarditis, hepatitis and amnionitis may develop as complications. In the pre-antibiotic era a case fatality rate of about 10% was reported; the rate is much lower nowadays with effective treatment.

LABORATORY DIAGNOSIS

An acute febrile illness associated with asymmetric arthropathy, a maculopapular rash involving the extremities and a history of contact with rodents may point to the diagnosis.

S. moniliformis can be isolated in culture from the patient's blood during the acute phase of the illness and from the synovial fluid of those who develop arthritis. Growth occurs in serum broth as a characteristic granular sediment, appearing like 'cotton wool balls' that do not disintegrate on shaking.

Mice are highly susceptible to intraperitoneal inoculation of infected blood or joint fluid, as a result of which they develop either a rapidly fatal generalized condition or a more chronic disease with swelling of the feet and legs.

Specific agglutinins may be detected in the patient's serum as early as 10 days or as late as several weeks after the rat bite. As agglutinins also occur in healthy individuals, and at least a four-fold rise in titre is needed for diagnosis, the test is no longer recommended.

ELISA, where available, is the method of choice for detecting antibodies. A PCR test may also be used. A false-positive reaction in the VDRL test (Venereal Disease Research Laboratory slide test; see p. 363) is seen in about 25% of patients.

TREATMENT

Penicillin, clarithromycin or oral tetracycline is usually effective.

KEY POINTS

- Brucellae are highly infectious coccobacilli that cause a septicaemic illness, undulant fever. Most human disease is caused by *Brucella melitensis, B. abortus* or *B. suis.*
- The disease is a typical zoonosis most commonly acquired from infected animals, or from infected meat or dairy products.
- Brucellosis is diagnosed by isolation of the organism from blood; alternatively serology or polymerase chain reaction tests can be used.
- Brucellosis is treated with a tetracycline, usually in combination with an aminoglycoside or rifampicin.
- *Bartonella bacilliformis* is a highly infectious agent causing the sandfly disseminated diseases, Oroya fever

(Carrion's disease) and verruga peruana in parts of South America.
- The organism infects blood cells and can be diagnosed in stained blood or tissue aspirates. Alternatively, PCR methods are used.
- Other bartonellae cause trench fever and cat scratch disease.
- Chloramphenicol, macrolides, aminoglycosides and tetracyclines are used in the treatment of bartonella infections.
- *Streptobacillus moniliformis* is one of the causes of a septicaemic illness, rat bite fever. Treatment with penicillin or a tetracycline is usually effective.

RECOMMENDED READING

Adal K A 1995 Bartonella: new species and new diseases. *Reviews in Medical Microbiology* 6: 155–164

Boot R, Oosterhuis A,Thuis H C 2002 PCR for the detection of *Streptobacillus moniliformis. Laboratory Animals* 36: 200–208

Corbel M J, Beeching N J 2003 Brucellosis. In Kasper D L, Braunwald E, Fauci A (eds) *Harrison's Principles of Internal Medicine,*16th edn, pp. 914–917. McGraw Hill, New York

Graves M H, Janda J M 2001 Rat bite fever (*Streptobacillus moniliformis*): a potential emerging disease. *International Journal of infectious Diseases* 5: 151–155

Jensen W A, Fall M Z, Rooney J, Kordick D L, Breischwerdt E B 2000 Rapid identification and differentiation of *Bartonella* species using a single step PCR assay. *Journal of Clinical Microbiology* 38: 1717–1722

Lopez-Goni I, Moriyon I (eds) 2004 *Brucella. Molecular and Cellular Biology.* Horizon Bioscience, Wymondham

Madkour M M 2001 *Madkour's Brucellosis.* Springer, Berlin

Maguina C, Garcia P, Gotuzzo E, Spach D 2001 Bartonellosis (Carrion's disease) in the modern era. *Clinical Infectious Diseases* 33: 772–779

Murray P R, Corbel M J 2005 Brucella. In Borriello SP, Murray P R, Funke G (eds) *Topley and Wilson's Microbiology and Microbial Infections,* 10th edn, Vol. 2, pp. 1719–1751. Hodder Arnold, London

Wullenweber M 1995 *Streptobacillus moniliformis* – a zoonotic pathogen. Taxonomic considerations, host species, diagnosis, therapy, geographical distribution. *Laboratory Animals* 29: 1–15

Yersinia, pasteurella and francisella

Plague; pseudotuberculosis; mesenteric adenitis; pasteurellosis; tularaemia

M. J. Corbel

The organisms within these three genera are animal pathogens that, under certain conditions, are transmissible to man, either directly, or indirectly through food and water or via insect vectors. They are Gram-negative coccobacilli, formerly contained within one genus, *Pasteurella*. Molecular genetics has indicated a completely separate identity for the three genera, each with its own disease manifestations in man and animals:

1. *Yersinia* belongs to the Enterobacteriaceae and includes many non-pathogenic species.
2. *Pasteurella* is closely related to the *Actinobacillus–Haemophilus* group.
3. *Francisella* is distantly related to *Legionella*.

YERSINIA

YERSINIA PESTIS

Yersinia pestis, the *plague bacillus*, is essentially a parasite of rodents. In certain parts of the world, burrowing animals such as ground squirrels, gerbils and voles act as reservoirs of infection that may be transmitted by fleas to susceptible animals such as bandicoots, marmots, squirrels and rats. The animals suffer from outbreaks of plague, and their fleas may transmit the infection to man, giving rise to sporadic disease referred to as *wild* or *sylvatic plague*. Farmers or trappers who come into contact with infected animals are at risk.

More serious for man is *urban plague*, resulting from the spread of infection among rats, especially the black rat, *Rattus rattus*, which used to flourish around human habitation. Outbreaks of human plague, following epidemics in rats, have in the past sometimes developed into pandemics.

Description

Y. pestis is a Gram-negative, non-sporing, non-motile, short coccobacillus. It occurs singly, in pairs or, when in liquid culture, in chains. Pleomorphism is marked, especially in old cultures in which pear-shaped or globular cells, suggestive of yeast cells (*involution forms*), may be seen. In smears from exudates and in cultures grown at 37°C they are frequently capsulate. In smears from tissues stained by methylene blue or Giemsa stain they show characteristic bipolar staining ('*safety pin*' appearance).

Y. pestis grows aerobically or anaerobically at 0–37°C (optimum 27°C), although small inocula may not grow aerobically in ordinary culture media. Small, slightly viscid, translucent, non-haemolytic colonies develop on blood agar within 24 h. Growth occurs on MacConkey's medium, but tends to autolyse after 2–3 days.

It is killed at 55°C in 5 min and by 0.5% phenol in 15 min. It is sensitive to drying but may remain viable in moist culture for many months, especially at low temperatures.

Pathogenesis

The heat-stable somatic antigen complex of *Y. pestis* comprises a rough-type lipopolysaccharide (LPS), which has endotoxin activity and is believed to contribute to the terminal toxaemia of plague. The heat-labile Fraction 1 (F1) protein capsular antigen helps the organism to resist phagocytosis and is a protective immunogen. This, and many other proteins associated with pathogenicity, are encoded by three plasmids. The largest of these contains genes activated at low calcium concentration that express various outer membrane and secreted proteins with a variety of functions, such as inhibition of phagocytosis and intracellular killing. The V antigen, part of the type III secretion system, is an important protective antigen. *Y. pestis* also produces a plasminogen activator and fibrinolysin, which may play a critical role in the initial stages of infection. A pathogenicity island encodes other proteins associated with virulence including cell surface adhesion and iron acquisition factors, some common to *Y. pseudotuberculosis* and *Y. enterocolitica*.

Three severe forms of human plague are described: bubonic, pneumonic and septicaemic plague. All may occur at different stages in the same patient. The disease may also present as pharyngitis or meningitis.

Bubonic plague

The transfer of *Y. pestis* from rats to man through the bites of infected fleas may occasionally result in a localized infection, known as *pestis minor*, with mild constitutional symptoms. More often the lymph nodes draining the area of the flea bite become affected, and the resulting adenitis produces intensely painful swellings or *buboes* in the inguinal, axillary or cervical regions, depending on the position of the bite. From these primary buboes the plague bacilli may spread to all parts of the body. Complications such as bronchopneumonia, septicaemia or meningitis may follow. In the absence of adequate antibiotic therapy administered early in the course of the disease, the case fatality rate may exceed 50%.

Pneumonic plague

This can develop in patients presenting with bubonic or septicaemic plague. It may also be acquired as a primary infection by inhalation of droplets infected with *Y. pestis*, usually from an individual with pneumonic disease or as a result of exposure to aerosols generated from cultures. A severe bronchopneumonia develops. As disease progresses, the sputum becomes thin and blood stained; numerous plague bacilli are demonstrable in stained films or on culture of the sputum. This type of plague is highly contagious and is almost invariably fatal unless treated very early.

Septicaemic plague

This may occur as a primary infection or as a complication of bubonic or pneumonic plague. The bacilli spread rapidly throughout the body and the outcome is almost invariably fatal, even in treated cases. Purpura may develop in the skin ('*Black Death*'), and disseminated intravascular coagulation is usually present. It should be noted that bacteraemia can occur in bubonic or pneumonic plague but is usually intermittent in the early stages.

Pointers to the disease include a sudden onset of high fever accompanied by prostration in individuals recently returned from an endemic area, or with a history of occupational exposure.

Laboratory diagnosis

Pneumonic plague is easily acquired in the laboratory by inhalation of aerosols generated from *Y. pestis* cultures. These and clinical specimens suspected of containing the organism should be handled only under containment conditions appropriate for class 3 pathogens. Laboratory animals used for diagnostic tests must be housed under insect-free containment conditions.

Plague is confirmed by demonstrating the bacilli in fluid from buboes or local skin lesions in the case of bubonic plague, in the sputum in pneumonic plague, and in blood films and by blood culture when septicaemic plague is suspected. Blood culture may be intermittently positive in all forms of the disease. Post mortem, the bacilli can usually be isolated from a wide range of tissues, especially spleen, lung and lymph nodes.

Smears of exudate or sputum are stained with methylene blue, Wayson stain (a mixture of basic fuchsin, methylene blue and phenol) or Giemsa stain. Characteristic bipolar-stained coccobacilli are confirmed as *Y. pestis* by culturing samples on blood agar and incubating at 27°C. If exudate is inoculated subcutaneously into guinea-pigs or white rats, or on to their nasal mucosa, infection follows and the animals die within 2–5 days. The bacilli may then be isolated from the blood or from smears of spleen tissue taken post mortem.

Characteristic colonies growing on blood agar plates are identified presumptively by various cultural and biological tests, by demonstrating chain formation in broth culture, and by 'stalactite' growth from drops of oil layered on the surface of fluid medium. Demonstration of the F1 capsular antigen by immunospecific staining confirms the presence of *Y. pestis*.

Serology is most likely to be useful in the convalescent stage. Tests used include a complement fixation test and a haemagglutination test with tanned sheep red cells to which the capsular F1 antigen has been adsorbed. In the latter case a rising titre or a single titre of at least 16 is considered significant. An enzyme-linked immunosorbent assay (ELISA) with F1 antigen is likely to become the method of choice.

A polymerase chain reaction (PCR), with primers based on F1 gene sequences, offers a rapid and less hazardous means of diagnosis than culture.

Treatment

Y. pestis is sensitive to many antibiotics, including aminoglycosides, fluoroquinolones, chloramphenicol, co-trimoxazole and tetracyclines, but not penicillin.

When plague is suspected, patients should be isolated and respiratory precautions observed for at least the first 48 h of treatment. Antibiotic therapy should be started without waiting for confirmation of the diagnosis.

Intramuscular streptomycin or intravenous gentamicin is highly effective. Chloramphenicol (given intravenously for the first 4 days) is recommended in patients with meningitic symptoms. Tetracycline may be adequate in uncomplicated bubonic plague if given in large doses within

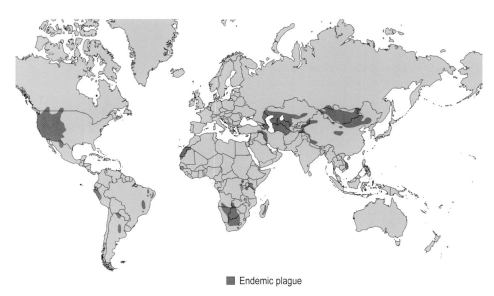

Fig. 35.1   Areas where endemic plague is known to have persisted.

48 h of onset, and continued for 10 days. Experience with other antibiotics is limited, but there are indications that ciprofloxacin is effective.

Although monotherapy is usually adequate, strains carrying antibiotic resistance plasmids have been reported, and combined therapy may be advisable until the sensitivity of the strain is known.

Plague is a toxigenic infection and antibiotics will not prevent death once the bacteraemia has exceeded a certain threshold. Most patients treated within 18 h of onset can be expected to survive.

Epidemiology

Plague was introduced into Europe from Asia in the thirteenth century and led to the great pandemic known as the Black Death, when about a quarter of the population of Europe succumbed to the disease. It was during a major outbreak of plague in Hong Kong in 1894 that Yersin first described the plague bacillus.

Plague largely disappeared from Europe in the seventeenth century, perhaps because the black rat was displaced by the spread of the brown (sewer) rat, *Rattus norvegicus*, which is susceptible to plague but does not commonly frequent human dwellings. Improvements in housing may also have played an important part in the elimination of plague from Europe.

The bacilli are transmitted between animals and from animals to man by fleas, notably, but not exclusively, *Xenopsylla cheopsis*, an ectoparasite of rats. In cool humid weather fleas multiply and plague spreads readily among susceptible rats. In hot dry weather, on the other hand, the fleas die out, limiting the spread of infection.

When a flea feeds on the blood of a sick animal, plague bacilli are sucked into the insect's midgut, where they multiply to produce a biofilm that may block the proventriculus, a process promoted by secreted proteins. When the host animal dies, the flea seeks an alternative host, which may be another rodent or human being. Because the 'blocked' flea is unable to suck readily, some of the infected midgut content is regurgitated and injected into the bite wound of the new victim.

When the epizootic among rats has reached a stage at which the number of susceptible animals has greatly decreased through death or immunity, it tends to die out, as does any human epidemic associated with it.

Domestic cats may become infected through contact with rodents. The animals may develop atypical disease and then transmit the infection to their owners or to veterinarians by the percutaneous or respiratory routes.

The sputum of persons suffering from pneumonic plague contains large numbers of plague bacilli, and under favourable conditions the disease spreads rapidly through the community by droplet infection, independently of rodents or fleas. Close contact is required and epidemics are most likely to occur when overcrowding in insanitary accommodation allows the infected droplets to spread readily from person to person. Cool, humid conditions favour transmission.

Endemic foci of wild rodent plague persist in many rural parts of the world, including North and South America, Africa and many parts of Asia (Fig. 35.1). Constant surveillance must be maintained to prevent its spread to urban populations, especially in areas where living conditions are below standard.

Y. pestis has been used as a biological warfare agent. Its potential application in bio-terrorism is of major concern.

Control

Bubonic plague

Periodic surveys are advocated in endemic areas to determine the prevalence of rodents and fleas so that control measures can be taken. Rats may be destroyed by rat poison and fleas by the liberal application of insecticide to rat runs.

Other control measures include the construction of rat-proof dwelling houses and buildings such as warehouses in dockland areas. The fumigation of ships and measures to prevent rats gaining access to ships and aircraft help to prevent the spread of plague from one country to another.

Pneumonic plague

Patients suffering from pneumonic plague should be isolated, with full respiratory precautions. Overcrowding of houses and other accommodation should be avoided. Co-trimoxazole, ciprofloxacin or tetracyclines administered to immediate contacts may afford some degree of protection.

Vaccination

Vaccines prepared from killed virulent strains of *Y. pestis* confer significant protection against bubonic but not pneumonic plague. Live vaccines prepared from avirulent strains are used in some countries, but can cause severe reactions. Neither type can be relied upon to confer long-term immunity, and revaccination is necessary at 6-month intervals if exposure to infection continues. Efficacy data for improved vaccines based on recombinant F1 and V antigens are not yet available.

YERSINIA PSEUDOTUBERCULOSIS

Y. pseudotuberculosis can cause disease in many species of wild and domesticated animals and birds. Although presentation can vary widely, it typically causes a fatal septicaemia, often accompanied by formation of small whitish nodules in the viscera ('*pseudotuberculosis*').

The infection is indirectly transferable to man, usually through contaminated food or water, resulting in a variety of presentations ranging in severity from subclinical to severe. Gastro-intestinal manifestations are common;

acute ileitis and mesenteric lymphadenitis are the most characteristic. Post-infectious immunologically mediated sequelae are not uncommon.

Description

Y. pseudotuberculosis is a small, ovoid, Gram-negative bacillus, with a tendency to bipolar staining. Genetically it is very similar to *Y. pestis*, which is probably a rough variant that has acquired additional plasmids encoding virulence factors. Initial growth may be best under anaerobic conditions. Isolation can be improved by 'cold enrichment' in buffered saline incubated at 4°C with periodic subculture for up to 6 weeks.

The organisms may be differentiated from *Y. pestis* by:

- motility when grown at 22°C
- ability to produce urease
- lack of the F1 antigen as shown by immunospecific staining or PCR.

There are eight major O serotypes, several of which can be separated into subtypes based on thermo-stable LPS somatic antigens. Unlike *Y. pestis* LPS, these are of smooth type and their specificity is determined by the O chain structure. The core regions are common to all serotypes and to *Y. pestis*. Thermo-labile flagellar antigens are present in cultures grown at 18–26°C. Many other protein antigens are shared with *Y. pestis* and *Y. enterocolitica*.

Pathogenesis

Like other yersiniae, *Y. pseudotuberculosis* carries a plasmid encoding factors essential for pathogenicity including a type III secretion system. At least one enterotoxin is also produced, as well as *invasin* and iron-regulated proteins encoded by a chromosomal pathogenicity island.

Infection may be subclinical, but occasionally results in a severe typhoid-like illness with fever, purpura and enlargement of the liver and spleen, which is usually fatal. More frequently it causes mesenteric lymphadenitis and terminal ileitis, usually accompanied by fever, diarrhoea and pain simulating acute or subacute appendicitis. All age groups may be attacked but young males aged 5–15 years seem to be more frequently affected. Recovery is usually uneventful, although immunological sequelae such as erythema nodosum or reactive arthritis develop in some patients.

Laboratory diagnosis

Infection in man is confirmed by isolation of the organism in culture from blood, local lesions or mesenteric nodes, particularly the ileocaecal nodes.

Specific serum antibodies are detected and measured by tube or micro-agglutination tests performed during the acute phase of the illness with smooth suspensions of strains of serotypes I–VI grown at 22°C. ELISA or haemagglutination of red cells sensitized with LPS can also be used. The antibodies decline rapidly and reach low levels within 3–5 months.

Treatment

Unlike *Y. pestis*, *Y. pseudotuberculosis* is usually sensitive in vitro to penicillins; it is also usually sensitive to aminoglycosides, chloramphenicol, tetracyclines, co-trimoxazole and fluoroquinolones.

Ileitis and mesenteric adenitis are usually self-limiting. Septicaemia demands parenteral treatment with ampicillin, chloramphenicol, gentamicin or tetracycline.

Epidemiology

Many animal species suffer from the infection, but there is little proof of direct transmission to man. Most human infections probably result from the ingestion of contaminated water, vegetables or other food.

About 90% of all human cases in Australia, Europe and North America are attributed to strains of serotype I, followed by serotypes II and III, whereas in Japan serotypes IV and V predominate.

YERSINIA ENTEROCOLITICA

By far the commonest manifestation of *Y. enterocolitica* infection is acute enteritis, which may simulate acute appendicitis. Like many environmental species of *Yersinia*, it may also occasionally cause opportunist infection in compromised patients, occasionally presenting as a plague-like syndrome with bubo formation or fulminant septicaemia.

Description

Morphologically and culturally, *Y. enterocolitica* resembles *Y. pestis* and *Y. pseudotuberculosis* but grows more readily; it differs from them antigenically, biochemically and genetically.

At least 54 different O antigens and 19 H factors have been identified, so that a large number of serotypes are recognized. Serotypes 3, 8 and 9 account for most human infections; other serotypes are probably non-pathogenic in immunocompetent individuals.

Pathogenesis

Y. enterocolitica infects primarily the lymphoid tissue of the small intestine and ileo-caecal junction. It carries a low calcium response plasmid and pathogenicity island encoding factors similar to those of *Y. pseudotuberculosis*. It causes mild and occasionally severe enteritis, mesenteric lymphadenitis and terminal ileitis. Septicaemia, which is often fatal, is most common in the elderly or in patients with predisposing conditions such as cirrhosis, iron overload or immunosuppression. Pneumonia and meningitis are rare presentations. Post-infectious complications include erythema nodosum, polyarthritis, Reiter's syndrome and thyroiditis. In young children the infection may produce fever, diarrhoea, abdominal pain and vomiting. The symptoms may last for several weeks.

Laboratory diagnosis

The organism is isolated from blood, lymph nodes or other tissues on blood agar or MacConkey's agar. Isolation from contaminated sources such as faeces is best done by cold enrichment in buffered saline incubated at 4°C for up to 6 weeks, followed by plating on a selective medium. Identity is confirmed by biochemical tests and motility.

The serotype may be determined by slide agglutination with specific rabbit antisera. Serum antibodies are measured by agglutination tests against appropriate O antigens. A significant rise in the titre to 160 or more over a 10-day period indicates acute infection. ELISA may also be used. Cross-reactions occur between serotype O9 and smooth *Brucella* strains. These are very difficult to differentiate. PCR may be of value but is difficult to apply to highly contaminated materials such as faeces.

Treatment

Y. enterocolitica is sensitive to many antibiotics, including aminoglycosides, chloramphenicol, co-trimoxazole, fluoroquinolones and tetracyclines, but is resistant to penicillin. Sensitivity to other β-lactam antibiotics is variable.

Uncomplicated gastro-intestinal infection is usually self-limiting and treatment is indicated only in severe cases. Tetracycline is probably the drug of choice. Invasive infections such as septicaemia require intensive parenteral antibiotic treatment.

Epidemiology

Y. enterocolitica has been isolated from caseous abscesses resembling those of pseudotuberculosis, from blood and infected wounds, and from the intestinal contents of apparently healthy animals of many species throughout the world. Pigs carry pathogenic serotypes quite frequently, cattle, sheep and goats less so.

Serotypes 3 and 9 account for most human infections in Europe, whereas serotype 8 is most common in the USA. Human disease usually results from ingestion of contaminated food or from contact with the environment. Raw pork, milk and drinking water have been implicated as sources. Person-to-person transmission also occurs.

Blood transfusion is a significant hazard as the organism can grow in refrigerated contaminated blood donations. Flies are believed to play a role in transmission by contaminating food, and infection has been demonstrated in fleas and lice. However, enteric infection is the usual route of transmission and preventive measures are those appropriate for food-borne disease.

PASTEURELLA

PASTEURELLA MULTOCIDA

Pasteurella multocida (formerly *P. septica*) is a commensal or opportunist pathogen of many species of domestic and wild animals and birds. Human beings occasionally become infected, especially following animal bites. Closely related bacteria formerly classified as *Pasteurella*, and now renamed as *Mannheimia, Gallibacterium* and *Photobacterium*, are rarely implicated as human pathogens.

Description

P. multocida organisms are aerobic and facultatively anaerobic coccobacilli, which are appreciably smaller than those of *Yersinia* species, although they are often pleomorphic in culture. They are Gram-negative, non-motile, non-sporing and capsulate in culture at the optimal growth temperature of 37°C. In smears of blood or tissue stained with methylene blue they show bipolar staining. *P. multocida* does not grow on MacConkey's medium.

Five capsular antigens A, B, D, E and F (C is not valid) and at least 11 somatic LPS antigens have been identified. The expression of the capsule is affected by cultural conditions and is lost in rough strains, which also fail to express smooth type O antigens.

The organisms are killed in a few minutes at 55°C and by phenol (0.5%) in 15 min. They may survive and remain virulent in dried blood for about 3 weeks, and in culture or infected tissues for many months if kept frozen.

Pathogenesis

P. multocida can be extremely virulent to many species of animals and birds, causing *fowl cholera* and haemorrhagic septicaemia, which are usually fatal. It also causes respiratory infections and contributes to the pathogenesis of atrophic rhinitis in pigs. Carriage of the organism is usually asymptomatic but stress may provoke fatal systemic infection.

The capsule is essential for full virulence, at least in mice and rabbits, and is the major protective antigen. Iridescent smooth strains show the greatest pathogenicity; mucoid strains are of reduced virulence and rough strains are avirulent. A dermo-necrotic protein toxin, a cytotoxin and a neuraminidase probably account for many of the local manifestations of infection, but the bacteria also contain LPS with endotoxin activity.

Human infections usually present as a local abscess at the site of a cat or dog bite, with cellulitis, adenitis and, sometimes, osteomyelitis. *P. multocida* is also implicated in infections of the respiratory system such as pleurisy, pneumonia, empyema, bronchitis, bronchiectasis and nasal sinusitis.

Rare manifestations of disease include meningitis or cerebral abscess (usually following head injury), endocarditis, pericarditis or septicaemia, and infections of the eye, liver, kidney, intestine and genital tract.

A history of a recent animal bite or of occupational exposure are indicators for suspecting a *Pasteurella* infection. The organisms may also be carried commensally in the respiratory tract and can cause infection after surgical operation or cranial fracture.

Laboratory diagnosis

Material from bite wounds, blood cultures, cerebrospinal fluid (in cases of meningitis) or respiratory secretions (in suppurative chest infections) are cultured on blood agar. The organisms are identified by various cultural and biochemical tests. Serology is unhelpful. PCR is potentially useful but rarely available.

Treatment

Infections usually respond to penicillin. Tetracycline, erythromycin or co-trimoxazole are suitable alternatives. In cases of osteomyelitis following dog or cat bites, antibiotic therapy must be continued for at least 8 weeks.

Epidemiology

P. multocida is carried in the nasopharyngeal region of many species of wild and domestic animals. In human infections following animal bites, the organism passes directly to the person in the animal's saliva. Cat bites are particularly hazardous. Human beings may also become infected through breathing droplets generated by the coughing of animals suffering from respiratory infection. Pig farmers may be particularly at risk.

The disease in farm animals can be prevented by vaccination with preparations derived from killed capsulate bacteria. This is not practicable for human infections because of their rarity.

OTHER PASTEURELLA SPECIES

Pasteurella spp. other than *P. multocida* cause human disease very infrequently.

- *P. caballi* causes respiratory and genital infections in horses and has caused bite wound infections in man.
- *P. dagmatis* has been associated with bite wounds and endocarditis.
- *P. (Mannheimia) haemolytica* causes pneumonia and haemorrhagic septicaemia in sheep, buffalo and cattle, and various diseases in poultry and other domesticated animals. It has been isolated from human cases of endocarditis, septicaemia and wound infection. It differs from *P. multocida* in forming haemolytic colonies on blood agar and by its ability to grow on MacConkey's medium.
- *P. (Actinobacillus) pneumotropica* is frequently isolated from the respiratory tract of laboratory animals. It has occasionally been isolated from human cases of septicaemia, upper respiratory tract infections and from animal bite wounds.
- *P. stomatis* has been isolated from cat bite wounds.
- *P. volantium* has been isolated from the human oropharynx.

FRANCISELLA

FRANCISELLA TULARENSIS

Francisella tularensis produces *tularaemia* in man and certain small mammals, notably rabbits, hares, beavers and various rodent species. It occasionally causes large epizootics in lemmings and other small rodents. It can be transmitted by direct contact, by biting flies, mosquitoes and ticks, by contaminated water or meat, or by aerosols.

F. novicida (probably a subtype of *F. tularensis*) and *F. philomiragia* have been reported from North America as rare causes of human disease.

Description

When first isolated from infected tissue, *F. tularensis* is a very small, non-motile, non-sporing, capsulate, Gram-negative coccobacillus. In culture larger, pleomorphic,

even filamentous, forms are present. It stains poorly with methylene blue but carbol fuchsin (10%) produces characteristic bipolar staining. Two biovars are recognized:

1. Type A (formerly *F. tularensis tularensis* or *F. tularensis nearctica*) is found only in North America, is often transmitted by ticks and is highly pathogenic.
2. Type B (formerly *F. tularensis palaearctica* or *F. tularensis holarctica*) occurs in Europe, Asia and North America, is transmitted by mosquitoes and is much less virulent.

F. tularensis is strictly aerobic. It will not grow on ordinary nutrient media, but grows well on blood agar containing 2.5% glucose and 0.1% cysteine hydrochloride. *F. novicida* and *F. philomiragia* are less fastidious.

F. tularensis is killed by moist heat at 55°C in 10 min, but may remain viable for many years in cultures maintained at 10°C, and for many days in moist soil and in water polluted by infected animals.

Pathogenesis

Little is known about mechanisms of pathogenicity. A carbohydrate capsule is essential for virulence. A smooth type LPS is also present in the outer membrane, but apparently has low endotoxin activity.

In animals suffering from tularaemia the bacteria are present in large numbers within the cells of the liver and spleen, including macrophages.

Most human cases are sporadic, although occasional large outbreaks have been reported. After an acute onset with fever, rigors and headache, the disease develops manifestations that vary according to the route of entry of infection.

Following cutaneous inoculation through direct contact with infected animals or a fly or tick bite, a small punched-out skin ulcer develops at the point of entry, accompanied by enlargement of the draining lymph nodes even to the extent of bubo formation (*ulceroglandular form*). If entry is via the conjunctiva a similar syndrome will develop involving the eye and pre-auricular nodes (*oculoglandular form*). A glandular form without ulceration also occurs. Inhalation of infected dust or droplets, or ingestion of contaminated meat or water, is more likely to lead to pulmonary or typhoidal disease, respectively. Either can be preceded or accompanied by painful pharyngitis.

Type A strains cause severe and, in the pre-antibiotic era, often fatal disease. Disease caused by type B strains is much less severe and associated with very low mortality rates, but can still cause prolonged disability.

Laboratory diagnosis

F. tularensis is extremely dangerous to handle in the laboratory and category 3 containment is required for all manipulations and animal work. All suspect samples should be labelled 'High risk'.

Human infections are usually diagnosed by inoculating tissue samples or the discharge from local lesions on to glucose–cysteine blood agar or cystine heart agar, and identifying any characteristic small mucoid colonies. Alternatively, the exudate may be inoculated into guinea-pigs or mice and the liver and spleen of the infected animals cultured post mortem. PCR methods are also available.

Serology is most likely to be positive after 3 weeks. Rising *F. tularensis* antibody titres or individual agglutinin titres of 160 are diagnostic. ELISA with confirmatory western blotting is now replacing agglutination as the preferred method. Serum from cases of brucellosis may cross-react with *F. tularensis* and vice versa, usually to relatively low titre. Western blotting permits differentiation of cross-reactions. Antigen for an intradermal test formerly used is no longer available.

Treatment

F. tularensis is sensitive to aminoglycosides, chloramphenicol, fluoroquinolones and tetracyclines, but resistant to most β-lactam antibiotics. Streptomycin and gentamicin are the antibiotics of choice in tularaemia and are usually curative. Treatment should be continued for at least 10 days (14 days if ciprofloxacin is used). Tetracyclines or chloramphenicol in high dosage are also effective, but relapse may occur with these bacteriostatic agents unless treatment is prolonged.

Epidemiology

Tularaemia has a worldwide distribution, but occurs mainly in the northern hemisphere. Cases have been reported from North America, from several European countries, including Scandinavia, and from Asia. It has not so far been identified in the UK.

It is a typical zoonosis, spread mainly by insects or ticks among lagomorphs and rodents. It is transmitted to humans through:

- handling of infected animals, such as rabbits or hares
- tick, mosquito or fly bites
- inhalation of contaminated dust (e.g. during harvesting or mowing)

- ingestion of contaminated water (as a result of pollution with the carcasses or excreta of infected rodents) or meat.

The organism is highly infectious, with a minimum infectious dose of about ten viable bacteria for the most virulent strains. Laboratory workers are especially at risk through handling infected laboratory animals or cultures of the organism. Person-to-person transmission of infection apparently does not occur. *F. tularensis* has been developed as a biological warfare agent and has potential application in bio-terrorism. A vaccine based on the live-attenuated LVS strain confers some protection.

KEY POINTS

- *Yersinia pestis* is the cause of human plague, transmitted to humans from rats and other rodents by their fleas. Pneumonic plague is transmitted from person to person by droplet infection.
- There are three main types of disease: bubonic, pneumonic and septicaemic plague. All are highly fatal without prompt treatment.
- The organism is readily cultured, but a polymerase chain reaction method is preferred for diagnosis as the organism is hazardous to handle.
- Aminoglycosides and chloramphenicol are commonly used in the treatment of plague.
- *Y. pseudotuberculosis* is an animal pathogen that occasionally causes human infection, which may be subclinical or severe.
- *Y. enterocolitica* causes a usually mild enteritis, but can give rise to a septicaemia, which may be fatal if untreated.
- *Pasteurella multocida* sometimes infects man, usually through animal bites.
- Pasteurella infection usually responds to penicillin or other antibiotics.
- *Francisella tularensis* is the cause of tularaemia, a febrile illness that can be severe and life threatening. It is usually acquired from infected animals.
- Streptomycin or gentamicin is the antibiotic of choice.

RECOMMENDED READING

Abdel-Haq N M, Asmar B I, Abrahamson W M, Brown W J 2000 *Yersinia enterocolitica* infection in children. *The Pediatric Infectious Disease Journal* 19: 954–958

Agger W A 2005 Tularemia, lawn mowers and rabbits' nests. *Journal of Clinical Microbiology* 43: 4304–4305

Bottone E J 1999 *Yersinia enterocolitica*; overview and epidemiologic correlates. *Microbes and Infection* 1: 323–333

Butler T 2000 Plague. In Ledingham J E G, Warrell D A (eds) *Concise Oxford Textbook of Medicine*, pp. 1625–1628. Oxford University Press, Oxford

Dennis D T, Campbell G L 2005 Plague and other yersinia infections. In Kasper D L, Braunwald E, Fauci A S, Hauser S L, Longo D L, Jameson J L (eds) *Harrison's Principles of Internal Medicine*, 16th edn, pp. 921–929. McGraw-Hill, New York

Janda W M, Mutters R 2005 *Pasteurella, Mannheimia, Actinobacillus, Eikenella, Kingella, Capnocytophaga* and other miscellaneous Gram-negative rods. In Borriello S P, Murray P R, Funke G (eds) *Topley and Wilson's Microbiology and Microbial Infections*, 10th edn, Vol. 2, pp. 1648–1691. Hodder Arnold, London

Schmitt P, Splettstosser W, Porsch-Ozcurumez M, Finke E J, Grunow R 2005 A novel ELISA and a confirmatory western blot useful for diagnosis and epidemiological studies of tularaemia. *Epidemiology and Infection* 133: 759–766

Wanger A 2005 Yersinia. In Borriello S P, Murray P R, Funke G (eds) *Topley and Wilson's Microbiology and Microbial Infections*, 10th edn, Vol. 2, pp.1458–1473, Hodder Arnold, London

Wilson W J, Erler A M, Nasarabadi S L, Skowronski E W, Imbro P M 2005 A multiplexed PCR-coupled liquid bead array for the simultaneous detection of four biothreat agents. *Molecular Cell Probes* 19: 137–142

36 Non-sporing anaerobes

Wound infection; periodontal disease; abscess; normal flora

R. P. Allaker

The significance of obligate anaerobes in general and of non-sporing anaerobes in particular is increasingly recognized. This heightened awareness of the important role that such organisms play, both as part of the normal microbial flora of the body and in a wide variety of infections, has come about largely through the application of greatly improved laboratory techniques for the isolation and cultivation of anaerobic bacteria, and the pioneering efforts of 'anaerobe enthusiasts' in various parts of the world.

A bewildering range of anaerobes is found in the mouth and oropharynx, gastro-intestinal tract and female genital tract of healthy individuals as part of the commensal flora. These include Gram-positive and Gram-negative cocci, rods and filaments, as well as a number of spiral forms (Table 36.1). Most infections with these organisms are of endogenous origin, except in the case of animal and human bite wounds, where the infecting organisms, usually mixed, are derived from the mouth of the aggressor.

The flora of the lower intestinal tract, in particular, harbours vast numbers of anaerobes; quantitative studies on the bacterial flora of human faeces (Table 36.2) reveal a total content of over 10^{10} anaerobes per gram of faeces.

Many of the bacteria isolated from anaerobic infections are opportunist pathogens. Such organisms are particularly likely to set up infections in damaged and necrotic tissue, when they are translocated to sites other than their normal habitat, or in a host that is compromised or debilitated in a way that leads to impairment of immunological or other defence mechanisms. Anaerobic infections of the head, neck and respiratory tract are often associated with organisms found in the mouth, whereas infections in the abdominal and pelvic regions are more commonly associated with gut bacteria.

FEATURES OF ANAEROBIC INFECTIONS

Clinical signs

A common, but not invariable, feature is the production of a foul or putrid odour. Foul-smelling pus or discharge should always alert the clinician to the likelihood that anaerobes are present, as no other organisms produce this effect, but the absence of this sign does not necessarily exclude the involvement of anaerobic bacteria. Other clues to the clinical diagnosis are listed in Box 36.1.

Polymicrobial flora

Infections involving non-clostridial anaerobes are often polymicrobial. The composition of these mixed infections varies according to the site affected. The complexity may vary from two or three species up to a dozen or more, and may include strict anaerobes, facultatively anaerobic and micro-aerophilic organisms. Such combinations frequently comprise mixtures of Gram-negative rods (such as *Bacteroides, Prevotella* and *Fusobacterium* species) and Gram-positive cocci (peptostreptococci or streptococci, or both). In most cases, with the occasional exception of actinomycosis, it is not possible to predict accurately which organisms are present from the clinical presentation, although the detection of red fluorescing pus under ultraviolet light usually indicates the involvement of one of the black-pigmented *Porphyromonas* species.

LABORATORY DIAGNOSIS

When anaerobic infection is suspected, it is important that adequate clinical specimens are collected and transported as soon as possible to the bacteriology laboratory, preferably under reducing conditions. After direct microscopical examination of the material, appropriate culture media should be inoculated for incubation under good anaerobic conditions, either in an anaerobic cabinet or in anaerobic jars. In some laboratories, gas–liquid chromatography is carried out directly on pus and other clinical specimens in order to detect metabolic products, such as butyric and propionic acids, that are characteristic of certain anaerobes. As many anaerobes are relatively slow growing, it is essential that cultures are incubated for several days before

Table 36.1 Anaerobic bacteria found as part of normal flora in man[a]

	Skin	Mouth	Gastro-intestinal tract	Genito-urinary tract
Gram-positive bacilli				
Actinomyces	–	+	+	+
Bifidobacterium	–	+	+	+
Clostridium	–	–	+	+
Eubacterium	–	+	+	+
Lactobacillus	–	+	+	+
Propionibacterium	+	+	+	+
Gram-positive cocci				
Coprococcus	–	–	+	–
Gemmiger	–	–	+	–
Peptococcus	+	–	+	+
Peptostreptococcus	+	+	+	+
Ruminococcus	–	–	+	–
Sarcina	–	–	+	–
Streptococcus	+	+	+	+
Gram-negative bacilli				
Anaerobiospirillum	–	+	+	(?)
Anaerorhabdus	–	–	+	(?)
Bacteroides	–	+	+	+
Bilophila	–	–	+	+
Butyrivibrio	–	–	+	–
Centipeda	–	+	+	(?)
Desulfomonas	–	–	+	–
Fusobacterium	–	+	+	+
Leptotrichia	–	+	+	–
Mitsuokella	–	+	+	(?)
Porphyromonas	–	+	+	+
Prevotella	–	+	+	+
Selenomonas	–	+	+	–
Succinimonas	–	–	+	–
Succinivibrio	–	–	+	–
Wolinella	–	+	+	–
Gram-negative cocci				
Acidaminococcus	–	–	+	+
Megasphaera	–	–	+	–
Veillonella	–	+	+	+
Spirochaetes				
Treponema	–	+	+	+
Other spiral forms	–	+	+	+

–, Not usually found; +, commonly present; (?), presence uncertain (further data required).
[a]Data from various sources.

being discarded. In mixed infections, fast-growing aerobic or facultatively anaerobic organisms are often detected within 24 h, whereas some anaerobes may require incubation for 7–10 days before their colonies can be recognized.

GRAM–NEGATIVE BACILLI

Gram-negative, anaerobic, non-spore-forming, non-motile rods were previously classified within three genera, *Bacteroides, Fusobacterium* and *Leptotrichia*, which constituted the family Bacteroidaceae. The assignment of species was based largely on the profile of metabolic end-products produced after growth in a carbohydrate and protein hydrolysate-rich medium. Thus, *Leptotrichia buccalis* produces mainly lactic acid, whereas *Fusobacterium* species produce copious amounts of butyric acid and lower levels of acetic and propionic acids. Organisms that possessed other metabolic end-product patterns were placed in the

Table 36.2 The bacterial flora of faeces of English subjects[a]

Bacterial group	Mean bacterial count[b]
Gram-negative anaerobic rods	9.8
Bifidobacterium spp.	9.8
Clostridium spp.	5.0
Veillonella spp.	4.2
Lactobacillus spp.	6.5
Bacillus spp.	3.7
Enterobacteria	7.9
Streptococcus spp.	7.1
Enterococcus spp.	5.8
Total anaerobes	10.1
Total aerobes	8.0

[a]From Hill M J, Drasar B S, Hawksworth G et al 1971 Bacteria and aetiology of cancer of large bowel. Lancet i: 95–100.
[b]Log_{10} viable organisms per gram of faeces.

BOX 36.1 Some clinical signs and indicators of non-clostridial anaerobic infections[a]

- Presence of foul-smelling pus, discharge or lesion
- Production of a large amount of pus (abscess formation)
- Proximity of lesion to mucosal surface or portal of entry
- Failure to isolate organisms from pus ('sterile' pus)
- Infection associated with necrotic tissue
- Deep abscesses
- Gas formation in tissues
- Failure to respond to conventional antimicrobial therapy
- Pus that shows red fluorescence under ultraviolet light
- Detection of 'sulphur granules' in pus (actinomycosis)
- Infection of human or animal bite wound
- Gram-negative bacteraemia
- Septic thrombophlebitis

[a]Adapted from Finegold and George (1989).

genus *Bacteroides*, which has since undergone extensive taxonomic revision.

Morphologically, *Leptotrichia* and some *Fusobacterium* species tend to form long filamentous rods, often with pointed ends. Such filaments are sometimes described as *fusiform* or spindle shaped. *Bacteroides* species are usually seen as shorter rods or coccobacilli, but may be pleomorphic.

Fusobacterium spp.

Fusobacteria colonize the mucous membranes of human beings and animals, and are generally regarded as commensals of the upper respiratory and gastro-intestinal tracts. Species such as *F. nucleatum*, *F. periodonticum* and *F. naviforme* are generally isolated from the oral cavity and are often associated with infections of this and related sites. *F. nucleatum*, the most studied species, is frequently recovered from mixed infections of the head and neck region, including dental abscesses and the central nervous system, and is also quite commonly isolated from transtracheal aspirates and pleural fluid. Various subspecies are recognized. *F. nucleatum* subspecies *nucleatum* is an important periodontal pathogen, particularly during the period when quiescent periodontitis becomes active.

F. necrophorum is an important animal pathogen. It is associated with human necrobacillosis and infections similar to those caused by *F. nucleatum* in humans, but is isolated less often.

F. mortiferum, *F. necrogenes*, *F. gonidiaformans* and *F. varium* are generally isolated from the gastro-intestinal and urogenital tracts of man and animals. These species, together with *F. nucleatum*, are often associated with mixed intra-abdominal infections, perirectal abscesses, osteomyelitis, decubitus, and other ulcers and various soft tissue infections. *F. ulcerans* was originally isolated from tropical ulcers but may be found in other sites.

Leptotrichia buccalis

This species was originally classified in the genus *Fusobacterium* and shares a number of properties with the fusobacteria. Although normally considered to be an oral species, it also occurs outside the oral cavity, but has not been widely studied. The association of *L. buccalis* with disease is not clear-cut, although it has been reported in acute necrotizing ulcerative gingivitis (*Vincent's gingivitis*), together with *Treponema*, *Porphyromonas* and *Fusobacterium* species. Some isolates described as *L. buccalis* probably represent separate species within the genus.

Bacteroides, Porphyromonas and Prevotella species

Bacteria once thought of as typical members of the genus *Bacteroides*, especially those isolated from human beings, form three broad groups according to whether they are asaccharolytic, moderately saccharolytic or strongly saccharolytic (Table 36.3):

1. The asaccharolytic, pigmented species are classified in the genus *Porphyromonas*, which includes the important periodontal pathogen *P. gingivalis*.

Table 36.3 Current taxonomic status of *Bacteroides*, *Porphyromonas* and *Prevotella* species

Group	Special
Saccharolytic (*B. fragilis* and related species)	*B. fragilis, B. caccae, B. distasonis, B. eggerthii, B. merdae, B. ovatus, B. stercoris, B. thetaiotaomicron, B. uniformis, B. vulgatus*
Moderately saccharolytic (*Prevotella* spp.)	*Prev. melaninogenica, Prev. bivia, Prev. buccae, Prev. buccalis, Prev. corporis, Prev. denticola, Prev. disiens, Prev. enoeca, Prev. heparinolytica, Prev. intermedia, Prev. loescheii, Prev. nigrescens, Prev. pallens, Prev. oralis, Prev. oris, Prev. oulora, Prev. tannerae, Prev. veroralis, Prev. zoogleoformans, Prev. salivae, Prev. shahii, Prev. multiformis, Prev. marshii, Prev. baroniae*
Asaccharolytic (*Porphyromonas* spp.)	*P. asaccharolytica, P. catoniae, P. gingivalis, P. endodontalis, P. uenonis*
Other (uncertain taxonomic status)	*B. splanchnicus, B. forsythus*

2. The moderately saccharolytic species that are inhibited by 20% bile and are largely indigenous to the oral cavity are assigned to the genus *Prevotella*.
3. The genus *Bacteroides* is now restricted to *B. fragilis* and related species that are saccharolytic and grow in 20% bile.

To add to the taxonomic complexity, many other former *Bacteroides* species that are usually isolated from non-human sources have undergone reclassification. *B. gracilis* and *B. ureolyticus* now belong to the genus *Campylobacter*, and the former *B. ochraceus* now belongs to the genus *Capnocytophaga*. *B. forsythus*, isolated from deep periodontal pockets, has been reclassified as *Tannerella forsythensis*.

Infections with *Bacteroides*, *Porphyromonas* and *Prevotella* species

Bacteroides species and related Gram-negative rods are, together with anaerobic cocci, the commonest cause of non-clostridial anaerobic infections in man. Organisms of the *B. fragilis* group are particularly significant, as they are the most commonly isolated and tend to be more resistant to antimicrobial agents than most anaerobes.

B. fragilis itself is substantially outnumbered by other *Bacteroides* species in the normal bowel microflora, but is often associated with intra-abdominal and soft tissue infections below the waist. *B. fragilis* is also the most common anaerobe found in bacteraemia, and has even occasionally been reported from head and neck infections, despite its apparent absence from the normal flora of the mouth. Species of the *B. fragilis* group account for about a quarter of all anaerobes isolated from clinical specimens.

Black-pigmented species, including those from the genera *Porphyromonas* and *Prevotella*, occur in abscesses and soft tissue infections in various parts of the body. They are rarely isolated in pure culture. *P. gingivalis* is associated with chronic adult periodontitis and *P. endodontalis* with dental root canal (endodontic) infections.

GRAM–POSITIVE ANAEROBIC COCCI

Most clinically significant species of strictly anaerobic Gram-positive cocci (as opposed to facultatively anaerobic, micro-aerophilic or carbon dioxide-dependent cocci, which may also be isolated on primary anaerobic culture plates) were formerly regarded as belonging to the genus *Peptostreptococcus*. However, several other genera are now recognized (Box 36.2). Distinguishing between these species depends upon a variety of physiological and biochemical tests, including the analysis of metabolic end-products by gas–liquid chromatography. Many species do not ferment carbohydrate substrates, so that some commonly used identification tests are of little value. As it is not always easy to identify clinical isolates precisely, they are often described simply as 'anaerobic cocci'.

Gram-positive anaerobic cocci comprise part of the normal microbial flora of the mouth, gastro-intestinal tract, genito-urinary tract and skin. They are often isolated from clinical specimens. Other genera, including *Coprococcus*, *Gemmiger*, *Ruminococcus* and *Sarcina*, are found as part of the flora of the bowel (see Table 36.1), but are not usually considered to be significant in infections.

Infections with anaerobic cocci

Anaerobic cocci are isolated from infections in various parts of the body, particularly from abscesses (Box 36.3). They are often found in association with other anaerobes, facultatively anaerobic or aerobic organisms. As with all mixed infections, it is difficult to assess the contribution of each individual organism to the pathogenic process. However, there is sufficient evidence from both clinical

Table 36.4 Some characteristics of anaerobic non-sporing Gram-positive rods

Genus	Common sites	Acid end-products
Propionibacterium	Skin, mouth, gut, vagina	Propionic acid
Bifidobacterium	Gut, mouth, vagina	Acetic and lactic acids
Lactobacillus	Mouth, gut, vagina	Lactic acid (major end-product)
Actinomyces	Mouth, gut, vagina	Succinic, lactic and acetic acids
Eubacterium[a]	Mouth, gut, vagina	Butyric and other acids

[a]Taxonomy currently undergoing revision; includes several different genera.

and experimental studies to confirm the pathogenic potential of the anaerobic cocci.

GRAM–NEGATIVE ANAEROBIC COCCI

Among genera recorded as part of the normal flora of the gastro-intestinal tract (see Table 36.1), only *Veillonella* is found regularly at other sites. In the mouth, for example, this genus is a regular component of supragingival dental plaque and of the tongue microflora. Veillonellae are able to use some of the lactic acid produced by bacteria such as streptococci and lactobacilli that potentially induce dental caries.

The role of *Veillonella* species and other anaerobic Gram-negative cocci in disease, if any, has not been established clearly, although they may be isolated from a variety of clinical conditions. In general, they are regarded as a minor component of mixed anaerobic infections, and antimicrobial chemotherapy is not generally directed specifically against them.

NON–SPORING GRAM–POSITIVE RODS

The spore-forming genus *Clostridium* is well known for its involvement in serious infections (see Ch. 22). The role of anaerobic non-sporing Gram-positive rods, on the other hand, is less well understood, although they are present in significant numbers in the normal flora of the mouth, skin, gastro-intestinal and female genito-urinary tracts, and are isolated from a variety of types of infection. The main genera and some of their characteristics are listed in Table 36.4. As with some other groups of anaerobes, the use of acid end-product analysis by gas–liquid chromatography is an important step in the identification of these bacteria.

Infections with Gram–positive rods

Any of these bacteria can occur as components of mixed anaerobic infections, and *Actinomyces* species can undoubtedly adopt a pathogenic role. Most cases of actinomycosis are caused by *Actinomyces israelii* and are cervicofacial, although the disease can also occur in the thorax, abdomen and female genital tract (see Ch. 20). *Actinomyces* species are not themselves strict anaerobes, but *A. israelii* requires good anaerobic conditions for primary isolation, and plates should be incubated for 7–10 days.

Propionibacterium propionicus is morphologically and biochemically very similar to *A. israelii*. It is particularly associated with infection of the tear duct in the condition called *lachrymal canaliculitis*. The significance of other genera in infections is not clear. Some species are found in acne; they are also isolated occasionally in infective endocarditis and in infections associated with implanted prostheses. *Eubacterium* species (possibly mistaken for *Actinomyces* species in some reports) are a large group of up to 45 currently recognized species. Reclassification of many existing species into new genera, including *Slackia* and *Eggerthella*, is currently being undertaken. These bacteria may play a role in infections around intra-uterine devices; others, for example *Slackia exigua*, may be involved in human periodontal disease. Similarly, there is only limited evidence for the pathogenicity of *Bifidobacterium* species, although *Bif. dentium* has been isolated occasionally from pulmonary infections.

SPIRAL–SHAPED MOTILE ORGANISMS

Several *Treponema* species are found in the mouth and elsewhere in the body (see Table 36.1). They are thought to be an important component of the mixed anaerobic infection associated with acute necrotizing ulcerative gingivitis along with fusobacteria and *Prev. intermedia*, and may also contribute to other forms of periodontal disease. The proportion of motile spiral organisms seen by dark-ground microscopy in samples from the gingival pocket increases markedly when there is evidence of periodontal destruction.

Motile, spiral-shaped, Gram-negative anaerobes of the genus *Anaerobiospirillum* have been isolated from patients with diarrhoea and from bacteraemia. Although comparatively rarely isolated from man, they can cause serious infections. The distribution and normal habitat of this and other morphologically similar organisms are not well understood. In some cases the source of infection may be domestic animals and pets.

TREATMENT

In many infections caused by anaerobes the most important aspect of treatment is surgical. This often involves drainage of pus from abscesses, but may also include debridement, curettage and removal of necrotic tissue. For minor infections surgical drainage alone may be sufficient, but in many cases antimicrobial chemotherapy is also indicated. The main groups of agents used are the penicillins and the nitroimidazoles, particularly metronidazole. Other agents with good anti-anaerobe activity include chloramphenicol, clindamycin and cefoxitin, but resistant strains occur.

Metronidazole is effective against virtually all obligate anaerobes, including *Bacteroides, Porphyromonas, Prevotella* and *Fusobacterium* species, but not against facultatively anaerobic or micro-aerophilic bacteria such as actinomyces and streptococci. Resistance to metronidazole is still relatively uncommon.

Most anaerobic species are sensitive to benzyl-penicillin, but members of the *B. fragilis* group are usually resistant. Such resistance is associated with β-lactamase production and these organisms are usually susceptible to combinations of penicillins with β-lactamase inhibitors (e.g. co-amoxiclav) and to carbapenems such as imipenem.

KEY POINTS

- Non-sporing anaerobes are found as part of normal flora in health.
- Most infections with anaerobes are of endogenous origin and are often polymicrobial. They act as opportunistic pathogens at damaged and necrotic tissue sites.
- Production of putrid odour is a common feature of infection.
- *Fusobacterium nucleatum* is often recovered from head and neck infections.
- Anaerobic Gram-negative rods, especially *Bacteroides fragilis* and anaerobic Gram-positive cocci, are the commonest cause of non-clostridial anaerobic infections.
- Black-pigmented *Porphyromonas* and *Prevotella* species occur in abscesses and soft tissue infections in various parts of the body.
- Penicillins and nitroimidazoles, especially metronidazole, are the main agents used for treatment.

RECOMMENDED READING

Allaker R P, Young K A, Langlois T, de Rosayro R, Hardie J M 1997 Dental plaque flora of the dog with reference to fastidious and anaerobic bacteria associated with bites. *Journal of Veterinary Dentistry* 14: 127–130

Bolstad A I, Jensen H B, Bakken V 1996 Taxonomy, biology and periodontal aspects of *Fusobacterium nucleatum. Clinical Microbiology Reviews* 9: 55–71

Borriello S P, Murray P R, Funke G (eds) 2005 *Topley and Wilson's Microbiology and Microbial Infections*, Vol. 2. *Bacteriology*. Hodder Arnold, London

Duerden B I, Drasar B S (eds) 1991 *Anaerobes in Human Disease*. Edward Arnold, London

Finegold S M, George W L (eds) 1989 *Anaerobic Infections in Humans*. Academic Press, San Diego

Fuller R, Perdigon G (eds) 2003 *Gut Flora, Nutrition, Immunity, and Health*. Blackwell Publishing, Oxford

Marsh P, Martin M V 1999 *Oral Microbiology*. Wright, Oxford

Wilson M 2005 *Microbial Inhabitants of Humans*. Cambridge University Press, Cambridge

Internet sites

List of Prokaryotic Names with Standing in Nomenclature. http://www.bacterio.cict.fr/

37 Treponema and borrelia

Syphilis; yaws; relapsing fever; Lyme disease

A. Cockayne

Members of the genera *Treponema* and *Borrelia* are spirochaetes. Human diseases caused by these bacteria (Table 37.1) include syphilis, which has been known for thousands of years, and infections such as Lyme disease, the prevalence and geographical distribution of which are still being evaluated.

Treponemal infections may be spread from person to person by intimate physical contact, by contact with infectious body fluids or, in some instances, by fomites. The treponemes that infect man are obligate human parasites, and no other natural hosts are known. In contrast, borreliae are transmitted to man by infected ticks or lice. The borreliae that cause Lyme disease and endemic relapsing fever also infect many other animal species, which act as reservoirs of infection; man is an unfortunate incidental host in the natural history of these pathogens.

Characteristically, treponemal and borrelial infections occur in several distinct clinical stages. These may be separated by periods of remission, and each stage may have a particular associated pathology. Commonly the causative organism is detectable in early lesions but is much more difficult to identify in later disease. The pathogen spreads from the initial site of infection to many organs via the bloodstream and, despite a vigorous immune response, in some untreated cases these infections may be progressive, destructive and, in some instances (e.g. tertiary syphilis), fatal. In other cases only the early symptoms are apparent and the later pathology is not seen.

Antigenic variation contributes to bacterial virulence for the relapsing fever borreliae but in general the pathogenic mechanisms employed by spirochaetes are poorly understood. No extracellular toxins have yet been identified, and the mechanisms that enable these organisms to persist in tissues despite vigorous immune responses remain unclear. *Borrelia burgdorferi* and other borreliae can vary their surface lipoproteins to avoid the immune system. The paucity of exposed antigenic proteins on the surface of *Treponema pallidum* may contribute to immune evasion. It is also likely that the later manifestations of the treponemal and some borrelial infections involve auto-immune phenomena.

In addition to the pathogenic species, many other spirochaetes form part of the normal bacterial flora of the mouth, gut and genital tract. Morphological and antigenic similarities between pathogenic and commensal spirochaetes may cause problems in the clinical and serological diagnosis.

Table 37.1 Principal human diseases caused by spirochaetes

Organism	Disease	Distribution	Primary mode of transmission	Animal reservoirs
T. pallidum	Syphilis	Worldwide	Sexual–congenital	None
T. pertenue	Yaws	Tropics and subtropics	Direct contact	None
T. endemicum	Bejel	Arid, subtropical or temperate areas	Mouth-to-mouth via utensils	None
T. carateum	Pinta	Arid, tropical Americas	Skin-to-skin contact	None
B. recurrentis	Epidemic relapsing fever	Central, East Africa; South American Andes	Louse bites	None
Borrelia spp.	Endemic relapsing fever	Worldwide[a]	Tick bites	Yes
B. burgdorferi sensu lato	Lyme disease	Worldwide[a]	Tick bites	Yes

[a]Distribution governed by presence of tick vectors.

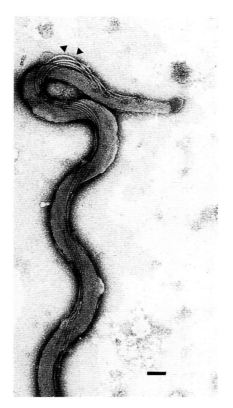

Fig. 37.1 Electron micrograph of *T. pallidum*. The flagella (arrowheads) are inserted at the tip and follow the helical contour of the bacterial cell enclosed within the outer membrane. Bar = 0.1 µm. (Photograph courtesy of Professor C. W. Penn.)

DESCRIPTION

Spirochaetes are slender unicellular helical or spiral rods (Fig. 37.1) with a number of distinctive ultrastructural features used in the differentiation of the genera (Fig. 37.2). The cytoplasm is surrounded by a cytoplasmic membrane, and a peptidoglycan layer contributes to cell rigidity and shape. In *Treponema* species, fine cytoplasmic filaments are visible in the bacterial cytoplasm (Fig. 37.3), but these are absent in *Borrelia* species. Members of both genera are actively motile; several flagella are attached at each pole of the cell and wrap around the bacterial cell body. In contrast to other motile bacteria, these flagella do not protrude into the surrounding medium but are enclosed within the bacterial outer membrane. Treponemal flagella are complex, comprising a sheath and core (Fig. 37.4), whereas those of *Borrelia* species are simpler and similar to the flagella of other bacteria. The spirochaetal outer membrane is unusually lipid rich and, at least in some treponemes, appears to be protein deficient and to lack lipopolysaccharide. This may account for the

susceptibility of these organisms to killing by detergents and desiccation.

Although the treponemes are distantly related to Gram-negative bacteria, they do not stain by Gram's method, and modified staining procedures are used. Moreover, the pathogenic treponemes cannot be cultivated in laboratory media and are maintained by subculture in susceptible animals. In contrast, borreliae stain Gram negative, and many pathogenic species can be cultured in vitro in enriched, serum-containing, media.

TREPONEMA

Treponema species pathogenic for man include the causative agents of venereal *syphilis* and the non-venereal treponematoses, *yaws*, *bejel* and *pinta*.

The spirochaetes causing these different infections are micro-aerobic and morphologically identical: tightly coiled helical rods, 5–15 µm long and 0.1–0.5 µm diameter. They show only subtle antigenic differences and are characterized primarily by the clinical syndromes they cause and minor differences in the pathology induced in experimental animals.

TREPONEMA PALLIDUM SSP. PALLIDUM (T. PALLIDUM)

T. pallidum, the causative agent of syphilis, was first isolated from syphilitic lesions in 1905. Infection is usually acquired by sexual contact with infected individuals and is commonest in the most sexually active age group of 15–30-year-olds. *Congenital syphilis* usually occurs following vertical transmission of *T. pallidum* from the infected mother to the fetus in utero, but neonates may also be infected during passage through the infected birth canal at delivery. Infection in utero may have serious consequences for the fetus. Rarely, syphilis has been acquired by transfusion of infected fresh human blood.

Pathogenesis

Untreated syphilis may be a progressive disease with *primary*, *secondary*, *latent* and *tertiary* stages. *T. pallidum* enters tissues by penetration of intact mucosae or through abraded skin.

The bacteria rapidly enter the lymphatics, are widely disseminated via the bloodstream and may lodge in any organ. The exact infectious dose for man is not known, but in experimental animals fewer than ten organisms are sufficient to initiate infection. The bacteria multiply at the initial entry site forming a *chancre*, a lesion characteristic of primary syphilis, after an average incubation period of 3 weeks.

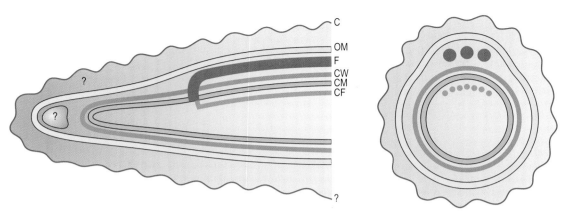

Fig. 37.2 Schematic representation of the structure of *T. pallidum* in longitudinal and cross-section: C, postulated capsular layer; OM, outer membrane; F, flagellum; CW, cell wall peptidoglycan; CM, cytoplasmic membrane; CF, cytoplasmic filaments (absent in borreliae). Areas of uncertainty (indicated by question marks) include the existence and form of the capsule, the continuity or otherwise of the outer membrane over the tip of the organism, the nature and form of the tip structure, and the exact juxtaposition of the ends of the cytoplasmic filaments with the bacterial flagellar basal bodies. (After Strugnell et al 1990 *Critical Reviews in Microbiology* 17: 231–250.)

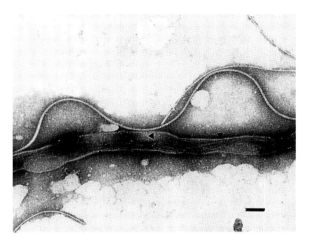

Fig. 37.3 Electron micrograph of a detergent- and protease-treated *T. pallidum* cell showing cytoplasmic filaments (arrowheads) in the bacterial cytoplasm. Bar = 0.1 μm.

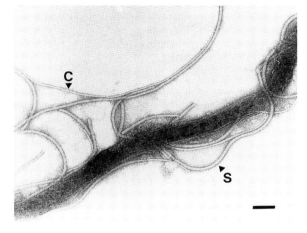

Fig. 37.4 Electron micrograph of a detergent-treated *T. pallidum* cell showing the complex structure of the treponemal flagellum. Both sheathed (S) flagella and the thinner flagellar cores (C) are visible. Bar = 0.1 μm.

The chancre is painless and most frequently on the external genitalia, but it may occur on the cervix, peri-anal area, in the mouth or anal canal. Chancres usually occur singly, but in immunocompromised individuals, such as those infected with the human immunodeficiency virus (HIV), multiple or persistent chancres may develop. The chancre usually heals spontaneously within 3–6 weeks, and 2–12 weeks later the symptoms of secondary syphilis develop. These are highly variable and widespread but most commonly involve the skin where macular or pustular lesions develop, particularly on the trunk and extremities. The lesions of secondary syphilis are highly infectious.

These lesions gradually resolve and a period of latent infection is entered, in which no clinical manifestations are evident, but serological evidence of infection persists. Relapse of the lesions of secondary syphilis is common, and latent syphilis is classified as early (high likelihood of relapse) or late (recurrence unlikely). Individuals with late latent syphilis are not generally considered infectious, but may still transmit infection to the fetus during pregnancy and their blood may remain infectious.

Late or tertiary syphilis may develop decades after the primary infection. It is a slowly progressive, destructive, inflammatory disease that may affect any organ. The three most common forms are *neurosyphilis*, *cardiovascular syphilis* and *gummatous syphilis* – a rare granulomatous lesion of the skeleton, skin or mucocutaneous tissues. Isolation of *T. pallidum* from patients with late syphilis is usually impossible, and much of the observed pathology may be due to auto-immune phenomena.

TREPONEMA PALLIDUM SSP. PERTENUE (T. PERTENUE)

T. pertenue is the causative agent of *yaws*, a disease that is endemic among rural populations in tropical and subtropical countries such as Africa, South America, South-East Asia and Oceania. Eradication programmes sponsored by the World Health Organization reduced the number of cases to fewer than 2 million in the 1970s, but termination of the programmes led to a resurgence of pockets of disease, particularly in West Africa.

Infection with *T. pertenue* occurs non-venereally following contact of traumatized skin with exudate from early yaws lesions. Infection is usually acquired before puberty. The incubation period is 3–5 weeks, and the initial lesions usually occur on the legs. The papular lesions enlarge, erode and usually heal spontaneously within 6 months. Eruption of similar lesions occurs weeks to months later, and relapse is common. Secondary lesions may involve bones, particularly the fingers, long bones and the jaw. Late yaws is characterized by cutaneous plaques and ulcers, and thickening of the skin on the palms and soles of the feet. Gummatous lesions may also develop. In contrast to syphilis, neurological and cardiovascular damage does not occur. As affected individuals acquire infection early in life, they are essentially non-infectious at childbearing age, and congenital yaws is unknown.

TREPONEMA PALLIDUM SSP. ENDEMICUM (T. ENDEMICUM)

This organism causes a non-venereal, syphilis-like disease called *endemic syphilis* or *bejel*. Bejel is endemic in Africa, western Asia and Australia, and affects mainly children in rural populations where living conditions and personal hygiene are poor. Transmission is by direct person-to-person contact and by sharing of contaminated eating or drinking utensils.

The initial lesion is usually oral and may not be detected. Secondary lesions include oropharyngeal mucous patches, condyloma lata and periostitis. Late lesions involve gummata in the skin, nasopharynx and bones. As in yaws,

the cardiovascular and central nervous systems are not involved, and congenital infection is rare because of the early age of infection.

TREPONEMA CARATEUM

Unlike the other treponematoses, the manifestations of *pinta*, caused by *T. carateum*, are confined to the skin. Although these lesions are non-destructive, they cause disfigurement with associated social problems for infected individuals. Pinta is probably the oldest human treponemal infection, with distribution now restricted to arid rural inland regions of Mexico, Central America and Colombia.

Spread of infection is by direct contact with infectious lesions. After a 7–21-day incubation period, small erythematous pruritic primary lesions develop, most commonly on the extremities, face, neck, chest or abdomen. The primary lesions enlarge and coalesce, and once healed may leave areas of hypopigmentation. Disseminated secondary lesions appear 3–12 months later and may become dyschromic. Recurrence of lesions is common for up to 10 years after the initial infection. The depigmented lesions are characteristic of the later stages of pinta but do not cause any serious harm.

OTHER TREPONEMA SPECIES

Treponemes are implicated in several other human infections. The microbial flora associated with these conditions is complex, and the exact role of the spirochaetes in the aetiology of infection remains to be determined. Moreover, the taxonomic status of some of these organisms is uncertain.

Oral infections

T. denticola, *T. socranskii* and *T. pectinovorum* form part of the normal flora, and the numbers of these organisms increase in acute necrotizing ulcerative gingivitis and chronic adult periodontal disease. *T. vincentii* (or Vincent's spirillum) is similarly associated with ulceromembranous gingivitis or pharyngitis, *Vincent's angina*. Several spirochaetes appear to be involved in the aetiology of a similar condition called *trench mouth*.

Gastro-intestinal infections

Several as yet unidentified weakly haemolytic spirochaetes have been implicated in the aetiology of persistent diarrhoea and rectal bleeding in certain human populations. A morphologically similar but genetically distinct organism,

Brachyspira (formerly *Serpulina*) *hyodysenteriae*, is the cause of swine dysentery.

Skin lesions

Tropical ulcer is a chronic skin condition in which spirochaetes of unknown identity have been implicated, usually in association with fusiform bacteria (see p. 354) and other organisms.

LABORATORY DIAGNOSIS

The inability to grow most pathogenic treponemes in vitro, coupled with the transitory nature of many of the lesions, makes diagnosis of treponemal infection impossible by routine bacteriological methods. Although spirochaetes are detectable by microscopy in primary and secondary lesions, diagnosis is based primarily on clinical observations and confirmed by serological tests. For practical purposes, the serological responses to all these pathogens are identical, and only their use in the serodiagnosis of syphilis is considered here.

Direct microscopy

Treponemes can be visualized directly in freshly collected exudate from primary or secondary lesions by dark-ground or phase-contrast microscopy. Although this method allows a rapid definitive diagnosis to be made, it is rather insensitive because primary lesions may contain relatively few bacteria. In addition, care must be taken to differentiate between pathogenic and commensal spirochaetes, which may occasionally contaminate such material. More sensitive and specific results may be obtained using fixed material in an immunofluorescence assay with an anti-treponemal antibody.

Serological tests

Infection with *T. pallidum* results in the rapid production of two types of antibody:

1. specific antibodies directed primarily at polypeptide antigens of the bacterium
2. non-specific antibodies (reagin antibodies) that react with a non-treponemal antigen called *cardiolipin*.

The mechanism of induction of non-specific antibodies remains unclear. Cardiolipin is a phospholipid extracted from beef heart, and it is possible that a similar substance, present in the treponemal cell or released from host cells damaged by the bacterium, may stimulate antibody production.

Historically, assays for non-specific antibody were used as routine screening tests for evidence of syphilis because of their low cost and technical simplicity. Enzyme immunosorbent assays that detect specific treponemal antibodies are increasingly replacing the older tests for screening purposes.

Non-specific serological tests for syphilis

The *rapid plasma reagin* (RPR) test, which has now largely superseded the earlier *Venereal Disease Research Laboratory* (VDRL) test, is a non-specific serological test for syphilis that uses cardiolipin as antigen. Immunoglobulin (Ig) M or IgG antibody present in positive sera causes a suspension of this lipoidal antigen to flocculate, and the result can be read rapidly by eye. Both of these assays may be used as screening tests, and are positive in approximately 70% of primary and 99% of secondary syphilitics, but are negative in individuals with late syphilis. These tests can be used quantitatively, and increases in antibody titres with time may be used to confirm a diagnosis of congenital syphilis. The RPR assay is not suitable for use with cerebrospinal fluid.

As a positive result in these tests usually indicates active infection, they can also be used to monitor the efficacy of antibacterial therapy.

Tests for specific antibody

Fluorescent treponemal antibody absorption (FTA-Abs) test. This is an indirect immunofluorescence assay in which *T. pallidum* is used as an antigen. Acetone-fixed treponemes are incubated with heat-treated sera, and bound antibody is detected with a fluorescein-labelled conjugate and ultraviolet microscopy. The serum is first absorbed with a suspension of a non-pathogenic treponeme, which removes non-specific cross-reactive antibodies that may be directed against commensal spirochaetes. The FTA-Abs test is positive in approximately 80, 100 and 95% of primary, secondary and late syphilitics, respectively, and, unlike the RPR and VDRL tests, remains positive following successful therapy.

T. pallidum haemagglutination assay (TPHA). In this test, *T. pallidum* antigen is coated on to the surface of red blood cells, and specific antibody in test sera causes haemagglutination. As in the FTA-Abs assay, sera are pre-absorbed with a non-pathogenic treponeme to remove antibody against commensal spirochaetes. The TPHA test is less sensitive than the FTA-Abs test in primary syphilis (positive in 65%), but both give similar results for secondary and late syphilis; the TPHA also remains positive for life following infection. This assay can be used to detect localized production of anti-treponemal antibodies in cerebrospinal fluid, a marker of neurosyphilis. The *T. pallidum particle agglutination (TPPA)* test works on the same principle as the TPHA, but treponemal antigen is coated on to coloured gelatin particles rather than red blood cells.

Enzyme immunoassay. In these tests monoclonal anti-*T. pallidum* antibodies are used to detect antibody responses to individual treponemal antigens. This allows rapid screening of large numbers of samples with potentially enhanced specificity. Assays that detect either IgM or IgG are available. Positive results should be confirmed by a second specific test such as the TPHA.

Problems in the serological diagnosis of syphilis

Occasionally, both the non-specific and specific tests produce false-positive results. The RPR and VDRL assays may give a transient positive result following any strong immunological stimulus such as acute bacterial or viral infection or after immunization. More persistent false-positive results occur in individuals with auto-immune or connective tissue disease, in drug abusers and in individuals with hypergammaglobulinaemia. False-positive results usually become apparent when negative results are found in specific serological tests, but in some cases FTA-Abs results may also be positive or borderline.

Rarely, the FTA-Abs test may be positive and the non-specific VDRL test negative. Lyme disease (see below) induces antibodies that react in the FTA-Abs but not in the VDRL assay. Other spirochaetal diseases such as relapsing fever, yaws, pinta and leptospirosis may give positive results in both specific and non-specific tests. Of particular difficulty is the differential diagnosis of syphilis and yaws in immigrants from areas in which yaws is endemic.

Some of the newer enzyme immunoassays are less sensitive in cases of primary syphilis.

Direct detection of spirochaetal deoxyribonucleic acid (DNA) in clinical material by molecular methods, such as the polymerase chain reaction (PCR), may have a future role in confirming a diagnosis of syphilis in difficult or atypical cases.

TREATMENT

All the pathogenic treponemes are sensitive to benzylpenicillin, and prolonged high-dose therapy with procaine penicillin has been the traditional method of treatment for primary and secondary syphilis. So far there have been no reports of penicillin resistance. If penicillin allergy is a problem, erythromycin, tetracycline or chloramphenicol may be used. There are reports of treatment failure with erythromycin, and an erythromycin-resistant variant of *T. pallidum* has been isolated. In late syphilis, aqueous benzylpenicillin is used, as this penetrates better into the central nervous system. In neurosyphilis, successful eradication of the organism may not result in a clinical cure. More aggressive and prolonged antibiotic therapy may be required in HIV-positive patients with syphilis owing to impaired immune function.

Antibiotic therapy of syphilitics, particularly with penicillin, characteristically induces a systemic response called the *Jarisch–Herxheimer reaction*. This is characterized by the rapid onset (within 2 h) of fever, chills, myalgia, tachycardia, hyperventilation, vasodilatation and hypotension. The response is thought to be due to release of an endogenous pyrogen from the spirochaetes.

EPIDEMIOLOGY AND CONTROL

Syphilis

The widespread introduction of antibiotic therapy shortly after the Second World War produced a dramatic decrease in the incidence of syphilis, but the disease remained endemic within the general population. Geographically localized outbreaks of syphilis, associated with specific recreational activities and lifestyles, now occur in countries such as the UK.

In the mid-1980s most cases of syphilis in developed countries occurred in male homosexuals. The advent of HIV and the acquired immune deficiency syndrome (AIDS) in the 1980s reduced the incidence among this group owing to changes in sexual practices. The early 1990s saw a resurgence of syphilis among the heterosexual population in the USA, resulting in an increased incidence among women and in the number of cases of congenital syphilis. Subsequent changes in sexual practices among homosexual males have reversed this trend: significantly more cases of primary and secondary syphilis now occur in men than women, whereas the incidence of congenital syphilis has fallen progressively from 2435 cases in 1994 to 413 in 2003. Total numbers of cases of syphilis in the USA fell to an all-time low of 32 976 in 2000, but steadily increased, principally in the male homosexual population, to 35 655 in 2003.

In the UK ongoing outbreaks of syphilis in various regions have seen the number of cases rise five-fold between 2000 and 2004, when 2252 cases of infectious syphilis were diagnosed. Although the number of cases in the male homosexual population continues to rise, the largest increase has occurred in heterosexual males and females. Syphilis therefore continues to pose major public health issues. The increased possibility of congenital infection and the acquisition of syphilis by blood transfusion mean that screening programmes of all pregnant women and blood donations are still required.

Control of syphilis is achieved by treating index cases and any known contacts. Treatment of contacts is important as some may be incubating the infection even if they have no overt signs of disease.

Control of the disease may have additional benefits: primary syphilis increases the risk of HIV infection two- to five-fold, presumably by permitting easier access of the virus through damaged skin or mucosal membranes.

Other treponematoses

The incidence of the other treponematoses is influenced primarily by socio-economic factors. Prevention and control involve treatment of individuals with active or latent disease and contacts, and improvement of living conditions and personal hygiene.

BORRELIA

The two principal human diseases associated with borreliae are *relapsing fever*, caused by *Borrelia recurrentis* and several other *Borrelia* species, and *Lyme disease* or *Lyme borreliosis*, a multi-system infection caused by *B. burgdorferi* sensu lato. The bacteria causing these infections are morphologically similar helical rods, 8–30 μm long and 0.2–0.5 μm in diameter, with three to ten loose spirals. Antigenic and genetic differences are used to differentiate the species.

RELAPSING FEVERS

Relapsing fevers are characterized clinically by recurrent periods of fever and spirochaetaemia.

Endemic or *tick-borne relapsing fever* is a zoonosis caused by several *Borrelia* species, including *B. duttoni*, *B. hermsii, B. parkeri* and *B. turicatae*, and is transmitted to man by soft-bodied *Ornithodoros* ticks. The natural hosts for these organisms include rodents and other small mammals on which the ticks normally feed. The disease occurs worldwide, reflecting the distribution of the tick vector.

Epidemic or *louse-borne relapsing fever* is caused by *B. recurrentis*, an obligate human pathogen transmitted from person to person by the body louse, *Pediculus humanus*. The incidence is influenced by socio-economic factors such as lack of personal hygiene, and, historically, increases during periods of war, famine and other social upheaval. The disease still occurs in central and eastern Africa and in the South American Andes.

The spirochaetes causing the two forms of relapsing fever differ in their mode of growth in the arthropod vector, and this influences the way in which human infection is initiated. *B. recurrentis* grows in the haemo-lymph of the louse but does not invade tissues. As a result the louse faeces are not infectious and the bacterium is not transferred through eggs to the progeny. Human infection occurs when bacteria released from crushed lice gain entry to tissues through damaged or intact skin, or mucous membranes. Spirochaetes causing tick-borne relapsing fever invade all the tissues of the tick, including the salivary glands, genitalia and excretory system. Infection occurs when saliva or excrement is released during feeding. Transovarial transmission to the tick progeny maintains the spirochaete in the tick population.

Pathogenesis

In both forms of relapsing fever, acute symptoms, including high fever, rigors, headache, myalgia, arthralgia, photophobia and cough, develop about 1 week after infection. A skin rash may occur, and there is central nervous system involvement in up to 30% of cases. During the acute phase there may be up to 10^5 spirochaetes per cubic millimetre of blood. The primary illness resolves within 3–6 days, and terminates abruptly with hypotension and shock, which may be fatal. Relapse of fever occurs 7–10 days later, and several relapses may take place.

Each episode of spirochaetaemia is terminated by the development of specific anti-spirochaete antibody. Subsequent febrile episodes are caused by borreliae that differ antigenically, particularly in outer membrane protein composition, from those causing earlier attacks. As the cycle of fever and relapse continues, the borreliae tend to revert back to the antigenic types that caused the original spirochaetaemia, and ultimate clearance of the infection appears to be due to antibody-mediated killing.

In general, louse-borne relapsing fever has longer febrile and afebrile periods than tick-borne infection, but fewer relapses. The case fatality rate varies from 4 to 40% for louse-borne infection and from 2 to 5% for tick-borne relapsing fever, with myocarditis, cerebral haemorrhage and liver failure the most common causes of death.

Laboratory diagnosis

Definitive diagnosis of relapsing fevers is made by detection of borreliae in peripheral blood samples. Thick or thin blood smears are stained with Giemsa or other stains such as acridine orange.

Although antibodies to the borreliae are produced during infection, serological tests are complicated by antigenic variation and the tendency to relapse. Serological tests for syphilis are positive in 5–10% of cases.

Treatment

Tetracycline, chloramphenicol, penicillin and erythromycin have been used successfully. As in the treatment of syphilis, antibiotics may elicit a Jarisch–Herxheimer reaction.

Prevention of infection involves avoidance or eradication of the insect vector. Insecticides can be used to eradicate ticks from human dwellings, but elimination from the environment is not feasible. Prevention of louse-borne infection involves maintenance of good personal hygiene, and delousing if necessary.

LYME DISEASE

Lyme disease, originally called *Lyme arthritis*, was recognized as an infectious condition in 1975 following an epidemiological investigation of a cluster of cases of suspected juvenile rheumatoid arthritis that occurred in Lyme, Connecticut, USA. A common factor in these cases was a previous history of insect bite, and the infectious agent, *B. burgdorferi*, was subsequently isolated from an *Ixodes* tick. Retrospective serological data suggest that Lyme disease was endemic in the USA as early as 1962, and the clinical manifestations of this infection have been known in Europe, including the UK, since the early 1900s. Lyme disease has also been reported in Scandinavia, eastern Europe, China, Japan and Australia.

The natural hosts for *B. burgdorferi* are wild and domesticated animals, including mice and other rodents, deer, sheep, cattle, horses and dogs. The larger animal hosts such as deer are probably more important in maintaining the size of tick populations rather than acting as a major source of *B. burgdorferi*. Infection in these animals may be inapparent, although clinical infection has been observed in cattle, horses and dogs.

B. burgdorferi is transmitted to man by ixodid ticks that become infected while feeding on infected animals. The principal vectors in the USA are *Ixodes dammini* and *I. pacificus*, and in Europe, *I. ricinus*. The life cycle of these ticks involves larval, nymph and adult stages, all of which are capable of transmitting infection, although the nymphal stage is most commonly implicated. In areas endemic for Lyme disease, 2–50% of ticks may carry *B. burgdorferi*. The bacterium grows primarily in the midgut of the tick, and transmission to man occurs during regurgitation of the gut contents during the blood meal. Transmission efficiency appears to be relatively low, but increases with the duration of feeding.

Although there is general similarity, clinical manifestations may differ in the USA and Europe. This variation is due in part to significant differences in the bacterial strains causing infection in the two continents, and has resulted in the division of *B. burgdorferi* into three distinct genospecies:

- *B. burgdorferi* sensu stricto is the sole cause of Lyme disease in the USA where Lyme arthritis is a common complication of infection.
- *B. afzeli* and *B. garinii* are responsible for most Lyme disease in Europe, and are associated with chronic skin and neurological symptoms, respectively.

Several other genetically distinct isolates of *B. burgdorferi* have been identified in ticks, but their importance in human infection has yet to be established.

Lyme disease may be a progressive illness, and is divided into three stages:

- *Stage 1* is characterized by a spreading annular rash, *erythema chronicum migrans* (ECM), which occurs at the site of the tick bite 3–22 days after infection. Lesions may contain very small numbers of bacteria, and the disproportionate intensity of the pathology seen may be due to stimulation of cytokines such as tumour necrosis factor- and secondary mediators. The bacterium also spreads to various other organs. In the USA, secondary lesions similar to those of ECM are common. Malaise, fatigue, headache, rigors and neck stiffness may also be apparent. ECM and secondary lesions fade within 3–4 weeks.
- *Stage 2* develops in some patients after several weeks or months. These patients exhibit cardiac or neurological abnormalities, musculoskeletal symptoms or intermittent arthritis.
- *Stage 3* may ensue months to years later, when patients present with chronic skin, nervous system or joint abnormalities.

Congenital infection may occur with serious, potentially fatal, consequences for the fetus.

Laboratory diagnosis

Once a clinical diagnosis has been made, culture of the spirochaete from suitable biopsy material provides a definitive diagnosis, but this is a lengthy, specialized technique that is not widely available. As the organism is also difficult to detect in histological sections, serological tests are used routinely for the confirmation of Lyme disease, although PCR techniques have been used in some laboratories.

Specific IgM antibodies develop within 3–6 weeks of infection. The earliest response appears to be against the bacterial flagellum and later against outer surface proteins. Subsequently, IgG antibodies are produced, and the highest titre is detectable months or years after infection.

An indirect immunofluorescence test is available, but enzyme-linked immunosorbent assay (ELISA) is now widely used. Immunoblotting with a panel of carefully selected recombinant antigens is used to confirm serological results. Serological diagnosis of early Lyme disease may still pose problems, as antibodies to the bacterium are slow to develop in some individuals and the formation of immune complexes may affect the test results. Antibodies that cross-react with *B. burgdorferi* may be produced after infection with other spirochaetes, and sera from patients with Lyme disease may give a positive FTA-Abs test, although the VDRL test is negative.

Serological evidence of infection may be detectable in the apparent absence of overt disease. The significance of these findings is unclear but it is possible that such individuals may develop late complications of Lyme disease.

Treatment

Penicillins, the newer macrolides, cephalosporins and tetracyclines have all been used successfully. Reports suggest that treatment with tetracyclines (doxycycline) produces fewer late complications than penicillin therapy. About 15% of patients experience a Jarisch–Herxheimer reaction after antibiotic therapy. Despite antibiotic treatment, some patients suffer from minor late complications of the disease, which may be mediated immunologically and may require supplementary immunosuppresive therapy, or may indicate low-level persistence of the organisms.

Antibiotic therapy may reduce or abolish the antibody response, and this may interfere with the serological confirmation of infection.

Epidemiology and control

The geographical distribution of Lyme disease is governed by that of the tick vector and its associated animal hosts. Forestry workers and farmers are particularly at risk, but infection is also increasingly associated with recreational activities. In the UK, Lyme disease occurs in areas that support large populations of wild or domesticated animals on which ixodid ticks feed. Infection may also be acquired after travel to countries where Lyme disease is endemic. It is difficult to assess accurately the true incidence of Lyme disease, because infection may be mild or asymptomatic and consequently not detected. In 2004 approximately 250 cases were confirmed in England and Wales (20–25% of which may have been acquired abroad) and around 450 cases in Scotland. By comparison, approximately 22 000 cases were reported in the USA in 2003.

Prevention of infection involves avoidance of endemic areas and education of the public regarding the possible risks of infection in these localities. Eradication of the tick vectors or mammalian hosts from such areas is not feasible. A vaccine to protect residents and visitors in areas in which Lyme disease is endemic would be useful, but attempts to develop effective and safe recombinant vaccines against *B. burgdorferi* have so far been unsuccessful.

KEY POINTS

- The genus *Treponema* includes the agents of syphilis (*T. pallidum*), yaws (*T. pertenue*), bejel (*T. endemicum*) and pinta (*T. carateum*). All are essentially morphologically and antigenically identical spirochaetes, which cannot be cultivated in vitro.
- These diseases, if untreated, characteristically progress to chronic disease through distinct early and late stages and pathologies whose appearance is separated by latent periods of variable length.
- Diagnosis of syphilis, which is primarily sexually transmitted, is made by clinical observation and confirmed serologically.
- All of these organisms are sensitive to benzylpenicillin, which can be used to treat the early stages of disease.
- The genus *Borrelia* includes agents of Lyme disease (*B. burgdorferi* sensu lato) and relapsing fevers (*B. recurrentis* and others), all of which are transmitted to man by ticks or lice.
- Lyme disease, if untreated, may progress through distinct clinical phases (stages 1–3), resulting in later pathology (e.g. arthritis, neurological damage) in some individuals.
- Relapsing fevers are characterized by recurring periods of fever and remission associated with antigenic variation of the borreliae. Infection may be fatal.
- Lyme disease is diagnosed clinically and confirmed serologically. Doxycycline may be used to treat the early stages.

RECOMMENDED READING

Antal G M, Lukehart S A, Meheus A Z 2002 The endemic treponematoses. *Microbes and Infection* 4: 83–94

Cullen P A, Haake D A, Adler B 2004 Outer membrane proteins of pathogenic spirochetes. *FEMS Microbiology Reviews* 28: 291–318

Edwards A M, Dymock D, Jenkinson H F 2003 From tooth to hoof: treponemes in tissue-destructive diseases. *Journal of Applied Microbiology* 94: 767–780

Egglestone S I, Turner A J 2000 Serological diagnosis of syphilis: PHLS Syphilis Serology Working Group. *Communicable Disease and Public Health* 3: 158–162

Genc M, Ledger W J 2000 Syphilis in pregnancy. *Sexually Transmitted Infections* 76: 73–79

Holt S C 1978 Anatomy and chemistry of spirochetes. *Microbiological Reviews* 42: 114–160

Steere A C, Coburn J, Glickstein L 2004 The emergence of Lyme disease. *Journal of Clinical Investigation* 113: 1093–1101

Wilske B 2003 Diagnosis of Lyme borreliosis in Europe. *Vector Borne Zoonotic Disease* 3: 215–227

Zeltzer R, Kurban A K 2004 Syphilis. *Clinics in Dermatology* 22: 461–468

Internet sites

Centers for Disease Control and Prevention. Syphilis. http://www.cdc.gov/std/syphilis/default.htm

Centers for Disease Control and Prevention. Lyme disease. http://www.cdc.gov/ncidod/dvbid/lyme/index.htm

38 Leptospira

Leptospirosis; Weil's disease

T. J. Coleman and A. S. Johnson

The recognition of human leptospirosis as a distinct clinical entity is usually attributed to Adolf Weil of the University of Heidelberg in 1886, although the disease had been described in animals since the mid 19th century. The term *Weil's disease* acknowledges Weil's observations in differentiating what was later proven to be a leptospiral infection from other forms of infective jaundice.

In 1914, Ryokichi Inada and his colleagues in Kyushu, Japan, observed spiral organisms in the livers of guinea-pigs inoculated with blood taken from Japanese miners with infectious jaundice, presumed to be Weil's disease. They named the organisms *Spirochaeta icterohaemorrhagiae*, reflecting their spiral shape and the fact that human infections were associated with jaundice (icterus) and haemorrhage. In Europe, similar organisms were demonstrated in some cases of jaundice in German soldiers involved in the First World War. In 1917, another Japanese scientist, Hideyo Noguchi, recognized that the organisms associated with Weil's disease differed from other known spirochaetes and proposed the genus name *Leptospira*, meaning a 'slender coil'.

More than 200 different pathogenic strains, referred to as *serovars*, are currently recognized. Genotypic classification has defined 17 genomospecies of *Leptospira*, but the phenotypic classification system is still most widely used as it provides more useful information to microbiologists, clinicians and epidemiologists.

Leptospirosis is a zoonosis and has one of the widest geographical distributions of any zoonotic disease. The highest incidence is found in tropical and subtropical parts of the world. Probably every mammal has the potential to become a carrier of some serovar of leptospira. These carriers harbour leptospires in their kidneys and excrete the bacteria into the environment when they urinate. This enables spread among their own kind and to other species, including human beings, who may directly or indirectly come into contact with their urine.

In man, the disease varies in severity from a mild self-limiting illness to the fulminating and potentially fatal disease described by Weil. Fortunately, full recovery without long-term morbidity is the most frequent outcome.

DESCRIPTION

Classification

The family Leptospiraceae belongs to the order Spirochaetales and can be subdivided into three morphologically indistinguishable genera: *Leptospira, Leptonema* and *Turneria*. Only *Leptospira* spp. are considered to be pathogenic for animals and man.

For practical purposes leptospires can be divided into two species:

1. *Leptospira interrogans* comprising those leptospires able to cause the disease *leptospirosis*.
2. *L. biflexa* contains all strains that are found in the environment but are believed to be harmless to animals and man. Unlike *L. interrogans*, they can grow at low ambient temperatures (11–13°C).

L. interrogans is further divided into 24 serogroups, which, in turn, can be split into more than 200 serovars based on differences in the outer protein envelope of the bacteria. Similarly, *L. biflexa* can be subdivided into 38 serogroups containing more than 60 serovars.

The accepted nomenclature is generic name, followed by species name, followed by serovar, followed by strain (if appropriate). For example:

- *Leptospira* (generic name) *interrogans* (species name), serovar *icterohaemorrhagiae*
- *Leptospira* (generic name) *interrogans* (species name), serovar *hardjo*, strain hardjoprajitno.

For simplicity the names are often abbreviated to only genus and serovar, so that the examples given above are shortened to *L. icterohaemorrhagiae* and *L. hardjo*, respectively.

The genetic classification of *Leptospira* does not correlate with the phenotypic classification. Of the 17 genomospecies identified so far, some include both pathogenic and non-pathogenic strains. The complete DNA sequence of one strain (*L. interrogans* serovar Lai) has been determined and this should aid studies into pathogenesis.

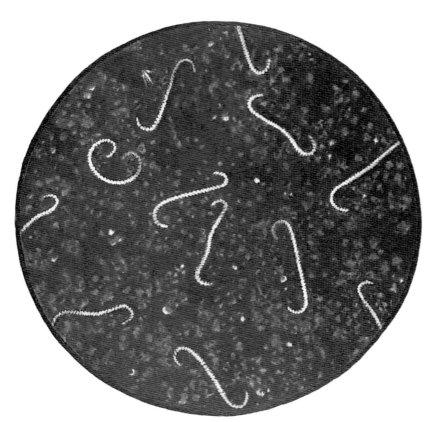

Fig. 38.1 Appearance of living leptospires as seen by dark-ground microscopy. Note the very fine coils and characteristic hooked ends. (From an original painting by Dr Cranston Low, in Low R C, Dodds T C 1947 *Atlas of Bacteriology*. Livingstone, Edinburgh.)

The organism

Leptospires range between about 6 and 20 μm in length, but are only about 0.1 μm in diameter, which allows them to pass through filters that retain most other bacteria. They are Gram negative, but take up conventional stains poorly. They can be visualized by Giemsa staining, silver deposition, fluorescent antibody methods or electron microscopy. They are best viewed by dark-ground microscopy: usually one or both ends appear hooked and they rotate rapidly around their long axis. They have many closely set primary coils that are often difficult to see in living bacteria (Fig. 38.1).

An envelope composed of three to five layers of protein, polysaccharide and lipid covers the bacteria and is the main target for the host immune response. Within the outer sheath is the protoplasmic cylinder, bounded by a cell membrane and cell wall. Leptospires have two endoflagella with their free ends towards the middle of the bacteria. They lie in the periplasmic space between the cell wall and the outer envelope, and are wrapped around the cell wall. Each flagellum is attached to a basal body located at either end of the cell. The flagella are similar in structure to those of other bacteria and are responsible for motility, but the mechanism involved in their rapid movement is incompletely understood.

Leptospires are killed rapidly by desiccation, extremes of pH (e.g. gastric acid) and antibacterial substances that occur naturally in human and bovine milks. They are susceptible to low concentrations of chlorine and are killed by temperatures above 40°C (after about 10 min at 50°C and within 10 s at 60°C).

Metabolism

Leptospires require aerobic or micro-aerophilic conditions for growth. Adequate sources of nitrogen, phosphate, calcium, magnesium and iron (as a haem compound or ferric ions) are essential. They can use fatty acids as their major energy source, but are unable to synthesize long-chain fatty acids with 15 or more carbon atoms. Members of the interrogans group require the presence of unsaturated fatty acids to utilize saturated fatty acids. Vitamins B_1 (thiamin) and B_{12} (cyanocobalamin) are also essential and the addition of biotin is needed for the growth of some strains. These components are provided in Ellinghausen–McCullough–Johnson–Harris (EMJH) medium.

Optimal growth of pathogenic species in culture takes place at 28–30°C at pH 7.2–7.6; supplementation of media with 0.1% agar enhances primary isolation. Growth is slow, often taking 3–4 weeks before colonies are visible. Culture media do not generally contain selective agents as leptospires may be sensitive to them, so great care must be taken to avoid bacterial or fungal contamination at the time of inoculation and during the prolonged incubation period.

PATHOGENESIS

The pathogenesis of leptospirosis is incompletely understood, but a vasculitis resulting in damage to the endothelial cells of small blood vessels is probably the main underlying pathology.

Infection is acquired by direct or indirect contact with infected urine, tissues or secretions. Ingestion or inhalation of leptospires is not thought to pose a risk and human-to-human spread is very rare. Leptospires generally gain entry through small areas of damage on the skin or via mucous membranes. It is possible that they may also pass through waterlogged skin, although this is probably not a major route of infection.

The term 'leptospirosis' should be used to describe all infections in both man and animals, regardless of the clinical presentation or strain of *Leptospira* involved. There are no serovar-specific disease patterns, although some serovars tend to cause more severe disease than others. In the past many names – epidemic pulmonary haemorrhagic fever, cane cutter's disease, Fort Bragg fever, Weil's disease, autumnal fever, etc. – were used to describe the particular clinical presentation or to reflect occupational, geographical, seasonal or other epidemiological features of leptospiral disease. Because of this, the full range of disease presentations was not appreciated and, even now, leptospiral infection may not be suspected unless the patient has the classically severe disease involving the liver and kidneys described originally by Weil. Some reports suggest that human infection with some serovars can, in rare cases, cause abortion.

Clinical features

Typically, acute symptoms develop 7–12 days after infection, although rarely the incubation period can be as short as 2–3 days or as long as 30 days. The infection presents with an influenza-like illness characterized by the sudden onset of headache, muscular pain, especially in the muscles of the lower back and calf, fever and occasionally rigors. Conjunctival suffusion and a skin rash may be seen in some cases.

During a bacteraemic phase lasting 7–8 days after the onset of symptoms, the leptospires spread via the blood to many tissues, including the brain. In severe cases the illness often follows a biphasic course: the bacteraemic phase is followed by an 'immune' phase, with the appearance of antibody and the disappearance of recoverable leptospires from the blood. In this phase patients may show signs of recovery for a couple of days before the fever, rigors, severe headaches and meningism return. Bleeding may occur, together with signs and symptoms of jaundice and renal impairment. Typically, bilirubin concentrations are markedly raised, but other liver function test results may be only moderately increased.

In some cases of leptospirosis, pulmonary manifestations of infection are predominant. Patients can present with cough, shortness of breath or haemoptysis. In severe cases, adult respiratory distress syndrome and pulmonary haemorrhage can supervene and lead to death.

In severe fulminating disease the patient may die within the first few days of illness, but with appropriate treatment the prognosis is usually good. Many deaths throughout the world are due to the failure to provide adequate supportive management, especially in relation to the maintenance of renal function. Generally, patients are well within 2–6 weeks but some require up to 3 months to recover fully. In a few patients, symptoms persist for many months, but neither long-term carriage of leptospires nor chronic disease has been conclusively demonstrated in man.

After infection, immunity develops against the infecting strain, but may not fully protect against infection with unrelated strains.

LABORATORY DIAGNOSIS

The initial diagnosis must rely on the medical history and clinical findings backed up with details of possible occupational or recreational exposure. A rapid diagnostic method using the polymerase chain reaction (PCR) has been developed, but is not yet available routinely. Until it is, a high index of clinical suspicion is essential.

Examination of blood and urine

In theory, leptospirosis can be diagnosed by dark-ground microscopy of blood taken during the first week of illness or, much less reliably, in urine during the second week. Dark-ground microscopy of blood is technically demanding as Brownian movement of collagen fibrils, red blood cell membranes and other artifacts can resemble viable leptospires. Examination of urine is seldom worthwhile as a method of early diagnosis of infection in man. Culture of blood may be useful in severe fulminating disease. Identification of infecting serovars requires specialized techniques that are available only in national reference laboratories.

Serology

In practice the detection of specific antibody offers the earliest confirmation of infection in most cases. Antibodies can usually be demonstrated by the sixth day after symptoms have developed, although their detection may be delayed if antibiotics were administered early in the course of the illness.

The microscopic agglutination test is generally accepted as the 'gold standard'. In addition to detecting genus-specific antibody, it can indicate the likely infecting serogroup or serovar. Doubling dilutions of patient's serum are titrated against pools of reference serovars representing the most common serogroups. The reference antigens may be either live or formolized leptospires. Tests with killed leptospires are more sensitive but less specific; they are also safer to handle and have a longer 'shelf-life'. After incubation, tests are read by low-power, dark-ground microscopy; 50% agglutination of the leptospires by the patient's or control serum represents a positive result. Sera collected soon after the onset of symptoms often show cross-reactivity to different serogroups in the microscopic agglutination test. In contrast, sera obtained during the convalescent phase of the illness generally show a significantly higher titre to the infecting serogroup or serovar.

Most other tests give no indication of the infecting serovar. Several enzyme-linked immunosorbent assay (ELISA) kits offer ease of use together with good sensitivity and specificity. Most detect IgM antibody, which, paradoxically, may remain detectable for several years, whereas IgG antibody may not be identified at all or for only a few months after infection. The complement fixation test is less sensitive and prone to cross-reactions; it is no longer recommended.

Molecular diagnostic techniques offer the potential for more rapid diagnosis of leptospirosis than the currently available serological methods. This is of particular value in the critically ill patient. PCR methods under development should allow rapid laboratory diagnosis.

Reference laboratories can advise on the examination of cerebrospinal fluid or other tissue, including those taken at post mortem.

TREATMENT

Antibiotics offer some benefit if started within 4 days of the onset of illness, and preferably within 24–48 h. In severe illness, intravenous benzylpenicillin is the drug of choice. For milder infections a 7–10-day course of oral amoxicillin is appropriate. Patients allergic to penicillins can be treated with erythromycin.

The value of antibiotic treatment is probably over-estimated and few trials have been conducted. However, supportive management to maintain tissue and organ function, such as the temporary maintenance of renal function by dialysis, may be life-saving.

EPIDEMIOLOGY

Animals that acquire infection may not develop discernible disease, but become long-term carriers – so-called *maintenance hosts*. Many rodents fall into this category. For example, rats acquiring inapparent infection with *L. icterohaemorrhagiae* may carry the bacteria in the convoluted tubules of the kidney (possibly life-long), resulting in chronic excretion of viable leptospires in their urine. Similarly, cattle may become a maintenance host for *L. hardjo*, dogs for *L. canicola*, and pigs for *L. pomona* or *L. bratislava*. The reasons for this tolerance are unclear, as infection with other serovars may cause illness of varying severity followed by the transient shedding of the leptospires in the urine for only a few weeks. Moreover, an animal may become a long-term maintenance host for one serovar and yet develop disease and transient carriage after infection with another.

Changes in industrial, agricultural and social practices may result in the rapid change of both the density and type of animal populations in an area, with subsequent change in the predominant serovars of *Leptospira* causing disease in people and animals.

Viable leptospires are present in the semen of infected animals; in rodents a significant increase in the carriage of leptospires is seen once sexual maturity has been reached. Spread across the placenta occurs in several animal species, leading to infection and possibly death of the fetus.

Outside the animal host, leptospiral survival is favoured by warm, moist conditions at neutral or slightly alkaline pH. This no doubt contributes to the seasonal pattern of human infections, which peak in the summer months in both hemispheres. Even small reductions below pH 7.0 markedly reduce the survival of leptospires. The anaerobic conditions and low pH of raw sewage explains their short survival time compared with that in aerated sewage. Salt water is also relatively toxic to leptospires. They do not survive well in undiluted cow's milk and therefore drinking unpasteurized milk poses minimal risk. However, they will survive in water at pH 7.0 or in damp soil for up to 1 month. If the soil is saturated with urine they may survive for up to 6 months, indicating the potential for long-term exposure to an infection risk even if the reservoir host has been removed for some time.

Epidemiology in the UK

In the UK, dockside fish workers, sewer workers and coal miners once accounted for almost half of all

Fig. 38.2 Leptospirosis acquired overseas, 1990–2004. (Data kindly provided by Mrs W J Zochowski, Leptospira Reference Laboratory, Hereford, UK.)

reported cases. Since 1980 improved health and safety measures adopted in these industries have resulted in a very marked change, and farm workers now represent the group at greatest occupational risk. Most cases are associated with dairy farms. In 'herring-bone' milking parlours the head of the operator is at about the same height as the udder and is, therefore, in danger of direct urine contamination!

Infections related to exposure to surface waters have also shown a significant rise. Here the predominant serovar is *icterohaemorrhagiae*. The increase is almost certainly due to the greater recreational use of surface waters for activities such as canoeing, windsurfing, fishing and pot-holing, and the use of rivers for the swimming section of triathlon competitions.

There has also been an increase in cases of lepto-spirosis acquired abroad, particularly among travellers on adventure holidays with water contact. In England and Wales between 1990 and 2004, leptospirosis was acquired by 111 travellers, 60 of whom had been to South-East Asia, compared with 194 indigenous cases (Fig. 38.2).

Men and women are equally susceptible, but reported cases in the UK show a marked male preponderance of about 20:1, reflecting differences in occupational and recreational exposure.

Deaths from leptospirosis have declined markedly, although about 4% of patients still succumb each year. This percentage is similar in other developed countries, but in some countries with limited facilities for medical care death may occur in 25% or more cases.

CONTROL

With more than 200 known strains, each able to infect a wide range of animals that may become long-term carriers capable of infecting others, together with the organisms' ability to survive for long periods in the environment, the complete prevention or eradication of leptospirosis is impossible.

Mass immunization of domestic livestock will prevent clinical disease in the animals and reduce the risk of human acquisition of infection. In the UK, immunization and treatment of infected farm and companion animals may have been an important contributory factor in the reduction in *L. hardjo* infection and the apparent disappearance of the human cases of *L. canicola* infection acquired from dogs.

To be fully effective, a vaccine should not only protect against disease in the animal but also prevent the establishment of the carrier state and the shedding of viable leptospires in the urine. It is also important that the vaccine contains the strains that predominate in the locality, as protection will be optimal only against the vaccine components. Current vaccines protect for only 1–2 years and the economics of farming may influence a farmer's decision as to whether or not to immunize cattle. No human vaccine is licensed for use in the UK.

Awareness of leptospirosis through the education of doctors, employers and the general public has helped to develop safer practices or procedures in the workplace and during recreational pursuits. This awareness should include consideration of leptospirosis in the differential diagnosis of fever in the returning tourist. Measures to reduce rodent populations in the vicinity of human activity, such as removing rubbish, especially waste food, and prevention of the access of rats into buildings is most important.

Simple measures to reduce the risks of acquiring infection include:

- covering cuts and abrasions with waterproof plasters
- wearing protective footwear before exposure to surface waters
- showering promptly afterwards if immersion takes place.

In parts of the world where the prevalence of human infection in certain groups is high, selective human immunization schemes may be of benefit if a suitable vaccine is available. Antimicrobial prophylaxis with doxycycline may be of value in high-risk exposure situations in which prompt medical help is unavailable.

KEY POINTS

- The term 'leptospirosis' is used to describe all infections in man and animals, regardless of the clinical presentation or strain of *Leptospira* involved. Infections are uncommon in the UK and associated predominantly with occupational or recreational exposure.
- Antibiotics offer some benefit if started within 4 days of the onset of illness, and preferably within 24–48 h. In severe illness, intravenous benzylpenicillin is the drug of choice.
- Animals that acquire infection may not develop discernible disease, but become long-term carriers, so-called *maintenance hosts*.
- More than 200 strains are known and each is able to infect a range of animals that may become long-term carriers capable of infecting others. The organisms can also survive for long periods in the environment. Together, these features make control of leptospirosis a substantial challenge.

RECOMMENDED READING

Alston J M, Broom J C 1958 *Leptospirosis in Man and Animals.* Livingstone, Edinburgh

Faine S, Adler B, Bolin C, Perolat P 1999 *Leptospira and Leptospirosis*, 2nd edn. MediSci, Melbourne

Kmety E, Dikken H 1993 *Classification of the Species* Leptospira interrogans *and History of its Serovars.* University Press, Groningen

Levett P N 2001 *Leptospirosis. Clinical Microbiology Reviews* 14: 296–326

Levett P N 2005 Leptospirosis. In Mandell G L, Bennett J E, Dolin R (eds) *Principles and Practice of Infectious Diseases*, 6th edn, pp. 2789–2794. Elsevier Churchill Livingstone, Edinburgh

Internet sites

eMedicine. Leptospirosis in humans. http://www.emedicine.com/emerg/topic856.htm

Centers for Disease Control and Prevention. Leptospirosis. http://www.cdc.gov/ncidod/dbmd/diseaseinfo/leptospirosis_g.htm

39

Chlamydia

Genital and ocular infections; infertility; atypical pneumonia

M. E. Ward

Chlamydiae are obligate intracellular bacterial pathogens of eukaryotic cells with a characteristic dimorphic growth cycle quite distinct from that of other bacteria. They are widely distributed in nature and two genera, *Chlamydia* and *Chlamydophila*, are responsible for a variety of human infections, especially ocular, genito-urinary and respiratory diseases (Table 39.1). There is some evidence that they may be involved in atherosclerosis and, possibly, other chronic diseases.

Among related organisms, *Simkania negevensis* has occasionally been associated with human respiratory tract disease. *Parachlamydia* species commonly occur in symbiotic association with freshwater amoebae and with pathogenic *Acanthamoeba* species. They are sometimes found in nasal swabs, soft contact lenses, humidifier water, etc., but are rarely sought.

DESCRIPTION

Chlamydiae are obligate intracellular bacteria that grow within characteristic cytoplasmic vacuoles. They stain poorly with Gram's stain, but have the typical lipopolysaccharide (LPS) of Gram-negative bacteria. They have a characteristic dimorphic growth cycle, in which infection is initiated by environmentally resistant, metabolically inert, infectious structures called *elementary bodies*, whereas the larger, more fragile, *reticulate bodies* are responsible for intracellular replication. The species are differentiated on the basis of:

- growth characteristics and host range
- characteristic morphology or staining properties of their inclusions: *C. trachomatis* in particular produces inclusions containing a glycogen-like carbohydrate that stains blue with iodine
- deoxyribonucleic acid (DNA) and ribosomal ribonucleic acid (RNA) sequences
- the presence of plasmids or (in *Chlamydophila*) bacteriophages

- serology based on surface antigens, notably the major outer membrane protein.

Classification

Chlamydia species

C. trachomatis, the most important cause of human disease, is divided into two biovars which reflect fundamental differences in their invasiveness for cell culture, and in their involvement in human disease.

- The oculogenital biovar causes minimally invasive *trachoma, inclusion conjunctivitis* (the so-called *TRIC agents*), oculogenital infections and reactive arthritis.
- The LGV biovar causes *lymphogranuloma venereum*, a more invasive genital tract infection associated with lymphoid pathology.

Both biovars can be divided into serovars (see Table 39.1) on the basis of epitopes carried on their major outer membrane protein.

- There are 15 serovars, designated A–K, in the oculogenital biovar.
- Trachoma, a potentially blinding eye infection of the developing world, is caused by serovars A, B, Ba or C.
- Chlamydial genital tract infection (or conjunctivitis secondary to chlamydial genital tract infection) is caused by serovars D–K.
- There are three serovars, designated L1, L2 and L3, within the LGV biovar.

There are geographical differences in the distribution of these serovars and it is not uncommon for an individual with genital infection to be infected simultaneously with more than one serovar. There is little difference in virulence between strains of the same serovar. Single strains of mixed serovar rarely arise from recombination events within the gene encoding the major outer membrane protein. In persons frequently exposed to infection these recombinants

Table 39.1 Human infections caused by chlamydiae

Site of infection	Disease	Organism (serovars)
Eye	Trachoma	*C. trachomatis* (A, B, Ba, C)
	Inclusion conjunctivitis	*C. trachomatis* (D–K)
	Ophthalmia neonatorum	*C. trachomatis* (D–K)
	Contact lens-associated	*Parachlamydia* spp.
Genital tract		
Male	Non-specific urethritis, proctitis, epididymitis	*C. trachomatis* (D–K)
Female	Cervicitis, urethritis, endometritis, salpingitis, PID, perihepatitis, peri-appendicitis, infertility with tubal occlusion	*C. trachomatis* (D–K)
	Abortion, premature birth	*C. trachomatis* (D–K)[a]
	Sheep-related abortion	*Ch. abortus*
Male and female	Lymphogranuloma venereum	*C. trachomatis* (L1–L3)
Respiratory tract	Neonatal atypical pneumonia	*C. trachomatis* (D–K)
	Pharyngitis, bronchitis, pneumonia	*Ch. pneumoniae*
		Simkania negevensis[a]
	Pneumonia	*Ch. abortus*
	Psittacosis, ornithosis	*Ch. psittaci*
Chronic diseases	Atherosclerosis, coronary disease	*Ch. pneumoniae*[a]
	Stroke, multiple sclerosis, sarcoidosis, Alzheimer's disease	*Ch. pneumoniae*[b]

PID, pelvic inflammatory disease.
[a]Unproven association; [b]weak association.

may enable chlamydiae to avoid the established immune response.

Other species of *Chlamydia* include *C. muridarum*, the former mouse pneumonitis agent, which provides a useful mouse model of human chlamydial infection, and *C. suis*, which is associated with widespread respiratory, intestinal and respiratory tract infection in pigs.

Chlamydophila species

Members of the genus *Chlamydophila* infect birds and mammals as well as human beings:

- *Ch. psittaci* causes respiratory, gut and systemic infections in birds, particularly in psittacine and ornithine birds (budgerigars, parrots), and may lead to severe and sometimes fatal pneumonia in man (*psittacosis* or *ornithosis*).
- *Ch. pneumoniae*, the former TWAR (Taiwan acute respiratory) agent, is a common cause of mild to severe acute respiratory disease in man and is also associated with chronic human disease.
- *Ch. abortus* causes abortion in sheep and, rarely, in pregnant women exposed to infected sheep. Sheep and goats often carry a related organism, *Ch. pecorum*, in their intestines.

- *Ch. caviae* causes oculogenital infection in guinea-pigs and is an important experimental model of human infection.

Structure

Elementary bodies

Chlamydial elementary bodies are small electron-dense structures about 300–350 nm in diameter. They are generally round, although in some strains of *Ch. pneumoniae* they are pear shaped. The elementary body is the only infectious stage of the chlamydial developmental cycle and is the primary target of efforts to prevent chlamydial infection.

The elementary body may be considered as a tough, rigid, 'spore-like' body whose purpose is to permit survival outside the host cell. In electron micrographs the most obvious feature is the electron-dense core of DNA, which is tightly compacted on to chlamydial histone protein (Fig. 39.1). Short projections of unknown function radiate from the surface of the elementary body.

The elementary body is thought to be metabolically inert until it attaches to, and is ingested by, a susceptible host cell. Its rigid cell wall contains only small amounts of peptidoglycan and derives its strength mainly from sulphur bridges in the cysteine- and methionine-rich proteins of the outer envelope.

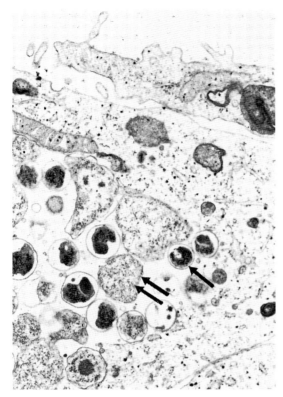

Fig. 39.1 Electron micrograph of a thin section of chlamydial inclusion showing small elementary body (single arrow) and reticulate body (double arrows) (original magnification ×15 000. (Courtesy of Dr Douglas R. Anderson, Miami.)

Reticulate bodies

Typically, reticulate bodies have a diameter of around 1 μm or more. They are non-infectious. They are metabolically active, so their cytoplasm is rich in ribosomes, which are required for protein synthesis. Their nucleic acid, unlike that of the elementary body, is diffuse and fibrillar.

Reticulate bodies are bounded by an inner cytoplasmic membrane and an external outer envelope covered with projections and rosettes that are similar to, but more numerous than, those seen on elementary bodies. Reticulate bodies are often packed around the edge of the intracellular inclusion, where they grow in close apposition to the inclusion membrane. The surface projections apparently penetrate the inclusion membrane and they may be part of a bacterial secretion mechanism, injecting chlamydial proteins to subvert the host cell. The reticulate bodies divide by binary fission in typical bacterial fashion. They eventually switch on histone synthesis and differentiate into one or more elementary bodies. A summary of some features of chlamydial elementary and reticulate bodies is shown in Table 39.2.

Nucleic acid

Chlamydiae have one of the smallest genomes among bacteria, with approximately 1.2 Mb of sequence coding for roughly 700 proteins. Gene sequencing has revealed a number of surprising features.

- Genes encoding metabolic pathways for the production of adenosine triphosphate (ATP). It had formerly been assumed that high-energy molecules for biosynthesis had to be provided by the host cell.
- Both *C. trachomatis* and *Ch. pneumoniae* have the full complement of molecules required for biosynthesis of peptidoglycan, the site of penicillin action and the main strengthening component of bacterial cell walls.
- The existence of a family of polymorphic membrane proteins, including a clostridia-like cytotoxin that may play a key role in pathogenicity.
- A surprisingly large number of chlamydial genes with apparent origins from higher organisms, including animals and plants. This may have implications for understanding the evolution of eukaryotic cells and their organelles.
- The absence of the *FtsZ* gene, previously considered essential for bacterial cell division.
- A family of genes encoding proteins with phospholipase D-like activity and probably involved in the interaction of chlamydiae with lipid-containing host cell organelles.

Table 39.2 Comparison of chlamydial elementary bodies and reticulate bodies

Characteristic	Elementary body	Reticulate body
Size	0.2–0.3 μm	1 μm
Morphology	Electron-dense core, rigid	Fragile, pleomorphic
Infectivity to host	Infectious	Non-infectious
RNA : DNA ratio	1:1 (condensed DNA core)	3:1 (increased ribosomes)
Metabolic activity	Relatively inactive	Active, replicating stage
Trypsin digestion	Resistant	Sensitive
Projections and rosettes	Few	More

Within *C. trachomatis* serovars, genomic sequencing indicates few major genetic differences with the exception of a region of hypervariable sequence, the plasticity zone. Within this zone differences in genes encoding tryptophan synthetase and a clostridia-like cytotoxin are thought to be important determinants of biovar-associated differences in tissue tropism and invasion.

Virtually all strains of *C. trachomatis* and several other chlamydiae carry a cryptic plasmid, whose function is unknown. However, as multiple copies of the plasmid are almost always present in *C. trachomatis* it is a particularly useful target for the detection and diagnosis of chlamydial infections by DNA amplification methods. There is no evidence that these plasmids carry antibiotic resistance genes.

Protein

In most chlamydiae the major outer membrane protein at the surface of the elementary bodies is immunodominant, and relatively high amounts of monoclonal antibody to the serovar-specific regions of the protein neutralize chlamydial infection in tissue culture cells. Furthermore, the major outer membrane protein forms a pore in the chlamydial surface, permitting the controlled ingress of molecules. One model of chlamydial infection is that, after entry of the elementary body into the cell, reduction of disulphide bridges permits the entry of nutrients into the elementary body and triggers differentiation.

In *Ch. pneumoniae*, the major outer membrane protein is not immunodominant and it may be overlaid by one of the other membrane proteins that this organism possesses, and which are likely to play an important role in pathogenicity.

Chlamydiae have at least two heat shock proteins, close relatives of which are found in other bacteria and in man. These proteins rescue and chaperone proteins that have been exposed to stress; exposure to active oxygen radicals inside cells might be such a stress for chlamydiae. Cross-reactivity of antibodies to human or bacterial heat shock proteins with the chlamydial protein may interfere with serological tests measuring antibody to whole chlamydiae, such as the micro-immunofluorescence test. Auto-immunity caused by the immune response to chlamydial heat shock protein may contribute to the pathogenesis of chronic chlamydial infection.

Lipopolysaccharide

Chlamydiae have a typical 'rough' Gram-negative LPS with weak endotoxin activity that carries a chlamydia-specific antigenic determinant detected by some laboratory diagnostic tests.

Growth

Laboratory propagation

Chlamydiae may be grown by centrifugation on to a monolayer of tissue culture cells, usually HeLa or McCoy cells, and incubation in special media for 48–72 h. The presence of chlamydial inclusions is determined by microscopy in conjunction with a suitable staining method, preferably fluorescence microscopy with labelled monoclonal antibody.

Newer laboratory tests have replaced the need for tissue culture in the laboratory confirmation of chlamydial infection.

Growth cycle

The chlamydial growth cycle (Fig. 39.2) is initiated by the attachment of an infectious elementary body to the host cell and entry into the cell, which follows closely. Typically, chlamydiae attach to the host cell near the base of microvilli, from where they are actively enclosed in tight endocytic vesicles. The process of chlamydial entry is relatively efficient, but ill understood. Electron microscopy suggests two possible mechanisms:

1. a sequential, zipper-like, microfilament-dependent process of phagocytosis that requires direct circumferential contact between bacterial adhesins and unknown host cell receptors
2. uptake by receptor-mediated endocytosis into clathrin-coated vesicles.

Intrinsic properties of the chlamydial cell wall initially delay normal fusion with lysosomes. Subsequently, the vacuole translocates to the perinuclear region, and chlamydial gene products modify the vacuole so that it intercepts and fuses with a subset of sphingomyelin-containing exocytic vesicles from the host cell Golgi apparatus. The chlamydial vacuole also grows, in part, by acquiring and subsequently modifying glycerophospholipids from the host cell. The result is that chlamydiae have a phospholipid composition much closer to that of the host cell than to other bacteria.

After the differentiation of elementary bodies to reticulate bodies, the latter divide by binary fission within the expanding endosome with a doubling time of about 2 h. After 18–24 h the DNA begins to re-condense, and reticulate bodies are transformed into one or more elementary bodies. The growth cycle is asynchronous, so that reticulate bodies, elementary bodies and various intermediate forms may be found in the same inclusion.

This process of replication, followed by the differentiation of reticulate bodies into elementary bodies, requires amino acids such as cysteine and tryptophan, deprivation of which (e.g. as a result of the action of interferon-γ) halts

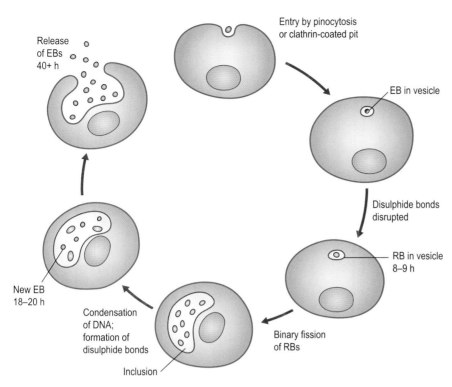

Fig. 39.2 The growth cycle of chlamydia.

Entry by pinocytosis or clathrin-coated pit

Release of EBs 40+ h

EB in vesicle

Disulphide bonds disrupted

RB in vesicle 8–9 h

New EB 18–20 h

Condensation of DNA; formation of disulphide bonds

Inclusion

Binary fission of RBs

EB = elementary body; RB = reticulate body

chlamydial growth. This can lead to persistent quiescent infection, as the amino acid-starved pathogen may lie dormant, transferring with host cell division, until the necessary amino acid becomes available again.

PATHOGENESIS

Comparatively little is known about how chlamydiae produce disease; what is known relates to *C. trachomatis*. The most severe sequelae of *C. trachomatis* infection (visual loss in trachoma, ectopic pregnancy or infertility in pelvic inflammatory disease) are caused by fibrosis and scarring due to the repair of tissue damaged by chlamydia-induced inflammation. There is little evidence for major differences in the basic virulence of *C. trachomatis* other than the differences between the LGV and trachoma biovars, and differences in susceptibility to interferon-γ. There are several apparent major determinants of severe disease.

- Socio-environmental factors, particularly in trachoma, affecting the number of infecting organisms, which in turn influences the magnitude of the initial inflammatory response and the likelihood of repeated infection. The more severe the inflammatory response,

the more likely there is to be significant fibrosis and damage to tissue.
- Host genotypic factors that regulate the immune response and associated repair mechanisms.
- The potential for self-induced (probably cell-mediated) immune damage to the host arising from chlamydial heat shock proteins or other chlamydial antigens that are similar to, or mimic, host components.
- The ability of chlamydiae to become sequestered at immunologically privileged or inaccessible sites (e.g. coronary arteries for *Ch. pneumoniae*; the joints for reactive arthritis due to *C. trachomatis*).

Clinical features

Ocular infection

Trachoma. Trachoma has been known for thousands of years. It is caused by several *C. trachomatis* serovars (see Table 39.1) that are spread by eye-seeking flies, fingers and contaminated articles. Active trachoma, characterized by the presence of lymphoid follicles on the conjunctiva and intermittent shedding of chlamydiae, is primarily a disease of children.

By contrast, blindness occurs mainly in adults. It is caused by conjunctival scarring leading to distortion of the eyelids (*entropion*) and abrasion of the cornea by the eye lashes (*trichiasis*). Adults with trachoma rarely shed viable chlamydiae from their eyes, although it is possible to demonstrate the presence of chlamydial antigen or chlamydial nucleic acid. Ocular damage is probably largely determined by:

- the cell-mediated immune response
- the severity of the original infection in childhood
- the frequency of re-infection
- the prevalence of other agents of bacterial conjunctivitis, such as *Moraxella* spp., that can exacerbate visual loss
- host genotypic factors that influence the balance between protective T helper 1 and T helper 2 cell-mediated immune responses.

Trachoma is generally a clinical diagnosis, which has been simplified by a grading scheme introduced by the World Health Organization.

Adult inclusion conjunctivitis. Inclusion conjunctivitis (*paratrachoma*) occurs worldwide and is most prevalent in sexually active young people, being spread from genitalia to the eye. In the acute stage it presents as a follicular conjunctivitis, often with a mucopurulent discharge, which persists if untreated. Patients occasionally develop scarring or pannus formation as in trachoma. However, the disease is much milder, is usually self-limiting and rarely causes visual loss, presumably because, unlike trachoma, repeated infection is less common.

Chlamydial ophthalmia neonatorum. This condition develops in infants around 14 days after birth. Follicles are rarely seen, reflecting the infant's immunological immaturity. The disease presents as a swelling of the eyelids and orbit, hyperaemia and a purulent infiltration of the conjunctiva (*inclusion blennorrhoea*), which does not respond to cleansing of the eye or to chloramphenicol eye ointment.

The organisms are acquired from the mother during birth; about half of infants born through a chlamydia-infected cervix develop ocular infection. These babies often have nasopharyngeal infection also, either from the eyes via the lachrymal ducts or by direct acquisition at birth. The diagnosis is confirmed by laboratory examination of a conjunctival swab obtained after removing pus from the eye. A smear stained with *C. trachomatis*-specific fluorescent monoclonal antibody often shows spectacular numbers of chlamydial elementary bodies, looking like the star-spangled sky at night.

If untreated the infection usually resolves, but a substantial proportion of these infants develop chlamydial pneumonia about 6 weeks after birth. This is a mild atypi-cal pneumonia, with insidious onset and surprisingly marked radiographic changes.

Genital infection

Infection in men. *C. trachomatis* serovars D–K are responsible for about 30% of cases of non-specific urethritis in men. This is one of the commonest sexually transmitted infections worldwide and repeat infections are common.

The infection is often asymptomatic, with infected men serving as a reservoir of infection in the community. In symptomatic patients, varying amounts of mucopurulent discharge are produced. Occasionally this progresses to epididymitis or prostatitis, especially in those aged less than 35 years. It is likely that chronic chlamydial epididymitis may eventually lead to occlusion of the tube and infertility due to azoospermia.

Anal intercourse may cause chlamydial proctitis in either sex. Associated symptoms include rectal pain and bleeding, mucopurulent discharge and diarrhoea.

Infection in women. In symptomatic women, *C. trachomatis* serovars D–K cause mucopurulent cervicitis and urethritis. However, many women harbour the organism asymptomatically in their cervix. Such patients are not only a risk to their sexual partners or offspring, but also to themselves, as ascending infection frequently occurs. This results first in an endometritis, in which chlamydiae survive monthly menstrual shedding of the uterine lining, followed by infection of the fallopian tubes to cause acute salpingitis. If untreated, this is likely to produce scarring and tubal occlusion which, if bilateral, leads to infertility. Alternatively, tubal function may be impaired rather than blocked, leading to implantation of the fertilized ovum in the oviduct rather than the womb, resulting in ectopic pregnancy.

Collectively, endometritis and salpingitis are known as *pelvic inflammatory disease*, which, in most developed countries, is largely caused by *C. trachomatis*. Chlamydial pelvic infection may lead to further abdominal involvement and the formation of pelvic adhesions. Perihepatitis (*Fitz-Hugh–Curtis syndrome*) and even peri-appendicitis may result.

Chlamydial pelvic inflammatory disease is a substantial and seldom recognized cause of pelvic pain in young women, leading to greatly increased likelihood of hysterectomy or other surgical interventions. Pain, costly investigations for infertility, and surgery can be avoided by screening young women aged 15–25 years for chlamydial infection. This is particularly important in women seeking a termination of pregnancy or undergoing surgery or instrumentation of the genital tract.

Lymphogranuloma venereum. Genital tract infection with *C. trachomatis* serovars L1–L3 may present as lymphogranuloma venereum. Commonest in the tropics, this condition is occasionally seen in developed countries. It usually begins with a genital ulcer followed by lymphadenopathy of the regional lymph nodes. Buboes are seen in men and, should the infection persist, can spread to the gastro-intestinal and genito-urinary tracts, causing strictures and, in rare cases, peno-scrotal elephantiasis.

Infection in pregnancy

Ch. abortus is commonly associated with infectious abortion in sheep. Pregnant women exposed to infection from aborting sheep occasionally suffer miscarriage or intra-uterine death. The organism seems to have a predilection for the placenta, causing placentitis. Patients develop a febrile prodrome a few days to 2 weeks after contact with sheep. There is also a potential hazard to nursing or medical staff dealing with unsuspected cases due to *Ch. abortus* because of the heavy load of the organism in the placenta.

C. trachomatis has also been isolated from abortion products, but its role in abortion, stillbirth, premature birth and premature rupture of the membranes is uncertain.

Respiratory infections

Chlamydophila pneumoniae. This organism causes pneumonia, pharyngitis, bronchitis, otitis and sinusitis with an incubation period of about 21 days. It is suspected to be a significant cause of acute exacerbations of asthma. *Ch. pneumoniae* is one of the commonest causes of community-acquired pneumonia, but is seldom identified as the causal agent because laboratory tests for its diagnosis are not widely used. The organism is a chronic, often insidious, respiratory pathogen to which there appears to be little immunity. Clinical reactivation of existing infection and re-infection are probably common, although the two are difficult to distinguish. Sero-epidemiological studies indicate that some 60–80% of people worldwide become infected with *Ch. pneumoniae* during their life, at an incidence of 1–2% per year.

Chlamydophila psittaci. Avian strains of *Ch. psittaci* cause psittacosis in man. The disease is more correctly called ornithosis, as infection can be acquired from birds other than psittacines. The incubation period is about 10 days, and the illness ranges from an 'influenza-like' syndrome, with general malaise, fever, anorexia, rigors, sore throat, headache and photophobia, to a severe illness with delirium and pneumonia. The illness may resemble bronchopneumonia, but the bronchioles and larger bronchi are involved as a secondary event and sputum is scanty. The organism is blood-borne through the body, and there may be meningo-encephalitis, arthritis, pericarditis or myocarditis, or a predominantly typhoidal state with enlarged liver and spleen, and even a rash resembling that of enteric fever. Endocarditis resembling that complicating Q fever (see p. 392) has been described.

Other diseases

Prospective serological studies initially indicated that *Ch. pneumoniae* was associated with the subsequent development of coronary heart disease, but this association has not been confirmed. However, *Ch. pneumoniae* is shed from the lungs into the bloodstream, and *Ch. pneumoniae* antigen or nucleic acid is commonly found in atherosclerotic plaque. In cell culture *Ch. pneumoniae* can bring about the conversion of macrophages into foam cells, and in experimental animals the organism may exacerbate formation of atheromatous plaque. Taken together, it seems plausible that *Ch. pneumoniae*, in association with known risk factors, might exacerbate coronary artery disease. However, in two large prospective trials anti-chlamydial antibiotics had no significant beneficial effect for the prevention of coronary heart disease. Thus, the use of antibiotics to prevent coronary artery disease is presently not justified.

Ch. pneumoniae has been suspected of playing a role in other chronic human diseases, including Alzheimer's disease, multiple sclerosis, stroke and sarcoidosis, but the evidence is even more slender than for heart disease.

Infection of animals

Chlamydiae infect animal cells ranging from amoebae to man. Non-human infections are important economically and as a source of human infection. *Ch. abortus* is an important cause of sheep abortion.

Ch. psittaci is an important cause of infections in a wide range of birds and the organism is shed in nasal secretions and droppings. The nasal secretions contaminate the feathers, where they dry and produce a highly infectious dust in which the organism can survive for months. This may give rise to severe pneumonia in man, called ornithosis or psittacosis depending on the bird species from which the infection was derived. The agricultural economy is also affected as large outbreaks of ornithosis have been reported in turkeys, geese and ducks. There are import controls in many countries to restrict the movement of birds, which are rendered more infectious by travel-induced stress.

Immunological response

Studies in animal models indicate that there are both humoral and cell-mediated immune responses to chlamydial infection.

Initial host response

Chlamydial infection generates a cytokine response by direct infection of the columnar or non-squamous epithelia that they often infect, and by interaction with cells of the immune system, particularly macrophages.

Cytokines generated by direct interaction with epithelial cells help to generate and sustain an inflammatory response. Interleukin (IL)-1α is particularly important, stimulating additional production of pro-inflammatory cytokines by adjacent uninfected cells. IL-8 attracts polymorphonuclear leucocytes, which can kill chlamydiae, whereas IL-6, generated by either epithelial cells or interaction of chlamydiae with T lymphocytes, is probably important, together with IL-12, for sustaining the protective T helper 1 cell-mediated immune response.

Tumour necrosis factor-α (TNF-α), produced by macrophages in response to chlamydiae, may play an important role in tissue damage. Some virulent strains of *Ch. psittaci* can replicate in macrophages, but it is unclear whether this is significant in the case of *C. trachomatis*. However, it is likely that chlamydial antigens, particularly LPS, persist in macrophages, perhaps acting as a long-term inflammatory stimulus.

Protective immunity

Multiple infections with *C. trachomatis* are common, suggesting that there is little natural immunity to chlamydial infections. Nevertheless, there is evidence that acquired immunity develops.

- Active trachoma is a childhood disease, suggesting that repeated exposure to high levels of re-infection in a hyper-endemic area eventually leads to immunity.
- Genital tract isolation of *C. trachomatis* decreases significantly with age independently of diminished sexual activity, suggesting that exposure leads to moderate protection. Moreover, genital tract infection is relatively uncommon in groups repeatedly exposed to infection, such as commercial sex workers.
- *C. trachomatis* infections are generally limited to the superficial mucosae and shedding tends to resolve if infection is untreated. However, protection is short-lived.
- Vaccination against trachoma with crude, whole preparations of *C. trachomatis* achieved some short-term protection in volunteers and in field trials. Limited protection has been achieved in experimental animals with various vaccines.

As intracellular pathogens, chlamydiae are, for most of their life cycle, inaccessible to antibody. The simplest model for immunity to chlamydial infections is that neutralizing antibody can protect against initial colonization by blocking attachment and invasion of elementary bodies, but thereafter cell-mediated immune responses are necessary to control established infection.

For *C. trachomatis*, the main neutralizing antibody response appears to be directed against the surface-exposed epitopes of the major outer membrane protein. Antibody to these epitopes in genital or ocular secretions is probably responsible for the short-term serovar-specific immunity to this organism that was observed in trachoma vaccine field trials and in experimental studies in primates. However, relatively high levels of antibody must be sustained at mucosal surfaces for effective neutralization. This is difficult to achieve.

It is likely that antibody is relatively unimportant compared with cell-mediated immunity for the eradication of established chlamydial infection. Experiments in guinea-pigs infected with *Ch. caviae* and mice infected with *C. muridarum* suggest that, as with other intracellular bacteria, the ability to mount a T helper 1 response is critical, and interferon-γ is the key protective cytokine. It is possible that the balance of cell-mediated immunity to chlamydiae and chlamydial replication and antigen production may account for the fact that chlamydial infections tend to be chronic, insidious and characterized by occasional periods of intermittent activity and shedding.

LABORATORY DIAGNOSIS

Cultivation

C. trachomatis can be demonstrated by characteristic iodine-staining inclusions in McCoy cell tissue culture. However, cell culture is tedious, insensitive, time consuming and expensive, and necessitates special transit of specimens to the laboratory to ensure chlamydial viability is sustained. It has been superseded almost entirely by diagnostic methods that are not dependent on viability.

Ch. pneumoniae is usually even more difficult to grow but, apart from the availability of a commercial fluorescent monoclonal antibody, there is no commonly accepted laboratory diagnostic method. Some laboratories have developed their own in-house polymerase chain reaction (PCR) or other nucleic acid-based tests, of variable quality.

Antigen detection

In smears of infected exudate from patients, *C. trachomatis* elementary bodies may be identified with fluorescein-labelled monoclonal antibodies and fluorescence microscopy. This is a sensitive and specific method, but is technically demanding and appropriate only for small numbers of specimens.

Chlamydial antigen, usually LPS, may also be detected by enzyme immuno-assay, and a number of tests are available commercially. The better methods have excellent specificity, but a sensitivity of only 70–80%. These methods can be automated and are appropriate for screening large numbers of samples. However, positive results need to be confirmed by other tests.

Nucleic acid detection

Commercial nucleic acid amplification based tests for *C. trachomatis* (but not other chlamydiae) are available. The best tests have sensitivity and specificity approaching 100% and are now the methods of choice for appropriate clinical specimens. However, sampling mathematics means that very low numbers of chlamydial elementary bodies in infected patients may still be missed. *C. trachomatis* DNA can be reliably detected in urine or vaginal tampons, reducing the need for direct sampling of the female genital tract. Amplification tests are not always licensed for this kind of sample and it is necessary to include controls to avoid false-negative reactions due to the presence of inhibitors in the specimen.

Serology

Serological tests have limited predictive value. The micro-immunofluorescence test is based on crude, whole, chlamydial antigens, antibody to which may be cross-reactive with other bacterial antigens. Patients with pelvic inflammatory disease often have high levels of antibody to *C. trachomatis*, but the converse is often not true. For *Ch. pneumoniae*, serological findings do not correlate well with the detection of the organism by PCR.

Serological tests are useful in the diagnosis of *C. trachomatis* pneumonia in the newborn and can be indicative in the investigation of pyrexia of unknown origin due to *Ch. psittaci*. Serology can be particularly helpful if a rising antibody titre can be demonstrated in acute versus convalescent sera.

TREATMENT

The antibiotic of choice is doxycycline or azithromycin in adults and erythromycin in babies. Penicillins inhibit, but do not kill, chlamydia and should not be used. Because chlamydiae have a prolonged replication cycle and may not be eradicated by short courses of antibiotics, treatment must be given for a minimum of 7 days. Many authorities advocate 3 weeks of treatment, especially for ascending and complicated genital infections in women.

Azithromycin provides sustained tissue levels and is often curative after a single dose. This is important where patient compliance is likely to be an issue. To prevent re-infection, the sexual partners of patients with chlamydial genital tract infection must also be treated.

In managing neonatal chlamydial conjunctivitis it is important to remember that the nasopharynx is probably infected and that atypical pneumonia may ensue; thus systemic antibiotics, preferably oral erythromycin, should be used. Chloramphenicol eye drops, often used to treat neonatal conjunctivitis, are not effective against chlamydiae.

EPIDEMIOLOGY

For *C. trachomatis*, the main routes of transmission are flies or contaminated secretions (trachoma) or sexual intercourse (genital tract infection). Trachoma is a disease of poverty and is now rare in industrially developed countries. Elsewhere, blindness affects 6–9 million adults, particularly women who care for infected children. Although the incidence of active trachoma in children is decreasing, it is likely that trachoma-induced blindness will increase for some years as populations in the developing world increase their life expectancy.

Ch. pneumoniae is uncommon in childhood, but ultimately most people become infected, with the main route of infection being aerosol droplets. In Scandinavia there have been some spectacular outbreaks of pneumonia caused by this organism in military camps and student populations.

Abortion in ewes caused by *Ch. abortus* can occur in one-third of a previously uninfected flock, leading to a continuing 1–2% abortion rate. The organism is found in sheep droppings, in the milk, and on the placenta and fleece of sheep. Aerosols can be produced and are a hazard to shepherds, who may develop a respiratory infection, as well as to pregnant women.

CONTROL

Trachoma

Prevention of trachoma relies on several approaches.

- Antibiotic prophylaxis with oral azithromycin, which, although more expensive, is as effective as tetracycline eye ointment and substantially more convenient. Long-term efficacy is still under evaluation.
- Education and face-washing campaigns, which are effective but difficult to sustain.
- Environmental hygiene improvements such as the introduction of clean water.

Genital infection

Genital infections can be insidious and asymptomatic. Community screening programmes for *C. trachomatis* infection in young women have therefore been instituted in some countries. It is essential that infected partners be traced and treated, even if clinically normal. In cases of neonatal infection the mother and her partner(s) should be examined and treated.

Animal contact

The control of importation of psittacine birds has reduced the risk of *Ch. psittaci* infection in pet owners and bird fanciers in many countries. Control of *Ch. abortus* infection may mean avoidance of contact with well known sources of infection, for instance sheep at lambing, milking and shearing. Pregnant women are particularly at risk from such infection and should avoid contact with sheep during pregnancy.

Immunization

There are no vaccines available against human chlamydial infections, even though it is clear that T helper 1 cell-mediated immune responses can be protective. A major problem with most of the human infections is that it is unclear how best to sustain immune responses at mucosal surfaces.

KEY POINTS

- Chlamydiae are obligate intracellular bacterial pathogens with a characteristic two-step growth cycle.
- *Chlamydia trachomatis* is a major cause of preventable blindness (trachoma) and genital tract infection.
- *Chlamydophila pneumoniae* is an important cause of respiratory tract infection.
- Chlamydial infections are frequently asymptomatic and persistent.
- Serious sequelae of chlamydial infection (blindness, pelvic inflammatory disease, infertility) are caused by immune response-driven scarring and fibrosis.
- Diagnosis requires laboratory tests, preferably those based on detection of chlamydial nucleic acid.
- Treatment is with doxycycline, erythromycin or azithromycin.
- Don't forget to treat the sexual partners of patients with chlamydial genital tract infection.
- Prevention depends on interrupting the transmission chain: there are no effective vaccines.

RECOMMENDED READING

Andraws R, Berger J S, Brown, D L 2005 Effects of antibiotic therapy on outcomes of patients with coronary artery disease. A meta-analysis of randomized controlled trials. *Journal of the American Medical Association* 293: 2641–2647

Beatty W L, Morrison R P, Byrne G I 1994 Persistent chlamydiae: from cell culture to a paradigm for chlamydia pathogenesis. *Microbiological Reviews* 58: 686–699

Brunham R C, Peeling R W 1994 *Chlamydia trachomatis* antigens: role in immunity and pathogenesis. *Infectious Agents and Disease* 3: 218–233

Carlson J H, Porcella S F, McClarty G, Caldwell H D 2005 Comparative genomic analysis of *Chlamydia trachomatis* oculotropic and genitotropic strains. *Infection and Immunity* 73: 6407–6418

Centers for Disease Control and Prevention 2002 Screening tests to detect *Chlamydia trachomatis* and *Neisseria gonorrhoeae* infections – 2002. *Morbidity and Mortality Weekly Report* 51 (RR-15). Online. Available: http://www.cdc.gov/mmwr/PDF/rr/rr5115.pdf

Danesh J, Whincup P, Walker M et al 2000 *Chlamydia pneumoniae* IgG titres and coronary heart disease: prospective study and meta-analysis. *British Medical Journal* 321: 208–212

Everett K D E, Bush R M 2001 Molecular evolution of the Chlamydiaceae. *International Journal of Systematic and Evolutionary Microbiology* 51: 203–220

Stephens R S (ed.) 1999 *Chlamydia: Intracellular Biology, Pathogenesis and Immunity*. American Society of Microbiology, Washington, DC

Internet sites

The comprehensive reference and education site to *Chlamydia* and the chlamydiae. http://www.chlamydiae.com

40

Rickettsia, orientia, ehrlichia, anaplasma and coxiella

Typhus; spotted fevers; scrub typhus; ehrlichioses; Q fever

D. H. Walker and Xue-Jie Yu

Few diseases have had a greater impact on the course of human history than epidemic typhus. Hans Zinsser's classic book *Rats, Lice and History* provides a graphic account of how *Rickettsia prowazekii*, the aetiological agent of this louse-borne disease, has caused millions of deaths and much human suffering in conditions of famine, poverty and war. Epidemic typhus now occurs mainly in poor populations in developing countries as world conditions have improved, but various other rickettsial diseases are still widely distributed.

The rickettsiae (*Rickettsia* and *Orientia* species) and anaplasmas (*Anaplasma, Ehrlichia* and *Neorickettsia* species) of medical importance are obligate intracellular bacteria that are transmitted by arthropod vectors. Molecular studies reveal that *Rickettsia* species, *Orientia tsutsugamushi*, *Ehrlichia* species and *Anaplasma* species evolved from a common ancestor closely related to mitochondria.

Coxiella burnetii, an obligate intracellular bacterium that can be isolated from arthropods, is more closely related to *Legionella pneumophila*, but is conveniently considered here alongside the rickettsia group.

RICKETTSIA AND ORIENTIA

DESCRIPTION

The genera *Rickettsia* and *Orientia* include organisms responsible for numerous diseases in many parts of the world (Table 40.1). The pioneering research of Ricketts and others in the early twentieth century demonstrated the rickettsial aetiology of Rocky Mountain spotted fever. Several other diseases, including epidemic and murine typhus, were later shown to be rickettsial infections. Previously unrecognized spotted fevers caused by *R. japonica, R. africae* and *R. honei* were discovered in the 1980s and 1990s in Japan, Africa and Australia, indicating that much remains to be learned about these organisms. In addition to species known to be associated with human disease, a number of presumably non-pathogenic rickettsiae have been isolated, primarily from arthropods, including even herbivorous insects, and are poorly understood.

Rickettsiae are small ($0.3–0.5 \times 0.8–1.0$ μm) Gram-negative bacilli. They are obligate intracellular bacteria that reside in the cytosol of host cells (Fig. 40.1). All are associated with a flea, louse, mite or tick vector. Species pathogenic for man parasitize endothelial cells almost exclusively. Rickettsiae have a small genome (approximately 1 Mb) and lack genes encoding many essential enzymes. Thus they depend on the host for nutrition and building blocks, and are yet to be cultivated outside eukaryotic cells. Like the chlamydiae (see Ch. 39), rickettsiae have a typical Gram-negative bacterial cell wall, including a bilayered outer membrane that contains lipopolysaccharide, and are energy parasites that transport adenosine triphosphate (ATP) with a unique translocase.

The genus *Rickettsia* is divided into two antigenically distinct groups based on their lipopolysaccharide: the typhus group and the spotted fever group. The immunodominant rickettsial outer membrane protein A (OmpA) exists only in the spotted fever group rickettsiae. Another major outer membrane protein, OmpB, exists in all *Rickettsia* species. OmpA and OmpB both contain cross-reactive and species-specific epitopes.

The scrub typhus rickettsiae are antigenically distinct and appear to be fundamentally different. They have been classified into a related but distinct genus as *Orientia tsutsugamushi* (see Table 40.1). The cell wall lacks lipopolysaccharide, peptidoglycan and a slime layer, and appears to derive its structural integrity from proteins linked by disulphide bonds. *O. tsutsugamushi* exhibits multiple major antigenic proteins with both strain-specific and cross-reactive epitopes.

PATHOGENESIS

Invasion and destruction of target cells

Rickettsiae normally enter the body through the bite or faeces of an infected arthropod vector. They are disseminated

Table 40.1 Human diseases caused by *Rickettsia* and *Orientia* species

Species	Disease	Geographical distribution	Mode of transmission	Primary vectors	Main vertebrate hosts
Typhus group					
R. prowazekii	Epidemic typhus	Extant foci in Africa, North and South America	Louse faeces	*Pediculus humanus corporis*	Man, flying squirrels
R. typhi	Murine typhus	Primarily tropics and subtropics	Flea faeces	*Xenopsylla cheopis* and other fleas	Rodents and other small mammals
Spotted fever group					
R. akari	Rickettsialpox	USA, Ukraine, Croatia, Korea	Bite of mouse mite	*Liponyssoides sanguineus*	House mice; possibly other rodents
R. australis	Queensland tick typhus	Australia	Bite of tick	*Ixodes holocyclus*	Unknown
R. conorii	Boutonneuse fever	Europe, Africa, Middle East, India	Bite of tick	*Rhipicephalus*	Unknown
R. japonica	Japanese spotted fever	Japan and north-eastern Asia	Bite of tick	*Dermacentor, Haemaphysalis, Ixodes*	Unknown
R. rickettsii	Rocky Mountain spotted fever	North and South America	Bite of tick	*Dermacentor, Rhipicephalus sanguineus, Amblyomma cajennense*	Rodents, dogs and other small mammals
R. africae	African tick bite fever	Africa and West Indies	Bite of tick	*Amblyomma hebraeum, A. variegatum*	
R. parkeri	American tick bite fever	North and South America	Bite of tick	*Amblyomma maculatum, A. americanum, A. triste*	Unknown
R. sibirica	North Asian tick typhus	Northern Asia	Bite of tick	*Dermacentor, Haemaphysalis, etc.*	Rodents and other small mammals
R. honei	Flinders Island spotted fever	Australia and South-East Asia	Bite of tick	*Aponomma hydrosauri*	Unknown
R. slovaca	Tick-borne lymphadenopathy	Eurasia	Bite of tick	*Dermacentor marginatum, D. reticularis*	Unknown
R. felis	Flea-borne spotted fever	Worldwide	Flea; undetermined mechanism	*Ctenocephalides felis*	Opossums
Scrub typhus group					
Orientia tsutsugamushi	Scrub typhus	Asia, Australia, islands of south-west Pacific and Indian oceans	Bite of larval mite	*Leptotrombidium* spp.	Rodents (especially rats)

through the bloodstream, attach to endothelial cells by OmpA and OmpB, enter endothelial cells by induced phagocytosis, escape from the phagosome, multiply intracellularly and eventually destroy their host cells. Observations in cell culture systems suggest that spotted fever and typhus group rickettsiae destroy the host cell by different mechanisms. After infection with *R. prowazekii* or *R. typhi*, the rickettsiae continue to multiply until the cell is packed with organisms (Fig. 40.2) and then bursts, possibly as a result of membranolytic activity; before lysis, host cells have a normal ultrastructural appearance.

Spotted fever group rickettsiae seldom accumulate in large numbers and do not burst the host cells, but stimulate polymerization of F-actin tails which propel them through the cytoplasm and into filopodia, from which they escape the cell or spread into an adjacent cell (Fig. 40.3). A protein, RickA, acts as a critical regulator of actin-based movement in spotted fever group, but not in typhus rickettsiae. Infected cells exhibit signs of membrane damage associated with an influx of water, which is sequestered within cisternae of dilated rough endoplasmic reticulum (see Fig. 40.1). Rickettsiae damage host cell membranes at least in part by stimulating production of free oxygen radicals by endothelial cells.

Scrub typhus rickettsiae also escape from the phagosome, reside free in the cytosol, and are released

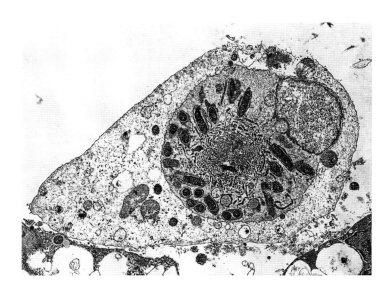

Fig. 40.1 Electron micrograph of a cell infected with *R. rickettsii*, showing the dilatation of the endoplasmic reticulum of host cells that occurs as a result of injury associated with infection by spotted fever group rickettsiae.

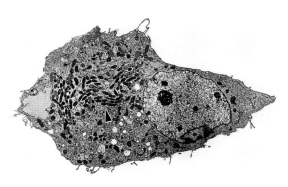

Fig. 40.2 Electron micrograph of a cell infected with *R. prowazekii*. The rickettsiae continue to multiply within the cell until it is completely packed with organisms and bursts. In contrast to cells infected with spotted fever group rickettsiae, the ultrastructural appearance of cells infected with typhus group rickettsiae remains normal until the cell lyses. The region of the cell containing rickettsiae is indicated by the arrowhead.

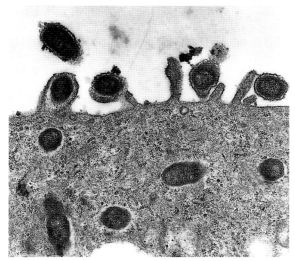

Fig. 40.3 Electron micrograph of *R. conorii* escaping from a host cell. Note the location of the rickettsiae within host cell filopodia.

from host cells soon after infecting them, but little is known about the mechanism(s) by which these organisms damage cells.

Pathological lesions

Rickettsial pathogenesis is due primarily to destruction of endothelial cells by replicating bacteria. All members of the genera *Rickettsia* and *Orientia* cause widespread microvascular injury leading to the destruction of infected endothelial cells. The pathological manifestations result more from direct rickettsial injury than from immunopathological mechanisms mediated via cytokines and cytotoxic T lympho-

cytes, inflammation, intravascular coagulation or endotoxin. Interference with normal circulation and increased vascular permeability following damage of blood vessels can cause life-threatening encephalitis and non-cardiogenic pulmonary oedema.

CLINICAL ASPECTS OF RICKETTSIAL DISEASES

Epidemic typhus

Headache and fever develop 6–15 days after being exposed to *R. prowazekii*. A macular rash, often noted 4–7 days after the patient becomes ill, first appears on the trunk and axillary folds and then spreads to the extremities. In mild cases the rash may begin to fade after 1–2 days, but in more severe cases it may last much longer and become haemorrhagic. Severe cases may also develop pronounced hypotension and renal dysfunction. The mental state of the patient may progress from dullness to stupor and even coma. Although the prognosis is grave for comatose patients, prompt appropriate treatment may be life-saving.

Individuals who survive a primary infection of louse-borne typhus may develop a relatively milder reactivation of latent infection many years later. This is referred to as recrudescent typhus or *Brill–Zinsser disease*. Such individuals are nevertheless immune to a second louse-borne infection.

Flea-borne fevers

Patients infected with *R. typhi* develop symptoms similar to those of epidemic typhus. Fatal cases are uncommon but occasionally occur, particularly in the elderly. Although murine typhus is much milder than epidemic typhus, it is still severe enough to require several months of convalescence. *R. felis* is maintained transovarially in cat fleas and causes a similar illness.

Tick-borne spotted fever

There are many clinical similarities among the tick-borne rickettsioses of the spotted fever group. The most severe is Rocky Mountain spotted fever, for which the average annual case-fatality rate for 1981–1998 was 3.3%. Risk factors for fatal infection include older age and delayed tetracycline treatment.

Patients become ill within 2 weeks of being bitten by an infected tick. Early symptoms include fever and severe headache and myalgia, often accompanied by anorexia, vomiting, abdominal pain, diarrhoea, photophobia and cough. An eschar ('*tache noir*') frequently occurs at the site of the tick bite in all spotted fever group infections except Rocky Mountain spotted fever. A maculopapular rash usually develops within 3–5 days. The rash of spotted fever usually develops first on the extremities rather than on the trunk. Absence of a rash does not exclude rickettsial infection, because a disproportionate number of fatal cases of Rocky Mountain spotted fever are of the 'spotless' variety. Spotted fever group rickettsiae are found within the endothelial cells and less often macrophages of vertebrate hosts, but *R. rickettsii* can also invade vascular smooth muscle.

Vascular damage in severe cases may result in hae-morrhagic rash, hypovolaemia, hypotensive shock, non-cardiogenic pulmonary oedema and impairment of central nervous system function. A fulminant form of Rocky Mountain spotted fever sometimes kills the patient within 5 days of the onset of symptoms; this form of the disease is more common in black males who are deficient in glucose-6-phosphate dehydrogenase, and may be related to haemolysis in these patients. Infection confers long-lasting immunity.

Rickettsialpox

The clinical course of rickettsialpox is similar to that of other spotted fever group infections and includes development of fever, headache and an eschar at the site where the infected mite fed. The rash is initially maculopapular but often becomes vesicular. Fever lasts for about a week, and patients usually recover uneventfully.

Scrub typhus

Human infection with scrub typhus rickettsiae may be mild or fatal, depending on host factors and presumably the virulence of the infecting strain. Symptoms develop 6–18 days after being bitten by infected mite larvae (chiggers). The disease is characterized by sudden onset of fever, headache, and myalgia. A maculopapular rash often develops 2–3 days after onset of the illness. An eschar is often apparent at the site of the bite with enlargement of local lymph nodes. Progression of the disease may be accompanied by interstitial pneumonitis, generalized lymphadenopathy and splenomegaly. Death may result from encephalitis, respiratory failure and circulatory failure. Patients who survive generally become afebrile after 2–3 weeks, or sooner if treated appropriately. Scrub typhus confers only transient immunity, and re-infection may occur with heterologous or homologous strains.

LABORATORY DIAGNOSIS

Timely and accurate diagnosis of rickettsial disease followed by administration of an appropriate antibiotic may mean the difference between death of the patient and uneventful recovery. The lack of widely available, reliable diagnostic tests that can detect the disease in its early stages remains a problem, particularly as symptoms are often non-specific. The rash may appear at a late stage in the infection or not at all, and may resemble exanthemata of many other diseases. The presence or significance of an eschar, if present, is also commonly overlooked.

Serological methods

Rickettsial diseases are usually acute and short-lasting. Antibodies appear in the second week of illness, when the patient is usually on the way to recovery. Death may occur before detectable levels of antibody are present. Serology is therefore not suitable for early diagnosis of rickettsial infections and is used mainly to confirm the diagnosis for epidemiologic investigations.

The oldest and most widely used laboratory method is the *Weil–Felix test*, which relies on agglutination of the somatic antigens of non-motile *Proteus* species. The test has unacceptably low levels of sensitivity and specificity and is no longer recommended.

More reliable diagnostic tests include immunofluorescence and enzyme immunoassay. These tests are commercially available.

Isolation of rickettsiae

Isolation of the organism provides conclusive proof of rickettsial infection. However, it is seldom attempted because of lack of facilities or expertise and because of the presumed danger to laboratory personnel of handling rickettsia-infected tissues. Such dangers have been overemphasized in this era of antibiotics, but use of containment facilities is appropriate.

Rickettsiae can be isolated in cell culture, in laboratory animals such as mice or guinea-pigs, or in embryonated chicken eggs. Cell culture yields timely results and is widely used for isolation of rickettsiae from clinical samples. Rickettsiae can be detected in cell culture 48–72 h after inoculation by the shell vial assay.

Detection of rickettsiae in tissue

Skin biopsies from the centre of petechial lesions can be examined by immunohistochemistry. This approach is virtually 100% specific and has a sensitivity of 70%. Rickettsiae can be visualized for up to 48 h after the administration of antirickettsial drugs. This approach is particularly effective for diagnosing infection post mortem. A method has been developed for capturing detached, circulating endothelial cells by antibody-coated magnetic beads and immunocytological detection of intracellular rickettsiae.

Polymerase chain reaction (PCR)

Detection of rickettsial DNA by PCR is more rapid than isolation and allows specific identification, but the test is not generally available. It is also insensitive, particularly early in the course of the infection when therapeutic decisions should be made. Peripheral blood mononuclear cells, skin biopsy and tache noire specimens from the site of the bite can be used.

PCR amplification of highly conserved genes can be used for diagnosis of rickettsial infections. A single primer pair that amplifies all or most rickettsiae can be designed from the genes encoding 16S ribosomal ribonucleic acid (rRNA), a 17-kDa protein, citrate synthase (*gltA*) or *ompB*. A suitable approach is to use a conserved genus-specific primer pair to amplify a rickettsial gene and then to identify the species by restriction fragment length polymorphism analysis or DNA sequencing. Scrub typhus can be diagnosed by PCR amplification of the 56-kDa protein gene of *O. tsutsugamushi*.

TREATMENT

The drug of choice for treating rickettsial infections of all types and in all age groups is doxycycline. Chloramphenicol is an alternative, but is less effective and carries a higher risk of death. Both drugs are rickettsiostatic and allow the patient's immune system time to respond and control the infection. Sulphonamides should not be administered as they exacerbate rickettsial infections; β-lactam antibiotics are ineffective. Some new quinolones and macrolides have antirickettsial effects in vitro, but clinical experience in severe rickettsioses is lacking. Short courses of doxycycline do not cause significant dental staining in young children.

Owing to the difficulties of accurate diagnosis and the risks involved in misdiagnosis, empirical tetracycline therapy is appropriate for patients who have had a fever for 3 days or more and a history consistent with the epidemiological and clinical features of rickettsial disease. The case fatality rate is increased significantly if treatment is delayed for more than 5 days.

Intensive nursing care, management of fluids and electrolytes, and administration of red blood cells to patients who develop anaemia may be needed. Surgery may also be necessary to remove digits and extremities that develop ischaemic necrosis.

EPIDEMIOLOGY

Typhus group infections

Epidemic typhus

R. prowazekii is transmitted from person to person by the body louse, *Pediculus humanus corporis*; the organisms are present in the faeces of infected lice and enter through the bite wound or skin abrasions. *R. prowazekii* causes a fatal infection of the louse, which is therefore incapable of long-term maintenance of the rickettsiae, and human beings appear to be reservoirs of epidemic typhus. Patients who suffer a bout of recrudescent typhus (Brill–Zinsser disease) circulate sufficient rickettsiae in their blood to infect approximately 1–5% of lice that feed on them – enough to initiate new epidemics of the disease, such as those that have occurred in Africa, South America and Russia. *R. prowazekii* is maintained in an enzootic cycle in North America involving flying squirrels and their fleas and lice.

Murine typhus

Murine typhus is distributed worldwide, particularly in tropical and subtropical coastal regions and port facilities where large numbers of rats are found. This disease is maintained in an enzootic cycle involving rats and their fleas, which remain infected for life. Even the inefficient rate of transovarial transmission in fleas may play an important role in maintaining the rickettsiae in nature. Man is infected by the contamination of abraded skin, respiratory tract or conjunctiva with infective flea faeces, in which the rickettsiae can survive for as long as 100 days under optimal conditions of temperature and humidity. The disease is an occupational hazard of working in rat-infested areas such as markets or ports.

Spotted fever group infections

Tick-borne infections

Tick-borne rickettsiae of the spotted fever group are maintained in enzootic cycles involving ticks and their wild animal hosts. Ticks are the primary reservoirs of the rickettsiae, and maintain the organisms by both trans-stadial transmission (during moulting of larvae to nymph and thence to adult tick) and transovarial or vertical transmission. Some horizontal transmission (tick to rodent to tick) is likely to be essential to the survival of virulent rickettsiae in nature because these rickettsiae are somewhat pathogenic to ticks.

Man becomes infected following the bite of infected ticks or through contamination of abraded skin or mucous membranes. People place themselves at risk when they enter areas infested with infected ticks. Individuals may also become infected if they are bitten by ticks of domestic dogs or if partially fed ticks rupture during manual deticking of dogs. Rocky Mountain spotted fever is endemic in the Americas, especially in the south-eastern and south-central USA. In 2004, 1514 cases were reported.

Rickettsialpox

R. akari is maintained in an enzootic cycle that involves house mice (*Mus musculus*) and their mites. As with other spotted fever group infections, the arthropod vector is also the primary reservoir and can maintain the organism

by trans-stadial and transovarial transmission. Other rodents and their ectoparasites may be able to maintain the rickettsiae in rural areas, but their importance remains unknown. Rickettsialpox is primarily an urban disease associated with mice-infested buildings.

Scrub typhus

The nymphal and adult stages of the mites transmitting *O. tsutsugamushi* are free living and do not feed on animals. The parasitic larvae (chiggers) occur in habitats that have been disturbed by the loss or removal of the natural vegetation. The area becomes covered with scrub vegetation, which is the preferred habitat for chiggers and their mammalian hosts, and gives the disease its name. The disease is often localized because of the restricted habitat of the chiggers. Persons entering infected areas are at risk.

CONTROL

It is virtually impossible to eradicate rickettsial infections because of their enzootic nature. Measures aimed at reducing rodent or ectoparasite populations may help to reduce the risk of infection. In addition to delousing infested persons, their clothing and bedding should be decontaminated.

Persons entering areas endemic for spotted fever group infections should wear protective clothing treated with tick repellent. Individuals should also examine themselves carefully for ticks as soon as possible after returning from tick-infested areas. The probability of infection is decreased if the tick is removed soon after it attaches. Transmission may require up to 24 h of feeding, perhaps because starved ticks require a partial blood meal if they are to reactivate the virulence of the rickettsiae.

There is no safe, effective vaccine for any of the rickettsial diseases. The attenuated E strain of *R. prowazekii* induces protective immunity, but is unsuitable for general use because it causes a mild form of typhus in 10–15% of those inoculated and reverts to a virulent state after animal passage. Rocky Mountain spotted fever vaccines containing killed rickettsiae afford incomplete protection and are no longer available. A recombinant or attenuated vaccine that contains cross-protective antigens stimulating cellular immunity could protect against both typhus and spotted fever rickettsioses. Scrub typhus vaccines derived from killed rickettsiae do not prevent infection, and experimental recombinant subunit vaccines have not proved effective in animals.

Antimicrobial prophylaxis is not recommended for infections with *Rickettsia* species, as they are only rickettsiostatic and disease develops as soon as the antibiotic is

Table 40.2   Human diseases caused by *Ehrlichia*, *Anaplasma* and *Neorickettsia* species

Species	Disease	Geographical distribution	Means of transmission	Primary vectors
E. chaffeensis	Monocytic ehrlichiosis	North America, Thailand, Africa	Tick bite	*Amblyomma americanum*
A. phagocytophilum	Granulocytic anaplasmosis	USA, Eurasia	Tick bite	*Ixodes* spp.
E. ewingii	Ehrlichiosis ewingii	USA	Tick bite	*A. americanum*
N. sennetsu	Sennetsu ehrlichiosis	Japan, Malaysia	Unknown[a]	Unknown

[a]Possibly ingestion of fluke-infested fish.

discontinued. Prolonged prophylaxis with weekly doses of doxycycline is effective against scrub typhus, but is probably inappropriate except under exceptional circumstances, for instance during military operations.

EHRLICHIA AND ANAPLASMA

DESCRIPTION

Long recognized as the aetiological agents of veterinary diseases such as bovine and ovine tick-borne fever in the UK, the first human ehrlichial disease was recognized in 1954 when *Neorickettsia* (formerly *Ehrlichia*) *sennetsu* was identified as the cause of an illness resembling glandular fever in Japan (*sennetsu* means glandular fever in Japanese). *Ehrlichia chaffeensis*, *Anaplasma phagocytophilum* and *E. ewingii* later emerged as the causes of tick-borne diseases in the USA. *Ehrlichia* and *Anaplasma* species are transmitted through the bite of ticks. *N. sennetsu* is suspected to be ingested with raw fish (Table 40.2).

These organisms are small Gram-negative bacteria. They multiply within membrane-bound cytoplasmic vacuoles, usually in various phagocytes, and form characteristic microcolonies resembling mulberries, termed morulae (Latin *morum* = mulberry) (Fig. 40.4). Electron microscopy reveals two distinct morphological forms, larger reticulate and smaller dense-core cells, both of which divide by binary fission.

PATHOGENESIS

Anaplasma and *Ehrlichia* species can establish prolonged, even persistent, infections in vivo, and some species, including *E. chaffeensis*, kill heavily infected cells in vitro. Available evidence suggests that cytokine-associated immunopathological mechanisms are important. *A. phagocytophilum* and *E. chaffeensis* evade the host immune system by modulating cytokine production.

The tropism for phagocytes indicates that these organisms have evolved strategies for evading the microbicidal

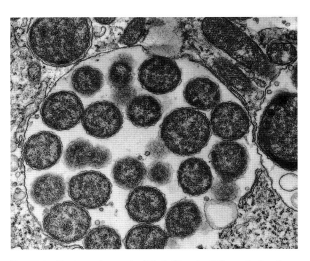

Fig. 40.4   Electron micrograph of *E. chaffeensis* within a cytoplasmic vacuole. (Micrograph by courtesy of Dr Vsevolod L. Popov.)

activities of the macrophage or granulocyte. Within phagocytes *A. phagocytophilum* and *E. chaffeensis* block the fusion of phagosome-containing bacteria with lysosomes to prevent killing by lysosomal enzymes. *A. phagocytophilum* prevents killing mediated by reactive oxygen species by lowering reduced nicotinamide adenine dinucleotide phosphate (NADPH) oxidase activity in neutrophils. To accommodate a slow generation time (about 8 h), *Anaplasma* and *Ehrlichia* prolong their host cells' lifespan by inhibiting apoptosis.

Antibody to ehrlichiae confers passive protection, and cellular immunity is crucial to recovery. Suppression of neutrophil function by *A. phagocytophilum* may predispose to opportunistic infection.

Clinical aspects of infection

Human monocytic ehrlichiosis

The disease begins 1–2 weeks after a bite of an infected tick. Clinical features frequently include fever, headache, myalgias, nausea, arthralgias and malaise. Other manifestations include cough, pharyngitis, lymphadenopathy, diarrhoea, vomiting, abdominal pain and changes in mental status. A fleeting or transient rash involving the extremities, trunk, face or, rarely, the palms and soles appears in 30–40% of patients about 5 days after onset. The rash may be petechial, macular, maculopapular or diffusely erythematous.

Cytopenia early in the course of the illness may provide presumptive clues to the diagnosis. Mild to moderate leucopenia is observed in approximately 60–70% of patients during the first week of illness, with the largest decreases occurring in the total lymphocyte count. Thrombocytopenia occurs in 70–90% of patients. Mildly or moderately raised hepatic transaminase levels are noted in most patients at some point during their illness.

About 50% of patients need admission to hospital. Those with severe disease may develop acute renal failure, metabolic acidosis, respiratory failure, profound hypotension, disseminated intravascular coagulopathy, myocardial dysfunction and meningo-encephalitis; about 3% die. Death is more common in elderly men and immunocompromised individuals, including those infected with human immunodeficiency virus (HIV).

Human granulocytic anaplasmosis

The incubation period is 7–10 days after a bite by an infected tick. Clinical signs and symptoms are similar to those of human monocytic ehrlichiosis, but the disease is less severe and the mortality rate is less than 1%. Rash and central nervous system involvement are rare.

Human granulocytic ehrlichiosis caused by *E. ewingii* elicits similar signs and symptoms. The infection has been observed mainly in immunocompromised patients with no fatalities.

LABORATORY DIAGNOSIS

Ehrlichia and *Anaplasma* species grow intracellularly, and isolation is difficult. Human infections are diagnosed mainly by demonstrating the development of specific antibodies during convalescence. Indirect immunofluorescence methods use cell culture-propagated organisms. *E. chaffeensis, A. phagocytophilum* and *E. ewingii* are detected diagnostically by PCR with specific primers to amplify the ehrlichial DNA. Human granulocytic anaplasmosis can often be diagnosed by identification of characteristic morulae in Giemsa-stained peripheral blood neutrophils; *E. chaffeensis* is seldom detected in monocytes in blood smears.

TREATMENT

Doxycycline is very effective in shortening the course of infection and reduces mortality. The effect of chloramphenicol on *E. chaffeensis* is controversial and use of the drug is not recommended.

EPIDEMIOLOGY

Deer and ticks are involved in the ecology of both human monocytic and granulocytic ehrlichioses and granulocytic anaplasmosis. Deer, canines, rodents and domestic ruminants are important reservoirs. Immature ticks obtain ehrlichiae from the blood of infected animals; the organisms are maintained trans-stadially but not transovarially, and are transmitted during a subsequent blood meal. Human infections are strongly associated with the season of tick activity, history of tick bite and the distribution of the tick vectors. Most cases of human monocytic ehrlichiosis are reported between March and November in the south-central and south-eastern USA, where *Amblyomma americanum* is prevalent. Human granulocytic anaplasmosis occurs from March to December in the northern states.

COXIELLA

DESCRIPTION

Query or Q fever was first identified as a distinct clinical entity in 1935 after an outbreak of typhoid-like illness among abattoir workers in Australia. Subsequent studies showed that the disease is widespread with an almost global distribution. The aetiological agent, *C. burnetii*, is an obligately intracellular prokaryote, genetically related to *Legionella* species.

C. burnetii is a pleomorphic coccobacillary bacterium. Ultrastructural studies reveal a Gram-negative cell wall. The organisms typically grow within the phagolysosome of macrophages of the vertebrate host (Fig. 40.5). Structurally distinct large and small cell variants have been described, suggesting that the organism has a developmental cycle. In acidic conditions, similar to those found within a phagolysosome, it actively metabolizes a variety of substrates and can accomplish significant levels of macromolecular synthesis. Prolonged cultivation in vitro results in phase variation due to structural changes in surface lipopolysaccharides analogous to the smooth to rough transitions observed in other bacteria. Phase I organisms are representative of strains in nature, whereas

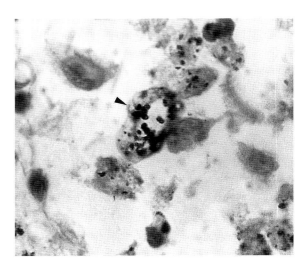

Fig. 40.5 Immunoperoxidase staining of *C. burnetii* in alveolar macrophages in a patient with Q fever. A cell containing many *Coxiella* organisms is indicated by the arrowhead. (Micrograph by courtesy of Dr J. Stephen Dumler.)

phase II organisms appear in laboratory cultures and are avirulent for laboratory animals. *C. burnetii*, unlike *Rickettsia* species, has been found to contain plasmids.

PATHOGENESIS

Human infection usually follows inhalation of aerosols containing *C. burnetii*. Entry into the lungs results in infection of the alveolar macrophages and a brief rickettsaemia. Most infections are subclinical, and only 2% of persons infected with *C. burnetii* are admitted to hospital. The incubation period for the acute form of the disease is usually about 2 weeks but can be longer.

Typical acute Q fever is a self-limiting flu-like syndrome with high fever (40°C), fatigue, headache and myalgia. The patient may also suffer pneumonitis, hepatic and bone marrow granulomata, and meningo-encephalitis. Chronic infections can develop, with the organism persisting in cardiac valves and possibly other foci. Endocarditis is rare, but potentially fatal, and may be accompanied by glomerulonephritis, osteomyelitis or central nervous system involvement.

Reactivation of latent infection may occur during pregnancy, and the organism is shed with the placenta or abortus.

LABORATORY DIAGNOSIS

Diagnosis relies on the demonstration of specific antibodies in an indirect immunofluorescence assay or enzyme immunoassay. Immunofluorescence assay titres peak at 4–8 weeks. PCR amplification has been used to detect *C. burnetii* DNA in clinical samples from acute and chronic Q fever patients.

Isolation of *C. burnetii* from patient specimens is a specialized procedure and is not generally recommended because of the extremely infectious nature of the organism.

TREATMENT

Most *C. burnetii* infections resolve without antibiotic treatment, but administration of doxycycline reduces the duration of fever in the acute infection and is definitely recommended in cases of chronic infection. In Q-fever endocarditis, long-term administration of a combination of two drugs among doxycycline, ciprofloxacin and rifampicin has been suggested. Alkalinization of the acidic phagolysosome with chloroquine has been suggested to achieve better bacterial killing.

C. burnetii may be recovered from some patients after months or even years of continuous treatment. In addition to antibiotic therapy, the haemodynamic status should be monitored. Valve replacement may be necessary in some cases of Q-fever endocarditis.

EPIDEMIOLOGY

Q fever has been found on all continents except Antarctica, but most cases are reported from the UK and France. Elsewhere, the disease often goes unreported, is misdiagnosed, or causes such a mild infection that treatment is not sought. The primary reservoirs of the disease are wild and domestic ungulates, including cattle, sheep and goats, rabbits, cats and dogs. Ticks can maintain *C. burnetii* by trans-stadial and transovarial transmission. Faeces of infected ticks contain very large numbers of *C. burnetii*, but arthropods are not an important source of infection.

C. burnetii may be the most infectious of all bacteria. Human infections generally follow inhalation of aerosols or direct contact with the organisms in the milk, urine, faeces or birth products of infected animals. The organism can survive on wool for 7–10 months, in skimmed milk for up to 40 months and in tick faeces for at least 1 year. Most individuals acquire the disease as an occupational hazard. Cases are common among abattoir workers and those associated with livestock rearing or dairy farming. Although Q fever normally occurs as isolated, sporadic cases, well documented outbreaks have been reported.

CONTROL

Elimination of infected reservoir hosts is probably impossible because of chronic infections among the animals and the ability of the organism to survive for long periods in the external environment. Exposure can be reduced by construction of separate facilities for animal parturition, destruction of suspect placental membranes, heat treatment of milk (74°C for 15 s) and efforts to reduce the tick population. Abattoir workers should take care while handling carcasses, especially in the removal and dissection of mammary glands and inner organs. Animal hides should be kept wet until the salting procedure begins. Appropriate containment procedures should be observed in laboratories working with this highly infectious organism.

Vaccines have been developed from formalin-killed whole cells. Vaccines derived from phase I organisms generate a much greater protective response than similar ones prepared from phase II organisms.

KEY POINTS

- Rickettsiae are small obligately intracellular bacteria that are associated with insects or ticks during at least part of their transmission cycle.
- *Rickettsia* and *Orientia* cause spotted fevers, typhus fevers and scrub typhus by infecting and damaging endothelial cells, resulting in increased vascular permeability, oedema, adult respiratory distress syndrome, meningoencephalitis and rash.
- *Ehrlichia* and *Anaplasma* reside in persistently infected vertebrate hosts such as deer, dogs or rodents, and are transmitted by feeding ticks.
- *Ehrlichia* and *Anaplasma* target human monocytes or granulocytes, where they grow to microcolonies in cytoplasmic vacuoles.
- The frequent absence of antibodies to rickettsiae, orientiae and ehrlichiae early in the course of illness hinders laboratory diagnosis, making clinico-epidemiologic suspicion of the diagnosis and empirical treatment, preferably with doxycycline, essential.
- *Coxiella burnetii* thrives in the acidic phagolysosome of macrophages, has a stable extracellular form and infects human beings who inhale aerosols from birth fluids or the placenta of infected animals.

RECOMMENDED READING

Anderson B, Friedman H, Bendinelli M (eds) 1997 *Rickettsial Infection and Immunity.* Plenum, New York

Audy J R 1968 *Red Mites and Typhus.* Athlone Press, London

Carlyon J A, Fikrig E 2003 Invasion and survival strategies of *Anaplasma phagocytophilum. Cellular Microbiology* 5: 743–754

Dalton M J, Clarke M J, Holman R C et al 1995 National surveillance for Rocky Mountain spotted fever, 1981–1992: epidemiologic summary and evaluation of risk factors for fatal outcome. *American Journal of Tropical Medicine and Hygiene* 52: 405–413

Dumler J S, Walker D H 2006 Ehrlichioses and anaplasmosis. In Guerrant R L, Walker D H, Weller P F (eds) *Tropical Infectious Diseases. Principles, Pathogens, and Practice,* 2nd edn, pp. 564–573. Elsevier Churchill Livingstone, Philadelphia

Hechemy K E, Avsic-Zupanc T, Childs J E, Raoult D A (eds) 2003 *Rickettsiology: Present and Future Directions.* New York Academy of Sciences, New York

Kawamura A, Tanaka H, Tamura A 1995 *Tsutsugamushi Disease.* University of Tokyo Press, Tokyo

Maurin M, Raoult D 1999 Q fever. *Clinical Microbiology Reviews* 12: 518–553

Paddock C D, Childs J E 2003 *Ehrlichia chaffeensis*: a prototypical emerging pathogen. *Clinical Microbiology Reviews* 16: 37–64

Raoult D, Brouqui P 1999 *Rickettsial Diseases at the Turn of the Third Millennium.* Elsevier, Amsterdam

Raoult D, Walker D H 2005 *Rickettsia prowazekii* (epidemic or louse-borne typhus). In Mandell G L, Bennett J C, Dolin R (eds) *Principles and Practice of Infectious Diseases,* 6th edn, Vol. 2, pp. 2303–2306. Elsevier Churchill Livingstone, Philadelphia

Treadwell T A, Holman R C, Clarke M J, Krebs J W, Paddock C D, Childs J E 2000 Rocky Mountain spotted fever in the United States, 1993–1996. *American Journal of Tropical Medicine and Hygiene* 63: 21–26

Walker D H, Raoult D H 2005 *Rickettsia rickettsii* and other spotted fever group rickettsiae (Rocky Mountain spotted fever and other spotted fevers). In Mandell G L, Bennett J C, Dolin R (eds) *Principles and Practice of Infectious Diseases,* 6th edn, Vol. 2, pp. 2287–2295. Elsevier Churchill Livingstone, Philadelphia

Watt G, Walker D H 2006 Scrub typhus. In Guerrant R L, Walker D H, Weller P F (eds) *Tropical Infectious Diseases. Principles, Pathogens, and Practice,* 2nd edn, pp. 557–563. Elsevier Churchill Livingstone, Philadelphia

Zhang J Z, Sinha M, Luxon B A, Yu X J 2004 Survival strategy of obligately intracellular *Ehrlichia*: novel modulation of immune response and host cell cycles. *Infection and Immunity* 72: 498–507

41 Mycoplasmas

Atypical pneumonia; genital tract infection

D. Taylor-Robinson

Mycoplasmas are the smallest prokaryotic organisms that can grow in cell-free culture. They are found in man, animals, plants, insects, soil and sewage. The first to be recognized, *Mycoplasma mycoides* ssp. *mycoides*, was isolated in 1898 from cattle with pleuropneumonia. As other pathogenic and saprophytic isolates accumulated from veterinary and human sources they became known as pleuropneumonia-like organisms (PPLO), a term long superseded by mycoplasmas. 'Mycoplasma' (Greek *mykes* = fungus; *plasma* = something moulded) refers to the filamentous (fungus-like) nature of the organisms of some species and to the plasticity of the outer membrane resulting in pleomorphism.

The term 'mycoplasma(s)' is often used, as here, in a trivial fashion to refer to any member of the class Mollicutes ('soft skins'), which embraces several families and numerous species including the spiroplasmas, which cause disease in plants and insects, and the acholeplasmas, which are ubiquitous in nature. The family Mycoplasmataceae contains the mycoplasmas that are of importance in human medicine. It is subdivided into two genera: *Mycoplasma*, of which there are at least 114 named species; and *Ureaplasma*, urea-hydrolysing organisms (trivially termed ureaplasmas), of which there are six species.

It has been proposed, but not yet confirmed, that several members of the genera *Haemobartonella* and *Eperythrozoon* should be transferred to the genus *Mycoplasma* on the basis of their 16S ribosomal RNA (rRNA) sequence similarities.

Mycoplasmas are distributed widely in nature, and various species cause economically important infections in cattle, goats, sheep, swine, other mammals, birds and cold-blooded animals (alligators, crocodiles, tortoises) as well as man. In addition, mycoplasmas are of concern to those who use cell cultures because of the problems of contamination.

Ureaplasmas were known originally as T strains or T mycoplasmas – 'T' for 'tiny' to describe the small size of the colonies in comparison with those produced by other mycoplasmas. Many animals are infected by ureaplasmas. Those of avian, bovine, canine and feline origin are antigenically distinct from the human strains and have been placed in separate species. Ureaplasmas of human origin, formerly classified as *Ureaplasma urealyticum*, comprise at least 14 serovars and fall into two groups differentiated in several ways, including genome size. Strains of small genome size include serovars 1, 3, 6 and 14, and now comprise a new species, *U. parvum*, whereas those of larger genome size include the other ten serovars and these remain as *U. urealyticum*.

DESCRIPTION

Reproduction

Mycoplasmas, like conventional bacteria, multiply by binary fission. However, cytoplasmic division may not always be synchronous with genome replication, resulting in the formation of multinucleate filaments and other shapes. Subsequent division of the cytoplasm by constriction of the membrane at sites between the genomes leads to chains of beads that later fragment to give single cells. Budding occurs when the cytoplasm is not divided equally between the daughter cells. The minimal reproductive unit of mycoplasmas is a roughly spherical cell about 200–250 nm in diameter. It is from organisms of this order of size, whatever the shape, that growth is initiated in cell-free medium; such organisms also make up, with larger forms (0.5–1.0 μm diameter), the substance of the characteristic agar-embedded colonies.

Morphology

Cell morphology varies with the species, environmental conditions and the stage of the growth cycle. Light microscopy reveals pleomorphic organisms, which may range from spherical through coccoid, coccobacillary, ring and dumb-bell forms, to short and long branching (Fig. 41.1), beaded or segmented filaments. The helical shape of spiroplasmas is characteristic but not always seen.

Several mycoplasmas, including *M. pneumoniae, M. genitalium* (Fig. 41.2) and *M. penetrans* of human origin,

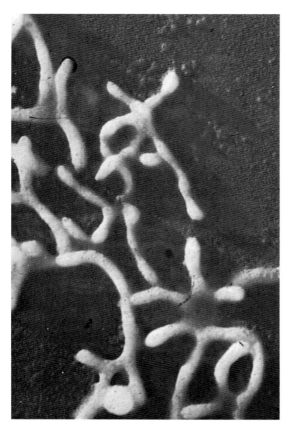

Fig. 41.1 Electron micrograph of *M. mycoides* ssp. *mycoides* (bovine origin), gold shadowed, to show branching filaments (original magnification ×28 000). (From Rodwell A W, Abbot A 1961 *Journal of General Microbiology* 25: 201–214.)

have specialized structures at one or both ends by which they attach to respiratory or genital tract mucosal surfaces. Sections of the terminal structures of the two former mycoplasmas may exhibit a dense rod-like core when viewed by electron microscopy.

Mycoplasmas, unlike conventional bacteria, do not have a rigid cell wall. This also differentiates them from bacterial L-forms for which the absence of the cell wall is a temporary reflection of environmental conditions. The mycoplasma cell is limited by a membrane, 7.5–10 nm wide, in which two electron-dense layers are separated by a translucent one (Fig. 41.3). Some species have an extramembranous layer, which, in the case of *M. mycoides* ssp. *mycoides*, for example, comprises galactan and has a dense capsular appearance. Some others, for example *M. gallisepticum* (avian), *M. genitalium*, *M. pneumoniae*, *M. pulmonis* (murine) and *Spiroplasma citri*, have surface spikes (sometimes described as a 'nap'), somewhat coarser than those seen on myxoviruses, which may contribute, through adhesin proteins, to attachment of the organisms to eukaryotic cells. *M. gallisepticum*, *M. genitalium* and *M. pneumoniae* attach to neuraminic acid receptors. Close adherence enables the mycoplasma to insert nucleases and other enzymes into the cell and to take from it the products of enzyme activity, such as nucleotides. Adherence to erythrocytes (*haemadsorption*), tissue culture cells, spermatozoa and other eukaryotic cells may be demonstrated with certain mycoplasmas (Fig. 41.4).

The electron-lucent part of the membrane comprises lipids with the long chains of the fatty acids arranged inwards and polar groups to the external and internal parts of the organisms; the electron-dense layers consist of

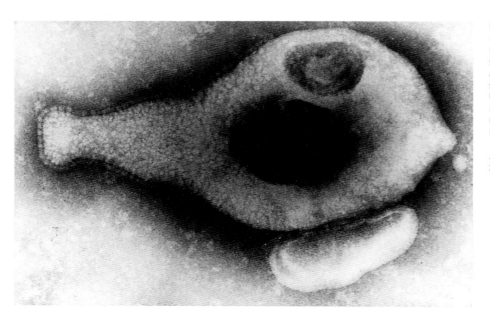

Fig. 41.2 Electron micrograph of *M. genitalium* (human origin), negatively stained, to show flask-shaped appearance and terminal specialized structure covered by extracellular 'nap' (original magnification ×120 000). (From Tully J G, Taylor-Robinson D, Rose D L et al 1983 *International Journal of Systematic Bacteriology* 33: 387–396.)

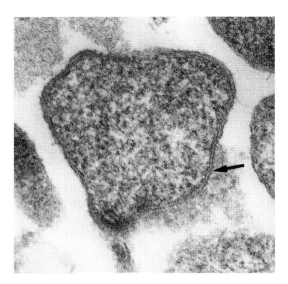

Fig. 41.3 Electron micrograph of *M. pulmonis* (murine origin); thin section illustrating trilaminar membrane (arrow) (original magnification ×75 000).

external osmotic pressures and in the absence of the rigid cell wall found in other bacteria. However, the presence of cholesterol renders the cell membrane susceptible to damage with agents such as saponin, digitonin and some polyene antibiotics (e.g. amphotericin B) that complex with sterols. As expected, mycoplasmas are completely resistant to antibiotics that influence bacterial cell wall synthesis and also to lysozyme.

The cytoplasm of mycoplasmas does not contain endoplasmic reticulum but is packed with ribosomes; fibrillar nuclear material (unbounded without nucleoli) is centrally placed or dispersed. Ribosomal protein synthesis is inhibited by antibiotics such as tetracyclines, aminoglycosides, erythromycin and chloramphenicol.

The genome of *M. genitalium* is the smallest known so far (580 kb) and was the first of any micro-organism to be fully sequenced. It is not much larger than that of a large poxvirus, an indication of the small amount of genetic information needed for a free-living existence and the reason why mycoplasmas have a paucity of biochemical activity and are nutritionally fastidious.

protein and carbohydrate. The mycoplasma membrane is the site of many metabolic reactions involving membrane-bound enzymes and transport mechanisms. Cholesterol, or carotenoid/carotenol, is interspersed between the phospholipid molecules and plays an important part in maintaining membrane integrity in the face of varying

Viruses and plasmids

Fourteen viruses have been identified: six in *Acholeplasma*, four in *Mycoplasma* and four in *Spiroplasma* species. They are rod-shaped (Fig. 41.5) or enveloped spheres that bud from the mycoplasma membrane surface, or are polyhedral with a tail. Those that have been examined in

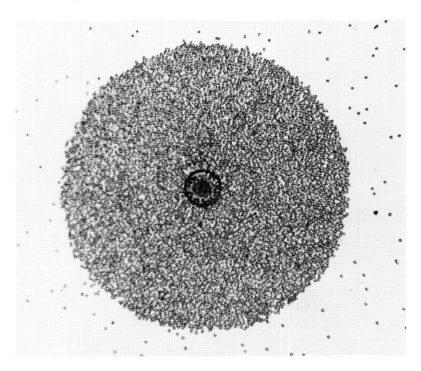

Fig. 41.4 Colony of *M. agalactiae* (caprine origin) showing adherent guinea-pig erythrocytes (phenomenon of haemadsorption).

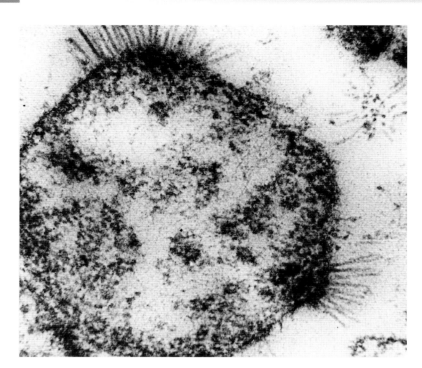

Fig. 41.5 Rod-shaped virus particles (75 × 7.5 nm) radiating from the surface of an *A. laidlawii* cell (original magnification ×100 000).

detail contain single- or double-stranded DNA in either a circular or linear form. There is evidence for integration of the viral genome into the mycoplasmal chromosome, especially in spiroplasmas, and this may provide a mechanism for the promotion of genetic diversity. The release of the viruses is continuous and is not accompanied by cell lysis. Plasmids, so far regarded as cryptic, have been detected in *Acholeplasma*, *Mycoplasma* and, most frequently, *Spiroplasma* species.

Cultivation

Mycoplasmas have limited biosynthetic abilities, so that they need a rich growth medium containing natural animal protein (usually blood serum) and, in most cases, a sterol component. Serum supplies not only cholesterol but also saturated and unsaturated fatty acids for membrane synthesis, components that the organisms cannot synthesize. A medium that has been widely used for isolation contains bovine heart infusion (PPLO broth) with fresh yeast extract and horse serum. However, these components vary in their ability to support growth and success may depend on use of different batches of the components, sera from other animal species or the addition of various supplements.

Cultivation of spiroplasmas is more difficult. A medium designated SP-4 has been helpful not only in their isolation, but also in the isolation of *M. genitalium* and

other fastidious mycoplasmas, such as *M. pneumoniae* and *M. fermentans*. Although various specific formulations have been described for the isolation of ureaplasmas, most mycoplasmal media containing urea may be suitable. The sterol-independent organisms also grow readily on these media, but serum, which often promotes growth, is not usually essential.

A strategy using Vero cell cultures has been described for isolating *M. genitalium*, which, otherwise, may be extremely difficult. The inoculated cell cultures are monitored for mycoplasmal growth by polymerase chain reaction (PCR) technology, followed by subculture to mycoplasmal medium. However, investigators should be aware of the dangers inherent in the widespread use of cell cultures (see below).

Most mycoplasmas are facultatively anaerobic but, because organisms from primary tissue specimens often grow only under anaerobic conditions, an atmosphere of 95% nitrogen and 5% carbon dioxide is preferred for primary isolation. The optimal temperature for most mycoplasmas and ureaplasmas from man or animals is 36–38°C, but is lower for acholeplasmas and spiroplasmas.

Colonial morphology

On agar, mycoplasmas often produce colonies that have a 'fried egg' appearance, with an opaque central zone of growth within the agar and a translucent peripheral zone

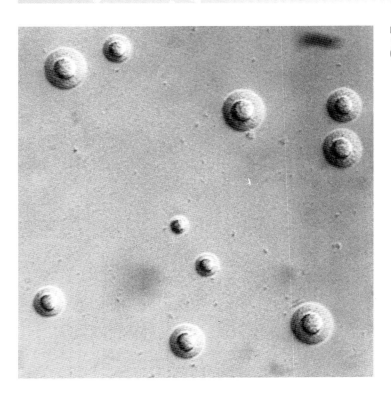

Fig. 41.6 Colonies of *M. hominis* (human origin) (up to 110 μm in diameter) with typical 'fried egg' appearance (oblique illumination; original magnification ×150).

on the surface (Fig. 41.6). However, some, such as *M. pneumoniae* on primary isolation, have just a spherical appearance. The size of the colonies varies widely: colonies of some bovine mycoplasmas and most acholeplasmas may exceed 2 mm in diameter, and are easily visible to the naked eye. Nevertheless, most require low-power microscopic magnification. Colonies of ureaplasmas are characteristically small (15–60 μm in diameter), mainly because they usually lack the peripheral zone of growth. Size and appearance also depend on the constituents and degree of hydration of the medium, the agar concentration, atmospheric condition and age of the culture.

Biochemical reactions

Most mycoplasmas use glucose (or other carbohydrates) or arginine as a major source of energy; a few use both, and others do not metabolize either substrate. In most mycoplasmas the respiratory pathways are flavin-terminated so that the haem compounds, cytochromes and catalase are absent. The unique and distinctive biochemical feature of ureaplasmas is the conversion of urea to ammonia by urease.

Mycoplasma, Ureaplasma and certain other species depend on sterol. They fail to grow in serum-free media and are inhibited by digitonin, distinguishing them from the species that do not require sterol. In addition, most *Mycoplasma* species and ureaplasmas produce hydrogen peroxide, which causes some lysis of guinea-pig or other erythrocytes when these are suspended in agar over developing colonies. *M. pneumoniae* produces clear zones of β-haemolysis. About one-third of all *Mycoplasma* species exhibit haemadsorption (see Fig. 41.4).

Antigenic properties

DNA analysis and sequencing has made a fundamental contribution to the taxonomy of mycoplasmas. Phylogeny-based rapid identification of some mycoplasmas and ureaplasmas has been described based on the amplification of part of the 16S rRNA gene by PCR and, in the future, it may play a more important role in simplifying species classification. For the present, identification and speciation rest largely on the use of specific antisera containing antibodies that reflect the antigenic composition of different mycoplasmas. The western blot technique is useful in assessing the importance of particular antigens, but gel diffusion and immuno-electrophoresis techniques are important for studying the antigenic structure and the relationships between them. For sero-epidemiological studies and for serological diagnosis, more sensitive tests, such as indirect haemagglutination, the enzyme-linked immunosorbent assay (ELISA) and micro-immunofluorescence, are used.

Table 41.1 Some properties of mycoplasmas of human origin

| Mycoplasma | Frequency of detection | | Metabolism of | Preferred pH | Haemadsorption[a] |
	Respiratory tract	Urogenital tract			
M. amphoriforme	? Rare	Not reported	Glucose	7.5	?
M. buccale	Rare	Not reported	Arginine	7.0	–
M. faucium	Rare	Not reported	Arginine	7.0	+
M. fermentans[d]	Common	Rare	Glucose and arginine	7.5	–
M. genitalium[d]	Rare	Rare[b]	Glucose	7.5	+
M. hominis[d,e]	Rare	Common	Arginine	7.0	–
M. lipophilum	Rare	Not reported	Arginine	7.0	–
A. laidlawii	Very rare	Not reported	Glucose	7.5	–
A. oculi	?	Not reported	Glucose	7.5	–
M. orale	Common	Not reported	Arginine	7.0	+
U. parvum[e]	Rare	Common	Urea	6.0 or less	+[c]
M. penetrans[d]	Not reported	Rare	Glucose and arginine	7.5	+
M. pirum[d]	Not reported	?	Glucose and arginine	7.5	?
M. pneumoniae	Rare[b]	Very rare	Glucose	7.5	+
M. primatum	Not reported	Rare	Arginine	7.0	–
M. salivarium	Common	Rare	Arginine	7.0	–
M. spermatophilum	Not reported	Rare	Arginine	7.0	–
U. urealyticum[e]	Rare	Common	Urea	6.0 or less	–

[a]Chick red blood cells.
[b]Rare, except in disease.
[c]Serovar 3 only.
[d]Reported in rectum of homosexual men.
[e]Reported in the anal canal of women attending a genito-urinary medicine clinic.

The way in which different antigens function is exemplified by considering *M. pneumoniae*. The glucose- and galactose-containing membrane glycolipids of the organism are haptens, which are antigenic only when bound to membrane protein. They induce antibodies that react in complement fixation, metabolism inhibition and growth inhibition tests. Glycolipids with fortuitously similar structure have been found in the human brain. Their cross-reactivity with antibodies to *M. pneumoniae* could feasibly account for the neurological manifestations of *M. pneumoniae* infection (see below). Furthermore, the ability of the organisms to alter the I antigen on erythrocytes sufficiently to stimulate anti-I antibodies (cold agglutinins) leads to an auto-immune response and damage to erythrocytes.

M. pneumoniae has two major surface proteins, including the P1 protein involved in attachment, that are recognized by the host; antibody to them is detected in convalescent sera and respiratory secretions. Variation in the P1 protein has led to the recognition of two subtypes of strains, which may vary epidemiologically and in pathogenicity.

Variable membrane lipoproteins of several mycoplasmas form an antigenic variation system that provides a means of escaping from the host immune response; the variations are restricted to a small number of protein antigens and

do not appear to alter the total cell protein profiles of the organisms.

PATHOGENESIS

Fourteen *Mycoplasma* species, two *Acholeplasma* species and two *Ureaplasma* species have been isolated from man (Table 41.1), mostly from the oropharynx. A newly recognized species, *M. amphoriforme*, has been recovered from the respiratory tract of patients with antibody deficiency and chronic bronchitis, but its relation to the disease is not yet clear. *M. fermentans*, *M. genitalium*, *M. hominis*, *M. pneumoniae* and *U. urealyticum* unequivocally cause disease or are strongly associated with disease.

Respiratory infections

Mycoplasmal pneumonia

Pneumonias not attributable to any of the common bacterial causes were labelled historically as *primary atypical pneumonias*. In one variety associated with the development of cold agglutinins, a filterable micro-organism, first called the *Eaton agent*, was isolated in embryonated eggs. Serious doubts about the possibility of the agent being a virus arose when its growth was found to be inhibited by chlortetracycline and gold salts, and cultivation on cell-free medium in the early 1960s finally clinched its mycoplasmal nature. The organism was subsequently named *M. pneumoniae*, and its importance as a cause of respiratory disease was confirmed by numerous studies based on isolation, serology, volunteer inoculation and vaccine protection.

Epidemiology. *M. pneumoniae* infection occurs worldwide. Although endemic in most areas, there is a preponderance of infection in late summer and early autumn in temperate climates; in some countries, such as the UK, epidemic peaks have been observed about every 4 years. Spread is fostered by close contact, for example in a family. Overall, *M. pneumoniae* may cause only about one-sixth of all cases of pneumonia, but in certain confined populations, such as military recruits, it has been responsible for almost half of the cases of pneumonia. Children are infected more often than adults, and the consequence of infection is also influenced by age. Thus, in school-aged children and teenagers, about a quarter of infections culminate in pneumonia, whereas in young adults fewer than 10% do so. Thereafter, pneumonia is even less frequent, although the severity tends to increase with the age of the patient.

Clinical features. *M. pneumoniae* infections often have an insidious onset, with malaise, myalgia, sore throat or headache overshadowing and preceding chest symptoms by 1–5 days. Cough, which starts around the third day, is characteristically dry, troublesome and sometimes paroxysmal, and becomes a prominent feature. However, patients usually do not appear seriously ill and few warrant admission to hospital. Physical signs such as rāles become apparent, frequently after radiographic evidence of pneumonia. Most often this amounts to patchy opacities, usually of one of the lower or middle lobes. About 20% of patients suffer bilateral pneumonia, but pleurisy and pleural effusions are unusual. The course of the disease is variable; cough, abnormal chest signs and radiographic changes may extend over several weeks and relapse is a feature. A prolonged paroxysmal cough simulating the features of whooping cough may occur in children. Very severe infections have been reported in adults, usually in those with immunodeficiency or sickle cell anaemia, although death is rare. Apart from being involved in some exacerbations of chronic bronchitis, it has been mooted that *M. pneumoniae* might not only exacerbate asthma, but sometimes be a primary cause in children.

Disease is limited usually to the respiratory tract. Extra-pulmonary manifestations include:

- Stevens–Johnson syndrome and other rashes
- arthralgia
- meningitis or encephalitis (and other neurological sequelae including, rarely, Bell's palsy)
- haemolytic anaemia
- myocarditis
- pericarditis.

Haemolytic anaemia with crisis is an auto-immune phenomenon brought about by cold agglutinins (anti-I antibodies). Some of the other complications, such as the neurological ones, may arise in a similar indirect way, although *M. pneumoniae* has been isolated from cerebrospinal fluid.

Respiratory infection in the newborn

M. hominis and ureaplasmas occasionally cause respiratory disease in the newborn, particularly in those of very low birth-weight, the infections being acquired in utero or at the time of delivery. The likelihood of death or chronic lung disease in infants of less than 1000 g with a ureaplasmal respiratory tract infection within 24 h of birth is double that of uninfected infants of similar birth-weight, or heavier infants. However, whether other bacteria that are associated with maternal bacterial vaginosis might also be involved is an issue that remains unresolved.

Other respiratory infections

M. hominis has produced sore throats when given orally to adult volunteers; however, it does not seem to cause naturally occurring sore throats in children or adults.

M. fermentans has been associated with adult respiratory distress syndrome with or without systemic disease, and with pneumonia in a few children with community-acquired disease. *M. genitalium* has been isolated, together with *M. pneumoniae*, from the respiratory secretions of a few adults but its role, if any, in respiratory disease seems small.

Urogenital infections

M. fermentans, M. genitalium, M. hominis, M. penetrans, M. pneumoniae, M. primatum, M. salivarium, M. spermatophilum and ureaplasmas have been isolated from the urogenital tract. *M. hominis* and ureaplasmas have been isolated most frequently, although use of the PCR technique has enabled some mycoplasmas, for example *M. fermentans* and *M. genitalium*, to be detected more often than would otherwise be the case.

Urogenital infections in men

Although *M. hominis* may be isolated from about 20% of men with non-gonococcal urethritis (NGU), it has not been incriminated as a cause. There is considerable evidence to implicate *U. urealyticum* as a cause of acute NGU, and the involvement of ureaplasmas as a cause of some cases of chronic NGU seems likely.

M. genitalium has been strongly implicated world-wide as a cause of acute and chronic NGU, but there is no evidence that this or other mycoplasmas are a cause of acute or chronic prostatitis.

There is a little evidence that ureaplasmas may occasionally cause acute epididymitis, but the role of *M. genitalium* has not been investigated. It is very doubtful that ureaplasmas have any role in male infertility. A possibility that *M. genitalium* organisms could have a role is based on their known ability to adhere to spermatozoa.

Reproductive tract disease in women

M. hominis organisms and, to a lesser extent, ureaplasmas are found in much larger numbers in the vagina of women who have bacterial vaginosis than they are in the vagina of healthy women and, with various other bacteria, may contribute to the development of the condition. These genital mycoplasmas do not cause cervicitis, but *M. genitalium* has been associated strongly with the disease.

Bacterial vaginosis, and hence the bacteria associated with it, including *M. hominis*, may have a role in the development of pelvic inflammatory disease. *M. hominis* has been isolated from the endometrium and fallopian tubes of about 10% of women with the disease, together with a specific antibody response. This suggests a causal role, but the significance of *M. hominis* is difficult to judge as several other micro-organisms are also present. Ureaplasmas have also been isolated directly from affected fallopian tubes, but the absence of antibody responses and failure to produce salpingitis in subhuman primates make them an even less likely cause of disease at this site.

The pathogenic potential of *M. genitalium* in pelvic inflammatory disease seems greater; it has been associated significantly with endometritis, and antibody responses to the mycoplasma in some patients with pelvic inflammatory disease and the production experimentally of salpingitis in subhuman primates suggest that it may have an important role. However, examination of laparoscopically derived specimens by PCR technology is needed to determine conclusively whether this is so.

The part that *M. hominis* is likely to play in infertility as a result of tubal damage is small. In contrast, serologically, *M. genitalium* has been associated strongly with tubal factor infertility.

Disease associated with pregnancy and the newborn

M. hominis and ureaplasmas have been isolated from the amniotic fluid of women with severe chorio-amnionitis who had pre-term labour. Similarly, ureaplasmas have been isolated from spontaneously aborted fetuses and stillborn or premature infants more frequently than from induced abortions or normal full-term infants.

Isolation of ureaplasmas from the internal organs of aborted fetuses, together with some serological responses and an apparently diminished occurrence following antibiotic therapy, suggest that these organisms may have a role in abortion. However, bacterial vaginosis is strongly associated with pre-term labour and late miscarriage, and, as *M. hominis* and ureaplasmas are part of the extensive microbial flora of bacterial vaginosis, it is possible that they act with a multitude of other micro-organisms and not independently. There is, so far, no evidence to suggest that *M. genitalium* is associated with pre-term labour. The same considerations as above relate to the role of genital mycoplasmas, particularly ureaplasmas, in causing low birth-weight in otherwise normal full-term infants. Bacterial vaginosis has been largely ignored in defining this association, and a study in which women given erythromycin in the third trimester delivered larger babies than those given a placebo has not been confirmed in later trials. Although there is obvious uncertainty surrounding this issue, premature infants are prone to meningitis caused by *M. hominis* or ureaplasmas within the first few days of life.

After an abortion, *M. hominis* has been isolated, apparently in pure culture, from the blood of about 10% of febrile women; half of them have exhibited an antibody response. This mycoplasma was not found in the blood of about 5–10% of women who aborted but remained afebrile, but it has been recovered from the blood of a similar proportion of women with postpartum fever. Observations of this kind have also been made for ureaplasmas and it seems that both micro-organisms induce fever, possibly by causing endometritis.

Urinary infection and calculi

M. hominis has been isolated from the upper urinary tract of patients with acute pyelonephritis and probably causes about 5% of such cases. Ureaplasmas do not seem to be involved in pyelonephritis. However, the fact that they produce urease, induce crystallization of struvite and calcium phosphates in urine in vitro, and produce calculi experimentally in animal models raises the question of whether they cause calculi in the human urinary tract. Ureaplasmas are found more often in the urine and calculi of patients with infective stones than in those with metabolic stones, suggesting that they may have a causal role.

Joint infections

Evidence that ureaplasmas are involved in the aetiology of sexually acquired reactive arthritis is based on synovial fluid mononuclear cell proliferation in response to specific ureaplasmal antigens. Ureaplasmas also have a role in arthritis in hypogammaglobulinaemic patients (see below). There are isolated reports of *M. genitalium* in the joints of patients with Reiter's disease or rheumatoid arthritis, but the significance of the observations is not clear. *M. fermentans* has been found by a PCR assay in the joints of about 20% of patients with rheumatoid arthritis and in those of patients with other chronic inflammatory rheumatic disorders, but not in the joints of patients with non-inflammatory arthritides. However, whether this mycoplasma is more than an innocent bystander remains to be determined.

Infections in immunocompromised patients

M. pneumoniae may cause severe pneumonia in immuno-deficient patients and may persist for many months in the respiratory tract of hypogammaglobulinaemic patients, despite apparently adequate treatment. The exact role of *M. amphoriforme* in immunodeficient patients with chronic bronchitis remains to be resolved. A few hypogammaglobulinaemic patients develop suppurative arthritis, and mycoplasmas are responsible for at least two-fifths of the cases. The mycoplasmas mainly involved have been *M. pneumoniae, M. salivarium* (usually regarded as non-pathogenic), *M. hominis* and, particularly, ureaplasmas. In some cases involving ureaplasmas, the arthritis has been associated with subcutaneous abscesses, persistent urethritis and chronic cystitis. Although sometimes responding to tetracyclines, the organisms and disease may persist for many months despite concomitant use of anti-inflammatory and γ-globulin replacement therapy. Administration of specific mycoplasmal antiserum pre-pared in an animal (e.g. sheep or rabbit) may aid clinical and microbiological recovery. A veterinary pleuromutilin antibiotic (Econor) has been used with some success to treat mycoplasmal respiratory infections in immuno-compromised subjects.

Septicaemia due to *M. hominis* has occurred after trauma and genito-urinary manipulations, and the mycoplasma has been found in brain abscesses and osteomyelitis. Haematogenous spread leading to septic arthritis, surgical wound infections and peritonitis seems to occur more often after organ transplantation and in other patients on immunosuppressive therapy. Particularly common are sternal wound infections caused by *M. hominis* in heart and lung transplant patients.

The suggestion that *M. fermentans* infection of the peripheral blood monocytes of human immunodeficiency virus (HIV)-positive subjects might lead to a more rapid development of the acquired immune deficiency syndrome has been investigated and seems highly unlikely. Similarly, nothing has emerged to support the notion that

M. penetrans is associated uniquely with HIV positivity and with Kaposi's sarcoma.

LABORATORY DIAGNOSIS

Mycoplasmal pneumonia

Because the clinical manifestations are not sufficiently distinct for definitive diagnosis, laboratory help is required. The difficulty and slowness of culturing *M. pneumoniae* means that serology is often relied upon in routine practice. The cold agglutinin test is non-specific and unreliable, and the more specific complement fixation test is often used. Because of cross-reactivity with *M. genitalium*, western blotting might be needed to check specificity in dubious cases. A four-fold or greater rise in antibody titre, with a peak at about 3–4 weeks, is indicative of a recent infection. Because paired sera are not always available or because of the delay in acquiring a second serum, a single antibody titre of 64–128, in a suggestive clinical setting, should be sufficient to institute therapy. However, the complement fixation test tends to detect immunoglobulin (Ig)M rather than IgG antibody and this accounts largely for any lack of sensitivity. Of a gamut of other tests, the enzyme immunoassay is sensitive and specific, and there are formats for testing paired sera or a single serum. The ImmunoCard test (Meridian Diagnostics) for IgM is simple and rapid and has proved reliable, especially in children. The Remel enzyme immunoassay detects IgM and IgG simultaneously and is sensitive and specific.

Attempted isolation of *M. pneumoniae* (and *M. fermentans*) requires the use of mycoplasmal SP4 broth medium supplemented with penicillin and glucose, and with phenol red as a pH indicator. After inoculation with sputum, throat washing, pharyngeal swab or other specimen, the medium is incubated at 37°C, and a colour change (red to yellow), which may take up to 3 weeks or longer, indicates fermentation of glucose due to multiplication of the organisms. The broth is then subcultured on agar medium to await the development of colonies, which are identified specifically by immuno-fluorescence with a specific antiserum or by other serological methods.

Isolation is not often attempted because of insensitivity and the time required. However, the PCR assay is rapid, sensitive and specific. If an isolate is required, a sensible approach is to test specimens by both the PCR assay and culture, and to continue the latter only for those specimens that prove to be PCR positive.

Diagnostic procedures for other mycoplasmas that might be found in the respiratory tract are considered below.

Urogenital infections

Material from urethral, cervical or vaginal swabs, or centrifuged deposit from urine, is added to separate vials of liquid mycoplasmal medium containing phenol red and 0.1% glucose, arginine or urea. *M. genitalium* metabolizes glucose and changes the colour of the medium from red to yellow. *M. fermentans* also metabolizes glucose but, in addition, converts arginine to ammonia. The latter conversion is also shown by *M. hominis* and *M. primatum*. The ureaplasmal urease also breaks down urea to ammonia. In each case, the pH of the medium increases, and the colour changes from yellow to red. The colour change produced by ureaplasmas occurs usually within 1–2 days, that due to *M. hominis* rarely after 4 days, whereas that for *M. genitalium* may take several months or not occur despite the organism being present. Subculture to agar medium usually results in the formation of characteristic colonies (see above, Cultivation).

On ordinary blood agar, *M. hominis*, but not ureaplasmas, produces non-haemolytic pinpoint colonies. *M. hominis* also multiplies in most routine blood culture media; the inhibitory effect of sodium polyanethol sulphonate, included as an anticoagulant, can be overcome by the addition of gelatin (1% w/v).

Serological tests are usually used for definitive identification of the respective species. Phylogeny-based rapid identification of urogenital mycoplasmas and ureaplasmas, based on the amplification of a part of the 16S rRNA gene by PCR, is also described. Kits designed to isolate and identify *M. hominis* and ureaplasmas, and to provide antimicrobial susceptibility profiles, are available commercially and are of particular value if the need to detect these micro-organisms arises infrequently.

DNA primers specific for several mycoplasmas including *M. fermentans, M. genitalium, U. urealyticum* and *U. parvum* have been developed and used for amplification by the PCR. This technique is much more sensitive than culture for detecting the two former mycoplasmas and is the only way of determining their presence reliably in clinical specimens.

Genital mycoplasmal infections stimulate antibody responses, but the various techniques to detect them are rarely used diagnostically.

TREATMENT

Mycoplasmal pneumonia

M. pneumoniae is sensitive to the tetracyclines and erythromycin in vitro and these antibiotics have been used widely in clinical practice, where they have sometimes proved less effective for treating pneumonia than in planned trials, probably because disease is often well established before treatment begins. Nevertheless, it is worthwhile administering a tetracycline or a macrolide to adults, and erythromycin to children and pregnant women.

Newer macrolides, such as clarithromycin and azithromycin, and the newer fluoroquinolones, are highly active in vitro and have a place in treatment. The fluoroquinolones seem to have a cidal effect, which may be useful as tetracyclines and macrolides only inhibit growth. Failure to kill the organisms, together with the fact that the organisms may become intracellular, probably explains persistence in the respiratory tract long after clinical recovery, as well as clinical relapse in some patients. Furthermore, a functioning immune system is important in eradication, so that in some hypogammaglobulinaemic patients the organisms may persist for months or years.

Antibiotic treatment of *M. pneumoniae* or other mycoplasma-induced infection should start as soon as possible, based on clinical suspicion, rather than waiting for laboratory confirmation. A course of at least 2 weeks is justified, particularly if supported by serological or other evidence of infection.

Urogenital infections

Treatment must take into account the fact that several different micro-organisms may be involved and that a precise microbiological diagnosis may not be attainable. Thus, patients with NGU should receive an antibiotic that is active against *Chlamydia trachomatis,* ureaplasmas and *M. genitalium.* Azithromycin, which is being used increasingly for chlamydial infections, is also active against a wide range of mycoplasmas, including *M. genitalium* and, to a lesser extent, ureaplasmas. This may be preferable to a tetracycline, because *M. genitalium* is less sensitive to tetracyclines and at least 10% of ureaplasmas are resistant.

A broad-spectrum antibiotic should also be included for the treatment of pelvic inflammatory disease to cover *C. trachomatis* and *M. hominis.* As about 20% of *M. hominis* strains are resistant to tetracyclines, other antibiotics such as clindamycin or fluoroquinolones may need to be considered.

Fever following abortion or childbirth often settles within a few days, but if it does not therapy with a broad-spectrum antibiotic should be started, keeping the above comments in mind.

MYCOPLASMAS AND CELL CULTURES

It is rare for primary cell cultures to become infected with mycoplasmas, but continuous cell lines do so frequently. The mycoplasmas most responsible are *M. arginini, M. fermentans, M. hyorhinis, M. orale, M. salivarium* and *Acholeplasma laidlawii.* The effects of contamination include those caused by mycoplasmal enzymes and toxins, and those resulting from metabolism of cell culture media components or from changes in pH. Despite the presence of up to 10^8 organisms per millilitre of culture fluid, there may be little effect on viral propagation, although it may decrease the yield. Occasionally, the yield may be increased as, for example, with vaccinia virus in *M. hominis*-infected cells.

Mycoplasmal culture methods and an indicator cell system with staining (e.g. Hoechst DNA dye) are used conventionally to detect contamination, but PCR methods have much greater sensitivity. Numerous procedures have been described to eliminate mycoplasmas from cell cultures, but none is consistently successful. Whenever possible it is easier to discard the cultures, replace them with mycoplasma-free cells and adhere to simple guidelines to prevent contamination. If it is imperative to save cells, treatment with an antibiotic that is likely to have mycoplasmacidal activity, such as a fluoroquinolone, identification of the contaminant and use of a specific antiserum, or a combination of these methods, is most likely to be successful in eradication.

KEY POINTS

- Mycoplasmas pathogenic for man include the genera *Mycoplasma* and *Ureaplasma*. They are found in the respiratory and genital tracts of man and many animal species, and sometimes invade the bloodstream to gain access to joints and other organs.
- Mycoplasmas are the smallest organisms that grow in cell-free bacteriological media. They have no peptidoglycan; hence they are not susceptible to β-lactam or glycopeptide antibiotics, and may assume a variety of shapes. Infections are treated with tetracyclines or macrolides.
- *M. pneumoniae* is a pathogen for the human respiratory tract. About a quarter of such infections in children result in pneumonia, whereas in young adults fewer than 10% do so.

- *M. genitalium* and *Ureaplasma urealyticum* are a cause of acute and chronic non-gonococcal urethritis in men.
- Other significant associations include: *M. hominis* and, to a lesser extent, ureaplasmas with bacterial vaginosis, and *M. hominis* and, probably, *M. genitalium* with pelvic inflammatory disease.
- *M. fermentans* has been detected in the joints of patients with chronic arthritides, although the significance of this is unknown. Other mycoplasmas, often ureaplasmas, are responsible for about two-fifths of the cases of suppurative arthritis occuring in patients with hypogammaglobulinaemia.

RECOMMENDED READING

Blanchard A, Browning G (eds) 2005 *Mycoplasmas: Molecular Biology, Pathogenicity and Strategies for Control.* Horizon Scientific Press, Wymondham, UK

Furr P M, Taylor-Robinson D, Webster A D B 1994 Mycoplasmas and ureaplasmas in patients with hypogammaglobulinaemia and their role in arthritis: microbiological observations over 20 years. *Annals of the Rheumatic Diseases* 53: 183–187

Maniloff J, McElhaney R N, Finch L R, Baseman J B (eds) 1992 *Mycoplasmas. Molecular Biology and Pathogenesis.* American Society for Microbiology, Washington, DC

Taylor-Robinson D 1996 Mycoplasmas and their role in human respiratory tract disease. In Myint S, Taylor-Robinson D (eds) *Viral and Other Infections of the Human Respiratory Tract*, pp. 319–339. Chapman and Hall, London

Taylor-Robinson D 2002 *Mycoplasma genitalium* – an update. *International Journal of STD & AIDS* 13: 145–151

Taylor-Robinson D, Bebear C 1997 Antibiotic susceptibilities of mycoplasmas and treatment of mycoplasmal infections. *Journal of Antimicrobial Chemotherapy* 40: 622–630

Taylor-Robinson D, Ainsworth J G, McCormack W M 1999 Genital mycoplasmas. In Holmes K K, Sparling P F, Mardh P-A et al (eds) *Sexually Transmitted Diseases*, 3rd edn, pp. 533–548. McGraw Hill, New York

Taylor-Robinson D, Gilroy C B, Jensen J S 2000 The biology of *Mycoplasma genitalium. Venereology* 13: 119–127

Tully J G, Razin S (eds) 1996 *Molecular and Diagnostic Procedures in Mycoplasmology. Diagnostic Procedures*, Vol. 2. Academic Press, London

Waites K B, Rikihisa Y, Taylor-Robinson D 2003 *Mycoplasma* and *Ureaplasma*. In Murray P R, Baron E J, Jorgensen J H, Pfaller M A, Yolken R H (eds) *Manual of Clinical Microbiology*, 8th edn, Vol. 1, pp. 972–990. American Society for Microbiology, Washington, DC

PART 4
VIRAL PATHOGENS AND ASSOCIATED DISEASES

42 Adenoviruses

Respiratory disease; conjunctivitis; gut infections

J. S. M. Peiris and C. R. Madeley

Adenoviruses were named from their original source, *adenoid tissue*, removed at operation and cultured as explants in vitro. Cellular outgrowth occurred readily, but this often deteriorated rapidly a week or 10 days later. The cause of the deterioration was found to be *adenovirus(es)* present in the original tissue and which replicated enthusiastically in the new cells growing from the explant.

This discovery initiated much research, which established that there were a considerable number of serotypes, mostly associated with mild *upper respiratory tract infections*. In addition, there were occasional serious (and even fatal) childhood *pneumonias* and infrequent, but readily transmissible, *eye infections*. Some of these infections, mostly in children and not always symptomatic, could persist for weeks or months. The focus of adenovirus research then shifted away from clinical virology with the discovery that some species could cause malignancies in laboratory rodents. As a result, virologists have thoroughly investigated adenovirus structure, replicative mechanisms and oncogenicity during the past 40 years.

Clinical interest revived in the middle 1970s with the discovery of two new serotypes (subsequently numbered types 40 and 41) linked (with several other hitherto unknown viruses) to that previously elusive entity 'viral gastro-enteritis'. This revival of clinical interest has been extended by the discovery in rapid succession of six new serotypes (numbered 42–47), most of them in patients with acquired immune deficiency syndrome (AIDS). Adenoviruses can cause serious disease in immuno-compromised patients (e.g. bone marrow transplant recipients), but they may also be present in the stools of congenitally immunodeficient children with little or no associated pathology. The apparently low pathogenic potential of adenoviruses has encouraged genetic manipulators to explore their possible use in gene therapy or tumour treatment. Human genes (up to 7 kb in size) have been inserted into replication-crippled adenoviruses and these new gene(s) carried into cells by adenovirus infection in the hope that their expression can correct defects caused by absent or defective genes. Alternatively, similarly modified adenoviruses have been used to target tumour cells (which are often of epithelial origin), either to carry lethal mutations into the cell or to induce the surface expression of target viral antigens, making the cell vulnerable to normal immune mechanisms.

Both of these approaches show promise, but formidable technical problems remain to be overcome before either gene therapy or oncolysis becomes routinely useful. These include difficulties in making the transferred genes persist and in overcoming pre-existing immunity in the patient to the serotype(s) of adenovirus used.

DESCRIPTION

Adenovirus virions provide a very good example of an *icosahedron*. Figure 42.1a shows a group of typical adenovirus particles, whereas Figures 42.1b–d compare a single virus particle, as seen by electron microscopy, with a model. The particles seen in Figure 42.1a do not show the apical fibres (they are rarely seen in situ by the electron microscope), but otherwise the particles in Figures 42.1a & b resemble the model closely.

The virion is 70–75 nm in size, depending on whether it is measured across the 'flats' or the apices, and there are probably minor variations between preparations and, possibly, serotypes. The surface has 252 visible surface 'knobs', the capsomers. Twelve apical capsomers, each surrounded by five others, are known as *pentons*; 60 peripentons surround the pentons and 180 other capsomers make up the major part of the faces (Fig. 42.1d). Except for the pentons, all of the other capsomers are surrounded by six others, and are therefore called *hexons*.

From each penton projects an apical *fibre*; these vary greatly in length, from 9 to 31 nm. Some avian types have double fibres, but human strains all have single ones.

Adenoviruses contain a single piece of double-stranded DNA, 33–45 kbp in size, which codes for a considerable number of proteins with molecular weights of 7500–400 000 Da. They include both structural and non-structural proteins. The ten structural proteins include three polypeptides that make up each hexon, one that

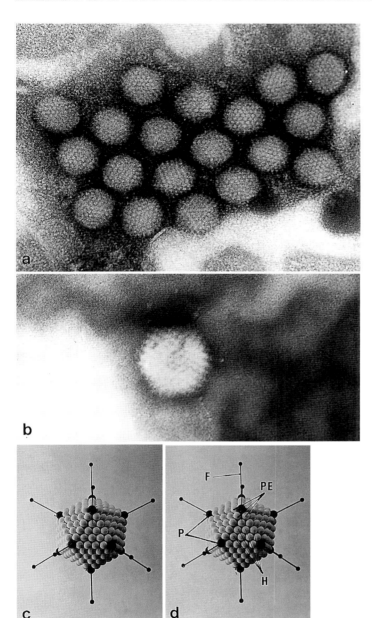

Fig. 42.1 a Group of adenovirus particles from a stool extract showing typical adenovirus morphology, although apical fibres are not visible. Individual capsomers can be seen as surface 'knobs'. These particles should be compared with the model seen in c. Negative contrast, 3% potassium phosphotungstate, pH 7.0 (magnification ×160 000). b Single adenovirus particle showing some of the apical fibres. This is an unusual finding. Negative contrast, 3% potassium phosphotungstate, pH 7.0 (magnification ×280 000). c,d Photograph of a model with pentons (P), peripentons (PE) and hexons (H) indicated. Note that the fibres (F) are attached to the penton.

forms the penton base and a glycoprotein for the fibre. These five lie on the surface of the particle, whereas the others are internal.

The pentons and fibres bear antigenic determinants that are 'type' specific, and antibody binding to them results in neutralization of virus infectivity. The hexons also have some type-specific antigenic determinants, but carry group-specific antigens as well. The same group antigens are found on all the mastadenoviruses (see

below), the genus to which the medically important human adenoviruses belong. These group-specific antigens provide a basis for tests to detect adenoviruses and antibodies to them (by immunofluorescence, enzyme immuno-assays, complement fixation, etc. – see Diagnosis, below).

Classification

The family Adenoviridae comprises two genera:

Table 42.1 Properties and classification of adenoviruses of species A–F

Species[a]	Serotypes	No. of *Smal* fragments[b]	Haemagglutination pattern[c]	Oncogenicity in newborn hamsters	Tissues most commonly infected
A	12, 18, 31	4–5	IV	High	Gut (no symptoms)
B	3, 7, 11, 14, 16, 21, 34, 35, 50	8–10	I	Low	Respiratory tract, kidney
C	1, 2, 5, 6	10–12	III	None	Respiratory tract, lymphoid tissue (tonsils and adenoids)
D	8–10, 13, 15, 17, 19, 20, 22–30, 32, 33, 36–39, 42–49, 51	14–18	II	None	Conjunctiva, gut, respiratory tract(?)
E	4	16–19	III	None	Conjunctiva, respiratory tract
F	40, 41	9–12	IV	None	Gut

[a]More than 50% DNA homology between members.
[b]Restriction endonuclease digestion. Some small fragments probably not included.
[c]I, complete agglutination of monkey erythrocytes; II, complete agglutination of rat erythrocytes; III, partial agglutination of rat erythrocytes; IV, agglutination of rat erythrocytes only after addition of heterotypic serum.
Adapted with permission from Wadell G 2000 In: Zuckerman A J, Banatvala J E, Pattison J R (eds) *Principles and Practice of Clinical Virology*, 4th edn, pp. 307–327. John Wiley, Chichester.

1. *Mastadenovirus*, whose members infect mammalian species, including humans
2. *Aviadenovirus*, whose members infect avian species.

These two genera are completely distinct antigenically. Human adenoviruses are further subdivided into six species (A–F) (Table 42.1), based on DNA homology; each species includes several serotypes. By definition, those with more than 50% homology are members of the same species, whereas those of different species have less than 20% homology.

Within each species, adenovirus serotypes are defined by cross-reactivity in neutralization tests. There are 47 recognized serotypes of human adenoviruses at present, of which types 40–47 have been recognized only within the past 35 years.

REPLICATION

The virus attaches to susceptible cells by the apical fibres and is then taken into the cell, losing both fibres and pentons in the process. It then passes to the nucleus, losing the peripentons at the nuclear membrane. Inside the nucleus the DNA is released and the process of replication is initiated (see Ch. 7). There are about 20 early proteins, most of which are not incorporated into new particles. The late proteins are produced in quantity in the cytoplasm, are mostly structural, and are later transported back to the nucleus where new virus particles are assembled, normally as crystalline arrays. The shutting down of host cell metabolism and the accumulation of thousands of new virions results in rupture (lysis) and death of the infected cell with release of the particles.

In cell cultures this process causes the cells to round up, swell and aggregate into clumps resembling bunches of grapes. The cells then disintegrate as they lyse.

CLINICAL FEATURES

Table 42.2 lists the more common associations of serotypes with disease. The great majority of infections with adenoviruses are probably undiagnosed and the full extent of pathogenesis is under-reported. In addition, virus infection and replication are not invariably associated with disease. For example, a wide variety of serotypes are isolated from faeces without evidence of gut pathology, and prolonged tonsillar carriage in children is common. However, some well recognized disease syndromes are caused by adenovirus infections.

Respiratory diseases

These are usually mild upper respiratory tract infections with fever, runny nose and cough. The majority are due to types 1–7, although higher serotypes may be involved sporadically. Types 1, 2, 5 and 6 are more commonly associated with endemic infections, whereas types 3, 4 and 7 are more epidemic. In children the associated clinical diagnoses include 'upper respiratory tract infection', 'increased secretions', 'wheezy', 'cold and cough' and 'failure to thrive'. Adenovirus infections can mimic whooping cough in some patients.

These infections are rarely serious but, occasionally and unpredictably, may progress to a pneumonia that is both extensive and frequently fatal. The majority of these pneumonias occur in young children. Unlike other viral

Table 42.2 Disease associated with adenovirus serotypes

Disease	Those at risk	Associated serotypes
Acute febrile pharyngitis:		
Endemic	Infants, young children	1, 2, 5, 6
Epidemic	Infants, young children	3, 4, 7
Pharyngoconjunctival fever	Older school-aged children	3, 7
Acute respiratory disease	Military recruits	4, 7, 14, 21
Pneumonia	Infants	1, 2, 3, 7
Follicular conjunctivitis	Any age	3, 4, 11
Epidemic keratoconjunctivitis	Adults	8, 19, 37
Haemorrhagic cystitis	Infants, young children	11, 21
Diarrhoea and vomiting	Infants, young children	40, 41
Intussusception	Infants	1, 2, 5
Disseminated infection	Immunocompromised (e.g. AIDS, renal, bone marrow and heart–lung transplant recipients)	5, 11, 34, 35, 43–51

Adapted with permission from Wadell G 2000 In: Zuckerman A J, Banatvala J E, Pattison J R (eds) *Principles and Practice of Clinical Virology*, 4th edn, pp. 307–327. John Wiley, Chichester.

respiratory infections, adenoviruses may be associated with a raised white cell count and enhanced levels of C-reactive protein, and therefore may be confused clinically with a bacterial infection.

In older children and young adults, a proportion of these infections will be labelled as 'colds', although epidemics of adenoviral infection with respiratory symptoms and fever are common in, for example, US military recruits, where violent exercise and close proximity combine to make the victims more vulnerable and to facilitate spread. These outbreaks can be severe in both numbers and extent of disease – severe enough, in fact, to warrant the development of a trial vaccine. Eye involvement is a common feature (see below), leading to such outbreaks being called pharyngoconjunctival fever. Types 3 and 7 are more often associated with these outbreaks, but other serotypes are found from time to time.

Eye infections

Adenoviruses have been associated with several outbreaks of conjunctivitis in the UK, frequently referred to as 'shipyard eye' because originally it was thought to be caused by steel swarf thrown up from welding and grinding. It was later shown that adenovirus was being transmitted through fluids, eyebaths and other instruments used to treat eye injuries in the shipyard First Aid Clinic, and which had become contaminated by virus from the index case. Use of properly sterilized instruments and single-dose preparations of eye ointment have now made this uncommon. Most such outbreaks were due to adenovirus type 8 (although types 19 and 37 may also be involved). This is not one of the easiest types to isolate, and sporadic cases may not be identified as readily as outbreaks, which inevitably attract a more concentrated effort at diagnosis. Eye infections with type 8 do not usually cause systemic symptoms.

Conjunctivitis caused by other serotypes may form part of outbreaks of pharyngoconjunctival fever. It is a follicular conjunctivitis resembling that caused by *Chlamydia*, from which it should be differentiated, but the associated and usually marked adenovirus respiratory symptoms provide a clue.

Gut infections

The common respiratory serotypes (1–7) are frequently isolated from faeces; in young children, the same serotype may be recovered from both ends of the child. There is little evidence to link such isolates with disease in the gut. However, when faecal extracts from children with diarrhoea were examined by electron microscopy, typical adenoviruses were seen in a proportion of cases, often in very large numbers. Surprisingly, these morphologically typical viruses could not readily be isolated in cell culture. It is now clear that these belonged to two hitherto unknown serotypes, 40 and 41, which are associated with a significant proportion of endemic cases of childhood diarrhoea and, in terms of numbers of cases, are second only to rotaviruses. They may contribute up to one-third of those cases in which a virus is found. As with the other viruses found in diarrhoeal faeces, such adenoviruses may also be present less frequently in the faeces of apparently normal babies, but there is no doubt of their pathogenic potential for the gut. How they differ from the other non-pathogenic types in the gut is not known, nor whether they cause respiratory infections.

The role of adenovirus(es) in mesenteric adenitis and intussusception is uncertain. Even when enlarged nodes are identified at laparotomy, it is not usual to excise one for diagnostic purposes, and direct evidence is lacking. Finding a coincidental adenovirus in faeces is not proof of involvement, nor is it clear how often adenitis precedes (and possibly initiates) intussusception. Where any such temporal association has been recorded it has involved the common serotypes 1, 2, 5 and 6.

Ten new serotypes (42–51) have been found, nine of them (43–51) in the faeces of patients with AIDS. Type 42 was isolated from the faeces of a normal child. Chronic diarrhoea is a feature of AIDS, although any causal role of these adenoviruses has not been proven. Interestingly, these new serotypes were isolated and identified in cell culture, suggesting that they are indeed 'new'. If so, both their origin and significance are unknown.

Other diseases

Infrequently there have been reports of adenovirus (mostly types 11 and 21) recovered from the urine of children with haemorrhagic cystitis. The finding of virus provides a (possible) retrospective cause. Adenoviruses may also be associated with haemorrhagic cystitis, hepatitis (in liver transplant patients) and pneumonitis in immuno-compromised patients. The serotypes involved are often uncommon ones (e.g. types 11, 34 and 35). Adenoviruses have been implicated as a major cause of systemic disease in bone marrow transplant recipients, replacing cytomegalovirus as the main viral infectious threat. In the newborn, adenoviruses may cause disseminated disease with a 'septic shock' form of presentation.

There are reports in the literature of recovery of adenoviruses from both the male and the female genital tracts. They may be sexually transmitted but are not the cause of a recognized sexually transmitted disease.

There is no good evidence of adenoviruses being involved in central nervous system disease, nor in human tumour production. Experimentally, adenoviruses may induce transformation of hamster cells in culture, and such transformed cells will initiate tumours in laboratory animals. There is no evidence that this can occur in man, although it has been diligently sought. Under laboratory conditions adenoviruses can also form hybrids with simian virus 40 (SV40), a papovavirus that contaminated early stocks of polio vaccine grown in monkey kidney cells. The hybrids, carrying part of the SV40 genome, are neither pathogenic nor oncogenic in man.

PATHOGENESIS

Adenoviruses mostly infect mucosal surfaces (respiratory tract, gut and eye), although it is clear that by no means all such infection leads to overt disease. Different serotypes appear to prefer different regions, perhaps due to the presence or absence of particular receptors. Infection of an individual cell will cause its death, but several studies have documented prolonged respiratory and gut excretion in healthy children, lasting for weeks or months. Such respiratory 'carriage' is probably in lymphoid tissue (tonsils and adenoids), and gut carriage may be in the equivalent Peyer's patches, although this has not been documented.

Most infections with adenoviruses, whatever the primary site, probably spread to include the gut. 'Respiratory' strains are frequently recovered from faeces and it seems improbable that this results solely from overflow from the upper respiratory tract. It is much more likely that faecal excretion follows a secondary gut infection, albeit asymptomatic in most cases.

The role of the nine recently discovered serotypes (43–51) in patients with AIDS is not known at present. All were recovered from faeces by culture and had not been identified before. They extend the unanswered questions about adenoviruses: Where do new types come from? Can they arise by mutation and/or recombination? For further discussion of this topic, see de Jong et al 1999.

LABORATORY DIAGNOSIS

Direct demonstration of 'virus'

Electron microscopy

Virus particles may be seen directly in stool extracts by electron microscopy, although this cannot identify serotypes. None the less, where virus is seen this is usually of type 40 or 41, particularly where large numbers are present (the level may reach more than 10^{10} particles per gram of faeces). The finding of virus in faeces by electron microscopy does not mean that virus is also present in the nasopharynx or eye(s).

Virus antigen

The presence of viral antigen in the nasopharynx may be identified by immunofluorescence with group-specific antibodies (polyclonal or monoclonal) directly on aspirates (*not* swabs), provided that they contain respiratory cells. The presence of such infected cells usually indicates a significant infection, in contrast to asymptomatic carriage. Alternatively, viral antigen may be detected by enzyme immuno-assays, although these detect virus antigen alone without indicating where it was located. Hence, they cannot distinguish between a significant presence in respiratory cells (indicating invasion of the mucosal surface) and mostly silent (and clinically insignificant) carriage, probably in the tonsils and adenoids.

Viral antigen in the stools can also be detected by a variety of commercial group-specific enzyme immuno-assays.

Viral DNA

Conserved parts of the adenoviral genomic DNA can be detected by molecular detection methods (e.g. the polymerase chain reaction) in a variety of clinical specimens. This may form part of a multiplex system to test for several respiratory viruses simultaneously, or in other conditions. This approach can be adapted to identify individual adenovirus serotypes, but is not routinely available.

Culture

Virus can be grown in cell culture from respiratory specimens (nasopharyngeal aspirates, and nose and throat swabs), eye swabs, faeces and, occasionally, urine. The speed of isolation is usually an indication of viral load in the specimens and can provide a pointer to the clinical significance of the finding. If it takes longer than 12 days it is less likely to be clinically significant, particularly with types 1, 2, 5 and 6.

Isolation of an adenovirus in cell culture from the faeces of patients with diarrhoea is by itself of little significance. As mentioned above, diarrhoea-causing adenoviruses are not readily isolated in culture, and cultivable adenoviruses in the stool are not usually those associated with diarrhoea. Nevertheless, the diarrhoea-associated adenoviruses go through partial replication in G293 cells (an adenovirus-transformed human embryo kidney cell line), and the antigens induced in these cells are type-specific. They may be identified by immunofluorescence using type-specific antibodies.

Serology

A rise in antibody levels indicates recent infection (although not its site or nature), but absence does not exclude it, especially in babies. Complement fixation is the test most frequently used; it provides only a group-specific diagnosis. Group- and type-specific enzyme immuno-assays have also been developed but are not used widely. Neutralization tests are both type-specific and more sensitive, but are not available as routine tests; neither is haemagglutination inhibition widely available. However, both tests may be used by reference laboratories or in research.

TREATMENT

As most adenoviral infections are not life threatening in healthy individuals, there is little demand for specific treatment and no antiviral drugs are available that are unequivocally effective for adenoviral infections. Ribavirin, ganciclovir, vidarabine and cidofovir have all been shown to have antiviral activity in vitro and there are anecdotal reports of their therapeutic use in immunocompromised patients, but with variable success.

EPIDEMIOLOGY

Adenoviruses are endemic. Types 1–7 spread readily between individuals, presumably by droplets and direct or indirect contact with infected secretions. Faecal–oral transmission can also occur and probably does in areas with poverty, poor hygiene and overcrowding. However, it is probable that types 40 and 41, which are widespread causes of diarrhoea even in highly developed countries, also spread via droplets.

Subtyping, which has shown, for example, eight subtypes of adenovirus type 7 (the prototype 7p and seven variants 7a–7g), also indicates geographical variation in the distribution. Such detailed analysis is not done routinely, however, and no information on the subtype distribution of types 40 and 41 is available.

It is probable that only a minority of adenovirus infections are diagnosed virologically.

CONTROL

For the same reasons as discussed above under Treatment, there is little demand for a vaccine. The value of a vaccine is questionable, as circulating antibody may not prevent re-infection, although volunteer studies indicate that re-infection rarely results in disease. Nevertheless, the problems with adenoviruses encountered by US armed forces in recruit camps led them to experiment with a live virus vaccine administered orally in enteric capsules. This provided adequate protection from disease and was licensed for use, but only in military personnel. The plethora of serotypes, their widespread presence and their generally benign outcome makes extension to the general public unlikely to be worthwhile.

A careful and rigorous attention to aseptic technique and single-dose vials of materials for use in the eye is the best approach to preventing outbreaks of adenovirus eye infections. As a component of pharyngoconjunctival fever, conjunctivitis is not preventable.

ADENOVIRUS–ASSOCIATED VIRUSES

The adenovirus-associated viruses (adeno satellite viruses) are members of the Parvoviridae. They are about 22 nm in diameter, appear to be more hexagonal than circular in outline, and contain insufficient single-stranded DNA to replicate on their own. They form a genus, *dependoviruses*, indicating their dependence on adenoviruses (or herpes simplex virus) to provide the missing functions.

True adeno-associated virus has not been implicated in clinical disease. However, as with any other virus found in faeces, large numbers of parvovirus-like particles have been seen in extracts of diarrhoeal faeces, sometimes (but not invariably) combined with smaller numbers of adenoviruses. Neither virus grows in cell culture, leaving the significance of these observations obscure (see Ch. 47).

KEY POINTS

- Adenoviruses, comprising 51 serotypes in six species (A–F), commonly infect man, and can persist in lymphoid tissue.
- They cause (mostly) mild respiratory infections, especially in children and military recruits, but can cause serious multi-organ disease in immunocompromised patients.
- Types 1, 2, 5 and 6 are mostly endemic; types 3, 4 and 7 may be associated with epidemic disease.
- Types 40 and 41 cause diarrhoea, but others are also found in faeces without associated disease.
- Type 8 (and also types 19 and 37) cause epidemic keratoconjunctivitis.
- Similar viruses infect both animals and birds.
- The use of adenoviruses as vectors for gene and cancer therapy, and for vaccines, is being explored but is not yet suitable for clinical use.

RECOMMENDED READING

Amalfitano A, Parks R J 2002 Separating fact from fiction: assessing the potential of modified adenovirus vectors for use in human gene therapy *Current Gene Therapy* 2: 111–133

De Jong J C, Wermenbol A G, Verweij-Uijterwaal M W et al 1999 Adenoviruses from human immunodeficiency virus-infected individuals, including two strains that represent new candidate serotypes Ad50 and Ad51 of species B1 and D, respectively. *Journal of Clinical Microbiology* 37: 3940–3945

Echavarria M 2004 Adenoviruses. In Zuckerman A J, Banatvala J E, Pattison J R, Griffiths P D, Schoub B D (eds) *Principles and Practice of Clinical Virology*, 5th edn, pp. 343–360. Wiley, Chichester

Mizaguchi H, Hayakawa T 2004 Targeted adenovirus vectors. *Human Gene Therapy* 15: 1034–1044

43 Herpesviruses

Herpes simplex; varicella and zoster; infectious mononucleosis; B cell lymphomas; cytomegalovirus disease; roseola infantum; Kaposi's sarcoma; herpes B

M. M. Ogilvie

The herpes viruses infect many animal species and share several features, including their structure, mode of replication and the capacity to establish life-long latent infections from which virus may be reactivated.

Latent infection. This has been defined as 'a type of persistent infection in which the viral genome is present but infectious virus is not produced except during intermittent episodes of reactivation'.

Reactivation. Reactivation from the latent state may be restricted to asymptomatic virus shedding.

Recurrence or recrudescence. These terms are used when reactivated virus produces clinically obvious disease.

At present eight human herpesviruses are recognized, and infection with each of the first seven has been shown to be common in all populations; studies with the eighth suggest it is an uncommon infection in developed countries.

1. Herpes simplex virus 1   HSV-1
2. Herpes simplex virus 2   HSV-2
3. Varicella–zoster virus   VZV
4. Epstein–Barr virus   EBV
5. Cytomegalovirus   CMV
6. Human herpesvirus 6   HHV-6 (HHV-6A and HHV-6B)
7. Human herpesvirus 7   HHV-7
8. Human herpesvirus 8   HHV-8

The herpes B virus of monkeys can be transmitted to man accidentally.

DESCRIPTION

Herpesviruses have a characteristic morphology visible on electron microscopy (Fig. 43.1). The icosahedral protein capsid (average diameter 100 nm) consists of 162 hollow hexagonal and pentagonal capsomeres, with an electron-dense core containing the DNA genome. Outside the capsid in mature particles is an amorphous proteinaceous layer, the *tegument*, surrounded by a lipid *envelope* derived from cell membranes. Projecting from the trilaminar lipid envelope are *spikes* of viral glycoproteins. Cryo-electron micrographs indicate that the capsid is organized into at least three layers, with viral DNA inserted in the innermost layer. The average enveloped particle is approximately 200 nm in diameter.

The genome of herpes virions is linear double-stranded DNA, varying in length from 125 to 248 kbp, with a base content ranging from 42 to 69 G + C mol% for the human herpesviruses. The presence of long and short unique regions bounded by repeated and inverted short segments allows recombination and isomeric forms in some cases (Fig. 43.2). Genes coding for viral glycoproteins, major capsid proteins, enzymes involved in DNA replication and some transcripts associated with latency have been identified. Conserved sequences appear in certain regions, and some genes show homology with regions of human chromosomes. Restriction endonuclease analysis permits epidemiological comparison of strains ('fingerprinting') within some herpes species.

Some herpesviruses are predominantly:

- neurotropic (HSV and VZV)
- lymphotropic (EBV, HHV-6 and HHV-7).

BIOLOGICAL CLASSIFICATION

The family Herpesviridae comprises three broad groups (subfamilies):

1. alphaherpesviruses (e.g. HSV, VZV and B virus); rapid growth, latency in sensory ganglia
2. betaherpesviruses (CMVs); slow growth, restricted host range
3. gammaherpesviruses (e.g. EBV); growth in lymphoblastoid cells.

Of the newer human herpesviruses, HHV-6 and HHV-7 have been classified with CMV as betaherpesviruses, although in a new genus, and HHV-8 is a gammaherpesvirus like EBV, but in a different genus.

The viruses are relatively thermo-labile and readily inactivated by lipid solvents such as alcohols and detergents.

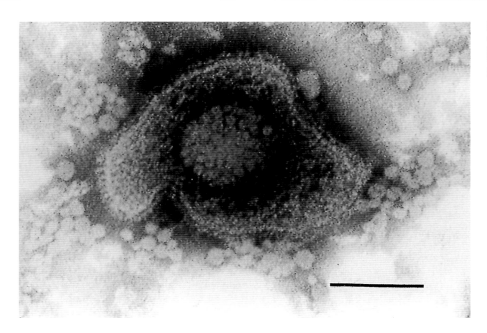

Fig. 43.1 Electron micrograph of HSV. Negative staining (2% phosphotungstic acid). Bar, 100 nm. (Prepared by Dr B. W. McBride.)

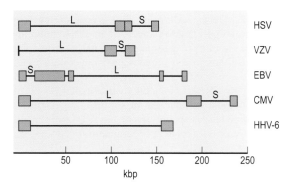

Fig. 43.2 Diagrammatic comparison of herpesvirus DNAs. Lines indicate long (L) and short (S) unique sequences, and repeated regions are boxed.

Viral glycoproteins are processed in the Golgi complex and incorporated into cell membranes, from which the viral envelope is acquired, usually from the inner layer of the nuclear membrane as the virus buds out from the nucleus. It then passes by way of membranous vacuoles to reach the cell surface. Productively infected cells generally do not survive (see first internet site reference for animations).

HERPES SIMPLEX VIRUS

HSV is ubiquitous, infecting the majority of the world's population early in life and persisting in a latent form from which reactivation with shedding of infectious virus occurs, thus maintaining the transmission chain.

DESCRIPTION

In contrast to other members of the group, HSV can be grown relatively easily in cells from a wide variety of animals, so that far more extensive studies have been undertaken with this virus.

There are two distinct types of HSV: type 1 (HSV-1) and type 2 (HSV-2). These two types are generally (but not exclusively) associated with different sites of infection in patients (see below); type 1 strains are associated primarily with the mouth, eye and central nervous system, whereas type 2 strains are found most often in the genital tract.

REPLICATION

After initial attachment to cell surface proteoglycans followed by specific binding to receptors, the envelope of herpes virions fuses with the cell membrane. The nucleocapsids cross the cytoplasm to the nuclear membrane; replication of viral DNA and assembly of capsids takes place within the nucleus. With HSV, it is known that tegument protein transactivates expression of the first set of genes. Between 65 and 100 viral proteins are synthesized, in an orderly sequence or cascade (see Ch. 7).

The viral capsid proteins migrate from the cytoplasm to the nucleus where capsid assembly occurs, and new viral DNA is inserted and located in the inner shell.

HSV glycoproteins

The envelope of HSV contains at least 11 glycoproteins. Three of the glycoproteins are essential for production of infectious virions: gB and gD, which are involved in adsorption to and penetration into cells, and gH, involved in fusion at entry and in the release of virus. Some of the glycoproteins have common antigenic determinants shared by HSV-1 and HSV-2 (gB and gD) whereas others have specific determinants for one type only (gG).

PATHOGENESIS

Primary infection

The typical lesion produced by HSV is the *vesicle*, a ballooning degeneration of intra-epithelial cells. The basal epithelium is usually intact, as vesicles penetrate the subepithelial layer only occasionally. The base of the vesicle contains multinucleate cells (*Tzanck cells*) and infected nuclei contain eosinophilic *inclusion bodies*. The roof of the vesicle breaks down and an ulcer forms. This happens rapidly on mucous membranes and non-keratinizing epithelia; on the skin the ulcer crusts over, forming a scab, and then heals. A mononuclear reaction is normal, with the vesicle fluid becoming cloudy and cellular infiltration in the subepithelial tissue. After resorption or loss of the vesicle fluid, the damaged epithelium is regenerated. Natural killer cells play a significant role in early defence by recognizing and destroying HSV-infected cells. Herpesvirus glycoproteins synthesized during virus growth are inserted in the cell membranes, and some are secreted into extracellular fluid. The host responds to all these foreign antigens, producing cytotoxic T cells (CD4+ and CD8+) and helper T lymphocytes (CD4+), which activate primed B cells to produce specific antibodies and are also involved in the induction of delayed hypersensitivity. The different glycoproteins have significant roles in generating these various cell responses; they probably all induce neutralizing antibody.

During the replication phase at the site of entry in the epithelium, virus particles enter through the sensory nerve endings that penetrate to the parabasal layer of the epithelium and are transported, probably as nucleocapsids, along the axon to the nerve body (neurone) in the sensory (dorsal root) ganglion by retrograde axonal flow. Virus replication in a neurone ends in neuronal cell death; however, in some ganglion cells a *latent infection* is established in which surviving neurones harbour the viral genome. Neurones other than those in sensory ganglia can be the site of herpes latency. It is not clear whether true latency occurs at epithelial sites; there is some evidence of persistence of virus at peripheral sites, but this may be due to a reactivation with a low level of virus replication.

Antibody does reduce the severity of infections, although it does not prevent recurrences. Neonates receiving maternal antibody transplacentally are protected against the worst effects of neonatal herpes. HSV-2 infection seems to protect against HSV-1, but previous HSV-1 infection only partly modifies HSV-2 disease.

Latent infection

Latent infection of sensory neurones is a feature of the neurotropic herpesviruses, HSV and VZV; HSV-1 latency is the best understood. Only a small proportion (about 1%) of cells in the affected ganglion carry the viral genome as free circular episomes – perhaps about 20 copies per infected cell. Very few virus genes are expressed in the latent state; in HSV some viral RNA transcripts (latency-associated transcripts) are found in the nuclei, but no virus-encoded proteins have been demonstrated in the cells, so these infected neurones are not recognized by the immune system. Latent herpes simplex genomes have been detected in post mortem studies on excised ganglia and other neuronal tissues. HSV-1 is regularly detected in:

- the trigeminal ganglion
- other sensory and autonomic ganglia (e.g. vagus)
- adrenal tissue and the brain.

HSV-2 latency in the sacral ganglia has been demonstrated. Either type may become latent in other ganglia.

Reactivation and recrudescence

Reactivation processes are still not understood. It is suggested that herpesvirus DNA passes along the nerve axon back to the nerve ending where infection of epithelial cells may occur. Not all reactivation will result in a visible lesion; there may be asymptomatic shedding of virus only detectable by culture or DNA detection methods.

The factors influencing the development of recrudescent lesions are not yet clearly identified. An increase of CD8+ suppressor cell activity is common at the time of recurrences. Some mediators (e.g. prostaglandins) and a temporary decrease in immune effector cell function, particularly delayed hypersensitivity, may enhance spread of the virus. Certainly the known triggers for recurrences are accompanied by a local increase in prostaglandin levels, and depression of cell-mediated immunity predisposes to herpes recurrence. The mechanism of reactivation is not known, but the event can be predicted accurately in those known to harbour latent virus. Thus, after bone marrow

Table 43.1 Types of herpes simplex virus (HSV) infection

Type of infection	Antibody status of patient
Primary – first HSV infection (any type at any site)	Seronegative
Latent – no symptoms	Seropositive
Recurrent – recrudescence of the latent HSV type(s)	Seropositive
Initial (non–primary) – first episode of the heterologous HSV in a seropositive patient	Seronegative for infecting type; seropositive for latent type
Re-infection (exogenous) – with a strain that differs from the latent HSV type	Seropositive

transplant or high-dose chemotherapy, herpes simplex recurs in 80% at a median interval of 18 days in the absence of prophylaxis.

In whatever way reactivation is achieved, it is a feature of HSV infection. It occurs naturally, and can be induced by a variety of stimuli such as:

- ultraviolet light (sunlight)
- fever
- trauma
- stress.

The interval between the stimulus and the appearance of a clinically obvious lesion is 2–5 days; this has been demonstrated regularly in patients undergoing neurological interference with their trigeminal ganglion, a common site of herpes latency.

CLINICAL FEATURES

Primary infection

Primary herpetic infection is the patient's first experience of HSV and therefore occurs in those with no antibody to the virus (Table 43.1). It usually involves the mucous membranes of the mouth, but may include the lips, skin of the face, nose or any other site, including the eye and genital tract.

Recurrence or recrudescence

Symptomatic recurrence is heralded by a prodrome in two-thirds of people, who experience pain or paraesthesia (tingling, warmth, itch) at the site, followed by erythema and a papule, usually within 24 h. Progression to a vesicle and ulcer, with subsequent crusting, takes 8–12 days, before natural healing. Because of their association with

febrile illness the lesions are popularly known as *cold sores* or *fever blisters*. The most common sites are at the mucocutaneous junction of the lip (seldom inside the mouth), on the chin or inside the nose. However, recurrent lesions can manifest at any site innervated by the affected neurone, determined by the site of initial infection (see Table 43.1).

Severe pain, extensive mucosal ulceration and delayed healing are features of recurrent herpes in severely compromised patients. The ulcers provide entry for other infections. Herpes simplex viraemia after reactivation is uncommon, even in the immunocompromised, but can lead to disseminated infection in internal organs. Some sufferers experience *erythema multiforme* following their recurrent herpes, associated with reaction to certain herpes antigens; this may take the form of the Stevens–Johnson syndrome. Prophylaxis to prevent the herpes recurring has proved useful.

Oral infection

Classically, the first infection presents as an acute, febrile *gingivostomatitis* in pre-school children. Vesicular lesions ulcerate rapidly and are present in the front of the mouth and on the tongue. Gingivitis is usually present. Vesicles may also develop on the lips and skin around the mouth, and cervical lymphadenopathy occurs. The child is miserable for 7–10 days in an untreated case before the lesions heal. However, the majority of primary infections go unrecognized, the episode being attributed to teething or mistaken for 'thrush' (candida infection). Herpetic stomatitis also occurs in older children and adults acquiring a primary infection. There may be an associated mononucleosis in the older patient; pharyngitis is also notable. Viraemia with dissemination of herpes to internal organs is rare except in pregnancy (primary infection, with hepatitis) or the neonate (see below) or the immunocompromised patient.

Skin infection

Herpetic whitlow

Hand infections with HSV are not uncommon. Three presentations may be seen:

1. The classical primary lesion on the fingers or thumb of the toddler with herpetic stomatitis, due to auto-inoculation.
2. Another classical and often primary infection is acquired by accidental inoculation in health-care workers. These infections may recur; the majority are HSV-1.
3. The commonest hand lesions are recurrent, associated with HSV-2 and genital herpes, and are seen

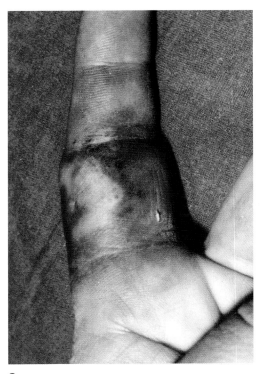
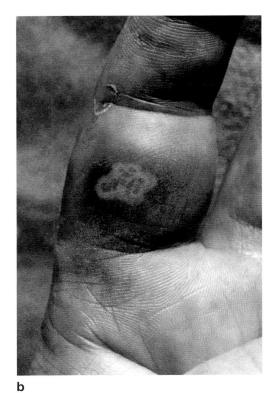

a   b

Fig. 43.3 Herpetic whitlow on left index finger, photographed during recurrences in **a** 1984 and **b** 1994.

in young adults. Pain and swelling occur, and the vesicles become pustular, but, if in well keratinized areas, do not always ulcerate. Associated lymphangitis is common. Primary lesions take up to 21 days to heal, recurrent ones 10 days. Figure 43.3 shows recurrent lesions at an interval of 10 years.

Eczema herpeticum

A severe form of cutaneous herpes may occur in children with atopic eczema – *eczema herpeticum* or *Kaposi's varicelliform eruption*. Vesicles resembling those of chickenpox may appear, mainly on already eczematous areas. Extensive ulceration results in protein loss and dehydration, and viraemia can lead to disseminated disease with severe, even fatal, consequences. A similar picture is seen occasionally in adults with pemphigus who develop herpes simplex. Patients with burns are also at risk. In each instance early recognition and antiviral therapy can be life-saving.

Eye infection

HSV infection of the eye may be initiated during a childhood primary infection or occur from transfer of virus from a cold sore. There may be conjunctivitis, or keratoconjunctivitis associated with corneal ulceration. Typically, branching (dendritic) corneal ulcers are found. If these recur untreated, the result is corneal scarring and impairment of vision. More extensive ulceration occurs if steroids have been used, but deeper infiltrates are common in long-standing cases and benefit from combined steroid and antiviral therapy. The presence of typical herpes vesicles on eyelid margins is a useful clinical guide but is not always seen. The majority of eye infections are with HSV-1, and most patients with recurring eye disease are aged over 50 years. More than half of the corneal grafting performed in the UK is for HSV corneal scarring, although the disease may recur in the graft. Acute retinal necrosis associated with HSV-1 or 2 has also been recognized.

Central nervous system infection

HSV may reach the brain in several ways. Viraemia has been detected during primary herpetic stomatitis, and infection may be carried within cells into the brain and meninges. Direct infection from the nasal mucosa along the olfactory tract is another possibility, but the most likely route is central spread from trigeminal ganglia.

HSV encephalitis

This is a rare condition, but is the commonest sporadic fatal encephalitis recognized in developed countries. The infection has a high mortality rate and significant morbidity in survivors of the acute necrotizing form. It presents:

- at any time of year
- at any age
- occasionally in the young, but more frequently in those aged 50–70 years.
- Some 70% of cases are in people with serological evidence and often a clinical history of previous herpes simplex.

Recurrent lesions are seldom apparent at the same time as encephalitis. A prodrome of fever and malaise is followed by headache and behavioural change, sometimes associated with a sudden focal episode such as a seizure, or paralysis; coma usually precedes death. The temporal lobe is most frequently affected, and virus replication in neurones, followed by the oedema associated with the inflammatory response, accounts for the haemorrhagic necrosis and space-occupying nature of this form of the disease. More diffuse, milder disease has been recorded. Brainstem encephalitis is another serious manifestation.

Clinical recognition leading to early specific antiviral therapy significantly reduces the 70% mortality rate and serious morbidity of untreated cases, but therapy must be started as soon as possible, before the patient progresses from a drowsy state into coma. Diagnostic confirmation used to be attempted by examination of brain biopsy tissue for virus antigens, virus particles and the characteristic histopathology. Brain biopsy is now seldom performed for suspected herpes encephalitis in the UK, but should tissue be available in the course of intracranial pressure monitoring it should be examined. Cerebrospinal fluid (CSF) collected in the acute stages of the disease should be sent to a laboratory offering HSV DNA amplification by the polymerase chain reaction (PCR). This is now established in several regional centres and, unlike culture or antigen detection, can demonstrate the presence of HSV DNA in most cases within the first 10 days after onset. The diagnosis can be confirmed later, after therapy, by the demonstration of intrathecal synthesis of HSV antibody. For this purpose CSF and peripheral blood taken on the same day are required.

HSV-1 is responsible for most cases of encephalitis where virus has been identified (outside the neonatal period; see below). HSV-2 encephalitis does occasionally occur in immunocompromised adults and may be seen more frequently in those infected with the human immunodeficiency virus (HIV).

HSV meningitis

Meningitis caused by HSV is much less serious. HSV-2 is almost always the cause, and reaches the CSF following radiculitis during genital herpes (see below). A lymphocytic reaction is seen in CSF, and there may be recurrent bouts of this meningitis (sometimes known as *Mollaret's meningitis*). HIV-infected patients may present with HSV meningitis.

Genital tract infection

Both types of HSV can infect the genital tract. Although the more common association has been with type 2 strains, type 1 infection is not infrequent, particularly in young women, where it may account for more than half of genital infections. Genital infection may be acquired by auto-inoculation from lesions elsewhere on the body, but most often results from intimate sexual contact, including orogenital contact. The lesions are vesicular at first but rapidly ulcerate.

- In the male, the glans and shaft of the penis are the most frequent sites of infection.
- In the female, the labia and vagina or cervix may be involved.
- In both sexes, lesions may spread to surrounding skin sites.

The incubation period is 2–20 days, with an average of 7 days. A primary infection is usually the most severe, especially in women. Fever and malaise are accompanied by regional lymphadenopathy and urethritis, and vaginal discharge may be present. The whole episode lasts for 3–4 weeks, and high titres of virus are shed. In some cases a lymphocytic meningitis develops, and urinary retention can also be a problem; these are manifestations of sacral radiculopathy. Where the infection is an initial (non-primary) genital herpes, the attack is generally less severe, but lasts for about 2 weeks.

Recurrent genital herpes

This can be as frequent as six or more episodes a year. Although the attacks are milder and shorter (around 7–10 days) than first episodes, the results are socially and psychologically distressing. Some patients experience prodromal symptoms in the distribution of the sacral nerves, but the patient is already infectious by this stage. Virus shedding from the genital tract is often asymptomatic. The patient or general practitioner may not recognize recurrent lesions as herpetic in origin, so that in many instances the risk of transmission to sexual partners is not apparent. HSV-1 genital infection recurs less often than HSV-2, and thus carries a better prognosis.

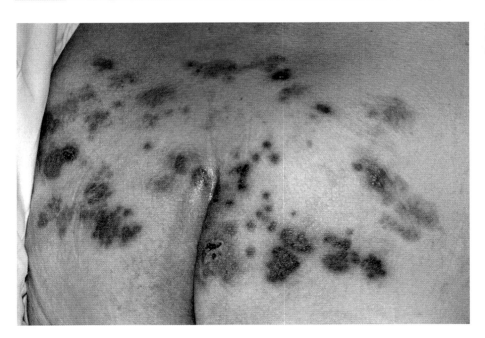

Fig. 43.4 HSV-2 recurrence over the sacral region in an elderly patient.

Either type is capable of transmission from mother to infant. Transplacental passage resulting in intra-uterine damage to the fetus has been recorded but is very rare, and probably limited to cases with substantial maternal viraemia. Ascending infection from the cervix may be more significant, especially when the membranes are ruptured for some time before delivery.

Genital herpes can be a significant problem in immuno-suppressed patients, and is seen as persistent, severe, peri-anal lesions with or without proctitis in many HIV-infected male homosexuals. Genital herpetic ulcers are known to increase the risk of transmission of infection with HIV. Figure 43.4 shows recurrent HSV-2 over the sacrum in an elderly person; the latent virus in the sacral ganglion travels to the skin dermatome served as well as the internal mucosal site; the patient had been nursed on a rubber ring.

Neonatal herpes

A rare but very serious infection, untreated neonatal herpes has a case fatality rate exceeding 60%, with half of the survivors severely damaged. Virus, commonly HSV-2, is acquired by passage through an infected genital tract. The greatest risk, with a 50% transmission rate (because there is more virus present and no antibody transfer), occurs when the mother has a primary infection at the time of delivery. With recurrent herpes at term, the transmission rate is only 8% or less. Neonatal infection usually presents about the sixth day post-partum.

Skin lesions may be absent, or few in number, and are commonly located on the presenting part (e.g. the scalp). Virus dissemination is the most serious complication, with signs of general sepsis, including fever, poor feeding and irritability. Pneumonia and jaundice develop, with or without signs of meningitis or encephalitis. Progressive liver failure with coagulopathy leads to death around the 16th day in the most serious disseminated form. Occasionally, infection is restricted to skin or mucosae, including the conjunctivae, or involves the central nervous system with or without dissemination to internal organs. Early high-dose antiviral therapy is the key to survival with minimal morbidity, although local recurrent lesions can be expected, especially at skin sites, in the first year.

The prevention of neonatal herpes is difficult, the vast majority of infections occurring in babies born to women with no past history of genital herpes and in whom the infection at term was either asymptomatic or unrecognized clinically. Routine pre-term screening for virus shedding, particularly as applied to those with a past history, does not predict the babies at risk. With the availability of type-specific antibody tests, susceptible women at risk of acquiring HSV-2 from partners can be identified. If suspicious lesions are seen during labour, swabs for virus culture and/or DNA amplification by PCR are taken. Caesarean section may reduce the risk of infection if performed in the early stages of labour. If the mother is known to have moderate levels of antibody, the baby is unlikely to develop the disease.

Some cases of neonatal herpes are acquired just after birth from contact with sources of HSV other than the mother's genital tract. The clinical presentation is similar, although the virus is likely to be type 1 from oral or skin lesions of attendants or relatives. In the case of suspected neonatal herpes without skin lesions, nasopharyngeal secretions, CSF and blood lymphocytes should be cultured.

LABORATORY DIAGNOSIS

Virus detection and serological studies both have their place in the diagnosis of infection with HSV. In all instances of acute infection, be it primary or recurrent, virus detection is the method of choice, as antibody responses are naturally later and much less informative. Indeed, in recurrent episodes the antibody titre may not vary. Sensitive assays for immunoglobulin (Ig) G antibody, including type-specific antibody, have an important place in prospective testing.

Herpes virus detection

Direct diagnosis of HSV infection is available, and should be sought in cases where there is any doubt as to the clinical diagnosis, or where rapid confirmation is required to support the choice of therapy or other management. Virus isolation, or PCR for DNA, is suitable for most common herpes infections, direct diagnosis being reserved for the atypical or serious situations.

Isolation of HSV is attempted in cultures of human diploid fibroblast cells. Growth is rapid, and within 24 h a *cytopathic effect* may be visible, presenting as rounded, ballooned cells in foci that later expand and eventually involve the whole cell sheet. Virus is released from infected cells into the culture fluid – hence the rapid spread of infection.

Herpes virions may be demonstrated by electron microscopy of vesicle fluid or tissue preparations. Detection of viral antigens in cells (immunofluorescence) provides a rapid diagnosis on cells scraped from the base of lesions. In the most serious infections easy access to the site of infection may not be possible. Detection of viral DNA by PCR in CSF or other samples is the most sensitive and specific method of diagnosis. The two HSV types can be differentiated by using either type-specific primers in the PCR, or common primers followed by analysis with restriction enzymes or hybridization probes for each type.

Antibody tests for HSV

Complement fixation tests are useful in the diagnosis of primary infections, when a significant change in antibody titre can be expected. The titre of antibody may not be high, especially in genital infection, and the test measures total antibody with no differentiation of type-specific antibodies. In the case of herpes encephalitis, a serological diagnosis depends on the demonstration of intrathecal synthesis of antibody to HSV. Complement fixation tests can be used, testing serum and CSF in parallel against HSV antigens and another unrelated antigen. In health there is no antibody detectable by such tests in the CSF. If the serum antibody:CSF antibody ratio is diminished from the normal 200:1 to 40:1 (or less), and the blood–CSF barrier integrity is confirmed by other antibody (or albumin) being excluded from the CSF, intrathecal antibody synthesis is demonstrated. It is also helpful to check for the possible transfer of antibody from blood to CSF by means of an IgG index performed on the same serum and CSF samples. Tests on serum alone cannot confirm herpes encephalitis.

Enzyme immuno-assays are much more sensitive and specific than complement fixation tests. When type-specific antigen preparations based on glycoprotein G are used, type-specific antibody can be detected. Therapy often results in delayed appearance of these type-specific antibodies, and late convalescent samples (at 6 weeks) should be included.

TREATMENT

Specific antiviral therapy has revolutionized the management of HSV infections over the past 25 years. Before the development of agents suitable for systemic use, topical application of the relatively non-selective idoxuridine was used successfully in the treatment of eye and skin infections. Aciclovir has a better therapeutic ratio and proven efficacy, when used early enough in appropriate dosage, for the whole range of acute HSV infections. Latency is not eradicated by this agent, which inhibits viral DNA synthesis. Prophylactic use of aciclovir is now used to prevent reactivation in the immunocompromised (e.g. transplant recipients). Long-term suppressive therapy with aciclovir has been particularly successful in the management of frequently recurring genital herpes and HSV-related erythema multiforme. Patient-initiated early treatment can also abort or modify recurrences.

Aciclovir is the most widely used antiherpes treatment, with an excellent safety record; it is available in preparations for topical, oral and intravenous use. Its use in pregnancy is no longer monitored as there is no adverse effect in the infants of treated women. Topical cream or ointment is suitable only for mild epithelial lesions, such as recurrent cold sores, genital herpes or corneal ulcers. Oral or intravenous therapy with aciclovir should be given for:

- any deeper lesions
- disease in immunocompromised hosts

- central nervous system and other systemic infection (intravenous therapy is required)
- any of the serious manifestations.

Dosage varies considerably, depending on the site of infection and whether the aim is suppression of recurrence or therapy of established disease. Because the level of aciclovir achieved in the CSF is only half that in plasma, the dosage for the treatment of encephalitis has to be twice that for other systemic disease. Therapy must be maintained until clinical signs indicate a favourable response. In serious systemic disease, or in the severely immunocompromised, therapy is continued for 2 weeks or longer. In the neonate 3 weeks' therapy at a higher (intravenous) dose should be followed by suppressive oral treatment for 6–12 months.

The poor bio-availability of oral aciclovir led to the development of a pro-drug, valaciclovir, which is rapidly converted into aciclovir, producing significantly higher plasma levels after oral dosage. Another effective antiherpes agent, penciclovir, is given in the form of an oral pro-drug, famciclovir. Both of these agents are licensed for treatment of genital herpes, and may be administered less frequently than aciclovir.

Resistance to aciclovir can develop. The commonest forms are HSV strains with deficient or altered thymidine kinase; hence they cannot phosphorylate aciclovir. This has not been a significant clinical problem but it must be remembered, particularly in severely immunocompromised hosts with acquired immune deficiency syndrome (AIDS) or following bone marrow transplant. A few resistant viruses with altered DNA polymerase have also been isolated associated with clinical disease. With widespread use of aciclovir, more resistant strains may arise, and monitoring of the antiviral sensitivity of herpes isolates might be necessary. Strains resistant to aciclovir are generally also resistant to famciclovir, and foscarnet may be used as it does not require activation by viral thymidine kinase.

EPIDEMIOLOGY

HSV is probably transferred by direct contact. Many children, especially in overcrowded conditions, acquire oral HSV-1 infections in the first years of life. Spread may not occur so readily in better social conditions, with the result that primary infection is delayed to young adulthood. This is the usual time of exposure to genital herpes and, as a result, primary infections may be HSV-2 or HSV-1.

Sensitive immuno-assays and type-specific assays have shown that:

- some 60–90% of adults have had an HSV-1 infection.
- many more adults have had HSV-2 infection than give a history of genital herpes.

Neonatal herpes is a rare complication in the UK; the British Paediatric Surveillance Unit reported 37 confirmed neonatal herpes infections over a period of 42 months. Eleven of the infants died within their first month, and several of the survivors suffered adverse sequelae. The rate of cases has increased in some populations as genital herpes has become more common. Herpes encephalitis outside of the neonatal period occurs sporadically at a rate of about 1 per 500 000 of the population per annum.

CONTROL

Transmission of herpes simplex can be reduced by:

- alleviating overcrowding
- practising simple hygiene
- education regarding the infectious stages.
- Sexual transmission may be significantly reduced by the use of condoms.

Reference has already been made to the prevention of neonatal infection, as has the use of prophylactic antiviral regimens to control predictable recurrence. Progress in understanding latency and reactivation will provide approaches to preventing reactivation. Protection from ultraviolet light and the use of inhibitors of prostaglandin synthesis may be useful in this context.

Experimental vaccines are under investigation, but none is licensed for use in the UK at present. Research into subunit vaccines based on the viral glycoproteins or other significant viral proteins may lead to an appropriate preparation to elicit the immune responses important in control of herpes simplex. Recent trials have shown limited protection against genital tract disease. A vaccine that stimulates cytotoxic T cell responses may be required.

VARICELLA–ZOSTER VIRUS

Infection with VZV presents in two forms:

1. The primary infection *varicella* (or chickenpox) is a generalized eruption.
2. The reactivated infection *zoster* (or shingles) is localized to one or a few dermatomes.

DESCRIPTION

The viruses isolated from varicella and zoster are identical. The virus has the morphology of all herpesviruses. Seventy genes code for 67 different proteins, including five families of glycoprotein genes. The glycoproteins gE, gB and gH are abundant in infected cells, and are present in the viral envelope (Fig. 43.5).

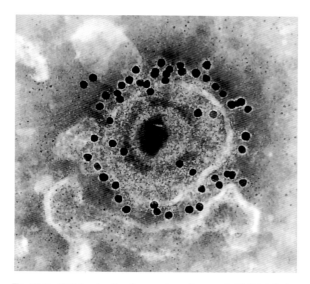

Fig. 43.5 VZV showing the virus envelope glycoprotein I (gE) labelled with monoclonal antibody and goat anti-mouse IgG conjugated with 15 nm colloidal gold (original magnification ×150 000. (Micrograph taken by C. Graham, supplied by Dr E. Dermott, Department of Microbiology and Immunobiology, Queen's University, Belfast.)

Virus replication takes place in the nucleus, and histological examination of infected epidermis reveals typical nuclear inclusions and multinucleate giant cells identical to those of herpes simplex. Human fibroblast cell cultures are most often used for isolation. The enveloped virions released from the nucleus remain closely attached to microvilli along the cell surface; studies of this 'cell associated' characteristic, with infection being passed from cell to cell, have been limited compared with studies of the lytic HSV. The typical cytopathic effect appears in cell cultures in 3 days to 2–3 weeks.

Until recently, only one antigenic type of VZV was known, although a limited identification of strains from different geographical sources is possible, and the vaccine virus (Oka) can be distinguished from wild-type VZV. A few virus isolates with mutations in glycoprotein E have been found. Antibodies to the three main glycoproteins all neutralize virus infectivity. One of these glycoproteins, gB, shares 49% amino acid identity with the gB of HSV, and this may account for the cross-reactive, anamnestic, antibody response that may be detected during infections with either virus.

PATHOGENESIS

Varicella

This is a disease predominantly of children, characterized by a vesicular skin eruption. Virus enters through the upper respiratory tract, or conjunctivae, and may multiply in local lymph tissue for a few days before entering the blood and being distributed throughout the body. After replication in reticulo-endothelial sites, a second viraemic stage precedes the appearance of the skin and mucosal lesions. An alternative model has been proposed, based on studies in a SCIDhu (severe combined immunodeficient, bearing human skin xenografts) mouse model. The work showed that activated human tonsillar T cells could be infected and, if given to the mice intravenously, homed rapidly to the skin grafts, where VZV could be detected by 7 days, although the typical epidermal vesicular lesion did not appear for 10–21 days.

In man, the VZV vesicles lie in the middle of the epidermis, and the fluid contains numerous free virus particles. Within 3 days the fluid becomes cloudy with the influx of leucocytes; fibrin and interferon are also present. The pustules then dry up, scabs form, and they desquamate. Lesions in all stages are present at any time while new ones are appearing. The clearance of virus-infected cells is dependent on functional cell-mediated immune mechanisms, cytotoxic T cells and antibody-dependent cell cytotoxicity in particular. Persons deficient in these responses and in interferon production have prolonged clear-vesicle phases and great difficulty in controlling the infection.

Zoster

The pathogenesis of zoster is not so well established as that of HSV recurrence. Nucleic acid probes have revealed VZV sequences in sensory ganglia, and explant cultures of ganglia produce some VZV proteins, although not fully infectious virus. The latent virus is found in neurones and satellite cells in sensory ganglia, and more than one region of the genome is transcribed, although the state of the latent VZV is not known. It seems likely that virus reaches the ganglion from the periphery by travelling up nerve axons, as HSV does, but there is also the possibility that during viraemia some virus enters ganglion cells. Another difference from HSV latency lies in the persistent VZV expression that has been detected in some mononuclear cells; this may have a role in VZV disease such as post-herpetic neuralgia (see below).

Reactivation of VZV as zoster can occur at any age in a person who has had a primary infection, which may or may not have been apparent clinically. The rate is greatest in persons aged 60 years or more and, as most primary infection takes place before the age of 20 years, there is usually a latent period of several decades. However, a much shorter latent period is seen in immunocompromised patients, and also in those who acquired primary infection in utero (see below). More than one episode of zoster is uncommon in any individual. The stimulus to reactivation

Table 43.2 Key differences between rashes of varicella and smallpox

Feature	Varicella	Smallpox
Characteristic lesions	Blisters – domed small superficial single vesicles with clear fluid	Several layers of cells cover multifocal vesicles; early umbilication seen
Distribution	Centripetal – most on trunk, neck, face, proximal parts of limbs	Centrifugal – face, forearms and palms, soles of feet
Appearance at any one time	Pleomorphic – papules, vesicles and crusts may all be seen in one area	Lesions all at same stage in one affected area

is not known, but virus travels from sensory ganglia to the peripheral site. The zoster is usually limited to one dermatome; in adults, this is most commonly in the thoracic or upper lumbar regions, or in the area supplied by the ophthalmic division of the trigeminal nerve. It is thought that this distribution is related to the density of the original varicella rash. There are associations with preceding trauma to the dermatome – injury or injections – with an interval of 2–3 weeks before the zoster appears. There is an associated suppression of specific cell-mediated responses in acute zoster, but rapid secondary antibody responses are usually found. Reactivation occurs more commonly in T cell immunodeficiency states.

Viraemia may occur in the course of zoster but is unusual, with pre-existing immunity normally being boosted rapidly. In the immunocompromised, however, viraemia leads to dissemination of zoster, either to internal organs or in a generalized manner similar to varicella.

CLINICAL FEATURES

Varicella

The incubation period averages 14–15 days but may range from 10 to 20 days. The patient is infectious for 2 days before and for up to 5 days after onset, while new vesicles are appearing.

The rash of varicella is usually centripetal, being most dense on the trunk and head. Initially macular, the rash rapidly evolves through papules to the characteristic clear vesicles ('dew drops'). A series of clinical pictures may be viewed through the second internet site reference (see below).

Presentations vary widely – from the clinically inapparent to only a few scattered lesions, or to a severe febrile illness with a widespread rash, especially in secondary cases in older members of a household. Normally a relatively mild infection in the young child, the complications of varicella are serious and cause significant morbidity, requiring hospital admission of normal children and adults.

In the past, and in preparation for any potential bioterrorist or accidental release of smallpox (see Ch. 44),

it was and is important to know the characteristic appearances of varicella as compared to a smallpox rash (Table 43.2).

Secondary bacterial infection of skin lesions is the commonest complication, mainly in the young child, and increases the amount of residual scarring. Thrombocytopenic purpura occurs, especially in immunocompromised hosts. A variety of organs may be affected, producing myocarditis, arthritis, glomerulonephritis and appendicitis. The two most frequent problems are related to the lungs and the central nervous system.

Pneumonia

Viral pneumonitis is a most serious complication, even in immunologically normal people. It may be subclinical, evident only on radiography, but has been recognized for the danger it brings, especially to smokers. Cough, dyspnoea, tachypnoea and chest pains begin a few days after the rash. Nodular infiltrates are seen in the lungs radiologically. Specific antiviral therapy must be used in sufficient dosage, and started at the first sign of pneumonia in an adult with chickenpox. Early investigations (radiography and gas exchange) are indicated in all smokers with chickenpox. Immunocompromised patients are at even greater risk of varicella pneumonia.

Central nervous system

Neurological complications include the common but benign cerebellar ataxia syndrome. Acute encephalitis is rare but more serious, and occurs mostly in immunocompromised patients. It may be confused with post-infectious encephalopathy, which, with other post-infectious manifestations such as transverse myelitis or Guillain–Barré syndrome, is immunologically mediated and not related to viral cytopathogenicity.

Varicella in pregnancy

Varicella virus can cross the placenta following maternal viraemia and infect the fetus. The infection may be more serious for the mother herself in pregnancy, with pneumonia the major problem. Two types of intra-uterine infection are noted:

1. *The fetal varicella syndrome* is a rare consequence of fetal infection in the first half of pregnancy. The baby has a characteristic scarring of the skin, hypoplasia of the limbs, and chorioretinitis. The maternal infection is usually varicella, but in one case at least it was disseminated herpes zoster. Fetal infection is not inevitable. Silent intra-uterine infection can also occur: no damage is seen, but the baby is born with latent VZV infection, having recovered from infection.

2. *Neonatal (congenital) varicella* occurs when varicella develops within the first 2 weeks of life after maternal varicella in late pregnancy. The interval between maternal viraemia and delivery is important. If the maternal rash begins 7 days or more before delivery, her antibody response will have developed and been transferred across the placenta so that the baby does not develop disease. However, the infant is at serious risk if varicella occurs 6 days or less before delivery (or up to 2 days after); this allows viraemic spread across the placenta, before antibody is made and transferred to the baby. Because the usual respiratory entry route has been bypassed, the incubation period is reduced to 10 days, on average. If infection arises, disseminated disease with pneumonitis and encephalitis may be found. The rate of transmission is 1 in 3, but there is no way of knowing which mother will transmit, and in this situation protection of the neonate with passive immunity (see below) and early antiviral therapy is indicated.

Zoster

This is the manifestation of reactivated VZV infection. It takes the form of a localized eruption, is unilateral and typically confined to one dermatome. Prodromal paraesthesia and pain in the area supplied by the affected sensory nerve are common before the skin lesions develop; these are identical to those of varicella, except in their distribution. The evolution of the rash is similar, with some new vesicles appearing while the earliest ones are crusting; however, the whole episode in the majority of cases is confined to the affected dermatome and heals in 1–3 weeks. Acute pain is not always a feature, but its presence should alert to the possibility of zoster, and a search for early lesions, perhaps internal, is indicated. Occasionally there are no skin lesions – '*zoster sine herpete*'.

Dissemination of zoster is indicated by lesions appearing in the skin at distant sites or, more seriously, by involvement of internal organs such as the lung and brain. In the immunocompromised, this results in severe disease, with occasional fatalities.

Post-herpetic neuralgia

This is the most common complication of zoster, a risk in 50% of patients aged over 60 years, and results in significant morbidity in around 20%. It is defined as intractable pain persisting for 1 month or more after the skin rash. Constant pain at the site, or stabbing pains or paraesthesiae, may continue for 1 year, or much longer in a number of individuals. This is an exhausting and disabling condition for which no satisfactory cure has been found; adequate early antiviral therapy may reduce the incidence.

Ophthalmic zoster

Involvement of the ophthalmic division of the trigeminal nerve occurs in up to one-quarter of zoster episodes, with ocular complications in more than half of the patients. Corneal ulceration, stromal keratitis and anterior uveitis may result in permanent scarring, so this complication may threaten sight when the nasociliary branch is involved. Ocular complications are reduced in patients given oral aciclovir early in ophthalmic zoster. Occasionally acute retinal necrosis is identified.

A contralateral hemiparesis due to granulomatous cerebral angiitis in the weeks following acute ophthalmic zoster is a recognized neurological complication. A more acute one, such as the *Ramsay–Hunt syndrome* (facial palsy with aural zoster vesicles), suggests that motor neurones can also be involved. Sympathetic ganglia may also be the site of latency, as indicated by cases in which the initial recrudescence has been in gastric mucosa, with subsequent dissemination.

Recurrent and chronic VZV

Immunodeficient patients, most particularly those with CD4$^+$ lymphopenia due to HIV infection, may develop recurrent and chronic infection. New lesions continue to appear, or re-appear after aciclovir therapy, often presenting an atypical hyperkeratotic appearance. Aciclovir-resistant VZV has been isolated in this situation.

LABORATORY DIAGNOSIS

Typical presentations of varicella or zoster seldom need laboratory confirmation; however, atypical presentations merit investigation, especially in the immunocompromised. Vesicular rashes due to enterovirus are sometimes confused with varicella and, in compromised patients, various vesicular lesions may be mistaken for zoster. Localized vesicular lesions other than those on the face or genitalia are commonly misdiagnosed as due to zoster (see Fig. 43.4);

many are due to recurrence of herpes simplex, and this is readily shown by antigen detection and virus isolation, or by PCR.

Virus detection

Early vesicular lesions provide the best diagnostic material. Vesicle fluid is collected in a capillary tube or aspirated with a fine needle and syringe. Direct examination by electron microscopy will reveal herpes particles; some of the fluid can be diluted in virus transport medium and inoculated into tissue culture for virus isolation, which takes between 5 days and 3 weeks. More rapid detection is possible with centrifugation-enhanced cultures ('shell vials'). If cells swabbed from the base of lesions are available, or biopsy tissue, virus antigens may be sought by immunofluorescence tests. VZV DNA amplification by PCR is used for the detection of VZV in CSF or aqueous humour, but more routinely now for all samples. The latter two methods (immunofluorescence testing and PCR) are the ones most used now for rapid diagnosis.

Serological diagnosis

Antibody testing with VZV antigens can confirm a diagnosis of varicella by demonstration of seroconversion or rising titres of antibody between acute and convalescent serum samples; complement fixation is still a useful test for this purpose. However, it is not sufficiently sensitive to determine past infection and, to assess immune status, assays need to be based on enzyme or radiolabelled methods, or immunofluorescent staining of infected cells. IgM to VZV is detectable by IgM capture systems in both varicella and zoster, appearing early in zoster. It is increasingly common to test for past infection (by measuring IgG antibody to VZV) in those who are or will become immunocompromised, or in women exposed antenatally to VZV, or health-care workers or other adults who are to be offered immunization.

TREATMENT

Aciclovir given intravenously is effective in the treatment of varicella and zoster in immunocompromised patients. Oral aciclovir can be used to accelerate healing and reduce new lesion formation in zoster in healthy patients if given early enough and may lower the rate of postherpetic neuralgia. VZV is not as sensitive to aciclovir as HSV, with 50% inhibitory dose (ID_{50}) values ranging from 4 to 17 μM aciclovir compared with 0.1–1.6 μM for HSV. For the intravenous preparation this means 10 mg/kg given every 8 h. Oral therapy has to be with a high-dose regimen of 800 mg five times daily. Newer preparations

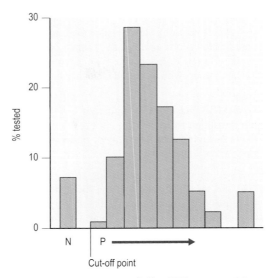

Fig. 43.6 Distribution of antibody (IgG) to VZV in a young adult population (southern England). The proportion confirmed as negative (N) was 7.8%; P →, increasingly positive result.

such as valaciclovir and famciclovir require less frequent dosing. Trials have shown that high-dose oral aciclovir shortens the course of varicella in healthy children and adults by 1 day if commenced within 24 h of the onset of the rash. The practical difficulties of achieving this rule out routine treatment for all cases of varicella in healthy children, but consideration should be given to treating all adults, and where possible adolescents and family contact cases, who are known to develop more extensive disease. Treatment of VZV infection is given primarily to all 'high risk of complication' groups:

- neonates (within the first 3 weeks of life)
- immunocompromised patients
- those with ophthalmic zoster
- healthy patients with varicella when there is an additional complicating factor such as smoking or pneumonia.

EPIDEMIOLOGY

Varicella is partly seasonal, being spread mainly by the respiratory route in winter and early spring. Some cases result from contact with zoster and occur sporadically at any season. Varicella is highly infectious to susceptible close contacts, as in a household; a past history of varicella is a good indicator of immunity. The majority of children contract varicella between the ages of 4 and 10 years in western countries, with around 8% of young adults remaining susceptible (Fig. 43.6). However, a much higher proportion of young adults remain susceptible in subtropical countries. The mortality rate from varicella is surprisingly high in otherwise healthy adults, particularly smokers, who develop pneumonia. Zoster is associated with decreased T cell function, and occurs with increased incidence in:

- old age
- the pre-AIDS phase of HIV infection
- organ transplant recipients
- patients receiving chemotherapy or radiotherapy for lymphoma or leukaemia.

In developed countries the increasing survival of people over 65 years of age means that the incidence of zoster will increase. This may be exacerbated by less natural boosting of immune responses as varicella decreases in countries offering childhood immunization for varicella.

CONTROL

Passive immunization

Passive immunity is partly protective for varicella, as seen in infants with maternal antibody or patients given *varicella–zoster immunoglobulin* (VZIG) within 72 h of exposure. VZIG (in the UK) or another similar high-titre antibody preparation is available for neonates, non-immune pregnant contacts or immunocompromised contacts. As most pregnant contacts will be immune, testing for antibody after exposure may prevent unnecessary use of VZIG. Current preparations seldom prevent infection, but do modify disease. VZIG does reduce the rate of transmission to the fetus.

Varicella vaccine

A live-attenuated varicella vaccine has been in use for some years in Japan and some European countries, and was approved in the USA for routine childhood immunization in 1995. This vaccine, given by intramuscular injection, has been found to be immunogenic in children with leukaemia in remission and in healthy children and adults; vaccinees have resisted infection on close exposure to varicella. Some symptoms are noted around 10 days post-vaccination, and vesicles appear at the site of injection in up to 5%. Immunization does not prevent latency developing; however, the incidence of zoster in vaccinees compared with that in the naturally infected is not increased. In the USA, surveillance has confirmed a significant reduction in the incidence of varicella, especially in the 1–4 years age group, and a substantial fall in the mortality and complication rates. There have been reports of 'breakthrough' epidemics locally, even in well vaccinated cohorts, and a booster dose may be

necessary. A large trial of vaccination in the elderly (those aged over 60 years) showed a 60% reduction in the incidence of zoster and post-herpetic neuralgia in those immunized with a high-titre preparation of Oka virus-based vaccine.

EPSTEIN–BARR VIRUS

In 1964, Epstein, Barr and Achong described herpesvirus particles in cells from a lymphoma in African children studied by Burkitt, who suspected a viral aetiology of the tumour. The link between that new herpesvirus – the *Epstein–Barr virus* – and a variety of lymphoproliferative diseases is now clear. EBV primary infection is:

- most often asymptomatic and occurs early in childhood
- the classical infectious mononucleosis (*glandular fever*) of adolescents in the developed world.

Man is the only natural host, but EBV infection can be transmitted to some subhuman primates.

DESCRIPTION

The characteristic morphology seen on electron microscopy placed EBV with the herpesviruses. This virus cannot be grown in human fibroblast or epithelial cell lines and there is no completely productive or permissive system for culture of EBV. This lymphotropic virus is classified as a gammaherpesvirus, genus lymphocryptovirus.

REPLICATION

The full replication (productive or lytic) cycle of EBV can take place in certain differentiated epithelial cells, although it is not known whether this is part of the natural history of EBV infection. The B lymphocyte is the principal and essential cell infected, through EBV receptors (the CD21 molecule) expressed on mature resting B lymphocytes. There are other receptors on cells of stratified squamous epithelium – in the oropharynx, salivary glands and ectocervix, for instance – and these epithelial cells can be infected through their basal aspect. Viral production is restricted to the differentiated cells of the granular layer and above, and virus is shed from the superficial cells. It is difficult to grow differentiating epithelia in culture, and most work on the lytic cycle of EBV has been done in lymphoblastoid cell lines, in which a proportion of the EBV-infected cells can be induced to produce virions.

The organization of the EBV genome differs from that of HSV, and some genes are present in one and not in the other. Approximately 80 proteins are encoded; some glycoproteins are known, including the major glycoprotein gp350/220, which mediates attachment to CD21, and gp85, which is involved in membrane fusion. The latent (non-productive) state of EBV infection is established in a subset of resting memory B lymphocytes, in which the EBV genome is maintained as multiple full-length circular episomes. Specific EBV-encoded small RNA species are found in all cells infected with the virus. A variable number of EBV genes are expressed in the different forms of the latent state. These are principally genes coding for one or more of six EBV nuclear antigens (EBNAs) and latent membrane proteins. Latent membrane protein 1 is a glycoprotein that may be involved in recognition by immune T cells. It also has a role in cell transformation. All of the latent genes are expressed in the resting memory B cells (unrestricted latency). More restricted gene expression is found in Burkitt's lymphoma and Hodgkin's disease cells.

The EBV nuclear antigen 1 (EBNA-1) maintains the replication of the EBV episome as the B cells divide. EBNA-1 is the only EBV protein produced in these cells, and has evolved to escape presentation to the immune sytem. EBNA-2 is necessary for transformation of B lymphocytes; there are two alleles of this gene, originally designated *EBNA-2A* and *EBNA-2B*. There are two EBV types, EBV-1 and EBV-2. The genes of the latent EBV are replicated by host cell DNA polymerase.

The full lytic cycle of EBV replication is accompanied by the production of virus structural antigens and assembly of virions. Although EBV does encode a viral thymidine kinase, aciclovir is phosphorylated by cellular thymidine kinase in cells producing EBV. The viral DNA polymerase is very sensitive to aciclovir triphosphate, and treatment with aciclovir reduces EBV production but has no effect on latency or the proliferation induced by the virus.

PATHOGENESIS

Virus in saliva initially infects oropharyngeal B lymphocytes in tonsillar crypts. This leads to activation of the cells, which progress to the local lymph nodes and on through the circulation, with the potential to enter a productive phase and release virus elsewhere in the body. Most shedding of virus takes place in the oral cavity, and virus can be detected regularly in the saliva of asymptomatic hosts, the amount increasing in immunosuppressed states.

Activated B lymphocytes secrete immunoglobulin, and EBV is a potent polyclonal activator of antibody production by B cells, independent of any accessory cells. IgM-producing lymphocytes predominate, and IgM antibody is

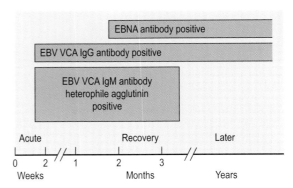

Fig. 43.7 Appearance and duration of diagnostic antibodies following primary EBV infection. EBNA, EBV nuclear antigen; VCA, viral capsid antigen.

found in high levels; however, detectable IgG (especially the IgG3 subclass), IgA and IgD are also found.

Recovery from primary EBV infection is associated with humoral and cellular responses; any delay in cellular control, or over-vigorous responses, will contribute to the severity of the infection. Thus, large initial infective doses result in high numbers of circulating infected B lymphocytes, followed by a marked T cell response. The polyclonal activation of B cells results in the transient appearance of antibodies (predominantly IgM), both autophile and heterophile. The cellular response is detected as large numbers of 'atypical lymphocytes' in the blood and infiltrating many tissues. These cells have been shown to be EBV-specific CD8+ cytotoxic T cells. The suppression is manifested as a general depression of immune responses. The cytotoxic elements carry the ability to kill EBV-infected cells, and this is not entirely human leucocyte antigen (HLA) restricted.

Antibody responses after EBV infection follow a characteristic pattern, with the initial IgM response to virus capsid antigens persisting for some months. The EBNA complex elicits antibodies in late convalescence only, perhaps after release from B cells lysed by the cytotoxic T cells (Fig. 43.7). Failure to produce antibody to EBNA is a feature of immunodeficiency states. This may be associated with increased levels of antibodies to EBV lytic cycle antigens (early antigen and viral capsid antigen), reflecting a high virus replication rate. High IgA levels to EBV capsid antigen are found in those at risk of developing nasopharyngeal carcinoma.

CLINICAL FEATURES

Primary infection with EBV is usually mild and unrecognized in the vast majority who acquire it in the first years of life.

Table 43.3 Diseases associated with EBV

Disease	Cells infected	Link
Infectious mononucleosis ('glandular fever')	Naive B lymphocytes	Causal; acute primary infection
Oral hairy leucoplakia (seen in AIDS)	Differentiated epithelium along edge of tongue	Causal; productive recurrence in immunocompromised host
Nasopharyngeal carcinoma (especially in South-East Asia and China)	Undifferentiated nasopharyngeal epithelium (long latent period, >30 years)	100% EBV-positive cells; co-factor(s) and genetic risk
Burkitt's lymphoma (endemic African form)	Monoclonal B cell tumour (short latent period, >5 years)	100% EBV-positive; immunocompromised by malaria
Immunoblastic lymphoma (post-transplant/AIDS; X-linked syndrome)	Activated B lymphocytes	70–100% EBV-positive cells; immunodeficiency states, genetic defect
Hodgkin's disease	Hodgkin–Reed–Sternberg cells (germinal-centre B lymphocytes)	30–90% EBV-positive cells in subsets of disease
Anaplastic gastric carcinoma	Epithelial cells	Strong link to the anaplastic subset
T cell lymphoma	T lymphocytes	30–100% EBV-positive cells in subsets of disease

Infectious mononucleosis

The disease known as *infectious mononucleosis*, or *glandular fever*, is a primary EBV infection seen predominantly in the 15–25-years age group. The incubation period is 30–50 days, and the onset is abrupt with a sore throat, cervical lymphadenopathy and fever, accompanied by malaise, headache, sweating and gastro-intestinal discomfort. Pharyngitis may be severe, accompanied by a greyish-white membrane and gross tonsillar enlargement. Lymphadenopathy becomes generalized, often with splenic enlargement and tenderness, mild hepatomegaly in some cases and clinical jaundice in 5–10%. Intermittent fevers with drenching sweats occur daily over 2 weeks. A faint transient morbilliform rash may be seen; a maculopapular rash may follow ampicillin administration, due to immune complexes with antibody to ampicillin. The illness can last for several weeks, and fatigue and lack of concentration are common in the aftermath.

Complications of glandular fever

Complications are rare but some are serious.

- Acute airway obstruction may occur as a result of the lymphoid enlargement and oedema; this merits emergency tracheostomy in some cases, but usually responds well to corticosteroids.
- Splenic rupture is rare.
- Neurological complications include meningitis, encephalitis and the Guillain–Barré syndrome.

Other EBV-associated disease, tumours and immunodeficiency

EBV is associated with an increasing number of diseases, including malignant tumours, some of which are listed in

Table 43.3. The role played by EBV in these conditions is not clear in all cases. Cellular immunodeficiency states lead to lack of T cell control over the B cell proliferative phase of EBV infection, and a risk of development of immunoblastic lymphomas. These are recognized in AIDS and post-transplant lymphoproliferative disease. In some situations, such as African Burkitt's lymphoma, EBV infection at an early age accompanied by chronic immunosuppression due to endemic malaria proceeds to a highly aggressive tumour. Hodgkin's disease is associated with EBV in 50% of cases, and the common undifferentiated nasopharyngeal carcinoma almost always contains EBV. The virus is referred to as the first human tumour virus. (see Ch. 7).

LABORATORY DIAGNOSIS

Infectious mononucleosis is accompanied by the production of heterophile agglutinins that can be detected by a rapid slide agglutination test or the *Paul–Bunnell test*. Agglutination of horse or sheep red cells by serum absorbed to exclude a natural antibody is the basis of this test. Atypical lymphocytes, accounting for 20% of the lymphocytosis common in this condition, are seen in blood films. Definitive diagnosis requires the demonstration of IgM antibody to the EBV viral capsid antigen, or seroconversion (IgG antibody). These tests, using indirect immunofluorescence or enzyme immunoassay, are generally available. Other serological tests are applicable in special situations.

Culture of EBV, from saliva or throat washings, is a research technique. Tissue sections can be stained for EBNA or other EBV proteins (Fig. 43.8), or probed for EBV early RNA species; these approaches, and PCR for EBV, are increasingly important in the diagnosis of

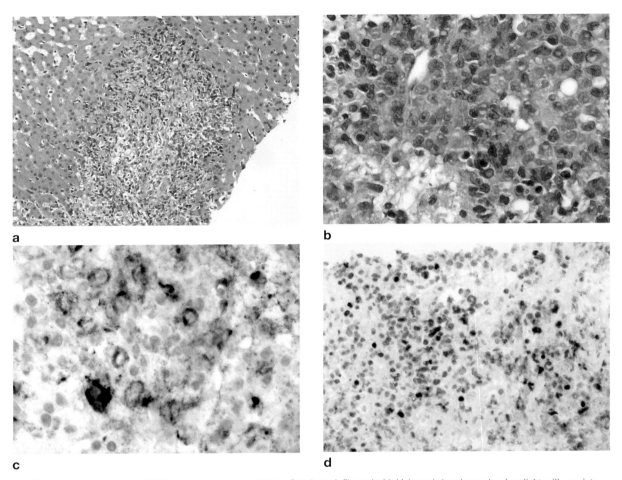

Fig. 43.8 Liver needle biopsy of PTLD in a liver transplant recipient. **a** Portal area infiltrated with high-grade lymphoma showing slight spill-over into adjacent and relatively normal-looking liver tissue. **b** High-power view of the neoplastic infiltrate showing large, atypical, lymphoid blast cells. **c** EBV latent membrane protein immunohistochemistry decorates many of the neoplastic lymphoid cells. **d** EBNA-2 immunohistochemistry also highlights many tumour cell nuclei (brown is positive). (Courtesy of Dr C. O. C. Bellamy, Department of Pathology, University of Edinburgh.)

disease in immunodeficiency states. The role of monitoring post-transplant levels of EBV DNA in the prediction of lymphoproliferative disease is still being assessed.

TREATMENT

Aciclovir therapy does reduce EBV shedding in acute infections, and there may be a place for it in patients suffering from, or at risk of, complications associated with an ongoing viral lytic cycle. Immunotherapy for EBV lymphoproliferative conditions with donor-derived or HLA-matched cytotoxic T cells is a possibility in some situations. Reducing immunosuppression in transplant recipients is sometimes an option. Once a tumour has developed, use of rituximab, a monoclonal antibody against CD20 (a marker

for the B cell immunoblasts), may reduce the size of the tumour.

EPIDEMIOLOGY

Infection with EBV is transmitted by saliva, and requires intimate oral contact. A potential role for sexual transmission has been proposed. Transmission has rarely been reported following transfusion of fresh blood to seronegative recipients. Infection is widespread, with most of the population infected from early in life, even in developed countries. Increased numbers of severely immunocompromised hosts are at risk of developing EBV-associated malignant lymphomas, including recipients of solid organ or bone marrow transplants.

CONTROL

A subunit vaccine based on the major membrane glyco-protein gp350/220 has been undergoing trials, and has been shown to protect marmosets against tumour-inducing doses of EBV. A peptide vaccine to stimulate cytotoxic T cell responses is under trial. Screening for IgA to EBV capsid antigen is used in populations at risk of nasopharyngeal cancer to detect preclinical cases.

CYTOMEGALOVIRUS

There are CMVs specific to other animals; the full name for the virus that infects man is human cytomegalovirus. This will be used only in descriptions of features specific to the human virus; otherwise CMV will be used. The name 'cytomegalovirus' was chosen because of the swollen state of infected cells in culture and in tissues. Nuclei of productively infected cells contain a large inclusion body, giving a typical 'owl's eye' appearance.

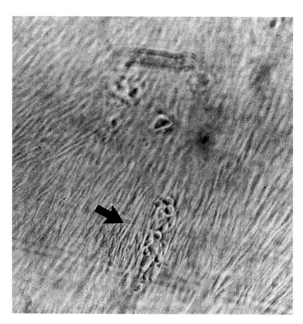

Fig. 43.9 Focus of CMV infection (arrowed) in a tissue culture monolayer of human embryo fibroblasts.

DESCRIPTION

The CMVs have the same general structure as other members of the herpes group.

Human fibroblast cells are required for isolation of human CMV in vitro, but in vivo the virus replicates in epithelial cells in:

- salivary glands
- the kidney
- the respiratory tract.

CMV remains highly cell associated, and is sensitive to freezing and thawing. Virus shed in urine is stable at 4°C for many days.

REPLICATION

The temporal regulation of viral protein synthesis in the growth cycle is more obvious in laboratory culture of the slower-growing CMV than with HSV. Immediate early (non-structural) protein (p72) appears in nuclei within 16 h of inoculation, whereas late (virion-structural) proteins are produced after DNA synthesis; the typical cytopathic effect is often not recognizable for 5–21 days. Foci of swollen cells expand slowly as infection passes from cell to cell (Fig. 43.9). Passage and storage of virus are best achieved by trypsinization and passage as infected cells.

Human CMV does not produce a virus-specified thymidine kinase; the protein kinase product of gene *UL97* carries out initial phosphorylation of ganciclovir in CMV-infected cells, and cellular kinases produce the triphosphate, which inhibits CMV DNA polymerase.

There are several families of glycoproteins in CMV, and these are important antigenic targets. Most neutralizing antibody is directed against gB.

PATHOGENESIS

Primary infection with CMV may be acquired at any time, possibly from conception onwards.

CMV persists in the host for life. Reactivation is common, and virus is shed in body secretions such as urine, saliva, semen, breast milk and cervical fluid. Mononuclear cells carry the latent virus genome: viral RNA transcripts of early genes have been detected in them. Bone marrow progenitor cells of the myeloid line may be the site of latency. Once their descendants have been activated to differentiate into tissue macrophages, the virus can enter the replication cycle. Recurrent infections may follow reactivation of latent (endogenous) virus, or re-infection with another (exogenous) strain. Isolates can be distinguished by restriction endonuclease analysis, and by variations in some envelope glycoproteins (gB and gH). Endothelial giant cells (multinucleate cells) have been found in the circulation during disseminated CMV infection. These cells are fully permissive for CMV replication.

Intra-uterine infection

Maternal viraemia may result in fetal infection in approximately one in three cases of primary CMV during pregnancy, and may lead to disease in the fetus. Infection may be acquired in utero when the mother has a reactivation, but this rarely results in disease. Transplacental infection is probably carried by infected cells, and transmission is associated with a high viraemic load.

Perinatal infection

This is predominantly acquired from infected maternal genital tract secretions or from breast-feeding (3–5% of pregnant women in Europe reactivate CMV). Rarely, perinatal blood transfusion was the source, but as leuco-depletion (removal of the white cells) is now practised this risk has been reduced.

Postnatal infection

This can be acquired in many ways. Saliva containing CMV is distributed profusely among young children, and at older ages by intimate kissing. Semen can have high titres of virus, and may be a source of sexual transmission or artifical insemination-associated infection. Whole blood transfusion used to be (and donated organs remain) an important source of CMV. It is not known which donors are most likely to transmit infection, so all antibody-positive ('seropositive') cell and organ donors are considered to be potential transmitters, as the presence of antibody implies the presence of persistent virus.

Host responses

The host response to primary CMV includes IgM, IgG and T cell responses. Some of the T cell responses may contribute to immunopathology by reacting with HLA molecules induced by CMV. CMV early genes transactivate other viral and cellular genes, and this may be an important interaction with HIV, leading to the production of HIV from latently infected cells. Because CMV infects mononuclear cells, there is a degree of immunosuppression associated with the acute infection. Cell-mediated responses are crucial to the control of CMV, as shown by the serious consequences of disseminated infections in those deficient in effector cell functions. The incubation period for primary infection is 4–6 weeks; reactivation, after transplantation for instance, appears from 3 weeks onwards.

CLINICAL FEATURES

Congenital CMV infection

CMV infection is asymptomatic in 95% of infected babies, but around 15% of these show later sensorineural deafness or intellectual impairment. Progression of the persistent CMV infection may be involved. All congenitally infected infants excrete abundant virus in the urine during the first year. The 5% of symptomatic infants have *cytomegalic inclusion body disease* with:

- growth retardation
- hepatosplenomegaly
- jaundice
- thrombocytopenia
- central nervous system involvement – a significant problem with CMV; microcephaly, encephalitis and retinitis are noted at birth. Some changes may even be detectable on ultrasonography in utero.

Mononucleosis

Postnatal infection with CMV is seldom recognized clinically, unless virus is isolated. Respiratory tract infection is common in infancy. A *mononucleosis syndrome* is seen occasionally, especially in young adults or when CMV is acquired from blood transfusion. Hepatitis, fever and atypical lymphocytosis are noted, but pharyngitis and lymphadenopathy are unusual and heterophile agglutinins are not found. This syndrome is also seen in some HIV-infected patients before the development of AIDS, and should prompt HIV-related investigations.

Infection in the compromised patient

Immunocompromised patients may develop symptoms as the result of primary or recurrent CMV infection. Dissemination of the virus in the blood, as indicated by a hectic fever, is a bad prognostic sign. The complications of CMV infection in cellular immunodeficiency include:

- pneumonitis – has a high mortality rate in receipients of bone marrow allografts
- encephalitis – often fatal
- retinitis – may occur on its own (10–40% of patients with AIDS)
- oesophagitis/colitis – 5–10% of patients with AIDS
- hepatitis
- pancreatitis or adrenalitis.

CMV infection in transplant recipients is a significant cause of morbidity and loss of graft. The mortality rate is high, particularly in allogeneic marrow recipients who develop an immunopathological pneumonitis associated

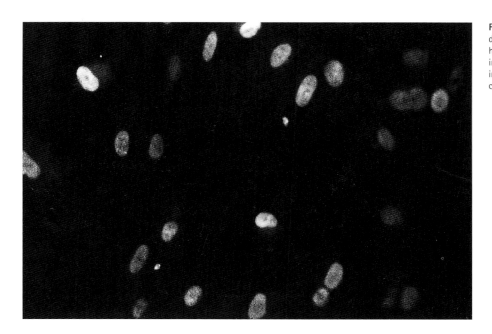

Fig. 43.10 CMV early antigen demonstrated in nuclei of human embryo fibroblasts by immunofluorescence after incubation for 24 h following centrifuge-assisted inoculation.

with graft-versus-host disease. Transplant protocols all include prophylactic or pre-emptive therapy for the prevention of CMV disease.

Retinitis due to CMV recurrence is a feature of late-stage AIDS. Early recognition, antiviral treatment and maintenance therapy are important in slowing the progression towards blindness.

LABORATORY DIAGNOSIS

Detection of CMV is the objective, if possible, showing its presence at the site of disease. Samples should include urine, saliva, broncho-alveolar lavage fluid or biopsy tissue if available, and peripheral blood collected in anticoagulant. In the neonate, urine samples taken in the first 2 weeks of life are sufficient for the diagnosis of congenital infection.

Virus detection

Rapid diagnostic methods can detect viral DNA in the samples with PCR or a hybridization assay; or CMV early antigen in cell cultures 24 h post-inoculation (Fig. 43.10). Conventional culture is used to isolate virus. High titres of CMV produce a cytopathic effect very quickly, but most cultures require 2–3 weeks. CMV isolation from urine does not prove that the virus is the cause of the disease being investigated. To demonstrate congenital infection, virus must be shown in a sample taken within the first 2 weeks of life, as later samples may be positive due to virus acquired in the postnatal period. Confirmation that CMV is related to a disease process comes from showing (by immunohistochemistry) that the virus is replicating in the affected tissues, and in some cases that a primary infection has occurred. The demonstration of CMV viraemia is a significant prognostic finding, by detection of DNA (by PCR) or leucocytes containing CMV late phosphoprotein antigen (pp65); above a certain level it is predictive of clinical disease. The quantitative PCR assay has become important in monitoring for CMV after transplantation.

Serology for CMV

Complement fixation tests are adequate for showing seroconversion after primary infection in competent hosts. To screen for 'seropositive' status a more sensitive assay, such as enzyme immuno-assay for CMV IgG or total antibody or latex agglutination assay, is appropriate. These tests can be done urgently for donor–recipient assessment. CMV IgM is found after primary or secondary infection, but it may not be possible to detect IgM in the neonate or immunocompromised patient.

TREATMENT

Antiviral agents for CMV infections are available but serious side effects limit their use to life- or sight-threatening complications. Ganciclovir is the agent most often used, given intravenously twice a day. Marrow toxicity results in neutropenia, and there can be long-term loss of spermatogenesis. Clinically, treatment has been successful in CMV hepatitis, colitis and encephalitis, and progression of CMV retinitis in AIDS has been controlled with prolonged maintenance therapy, which may now be given orally. Ganciclovir-resistant virus has been found; usually a mutation in the protein kinase gene is responsible. Foscarnet (phosphonoformate) is an alternative agent that does not require phosphorylation. Aciclovir is not effective therapy for CMV, but some benefit from high-dose prophylaxis of CMV in transplant recipients has been noted. Oral ganciclovir has been used for prophylaxis in transplant recipients for several years, and now an oral pro-drug, valganciclovir, is used to provide systemic levels of ganciclovir virtually equivalent to those achieved by intravenous therapy. Cidofovir is an alternative antiviral agent used locally in the eye for CMV retinitis.

EPIDEMIOLOGY

Primary CMV infection is acquired by 40–60% of persons with limited exposure by mid-adult life, and by more than 90% of those with multiple intimate exposures. Fewer than 5% of units of whole blood from seropositive donors result in transmission to seronegative recipients, whereas 80% of kidneys transmit infection from seropositive donors. The full extent of congenital CMV disease is not known, but this infection occurs in approximately 3 of 1000 live births in the UK, and is thought to be the commonest viral cause of congenital infection.

CONTROL

Screening of organ donors and recipients is done to avoid, where possible, a seronegative recipient receiving an organ from a seropositive donor. This reduces morbidity and mortality significantly for all forms of allogeneic transplant. Blood donor screening to select CMV-seronegative units for support of seronegative patients in transplant programmes and for premature babies is less important now with the use of leucodepletion. Antiviral prophylaxis (starting from time of transplant) or pre-emptive treatment (starting from time of virus detection) are routine procedures in solid organ and stem cell transplant programmes, for those at risk of CMV disease.

No CMV vaccine is licensed for use. Experimental live-attenuated vaccines have been tried, but hopes rest on subunit vaccines or on a combined approach.

HUMAN HERPESVIRUSES 6 AND 7

DESCRIPTION

The existence of further herpesviruses infecting man was not suspected until one was isolated in 1986 from the blood of patients with lymphoproliferative disorders, some of whom had AIDS. Electron microscopy revealed a virus with characteristic herpes group features. DNA sequence studies showed this virus, now officially named human herpesvirus 6 (HHV-6), to be distinct from the then five known human herpesviruses, but closer to human CMV, with which it has some homology. Two variants have been identified on sequence analysis, HHV-6A and HHV-6B; they are sufficiently different that they might have been allotted separate species (like HSV-1 and HSV-2). Although first isolated from B cells, HHV-6 was shown to infect CD4$^+$ T cells preferentially. HHV-6B is the cause of a common disease of infancy called exanthem subitum or roseola infantum, whereas variant 6A has no clear disease association. The cell receptor for HHV-6 is CD46, a complement regulatory glycoprotein present on all nucleated cells. Bone marrow progenitor cells contain truly latent HHV-6; low-level persistent infection is found in salivary glands and brain cells.

In 1990 another new human herpesvirus was isolated in similar circumstances, and has been shown to be sufficiently distinct to be called HHV-7. Primary infection with HHV-7 has been identified in some cases of roseola infantum also. A new genus *Roseolovirus* has been established for these viruses, which are members of the subfamily betaherpesviruses, distantly related to human CMV, but with much smaller genomes.

PATHOGENESIS AND CLINICAL FEATURES

Both HHV-6B and HHV-7 are shed persistently in saliva, and both have been found in female genital secretions. Only HHV-7 is found in breast milk. It is known that HHV-6 can be integrated in the human chromosome, and can be inherited from either parent. Such congenital infection leads to a persistently high level of viral DNA within cells in blood. In most people only low numbers of copies of HHV-6 or HHV-7 are found in circulating cells.

Roseola infantum (exanthem subitum)

This disease was long considered to be an infection caused by a virus, and transmission by blood was confirmed experimentally years before HHV-6 was isolated. The infection is extremely common in the first years of life, presenting between 6 months and 3 years of age with a sudden onset of fever (39.4–39.7°C). The acute febrile illness due to these viruses accounts for 20% of all fevers of early childhood. The child is not usually ill, remaining alert and playful, with some throat congestion and cervical lymphadenopathy. Sometimes more pronounced respiratory symptoms occur with febrile convulsions. Fever usually persists for 3 days, when the temperature falls suddenly; a widespread macular rash appears in around one-tenth of cases. HHV-6B has been isolated from peripheral blood in the acute febrile phase, in patients who subsequently produced a rash and in many who did not. Seroconversion is noted in convalescence, confirming a primary infection with HHV-6. Some cases are associated with HHV-7 infection, but no other disease association has yet been found for HHV-7.

Neurological disease

A British survey has confirmed the importance of roseoloviruses in neurological illness (fits and encephalitis) in early childhood. Diagnosis is difficult as viral DNA is not always detected in CSF, but seroconversion to one of the viruses may confirm a primary infection. HHV-6 strains appear to remain thereafter in brain, and these viruses are now considered to be commensal there. No association with demyelinating diseases, particularly multiple sclerosis, has been shown.

Other associations

In the rare adult case, hepatitis and lymphadenopathy have been found, and sometimes a heterophile agglutinin-negative mononucleosis. Reactivation in immunocompromised hosts may be diagnosed by viral load or on the basis of serology, but this is difficult to interpret. HHV-6 and HHV-7 reactivate in the first weeks after transplant and appear to make CMV disease worse, but the full extent of disease in the transplant recipient has still to be established.

Both HHV-6 and HHV-7 infect T lymphocytes; indeed, HHV-7 uses the same receptor (CD4) as HIV. The potential significance of these interactions has still to be established.

LABORATORY DIAGNOSIS

Isolation of HHV-6 or HHV-7 involves co-cultivation of peripheral blood lymphocytes with mitogen-activated cord blood lymphocytes. Very large, refractile, multinucleate cells are produced in culture, and many intact enveloped virions are released into the culture medium. These viruses may be isolated from saliva, particularly HHV-7. Viral

DNA detection (PCR) is also used in diagnosis, with a multiplex assay capable of distinguishing HHV-6A from HHV-6B and HHV-7.

Laboratory diagnosis is currently available only from specialist laboratories. Antibody tests are becoming more widely used, but results can be confusing. Antibody avidity tests can help to establish whether a primary infection has occurred, as the binding of recently acquired low-avidity IgG is easily disrupted by protein-denaturing agents. Both this method and PCR have proven useful in establishing the correct diagnosis of HHV-6- and HHV-7-associated roseola infantum in a substantial proportion of suspected cases of infant measles.

TREATMENT

HHV-6 and HHV-7 are sensitive to ganciclovir, but not to aciclovir. Foscarnet and cidofovir may also be used in serious cases, for example of encephalitis in immunocompromised patients. Treatment is seldom indicated clinically.

EPIDEMIOLOGY

Antibody analysis to date has been based mainly on immunofluorescence studies with cells infected with either HHV-6 or HHV-7 as antigen. High antibody titres to HHV-6 are found in young children in the first 4 years of life, reflecting recent primary infections in this group. Primary infection, with viraemia and seroconversion, has also been detected in seronegative transplant recipients of liver or kidney from seropositive donors. Possible transmission by blood transfusion has not been excluded. Infection with HHV-7 occurs, on average, a few years later. Older persons have lower levels of antibody, but more sensitive enzyme immuno-assay tests reveal almost universal seropositivity. The full spectrum of disease associated with these viruses and the extent of asymptomatic shedding remain to be established.

HUMAN HERPESVIRUS 8 (KAPOSI'S SARCOMA-RELATED HERPESVIRUS)

Sequences of DNA representing a completely new human herpesvirus were identified in 1994 in tissues from the epidemic form of Kaposi's sarcoma in patients with AIDS. The DNA fragments unique to the tumour tissue were found to have some homology with the gammaherpesvirus subfamily, including EBV. This is a new virus, officially named HHV-8, but also referred to as *Kaposi's sarcoma-associated herpes virus*, and it is the latest human tumour virus.

The eighth human herpesvirus is classified in the rhadinovirus genus of gamma2herpesviruses, along with known viruses of non-human primates. The genome is 165–170 bp and encodes around 95 proteins, including a group that are homologues of human proteins involved in cell growth regulation and cytokine production. Like other herpesviruses, HHV-8 attaches to cells first via heparin sulphate through an envelope glycoprotein (K8.1). HHV-8 infects dividing B cells – $CD45^+$. It then proceeds to either a lytic replication cycle, releasing many infectious virions, or enters latency, expressing only the latency-associated nuclear antigens (LANAs). LANA-2 suppresses apoptosis via inhibition of p53-mediated transcription. Recent work has shown that valproate (used as an anti-epileptic drug) can stimulate replication in cell lines in which the virus is latent, and leads on to apoptosis in the presence of anti-herpes antiviral agents capable of halting viral DNA replication, so that no infectious virus is produced.

Kaposi's sarcoma

Whether the new virus is the cause of Kaposi's sarcoma has not yet been established; the evidence is that HHV-8 is an essential agent but not the only factor. The mucocutaneous neoplasm (Kaposi described it as 'a multifocal pigmented sarcoma') was the commonest tumour in one group of HIV-infected individuals before the era of highly active retroviral therapy (HAART). The classic form was described as a rare finding in elderly patients of southern Mediterranean descent. A more aggressive endemic form is seen in Africa and in some post-transplant patients, but the aggressive epidemic form has been a feature of HIV-infected homosexual male patients, and, in fact, was one of the heralds of the AIDS epidemic. A transmissible agent was long suspected in Kaposi's sarcoma on epidemiological grounds, transmitted sexually but rarely by blood. Endothelial cells of vascular or lymphatic origin are involved; they have a characteristic spindle shape and are arranged in bundles. Although occurring at multiple sites in skin, lymph glands and the gastro-intestinal tract, the tumour itself does not lead to death. Local radiotherapy and systemic chemotherapy have been used in treatment. Lesions on the exposed parts are blue–pink or red–brown plaques, sometimes raised, and are a cause of considerable concern for the affected individual.

'Body cavity–associated lymphomas'

HHV-8 genomes have been found in lymphoma cells from AIDS-related B cell lymphomas and similar conditions. These are termed body cavity lymphomas and are found primarily in immunodeficient HIV-infected patients. Primary effusion lymphoma is the best known, and all of its cells contain HHV-8. Cell lines established from this condition can be induced to produce viral proteins and particles, and have been used as a basis for diagnostic tests. A subset of another condition, multicentric Castleman's disease, is also strongly associated with HHV-8.

LABORATORY DIAGNOSIS

DNA amplification by PCR is used to detect HHV-8. As mentioned above, lytic growth of HHV-8 can be induced in latently infected B cell lymphoma lines and, as a result, reagents for antibody testing can be produced – both infected cell lysates for immunoblots, and antigen-containing cells for immunofluorescence. Recombinant proteins have been synthesized and are useful for enzyme immuno-assays. Now that reagents free of EBV are available, the specific epidemiology of infection with this new herpesvirus is being studied.

EPIDEMIOLOGY

Molecular sequence data have revealed five variants of HHV-8, termed groups A–E. Group B predominates in Africa, whereas in Europe and North America groups A and C are more common. Serological surveys have shown that, unlike most of the human herpesviruses, HHV-8 infection is not common, at least in developed countries (fewer than 5% of blood donors in North America and northern Europe have antibodies; 1–2% of UK donors). Rates of seropositivity are higher in risk groups for Kaposi's sarcoma, with 30% seropositivity in men with male sex partners. Confirmation of a high seropositivity rate (up to 40%) in children in sub-Saharan African countries and studies on transmission have shown a link between mothers and their children and between siblings. Transmission from saliva is considered the most likely route, and the high rates in communities where crowding in childhood is seen support this. In Europe there is increased seropositivity around the Mediterranean, and the presence of HHV-8 DNA in elderly Italians is highly predictive of the development of Kaposi's sarcoma following organ transplantation or HIV infection.

TREATMENT AND CONTROL

Ganciclovir has activity against HHV-8, and its use in CMV prophylaxis after organ transplantation may contribute to reducing the risk of reactivation and disease due to HHV-

8 in seropositive recipients, or in seronegative recipients who have a seropositive donor. Tumours associated with HHV-8 infection may regress if it is possible to reduce the dosage of immunosuppressive drugs (post-transplant) or to improve immune capability with HAART (in patients with AIDS). The anti-CD20 monoclonal antibody may also have some effect in patients with body cavity lymphomas.

CERCOPITHECINE HERPESVIRUS 1 (B VIRUS)

DESCRIPTION

This virus, antigenically related to HSV, commonly infects Old World (Asiatic) macaque monkeys, causing a mild vesicular eruption on the tongue and buccal mucosa analogous to primary herpetic stomatitis in man. The infection rate in monkeys increases markedly if they are kept in crowded conditions and, although relatively benign in the monkey, this virus is highly pathogenic for man.

Human infection with B virus is rare. It is usually acquired from a bite or from handling infected animals without appropriate protective wear. In one instance the wife of an infected monkey handler became infected through contact with her husband's vesicles, and B virus has also been transmitted in the laboratory from infected monkey cell cultures. Within 5–20 days of exposure, local inflammation may appear at the site of entry, usually on the skin, accompanied by some itching, numbness and vesicular lesions. Ascending myelitis or acute encephalomyelitis may follow, occasionally as long as 5 weeks after exposure, and may not always be recognized. Delay in specific therapy leads to a high mortality rate and serious neurological sequelae in survivors.

DIAGNOSIS AND TREATMENT

Diagnosis by isolation of the virus from blood, vesicle fluid, conjunctival swabs or CSF is not always possible. Herpes virions may be detectable on electron microscopy of vesicle fluid. Definitive identification of the virus is available in specialist reference laboratories, using PCR for DNA or monoclonal antibodies. Demonstration of specific antibodies is complicated by cross-reacting HSV antibody, but new specific antigens based on recombinant glycoproteins offer improved serology. B virus is not as sensitive to aciclovir as HSV, requiring concentrations equivalent to those used for VZV. Treatment needs to be given promptly to be effective, and very high-dose intravenous aciclovir for 14 days or longer is recommended. Ganciclovir should be used when there is evidence of central nervous system involvement. Because of the small number of cases, the best therapeutic regimen is not well established.

PREVENTION OF B VIRUS INFECTION

Guidelines have been issued for the protection of those handling monkeys or monkey tissues. These include:

- recommendations for training to prevent exposure
- safe handling and protective wear procedures
- care of wounds
- information regarding the risks and nature of the infection.

Prophylaxis in the event of possible exposure involves prompt, rigorous wound washing and cleansing with 10% iodine in alcohol. A course of high-dose oral aciclovir (800 mg five times daily for 2 weeks) or valaciclovir is reserved for prophylaxis of high-risk exposures, namely those on head, neck or torso, deep wounds, and where the source animal has lesions, recent stress or illness, or the material is of central nervous system or mucosal origin. After prophylaxis, a prolonged observation period is recommended as the onset of infection may be delayed.

KEY POINTS

- All herpesviruses establish a latent (non-replicating) state from which they may be reactivated under certain conditions.
- Cell-mediated immunity, especially the action of cytotoxic T cells, is essential in the control of herpesvirus infections.
- Acute necrotizing sporadic viral encephalitis caused by HSV is a medical emergency requiring urgent treatment with intravenous aciclovir.
- Genital herpes simplex may be due to HSV-1 or HSV-2, and the individuals are often unaware that they have this recurring infection.
- A live-attenuated vaccine can protect against chickenpox (varicella); it may boost immunity in the elderly with a reduction in the symptoms of zoster (shingles), which arises from reactivated VZV travelling back down the nerve.
- Glandular fever is a clinical presentation of acute EBV infection mainly in the 15–25 years age group; most people have a silent infection earlier.
- EBV is a recognized human tumour virus associated with relatively rare lymphoproliferative tumours in certain compromised individuals or situations.
- CMV is commonly acquired without symptoms; however, in the immunocompromised CMV disease is serious and merits prophylaxis.
- Less well known herpesviruses (human herpesviruses 6 and 7) cause common childhood infections, whereas HHV-8 is associated with Kaposi's sarcoma.

RECOMMENDED READING

Thorley-Lawson D A 2005 EBV the prototypical human tumor virus – just how bad is it? *Journal of Allergy and Clinical Immunology* 116: 251–261

Ward K N 2005 The natural history and laboratory diagnosis of human herpesviruses-6 and -7 infections in the immunocompetent. *Journal of Clinical Virology* 32: 183–193

Zuckerman A J, Banatvala J E, Pattison J R, Griffiths P D, Schoub B D (eds) 2004 Herpesviridae. In *Principles and Practice of Clinical Virology*, Section 2, 5th edn, pp. 23 –198. Wiley, Chichester (separate chapters on the human herrpesviruses and an overview of the family)

Internet sites

The Homepage of Dr Edward K. Wagner. Herpesviruses. http://darwin.bio.uci.edu/~faculty/wagner/ (for details and animations showing all stages of HSV infection)

Immunization Action Coalition. Vaccine information. http://vaccineinformation.org/varicel/photos.asp (for a selection of clinical photographs of varicella)

44 Poxviruses

Smallpox; molluscum contagiosum; parapoxvirus infections

T. H. Pennington

The world's last naturally occurring case of smallpox was recorded in Merca, southern Somalia, in October 1977. This momentous event marked the end of a long campaign against smallpox, which in its 'modern' phase started with the introduction of vaccination by Edward Jenner at the end of the eighteenth century. With the eradication of smallpox, the importance of poxviruses in medical practice may appear to be much diminished, as the other naturally occurring viruses in this family that infect man nearly always cause only self-limiting and trivial skin lesions. Smallpox caused a generalized infection with high mortality rate, and fell into that small group of viruses whose infections are commonly severe and frequently fatal. Notwithstanding their minor role today as human pathogens, poxviruses retain their importance, and their place in this book, for three reasons.

Firstly, the successful smallpox eradication campaign is important in its own right as a major achievement. It is also important because it highlights the principles and problems associated with projects that aim to control infections by eradicating the pathogen. Smallpox was the first disease to fall to this approach and remains the only successful example to date.

Secondly, work on the molecular biology of poxviruses has led to the identification of distinctive properties and to the development of techniques and approaches that have made it possible to move significantly towards constructing single-dose vaccines that protect simultaneously against a wide range of diseases. Genes coding for foreign non-poxvirus antigens have been inserted into the poxvirus genome so that the antigens are expressed during virus infection and induce immunity.

Thirdly, no account of virus diseases in general and poxvirus infections in particular is complete without consideration being given to the events that followed the introduction of myxomatosis into Australia – by far the best studied example to date of the evolution of a new virus disease.

DESCRIPTION

Classification

A large number of different viruses belonging to the family Poxviridae have been described. They infect a wide range of vertebrate and invertebrate hosts. The subfamily Chordopoxvirinae contains all of the viruses that infect vertebrates; it is divided into six genera, each containing related viruses, which generally infect related hosts. Thus, members of the genus *Leporipoxvirus* infect rabbits and squirrels, *Avipoxvirus* members infect birds, *Capripoxvirus* members infect goats and sheep, *Suipoxvirus* members infect swine, and *Parapoxvirus* members infect cattle and sheep. Some viruses, including that of molluscum contagiosum, which infects man, remain unclassified. By far the most intensively studied poxvirus is vaccinia virus, the Jennerian smallpox vaccine virus. This virus has been placed in the genus *Orthopoxvirus*, together with smallpox virus and some viruses that infect cattle and mice.

The virion

Poxviruses are the largest animal viruses. Their virions are big enough to be seen as dots by light microscopy after special staining procedures. They are much more complex than those of any other viruses (Figs 44.1 and 44.2). They are also distinctive in that they do not show any discernible symmetry. The core contains the DNA genome and 15 or more enzymes that make up a transcriptional system whose role is to synthesize biologically active polyadenylated, capped and methylated virus messenger RNA (mRNA) molecules early in infection. The core has a 9-nm thick membrane, with a regular subunit structure. Within the virion, the core assumes a dumb-bell shape because of the large lateral bodies. The core and lateral bodies are enclosed in a protein shell about 12 nm thick (the outer membrane), the surface of which consists of

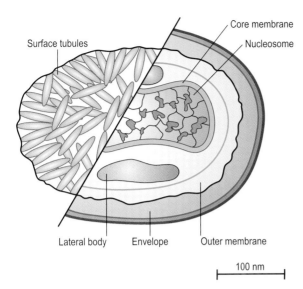

Fig. 44.1 Structure of the vaccinia virion. Right-hand side, section of enveloped virion; left-hand side, surface structure of non-enveloped particle. (From Fenner et al 1988 *Smallpox and its Eradication*. World Health Organization, Geneva.)

irregularly arranged tubules, which in turn consist of a small globular subunit. Virions released naturally from the cell are enclosed within an envelope that contains host cell lipids and several virus-specified polypeptides, including haemagglutinin; they are infectious. Most virions remain cell associated and are released by cellular disruption. These particles lack an envelope, so that the outer membrane constitutes their surface; they are also infectious. More than 100 different polypeptides have been identified in purified virions.

The genome

Their DNA genomes range in mass from 85 MDa (para-poxviruses) to 185 MDa (avipoxviruses). The vaccinia virus genome has 186 000 base pairs (123 MDa). The poxvirus genome is distinctive in that covalent links join the two DNA strands at both ends of the molecule, the genome thus being a single uninterrupted molecule that is folded to form a linear duplex structure. The occurrence of inverted terminal sequence repetitions is also a characteristic feature, identical sequences being present at each end of the genome.

REPLICATION

Poxviruses are unique among human DNA viruses in that virus RNA and DNA synthesis takes place in the cytoplasm of the infected cell and they code for all the proteins needed for their replication. They have early and late phases of protein synthesis (see Ch. 7). Virion assembly takes place in the cytoplasm and proceeds in a series of steps which include the formation of spherical immature particles. The virus causes irreversible inhibition of host protein synthesis due to the functional inactivation and degradation of host cytoplasmic RNA molecules, leading to the death of infected cells.

CLINICAL FEATURES

Smallpox virus had no animal reservoir and spread from person to person by the respiratory route. After infecting mucosal cells in the upper respiratory tract without producing symptoms, it spread to the regional lymph nodes and, after a transient viraemia, infected cells throughout the body. Multiplication of virus in these cells led to a second and more intense viraemia, which heralded the onset of clinical illness. During the first few days of fever the virus multiplied in skin epithelial cells, leading to the development of focal lesions and the characteristic rash. Macules progressed to papules, vesicles and pustules, leaving permanent pockmarks, particularly on the face. Two kinds of smallpox were common in the first half of the twentieth century:

1. variola major, or classical smallpox
2. variola minor, or alastrim.

Variola major had case fatality rates varying from 10% to 50% in the unvaccinated. Variola minor caused a much milder disease and had case fatality rates of less than 1%. The viruses are very similar but can be distinguished in the laboratory by restriction enzyme fragment length polymorphisms of their genomes.

CONTROL OF SMALLPOX

Before vaccination

Before the introduction of vaccination, the control of smallpox relied on two approaches, variolation and isolation. Variolators aimed to induce immunity equivalent to that after natural infection. Susceptible individuals were deliberately infected with smallpox pus or scabs by scratching the skin or by nasal insufflation. Although the virus was not attenuated, infections had lower case fatality rates (estimated to be 0.5–2%) and were less likely to cause permanent pockmarks than those acquired naturally. Variolation was first recorded in China nearly 1000 years ago, and was practised in many parts of the world. In Afghanistan, Pakistan and Ethiopia

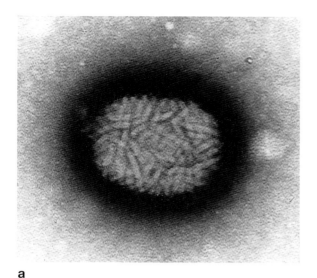

a

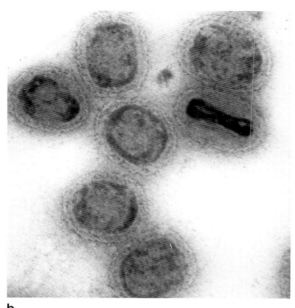

b

Fig. 44.2 Electron micrographs of poxviruses. **a** Molluscum contagiosum virus (MCV), ×15 000. **b** MCV showing internal structure, ×75 000. (Prepared by N. Atack.) **c** Parapoxvirus: orf. Bar, 100 nm. (Courtesy of Dr D. W. Gregory.)

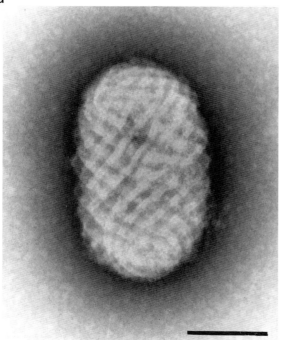

c

the activities of variolators caused problems towards the end of the smallpox eradication programme in the 1970s because they spread virus in a way that evaded the measures erected to control natural virus transmission.

Vaccination

Edward Jenner vaccinated James Phipps with cowpox virus on 14 May 1796 and challenged him by variolation some months later. He repeated this 'trial', as he called it, in other children, and the description of these events in his 'Inquiry' in 1798 led to the rapid worldwide acceptance of vaccination. Introduction of the vaccine virus into the epidermis led to the development of a local lesion and the induction of a strong immunity to infection with smallpox virus that lasted for several years. Although the essentials of *Jennerian vaccination* remained unchanged for the rest of its history, early vaccinators developed their own

Table 44.1 Features of smallpox that facilitated its eradication

Feature	Importance
Disease severe	Ensured strong public and governmental support for eradication programme
Detection of cases easy because of characteristic rash and subsequent development of facial pockmarks	Facilitated containment of outbreaks and audit of success of programme
Slow spread and poor transmissibility	Facilitated containment of outbreaks by vaccination and isolation
Transmission by subclinical cases not important	Meant that control of spread by isolation of cases was an effective procedure
No carrier state in humans	Meant that control of spread by isolation of cases was an effective procedure
No animal reservoir	Meant that control of spread by isolation of cases was an effective procedure
Vaccine technically simple to produce in large amounts, in high quality and at low cost (in skin of ungulates)	Meant that vaccine availability was not an important constraint in the eradication programme
Vaccine delivery simple and optimized by use of re-usable, cheap, specially designed needles to deliver a standard amount of vaccine to scratches in skin	Meant that failure at the point of vaccination was not an important constraint in the eradication programme
Freeze-dried vaccine stocks were heat-stable with a very long shelf-life in the tropics	Meant that vaccine viability under adverse environmental conditions was not an important constraint in the eradication programme

vaccine viruses, which became known as vaccinia. The origin of these viruses is obscure, and modern vaccinia viruses form a distinct species of orthopoxvirus, related to but very clearly distinct from the viruses of both cowpox and smallpox.

The eradication campaign

Smallpox was brought under control by:

- routine vaccination of children – compulsory in some countries
- outbreak control by isolation and selective vaccination.

This was achieved gradually in Europe, the former USSR, North and Central America, and Japan, and the virus had been eradicated from all these areas by the mid-1950s. In 1959 this achievement prompted the World Health Organization (WHO) to adopt the global eradication of smallpox as a major goal. At this time 60% of the world's population lived in areas where smallpox was endemic. A slow reduction in disease was maintained for the next few years, but epidemics continued to be frequent. Consequently the WHO initiated its Intensified Smallpox Eradication Programme. This started on 1 January 1967 when the disease was reported in 31 countries. It had the goal of eradication within 10 years. The goal was achieved in 10 years, 9 months and 26 days.

From a starting point of 10–15 million cases annually, and against a background of civil strife, famine and floods, success came because of a major international collaborative effort – aided by some virus-specific factors (Table 44.1). At the beginning of 1976, smallpox occurred only in Ethiopia.

Transmission was interrupted there in August of that year, although an importation of virus into Somalia and adjacent countries had occurred by then. This was the last outbreak. The last case was recorded on 26 October 1977. In the final years of the programme its emphasis moved from mass vaccination to a strategy of surveillance and containment. This strategy rapidly interrupted transmission because:

- cases were easy to detect owing to the characteristic rash.
- patients usually transmitted disease to only a few people – and only to those in close face-to-face contact.
- only persons with a rash transmitted infection.

The WHO Global Commission for the Eradication of Smallpox formally certified that smallpox had been eradicated from the world on 9 December 1979.

Smallpox and bio-terrorism

After eradication, virus strains were kept in government laboratories in Atlanta and Novosibirsk Region, Russia. The possibility that virus could fall into unauthorized hands is small, but vaccine stocks have been ordered and contingency plans prepared. The experience from outbreaks in the past suggests that outbreaks could be controlled rapidly. There were 49 importations into Europe after 1949. The average number of cases was 14, justifying the characterization of the spread as slow and plodding. In some outbreaks, public pressure for protection led to unnecessary vaccinations, in turn leading to more deaths from vaccination (see below) than from smallpox. In spite of frequent misdiagnoses of early cases as chickenpox (see Ch. 43), the outbreaks were rapidly controlled. Fortunately smallpox spreads slowly.

OTHER HUMAN POXVIRUS INFECTIONS

Molluscum contagiosum

The lesions of this mild disease are small, copper-coloured, warty papules that occur on the trunk, buttocks, arms and face. It is spread by direct contact or by fomites. The lesion consists of a mass of hypertrophied epidermis that extends into the dermis and protrudes above the skin. In the epithelial cells very large hyaline acidophilic granular masses can be observed. They crowd the host cell nucleus to one side, eventually filling the whole cell. When material from the lesions is crushed, some of the inclusions burst open, liberating large numbers of virions. These have the size, internal structure and morphology of vaccinia virus. The infection has been transmitted experimentally to human subjects, but the virus has not been grown in cultured cells. The development of immunity is slow and uncertain. Lesions can persist for as long as 2 years, and re-infection is common.

Monkeypox

This has been implicated occasionally in a smallpox-like condition in equatorial Africa. It may be fatal in unvaccinated individuals, but is less transmissible from person to person than smallpox.

Parapoxvirus infections

The virions of parapoxviruses are characterized by a criss-cross pattern of tubes in the outer membrane (see Fig. 44.2c) and genomes that are considerably smaller (85 MDa) than those of other poxviruses. They infect ungulates and cause the human occupational diseases of *orf* and *milker's nodes*. The lesions of orf (which causes a disease in sheep known as contagious pustular dermatitis) are often large and granulomatous. Erythema multiforme is a relatively frequent complication. The lesions of milker's nodes are highly vascular, hemispherical papules and nodules. Both diseases are:

- self-limiting
- commonest on the hands
- contracted by contact with infected sheep (orf) or cows (milker's nodes)
- occupational diseases, mainly seen in farm workers such as shepherds, slaughterhouse workers or butchers.

VACCINIA VIRUS AS A VACCINE VECTOR

The vaccinia virus genome can accommodate sizeable losses of DNA in certain regions, to the extent that as many as 25 000 base pairs can be lost without lethal effect. By replacing this non-essential DNA with foreign genes it has been possible to construct novel *recombinant virus strains*, which express the foreign genes when they infect cells. Recombinant vaccinia strains containing as many as four foreign genes, coding for combinations of bacterial, viral and protozoal antigens, have been constructed.

The advantages of this ingenious way of developing new vaccines are:

- they are applicable to many different antigens
- the possibility of constructing multivalent vaccines that could give protection against several diseases after a single 'shot'
- stimulates cell-mediated immunity
- the ease of administration
- cheapness of vaccinia as a vaccine.

A strain that expresses the rabies glycoprotein antigen is being used in Europe to protect foxes. Serious disadvantages remain for human diseases, however. These are primarily those associated with vaccinia virus itself, as its use was associated with a number of serious complications. The most important of these were:

- *progressive vaccinia*, a fatal infection that occurred in immunodeficient individuals
- *eczema vaccinatum*, a serious spreading infection that occurred in eczematous individuals
- *post-vaccinal encephalitis*, which, although rare, was severe and occurred in normal healthy individuals.

These disadvantages preclude the use of vaccinia as a vector for foreign antigens in man, and work is being done on the modification of its virulence to circumvent these problems. So far, studies on virulence genes have shown that poxviruses code for an impressive array of factors that interfere with host defences. These include proteins that bind complement components, act as receptors for interleukin-1β, interferon and tumour necrosis factor, and synthesize steroids. These factors are all exported from infected cells. Factors that act inside cells include proteins that block the action of interferon, inhibit the post-translational modification of interleukin-1β, and prevent the synthesis of a neutrophil chemotactic factor. It is possible that abrogation of virulence factors such as these may lead to safer vaccines.

MYXOMATOSIS: AN EVOLVING DISEASE

As a rule, virus infections are mild and self-limiting. Viruses are obligate parasites and it is not in their interests to cause the extinction or massive reductions in the size of host populations. It is reasonable to suppose that the type of disease caused by a virus reflects the outcome of a process in which host and virus have co-evolved to levels of resistance and virulence optimal for the maintenance of their respective population numbers. The high mortality rate of classical smallpox is considered by some to have been a major factor in the restriction of human population size, and this has been used to support the hypothesis that the association between smallpox and man has been established – in evolutionary terms – only in recent times, co-evolution of the relationship being a long way from equilibrium. It is impossible to test this hypothesis directly for variola, but the relationship between another poxvirus, myxoma virus, and the rabbit in Australia, has provided a dramatic example. This is the only example of co-evolution where changes in an animal host and virus and the evolution of a disease have been studied in real time.

Myxoma virus is South American. It causes a benign local fibroma in its natural host, the rabbit *Sylvilagus brasiliensis*, but causes *myxomatosis* in the European rabbit, *Oryctolagus cuniculus*. This is a generalized infection with a very high mortality rate. Field trials to test its efficacy as a measure for controlling the European rabbit were carried out in Australia in 1950. The virus escaped and caused enormous epidemics in the years that followed. The original virus caused infections with a case mortality rate greater than 99%, and rabbits survived for less than 13 days. Within 3 years virus isolates from epidemics had become much less virulent, causing infections with mortality rates of 70–95% and survival times of 17–28 days. Changes in the resistance of the rabbit also occurred, with mortality rates from infection falling (from 90% to 25% after challenge with strains of virus with modified virulence, for example) and symptomatology becoming less severe. In myxomatosis, natural selection favoured virus strains with intermediate virulence because such strains are transmitted more effectively than highly virulent strains, which kill their hosts too quickly, and non-virulent strains, which are poorly transmitted.

KEY POINTS

- Poxviruses are large cytoplasmic DNA viruses that infect a wide range of species.
- Smallpox is a generalized disease with a vesicular rash and associated with significant mortality.
- Live vaccine is used to eradicate disease, but is not risk free.
- Eradication was successful because only human beings are infected, the disease is easily recognized, transmitted to only a few contacts, and there is no carrier state. Bio-terrorism threatens the re-introduction of smallpox.
- Other human poxvirus infections are of minor significance and include molluscum contagiosum, monkeypox and orf.

RECOMMENDED READING

Baxby D 1981 *Jenner's Smallpox Vaccine*. Heinemann, London

Fenner F, Ratcliffe F N 1965 *Myxomatosis*. Cambridge University Press, Cambridge

Fenner F, Henderson D A, Arita I, Jezek Z, Ladnyi I D 1988 *Smallpox and its Eradication*. World Health Organization, Geneva

Fields B N, Knipe D M, Howley P M (eds) 1996 *Virology*, 3rd edn. Lippincott-Raven, Philadelphia

Jenner E 1798 *An Inquiry into the Causes and Effects of the Variolae Vaccinae*. Sampson Low, London

45 Papillomaviruses and polyomaviruses

Warts: warts and cancers; progressive multifocal leuco-encephalopathy

H. A. Cubie

The *papillomavirus, p*olyomavirus and the *v*acuolating virus of rhesus monkeys (simian virus 40; SV40) are small DNA viruses which used to be classified in two genera (*Papillomavirus* and *Polyomavirus*) of the family Papovaviridae, the name being derived from the first two letters of the names of each of the members. All are small DNA viruses with icosahedral capsid structure, but the molecular biology and pathogenesis of the two groups differ considerably, and the two groups are now regarded as separate families: the Papillomaviridae and the Polyomaviridae.

PAPILLOMAVIRUSES

Papillomaviruses are widely distributed in nature and are found in more than 20 mammalian species as well as in birds and reptiles. They are species-specific DNA viruses that infect squamous epithelia and mucous membranes, and are responsible for many varieties of *warts* and fibropapillomata. In 1933, Shope found that the cottontail rabbit papillomavirus caused benign warts, with 25% of the lesions undergoing malignant change within a year. Viral DNA remains detectable within the tumours. In domestic rabbits, however, up to 75% of lesions progressed to cancer. Application of tar to the lesions led to more rapid malignant change. Bovine papillomavirus (BPV) causes tumours of the alimentary canal and bladder in cattle fed a diet containing bracken fern. The tumours are preceded by infection with BPV types 4 and 2, respectively, but the virus is not detected after malignant transformation. These studies show that malignant progression can be influenced by both genetic composition and exposure to chemicals. Human papillomavirus (HPV) also causes warts and tumours.

Genome organization and replication

The papillomaviruses have:

- a diameter of 52–55 nm
- an icosahedral capsid composed of 72 capsomeres (Fig. 45.1)

- a supercoiled double-stranded DNA genome, with a single coding strand, a molecular weight of approximately 5×10^6 Da and consisting of about 7900 base pairs.

All papillomaviruses are dependent on epithelial differentiation for completion of their life cycle. Although viral DNA can be found in the basal cells, whole virions are found only in terminally differentiated keratinocytes, with expression of viral gene products being closely regulated as infected cells move from the basal to superficial layers.

The genome is divided into an early region with two large (E1 and E2) and several smaller (E4–E7) open reading frames and a late region with two large genes (L1 and L2) that code for the capsid proteins (Fig. 45.2). The coding sequences are distributed in all three open reading frames with considerable overlap along a single strand of DNA. Complex splicing events allow many different transcripts to be produced, with the resulting proteins having different biological roles. The gene functions are summarized in Table 45.1. Unravelling the functions of these genes has contributed significantly to our understanding of the pathogenicity of different HPV types and particularly to the transforming ability of the high-risk types.

HUMAN PAPILLOMAVIRUSES (HPV)

Classification

The existence of different HPV types at different sites was first recognized in the late 1960s, with differentiation of cutaneous and genital warts. Present data suggest that more than 200 HPV types exist, with the genome of approximately 100 types being fully sequenced. An HPV type is defined as a complete genome, where the sequence of the L1 gene is at least 10% dissimilar to that of any other type. Subtypes vary by 2–10% in L1 sequence, whereas variants have less than 2% variation in L1. Eight genera of HPVs have been agreed, with the

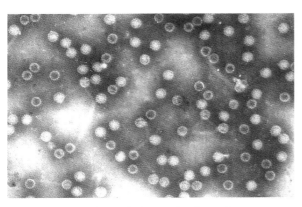

Fig. 45.1 Electron micrograph of a wart virus from a plantar wart (phosphotungstic acid stain).

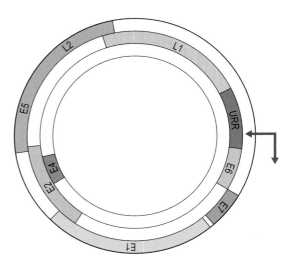

Fig. 45.2 Simplified circular map of HPV-16 genome, identifying early (E) and late (L) open reading frames. Arrow at origin. URR, upstream regulatory region.

Table 45.1 Papillomavirus gene functions

Gene	Function	Comments
URR (upstream regulatory region)	Regulation of gene function and initiator of viral replication	Both positive and negative transcriptional control elements
E1	Episomal maintenance	Frequently disrupted by integration
E2	Regulation of transcription and replication. DNA-binding protein; very similar in BPV and HPV	Regulates transcription and viral replication in association with E1. Frequently disrupted by integration
E4	Virion maturation, disrupts cytoskeleton	May facilitate release of virions from differentiated keratinocytes
E5	Transforming function – alters signalling from growth factor receptors	Often lost or not expressed after integration. Not found in all papillomaviruses
E6	Transforming function – binds to p53 and other pro-apoptotic proteins	Co-operates with E7 to stimulate cells into S phase, retard cell differentiation and increase efficiency of transformation
E7	Major transforming gene – binds to members of pocket protein family such as Rb	Transforms cells independently of E6
L1	Production of major capsid proteins	Group- and type-specific determinants
L2	Production of minor capsid proteins	Type-specific determinants

majority of HPVs falling into the Alpha genus. Different HPVs have evolved to fill different biological niches, with the particular tissue infected and the appearance of the lesions produced being associated with specific virus types (Table 45.2).

Types can be categorized as low, intermediate or high risk according to the extent of their oncogenic potential. The tissue tropisms and relatedness of HPVs can be seen in phylogenetic trees produced by computer algorithms of aligned sequences. Cutaneous types form one clear

Table 45.2 Classification of human papillomaviruses by genus, according to International Council on Taxonomy of Viruses

Genus	Type	Properties
Alpha	HPV-2, -27, -57	Common skin warts, also frequently in genital warts in children
	HPV-18, -39, -45, -59	Mucosal lesions, high risk, more frequent in adenocarcinomas
	HPV-16, -31, -33, -35, -52, -58, -67	Mucosal lesions, high risk, more frequent in squamous carcinomas
	HPV-6, -11, -13, -44, -74	Benign mucosal lesions – genital warts, laryngeal papillomas
Beta	HPV-5, -8	Cutaneous benign and malignant lesions in immunosuppressed patients
Gamma	HPV-4, -65	Cutaneous benign lesions
Mu	HPV-1, -63	Cutaneous benign lesions, frequently on feet

Adapted from Bernard SU 2005 The clinical importance of the nomenclature, evolution and taxonomy of human papillomaviruses. *Journal of Clinical Virology* 32S: S1–S6.

branch of the tree, and genital types with greatest malignant potential form a second branch (Fig. 45.3).

Transformation

Transfection of viral DNA into cultured cells has yielded much information on the genes responsible for the tumorigenic potential of HPV. Integration of the viral genome results in disruption of E1 and E2, with subsequent increase in the expression of E6 and E7. The immortalizing properties of high-risk HPV types such as HPV-16 and HPV-18 are due largely to expression of E7, but co-operation between E6 and E7 is necessary for efficient transformation.

Both E6 and E7 stimulate cellular proliferation. E7 mediates entry of infected cells into S phase by binding to the tumour suppressor gene product, Rb, which normally prevents S-phase entry. Thus, binding of E7 to Rb permits continued cell growth, thus preventing differentiation. E6 prevents induction of apoptosis in response to this unscheduled S-phase entry by binding to the tumour suppressor gene product p53 and causing its rapid degradation. The function of p53 is to induce cellular arrest to allow repair of DNA damage. Only E6 from high-risk HPV types can overcome the growth arrest function of p53. In-vitro studies with BPV and HPV show several E5 interactions with growth factor receptors, resulting in their sustained activation and a clear contribution to the transformed state, but not all papillomaviruses possess an *E5* gene, so it cannot be vital for transformation.

Clinical features and pathogenesis of HPV infections

Cutaneous warts

Cutaneous warts commonly infect the keratinized epithelium of the hands and feet, producing typical

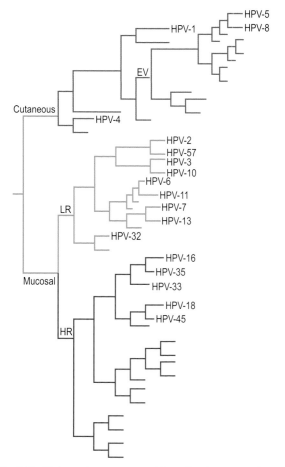

Fig. 45.3 Phylogenetic tree constructed from aligned sequences of part of the *E6* gene of HPV. LR, low risk; HR, high risk (of malignancy). (Adapted from van Ranst M, Tachezy R, Burk R D 1996 Human papillomaviruses: a neverending story? In Lacy C (ed.) Papilloma Reviews: Current Research on Papillomaviruses, pp. 1–19. Leeds University Press, Leeds.)

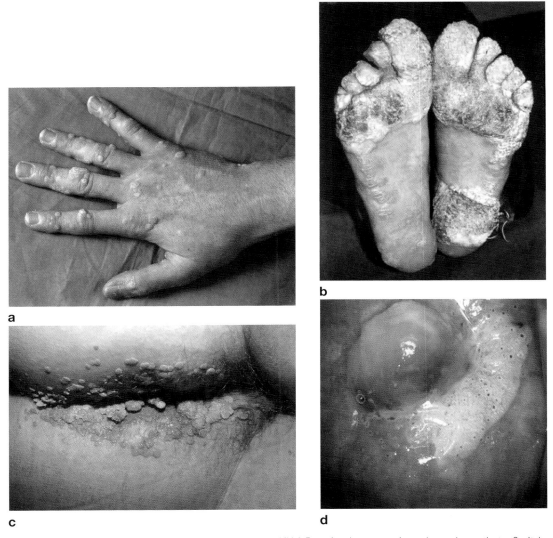

Fig. 45.4 Typical clinical presentations of HPV. **a**. Common hand warts on child. **b** Extensive plantar warts in renal transplant patient. **c** Genital warts (condylomata acuminata). **d** Cervical flat wart after application of acetic acid. (Courtesy of Dr M. H. Bunney and Dr E. C. Benton.)

warts frequently seen in young children and adolescents (Fig. 45.4a). Table 45.3 gives the grouping of HPV types by their clinical manifestations. Typically, HPV-1 and HPV-4 are commonly found on the feet, whereas HPV-3 and HPV-10 are associated with flat warts and HPV types 2, 4 and 7 with common warts.

Histologically, the lesions are benign with hypertrophy of all layers of the dermis and hyperkeratosis of the horny layer (Fig. 45.5). They usually disappear spontaneously but occasionally may be resistant to treatment. Regrowth of the lesions after treatment is probably due to persistence of the virus in the skin surrounding the original wart.

The incidence of viral infection greatly exceeds that of associated disease, indicating that host factors contribute to the control and outcome of infection. Both humoral and cell-mediated immune responses develop, with regression largely resulting from T helper (T_H) type 1-driven cytotoxic T cells and protection from subsequent infection with the same HPV type from T_H2 stimulation of B cells. Patients with primary or secondary cell-mediated immunodeficiency have an increased risk of developing warts, which can be extensive (Fig. 45.4b), persistent or recurrent, and likely to progress. For example, in people suffering from the rare genetic skin

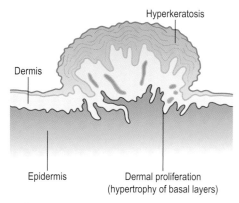

Fig. 45.5 Diagrammatic representation of the histological appearance of a wart.

Table 45.3	HPV types associated with specific lesions
Lesion	HPV type
Cutaneous sites	
Deep plantar warts (verruca plantaris)	1, 4
Common warts (verruca vulgaris)	2, 4, 7
Plane or flat warts	3, 10
Epidermodysplasia verruciformis-like macular lesions	5, 8
Mucocutaneous sites	
Genital warts (condyloma acuminatum)	6, 11
Flat condyloma and intra-epithelial neoplasia	16, 18, 31, 33, 45
Laryngeal papillomata	6, 11
Oral papillomata	2, 6, 11, 16, 18, 57
Oral hyperplasia	13, 32

disorder *epidermodysplasia verruciformis*, in which there is a selective depletion of specific T cell clones, large plane warts associated with virus types such as HPV-5 and HPV-8 develop and persist for life. Similarly, up to 40% of renal allograft recipients develop cutaneous warts within a year of receiving the graft, rising to more than 90% in those with graft survival for longer than 15 years. Epidermodysplasia verruciformis EV-associated HPV types are found in a variety of lesions, including a high proportion of psoriatic skin lesions but also in normal skin and hair follicles in people with no immune defects, suggesting that the viruses are widespread and that development of visible lesions is well controlled by the immune system of healthy individuals.

Anogenital warts

Genital HPV infections are very common in the general population and a huge increase in incidence has been reported in recent years. Anogenital warts (also known as *condylomata acuminata;* Fig. 45.4c) are found predominantly in sexually active adults and are the most common clinical manifestation. In women they are found on the vulva, within the vagina and on the cervix.

Vulvar and vaginal warts are usually plainly visible, but on the cervix they may be indistinguishable from the normal mucosa without the aid of a colposcope to magnify the cervical epithelium. The application of 5% acetic acid causes whitening of epithelium in which there is a high concentration of nuclear material and reveals subclinical lesions as areas of densely white whorled epithelium known as flat or non-condylomatous warts (Fig. 45.4d). Subclinical lesions may become clinically apparent if the immune response is disturbed, as in pregnancy.

In men, the most common sites for lesions are the shaft of the penis, peri-anal skin and the anal canal. Subclinical infection may also occur in men and is visible colposcopically only when the penis is painted with 5% acetic acid.

About 90% of genital warts are associated with HPV types 6 and 11. High-risk types such as HPV-16 and HPV-18 are more likely to be associated with subclinical or latent infections, which can be detected only in the laboratory, usually by looking for HPV DNA. From multiple international studies, the median prevalence of high-risk HPV is 15% among women with normal cytology in superficial cells from cervical scrapes. The peak prevalence is found in young women aged under 25 years, and declines with age. HPV-16/18 is more commonly found than HPV-6/11, and persistent latent infection carries a much greater risk of progression to cancer.

Orolaryngeal lesions

Recurrent respiratory papillomatosis (RRP) This is a rare condition characterized by the presence of benign squamous papillomata on the mucosa of the respiratory tract, most commonly on the larynx. It has a bimodal age distribution with peaks of incidence in children under 5 years of age and adults after the age of 15 years, and is caused by infection of the respiratory mucosa with HPV types 6 and 11. Children acquire the disease by passage through an infected birth canal, whereas adults acquire the disease from orogenital contact with an infected sexual partner. The transmission rate is low. The disease presents with hoarseness of voice or, in children, with an abnormal cry. As the lesions grow they may cause stridor and upper airway obstruction, which can be life-threatening and require emergency surgery. Recurrence after treatment is common, and extension of the disease to the bronchial tree may occur. Malignant conversion of laryngeal papillomata has been described in the past, usually after radiotherapy to treat the initial lesion.

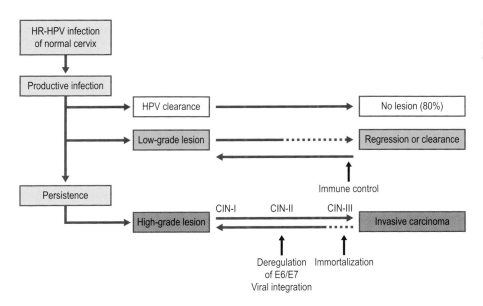

Fig. 45.6 HPV infection and pathogenesis of cervical cancer. HR, high risk; CIN, cervical intra-epithelial neoplasia.

Oral papillomatosis. A variety of papillomata and benign lesions occur on the oral mucosa and tongue. Several HPVs, including types 2, 7, 13 and 32, have been found. Multiple lesions may develop on the buccal mucosa, a condition known as oral florid papillomatosis. Subclinical lesions can be detected on the oral mucosa of normal adults after the application of acetic acid. The virus types here are those found more commonly in the genital tract, and infection is acquired during orogenital contact with an infected sexual partner.

HPV and cancer

Premalignant lesions of the genital tract

Malignant disease of the cervix is preceded by neoplastic change in the surface epithelium, a condition known as *cervical intra-epithelial neoplasia* (CIN). A similar pattern of events takes place in other sites in the genital tract of both men and women. The initial transforming event takes place in the deepest layer of the epithelium, the germinal layer, and abnormal cells spread through the surface layers. This condition increases in severity from:

- CIN-I (low-grade squamous intra-epithelial lesions (LSIL in American classification), to
- CIN-II and CIN-III (high-grade or HSIL), in which abnormal cells including mitotic figures can be found in all layers of the epithelium with loss of stratification and differentiation.

Untreated CIN-II/III can progress to invasive cancer in a large proportion of affected individuals, whereas CIN-I lesions are unlikely to progress. Regression may occur spontaneously at all stages (Fig. 45.6). HPV DNA can be detected in all grades of the premalignant lesions of the female and male genital tract.

- HPV types 6 and 11, and other low-risk types, are found most commonly in low-grade disease, whereas
- HPV types 16 and 18, and other high-risk types, are more commonly associated with lesions of greater severity and with invasive cancer.
- Infection with multiple types is common but is not associated with an increased risk of progression.

Squamous cell carcinoma

The association of wart viruses with invasive cancers of the skin, larynx and genital tract is well documented. Malignant conversion of skin warts occurs in about one-third of patients with epidermodysplasia verruciformis at a relatively young age on skin exposed to sunlight, with HPV type 5 or 8 in more than 90% of squamous cancers. Malignant conversion of cutaneous warts in renal allograft recipients also occurs. Squamous cell carcinoma of the aerodigestive tract has been shown to harbour HPV to varying extents in different studies and in different tumour types, with more than 20 HPV types found. More than half of all carcinomas of the tonsils harbour HPV, with HPV-16 present in 84%, although the nature of the association is not fully established.

The commonest association with invasive cancer, however, is with tumours of the anogenital tract. More than 99% of invasive cervical cancers contain HPV DNA. The proportion of cancers containing different HPV types varies around the world, and the incidence shows wide demographic, ethnic and socio-economic variation.

- In Indonesia HPV-18 predominates.
- In Europe HPV-16 is by far the most common.
- HPV-16 is the cause of more than 50% of invasive cervical cancers worldwide.
- Cervical cancer is the second most common cancer of women worldwide with a mean incidence of 35 per 100 000.

HPV-18 and the related HPV-45 are found more frequently in the more aggressive adenocarcinomas. In-situ DNA hybridization of tumour biopsies has shown that HPV DNA is unevenly distributed throughout the tumour, suggesting that these cancers are polyclonal in origin.

In most animal cancers associated with papillomavirus there is a co-factor. In man, smoking, seminal fluid factors, immune status and the genetic background of the individual all act as co-factors in HPV-associated genital tract malignancies.

- Women who smoke are at increased risk of CIN and cancer of the cervix.
- Women who are immunosuppressed, whether due to drug treatment (transplant recipients) or secondary to haematological malignancies (leukaemias) or to infection with human immunodeficiency virus (HIV), are also at increased risk.

The incidence of anal cancer in men who have sex with men and in HIV-infected men is similar to that of cervical cancer. It is thought that the presence of a cellular transformation zone makes these sites more susceptible to the oncogenic effects of high-risk HPV infection. In contrast, vaginal, vulval and penile cancers are extremely rare, and these areas lack a transformation zone. In the development of HPV-related carcinomas at other sites, exposure to ultraviolet or ionizing radiation, smoking, chewing of betel-quid, immunosuppression and genetic make-up are all likely to be important.

Transmission and epidemiology

Clinical studies of the incidence of skin warts show that infection is common in early childhood and is acquired by direct contact with an infected person. HPV is a stable, hardy virus, so fomites are also important, as shown by outbreaks of hand warts from the use of gymnastic apparatus and plantar warts acquired from swimming pool surrounds. Mild shearing trauma is necessary to allow the virus to reach the basal layers of the skin.

Transmission of genital warts occurs mainly during sexual activity and there is a strong association between increasing numbers of sexual partners and the prevalence of genital HPV infection. The major risk factors in the development of genital HPV disease include:

- sexual activity
- smoking
- long-term oral contraceptive use
- infection with other sexually transmitted infections.

It is, however, extremely difficult to assess the relative interdependence of all risk factors and epidemiological studies have not been able to identify consistently which risk factors are most strongly associated with progression to cancer.

Genital wart types can be acquired non-sexually. Vertical transmission of low- and high-risk HPV types from mother to infant at birth and perinatal transmission within the first 6 weeks of life have both been clearly demonstrated. The high rates and persistence of buccal infection with HPV-16 in primary schoolchildren could also be explained by horizontal transmission. Furthermore, genital HPV DNA has been detected in virgins and antibodies to HPV-16 have been demonstrated in up to 10% of young schoolgirls.

Laboratory diagnosis

Morphological identification

HPV infection may be readily diagnosed when there are typical clinical lesions. Subclinical infection requires laboratory confirmation using cytological, histological, immunocytochemical and molecular detection of nucleic acid. HPV infection can be recognized morphologically in cervical smears by the presence of vacuolated cells with atypical enlarged hyperchromatic nuclei described as *koilocytes*. However, koilocytes are not always present, are not sufficiently specific for HPV, and cannot differentiate between low- and high-risk HPV types. Histological examination of a biopsy taken from a lesion identified at colposcopy will show more specific features of HPV infection, including papillomatosis, hyperkeratinization of the surface layer, hypertrophy of the basal layers and disorganization of the epidermal structure.

Molecular HPV testing is more sensitive than cytology for detection of high-grade CIN, but it has a lower specificity, especially in younger women in whom transient infection is common. This provides a role for HPV screening within cervical screening programmes for more effective identification of women with lesions likely to progress to cervical cancer. HPV testing could be used in association with cytology or, indeed, instead of cytological screening in countries with no cervical screening programmes. Large-scale trials are ongoing in several countries to evaluate further the effectiveness of HPV testing.

Molecular methods

The first molecular method used for HPV detection was in-situ hybridization on tissue sections or exfoliated cells; it has the advantage of providing histological information at the same time as HPV detection. In-situ hybridization lacks sensitivity, and hybridization in solution using a cocktail of high-risk HPV probes followed by immuno-chemical signal amplification (Hybrid Capture® assay) has proved more robust and been used in several large cervical screening studies. Amplification of HPV DNA by polymerase chain reaction (PCR) can also be used with a variety of consensus or degenerate primers to provide a generic screening test. For diagnosis of persistence or follow-up after treatment of cervical disease, type-specific HPV tests are required. Currently these are based on reverse hybridization of PCR product to multiple probes immobilized on membrane strips (line blot assays).

Serology

The lack of native antigen hampered the development of serological assays for many years. Studies reporting the detection of antibody to recombinant proteins or synthetic peptides from the E6 or E7 gene in women with high-grade cervical disease suggest there may be a correlation between antibody presence and poor prognosis. Recent production of virus-like particles in eukaryotic expression systems using baculovirus or vaccinia recombinants has allowed the detection of antibodies to specific HPV types. Such assays are useful for showing evidence of past exposure rather than current infection, and are appropriate for epidemiological studies. They are also essential in studies of the effectiveness of HPV vaccination where responsiveness to each HPV component in the vaccine should be monitored.

Treatment and control

In most immunocompetent people, warts are a cosmetic nuisance and will eventually disappear spontaneously. Warts may be destroyed by cryotherapy with dry ice or liquid nitrogen, but simple topical treatment with salicylic acid at home is often successful. Genital warts are more problematic. They are a common sexually transmitted infection and represent a significant cost to health-care resources. In pregnant women, vaginal warts may occasionally grow to such a size that the birth canal is obstructed and surgical removal is required. Prevention of spread of wart viruses can be achieved by avoiding contact with affected individuals. Thus, the use of condoms will diminish the risk of spread of genital warts. Interferon, photodynamic therapy and indole-3-carbinol have been used for treatment of recurrent laryngeal warts after the reduction of tumour load by cautery or excision.

Many treatments can be used for genital warts including:

- antiproliferative agents such as podophyllin or 5-fluorouracil, although close monitoring is required
- destructive therapies such as trichloracetic acid, liquid nitrogen or surgical excision
- immunomodulators such as imiquimod, which activates monocytes/macrophages and causes direct release of interferon-α (self-applied topical treatment).

All of these treatments will remove the lesions but do not always eradicate the virus from surrounding normal epithelium. Incomplete treatment is the commonest cause of recurrence of warts; all of a patient's warts must have disappeared with restoration of normal skin texture before a cure is considered. This is also true for treatment of CIN. Long-term follow-up represents a significant cost to health care.

Vaccines

Vaccines are available which can prevent the development of papillomata and cancers in cattle and rabbits, consisting principally of virus-like particles of L1 or L1 with L2 proteins. In man, similar prophylactic vaccines containing L1 of HPV-16 and 18 or of HPV-6, 11, 16 and 18 have been under trial for over 5 years. Both proved to be safe and highly efficacious and the first license was granted in 2006. They provide almost complete protection from cervical disease due to HPV-16 and 18 (both vaccines) and from genital warts (quadrivalent vaccine). Therapeutic vaccines containing viral oncoproteins have been used to stimulate cell-mediated immunity, and again trials are under way to assess effectiveness. While protection from infection is now possible, it will be many years before the ability of vaccine to prevent or halt cervical cancer development can be assessed.

POLYOMAVIRUSES

The term *polyomaviruses* literally stands for many (poly) tumour (oma) viruses. The viruses are species specific and, although tumour induction is well described in experimental animals, there is to date no association with any human tumour. Members include the mouse polyomavirus, simian virus (SV) 40 of monkeys and two viruses of man, grouped as *Polyomavirus hominis* (huPyV) type 1 (JC virus; JCV) and type 2 (BK virus; BKV), both named after the initials of the people from whom they were first isolated. Each type contains different genotypes.

Table 45.4 Polyomavirus-associated infection and disease

	BK virus		JC virus	
	Prevalence of infection	Associated disease	Prevalence of infection	Associated disease
Healthy child	50% seropositive by age 3 years	Upper respiratory tract infection; mild pyrexia	5% seropositive pre-school	None
Healthy adult	90–100% seropositive	Transient cystitis	80% seropositive	None
Immunocompromised patients Urogenital		Ureteric stenosis; haemorrhagic cystitis; tubular nephritis		Ureteric stenosis (occasional)
Lung		Interstitial pneumonitis		None
Central nervous system		Subacute meningoencephalitis		PML

Adapted from Dorries et al (2004). PML, progressive multifocal leuco-encephalopathy.

The virions are 42–45 nm in size with a 72-capsomere icosahedral capsid. The genome has a molecular weight of 3.4×10^6 Da and is approximately 5100 bp in length. Like the papillomaviruses, it is a double-stranded supercoiled loop of DNA, but both strands code for virus proteins. Capsids contain three structural proteins, VP1 (major), VP2 and VP3 (minor), and each capsomere contains five molecules of VP1 and one of VP2 or VP3.

Replication and transformation

There are three non-capsid regulatory proteins: large T (*Tag*), small t and agnoprotein. These are the first antigens to appear after infection; they accumulate in the nucleus, stimulate cellular growth and are important for replication. This is followed by a switch to transcription of the late region and production of the structural proteins (VP1, VP2 and VP3). The growth cycle in culture is 36–44 h, and the release of mature virus particles follows lysis of the cell. The structural proteins determine host range and infectivity. BKV can be grown with difficulty in human embryonic kidney cells, whereas JCV, which is not only species but also tissue specific, replicates only in human embryo glial cell cultures.

The early proteins are associated with immortalization and transformation. Large T antigen binds to both Rb and p53, and prevents the induction of cell death. Only murine and hamster polyomaviruses appear to be oncogenic in their natural hosts. Both BKV and JCV have been shown to be oncogenic for newborn hamsters, but have not been found consistently in any human tumours.

It would appear that interaction between polyomaviruses and other viruses can occur. For example, the HIV-1

Tat gene and possibly cytomegalovirus (CMV) immediate early gene *IE2* induce the JCV promoter, whereas BKV *Tag* can induce the expression of CMV early genes. It is not yet known how these effects contribute to pathogenesis.

Clinical features and pathogenesis

Serological studies show that infection is common, but primary polyomavirus infection is a rare diagnosis. BKV seroconversion has been associated with tonsillitis, acute upper respiratory tract infection, mild pyrexia and transient cystitis (Table 45.4). Reactivation is common, particularly in organ transplant patients, with up to 40% of renal allograft recipients and bone marrow transplant patients excreting polyomavirus in the early months after transplantation. Reactivations of BKV and JCV infections are also quite common in pregnancy, with viruria being detected in 3–7% of women, especially during the second and third trimesters. Although there are no specific associated symptoms, women with persistent viruria have more illnesses and their babies are more often premature or jaundiced, but there is no evidence that polyomaviruses are transmitted transplacentally.

Thus disease is associated with persistence and reactivation of virus in immunocompromised people.

Progressive multifocal leuco-encephalopathy (PML)

This condition was first described in 1958 in patients with *Hodgkin's lymphoma* and *chronic lymphocytic leukaemia*, and is associated with the reactivation of JCV in the brain. Over the past 10 years, JCV has been found almost exclusively in patients with acquired immune deficiency syndrome (AIDS).

Patients with PML have multiple foci of demyelination, usually in the cerebral hemispheres, but occasionally elsewhere in the central nervous system. A unique combination of pathological changes in PML is identifiable histologically. The peripheral zone surrounding areas of demyelination contains distinctively altered oligodendrocytes. These are swollen, with hyperchromatic nuclei and occasional basophilic inclusions pathognomonic of PML. The cells contain many virions detectable by electron microscopy. Replication of the virus occurs in the nucleus, causing cell destruction and breakdown of the myelin sheath. In the centre of the lesion, oligodendrocytes are absent and, in contrast, the astrocytes present look more neoplastic and do not contain virions. Some astrocytes may be transformed but tumour formation is absent. Malignant gliomas are occasionally found near areas of demyelination. Genotypes 1 and 2 of JCV are consistently associated with PML, although low levels of JCV DNA have been reported in patients without clinical evidence of PML. Whether this indicates a latent state or a preclinical stage of PML is currently unknown.

The clinical features depend on the areas affected. Symptoms include visual, mental and speech impairment, hemiplegia, loss of memory and dementia. Death usually occurs within 6 months of the first signs of the disease. Similar symptoms can have other causes and differential diagnosis is essential. Use of highly active antiretroviral therapy (HAART) has prolonged survival in HIV-infected patients with PML, such that 'active' and 'inactive' disease states have been proposed. Patients with the former are likely to die within a few months, whereas those with the latter may be stable over years of follow-up.

Polyomavirus infection and transplant recipients

In renal transplant recipients, polyomavirus infection appears to be as common as CMV infection. Most infections appear to be due to reactivation of BKV, although primary infection with JCV has been recognized with the donor kidney as the likely source of virus. Half of the infections occur between 4 and 8 weeks after transplantation, but subclinical reactivations of virus occurring up to 12 months after transplantation are not uncommon. Interstitial tubular nephritis is the most common disease and is associated with a transient decrease in graft function. If this is wrongly interpreted as being a rejection episode and managed by further immunosuppression, complications may occur. Polyomavirus infection of the transplanted ureter results in cell proliferation, leading to ureteric stenosis and occasionally to obstruction. Obstruction can occur for up to 300 days after surgery and may not be recognized until nephrectomy or autopsy is performed.

In bone marrow transplant recipients, 38% of a study series in the USA were found to be excreting polyomavirus in the urine, with the highest incidence in patients with acute myeloid leukaemia. Excretion of BKV appears to be much more common than that of JCV. Although not associated with graft-versus-host disease, some patients had a transient post-transplant hepatitis. Late-onset BKV viruria has been linked to severe haemorrhagic cystitis in up to 25% of patients.

Recently cases of PML have also been found in patients after solid organ or cell transplants.

Laboratory diagnosis

Detection of virus particles by electron microscopic examination of negatively stained urine deposits or brain homogenates is possible, and viral antigen can be detected by means of indirect immunofluorescence on exfoliated cells from urine or in acetone-fixed brain smears. The 'gold standard' for diagnosis of PML depends on a combination of neuro-imaging and biopsy. Immunohistochemical methods for detecting viral antigens or DNA by in-situ hybridization, together with histological markers such as astrocyte morphology and lipid-laden macrophages, are used for diagnosis of active PML. There is a need for less invasive methods and nucleic acid amplification, particularly nested PCR, is required for detection of JCV DNA in cerebrospinal fluid. When this is combined with intrathecal antibody detection, it may provide a sensitive non-invasive diagnostic procedure. However, standardization is required and, as yet, none of the available clinical, radiological or virological methods can be used as stand-alone diagnostic techniques.

Antibodies to polyomaviruses can be measured using haemagglutination inhibition (HAI) and enzyme-linked immunosorbent assay (ELISA). HAI uses whole virions and gives a species-specific result, whereas in ELISA disrupted virions or purified VP1 can be used to give a more sensitive and type-specific result. Rising titres and the presence of immunoglobulin M are diagnostic of recent infection. Failure to detect antibody to JCV nearly always excludes a diagnosis of PML.

Transmission and epidemiology

Very little is known about the transmission of polyomaviruses. Molecular detection and virus isolation of BKV suggest both respiratory and sexual routes are possible, although oral–faecal transmission has also been suggested and studies of sewage in Europe show faecal contamination of fresh water by JCV. Urinary excretion was shown to be the route of transmission in Japanese families. Isolation of both viruses has been principally from immunocompromised individuals, with BKV from urine or kidney and JCV from the brain, urine and renal tract. Detection by PCR shows a wider distribution, with BKV DNA found in tonsils, lung, lymph nodes and spleen, and JCV genome in lung, liver, spleen, lymph nodes and leucocytes.

Serological studies show that BKV infection is a common event in early childhood. By 3 years of age 50% of children have antibody, rising to almost 100% by the age of 10 years. JCV circulates independently of BKV in the community, and acquisition of antibodies is slower and more variable. In the UK, 5% of the pre-school population have antibody, with this proportion rising to 50% in adolescence and to 80% in adults (see Table 45.4). In Japan, infection is more rapid and widespread, with two-thirds of children seropositive by the age of 6 years, rising to 90% of adults.

Treatment

There is no established treatment for PML, although many antiviral agents have been tried. Nucleoside analogues such as cytosine arabinoside (ARA-C, cytarabine) may allow some improvement, especially if given intrathecally for a long time. Current efforts are focusing on targeted intraparenchymal delivery of drugs. Cidofovir in combination with HAART may also be useful. Use of HAART in HIV-associated PML leads to stabilization and a decrease in JCV viral load in cerebrospinal fluid, with a significant increase in the survival of patients on regimens that include a protease inhibitor.

Reducing the level of immunosuppression is likely to be beneficial, but may not be possible because of the underlying disease. Immunomodulators such as interferon and interleukin-2 have been tried without great success.

Few large-scale trials have been undertaken, but the increased incidence of polyomavirus diseases make such studies essential.

KEY POINTS

- Papillomaviruses are ubiquitous, found in most mammalian species as well as in birds and reptiles. More than 100 types of HPV are recognized.
- Some HPVs are associated with skin lesions such as common hand warts and plantar verrucas.
- More than 30 HPV types infect the genital mucosa and are associated with a variety of lesions ranging from benign genital warts to invasive cervical cancers.
- *E6* and *E7* are transforming genes of HPV whose protein products bind to cellular tumour suppressors p53 and Rb respectively.
- HPV types can be grouped into high-risk and low-risk types, dependent on their oncogenic potential.
- HPV-16 is the most common high-risk HPV worldwide, and is associated with more than 50% of cervical cancers.
- BKV and JCV are human polyomaviruses associated with disease in immunosuppressed individuals.

RECOMMENDED READING

Best J M, Cubie H A (eds) 2005 Human papillomaviruses. *Journal of Clinical Virology* 32(Suppl 1)

Bunney M H, Benton E C, Cubie H A 1992 *Viral Warts – Biology and Treatment*, 2nd edn. Oxford University Press, Oxford

Dorries K 2004 Human polyomaviruses. In Zuckerman A J, Banatvala J, Pattison J R, Griffiths P D, Schoub B D (eds) *Principles and Practice of Clinical Virology*, 5th edn, pp. 675–702. John Wiley, Chichester

Khalili K, Stoner G L (eds) 2001 *Human Polyomaviruses. Molecular and Clinical Perspectives*. Wiley-Liss, New York

McCance D 2004 Papillomaviruses. In Zuckerman A J, Banatvala J, Pattison J R, Griffiths P D, Schoub B D (eds) *Principles and Practice of Clinical Virology*, 5th edn, pp. 661–674. John Wiley, Chichester

Zur Hausen H 1998 Papovaviruses. In Collier L, Balows A, Sussman M (eds) *Topley and Wilson's Microbiology and Microbial Infections*, Vol. 1: *Virology*, 9th edn, pp. 291–308. Arnold, London

46

Hepadnaviruses

Hepatitis B infection; deltavirus infection

P. Simmonds and J. F. Peutherer

The family Hepadnaviridae includes the human hepatitis B virus (HBV) and the woodchuck, ground squirrel and Pekin duck viruses; others have been identified, but these are the best known and studied. The viruses show similarities in the structure of their virions and associated particles, the size, nature and replication of the DNA genome and their ability to cause both acute and chronic infections in their natural hosts. HBV is a major cause of chronic liver disease and hepatocellular carcinoma; worldwide it is estimated to cause more than 1 million deaths each year.

HEPATITIS B VIRUS

PROPERTIES

Structure

Three different particles can be seen in the blood in HBV infection (Figs 46.1 & 46.2). The predominant form is a small, spherical particle with a diameter of 22 nm. Filaments are also present with a diameter of about 22 nm. Both types of particle are composed of lipid, protein and carbohydrate; they are not infectious and consist solely of surplus virion envelope. The particles carry the hepatitis B surface antigen (HBsAg). The third type of particle, the virion or *Dane particle*, has a diameter of 42 nm; enclosed within the envelope is the core (27 nm), which contains the viral DNA and polymerase within a shell composed of hepatitis B core antigen (HBcAg). There may be as many as 10^{13} of the small particles and filaments per millilitre. The virions are present in much smaller numbers, usually by a factor of 10^3 or more, and the proportion varies considerably in different stages of the disease. The viral DNA is about 3200 nucleotides long and is circular (Fig. 46.3). The long strand is complete, but there is a gap of variable length of about 1000 nucleotides in the complementary strand; this can be closed via the action of the virion polymerase when virus replication starts.

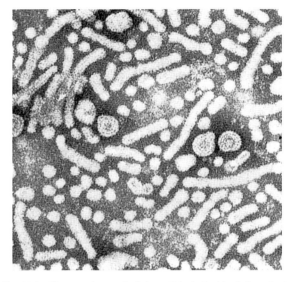

Fig. 46.1 Electron micrograph of the particles in the blood of a patient infected with HBV (original magnification 130 000). (Courtesy of Dr A. Keen, University of Cape Town.)

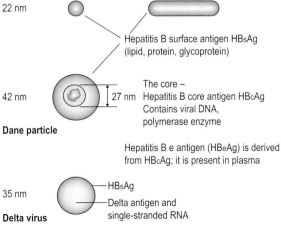

22 nm

Hepatitis B surface antigen HBsAg (lipid, protein, glycoprotein)

42 nm — 27 nm

Dane particle

The core –
Hepatitis B core antigen HBcAg
Contains viral DNA, polymerase enzyme

Hepatitis B e antigen (HBeAg) is derived from HBcAg; it is present in plasma

35 nm

HBsAg
Delta antigen and single-stranded RNA

Delta virus

Fig. 46.2 Particles and antigens of HBV and delta virus.

There are four overlapping genes coding for the core, surface and polymerase proteins, and an X protein that may act as an activator of transcription. The hepatitis B e antigen (HBeAg) is translated from the HBcAg gene using an upstream initiating codon. The protein is secreted from infected cells into plasma, especially at times when there is active viral replication reflected by the release of Dane particles into the circulation. The surface antigen gene is transcribed to produce three messenger RNAs (mRNAs): L, M and S. These are translated to give three proteins, each containing the S protein. The product of the M mRNA consists of the S and pre-S_2 proteins. The protein from the L mRNA comprises pre-S_1, pre-S_2 and S. The L product is present only in the virion, and the M and S proteins are found in each type of particle.

Genetic variation

HBV can be classified into at least eight genotypes, A–H, by comparison of their nucleotide sequences (Fig. 46.4). Some genotypes are distributed worldwide, whereas others have a more restricted geographical distribution; genotype E is found predominantly in sub-Saharan Africa, whereas genotypes B and C are found in the Far East, and genotypes F and H in Central and South America. In western countries, there are differences in genotype frequencies between risk groups for HBV infection, although this is not obviously linked to transmissibility or disease causation. There are antigenic differences in the HBsAg protein. HBsAg contains a common 'a' antigen and two sets of mutually exclusive determinants, 'd' or 'y' and 'w' or 'r', giving the four main types – adw, adr, ayw and ayr, the latter being associated with different HBV genotypes. Fortunately, the invariant 'a' determinant is the main target of the protective antibody response to infection, and immunity induced by infection or immunization with one HBV genotype cross-protects against infection with others.

HBV infection also occurs in the wild in a number of non-human primate species, such as chimpanzees, orang-utans and gibbons. Each of these species harbours variants of HBV distinct from those found in man (see Fig. 46.4). Whether primate HBV infection contributes to the pool of human carriers in areas of high endemicity, such as sub-Saharan Africa, is currently unclear. There are no known examples of primate to human transmission of HBV and almost invariably genotypes are distinct in human and non-human primates.There is little current evidence for the converse possibility that HBV infection of primates resulted from human contact.

Stability

It is difficult to assess the stability of HBV owing to the lack of a suitable laboratory culture system. Indirect evidence

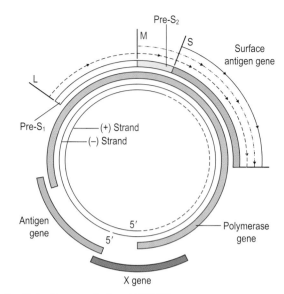

Fig. 46.3 Gene organization of HBV DNA.

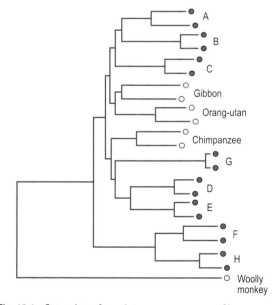

Fig. 46.4 Comparison of complete genome sequences of human genetic variants (A–H; blue circles) and those found in primates infected in the wild (chimpanzees, gibbons, orang-utans, woolly monkeys; unfilled circles). Human genotypes A–H cluster into a number of separate groups, each distinct from HBV variants found in different species.

has been obtained from the study of recipients of blood products treated in various ways and from chimpanzee inoculation experiments. Thus it was established that:

- heating to 60°C for 10 h inactivates virus by a factor of 100–1000-fold

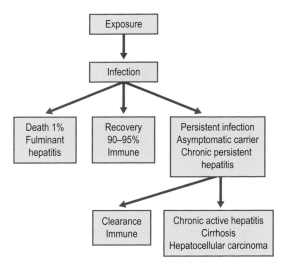

Fig. 46.5 Outcome of hepatitis B infection.

- treatment with hypochlorite (10 000 ppm available chlorine) or 2% glutaraldehyde for 10 min will inactivate virus 100 000-fold.

Studies based on the survival of HBsAg show that this is much more resistant to destruction.

Replication

Replication of viral nucleic acid starts within the hepatocyte nucleus where viral DNA can be free, extrachromosomal or integrated at various sites within the host chromosomes. However, integration is not essential for viral replication. There are some parallels between the hepadnaviruses and the retroviruses, in that:

- both synthesize DNA from an RNA template
- there is amino acid sequence homology between the enzymes involved.

To replicate hepadnavirus DNA, a full-length RNA copy is enclosed in core protein in the hepatocyte nucleus. This is copied to DNA by the polymerase, the RNA is destroyed, and the DNA is copied to form double-stranded DNA as the virion matures (see Ch. 7).

CLINICAL FEATURES

The possible outcomes of infection with HBV are summarized in Figure 46.5. The incubation period varies widely, from 40 days to 6 months, but is often about 2–3 months. A dose-related effect has been observed, as shorter incubation periods have been associated with the inoculation of large doses of virus (as in the transfusion of

infected blood). A prodromal illness occurs in some patients, who complain of malaise and anorexia accompanied by weakness and myalgia. Arthralgia also occurs and may be accompanied by an urticarial or maculopapular rash. These features may be related to circulating immune complexes containing HBsAg, which have been implicated in the rarer complications of polyarteritis nodosa and glomerulonephritis. Complexes are usually present in the plasma in cases of fulminant hepatitis B. In an acute case, hepatocellular damage is detectable biochemically before the onset of jaundice and persists after it has resolved. The patient usually begins to feel better when the jaundice appears, accompanied by pale stools and dark urine. Carriers of HBV are initially symptom-free, and many remain so for many years. As many as 25% of carriers will develop the clinical features of chronic hepatitis and cirrhosis, and eventually hepatocellular carcinoma.

PATHOLOGY

All types of viral hepatitis produce similar changes at the histological level. In the acute stage there are signs of inflammation in the portal triads; the infiltrate is mainly lymphocytic. In the liver parenchyma, single cells show ballooning and form acidophilic (*Councilman*) bodies as they die. In healthy carriers, the inflammatory response is mild, and the affected hepatocytes are pale staining and glassy.

In chronic hepatitis, damage extends out from the portal tracts, giving the piecemeal necrosis appearance. Some lobular inflammation is also seen. As the disease progresses fibrosis develops and, eventually, cirrhosis.

Pathogenesis

Acute disease

HBV replicates in the hepatocytes, reflected in the detection of viral DNA and HBcAg in the nucleus and HBsAg in the cytoplasm and at the hepatocyte membrane. During the incubation period, high levels of virus are present before the host immune response develops and controls the virus. During replication, HBcAg and HBeAg are also present at the cytoplasmic membrane. These antigens induce both B and T cell responses; damage to the hepatocyte can result from antibody-dependent, natural killer (NK) and cytotoxic T cell action. Expression of major histocompatibility complex (MHC) class I antigens is poor in hepatocytes but can be enhanced as interferons are produced in response to the infection. This in turn leads to increased antigen recognition and lysis of the infected hepatocytes. HBV is non-cytocidal without the assistance of the host's immune system, and the disease is usually milder in the immunocompromised. A similar sequence of events occurs in the silent case except that the liver damage is less severe.

Persistence of HBV

Persistence of HBV is indicated by the continued presence of HBsAg and HBV DNA in the blood for more than 6 months. This occurs:

- in 5–10% of adult cases
- in 30% of childhood cases (<6 years)
- in 90% of newborn infections
- more frequently in males
- more often in the immunocompromised.

It is not yet clear what determines that an individual will progress to the carrier state. There may be genetic factors, but the absence, or relative inefficiency, of the immune response is important, as shown by the increased likelihood of the carrier state in the very young and the immunocompromised. In the neonate, infection occurs in the presence of maternal immunoglobulin (Ig) G anti-HBc and tolerance to HBeAg, which can cross the placenta. This will have the effect of masking HBcAg on hepatocyte membranes and thus will prevent its recognition by cytoxic T cells and other immune mechanisms that could lead to clearance of the virus.

Carriers may continue to replicate virus to high levels (HbeAg positive, DNA positive) without evidence of liver damage. This has been called the early replicative phase of chronic infection. It ends with the disappearance of HBeAg and the appearance of anti-HBe. The change happens at a variable time after infection, but each year 5–20% of patients go through this transition, usually associated with a period of liver cell damage. This is often, but not always, accompanied by the disappearance of HBV DNA from the blood, signalling a transition from high to low infectivity carrier status. A carrier may undergo several episodes of hepatitis. Eventually, HBV may disappear in 1–2% of carriers each year. This is the picture in most Caucasian populations infected in adolescence or as adults.

Chronic liver disease and hepatocellular carcinoma

It is estimated that about 1 million people die each year from the the chronic effects of HBV infection. Chronic liver damage results from continuing, immune-mediated destruction of hepatocytes expressing viral antigens. In addition, auto-immune reactions may contribute to the damage as immune responses are induced to various liver-specific antigens. There may be a role for the HBeAg-negative mutants that arise during the transition

from HbeAg positive to anti-Hbe positive. In Asian and Mediterranean patients cirrhosis and hepatocellular carcinoma can arise after the loss of HbeAg and the clearance of HBsAg from the blood, although viral DNA may still be present.

The rate of progression to cirrhosis and carcinoma varies according to:

- the age of infection and stage reached
- the state of the patient's immune system
- geographic factors
- genetic factors.

Several of these may be related.

Hepatocellular carcinoma is almost always fatal and is one of the ten most common tumours in the world. There is considerable evidence that 80% are caused by chronic infection with HBV. Thus, the highest rates are found in areas where HBV is highly endemic and where infection occurs at a very early age. This is necessary as there may be an interval of 30–40 years between infection and tumour development, although shorter intervals are seen. The highest rates of carcinoma are found in Asian patients, who are usually infected early in life. Integrated viral DNA can be found in tumour cells but the site differs in different tumours. The administration of HBV vaccine to children in Taiwan has already caused a significant decrease in cases of hepatocellular carcinoma.

HBV with mutations in the surface, core and polymerase genes

HBsAg variants were first detected in children born to infected mothers and in patients given a new liver because of chronic HBV disease. In both situations the variants arose during infection in the presence of anti-HBs. The children were exposed to maternal virus during birth and were then given anti-HBs immunoglobulin and active immunization to reduce the risk of infection. This treatment is successful in most cases but, in some, mutants appear, allowing the virus to escape the neutralizing anti-HBs given to, or induced in, the baby. Escape mutants have been detected in many areas of the world, including Italy, Singapore, Japan, Brunei, USA and China after the start of vaccination campaigns.

In liver transplant recipients, anti-HBs is administered to protect the new liver from infection with virus already present. The variants have been seen to disappear when anti-HBs therapy is stopped. The phenomenon has been found in liver transplant patients in the USA, Germany and the UK.

The mutations affect the 'a' determinant of HBsAg, the principal target of anti-HBs. Mutant virus can be transmitted to new hosts; it is associated with the virus-positive carrier state. The widespread occurrence of HBsAg mutants would create considerable problems for vaccination programmes, as well as compromising many existing diagnostic assays for HBsAg that detect epitopes associated with the 'a' determinant. Other isolates with mutations in the pre-S_1 and pre-S_2 regions have been described. They can replicate and may be the only detectable form of virus in some cases. They appear during the chronic phase of infection.

HBcAg variants are known with mutations in the core promoter or pre-core coding regions of the HBcAg gene. These generally suppress the production of HBeAg, without affecting the synthesis of HBcAg and the assembly of complete virions. They are therefore often found in carriers who are HBeAg negative, but positive for HBV DNA by polymerase chain reaction (PCR). The mutants are rarely seen in patients with minimal liver disease, but can appear during or after the loss of HBeAg and the development of anti-HBe. In Greece, 90% of carriers are negative for HBeAg and all carry mutant genomes. Similar variants are found in the Far East and the Mediterranean region; such cases are increasingly recognized worldwide.

Polymerase variants can often be detected during therapy with nucleoside analogues and confer drug resistance. Their presence can affect the choice of drugs.

The frequency with which mutations appear in HBV is dependent on the high rate of replication of the virus and its dependence on replicating DNA via RNA and an RNA-dependent DNA polymerase. In highly viraemic carriers as many as 10^{10} mutant genomes may arise each day. Different variants may be present in the blood, liver and peripheral blood mononuclear cells of a carrier. Most mutants are defective, but some may explain treatment failure.

LABORATORY DIAGNOSIS

The virology laboratory can test for a wide range of HBV antigens and antibodies, using radio-immuno-assays and enzyme-linked immunosorbent assays (ELISAs), and for HBV DNA by PCR. The standard screening test is for HBsAg, which, if present in the serum, indicates that the patient is infected with HBV, either as a recent acute infection or as a carrier.

Acute infection

In an acute case (Fig. 46.6), HBsAg is present for some weeks before the onset of symptoms and is at maximum titre at the height of liver damage. In most cases HBsAg cannot be detected beyond 3–4 months. Over this period, HBV DNA can be detected in plasma by PCR. In a few patients, antigenaemia is of short duration and may not be detectable at the onset of symptoms. In such cases

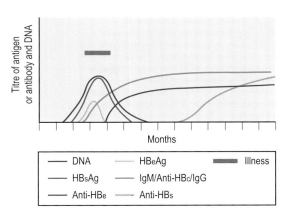

Fig. 46.6 Hepatitis antigens, antibodies and DNA in a patient recovering from acute infection.

the presence of anti-HBc, especially IgM anti-HBc, and anti-HBe may be the only indications of recent HBV infection. If successive serum samples are examined, the development of anti-HBs will confirm primary infection, although there is considerable variation in the appearance of this antibody.

Typical responses shown by at least three-quarters of cases are outlined in Figure 46.6. HBeAg is produced when virus is replicating and thus is usually found soon after HBsAg. IgM anti-HBc is the next response to be detected. HBeAg is correlated strongly with the detection of viral DNA, virions and viral polymerase in the serum. IgM anti-HBc is a transient response and, if present in high titre, indicates a recent acute infection. The response declines with time and is usually absent by 6 months. Tests for HBeAg and anti-HBe during an acute infection can be helpful in that the disappearance of HBeAg and replacement with anti-HBe indicates that the patient is responding to the infection and will clear HBsAg. Some 90% of patients develop anti-HBs; in a few cases it is present at the same time as HBsAg, but usually there is a gap of up to 6 months before anti-HBs is detected. Once present, the patient is immune to further infection with HBV.

Chronic infection

As illustrated in Figure 46.7, a carrier of HBV is almost always HBeAg-positive and HBV DNA-positive beyond 6 months. During this time, IgM anti-HBc disappears and is replaced with IgG anti-HBc. This is typical of the early replicative phase of chronic infection. It may be succeeded by the loss of HBeAg and the appearance of anti-HBe (Fig. 46.8); the change happens at a variable time. Many patients continue to be positive for HBV DNA.

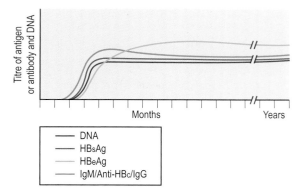

Fig. 46.7 Sequence of laboratory results in a patient who becomes a carrier of HBV.

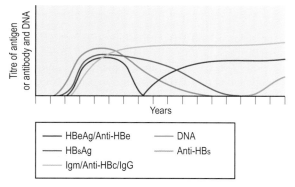

Fig. 46.8 Sequence of laboratory results in a hepatitis B carrier who loses HBeAg.

TREATMENT

Much effort has been focused on the treatment of the chronic carrier with the aims of:

- reducing the possibility of transmission to others
- preventing progression of liver disease and improving liver histology
- preventing the development of hepatocellular carcinoma.

Historically, α-interferon (α-IFN) has been the principal drug used for the treatment of chronic HBV infection. α-IFN inhibits the packaging of RNA in nucleocapsids during virus assembly and thus has a directly antiviral action. It also upregulates the expression of HBsAg and other viral proteins on the surface of the infected cell, and therefore enhances cytotoxic T cell responses. A course of peginterferon (see Ch. 5) for 4 months can achieve a significant reduction in the levels of circulating virus and normalization of alanine aminotransferase levels in

approximately one-third of those treated. Conversion from HBeAg positivity to anti-HBe-positive status occurs in approximately 20%, generally followed by the clearance of HBsAg in subsequent years. Pretreatment variables favouring response are:

- recent infection
- adult patient
- female
- Caucasian
- no serological evidence of human immunodeficiency virus (HIV) infection
- increased serum ALT levels
- HBeAg positive
- low HBV DNA concentration
- hepatitis D virus negative
- therapy maintained for more than 4 months.

The incidence of development of cirrhosis of the liver and of carcinoma is lower in those receiving α-IFN therapy. Side effects are common and include influenza-like symptoms, mild exacerbation of hepatitis (often a favourable indication of subsequent response) and granulocytopenia.

Antiviral drugs that inhibit the replication of HBV have also been developed. These compounds are nucleoside analogues that, after phosphorylation in the cell, are recognized by the HBV-encoded RNA polymerase/reverse transcriptase.

The most widely used nucleoside analogue is lamivudine, a potent inhibitor of HBV DNA synthesis that is effective at low concentrations in vitro. At a dose of 100–300 mg daily, lamivudine leads to a marked reduction or elimination of detectable HBV DNA in plasma and normalization of alanine aminotransferase levels in approximately 40% of individuals. Over a treatment period of 1 year, about 30% of those who are HBeAg positive become anti-HBe positive. In some patients withdrawal of treatment leads to the reappearance of hepatic inflammation and the return of HBV DNA to pretreatment levels, except in those achieving HBeAg seroconversion. Although toxicity is low, and in the long term the drug is generally tolerated well, prolonged administration is complicated by the emergence of antiviral resistance, manifested by the reappearance of HBV DNA in plasma and the development of abnormal transferase levels. Resistance may be associated with mutations in the catalytic core of the HBV polymerase enzyme. Resistance is more frequent in those who are immunosuppressed, a group that includes patients receiving liver transplants.

The other licensed antiviral in clinical use is adefovir. This is an orally administered nucleoside analogue of deoxyadenosine monophosphate (dAMP) that has demonstrated efficacy in clearing virus (20–50% response rates) and normalizing liver function (50–70% response rates). Resistance to adefovir emerges less frequently than to lamivudine, a potential advantage to patients on long-term therapy. Additionally, adefovir is effective against lamivudine-resistant mutants. Conversely, mutations that arise in response to adefovir therapy are sensitive to lamivudine. These observations led to the investigation of the efficacy of lamividune–adefovir combination therapy; although overall efficacy was no different from that with each drug alone, the frequency of antiviral resistance to lamivudine was substantially lower with the combined therapy.

A number of other nucleoside analogues are in phase II or III trials. These include entecavir, emtricitabine, telbivudine and clevudine. These work in a similar way to lamivudine and can often achieve comparable or improved results in reduction or clearance of HBV DNA and normalization of alanine aminotransferase. As for adefovir, many of the new compounds are active against HBV mutants resistant to lamivudine, enhancing prospects for the development of combination therapy, as in the treatment of HIV infection.

For end-stage liver disease, liver transplantation is necessary. Attempts have been made to prevent reinfection of the graft with HBV through the administration of anti-HBs immunoglobulin, and more recently this has been combined with a nucleoside analogue such as lamivudine. Used singly, anti-HBs immunoglobulin and lamivudine are rarely effective, but recent studies have indicated that their combined use may prevent HBV recurrence after transplantation, and give improved graft function and significantly improved patient survival.

EPIDEMIOLOGY

HBV is present in the blood as well as in body fluids such as semen, vaginal secretions and saliva, although the concentration is only about 1 in 1000 of that in blood. Even so, this may still represent a large number of infectious virions. The presence of HBV in blood underlines the original association of infection with blood transfusion or the use of blood products (post-transfusion 'serum' hepatitis) and infections associated with needlestick injuries. Sexual transmission is also recognized, as is that which occurs between family members, siblings, peers and residents in institutions for the learning impaired. In these circumstances there will be frequent contact with blood and saliva; the virus gains entry through cuts and abrasions or across mucous membranes. Biting and scratching are also important. Vertical transmission from mother to child is one of the most important routes. Transmission occurs when maternal blood contaminates the mucous membranes of the baby during birth. Transplacental infection is thought to be quite rare.

The prevalence of HBV infection varies widely in different parts of the world.

- *Highest rates*: HBsAg 10–15%; anti-HBs 70–90% in South-East Asia, China, equatorial Africa, Oceania and South America. Vertical and horizontal transmission are both common.
- *Intermediate rates*: HBsAg 5%; anti-HBs 30–40% in Eastern Europe, around the Mediterranean, South America and the Middle East.
- *Lowest rates*: HBsAg 0.1–0.5%; anti-HBs not more than 5% in western Europe, North America and Australia.

Overall, it is estimated that one-third of the world's population has been infected with HBV and that there are 300 to 400 million carriers worldwide. In areas of low endemicity the risk of infection varies widely in different groups according to behaviour. Most cases occur in parenteral drug injectors who share scarce needles and syringes, and by sexual transmission by both homosexual and heterosexual intercourse. In about one-third to one-quarter of cases it may be impossible to identify a source. Screening of all blood donations has virtually eliminated transmission by transfusion and blood products. Some groups of patients are at increased risk of infection and of becoming carriers; these include patients on maintenance haemodialysis and in long-stay homes for those with learning difficulties.

Health-care personnel and laboratory workers are at risk from patients, although the degree of risk varies with the place and nature of their work, the care with which it is performed and their immune status. Surgery, dental surgery, obstetrics and gynaecology often involve working with sharp instruments, often in restricted spaces. Operators may injure themselves and inoculate patient's blood. The spilling of patient's blood will pose a threat only if there is contamination of unprotected skin, with abrasions, or mucous membranes.

Patients are also at risk from staff, and episodes have been identified in which a surgeon, gynaecologist, dentist or other staff member has transmitted the virus to patients during invasive procedures, especially in difficult operations where the operator's hands are hidden and needles and instruments are guided by touch. Apart from small outbreaks linked to a common source, such as surgery and haemodialysis, epidemics are rare. In the past, before specific tests were available, a few outbreaks were described following the use of blood products prepared in bulk from large numbers of donations. A large number of cases occurred in the US armed forces during the Second World War owing to the use of yellow fever vaccine stabilized with infectious human plasma.

CONTROL

Broadly there are two approaches to the prevention of infection with HBV. The possibility of transmission can be reduced or removed by modifying risk behaviour. The measures include avoiding unprotected sexual contact by the use of condoms, and sharing needles by injecting drug misusers. Implementation of sensible control of infection policies can reduce the risks considerably to health-care workers and patients. It is essential, of course, that blood for transfusion is screened. However, there are limits to these approaches, and immunization, both passive and active, offers many advantages, not least in situations where the prevention of infection is otherwise difficult or impossible.

Passive immunization

Hyperimmune hepatitis B immunoglobulin (HBIG) is prepared from donors with high titres of anti-HBs. Doses of 300–500 IU in 3 ml are given intramuscularly. The use of HBIG is still necessary in the following situations:

- after accidental exposure if the victim is not vaccinated, or did not respond to the vaccine
- to babies born to infected mothers (with active immunization)
- to prevent infection of a new liver transplanted for chronic HBV.

HBIG must be given as soon as possible after an accident and preferably within 48 h; a second dose is given 4 weeks later to those who do not respond to current vaccines. Such a regimen does not give absolute protection, but an efficacy of 76% has been reported. In some such cases, a transient viraemia may occur. If the victim has not been vaccinated, HBIG should be used and a course of active immunization started, injecting the two materials into different body sites. A course of 6-monthly injections of HBIG can reduce the infection rate in babies by 70%, but even greater protection is provided by combining HBIG with active immunization, starting at birth. The protective efficacy of this combined treatment is 90%. In transplant recipients, it may be more effective to combine HBIG with lamivudine owing to the appearance of HBsAg mutants.

Active immunization

Currently available vaccines are produced by cloning the surface antigen gene in yeast cells. The product is particulate and resembles the small particles seen in patients, although it is not glycosylated. The vaccine is administered with alum as adjuvant and injected intramuscularly; care should be taken to avoid injection into fat as this can produce poorer seroconversion rates. For this reason, injection into the deltoid muscle of the upper arm is recommended. The vaccine is free from major side effects; local swelling and reddening may occur in up to one in five recipients, with a slight fever in only a few cases.

Three doses of vaccine are given at 0, 1 and 6 months. Shorter schedules may be appropriate in some circumstances. The seroconversion rate is influenced by a number of factors, the most important of which are the age and sex of the vaccinee. Rates in excess of 95% are seen in young women, whereas the rate may drop to 80% in older men. Immunosuppressed patients show even lower rates, for example only 50–60% in patients on maintenance dialysis. The level of the anti-HBs should be greater than 10 mIU/ml. If vaccine is used selectively to protect particular groups in the population such as health-care workers, it is reasonable to check the response within a few months of the third dose of vaccine, although this is not appropriate when the vaccine is given as part of a schedule applied to a population.

The duration of the response to vaccine is variable and dependent on the titre of anti-HBs after completion of the course. If the response is poor, a booster dose should be given. Low or non-responders need to be identified and told that they are not protected and that they must seek prophylaxis by passive immunization if they suffer accidental exposure. Those who are known to have responded can be given a booster if they are exposed to the virus, although the need for this depends on the titre of anti-HBs achieved and the time since completion of the course. Individuals immunized at birth show a drop in positivity rates over 15 years; booster doses of vaccine may be needed.

Alternative vaccines are under development to improve the seroconversion rate, especially in those who do not respond to current preparations, and the ease of administration by reducing the number of doses needed and by combining with other vaccines. Incorporation of the pre-S$_1$ and pre-S$_2$ proteins is one approach. Alternatives are the use of synthetic peptides and hybrid virus vaccines.

Who should be immunized?

The aim of immunization is to prevent transmission of HBV by the routes described and to eradicate the virus from the population of the world. This will not be achieved quickly or easily. Carriers are an important reservoir and can transmit the virus over many years.

Transmission from mother to child is an important route and requires intervention at the birth, or within 12 h, to protect the child. As most babies infected at birth will become long-term carriers, it is essential to target this group. Passive and active immunization are combined to give 90% protection; active immunization alone gives 70–85% protection. Intervention at this early stage is expensive and should be reserved for babies born to carrier mothers identified by antenatal screening.

Herd immunity can be provided by the incorporation of HBV vaccine into the schedules of routine immunizations; a polyvalent vaccine containing HBV and pertussis, diphtheria and tetanus antigens is immunogenic, as is a combined hepatitis A and B vaccine.

Horizontal transmission occurs at later ages; this can be tackled by immunization of adolescents before they become sexually active or adopt at-risk lifestyles. The strategy is:

- antenatal screening to identify infected mothers and protect neonates
- universal infant vaccination
- vaccination of all adolescents.

Vaccine used in this way has been shown to reduce the number of cases of acute hepatitis B and the carrier rate in the population.

In all populations medical staff are at special risk of infection because of the nature of their work; they should be immunized and their responses checked. This will also protect patients against infection from staff. Until the vaccine has been used routinely in a population for some years, it is sensible to protect other groups at special risk:

- sexual partners of known HBV positives
- those with frequent unprotected sexual exposure
- parenteral drug misusers (most have been infected by the time identified)
- ambulance, police and fire service staff and military personnel
- patients with learning difficulties living in long-stay homes
- patients needing frequent transfusions and/or blood products
- patients on maintenance dialysis.

The selective use of vaccine in health-care staff in the UK has led to a significant reduction in the number of cases of acute infection in this group.

THE DELTA AGENT (HEPATITIS D VIRUS)

The delta (δ) antigen was first identified in Italian patients infected with HBV. Initially, it was thought to be an antigen of HBV, but it is part of another virus that cannot replicate without assistance from HBV (or another hepadnavirus). The virus is a small (35–37 nm), enveloped particle containing a single, small, circular molecule of RNA of 1.7 kilobase pairs. The internal protein is the δ antigen. The envelope of the virus is the same as that of HBV (see Fig. 46.2). The origin of the virus is unknown and it has no homology with HBV. The closest relatives are the satellite viruses of plants.

Analysis of the RNA from different isolates has shown that there are three distinct genotypes.

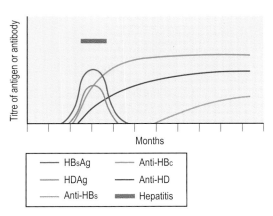

Fig. 46.9 Test results in a patient infected simultaneously with HBV and hepatitis D virus.

CLINICAL FEATURES AND PATHOGENESIS

Hepatitis D virus (HDV) can only infect simultaneously with HBV or as a superinfection of a chronic carrier of HBV. The symptoms are similar to those of acute and chronic hepatitis B. The presence of HDV may increase the severity of the clinical features compared with those seen with HBV alone. This is reflected in a 10% risk of fulminant hepatitis with simultaneous infection, and a 20% risk in superinfections.

In about 10% of cases, co-infection results in a biphasic illness. HDV superinfection can cause an acute hepatitis and may lead to a persistent state. The risk of progression of liver damage is also increased.

DIAGNOSIS

Tests are available for δ antigen and antibody. The sequence of appearance of the various markers in a patient co-infected with HBV and HDV is shown in Figure 46.9. The initial antibody to HDV is of the IgM class. In cases of superinfection, the test results are as illustrated in Figure 46.10. During the episode there may be a drop in the HBsAg titre, which, although usually still detectable, may disappear temporarily in a few cases. This can cause some confusion if the episode is the first presentation of the patient.

EPIDEMIOLOGY

HDV is not a new virus as there is evidence of infection in the 1930s. It has been estimated that there may be 25 million carriers in the world, but there is considerable geographical variation. The three genotypes are found as follows:

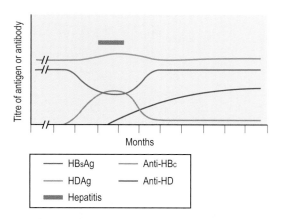

Fig. 46.10 Test results in a hepatitis B carrier superinfected with HDV.

- Type I: Europe, North America, North Africa, Middle East and East Asia
- Type II: Japan, Taiwan and Okinawa
- Type III: The north of South America.

In some populations HDV is associated with drug misusers, whereas in others it is present throughout the population. Sexual transmission is recognized, but is not so easy as via blood, as may occur in drug misuse and transfusion. In several episodes HDV has spread rapidly among carriers, causing hepatitis, death from fulminant hepatitis and chronic illness. The virus is rare in South-East Asia and China.

TREATMENT AND CONTROL

There is no treatment for HDV infection other than for the co-existing HBV infection. The same general control measures as for HBV are also relevant. HBV vaccine will prevent HDV infection, but there is no means of protecting existing HBV carriers against the consequences of HDV superinfection.

KEY POINTS

- Hepatitis B virus (HBV) causes acute hepatitis and chronic infection leading to chronic liver disease and hepatocellular carcinoma.
- HBV is a partially double-stranded DNA virus, carrying a reverse transcriptase-like enzyme to replicate viral DNA on an RNA intermediate.
- HBV is transmitted in blood, by sexual intercourse and from mother to child.
- There is considerable geographical variation in exposure; the highest rates are in the East, Equatorial Africa, Oceania and South America (about 10% carry virus and most people have been infected).
- Injecting drug misuse, unprotected sexual intercourse and blood from unscreened donors carry risk; mother-to-child transmission is important.
- Effective vaccine is available, given in three doses over 6 months. Wide use in young children will reduce the carrier rate and long-term sequelae, including carcinoma.
- Carriers can be treated with interferon and nucleoside analogues to reduce viral replication.

RECOMMENDED READING

Banatvala J E, van Damme P 2003 Hepatitis B vaccine – do we need boosters? *Journal of Viral Hepatitis* **10**: 1–6
Centers for Disease Control and Prevention 2001 Updated US Public Health Service guidelines for the management of occupational exposures to HBV, HCV and HIV and recommendations for postexposure prophylaxis. *MMWR. Morbidity and Mortality Weekly Report* **50**: 1–42
Fields B N, Knipe D M, Howley P M (eds) 1996 *Virology*, 3rd edn. Lippincott–Raven, Philadelphia
Karayiannis P, Main J, Thomas H C 2004 Hepatitis vaccines. *British Medical Bulletin* **70**: 29–49

Zuckerman A J, Banatvala J E, Pattison J R, Griffiths PD, Schoub BD (eds) 2004 *Principles and Practice of Clinical Virology*, 5th edn. Wiley, Chichester

Internet sites

Health Protection Agency. Hepatitis B. http://www.hpa.org.uk/infections/topics_az/hepatitis_b/menu.htm

47 Parvoviruses

B19 infection; erythema infectiosum

P. J. Molyneaux

Parvoviruses (*parvum* means small) have been isolated from a wide range of organisms, from arthropods to human beings. The family Parvoviridae is divided into two subfamilies: the Parvovirinae and the Densovirinae. The latter group infects only invertebrates. The Parvovirinae contain five genera: *Parvovirus, Dependovirus, Erythrovirus, Amdovirus and Bocavirus* (Table 47.1). They are differentiated by their replication:

- autonomous (*Parvovirus*)
- requirement for a helper virus (*Dependovirus*)
- replication in erythroid progenitor cells (*Erythrovirus*).

Only B19 within the genus *Erythrovirus* is known to be a human pathogen; erythroviruses distinct from B19 may yet be linked with human disease. Although adeno-associated viruses within the genus *Dependovirus* can be detected in human beings at various sites, they have not been linked definitely with human disease (see pp. 313–314).

DESCRIPTION

The B19 virion (Fig. 47.1) is 20–25 nm in diameter, unenveloped, and contains a single strand of DNA (5.9 kilobase pairs). See Chapter 7 for details of replication. There are two capsid proteins, VP1 and VP2. B19 is genetically and antigenically stable with only one serotype, and is extremely resistant to lipid solvents, acid, alkali and high salt concentrations. It is relatively heat resistant as infectivity from high-titre virus can persist after 80°C for 72 h in clotting factor concentrates. The name B19 was given because it was first found in the blood of an asymptomatic blood donor (coded 19 in panel B) where it caused a false-positive result in an early test for hepatitis B surface antigen.

PATHOGENESIS AND CLINICAL FEATURES

The clinical manifestations of B19 infection are due to either the direct effect of viral replication, which causes cell death, or the subsequent immune response. B19 attaches to cells with the blood group P antigen (globoside), and individuals who lack P antigen (p phenotype) are not susceptible to B19 infection. P antigen is present on mature erythrocytes, erythroid progenitors, megakaryocytes, vascular endothelium, placental cells, and fetal myocardial and liver cells. For B19 to replicate, it requires actively dividing cells; hence the cells most affected by B19 are the rapidly dividing red cell precursors in the bone marrow. The outcomes of infection depend on the immune competence and any red cell-related haematological dysfunction in the infected person. Development of antibody is needed to control B19 replication; capsid protein VP1 is the target for neutralizing antibody. After recovery, immunity is life-long in normal individuals. Persistent and sometimes relapsing infection occurs in the immunocompromised as the little antibody they may produce in response to infection is not capable of neutralizing the virus and controlling the infection. Persistent infection can also develop in the fetus, the maternal viraemia giving ample opportunity for infection of the placenta and fetus. Re-infection may occur, but is associated with illness only in the immunocompromised.

Volunteer studies in humans

Experimental infection of healthy volunteers has revealed the steps in the pathogenesis of B19 infection (Fig. 47.2). The virus is infectious when given in nasal drops. One week later there is an intense viraemia with virus also present in respiratory secretions, but not in faeces or urine. At this time, a febrile influenza-like illness may occur, thought to be cytokine induced. The intense viraemia (up to 10^{11} particles per millilitre) lasts for only a few days

Table 47.1 Taxonomic organization of Parvoviridae, their natural hosts and the diseases they cause

Subfamily	Genus	Virus	Host	Disease
Densovirinae	Three genera	Densonucleosis viruses	Arthropods	Many fatal diseases
Parvovirinae	Parvovirus	Minute virus of mice	Mice	Subclinical
		Feline panleucopenia virus	Cat	Enteritis, leucopenia, cerebellar ataxia
		Canine parvovirus	Dog, fox	Enteritis, myocarditis
		Porcine parvovirus	Pigs	Reproductive failure
	Amdovirus	Aleutian disease virus	Mink	Pneumonitis
		Mink enteritis virus	Mink	Enteritis
	Dependovirus	Adeno-associated viruses	Man and others	Unknown
	Erythrovirus	Human parvovirus B19	Man	Respiratory tract illness, aplastic crisis, erythema infectiosum/fifth disease, fetal hydrops
		Simian parvovirus	Monkey	Anaemia
	Bocavirus	Human bocavirus	Man	Unknown

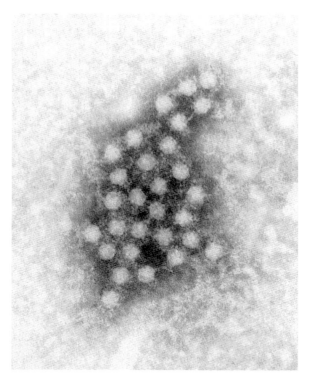

Fig. 47.1 B19 particles in an immune electron microscopy preparation of serum from a child with a petechial rash and arthritis (original magnification ×200 000). (Courtesy of Dr Hazel Appleton, Health Protection Agency, London.)

before there is a brisk antibody response, initially of the immunoglobulin (Ig) M class, but followed rapidly by the appearance of IgG antibody. B19 viraemia can sometimes be detected for months after infection by polymerase chain reaction (PCR).

Haematological changes take place in the second week after inoculation. At 10 days after inoculation erythroid precursors are absent from the bone marrow and reticulocytes disappear from the peripheral blood; there is a small subclinical fall in the haemoglobin level. Lymphocyte, neutrophil and platelet counts sometimes fall transiently, but this is not due to lack of precursors in the bone marrow. Haematological changes are the direct result of virus-induced cell death of erythroid progenitor cells in bone marrow with interruption of erythrocyte production. No effect on the precursor cells of the myeloid series is observed.

The *rash* and *arthralgia* associated with B19 infection occur in infected volunteers during the third week after inoculation. As they follow the disappearance of the viraemia and occur when there is an easily detectable immune response, both are assumed to be immune mediated.

Clinical disease caused by B19

There is a spectrum of clinical consequences of B19 infection (Table 47.2).

Minor illness or subclinical infection

Asymptomatic or mild infection can occur at any age. In children this accounts for about half of all infections. A non-specific febrile respiratory tract illness is common at the viraemic phase; it is usually mild, but may mimic influenza.

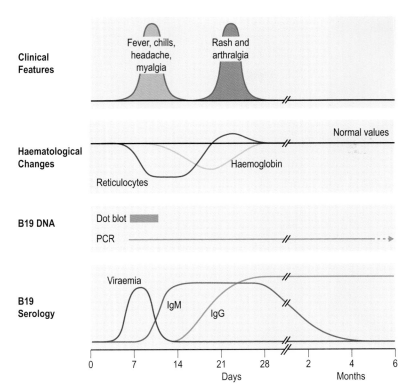

Fig. 47.2 Virological, haematological and clinical events during B19 virus infection of volunteers. (Reproduced with permission of Wiley, Chichester).

Table 47.2 Spectrum of disease due to B19 related to host factors

Disease	Host	Treatment
Asymptomatic	Normal children and adults	
Respiratory tract illness	Normal children and adults	
Rash illness	Normal children and adults	
Erythema infectiosum, fifth disease, slapped cheek syndrome	Normal children	
Arthralgia	Normal adults	
Transient aplastic crisis	Patients with increased erythropoiesis	Blood transfusion
Persistent anaemia	Immunocompromised patients	Intravenous immunoglobulin
Congenital anaemia or hydrops	Fetus <20 weeks	Intra-uterine transfusion

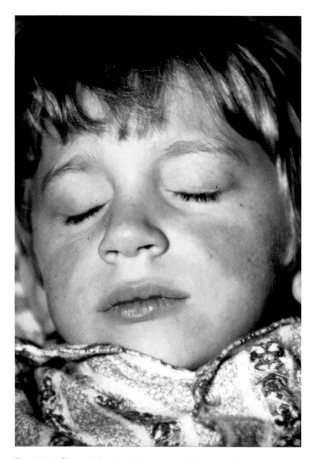

Fig. 47.3 Slapped cheek syndrome. Note the intense cheek erythema and circumoral pallor. (Courtesy of Dr Ken Mutton, Medical Microbiology, Manchester Royal Infirmary.)

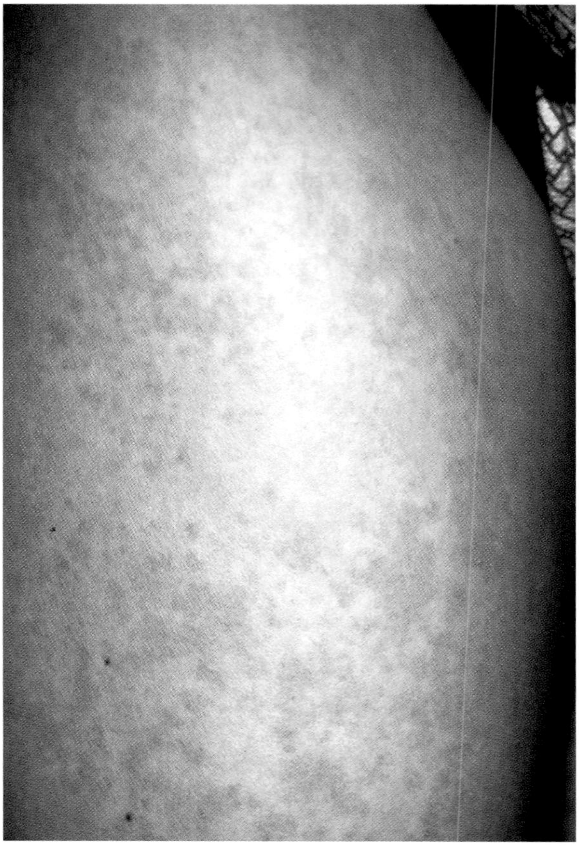

Fig. 47.4 Late stage of B19 rash with central clearing showing lacy appearance.

Rash illness

In healthy individuals, B19 can cause erythematous, macular or maculopapular or, less commonly, purpuric rashes. In its most distinct form the erythematous rash is called *erythema infectiosum* or *slapped cheek syndrome*, so called because of the intense erythema of the cheeks (Fig. 47.3). Rash illness is commonest in children aged 4–11 years, and is sometimes called *fifth disease*, as it was the fifth of six erythematous rash illnesses of childhood in an old classification. Classically, it starts with facial erythema, followed by a maculopapular rash of the trunk and limbs. It lasts only a day or two, although transient recurrences may occur over 1–3 weeks. It may be exacerbated when the individual is hot, for example after exercise or a hot bath, and may be itchy. As it spreads out, there may be central clearing of the erythema which can give a lacy or reticular appearance (Fig. 47.4). There may be associated lymphadenopathy and joint symptoms.

Even during a community outbreak of slapped cheek syndrome, clinical diagnosis is not wholly reliable as the illness can be very similar to rubella (where arthropathy and lymphadenopathy may also occur). Cheek erythema is not always prominent, the rash may not appear lacy, and it may be on the palms and soles; rarely there are vesicles. A purpuric rash limited to the hands and feet in a gloves-and-socks distribution with pain, pruritus and oedema can also occur. In the absence of virology tests the most frequent clinical diagnoses made are rubella, allergy and 'viral illness'. Where the rash is purpuric, the platelet count is usually normal, but a transient thrombocytopenia may occur.

Joint disease

Symptoms and signs of joint involvement may occur with or without rash illness, and B19 infection should be considered in the differential diagnosis of acute arthritis. This is more common in adults, especially women, of whom 80% have joint symptoms, compared with about 10% in childhood. Like the rash, the arthropathy is very similar to that seen with rubella, being a symmetrical arthralgia or arthritis involving mainly the small joints of the hands, although feet, wrists, knees, ankles and other joints may be affected. In children, the arthropathy may be less symmetrical. Arthropathy usually resolves within 2–3 weeks, but may occasionally persist or recur for months and very rarely for years. Some of these patients may be classified clinically as having early benign rheumatoid arthritis. However, B19 arthropathy is not destructive and, if rheumatoid factor is detected, it does not persist and B19 virus is not linked to rheumatoid arthritis.

Transient aplastic crisis

This is an acute transient event seen in those with various underlying haematological problems:

- decreased red cell survival (e.g. haemolytic disorders)
- where the bone marrow is 'stressed', for example because of haemmorhage or iron deficiency anaemia.

There is a virtual absence of red blood cell precursors in the bone marrow at the beginning of the crisis, followed by disappearance of reticulocytes from the peripheral blood and a subsequent fall in haemoglobin concentration. The cessation of erythropoiesis lasts for 5–7 days, and patients present with symptoms of acute anaemia, which can be life-threatening. Blood transfusion is required, but after a week or so specific antibody production controls viral replication, the bone marrow recovers rapidly, there is a reticulocytosis and the haemoglobin concentration returns to steady-state values.

Throughout the world, B19 infection is responsible for 90% of cases of transient aplastic crisis in those with underlying haemolytic disorders, most commonly in children with, for example, sickle cell anaemia, hereditary spherocytosis or thalassaemia.

Persistent infection in the immunocompromised

The inability to mount an effective neutralizing antibody response to B19 results in persistent infection. This has been described in patients with:

- congenital immunodeficiency
- leukaemia
- acquired immune deficiency syndrome (AIDS)
- transplant recipients.

The bone marrow picture is typical of that seen in transient aplastic crisis with absent red cell precursors, but here results in *pure red cell aplasia*, ongoing viraemia and a non-specific febrile illness presenting with symptoms of chronic anaemia or a remitting and relapsing anaemia, without rash or joint disease.

B19 in pregnancy

Although B19 infection can cause fetal loss, most pregnancies result in the birth of a normal baby. There is no evidence of birth defects or developmental abnormalities in children exposed to B19 in utero.

Transplacental infection can occur during acute maternal infection, whether or not symptoms of B19 infection occur in the mother, before maternal antibody has developed and crossed the placenta to protect the fetus. Infection in the first 20 weeks can lead to intra-uterine death (increased risk 9%) and *non-immunological fetal hydrops* (risk 3%), of which about half die and are included in the 9%. This is in contrast to hydrops, which occurs following Rhesus incompatibility and is immune mediated. The greatest risk of fetal loss is during the second trimester.

These effects are due to the immaturity of the fetal immune response and the shorter survival of fetal red cells. The large increase in red cell mass (in both the bone marrow and extramedullary sites of erythropoiesis) during the second trimester leads to the increased risk of developing anaemia at this time. The anaemia persists and becomes chronic because of ongoing viral replication. Infection of fetal myocardial cells can also occur, causing myocarditis. Both the anaemia and the myocarditis contribute to the development of cardiac failure, leading to the development of ascites and fetal hydrops (Fig. 47.5). Fetal loss is usually 4–6 weeks after maternal infection, but can be as long as 12 weeks later. Overall, B19 probably accounts for 10% of cases of fetal hydrops. A more effective fetal immune response reduces the risk of fetal loss in the third trimester; very rarely chronic anaemia in the newborn has followed intra-uterine infection.

Other

In healthy people, case reports suggest that the manifestations of B19 may on occasion be very wide, for example meningitis, hepatitis, haemophagocytic syndrome. However, as both IgM and DNA may persist for months, some associations may be casual rather than causal.

LABORATORY DIAGNOSIS OF B19 INFECTION

Detection of virus

During pure red cell aplasia, and the acute illness of transient aplastic crisis, viraemia and throat virus excretion coincide

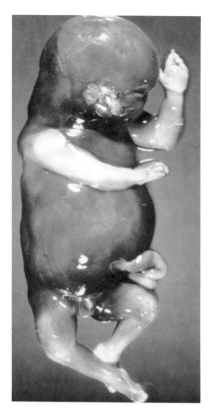

Fig. 47.5 Fetal hydrops due to B19. Note pallor of limbs. In-utero autolysis of fetus disguises pallor of body.

with haematological changes, so detection of the viral genome in serum or bone marrow by DNA hybridization or PCR is the method of choice. Serum is the usual specimen examined; it contains high concentrations of virus and, if virus is not detected, it can serve as an acute-phase serum for antibody assays. While awaiting B19 detection results in those with anaemia, the reticulocyte count is a quick test that can be helpful as a low or absent count is consistent with current B19 infection, whereas a normal or raised count excludes the diagnosis. Hybridization or PCR are also used to investigate fetal samples. Virus culture is not appropriate as B19 does not grow in routine laboratory cell cultures.

Antibody detection

In most cases B19-specific IgM is detectable within a day or two of the onset of the rash, although it may be up to 2 weeks before antibody is present in convincing amounts. Once B19-specific IgM has appeared in the serum, it rapidly reaches peak concentrations and is usually detectable for 2–3 months. High concentrations of IgG antibody are generally found by 1 month after infection, and IgG usually remains detectable for life. Diagnosis of recent infection can be made by demonstrating seroconversion or increasing amounts of IgG antibody.

Investigations during pregnancy and infection in the fetus

At booking, all pregnant women should be advised to inform their general practitioner, midwife or obstetrician as soon as they develop a rash, unexplained arthropathy or are in contact with someone with a possible viral rash. Recommendations for investigating a pregnant woman who has had significant contact with possible B19 are summarized in Figure 47.6. Testing of a stored blood (e.g. booking sample) taken prior to the contact can be done, but if this is IgG negative a current sample should also be tested for IgG and IgM. Where maternal infection is proven, whether or not symptoms have occurred, obstetric referral for investigations (e.g. fetal ultrasonography) should be arranged without delay as deterioration from early hydrops to intra-uterine death can be rapid. Although fetal loss has been documented only with infection up to 20 weeks' gestation, investigation of rash or contact is recommended at any gestation and is also indicated for non-immune fetal hydrops.

The diagnosis of B19 infection in a fetus is complex. When fetal hydrops is found in an asymptomatic pregnant woman, maternal infection is likely to have occurred some weeks previously and B19 IgM may not be detectable in the mother or fetus. If available, a stored blood sample should be tested for IgG and IgM, looking for IgG seroconversion. Where fetal diagnosis is needed during pregnancy, testing of amniotic fluid or fetal blood for B19 DNA is appropriate. In placenta or fetal tissues taken at autopsy, diagnosis can be made by PCR or in-situ hybridization for B19 DNA on formalin-fixed, paraffin-embedded tissue sections (Fig. 47.7) or by demonstrating characteristic intranuclear inclusions in erythroid precursors (Fig. 47.8).

EPIDEMIOLOGY

B19 infection is common and worldwide in distribution. Serological studies indicate that infection is most commonly acquired between 4 and 10 years of age, and by 15 years approximately 50% of children have antibody. Infection occurs throughout adulthood and up to 90% of elderly people are seropositive.

B19 virus infections are endemic throughout the year in temperate climates, with a seasonal increase in frequency in late winter, spring and early summer months. There are also longer-term cycles of infection with a periodicity of

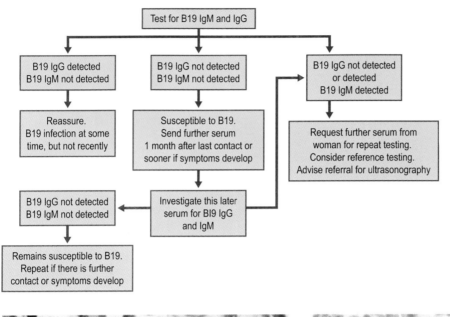

Fig. 47.6 Investigation for B19 in a pregnant woman with significant exposure to rash illness. (Reproduced with modifications with permission of Health Protection Agency, London.)

```
Test for B19 IgM and IgG
```

```
B19 IgG detected        B19 IgG not detected        B19 IgG not detected
B19 IgM not detected     B19 IgM not detected         or detected
                                                      B19 IgM detected
```

```
Reassure.               Susceptible to B19.          Request further serum from
B19 infection at some    Send further serum           woman for repeat testing.
time, but not recently   1 month after last contact or  Consider reference testing.
                         sooner if symptoms develop    Advise referral for ultrasonography
```

```
B19 IgG not detected     Investigate this later
B19 IgM not detected     serum for BI9 IgG
                         and IgM
```

```
Remains susceptible to B19.
Repeat if there is further
contact or symptoms develop
```

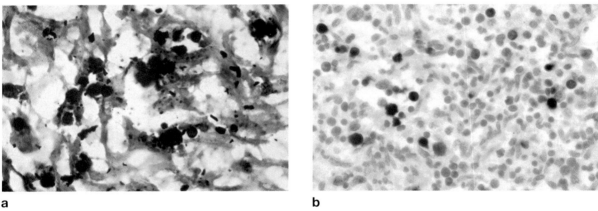

Fig. 47.7 In-situ hybridization using a pool of B19 oligonucleotide probes showing enlarged B19-infected erythroblasts in **a** fetal heart and **b** fetal liver from a fatal case of fetal hydrops. (Courtesy of Dr Heather Cubie, Specialist Virology Centre, Royal Infirmary, Edinburgh.)

about 4–5 years. Most infections are transmitted by close contact by the respiratory route, with a seroconversion rate of 20–30% for day-care personnel in close contact with infected children. Asymptomatic blood donors with high-titre viraemia enable transmission by blood or blood products, and the stability and resistance of B19 enables transmission by heat-treated factor VIII and IX to haemophiliacs.

MANAGEMENT

Most cases of B19 infection are mild and self-limiting, and specific treatment is not required other than symptomatic

relief for joint disease. No antiviral drug is available. However, there are three situations (see Table 47.2) in which severe anaemia occurs as a consequence of B19 infection, and in each of these blood transfusion or intravenous immunoglobulin is indicated. In transient aplastic crisis, blood transfusion tides patients over the relatively short period of erythroid aplasia before the immune response rapidly clears the virus infection. In the immunosuppressed there is a failure to produce neutralizing antibody, so management consists of blood transfusion plus a course of immunoglobulin over 5–10 days. This leads to disappearance of the viraemia, sometimes accompanied by the development of rash illness. If the viraemia recurs, further immunoglobulin is usually necessary. With infection during

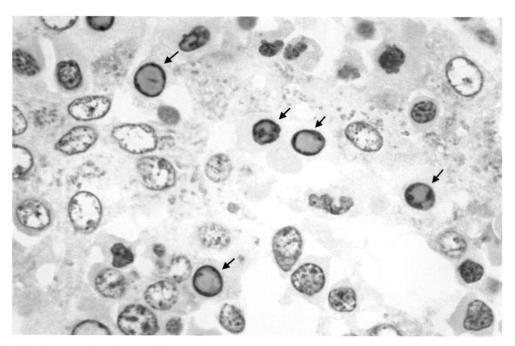

Fig. 47.8 Fetal liver (haematoxylin and eosin stain) from fetal hydrops showing characteristic B19-infected erythroblasts (arrows) with intranuclear inclusions and marginated chromatin. (Courtesy of Dr Elizabeth Gray, Pathology, Aberdeen Royal Infirmary.)

pregnancy, termination of pregnancy is not indicated as a fetus that survives B19 infection has a good prognosis. Although spontaneous resolution of hydrops due to B19 with the birth of a healthy baby has been documented, the overall fetal mortality rate is decreased with intra-uterine transfusion, which may need to be repeated.

Prevention of disease by isolating susceptible individuals is generally impractical as infections may be subclinical and symptomatic individuals are infectious before distinctive symptoms appear. In addition, as the rash and arthralgia of B19 infection are caused by immune reactions, not viraemia, those with rash or arthropathy are not infectious. However, susceptible pregnant health-care workers should not care for those with transient aplastic crisis or pure red cell aplasia; those with the disease are likely to be highly infectious. Theoretically, susceptible individuals at risk of significant anaemia (e.g. immunocompromised children) could be protected temporarily by the administration of human immunoglobulin, but this has not been tried. Good hand hygiene should be encouraged.

There is as yet no licensed vaccine against B19 but, by analogy with the animal viruses where immunization against feline and canine parvoviruses is routine in veterinary practice, such a strategy would be expected to be very effective in man. Recombinant B19 virus-like particles can be produced that induce neutralizing antibodies and so are potentially suitable candidates for a human vaccine. At present, screening of at-risk groups is not recommended as vaccine and prophylaxis are not available.

KEY POINTS

- Parvoviruses, including B19, are the smallest viruses known to infect and cause human disease.
- B19 replicates in erythroid precursor cells, causing cell death.
- Spread is by respiratory, blood and transplacental routes.
- In healthy individuals, B19 causes immune-mediated rashes (fifth disease, slapped cheek syndrome) and arthropathy.
- In patients with underlying haematological disease, B19 may result in anaemia or a transient aplastic crisis.
- In immunocompromised patients, chronic anaemia or pure red cell aplasia may ensue.
- In pregnancy, maternal infection at 0–20 weeks' gestation may lead to fetal loss or non-immunological fetal hydrops.
- There is no specific antiviral therapy, and no available vaccine for B19.

RECOMMENDED READING

Abdel-Fattah S A, Soothill P W 2001 Parvovirus B19 infection in pregnancy. In MacLean A B, Regan L, Carrington D (eds) *Infection and Pregnancy*, pp. 271–282. Royal College of Obstetricians and Gynaecologists Press, London

Allander T, Tammi M T, Eriksson M et al 2005 Cloning of a human parvovirus by molecular screening of respiratory tract samples. *Proceedings of the National Academy of Sciences of the USA* 102: 12891–12896

Bloom M E, Young N S 2001 Parvoviruses. In Knipe D M, Howley P M (eds) *Virology*, 4th edn, pp. 2361–2379. Lippencott Raven, Philadelphia

Brown K E 2004 Human parvoviruses. In Zuckerman A J, Banatvala J E, Pattison J R, Griffiths P D, Schoub B D (eds) *Principles and Practice of Clinical Virology*, 5th edn, pp. 703–720. Wiley, Chichester

Faisst S 2000 *Parvoviruses, from Molecular Biology to Pathology. Contributions to Microbiology*, Vol. 4. Karger, Basel

Jones M S, Kapoor A, Lukashov V L et al 2005 New DNA viruses identified in patients with acute viral infection syndrome. *Journal of Virology* 79: 8230–8236

Morgan-Capner P, Crowcroft N S 2002 Guidelines on the management of, and exposure to, rash illness in pregnancy (including consideration of relevant antenatal screening programmes in pregnancy). *Communicable Disease and Public Health* 5: 59–71

Young N S, Brown K E 2004 Parvovirus B19. *New England Journal of Medicine* 350: 586–597

Internet site

Health Protection Agency. Guidance on the management of rash illness and exposure to rash illness in pregnancy. http://www.hpa.org. uk/infections/topics_az/pregnancy/rashes/default.htm

48 Picornaviruses

Meningitis; paralysis; rashes; intercostal myositis; myocarditis; infectious hepatitis; common cold

S. M. Burns

The family Picornaviridae comprises small (pico) RNA viruses with a diameter of 27–30 nm. It is subdivided into nine genera of which the *Enterovirus, Rhinovirus* and *Hepatovirus* genera are of considerable human importance. Viruses in the genus *Enterovirus* infect via the gut, and include the echoviruses, coxsackieviruses and polioviruses. The virus of hepatitis A belongs to the genus *Hepatovirus*. The members of the other genus, *Rhinovirus*, are the most important causes of the common cold. The properties of enteroviruses and rhinoviruses are compared in Table 48.1.

ENTEROVIRUSES

DESCRIPTION

The polioviruses, coxsackieviruses and echoviruses are described as enteroviruses because they are all found in the intestines and are excreted in the faeces. New enteroviruses continue to be isolated, but they are no longer subdivided into the three named groups. Instead they are called enterovirus and are numbered, for example, enterovirus 71. Hepatitis A, previously classified as enterovirus 72, has now been assigned its own genus, *Hepatovirus*.

Composition

The RNA is single stranded and of positive sense, and can be translated directly by host ribosomes. It is surrounded by a capsid consisting of a protein shell arranged in icosahedral symmetry around the RNA molecule, the complete sequence of which is now known for several strains. Four major peptides are recognized in the shell: viral protein (VP) 1, VP2, VP3 and VP4, and are formed from a single precursor protein, VP0, by proteolytic cleavage. Specific neutralizing antibodies are considered to be the major mechanism of protection against infection, and a major antigenic site on the VP1 protein of polio type 3 has been identified.

In addition to the properties listed in Table 48.1, the three groups of viruses and those designated as enteroviruses 68–71 have a number of features in common.

- They attach to cells of the intestinal tract at specific receptor sites and replicate inside the intestinal cells.
- They commonly cause asymptomatic immunizing infections, which protect against future infections with the same virus.
- They occasionally cause infection of the central nervous system and other target organs.
- They are commoner in children than in adults.
- In temperate climates they usually cause infections in the summer and autumn.

PROPERTIES OF ENTEROVIRUSES

In the presence of moist organic material enteroviruses:

- survive at room temperature for weeks
- survive at 4°C for months
- are killed at 50–55°C.

They are rapidly inactivated by 0.3% formaldehyde, 0.1 M hydrochloric acid and solutions giving a free residual chlorine concentration of 0.3–0.5 ppm. However, chemical inactivation is ineffective when organic matter is present; e.g. in swimming baths.

Polioviruses

There are three types of poliovirus, identified by neutralization tests; their RNA molecules differ by 50% in hybridization studies. Two outstanding characteristics of the viruses are their affinity for nervous tissue and the narrow host range. Only human beings and primates are susceptible. Cynomologus and rhesus monkeys can be infected by the oral route, and develop paralysis; in chimpanzees the infection is often asymptomatic.

Table 48.1 Some properties of picornaviruses

Property	Enteroviruses	Rhinoviruses
Size (nm)	22–30	30
Capsid		
Form	Icosahedral	Icosahedral
Polypeptides	VP1, VP2, VP3, VP4	VP1, VP2, VP3, VP4
RNA type	Single-stranded, positive-sense	Single-stranded, positive-sense
RNA molecular weight (Da)	2×10^6–2.6×10^6	2.6×10^6
Acid	Stable (pH 3–9)	Labile (pH 3–5)
Optimal temperature for growth (°C)	37	33–34
Density in caesium chloride (g/ml)	1.34	1.39–1.42

Under the influence of cortisone, monkeys become more susceptible to small parenteral doses of the virus.

The prototype strains are:

- type 1, the Brunhilde and Mahoney strains
- type 2, which includes the rodent-adapted strains, the Lansing and MEFI strains
- type 3, the Leon and Saukett strains.

The three types are antigenically distinct, but overlap occurs in neutralization tests. Type 1 is the common epidemic type, type 2 is usually associated with endemic infections, but type 3 has caused some epidemics. The size, chemical and physical properties, and the resistance of the three types are all identical, and so their antigenic properties provide one of the main methods of differentiating them.

Echoviruses

Echoviruses were identified by cell culture of the faeces of patients suffering from paralytic and non-paralytic illness. There are over 30 serotypes of the *enteric cytopathic human orphan* viruses and, true to their name, there is still no clear association of some types with specific disease. Echoviruses 22 and 23 have a very different RNA sequence and have been classified into a separate genus, *Parechovirus*. Most do not infect laboratory animals. Several types can agglutinate human group O red cells.

Antigenic characters

Thirty-three distinct antigenic types have so far been distinguished by neutralization tests in cell cultures. Some cross-reactions occur. Antigenic variation is known to occur in types 4, 6 and 9, and may be a common occurrence under natural conditions.

Coxsackieviruses

This group (of 30 serotypes) was named after the place in New York State where the first member was isolated.

Table 48.2 Features of coxsackievirus infection in the laboratory

	Types	Growth in MK cell culture	Effect in suckling mice
Coxsackie A virus	1–24[a]	+	Paralysis
Coxsackie B virus	1–6	+	Spasticity

MK, monkey kidney.
[a]Coxsackievirus A23 now classified as echovirus 9.

Coxsackieviruses are pathogenic for newborn mice and hamsters. Two groups, A and B, of viruses were recognized according to the histological nature and the types of lesions they produce in mice (Table 48.2).

Growth in culture and mice

All group B coxsackieviruses, but only a few group A viruses, grow readily in monkey kidney cell cultures.

The other type A viruses were isolated by inoculation of suckling mice; they are now detected by reverse transcription–polymerase chain reaction (RT-PCR). Type A7, which has caused paralytic disease, is identical to the Russian strain AbIV, thought at one time to be a fourth type of poliovirus. Mouse inoculation has not been used as a diagnostic test for years. It has been replaced by RNA detection using RT-PCR.

Antigenic characters

Thirty antigenic types were defined by cross-neutralization tests in mice or cell culture, and by cross-complement fixation reactions. Twenty-three have the features of group A and six have those of group B. Each of the six group B types is subject to antigenic variation, and sera from convalescent cases may show heterotypic responses. Coxsackievirus A23 is now been reclassified as echovirus 9.

Other enteroviruses

Detailed examination of the characteristics of the viruses in cell culture and their ability to infect laboratory animals, which had been the basis of classification into polioviruses, coxsackieviruses or echoviruses, showed that these were not reliable features. Since 1970 new isolates are simply called enterovirus and given the next available consecutive number. Thus, enterovirus types 68, 69, 70, 71 and 72 are now recognized as additions to the family. The original names are still used.

REPLICATION

Knowledge of enterovirus replication is based on studies of poliovirus. The different stages of the cycle are described in Chapter 7. The end of replication is signalled by lysis of the host cell, with the release of all the new virus particles.

CLINICAL FEATURES AND PATHOGENESIS

The enteroviruses are associated with a wide variety of clinical presentations, although the majority of infections are asymptomatic. The sites affected are:

- the central nervous system (meningitis, paralysis), particularly in children
- the skin (rashes)
- muscle (intercostal myositis, myocarditis).

In neonates infection can be severe, extensive and occasionally fatal. There is no strong evidence that enteroviruses have a role in the development of the chronic fatigue syndrome.

There is a possible link between enterovirus infections and the onset of insulin-dependent (type 1) diabetes mellitus, but further research is required to confirm this association.

Polioviruses

There are three types of poliovirus infection:

1. Asymptomatic infection or a mild, transient 'influenza-like' illness; this is the most likely outcome, with only 1% of infections resulting in recognizable clinical illness. The virus is excreted in the faeces for a limited time, and an immunological response develops that protects against re-infection with the same strain.

2. Infections with the same symptoms as above and evidence of involvement of the central nervous system with headache, neck stiffness and back pain (meningitis). Rapid and complete recovery in less than 10 days is usual.

3. Paralytic poliomyelitis in which the patient develops paralysis; this is the most dramatic form of the infection but is very uncommon, occurring in 1 in 1000 poliovirus infections in children, although 1 in 75 adults may be affected. The paralysis is usually flaccid due to destruction of lower motor neurones, although invasion of the brainstem by the virus can lead to incoordination of muscle groups and painful spasms. Paralysis occurs early in the illness but the extent is variable. Often the paralysis is greatest initially; some function may return over the next 6 months. Damage to nerve cells in the brainstem (bulbar paralysis) can lead to the inability to swallow and breathe.

The time between infection and the development of symptoms is usually 14 days but can range from 3 to 21 days. During this period several factors are known to have an adverse effect on the outcome of infection.

- Severe muscular activity can lead to paralysis of the limbs used, possibly due to increased vascularity, either in the limb or the appropriate area of the spinal cord, allowing increased access of virus to nerve endings.
- Pregnant women in the third trimester of pregnancy can have severe disease, but there is no firm evidence of congenital defects in infants born to mothers with poliomyelitis. Maternal infection acquired late in pregnancy may lead to perinatal infection and disease of the newborn.
- Injection of adjuvant-containing vaccines, and irritant substances such as heavy metals, can result in paralysis occurring in the limbs that received the inoculation. Paralysis develops (incidence 1 in 37 000 injections) when poliovirus is contracted within 1 month of receiving the inoculation.
- Patients who have had their tonsils removed have a higher chance of developing bulbar poliomyelitis. This has been attributed to the reduction of secretory immunoglobulin (Ig) A in the pharynx, and thus reduced neutralization of the virus.

Paralysis is a rare complication of oral polio vaccine or following infection from contact with a vaccinee excreting virus.

Non-polio enteroviruses have also been associated with central nervous disease and paralysis (e.g. echovirus 4 and enterovirus 71); as with polio, paralysis is a rare manifestation of infection.

Echoviruses

Most echovirus infections cause few or no clinical symptoms, but some have been associated with clinical syndromes.

The main clinical syndromes associated with echoviruses are:

- aseptic meningitis
- paralysis
- rash
- respiratory disease.

Less frequently, echoviruses cause:

- pericarditis and myocarditis
- neonatal infection.

Infection can be widespread in a community, although only a few suffer from clinical illness. Symptoms occur following a short incubation period of 3–5 days (simple fever, upper respiratory symptoms or diarrhoea). Non-specific rashes of fleeting duration have been reported.

The onset of meningitis is abrupt, with severe headache and vomiting. Symptoms are self-limiting, and after a variable convalescent period a full recovery is made, although rare cases of paralysis do occur.

Most of the echovirus types have been associated with sporadic cases of aseptic meningitis or one of the other disease patterns already mentioned. A number of types, notably 4, 6, 9, 16, 20, 28 and 30, have considerable epidemic potential, and the clinical features are very varied. As examples, echovirus 9 epidemics have been common in Europe and North America as large outbreaks of a biphasic fever, with a sore throat and a rash on the face, neck and chest and, less commonly, on the lower trunk and extremities. A minority of patients show clinical signs of meningitis, but many without distinct clinical signs have a pleocytosis in the cerebrospinal fluid (CSF). In 2002, echovirus 13 was isolated from patients with asceptic meningitis in Korea.

Echovirus 16 epidemics have been called *Boston fever* after the city where the illness was first reported. Clinically, the infection starts with a sharp fever, abdominal pains and a mild sore throat. After 24–48 h, a pink discrete macular or maculopapular rash, mostly on the face, chest and back, appears in children. Echovirus 4, 6 and 30 epidemics have been associated with considerable numbers of cases of meningitis in children and adults. Echoviruses 6, 9 and 11 have caused severe and fatal infections in newborn infants. The virus is probably transmitted during birth or postnatally from the mother or nursery attendants, who are asymptomatic. Circulatory collapse, hepatitis and meningo-encephalitis develop and infection can spread rapidly to other infants in the nursery or special care baby unit. The type of virus circulating in the community at the time is usually implicated.

Echovirus 18 has been recovered from the faeces of many infants in an outbreak of diarrhoea. The children had

Table 48.3 Features of human coxsackievirus infection

Coxsackievirus group	Type	Clinical illness
A	1–24	Aseptic meningitis Febrile illness Herpangina Hand, foot and mouth disease
B	1–6	Neonatal disease Bornholm disease Myocarditis, hepatitis Meningitis

no rash and no severe involvement of the central nervous system or meningeal reaction. Several echoviruses, including types 1, 11, 19, 20 and 22, have been isolated from cases of respiratory illnesses in children, in whom symptoms include pneumonia, bronchiolitis and upper respiratory tract illness.

Coxsackieviruses

Group A viruses

These viruses give rise to a number of different illnesses (Table 48.3). Aseptic meningitis, indistinguishable clinically from that caused by other enteroviruses, is caused by a number of types (e.g. 2, 4, 7, 9 or 23). Type A7 has given rise to epidemics of paralytic disease in Scotland, the former USSR and elsewhere. Herpangina is an acute feverish disease, usually in young children, characterized by lesions in the mouth consisting of papules on the anterior pillars of the fauces; these papules become vesicles and finally shallow ulcers with a greyish base and punched-out edge. There are usually small numbers of lesions. A fine macular rash (rubella-like) is a feature of some Coxsackie infections. Outbreaks may be seen in nurseries and schools. Hand, foot and mouth disease presents as a painful stomatitis with a vesicular rash on the hands and feet. Typically, it lasts about a week; most cases are seen in the summer in children aged 1–10 years; cases can occur in clusters, and in families. The viruses usually implicated are types 5 and 16.

Coxsackievirus A21 has caused epidemics of colds in camps of military recruits.

Coxsackievirus A24 caused a major outbreak of 6000 cases of acute haemorrhagic conjunctivitis in French Guyana in 2003. Molecular typing confirmed that this virus had been introduced from South-East Asia.

Table 48.4 Illness associated with enteroviruses 68–72

Enterovirus type	Clinical illness
68	Pneumonia and bronchiolitis
69	Isolated from an ill person in Mexico
70	Acute haemorrhagic conjunctivitis
70, 71	Paralysis, meningo-encephalitis
71	Hand, foot and mouth disease
72[a]	Hepatitis A

[a]Reclassified as *Hepatovirus*.

Group B viruses

Epidemic myalgia or *Bornholm disease*, first described on the Danish Island of Bornholm, is characterized by fever and the sudden onset of agonizing stitch-like pains in the muscles of the chest, epigastrium or hypochondrium. Although the disease is most frequently recognized in its epidemic form, many sporadic cases occur. Pleurisy and pericarditis may complicate epidemic myalgia, although most cases recover within a week.

In newborn infants, severe and often fatal *myocarditis* has been reported, and the virus can be found in high concentrations in the myocardium at autopsy. Epidemics have occurred in nurseries when there is evidence of group B virus activity in a community. The baby is highly susceptible to the virus, as is the infant mouse. Myocarditis and pericarditis can occur in children and adults, and virus has been isolated from pericardial fluid.

Coxsackie B viruses are major causes of human *myo-pericarditis*, but this is a difficult diagnosis to confirm. Although severe in the neonate, it tends to follow a more benign course in the adult. The initial symptoms are often of an upper respiratory or 'influenza-like' illness followed 7–10 days later by clinical heart disease. Chest pain is a feature and electrocardiographic abnormalities such as tachycardia and arrhythmias have been found. On clinical examination, murmurs, rubs and, occasionally, pericardial effusions are detected.

Aseptic meningitis, sometimes with paralysis, is a common manifestation of infection due to group B virus. Occasionally a rash is present.

Enteroviruses 68–71 (Table 48.4)

Enterovirus 68

This has been reported to cause pneumonia.

Enterovirus 69

This virus has not yet been associated with significant clinical illness.

Enterovirus 70

In 1969, outbreaks of *acute haemorrhagic conjunctivitis* spread throughout Africa and Asia; more recently the disease has occurred in Mexico. This condition is highly infectious and has a short incubation period of 24–48 h. Attack rates are very high where there is overcrowding and poor sanitation. Although subconjunctival haemorrhage is a complication, most patients recover within 7 days. Neurological complications can occur, including a polio-like paralytic illness.

Enterovirus 71

Like other enteroviruses, enterovirus 71 produces infections that are usually inapparent. This virus was first recognized in 1969 in California, when it was isolated from the faeces of a 9-month-old infant suffering from encephalitis. Since then it has been isolated from patients with a spectrum of illness. During an epidemic in Bulgaria in 1975, there was a high incidence of paralytic disease, mostly in children under 5 years of age. A large outbreak occurred in Malaysia in 1997 where 2140 children were affected and 27 cases of fatal myocarditis were reported. This was followed by a related large outbreak in Taiwan in 1998, which was characterized by hand, foot and mouth disease, and caused fatalities and severe handicap.

PATHOGENESIS

All enterovirus infections follow a similar pattern, as illustrated in Figure 48.1 for poliovirus, with differences in the target organs (e.g. central nervous system, skin, heart or muscle). Due to the potential severity of infection and the availability of an animal model, most is known about poliomyelitis.

Virus is ingested and, after multiplication in the lymphoid tissue of the tonsil or Peyer's patches and the local lymph nodes, it enters the blood and then the central nervous system. The paralytic effect of poliovirus is due to destruction of motor neurone cells in the anterior horns of the spinal cord or bulbar regions. Localization in the motor neurones is probably due to the presence of receptors for the poliovirus. Once within the brain or cord, virus can spread to neighbouring cells or via the CSF.

The viraemic phase marks the end of the incubation period, and is manifest in the patient by fever and generalized symptoms; it is followed by a period of about 48 h of relative well-being (the disease is biphasic) while the virus is invading nerve tissue and then, in serious cases, the signs of paralysis appear. If antibody is present in the blood, virus can be prevented from reaching the central nervous system.

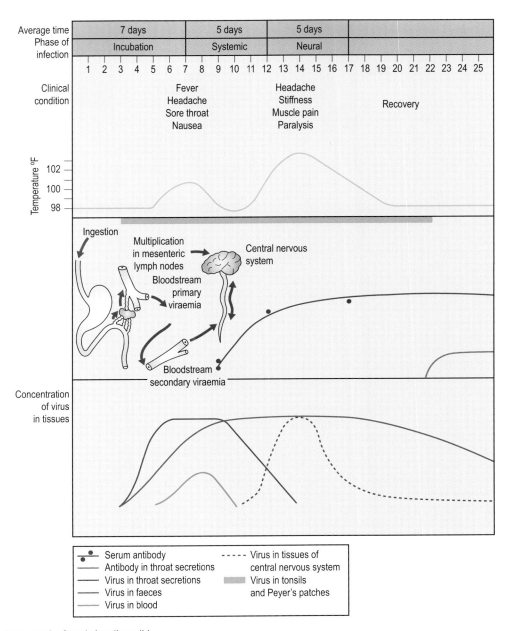

Fig. 48.1 Pathogenesis of paralytic poliomyelitis.

Enteroviruses are lytic, destroying the infected cell within a few hours to days. They are non-enveloped, so are unlikely to induce cytotoxic T cell responses. However, cell damage will trigger an inflammatory response; the resultant oedema may affect neurones other than those infected with the virus. As the oedema resolves, function will recover in these cells, thereby explaining the apparent improvement in the degree of paralysis in the weeks to months after the acute stage.

LABORATORY DIAGNOSIS (Table 48.5)

Culture

Virus isolation was the most useful method for establishing a diagnosis but has been superseded by RT-PCR for RNA detection in CSF. Examination of CSF is an essential part of the laboratory diagnosis of meningitis. Faecal samples should also be submitted, although isolation can be made

Table 48.5 Laboratory diagnosis of picornaviruses

	Sample	Timing	Laboratory method
Enterovirus	CSF	Acute <5 days	Cell culture, PCR
	Faeces		Cell culture
	Blood		IgM[a]
Poliovirus	CSF	Acute <5 days	Cell culture, PCR
	Faeces		Cell culture
Rhinovirus	Respiratory secretions	Acute <5 days	Cell culture, PCR

[a]Including hepatovirus

from rectal and throat swabs. As virus excretion can be intermittent, two specimens should be collected on successive days, as early as possible in the illness, ideally within 5 days of the onset.

In the UK, poliovirus isolates are likely to be attenuated vaccine strains, which can be differentiated from wild-type virus by several genetic markers.

In paralytic poliomyelitis, the virus can be found in the faeces for a few days preceding the onset of acute symptoms, and is present in more than 80% of cases in the faeces during the first 4 days. Only a few patients continue to excrete the virus after the 12th week. Prolonged excretion occurs in immunocompromised hosts. The virus can be isolated from the oropharynx of many patients for a few days before and after the onset of the illness. Isolation from the CSF is seldom successful in cases of paralytic poliomyelitis.

Echoviruses, coxsackie B viruses and coxsackie A9 virus are readily isolated from nose and throat swabs, stools or CSF; they are present in 80% of cases in the faeces for at least 2 weeks after the onset. Identification is by neutralization, now carried out in reference laboratories. Isolation of virus from the CSF is a significant finding. However, the relevance of virus isolation from faeces is not always clear, as enteroviruses can be excreted for some time after asymptomatic infection.

Polymerase chain reaction

Identification of conserved sequences within the 5′ non-coding region of the enterovirus genome has resulted in the development of PCR primers that allow the detection of most enteroviruses. Many studies have confirmed that enterovirus RT-PCR is more sensitive than culture for the detection of enteroviruses. The current 'gold standard' of diagnosis is to detect the enterovirus genome in CSF by PCR.

Serological tests

Although specific IgM and IgG assays for enteroviruses are available, their use is limited by the heterotypic and anamnestic responses associated with infections. IgM tests for hepatovirus (hepatitis A virus) have been developed (see p. 485).

TREATMENT

Treatment is symptomatic. A number of antiviral agents were investigated many years ago for the treatment of paralytic poliomyelitis, but were of no practical value owing to the rapid emergence of drug resistance.

EPIDEMIOLOGY AND TRANSMISSION

The only natural source of poliovirus is man. The virus is spread from person to person, and no intermediate host is known. Echovirus and coxsackievirus have a similar epidemiology; no vaccines are available.

Transmission

All enteroviruses are excreted in large numbers from the gut and are ingested to cause infection. This can be achieved in the following ways:

- by direct transfer on fingers (faecal–oral transfer)
- on eating utensils
- through contamination of food or drink
- only rarely by entry through the conjunctiva.

Outbreaks are often seen in closed communities and schools. In the acute phase of infection, virus is present in the throat and, although droplet spread could occur, the faecal–oral route is the usual means of transmission. After the acute phase, it is the only possible route. Cases are most infectious late in the incubation period, but infection can be transmitted at any time during virus excretion in the faeces.

Infection rates can reach 100% in closed communities and households, particularly where children are present. Social factors, such as standards of hygiene and overcrowding, are also important. Sewage can contain polioviruses, particularly when there is infection in a community. Enteroviruses can survive for several months in river water, but are unlikely to survive in chlorine-treated water or swimming pools where the recommended level of chlorination (without protein contamination) is achieved. Flies and cockroaches have been found to harbour viruses, but their role in transmission is minimal.

PREVENTION AND CONTROL

After natural infection, immunity is permanent. Virus-neutralizing antibodies are formed early during the disease (often before the seventh day) and persist for several decades. Secretory IgA is produced in the gut. There is some doubt about the duration of immunity to some of the coxsackieviruses. In the clinical setting, standard infection control precautions are critical in preventing spread.

Immunization

There are two types of polio vaccine: inactivated and live-attenuated.

Inactivated polio vaccine (Salk vaccine)

This was introduced in 1956 for routine immunization. The vaccine contains strains of the three types of virus grown in monkey kidney cell culture and inactivated by exposure to formaldehyde. The batches of vaccine are tested for the presence of residual live poliovirus and must be free of contaminants. Inactivated vaccines are used in Sweden, Finland, Iceland and Holland. With acceptance rates in excess of 90%, these countries have virtually eliminated poliovirus. The circulation of poliovirus in the community has been reduced dramatically despite the fact that inactivated vaccine does not induce much secretory IgA in the alimentary tract. A high rate of immunization is necessary and antibody levels need to be maintained because it has been shown in children that the outcome of exposure to virus is directly related to the level of antibody at the time of exposure. An outbreak in Finland was due to a poorly immunogenic type 3 component in the vaccine. The vaccine should be administered by deep subcutaneous or intramuscular injection, and is not associated with local or general reactions. A course of three injections given with intervals of 6–8 weeks between the first and second doses and 4–6 months between the second and third doses produces long-lasting immunity to all three poliovirus types. Inactivated vaccine is recommended for immunocompromised individuals and their contacts, and others for whom a live vaccine is contra-indicated.

Inactivated polio vaccine may be used simultaneously with the triple vaccine for diphtheria, pertussis and tetanus. Booster doses of polio vaccine can be given at the same time as diphtheria/tetanus and also with the combined mumps, measles, rubella (MMR) vaccine.

Live-attenuated polio vaccines (Sabin vaccine)

This replaced the Salk vaccine for routine use in the UK in 1962 and is used extensively in many other countries. It contains live but attenuated strains of poliovirus types 1, 2 and 3, grown in cultures of either monkey kidney cells or human diploid cells. The strains, developed by Sabin in 1959, were obtained by growing less virulent polioviruses and, after passage, selecting strains that had lost their neurovirulence.

The vaccine is administered orally and parallels natural infection, with stimulation of circulating IgG and local secretory IgA in the pharynx and alimentary tract, thereby producing local resistance to subsequent infection with wild poliomyelitis viruses. Herd immunity is important in preventing the circulation of the wild-type virus and high levels of immunization uptake are necessary. The wide circulation of vaccine virus, which helps to maintain immunity in the community, aids this. However, vaccine strains are not completely stable and studies of sequential isolates of virus from vaccine recipients show that changes can be detected very rapidly. There is therefore the theoretical possibility that vaccine virus, attenuated by serial passage in culture, could revert to neurovirulence with multiple rounds of replication in the vaccinee and after transmission to contacts. Vaccine-associated poliomyelitis has been reported in oral vaccine recipients at a rate of 1 in 2 million doses, and it has also been seen in contacts of recipients. It is not possible to predict who will be affected, although the extended replication, which occurs in the immunocompromised, should be avoided, as the rate of vaccine-associated poliomyelitis in such patients is 10 000 times greater than in people with normal immunity. Non-immunized parents and household contacts of children receiving primary immunization should be immunized against poliomyelitis at the same time as the children.

Until 2004, oral polio vaccine was recommended for infants from 2 months of age. Oral vaccine is no longer available for routine use and since 2004 has only been available for outbreak control. The primary course consists of three separate doses given at the same time as diphtheria/tetanus/pertussis vaccine (see Ch. 69). Each dose contains all three strains. In infants, three drops are dropped from a spoon directly into the mouth, which may be open in response to the simultaneous administration intramuscularly or subcutaneously of diphtheria/tetanus/pertussis vaccine. Breast-feeding does not interfere with the antibody response and should continue. A reinforcing dose is given to children at school entry and prior to leaving, but it is not necessary for adults unless they are at special risk, for instance through travel or occupational exposure. The effectiveness of live vaccine is shown by the experience in Scotland, where the last notified case was in 1994 and was associated with vaccine.

When a case of paralytic poliomyelitis is diagnosed, a dose of oral vaccine should be given to all persons in the immediate neighbourhood of the case who are immunocompetent, whether or not they have a history of previous vaccination against poliomyelitis. This should be followed by completion of the primary course in those not immunized. If the source of the outbreak is uncertain, it should be assumed to be a 'wild-type' virus until proven otherwise.

Poliovirus immunization in the tropics

Serological studies in Africa have shown that children are infected with two or three types of poliovirus by the age of 5 years. The rate of paralytic disease is low in young children, but significant numbers occur as the infection is widespread. The disease is known as *infantile paralysis*. Through the expanded programme on immunization, the World Health Organization has increased the rate of immunization, chiefly with the use of oral vaccine. Although this is cheaper than inactivated vaccine, it must be stored at 0–4°C and thus requires the existence of an effective cold chain to ensure successful immunization. Apart from inactivation of the vaccine, other explanations for low rates of seroconversion include interference from other enteroviruses, malnutrition and the presence of inhibiting factors in the gastro-intestinal contents.

Global eradication

The World Health Organization had set a target date of the year 2000 for the global eradication of poliomyelitis. Although, overall, the number of cases has fallen by more than 95%, 30 countries in southern Asia and West and Central Africa have recent reports of cases, and a revised deadline has had to be set. The eradication is being attempted with annual national immunization days to ensure that each child receives an adequate number of doses of oral polio vaccine. Inactivated vaccine has also been shown to provide excellent mucosal protection and with its use regions such as Scandinavia have eliminated wild virus. No decision will be made to discontinue immunization against polio until all countries are free of wild poliovirus infections. In countries where no wild poliovirus infection is reported, surveillance systems are introduced that are effective at detecting wild poliovirus infections. All cases of paralytic illness are investigated. Only then will a country be certified poliovirus-free.

Prospects for the future

Routine and mass administration of oral polio vaccine since 1961 has prevented many millions of cases of paralytic poliomyelitis; however, when the incidence of poliomyelitis falls dramatically, as in the USA for example, the proportion of paralytic cases attributable to vaccine, although very low, becomes increasingly significant. Recent work has identified the amino acid sequence of antigenic sites, which are important for neutralization of the virus. Information concerning the viral factors necessary for virulence is also available. Therefore, it may be possible to develop modified vaccine viruses that cannot revert to neurovirulence.

New drugs with some efficacy are under development. Pleconaril is one such drug. It inhibits the uncoating and attachment of virus to cell receptors and shows early promise against enteroviruses and rhinoviruses.

HEPATOVIRUSES

HEPATITIS A VIRUS

Hepatitis A has a worldwide distribution and occurs in epidemics as well as sporadically. The virus was first detected by electron microscopical examination of faeces.

Hepatitis A virus is the causative agent of *infectious hepatitis*. It has similar polypeptides to the four major polypeptides of the enterovirus family and shares the same properties of resistance to physical and chemical agents. Recently, it has been adapted to cell culture; it will grow only in cells of primate origin.

Clinical features

Although the incidence has fallen in the past decade, hepatitis A is still responsible for almost 60% of acute viral hepatitis in the USA. The illness is usually mild and occurs after an incubation period of 3–6 weeks. There is a prodrome of malaise, muscle pain and headache, and there may be a low-grade fever. The symptoms improve and disappear as jaundice develops. Serological tests show that many patients have a subclinical illness, but fulminating hepatitis and liver failure are rare. There are no carriers of the virus. Infection is mildest in young children, often accompanied only by nausea and malaise. Of children aged under 3 years, only 5% develop jaundice, but this rises progressively to more than 50% in adults. The fatality rate also rises with age, to 2% in adults. Some patients develop diarrhoea and some appear to have a relapse a few weeks after the onset. Arthritis and aplastic anaemia are rare complications. Treatment is supportive (e.g. replacement of fluid). There are no specific antiviral drugs. Severe cases where there is liver failure can be transplanted.

Pathogenesis

Like the enteroviruses, hepatitis A virus probably infects cells in the gut initially and then spreads to the liver via the blood. The histopathological appearance is similar to that of hepatitis B, with periportal necrosis and infiltration of mononuclear cells. Viral antigens are seen in the cytoplasm of the hepatocytes. Virus is excreted via the bile into the gut 1–2 weeks before the onset of jaundice, and excretion then declines rapidly over the next 5–7 days. Virus is also present in the urine of clinical and subclinical cases during the same period.

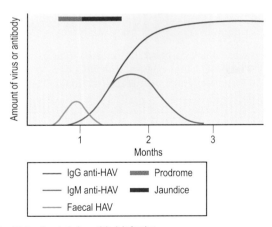

Fig. 48.2 Events in hepatitis A infection.

Legend:
— IgG anti-HAV — Prodrome
— IgM anti-HAV — Jaundice
— Faecal HAV

Laboratory diagnosis

Although the virus has been grown in cell culture, it is not possible to do this routinely from the faeces of patients. Diagnosis relies on the demonstration of specific IgM antibody, which develops very early in the course of infection and is generally present by the time the patient is investigated. It is detectable in the serum for 2–6 months after the onset of symptoms (Fig. 48.2). IgG antibody usually persists for many years and is a useful indicator of immunity.

Epidemiology

Only one major type of hepatitis A virus has been recognized. Molecular typing is now available to investigate outbreaks and link sporadic cases with common source outbreaks. Serological tests have made it possible to study the rate of infection in different populations throughout the world. Such studies confirm that, as the virus is spread by the faecal–oral route, it is prevalent in countries where sanitation is poor, and children are infected early in childhood. Serological studies also show that, even in developed countries, more than half of the population has been infected with hepatitis A virus. However, with increasing use of vaccine, lower rates are seen. This can result in increased numbers of acute cases in young adults and older patients who may have a history of recent travel to an endemic area.

Outbreaks of hepatitis A virus infection have been associated with food. Shellfish have often been incriminated, particularly when they are harvested from coastlines adjacent to sewage outlets. The mussel, for example, is a filter feeder and can concentrate the virus. Shellfish are eaten raw or partially cooked, thus protecting the virus. Contaminated raspberries were incriminated in another notorious out-

break in which uncooked frozen raspberries were eaten many months after picking. Infection is assumed to have come from an infected raspberry picker. The infectious dose is low – less than 100 virus particles. Because there is only a transient viraemic phase, only a few infections have been recognized after blood transfusion.

Prevention and control

A highly effective and immunogenic vaccine was licensed for use in the UK in 1991. It is a formaldehyde-inactivated vaccine prepared from virus grown in human diploid cells. A primary course of two doses (or, more recently, a single dose) of vaccine given intramuscularly produces good levels of neutralizing antibody that are known to persist for at least 10 years. The vaccine is an alternative to human normal immunoglobulin for frequent travellers and those whose occupation takes them to endemic areas abroad. Recent formulations combine hepatitis A and B antigens in the same vaccine.

RHINOVIRUSES

Although other viruses can give a similar illness, the viruses of this genus are responsible for the most frequent of all human infections, the 'common cold'. Most people suffer from two to four colds every year and, although the primary infection is not a severe one, the symptoms of secondary bacterial infection often follow and may be more severe than those of the original cold. Sinusitis and otitis media are quite common. Recently the rhinoviruses have been associated with acute exacerbations of asthma. These viruses are of major economic significance because they cause the loss of many million man-hours of work, accounting for 48% of upper respiratory tract infections.

PROPERTIES

As the name rhinovirus implies, the genus is associated with the nose. These viruses can be distinguished from the enteroviruses by their acid lability; consequently, they do not infect the intestinal tract. There are more than 100 serotypes of rhinoviruses; all are fastidious in cell culture. Electron microscopy cannot differentiate them from other family members (Fig. 48.3). Some primates may be susceptible to human viruses, and there are related viruses in cattle, cats and horses. The genomes of some rhinoviruses have 45–60% homology with polioviruses in hybridization tests.

The capsid of rhinoviruses appears to be less rigid than that of the enteroviruses. This loose packing is consistent with its greater buoyant density and sensitivity to acid.

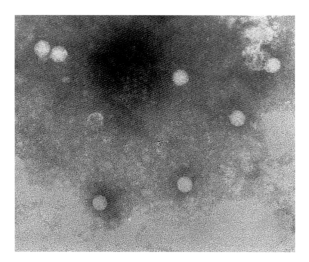

Fig. 48.3 Human rhinovirus. This virus is indistinguishable in appearance from other picornaviruses. Approximate size 25–30 nm. (From Madeley C R, Field A M 1988 *Virus Morphology*, 2nd edn. Churchill Livingstone, Edinburgh.)

Cultivation

Rhinoviruses show a distinct preference for cells of human origin, especially fetal lung and kidney. They are divided into major (90%) and minor (10%) groups according to their cell receptors.

Stability

Inactivation of rhinoviruses occurs below pH 6.0 and is more rapid the lower the pH. Complete inactivation occurs at pH 3.0.

Some rhinoviruses may survive heating at 50°C for 1 h. They are relatively stable in the range of 20–37°C and can survive on environmental surfaces such as doorknobs for several days. They can be preserved at −70°C.

Rhinoviruses are resistant to 20% ether and 5% chloroform, but are sensitive to aldehydes and hypochlorites.

REPLICATION

Most rhinoviruses attach to the same cell receptors on HeLa cells (90%). The minor group (10%) attaches to a low-density lipoprotein receptor and its related protein. The viruses replicate in the cytoplasm of infected cells to give a cytopathic effect that coincides with release of the virus in the same way as other picornaviruses. If the infected cultures are incubated at 37°C, the yield is reduced to 30–50% of that at 33°C.

CLINICAL FEATURES AND PATHOGENESIS

The typical illness is generally referred to as a common cold. The onset, after contact with infection, is usually within 2–3 days, sometimes as long as 7 days. The symptoms are:

- clear watery nasal discharge, which often becomes mucoid or purulent due to secondary bacterial infection
- sneezing and coughing
- sore throat
- headache and malaise.

The symptoms are most severe for 2–3 days when nasal virus titres are maximal and, although recovery is usually complete within a week, symptoms can persist for 2 weeks or longer. The ratio of symptomatic to asymptomatic infection is about 3:1, and the illness is generally worse in cigarette smokers. Rhinoviruses have frequently been isolated from patients during acute exacerbation of chronic obstructive airway disease, and are the most common viruses to be associated with wheeze in pre-school children.

It should be remembered that all respiratory viruses may cause the symptoms of the common cold.

In organ cultures, it has been shown that the virus settles on the ciliated nasal epithelial cells, enters, infects and spreads from cell to cell in the epithelium. The cilia become immobilized, and both cilia and cell degenerate as the virus replicates. Bacterial invasion of the damaged epithelium can then occur. Interferon is usually detectable shortly after the peak of virus shedding and probably plays a part in recovery. When specific antibody is first detected in nasal secretions, virus shedding ceases, suggesting that this may be the main factor leading to recovery. Little is known of the importance of cell-mediated immunity. The symptoms probably relate to the local inflammatory response and interferon release. Rhinoviruses have been recovered in pure culture from sinus fluids collected from patients with acute sinusitis, but secondary bacterial infection is thought to be the usual cause.

Lower respiratory tract infection may occur on some occasions since:

- patients with colds may also have lower respiratory tract symptoms with abnormal lung function
- children who develop colds may develop wheeze
- adults with colds may suffer exacerbation of chronic obstructive airway disease.

Immunity

After the acute illness, neutralizing antibody can be detected in both serum and nasal secretions. It can continue to rise in titre for 4–5 weeks after infection and may persist for up to 4 years, although some infections may provoke only a poor response, leaving the patient susceptible to the same serotype after a few weeks or months. New virus types continue to emerge by a process of immune selection and random mutation.

LABORATORY DIAGNOSIS

Culture

Nose and throat swabs in virus transport medium are the specimens of choice for the recovery of virus from all age groups. Nasopharyngeal aspirates are excellent specimens from children. Specimens should be taken as early in the illness as possible, preferably within the first 3 days.

Cell cultures of human origin, such as MRC5 or WI38, are preferred for the isolation of rhinoviruses. Organ cultures are not used routinely. Cultures are incubated at 33°C and observed microscopically for a cytopathic effect.

The majority of isolates are apparent within 2 weeks of inoculation, although some may take longer. Identification of an isolate as a rhinovirus may be made by considering the cells in which the cytopathic effect develops, the appearance of the cytopathic effect, and the demonstration of acid lability.

Serology

Serological methods cannot be used in the routine diagnosis of rhinovirus infections because of the multiplicity of serotypes and the lack of a common antigen.

Nucleic acid detection

Culture is slow and cumbersome and serology inadequate, so nucleic acid detection is used for diagnosis.

EPIDEMIOLOGY AND TRANSMISSION

Rhinoviruses can be isolated from patients with respiratory illnesses throughout the year, but in temperate climates the incidence of colds due to rhinoviruses increases in the autumn and spring and is lowest in the summer months. In the tropics the peak incidence occurs in the rainy season. Deliberate exposure of volunteers to wet and chilling does not cause colds. Rhinoviruses may be transmitted by inhalation of droplets expelled from the nose of a patient and also by hand-to-surface contact. During the acute phase of the illness high concentrations of virus are present in nasal secretions and may contaminate the fingers, and thereafter the contaminated fingers may touch the eye or nasal mucosa. The incidence of rhinovirus infections is highest in pre-school children, who often introduce colds to the home. People who are in contact with young children are at increased risk of infection.

Colds are mostly trivial and an inconvenience.

TREATMENT AND CONTROL

Although inactivated vaccines can be produced, there remains the considerable problem of deciding on the antigenic composition. Much effort has been devoted to the development of suitable antiviral therapy. Pleconaril is one such drug showing activity against rhinoviruses and enteroviruses (see above). Isolation of the infected person, although perhaps desirable, is not a practical method of preventing the spread of infection. Good infection control practices, including hand washing, will reduce spread of infection in the hospital setting.

KEY POINTS

- Picornaviruses pathogenic for man belong to the *Enterovirus*, *Rhinovirus* and *Hepatovirus* genera.
- Enteroviruses cause a wide range of diseases, including meningitis, skin rashes and myocarditis.
- Rhinoviruses are responsible for the most frequent of all human infections – the 'common cold'.
- There are effective vaccines against two members – hepatitis A and poliovirus.
- Enteroviruses are among the most stable viruses.
- Diagnosis of enterovirus infection now relies on nucleic acid detection.
- New enteroviruses regularly emerge and can cause major epidemics of disease.

RECOMMENDED READING

Archimbaud C, Mirand A, Chamber M et al 2004 Improved diagnosis on a daily basis of enterovirus meningitis using a one-step real-time RT-PCR assay. *Journal of Medical Virology* 74: 604–611

Cuthbert J A 2001 Hepatitis A: old and new. *Clinical Microbiology Reviews* 14: 38–58

Department of Health 2006 *Immunisation Against Infectious Disease.* HMSO, London

Dowdle W R, De Gourville E, Kew O M, Pallansch M A, Wood D J 2003 Polio eradication: the OPV paradox. *Reviews in Medical Microbiology* 13: 277–291

Grist N R, Bell E J 1984 Paralytic poliomyelitis and non-polio enteroviruses. Studies in Scotland. *Reviews of Infectious Diseases* 6(Suppl 2): S385–S386

John T J 2000 The final stages of the eradication of polio. *New England Journal of Medicine* 343: 806–807

Scheltinga S A, Templeton K E, Beersma M F, Claas E C 2005 Diagnosis of human metapneumovirus and rhinovirus in patients with respiratory tract infections by an internally controlled multiplex real-time RNA PCR. *Journal of Clinical Virology* 33: 306–311

Taylor K W 2004 Conference report: CIBA Foundation meeting on enteroviruses and early childhood diabetes. *Diabetic Medicine* 13: 910–911

Internet sites

Wiley InterScience. http://www3.interscience.wiley.com

Global Polio Eradication Initiative. http://www.polioeradication.org

49 Orthomyxoviruses

Influenza

S. Sutherland

The family Orthomyxoviridae comprises four genera: influenza A, B and C viruses and thogotoviruses.

Influenza A viruses can infect a variety of different host species, a fact that is of great importance in determining their ability to cause pandemic infection in man. Influenza B infects only human beings, and influenza C, although assumed to be primarily a human infection, has been isolated from pigs in China. The thogotoviruses form a newly discovered fourth genus of the orthomyxovirus family and are found in mosquitoes, ticks and the banded mongoose.

Influenza virus type A was the first to be isolated in 1933 by intranasal inoculation of the ferret. Thereafter, type B was isolated along with type A in cell culture in 1940. One of the most prominent features of the influenza viruses is their ability to change antigenically either gradually (*antigenic drift*) or suddenly (*antigenic shift*). Only influenza A virus has the potential to shift, whereas A, B and C may drift antigenically, although only very minor changes have been demonstrated in influenza C.

THE VIRUSES

The virions are spherical, 80–120 nm in diameter, but may be filamentous, sometimes up to several micrometres in length (Fig. 49.1). They have a helical nucleocapsid comprising, in A and B, eight segments of single-stranded RNA. Also present within the virion is the viral RNA-dependent RNA polymerase; this is essential for infectivity as the virion RNA is of negative sense and therefore has to be transcribed to produce viral messenger RNA (mRNA). The nucleocapsid is surrounded by an M1 protein shell, immediately exterior to which is a lipid envelope derived from the host cell. The M2 protein projects through the envelope to form ion channels, which allow pH changes in the endosome. Two types of spike (Fig. 49.2) project from the envelope, the *haemagglutinin* (H) and the *neuraminidase* (N). The haemagglutinin, so called because the virus agglutinates certain species of erythrocyte, is about 10 nm in length and consists of trimers of identical glycoprotein subunits, each consisting of two polypeptide chains, HA1 and HA2. The two polypeptides are joined in each subunit by a linkage site that may be a single base, arginine, or multibasic. Influenza viruses bind to cells by the haemagglutinin interacting with cell membrane receptors containing *N*-acetylneuraminic acid (sialic acid). The epitopes involved in receptor binding show great variability because of mutations in the RNA causing amino acid substitutions

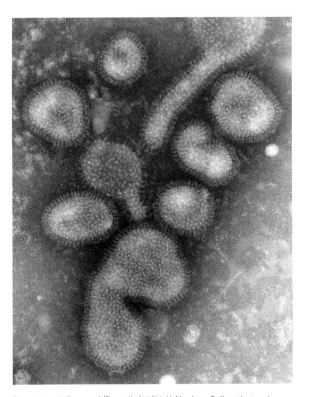

Fig. 49.1 Influenza A/Dunedin/27/83 H_1N_1 virus. Spikes that make up the fringe are also seen in the end-on view, covering the surface of the virus particles and giving them a 'spotty' appearance. Approximate size 80–120 nm. (From Madeley C R, Field A M 1988 *Virus Morphology*, 2nd edn. Churchill Livingstone, Edinburgh.)

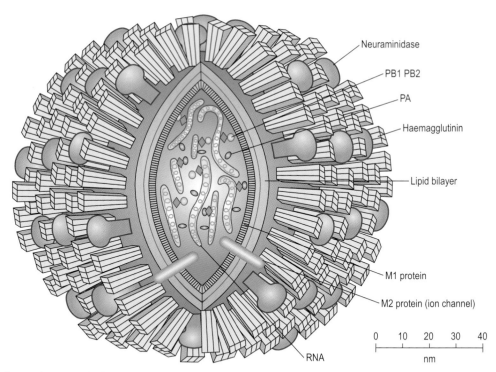

Fig. 49.2   Influenza virus particle in diagrammatic form showing projections and internal composition.

at several sites on the HA1 molecule. These changes can be located in the three-dimensional structure of the molecule and are found at only a few well defined areas close to the attachment site.

Between the H spikes are the mushroom-shaped protrusions of neuraminidase. The head is box-shaped and is assembled from four roughly spherical subunits attached to the stalk containing the hydrophobic region by which it is embedded in the viral envelope. The enzyme catalyses the cleavage of sialic acid and an adjacent sugar residue in glycoproteins found in mucus. This action allows the virus to permeate mucin and escape from the so-called 'non-specific' inhibitors. Neuraminidase activity is also thought to be important in the final stages of release of new virus particles from infected cells. Sialic acid is always present in newly synthesized virions, and its removal by neuraminidase prevents the new virus particles agglutinating, thus increasing the number of free virus particles and hence spread of the virus from the original site of infection. The viral genes and their functions are shown in Table 49.1.

Nomenclature

The World Health Organization (WHO) system of nomenclature includes the host of origin, geographical

origin, strain number and year of isolation; then, in parentheses, the antigenic description of the haemagglutinin and neuraminidase is given, for example A/swine/Iowa/3/70 (H_1N_1). If isolated from a human host the origin is not given: A/Scotland/42/89 (H_3N_2), for example. There are 16 different H antigens and nine N antigens that can occur in any combination. So far only H_1, H_2 and H_3, and N_1 and N_2, have been found in human epidemic strains, the rest in animals and birds.

Physical characteristics

The influenza virus withstands slow drying at room temperature; it has been demonstrated in dust beyond 2 weeks. Virus can survive in cold sea-water for several weeks. It can be preserved for long periods at −70°C, and remains viable indefinitely when freeze-dried.

Exposure to heat for 30 min at 56°C is sufficient to inactivate most strains; the few that survive this treatment are killed after 90 min. The viruses are inactivated by a variety of substances, such as 20% ether in the cold, phenol, formaldehyde, salts of heavy metals, detergents, soaps, halogens and many others.

Table 49.1 Influenza virus proteins

Gene segment	Proteins	Molecular weight ($\times 10^3$)	Location in virion	Function	Comments
1	PB2	0.87	Internal	RNA transcription	Polymerase proteins; highly conserved
2	PB1	0.96			
3	PA	0.85			
4	HA1	47.5	Spikes	Binding to cell receptors; envelope fusion to endosome	Glycoprotein; haemagglutinin varies antigenically
	HA2	28.8			
5	NA	48–63	Spikes	Neuraminidase	Glycoprotein enzyme activity
6	NP	50–60	Internal	Subunit of nucleocapsid	Nucleoprotein helical arrangement; type-specific
7	M1	0.28	Beneath lipid bilayer of envelope	Major structural component	Involved in assembly and budding
	M2	11–15	Transmembrane	Ion channel	Changes pH in endosome; blocked by amantadine
8	NS1	0.25	Internal	mRNA transport inhibition; interferon antagonist	Non-structural proteins; potential target for attenuation of live vaccine
	NS2	0.13		Unknown	

REPLICATION

Human influenza viruses recognize receptors that contain *N*-acetylneuraminic acid (sialic acid) attached to the penultimate sugar via an $\alpha(2,6)$-linkage, whereas avian strains prefer receptors with an $\alpha(2,3)$-linkage; both types are present in pigs. Attachment and entry to cells, and the rest of the replication cycle, is described in Chapter 7.

PATHOGENESIS

The pathogenicity of influenza viruses is multifactorial and may involve viral, host and environmental factors. Work with avian strains has identified host receptors and proteases as determinants of tropism and hence of pathogenicity. The sensitivity of strains with monobasic links at the HA1/HA2 junction to the trypsin-like proteases that are active extracellularly tends to limit spread of virus locally in the respiratory tract in man. Human strains are almost always of this type, whereas avian strains with multibasic links are associated with systemic spread in birds and animals, and are highly pathogenic. The neuraminidase type may also determine pathogenicity. One neuraminidase binds plasminogen, which could act as an ubiquitous protease even for monobasic strains. Receptor preference also determines tropism, and the length of the NS1 non-structural protein may also influence pathogenicity by its role in interferon antagonism.

Inhaled virus is deposited on the mucous membrane lining the respiratory tract or directly into the alveoli, the level depending on the size of the droplets inhaled. At the mucous membrane it is exposed to mucoproteins containing sialic acid that can bind to the virus, thus blocking its attachment to respiratory tract epithelial cells. However, the action of neuraminidase allows the virus to break this bond. Specific local secretory immunoglobulin (Ig) A antibodies, if present from a previous infection, may neutralize the virus before attachment occurs, provided the antibody corresponds to the infecting virus type. If not prevented by one of these mechanisms, virus attaches to the surface of a respiratory epithelial cell and the intracellular replication cycle is initiated.

The major site of infection is the ciliated columnar epithelial cell. New viruses bud from the apical membrane, the cilia are lost and viruses spread to other areas of the respiratory tract. The cell damage initiates an acute inflammatory response with oedema and the attraction of phagocytic cells. The earliest response is the synthesis and release of interferons from the infected cells: these can diffuse to and protect both adjacent and more distant cells before the virus arrives. It appears that interferons released in this way cause many of the systemic features of the 'flu-like' syndrome. Although viral components are absorbed and trigger the immune system, the virus itself is confined to the epithelium of the respiratory tract. Specific antibody helps to limit the extracellular spread of the virus, whereas T cell responses are directed against the viral glycoproteins on the surface of infected cells, leading to their destruction by cytotoxic T cells and also by antibody-dependent cell cytotoxicity.

CLINICAL FEATURES

Influenza A

In classical influenza:

- the incubation period is short, 2 days, but may vary from 1 to 4 days
- the illness is characterized by a sudden onset of systemic symptoms such as chills, fever, headache, myalgia and anorexia
- respiratory symptoms are also common but take second place to the systemic effects, especially early in the illness.

Many patients have both upper and lower respiratory tract infection, often with a troublesome, dry cough. The main physical finding is pyrexia, which rises rapidly to a peak of 38–41°C within 12 h of onset. Fever usually lasts for 3 days, but may be present for 1–5 days. During the second and third days of the illness the systemic effects diminish, and by the fourth day the respiratory symptoms and signs are predominant. In adults, systemic illness without respiratory symptoms is common. Some symptoms are age-specific, for example febrile convulsions and otitis media in children and dyspnoea in the elderly. About one-third of patients suffer only a common cold-like illness, and it has been shown that as many as 20% of cases are subclinical. A long convalescence is common, and cough, lassitude and malaise may last for 1–2 weeks after the disappearance of other manifestations. Many other respiratory viruses can cause typical influenza-like illnesses, although the severity of the systemic symptoms is usually greatest with influenza virus. Conjunctivitis was a feature of H_7N_7 cases in the Netherlands.

Influenza B

Symptoms closely resemble those associated with influenza A infections, consisting of a 3-day febrile illness with predominantly systemic symptoms. Overall, the infection is somewhat milder and in children vomiting may be a prominent symptom.

Influenza C

Clinically, influenza C causes an afebrile upper respiratory tract infection usually confined to young children; outbreaks are not recognized.

Complications of influenza

- Primary *influenza pneumonia*, especially in young adults during an outbreak, can be fatal after a very short illness of sometimes less than 1 day. A similar rapid illness can occur in the elderly.
- More commonly a bacterial pneumonia caused by *Staphylococcus aureus*, *Streptococcus pneumoniae* or *Haemophilis influenzae* occurs late in the course of the illness, often after a period of improvement, resulting in a classical biphasic fever pattern.

The incidence of chest complications is related to the age of the patient, increasing progressively after the age of 60 years. Severe infections and sudden death can occur, especially if there is some underlying disease, such as cerebrovascular, cardiovascular or chronic respiratory disease. The 1918 pandemic killed 20 to 30 million people; there were no antibiotics to treat the complications at that time. The genome has been reconstructed from material recovered from the body of a soldier who died from influenza; it will be of great interest to compare the genetic composition with that of later strains.

In the immunocompromised, symptoms may last longer and viral excretion may go on for weeks to months. Excess mortality was reported in pregnant women during the 1918 and 1957 pandemics, and even in non-pandemic outbreaks an increase in hospital admission due to cardiorespiratory disease is seen in the second and third trimesters. Between 1997, when the first human cases were recognized, and 2006, 151 of 256 people identified as being infected with the avian virus H_5N_1 died.

Immunity

After an attack of influenza the ensuing immunity to the particular subtype of infecting virus is of long duration. It is related to the amount of local antibody (IgA) in the mucous secretions of the respiratory tract together with the specific IgG serum antibody concentration. Immunity to infection, especially with type A, is subtype specific, giving little or no protection against subtypes possessing immunologically distinct H or N proteins. Once recovered from an initial influenza infection, exposure to later strains will boost IgG levels to the earlier strains, the so-called *original antigenic sin*.

LABORATORY DIAGNOSIS

Virus isolation and detection

For primary isolation the most suitable cells are primary monkey kidney or human embryo kidney cells, but as these tissues are scarce most laboratories now use secondary baboon kidney cells or Madin–Darby canine kidney cells. The presence of virus may be detected in baboon kidney cell cultures incubated at 33°C as early as 18 h after inoculation, by haemadsorption with human group O, fowl or guinea-pig red blood cells. Usually, if viable viruses are present in the clinical specimen, isolation is made within 7 days. The virus may then be identified by immunofluorescence.

Rapid diagnosis

Since the advent of antiviral agents with activity against influenza viruses, rapid diagnosis of respiratory infections has increased in importance, particularly in the hospital or institutional context, such as homes for the elderly, so that attempts can be made to limit the spread of the virus by a combination of patient isolation, vaccination and chemoprophylaxis.

Immunofluorescent detection of influenza antigens in respiratory specimens, either directly or after amplification in cell culture for 24 h, has speeded up diagnosis considerably. The best specimens are nasal aspirates or nasal washes, but nasal or throat swabs taken vigorously to obtain cells and put into viral transport medium can be satisfactory if taken in the first few days of illness.

It is important in the early stages of an outbreak or in sporadic cases that viruses should be isolated and analysed antigenically for epidemiology and vaccine production. For patient management and control of infection, rapid antigen detection is more useful. Detection of influenza RNA by reverse transcriptase–polymerase chain reaction is more sensitive than antigen detection but is not widely available.

Serology

Serological confirmation of the clinical diagnosis is obtained when a four-fold or greater rise in antibody titre can be shown to any one type of virus. Complement fixation tests are still widely used. This test uses nucleocapsid antigens that are type-specific and can distinguish A from B and C infections. Strain differences can be demonstrated by means of haemagglutination inhibition and neutralization assays or strain-specific complement fixation tests. In patients who are normally healthy a high titre found during an outbreak can, with some assurance, be related to a recent illness.

EPIDEMIOLOGY

Influenza A

Epidemics that were probably caused by influenza viruses have been described for more than 2000 years. The epidemics occur frequently at irregular intervals and were thought at one time to be under the influence of the stars, hence the term *influenza* (Italian for influence). It is well recognized that epidemics vary in severity. Occasionally infection spreads to infect throughout the world; such *pandemics* have been recognized at varying intervals. The

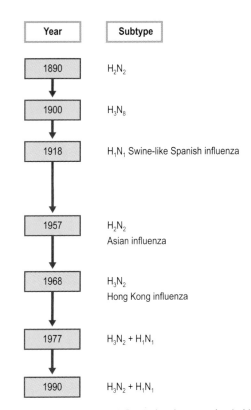

Fig. 49.3 Sequential changes in influenza A antigens associated with antigenic shift. Antigenic drift occurs after the appearance of a new subtype.

great pandemic of 1918–1919 was particularly severe, killing 20 to 40 million people as it spread over a few years. Modern travel means that pandemics can now build up very quickly, within 1–2 years. Between the pandemics of 1918, 1957 (Asian influenza) and 1968 (Hong Kong influenza) there are more localized epidemics (Fig. 49.3). Isolation studies from 1933 onwards, and analysis of isolates, have given an understanding of how the epidemic and pandemic behaviour relates to changes in the virus. In addition, it is possible to deduce which viral antigens circulated before virus isolation became possible; sera from older people show which antigens they encountered early in life, back to the end of the nineteenth century.

The major pandemics are associated with *antigenic shifts* – when the viral H or N (or both) is changed. This is too extensive to be the result of mutation, and analysis of the viral RNA indicates that shift results from the acquisition of a complete new RNA segment 4 and/or 5. In the laboratory it is easy to show that, if a cell is infected with two different strains, the progeny will include viruses whose RNA molecules are derived from each parent. Thus, a 'new' virus can result from the process of *reassortment*. All of the H and N antigenic subtypes are found in aquatic birds (both seabirds and ducks), but in these animals the viruses vary little. The genetic reassortment may take place in pigs that have receptors for both human and avian strains, or in humans infected simultaneously with human and avian viruses. The conditions in South-East Asia where there are high densities of people, poultry and pigs favour such an outcome.

Until 1977, when H_1N_1 reappeared, it was the rule that when a 'new' virus appeared the 'old' one disappeared, but since that time two subtypes have been circulating concurrently, namely H_3N_2 and H_1N_1. The latter antigens had not been found since the 1950s and, as they were antigenically very similar to viruses from the 1957 pandemic, may have reappeared from a frozen source. There is no evidence of latent or persistent infection of man. The epidemics arising between pandemics are associated with antigenic drift in the H antigen. Amino acid changes found near the receptor-binding site allow the virus to infect despite the presence of antibody to previous strains.

High rates of infection are found in pre-school children; thereafter the rates are lower, even in the elderly. However, elderly patients in residential homes, if not protected, are at particular risk from acute illness and sudden death. Thus, even if the attack rate is low, the case fatality rate is high. Staff, visitors or other patients may introduce the virus.

The future

There is considerable evidence that avian strains can infect human beings who are in close contact with infected birds. Since 1997, avian viruses with H_9N_2, H_7N_2, H_7N_7, H_9N_2 and H_7N_3 antigens have caused mainly mild infections, some more severe and a few deaths. Conjunctivitis was the main feature in the H_7N_7 outbreak in the Netherlands among those in close contact with the birds; there was one death.

More persistently, H_5N_1 has been the cause of several large outbreaks of infection in poultry and has regularly infected human beings in contact with the birds, with some deaths. South-East Asia has been the main focus. Millions of birds have been culled to contain the virus. In 1977 in Hong Kong there were 18 human cases with six deaths. Again in 2003, there was one death in Hong Kong after travel in China. Since 2004 there have been large outbreaks in poultry in several countries, including Thailand, Vietnam, Cambodia, South Korea, Japan, China, Indonesia, Malaysia, Laos and, most recently further to the west, Turkey. Migratory birds are infected and could spread the virus even further. The virus is not efficient at infecting man and spreading to human contacts. The risk is that virus may mutate or, more likely, as a result of reassortment following dual infection in a host, that virus adapted to man will acquire the H_5N_1 surface antigens; a pandemic could result.

Influenza B

Influenza B viruses do not undergo antigenic shift as there is no animal reservoir and, although epidemics do occur at 3–6-year intervals, they never reach pandemic proportions and their extent is usually limited to small communities such as boarding schools or residences for the elderly. The antigenic changes result from mutation, as do those seen in influenza A after the appearance of 'new' virus strains; the changes are the cause of antigenic drift.

Influenza C

Influenza C virus is not associated with epidemics but gives rise only to mild upper respiratory tract infections, especially in young children. Most infections are asymptomatic.

TREATMENT

Oral amantadine hydrochloride was introduced in the early 1980s, followed by a derivative, rimantadine. These drugs work by blocking the ion channels in the envelope, thus preventing the pH changes that precede the membrane fusion step essential for nucleocapsid release. Unfortunately, these compounds have activity only against influenza virus type A, not B, C or other respiratory viruses. Therefore, it is essential to know which virus is responsible for the illness or outbreak. A clinical diagnosis can be made fairly confidently in a typical case seen during an epidemic due to a known type, but is impossible in sporadic or milder cases. Amantadine is effective when given prophylactically, and also therapeutically in patients treated within 24 h of onset of illness. Viruses resistant to amantadine and rimantadine may appear within a few days of drug administration; however, resistant strains show no increased pathogenicity or transmissibility.

More recently two neuraminidase inhibitors, zanamivir and oseltamivir, have received approval for therapeutic use in both influenza A and B infections. They can reduce the duration of symptoms by 1–3 days if given within 36 h of the onset of illness. Zanamivir has poor bio-availability and is administered by inhalation of a dry powder twice daily for 5 days. Oseltamivir is given by mouth as a pro-drug and has excellent bio-availability, although nausea and vomiting may occur in some patients. Twice-daily dosage for 5–7 days has been used in those with normal renal function; once-daily dosage is recommended when renal function is impaired.

Resistance to both of these drugs has been demonstrated and may be neuraminidase-independent or

-dependent. Both are effective against amantadine- or rimantadine-resistant strains.

CONTROL

The aim of immunization is to produce haemagglutination inhibiting or neutralizing antibody in all vaccinees. This protects them against infection, but only with strains closely related to those in the vaccine. Subunit influenza vaccine for intramuscular injection has been widely available for many years. In the UK these vaccines contain the H and N subunits from two type A strains and one type B strain. The strains are updated annually on the recommendation of the WHO and, in the UK, are recommended for use in people aged over 65 years, and in those of any age who suffer from chronic cardiorespiratory problems, diabetes, renal failure or an immunosuppressive illness. In interpandemic years in the elderly given annual boosting, the mortality rate can be reduced by 75% and the rate of hospital admission with complications by about 50%.

Intervention during a pandemic is likely to be more difficult owing to the shortage of appropriate vaccine, as the interval from isolation of a potential pandemic strain to release of an inactivated vaccine made by current methods using eggs is seldom less than 6 months. Such vaccines rely on adequate quantities of virus with the appropriate H and N antigens being produced in the allantoic cavity of embryonated hens' eggs inoculated with seed virus. Reassortment of two strains, one a high-yielding laboratory-adapted strain and the other containing the required H and N antigens, is employed to produce an appropriate inoculum for growth in eggs so that vaccine is prepared as quickly as possible. Separated whole virus particles are inactivated by either formalin or β-propiolactone. Whole virus vaccine should not be given to those who are allergic to egg protein. The H and N antigens may be separated from the whole virus by treatment with detergent, and these subunit or split-virus vaccines are better tolerated, especially by young children. Other types of vaccines have been tried, such as cold-adapted live-attenuated vaccines given intranasally. These vaccines have been generally effective in provoking a good local (IgA) antibody response but are not widely used at present. Genetic techniques that assemble appropriate vaccine strains by cloning genes have been developed; they are not yet licensed for widespread use but may well be a successful innovation to speed up vaccine production.

During a pandemic there is a major role for the prophylactic use of anti-influenza drugs such as rimantidine, oseltamivir and zanamivir in those at high risk, in healthcare staff and in staff in long-term care facilities who have not been protected by an appropriate vaccine. Early studies have identified protection rates of 60–70%.

The WHO Global Influenza Programme with its network of reference laboratories plays a very important role in monitoring the evolution of influenza viruses, selecting and developing prototype pandemic vaccine strains, and developing and updating WHO diagnostic reagents. Recent changes in International Health Regulations have increased the obligation in countries world-wide to have the capacity for preventive measures and to be able to detect and respond to infections of international concern. Aware of the ubiquity and speed of travel, the world is preparing to deal with a potential avian influenza pandemic.

KEY POINTS

- Influenza viruses cause respiratory infections world-wide; there are three types: A, B and C.
- Influenza A is found in aquatic birds, poultry and pigs, and these play important roles in the epidemiology of human infections.
- Influenza viruses are segmented RNA viruses that regularly undergo genetic change.
- The surface proteins, the haemagglutinin and the neuraminidase, can show gradual change (*antigenic drift*) and sudden major change (*antigenic shift*).
- In man, influenza A results in high morbidity and mortality rates in the winter months in temperate zones, but throughout the year in more tropical climates.
- The disease may be pandemic, epidemic or sporadic.
- Since 1997, avian influenza has been a major problem in the Far East, with strains from domestic poultry occasionally infecting man directly, resulting in high mortality rates; virus could become adapted to man.
- Vaccines have been available for many years, recommended for annual boosting of high-risk groups in interpandemic periods to reduce mortality, morbidity and hospital admission rates.

RECOMMENDED READING

Cox N J, Subbarao K 1999 Influenza. *Lancet* 354: 1277–1282

Fields B N, Knipe D M, Howley P M (eds) 1996 *Virology*, 3rd edn. Lippincott-Raven, Philadelphia

Gubareva L V, Kaiser L, Hayden F G 2000 Influenza virus neuraminidase inhibitors. *Lancet* 355: 827–835

Nichol K L, Margolis K L, Wuorenma J, Von Sternberg T 1994 The efficacy and cost effectiveness of vaccination against influenza among elderly persons living in the community. *New England Journal of Medicine* 331: 778–784

Stephenson I, Nicholson K G, Wood J D, Zambon M C, Katz J M 2004 Confronting the avian influenza threat: vaccine development for a potential pandemic. *The Lancet Infectious Diseases* 4: 499–509

Internet site

World Health Organization. Avian influenza. http://www.who.int/csr/disease/avian_influenza/en/

50 Paramyxoviruses

Respiratory infections; mumps; measles; Hendra/Nipah disease

J. S. M. Peiris and C. R. Madeley

The paramyxoviruses are a family of enveloped viruses containing single-stranded RNA as a single piece. They resemble the orthomyxoviruses in both morphology and their affinity for sialic acid receptors on mammalian cells, but they are larger and more fragile (Fig. 50.1a).

Within the family Paramyxoviridae there are four genera, each with several members, that cause disease in man or animals (Table 50.1).

STRUCTURE AND REPLICATION

Originally, these viruses were classified together because they were thought to be similar in structure and function.

Neither property is constant throughout the family but there are similarities.

Structure

Parainfluenza, mumps and measles viruses are identical when seen in the electron microscope (and as described below), whereas respiratory syncytial virus and metapneumovirus, although very similar, have slightly longer surface spikes and are more difficult to visualize.

Functionally there are other differences. Parainfluenza viruses (1–4a,b), Newcastle disease virus (NDV) and mumps virus have a surface haemagglutinin and neuraminidase on the same spike; measles virus has haemagglutinin but

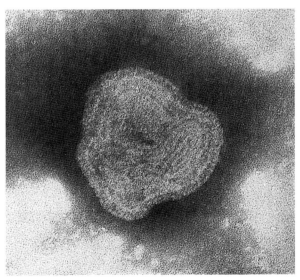

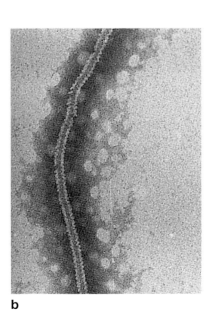

Fig. 50.1 a Electron micrograph of a typical paramyxovirus. b Separate internal helical nucleocapsid. Negative contrast, 3% potassium phosphotungstate, pH 7.0, magnification ×200 000.

Table 50.1 Classification and important pathogens of the paramyxoviruses

Genus	Human viruses	Animal viruses

Paramyxovirus	Paramyxovirus
	    Parainfluenza viruses types 1, 3
	Rubulavirus
	    Mumps virus
	    Parainfluenza viruses types 2, 4a, 4b
		Newcastle disease virus (NDV) (poultry), simian virus 5

Morbillivirus	Measles virus	Canine distemper virus, rinderpest virus, equine morbillivirus, morbilliviruses of seals, dolphins and porpoises
Pneumovirus	Respiratory syncytial (RS) virus Human metapneumovirus (hMPV)	Turkey rhino-tracheitis virus (avian metapneumovirus)
Henipavirus	(Hendra virus)[a] (Nipah virus)[a]	Hendra virus Nipah virus

[a]These viruses cause disease in animals but can cause serious human disease (see text).

---

no neuraminidase activity; and pneumovirus has neither. In addition, measles virus has a haemolysin not possessed by the others, whereas respiratory syncytial (RS) virus has a large surface glycoprotein, G, which has a similar cell-attaching function as a haemagglutinin. Other spikes carry fusion (F) proteins; all envelopes have matrix (M) proteins. The RNA is complexed with protein to form the nucleocapsid.

Replication

The replication of paramyxoviruses follows a common theme. After attachment, the F protein fuses the viral envelope to the cell membrane, releasing the nucleocapsid into the cell. The negative-sense genome cannot act as messenger RNA (mRNA), necessitating the virus to carry its own RNA-dependent RNA polymerase within the virion. This polymerase produces subgenomic-sized mRNA transcripts, which are translated to produce some of the early virus-specific polypeptides. These include a second RNA polymerase, which copies the genome into full-length positive complementary strands that are, in turn, copied back into negative strands for transcription of later mRNA (coding for structural proteins) and for incorporation into new virions. The virus buds off from the cell, which is able to adsorb red blood cells (haemadsorption) owing to the presence of the viral haemagglutinin (see Ch. 7).

In the 1990s, two new paramyxoviruses, Hendra and Nipah, were discovered in Australia and South-East Asia; they are animal viruses occasionally transmitted to man. As they are genetically distinct from the other members of this family, they are classified as a separate genus. Their ecology is still being investigated and their full significance as human pathogens is not yet clear.

PARAINFLUENZA VIRUSES

DESCRIPTION

Parainfluenza viruses, mumps virus, NDV and simian virus 5 are indistinguishable in the electron microscope. The surface projections are 12–14 nm long and 2–4 nm wide.

Within the virion the genome has a basic length of about 1 μm; the helical nucleoprotein has a herring bone or 'zipper'-like appearance (Fig. 50.1b). This is more easily recognized in the electron microscope than the complete particle; it is 15–19 nm wide with a pitch of about 6 nm.

CLASSIFICATION

There are four types of parainfluenza viruses (1–4) that are antigenically distinct. Nevertheless, there are conserved antigenic epitopes on the paramyxovirus envelope proteins, which cause the serological cross-reactions that are found between the parainfluenza viruses, mumps virus and simian virus 5. These cross-reactions make the results of serological tests virtually uninterpretable. Infection with one of these viruses results in a boost of antibody titres to previously experienced paramyxoviruses as well as to the currently infecting virus (original antigenic sin). Type 4 has two subtypes, 4a and 4b, which can be distinguished only by neutralization or haemadsorption inhibition tests. The paramyxoviruses have been subdivided into two genera, *Paramyxoviruses* and *Rubulaviruses* (see Table 50.1), but the distinction is not important for clinicians.

NDV is a typical paramyxovirus. It infects chickens and other domestic birds. The severity of infection varies considerably from inapparent to fatal. Because some strains can cause major outbreaks with high mortality, an effective live vaccine based on avirulent strains has been developed. Simian virus 5 is often present in normal uninfected monkey kidney cell cultures but does not appear to reduce their sensitivity to other viruses and does not cause human illness. Other parainfluenza viruses are natural pathogens for cattle and other domestic species; they are not known to infect man.

CLINICAL FEATURES AND PATHOGENESIS

The parainfluenza viruses are mostly associated with:

- *croup*, a harsh brassy cough in children familiar to many parents as a middle-of-the-night irritant caused by a combination of tracheitis and laryngitis
- minor upper respiratory tract illness
- some cases of *bronchiolitis* and 'failure to thrive'.

They are responsible for 6–9% of respiratory infections for which a virus cause can be identified. The incubation period is 3–6 days, during which the virus spreads locally within the respiratory tract.

Many infections occur in infants in the presence of circulating maternal antibody that does not appear to be protective or to make the illness worse.

NDV may cause a mild conjunctivitis in man, usually as a result of a laboratory accident.

LABORATORY DIAGNOSIS

Rapid diagnosis may be made by immunofluorescent staining of exfoliated respiratory cells separated from well taken nasopharyngeal secretions. There are monoclonal antibody reagents for immunofluorescent detection of parainfluenza virus types 1, 2 and 3, but reagents for type 4 are less widely available. Although these monoclonal antibody reagents may not provide the detection sensitivity associated with well made and characterized polyclonal reagents, their wide commercial availability is an advantage.

Similar antiviral antibodies may also be used in enzyme immuno-assays to identify viral antigen in specimens from the patient. They have the advantage of requiring only antigen to be present in the specimen, whereas other assays require intact infected cells (for immunofluorescence) or infective virus (for culture). However, enzyme immuno-assays give no information on the quality of the specimen, nor the extent of infection.

Molecular methods (e.g. reverse transcription–polymerase chain reaction) have been developed for detection of parainfluenza virus RNA for diagnosis; some

of these assays have been incorporated into multiplex nucleic acid amplification tests to a number of respiratory viruses within a single reaction tube. Virus may be isolated in monkey kidney cells. Visible cytopathic effects in the cell sheets are minimal and it is usually necessary to show infection of the cells by the haemadsorption of 1% guinea-pig or human group O red blood cells. The infecting virus can be typed (or subtyped) by reacting the cell cultures with type-specific antisera before adding the red cells. The appropriate antibody inhibits the haemadsorption. The typing can be confirmed by a neutralization test or by immunofluorescence, but this is not usually necessary.

Serological assessment is not used routinely in diagnosis. Commonly available tests, such as complement fixation, are difficult to interpret because of cross-reactions between parainfluenza viruses with mumps virus and, possibly, simian virus 5 (see above). Type-specific antibodies may be detected by neutralization or haemadsorption inhibition but these tests are too complex for routine use.

EPIDEMIOLOGY AND TRANSMISSION

Epidemiologically, parainfluenza type 1 infections are more frequent in the winter whereas type 3 is a summer infection, with small epidemics appearing reliably each year. Type 2 and 4a and 4b infections are more infrequent, in Newcastle upon Tyne at least, although elsewhere in Britain type 2 can be another summer visitor. Type 4 infections are underdiagnosed because of a lack of suitable reagents, and reported figures are too low for epidemiological patterns to be clear. The reasons underlying these individual epidemiological patterns are unknown.

Numerically, parainfluenza infections are far fewer than those due to RS virus, and most diagnosed infections are in pre-school and primary school children. Fatalities are very rare, and reinfections occur.

The viruses are present in respiratory secretions and are expelled during coughing and sneezing. Infection is acquired by inhalation of infected droplets and by person-to-person contact.

No vaccine is yet available for routine use.

MUMPS VIRUS

DESCRIPTION

Mumps virus is a typical paramyxovirus, indistinguishable in appearance from parainfluenza viruses, measles virus and NDV, with a similar ribonucleoprotein, which may be the only virus-like material seen on electron microscopy. There is only one serotype, although monoclonal antibodies have shown minor variations in the various surface antigenic epitopes.

CLINICAL FEATURES AND PATHOGENESIS

Mumps is an 'iceberg' disease which, although common as a childhood infection, is often subclinical. Although the salivary glands are often involved, inapparent or minor infections are more common. Difficulty in recovering the virus and ambiguity in serology mean that infections are rarely confirmed by the laboratory.

Infection is probably acquired by inhalation of droplets into the respiratory tract. The incubation period is 14–18 days and is followed by a generalized illness with localization in the salivary glands, usually the parotids. The generalized phase is the usual 'flu-like' illness with fever and malaise, followed by developing pain in the parotid glands, which then swell rapidly. Much of the swelling is due to blockage of the efferent duct of the parotid gland, and sucking a lemon in front of a sufferer is a refined form of torture!

Neurological involvement is common in mumps (more than 50% of infections), although the majority of cases are not clinically apparent. However, clinical meningitis remains the most common serious complication of mumps, occurring in 1–10% of patients with mumps parotitis. Meningitis (like any other complication of mumps) can occur before, during, after or in the absence of salivary gland involvement. Before the widespread use of the MMR (mumps, measles, rubella) vaccine, mumps virus and the enteroviruses accounted for most of the cases of aseptic meningitis in the UK. Mumps meningitis is rarely fatal and complete recovery is usual. Meningo-encephalitis has been described, but is much rarer, carries a poorer prognosis and may result in long-term neurological sequelae or death. Deafness and tinnitus have also been described, but are very rare.

In prepubertal children the acute illness usually subsides in 4–5 days, with complete recovery. The best known complication, in postpubertal males, is *orchitis*. This, although painful and causing softening and atrophy of the affected testicle, is usually unilateral and rarely causes sterility. *Oophoritis* also occurs in girls, and should be distinguished from a ruptured cyst or acute appendicitis. Both orchitis and oophoritis usually develop as the parotitis resolves, and a history of previous parotid pain and swelling usually provides the clue.

The role of mumps in pancreatitis is difficult to establish. There may be abdominal pain in acute mumps but the levels of serum amylase do not correlate with the clinical picture. High levels may provide supportive evidence but are not diagnostic. Although uncomfortable, it is not fatal.

LABORATORY DIAGNOSIS

Detection and isolation

Typical mumps does not usually require laboratory confirmation, but mild cases with little parotid swelling may not be noticed until complications develop. By this time virus may not be plentiful in the oropharynx. Direct demonstration of infected cells by immunofluorescence on secretions is very rarely successful, and the diagnosis depends on growing the virus from throat swabs, from saliva from affected glands or from the cerebrospinal fluid (CSF), although the virus may take up to a week to grow. Virus may be present in urine but not reliably so. However, when found in CSF, it is diagnostic of mumps meningitis.

The detection of ribonucleoprotein helix in CSF by electron microscopy is diagnostic of mumps because the only other similar virus found in the central nervous system (measles virus) does not reach levels detectable by electron microscopy in CSF. However, the small quantities of CSF usually taken make electron microscopic diagnosis of mumps meningitis impractical.

Growth of virus in cell cultures (usually monkey kidney or HEp2 cells) produces little cytopathic effect, but the virus can be detected by haemadsorption, which can be inhibited by specific antiserum to confirm the presence of mumps virus.

Serology

The diagnosis can be made serologically by showing a rise in antibody titre between acute and convalescent specimens. Complement fixation is the usual test, using soluble (S) and viral (V) antigens. Antibodies to the S antigen are said to develop early, within 1 week of onset, followed by anti-V antibodies, which persist longer. However, these patterns are not invariable, and only a rise in titre to both antigens can be relied on. A rise to only one antigen in the absence of typical illness is difficult to interpret and may reflect infection with other paramyxoviruses. Low rises or low static titres in cases of meningitis are difficult to interpret, and only isolation of the virus provides unequivocal evidence.

Other serological tests have been used, based on enzyme immuno-assays and single radial haemolysis. They have not been employed as widely as complement fixation, being used more often in serological surveys or to complement other assays. Neutralization and haemagglutination inhibition tests are more complex and do not offer any advantages in routine diagnosis.

EPIDEMIOLOGY

Mumps is a world-wide disease, with man the only known reservoir. Most infections are in children of school age, with those in adults being more severe and more likely to involve complications. Although epidemics occur, mumps is less infectious than measles or chickenpox. Infection appears to confer life-long immunity, and second infections do not occur.

CONTROL

Some, but not very reliable, protection can be given by passive immunization, which may prevent severe orchitis even when given at the stage of parotitis.

Mumps vaccine, based on the *Jeryl Lynn* or *Urabe* strains, has been available as a monovalent vaccine for some time, particularly in the USA, but is now incorporated into the triple MMR vaccine. All three components are live-attenuated viruses, and the mumps component induces good antibody levels, lasting long enough to suggest that the recipients will not become susceptible as adults. A few cases of mild post-vaccine meningitis have been described but have not caused serious concern.

MEASLES VIRUS

DESCRIPTION

Measles virus, a *morbillivirus*, is morphologically indistinguishable from other members of the group. The ribonucleoprotein helix is readily released from the virion and may, as with the others, be the only identifiable virus structure seen on electron microscopy. The virion structure differs from other paramyxoviruses since the:

- spikes carry a haemagglutinin but not a neuraminidase function
- F protein is also a haemolysin.

There is only one serotype of measles virus and no subtypes have yet been recognized, although monoclonal antibodies show that there may be differences between wild and cultivated strains.

Human morbilliviruses are related to a number of animal strains. *Canine distemper* and *rinderpest* in cattle are well known relatives, but in the past few years other similar viruses have been isolated from seals (of several species), dolphins and porpoises, and an equine morbillivirus has reappeared that has apparently been transmitted to man in contact, fatally in one case. All are distinct but can cause serious illness in their natural species, although survivors develop solid immunity. There is partial cross-protection in ferrets between measles and canine distemper viruses.

CLINICAL FEATURES AND PATHOGENESIS

Measles is an acute febrile illness, mostly in childhood, after an incubation period of 10–12 days. The onset is 'flu-like', with high fever, cough and conjunctivitis. *Koplik's spots* (red spots with a bluish-white centre on the buccal mucosa) may be present at this stage. After 1–2 days the acute symptoms decline, with the appearance of a widespread maculopapular rash. Viral antigen, but not infectious virus, may be found in the spots. The rash can be inhibited by local injections of immune serum, but does not appear at all in those who are severely immunocompromised ('spotless measles' – usually rapidly fatal), and this has been thought to point to an immunopathological (T cell-mediated) mechanism.

Over the next 10–14 days recovery is usually complete as the rash fades, with considerable desquamation. Complications include:

- giant cell pneumonia, more common in adults
- otitis media
- post-measles encephalitis.

The pneumonia is due to direct invasion with virus, but the role of virus in the other two complications is uncertain. Measles encephalitis can cause severe and permanent mental impairment in those it does not kill. It is rare but disastrous.

The mortality rate associated with uncomplicated measles in immunocompetent, well nourished children is low but rises rapidly with malnourishment (marked in Africa), in the immunocompromised and, to a much lesser extent, with age. The virus has also been devastating in isolated populations into which it was introduced as a 'new' disease.

One further complication of measles is *subacute sclerosing panencephalitis* (SSPE), which occurs in children or early adolescents who have had measles early in life, usually when less than 2 years of age. It is a progressive and inevitably fatal degenerative disease. Within infected cells is a defective form of measles virus, which, because it is unable to induce the production of a functional M protein, is not released as complete virus from the cells. Patients deteriorate over several years, losing intellectual capacity before motor activities. Oligoclonal antibodies to measles virus proteins appear in the CSF, but the virus cannot be cultivated unless it is 'rescued' by co-cultivating neuronal cells with a susceptible cell type.

The virus has been linked with multiple sclerosis, Paget's disease of bone and Crohn's disease. In each case, tubular structures resembling measles nucleocapsids have been seen by thin-section electron microscopy, and immunofluorescence has been used to demonstrate measles 'antigens' in biopsy material. Serum from about 50% of adults aged over 50 years, however, will fix complement with measles antigen, although the individuals give no history suggestive of recent measles, and it is possible that auto-antibodies to a measles-like protein can be induced with age. If so, its significance is unknown but would be a factor in assessing measles involvement in older patients with chronic diseases. The evidence linking measles virus to the aetiology of these diseases is not compelling.

LABORATORY DIAGNOSIS

Detection and isolation

Most cases of measles are diagnosed clinically, usually in the patient's home or in a general practice setting. However, the widespread use of vaccine has made the disease rarer in these areas, and consequently fewer clinicians are familiar with it. Direct virological confirmation is often not attempted and is difficult to do in general practice. In hospital and particularly in immunocompromised patients, in whom the disease is often rashless, the diagnosis may be made rapidly by immunofluorescence on exfoliated respiratory cells in well-taken nasopharyngeal secretions. The presence of a large number of giant cells, particularly in patients on cytotoxic drugs, is a bad prognostic sign.

Other immuno-assays have been developed but give no feedback on the extent of the infection. Otherwise the virus may be grown in human fibroblasts, primary monkey cells and Vero cells, although it does not grow readily.

Serology

Measles induces a good antibody response, and a rise in complement-fixing antibody is diagnostic. However, complement-fixing antibody in older patients is often detected and its significance is unknown. Other tests for antibody have been developed but are not used widely. As measles vaccine (see below) is used more widely, clinical measles is becoming rarer and the need for laboratory confirmation of probable cases is rising. Antibody tests for measles-specific immunoglobulin (Ig) M and IgG on specimens of saliva are now available in some countries and avoid the need for venesection.

EPIDEMIOLOGY

Transmission is from person to person, probably by respiratory droplets, but the associated conjunctivitis

may also be a source. Despite some anecdotal evidence, there is no good evidence that distemper in dogs can be a source of measles in humans.

Measles epidemics occur every 2 years in developed countries in the absence of widespread use of the vaccine. This periodicity is absent in isolated populations too small to maintain transmission (<400 000), in poverty and overcrowding, and following the widespread use of vaccine. The disease is ubiquitous throughout the world and, although a candidate for eradication, this may be difficult to achieve.

In tropical areas, particularly Africa, children become infected under the age of 1 year, and the mortality rate rises in consequence, reaching as high a value as 42% in children under 4 years of age. Malnutrition is one of the main underlying causes of this excess mortality. The attack rate is also very high in isolated populations that have not experienced the disease for some years. In the Faroe Islands in the 1840s, three-quarters of the population were infected, although the mortality rate was low. Most of those who were not infected were aged over 65 years, the interval since the last time the disease had been present in the islands, and confirms that infection gives prolonged immunity.

CONTROL

The first measles vaccine was a formalin-inactivated one. Although inducing circulating antibody, it was found that vaccinees exposed to natural measles were likely to develop atypical disease. The rash was more peripheral, involving the palms and soles, and pneumonia was common. It was later recognized that the vaccine had failed to induce adequate levels of antibody to the haemolytic F protein, and the immunity induced did not inhibit cell-to-cell spread of the virus. Consequently it was withdrawn and replaced with a live-attenuated vaccine, containing the *Edmonston B* or *Schwarz* strains, which have given a seroconversion rate of over 90%. So far (over about 30 years) the immunity induced by the vaccine has persisted and may be life-long.

Measles vaccine is now combined with those against mumps and rubella to form the MMR vaccine. This combination of three attenuated viruses has been shown to induce good immunity to all three. Introduced initially in the USA, it is now the preferred vaccine in the UK for administration to children aged between 12 and 18 months.

The attenuated measles vaccine, alone or in combination with mumps and rubella, has been shown to be effective and safe. Unfortunately, in the UK, vaccine uptake has fallen recently due to fears over its safety, particularly as a possible cause of autism. These fears have now been confirmed as unsubstantiated.

Eradication has been attempted in the USA; the number of cases was reduced from over 500 000 per year to about 2000 (a reduction of over 99%), but outbreaks in immigrants and high-school students have emphasized the problems of preventing imported cases and of keeping up a high level of immunization.

Immunization in the high endemic regions of Africa still presents problems, however. Because many infants are infected before their first birthday, the vaccine has to be given to babies aged under 6 months to have any effect. Passively transferred maternal antibody often interferes with the immune response to a live vaccine, and such early immunization does not always produce adequate immunity. A second dose at 12–13 months is then probably necessary, but adds to the cost and the logistic difficulties. Solutions to both will have to be found before progress is made towards measles eradication in the developing world.

RESPIRATORY SYNCYTIAL VIRUS

DESCRIPTION

Superficially, RS virus resembles other paramyxoviruses, with a similar pleomorphic envelope studded with surface spikes that may be seen more clearly on electron microscopy than those on the parainfluenza viruses, mumps and measles. The spikes may also be slightly longer, but neither the complete virus particles nor the nucleoprotein helix (with a diameter of 17 nm) are easy to visualize in the electron microscope. However, individual particles are generally larger than other paramyxoviruses (Fig. 50.2).

RS virus is placed in a separate genus – *Pneumovirus* – because of these minor physical differences and the lack of haemagglutinin, haemolysin or neuraminidase. It has a G lipoprotein, a receptor for cell attachment (but not to red blood cells), which differs in chemical composition from the HN protein of other paramyxoviruses. The F protein induces the syncytia in cell cultures from which the virus gets its name, and is probably responsible for both virus penetration and spread in the host. The virus is relatively fragile and may not survive even snap-freezing to −70°C. Specimens for isolation should not be frozen.

For most purposes there is only one serotype, although the advent of monoclonal antibodies has confirmed that there are two subtypes, A and B. In Newcastle upon Tyne, strains of subgroup A have been prevalent every year since 1974, but subgroup B strains have been more erratic and have not been isolated every winter. The reason for this is unknown. Analysis of the genome has revealed further minor differences that appear to be unimportant in diagnosis (i.e. do not represent major antigenic variations)

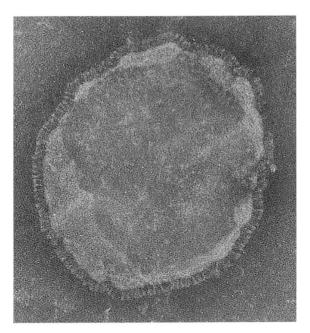

Fig. 50.2 Electron micrograph of respiratory syncytial virus. Human metapneumovirus is indistinguishable in appearance. Negative contrast, 3% potassium phosphotungstate, pH 7.0, magnification ×200 000.

but which have allowed various strain lineages to be recognized.

RS virus is also a significant pathogen in cattle and infects chimpanzees readily – early isolates were termed *chimpanzee coryza agent*. Both goats and sheep may be infected naturally, and there is evidence that several other domestic and rodent species are susceptible, either naturally or after some adaptation.

CLINICAL FEATURES AND PATHOGENESIS

The most serious illness caused by RS virus is *bronchiolitis* in young babies, in whom the bronchiolar inflammation acts as a one-way valve leading to hyperinflation of the lungs (very characteristic on radiography), but it is also associated with minor upper tract infections and non-specific 'failure to thrive'. The peak incidence is in those under 1 year of age, and there are (possibly as a consequence) annual winter epidemics. This infection is potentially life-threatening, particularly in those with bronchopulmonary dysplasia or congenital heart defects, or in those who are immunosuppressed or immunodeficient. In normal babies it is rarely fatal where medical staff have the experience and facilities for appropriate management. RS virus has been recovered from some victims of the *sudden infant death syndrome* (SIDS). Although it may have contributed to the death, other factor(s) are also significant.

The main clinical feature is bronchiolitis, but the upper respiratory tract remains infected, and this makes it possible to confirm the diagnosis. If RS virus is present in the nasopharynx and there is clinical evidence of lower respiratory tract involvement, RS virus is likely to be responsible.

Recovery is apparently complete, although it has been suggested that the infection predisposes to chronic respiratory tract disease (asthma, bronchiectasis, etc.). This has yet to be confirmed, although studies are in progress.

The sequelae to the use of an inactivated vaccine (see below) have led to the suggestion that some of the severity of bronchiolitis is due to hypersensitivity induced by an earlier infection. Studies in various centres have neither confirmed this theory nor fully excluded it.

In older children and adults, the virus causes only minor infections, possibly because their air passages are larger. Re-infections are common and in adults may cause no more than a cold. However, the failure of acute infections to immunize, even in the immunocompetent, highlights the difficulty of producing a vaccine.

There have been reports of severe illness, with some fatalities, in old people's homes as well as in the elderly living in the community. The under-recognition of RS virus in these groups may be due to the difficulty in confirming a virological diagnosis in adults and the elderly (see below). An associated neurological syndrome has also been described, but whether RS virus is the cause is unproven.

LABORATORY DIAGNOSIS

Detection and isolation

During the acute phase of illness virus may be readily demonstrated in nasopharyngeal secretions (which are usually copious) by immunofluorescence, enzyme immuno-assays or culture.

Rapid diagnosis in less than 1 h with commercially available, directly conjugated, monoclonal antibodies can be made reliably by immunofluorescence, provided an adequate number of desquamated respiratory cells is collected in the secretions. Generally, similar results can be obtained with enzyme immuno-assays. Such assays may be less sensitive compared with culture, but the virus grows slowly and positive results will come too late to influence management. Although antigen detection and culture methods are good for diagnosing RS virus infections in infants and young children, they are less reliable in adults and the elderly because the levels of virus or antigen are lower in adult secretions compared with those in infants.

As with other respiratory viruses, molecular methods such as reverse transcription–polymerase chain reaction, either for a single virus or multiplexed to detect a panel of viruses (see under Parainfluenza viruses) can be used.

Serology

Serological assessment using complement fixation is generally not helpful. Many of the patients are too young to respond reliably and even adults do not always produce a detectable rise in serum antibody levels. However, immuno-assays for the G and F proteins may offer more reliable serological tests, especially in adults where other options are limited.

TREATMENT

Appropriate management includes use of oxygen, if indicated, and tube feeding to maintain energy intake if the baby has difficulty in suckling. Most babies can be managed symptomatically by these measures. The only specific antiviral drug available for chemotherapy is ribavirin. In vivo there is some evidence of efficacy for ribavirin when given as a small-particle aerosol, although it is apparently not effective when given by intravenous infusion. The evidence of efficacy by any route is poor. This may be because the most affected parts of the lungs are also the least well aerated and therefore least accessible to the aerosolized drug. The drug is expensive and its recommended use is confined to those babies who are at high risk from rampant RS virus because they have congenital heart or lung abnormalities. Even in these babies the virus may have caused damage before the drug can be given.

Hyperimmune RS virus immunoglobulin and human-ized monoclonal antibodies have become available for the prevention and/or treatment of RS infections. These preparations are very expensive and their use may be difficult to justify except in groups at very high risk, such as very premature babies or those with pre-existing bronchopulmonary dysplasia.

EPIDEMIOLOGY

In temperate climates in both the northern and southern hemispheres, RS virus causes a substantial winter epidemic every year. In the Newcastle upon Tyne/Tyne and Wear conurbation (total population about 1 million), the annual number of virologically confirmed diagnoses regularly exceeds 500. Similar figures are obtained elsewhere where adequate facilities for diagnosis exist. Most infected babies reach hospital, and the failure to recover RS virus from babies with 'colds' attending well baby clinics in Newcastle upon Tyne has suggested that comparatively few cases go unrecognized.

Why RS virus induces an epidemic every October–November is unknown. In other regions of the world the pattern can be markedly different. In tropical regions there is an equally clear epidemic, but it occurs in the hot and humid months of the summer. Therefore, RS virus epidemics do not always correspond to the same short-term climatic factors (temperature, humidity, etc.), and sporadic cases occur anyway throughout the year. Moreover, this apparent climatic paradox is found with other respiratory viruses as well. The virus is distributed all over the world, but its activities in tropical, overcrowded and poor areas are under-recorded so far.

The significance, if any, of subtypes in explaining these RS virus phenomena is still under investigation.

CONTROL

A formalin-inactivated, crude, whole-virus vaccine was tried in the 1960s. It induced good levels of circulating antibody but failed to protect the recipients, who actually became more ill than placebo controls when subsequently exposed to RS virus. As with measles, this may have been due to the vaccine failing to induce the right protective antibodies. Subsequently, several live vaccines based on cold adaptation, temperature-sensitive mutants or administration by a different route (intramuscularly) were tried but none has proved satisfactory.

A major obstacle in developing a good vaccine is the fact that the peak of disease occurs within the first year of life, and thus a safe vaccine immunogenic to such young and immunologically immature recipients is difficult to prepare. The increasing recognition that RS virus causes morbidity in the elderly may stimulate the preparation of an adult vaccine – possibly an easier challenge to meet. In the future, genetic engineering offers new prospects, but at present (2006) there is no satisfactory vaccine.

HUMAN METAPNEUMOVIRUS

DESCRIPTION

There have been a number of illnesses typically viral in nature from which no virus had hitherto been found. Part of this diagnostic gap has been filled with the discovery in the Netherlands in 2001 of a virus very similar in appearance on electron microscopy to RS virus (see Fig. 50.2), but differing in fine detail. This is human metapneumovirus (hMPV). It has a single-stranded negative-sense RNA genome of 13.4 kilobases, similar to RS virus, although slightly smaller. The order of the genes is different, resembling more closely that of an avian virus, turkey rhinotracheitis virus. The genes, however, code for similar structural and non-structural proteins. Genetic analysis indicates that at least two separate lineages circulate world-wide; antigenic studies confirm that there may be more than one serotype, although further work is needed to confirm this.

Like RS virus, hMPV appears to lack haemagglutinin, haemolysin and neuraminidase. It can be cultured in tertiary monkey kidney cells, but grows slowly, and these cells are becoming less available.

CLINICAL FEATURES AND PATHOGENESIS

Clinically hMPV causes a very similar spectrum of disease to RS virus. It affects mostly young children in whom it causes a bronchiolitis with fever of up to 39°C, wheezing, crepitations, and changes on chest radiography. The proportion of respiratory patients infected with hMPV has been reported to be about 12%, but there is yearly variation in the activity of this virus.

LABORATORY DIAGNOSIS

Routine diagnosis is not yet widely available. Molecular detection using reverse transcriptase–polymerase chain reaction is most often used. Culture is relatively less sensitive and commercially available rapid antigen tests and serology are still being developed.

TREATMENT AND CONTROL

As yet, there is no specific drug treatment, although animal models are being developed with the aim of exploring both treatment and vaccine prevention. It is likely that very similar obstacles to those found with RS virus will present similar difficulties.

EPIDEMIOLOGY

The epidemiology of hMPV is still being researched. Serological studies in the Netherlands have shown that the virus has been circulating there for more than 50 years, indicating that this is not a 'new' human virus. The seasonality of infections, where data are available, has shown similar patterns to those of other respiratory viruses, and RS virus in particular.

NIPAH AND HENDRA VIRUSES

During 1998–1999, an outbreak of respiratory disease in pigs was associated with encephalitis in humans in Malaysia. There were more than 200 human cases, with 105 deaths. The causative agent was found to be a paramyxovirus given the name Nipah. Another episode of 13 cases of human Nipah disease occurred in Bangladesh in 2001–2003, without obvious involvement of pigs as an intermediate host; pigs are rare, in any case, in that country. It is distinct genetically from all the other paramyxoviruses and most closely related to Hendra, another recently (1994) discovered paramyxovirus causing epidemic fatal respiratory disease in horses that can be transmitted to man, resulting in at least one fatal infection.

These new viruses are now officially classified as paramyxoviruses, but in a separate genus within the *Paramyxoviridae*. Fruit bats appear to be the natural reservoir of both viruses, with transmission to mammals (including man) an exceptional event. Nevertheless, these discoveries underline the fact that new pathogens capable of causing human disease continue to emerge.

KEY POINTS

- Respiratory syncytial virus, parainfluenza viruses and human metapneumovirus are important and frequent causes of respiratory tract infections, especially in children.
- Seasonality of respiratory viruses is variable depending on the geographical (i.e. climatic) region.
- Measles remains a serious disease, especially in the immunocompromised in whom it may be 'spotless' and often fatal, and in less developed countries where it contributes to significant morbidity and mortality.
- Vaccines are available for measles and mumps viruses, but not for other paramyxoviruses.
- There are no reliable antiviral drugs for any of the paramyxoviruses.

RECOMMENDED READING

Peiris J S M, Tang W-H, Chan K-H et al 2003 Children with respiratory disease associated with metapneumovirus in Hong Kong. *Emerging Infectious Diseases* 9: 628–633

Van den Hoogen B G, de Jong J C, Groen J et al 2001 A newly discovered human pneumovirus isolated from young children with respiratory tract disease. *Nature Medicine* 7: 719–724

Mandell G L, Cheatham O R, Bennett J E, Dolin R (eds) 2005 *Principles and Practice of Infectious Disease*, 6th edn. Churchill Livingstone, Philadelphia

Zuckerman A J, Banatvala J E, Pattison J R Griffiths P D, Schoub B D (eds) 2004 *Principles and Practice of Clinical Virology*, 5th edn, Chs 6, 7, 11 & 13. Wiley, Chichester

51

Arboviruses: alphaviruses, flaviviruses and bunyaviruses

Encephalitis; yellow fever; dengue; haemorrhagic fever; miscellaneous tropical fevers; undifferentiated fever

A. D. T. Barrett and S. C. Weaver

The name 'arbo' (arthropod-borne) virus has been used for many years to denote viruses transmitted biologically by arthropod (mainly insect) vectors. However, it is now recognized that there are many different taxa of arboviruses. Indeed, there are more than 500 individual arbovirus species that are now officially classified in six virus families. Many arboviruses are highly pathogenic and are classified at biosafety level 3 or 4. As there are many similarities in their transmission cycles and in the diseases that they cause, they are considered together in this chapter.

Arboviruses were defined by a World Health Organization Scientific Group as 'viruses that are maintained in nature principally, or to an important extent, through biological transmission between susceptible vertebrate hosts by haemotophagous arthropods or through transovarian and possible venereal transmission in arthropods; the viruses multiply and produce viraemia in the vertebrates, multiply in the tissues of arthropods, and are passed on to new vertebrates by the bites of arthropods after a period of extrinsic incubation.'

Certain viruses within the six families containing arboviruses are not transmitted by arthropods, but are maintained in nature within rodent reservoirs that may transmit infection directly to man. These include the *Hantavirus* genus of the family *Bunyaviridae*.

DESCRIPTION

Classification

Arboviruses are classified within six families (Table 51.1). Most are members of the families *Togaviridae*, *Flaviviridae* and *Bunyaviridae*; some are assigned to the families *Reoviridae* (genera *Coltivirus* [e.g. Colorado tick fever virus] and *Orbivirus* [e.g. Bluetongue viruses]), *Orthomyxoviridae* (e.g. Thogoto virus) and *Rhabdoviridae* (members of the genera *Vesiculovirus* [e.g. vesicular stomatitis virus] and *Lyssavirus*). Within the *Togaviridae*, only one (*Alphavirus*) of the two genera contains arthropod-borne viruses; the other genus, *Rubivirus*, contains rubella virus, which is not arthropod-borne, as its sole member (see Ch. 52). The *Flaviviridae* contains three genera (*Flavivirus*, *Pestivirus* and *Hepacivirus*), but only the *Flavivirus* genus contains arthropod-borne viruses. Pestiviruses infect only vertebrate animals (e.g. bovine viral diarrhoea virus), and hepatitis C virus is described in Chapter 52. The Bunyaviridae consists of five genera (*Orthobunyavirus*, *Hantavirus*, *Nairovirus*, *Phlebovirus* and *Tospovirus*), containing a total of over 300 species, and is the largest virus family. The *Tospovirus* genus contains plant viruses that are transmitted by vectors (thrips), whereas the *Hantavirus* genus contains viruses that are transmitted by rodents rather than arthropods.

Many arboviruses show close relationships with other arboviruses. The classification of arboviruses into individual species has been made on the basis of neutralization or other serological tests. Clusters of viruses that show antigenic overlap are termed serogroups or antigenic complexes. The molecular properties and nucleotide sequences of genomes of viruses have increasingly contributed to the classification of viruses. Table 51.2 lists some of the important members.

Properties

Arboviruses share common biological attributes (see Table 51.1):

1. Most induce fatal encephalitis 1–10 days after intracerebral inoculation of mice aged less than 48 h; some also induce fatal encephalitis after intracerebral inoculation of weaned mice aged 3–4 weeks.

2. Haemagglutination. Most arboviruses can agglutinate erythrocytes, an ability that is inhibited by antiserum against viruses within the same serogroup. Seroreactivity against viruses from dissimilar serogroups is generally weak.

3. Many arboviruses multiply in continuous polyploid tissue cultures of mammalian cells incubated at 37°C, such as grivet monkey kidney (Vero) and baby hamster kidney (BHK).

Table 51.1 Characteristic properties of arboviruses

Property	Arbovirus family (principal genus)					
	Togaviridae (*Alphavirus*)	Flaviviridae (*Flavivirus*)	Bunyaviridae (*Orthobunyavirus*)[a]	Rhabdoviridae (*Rhabdovirus*)[b]	Reoviridae (*Reovirus*)[c]	Orthomyxoviridae
Symmetry[d]	Cubic	Cubic	Helical	Bullet-shaped	Cubic	Cubic
Total diameter (nm)	70	40–60	80–100	180 × 85	60–80	15–120
Nucleic acid	(+)ssRNA	(+)ssRNA	(−)ssRNA and antisense	(−)ssRNA	dsRNA	(−)ssRNA
Molecular weight (×10^6) (Da)	4.2–4.4	4.2–4.4	0.3–3.1	3.5–4.6	0.2–3.0	
No. of molecules	1	1	3	1	10–12	6–7
No. of viruses	29	68	318	63	77	2
Inactivation by diethyl ether or sodium deoxycholate	+	+	+	+	−	+

ssRNA, single-stranded RNA; dsRNA, double-stranded RNA.
[a]Other important genera: *Nairovirus*, *Phlebovirus* (arthropod-borne), *Hantavirus* (not arthropod-borne).
[b]See Chapter 58.
[c]See Chapter 54.
[d]All have enveloped virions (except Reoviridae).

Table 51.2 Some important arboviruses

Family and genus	No. of members	Some important members	Comments
Togaviridae			
Alphavirus	29	Western equine encephalitis virus Eastern equine encephalitis virus Venezuelan equine encephalitis virus Chikungunya virus Ross River virus	Mosquito-borne
Flaviviridae			
Flavivirus	68	St Louis encephalitis virus Japanese encephalitis virus Murray Valley encephalitis virus Yellow fever virus Dengue virus West Nile virus	Mosquito-borne
		Louping ill virus Powassan virus Tick-borne encephalitis virus Kyasanur Forest virus Omsk haemorrhagic fever virus	Tick-borne
Bunyaviridae	**318**		
Orthobunyavirus	172	La Crosse virus Snowshoe hare virus Oropouche virus	
Phlebovirus	51	Rift Valley fever virus Punta Toro virus Sandfly fever virus Toscana virus	California (CAL) serogroup
Nairovirus	34	Crimean–Congo haemorrhagic fever virus	
Hantavirus	15	Sin Nombre virus (not arthropod-borne)	

4. Many arboviruses, such as dengue virus and Ross River virus, multiply in continuous tissue cultures of mosquito cells when incubated at 34°C or lower temperatures; *Aedes albopictus* C6/36 mosquito cells are often used. In general, mosquito-borne viruses do not replicate in tick cell cultures and vice versa.

5. Mosquito-borne arboviruses multiply after oral feeding or intrathoracic injection of several *Aedes* and *Culex* mosquito species after incubation at 4–28°C (depending on the mosquito species). Intrathoracic susceptibility of mosquitoes to dengue and California serogroup agents is 10 to 100 times higher than that of mammalian tissue cultures or suckling mice. Some arboviruses, including members of the *Orthobunyavirus* and *Phlebovirus* genera of the Bunyaviridae, *Flavivirus* genus of the *Flaviviridae*, *Alphavirus* genus of the family *Togaviridae* and some members of the *Vesiculovirus* genus of the family *Rhabdoviridae*, are also transmitted transovarially by vectors. Ticks do not normally transmit mosquito-borne arboviruses; mosquitoes do not normally transmit tick-borne arboviruses. Sandfly-borne viruses are transmitted only by sandflies (*Phlebotomus* spp. and *Lutzomyia* spp.).

6. Tick-borne arboviruses multiply after oral feeding to larval or nymphal ixodid ticks (hard ticks of the genera *Dermacentor* and *Ixodes*). The virus is transferred trans-stadially to the next developmental stage (nymph or adult, respectively), which then transmits virus by biting susceptible vertebrates.

REPLICATION

The replication of the various arboviruses differs significantly and is one of the major criteria used in their classification (see Ch. 7 for a general account).

Alphaviruses

Alphaviruses enter cells by receptor-mediated endocytosis but co-receptors or factors probably also contribute to entry and cell specifity. The virion fuses with an endosome membrane via the E1 envelope glycoprotein. The nucleocapsid is released and binds to ribosomes, and the non-structural proteins are translated directly from the genomic RNA. RNA replication occurs in complexes comprised of the non-structural proteins and cellular proteins, which are associated with cytoplasmic membranes. Genomic RNA is messenger sense (positive strand) and serves as a template for full-length negative-sense RNA synthesis; these negative-sense RNAs are a template for the production of positive-sense genomic RNA, as well as a subgenomic mRNA designated 26S that encodes the structural proteins. Both of these RNAs are capped at the 5′ end and polyadenylated at the 3′ end.

Regulation of positive- versus negative-strand synthesis occurs via changes in the non-structural protease activity mediated by different cleavage patterns of the non-structural polyprotein. The 26S message is translated to yield a polyprotein comprised of the capsid and envelope glycoproteins. The capsid is cleaved co-translationally in the cytoplasm via its own protease activity, and the remaining polyprotein enters the endoplasmic reticulum, where it is processed through the secretory pathway to yield glycosylated E2 and E1 protein heterodimers in the plasma membrane. Genomic RNA combines in the cytoplasm with 240 copies of the capsid protein to form a nucleocapsid, and nucleocapsids interact with the cytoplasmic tail of the E2 envelope protein to mediate budding, whereby 240 E2/E1 protein heterodimers and a portion of the plasma membrane are incorporated into the mature virion.

Flaviviruses

Morphologically, flaviviruses, although smaller; are similar to alphaviruses; however, their molecular biology differs from that of the alphaviruses. Flavivirus virions have three structural proteins; the viral RNA genome is encapsidated by a small core protein and there are two proteins, the membrane (M) and envelope (E), on the outside of the virus particle. The E protein is the major protein of the virus. It is normally glycosylated, has haemagglutination activity, and is the target of neutralizing antibodies. The NS3 protein (see below) contains the majority of T cell epitopes. The virus genome is one single-stranded, positive-sense RNA molecule. Flaviviruses do not have a poly A tail at the 3′ terminus of the genome. The 5′ one-third of the genome encodes the three structural protein, and the remaining two-thirds encodes seven non-structural (NS) proteins involved in virus replication, including an RNA-dependent RNA polymerase. Flaviviruses replicate in the cytoplasm of cells. The input virion RNA is translated as a single open reading frame to generate a polyprotein precursor that is rapidly co- and post-translationally processed by viral and cellular proteases to yield the structural and non-structural proteins. Flavivirus particles assemble by budding through Golgi vesicles and contain prM, a precursor to M protein, as a chaperone for the E protein. Mature virions are produced at the cell surface where the 'pr' portion of prM is cleaved by the cell enzyme furin to yield the mature M protein found in virions.

Bunyaviruses

The *Bunyaviridae* are icosahedral enveloped viruses, with a diameter of 100–120 nm. They have a tripartite RNA genome of negative sense. The three pieces are the large (L), medium (M) and small (S) segments. The L

RNA encodes the RNA-dependent RNA polymerase carried in the virion. The L protein of Crimean–Congo haemorrhagic fever virus encodes an ovarian tumour-like cysteine protease motif, suggesting that the L polyprotein is cleaved autoproteolytically. The M RNA encodes two glycoproteins, termed G1 and G2, now called Gc and Gn, that are found on the surface of virions. The S RNA encodes a nucleocapsid (N) protein.

The *Orthobunyavirus* genus also contains a nonstructural protein, NSs, which is an interferon antagonist that also inhibits host cell protein synthesis. The tripartite genome enables bunyaviruses to undergo genetic reassortment. Reassortment has been shown to take place in nature between closely related bunyaviruses and is a process that contributes to genetic variation and evolution. The *NSm* gene of the *Tospovirus* genus and the *NSs* gene of the *Phlebovirus* and *Tospovirus* genera are encoded as genes in the positive-sense orientation. Thus, the S- and M-RNA segments of tospoviruses and the S-RNA of phleboviruses are termed ambisense RNAs.

Bunyaviruses replicate in the cytoplasm of cells and assemble by budding through Golgi vesicles.

PATHOGENESIS

Natural vertebrate infection by arboviruses is initiated when mosquitoes or other arthropods deposit saliva in extravascular tissues and blood vessels while blood-feeding. In some alphaviruses, murine model systems using needle inoculations indicate that the initial site of replication is the Langerhans cell. Alphavirus replication appears to stimulate both the migratory response of the Langerhans cell to the lymph nodes and the accumulation of leucocytes in the draining lymph node, where local replication produces viraemia. Arboviruses induce high titres of viraemia in many susceptible vertebrates 1–4 days after parenteral inoculation or following bites by infected arthropods; viraemia persists for several days and is a source of infective blood meals for other biting arthropods. Invasion of the central nervous system (CNS) via the olfactory nervous tract may ensue in some infections, whereas other viruses cross the blood–brain barrier. In alphavirus infections accompanied by rash and arthritis, virus replication and necrosis occur in the epidermis and possibly the muscles, tendons and connective tissue. Infection of macrophages may mediate musculoskeletal pathology via suppression of cytokine induction. Wild bird and mammal reservoir hosts regularly exhibit viraemia without symptoms.

Antibodies are first detected when the fever subsides, usually within 5–10 days after infection, and may persist for many years. Antibodies are of the immunoglobulin (Ig) M class for 1–7 weeks after infection; subsequently they are of the IgG class.

Arboviruses cause a spectrum of disease ranging from inapparent infection to acute encephalitis, and some have high rates of inapparent infection. Within the CNS, arboviruses multiply in and induce necrosis of neurones, which in turn become surrounded by microglia, forming glial knots. There is also evidence of apoptosis for some virus infections. An age dependence of CNS disease has been observed for many arboviruses, and animal model systems indicate that age-dependent apoptosis of neurones may explain this phenomenon. Perivascular cuffing with mononuclear cells affects many cerebral blood vessels. Usually there is concomitant meningitis with accumulation of mononuclear cells in the subarachnoid space and hyperaemia of adjacent capillaries. Invasion of the CNS appears to be a critical determinant in the pathogenesis and is due in part to the level of viraemia. The role of the immune system is not clear, although it may be involved in the pathogenesis of dengue haemorrhagic fever and dengue shock syndrome, seen in young children who are experiencing a second dengue virus infection. Antigen–antibody complex formation may underlie the syndrome, which is associated with increased capillary permeability and shock, often with haemorrhage. It is known that the uptake of virus into macrophages is enhanced in the presence of antibody, as the virus–antibody complexes bind to Fc receptors, so there is likely to be a great increase in the uptake of virus and hence the release of virus from macrophages.

Some arboviruses including alphaviruses (e.g. Ross River virus), flaviviruses (e.g. West Nile virus) and bunyaviruses (e.g. Bunyamwera virus) can disrupt the activation of specific innate immunity antiviral pathways, including interferon induction, to enhance their replication.

CLINICAL FEATURES

The tissue tropism of arboviruses can be divided into three categories:

1. the CNS (e.g. encephalitis, aseptic meningitis)
2. visceral organs (e.g. hepatitis and haemorrhagic fevers)
3. febrile infections.

Human arbovirus infections become clinically manifest according to the target organ principally infected (Table 51.3). These include the following syndromes.

Encephalitis

For many arboviruses, encephalitis is most common in children and/or the elderly. Illness leading to encephalitis typically has an abrupt onset within 1 week of infection, with headache, fever, myalgia and dysthesias, and sometimes lethargy, chills, dizziness, nausea, vomiting and

Table 51.3   Clinical syndromes associated with selected arboviruses and their geographical distribution

Syndrome	Genus	Serogroup (vector)	Causative arbovirus serotype		Geographical distribution
Encephalitis or aseptic meningitis	*Alphavirus*	(mosquito)	EEE	Eastern equine encephalitis	Eastern Canada, USA, Caribbean
			VEE	Venezuelan equine encephalitis	North, Central and South America
			WEE	Western equine encephalitis	North, Central and South America, Caribbean
	Flavivirus	(mosquito)	JE	Japanese encephalitis	Orient (Japan to Malaysia)
			MVE	Murray Valley encephalitis	Australia
			SLE	St Louis encephalitis	Canada, USA, Central America
		(tick)	LI	Louping ill	Scotland, Northern Ireland
			POW	Powassan	Canada, Northern USA
			TBE	Tick-borne encephalitis complex	Central and northern Europe, Siberia
	Orthobunyavirus	CAL (mosquito)	LAC	La Crosse	USA
			SSH	Snowshoe hare	Canada
Yellow fever	*Flavivirus*	(mosquito)	YF	Yellow fever	Tropical Africa, Caribbean, tropical South America
Dengue	*Flavivirus*	(mosquito)	DEN	Dengue (four types)	Entire tropical zone
Haemorrhagic fever	*Alphavirus*	(mosquito)	CHIK	Chikungunya	East Africa, India, South-East Asia
	Flavivirus	(mosquito)	DEN	Dengue (four types)	India, Philippines, South-East Asia, Oceania
		(tick)	KFD	Kyasanur Forest disease	India
			OMSK	Omsk haemorrhagic fever	Siberia
	Nairovirus	CCHF (tick)	CCHF	Crimean–Congo haemorrhagic fever	Central and southern Africa, Asia, Europe
Miscellaneous tropical fevers	*Alphavirus*	(mosquito)	CHIK	Chikungunya	East Africa, India, South-East Asia
			RR	Ross River[a]	Australia, Oceania
	Flavivirus	(mosquito)	ILH	Ilheus	Caribbean, South America
			WN	West Nile	Central and northern Africa, Europe, North, Central and South America, Caribbean
	Orthobunyavirus	SIM (mosquito)	ORO	Oropouche	Caribbean, South America
	Phlebovirus	PHL (mosquito)	RVF	Rift Valley fever	Northern, eastern and southern Africa
		PHL (sandfly)	PT	Punta Toro	Central America
		PHL (sandfly)	SFN	Sandfly fever – Naples	Mediterranean
	Vesiculovirus	VS (sandfly)	VSI	Vesicular stomatitis – Indiana	USA, Central America
Undifferentiated fever	*Coltivirus*	CTF (tick)	CTF	Colorado tick fever	Western USA

[a]Ross River virus infections frequently exhibit polyarthralgia.

prostration. Inflammation of the throat, cervical lymphadenitis and abdominal tenderness are also common. Signs and symptoms usually subside after several days, but may recrudesce later. Progression to encephalitis may occur rapidly, or a prodromal illness may last for 1 week or more. Severe CNS disease is accompanied by neck stiffness, motor weakness and paralysis, meningismus, cranial nerve palsy, confusion, convulsions and somnolescence, leading to coma. Rigidity or weakness of the limbs may occur along with reduction in reflexes. White blood cells, predominantly lymphocytes, and raised glucose levels may occur in the cerebrospinal fluid, which can exhibit increased pressure. Peripheral blood cell counts may be raised, with a left shift. During most outbreaks of arboviral encephalitis, a proportion of patients develop aseptic meningitis alone, without significant neuronal involvement, whereas up to 50% of patients recover from acute encephalitis to suffer from neuropsychiatric sequelae that may last from months to years. These sequelae vary from physiological impairment to mental disorder.

Yellow fever

Yellow fever is caused by a mosquito-borne flavivirus that is found in tropical South America and Africa. The disease has an incubation period of 3–6 days, characterized by the sudden onset of headache and fever (temperatures may exceed 39°C), with generalized myalgia, nausea and vomiting. Jaundice may appear by the third day of illness, but frequently is mild or absent. Haematemesis, melaena, epistaxis and bleeding gums may also be noted. Albuminuria and oliguria may also begin suddenly during the first week of illness. In severe cases, death may occur 3–6 days after the onset of illness, and mid-zonal necrosis is observed in the liver. The case fatality rate is estimated as 20–50%. Patients with mild infection may develop fever, headache and myalgia alone, without gastro-intestinal upsets, jaundice or albuminuria. In such cases, diagnosis is made by isolation of virus from blood or by demonstration of rising levels of antibody.

Dengue

Dengue is caused by four serologically related flaviviruses called dengue-1, -2, -3 and -4. These viruses are found in most tropical parts of the world. Dengue presents as an acute febrile illness with chills, headache, retro-ocular pain, body aches and arthralgia in more than 90% of apparent cases, with nausea or vomiting and a maculopapular rash resembling measles lasting for 2–7 days in about 60% of cases. Illness persists for 7 days, fever remitting after 3–5 days, followed by relapse ('saddleback fever') and pains in the bones, muscles and joints sufficiently severe to earn the epithet 'breakbone fever'. Rash occurs more commonly in patients aged less than 14 years. Complete recovery is the rule. The incubation period is 5–11 days.

Dengue haemorrhagic fever

This is a less common manifestation of dengue, with about 500 000 cases per year and a case fatality rate of 15%, mainly affecting children. It is occasionally accompanied by a shock syndrome, known as dengue shock syndrome, with a case fatality rate of 50%. These two severe forms of dengue are observed in patients who undergo successive infection with two different dengue viruses (e.g. a primary dengue-1 infection followed by a secondary infection with dengue-2 virus). After an acute onset, fever of 40°C,

accompanied by vomiting and anorexia, enlarged liver and petechiae persists for 5–10 days. This is followed by a complete recovery unless shock supervenes, as occurs in 7–10% of patients 2–7 days after onset, usually accompanied by haematemesis and melaena. Although dengue haemorrhagic fever with shock syndrome was initially recognized in South-East Asia in 1951, it has since occurred in the south-west Pacific, Caribbean and South American countries.

Miscellaneous tropical fevers

These comprise an increased temperature to above 39°C with any combination of headache, myalgia, malaise, nausea or vomiting, and sometimes accompanied by maculopapular rash or polyarthralgia, that is, a dengue-like syndrome but without haemorrhagic manifestations or the shock syndrome. Tropical fevers arise from infection with a wide variety of arboviruses (see Table 51.3) and clinical diagnosis is impossible without access to serological tests for diagnosis. Many of the Old World alphaviruses such as Ross River, chikungunya, o'nyong-nyong and Sindbis viruses cause an arthritic syndrome accompanied by rash, which can persist for several months. Chronic or relapsing arthritic symptoms may be caused by inflammation associated with periodic increase in replication within persistently infected synovial macrophages which persist despite neutralizing antibodies and antiviral cytokine response. Inhibition of cytokine responses by virus–antibody complexes binding to Fc receptors, and interleukin-10 induction, may facilitate persistence.

Undifferentiated fever

A US example is Colorado tick fever (genus Coltivirus) in which symptoms of chilliness, headaches, retro-orbital pain and generalized aches, especially in the back and limbs, appear 3–6 days after bites by infected Dermacentor andersoni ticks in wooded areas of the Rocky Mountain region. Fever of 39–40°C often shows a biphasic course, with eventual defervescence within 1 week, followed by complete recovery.

Hantavirus pulmonary syndrome

This syndrome, due to the hantavirus Sin Nombre, was first recognized in south-western USA during 1993. Sudden onset of fever, myalgia, headache, cough, and nausea or vomiting is accompanied by rapid respirations exceeding 20 per minute, temperature above 38°C and hypotension. Extensive interstitial and alveolar infiltrates are observed in the lungs, accompanied by reduced oxygen saturation. Fatal cases develop progressive pulmonary oedema with hypoxia and severe hypotension, and die 2–16 days after onset of symptoms; the case fatality rate may exceed 50%. Patients with non-fatal infection usually recover within 1–3 weeks.

General principles

Certain principles apply to human illnesses induced by arboviruses:

1. For each syndrome, several different virus families may be involved, as in encephalitis due to western equine encephalitis (WEE), Japanese encephalitis or La Crosse virus infections.
2. There is a high proportion of subclinical infections to clinical illnesses, for example from 64:1 to 209:1 for St Louis encephalitis (SLE) virus, but this proportion is considerably lower for dengue, Venezuelan equine encephalitis and miscellaneous tropical fevers.
3. There may be localization of a particular virus within a particular geographical zone. However, dengue is distributed widely throughout tropical regions of each continent and in Oceania.
4. Virus transmission by the bite of culicine mosquitoes or ixodid ticks occurs after ingestion of relatively small titres (0.1–10 plaque-forming units) of virus in blood meals from naturally infected viraemic vertebrates after extrinsic incubation periods from less than 1 to 3 weeks at usual summertime temperatures.
5. Viraemia develops asymptomatically in birds (some alphaviruses and mosquito-borne flaviviruses) and mammals (some alphaviruses, bunyaviruses, tick-borne flaviviruses and orbiviruses) at titres sufficient to infect engorging mosquitoes or ticks. However, some vertebrates are dead-end hosts owing to low viraemias that prevent infection of engorging vectors (e.g. human beings infected with Japanese encephalitis and West Nile viruses).

LABORATORY DIAGNOSIS

Diagnosis of arbovirus infections depends on:

- the isolation of virus from blood, cerebrospinal fluid or tissues
- detection of arbovirus-specific RNA in blood, cerebrospinal fluid or tissues
- serology.

Similarly, enzyme-linked immunosorbent assays (ELISAs) can be used to detect serum antibodies from patients. However, these assays are dependent on availability of virus and/or antigens for use in ELISAs. One procedure that is commonly used to identify arbovirus infections is indirect immunofluorescence as this can give a result in a few hours.

Virus isolation

Virus isolation is an important diagnostic tool, but is normally a very slow process and unlikely to generate results until after an acute virus infection. The causative virus can be isolated from blood collected during the initial 3-day febrile illness when viraemia titres are at a maximum. Some arboviruses can also be isolated from cerebrospinal fluid or brain biopsy, or from brain at autopsy of fatal encephalitis cases. Liver may yield virus isolation from fatal cases of yellow fever. The isolation procedures are as follows:

1. Inoculate mammalian tissue cultures (e.g. Vero or BHK cells) and incubate at 37°C for 14 days or until a cytopathic effect is evident, whichever is sooner. Identify the virus by neutralization tests, ELISA, haemagglutination inhibition tests or indirect immunofluorescence.
2. For mosquito-borne viruses, inoculate continuous cultures of mosquito cells (e.g. *Aedes albopictus* C6/36) and incubate at 28°C for up to 14 days; examine cells for arbovirus antigen by immunofluorescence.
3. Inoculate acute-phase blood from patients with dengue intrathoracically into mosquitoes and incubate at 30°C for 14 days. Detect viral antigen in head squashes by direct immunofluorescence.
4. Inoculate acute-phase blood, cerebrospinal fluid, or brain or liver suspensions intracerebrally into mice aged less than 48 h and examine daily for up to 14 days for development of encephalitis. Perform preliminary virus indentification by haemagglutination inhibition tests on sucrose–acetone extracts of brain suspension, followed by final virus identification by neutralization tests in suckling mice or by plaque reduction in mammalian tissue cultures (e.g. Vero or BHK cells).

Arbovirus–specific RNA detection

Detection of virus-specific RNA, by reverse transcriptase–polymerase chain reaction (RT-PCR) and the more sensitive real-time RT-PCR, is a rapid approach to diagnosis but is dependent on the availability of oligonucleotide primers that will specifically amplify a given virus. Given the large number of arboviruses, there must be significant differential diagnosis before selection of appropriate primers. Genus-reactive primers have been described for alphaviruses, flaviviruses and bunyaviruses.

After amplification of extracted RNA, the product is analysed by restriction enzyme digestion and/or determination of the nucleotide sequence of the PCR product and comparison with sequences in Genbank or other nucleotide sequence databases.

Serology

Serological testing is often the only available means of laboratory diagnosis of encephalitis. The detection of a four-fold or greater rise of antibody titre by haemagglutination inhibition tests or ELISAs on paired sera collected during the initial week and several days later may provide good, but not definitive, evidence of concurrent infection. Antibodies detected by complement fixation testing first appear 2 weeks or more after onset and become undetectable by 3 years. Haemagglutination inhibition and complement fixation test results are usually not as specific as those obtained by neutralization. Virus-specific IgM antibody may be detected within 1 day of onset of clinical symptoms using an IgM capture ELISA. IgM antibody wanes 6 weeks after onset and is replaced by IgG antibodies. IgM antibodies are indicative of a recent infection.

TREATMENT

Currently, no specific anti-arboviral therapeutic agent is available. Patients with encephalitis are managed supportively, using anticonvulsants as required, and ice packs are applied when indicated to reduce hyperthermia. Raised intracranial pressure can also be treated, and airway protection may be needed in unconscious patients, with hyperventilation accompanied by anaesthesia and sedation. Brain swelling can be minimized by regulating serum sodium levels and osmolarity. Nosocomial infections, especially pneumonia, should be prevented and treated aggressively when they occur. Similarly, in dengue and haemorrhagic fevers, supportive measures may include intravenous infusions to replace fluids and electrolytes. Ribavirin shows activity against some bunyaviruses and alphaviruses (and in combination with interferon-α), but has not been well evaluated and is not therapeutic in flavivirus infections.

EPIDEMIOLOGY

Natural cycles

Arboviruses are maintained in natural transmission cycles involving reservoir hosts and arthropod vectors, typically:

- ticks
- mosquitoes
- other biting flies.

Except for dengue, arboviruses are zoonotic pathogens that utilize wild animals as reservoir hosts. Many arboviruses also use non-human animals as amplification hosts during epidemics, and human beings are often tangentially infected, *dead-end hosts* during these outbreaks. Mosquitoes or other arthropods become

infected by engorging on a viraemic vertebrate host. In mosquitoes, infection begins in the midgut epithelium and spreads to the haemocele or open body cavity, where it may disseminate to other tissues and organs including the salivary glands. Finally, this extrinsic incubation period is completed when replication in salivary gland acinar cells leads to virus release into the apical cavities and salivary ducts. Transmission may then occur upon a subsequent blood meal, when mosquitoes deposit saliva in extravascular tissues while probing to locate a venule, or intravascularly. Infection of the vertebrate then leads to viraemia and the opportunity for infection of additional vectors. Nonviraemic transmission, whereby an infected vector transmits to an uninfected vector feeding at the same time on the same host before replicative viraemia can occur, has been described for tick-borne viruses.

Examples of natural cycles are:

- human–mosquito cycle, as in dengue and urban yellow fever (Fig. 51.1)
- mosquito–bird cycle, as in SLE and West Nile viruses (Fig. 51.2)
- mosquito–mammal cycle, as in Venezuelan equine encephalitis (VEE) virus (Fig. 51.3).

Epidemiological aspects of arbovirus infections with major impact on human health are described according to syndrome and infecting serotype (see also Table 51.2).

Alphaviruses

Western equine encephalitis virus

WEE virus was first isolated from the brain of a horse with encephalitis. It has caused periodic outbreaks of equine and human encephalitis in the western half of North America, as well as in Brazil and Argentina. The mean incidence of WEE cases throughout the USA between 1966 and 1985 was 17 per year, constituting 7% of the total reported cases of arbovirus encephalitis. In North America, cases occur only during June to September, when mosquitoes are abundant. Outbreaks of infection occur at intervals of a few years and are preceded by epizootic peaks of encephalitis among horses. The principal mosquito vector species in Canada and the western USA is *Culex tarsalis*. Ducks and other water birds constitute the principal natural reservoirs. Human beings and equines are considered dead-end hosts because they produce little viraemia. Although all age groups may be involved, encephalitis due to WEE virus and California serogroup agents commonly affects children, in whom the disease is often severe; SLE virus is seen most often in persons aged 55 years or more.

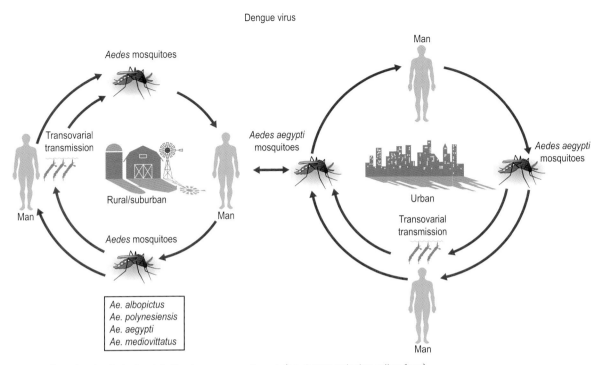

Dengue virus

Fig. 51.1 Natural cycle of arbovirus infection: human–mosquito cycle (e.g. dengue and urban yellow fever).

Eastern equine encephalitis virus

Eastern equine encephalitis (EEE) virus was first isolated in 1933 from the brain of a horse with encephalitis. Human and horse cases of encephalitis have occurred repeatedly during the summer in Atlantic coastal areas extending from Massachusetts and New Jersey to Florida and Texas and north-eastern Mexico, and in a few inland locations such as Michigan and Wisconsin. Occasionally, disease extends northwards into the Province of Quebec, Canada. During the 20-year period from 1966 to 1985 in the USA, EEE virus caused a mean of four cases of encephalitis per year, constituting 2% of the total cases of arbovirus encephalitis. Illness affected mainly those aged less than 14 years and more than 55 years, with overall case fatality rates as high as 69%. Surviving patients usually have severe neurological sequelae. Most human cases occur during late August and September, about 3 weeks after the peak of horse cases. Distinct forms of EEE virus occur throughout South and Central America, but are rarely associated with human disease. In North America, EEE virus is maintained enzootically in hardwood swamp habitats, where the mosquito *Culiseta melanura* transmits among passerine birds. During years of hyper-enzootic transmission, horses and human beings residing near the swamp habitats become infected via bridge vectors such as *Aedes* spp. (*Cs. melanura* feeds primarily on birds).

Equines and man are considered dead-end hosts because they develop little viraemia, and outbreaks are therefore confined to regions of enzootic activity.

Venezuelan equine encephalitis virus

VEE virus was first isolated in 1938 from a horse in Venezuela. Periodic, sometimes widespread, outbreaks involving up to hundreds of thousands of human beings and equines have occurred since, primarily in northern South America, with one outbreak extending as far north as Texas in the USA in 1971. During outbreaks, VEE virus is transmitted among equines by a variety of mosquitoes such as *Aedes* and *Psorophora* spp. Equines are extremely effective amplifying hosts because they develop high-titred viraemia and are attractive to large numbers of mosquitoes. Antigenically related viruses occur in sylvatic and swamp habitats through much of the Neotropics and subtropics, as far north as Florida, USA, and as far south as northern Argentina. These viruses can cause febrile illness and encephalitis in human beings that enter these foci, and one antigenic subtype (designated ID) can mutate to become capable of initiating widespread outbreaks. Rates of encephalitis in apparent infections (generally 5–15% of symptomatic cases) and mortality (around 0.5%) are generally lower than for EEE virus, with most fatal cases

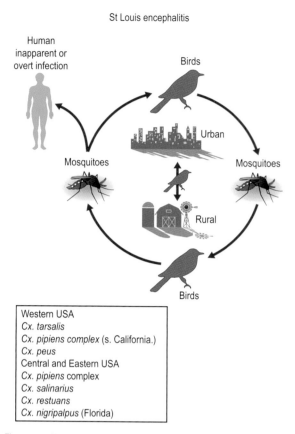

Fig. 51.2 Natural cycle of arbovirus infection: bird–mosquito cycle (e.g. St Louis encephalitis).

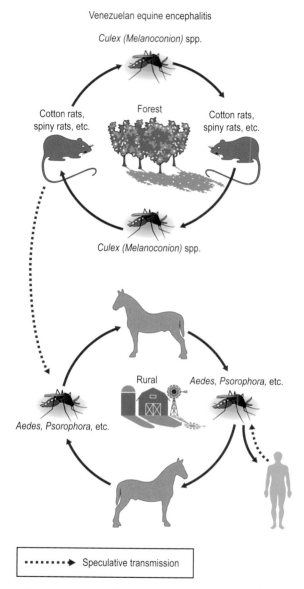

Speculative transmission

Fig. 51.3 Natural cycle of infection: mammal–mosquito cycle (e.g. Venezuelan equine encephalitis).

occurring in young children. Neurological sequelae are common. VEE attack rates can exceed 50%.

Ross River virus

Ross River virus infection was first described in 1928 in Australia, where the disease is known as epidemic polyarthritis. Since that time numerous outbreaks have occurred in Australia and the South Pacific. In Australia, epidemic polyarthritis occurs primarily during the summer and autumn as sporadic cases and small outbreaks; the geographical distribution has expanded. An average of 5000 cases are reported each year in Australia, mostly involving vacationers and others travelling in rural areas. Ross River virus is maintained in a zoonotic vertebrate–mosquito cycle, with *Culex annulirostris* and *Aedes vigilax* serving as the principal vectors in Australia, and flying foxes and marsupials implicated as reservoir hosts. Human infections have also been documented in New Guinea, the Solomon Islands, New Caledonia, Fiji, American Samoa and the Cook Islands. During 1979–1980 a large, explosive epidemic of Ross River polyarthritis swept across the South Pacific, with 40–60% of the population affected on some islands. *Aedes polynesiensis* was implicated as the vector, and epidemiological studies suggested a man–mosquito–man transmission cycle.

Chikungunya virus

Chikungunya (CHIK) virus was first isolated during a 1952 epidemic in Tanzania. In Swahili, 'chikungunya' means 'that which bends up', and refers to the posture of patients with severe joint pains. CHIK virus has probably occurred sporadically in India and South-East Asia for at least 200 years, and also occurs in most of sub-Saharan Africa, Indonesia and the Philippines. Its origin has been traced to East Africa. In Africa, a sylvatic transmission cycle occurs between wild primates and arboreal *Aedes* mosquitoes, while urban CHIK virus epidemics in India and South-East Asia, causing hundreds of thousands of cases, involve *Ae. aegypti* transmission in a man–mosquito–man cycle. Unlike dengue, which is endemic in many of these Asian cities, CHIK virus may disappear and reappear at irregular intervals. However, because it cannot be distinguished clinically from dengue fever, CHIK virus infection may be overlooked in the absence of appropriate surveillance. The mechanism of virus maintenance in Asia during inter-epidemic periods or reintroduction is unknown. Since March 2005 epidemics have taken place in Asia and Africa involving more than one million people with some fatalities.

O'nyong-nyong virus

O'nyong-nyong (ONN) virus, derived from the description by the Acholi tribe, meaning 'joint breaker', was first isolated during a 1959–1962 epidemic affecting 2 million people in Uganda, Kenya, Tanzania, Mozambique, Malawi and Senegal. Another major outbreak involving an estimated 1 million cases occurred in 1996 in areas of Uganda and northern Tanzania. Attack rates are generally high and all age groups are affected during ONN virus epidemics. An antigenically closely related virus, Igbo-Ora, isolated from febrile patients in Nigeria, has been shown to be a strain of ONN virus. The transmission cycle of ONN virus involves *Anopheles funestus* and *Anopheles gambiae* mosquitoes; ONN virus is the only known alphavirus with *Anopheles* vectors. The reservoir and amplification hosts are unknown.

Flaviviruses: mosquito–borne

St Louis encephalitis virus

SLE virus was first isolated from the brain of a person dying from acute encephalitis in St Louis, Missouri. SLE virus activity is widely distributed throughout the USA, from the Ohio and Mississippi valleys extending westwards through Colorado to California and Washington, northwards into the Canadian Provinces of Ontario, Manitoba and Saskatchewan, eastwards to Ohio, and southwards to Florida and Texas. As with WEE, SLE became prevalent in arid areas of the north-western and south-western USA after the development of irrigation farming led to the breeding of massive populations of the principal mosquito vector *Cx. tarsalis* in semipermanent collections of water in grassy locations. During the largest outbreak of the past quarter-century, in 1975, when 1815 (86%) of 2113 confirmed cases of arbovirus encephalitis were due to SLE virus, the Chicago metropolitan area was heavily affected for the first time. Substantial outbreaks affected the Houston, Texas, metropolitan area in 1964 and 1986, where attack and case fatality rates were highest among persons over 55 years of age. Los Angeles was affected first, with 26 human cases of encephalitis in 1984. Mosquito vectors in California, the Rocky Mountains and plains states comprise mainly *Cx. tarsalis*, but *Cx. pipiens* and *Cx. quinquefasciatus* are important along the Mississippi Valley and eastwards. The salt marsh mosquitoes *Cx. restuans* and *Ae. sollicitans* may be important vectors in some localities.

Japanese encephalitis virus

Japanese encephalitis virus was first isolated from the brain of a patient with fatal encephalitis in Tokyo in 1935. It continues to cause epidemics of encephalitis, affecting children, particularly in India, Korea, China, South-East Asia and Indonesia, with case fatality rates often exceeding 20%. The virus has recently emerged in the Torres Straits and the tip of northern Australia. Approximately 50 000 cases occur each year, of which 15 000 are fatal. The ratio of apparent to inapparent infection is 1:50–400, depending on the geographical area. *Cx. tritaeniorhynchus* mosquitoes are the principal vectors, and maximal virus isolation rates from mosquitoes occur during late July, simultaneously with human and equine epidemics. Important vertebrate reservoirs are black-crowned night herons and other water birds, and pigs are considered to be an amplifying host. In Malaysia, *Cx. gelidus* is an important vector as well as *Cx. tritaeniorhynchus*. In northern Thailand cases of encephalitis due to Japanese encephalitis virus have been diagnosed every year since the late 1960s, with most occurring during the rainy season in June, July and August. The case fatality rate of virologically confirmed cases is 33%, and about one in every 300 human beings infected with the virus develops encephalitis.

West Nile virus

West Nile virus was first isolated from a febrile human being in the West Nile district of Uganda in 1937. The virus has a wide geographical distribution, including southern Europe, Africa, central and south Asia, and Oceania. Genetic studies have shown that the related Kunjin virus, found in Australasia, is a subtype of West Nile virus. Although the virus infects a wide variety of animals, including horses, cattle and man, the major vertebrate hosts are wild birds, and it is thought that migratory birds are important in the spread of the disease owing to long-term high-titre viraemia. Both mosquitoes and ticks have been reported as vectors; the principal vectors are considered to be mosquitoes of the *Culex* genus. Both sylvatic and urban transmission cycles have been reported, with *Cx. pipiens* implicated as the major urban vector. West Nile virus usually causes a febrile illness, but encephalitis is seen in patients over 50 years of age. Epidemics vary in size, with the largest reported in Israel and South Africa, including an epidemic of 3000 clinical cases in South Africa in 1974 and over 400 cases in Russia in 1998, and an equine epidemic in North Africa in 1998. In the late summer of 1999, the first outbreak of West Nile virus infection was reported in the western hemisphere, in New York. A total of 62 clinical cases was reported, mostly in the elderly, including seven deaths.

Since then the virus has spread across the USA with at least 4161 human cases (277 fatalities) and more than 14 000 equine cases reported in 2002. This is the largest recorded epidemic of arboviral meningo-encephalitis in the western hemisphere. Overall, in the USA there had been more than 24 000 human cases and over 900 deaths by the end of 2006. As of 2006, the distribution of the virus has expanded to include the continental USA and seven Canadian provinces, as well as Mexico, El Salvador and many of the Caribbean Islands. During 2006 the virus has spread as far south as Argentina, causing disease in equines. Significantly, the virus has infected at least 308 species of birds, 30 other vertebrate species and 60 species of mosquitoes. In addition, cases of West Nile virus infection have been imported into European countries from endemic areas. Genetic studies have identified two West Nile virus-related viruses, Rus98 from Russia and Rabensburg virus from the Czech Republic. It remains to be seen whether these viruses are veterinary or public health problems.

Usultu virus

Usultu virus, isolated in Africa, is a member of the Japanese encephalitis serogroup of the *Flavivirus* genus and was not known to cause disease. In 2001 the virus emerged in Austria, in Vienna and surrounding districts, and was responsible for a large epizootic with high mortality in birds, especially Eurasian blackbirds (*Turdus merula*) and owls. The virus survived the winter of 2001 and caused a significant outbreak of avian mortality in 2002.

Murray Valley encephalitis virus

Murray Valley encephalitis (MVE) virus was first isolated from the brain of a patient with fatal encephalitis at Mooroopna, Victoria, Australia. MVE virus caused epidemics of encephalitis in irrigated farming regions of the Murray–Darling River basin during the summer months (January to March) of 1951 and 1974, with case fatality rates approaching 40%. MVE virus was isolated from *Cx. annulirostris* mosquitoes only during 1974 but not during intervening years, which suggests epidemic introduction of virus into this dry temperate region. In the tropical irrigated Ord River region of Western Australia, the virus is endemic and encephalitis occurred in 1974, 1978, 1981 and 1986. Another endemic focus exists in the Gulf of Carpentaria region of Queensland. Natural cycles of transmission of MVE virus involve *Cx. annulirostris* as the principal mosquito vector and water birds as reservoirs.

Yellow fever virus

Yellow fever virus was first isolated in Ghana from the blood of a male patient with fever, headache, backache and prostration. Yellow fever virus is found mostly in tropical Africa and tropical South America. There are two epidemiological patterns. In the sylvatic cycle, virus is transmitted among monkeys by mosquitoes (*Haemagogus* and *Sabethes* species in South America and *Aedes* species in Africa). Humans are infected incidentally when entering the area, for example to work as foresters. In the urban cycle, person-to-person transmission is via *Ae. aegypti*, which breeds close to human habitation in water, pits and scrap containers such as oil drums. No urban yellow fever has been reported in South America since 1954.

In the Americas, the majority of cases of yellow fever are reported in Peru and Bolivia, and involve males aged 15–45 years who are agricultural and forest workers. During 1995 there were at least 490 cases in Peru, representing the largest epidemic in South America in 40 years. Yellow fever usually occurs from December to May, and peaks during March and April, when populations of *Haemagogus* mosquitoes are highest during the rainy season.

Yellow fever is endemic in many parts of West, Central and East Africa, principally between latitudes 15°N and 15°S, extending northwards into Ethiopia and Sudan. Continuing activity has been encountered in Nigeria, with occasional outbreaks in other parts of West Africa. There are relatively few outbreaks in East Africa. The last major outbreaks were in Kenya in 1993, with 54 cases (fatality rate 50%), and in Sudan in 2003 and 2005. *Ae. africanus* is the principal vector in Africa.

Dengue viruses

Dengue infection is endemic in all tropical regions between latitudes 23.5°N and 23.5°S. Four serologically related viruses termed dengue-1, -2, -3 and -4 cause the disease *dengue*. The febrile clinical symptoms associated with dengue are similar to those of other arboviruses from other families; this has resulted in confusion in the diagnosis of dengue. In particular, the high incidence of dengue has resulted in the misdiagnosis of some arbovirus outbreaks. Serological studies are needed to differentiate these.

Dengue viruses are thought to have originated in a sylvatic cycle involving arboreal mosquitoes and monkeys. However, endemic viruses have evolved and are now maintained in nature by a cycle involving man as both reservoir and definitive host, and domestic mosquitoes, principally *Ae. aegypti*, as vectors. However, other mosquito species can also transmit the viruses.

Dengue virus has caused numerous outbreaks throughout the south-west Pacific region since it was first reconized among servicemen during the Second World War, when the average monthly attack rate of 54 per 1000 peaked to 197 per 1000 during the wet season. Subsequently, in the south-west Pacific, multiple dengue viruses have affected Tahiti, with haemorrhagic dengue first encountered in 1971. In addition to *Ae. aegypti* other species have been implicated as vectors, including *Ae. scutellaris hebrideus* in New Guinea, *Ae. polynesiensis* in Tahiti and *Ae. cooki* in Niue. Cases of imported dengue occurred in Hawaii during 1943 and 1944, probably by transport of dengue-infected mosquitoes on ships returning from dengue-endemic Pacific islands. Hawaii was dengue-free from 1944 to 2001 until, in 2001–2002, there was a dengue-1 virus outbreak involving at least 122 cases and *Ae. albopictus* as the vector.

Dengue virus activity continues in many tropical countries of the Pacific rim, including northern Australia. Most South-East Asian countries, including Indonesia, Malaysia, Thailand, Vietnam, China, the Philippines and India, experience repeated epidemics of dengue caused by all four viruses, with most cases occurring between June and November. Mostly children are affected; in some outbreaks up to 25% may develop haemorrhagic fever. Most epidemics occur in urban areas and villages where *Ae. aegypti* is abundant, but not in rural environments. Up to 100 million infections may occur each year, mostly in children; this is a major public health problem.

Dengue is endemic throughout tropical Africa, including Nigeria in the west and Mozambique in the east, and also in Middle Eastern countries such as Saudi Arabia. Dengue disease is seldom severe in Africa and dengue haemorrhagic fever is rare.

Caribbean countries have been involved in epidemic waves of dengue since 1827, with little evidence of clinical dengue during interepidemic periods. All four dengue viruses have been implicated in outbreaks affecting residents of Caribbean islands and adjacent areas of Central and South America. Each year cases of dengue are imported into continental USA and Europe following visits to dengue-endemic countries. Although most infections in the USA occur among travellers returning from the Caribbean, small numbers of cases (less than 30 per year) of indigenous dengue occur among residents of southern Texas who live near the border with Mexico. The semi-tropical climate in this area allows *Ae. aegypti* to flourish during the nine warmer months of each year. In 1985, *Ae. albopictus* was introduced from Asia into the Americas via used motor tyre casings that had been imported for retreading into Texas, from Japan, South Korea and several South-East Asian countries. The mosquito rapidly established itself in Texas and spread into mid-western, north-eastern and north-western states of the USA, and also into Mexico and other countries. A peri-domestic mosquito, *Ae. albopictus*, was first identified as a dengue vector in Malaysia during the 1960s, and it transmits dengue both in the human–mosquito cycle and by transovarial transfer. However, to date, there have been few reports of dengue virus transmission in the Americas by *Ae. albopictus*.

Flaviviruses: tick-borne

Powassan virus

Powassan virus, the sole North American tick-borne flavivirus, was first isolated from the brain of a fatal human case of encephalitis in Powassan, Ontario, Canada. To date, it has caused approximately 30 cases of encephalitis and four deaths among residents of forested areas of Ontario, Quebec and Nova Scotia in Canada, as well as Massachusetts, New York State and Pennsylvania in the USA. All of these cases occurred between May and October. Principal tick vector species in Ontario are *Ixodes cookei*, which feeds on groundhogs, and *I. marxi*, which feeds on tree squirrels; both of these mammals serve as reservoirs.

Tick-borne encephalitis viruses

Tick-borne encephalitis (TBE) virus is used to describe a serocomplex of related viruses that are transmitted by ticks and cause similar diseases. These include central European (also known as the western subtype) TBE virus that causes central European encephalitis and Russian spring–summer encephalitis (also known as the Far Eastern subtype of TBE virus). In addition, a third subtype, Siberian TBE virus, has been described mainly on the basis of genetic studies. Russian spring–summer encephalitis occurs in eastern Europe and parts of Asia, including northern Japan, and causes a more severe disease than the central European encephalitis found in western Europe and Scandinavia. The case fatality rate of central European encephalitis is usually below 10%, whereas it can reach 30% for Russian spring–summer encephalitis. In Britain and Ireland and some parts of France and Scandinavia, a mild form of tick-borne encephalitis is caused by a related virus called 'louping ill'. The latter virus also infects sheep and grouse and gets its name from the leaping gait in infected sheep; the virus rarely infects man.

Human infections with TBE virus may range in severity from mild biphasic meningo-encephalitis, which is characteristic of the central European TBE virus and louping ill virus, to a severe form of polio-encephalomyelitis that is characteristic of Russian spring–summer encephalitis virus.

Natural cycles involve *I. ricinus* ticks as vectors, and mice, shrews and other small rodents as reservoirs, with infection transferred tangentially to sheep or other farm animals, and also to human beings. In Siberia, *I. persulcatus* ticks serve as vectors. Recently, additional viruses have been described from Spain, Turkey and Bulgaria that are related to louping ill virus and are members of the TBE virus complex and cause TBE virus-like disease symptoms in sheep and other animals.

The TBE virus complex contains three viruses associated with haemorrhagic fever: Omsk haemorrhagic fever (OHF) and Kyasanur Forest disease (KFD) and Alkhumra viruses. OHF was identified during the Second World War in Omsk and the virus causes occasional outbreaks in Russia, whereas KFD is found only in India and causes regular outbreaks of haemorrhagic fever. Alkhumra virus was isolated in 1995, and subsequently in 2001 in Saudia Arabia; it caused haemorrhagic fever with a fatal outcome in 25% of patients. Genetic studies suggest that Alkhumra virus is a subtype of KFD virus.

Bunyaviruses: *Bunyavirus* genus

California (CAL) serogroup

There are at least 14 viruses related antigenically to the prototype California encephalitis virus. In the USA, encephalitis and aseptic meningitis arise commonly from infections with La Crosse virus, first isolated from the brain of a patient with fatal encephalitis at La Crosse, Wisconsin. Other viruses occasionally associated with aseptic meningitis are snowshoe hare virus, isolated from the blood of a snowshoe hare in Montana, and Jamestown Canyon (JC) virus, isolated from *Culiseta inornata* mosquitoes collected at Jamestown Canyon, Colorado. In central Europe, febrile illness, sometimes with aseptic meningitis, arises from infection with Tahyna virus, isolated from *Ae. caspius* mosquitoes collected near Tahyna, in the former Czechoslovakia.

Currently, CAL serogroup viruses are the commonest arboviruses associated with encephalitis in the USA. The highest attack rates occur in states adjoining the Great Lakes, affecting mainly children aged less than 15 years. Abundant tree holes in wooded areas provide optimal breeding sites for the principal mosquito vector *Ae. triseriatus*, but rainwater collected in disused motor tyres is a suitable breeding ground for *Ae. triseriatus* in suburban locations. As adult mosquitoes die in winter, CAL serogroup viruses survive through transovarial transmission. Principal vertebrate reservoirs are tree squirrels and chipmunks.

Oropouche (ORO)

This virus is the only human pathogen in the Simbu serogroup; it was isolated in Trinidad in 1955. Outbreaks of oropouche fever, a febrile illness, have occurred in Brazil since 1961, involving urban transmission by the midge *Culicoides paraensis*. An outbreak was reported in Panama in 1989, and cases of oropouche fever have been reported in Peru since 1992.

Garissa virus

In 1997–1998 Garrisa virus was described as a reassortant bunyavirus. It was isolated from patients with acute haemorrhagic fever during an epidemic of Rift Valley fever in Kenya and southern Somalia. Acute sera from patints with haemorrhagic fever yielded either virus isolation or PCR evidence of infection. Initial studies indicated that the sequences of the L and S RNA segments were nearly identical to those of Bunyamwera virus, whereas the sequence of the M segment was very different (33% nucleotide and 28% amino acid differences from Bunyamwera virus). Very recent studies have shown that Garissa and Ngari virus M segment are nearly identical in sequence, whereas the L and S segments of Bunyamwera virus are similar to those of Ngari virus. These data indicate that Garissa virus is not a reassortant but an isolate of Ngari virus, which is a reassortant of Bunyamwera virus. Previously Ngari virus had not been considered a cause of haemorrhagic fever; further studies are required to investigate the pathogenesis and epidemiology of Ngari virus.

Bunyaviruses: *Phlebovirus* genus

Rift Valley fever virus

Rift Valley fever virus was first isolated in 1930 from sheep during an epizootic causing abortion and death in the Rift Valley near Lake Niavasha, Kenya, but is present from South Africa to Egypt. The virus infects many large domestic animals and a wide variety of mosquito species, with sheep, cattle, buffaloes and rodents as reservoirs. The virus is epizootic with long inter-epizootic periods.

Outbreaks of Rift Valley fever occurred in Egypt during 1977 involving an estimated 200 000 human cases and 600 deaths. Subsequently, Rift Valley fever has been reported during 1998–1999 in Mauritania and Senegal, and in Yemen and south-west Saudi Arabia in 2000–2001, involving over 2000 clinical cases and a mortality rate of 14%. Genetic studies of the virus are consistent with the theory that infection came from East Africa.

Sandfly fever group

These viruses are distributed throughout the European and North African countries surrounding the Mediterranean Sea, extending eastward through Israel and Iran to West Pakistan and central India. Sandfly fever (Naples and Sicilian) viruses were first isolated from the sera of US servicemen during an outbreak in the Second World War.

Epidemics of dengue-like fever occur during the sandfly season (June to September) and affect mainly visitors rather than residents. Natural vectors are *Phlebotomus papatasi* and other phlebotomine sandflies. Isolation of virus from male sandflies collected during July suggests transovarial transfer of virus. The natural cycle of sandfly fever appears to involve solely human beings, as definitive host and reservoir, with sandflies as vectors.

Toscana virus

Toscana virus was first isolated from sandflies collected in Tuscany, Italy; it is transmitted by *Phlebotomus perniciosus* sandflies, and can be transferred transovarially. Natural reservoirs of Toscana virus appear to be small rodents (*Apodemus sylvaticus*).

Bunyaviruses: *Nairovirus* genus

Crimean–Congo haemorrhagic fever virus

Crimean–Congo haemorrhagic fever (CCHF) virus was originally described as two separate viruses: Crimean haemorrhagic fever virus, isolated from the serum of a man with fatal haemorrhagic fever near Samarkand, Uzbekistan, and Congo virus isolated from the serum of a child with fever and arthralgia in Zaire. Antigenic and genetic studies show that the two viruses are identical. CCHF virus is distributed widely throughout tropical Africa, from Mauritania to Uganda and Kenya, the Middle East and West Pakistan, and southwards to South Africa. It is also found in Asia, including parts of China. The geographical distribution of CCHF virus corresponds to that of *Hyalomma* sp. ticks, from which the virus can be isolated. Human infection is rare but mortality rates of up to 50% have been reported.

Bunyaviruses: *Hantavirus* genus

Unlike the other genera in the *Bunyaviridae*, members of the *Hantavirus* genus are rodent-associated viruses. They are zoonotic viruses of rodents (mainly mice or voles) that excrete virus in urine for prolonged periods. Virus is transmitted to man by contact with aerosols of rodent urine.

Hantaan and Puumala viruses

These viruses induce either (1) a severe illness, termed *haemorrhagic fever* with *renal syndrome*, due to Hantaan virus in Japan, Korea, China and Siberia and Puumala virus in Scandinavia, or (2) a mild illness, termed *nephropathia epidemica*, due to Puumala virus in Scotland, France, Belgium and Germany, the Balkans and Greece. Hantaan virus was first identified in soldiers serving in Korea in 1951 and named after the area where it was detected. Principal vertebrate reservoirs comprise *Apodemus agrarius* rodents in Asia and *Clethrionomys glareolus* (bank vole) in Europe.

Sin Nombre virus

Sin Nombre virus induces hantavirus pulmonary syndrome, a severe acute respiratory illness with a case fatality rate exceeding 50%. Initially encountered in May 1993 in the Four Corners region of the USA (Arizona, Colorado, New Mexico and Utah), cases have since occurred elsewhere in the USA, Canada and many countries in South America. The principal rodent reservoir is considered to be *Peromyscus maniculatus* (deer mouse), but each hantavirus pulmonary syndrome-causing virus occupies a geographical region defined by the rodent species that carries it. Thus, Black Creek Canal virus found in cotton rats (*Sigmadon hispidus*) in Florida is related to viruses found in South America, with a distinct rodent host.

This was the first occasion on which molecular methods identified an unknown arbovirus before the virus was isolated. Hantavirus group antibodies were detected in sera from human cases and rodents. RNA from lung and liver of fatal human cases and seropositive rodents was amplified by RT-PCR; positive bands were detected using primers for Prospect Hill (North American) and Puumala virus-like hantaviruses, but not Hantaan and Seoul-like (Asian) hantaviruses. This was achieved within 3 weeks after death of the initial human cases. Several months later, Sin Nombre virus, genomically distinct from Prospect Hill and other hantaviruses, was isolated from tissue suspensions, initially from *P. maniculatus* and subsequently from man.

CONTROL

Strategies for the prevention of arbovirus infections depend on either vector control or active immunization with vaccine.

Vector control

This is possible for mosquito-borne viruses in urban and suburban localities; suppression of populations of vector mosquito species can halt virus transmission during epidemics. This can be achieved by:

1. Use of insecticides to kill adult mosquitoes ('adulticiding'), for instance aerial sprays of malathion; this may kill many other insect species. The use of persistent insecticides on the interiors of houses has decreased.
2. Elimination of breeding sites of domestic *Ae. aegypti* by removal of objects such as tin cans and motor tyres that could contain rainwater, both near human habitations and in public parks and drainage systems; this has prevented the occurrence of dengue in Singapore and urban yellow fever in metropolitan areas in Caribbean countries.
3. Chemical control of larvae (barricading) by the use of temephos granules or malathion in oil for even coverage of small breeding sites; this has reduced mosquito vector populations substantially in irrigated localities in California and elsewhere.
4. Biological control of larvae is attempted by micro-biological agents such as *Bacillus thuringiensis israeliensis*, larvivorous fish, flatworms or mermethid nematodes, or insect growth regulators such as the juvenile hormone that mimic methoprene.
5. Personal protection against bites by mosquitoes; involves a combination of wearing protective clothing, preferably impregnated with permethrin, screening of dwellings to prevent entry of mosquitoes and frequent application of mosquito repellants such as *N,N*-diethyl-*m*-toluamide (DEET) or picaridin to exposed skin areas. For tick-borne viruses, protective clothing should be worn outdoors, followed by rigorous inspection to remove attached ticks from the skin.

Vaccines

To date relatively few vaccines have been developed to control arbovirus diseases.

Alphaviruses

There are no commercially licensed vaccines for use in man. However, formalin-inactivated vaccines have been developed for EEE and WEE viruses. These are used to immunize horses, researchers who work with the viruses, and military personnel. A live VEE vaccine, known as TC-83, is also administered to horses and researchers. TC-83 is associated with significant adverse reactions and lack of seroconversion in many human vaccinees, and is not suitable for use in the general population. There is also an experimental killed Ross River virus vaccine.

Flaviviruses

Vaccines have been licensed against yellow fever, Japanese encephalitis and TBE viruses. A live yellow fever vaccine, 17D, was developed in the 1930s by 176 passages of wild-type strain Asibi through chicken tissue. One dose of vaccine administered subcutaneously and containing 5000–200 000 plaque-forming units of virus gives protective immunity 10 days after immunization; immunity lasts for at least 10 years. The World Health Organization recommends immunization every 10 years to maintain immunity; the vaccine can be given to children over 9 months of age. Immunization is contra-indicated in immunocompromised individuals and pregnant women. More than 400 million doses of vaccine have been administered, with only 21 cases of post-vaccination encephalitis reported. It is one of the finest vaccines ever developed.

Both live and killed vaccines have been developed to control Japanese encephalitis. Formalin-inactivated vaccines were developed in the 1940s based on virus grown in mouse brain. Although these vaccines are still used, there is a risk of allergic reaction following immunization. Killed vaccines based on virus grown in Vero cell culture are being developed. Two doses of killed vaccine given 7–28 days apart are required for protective immunity. A booster is given at 1 year and subsequently every 3–4 years to maintain immunity. A live vaccine has been developed: SA14-14-2 was generated in the People's Republic of China by 126 passages of wild-type strain SA14 in primary hamster kidney cell culture. It is given as two doses and has been administered to more than 150 million people in China without any reports of adverse reactions. To date the vaccine has been used only in Nepal, India and South Korea outside China. In addition, a live vaccine is used to immunize pigs.

A formalin-inactivated TBE vaccine using virus grown in chicken eggs was developed in the 1970s to control central European tick-borne encephalitis virus. The vaccine was improved by transferring manufacture to primary chick embryo fibroblast cell culture. Two doses are given, 2 weeks to 3 months apart, followed by a booster given 9 months to 1 year later. Boosters are recommended every 3 years. The vaccine has proved to be very effective, with few adverse reactions, and has resulted in the near elimination of tick-borne encephalitis in Austria. The effectiveness of the vaccine against Russian spring–summer encephalitis is uncertain.

There are no vaccines available to prevent dengue. However, a number of candidate live vaccines are currently undergoing human trials.

experimental live and killed vaccines have been developed against Rift Valley fever. In addition, experimental deoxyribonucleic acid (DNA)-based vaccines have been developed against several hantaviruses and show promise in preclinical studies.

Bunyaviruses

Although there are no commercially available vaccines against diseases caused by bunyaviruses, a number of

KEY POINTS

- Arboviruses are transmitted biologically by arthropod vectors (mosquitoes, ticks and biting flies).
- There are three major types of clinical disease caused by these viruses: central nervous system, visceral organs/haemorrhagic fever and febrile infections, with many progressing from the latter to the former syndromes. Many arboviruses are highly pathogenic and require a high level of biocontainment.
- There are more than 500 recognized arbovirus species classified in six virus families: Togaviridae, Flaviviridae, Bunyaviridae, Rhabdoviridae, Reoviridae and Orthomyxoviridae.
- In the UK arboviruses are overwhelmingly of concern to travellers. Yellow fever and dengue viruses are two

- widely distributed flaviviruses of global concern, but numerous other agents must be considered in specific locations.
- The *Hantavirus* genus of the Bunyaviridae contains many pathogens associated with haemorrhagic fever with renal syndrome or hantavirus pulmonary syndrome. Hantaviruses are transmitted by rodents, not arthropods.
- Commercial vaccines are available only for yellow fever, Japanese encephalitis and tick-borne encephalitis, but a number of experimental vaccines show promise.
- The only effective antiviral treatment is ribavirin, which has efficacy only against selected bunyaviruses and alphaviruses.

RECOMMENDED READING

Barrett A D T 2001 Japanese encephalitis. In Service M W (ed.) *The Encyclopedia of Arthropod-Transmitted Infections*, pp. 239–246. CAB International, Wallingford, UK

Elliott R M (ed.) 1996 *The Bunyaviridae*. Plenum Press, New York

Karabatsos N 1985 *International Catalogue of Arboviruses Including Certain Other Viruses of Vertebrates*, 3rd edn. American Society for Tropical Medicine and Hygiene, San Antonio

Ksiazek T G, Peters C J, Rollin P E et al 1995 Identification of a new North American hantavirus that causes acute pulmonary insufficiency. *American Journal of Tropical Medicine and Hygiene* 52: 117–123

Monath T P (ed.) 1988 *The Arboviruses: Ecology and Epidemiology*, Vols 1–5. CRC Press, Boca Raton

Weaver S C 2001 Eastern equine encephalitis. In Service M W (ed.) *The Encyclopedia of Arthropod-Transmitted Infections*, pp. 151–159. CAB International, Wallingford, UK

Weaver S C, Barrett A D 2004 Transmission cycles, host range, evolution and emergence of arboviral disease. *Nature Reviews Microbiology* 2: 789–801

Weaver S C, Ferro C, Barrera R, Boshell J, Navarro J C 2001 Venezuelan equine encephalitis. *Annual Reviews of Entomology* 49; 141–174

US Department of Health and Human Services 1999 *Biosafety in Microbiological and Biomedical Laboratories*, 4th edn. US Government Printing Office, Washington, DC

Togavirus and hepacivirus
Rubella; hepatitis C and E viruses

P. Morgan-Capner, P. Simmonds and J. F. Peutherer

RUBELLA

Rubella (German measles) was first described in the eighteenth century and was considered a mild illness with only occasional complications. However, in 1941, an Australian ophthalmologist, Sir Norman Gregg, described the association between maternal rubella in pregnancy and congenital abnormalities in the infant. He noted an increased incidence of congenital cataracts in children. When questioned, the mothers gave a history of having had rubella in early pregnancy when there had been an epidemic in Australia. Since then the importance of maternal rubella for the fetus has been confirmed by many studies.

DESCRIPTION

Rubella virus was first isolated in cell culture in 1962. It is a single-stranded RNA virus with an envelope, and is classified as a togavirus, being the only member of the genus *Rubivirus*. Virions are pleomorphic in appearance and 60–70 nm in diameter with a nucleocapsid of icosahedral symmetry. Rubella is inactivated by many chemical agents. The single-stranded RNA is infective, and replication occurs in the cytoplasm of infected cells. Virions acquire the envelope by budding from cell membranes either into intracellular vesicles or to the exterior. There are three major virion polypeptides: C and the envelope glycoproteins E1 and E2. The virus envelope carries a haemagglutinin (E1), present as 5–6-nm projections, which agglutinates the erythrocytes of 1-day-old chicks, pigeons, sheep and human beings, a characteristic that is utilized in the haemagglutination inhibition test for specific antibodies. Experimentally, infection can be transmitted to rhesus monkeys, rabbits and some other animals, but man is the only naturally infected species. Virus can be isolated in a range of primary and continuous cell lines (e.g. Vero, RK13 and baby hamster kidney-21 cells), but in only some cell lines,

such as RK13, does a cytopathic effect occur. In other cell lines the presence of virus must be demonstrated by immunofluorescence with specific antibody or by resistance to superinfection with another unrelated virus such as echovirus 11. Only one antigenic type of rubella virus is recognized, although minor differences can be found between strains.

CLINICAL FEATURES

Postnatal rubella

The incubation period for postnatal primary rubella is 12–21 days, with an average of 16–17 days. Virus may be excreted in the throat for up to a week before and after the rash, and this covers the period of infectivity. Infection has several characteristic clinical features.

- A macular rash (Fig. 52.1), which usually appears first on the face and then spreads to the trunk and limbs. Particularly in childhood, the rash may be fleeting and perhaps 50% of infections in children are asymptomatic. In adults, asymptomatic rubella is less common.
- General features such as minor pyrexia, malaise and lymphadenopathy also occur, with the suboccipital nodes being those most commonly enlarged and tender.
- Arthralgia is uncommon in children but may occur in up to 60% of women. The joints commonly involved are the fingers, wrists, ankles and knees, and, although arthralgia usually lasts for only a few days, it may occasionally persist for some months.
- Encephalitis and thrombocytopenia are rare complications of rubella and recovery is usually complete.

Rubella presents little danger to the immunocompromised patient, in whom the clinical features are similar to those seen in normal individuals.

Rubella reinfection is diagnosed when an antibody response is demonstrated in someone who has had natural

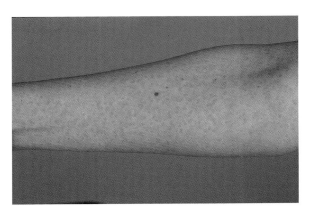

Fig. 52.1 Rubella rash.

rubella or has been immunized successfully. Such reinfections are seldom clinically apparent and are usually found when someone is investigated after contact with rubella. It is very uncommon for fetal infection and damage to occur in subclinical re-infection. It can be difficult, however, to distinguish serologically between subclinical primary rubella, which is of major risk to the fetus, and re-infection. The rare re-infection that is clinically apparent must be assumed to present a risk to the fetus similar to that of primary rubella.

Rubella is notoriously difficult to diagnose clinically, as other virus infections, such as some enteroviruses and human parvovirus B19, can present with identical clinical features. For correct diagnosis, laboratory investigation is essential; a history of rubella is an unreliable indicator of immunity unless it has been confirmed serologically.

Congenital rubella

If the fetus is infected during a primary maternal infection, a wide spectrum of abnormalities may occur. The classical congenital rubella syndrome (CRS) triad consists of abnormalities of the eyes, ears and heart:

1. Abnormalities of the eyes, which may be bilateral or unilateral, include cataracts, micro-ophthalmia, glaucoma and pigmentary retinopathy, which may result in blindness.
2. Bilateral or unilateral sensorineural deafness may be present at birth, although it may not be detected until later in life; it may increase in severity as the child gets older.
3. There are many possible heart defects, with patent ductus arteriosus, pulmonary artery and valvular stenosis, and ventricular septal defect being most common.

The baby often has a low birth-weight due to intra-uterine growth retardation. A purpuric rash due to

thrombocytopenia may be present at birth but this usually resolves, as does hepatosplenomegaly. Microcephaly, psychomotor retardation and behavioural disorders are manifestations of central nervous system involvement. Rarely, a persistent infection of the central nervous system occurs called *progressive rubella subacute panencephalitis*, which is similar clinically to the *subacute sclerosing panencephalitis* due to measles virus. Other problems that may not be present at birth but that may present later in life include pneumonitis, diabetes mellitus, growth hormone deficiency and abnormalities of thyroid function.

The fetus may be so severely affected that intra-uterine death with abortion or stillbirth occurs. Many infected babies, however, are born with no abnormalities, and the risk to the fetus depends on the gestational period in which primary rubella occurs.

- In the first trimester the risk is considerable, as more than 70% of babies will be affected and have some or many of the abnormalities described above.
- In the fourth month of pregnancy the risk reduces to approximately 20%, and the only abnormality likely to be seen is sensorineural deafness.
- After the 16th week of pregnancy, although fetal infection still occurs, congenital abnormalities are very infrequent and no more likely to occur than in an apparently uncomplicated pregnancy.
- Before conception the fetus is unlikely to be harmed.

PATHOGENESIS

Rubella virus is transmitted by the air-borne route. Infection is established in the upper respiratory tract; towards the end of the incubation period, a viraemia occurs and seeds the target organs such as the skin and joints. Most of the clinical features are probably a consequence of the host's immune response to the virus; for example, virus can be demonstrated not only in individual lesions of the macular rash but also in unaffected skin. During the viraemia the virus is able to infect the placenta and cross this barrier to infect the differentiating cells of the fetus. If such fetal infection occurs in early pregnancy, a persistent infection is likely.

The congenital abnormalities arise from a number of effects of the virus infection. Cell division is slowed, cell differentiation is disordered, and damage to small blood vessels may occur. Such effects may lead to the abnormalities seen at birth, but the persistence of infection may result in the clinical problems presenting later in life, either due to direct damage, such as late-onset deafness, or because of immunopathological mechanisms, such as pneumonitis. Virus can persist for many years in babies

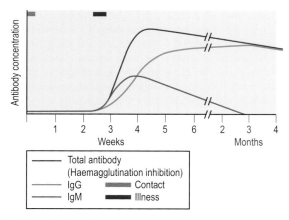

Fig. 52.2 Serological response in primary rubella.

infected during early gestation, but persistent infection is rare in later pregnancy, when the babies are not damaged.

DIAGNOSIS

Postnatal rubella

Laboratory investigation is required if a diagnosis of rubella has important implications, such as in the pregnant patient. As subclinical rubella may occur, pregnant women should also be investigated if they are exposed to someone with possible rubella. Investigation by virus isolation is not indicated as it is unreliable and time consuming. Serological diagnosis is the method of choice, using techniques to detect total rubella antibody or rubella-specific immunoglobulin (Ig) G, and rubella-specific IgM. In primary rubella, specific antibody becomes detectable about the time of the rash, although there may be a delay of 7–10 days, and rapidly increases in concentration (Fig. 52.2). Specific IgM usually, but not always, precedes specific IgG by 1–2 days but, unlike specific IgG, which persists for life, specific IgM is usually detectable for only 1–3 months. Thus, seroconversion may be demonstrated if a serum is obtained soon after contact, prior to the illness, and is examined in parallel with a sample collected at or soon after development of the rash; primary infection is confirmed by the development of specific IgM. If serum is collected after the rash or more than 12–14 days after contact, any total rubella antibody or specific IgG may have resulted from recent infection or infection many years previously. Tests for specific IgM must be used to determine whether the infection was recent.

Tests for total rubella antibody or rubella-specific IgG are also used for serological screening to ascertain susceptibility and whether rubella immunization is indicated. Many assays of appropriate sensitivity and specificity are available. False-positive results must be avoided as a susceptible woman would not be immunized and may not be investigated if contact or illness occurs.

Congenital rubella

The majority of babies with congenital rubella syndrome excrete large amounts of virus for the first few months of life and, indeed, are highly infectious for their attendants. Thus, virus isolation in cell culture from throat swab or urine is indicated, although many laboratories may not have appropriate cell lines readily available. A more sensitive technique in the first 3–6 months of life is serological testing for specific IgM. Maternal IgM does not cross the placenta, so the detection of specific IgM is diagnostic of intra-uterine infection. As specific IgG crosses the placenta from the mother, its detection in the early months of life is of no help in diagnosis. Maternal IgG has a half-life in the infant of 3–4 weeks, however, so persistence to the age of 9–12 months is diagnostic of congenital rubella. It is uncommon for the specific IgM to persist to 1 year of age. Occasionally, if an older child presents with deafness, a diagnosis of possible congenital rubella may be suspected. It is impossible, however, to discriminate specific IgG resulting from intra-uterine infection from that persisting after postnatal infection or immunization, so the reliable diagnosis of congenital infection in older children is impossible.

EPIDEMIOLOGY

Rubella has a worldwide distribution, and infection is endemic in all countries that have not had a highly successful infant immunization policy. Outbreaks usually occur in spring and early summer, with major epidemics occurring every 4–8 years. Infection is common in childhood. Currently, only 2–3% of young adult females are susceptible in the UK due to immunization; many of those who are susceptible come from areas of the world, such as the Indian subcontinent, where immunization is uncommon.

CONTROL

Passive prophylaxis

There is little evidence that administering normal human immunoglobulin after contact reduces the risk of maternal rubella and fetal infection, although it may attenuate the illness.

Active prophylaxis

Attenuated live rubella virus vaccines have been widely available since the early 1970s. They are safe, although occasional fleeting rashes and arthralgia occur. Seroconversion occurs in more than 95% of susceptible vaccinees, and protection persists for more than 20 years. Although vaccine virus may be isolated from the throat of vaccinees, there is no evidence of transmission to susceptible contacts. As it is a live virus vaccine, administration in pregnancy is contra-indicated, and pregnancy should be avoided for the month following immunization. Although the vaccine virus has been shown to infect the fetus, there is no evidence of teratogenicity, so if a susceptible pregnant woman is inadvertently immunized she should be reassured that the risk to her baby is very remote. It is advisable, however, that all women of childbearing age are screened for rubella antibody before immunization so that only susceptible women are offered vaccine.

The objective of rubella immunization is to eradicate congenital rubella. Two possible approaches are possible. In the UK it was decided in the early 1970s to target the vaccine at girls aged 11–14 years and susceptible adult women, identified by screening all pregnant women and women attending such facilities as family planning clinics, occupational health services and on the initiative of general practitioners. Such a policy was adopted because of uncertainties about the duration of protection and the belief that women would have better protection from having had natural rubella; 50% of adolescent girls would have had natural rubella. It was also thought that vaccine-induced protection would be boosted by occasional exposure to natural rubella. It became apparent in the late 1980s that such a policy would never achieve eradication as 10 to 20 cases of congenital rubella and 100 to 200 terminations of pregnancy still occurred each year in the UK. In the second approach, which was adopted in 1988, all children were immunized in an attempt to eradicate rubella from the community and so avoid exposure of any susceptible pregnant woman. Thus, the immunization schedule was augmented by offering rubella vaccine together with mumps and measles vaccine (MMR vaccine) to all children at 15 months of age. For the first few years, MMR vaccine was also offered to children at school entry as a 'catch-up' programme. In the late 1990s, a second dose of MMR vaccine was introduced at 3–5 years of age to 'boost' immunity in those who had received one dose, and to reach some of the children who had missed a first dose. This strategy has now almost eliminated congenital rubella in the UK, as it has done in the USA. To achieve and maintain such success, however, there must be at least a 90% uptake of vaccine in infancy. Concern in recent years in the UK about the safety of MMR vaccine has led to a fall in immunization rates and may lead to a resurgence of rubella.

HEPATITIS C VIRUS

Hepatitis C virus (HCV), discovered in 1989, is the cause of post-transfusion non-A, non-B hepatitis. Development of serological screening assays for HCV infection has virtually eliminated the transmission of HCV by blood transfusion, and indeed of transfusion-associated hepatitis.

Infection with HCV is widespread throughout the world, and is particularly associated with risk groups for parenteral (blood-borne) exposure. Among these, drug users sharing needles are numerically the most significant in western countries. HCV infection is frequently persistent, and leads to the development of significant liver disease, such as cirrhosis and hepatocellular carcinoma, only after a long asymptomatic carrier phase. HCV is currently the subject of intensive efforts to develop effective antiviral treatment for chronic infection, and protective vaccines.

PROPERTIES

Structure

HCV is a small, enveloped virus with a single-stranded RNA genome of positive (coding) polarity (Fig. 52.3a). HCV has been visualized in the plasma of HCV-infected individuals as small (50 nm) round particles. The surface of the virus particle contains a number of small surface projections thought to be formed from complexes of the virally encoded envelope glycoproteins, E1 and E2. The RNA genome is approximately 9400 bases in length, of which over 98% contains protein-coding sequence. In common with many other small RNA viruses, the gene sequences of HCV are translated in a single block to produce a large (more than 3000 amino acids) polyprotein. During and after translation, proteases cleave this precursor into a total of nine mature proteins, which are involved in virus replication (NS2–NS5B) or form structural components of virus particle (core, E1 and E2).

Replication

Information on the replication of HCV is limited because it is problematic to culture the virus in vitro. However, investigations of individual HCV-encoded proteins and comparison with related viruses have allowed the functions of most to be ascertained (Fig. 52.3b).

The principal targets of HCV replication in vivo are hepatocytes, although there is evidence that it may also be able to replicate in haemopoietic cells such as B lymphocytes, and stem cells in bone marrow. Entry of HCV into cells has been shown to be dependent on expression of CD81, but other factors are additionally required. After entry, the RNA genome is translated, a process that generates the RNA replicating and unwinding enzymes such as NS5B, NS5A and NS3 required for genome replication. The genomic RNA is transcribed into a full-length negative-

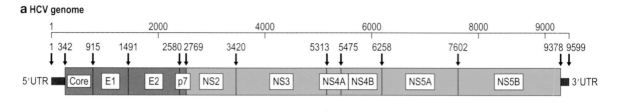

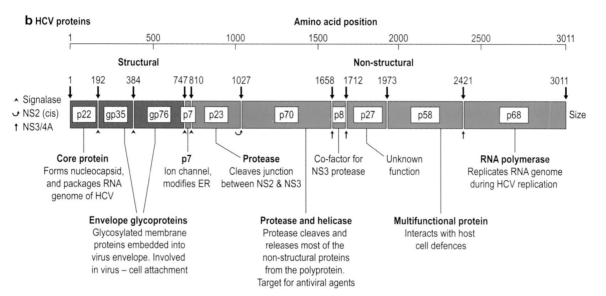

Fig. 52.3 Organization of HCV genome, showing a the structural (core, E1 and E2) and non-structural (NS2–NS5B) genes, and b the proteins produced from them by proteolytic cleavage of the translated polyprotein. The properties and functions, where known, of the HCV proteins are summarized below. ER, endoplasmic reticulum; UTR, untranslated terminal region.

polarity copy, which in turn acts as the template for the production of multiple positive-stranded RNA copies. These can be used for further rounds of transcription, or for the production of further viral proteins. Factors that regulate negative- and positive-strand synthesis, and the use of the latter transcripts for translation, remain undetermined and are likely to be complex.

Translation of the HCV polyprotein takes place in membrane compartments associated with the cellular endoplasmic reticulum. Although the core protein remains within the cytoplasm after cleavage from E1, E1 and E2 are embedded in the endoplasmic reticulum membrane, and their extracellular domains are glycosylated. Understanding of the subsequent stages awaits further studies.

Classification

The *Flaviviridae* are currently divided into three named genera (see Ch. 51); the *Hepacivirus* genus contains HCV and GB virus B, originally isolated from a captive tamarind, in which it causes acute hepatitis and liver

disease similar to that of HCV in man. It is currently under evaluation as a possible experimental model for HCV vaccines and antiviral treatment.

GB virus C (or hepatitis G virus) is in the currently unnamed fourth genus of the *Flaviviridae*. This virus is widely distributed in man and infection appears to be entirely asymptomatic, despite persistence of the virus in a significant proportion of those it infects. Its genome shows a number of similarities to that of HCV, but it lacks a protein corresponding to the core protein of HCV that forms the nucleocapsid. Viruses similar to GB virus C are widely distributed in a range of non-human primate species (GB virus A).

Virus stability

HCV is inactivated by exposure to chloroform, ether and other organic solvents, and by detergents. The effectiveness of a number of virus-inactivating procedures has been demonstrated by studies of the infectivity of products manufactured from plasma, such as the factor

VIII and IX concentrates used to treat haemophiliacs. Thus, the infectivity HCV can be destroyed by:

- dry heat treatment at 80°C
- wet heat treatment at 60°C
- organic solvents (*n*-heptane)
- detergents.

DIAGNOSIS

Chronic infection with HCV is associated with the presence of both plasma viraemia and antibody to HCV. Methods to detect antibody and HCV directly have been used for diagnosis.

Serological diagnosis

Tests for antibody are used for screening and diagnostic testing. Antibody tests for HCV are based upon cloned HCV RNA sequences of HCV genotype 1a, originally derived from an experimentally infected chimpanzee. Recombinant proteins expressed from these clones are the basis for almost all assays for antibody to HCV. Although the original first-generation assay was restricted to antigens expressed from the NS3 and NS4A regions of the genome, subsequent assays have incorporated additional antigens from the core, NS3 and NS5 regions to produce assays of greater sensitivity and specificity for antibody to HCV. The incorporation of antigens from the core and NS3 regions in these second- and third-generation assays gives a greater sensitivity for antibody elicited by infection with non-genotype 1 infections.

Testing for antibody to HCV is performed as a two-stage procedure. An enzyme-linked immunosorbent assay (ELISA) format is used for initial testing of serum or plasma from patients or blood donors. Repeatedly reactive samples are then retested by a second ELISA in a different format or by a supplementary assay such as the Ortho Recombinant Immunoblot Assay, which contains several separate HCV antigens. Currently used serological tests for anti-HCV are now highly sensitive and specific in most patient groups, although individuals who are immunosuppressed, such as those co-infected with HIV, renal dialysis or transplant patients, and those with congenital immunodeficiencies can produce false-negative serological test results. In these cases, direct detection methods for HCV, such as the polymerase chain reaction (PCR) to detect viral RNA, should be considered.

Direct detection methods

Current serological tests can detect the vast majority of chronic, established infections, but there remains a considerable window period in HCV infection between exposure and development of antibody detectable by the best current ELISAs (Fig. 52.4). For this reason, further direct detection methods for the detection of HCV antigens or RNA sequences are required for the effective diagnosis of HCV infection in acute hepatitis, or for the diagnosis of HCV in immunosuppressed individuals. More recently, genome detection methods have been adopted in addition to serological tests for the routine screening of blood donors for HCV in most western countries.

Direct detection methods are based on the detection of HCV RNA sequences by PCR or other nucleic acid amplification methods, or of viral antigens by ELISA. The most commonly used method for direct detection of HCV is the PCR. Diagnostic PCR methods are commercially available; they are capable of high sensitivity and specificity for the detection of HCV RNA sequences in plasma (or liver biopsy) specimens. Alternative, non-PCR-based methods, such as transcript-mediated amplification, have recently been developed and provide comparable sensitivity to PCR. An alternative to nucleic acid detection is the use of ELISA-based methods for the detection of HCV core protein in plasma, and more recently combined antibody–antigen detection methods.

Direct virus detection is the only reliable method of following the effect of treatment with antiviral drugs in patients with chronic infection, as normalization of biochemical liver function tests is not always associated with virus clearance. Similarly, chronic HCV infection with associated disease is known in patients with normal liver function test results. Quantification of the virus genome titre is now increasingly possible with commercial tests based on PCR, other amplification methods or hybridization, and has been used as a predictor for response to antiviral therapy (see Treatment).

EPIDEMIOLOGY

Genetic variation

Nucleotide sequences of HCV frequently show substantial differences from one another. This has led to the current genotypic classification of HCV, in which variants from various geographical locations can be classified into six main genotypes and a number of subtypes (Fig. 52.5). Genotypes show approximately 30% sequence divergence from each other, differences that greatly modify their antigenic properties and their biology (see Treatment).

Some genotypes of HCV (types 1a, 2a and 2b) show a broad worldwide distribution, whereas others, such as

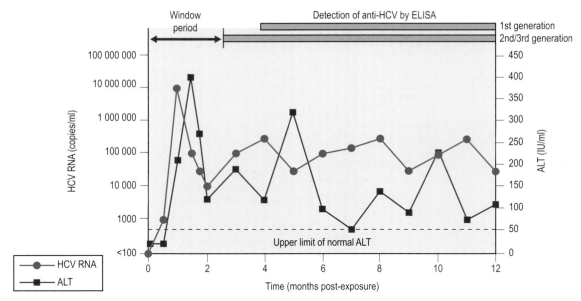

Fig. 52.4 Virological and biochemical markers of acute HCV infection. HCV RNA and abnormal alanine aminotransferase (ALT) levels appear approximately 50 days after exposure to HCV in a typical individual. The subsequent development of chronic hepatitis is indicated by persistent viraemia and fluctuating abnormal levels of ALT. Antibody to HCV first appears after the onset of acute hepatitis, in this example leading to a 'window period' of around 100–150 days for the first-generation serological assay (containing only NS3 and NS4 proteins) and approximately 60–80 days for second- and third-generation assays (containing additional NS3, (NS5) and core proteins).

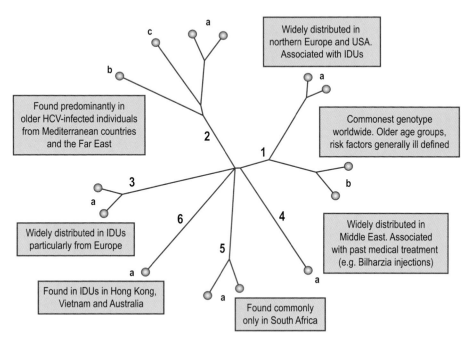

Fig. 52.5 Comparison of complete genome sequences of the common genotypes of HCV plotted as a phylogenetic tree, showing the six main genetic groups (genotypes 1–6) and a number of the commoner subtypes (a–c). The distribution of the different genotypes in the principal risk groups for HCV infection is indicated. IDU, injecting drug user.

types 5a and 6a, are found only in specific geographical regions (South Africa and South-East Asia, respectively). HCV infection in blood donors and patients with chronic hepatitis from countries in western Europe and the USA frequently involves genotypes 1a, 1b, 2a, 2b and 3a. The relative frequencies of each may vary geographically, such as the trend for more frequent infection with type 1b in southern and eastern Europe, and the association of genotype 1a and 3a infection with infection through drug use. HCV genotypes can be identified by analysis of sequences from the 5′-untranslated terminal region (UTR) or from coding regions. Methods for rapid genotyping of HCV have been developed and play a role in the pretreatment assessment of patients receiving antiviral treatment (see Treatment).

Transmission

In western Europe, Australia and North America, most HCV-infected patients have a history of parenteral exposure to the virus, and the majority are (or have been) injecting drug users. The seroconversion rate of injecting drug users has been estimated at 20% per year, and so long-term drug users are almost invariably HCV infected. Drug use was uncommon before the 1960s and so drug users tend to be younger than patients infected through other routes such as transfusion. Most drug users have an asymptomatic infection with no history of jaundice but have chronic hepatitis; few have overt clinical signs or symptoms of liver disease or liver failure. Other blood-borne routes of HCV transmission include:

- blood transfusion before 1991 (when universal donor screening was initiated)
- recipients of pooled plasma products, such as factor VIII, and immunoglobulin manufactured before 1986 (when virus inactivation procedures for plasma-derived blood products became widely used)
- transplant recipients
- haemodialysis patients
- health-care workers from needlestick injuries
- tattooing and acupuncture
- in countries of high prevalence, the use of unsterilized needles for cultural rituals, medical treatment or vaccination programmes.

The lowest frequencies of HCV infection are found in Scandinavian and other northern European countries such as the UK (0.3–0.4%), with slightly higher prevalences in North America (1%) and Australia. Prevalence is intermediate in eastern and southern European countries, even higher in Japan, and most prevalent in the Middle East; frequencies of HCV infection of up 30% have been recorded in areas of Egypt. In this latter case, bilharzia treatment using re-usable and unsterile needles in the 1960s has been identified as the main source of infection.

There is little evidence for non-parenteral transmission of HCV, for example by sexual contact. The frequency of mother-to-child transmission of HCV is between 3% and 10% in most studies. Transmission generally occurs at birth, presumably through contact with blood, with some evidence that elective caesarean section may prevent transmission.

CLINICAL FEATURES

HCV infection causes an indolent and slowly progressive liver disease that is asymptomatic until the development of decompensated liver disease, and often liver cancer.

Acute hepatitis

Exposure to HCV usually results in an asymptomatic infection without jaundice, followed, in most cases, by a chronic carrier state. There is an interval of around 8 weeks to the development of abnormal liver function test results (such as alanine aminotransferase [ALT] levels), although viraemia can be detected earlier (see Fig. 52.4). Clinically, hepatitis caused by HCV is indistinguishable from that caused by other hepatitis viruses; jaundice may develop, but more usually symptoms are non-specific (e.g. fatigue, anorexia and nausea). Viraemia can be detected in the early stages of acute hepatitis, appearing at the same time or slightly earlier than abnormal ALT levels, whereas seroconversion for antibody may be delayed for several weeks or months after the onset of hepatitis. Histological features of acute HCV are similar to those associated with acute hepatitis A and B virus infections; liver biopsy is rarely indicated to make a diagnosis of acute HCV infection.

Chronic hepatitis

The frequency of chronic infection following exposure to HCV is approximately 50%. Persistent infection with HCV is generally associated with persistent and progressive hepatitis, with fluctuating or continuously abnormal ALT levels. Although viraemia is invariably detected in patients with chronic hepatitis, there is no correlation between the level of viraemia and the severity of liver disease, ALT levels, or other biochemical abnormalities associated with hepatitis. HCV infection causes a range of histological changes in the liver, although few permit a specific diagnosis of HCV infection to be made. These include:

- lymphoid follicles within the portal tracts
- a dense periportal inflammatory process
- bile duct damage, lobular hepatitis, with lymphocyte infiltration within sinusoids surrounding the hepatocytes.

Liver histology in HCV infection is commonly classified as chronic persistent and chronic active hepatitis with or without cirrhosis, although more informative scoring systems such as the Knodell score have also been devised.

The percentage of chronically infected individuals who develop cirrhosis and liver failure can only be estimated approximately (Fig. 52.6). When chronic hepatitis does progress to clinically significant

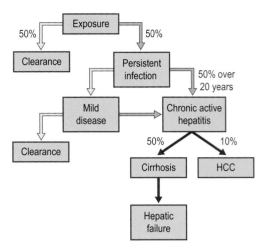

Fig. 52.6 Natural course of HCV infection, showing the approximate frequencies and time-course of persistence, and of progression to clinically significant disease. HCC, hepatocellular carcinoma.

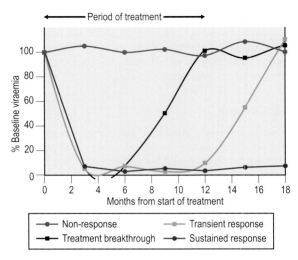

Fig. 52.7 Outcomes of α-interferon therapy in chronic hepatitis C. Non-responders are patients whose RNA levels remain unaltered during or after cessation of treatment. Among those initially responding to α-interferon and showing virus clearance, a proportion will relapse during therapy (treatment breakthrough) and some will relapse once therapy has stopped (transient response). The desired outcome is a complete response, where individuals remain non-viraemic for 12 months or longer after the cessation of treatment. HCV RNA levels parallel replication of HCV in the liver and the resulting inflammatory damage. Levels of the liver enzyme ALT and other transaminases normalize along with virus clearance.

liver damage, progression is almost invariably very slow, although faster progression may be observed in the immunosuppressed. Particularly aggressive HCV-associated liver disease has been observed in immunosuppressed organ transplant recipients, and in patients with inherited immunodeficiency states. Cirrhosis is rarely observed within 10 years of infection, and only 20% of infected patients have cirrhosis after 20 years of follow-up. Cirrhosis may be complicated by liver failure (decompensated cirrhosis), manifested as jaundice, portal hypertension and variceal bleeding; these manifestations of liver failure are shared with other forms of cirrhosis. Hepatocellular carcinoma frequently complicates chronic hepatitis C, although it is rare within 15 years of initial infection. In many western countries, such as Spain and Italy, and in Japan, HCV infection is found in 60–90% of patients with hepatocellular carcinoma.

Extrahepatic manifestations

In a minority of infected patients, HCV may be responsible for extrahepatic clinical manifestations and disease. These include certain types of vasculitis and glomerulonephritis caused by immune complex deposition. Associations between HCV infection and Sjögren's syndrome, essential mixed cryoglobulinaemia and membranoproliferative glomerulonephritis type 1 have been suggested.

TREATMENT AND CONTROL

α-Interferon combined with ribavirin is the standard treatment for chronic hepatitis associated with HCV

infection. Typically, 6 megaunits three times a week, and ribavirin are used for 6–12 months. Tests for HCV RNA, serum transaminase and ALT levels serve as surrogate markers of response to treatment. Three patterns of response are observed (Fig. 52.7). Depending on genotype (see below), approximately 20–40% of treated patients have persistently raised ALT levels and remain RNA positive despite treatment, whereas the remainder respond with clearance of viraemia and normalization of ALT. A proportion of this latter group relapse during therapy or when it is withdrawn. The treatment objective is a sustained response with persistently normal ALT levels and RNA clearance after treatment is concluded; this is achieved in 50–80% of treated patients. To improve response rates, and to avoid the unnecessary treatment of potential non-responders, clinical and virological features associated with a sustained response have been defined.

A variety of factors that correlate with severity of liver disease are predictive of the response to antiviral treatment. Historically, much information related to interferon monotherapy, although it is likely that most factors will be similar for interferon and ribavirin combined therapy. Factors that have been shown to influence response to interferon treatment include:

• patient age
• duration of infection

- presence of cirrhosis before treatment; a sustained response is seldom observed for patients with cirrhosis and portal hypertension
- HCV genotype
- Pre-treatment level of circulating viral RNA in plasma.

A consistent finding is the greatly increased rate of long-term response in patients infected with genotypes 2a, 2b and 3a, compared with type 1. Using standard combination therapy, sustained clearance of viraemia is typically observed in 50% of patients infected with type 1, and in 80–85% of those with type 2 and 3 infections. Prolonging combination treatment to 12 months significantly improves the frequency of response in those infected with type 1, but not non-type 1, genotypes. The mechanism by which different genotypes might differ in responsiveness to treatment remains obscure, particularly as it is not clear whether interferon has a directly antiviral action, or whether it acts as an immunomodulatory agent and enhances immune responses to HCV infecting cells in the liver. It is unlikely that the greater response rates achieved with types 2 and 3 are simply secondary to differences in disease severity. Pretreatment assessment of genotype allows for more appropriate patient selection and is widely used to calculate the dose and duration needed to obtain a sustained clearance of viraemia.

In the future, a new generation of drugs with specific antiviral actions against HCV are likely to be incorporated into or replace interferon treatment. Two classes of antivirals are currently in clinical trials, inhibitors of the viral NS3/4B protease (such as BILN2061) and RNA polymerase. Analogous to the antiretrovirals used for human immunodeficiency virus (HIV) therapy, polymerase inhibitors are either nucleoside analogues that cause chain termination of RNA transcripts, or non-nucleoside protein-binding inhibitors of RNA polymerase. Also analogous to HIV therapy is the rapid development of antiviral resistance to each class of anti-HCV drugs through mutations in the NS3 and NS5B proteins. Effective treatment may depend on the use of multiple drug combinations to suppress virus replication sufficiently to prevent resistance developing. Early evidence suggests major genotype-associated differences in drug efficacy. The protease and non-nucleoside RNA polymerase inhibitors bind with greater affinity to and show greater inhibition of genotype 1 enzymes than those of other genotypes. This is opposite to the genotype-associated differences in interferon response described above. How this problem will be resolved in the future is currently unclear.

Liver transplantation is indicated for patients with decompensated HCV cirrhosis, and for some patients with liver carcinoma complicating HCV infection. Liver transplantation does not cure HCV infection, and reinfection of the graft is probably inevitable.

Prevention

Screening of blood donors has proved effective in preventing the transmission of HCV infection through blood transfusion. A combination of blood donor screening and virus inactivation has virtually eliminated HCV transmission by blood products such as clotting factor concentrates and immunoglobulins.

The main continuing risks for HCV transmission are injecting drug abuse and the use of unsterile needles for medical and dental procedures, tattooing and other percutaneous exposures. Much of this could be prevented by education, greater availability of disposable needles, and for drug abusers by needle exchange programmes. Many of the public health measures adopted to prevent transmission of HIV by parenteral routes will assist efforts at controlling HCV.

Immunization

The development of a vaccine for HCV will be difficult because of viral heterogeneity and the lack of an appropriate animal model. Despite these difficulties, encouraging results have been obtained using recombinant envelope proteins (E1 and E2), which induce a short-lived specific anti-E1 and E2 response in immunized chimpanzees, and transient protection from challenge with the same virus strain. There is, however, little prospect of an effective vaccine for human use in the coming decade.

Infection with HCV is a growing medical problem worldwide. A combination of public health preventive measures, improved diagnosis, screening, antiviral treatment and immunization will undoubtedly all be required to combat its spread in the future.

HEPATITIS E VIRUS

Hepatitis E was first recognized in 1980, although it was difficult to diagnose as tests were not available for the other causes of hepatitis. Since then hepatitis E virus (HEV) has been identified and tests developed.

DESCRIPTION

The virion is non-enveloped, with a genome of single-stranded RNA of positive sense. It is classified as a *Hepevirus*. Virion structure and genome organization resemble those of the caliciviruses; there is also some similarity with rubella and other togaviruses. There is a single serotype with strong cross-reactions among the four known human species (HEV-1, -2, -3 and -4) and also with other unidentified viruses of mammals as well as an avian virus.

CLINICAL FEATURES

The disease has an incubation period of 15–60 days (average 40 days). The clinical features are those of acute hepatitis with anorexia, fever, abdominal pain, nausea and vomiting. The urine darkens and the stools lighten. Liver function tests show hepatocellular damage, and jaundice develops.

Most clinical cases occur in those aged 15–40 years. Infections in younger patients are mostly mild and anicteric.

The virus is acquired by the oral route; virus excretion can be detected about 4 weeks after infection and persists for about 2 weeks. Evidence of liver damage can be found after about 3 months. The overall mortality rate is 1–3%, but can reach 15–25% in the third trimester of pregnancy.

There is no evidence of a chronic carrier state.

LABORATORY DIAGNOSIS

Antibodies to the virus can be measured and show the usual profile of IgM and IgG responses. Reverse transcriptase–PCR can be used to detect viral RNA, although it is not widely available.

EPIDEMIOLOGY AND TRANSMISSION

The virus is acquired orally and outbreaks, often involving many thousands of cases, are linked to faecal contamination of water supplies or food. Unlike hepatitis A, there is little evidence of person-to-person transmission. Several large water-borne outbreaks have been recognized in Africa, India, China and Mexico. Infection is endemic in undeveloped areas of the world, and HEV is the cause of more than 50% of cases of sporadic hepatitis. In the developed world the prevalence of antibody to the virus is less than 2%, and cases are seen in travellers returning from endemic areas.

Antibody studies suggest that several animal species, including primates, pigs, cows, sheep, goats and rodents, may be infected with HEV or related viruses. Strains from swine are related to human strains 3 and 4. The virus may be maintained in endemic areas by person-to-person transmission or in a non-human reservoir.

TREATMENT AND CONTROL

There is no specific therapy. Prevention requires the provision of safe clean water and raising standards of food hygiene. Travellers should take the same precautions as recommended for protection from any faecal–oral infection.

Passive immunization is not available. Studies to develop a vaccine are under way.

KEY POINTS

- Rubella (German measles) is caused by infection with a single-stranded, enveloped RNA virus of a single antigenic type. It is characterized by a widespread macular rash.
- In the first 3 months of pregnancy, rubella can infect the fetus, causing the congenital rubella syndrome in 70% of cases and leading to abnormalities of the eye, ear and heart, as well as a range of other problems including mental handicap and purpura.
- Clinical diagnosis is unreliable. Serological detection of specific IgM is the investigation of choice.
- Rubella is now uncommon in the UK owing to widespread immunization with a live-attenuated vaccine included in the MMR preparation.
- Hepatitis C virus (HCV) is principally transmitted by parenteral routes. The development of serological and PCR-based methods for the diagnosis of HCV infection is of major value in blood donor screening and investigation of clinical hepatitis.
- HCV infection is persistent in a large proportion of those exposed. Significant liver disease develops slowly (20–30 years) and can lead to cirrhosis, end-stage liver failure and hepatocellular carcinoma.
- Chronic HCV infection can be successfully treated in a proportion of individuals by α-interferon–ribavirin combination therapy.
- Hepatitis E virus produces an acute infection prevalent in South-East Asia, north and central Africa, India and Central America, transmitted by faecal contamination, often affecting water supplies and caused by a single-stranded, non-enveloped RNA virus.
- Most cases are recognized in those aged 15–40 years. Large outbreaks occur; a severe form with fulminant hepatitis may be seen in the third trimester of pregnancy.

RECOMMENDED READING

Rubella

Best J M, Banatvala J E 2004 Rubella. In Zuckerman A J, Banatvala J E, Pattison J R et al (eds) *Principles and Practice of Clinical Virology*, 5th edn, pp. 427–457. Wiley, Chichester

Miller E, Cradock-Watson J E, Pollock T M 1982 Consequences of confirmed maternal rubella at successive stages of pregnancy. *Lancet* ii: 781–784

Hepatitis C

Heathcote J, Main J 2005 Treatment of hepatitis C. *Journal of Viral Hepatitis* 12: 223–235

Inchauspe G, Feinstone S 2003 Development of a hepatitis C virus vaccine. *Clinics in Liver Disease* 7: 243–259, xi

Lindenbach B D, Rice C M 2005 Unravelling hepatitis C virus replication from genome to function. *Nature* 436: 933–938 (and other articles in that issue)

Pawlotsky J M 2005 Current and future concepts in hepatitis C therapy. *Seminars in Liver Disease* 25: 72–83

Shepard C W, Finelli L, Alter M J 2005 Global epidemiology of hepatitis C virus infection. *The Lancet Infectious Diseases* 5: 558–567

Internet sites

Health Protection Agency. Rubella (German Measles). http://www.hpa. org.uk/infections/topics_az/rubella/menu.htm

53 Arenaviruses and filoviruses

Lassa, Junin, Machupo, Sabia and Flexal virus haemorrhagic fevers; Marburg and Ebola fevers

C. A. Hart

ARENAVIRUSES

PROPERTIES

Members of the family Arenaviridae have a single-stranded ambisense RNA genome. The genome has two segments, L (large) and S (small), of 7200 and 3400 nucleotides, respectively. The virions are spherical enveloped particles with diameters ranging from 90 to 100 nm on cryoelectron microscopy, enclosing a helical nucleocapsid (Fig. 53.1a). The lipid envelope is derived from the host plasma membrane and T-shaped glycoprotein spikes extend 7–10 nm from its surface. Virions are relatively unstable and infectivity is abolished by ultraviolet or gamma irradiation, heating to 56°C, exposure to detergents or other lipid solvents, and pH outside the range 5.5–8.5.

The virions contain not only virus genome but also host ribosomes (both 28S and 18S ribosomal RNA) which give the virus its characteristic grainy morphology (Fig. 53.1b) and the family name (arena is Latin for sand). The S segment is always more abundant and encodes the nucleoprotein and the two glycoproteins (GP-1 and GP-2). The L segment encodes the viral RNA polymerase, L (at the 3′ end), and a zinc-binding protein, Z (at the 5′ end). For each segment, the two proteins are encoded on non-overlapping reading frames of opposite polarities and the intergenic spaces are predicted to form hairpins.

The human arenaviruses are grouped as New World and Old World viruses (Table 53.1). Lymphocytic choriomeningitis (LCM) virus is a rare cause of human disease. Although classed as an Old World virus, arenavirus has a worldwide distribution as the mouse has moved with man. The phylogeny of the arenaviruses is usually determined by comparison of the nucleoprotein gene sequence portion of the 3′ terminus. This confirms the distinction between Old and New World viruses and shows, for example, that Junin and Machupo viruses are closely related, and that Sabia virus is distinct from all other New World arenaviruses but shares a common

ancestor. The New World arenaviruses are divided into three clades (A–C); Junin, Machupo and Sabia viruses, for example, are in clade B, and Whitewater Arroyo virus is in clade A. Although these viruses have segmented genomes, there is no evidence of reassortment with Guanarito and Pirital viruses in Venezuela.

REPLICATION

Arenaviruses can replicate in a number of mammalian hosts and in most tissues. Growth is restricted in terminally differentiated cells such as lymphocytes or macrophages. LCM virus has been shown to bind by GP-1 and GP-2 to a surface glycoprotein (α-dystroglycan) on rodent fibroblasts but its structure and normal functions are unknown. After the endocytotic vesicles have become acidified, causing conformational changes in GP-1 and GP-2, the viral envelope fuses with the vesicle membrane and the nucleocapsid is released. Viral transcription and replication take place in the cytoplasm but the nucleus may also be involved, perhaps in providing capped cellular messenger RNA (mRNA) for priming arenavirus transcripts. The nucleoprotein and L segment mRNAs can be transcribed from genomic RNA, whereas GP and Z mRNAs are transcribed only from antisense transcripts of the genome. The stem-loop hairpins act to separate genes on both L and S segments, and there are transcription terminations on the distal part of each loop.

The L protein (RNA polymerase, molecular weight 250 kDa) catalyses the production of both transcripts and new genome, and is not in the virion. The Z protein is involved in viral transcription and as a structural protein, perhaps linking the glycoproteins to the nucleocapsid. The nucleoprotein (63 kDa) is the most abundant protein with 1500 to 2000 copies per virion. The surface glycoproteins GP-1 and GP-2 are produced as a precursor protein (GP-C) on the rough endoplasmic reticulum. The protein is glycosylated and cleaved to GP-1 (43 kDa) and GP-2 (35 kDa) in the Golgi apparatus and delivered to the cell

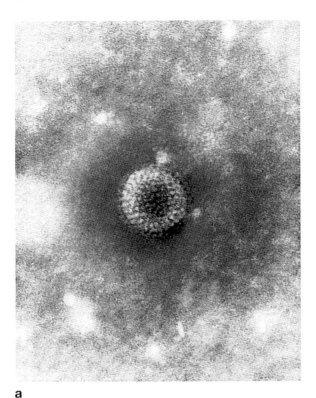
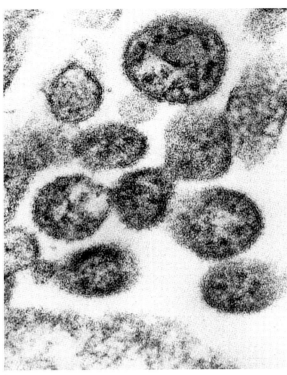

Fig. 53.1 a Lassa fever virus. b Lassa fever virus cross-section showing granular appearance of cell ribosomes.

surface. Together they form the T-shaped spikes and there are about 650 copies of each per virion.

The clade C arenaviruses (Parana, Ampari, Latino) use α-dystroglycan as a receptor, but the receptor for clades A and B is unknown. There is little information on arenavirus assembly and release. Arenaviruses have no matrix protein; the hydrophilic tail of GP-2 and the Z protein interact with the nucleoprotein. The assembled virus, together with host ribosomes, is released by budding. Arenaviruses can establish persistent infection in cell lines.

CLINICAL FEATURES AND PATHOGENESIS

Lymphocytic choriomeningitis virus

Most human infections are acquired by contact with laboratory mice or hamsters. Recent fatal infections have occurred after transplantation of infected lung, liver and kidneys from a woman who died with LCM virus infection. Corneal transplants from the same donor did not result in disease. A pet hamster was the original source of infection.

LCM is rare and may present as:

- an undifferentiated febrile illness
- aseptic meningitis
- encephalitis.

The incubation period is 1–2 weeks and the illness is of short duration. It can vary from a mild disease with headache and fever to neck stiffness, myalgia and photophobia. Long-term effects including persistent headache, paralysis and psychological changes have been described. Although persistent infection and T cell reactivity to epitopes on GP-2 associated with disease manifestation have been described in mice, there is no evidence of either in human infections.

Lassa fever

The incubation period is 1–3 weeks with a gradual onset of fever, headache, and muscle and joint pain. Pharyngitis with a non-productive cough is a common feature. In severe cases there is vomiting, diarrhoea and a raised haematocrit. A few days later the patient becomes increasingly febrile and has abdominal and retrosternal pain. The patient is lethargic with oedema of the face and

Table 53.1 Old and New World arenaviruses

	Year first isolated	Human disease	Rodent host	Distribution
Old World				
Lymphocytic choriomeningitis	1934	Mild to severe meningitis	*Mus musculus*	Worldwide
Lassa	1975	Asymptomatic to severe VHF	*Mastomys natalensis*	West Africa
Ippy	1970	Not known	*Mas. natalensis*	Central African Republic
Mopeia	1977	Not known	*Mas. natalensis*	South Central Africa
Mobala	1983	Not known	*Praomys jacksoni*	Central African Republic
New World				
Junin	1958	Argentinian VHF	*Calomys musculinus, C. laucha, Akodon azari*	Argentina
Machupo	1965	Bolivian VHF	*C. callosus, C. laucha*	Bolivia
'Tacaribe' complex	1963	Not known	*Artibeus bato*	West Indies
Ampari	1966	Not known	*Oryzomys capito, Neacomys guianae*	Brazil
Parana	1970	Not known	*O. buccinatus*	Paraguay, USA
Tamiami	1970	Asymptomatic	*Sigmodon hispidus*	USA
Pichinde	1971	Not known	*O. albigularis, Thomasomys fuscatus*	Columbia
Latino	1973	Not known	*C. callosus, Oryzomys* spp., *Zygodontomys brevicauda*	Bolivia
Flexal	1977	VHF[a]	*Neacomys* spp.	
Guanarito	1989	Venezuelan VHF	*S. alstoni*	Venezuela
Sabia	1994	Brazilian VHF	Not known	Brazil
Pirital	1997	Not known	*S. alstoni*	Venezuela
Whitewater Arroyo	2000	VHF, severe	*Neotoma albigula, N. mexicana*	USA (California)

VHF, viral haemorrhagic fever.
[a]Moderate laboratory-associated infection described.

neck and enlarged lymph nodes. Oedema and bleeding may occur together or independently. Recovery takes 1–3 weeks.

In fatal cases the fever is maintained and rapid deterioration occurs over the first 2 weeks, associated with:

- hypovolaemia, hypotension, pleural effusion, ascites and anuria
- bleeding from the gums, nose, intestine or vagina, linked to platelet dysfunction
- acute neurological changes varying from unilateral or bilateral deafness (in one-third of patients), to signs of encephalopathy including generalized seizures, dystonia and neuropsychiatric changes.

Infection in pregnancy (third trimester) is particularly severe with a 20% mortality rate and an 87% rate of fetal or neonatal loss. In children, infection is associated with a 'swollen baby' syndrome of widespread oedema, abdominal distension and bleeding. Blood platelet and lymphocyte counts fall early in the illness. Endothelial cell damage leads to extravascular fluid loss and thus oedema and, together with platelet loss and dysfunction, to the haemorrhagic manifestations. Lassa virus antibodies are produced but this may not

be associated with viral clearance. Viraemia can persist for weeks and reaches high levels in severe and fatal cases.

South American haemorrhagic fever

Argentinian, Bolivian, Venezuelan and Brazilian haemorrhagic fevers are caused by Junin, Machupo, Guanarito and Sabia viruses, respectively. Flexal and Tacaribe viruses have caused laboratory-acquired haemorrhagic fever, and in 2000 a variant (approximately 89% similar) of Whitewater Arroyo virus caused three fatal cases of viral haemorrhagic fever in California, USA. The incubation is 1–2 weeks and illness begins with a 'flu-like' prodrome. In severe disease petechiae develop and there can be bleeding from the gastrointestinal tract. Leakage of fluid through damaged vascular endothelium leads to hypotension, oliguria and hypovolaemic shock; encephalopathy occurs in many patients. The mortality rate varies between outbreaks in the range of 5–30%. Virus disseminates throughout the body and there is leukopenia and thrombocytopenia. Haemorrhage results from both capillary damage and platelet dysfunction. Raised levels of tumor necrosis factor (TNF)-α and thrombopoietin are proportional to disease severity. There is often bone marrow hypoplasia, perhaps related to low levels of erythropoietin.

DIAGNOSIS

Diagnosis depends upon an initial clinical suspicion of infection and involves obtaining a history of potential contact from travel or contact with infective material. Specific diagnosis depends on detection of the virus, its antigens or genome. A specific immune response may provide retrospective diagnosis or information on the population prevalence of infection.

Isolation

Arenaviruses can be isolated from:

- blood or serum in acutely infected patients, or for up to 2 (Junin, Machupo) to 4 (Lassa) weeks subsequently in severe cases
- throat swabs, breast milk, cerebrospinal fluid, urine or a variety of tissues taken by biopsy or at autopsy.

Cell culture (baby hamster kidney, Vero cells) is the most convenient and efficient mode of isolation, although animals (suckling mice, guinea-pigs, hamsters) have been used. The arenaviruses that cause viral haemorrhagic fever are all in hazard group 4, and culture should not be attempted, except in designated high-security laboratories.

Antigen detection

Antigen detection from blood by immunofluorescence or enzyme-linked immunosorbent assay (ELISA) is useful for early diagnosis and for providing information on prognosis. Early immunofluorescence assay positivity is correlated with risk of death.

Genome detection

Genome detection is by reverse transcriptase–polymerase chain reaction (RT-PCR) amplification with or without confirmation by deoxyribonucleic acid (DNA) hybridization. A strategy involving overlapping primers to amplify the whole of the S segment followed by restriction endonuclease digestion of amplicons has been shown to detect Old and New World arenaviruses, providing both diagnosis and identification of the virus. RT-PCR is highly sensitive and specific, and is particularly valuable for early diagnosis. As little as 0.01 plaque-forming units of Junin virus, for example, can be detected in blood using such a strategy.

Serology

Antibody detection by ELISA is the most useful test and can be adapted to detect a specific immunoglobulin (Ig) M response. In human cases of LCM virus infection, the cerebrospinal fluid shows changes typical of viral meningitis. LCM virus can be grown from cerebrospinal fluid, blood or brain tissue.

TREATMENT

Immunotherapy, by infusion of convalescent plasma, is beneficial, especially if given early in the infection when it can reduce the mortality rate from 16% to 1% in Junin virus haemorrhagic fever. However, the efficacy is related to the concentration of neutralizing antibodies and approximately 10% of those receiving immunotherapy develop a neurological syndrome 4–6 weeks later.

Ribavirin given early in the infection (within 6 days) is effective in both animal models and human infection with Lassa fever virus. There is anecdotal evidence of benefit in treating Junin and Sabia virus infections. Intravenous rather than oral ribavirin should be given.

Fluid, electrolyte and osmotic balance must be maintained and evidence of shock sought, so as to allow rapid resuscitation.

EPIDEMIOLOGY AND TRANSMISSION

All arenavirus infections are zoonotic and, in common with many zoonoses, the animal reservoir host is largely unaffected. In general, arenaviruses persistently infect only two of the rodent families:

1. The *Muridae* (house mice, *Mastomys* and *Praomys*) inhabit the same ecosystem as man.
2. The *Cricetidae* (voles, deer-mice, gerbils) inhabit open grasslands and it is only when they invade human territory, or vice versa, that human infections occur.

Thus the peak period for Junin infections is harvest time in Argentina. Infection in the reservoir host probably occurs in utero or shortly after birth, and becomes persistent with excretion of the virus in saliva and urine. Human infection usually occurs via inoculation into cuts and grazes or inhalation of dust contaminated with urine or saliva from infected rodents. Tacaribe virus is the only virus isolated from outside these two rodent families, being excreted by the fruit-bat *Artibeus*. The reservoir hosts for the different arenaviruses are shown in Table 53.1.

Lassa fever virus is a persistent infection of the multi-mammate rat, *Mastomys natalensis*. The first case to be described was in a nurse who may have become infected from patients in 1969, in Nigeria. Infection also occurred in a laboratory worker who made the first isolation of the virus. Thus, person-to-person transmission and laboratory-acquired infection do occur. Subsequently there have been a number of outbreaks with case-to-case transmission and mortality rates in excess of 40%. It is estimated that Lassa fever causes 5000 deaths each year in West Africa. However, in these endemic areas serosurveys show that infection is widespread and is mild in most cases. Person-to-person transmission is associated with blood and body fluid contact and inoculation injuries rather than by the respiratory route.

Argentinian haemorrhagic fever (Junin) is found in the provinces of Buenos Aires, Cordoba, Santa Fe and La Pampa. It is seasonal (from April to June), but cases or epidemics occur every year, with 100–4000 cases being reported annually. In untreated individuals the mortality rate ranges from 15% to 30%. Nosocomial infections and necropsy and laboratory-acquired infections have been described for many of the New World arenaviruses.

CONTROL

An effective Junin virus vaccine is in use in Argentina. A vaccinia-recombinant Lassa fever vaccine expressing each of the structural proteins has been shown to be highly effective in macaques. Prophylaxis with ribavirin in contacts has also shown some benefit. It is not possible at present to control the natural reservoirs, but this might be a useful approach. Protection against contamination with rodent excreta is important; this should be possible in laboratories to protect animal handlers against LCM virus infection, but is a very difficult task in the endemic areas of Africa and South America. Suspected and confirmed cases must be cared for in strict isolation in designated hospitals to prevent exposure of the attendant staff to the high levels of virus present in acutely ill patients, especially those with Lassa fever.

FILOVIRUSES

PROPERTIES

Marburg and *Ebola* viruses are enveloped viruses with a single-stranded, unsegmented, helical, negative-sense RNA genome. Virions are approximately 80 nm in diameter but vary in length up to 14 000 nm. Marburg is more uniform, with a mean length of 865 nm, but Ebola is more variable (mean 1250 nm). Virions usually appear as long threads (filo is Latin for thread), but they can be U-

Table 53.2 Differences between Marburg and Ebola viruses

	Ebola	Marburg
Antigenic cross-reactivity	No	No
Subtypes	4	1
Glycoprotein (kDa)	140	170
Terminal sialylation of carbohydrates	Yes	No
Secretory GP	Yes	No
Nucleoprotein (kDa)	105	95
Transcriptional editing	Yes	No
Gene overlaps	2–3	1
Mean virion size (nm)	80 × 1250	80 × 865
Peak infectivity virion length (nm)	805	665
Secondary spread common	Yes	No

shaped, branched or shaped like the number 6 (Fig. 53.2). The genome is approximately 19 kilobases long. The non-coding leader (3′) and trailer (5′) ends of the genome are highly conserved among the different filoviruses. There are seven structural genes encoding (from 3′ to 5′), the nucleoprotein, a rhabdovirus phosphoprotein analogue (VP-35), a matrix-like protein (VP-40), a surface glycoprotein, a minor nucleocapsid protein (VP-30), a membrane-associated hydrophobic protein (VP-24) and the viral RNA polymerase (L). Some genes are separated by intergenic regions, but others have short overlaps between them and the next gene. There is one overlap (VP30–VP24) in the Marburg virus genome, and three in Ebola virus (Zaire).

Embedded in the viral envelopes are glycoprotein spikes (homotrimers of the 120–160-kDa glycoprotein in the case of Marburg virus) that protrude 7 nm above the virion surface. Ebola virus also secretes a non-structural glycoprotein which has approximately 300 amino-terminal amino acids in common with glycoprotein but a different carboxy-terminus. The differences between Ebola and Marburg viruses are summarized in Table 53.2.

Human filovirus disease was first described in 1967 in Marburg (Germany) and shortly afterwards, it is presumed as part of the same outbreak, in Frankfurt and Belgrade. In each case infection was related directly or indirectly to blood or tissues of vervet monkeys (*Cercopithecus aethiops*) imported to laboratories from Uganda via London. There have been six subsequent episodes involving one or more cases of Marburg virus infection, the largest in 2005 in Angola with an 84% mortality rate (Table 53.3).

Outbreaks of severe haemorrhagic fever occurred almost simultaneously in Nzara, Sudan (June 1976) and Yambuku, Zaire (August 1976). The filovirus responsible was named Ebola after a small nearby river in Zaire. The two outbreaks were not linked epidemiologically and it was subsequently demonstrated that two different subtypes

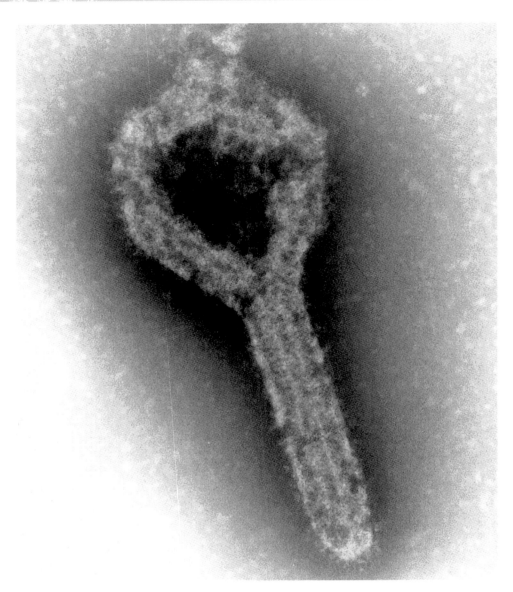

Fig. 53.2 Ebola virus.

Ebola (Sudan) and Ebola (Zaire) were responsible. In 1989, cynomologus macaques imported from the Philippines and held in a quarantine facility in Reston, Virginia, died from a similar haemorrhagic fever. Ebola (Reston) was isolated from the macaques. Four employees of the facility became infected asymptomatically. In 1994 a Swiss zoologist who investigated a troup of chimpanzees that had died from haemorrhagic fever himself developed viral haemorrhagic fever. This was due to a new subtype of Ebola (Côte d'Ivoire). In 1994, two outbreaks of viral haemorrhagic fever occurred in Gabon, one in hunters who had butchered a chimpanzee. The Ebola

virus that was isolated was related to, but distinct from, Ebola (Zaire). Phylogenetic analysis using the entire glycoprotein gene shows that Ebola and Marburg viruses are 72% different and that the four Ebola subtypes (Sudan, Zaire, Reston and Côte d'Ivoire) have 47% nucleotide sequence differences from one another.

REPLICATION

Filoviruses can be grown in a variety of cell lines, including Vero, but culture requires category 4 containment. Although

Table 53.3 Filovirus infections

Place	Year	Virus/Subtype	No. of cases	Mortality rate (%)	Epidemiology
Germany/Yugoslavia	1967	Marburg	32	23	Imported vervet monkeys
Zimbabwe/South Africa	1975	Marburg	3	33	Index infected in Zimbabwe; two secondary cases in companion and nurse
Southern Sudan	1976	Ebola/Sudan	284	53	Origin in cotton factory (bat infested); nosocomial spread
Northern Zaire	1976	Ebola/Zaire	318	88	Unknown origin; nosocomial spread by needlestick, etc.
England	1976	Ebola/Sudan	1	0	Laboratory infection by needlestick
Tandala, Zaire	1977	Ebola/Zaire	1	100	Sporadic case
Southern Sudan	1979	Ebola/Zaire	34	65	Nzara; same site as in 1976
Kenya	1980	Marburg	2	50	Index travelled in area that was source of vervet monkeys in 1967; secondary infection in doctor who survived
Kenya	1987	Marburg	1	100	Traveller to Mount Elgon Cave (bat infested)
USA	1989	Ebola/Reston	4	0	Monkeys imported from Philippines fatally infected; four asymptomatic human infections
Russia	1990	Marburg	1	0	Laboratory acquired
USA	1990	Ebola/Reston	0	0	Imported monkeys from Philippines to same quarantine facilities in Virginia and Texas; no human infections
Italy	1992	Ebola/Reston	0	0	Imported monkeys from Philippines; no human infections
Gabon	1994	Ebola/?Zaire	44	63	Unknown origin, identified retrospectively
Côte d'Ivoire	1994	Ebola/Côte d'Ivoire	1	0	Conducted autopsy on dead chimpanzee
Liberia	1994	Ebola/Côte d'Ivoire	1	0	Serological diagnosis only
Kikwit, Zaire	1995	Ebola/Zaire	317	78	Source unknown; secondary family and nosocomial cases
Gabon	1996	Ebola/?Zaire	37	57	Butchering dead chimpanzee plus family spread
Gabon	1996	Ebola/?Zaire	60	75	Index case a hunter; secondary spread to close contacts
South Africa	1996	Ebola/Zaire	2	50	Doctor treating patients in Gabon flew to South Africa, infected nurse who died
USA	1996	Ebola/Reston	0	0	Imported monkeys from same facility in Philippines
Philippines	1996	Ebola/Reston	0	0	Quarantine contact
Democratic Republic of Congo	1999	Marburg	154	83	Community outbreak lasting more than 2 years
Gulu, Uganda	2000	Ebola/?Zaire	400	40	Community and nosocomially acquired
Democratic Republic of Congo	2005	Ebola/Zaire	12	83	Eating a chimpanzee
Angola	2005	Marburg	401	84	Outbreak in an urban setting

the asialoglycoprotein on hepatocytes is known to be a receptor for Marburg virus, this protein is not expressed on many of the other cells infectable by filoviruses. There is evidence that C-type lectins on monocytes and macrophages are receptors for viral entry into these cells; the mode of entry is unknown. Filovirus replication takes place in the cytoplasm and large inclusions are formed. Mature virus is released as nucleocapsids bud through areas of plasma membrane rich in GP. It is thought that the process is orchestrated by VP-40, the matrix-like protein, and/or VP-24. In general, the replication rate of Ebola (Sudan and Reston) is slower than that of the others, with a less dramatic cytopathic effect.

CLINICAL FEATURES AND PATHOGENESIS

After an incubation period of 4–10 days there is a rapid onset of:

- fever
- malaise
- myalgia
- severe frontal headaches
- bradycardia and conjunctivitis occur early in the disease.

There is rapid progression with:

- nausea
- vomiting
- abdominal pain and diarrhoea
- haematemasis and melaena.

Within 5–7 days of onset there will be frank haemorrhagic manifestations including petechiae, ecchymoses and uncontrollable bleeding from venepuncture sites. A maculopapular rash appears around day 5 and is followed by desquamation. Death from shock usually occurs 6–9 days after onset, with a mortality rate of 50–80% for Ebola virus infection and a slightly lower rate for Marburg virus (see Table 53.3). Infection of pregnant women usually results in abortion with fatal infection of the neonate. Recovery in survivors is slow, with weight loss, prostration and amnesia for the period of acute infection.

The filoviruses replicate in the liver and most other tissues of the body. The cause of death is related to both the haemorrhagic diathesis and increased vascular permeability, which results in fluid and blood loss into the extravascular spaces resulting in shock. The haemorrhagic diathesis is related to endothelial cell damage and probably also to platelet dysfunction. The endothelial damage seems to be mediated by the membrane glycoprotein, since transfection of glycoprotein from Ebola virus (Zaire) but not Ebola virus (Reston) into human endothelial cells causes their complete disruption. During filovirus haemorrhagic fever there is a marked lymphopenia, but a neutorophilia with a predominance of band forms. The secretory glycoprotein (sGP) appears to bind to the Fcγ receptor (CD16b) on neutrophils and to inhibit some neutrophil activities. VP-35 causes decreased secretion of interferon-α. Interestingly, asymptomatic Ebola virus infection in man is linked to a strong inflammatory response characterized by high levels of circulating chemokines and cytokines. Although the glycoprotein is a major target for an immune response, it has an external peptide domain (highly conserved) close to the transmembrane domain that has homology to an immunosuppressive domain in retroviruses. Humoral antibodies are produced within 10–14 days, but are usually non-neutralizing. Persistent infection has been demonstrated in convalescent patients.

DIAGNOSIS

Virus can be isolated from blood by cell culture. Inclusions develop and can be demonstrated by immunofluorescence. Virus can be visualized by electron microscopy, either following culture or directly in blood in the early stages of infection. ELISA techniques are available for the detection of either viral antigen or specific IgM. An RT-PCR test amplifying a conserved region of the L gene has proved to be the most sensitive and specific method for rapid diagnosis.

TREATMENT

Supportive treatment is all that can be offered to patients. Ribavirin has no effect on filovirus replication in vitro. Neither passive immunization nor interferon therapy appears effective. In a mouse model, inhibitors of *S*-adenosyl-L-homocysteine hydrolase were very effective, even in immunosuppressed animals.

EPIDEMIOLOGY AND TRANSMISSION

Both Ebola and Marburg viruses are undoubtedly zoonotic, and disease appears to be more severe in primary than in secondary cases. The nature of the reservoir host is unclear. Although a number of cases have been acquired from chimpanzees and monkeys, these animals are unlikely to be the definitive host, since they also develop severe or fatal infection. Persistent infection in bats has been achieved experimentally and contact with bats was a risk factor for infection in Sudan and Mount Elgon. A recent study has demonstrated the presence of Ebola virus genome and, in one case, nucleocapsid in the organs of small rodents in the Central African Republic, although virus was not grown. Species included *Mus setulosus*, *Praomys* spp., and a shrew *Sylvisorex ollula*. Transmission appears to be via contamination with blood-stained body fluids or tissues. Although air-borne spread has been demonstrated, for example with Ebola virus (Reston) in monkeys, this does not seem to play a major role in human spread. Nosocomial spread is common, particularly affecting nurses and doctors. Transmission by sexual intercourse has been described from one patient 83 days after initial infection, to the man's wife.

CONTROL

In hospital, spread is controlled by patient isolation, use of protective clothing and strict attention to safe disposal of needles, syringes, and blood and other body fluids. Filovirus infectivity is destroyed by:

- heating at 60°C for 30 min
- ultraviolet and gamma irradiation; formalin (1%)
- lipid solvents

- β-propiolactone
- hypochlorite or phenolic disinfectants.

There is no safe and effective vaccine, although a DNA vaccination strategy (glycoprotein genes) has proved highly effective in preventing Ebola virus infection in monkey models, as has a recombinant vesicular stomatitis virus vector expressing Ebola and Marburg virus glycoproteins.

KEY POINTS

- Members of the Arenavirus and Filovirus families can cause a spectrum of illness including viral haemorrhagic fever (e.g. Lassa, Junin and Machupo viruses [Arenaviruses] and Marburg and Ebola viruses [Filoviruses]).
- Viral haemorrhagic fever presents as fever, headache, and muscle and joint pain; pharyngitis; and later nausea, vomiting and diarrhoea. There follows increasing fever, abdominal pain, oedema and enlarged lymph nodes with bleeding from gums, nose, vagina and into the gut, owing to thrombocytopenia.
- Arenaviruses have rodent hosts; the hosts of filoviruses are unknown.
- Close contact, as in hospitals, has resulted in transmission of virus to medical staff.
- Infusion of convalescent plasma and ribavirin may be of some value in treating arenavirus, but not filovirus, infections.

RECOMMENDED READING

Fields B N, Knipe D M, Howley P M (eds) 1996 *Virology*, 3rd edn. Plenum Press, New York

Richman D D, Whitley R J, Hayden F G (eds) 1997 *Clinical Virology*. Churchill Livingstone, New York

54 Reoviruses

Gastro-enteritis

N. A. Cunliffe and O. Nakagomi

The family *Reoviridae* comprises a diverse collection of viruses that infect man, many mammals, other vertebrates (including birds), plants and insects. The genus *Rotavirus* is the most important cause of severe infantile gastro-enteritis throughout the world, and is a recognized aetiological agent of diarrhoea in the young of many animal species. Viruses in the genera *Orbivirus* and *Coltivirus* infect various species of insects and can be transmitted to vertebrates including man, and can thus be described as arboviruses (see Ch. 51).

CLASSIFICATION

Four of the nine genera of the Reoviridae family infect man:

- *Orthoreovirus* – reovirus types 1, 2 and 3
- *Orbivirus* and *Coltivirus* – various serogroups
- *Rotavirus* – seven serogroups (A–G) and multiple serotypes within group A rotavirus.

The mature viruses have icosahedral symmetry and are triple-layered (the double-shelled capsid surrounds a core); they do not possess an envelope and measure 75–80 nm in diameter. The main features are listed in Table 54.1. The RNA genome consists of 10 to 12 segments of double-stranded RNA.

REPLICATION

The virus replicates in the cytoplasm of infected cells. Attachment is via a specific protein component of the outer capsid (e.g. σ1 of orthoreoviruses or viral protein (VP) 4 of rotaviruses) to a cellular receptor (junctional adhesion molecule-A and sialoglycoproteins for orthoreoviruses, and sialoglycoproteins and integrins for rotaviruses). The viral protein–receptor interaction leads to a conformational change in the viral capsid, which enables orthoreoviruses to enter the cell by receptor-mediated endocytosis. The mechanism of cell entry for rotaviruses has not yet been determined. In the cytoplasm the outer capsid is removed, but the viral genomic RNA remains in the inner capsid of the double-layered particle (subviral particle). RNA is transcribed by the virion-associated, RNA-dependent RNA polymerase complex to produce messenger RNA (mRNA) molecules from each genomic RNA segment. The mRNA molecules, which are capped but not polyadenylated, leave the subviral particle through channels penetrating the outer and inner capsids, and are then translated to generate the various viral proteins. They also act as templates for RNA replication, which is completed by the RNA polymerase complex, and packaged within new core structures. Assembly takes place in the cytoplasm, and aggregates of core particles and subviral particles may be detected histologically as inclusion bodies.

Mature, triple-layered virions are formed in the endoplasmic reticulum and released after cell lysis. Viral morphogenesis in the endoplasmic reticulum is unique in rotaviruses. The double-layered particles that are formed in the viroplasm bud into the lumen of the endoplasmic reticulum mediated by non-structural protein 4 (NSP4), and the particles transiently acquire an envelope, which is lost in the final steps of maturation.

ASSOCIATION WITH CLINICAL ILLNESS

The first human isolates of the Reoviridae were recovered from both respiratory secretions and faeces, but could not be associated with disease – hence the name *reo*virus (*r*espiratory *e*nteric *o*rphan). However, orthoreoviruses cause systemic disease (meningitis and encephalitis) in mice, and this virus–host system has been used extensively to study the different steps and mechanisms of viral pathogenesis. Although most of the human population is exposed to and develops antibodies against orthoreoviruses from an early age, there is still no clear link between these viruses and illness.

The orbiviruses include several serogroups that infect man and are transmitted by a variety of insects (ticks, midges

Table 54.1 Classification and features of genera of Reoviridae that infect man

	Orthoreovirus	Rotavirus	Orbivirus	Coltivirus
Size (nm)	80	75	80	80
No. of RNA segments	10	11	10	12
Molecular weight of genome (Da)	15.5×10^6	12.2×10^6	12.7×10^6	$\approx 18 \times 10^6$
Host range	Man, other vertebrates	Man, other vertebrates	Man, other vertebrates, arthropods	Man, other vertebrates, arthropods

and mosquitoes). In the blood the virus is associated with erythrocytes, and thus is hidden from immune responses. Clinically, patients experience a febrile illness, with rashes in 10% and leucopenia in two-thirds of cases. Infections of the central nervous system leading to meningitis or encephalitis are seen in 3–7% of laboratory-confirmed cases.

The coltiviruses are also transmitted by insects (ticks, mosquitoes), and rodents are considered to be the main animal reservoir. Infection is spread to man by insect bite. A febrile illness develops with gastro-intestinal symptoms (20%), rash (10%), and meningitis or encephalitis (3–7%) (see Ch. 51).

Rotaviruses are the main cause of acute gastro-enteritis in infants and young children, as well as in the young of many animal species.

ROTAVIRUSES

The cardinal disease syndrome is acute gastro-enteritis, which is usually mild to moderately severe among children in developed countries but can be very severe and associated with high mortality rates in developing countries. Rotaviruses also cause diarrhoea in the young of a wide variety of birds and mammals including cattle, sheep, goat, horses, pigs, dogs, cats and mice, and also elks, rabbits, monkeys and many others.

Description and classification

Morphology

The virion measures 75 nm in diameter and has characteristic sharp-edged, double-shelled capsids, which in electron micrographs look like spokes grouped around the hub of a wheel (the Latin word, *rota*, means wheel); this appearance is diagnostic (Fig. 54.1). Cryo-electron microscopy has shown that the triple-layered particle is penetrated by 132 large channels and that the virion has 60 spikes on its surface that consist of trimers of VP4 (Fig. 54.2) (see below).

Genome and gene-coding assignments

The 11 segments of double-stranded RNA can be extracted easily from rotaviruses and separated by polyacrylamide gel electrophoresis. This has been used to establish so-called 'electropherotypes' of rotavirus isolates (Fig. 54.3). As well as 'long' and 'short' electropherotypes (differing in the rates of relative migration of RNA segments 10 and 11), various minor differences in the migration of corresponding segments have been recognized and utilized extensively in epidemiological studies (see Fig. 54.3).

Generally, each genome segment codes for only one virus-specific protein (VP). The gene-coding assignments have been established (Fig. 54.4). RNA segments 1, 2 and 3 code for the inner core proteins VP1, VP2 and VP3, respectively; VP2 is the main scaffolding protein. RNA segment 6 codes for the middle layer protein, VP6, which is the single most abundant rotavirus protein and which interacts with the core protein VP2 and the outer capsid proteins VP7 and VP4 (see below). VP6 carries epitopes specifying groups and subgroups. To date, seven different groups (A–G) have been identified; for groups A–E a complete lack of serological cross-reactivity has been shown. Only groups A, B and C have been associated with human illness.

- Group A rotavirus, or *Rotavirus A*, is responsible for the vast majority of human rotavirus infections (mostly children).
- Group B rotavirus, or *Rotavirus B*, has caused outbreaks of gastro-enteritis in adults in China, India and Bangladesh.
- Group C rotavirus, or *Rotavirus C*, causes occasional episodes of gastro-enteritis in children.

The outer capsid (third layer) is formed by two proteins: VP7, a glycoprotein (encoded by RNA 7, 8 or 9, depending on the strain), and VP4 (encoded by RNA 4). Each of these surface proteins carries neutralization-specific epitopes that define the virus serotype. VP4 is cleaved post-translationally into the VP5* and VP8* subunits; proteolytic cleavage is essential for infectivity. Six nonstructural proteins are coded for by RNAs 5 (NSP1), 7, 8 or 9 (NSP2 and NSP3, depending on the strain),

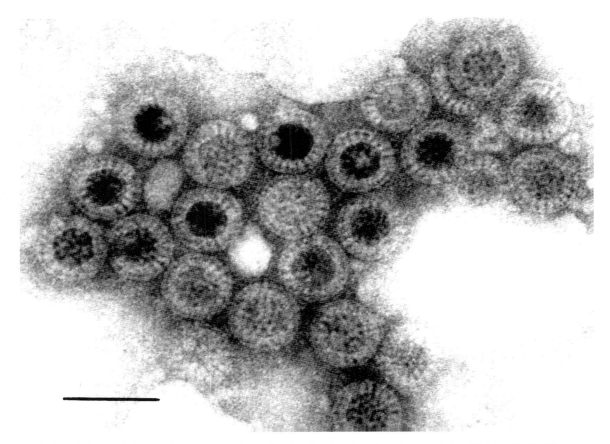

Fig. 54.1 Negatively stained electron micrograph of rotavirus particles in a faecal specimen. Potassium phosphotungstic acid stain. Bar, 100 nm.

10 (NSP4) and 11 (NSP5 + NSP6); they have various functions during replication, mainly in morphogenesis.

Antigenic and genetic diversity

Among group A rotaviruses, two main subgroups (I and II), 15 different VP7-specific serotypes (G types, derived from glycoprotein) and 26 different VP4-specific types (P types, derived from protease-sensitive protein) have been distinguished. Reverse transcriptase–polymerase chain reaction (RT-PCR) with gene- and type-specific primers has been widely applied as a reliable typing procedure. Although the correlation between G serotypes (determined by neutralization assays) and G genotypes (determined by molecular methods) is absolute, not all P types have as yet been confirmed as serotypes, and therefore the P-serotype designation differs from the P-genotype designation (which is included in a square bracket). For example, strain Wa is designated as P1A[8]G1, strain DS-1 as P1B[4]G2, etc. Most, but not all, human group A

rotaviruses are of long electropherotype, subgroup II and serotype P1A[8]G1, P1A[8]G3 or P1A[8]G4, or of short electropherotype, subgroup I and serotype P1B[4]G2.

At the molecular level, several factors have been identified that can explain the genomic and antigenic variability of co-circulating rotavirus strains.

- Like other viruses that depend on virion-associated, RNA-dependent RNA polymerases for their replication, rotavirus genomes undergo frequent point mutations that accumulate over time and give rise to multiple lineages and sublineages (when this occurs in the neutralization proteins, *antigenic drift* will result).
- Rotaviruses, like other segmented RNA viruses, undergo extensive reassortment in doubly infected cells. This has been shown to occur both in vitro and in vivo. If RNA segments coding for serotype-specific proteins are involved, major antigenic changes can result (*antigenic shift*).
- Rotaviruses may be transferred to man from animal species either as whole virions or by reassortment of

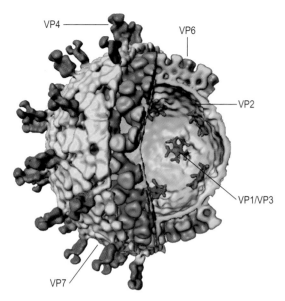

Fig. 54.2 Three-dimensional structure of mature rotavirus particle derived from cryo-electron microscopy images and computer image processing. The cut-away representation of the reconstruction shows the triple-layered icosahedral architecture of the virion and the locations of the various structural proteins. The outer (VP7 and VP4), middle (VP6) and inner (VP2) layers are shown in orange, blue and green, respectively. The flower-like structures inside the VP2 layer represent VP1–VP3 complexes. (Courtesy of B. V. V. Prasad, Baylor College of Medicine, Houston, Texas, USA.)

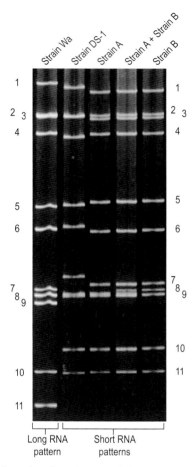

Fig. 54.3 Separation of rotavirus genomic RNA into 11 bands by polyacrylamide gel electrophoresis. Two RNA patterns, long and short, are represented by prototype strains Wa and DS-1, respectively. Strains A and B, both of short electropherotype pattern, differ only in the migration profile of RNA segment 8. This difference is clearly demonstrated by using the technique of co-electrophoresis (both RNAs are run in the same lane).

genome segments. Thus, the genome of human group A rotavirus isolates is closely related to that of animal rotaviruses (e.g. cat, cow, pig), either as a whole or in part.

- Rotaviruses that establish chronic infections in immunodeficient hosts may undergo various forms of genome rearrangement, resulting in highly atypical RNA profiles.

Pathogenesis and immunity

Rotaviruses replicate exclusively in the differentiated epithelial cells at the tips of the small intestinal villi. Progeny virus is produced after 10–12 h, and released in large numbers into the intestinal lumen ready to infect other cells. Biopsies show atrophy of the villi and mononuclear cell infiltrates in the lamina propria. The cellular damage leads to malabsorption of nutrients, electrolytes and water, resulting in diarrhoea with vomiting followed by dehydration. Additional mechanisms that may contribute to the pathogenesis of rotavirus diarrhoea include:

1. stimulation of the enteric nervous system leading to increased paracellular permeability
2. rotavirus NSP4, which functions as a viral enterotoxin.

Infection is followed by a mucosal humoral and cell-mediated immune response, and the virus is normally cleared within 1 week. Rotavirus-specific immunoglobulin (Ig) A antibodies on the enteric mucosal surface are thought to mediate protective immunity. Infection with one serotype provides serotype-specific (homotypic) protection, and repeated infections lead to partial cross-serotype (heterotypic) protection. In the immunodeficient host, however, a persistent infection may occur with severe chronic diarrhoea associated with rotavirus excretion that can last for many months. Extra-intestinal spread of rotaviruses in man has been documented, with occasional reports of infection in the liver and central nervous system. Recently, rotavirus antigen has been detected in acute-phase serum from immunocompetent infants as

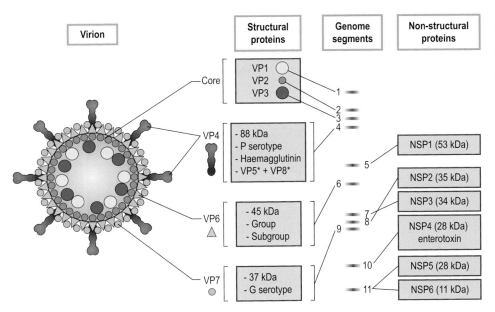

Fig. 54.4 Schematic diagram showing the relationships between the structure of a rotavirus virion and the genomic double-stranded RNA segments.

well as from experimentally infected animals; the clinical significance of this finding is being pursued.

Clinical features

The onset of symptoms is abrupt after a short incubation period of 1–2 days. Fever, vomiting and watery diarrhoea are seen in the majority of infected children, lasting for 2–6 days. If body fluids are not replaced, dehydration ensues that may range in severity from mild to life threatening. Multiple rotavirus infections commonly occur during infancy and early childhood; the first rotavirus infection typically results in the most severe disease outcome, with subsequent infections generally associated with milder disease or even asymptomatic infection. There is little evidence that illness severity is related to virus serotype. Rotavirus infection can cause gastro-enteritis in older children and adults, although severe disease is less common. Outbreaks of gastro-enteritis due to rotavirus infection have been observed in the elderly, in whom severe dehydration can result.

Laboratory diagnosis

At the peak of infection, as many as 10^{11} virus particles per millilitre of faeces are present, and can be detected by a variety of methods. Electron microscopy will easily detect the characteristic virus particles (see Fig. 54.1). In the majority of cases there are sufficient numbers of virions in faeces to allow identification of RNA profiles (see Fig. 54.3). Antigen detection tests, targeted on the

middle-layer protein VP6, include latex agglutination assays, and immunochromatographic assays have been developed that can be used at the bedside. However, enzyme-linked immunosorbent assay is the most sensitive and commonly used method.

Rotaviruses can be propagated in cultures of monkey kidney cells with trypsin incorporated in the culture medium. The diagnosis of rotavirus infection can also be made by demonstration of a four-fold rise in antibody titre with serological assays, complement fixation tests and virus neutralization assays. However, cell culture and serology are research tools.

Epidemiology

Rotavirus infections occur worldwide, but the vast majority of deaths occur in children in developing countries (Fig. 54.5). Thus, an estimated 2 million children under the age of 5 years die from diarrhoeal disease in developing countries each year, and rotavirus accounts for about 40% of these deaths (approximately 702 000 rotavirus-associated deaths per year). In industrialized countries rotavirus accounts for 40–50% of hospital admissions due to diarrhoeal disease. Most symptomatic infections are seen in children between 6 months and 2 years of age in industrialized countries, although in developing countries symptomatic rotavirus infection below 6 months of age is common. By the age of 5 years, virtually all children have been infected with rotavirus. In temperate countries rotavirus infections display marked seasonality, with distinct peaks during the winter months

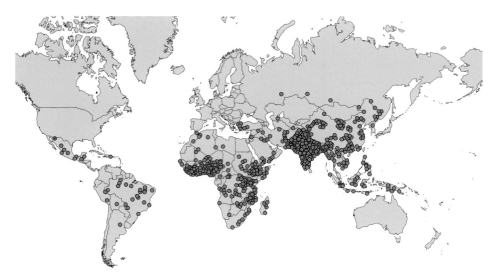

Fig. 54.5 Map showing the global distribution of rotavirus deaths in children aged less than 5 years. Each dot represents 1000 deaths. (Redrawn from Parashar U D, Hummelman E G, Bresee J S, Miller M A, Glass R I 2003 Global illness and deaths caused by rotavirus disease in children. *Emerging Infectious Diseases* 9: 565–572.)

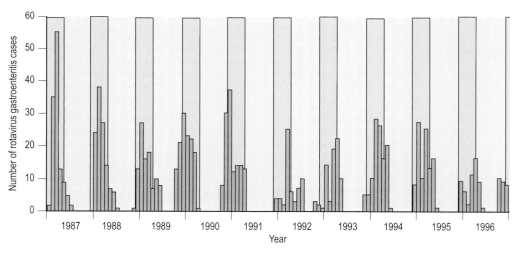

Fig. 54.6 Rotavirus gastro-enteritis in children admitted to hospital in Akita, Japan, 1987–1996. The light blue shaded area represents the winter season (December to March).

and few infections identified outside this period (Fig. 54.6). In contrast, rotavirus infections occur year-round in most tropical countries. Transmission of rotavirus within families to siblings and parents is well recognized. The release of enormous numbers of virions during the acute stage contributes to the easy transmission of the virus. Only a few virus particles are sufficient to cause disease in the susceptible host.

Among group A rotaviruses, four serotype combinations comprise approximately 75% of rotaviruses detected in man, including P[8]G1, P[4]G2, P[8]G3 and P[8]G4 (Fig. 54.7). However, serotype G9 emerged as a globally important serotype around the turn of the century, and tropical regions may harbour rotavirus serotypes that are uncommon in temperate regions (e.g. serotype G5 in Brazil and G8 in Malawi).

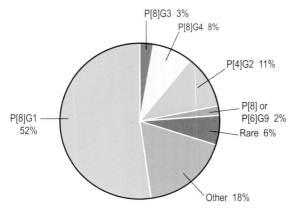

P[8]G3 3%
P[8]G4 8%
P[4]G2 11%
P[8] or P[6]G9 2%
Rare 6%
P[8]G1 52%
Other 18%

Fig. 54.7 Global distribution of combined P and G types, 1994–2003. (Adapted from Gentsch et al 2005.)

Treatment

Although no specific anti-rotaviral treatment is available routinely, probiotic therapy (e.g. with *Lactobacillus* GG) has been shown in clinical trials to shorten the duration of symptoms of gastro-enteritis. In selected clinical situations, anti-rotaviral immunoglobulin therapy has been used as prophylaxis against, and as treatment of, rotavirus gastro-enteritis. However, the mainstay of therapy consists of oral rehydration with fluids of specified electrolyte and glucose composition (Table 30.1, p. 312). Intravenous rehydration therapy is reserved for patients with severe dehydration, shock or reduced level of consciousness.

Control

Attention to hygienic measures such as handwashing, safe disposal of faeces and disinfection of contaminated surfaces is essential in reducing the risk of transmission. However, universal vaccination of infants is the most important preventive strategy. The rotavirus vaccines that have been developed are live-attenuated, orally administerable strains. Attenuation of virulence has been achieved either by repeated passage in cell culture or by substitution through genetic reassortment of serotype-determining gene segment(s) of a human rotavirus into the backbone of an animal rotavirus (which is both naturally attenuated to man and attenuated through repeated cell culture passage).

The first licensed rotavirus vaccine, a Rhesus monkey rotavirus-based tetravalent human reassortant vaccine (RotaShield), was withdrawn after this live oral vaccine was associated with the development of intestinal intussusception in approximately 1 in 10 000 vaccine recipients in the USA. Two further live-attenuated oral rotavirus vaccines have completed phase III clinical trials and appear to be safe with respect to intussusception. These are the monovalent P1A[8]G1 human rotavirus vaccine Rotarix, and the polyvalent human–bovine reassortant rotavirus vaccine RotaTeq, which includes the world's most common human serotypes G1, G2, G3, G4 and P[8] on a bovine rotavirus background. Both vaccines are highly (more than 90%) effective in preventing severe disease outcomes where the predominant circulating serotype is P1A[8]G1. It is not known to what extent vaccination with one rotavirus serotype cross-protects against other serotypes, and the efficacy of both vaccines in settings with a wide diversity of rotavirus serotypes remains to be demonstrated. However, it is likely that both Rotarix and RotaTeq will be licensed for global use within the next few years.

KEY POINTS

- Group A rotaviruses are the major cause of acute gastro-enteritis in infants and young children; in developing countries they cause 702 000 deaths per year in children under the age of 5 years.
- Diagnosis of rotavirus gastro-enteritis is usually made by detecting the characteristic wheel-shaped virus particles in faeces by electron microscopy, or by the detection of viral antigen (VP6) using a variety of immunological methods.
- A distinct winter seasonality of rotavirus infection is evident in temperate countries; infection is year-round in tropical countries.
- Rehydration and restoration of electrolyte balance are the primary aims of treatment of acute rotavirus gastro-enteritis.
- The licensing of two live oral rotavirus vaccines is imminent and, if incorporated into childhood immunization schedules, could substantially reduce the global rotavirus disease burden.

RECOMMENDED READING

Blutt S E, Kirkwood C D, Parreno V et al 2003 Rotavirus antigenaemia and viraemia: a common event? *Lancet* 362: 1445–1449

Cunliffe N A, Nakagomi O 2005 A critical time for rotavirus vaccines: a review. *Expert Review of Vaccines* 4: 521–532

Dormitzer P R 2005 Rotaviruses. In Mandell G L, Bennett J E, Dolin R (eds) *Principles and Practice of Infectious Diseases*, 6th edn, pp. 1902–1913. Elsevier, Philadelphia

Estes M K 2001 Rotaviruses and their replication. In Knipe D M, Howley P M (eds) *Fields Virology*, 4th edn, pp. 1747–1785. Lippincott Williams & Wilkins, Philadelphia

Gentsch J R, Laird A R, Bielfelt B et al 2005 Serotype diversity and reassortment between human and animal rotavirus strains:

implications for rotavirus vaccine programs. *Journal of Infectious Diseases* 192: S146–S159

Kapikian A Z, Hoshino Y, Chanock R M 2001 Rotaviruses. In Knipe D M, Howley P M (eds) *Fields Virology*, 4th edn, pp. 1787–1833. Lippincott Williams & Wilkins, Philadelphia

55 Retroviruses

Acquired immune deficiency syndrome; lymphoma

P. Simmonds and J. F. Peutherer

The family Retroviridae contains many viruses from widely different host species. They have been studied for many years, initially because a wide variety of tumours are caused by the oncovirus genus, including leukaemias and lymphomas, sarcomas, breast and brain tumours, auto-immune disease and blood disorders. The host species include birds, mice, cattle, pigs and several primates. The first human retrovirus was isolated from T cells of patients with T cell leukaemia – human T lymphotropic virus type I or HTLV-I – in 1980. The acquired immune deficiency syndrome (AIDS) is caused by a retrovirus, also with a predilection for T cells but differing significantly from HTLV-I. It is known as human immunodeficiency virus type 1 or HIV-1. Infection with this virus has become pandemic and is a major cause of morbidity and mortality in sub-Saharan Africa and other developing countries. HIV-2 infection is restricted largely to West Africa and is less pathogenic.

DESCRIPTION

All retroviruses have an outer envelope of lipid and viral proteins; the envelope encloses the core, consisting of other viral proteins, within which lie two molecules of viral RNA and the enzyme reverse transcriptase, an RNA-dependent DNA polymerase. The virions have a diameter of about 100 nm (Fig. 55.1) and, in thin section, differences can be seen in the appearance of the core; those with a central condensed structure are known as type C particles, whereas those with an eccentric bar structure are type D particles.

Virus stability

HIV is inactivated by:

- heat, in an autoclave or hot-air oven
- glutaraldehyde (2%)
- hypochlorite (10 000 ppm); 1 in 10 dilution of domestic bleach
- other disinfectants, including alcohols.

The chemicals will inactivate at least 10^5 units of virus within a few minutes, but disinfectants are inactivated in the presence of organic material.

HIV can survive for up to 15 days at room temperature and for 10–15 days at 37°C. At temperatures greater than 60°C, virus is inactivated 100-fold each hour.

CLASSIFICATION

The Retroviridae were originally divided into the sub-families *Oncovirinae, Spumavirinae* and *Lentivirinae*, based on their biological properties and appearances in cell cultures. However, nucleotide sequencing of a large number of human and animal retroviruses has shown that viruses referred to as oncoviruses belong to several distinct groups. There are five groups in the new classification (Fig. 55.2).

The human viruses HTLV-I and HTLV-II are related to the simian viruses STLV-I and -II, which are widely distributed in Old and New World monkeys. Like HTLV-I, they cause lymphomas in some primates. STLV-I shows approximately 90% similarity to HTLV-I at the sequence level. A more distantly related virus is found in cows (bovine leukaemia virus).

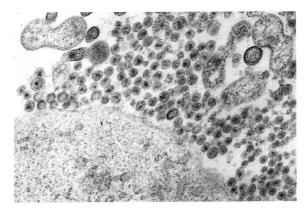

Fig. 55.1 Electron micrograph of HIV. Thin section of infected T lymphocyte. There are numerous virions lying outside the cell membrane.

The spumaviruses have been detected in various species, including cats, cattle and primates, but are not associated with disease. The name is derived from the foamy (vacuolated) appearance of infected cells in culture. There is no evidence for widespread or pathogenic human infection with spumaviruses.

The lentiviruses are associated with slowly progressive diseases. Visna and maedi of sheep were the first to be recognized; they are different clinical presentations of the same virus. The genus includes a virus causing arthritis and encephalitis in goats, equine infectious anaemia virus, and feline, bovine and simian viruses. HIV-1 and -2 are more closely related to lentiviruses that infect Old World primates. HIV-2 is almost identical to simian immunodeficiency virus (SIV_{smm}) found in sooty mangabeys, and it is likely that human infection originated through cross-species transmission. Similarly, HIV-1 corresponds closely to SIV_{cpz} variants that infect chimpanzees in Central Africa, the probable source of the human virus.

The genome organization is similar for all retroviruses as their genomes contain in the same order the genes *gag*, *pol* and *env*, coding for the three groups of structural proteins (Figs 55.3 & 55.4). The long terminal repeat sequences (LTR) at both ends of the genome contain promoter and enhancer sequences. There are important differences between the types in the genes involved in the regulation of replication. In HIV, *tat* codes for a protein that has a stimulating effect on the synthesis of all viral proteins by binding to a region in the LTR that promotes transcription of viral messenger RNAs (mRNAs). The *rev* gene product has a regulatory effect, switching on viral protein synthesis by favouring the production of full-length RNA molecules rather than the spliced RNA from the regulatory genes. The product of other transactivating genes, such as that of cytomegalovirus, may also act on the same sequence within the LTR of HIV. The four proteins coded for by the *gag* gene of HIV are all found in the virion (see Fig. 55.4). The *pol* gene products are a protease, endonuclease, integrase and reverse transcriptase; all required during replication. The *env* gene codes for a large protein that is glycosylated and cleaved to gp41, the transmembrane protein and gp120, present on the envelope as a trimer with many glycosylation sites.

Both HIV-1 and HIV-2 show considerable sequence variability, allowing their classification into a number

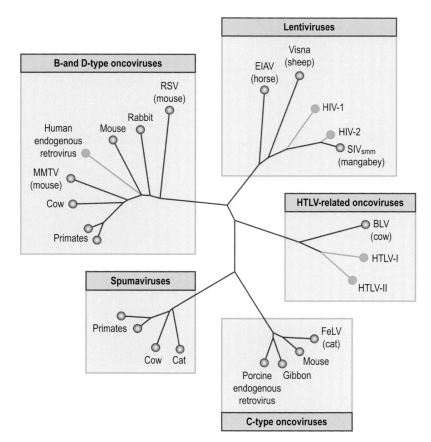

Fig. 55.2 Tree of sequences from the retroviral *pol* gene showing relatedness of retroviruses infecting man and a range of animal species. Sequences form five main groups, in which human retroviruses (HTLV, HIV and human retrovirus 5; shown in blue) are found in three. Viruses previously classified as oncoviruses comprise three genetically distinct groups (C type, B and D type, and HTLV-related). For animal retroviruses (red), the host species and names of familiar viruses are indicated (BLV, bovine leukaemia virus; EIAV, equine infectious anaemia virus; FeLV, feline leukaemia virus; MMTV, Moloney murine tumour virus; RSV, Rous sarcoma virus; SIV, simian immunodeficiency virus). The tree includes sequences of endogenous retroviruses found in the germline of the host (man and pig).

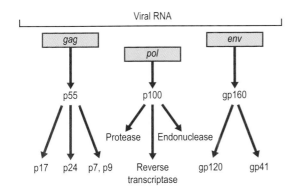

Fig. 55.3 The genomic organization of HIV structural genes and their protein products.

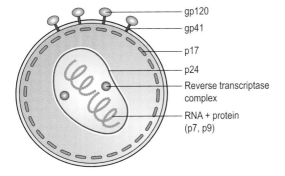

Fig. 55.4 Diagram of HIV to show location of structural proteins.

of subtypes with marked differences in geographical distribution and association with risk groups. HIV variants are currently classified into eight subtypes (A–H) differing by 20–30% in nucleotide sequence. Subtype B is most frequently found in western countries, whereas other genotypes such as C (South Africa, parts of Asia), E (Thailand) and F (South America) are the main subtypes responsible for the recent epidemic spread. HIV-1 diversity is greatest in sub-Saharan African countries such as the Congo, where there is wide co-circulation of most of the current subtypes.

REPLICATION

Retroviruses replicate and produce viral RNA from a DNA copy of the virion RNA (hence their name).

The best studied method of attachment of HIV to cells is by the interaction of the external envelope glycoprotein gp120 with part of the CD4 molecule of T helper lymphocytes and other cells; the HIV envelope then interacts with a second (co-) receptor. These include the chemokine receptors, CCR5 and CXCR4 expressed on a wide range of lymphoid and non-lymphoid cells, whose ligands are chemotactic cytokines such as macrophage inflammatory protein-1α. After this second binding step, entry of the virus occurs by fusion of the viral envelope with the cellular membrane, which requires exposure of a hydrophobic domain in gp41. Once the RNA is released into the cytoplasm, the reverse transcriptase acts to form the double-stranded DNA copy, which is circularized, enters the nucleus and is spliced into the host cell DNA (see Ch. 7). Once inserted into the host DNA, infection with HIV is permanent. The virus may stay latent or enter a productive cycle. Transcription of mRNA from the provirus is by the host RNA polymerase to produce viral mRNA and RNA. Proteins are synthesized and processed to form the virion components (see Figs 55.3 & 55.4). Virions are assembled at the cell membrane where envelope and core proteins have located. The internal structure of the virion matures as the virion buds from the cell. In the productive growth cycle the host cell is frequently destroyed.

CLINICAL FEATURES

HTLV–I infection

Adult T cell leukaemia/lymphoma (ATL) was first recognized in Japan. It is an acute T cell proliferative malignancy; the features are leukaemia, generalized lymphadenopathy and hepatosplenomegaly with bone marrow and skin involvement. Other less dramatic forms exist, including a variant that runs a slow course, associated with adenopathy and splenomegaly. A non-Hodgkin T cell lymphoma has been identified. The T cells involved carry the CD4 antigen. HTLV-I is also the cause of *tropical spastic paraparesis*, a slowly progressive myelopathy with spastic or ataxic features. Pathologically, areas of demyelination with lymphocytic inflammation and perivascular cuffing are seen. Males are at greater risk than females of developing ATL. A form of uveitis is also recognized.

No acute disease is apparent at the time of seroconversion. The period of latency until ATL arises lasts for many years, often decades.

During the latent period viral proteins are expressed, as there are steady high antibody titres to various proteins, particularly the *gag* proteins. The virus is genetically stable and little cell-free virus is produced. However, during the latent period, virus is present as integrated provirus and is replicated with the cellular DNA as the cell divides. The tumour cells contain monoclonally integrated HTLV-I provirus at random sites. There are no transforming genes. The T cell proliferation is the result of the action of the

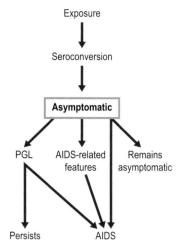

Exposure

↓

Seroconversion

↓

Asymptomatic

PGL   AIDS-related   Remains
features   asymptomatic

Persists   AIDS

Fig. 55.5 Stages of infection with HIV. PGL, persistent generalized lymphadenopathy.

Table 55.1 Classification of HIV infection and AIDS (Communicable Disease Center, USA)

Classification	Clinical picture
Group I	Seroconversion illness
Group II	Asymptomatic
Group III	Persistent generalized lymphadenopathy
Group IV	
A	Constitutional disease
B	Neurological disease
C	Secondary infectious disease
D	Secondary cancers
E	Other conditions

viral *tax* gene, which can activate transcription of cellular genes including those for interleukin-2 and its receptor, and cause cell proliferation. It is not known what triggers this effect after the long latency in the 1–4% of those infected who develop disease. Antibody to the tax protein can block the stimulation of cell division; loss or decay of immune control may be important.

HTLV-II is not linked to a particular disease, although the first isolation was from a patient with a rare T hairy-cell leukaemia.

HIV and AIDS

The different stages of HIV infection are summarized in Figure 55.5 and Table 55.1.

The acute seroconversion illness resembles glandular fever, with adenopathy and influenza-like symptoms. Although most patients experience some symptoms, only 5–10% show the full picture. Even fewer have the rare encephalitic presentation.

Persistent generalized lymphadenopathy is present in 25–30% of patients who are otherwise asymptomatic (see Table 55.1). The enlarged lymph nodes are painless and symmetrical in distribution. The rate of progression of patients to AIDS is no greater than that in those without adenopathy.

The *acquired immune deficiency syndrome* (AIDS) presents in many ways, all due to loss of the ability to respond to infectious agents and to control tumours. The features classified as group IV include what was known as the *AIDS-related complex*. This label was applied to patients with constitutional symptoms of fever, weight loss and diarrhoea, and minor opportunistic infections. Without treatment, such patients progress rapidly to AIDS. The clinical features of AIDS are varied and reflect the specific agents involved; a diagnosis is made if the conditions listed in Box 55.1 are present.

Oral hairy leucoplakia appears to be unique to HIV-infected patients. The margins of the tongue show white ridges of fronds on the epithelium. An association with Epstein–Barr and papilloma viruses has been proposed.

Kaposi's sarcoma was one of the earliest diseases used to define AIDS. This rare tumour had been known for many years; it usually occurred at a single site and was not aggressive. In patients with AIDS the tumour arises in many sites, including the skin, mouth, gut and eye. The tumours arise from endothelial cells of blood vessels, causing bluish-purple, raised, irregular lesions. The aetiological agent is human herpesvirus 8 (see Ch. 43). The tumours were seen only in homosexual men; the incidence has now declined.

Pneumocystis carinii (*P. jirovecii*) pneumonia was the presenting feature in many of the first patients. This opportunist pathogen was known to cause infections in the immunocompromised, but the diagnosis in young men with no explanation for their immunosuppression was the first clue to the recognition of AIDS.

Toxoplasma gondii infections can manifest at various sites, but are always associated with immunocompromised patients. The brain is an important site.

HIV dementia develops in 25% of patients with AIDS and is marked by a gradual loss of cognition, progressing to overt dementia. Brain scans show a loss of tissue, with widening of sulci and ventricles.

In developing countries many of the same infections are seen but there is an emphasis on local problems. *Mycobacterium tuberculosis* infections are an enormous problem in many regions, with the development of strains of the organism resistant to many antibiotics. Many patients show profound weight loss, perhaps accompanied by chronic diarrhoea; the term 'slim disease' has been given to this presentation.

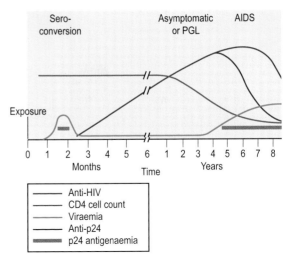

Fig. 55.6 Events in HIV infection. PGL, persistent generalized lymphadenopathy.

Paediatric patients with AIDS suffer from many of the same problems as adults. However, children infected early in life or at birth are at risk of recurring bacterial infections as they have not acquired immunity to micro-organisms. Lymphoid interstitial pneumonia and pulmonary lymphoid hyperplasia are presentations seen only in young children.

PATHOGENESIS OF HIV INFECTION AND AIDS

The major virological and immunological features of the acute and persistent stages are shown in Figure 55.6.

The incubation period in the acute stage is 1–2 months. This is preceded by a period of intense, unrestrained viral replication, reflected in the presence of high numbers of viral RNA genomes and p24 antigen in the circulation. After entering the body, virus is taken up by cells such as dendritic cells that express viral receptors. Within 24–48 h infected cells are present in the regional lymph nodes; virus can be detected in the blood and circulating lymphocytes by 5 days; the number of circulating CD4 positive (CD4+) lymphocytes is decreased. As the immune system responds, both p24 antigen and RNA copy number decrease, so that by 6–12 months p24 antigen is usually undetectable and the RNA load has stabilized at a lower level; in some it may be undetectable. The decline in RNA copy number is usually by at least 3–4 $\log_{10}$. Temporary increases in the RNA level can be seen during intercurrent infections, immunizations and pregnancy, when AIDS is diagnosed. CD4+ cell counts recover and remain more or less normal until progression to AIDS occurs, the counts then being less than 200/μl.

In peripheral blood, lymphoid tissue and other tissues such as brain where HIV replication occurs, HIV targets CD4+ cells and cells of the monocyte–macrophage lineage; the latter may act as a reservoir of virus. Macrophages are also important in carrying the virus into the central nervous system across the blood–brain barrier.

Destruction of CD4+ cells is caused by:

- viral replication
- syncytium formation via membrane gp120 binding to cell CD4 antigen
- cytotoxic T cell lysis of infected cells

- cytotoxic T cell lysis of CD4+ cells carrying gp120 released from infected cells
- natural killer cells
- antibody-dependent cell cytotoxicity.

The proportion of infected CD4+ cells and the titre of circulating virus rise as the infection progresses, until the patient becomes symptomatic. The RNA copy number will rise and reach high levels ($>10^5/mm^3$) and CD4+ cell counts drop below 200/µl.

Analysis of the viral genomes from a patient shows that there are several different viral genomes present at any time and that these change with time. Virus isolated in culture may be different from the predominant variants in the blood. Viruses isolated in the later stages of infection have been shown to grow more rapidly, to higher titres and to form syncytia (giant cells) more readily than virus isolated in the early stages. The switch to syncytium-inducing variants is accompanied by a change in co-receptor use from CCR5 to CXCR4. Regions of the envelope glycoproteins show most variation and this could affect the ability of antibody to react with the viruses. Although this could be relevant to the progression of the infection, it also has important implications for the development of vaccines.

Disease progression

There are host genetic differences influencing the risk of disease progression; for example, human leucocyte antigen (HLA) haplotype A1B8DR3 has been linked to rapid development and severe disease. There is also an age effect, with evidence of fast progression in some infants and in the elderly. Box 55.2 lists a number of laboratory markers that are associated with progression. The most useful in assessing the state of a patient's immune system is the absolute CD4+ cell count. Although this can vary, a downward trend is indicative of progression: when the count reaches 200/µl the patient is severely compromised and the diagnosis of AIDS is made even in the absence of an AIDS indicator disease. High viral loads (and p24 antigen) and cytopathic phenotype all correlate with progression.

The numbers progressing to AIDS have been studied over many years, although the use of specific therapy has changed the outcome. Untreated, from initial infection the results were:

- 5% within 3 years
- 20–25% by 6–7 years
- >50% by 10 years
- <5% asymptomatic for more than 10 years.

When it became possible to measure viral load, it was established that:

BOX 55.2 Laboratory markers associated with progression of HIV infection

- Declining number of CD4 lymphocytes
- Increasing proportion of infected CD4 cells
- Increasing titre of virus in plasma (HIV RNA copy number)
- p24 antigen in plasma
- Isolation of virus in culture – rapid growth, syncytium formation
- Loss of cutaneous hypersensitivity

- 13% of patients with a viral RNA copy number of less than 1500/ml will develop AIDS within 9 years
- 93% progression in 9 years with RNA copy number greater than 55 000/ml.

Paediatric infection

In most cases, infection is transmitted to the baby in the perinatal period when the child's immune system is immature. This results in a major difference from the picture seen in older children and adults in that the initial replicative phase is not limited by the immune response and high levels of viral RNA persist. RNA viral loads are often greater than $10^5/ml$ at 2 months. About 75% show a steady decline thereafter; however, by age 9–16 years a third are still asymptomatic and show little impairment of immune function. The other quarter of the children have high levels of viral RNA and develop early-onset disease, with death by 20–24 months without treatment. These babies may have been infected before birth by a mother with advanced disease. This group of rapid progressers can be identified by detection of virus by RNA and culture within 48 h of birth. Analysis of the child's RNA may show that it differs from that of the mother, suggesting that replication has occurred by the time of sample collection or that a minor maternal variant has been transmitted to the baby.

LABORATORY DIAGNOSIS

HIV infection

Both direct and indirect diagnostic methods are available. Infection is invariably persistent, and it is possible to diagnose infection through detection of antibodies to the virus (anti-HIV), or to detect the virus itself. Of these, the most sensitive assay is reverse transcriptase–polymerase chain reaction (RT-PCR) for viral nucleic acid; this method can detect a single copy of RNA or proviral DNA from infected cells.

Tests for anti-HIV

The main approach to the diagnosis of infection in patients and for screening populations (e.g. blood donors) has been by testing for anti-HIV. Many different assays are available, most using enzyme-linked immunosorbent assay (ELISA). All current tests use HIV antigens derived from cloned recombinant HIV *gag*, *pol* and *env* genes expressed in *Escherichia coli*, or synthetic peptides. Western or immunoblotting has been used extensively as a confirmatory assay. Most current assays can detect antibody to both HIV-1 and HIV-2 antigens. A first positive result must be confirmed by at least two other different assays with different viral antigens, and a second serum sample checked to confirm that the original sample was identified correctly. Most patients will seroconvert within 2–3 months (see Fig. 55.6), but some may take longer. Thus there is a window before antibody tests can detect infection. Rapid tests are now available to detect antibody in saliva.

Combination assays

Reliable and highly sensitive methods to detect anti-HIV and p24 antigen in a single ELISA are now available; these allow diagnosis irrespective of the stage of infection (before or after seroconversion). Although p24 antigen is less sensitive than RT-PCR for HIV RNA, the combination test is of value for diagnosis and for large-scale screening (e.g. blood donors).

PCR

Direct detection methods are required when serological tests are inappropriate, such as during the early acute stage and in infants who still carry maternal anti-HIV, and for monitoring progression.

Both HIV RNA and DNA sequences can be detected in blood. RNA sequences are found in extracellular virus particles in plasma, and the RNA can be accurately quantified as the number of RNA copies to indicate the extent of virus replication in the patient. Measurement of plasma virus load is now essential for monitoring disease progression and the response to antiviral therapy. A number of commercial assays have been developed to provide accurate and standardized viral load measurements in clinical laboratories.

HIV proviral DNA is present in infected cells and can be detected in peripheral blood mononuclear cells. This method is used principally to diagnose infection in infants born to HIV-infected mothers. As infection is often acquired perinatally, a negative test result at 3 months or later is required to exclude infection. In horizontally transmitted infections, detection of proviral DNA can be used to diagnose an acute infection before seroconversion, although detection of plasma viral RNA is used more often.

Virus isolation

Isolation of HIV is slow, taking 3–6 weeks. The usual sample is blood, from which the lymphocytes are separated and co-cultured with stimulated donor lymphocytes. Virus presence is detected by assays for reverse transcriptase and p24 antigen in the culture fluids. With the advent of PCR, there are now few, if any, diagnostic uses of virus isolation, apart from phenotypic measurement of antiviral resistance.

HTLV-I

Assays for the detection of antibody to HTLV-I are available. As with HIV, confirmation by other assays or immunoblotting must be attempted, although interpretation can be difficult. Confirmatory tests are able to distinguish between HTLV-I and HTLV-II. Detection of HTLV proviral sequences by PCR can also be used as a confirmatory test, and to distinguish HTLV-I and HTLV-II.

TREATMENT

HTLV-I infection

Interferon and inhibitors of reverse transcriptase may have a role, but further evaluation is needed.

HIV infection and AIDS

The development and assessment of clinical efficacy of antiretroviral drugs active against HIV has been an area of intense research over the past 15 years. Currently used multiple drug regimens (often referred to as *h*ighly *a*ctive *a*ntiretroviral *t*herapy, or HAART; now potent combination antiretroviral therapy) have achieved considerable success in halting the progression to AIDS, and have led to a major fall in AIDS-related deaths in industrialized countries with access to therapy.

The first antiretroviral drug to be developed, and still the most widely used, is a nucleoside analogue, azido-thymidine or zidovudine (see Ch. 5). Once it is phosphorylated, the drug inhibits reverse transcriptase (nucleoside reverse transcription inhibitor, or NRTI). Initial studies showed that zidovudine improved the symptoms and survival of patients with AIDS, although about 25% developed serious side effects, in some cases requiring frequent blood transfusions and cessation of therapy. Treated patients usually show reduced viral load and partial recovery of CD4 numbers. The clinical value of zidovudine monotherapy was limited by the development of drug resistance due to mutations in the *pol* gene that modify the substrate specificity of reverse transcriptase. Several nucleoside and non-nucleoside analogues are now licensed. Further approaches have also been exploited; viral protease inhibitors are now in use and inhibitors of the fusion of viral envelope with cell membranes have been discovered (see pp. 63–64).

To maximize effect and delay or prevent the emergence of viral resistance, at least two or preferably three drugs are prescribed together. The decision to start therapy in patients known to have been infected for some time can be made on the basis of:

1. clinical deterioration
2. CD4 count falling or already less than 200/µl, or less than 350/µl even if patient asymptomatic
3. high plasma viral load ($>10^5$/mm^3) or rising.

Treatment may be deferred in an asymptomatic patient who has a CD4 count above 350/µl and a viral load of less than 10^4/mm^3.

The aims of therapy are to:

- prolong life
- reduce clinical problems
- reduce infectivity.

These objectives can be achieved by:

- preserving or improving immune function: measure by sequential CD4 counts
- producing the maximum, lasting reduction in viral load – measure by sequential assays of viral RNA copy number in plasma
- regular monitoring of patient clinically and measuring CD4 count and viral load.

A patient who has not been treated before may be started on a combination of drugs such as:

1. a non-nucleoside reverse transcriptase inhibitor + two nucleoside reverse transcriptase inhibitors, e.g. efavirenz + lamivudine (or emtricitabine) + zidovudine (or tenofavir)
2. one or two protease inhibitors + two nucleoside reverse trancriptase inhibitors, e.g. lopinavir and ritonavir (co-formulation) + lamivudine (or emtricitabine) + zidovudine
3. three nucleoside reverse transcriptase inhibitors.

The management of treatment is difficult; advice on drug combinations and interactions and toxicity is available from many national bodies, including the National Institutes of Health in the USA (see internet site reference). The World Health Organization makes similar recommendations for patients in 'resource-limited settings', taking account of local conditions such as the need for a cold chain to deliver the drugs to the patients and how they are administered. Cost is an important consideration as therapy is expensive at many thousands of pounds per annum. Reduced-cost drugs are now available, but the number of patients needing them means that very large resources are needed.

A patient on combination therapy will have to adhere to a strict regimen. This may be a problem in very young and adolescent patients. The drugs chosen may have side effects in a particular patient or in pregnancy. There may be interactions with therapy for other conditions and infections such as hepatitis B and C viruses, herpesviruses, tuberculosis, toxoplasmosis and *P. carinii* (*P. jirovecii*). Therapy to prevent or treat these infections must be maintained.

A knowledge of previous antiretroviral therapy is also essential as drug resistance may have arisen. If indicated, it is possible to test the patient's virus population for drug resistance.

There is no clear case for treating a patient diagnosed during the acute phase of infection.

Monitoring progress

The CD4 count and the plasma viral load should be assayed when therapy is started and at 1-month and 3–4-month intervals thereafter. If there is a response, the RNA load will decrease within a few days, will drop by 1 $\log_{10}$ at 2–8 weeks and be less than 50 copies/ml by 4–6 months. If these objectives are not achieved, or the viral load increases after a time on therapy, or the clinical state deteriorates, a new combination regimen should be started. The resistance pattern of the patient's virus may be tested before selecting new drugs.

About 50% of those on long-term therapy will suffer a redistribution of body fat – the lipodystrophy syndrome. Many patients have been treated for years with a good quality of life as a result of these advances.

The prophylaxis of perinatal infection and accidental exposure is described below.

TRANSMISSION AND EPIDEMIOLOGY

HTLV–I and HTLV–II

The three lineages of HTLV-I strains are linked to Melanesia, Central Africa and various countries (the Cosmopolitan group). The latter includes viruses from Japan, North and West Africa and the Caribbean, which can be distinguished. HTLV-I and the simian virus, STLV-I, are closely related and it is proposed that human infection occurred many thousands of years ago in Africa and that the presence of the virus in many different parts of the world is related to the migration of ancient peoples. The slave trade may account for foci found in the West Indies and the southern USA.

Where it is found, the virus is endemic in certain communities. In parts of Japan, the prevalence of antibody can be 27%, with a rising trend from 7–8% in the 20–39-year age group to 52% in females and 32% in males by 80 years. In the Caribbean, the rates are in the range of 5–10%, with clusters in communities and families. In other regions, infection has been found in parenteral drug misusers and prostitutes.

The virus is cell associated in the host, so transmission will occur when infected cells are transferred. This can occur during sexual intercourse, via breast milk or blood transfusion and through sharing injecting equipment by drug misusers.

HTLV-II is transmitted by the same routes. The strains found in drug misusers in different countries are related.

Transmission of HIV-1 and HIV-2

The scale of the HIV-1 epidemic is revealed by co-ordinated surveillance by the United Nations AIDS Programme and the World Health Organization (WHO). For 2004, it is estimated that there were 39.4 million individuals with HIV infection worldwide, 4.9 million newly infected individuals and 3.1 million AIDS-associated deaths, of whom 510 000 were children under 15 years of age. Frequencies of HIV infection remain highest in sub-Saharan Africa (with a mean 7.4% overall population prevalence). HIV is spreading rapidly in parts of Asia with an estimated 1.2 million new cases in 2004. The WHO estimates that more than 40 million people have been infected with HIV and that 16 million have died.

Virus is present in the blood, semen, and cervical and vaginal secretions, and these sources are important in transmission. Virus may also be present in cerebrospinal fluid, saliva, tears and urine, but at lower titres than in blood. There is no epidemiological evidence that these are significant sources for transmission. Free virus is present at high titre during the early stage of infection and increases in titre in the blood in the later stages of the disease; there is evidence of a greater risk of transmission from such patients.

To transmit, virus has to reach susceptible cells at the point of entry (e.g. Langerhan's cells in mucous membranes) or after entering the circulation.

The three important routes of transmission of HIV are:

1. by unprotected, penetrative sexual intercourse
2. from mother to child
3. by blood and blood products.

Sexual intercourse

Heterosexual transfer of virus is the route by which the great majority of infections are spread, accounting for 90% of the global total, mostly in the developing world. Both sexes are affected equally. Overall the estimated risk of transmission from one unprotected exposure is 0.1–0.2% for vaginal intercourse. The probability of transfer is increased if either partner has ulcerative genital or other sexually transmitted disease. Any trauma during intercourse will also facilitate transfer, by allowing direct access of the virus to susceptible cells and the circulation. Sex workers are at high risk from their many partners and may be an important reservoir. Transmission may be more likely from male to female.

AIDS was first recognized in homosexual men in the USA. Most early studies established that unprotected anal intercourse was a particular risk, especially to the passive, receptive partner. The estimated risk from a single exposure is 0.1–0.3%.

Transmission during oral sexual contact has been documented, but is not a major route.

Mother to child

Most perinatal transmission occurs late in pregnancy or during birth. The most likely source is cells and virus in the cervix and vagina, as the baby passes through the birth canal. The risk of transmission varies from 13–32% in the developed world to 25–48% in the developing world. Prolonged and difficult labour is a factor. Breast milk is another possible source. It is difficult to be precise about the contribution of this route.

Blood and blood products

All blood for transfusion and the preparation of products such as factor VIII for haemophiliacs is screened by sensitive assays. This eliminates almost all the risk, but it is important to ask donors about possible exposure to risk. Preparation of blood products from large pools of donations was a major factor in contaminating the product as even one infected donation could introduce virus to all the material. Transplanted organs have been implicated in a few cases.

Intravenous drug misuse is a risk factor in about one-quarter of patients with AIDS in the USA and to a varying extent elsewhere in the world. The risk rises with the volume of blood injected and the frequency of sharing contaminated equipment. The withdrawal of blood before injection increases contamination. By sharing syringes, the virus can spread very rapidly so that most misusers in an area become infected in a few months. Those infected in this way can spread the virus to their sexual partners or children. Drug and sexual routes merge when misusers support their habit by prostitution.

Occupational exposure of health-care workers to infected patients has resulted in transmission in a relatively small number of cases. The route is via accidental penetrating injuries with needles and sharps contaminated with blood. The risk from a needlestick is 1 in 200–300; contamination of eyes and mucous membranes has a low risk of transmission. Transmission from health-care workers to patients has been suspected in only a few cases.

HIV-2 is transmitted by the same routes as HIV-1.

General

The majority of infected individuals have a recognized exposure to a known source of infection. In some this may be difficult to establish, but there is no evidence that HIV can spread by casual contact or inhalation.

Studies of people exposed to the virus have on many occasions shown that a few show no evidence of infection, and remain negative for anti-HIV. How these individuals are resistant to infection is of great interest in understanding protective immunity.

Epidemiology of HIV

The extent of spread of infection can be measured by the numbers of cases identified clinically and by serological testing. Much more evidence can be obtained from seroprevalence surveys of particular groups or the general population. Surveys have been performed on patients attending hospitals, antenatal clinics, sexually transmitted disease clinics and blood donors. Specific groups such as drug misusers and prostitutes can be targeted; non-invasive sampling (e.g. collecting saliva) may make these studies more feasible. Repeat testing over time will give an indication of the trend of infection in that population. Such studies are important in monitoring the effect of intervention strategies and forecasting the demand for health services.

HIV was isolated in the early 1980s, but the first identified cases date to the 1960s. During the 1970s, the virus began to spread widely in some populations and groups by the routes described above.

Regions of Africa have suffered the greatest epidemic spread of the virus, particularly in most of the countries of the sub-Saharan region. Globally, most infections are in this area. The viruses circulating here are of subtypes C, A, D and E. Of these, C has spread rapidly and now accounts for 50% of all infections. Most AIDS deaths have occurred in Africa. The virus continues to spread, and access to antiviral therapy is restricted. The social and economic consequences of this epidemic are devastating,

with the loss of parents and wage earners. The high infant mortality rate will have profound effects in the future.

In North and South America, Europe and Australia, at least 30–40% of cases are in gay men. Parenteral drug misusers are the other major risk group in these areas. Virus of subtype B is closely associated with these groups and accounts for at least 50% of all infections. The other cases are in the heterosexual partners of bisexual men, drug misusers, and men and women infected in other areas of the world. Infected blood caused some cases before screening was introduced. The estimated prevalence in the populations is 0.05–0.35%, with fewer than 0.01% in antenatal patients. The numbers of infections in risk groups can change as health education programmes are introduced; however, the success of programmes varies and advice may be ignored if the perception of risk changes.

HIV was introduced into South and South-East Asia later than in the rest of the world; infection is spreading rapidly. The earliest infections were in drug misusers, but this did not lead to wide spread outside the risk group. The situation changed with the introduction of subtype C virus into the general population and there is now rapid heterosexual transmission of this strain. Some 1–10% of antenatal patients may be infected. Without effective intervention large numbers of cases and deaths will occur, with all the expected human and socio-economic consequences. Throughout the region, subtypes A, B, C, D and E have been found.

As a result of the wide circulation of different subtypes in some populations, recombinant viruses have been identified. AC and AD recombinants have been found. Studies in Tanzania suggest that 15% of the virus population in the country is recombinant and that these viruses can be transmitted. A few recombinants have been isolated from areas in the East.

The existence of different subtypes of HIV, and recombinants, is important for two reasons. Firstly, assays for anti-HIV and viral nucleic acid must be able to recognize all types. Secondly, vaccine developers must take account of the various types and establish the spectrum of protection of candidate vaccines.

CONTROL

Until a vaccine is available, the emphasis in controlling the spread of infection must be on risk reduction. In future, antiviral therapy will play an important part in attempts to contain the spread of HIV-1.

Sexual transmission

The emphasis is on risk reduction by avoiding unprotected penetrative intercourse with partners of unknown status. Despite knowledge of the major routes of infection, there has been only limited success in reducing sexual transmission. Globally the problem is enormous and efforts are hampered by the poverty and lack of resources of the countries worst affected. The use of condoms and vaginal antiseptics could have an impact, but they need to be available and acceptable to the local population.

In the areas of the world with low levels of infection, early efforts to encourage safe practices had an effect on the spread of the virus among gay men in the Americas, Europe and Australia, but this was not always maintained as the perception of the risks changed as a result of declining rates of infection and, more recently, as the latest therapies appeared to be succeeding and prolonging survival. In addition, it is difficult to persuade the heterosexual majority that safe practices are relevant to them.

Mother to child transmission

This can be reduced by identifying infected mothers and giving specific therapy in the later stages of pregnancy and to the baby after birth. Zidovudine alone given to the mother before delivery and to the baby for 6 weeks can reduce transmission three-fold even if the mother has advanced disease.

If the mother is already being treated with standard combination therapy, the risk of neonatal infection is less than 2%. Transplacental transfer of the drugs is important and may be aided by phosphorylation of the drug in the placenta. The much more limited therapy available in resource-poor countries such as sub-Saharan Africa can also play a major preventive role. Thus, single doses of nevirapine given to the mother at the onset of labour and to the neonate within the first 3 days after birth reduce the rate of transmission by more than 50%.

As exposure to infected genital secretions is the source of the virus, avoiding prolonged rupture of the membranes can reduce risk. Caesarean section may have a similar effect, but is of limited applicability.

Breast-feeding is another possible route. However, studies of children protected by specific therapy at and after birth do not show that there is a significant extra risk of infection by this route. Even without therapy, the advantages of breast-feeding, if no other adequate nutrition is available, far outweigh any risk of infection Where alternative nutrition is available, the baby may not be breast-fed.

Exposure to blood

Drug injectors can avoid risk by not injecting, or can reduce risk by using only clean equipment. Screening of all blood donors should eliminate almost all possibility of transmission. Factor VIII and other blood products are heat treated, if possible, to inactivate HIV. All organ donors must be screened.

Occupational risk in the health-care setting can be controlled by the implementation of safe working practices to prevent accidental injury and contamination with blood and body fluids. The use of gloves, masks and eye protection is important in situations such as surgical procedures where bleeding and spattering are possible. The risk must be assessed in other situations. Safe disposal of used needles, scalpel blades and other sharps is an essential requirement. The sensitivity of HIV to heat and various disinfectants is described above.

If an accidental exposure occurs, any wound should be washed with soap and water, or mucous membranes flushed with water. The accident must be reported so that, if necessary, prophylaxis can be started as soon as possible. The risk must be assessed through knowledge of the circumstances:

- The HIV status of the source patient; if unknown, can the source be tested?
- Any particular risk of infection of the source patient.
- The nature of the exposure (e.g. penetrating injury or contamination of skin or mucous membranes).

The risk of infection from splashing on to mucous membranes or skin is hard to quantify, but is certainly less than with penetrating injuries. An intact skin is an effective barrier, but abrasions and diseases such as eczema may impair this protection.

If a sharp injury is reported the nature of the injury has to be assessed.

- Needlestick or cut with sharp instrument.
- Depth of penetration.
- Volume of blood involved.
- Whether blood vessel entered.

If there is an indication of risk, therapy must be started within 1–2 h, and not later than 48–72 h. If no professional advice is available, for instance at night, prophylaxis should be started and advice obtained. A decision should made about continuing with the drugs within 12–24 h. The victim should be involved in the decision, with discussion of the risks and the possible side effects of the drugs.

Zidovudine alone can reduce the transmission rate, but should now be combined with another reverse transcriptase inhibitor (e.g. lamivudine) and a protease inhibitor. The combination of drugs can be varied with knowledge of any drug resistance in the source. Therapy should be continued for 4 weeks and the victim followed with testing for virus for the next 6 months. A few cases of transmission have been seen in cases given prophylaxis.

Vaccines

Much effort has been devoted to the development of a vaccine to provide protection against infection after exposure (prophylactic vaccine) or to boost the immune system of those infected (therapeutic vaccine). Major problems arise because of the antigenic variability of HIV and the difficulty of developing immunogens that elicit protective rsponses to all variants. In addition, HIV may be transferred by blood-borne or mucosal routes, through transfer of free virus or infected cells. To protect, therefore, it is likely that both cell-mediated and humoral responses need to be stimulated. Whether an HIV vaccine could ever induce fully protective immunity is subject to some doubt, because the immune response, although highly active during acute infection, is never capable of fully clearing infection, and life-long persistence is the norm.

Most efforts have been directed to the development of vaccines containing the viral Env proteins gp160, gp120 or gp41 prepared by recombinant DNA cloning and expression, or synthetic peptides known to be important epitopes for induction of neutralizing antibodies. To date, human trials have shown no evidence of protection from infection by sexual transmission and injecting drug use. Some multiply exposed individuals who remain seronegative mount effective T cell responses that contain systemic spread. Vaccines that stimulate these locally acting immune responses are an aim of future research.

KEY POINTS

- Some retroviruses, including HTLV-I, the cause of human T-cell leukaemia, cause tumours in natural hosts.
- HIV-1 is the cause of the acquired immune deficiency syndrome (AIDS), a persistent infection leading to loss of CD4 T cells, immunodeficiency and many opportunistic infections.
- Disease status can be measured by sequential changes in CD4 count and titre of virus in plasma (HIV RNA copy number). Untreated, 5–10% of those infected develop AIDS each year; only 2% are asymptomatic after 12 years.
- HIV has a global distribution; it is spread by sexual intercourse, mother-to-child transmision and via blood and blood products. The greatest incidence is in sub-Saharan Africa and South-East Asia.
- Viral replication can be inhibited by drugs that block reverse transcription, proteases and membrane fusion.
- Drug combinations (two reverse transcriptase inhibitors and a protease inhibitor) reduce the appearance of drug resistance and, combined with control of opportunistic infections, substantially improve survival.

RECOMMENDED READING

Fields B N, Knipe D M, Howley P M (eds) 1996 *Virology*, 3rd edn. Lippincott-Raven, Philadelphia

UK Health Departments 1998 *Guidance for Clinical Health Care Workers; Protection Against Infection with Blood-Borne Viruses.* Recommendations of the Expert Advisory Group. HMSO, London

Winston A, Stebbing J 2005 New drugs for old. *Journal of HIV Therapy* 19: 11–16

Internet sites

World Health Organization. HIV infections. http://www.who.int/topics/hiv_infections/en/

US Department of Health and Human Services. AIDSinfo. http://www.aidsinfo.nih.gov

56 Caliciviruses and astroviruses
Diarrhoeal disease

W. D. Cubitt

The introduction of electron microscopy for the examination of faecal samples led, in the 1970s, to the discovery of a number of viruses that cause diarrhoeal disease in man and animals. The viruses discussed here are *noroviruses* and the *sapoviruses* that form two distinct genera of the Caliciviridae, and viruses belonging to the family Astroviridae. Epidemiological surveys, human volunteer experiments and laboratory investigations have confirmed that they cause diarrhoeal disease. Noroviruses are a major cause of food- and water-associated outbreaks of diarrhoea and vomiting, throughout the world. Sapoviruses are generally associated with sporadic cases of diarrhoea and vomiting, and astroviruses have been associated with extensive outbreaks of food-borne infection in Japan.

DESCRIPTION

The properties of the viruses are summarized in Table 56.1.

Morphology

Sapoviruses have a characteristic surface morphology (Fig. 56.1), formed by the 32 cups or 'calices'. Three distinct appearances can be observed depending on the orientation of the particles (Fig. 56.2). Factors such as freezing and thawing, the presence of proteolytic enzymes or incorrect staining can affect the appearance of the particles, which may then be indistinguishable from 'small round structured viruses'. The morphology may be masked also by the presence of antibodies. Noroviruses have a similar structure when studied by cryo-electron microscopy, but the tips of the capsomeres are bent and partially obscure the hollows, resulting in an amorphous surface structure with a ragged outline (Fig. 56.3). Complete virions measure 35–40 nm in diameter with a solid inner shell at a radius of 11.5–15.5 nm surrounding the RNA.

Astroviruses can be recognized by a five- or six-pointed star on their surface (Fig. 56.4), although this is generally evident on only a minority of particles in a preparation. The particles have a radius of 35 nm and are surrounded by a hair-like fringe.

Table 56.1 Properties of human caliciviruses (HuCVs) and astrovirus

	Sapovirus	Norovirus	Astrovirus
Nucleic acid	ssRNA	ssRNA	ssRNA
Protein	VP1	VP1	VP1, VP2, VP3?
Molecular weight	65 000 Da	60 000–70 000 Da	29 000–39 000 Da
Lipid	None	None	None
Buoyant density (g/cm³)	1.38–1.4	1.38–1.41	1.36–1.38
Morphology	See Figs 56.1 & 56.2	See Fig. 56.3	See Fig. 56.4
Diameter (nm)	30–35	30–38	28–35
Antigenic strains	≥5	≥12	8
Replication	Cytoplasm	Cytoplasm	Cytoplasm
Host range	Man	Man	Man
Transmission	Faecal–oral	Faecal–oral, air-borne, contaminated food and water	Faecal–oral, air-borne, contaminated food and water

ssRNA, single-stranded ribonucleic acid.

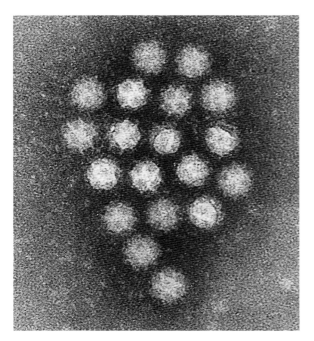

Fig. 56.1 Sapovirus, displaying characteristic cupped surface morphology. Original magnification ×300 000. (From Cubitt W D, Blacklow N R, Herrmann J E, Nowak N A, Nakata S, Chiba S 1987 Antigenic relationships between human caliciviruses and Norwalk virus. *Journal of Infectious Diseases* 156: 806–814.)

Physicochemical and physical properties

The properties of the noroviruses and sapoviruses are identical (see Table 56.1). Astroviruses have similar properties but appear to be more resistant to inactivation, withstanding acid (pH 3.0), 50°C for 1 h and 60°C for 5 min.

Genome organization

It is now known that morphologically typical human caliciviruses and many small round structured viruses are members of the family Caliciviridae. Based on differences in their genomic organization (Fig. 56.5a), human caliciviruses have been characterized within two genera of the Caliciviridae:

1. Sapoviruses (e.g. human calicivirus/Manchester)
2. Noroviruses (e.g. human calicivirus/Norwalk).

Non-structural proteins are encoded by open reading frame (ORF)1 and structural proteins by ORF2 and ORF3. Caliciviruses possess highly conserved regions within ORF1, encoding a helicase (2C), a protease (3C) and the RNA-dependent RNA polymerase 3D. These serve as suitable sites to direct primers for reverse transcriptase–polymerase chain reaction (RT-PCR). Primers have also been designed to amplify regions within ORF2 that encode the antigenic domains. Phylogenetic analyses of the human caliciviruses have shown that there are at least five distinct clades within the sapoviruses and two major clades, GI and GII, within the noroviruses. Recent studies have shown that recombinants of sapoviruses and noroviruses exist. This results in strains that have the non-structural genes of one strain and the structural gene of another (e.g. *Norovirus*/Harrow/Mexico).

Astroviruses have a unique genomic arrangement (Fig. 56.5b) and constitute a separate family, the Astroviridae.

Antigenic properties

At least four antigenically distinct strains of sapovirus have been identified: Saporo, Houston, London and Stockholm. The use of immune electron microscopy and enzyme immuno-assay has demonstrated that there are numerous strains of Norwalk-like viruses.

Eight strains of astrovirus can be identified with specific antisera. The use of a monoclonal antibody has shown that all serotypes share a group antigen.

Host range

In-vivo studies with strains of sapoviruses and noroviruses suggest that they are not readily transmitted to or from other species, although other caliciviruses have been identified in primates, domestic and farm animals, birds, fish, reptiles, amphibians and insects, and some are known to cross species barriers. There are two reports suggesting that primates can undergo subclinical infection when fed with Sapovirus or Norwalk virus. In both of these cases, animals showed a significant antibody response and were found to be excreting small numbers of virus particles in faeces. Similar experiments have not been conducted with human astroviruses.

Human volunteer studies with caliciviruses (Norwalk, Hawaii and Snow Mountain agents) and astroviruses have shown that some adults can be infected with faecal filtrates containing virus. Pre-existing antibody to Norwalk virus did not confer immunity to subsequent challenge. A person's ABO and secretor (functional fucosyltransferase enzyme [FUT-2]) phenotype determine whether or not they are susceptible to Norwalk virus infection. Approximately 20% of caucasians/Europeans do not secrete the enzyme and are resistant to infection.

Replication

Attempts to propagate enteric caliciviruses in vitro have been unsuccessful.

Astroviruses can be propagated readily in a human intestinal cell line, CaCo2, and in HT-29 cells provided

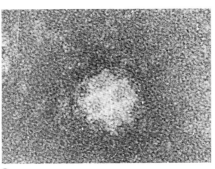

a

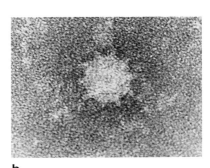

b

Fig. 56.2 Electron micrograph showing sapovirus morphology when viewed a–c along the two-, five- and three-fold axes of symmetry. Original magnification ×450 000. (From Cubitt W D, McSwiggan D A, Moore W 1979 Winter vomiting disease caused by calicivirus. *Journal of Clinical Pathology* 32: 786–793.)

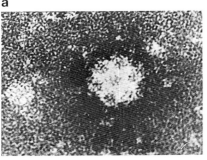

c

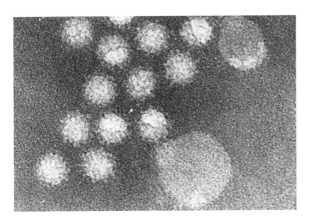

Fig. 56.3 Norovirus. Original magnification ×300 000. (From Cubitt W D, Blacklow N R, Herrmann J E, Nowak N A, Nakata S, Chiba S 1987 Antigenic relationships between human caliciviruses and Norwalk virus. *Journal of Infectious Diseases* 156: 806–814.)

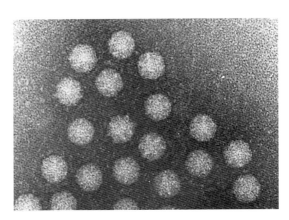

Fig. 56.4 Astroviruses, showing surface star. Original magnification ×280 000.

PATHOGENESIS AND CLINICAL FEATURES

trypsin is incorporated in the medium. Immunofluorescence studies show that replication occurs within the cytoplasm and that a protein is present in the nucleolus at an early stage in the replication cycle. Electron microscopic examination of thin sections of infected cells shows the presence of crystalline arrays of particles adjacent to cytoplasmic vacuoles.

Human volunteer studies with Hawaii and Norwalk viruses have shown that replication occurs in the jejunum. Light microscopy showed that the villi in the proximal part of the small intestine were broadened and blunted, and the enterocytes covering the damaged villi were cuboidal and vacuolated. At the same time the numbers of intra-epithelial lymphocytes and neutrophils were increased.

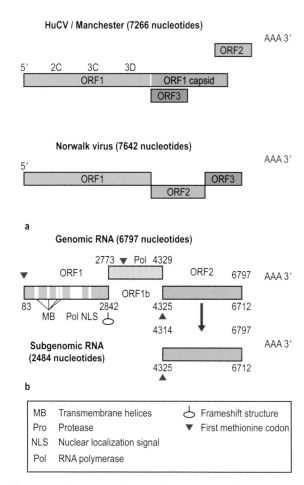

HuCV / Manchester (7266 nucleotides)

Norwalk virus (7642 nucleotides)

a

Genomic RNA (6797 nucleotides)

Subgenomic RNA (2484 nucleotides)

b

MB	Transmembrane helices	⏚	Frameshift structure
Pro	Protease	▼	First methionine codon
NLS	Nuclear localization signal		
Pol	RNA polymerase		

Fig. 56.5 a Genomic organization of the two genera of human caliciviruses: *Sapovirus* (Manchester strain) and *Norovirus* (Norwalk virus). b Genomic arrangement of astrovirus. HuCV, human calicivirus. (From Murphy et al 1995 by kind permission of Springer-Verlag.)

Electron microscopical studies showed that epithelial cells remained intact but the microvilli were disarranged and reduced in length. A similar histopathological picture has been found in calves infected experimentally with a bovine calicivirus ('Newbury agent 2').

Limited data on astrovirus infection in man come from the examination of duodenal and jejunal biopsies obtained from infants with symptoms of gastro-enteritis. Histopathological examination showed blunting of the villi, non-specific alterations in epithelial cells, and a mixed lamina propria inflammatory infiltrate. In the jejunum, viral antigen was present in surface epithelia near the villus tips (Fig. 56.6). Electron microscopic examination of jejunal enterocytes showed the presence of paracrystalline arrays of virus.

Clinical features

The clinical features of infection with calicivirus and astrovirus are shown in Table 56.2. The symptoms are similar, but vomiting, sometimes projectile, is more frequently reported in calicivirus infections. In some outbreaks of calicivirus infection involving adults, the illness resembles 'gastric flu' (i.e. diarrhoea, headache, fever, aching limbs and malaise). Patients of all age groups are affected by norovirus and in outbreaks attack rates are often greater than 70%.

The incubation period for caliciviruses is between 12 and 72 h, and slightly longer (3–4 days), for astrovirus. Illness typically lasts for 1–4 days, with excretion of detectable numbers of particles for the same period. Application of RT-PCR has shown that virus may continue to be shed for up to 2 weeks. Occasionally, symptoms persist for up to 2 weeks.

In patients with severe combined immune deficiency disease, persistent excretion of caliciviruses, astroviruses and rotaviruses can occur, either individually or simultaneously; in one report a patient was found to be excreting five different enteric viruses over a period of several weeks before he died.

Symptoms of illness are generally mild and seldom require admission to hospital. However, when outbreaks occur among debilitated elderly patients or infants with other underlying problems, intravenous rehydration may be necessary; fatalities are extremely rare.

LABORATORY DIAGNOSIS

Specimens required

Faecal samples should be collected as soon as possible and stored at 4°C. Paired serum samples should be obtained, the first taken as soon as possible after onset of symptoms and a further sample 10–14 days later. Blood samples from infants can be obtained by finger or heel pricks and dried on filter paper.

Laboratory tests

Until recently the only widely available test for the diagnosis of calicivirus and astrovirus infections was electron microscopy. All of the viruses are small and often difficult to recognize. The sensitivity of electron microscopy and virus particle recognition are increased by solid-phase immune electron microscopy (IEM) or conventional IEM, in which virus reacts with antibodies in a fluid phase, resulting in aggregates of particles, provided that the particles are not masked by excess antibody. IEM can also be used to measure antibody responses.

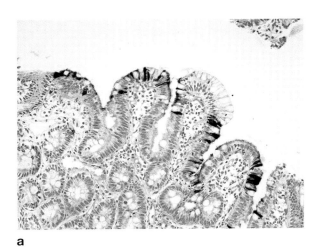
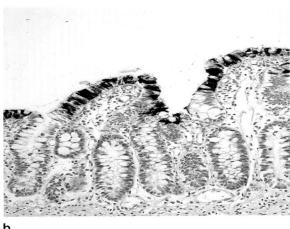

a   b

Fig. 56.6 a Duodenal and b jejunal biopsies from a bone marrow transplant recipient with astrovirus infection stained with anti-astrovirus monoclonal antibody, demonstrating progressively more extensive staining of surface epithelial cells near the villus tips. (From Sebire N J, Malone M, Shah N, Anderson G, Gaspar H B, Cubitt WD 2004 Pathology of astrovirus associated diarrhoea in a paediatric bone marrow transplant recipient. *Journal of Clinical Pathology* 57: 1001–1003.)

Table 56.2   Clinical features recorded in outbreaks

Virus	No. of cases	Vomiting	Diarrhoea	Fever	Abdominal pain	Nausea	Aching limbs	Headache
	Cases presenting with symptom (%)							
HuCV/Norwalk	30	66	83	47	70	100	73	83
HuCV/Snow Mountain	59	71	70	32	67	72	NS	68
HuCV/UK	181	52	66	65	60	NS	56	NS
HuCV/UK	9	100	22	NS	33	NS	NS	NS
HuCV/Sapporo	250	42	96	18	77	NS	NS	NS
Astrovirus	14	74	30	30	49	NS	NS	NS

HuCV, human calicivirus; NS, not stated.

Enzyme immuno-assays

Commercial assays are available to detect some strains of norovirus; human astroviruses share a group antigen and a commercial enzyme immuno-assay is available from Dako Ltd that detects all eight serotypes. An enzyme immuno-assay for sapovirus is not available.

RT-PCR

The success in obtaining the complete sequence of three caliciviruses (Norwalk and Southampton in 1991, and Manchester in 1995) has enabled primers to be designed, directed to the highly conserved 3-D region, the VP1 region or the area bridging the two regions. RT-PCR is now the most widely used method for diagnosing norovirus and sapovirus infections, but there is a lack of agreement about which primers are best, because of the extreme genetic diversity of these viruses. The predominant strain circulating in the community often changes from year to year.

RT-PCR has been applied for the diagnosis of astrovirus infection using primers directed to conserved regions within ORF2. Primers have been designed that are group-specific and others that can differentiate between the serotypes.

EPIDEMIOLOGY

Age distribution

Tests for antigen and sero-epidemiological surveys indicate that caliciviruses and astroviruses have a worldwide distribution. All age groups can be affected, but outbreaks

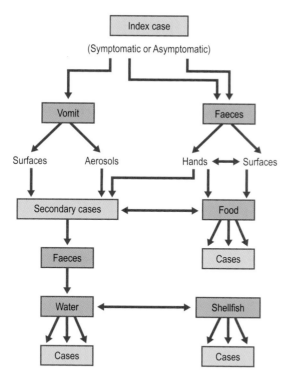

Fig. 56.7 Routes of transmission of viruses associated with gastro-enteritis.

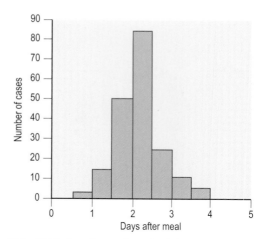

Fig. 56.8 Distribution of cases of norovirus infection following a meal, indicative of a point source of infection.

of sapovirus and astrovirus infection most commonly involve:

- infants
- schoolchildren
- the elderly.

In contrast, noroviruses cause extensive outbreaks, particularly among adults and the elderly, although there are increasing reports of episodes in children.

Modes of transmission

The various routes of transmission are shown in Figure 56.7. Epidemiological studies can identify two characteristic epidemic curves for an outbreak. In one (Fig. 56.8), when most cases appear at about the same time, this usually results from a *point source*, such as contaminated food or water. In the other (Fig. 56.9), cases occur in smaller numbers over a longer period, often with short intervals between the occurrence of new cases. This is characteristic of person-to-person spread.

Faecal–oral route

Affected individuals may excrete large numbers of particles in their faeces and/or vomit (>10⁹ particles per gram). Volunteer studies indicate that virus can remain viable for several years and that the infectious dose is 10–100 particles. Therefore, even minor contamination of hands, work surfaces, taps, carpets, etc. can be a major source of infection. This is illustrated by the number of people who can become ill when food is contaminated by a single handler (Table 56.3).

Respiratory route

There is strong epidemiological evidence that inhalation of aerosols of vomit or faecal material, from bed-linen or

Fig. 56.9 Distribution of cases of astrovirus infection in a geriatric ward, indicative of person-to-person spread.

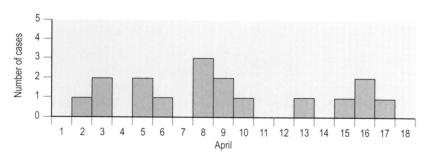

Table 56.3 Food- and water-borne infections with noroviruses

Source	Origin	Attack rate
Asymptomatic food handler; salads	USA	220/383 (57%)
Asymptomatic food handler; melon	UK	239/280 (85%)
Caterers; cold foods	Japan	835/3000 (28%)
Oysters	UK	500/1700 (29%)
Oysters	Japan	63/121 (52%)
Cockles and mussels	UK	>130/>300
Drinking water	USA	495/647 (76%)
Secondary home contacts		719/1740 (41%)

nappies, can result in infection. There is no evidence that virus replicates in the respiratory tract.

Cold foods

Cold foods that undergo extensive handling during preparation are a major source of outbreaks of norovirus infection. The foods involved, such as sandwiches, iced cakes, melons and salads, are not generally considered as potential vehicles of food poisoning; the true extent of the problem may not be recognized.

Shellfish

Numerous outbreaks of gastro-enteritis due to noroviruses and, occasionally, astroviruses have been caused by the consumption of bivalve shellfish (i.e. oysters, clams, mussels, cockles and scallops). These are harvested from estuarine or coastal waters polluted with faecal material, which is greatly concentrated by the filter-feeding bivalves. Methods to cleanse them prior to consumption (i.e. holding them in tanks of ultraviolet-irradiated, circulating filtered water), although successful in removing bacteria, are ineffective in freeing them of viruses. A further problem is that shellfish are frequently eaten raw or after minimal cooking.

Water

Outbreaks of norovirus and astrovirus infection associated with the consumption of untreated water, contaminated municipal drinking water or ice have been reported in the USA, UK and Australia. In the developing world, where water is often untreated, the problem is likely to be far greater.

Asymptomatic excretors

Epidemiological investigation and volunteer trials have shown that asymptomatic excretion of noroviruses and astroviruses is not uncommon. Such individuals may serve as an important reservoir of infection, particularly in situations such as hospitals or the catering industry.

TREATMENT

At present there is no specific treatment for infections with these agents. Severe dehydration in infants or elderly debilitated patients should be managed in hospital with parenteral fluid replacement.

CONTROL

Control of food–borne outbreaks

The following guidelines for the management of an outbreak associated with food have been proposed in the UK by the Public Health Laboratory Service Working Party on Viral Gastro-enteritis.

- Staff who develop or have had symptoms such as diarrhoea and/or vomiting should be excluded from work until 48 h after recovery.
- If kitchen or adjacent areas have been fouled (e.g. by vomitus) then (a) the area should be thoroughly cleaned and disinfected with a 10 000 ppm hypochlorite solution, and (b) all food to be eaten uncooked should be destroyed.
- The importance of hygienic practices, particularly hand-washing, should be reinforced.
- High-risk foods such as bivalve shellfish should be excluded from the kitchen, and other foods that require much handling (e.g. salads and sandwiches) should be bought in or obtained from other branches if at all possible.
- Unnecessary kitchen traffic should be stopped: the kitchen should not be used as a short cut for other staff, particularly during an outbreak.

Management and prevention of hospital outbreaks

- Ensure that both bacteriological and virological investigations are instigated at the same time.
- Whenever possible, affected patients should be isolated and infected nursing, medical and support staff excluded from work.
- All staff and patients on an affected ward should be screened as asymptomatic infections are common.
- Particular attention should be paid to hand-washing; 70–90% methanol or ethanol has been shown to be

effective against astroviruses and rotaviruses even in the presence of faeces.

- Bed-pan washers need to be examined to ensure they are working efficiently.
- In some outbreaks it may be necessary to close wards to new admissions until all patients have stopped excreting virus and no new cases have occurred for a period of 72 h.
- Staff movement from affected to unaffected wards should be restricted, group activities stopped and visits by children discouraged.

KEY POINTS

- Caliciviruses (noroviruses and sapoviruses) are common causes of diarrhoeal disease transmitted worldwide by the faecal–oral route and via fomites. Diagnosis is established by electron microscopic examination of faeces or by RT-PCR.
- They cause diarrhoea and vomiting with an incubation period of 12–72 h (3–4 days for astroviruses). Illness lasts for 1–4 days, and virus may be shed for up to 2 weeks.
- The very young and elderly are most at risk and may require rehydration. Prolonged excretion occurs in some immunocompromised patients.
- Cold foods are an important source, as are shellfish harvested from contaminated seawater.

RECOMMENDED READING

Brievogel B 2003 Diary of a Norwalk virus. Following an outbreak in long-term care. *Assisted Living Success* 3: 1–9

Caul E O 1996 Viral gastroenteritis: small round structured viruses, caliciviruses and astroviruses. Part II. The epidemiological perspective. *Journal of Clinical Pathology* 12: 959–964

Cowden J M, Wall P G, Adak C, Evans H, le Baigue S, Ross D 1995 Outbreaks of foodborne infectious intestinal disease in England and Wales; 1992 and 1993. *Public Health Laboratory Service Communicable Disease Report* 5: review 8

International Workshop on Human Caliciviruses 1999 *Journal of Infectious Diseases* 181(Suppl 2): S1–391

Kroneman A, Vennema H, van Duijinhoven Y, Duizer E, Koopmans M 2005 High number of Norovirus outbreaks associated with a GII.4 variant in the Netherlands and elsewhere: does this herald a worldwide increase? *Eurosurveillance* 10: 51–52

Moreno-Espinosa S, Farkas T, Jiang X 2004 Human caliciviruses and pediatric gastroenteritis. *Seminars in Pediatric Infectious Diseases* 4: 237–245

Murphy A F, Fauquet C M, Bishop D H L, et al (eds) 1995 *Virus Taxonomy: The Sixth Report of the International Committee on Taxonomy of Viruses*, p. 365. Springer-Verlag, Wien

Willcocks M M, Carter M J, Madeley C R 1992 Astroviruses. *Reviews in Medical Virology* 2: 97–106

Internet site

Institute for Animal Health. http://www.caliciviridae.com

57 Coronaviruses

Upper respiratory tract disease

J. M. Darville

Until the identification in 2003 of the agent responsible for severe acute respiratory syndrome (SARS) as a coronavirus (SARS-CoV), this group of viruses was considered to include only minor human pathogens. Coronaviruses were discovered in the early 1930s when an acute respiratory infection of domesticated chickens was shown to be caused by a virus now known as avian infectious bronchitis virus (IBV). In the early 1950s it was shown that colonies of mice maintained for laboratory research were endemically infected with a virus that in some circumstances caused outbreaks of fatal hepatitis. This virus was termed mouse hepatitis virus (MHV). Feline infectious peritonitis virus has been associated retrospectively with a case reported in the literature in 1912.

In the 1960s, research with human volunteers at the Common Cold Unit near Salisbury, UK, showed that colds could be induced by nasal washings that did not contain rhinoviruses. Subsequent in-vitro work using organ cultures revealed the presence of enveloped viruses in these washings. Electron microscopy studies demonstrated a unique morphology for these viruses, and comparative studies demonstrated a similarity to the previously described IBV and MHV. The term *coronavirus* was adopted for these agents in 1968, reflecting their morphology in the electron microscope after negative staining, and in 1975 the name Coronaviridae was accepted for the family. It is now evident that coronaviruses are widespread in nature, infecting a range of hosts with strong but variable tissue tropisms. Although highly species specific, they can cross species barriers to cause severe disease in the new host. In man they appear to be limited to infections of the respiratory and probably the enteric tracts, but their involvement in systemic disease in other animals suggests that they may behave likewise in human beings. Evidence for central nervous system (CNS) involvement in man is now emerging and research in this area is expanding.

PROPERTIES

Morphology and structure

The most studied coronaviruses are IBV, MHV and SARS-CoV. The particles are pleomorphic and enveloped, varying between 60 and 220 nm in diameter, although this measurement is affected by the choice of negative stain. Widely spaced club-shaped surface projections or peplomers (composed of surface, or S, protein) of approximately 20 nm in length are seen in all species, giving the particles their characteristic fringed appearance, reminiscent of the solar corona, when negatively stained (Fig. 57.1). It is from this unique appearance that the name of the virus (Latin, *corona* = crown) is derived. There is some variation in the morphology and spacing of the peplomers. Bovine coronavirus (BCoV) has a distinct inner fringe of short peplomers of haemagglutinin esterase, as well as the outer fringe. This double fringe is also seen in other coronaviruses of the same antigenic group, including human coronavirus (HCoV) OC43.

The genome is encoded in non-segmented single-stranded positive-sense RNA of approximately 30 kilobases, making these the largest known RNA virus genomes. In the virion this is complexed with nucleoprotein (N) in an extended helical nucleocapsid 9–11 nm in diameter. This is enclosed within a lipoprotein envelope in association with a transmembrane protein (M). Coronaviruses also have a small membrane minor protein required for virus replication. S protein is the major inducer of neutralizing antibody, although haemagglutinin esterase also induces it; monoclonal antibodies raised against M protein can neutralize infectivity in the presence of complement. Antigenic variation is a feature of the S protein, whereas the N protein is relatively conserved. In addition to the proteins described, other minor proteins have been reported but not consistently.

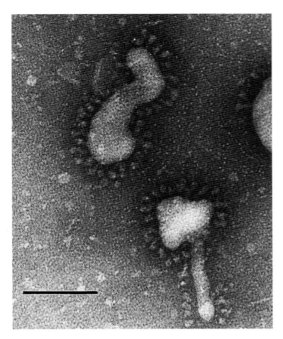

Fig. 57.1 Particles of HCoV serogroup 229E grown in human fibroblast cells and stained with 1.5% phosphotungstic acid. Bar, 100 nm.

Genetic recombination readily occurs between members of the same and of different coronavirus groups. Many coronaviruses have the ability to haemagglutinate, a property that has been used in their diagnosis.

Taxonomy

Viruses of the order Nidovirales have similarities in the organization and expression of their genomes. In particular, all produce nested messenger RNAs (Latin, *nidus* = nest). The order comprises the two families Coronaviridae and Arteriviridae. The former contains two genera, *Coronavirus* and *Torovirus*. Toroviruses are well recognized as animal pathogens, and have been detected in human infantile gastro-enteritis. The Arteriviridae contains one genus, *Arterivirus*, which contains no known human species. Coronaviruses are readily distinguishable from one another by their limited host range. Serological (N, M and S proteins) and phylogenetic studies have revealed relationships that allow the classification of coronaviruses into three or possibly four groups (Box 57.1). Some remain unclassified.

It is accepted that 229E and OC43 are two species of HCoV causing respiratory disease, rather than two strains of one species. Human enteric coronavirus (HECoV) is a likely third species. SARS-driven research has led to the discovery of two more coronaviruses implicated in human respiratory disease. These are HCoV-NL63, closely related to HCoV-229E in group 1, and CoV-HKU1, a

group 2 agent not demonstrably related to others in the group. Also recognized are coronaviruses of chickens (causing bronchitis), mice (hepatitis, gastro-enteritis and encephalitis), pigs (gastro-enteritis and encephalomyelitis), dogs, turkeys, cattle, horses (all gastro-enteritis), rats (pneumonia and swelling of salivary glands) and cats (peritonitis). In addition, coronavirus-like particles have been demonstrated in the faeces of cats, monkeys and rabbits, both with enteritis and in health. In all probability coronaviruses infect all higher animal species.

Cultivation

Some avian and mammalian coronaviruses can be cultivated readily in fertile eggs or in cell culture, but many, including HCoV, are difficult to grow. The traditional method of cultivating HCoV was in fetal tracheal organ culture. However, strains related to 229E may be isolated more readily in human diploid cells, and strains related to OC43 may be adapted to grow in them after initial isolation in organ culture. A wider range of cell lines capable of supporting coronavirus replication has now been developed. SARS-CoV grows readily in Vero cells.

Replication

Coronaviruses attach to glycoprotein receptors on host cells via their S (and haemagglutinin esterase) proteins.

After fusion of the viral envelope with the host cell membrane, the core penetrates the cell. An RNA-dependent RNA polymerase translated from the genomic RNA makes the negative strand template from which it then synthesizes a series of 3′ co-terminal nested genomic mRNAs. The viruses replicate in the cytoplasm with a growth cycle of 10–12 h. They bud not from the plasma membrane but from the rough endoplasmic reticulum (where the M protein localizes) into intracytoplasmic vesicles. These are transported via the Golgi apparatus to the plasma membrane through which they are released by exocytosis. Viral infection may result in cell lysis, but fusion of adjacent cells leading to the formation of syncytia can also occur and the virus may have a potential for persistence. The optimum temperature for the replication of HCoV is 33°C, reflecting its usual habitat in the upper respiratory tract.

Pathogenesis

It is presumed that the major pathogenic process in HCoV infection is the direct cytolysis of infected cells. However, the ciliostasis observed in organ culture may also be a contributory factor in vivo. As HCoV can cause disease on re-infection soon after primary infection, it has been speculated that the immune response has some role in causing or aggravating acute disease. Coronaviruses can cause disease by immunopathological mechanisms, as in the examples already given of CNS disease in mice and rats and peritoneal disease in cats. Furthermore, in SARS high levels of cytokines and infiltrates of mononuclear cells are present in the lung.

Transmission

Coronaviruses usually infect via the gut and/or respiratory tract. Different isolates of the same coronavirus type show different tropisms for the gut and respiratory tract. This tropism is determined by the S protein and by the type and distribution of receptors. All species may replicate in the respiratory tract to some extent. Once established in the gut or respiratory tract some viruses remain there, whereas others can spread to specific organs; infections of the liver and CNS are well recognized in animals. Coronaviruses can also infect macrophages and may persist in neural cells. HCoV infects the respiratory tract by the inhalation of droplets or aerosols generated by the coughs and sneezes of infected individuals. There is some evidence that fomites are a secondary factor in transmission.

CLINICAL FEATURES

Human coronavirus

The only significant condition known to follow HCoV infection is upper respiratory tract disease; it is estimated that coronaviruses cause up to 30% of 'common colds'. This figure could well be higher depending on the prevalence of unrecognized HCoVs. Statistically, when compared with rhinoviruses, coronaviruses cause more coryza and discharge but less pharyngitis and coughing. However, individual cases cannot be attributed to either virus group on clinical grounds.

In some cases infection is mild or even subclinical. In contrast, there is some suspicion that HCoV may cause severe lower respiratory tract infection in the very old and the very young, including premature infants. However, the association is tentative at present. There is some evidence that they may cause pneumonia in immunocompromised patients. Coronavirus infections of the upper respiratory tract have been linked with wheezing attacks, especially in asthmatic children, in whom the virus may persist.

Although the morbidity caused by HCoV infection is trivial to the individual, it is economically important as a major contributor to time lost from work and academic studies.

The incubation period is 2–4 days, and virus is detectable at the onset of symptoms and for 1–4 days thereafter. Symptoms outlast virus shedding and typically persist for a week, perhaps as a result of secondary bacterial infection.

After infection, humoral and local serological responses are detectable and cellular immune responses probably develop. These are presumed to be responsible for clearing the virus. Despite this, however, re-infection commonly occurs, and this can happen as little as 4 months after infection with the same serotype. Detectable antibody usually disappears after about 1 year.

CNS involvement

HCoVs are neuro-invasive. They can replicate in, persist in and activate neural cell lines of human origin and models of putative human CNS diseases have been developed. Molecular evidence also suggests that HCoVs often reach the CNS. The HCoVs have been implicated in the aetiology of human multiple sclerosis, but their role, if any, is unclear.

SARS coronavirus

This clinically distinct respiratory illness emerged in late 2002 in southern China, whence it spread to Hong Kong and thence to South-East Asia, North America and Europe in the space of a few weeks. Within a few months three groups had independently demonstrated that the illness was caused by a coronavirus new to the human population. The SARS coronavirus is very closely related to a coronavirus isolated from the palm civet, and in all probability the virus crossed to man from a captive wild animal in a street market.

Transmission of SARS-CoV from person to person requires close contact, such as that seen in families and health-care settings. This indicates that large droplets are most effective in transmission, although fomites are probably also involved. As the virus is shed in stools there may also be a role for faecal–respiratory spread. In hospitals the use of negative-pressure isolation, the wearing of gloves, gowns, masks and eye protection, and the rigorous observance of washing and disinfection procedures have proved successful in stopping outbreaks.

SARS has an incubation period of almost a week. It is often severe and sometimes fatal. The first non-specific signs and symptoms of infection include fever, malaise and myalgia; upper respiratory tract symptoms are rarely seen. At this stage there is a high viral load. Generally these symptoms abate before the patient develops pneumonia, for which oxygen may be required. Diarrhoea is also a common feature. By now the viral load has fallen and immunopathological mechanisms predominate. About 20% of patients progress to acute respiratory distress syndrome that requires mechanical ventilation. The mortality rate in SARS, originally thought to be exceptionally high, is about 5%, similar to that seen in other community-acquired pneumonias.

Treatment is essentially symptomatic. The use of ribavirin has been at best equivocal and may indeed be damaging through its toxic side effects. Steroid treatment may be of value in reducing the inflammatory component of the disease.

SARS-CoV has now disappeared from human populations but it, or a similar virus, could re-emerge.

Human enteric coronavirus

In 1975 simultaneous reports of the discovery of coronavirus-like particles in human faeces were made in Bristol, UK and southern India. It was soon shown that they could be propagated in human fetal intestinal organ cultures with a replication cycle indistinguishable from that of HCoV. The agent has now been adapted to grow in a range of cell lines and subjected to detailed analysis. Although a causative association of *human enteric coronavirus* (HECoV) with human enteric disease has yet to be proven, coronaviruses are significant agents of enteritis in other animals. Two serotypes of HECoV may exist, one of which is reported to cross-react serologically with HCoV strain OC43.

Other than the labour-intensive organ culture method described above, the only means of detecting HECoV is by electron microscopy. The introduction of other techniques awaits its confirmation as a pathogen.

HECoV appears to be endemic throughout the world, with a high prevalence in developing countries, where there may be some seasonal variation. In western countries the prevalence is high in travellers from third world countries and in low socio-economic groups, and is markedly higher in male homosexuals than in the normal population. Although it has not been proved that HECoV is spread by the enteric or faecal–oral route, there is strong circumstantial evidence that this is so; transmission by contaminated water may also be possible. The observed high prevalence among western male homosexuals may be explained by oral–anal–genital contact.

Non-human coronaviruses

In other animals coronaviruses regularly infect and cause disease beyond the mucosal surfaces. In infected adult mice, for example, hepatitis is sometimes a consequence of reactivation. In rats and mice, MHV strain JHM has been shown to cause demyelinating disease via an auto-immune mechanism. The cause of an encephalomyelitis in pigs has been shown to be a coronavirus, haemagglutinating encephalomyelitis virus. Feline infectious peritonitis virus infects most species and at all ages, and there is evidence for a carrier state. Although usually asymptomatic, the infection can cause severe peritonitis through antibody-dependent enhancement of infectivity, and is usually fatal.

LABORATORY DIAGNOSIS

There is no clinical requirement for laboratory diagnosis as human respiratory tract infections with coronaviruses appear to be mild. At present, therefore, laboratory methods have only research and epidemiological applications. This would change, however, should SARs-CoV or another coronavirus causing severe disease become established in man.

Virus isolation

Members of the OC43 group of HCoV will grow only on primary inoculation in organ culture. Virus growth may be detected by indirect means, such as passage to suckling mouse brain or by the demonstration of a haemagglutinin. Passage to diploid cell cultures may result in a cytopathic effect, whereas infection of human volunteers may result in clinically apparent respiratory disease. Premature inhibition of ciliary motion in organ cultures (i.e. ciliostasis) can be seen. Furthermore, increasing numbers of characteristic particles can be demonstrated by electron microscopy.

In contrast to the OC43 group, the 229E group can be cultivated directly in human diploid cells such as W138 or human embryo kidney. The cytopathic effect, often described as 'tatty', resembles the non-specific degeneration of aged cultures with individual cells rounding and detaching from the cell monolayer. However, overlays such as agarose or methylcellulose may reveal the development of plaques by some strains, whereas others produce syncytia.

Serology

Antibodies to HCoV have been detected by complement fixation, haemagglutination inhibition (some strains) and neutralization. The latter is always a complex procedure and especially so with these viruses. The preferred technique for detecting anti-HCoV antibody is now enzyme-linked immunosorbent assay (ELISA), although limited to research and epidemiology. Antibody is a marker of past infection but, as re-infections are well documented, such antibody cannot be regarded as reflecting immunity.

Rapid techniques

Although not yet of any practical value in monitoring HCoV disease, the methods developed for the rapid diagnosis of other agents have been applied to HCoV. ELISAs have been used to detect antigens in respiratory secretions and, as coronavirus antigens are associated with cell membranes, the virus can be detected using immuno-fluorescence techniques, like other respiratory viruses. Reliable reverse transcriptase–polymerase chain reaction techniques for detecting and quantifying coronavirus RNA are widely available. Molecular techniques were vital to the identification of SARS-CoV.

EPIDEMIOLOGY

Studies using both virus isolation and serology have shown that HCoV infections occur wherever in the world they have been sought. Similarly, HCoV activity can be widespread in the community in widely separated areas. Rates of infection are similar in all age groups and there are no other factors, such as sex or socio-economic status, that influence the frequency of infection. Coronaviruses thus behave much as other human respiratory viruses.

Infection with HCoV is markedly seasonal, with peaks of disease usually occurring in late winter or early spring. However, the exact timing of the peak can vary within the season, as it does with respiratory syncytial virus, and occasionally peaks of infection have been observed outside the usual season.

The two serogroups, 229E and OC43, display an unusual periodicity. Each group becomes prevalent every 2–3 years with only sporadic isolates being made of the non-dominant type within the same community. However, the cycles do overlap, so that in a single season 229E may predominate in one region and OC43 in another. This short-term cycling indicates that, despite the diversity that exists within the OC43 serogroups at least, there are functionally few serotypes of HCoV.

CONTROL

Treatment

The treatment of coronavirus infections, remains symptomatic only. Indeed, as human disease is almost invariably mild, HCoV infection ranks low on the list of candidates for specific antiviral chemotherapy. Although ribavirin has an in-vitro effect against coronaviruses it is not used in vivo, other than for SARS-CoV infection.

Prevention

Given the economic impact of the common cold, HCoV infections, like those with rhinoviruses, are obvious targets for control. Recombinant α-interferon has been used to prevent infection in volunteers in experimental studies. The ease of re-infection suggests that the development of conventional vaccines would be an unrewarding approach. Nevertheless, in the veterinary field, effective vaccines have been developed to protect animals of economic importance from outbreaks of infection, in which flocks or herds can be devastated. For example, both attenuated and inactivated vaccines are available to protect chickens from IBV, and live vaccines are used in sows to give their piglets lactogenic immunity to porcine transmissible gastro-enteritis virus. There is, however, a need for improved vaccines. Care is needed to avoid creating further problems through postulated immune-mediated pathogenic mechanisms, such as have been seen with some feline coronavirus vaccines and with an inactivated human measles virus vaccine.

KEY POINTS

- Coronaviruses cause upper respiratory tract infection and possibly enteric infection in man.
- They are widespread among mammals and birds, causing severe systemic infections in these animals.
- SARS transferred to man from the wild animal reservoir in 2002 and was then transmitted from person to person. The disease has now disappeared.
- No vaccines or antivirals are available for HCoV.

RECOMMENDED READING

Ashley C, Caul E O 1989 Human enteric coronaviruses. In Farthing M J G (ed.) *Viruses and the Gut (Proceedings of the Ninth BSG: SK & F International Workshop 1988)*, pp. 91–95. Smith, Kline and French Laboratories, Welwyn Garden City

Cavanagh D 2004 Coronaviruses and toroviruses. In Zuckerman A J, Banatvala J E, Pattison J R, Griffiths P, Schoub B (eds) *Principles and Practice of Clinical Virology*, 5th edn, pp. 379–397. John Wiley, Chichester

Lavi E, Schwartz T, Jin Y-P, Fu L 1999 Nidovirus infections: experimental model systems of human neurologic diseases. *Journal of Neuropathology and Experimental Neurology* 58: 1197–1206

Luby J P, Clinton R, Kurtz S 1999 Adaptation of human enteric coronavirus to growth in cell lines. *Journal of Clinical Virology* 12: 43–51

McIntosh K 1997 Coronaviruses. In Richman D D, Whitley R J, Hayden F G (eds) *Clinical Virology*, pp. 1123–1132. Churchill Livingstone, New York

Murphy F A, Gibbs E P J, Horzinek M C, Studdert M J 1999 1. Coronaviridae and 2. Arteriviridae. In *Veterinary Virology*, 3rd edn, pp. 495–508, 509–515. Academic Press, San Diego

Siddell S G (ed.) 1995 *The Coronaviridae*. Plenum Press, New York

St-Jean J R, Jacomy H, Desforges M et al 2004 Human respiratory coronavirus OC43: genetic stability and neuroinvasion. *Journal of Virology* 78: 8824–8834

The New England Journal of Medicine 2003; 348 contains several papers on SARS and the discovery of SARS-CoV.

Internet sites

National Center for Biotechnology Information. http://www.ncbi.nlm.nih.gov

Centers for Diseae Control and Prevention. Emerging infectious diseases. http://www.cdc.gov/ncidod/EID/

Wikipedia. Coronavirus. http://en.wikipedia.org/wiki/Coronavirus

www-micro.msb.le.ac.uk/3035/Coronaviruses.html

Virology Journal. All the Virology on the WWW. http://www.virology.net

58 Rhabdoviruses

Rabies

S. Sutherland

The hosts of viruses that constitute the family Rhabdoviridae include mammals, reptiles, birds, fish, insects and plants. By definition, members of the family must be enveloped, single-stranded RNA viruses with bullet-shaped or rod-shaped morphology. As with other virus families its members are unrelated serologically unless they form part of a distinct genus. About 80 viruses have vertebrate hosts, with many found in invertebrates also. Most of these rhabdoviruses have not yet been fully categorized, but two genera are known whose members have important roles in animal or human disease. The smaller genus, with which this chapter is concerned, has been designated *Lyssavirus*, the name being derived from the Greek word for madness or frenzy. This genus has rabies virus as its prototype. The larger genus, named *Vesiculovirus*, whose members are associated with the disease vesicular stomatitis, is particularly prevalent among horses, cattle and pigs in the Americas.

Rabies has been recognized in human beings and animals for many centuries, even before biblical times, as a distressing disease that develops rapidly into an acute encephalomyelitis, often frenzied initially, then subsiding into delirium, coma and death. A prominent feature in man is *hydrophobia* – fear of water and inability to swallow it. Figure 58.1 shows part of a case report from Edinburgh, dated 1747.

STRUCTURE

The genus *Lyssavirus* includes rabies virus and six other rabies-related viruses (Table 58.1).

Rabies virus, first established as a transmissible agent by Pasteur, is bullet shaped (Fig. 58.2), measures 75×80 nm and has helical symmetry. The non-segmental RNA genome has negative sense, a molecular weight of 4.5×10^6 Da and encodes five viral proteins: nucleoprotein, phosphoprotein, matrix protein, glycoprotein and polymerase. The nucleoprotein is closely attached to the single-stranded RNA genome. This nucleoprotein core together with the polymerase and phosphoprotein is enclosed in the matrix protein and covered by a lipoprotein membrane into which is inserted the glycoprotein (Fig. 58.3). The RNA-dependent RNA polymerase complex is necessary to initiate replication. Molecular analysis of the glycoprotein in virulent and non-virulent strains has shown an amino acid substitution at site 333 in the more pathogenic variants.

Rabies virus is rapidly inactivated at 60°C. It can remain viable for some days at 4°C, for longer periods when stored as infected brain tissue suspended in 50% glycerol at 4°C, and indefinitely when stored at or below −70°C or in a freeze-dried state. It is sensitive to sunlight, ultraviolet light, X-rays, lipid solvents, β-propiolactone, detergents and proteolytic enzymes.

REPLICATION

Virus replication will occur in all warm-blooded animals, mice being mainly used for primary isolation. Growth also occurs in the chick or duck embryo and in a range of cell cultures, including baby hamster kidney and mouse neuroblastoma cells, human diploid lung fibroblasts, chick embryo fibroblasts and Vero monkey kidney cells, although with minimal cytopathic effects. The last three are among cell lines used in vaccine production.

Virus attaches to cells by the glycoprotein. In neural tissue, attachment is at neuromuscular junctions by the acetylcholine receptors, neuronal cell adhesion molecules and the P75 neurotrophin receptor. However, although this may account for the localization and spread of the virus within the nervous tissue, there must be other receptors as the host cell range is broad and not confined to the central nervous system. Entry by endocytosis is followed by fusion of the viral envelope and cell membrane, and release of the nucleocapsid. Transcription of five messenger RNA species is catalysed by the polymerase-phosphoprotein complex, resulting in the three structural proteins and two others that form the polymerase complex. Viral RNA is replicated on a positive-strand template by a viral polymerase. Only negative strands are enclosed in new virions. The matrix

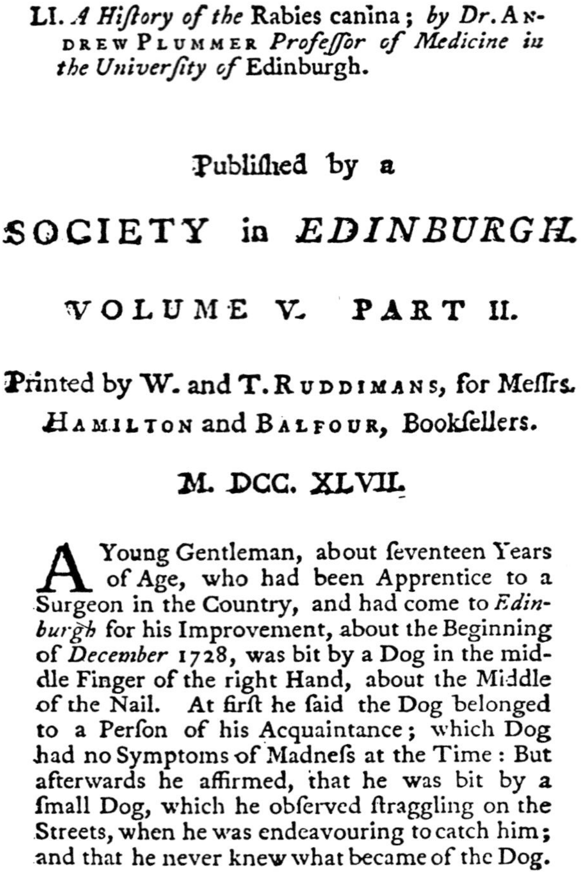

LI. *A Hiſtory of the* Rabies canina; *by Dr.* ANDREW PLUMMER *Profeſſor of Medicine in the Univerſity of* Edinburgh.

Publiſhed by a

SOCIETY in *EDINBURGH.*

VOLUME V. PART II.

Printed by W. and T. RUDDIMANS, for Meſſrs. HAMILTON and BALFOUR, Bookſellers.

M. DCC. XLVII.

A Young Gentleman, about ſeventeen Years of Age, who had been Apprentice to a Surgeon in the Country, and had come to *Edinburgh* for his Improvement, about the Beginning of *December* 1728, was bit by a Dog in the middle Finger of the right Hand, about the Middle of the Nail. At firſt he ſaid the Dog belonged to a Perſon of his Acquaintance; which Dog had no Symptoms of Madneſs at the Time: But afterwards he affirmed, that he was bit by a ſmall Dog, which he obſerved ſtraggling on the Streets, when he was endeavouring to catch him; and that he never knew what became of the Dog.

Fig. 58.1 Rabies case report, Edinburgh 1747. (Courtesy of Dr A. G. Dempster, University of Otago, New Zealand.)

Table 58.1 Lyssaviruses

Species	Serotype/genotype	Phylogroup	Source	Occurrence
Rabies (prototype)	1	I	Carnivores, cattle, man, bats (warm-blooded animals)	Worldwide
Lagos bat (rabies-like)	2	II	Fruit-eating bats, cats	West Africa
Mokola (rabies-like)	3	II	Shrews, man, cats, dogs, rodents	Africa
Duvenhage (rabies-like)	4	I	Man, insectivorous bats	South Africa, Europe
European bat lyssavirus 1	5	I	Bats, man	Europe
European bat lyssavirus 2	6	I	Bats, man	Europe
Australian bat lyssavirus	7	I	Fruit- and insect-eating bats, man	Australia

protein appears to be important in packaging the RNA and nucleoprotein and linking it to the envelope. Virions are formed by budding at the endoplasmic reticulum of the cell. The virus affects cell protein synthesis and the cell will die but, before this stage, it is possible to detect viral antigens by immunofluorescence or immunoprecipitation tests. Accumulation of cytoplasmic viral protein inclusions (*Negri bodies*) may be visible by light microscopy after appropriate staining. This has long been a useful diagnostic feature.

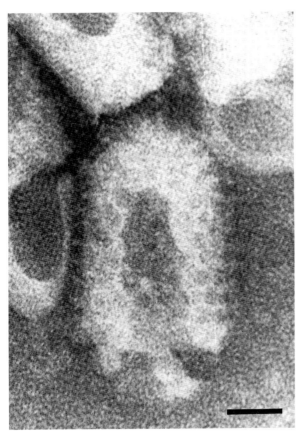

Fig. 58.2 Rabies virus particle. Bar, 30 nm. (Courtesy of Dr Joan Crick, Animal Virus Research Institute, Pirbright, UK.)

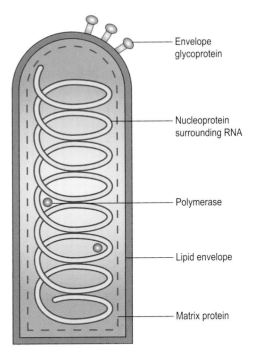

Fig. 58.3 Diagram of rabies virus.

Newly isolated strains from animals are capable of killing laboratory animals such as mice within 10–20 days after intracerebral inoculation. The appearance of eosinophilic Negri bodies, particularly in nerve cells of the hippocampus, brainstem or cerebellum, and specific rabies immunofluorescence confirm the diagnosis. Such strains, stemming from Pasteur, are called 'wild' or 'street' viruses. Serial animal passage may select out attenuated strains of lesser virulence called 'fixed' viruses. These viruses are usually no longer able to multiply when injected extraneurally, but reversion to greater virulence is possible by the use of an alternative host.

TRANSMISSION

Rabies virus does not penetrate intact skin; it enters the body from the saliva of a rabid animal by:

- abrasions or scratches on the skin
- mucous membranes exposed to saliva from licks
- most frequently via deep penetrating bite wounds.

The amount of virus excreted in saliva is variable. Bites through clothing may reduce the amount of virus reaching the wound so that not every person bitten by a rabid animal necessarily develops the disease. Uncommon routes include:

- inhalation while in a bat-infested cave
- aerosols released during centrifugation of infected materials in the laboratory
- ingestion of the flesh of rabid animals; high doses of virus would be necessary in such instances
- corneal transplants.

Despite the excretion of virus in saliva and conjunctival exudates, person-to-person spread has not been recorded. Transfer via infected corneal transplants has, however, been reported on at least eight occasions in five countries.

Transmission from a donor who died from encephalitis of unknown origin to four recipients of kidneys, liver and an arterial segment occurred in the USA in 2004. Subsequently a history of a bat bite in the donor was obtained.

CLINICAL FEATURES AND PATHOGENESIS

Rabies may present as:

- predominantly encephalitic disease – *furious rabies*
- paralytic illness – *dumb rabies*.

In man, about two-thirds develop the encephalitic form and die within 7 days; the rest initially present with paralysis then develop encephalitis, and death may not occur for 2–3 weeks. Survival is exceedingly rare.

The incubation period in man, mostly pinpointed from the time of a bite, can be very variable. It may be less than a week after head and neck wounds, when the virus site of entry is close to the brain, or range up to several years. Molecular studies have identified cases with incubation periods of 4 and 6 years. The average is between 1 and 3 months, with a shorter duration in children than adults. Initial virus replication is considered to occur in the tissues at the point of entry, persisting there for 48–72 h. The virus then gains access to the nerves via the motor end-plates. Once within the nerve fibres it is out of reach of any circulating antibody as it travels by anterograde axoplasmic flow along the axons towards the central nervous system. The two forms cannot be correlated with specific anatomical localization. The manifestations of illness are initially fever, malaise and headache, then symptoms related to the wound site – tingling, pain, lumbar weakness and ascending paralysis after leg bites and numbness, and pain with increasing shoulder weakness after hand or arm bites.

The prodromal symptoms include a profound sense of apprehension and feelings of irritation with paraesthesia at the wound sites. There are complaints of dry throat, cough and thirst, but patients will not drink. High fever, rigors, difficulty in swallowing and revulsion to water predominate, followed by bizarre behaviour, excitement, agitation, hallucinatory seizures, laryngeal spasms, choking and gagging, intermingled with lucid intervals. Spasms can be induced by tactile, auditory, visual or olfactory stimuli. This state, the furious form of rabies, gradually subsides into delirium, convulsions, coma and death. Sometimes only the dumb form is seen, with symmetrical ascending paralysis followed by coma and death.

The clinical course is usually short but causes much suffering as the patient often remains conscious. Once a diagnosis has been made immunoglobulins or antivirals are ineffective and the patient should be cared for with heavy sedation in a quiet, draught-free place.

The predominantly neurological mode of spread has some experimental support, as section of the main nerve trunk proximal to the inoculation site can prolong the incubation period, as can the use of drugs such as colchicine, which inhibit axonal flow. Experimentally, pathogenicity has been related to the capacity of a strain to induce cell fusion in neuroblastoma cells.

Despite an inflammatory reaction from the developing encephalomyelitis accompanied by considerable virus multiplication, observable damage to the nerve cells in the brain appears minimal, although if survival is sufficiently prolonged specific Negri bodies will appear.

Non-specific changes include a parenchymal microglial response and perivascular cuffing, with lymphocyte and plasma cell infiltration in the grey matter of the brainstem and spinal cord. From the brain the virus spreads via efferent nerves to most body tissues, including:

- salivary glands, with multiplication in the acinar cells and extrusion into the saliva
- conjunctival cells, with release into tears and exudates
- the kidneys, with excretion in the urine
- lactating glands and milk after pregnancy.

Virus has also been found in the suprarenal glands, pancreas, myocardium and at the base of hair follicles, as in the neck. Study by immunofluorescence of frozen sections of hair follicle biopsy from the nape of the neck may provide an initial diagnosis while the person is still alive. Virus excretion is intermittent and, although there is no evidence of risk to staff, the World Health Organization (WHO) recommends that staff caring for patients should adhere to barrier nursing techniques as for other infectious diseases.

LABORATORY DIAGNOSIS

The history may be so characteristic that early laboratory confirmation of rabies, often a problem in itself, is not requested. Difficulties can arise when information about exposure to a rabid animal is not elicited, as shown in the USA where more than a fifth of deaths due to rabies are in this category. It is worth remembering that when any visits, particularly to known rabies-endemic areas, have been made up to a year or more before a death that is ascribed to encephalitis, post-mortem examination of brain and cord tissue should include a search for rabies virus.

Because of risks from contact and handling, the British Advisory Committee on Dangerous Pathogens, in line with WHO safety recommendations, has classed rabies virus as a hazard group 3 pathogen on the basis that it can cause severe human disease and is dangerous for any person in contact, but effective prophylaxis is available. Regulations provide that high-risk diseases such as rabies and viral haemorrhagic fevers should be treated in secure isolation units, and the virological investigation of specimens from such patients is permitted only in specially designated laboratories. Similarly, for suspect rabid animals there is a designated veterinary laboratory. Stringent government regulations allow propagation of rabies virus only in designated category 4 laboratories.

Where rabies is endemic, wild animals or bats captured after biting incidents should be sent immediately for laboratory confirmation of rabies, but post-exposure treatment of persons bitten should not be delayed pending a laboratory diagnosis. Domesticated dogs and cats, particularly if previously vaccinated against rabies, may be observed in isolation for up to 10–14 days. If they survive for that time it is unlikely they were incubating rabies at the time of the incident but, if they succumb quickly, antirabies treatment of persons bitten should be started without waiting for laboratory confirmation.

Diagnostic methods

Virus detection is difficult as excretion of virus is intermittent and so negative results do not exclude rabies infection.

Virus identification. Identification of rabies antigen by specific immunofluorescence is rapid and sensitive.

- Ante mortem – in hair follicles. The WHO recommends examination of at least 20 thin sections of a frozen skin biopsy from the nape of the neck. Negative results, which are frequent, do not exclude rabies.
- Post mortem – in impression smears of the cut surface of the salivary gland, hippocampus, brainstem or cerebellum.

Virus isolation:

- Ante mortem – from saliva or cerebrospinal fluid.
- Post mortem – from salivary gland or brain tissue extract by mouse intracerebral inoculation or mouse neuroblastoma cell culture inoculation may be used, but cytopathic changes are minimal and viral antigen must be looked for by specific immunofluorescence techniques. However, this can shorten the time to diagnosis to 1–2 days, from 10–15 days for the mouse inoculation test.

Histological examination. This involves examination of fixed brain tissues by staining or immunofluorescence, including a search for Negri inclusions.

Serology. Fluorescence or enzyme-linked immuno-sorbent assay is carried out on serum or cerebrospinal fluid for evidence of specific antibody.

Detection of RNA. Amplification of DNA following reverse transcription of viral RNA is the most sensitive technique now in use and can be used on saliva to give the best ante-mortem results. When combined with the use of restriction enzymes, species-specific variants can be identified, allowing useful molecular epidemiology to trace the source of the virus in patients with no history of exposure.

EPIDEMIOLOGY

More than 99% of human deaths due to rabies occur in tropical and developing countries in Africa, Asia and South America. World surveys identify about 55 000 cases annually. The highest numbers are in India and 90% result from dog bites. About 50% of dog bite exposures are in children in poor rural areas where they are unlikely to receive post-exposure prophylaxis.

During 1946–2005 there were 23 human cases of rabies in the UK. All except one were in people infected in other countries. The remaining case in Scotland in 2002 was a bat handler who died from a European bat lyssavirus 2 (EBLV-2) infection. Of the 15 cases since 1975, none had received any post-exposure treatment.

Among animals, the disease may spread via two intercommunicating pathways, urban and sylvatic.

- The *urban* mode, more immediately dangerous for human beings, diffuses among domestic or scavenger dogs and cats, a situation prevailing in third world countries.
- The *sylvatic* mode, which affects small carnivores and mustelids, provides less opportunity for contact with man. Its main focus in different regions may be confined to separate species, for example foxes in continental Europe, Canada and northern parts of the USA, racoons along the eastern seaboard, skunks in the mid-western states and coyotes in the southern states of the USA, and the mongoose in the West Indies. Spread to other species or to man, although a potential threat, appears unusual. Perpetuation of the disease comes particularly from salivary excretion of the virus during the early stages of illness, combined with the bites of carnivores.

Dogs, and to a lesser extent cats, are the main sources of human infection so that control of such animals by surveillance, with elimination of strays and the initiation of vaccination programmes, as has occurred in the USA and many European countries, has considerably reduced the incidence of human rabies in these countries. In several European countries the main reservoir of virus for dogs and cats was the red fox, but the use of oral vaccines in bait since 1978 in Switzerland and subsequently in Belgium, Finland, France, Italy, Luxemburg and the Netherlands eradicated fox rabies by 2001.

As well as the involvement of terrestrial animals there is widespread rabies infection in numerous species of bats. This was first noted in Brazil in about 1916 when it was realized that both cattle and human beings could develop rabies after being bitten by blood-sucking vampire bats. A similar situation was then found to exist in many Central and South American countries, and also in the West Indies, with considerable mortality in cattle. Some human and animal infections may have followed the eating of infected carcasses.

Reports, initially from the USA in 1953, have shown that many species of insectivorous and fruit-eating bats may also harbour rabies viruses, with excretion and transfer of infection when they are sick. It is now believed that affected vampire bats, although thought to be mostly carriers, do ultimately succumb to the disease. Spread can occur readily in some bat colonies from the close contact among large numbers congregating together.

Most viruses isolated from bats in the western hemisphere resemble the classic serotype 1 rabies virus, although in cross-protection and neutralization tests, as would be expected, minor antigenic differences are found. As dog rabies is brought under control in the USA and Europe, a proportionate increase in bat rabies infections has been identified. Of 21 human rabies cases indigenous to the USA from 1990 to 1997, 19 (90%) were bat variants even though a history of bat bites was obtained in only one. Bites by bats are assumed to be the usual mode of transmission to man, although in two reported fatal cases in the USA, in 1956 and 1958, direct inhalation of virus while working in bat-infested caves in Texas led to the disease.

In continental Europe rabies-like viruses have also been recovered occasionally from bat species in Germany since 1954. Infected bats have also been identified in the former Yugoslavia, Turkey, the former USSR, Poland, Finland, the Czech Republic, Denmark, Holland, France, Switzerland, Spain and England. A bat virus closely related to rabies has been responsible for at least two human deaths in Australia, which is officially rabies-free.

In the UK the system of 6 months' quarantine for imported canines and felines was introduced in 1886, together with measures for the muzzling of dogs. By 1902 these steps had succeeded in freeing the country from rabies. The disease was re-introduced in 1918 via a dog brought back illegally from Europe. The disease was again eliminated from animals by 1922. Since that time the UK has been free from indigenous rabies despite two incidents in 1969 and 1970 when dogs released from quarantine developed rabies but, fortunately, there was no further spread. The quarantine barrier was extended to include most other imported animals, with an additional proviso that quarantined dogs and cats must be injected, under veterinary supervision, with an acceptable animal rabies vaccine on entry to quarantine and again after 1 month.

In September 1998 the Report of the Advisory Group on Quarantine recommended that, for some animals from certain designated countries, quarantine should be replaced by a system ensuring that an imported animal was electronically identifiable by microchip, vaccinated against rabies, blood-tested to confirm immunity, freedom from tapeworms and ticks, and certified as such. As a result of these recommendations, the Pet Travel Scheme was piloted from 28 February 2000. This allows pet dogs and cats from western Europe and assistance dogs from Australia to enter without quarantine provided they have complied with the above scheme. At present pets from other locations are still subject to quarantine and vaccination regulations, but in future the pet travel scheme is likely to be extended. In Britain, foxes have not been involved, but since 1996 bat viruses have been isolated from four Daubenton's bats in England and a recent survey demonstrated EBLV antibody in 4% of bats sampled.

IMMUNOPROPHYLAXIS

The risk of infection must first be assessed to take account of the type of exposure, the animal involved and whether rabies is known to be present in that species in the geographical area where the injury occurred. As there is no effective antiviral drug, post-exposure immunoprophylaxis must be started as soon as possible after the bite or other type of exposure. The following procedures are recommended in the UK and USA for those who have not previously had a full intramuscular vaccine course.

- Thorough wound cleaning with soap solution or a detergent and running water for 5 min followed by application of 40–70% alcohol or tincture or aqueous solutions of iodine or quaternary ammonium compounds. Scrubbing should be avoided and suturing delayed if possible.
- Passive treatment with antirabies immunoglobulin, preferably human, 20 IU/kg body-weight, given half in and around the wound and half in the gluteal muscle.
- Active immunization with inactivated whole virus vaccine (grown in cell culture), containing at least 2.5 IU/dose, given into the deltoid muscle. The complete course of post-exposure vaccination recommended in the UK comprises five intramuscular 1-ml doses on days 0, 3, 7, 14 and 30. The WHO also approves two alternative schedules: (a) the 2-1-1 regimen in which a 1-ml intramuscular dose given into each deltoid muscle on day 0 is followed by one dose on days 7 and 21; and (b) an intradermal 2-2-2-0-1-1 regimen in which two doses of 0.1 ml are given on days 0, 3 and 7 and one dose of 0.1 ml on days 30 and 90.

In individuals previously immunized against rabies, passive prophylaxis is not given but two booster doses of vaccine are given on days 0 and 3.

Britain usually offers vaccine only to persons bitten while abroad, but will make an exception for anyone bitten by a bat.

Vaccination

Pasteur introduced vaccination after exposure to rabies in 1885 on the basis that a long incubation period should allow time for immunity to develop before the onset of symptoms. His vaccine was a crude extract of rabbit spinal cord containing virus 'fixed' as a result of serial passage. A well publicized early success established the procedure, which is still in use despite various vicissitudes. A phenolized brain suspension formed the basis of the *Semple vaccine*, used in the UK from 1919 to1966, although there was a high rate of allergic encephalitis as a side effect.

A *suckling mouse brain vaccine* was developed containing much less myelin and causing less allergic encephalitis. It is widely used in Latin American countries. A non-neurogenic *duck embryo vaccine* was used in the UK from 1966 to 1976, but this vaccine has been superseded by *cell culture vaccines*, of which there are several types. Those available worldwide, though not necessarily used because of their high cost, include: diploid cell vaccine, purified chick embryo cell vaccine and Vero cell vaccine.

These all have good immunogenicity and safety. However, they are still not available to the poor in most endemic areas and do not protect against genotype 2 and 3 lyssaviruses.

The diploid cell vaccine is the only one licensed in the UK for both pre- and post-exposure prophylaxis and is the only vaccine recommended for intradermal administration. Severe reactions are rare after use, although up to 20% may report minor local effects, and a smaller proportion report systemic, influenza-like or sensitization effects. Intradermal vaccination is not recommended while antimalarials are being taken because these may interfere with the immune response. Immunosuppressed persons may show a poor response and should have antibody levels checked.

In the UK, stocks of vaccine and of human rabies immune globulin are held and distributed mainly through the Health Protection Agency in England and Wales and from designated centres in Scotland and Northern Ireland.

Several vaccine types and modes of delivery have been used in animals. Those used in dogs and cats are given intramuscularly, and require boosting every 1–3 years depending on the vaccine type. Attempts have been made to control the infection in wild animals by the use of live-attenuated vaccines delivered orally. By carefully selecting desirable baits, such as chicken heads for foxes, successful vaccination programmes have been carried out in several European countries, in parts of Canada and the USA.

Work continues on a number of recombinant and subunit vaccines, and on alternative antibody preparations for use in man.

CONTROL

Because rabies has a worldwide distribution its complete elimination would need the eradication of infection from all susceptible animal species. First steps in this direction have been made to control rabies in foxes in Europe and Canada and racoons in the USA. As most human exposure has resulted from contact with infected dogs and cats, vaccination of domestic dogs and cats combined with post-exposure prophylaxis for those exposed in specific incidents to suspect rabid animals, and pre-exposure for those who may come in contact with such animals in the course of their work, has reduced the number of human cases in many countries. Pre-exposure vaccination is recommended for the following groups:

- laboratory workers handling the virus
- those handling imported animals at animal quarantine centres, zoos, research centres and ports
- veterinarians and their technical staff
- animal health inspectors
- licensed bat handlers
- travellers to enzootic areas if work involves handling animals or patients with rabies
- those travelling more than a day's journey from modern medical treatment.

Pre-exposure immunization requires two injections of 1 ml vaccine, given into the deltoid muscle 4 weeks apart. A test for neutralizing antibody is advised 4 weeks later. A reinforcing dose is given at 12 months. A booster should be given after any potential exposure. Those at high risk of exposure should have periodic antibody tests and boosting as required every 6 months to 2 years.

Clear signs of case reduction have come from the developed countries that have applied this scheme, and now the need is to extend the procedure to other, particularly enzootic, regions where dogs are the major reservoir of the virus. Past attempts to control wildlife rabies by such draconian measures as shooting and gassing have had short-lived effects. Species vary in their susceptibility to rabies and live vaccine strains that may be used for one species may be unsuitable for another. Extension to canine and other species may become feasible when suitable oral vaccine strains are identified.

The WHO has introduced a computerized data management scheme, RABNET2, which supplies interactive maps of rabies data at global and country level.

KEY POINTS

- Rabies has the highest case-fatality rate of any infectious disease. Globally at least 55 000 cases occur annually.
- Most human infections result from bites by rabid dogs; about half occur in children under 15 years in poor rural parts of the developing world. The virus enters wounds and travels to neurones via nerve axons, causing a fatal encephalitis.
- There is no effective treatment, but immunoprophylaxis and vaccination given soon after injury can prevent the disease.

- Control of animal rabies by vaccination of dogs and some wild animals reduces the risk to human beings.
- Bat viruses are increasingly recognized as the cause of human rabies.
- Pre-exposure immunization is recommended for groups at high risk such as animal handlers, virus laboratory workers and travellers to remote endemic areas.

RECOMMENDED READING

Baer G M 1994 Rabies – an historical perspective. *Infectious Agents and Disease* 3: 168–180

Department of Health 2000 *Memorandum on Rabies: Prevention and Control.* Department of Health, London

Rupprecht C E, Dietzschold B, Koprowski H (eds) 1994 Lyssaviruses. *Current Topics in Microbiology and Immunology* 187

Smith J S 1989 Rabies virus epitopic variation: use in ecologic studies. *Advances in Virus Research* 36: 215–253

Winkler W G, Bogel K 1992 Control of rabies in wildlife. *Scientific American* 266: 86–92

Internet sites

Department of Health. Memorandum on Rabies: Prevention and Control. http://www.dh.gov.uk

World Health Organization. Human and animal rabies. http://www.who.int/rabies

World Health Organization. Rabnet. http://www.who.int/rabies/rabnet

World Health Organization. Immunization, vaccines and biologicals. Rabies vaccine. http://www.who.int/vaccines/en/rabies.shtml

Department for Environment, Food and Rural Affairs. Disease factsheet: Rabies. http://www.defra.gov.uk/animalh/diseases/notifiable/rabies/index.htm

59

Transmissible spongiform encephalopathies (prion diseases)

Scrapie; bovine spongiform encephalopathy; Creutzfeldt–Jakob disease; variant Creutzfeldt–Jakob disease

J. W. Ironside

INTRODUCTION

The transmissible spongiform encephalopathies (TSEs), or prion diseases, are a unique group of fatal neuro-degenerative disorders occurring in human beings and animals that take their name from two major characteristics:

1.   All are transmissible to a variety of mammals, either experimentally or by natural exposure. The precise nature of the transmissible agents involved is unknown (see below), but they possess physical and chemical properties that are quite distinct from those of conventional viruses and bacteria. One hypothesis concerning these agents suggests that they are composed entirely of protein, without any nucleic acid, for which the term *prion* (proteinaceous infectious particle) is used. No evidence of a conventional host immune reaction has been found in these diseases.

2.   The diseases caused by these agents are characterized in all species by spongiform change in the central nervous system (CNS). This consists of numerous small vacuoles (10–200 μm) that are formed within neuronal cell bodies and their processes (Fig. 59.1a). Neuronal death and a reactive proliferation of astrocytes and microglia also occur, and an abnormal form of the prion protein (PrP^{Sc}) accumulates within the CNS, both as diffuse deposits and in the form of amyloid plaques.

SCRAPIE

The first animal spongiform encephalopathy to be described was *scrapie*, an endemic disorder of sheep and goats that has been present in Europe for at least two centuries. Affected animals become ataxic and wasted, and often rub or scrape the fleece off the sides of their bodies, hence the name. The first experimental transmission of scrapie from affected to healthy sheep by intra-ocular injection of spinal cord homogenate was reported in 1936. Scrapie can be transmitted from ewe to lamb, but the precise mechanism and route of infection is unknown. Natural transmission is thought to occur at parturition, or by scarification or via the oral (alimentary) route. The placenta from an infected ewe is one known source of infection that can contaminate the farm environment. There is no epidemiological evidence that scrapie is pathogenic to man.

The transmissible agent: a prion?

The scrapie agent is by far the best characterized of the TSE agents. Although its precise nature is uncertain, it is subviral in size and notoriously resistant to inactivation by many physical and chemical agents, including:

- heat
- exposure to ionizing or ultraviolet radiation
- deoxyribonuclease (DNAase) and ribonuclease (RNAase)
- formaldehyde and glutaraldehyde.

The agent has been studied by experimental transmission to mice and hamsters; this has identified around 20 strains of scrapie. These strains are defined by their biological properties, particularly the disease incubation period and the nature and distribution of the pathology in the brain of the inoculated animals. Incubation periods range from 60 days to more than 2 years, which is close to the natural lifespan of mice and hamsters, and cases of asymptomatic infection have been recorded in inoculated animals who die from unrelated causes after a prolonged period. The incubation period is influenced by:

- the route of inoculation – intracerebral inoculation is the most efficient mode of infection; oral or parenteral routes are less efficient
- the dose of the injected inoculum and its infective titre.

Host genetic factors also influence the incubation period (susceptibility) of these diseases and it has recently been recognized that the major host gene involved is the prion protein gene.

Electron microscopical studies of extracts from scrapie-infected brain have revealed abnormal fibrillary

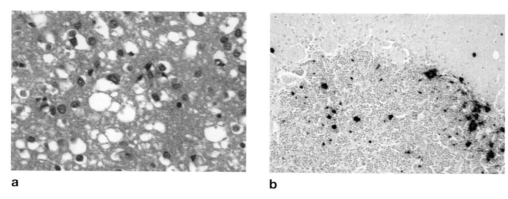

Fig. 59.1 a Spongiform change in the cerebral cortex in a case of sporadic CJD consists of numerous small cyst-like spaces that tend to coalesce in the neuropil and around neurones (centre). Haematoxylin and eosin stain, original magnification ×400. b In this case of sporadic CJD, prion protein (PrP) accumulation in the cerebellum has occurred in the form of numerous plaques that stain intensely on immunohistochemistry for PrP. Original magnification ×200.

structures named scrapie-associated fibrils, which comprise a modified form of the normal host prion protein (PrP^C). PrP^C is expressed in a variety of cells, including neurones in the CNS, where it may act as a copper-binding protein, and is thought to play a role in synaptic function. This protein has a molecular weight of 33–35 kDa and is highly conserved through a wide range of species, but its amino acid sequence varies from one species to another. This variation is thought to influence the species barrier, a factor that regulates the success of disease transmission from one species to another under experimental and natural conditions.

During scrapie infection, PrP^C appears to undergo a conformational change to convert to PrP^{Sc}. This abnormal isoform has a relatively high beta-pleated sheet conformation, which renders it partially resistant to digestion with proteinase K and allows it to aggregate as scrapie-associated fibrils in the brain. In the prion hypothesis, conversion of the PrP^C to PrP^{Sc} occurs by a direct interaction, which sparks off a catalytic conversion. This has been replicated experimentally in a cell-free system; it is not known whether this biochemical change alone accounts for all the biological properties of TSE agents.

Many of the physical and chemical features of TSE agents can be explained by the prion (protein-only) hypothesis. However, the mechanisms by which the various strains of scrapie agent have different biological properties are unclear; some research groups believe that a second (unidentified) molecule determines the strain of the infectious agent. Claims have been made that recombinant prion protein is infectious, but this remains a controversial area.

Pathogenesis

Transmission of scrapie to experimental hamsters and mice can be achieved by inoculation of infected brain tissue into a number of sites; the most rapid onset of disease occurs following intracerebral inoculation. After oral inoculation, spread of infection may occur along either the vagus nerve to the brainstem, or along sympathetic nerve fibres that connect via the splanchnic nerve complex to the spinal cord. Infection then spreads to the brain at a rate of around 1 mm/day, probably within neurones. Peripheral inoculation of the agent may be followed by a phase of accumulation and replication in the lymphoreticular system, mainly in follicular dendritic cells. In the spleen, there is a postulated 'neuro-immune' connection that allows access to the splanchnic plexus and then to the CNS. In the brain, the mechanisms through which spongiform change occurs are unknown, although dilatation of neuronal lysosomal, Golgi and endoplasmic reticulum structures has been suggested.

BOVINE SPONGIFORM ENCEPHALOPATHY AND OTHER ANIMAL PRION DISEASES

Scrapie-like diseases have been recorded in an increasing number of captive or domesticated animals (Table 59.1), the most significant of which is bovine spongiform encephalopathy (BSE). This was identified in 1985 in the UK and spread rapidly throughout the cattle population to affect around 1% of all adult cattle by 1992. It appears that this epidemic resulted from:

- oral ingestion of a TSE agent (possibly a novel strain of scrapie) via contaminated meat and bone meal in cattle feed
- changes in the rendering processes by which this feed was produced in the 1980s, allowing the BSE agent to persist as a contaminant

Table 59.1 Animal diseases caused by scrapie and related agents

Disease	Occurrence	Hosts
Scrapie	Common in many countries worldwide	Sheep and goats
Transmissible mink encephalopathy	Very rare, mostly in farms in the USA	Mink
Chronic wasting disease	Spreading from Colorado and Wyoming	Mule deer and elk
Bovine spongiform encephalopathy	Widespread epidemic in dairy cattle in the UK; smaller outbreaks in other countries	Domestic cattle
Unnamed spongiform encephalopathy	Small number of exotic ungulates in zoos	Nyala, gemsbok, greater kudu, Arabian oryx, eland
Feline spongiform encephalopathy	Small numbers of domestic cats in the UK and a few large cats in zoos	Domestic cats, cheetah, lion, puma, ocelot, tiger

- recycling of contaminated cattle carcasses in the meat and bone meal fed to cattle appears to have fuelled the epidemic.

The incidence of BSE is now declining as a result of a ban on the use of bovine organs as foodstuffs for other animals. Experimental studies on BSE indicate that it represents a single strain of the TSE agent. The BSE agent was also transmitted by foodstuff to domestic cats, large cats and exotic ungulates in zoos (see Table 59.1). There is good evidence that the BSE agent has also spread to man, causing a new disease known as variant Creutzfeldt–Jakob disease (CJD; see below).

HUMAN PRION DISEASES

The commonest form of human TSE was first described in the 1920s and is known as *Creutzfeldt–Jakob disease* after the authors of the early reports. Since then, an ever-widening spectrum of human TSE has been identified, with three main subgroups, comprising:

- idiopathic
- familial
- acquired.

The identification of the prion protein and sequencing of the human prion protein gene on chromosome 20 have greatly increased our understanding of human transmissible spongiform encephalopathies (Table 59.2).

DIAGNOSIS

The diagnosis of human prion diseases requires a range of clinical, biochemical, genetic and pathological studies. Careful studies of the clinical features (see below) by an experienced neurologist allows a presumptive diagnosis of CJD to be made with a high level of accuracy (at least 70%). Analysis of the prion protein gene is essential to

Table 59.2 Human spongiform encephalopathies classified by aetiology

Aetiology	Encephalopathy
Idiopathic disorders	Sporadic CJD (around 90% of all cases)
Familial disorders	Familial CJD (around 10% of all cases)
	Gerstmann–Sträussler–Scheinker syndrome
	Fatal familial insomnia
Transmitted from person to person	Kuru
	Iatrogenic CJD (less than 1% of all cases)
Transmitted from bovines to man	Variant CJD

CJD, Creutzfeldt–Jakob disease.

identify cases of familial CJD associated with a pathogenic mutation. Genetic studies have also demonstrated a naturally occurring polymorphism at codon 129 in the prion protein gene, which is important in determining disease susceptibility. At present, there is no form of screening test for human prion diseases and no specific treatment is available. A definitive diagnosis depends on examination of the brain at autopsy. Brain biopsy is not a routine investigation, as the procedure can compromise an already ill patient and the neurosurgical instruments used have to be destroyed if the diagnosis is confirmed, to prevent iatrogenic infection via subsequent neurosurgical procedures.

NEUROPATHOLOGY

It is recommended that all suspected CJD cases should be investigated by autopsy, with appropriate permission for retention and examination of the brain to allow confirmation of the clinical diagnosis and accurate subclassification of the case. The principal neuropathological features of human TSEs are:

a

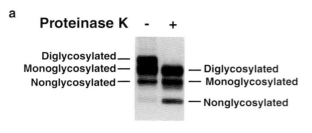

Proteinase K   -   +

Diglycosylated —
Monoglycosylated —
Nonglycosylated —

— Diglycosylated
— Monoglycosylated

— Nonglycosylated

b

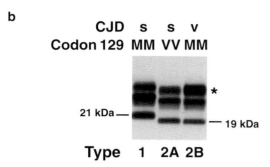

CJD   s   s   v
Codon 129   MM   VV   MM

      *

21 kDa —

— 19 kDa

Type   1   2A   2B

Fig. 59.2 Western blot analysis of prion protein (PrP) in post-mortem CJD brain. PrP from CJD brain occurs in three glycoforms resulting in diglycosylated, monoglycosylated or non-glycosylated PrP, which can be separated by western blotting into distinct bands. Proteinase K treatment (+) destroys the normal form of the prion protein, but in CJD it only partly denatures the disease-associated protein, which has an increased mobility (shown in **a**). Two main subtypes of the abnormal protein can be identified: **b** types 1 and 2 are found in sporadic CJD (s) irrespective of their genotype at the codon 129 genotype. Examples of two types (MM1 and VV2) are shown here. All cases of variant CJD (v) thus far tested are methionine homozygotes (MM) and have had a type 2 PrP characterized by the predominance of the diglycosylated glycoform (*), termed type 2B. The type 2 cases found in sporadic CJD, in which the monoglycosylated glycoform predominates, are termed type 2A.

- spongiform change (Fig. 59.1a)
- neuronal loss
- astrocytosis
- accumulation of PrP^{Sc}
- amyloid plaque formation (Fig. 59.1b).

A range of techniques is now available to detect PrP^{Sc} accumulation within the CNS. PrP^{Sc} can be detected by immunocytochemistry in the CNS in a variety of patterns, including amyloid plaques (see Fig. 59.1b), perivacuolar deposition around areas of spongiform change, perineuronal and axonal deposition. PrP^{Sc} can also be detected by western blot techniques, which allow further study of the PrP^{Sc} isotype in terms of its glycosylation and molecular weight following digestion with proteinase K (Fig. 59.2).

SPORADIC CJD

CJD occurs most commonly as a sporadic disorder, affecting around one person per million population per year.

Table 59.3 Prion protein gene codon 129 polymorphism in CJD

Codon 129 genotype	Methionine/ Methionine	Methionine/ Valine	Valine/ Valine
Normal population	37%	51%	12%
Sporadic CJD	75%	11%	14%
Variant CJD	100%	0%	0%

CJD, Creutzfeldt–Jakob disease.

CJD usually presents as a rapidly progressive dementia of less than 1 year's duration, often accompanied by other neurological abnormalities, including cerebellar dysfunction, pyramidal and extrapyramidal signs, cortical blindness and akinetic mutism. Many of these features evolve as the disease progresses, and electro-encephalography often shows characteristic diffuse, periodic, synchronous discharges. The peak incidence is in the seventh decade of life, but has been described in teenagers and even in the ninth decade. The disease is untreatable and invariably fatal; most patients survive for only 4 months after the onset of major symptoms.

A naturally occurring methionine/valine polymorphism at codon 129 in the prion protein gene is of major influence in determining susceptibility to sporadic CJD (Table 59.3). The mechanism for this influence is unclear. The cause of this disorder remains unknown; it might possibly reflect selective exposure to an unidentified ubiquitous agent, perhaps causing disease only in genetically susceptible individuals, or a somatic mutation involving the prion protein in the CNS (Fig. 59.3). No occupational or dietary risk factors exist for sporadic CJD. The global distribution of sporadic CJD is markedly different from that of scrapie, particularly in Australia, where scrapie has been eradicated for many decades, but sporadic CJD occurs at a similar frequency to that in the UK, where scrapie is endemic.

FAMILIAL PRION DISEASES

Around 10% of cases of CJD occur as inherited disorders; these are associated with mutations or insertions in the open reading frame of the human prion protein gene on chromosome 20 (see Fig. 59.3). In recent years, many new mutations and insertions have been identified in this gene in familial prion diseases. *Gerstmann–Sträussler–Scheinker syndrome* (GSS) is the best known example of an inherited human TSE. In this very rare disorder, which occurs as an autosomal dominant disease affecting middle-aged adults, cerebellar ataxia, nystagmus and gait abnormalities are prominent clinical features, with

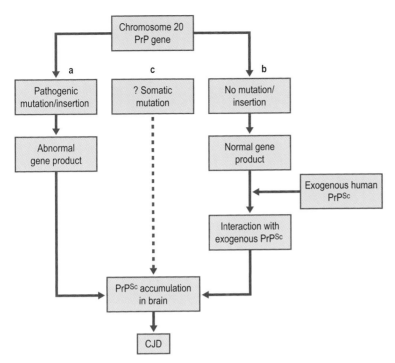

Fig. 59.3 Postulated mechanisms of PrP^Sc accumulation in CJD. In **a**, an inherited PrP gene mutation results in familial CJD. In **b**, exogenous human PrP^Sc interacts with the normal host precursor protein, resulting in PrP^Sc accumulation. This mechanism is thought to operate in iatrogenic CJD. The cause of sporadic CJD is unknown, but proponents of the prion hypothesis have suggested that a somatic mutation in a neurone (**c**) might initiate the process of PrP^Sc accumulation in the brain. This 'rare event' might also explain the low incidence of sporadic CJD.

dementia occurring only towards the end of the illness. The neuropathology is characteristic, with enormous multicentric PrP amyloid plaques present throughout the CNS and widespread loss of neurones, especially in the cerebellum and basal ganglia. However, these changes are not entirely specific; similar features may occur in both sporadic and other familial cases of CJD. GSS was the first human disease linked to a mutation in the PrP gene; a codon 102 proline-to-leucine mutation was first identified in 1989 and has subsequently been found in several GSS families, including the original kindred described in 1936. *Fatal familial insomnia* is an extremely rare inherited disorder, characterized clinically by disturbances of sleep and autonomic function with relative intellectual preservation in middle age. This disorder is associated with a unique PrP genotype (codon 178 asparagine, 129 methionine/methionine), and neuropathological changes mainly in the thalamus. The relationship between the genotype, neuropathology and clinical features of this enigmatic disorder await further study.

ACQUIRED PRION DISEASES

Iatrogenic CJD

CJD can occur by accidental transmission from one person to another (see Fig. 59.3). *Iatrogenic CJD* was first described in 1974 in a corneal graft recipient. Other examples of iatrogenic CJD have involved inoculation of contaminated material from a human CJD victim to another patient, for example following the implantation of inadequately decontaminated intracerebral electrodes, or via human dura mater grafts. In the UK, the commonest form of iatrogenic CJD occurs in recipients of human growth hormone derived from cadaveric pituitary glands, at an approximate incidence of 1 in 10 000 at-risk individuals, with an average age at death of 25 years and an incubation period of up to 20 years. This variant of CJD is unusual in its clinical features, in which a cerebellar syndrome predominates with prominent ataxia, gait and visual disturbances; gross cerebellar atrophy and a predominant accumulation of PrP within the cerebellum are characteristic neuropathological features.

Kuru

Kuru was a disease occurring in the Fore tribe of Papua New Guinea, apparently in association with practices related to ritualistic cannibalism. It is uncertain whether the brain was actually eaten, but it was handled after death, particularly by females and children, who went on to develop this disorder. The name *kuru* means shivering or trembling in the local dialect, and reflects the predominant clinical manifestations of a progressive cerebellar syndrome dementia. The incubation period was variable, ranging from around 5 to 40 years. There was no evidence of mother-to-child transmission. Its increasing prevalence made it the primary cause of death in most individuals within the tribes in the 1960s. The disease is now virtually extinct since cannibalistic practices were abandoned.

Variant CJD

The emergence of BSE as a new epidemic in cattle had potential implications for human disease, under the assumption that this disorder might be transmitted to man through the food chain. A National CJD Surveillance Project was established for the UK in May 1990, based at the Western General Hospital, Edinburgh. In 1996, the Unit identified a new form of human TSE, now known as *variant CJD*, which by 2006 had affected more than 160 people in the UK, 21 in France, four in Ireland, two in both Nertherlands and USA and one case in each of Canada, Japan, Portugal, Saudi Arabia and Spain.

- The age of onset is unusually young (mean 28 years), with a range from 12 to 74 years.
- The clinical illness is prolonged, with an average duration of 14 months (range 6–39 months).
- The clinical features are also unusual for CJD, with psychiatric and sensory symptoms at onset, followed by ataxia and myoclonus, with dementia only in the final stages of the illness.
- All patients with variant CJD so far are methionine homozygotes at codon 129 in the prion protein gene (see Table 59.3).

The neuropathological features are relatively uniform and consist of massive accumulations of PrP^{Sc} throughout the brain, with the formation of multiple amyloid plaques surrounded by spongiform change, known as florid plaques (Fig. 59.4). Variant CJD differs from other forms of human prion disease in that PrP^{Sc} can be detected in lymphoid tissues (in follicular dendritic cells) both during the clinical illness (Fig. 59.5) and in the late preclinical illness. This is consistent with variant CJD being acquired by the oral route (see below) and has implications for the possible transmission of the variant CJD agent by surgical instruments used on lymphoid tissues (e.g. in tonsillectomies). This finding could mean that the variant CJD agent is present in the blood of individuals in the preclinical incubation phase of the illness. Three cases of variant CJD transmission by blood transfusion have been reported in the UK. Many steps have been taken in the UK to prevent further transmission by this route, including filtration of whole blood to remove white cells (which may carry the highest levels of infectivity), sourcing of plasma products from overseas, and banning recipients

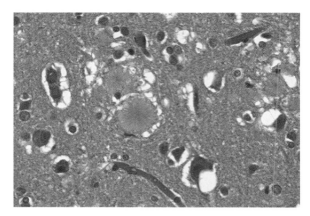

Fig. 59.4 The cerebral cortex in variant CJD contains numerous aggregates of PrP^{Sc} in the form of fibrillary amyloid plaques (centre), surrounded by spongiform change. Haematoxylin and eosin stain, original magnification ×200.

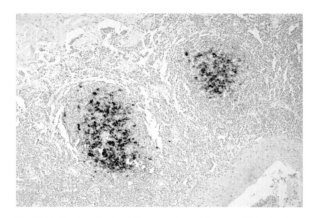

Fig. 59.5 Section of a tonsil from a patient with variant CJD stained to show the prion protein by immunohistochemistry. The dark stain demonstrates accumulation of the prion protein in the follicular dendritic cells within germinal centres of the tonsil. Accumulation of prion protein in lymphoid tissues does not occur in sporadic CJD. Original magnification ×100.

of blood transfusions from themselves becoming blood donors.

Experimental strain typing studies in mice have shown that the transmissible agent in variant CJD is identical to the BSE agent. Furthermore, biochemical analysis of the PrP^{Sc} from cases of variant CJD shows a similar banding pattern in western blot preparations to PrP^{Sc} from BSE in cattle, which is distinct from sporadic CJD (see Fig. 59.2). These findings provide strong support for the hypothesis that variant CJD is causally linked to BSE, and there is epidemiological evidence that dietary exposure through beef products is the most likely route of infection. Recently, a decrease in the incidence and death rate for variant CJD has been reported in the UK. However, because the incubation period for this new disease is unknown and the number of individuals exposed to the BSE agent uncertain, it is difficult to predict future numbers of variant CJD cases. Continuing surveillance is required.

KEY POINTS

- Prion diseases, or transmissible spongiform encephalopathies, are fatal neurodegenerative disorders caused by unconventional agents with very lengthy incubation periods. They are transmitted by inoculation or ingestion; the intracerebral route of transmission results in the shortest incubation period.
- Prion diseases occur in mammals, including scrapie in sheep, bovine spongiform encephalopathy (BSE) in cattle and Creutzfeldt–Jakob disease (CJD) in man.

- Prions are thought to lack nucleic acid, and appear to consist entirely of a modified host-encoded protein, the prion protein.
- Prion diseases do not elicit conventionally detectable immune responses. Asymptomatic or subclinical infections are very difficult to detect.
- Human prion diseases occur in sporadic, familial and acquired forms, all of which are rare diseases. In 1996, a new form of human prion disease known as variant CJD was identified, which results from infection with the BSE agent, probably via the oral route.

RECOMMENDED READING

Bruce M E, Will R G, Ironside J W et al 1997 Transmissions to mice indicate the 'new variant' CJD is caused by the BSE agent. *Nature* 389: 498–501

Chesebro B 2003 Introduction to the transmissible spongiform encephalopathies or prion diseases. *British Medical Bulletin* 66: 1–20

Ironside J W 1998 Prion diseases in man. *Journal of Pathology* 186: 227–234

Ironside J W, Head M W, Bell J E et al 2000 Laboratory diagnosis of variant Creutzfeldt–Jakob disease. *Histopathology* 37: 1–9

Ladogana A, Poupulo M, Croes E A et al 2005 Mortality from Creutzfeldt–Jakob disease and related disorders in Europe, Australia and Canada. *Neurology* 64: 1586–1591

Peden A H, Head M W, Ritchie D L et al 2004 Preclinical vCJD after blood transfusion in a PRNP codon 129 heterozgous patient. *Lancet* 364: 527–529

Prusiner S B 1998 Prions. *Proceedings of the National Academy of Sciences of the USA* 95: 13363–13383

Van Duijn C M, Delasnerie-Laupretre N, Masullo C et al 1998 Case-control study of risk factors of Creutzfeldt–Jakob disease in Europe during 1993–1995. *Lancet* 351: 1081–1085

Will R G, Ironside J W, Zeidler M et al 1996 A new variant of Creutzfeldt–Jakob disease in the UK. *Lancet* 347: 921–925

Internet sites

National Creutzfeldt–Jakob Disease Surveillance Unit. http://www.cjd.ed.ac.uk

Medical Research Council Prion Unit. http://www.prion.ucl.ac.uk

Department of Health. http://www.doh.gov.uk/cjd (search for CJD)

DEFRA. Bovine Spongiform Encephalopathy (BSE). http://www.defra.gov.uk/animalh/bse/index.html

PART 5

FUNGAL PATHOGENS, PARASITIC INFECTIONS AND MEDICAL ENTOMOLOGY

60 Fungi

Superficial, subcutaneous and systemic mycoses

D. W. Warnock

Fungi constitute a large, diverse group of heterotrophic organisms, most of which are found as saprophytes in the soil and on decomposing organic matter. They are eukaryotic, with a range of internal membrane systems, membrane-bound organelles, and a well defined cell wall composed largely of polysaccharides (glucan, mannan) and chitin. They show considerable variation in size and form, but can be divided into three main groups:

1. *Moulds (multicellular filamentous fungi)*, which are composed of branching filaments, termed hyphae, that grow by apical extension to form an intertwined mass, termed a mycelium. In most fungi the hyphae have regular cross-walls (septa) but in lower fungi these are usually absent. Moulds reproduce by means of spores produced, often in large numbers, by an asexual process (involving mitosis only) or as a result of sexual reproduction (involving meiosis, preceded by fusion of the nuclei of two haploid cells). Many fungi can produce more than one type of spore, depending on the growth conditions. The precise method of spore production and the type(s) of spore produced are unique to each individual fungal species. In some higher fungi the sexual spores are produced in macroscopic structures such as mushrooms and toadstools. In laboratory cultures, moulds produce mainly asexual spores.

2. *Yeasts*, which are predominantly unicellular and oval or round in shape. Most propagate by an asexual process called budding in which the cell develops a protuberance, which enlarges and eventually separates from the parent cell. Some yeasts produce chains of elongated cells (pseudohyphae) that resemble the hyphae of moulds; some species also produce true hyphae. A small number of yeasts reproduce by fission. Yeasts are neither a natural nor a formal taxonomic group, but are a growth form shown by a wide range of unrelated fungi.

3. *Dimorphic fungi*, which are capable of changing their growth to either a mycelial or yeast phase, depending on the growth conditions.

The fungal kingdom is divided into four divisions: the Chytridiomycota, Zygomycota, Ascomycota and Basidiomycota. One of these divisions, the Chytridiomycota, does not contain any human pathogens. Fungal classification is based primarily on the method of sexual spore production. In some fungi, however, asexual spore production has proved so successful as a means of rapid dispersal to new habitats that sexual reproduction has disappeared, or, at least, has not been discovered. Even in the absence of the sexual stage it is now often possible to assign these fungi to the divisions Ascomycota or Basidiomycota on the basis of ribosomal DNA sequences. There is no longer any separate formal grouping for those fungi that appear to be strictly asexual. None the less, mycologists continue to employ the asexual reproductive characteristics of moulds, at least for routine identification purposes. Yeasts are identified primarily according to their ability to ferment sugars and to assimilate carbon and nitrogen compounds.

FUNGAL DISEASES OF MAN

FUNGAL PATHOGENS

There are at least 100 000 named species of fungi. However, fewer than 500 have been recognized to cause disease (*mycosis*) in man or animals. Most are moulds, but there are a number of pathogenic yeasts and some are dimorphic. Dimorphic fungi usually change from a multicellular mould form in the natural environment to a budding single-celled yeast form when causing infection. In the laboratory, the tissue form can be induced by culture at 37°C on rich media such as blood agar, whereas the mould form develops when incubated at a lower temperature (25–30°C) on a less rich medium such as Sabouraud dextrose agar.

Some fungi, such as the systemic pathogens *Histoplasma capsulatum* and *Coccidioides immitis*, can establish an infection in all exposed individuals. Others, such as *Candida* and *Aspergillus* species, are opportunist pathogens that ordinarily cause disease only in a compromised host.

In some mycoses the form and severity of the infection depend on the degree of exposure to the fungus, the site and method of entry into the body, and the level of immunocompetence of the host.

Some fungi may cause serious, occasionally fatal, toxic effects in man, either following ingestion of poisonous toadstools or consumption of mouldy food that contains toxic secondary metabolites (*mycotoxins*). Allergic disease of the airways may result from inhalation of fungal spores.

EPIDEMIOLOGY

Most human fungal infections are caused by fungi that grow as saprophytes in the environment. Infection is acquired by inhalation, ingestion or traumatic implantation. Some yeasts are human commensals and cause endogenous infections when there is some imbalance in the host. Only dermatophyte infections are truly contagious.

Many fungal diseases have a worldwide distribution, but some are endemic to specific geographical regions, usually because the causal agents are saprophytes restricted in their distribution by soil and climatic conditions.

TYPES OF INFECTION

Superficial mycoses

Diseases of the skin, hair, nail and mucous membranes are the most common of all fungal infections and have a worldwide distribution.

- Dermatophytosis (ringworm) is a complex of diseases affecting the outermost keratinized tissues of hair, nail and the stratum corneum of the skin; it is caused by a group of closely related mould fungi called *dermatophytes*, which can digest keratin. Dermatophyte infections occur in both man and animals.
- Yeast infections affect the skin, nail and mucous membranes of the mouth and vagina, and are usually caused by commensal *Candida* species, notably *Candida albicans*. Infection is generally endogenous in origin but genital infection can be transmitted sexually. The yeast *Malassezia furfur*, a skin commensal, can cause an infection of the skin called *pityriasis versicolor*.

Subcutaneous mycoses

Mycoses of the dermis, subcutaneous tissues and adjacent bones that show slow localized spread occur mainly in the tropics and subtropics; they usually result from the traumatic inoculation of saprophytic fungi from soil or vegetation into the subcutaneous tissue. The principal subcutaneous mycoses are mycetoma, chromoblastomycosis and sporotrichosis.

Systemic mycoses

Deep-seated fungal infections generally result from the inhalation of air-borne spores produced by the causal moulds, present as saprophytes in the environment. Initially there is a pulmonary infection, but the organism may become disseminated to other organs. The organisms that cause systemic mycoses can be divided into two distinct groups: the true pathogens and the opportunistic pathogens. The first of these groups is comprised mostly of dimorphic fungi and infections occur mainly in the Americas. The principal diseases are:

- blastomycosis (caused by *Blastomyces dermatitidis*)
- coccidioidomycosis (*C. immitis*)
- histoplasmosis (*H. capsulatum*)
- paracoccidioidomycosis (*Paracoccidioides brasiliensis*).

Systemic mycoses caused by opportunistic pathogens such as *Aspergillus*, *Candida* and *Cryptococcus* species have a more widespread distribution. These infections are being seen with increasing frequency in patients compromised by disease or drug treatment. In transplant patients, for example, these fungi are among the most frequent causes of death due to infection.

INCIDENCE

The incidence of all the mycoses is related directly to factors that affect the degree of exposure to the causal fungi, such as living conditions, occupation and leisure activities.

- Dermatophytosis of the foot (*athlete's foot*), with associated infections of nails and groin, occurs most commonly in swimmers, sportspersons and industrial workers who use communal bathing facilities.
- Animal dermatophytosis is an occupational hazard for farmers, veterinarians and others closely associated with animals.
- Agricultural workers in warm climates who wear little protective clothing frequently contract subcutaneous infections following minor injuries from vegetation.
- In developed countries the incidence of infections due to opportunistic fungal pathogens has increased following major advances in health care that have led to increases in the size of the population at risk.
- In many developing countries the acquired immune deficiency syndrome (AIDS) pandemic has been

associated with a marked increase in the rate of opportunistic fungal infections. High mortality rates have been reported from countries where adequate treatment is often unavailable.

- The incidence of several of the systemic mycoses that are endemic in the Americas has increased as a result of urban development and changing land use in the endemic areas. Increased international travel and tourism has also led to a rise in the number of cases of disease among individuals who normally reside in countries far from the endemic areas.

DIAGNOSIS

Diagnosis of fungal infections is based on a combination of clinical observation and laboratory investigation.

Clinical investigation

Superficial and subcutaneous mycoses often produce characteristic lesions, but they may also closely resemble and be confused with other diseases. In addition, the appearance of lesions may be modified beyond recognition by previous therapy, for example with topical steroids.

The first indication that a patient may have a systemic mycosis is often their failure to respond to antibacterial antibiotics. As early diagnosis considerably increases the chances of successful treatment, it is important that the possibility of fungal involvement be considered from the outset, particularly in those known to be at risk of developing a fungal infection. Computed tomography is widely used to help diagnose *Aspergillus* infections and other invasive mycoses.

Laboratory diagnosis

Laboratory diagnosis depends on:

- recognition of the pathogen in tissue by microscopy
- isolation of the causal fungus in culture
- the use of serological tests
- detection of fungal DNA by the polymerase chain reaction (PCR).

It is important that the correct type of specimen, together with adequate clinical data, is sent to the laboratory so that the appropriate investigations can be carried out. Information on factors such as travel or residence abroad, animal contacts and the occupation of the patient enable the laboratory staff to direct their investigations towards a particular fungus or group of fungi when appropriate.

Types of specimen

Skin scales, nail clippings and scrapings of the scalp that include hair stubs and skin scales are the most suitable specimens for the diagnosis of ringworm; these are collected into folded paper squares for transport to the laboratory. Swabs should be taken from suspected *Candida* infections from the mucous membranes and preferably sent to the laboratory in 'clear' transport medium. For subcutaneous infections the most suitable specimens are scrapings and crusts, aspirated pus and biopsies. In suspected systemic infection, specimens should be taken from appropriate sites.

Direct microscopy

Most specimens can be examined satisfactorily in wet mounts after partial digestion of the tissue with 10–20% potassium hydroxide. Addition of Calcofluor white and subsequent examination by fluorescence microscopy enhances the detection of most fungi as the fluorescent hydroxide–Calcofluor binds to the fungal cell walls. Gram films may also be used for the diagnosis of yeast infections of mucous membranes. Giemsa staining of smears is advised for detection of the yeast cells of *H. capsulatum* because of their small size.

Histology

Invasive procedures are required to obtain specimens for histological examination. Although sometimes necessary to provide firm evidence of invasive disease, such procedures are often impracticable on patients who are already seriously ill. Haematoxylin and eosin staining is seldom of value for demonstrating fungi in tissue, and specific fungal stains such as periodic acid–Schiff and Grocott–Gomori methenamine–silver are widely used.

Culture

Most pathogenic fungi are easy to grow in culture. Sabouraud dextrose agar and 4% malt extract agar are most commonly used. These may be supplemented with chloramphenicol (50 mg/l) to minimize bacterial contamination and cycloheximide (500 mg/l) to reduce contamination with saprophytic fungi. Many fungal pathogens have an optimum growth temperature below 37°C. Consequently, cultures are incubated at 25–30°C and at 37°C. With some dimorphic pathogens, enriched media such as brain–heart infusion or blood agar are used to promote growth of the yeast phase.

Many fungi develop relatively slowly and cultures should be retained for at least 2–3 weeks (in some cases up to 6 weeks) before being discarded; yeasts usually grow within

1–5 days. Moulds are identified by their macroscopic and microscopic morphology. Yeasts are identified by sugar fermentation and their ability to assimilate carbon and nitrogen sources. Commercial kits are available for the identification of medically important yeasts.

Culture may provide unequivocal evidence of fungal infection when established pathogens are isolated or when fungi are recovered from normally sterile sites. However, when commensals such as *Candida* species are isolated, results must be interpreted according to the quantity of the fungus isolated, the source and clinical evidence.

Serology

The most common tests for fungal antibodies are:

- immunodiffusion
- countercurrent immuno-electrophoresis (CIE)
- whole cell agglutination
- complement fixation
- enzyme-linked immunosorbent assay (ELISA).

For antigen detection the following are used:

- latex particle agglutination
- ELISA.

Polymerase chain reaction

Detection of fungal DNA in clinical material, principally blood, serum, broncho-alveolar lavage fluid and sputum, is increasingly used for diagnosis.

TREATMENT

There are relatively few therapeutically useful antifungal agents compared with the large number of antibacterial agents that are available (see Chs 5 & 66). As fungi and human beings are both eukaryotes, most substances that kill or inhibit fungal pathogens are also toxic to the host. Antifungal agents vary considerably in their spectrum of activity (see Table 5.4, p. 61). Most exploit differences in the sterol composition of the fungal and mammalian cell membranes, although the echinocandins (caspofungin and micafungin) interfere with β-glucan synthesis in the fungal cell wall.

Most antifungal agents are available only for topical use. Relatively few can be administered systemically.

- Amphotericin B and the echinocandins are given parenterally because of poor absorption from the gastro-intestinal tract.
- Fluconazole, itraconazole, voriconazole and flucytosine are available for oral or parenteral administration.

- Terbinafine and griseofulvin are usually administered orally.
- Amphotericin B has long been the treatment of choice in life-threatening disease, despite its toxicity; liposomal and lipid complex formulations are less toxic but much more expensive.
- A combination of amphotericin B and flucytosine reduces the likelihood of the emergence of resistance to flucytosine. Combinations of azole drugs and amphotericin B are seldom used therapeutically.
- New antifungals are being evaluated for use in systemic mycoses. Voriconazole may become the drug of choice for aspergillosis, and the echinocandins appear highly effective in systemic candidiasis.

Antifungal prophylaxis is often used to help prevent opportunistic infections in patients undergoing solid-organ, blood or marrow transplants and in those with haematological malignancies. Oral or topical antifungals are also used to prevent recurrent vaginal candidosis.

Primary or acquired resistance is not a major problem. Resistance to azole antifungals is sometimes encountered, especially after prolonged fluconazole therapy of oropharyngeal *Candida* infections in persons with AIDS. Some yeast species (e.g. *Candida krusei, Candida glabrata*) are inherently resistant to triazoles such as fluconazole. Consequently, susceptibility testing is often carried out for fluconazole and for any drug that fails to produce the expected therapeutic response.

SUPERFICIAL INFECTIONS

DERMATOPHYTOSIS

Dermatophyte infections are common diseases of the stratum corneum of the skin, hair and nail; they are also referred to as ringworm or as *tinea*, a name that is qualified by the site affected, for example *tinea pedis* or *tinea capitis* for infections of the feet or scalp, respectively. These infections are caused by about 20 species of fungi that are grouped into three genera: *Trichophyton, Microsporum* and *Epidermophyton*. Some species are worldwide in distribution, whereas others are restricted to, or are more common in, particular parts of the world.

Many dermatophyte species produce two types of asexual spore: multi-celled macroconidia and single-celled microconidia. Classification into the three genera *Trichophyton, Microsporum* and *Epidermophyton* is based on the morphology of the macroconidia, although the identification of species is also based on the shape and disposition of the microconidia and the macroscopic appearance of the colonies.

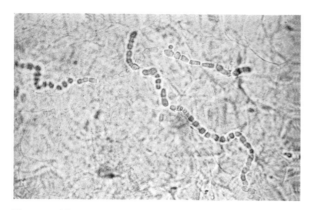

Fig. 60.1 Microscopical appearance of infected skin scrapings showing the development of arthroconidia. (Reproduced with permission from Richardson M D, Warnock D W, Campbell C K 1995 *Slide Atlas of Fungal Infection: Superficial Fungal Infections.* Blackwell Science, Oxford.)

The clinical appearances of dermatophyte infections are the result of a combination of direct tissue damage caused by the fungus and of the immune response of the host. The damage to tissue is due to a combination of mechanical pressure and enzymatic activities. In tissue the dermatophytes take the form of branching hyphae, which may eventually break up into arthroconidia, particularly in infected hair (Fig. 60.1).

Epidemiology

The dermatophytes can be divided into three groups depending on whether their normal habitat is the soil (geophilic species), animals (zoophilic species) or man (anthropophilic species). Members of all three groups can cause human infection, but their different natural reservoirs have important epidemiological implications in relation to the acquisition, site and spread of human disease.

Although geophilic dermatophytes occasionally cause infection in both animals and man, their normal habitat is the soil. Members of the anthropophilic and zoophilic groups are thought to have evolved from these and other keratinophilic soil-inhabiting fungi, different species having adapted to different natural hosts. Individual members of the zoophilic group are often associated with a particular animal host, for instance *M. canis* with cats and dogs and *T. verrucosum* with cattle. However, these organisms can also spread to man. The anthropophilic species are the most highly specialized group of dermatophytes. They rarely infect other animals and often show a strong preference for a particular body site, only occasionally being found in other regions. For instance, *M. audouinii* commonly infects scalp hair, whereas *E. floccosum* is usually found on the skin.

Infections are spread by direct or indirect contact with an infected individual or animal. The infective particle is usually a fragment of keratin containing viable fungus. Indirect transfer may occur via the floors of swimming pools and showers or on brushes, towels and animal grooming implements. Dermatophytes can remain viable for long periods of time and the interval between deposition and transfer may be considerable. In addition to exposure to the fungus, some abnormality of the epidermis, such as slight peeling or minor trauma, is probably necessary for the establishment of infection.

In industrialized countries, tinea capitis is relatively uncommon, and is caused by dermatophytes of both human and animal origin, although infections with the anthropophilic species *T. tonsurans* are on the increase among urban populations in Europe and the Americas. However, the use of communal bathing facilities has resulted in a considerable increase in the incidence of tinea pedis and associated nail and groin infections. These now comprise about 75% of all dermatophyte infections diagnosed in temperate zones.

In developing countries, particularly in warm climates, scalp, body and groin infections predominate, with *T. rubrum* and *T. violaceum* among the most common causes.

Clinical features

Lesions vary considerably according to the site of the infection and the species of fungus involved. Sometimes there is only dry scaling or hyperkeratosis, but more commonly there is irritation, erythema, oedema and some vesiculation. More inflammatory lesions with weeping vesicles, pustules and ulceration are usually caused by zoophilic species.

In skin infections of the body, face and scalp, spreading annular lesions with a raised, inflammatory border are usually produced (Fig. 60.2). Lesions in body folds, such as the groin, tend to spread outwards from the flexures. In tinea pedis, infection is often confined to the toe clefts, but it can spread to the sole; sometimes, painful secondary bacterial infection occurs in the toe clefts.

In nail infection, the nail becomes discoloured, thickened, raised and friable; most nail infections are due to *T. rubrum* and involve the toenails (Fig. 60.3).

In scalp infections there is scaling and hair loss, the extent of which depends on the causal fungus. Some zoophilic species give rise to a highly inflammatory, raised, suppurating lesion called a *kerion*; kerions may also occur in the beard area of adults. It is important that tinea capitis is recognized and treated promptly because it can lead to scarring and permanent hair loss.

In scalp infection the fungus invades the hair shaft and the hyphae then break up into chains of arthroconidia. In some species (e.g. *T. tonsurans*, *T. violaceum*) the

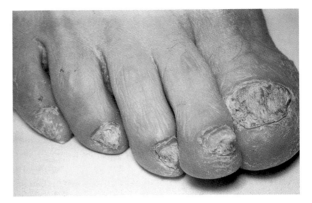

Fig. 60.3 Toenail infection due to *Trichophyton rubrum*. (Reproduced with permission from Richardson M D, Warnock D W, Campbell C K 1995 *Slide Atlas of Fungal Infection: Superficial Fungal Infections*. Blackwell Science, Oxford.)

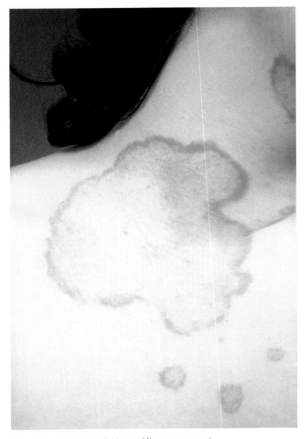

Fig. 60.2 Tinea corporis due to *Microsporum canis*.

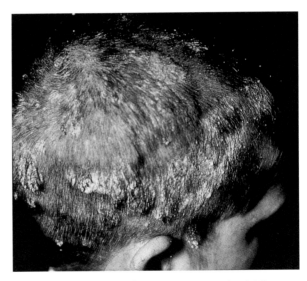

Fig. 60.4 Tinea capitis (favus) due to *Trichophyton schoenleinii*. (Reproduced with permission from Richardson M D, Warnock D W, Campbell C K 1995 *Slide Atlas of Fungal Infection: Superficial Fungal Infections*. Blackwell Science, Oxford.)

arthroconidia are retained within the hair shaft (endothrix invasion), whereas in others (e.g. *Microsporum* species, *T. verrucosum*) they are produced in a sheath surrounding the hair shaft (ectothrix invasion). The pattern of hair invasion affects the clinical appearance of the lesion.

- In endothrix infection the hair breaks off at, or just below, the mouth of the follicle to give what is described as a 'black dot' appearance.
- In ectothrix infection the hair usually breaks off 2–3 mm above the mouth of the follicle, leaving short stumps of hair.
- In favus, caused by *T. schoenleinii*, fungal growth within the hair is minimal. The hair remains intact, but intense fungal growth within and around the hair follicle produces a waxy, honeycomb-like crust on the scalp (Fig. 60.4).

Infections of the groin, hands and nails are nearly always secondary to infection of the feet and are usually (except in developing countries) caused by *T. rubrum*, *T. mentagrophytes* or *E. floccosum*. Mixed infections also occur.

Occasionally, patients with inflammatory infections develop a secondary rash known as an *id reaction*, which is thought to be an immunological reaction to fungal antigens. In patients with tinea pedis this takes the form of a vesicular eczema of the hands, whereas patients with tinea capitis (especially kerion) develop a follicular rash, usually on the trunk or limbs. These secondary lesions do not contain viable fungus and they disappear spontaneously when the infection subsides.

Laboratory diagnosis

Dermatophyte infections may be reliably diagnosed in the laboratory by direct microscopical examination and culture of skin, crusts, hair and nail.

Collection of samples

Skin, hair and nail samples are best collected into folded squares of black paper or card, which can be fastened with a paper clip. The use of paper allows the specimen to dry out, which helps reduce bacterial contamination and provides conditions under which specimens can be stored for 12 months or more without appreciable loss in viability of the fungus.

Nail samples should be collected by taking clippings from any discoloured, dystrophic or brittle parts of the nail and, importantly, by scraping material from underneath the nail. The sample should be taken from as far back as possible from the free edge of the nail.

Scales from skin lesions should be collected by scraping outwards with a blunt scalpel from the edges of the lesions, where most viable fungus is likely to be. Specimens from the scalp should include hair stubs, the contents of plugged follicles and skin scales. Infected hairs are usually easy to pluck from the scalp with forceps. Cut hairs are unsatisfactory because the focus of infection is usually below or near the surface of the scalp.

Wood's lamp

This is a source of long-wave ultraviolet light that can be used to detect fluorescence in infected hair. It is especially useful for the detection of inconspicuous scalp lesions, and to select infected hairs for laboratory investigation.

Hairbrush sampling

Adequate material from minimal lesions may be obtained by brushing the scalp with a sterilized plastic hairbrush or scalp massage pad; this is then used to inoculate an appropriate culture medium by pressing the brush or pad spines into the agar.

Processing of specimens

If there is insufficient material for both microscopy and culture, the sample should be used for culture, since this is generally the more sensitive procedure (except for nails).

The specimen should first be examined macroscopically; hair samples are examined under a Wood's lamp. Material from representative parts, and any fluorescent hairs, are divided up into 1–2-mm fragments with a sterile scalpel blade before microscopical examination and culture.

Direct microscopy

Microscopy of wet mounts of keratinous material in potassium hydroxide is simple and reliable. The preparation is allowed to stand for 15–20 min to digest and 'clear' the keratin. Dermatophytes are seen in skin and nail as branching hyphae, which often appear slightly greenish in colour and run across the outlines of the colourless host cells. With Calcofluor (see above) the cell outlines fluoresce white.

Culture

Small fragments of keratinous material are planted or scattered on Sabouraud dextrose or 4% malt extract agar and incubated at 28–30°C for up to 2 weeks; room temperature is adequate but the dermatophytes grow more slowly. Only *T. verrucosum* grows well at 37°C.

Identification is based on colonial appearance and colour, pigment production, and the micromorphology of any spores produced. Special tests exist for differentiating certain morphologically similar species. Thus, the ability of *T. mentagrophytes* to produce urease within 2–4 days distinguishes it from *T. rubrum*, and the ability to grow on rice grains distinguishes *M. canis* from *M. audouinii*.

Treatment and prevention

Topical therapy is satisfactory for most skin infections, although oral antifungals are required to treat infections of the nail and scalp, and severe or extensive skin infections.

Topical agents include azole compounds, terbinafine, amorolfine and ciclopirox olamine. Oral griseofulvin is useful for scalp, skin and fingernail infections, but gives poor results in toenail infections, even after 18 months' therapy. Terbinafine and itraconazole have largely replaced griseofulvin for the treatment of nail infections because of their much better cure rates and shorter periods of treatment (around 85% cure for toenails after 3 months' therapy).

Relatively little has been done to control the spread of dermatophyte infections. The prophylactic use of antifungal foot powder after bathing helps to reduce the spread of infection among swimmers. Foot-baths containing antiseptic solutions, which are commonplace in swimming pools, are of no value.

SUPERFICIAL CANDIDOSIS

Superficial *Candida* infections involving the skin, nails and mucous membranes of the mouth and vagina are very common throughout the world. *Candida albicans* accounts for 80–90% of cases, but other species, notably *C. glabrata*, *C. parapsilosis*, *C. tropicalis*, *C. krusei* and *C. guilliermondii*, may occur.

On Sabouraud dextrose agar *Candida* species grow predominantly in the yeast phase as round or oval cells, 3–8 mm in diameter. A mixture of yeast cells, pseudohyphae and true hyphae is found in vivo and under micro-aerophilic growth conditions on nutritionally poor media. *C. glabrata* does not form either hyphae or pseudohyphae.

Epidemiology

Candida species, usually *C. albicans*, are found in small numbers in the commensal flora (mouth, gastro-intestinal tract, vagina, skin) of about 20% of the normal population. The carriage rate tends to increase with age and is higher in the vagina during pregnancy. Commensal yeasts are more prevalent among patients in hospital. Yeast overgrowth and infection occur when the normal microbial flora of the body is altered or when host resistance to infection is lowered by disease.

In most cases, infection is derived from an individual's own endogenous reservoir. In some instances, however, transmission from person to person can occur; for example neonatal oral candidosis is more common in infants born of mothers with vaginal candidosis. The hands of health-care workers are another potential source of neonatal infection.

Individuals colonized with *Candida* species possess numerous non-specific and immunological defences to prevent infection. In superficial candidosis the non-specific inhibitory factors include inhibitors in serum, such as unsaturated transferrin, and epithelial proliferation. Specific defence largely depends on the development of active T cell-mediated immunity.

Both general and local predisposing factors are important in the development of oropharyngeal candidosis. Debilitated and immunosuppressed individuals, such as persons with diabetes mellitus, stem cell transplant recipients and those with human immunodeficiency virus (HIV) infection, are more susceptible to infection. Local factors, such as xerostomia and trauma from unhygienic or ill-fitting dentures, are also important. Local tissue damage is also a critical factor in the development of cutaneous forms of candidosis, Most infections occur in moist, occluded sites and follow maceration of tissue.

Candida vaginitis is more common during pregnancy. The lower prevalence of this infection after menopause emphasizes the hormonal dependence of the infection. Most cases of vaginal candidosis in older women are associated with uncontrolled diabetes mellitus or the use of exogenous oestrogen replacement treatment.

Clinical features

Mucosal infection

This is the commonest form of superficial candidosis. Discrete white patches develop on the mucosal surface,

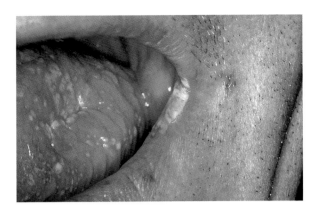

Fig. 60.5 Pseudomembranous oral candidosis with associated angular cheilitis.

and may eventually become confluent and form a curd-like pseudomembrane (Fig. 60.5).

In oropharyngeal candidosis white flecks appear on the buccal mucosa, tongue, and the hard and soft palate; although these are adherent, they can be removed. The surrounding mucosa is red and sore. This form of oropharyngeal candidosis occurs most frequently in infancy and old age, or in severely immunocompromised patients, including those with AIDS. Other forms of oral candidosis occur:

- lesions in the occluded area under the denture in those who wear dentures
- painful infection of the tongue in some individuals receiving antibiotic therapy
- chronic infection with extensive leucoplakia and infection of the angles of the mouth (angular cheilitis).

In vaginal candidosis, itching, soreness and a non-homogeneous white discharge accompany typical white lesions on the epithelial surfaces of the vulva, vagina and cervix. Sometimes the mucosa simply appears inflamed and friable. The perivulval skin may become sore and small satellite pustules may appear around the perineum and natal cleft. Some women suffer recurrent episodes.

Chronic, intractable oropharyngeal candidosis, which may extend to give oesophageal infection, is common in persons with HIV infection, although the use of combinations of antiretroviral drugs has reduced its incidence. The appearance of this infection can be the indicator of the transition from HIV-positive status to AIDS.

Skin and nail infection

Cutaneous candidosis is less common than dermatophytosis. The lesions usually develop in warm, moist sites such as the axillae, groin and submammary folds. In

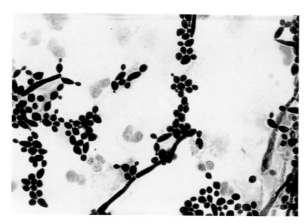

Fig. 60.6 Microscopical appearance of *Candida albicans* yeast cells and pseudohyphae in a Gram-stained vaginal smear.

infants, *Candida* species are often secondary invaders in napkin dermatitis. Infection of the finger webs, nail folds and nails is associated with frequent immersion of the hands in water.

Chronic mucocutaneous candidosis

This is a rare form of candidosis that usually becomes apparent in childhood and takes the form of a persistent, sometimes granulomatous, infection of the mouth, scalp, hands, feet and nails. In some cases, disfiguring hyperkeratotic lesions develop on the scalp and face. Some of those who develop this condition have subtle defects in lymphocyte function.

Laboratory diagnosis

Specimens of skin and nail are collected in the same way as for suspected dermatophytosis. For infections of the mouth or vagina, scrapings taken with a blunt scalpel or a spatula from areas with white plaques or erythema are better than swabs if the material is to be processed immediately. However, swabs are more convenient for transport to the laboratory, and they are better for collecting vaginal discharge. Swabs should first be moistened with sterile water or saline before taking the sample and should be sent to the laboratory in 'clear' transport medium.

In Gram-stained smears of mucous membrane samples the fungus is seen as budding Gram-positive yeast cells; pseudohyphae are usually present except in the case of *C. glabrata* (Fig. 60.6). Contrary to popular belief, the presence of *Candida* pseudohyphae in clinical material does not confirm infection with the organism, particularly as it may have developed in the period between collection and processing of the sample.

Candida species grow well on Sabouraud medium or on blood agar at 25–37°C; typical yeast colonies appear within 1–2 days. *C. albicans* isolates can be identified by the germ tube test: after incubation in serum at 37°C for 1.5–2 h, *C. albicans* produces short hyphae known as germ tubes. Other yeasts may be identified with one of the commercial kits, or by fermentation and assimilation tests.

Quantification of growth, especially in the case of vulvovaginal samples, may help the clinician to distinguish between commensal carriage and infection.

Treatment and prevention

Most superficial infections respond well to topical therapy with an imidazole. In oral candidosis, nystatin, amphotericin B or miconazole may be effective in lozenge or gel form. Most patients with vaginal candidosis can be treated successfully with a single application of a topical imidazole, or with oral fluconazole or itraconazole. Intermittent prophylaxis with an oral azole or vaginal pessaries is of benefit in controlling recurrent vaginal candidosis.

Treatment of chronic paronychia involves a combination of antifungal therapy, nail care and avoidance of prolonged exposure to water by use of protective gloves; patients should dry their hands carefully after washing. Regular application of an azole lotion or an azole given orally, sometimes in conjunction with a topical steroid and an antibacterial agent, is the most appropriate therapy but it may take several months to cure the condition; antifungal creams or ointments are less effective.

Oral therapy is essential for the treatment of intractable chronic *Candida* infections; treatment is given until remission is achieved but in some patients, for instance those with AIDS, relapse is common, and intermittent or prolonged therapy may be required. This may, however, lead to the development of resistance, as occasionally happens with fluconazole.

PITYRIASIS VERSICOLOR

This is a common, mild and often recurrent infection of the stratum corneum that produces a patchy discoloration of the skin caused by lipophilic yeasts of the genus *Malassezia*. These organisms require lipids for growth and special media containing Tween and lipid supplements have been developed.

On normal skin and in conditions such as dandruff and seborrhoeic dermatitis (in which its precise role is uncertain), *Malassezia* occurs as an oval or bottle-shaped yeast, which characteristically produces buds on a broad base. In pityriasis versicolor the organism produces predominantly round yeast cells and short hyphae.

Epidemiology

Malassezia species are common members of the normal skin flora and most infections are thought to be endogenous. The incidence of skin colonization rises from around 25% in children to almost 100% in adolescents and adults. Disease is probably related to host and environmental factors. It is very common in hot, humid tropical climates, where 30–40% of adults may be affected.

Clinical features

Small patches of well demarcated, non-inflammatory scaling are usually present on the upper trunk or neck; these may appear hypopigmented or hyperpigmented, depending on the degree of pigmentation of the surrounding skin. The lesions tend to spread and coalesce, and occasionally they spread to other sites.

Laboratory diagnosis

The diagnosis can be confirmed reliably by direct microscopy of skin scales; culture is unnecessary. Demonstration of clusters of the characteristic round yeast cells (5–8 µm in diameter) with short, stout hyphae, which may be curved and occasionally branched, is diagnostic.

Treatment

Pityriasis versicolor responds well to topical therapy with 2% selenium sulphide or azoles such as ketoconazole in cream, lotion or a shampoo. Oral azole therapy is sometimes used for recalcitrant or widespread infections. Relapse is common, particularly in hot climates.

OTHER SUPERFICIAL INFECTIONS

Skin and nail

Certain non-dermatophyte moulds may cause infection of skin and nail. It is important that these are recognized because they are often resistant to the agents used to treat dermatophytosis and superficial candidosis.

In the UK, about 5% of fungal nail infections are caused by non-dermatophyte moulds. *Scopulariopsis brevicaulis*, a ubiquitous saprophyte of soil, is the most common, although other saprophytic moulds such as *Fusarium*, *Aspergillus* and *Acremonium* species are also occasionally implicated. There is some debate as to whether these moulds are primary pathogens of nails – infection usually follows trauma and in many cases they are found in nails along with a dermatophyte.

Non-dermatophyte mould infections do not respond to existing antifungal agents. Attempts may be made to remove the nail with topical 40% urea paste.

Otomycosis

About 10–20% of chronic ear infections are due to fungi. The disease has a worldwide distribution but is more common in warm climates. Topical antibiotics and steroids are predisposing factors. The commonest causes are species of *Aspergillus*, in particular *A. niger*. The fungi are easy to see in material from swabs or scrapings and grow readily in culture.

Treatment with topical antifungals is usually successful, although relapse is common. Any concurrent bacterial infection or other underlying abnormality should also be treated.

Mycotic keratitis

Fungal infections of the cornea usually follow traumatic implantation of spores. Topical antibiotics and steroids are important predisposing factors. These infections occur most often in hot climates and are caused by common saprophytic moulds, in particular *Aspergillus* and *Fusarium* species. Culture results should be interpreted with care as these opportunist pathogens are also encountered as contaminants. Superficial swabs are of no value for laboratory investigation, and scrapings should be taken from the base or edge of the ulcer. The branched, septate hyphae may be rather sparse in potassium hydroxide mounts and some of the material should also be stained with periodic acid–Schiff or Grocott–Gomori methenamine–silver.

Management entails surgical debridement of infected tissue, discontinuation of topical corticosteroids, and topical or oral treatment with an antifungal drug. Topical treatment with natamycin is often successful, but oral treatment with an azole drug is required in patients with severe or worsening lesions. Even with intensive treatment, corneal perforation can occur.

SUBCUTANEOUS INFECTIONS

MYCETOMA

Mycetoma is a chronic granulomatous infection of the skin, subcutaneous tissues, fascia and bone that most often affects the foot or the hand. It may be caused by one of a number of different actinomycetes (actinomycetoma) (see Ch. 20) or moulds (eumycetoma). The disease is most prevalent in tropical and subtropical regions of Africa, Asia, and Central and South America.

A large number of organisms have been implicated in this disease, including species of *Madurella*, *Leptosphaeria*, *Acremonium*, *Pseudallescheria*, *Actinomadura*, *Nocardia* and *Streptomyces*. Within host tissues the organisms

Fig. 60.7 Microscopical appearance of *Madurella grisea* in a stained mycetoma grain. (Reproduced with permission from Richardson M D, Warnock D W, Campbell C K 1995 *Slide Atlas of Fungal Infection: Subcutaneous and Unusual Fungal Infections.* Blackwell Science, Oxford.)

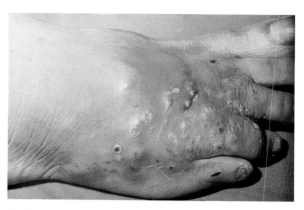

Fig. 60.8 Mycetoma of the foot.

develop to form compacted colonies (grains) 0.5–2 mm in diameter, the colour of which depends on the organism responsible; for example, in unstained preparations *Madurella* grains are black and *Actinomadura pelletieri* grains are red (Fig. 60.7).

Epidemiology

Infection follows traumatic inoculation of the organism into the subcutaneous tissue from soil or vegetation, usually on thorns or wood splinters. Consequently, the disease occurs most frequently in agricultural workers, in whom minor penetrating skin injuries are common.

Clinical features

Localized swollen lesions that develop multiple draining sinuses are usually found on the limbs, although infections occur on other parts of the body (Fig. 60.8). There is often a long period between the initial infection and formation of the characteristic lesions; spread from the site of origin is unusual but may occur, particularly from the foot up the long bones of the leg.

Laboratory diagnosis

The presence of grains in pus collected from draining sinuses or in biopsy material is diagnostic. The grains are visible to the naked eye and their colour may help to identify the causal agent. Grains should be crushed in potassium hydroxide and examined microscopically to differentiate between actinomycetoma and eumycetoma; material from actinomycetoma grains may be Gram stained to demonstrate the Gram-positive filaments. Samples should also be cultured, at both 25–30°C and 37°C, on brain–heart infusion agar or blood agar for actinomycetes

and on Sabouraud agar (without cycloheximide) for fungi. The fungi that cause eumycetoma are all septate moulds that appear in culture within 1–4 weeks, but their identification requires expert knowledge.

Treatment

The prognosis varies according to the causal agent, so it is important that the identity is established. Actinomycetoma responds well to medical treatment: combinations of streptomycin with co-trimoxazole or dapsone are often effective, but an average of 9 months' therapy is required. In eumycetoma, chemotherapy is ineffective and radical surgery is usually necessary. However, some antifungals have yet to be properly evaluated in this condition.

CHROMOBLASTOMYCOSIS

This disease, also known as *chromomycosis*, is a chronic, localized infection of the skin and subcutaneous tissues, characterized by slow-growing verrucous lesions usually involving the limbs. The disease is encountered mainly in the tropics, in Central and South America, and Madagascar. The principal causes are *Fonsecaea pedrosoi*, *F. compacta*, *Phialophora verrucosa* and *Cladophialophora carrionii*. Like mycetoma, infection follows traumatic inoculation of the organism into the skin or subcutaneous tissue, and is seen most often among those with outdoor occupations.

Laboratory diagnosis

The characteristic clusters of brown-pigmented, thick-walled fungal cells are relatively easy to see on microscopical examination of skin scrapings, crusts and pus. Culture on Sabouraud agar at 25–30°C yields slow-growing, greenish grey to black, compact, folded colonies. Cultures should be incubated for 4–6 weeks. Specific identification of these closely related fungi is usually left to a reference laboratory.

Treatment

There is no ideal treatment for this disease, but promising results have been obtained with terbinafine and itraconazole, both of which can be combined with flucytosine in difficult cases. Early, solitary lesions may be excised.

SPOROTRICHOSIS

Sporotrichosis is a chronic, pyogenic, granulomatous infection of the skin and subcutaneous tissues that may remain localized or show lymphatic spread. It is caused by *Sporothrix schenckii*, which is found in the soil and on plant materials, such as wood and sphagnum moss. The disease is worldwide in distribution, but occurs mainly in Central and South America, parts of the USA and Africa, and Australia; it is rare in Europe.

S. schenckii is a dimorphic fungus. In nature and in culture at 25–30°C, it develops as a mould with septate hyphae. The yeast phase is formed in tissue and in culture at 37°C, and is composed of spherical or cigar-shaped cells (1–3 × 3–10 μm).

Epidemiology

Minor trauma, such as abrasions or wounds due to wood splinters, is often sufficient to introduce the organism. Infection occurs mainly in adults, and is more common among individuals whose work or recreational activities bring them into contact with soil or plant materials, such as gardeners and florists.

Clinical features

Sporotrichosis presents most frequently as a nodular, ulcerating disease of the skin and subcutaneous tissues, with spread along local lymphatic channels (Fig. 60.9). Typically, the primary lesion is on the hand with secondary lesions extending up the arm. The primary lesion may remain localized or disseminate to involve the bones, joints, lungs and, in rare cases, the central nervous system. Disseminated disease usually occurs in debilitated or immunosuppressed individuals.

Laboratory diagnosis

Diagnosis is confirmed by isolation of the causative organism by culture of swabs from moist, ulcerated lesions or pus aspirated from subcutaneous nodules; biopsy specimens

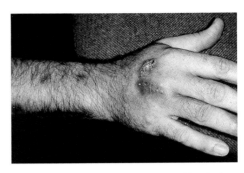

Fig. 60.9 Sporotrichosis of the hand showing local lymphatic spread.

may be necessary in some cases. Direct microscopy is of little value as so few of the small *S. schenckii* yeast cells are present in diseased tissue. The mycelial phase develops within 7–10 days on Sabouraud agar or blood agar at 25–30°C; the yeast phase develops in 2 days at 37°C. Identification depends on the micromorphology of the mould phase and its conversion to the yeast phase at 37°C.

Treatment

Prolonged therapy is usually required. For the lymphocutaneous form, treatment with potassium iodide or itraconazole is satisfactory. In disseminated disease, intravenous amphotericin B is required.

OTHER SUBCUTANEOUS MYCOSES

Phaeohyphomycosis is a general term used to describe solitary subcutaneous lesions caused by any brown-pigmented mould. If left untreated these lesions slowly increase in size to form a painless abscess. Diagnosis is often made at surgery, and treatment is by excision.

Several other fungi, including *Lacazia loboi*, *Basidiobolus ranarum* and *Conidiobolus coronatus*, occasionally cause subcutaneous infections, usually in the tropics. Surgical excision is often curative in *L. loboi* infections; antifungal therapy may be of use for the other infections, but the newer drugs have not been properly evaluated.

SYSTEMIC MYCOSES

COCCIDIOIDOMYCOSIS

This is primarily an infection of the lungs caused by *Coccidioides immitis*, a dimorphic fungus found in the soil of semi-arid regions of the western hemisphere. In the USA, the endemic region includes parts of California, Arizona, New Mexico and Texas. The endemic region

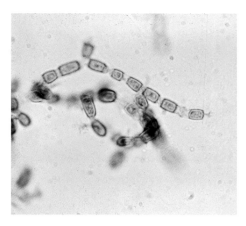

Fig. 60.10 Microscopical appearance of *Coccidioides immitis* arthroconidia.

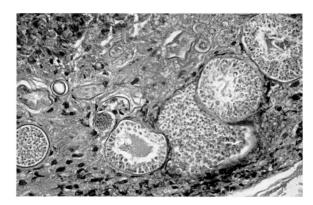

Fig. 60.11 Microscopical appearance of *Coccidioides immitis* spherules in tissue.

extends southwards into the desert regions of northern Mexico, and parts of Central and South America.

In culture and in soil *C. immitis* grows as a mould, producing large numbers of barrel-shaped arthroconidia (4 × 6 µm diameter), which are easily dispersed in wind currents (Fig. 60.10). In the lungs the arthroconidia form spherules (up to 120 µm in diameter) which contain numerous endospores (2–4 µm in diameter) (Fig. 60.11). Endospores are released by rupture of the spherule wall and develop to form new spherules in adjacent tissue or elsewhere in the body. In culture the mould colonies are initially moist and white but change within 5–12 days to become floccose and pale grey or brown.

Epidemiology

Infection is acquired by inhalation; the incubation period is 1–3 weeks. The major risk factor for infection is environmental exposure. Outbreaks have been associated with ground-disturbing activities, such as building construction and archaeological excavation, as well as with natural events that result in the generation of dust clouds, such as earthquakes and dust storms. The most serious disseminated forms of the disease are more common among those of black, Asian or Filipino race, and among pregnant women in the third trimester.

Clinical features

C. immitis causes a wide spectrum of disease, ranging from a transient pulmonary infection that resolves without treatment, to chronic pulmonary infection, or to more widespread disseminated disease. About 40% of newly infected persons develop an acute symptomatic and often severe influenza-like illness. However, most otherwise healthy persons recover without treatment,

their symptoms disappearing in a few weeks. In some cases primary infection may result in chronic, cavitating, pulmonary disease.

Fewer than 1% of infected individuals develop disseminated coccidioidomycosis. This is a progressive disease that usually develops within 3–12 months of the initial infection, although it can occur much later following reactivation of a quiescent infection in an immunosuppressed individual. The clinical manifestations range from a fulminant illness that is fatal within a few weeks if left untreated, to an indolent chronic disease that persists for months or years. One or more sites may be involved, but the skin, soft tissue, bones, joints and the central nervous system are most commonly affected. Meningitis is the most serious complication of coccidioidomycosis, occurring in 30–50% of patients with disseminated disease. Without therapy, it is almost always fatal.

Laboratory diagnosis

Microscopical examination of sputum, pus and biopsy material is helpful as the relatively large size and numbers of mature spherules present makes their detection and identification comparatively straightforward. Material for culture should be inoculated on to screw-capped slopes of Sabouraud agar and incubated at 25–30°C for 1–2 weeks. The fungus can be identified by its colonial morphology and the presence of numerous thick-walled arthroconidia formed in chains from alternate cells of the septate hyphae.

The arthroconidia are highly infectious and are a serious danger to laboratory staff. Consequently, Petri dishes should never be used for isolation of the organism and all procedures should be carried out in a biological safety cabinet under Category 3 containment. Preparations for microscopy should be made only after wetting the colony to reduce spore dispersal.

Serological tests play an important part in diagnosis. The precipitin test is most useful for detection of early primary infection or exacerbation of existing disease; precipitins appear 1–3 weeks after infection but are seldom detectable after 2–6 months, or in patients with disseminated coccidioidomycosis. The latex agglutination test gives similar results to the precipitin test, but is less specific. Complement-fixing antibodies appear 1–3 months after infection and persist for long periods in individuals with chronic or disseminated disease. In most cases the titre is proportional to the extent of infection; failure of the titre to fall during treatment of disseminated coccidioidomycosis is an ominous sign.

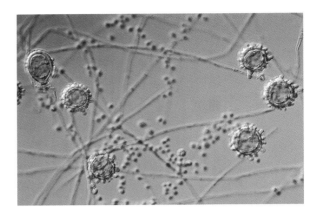

Fig. 60.12 Microscopical appearance of *Histoplasma capsulatum* microconidia and macroconidia.

Treatment

The historical standard of treatment is intravenous amphotericin B, but oral fluconazole is now used to treat many patients with skin, soft tissue, bone or joint infections. Itraconazole is also effective, but less well tolerated. Because oral fluconazole is so much more benign than intrathecal amphotericin B, it is now the drug of choice for coccidioidal meningitis.

HISTOPLASMOSIS

This disease is caused by *H. capsulatum*, a dimorphic fungus found in soil enriched with the droppings of birds and bats. Histoplasmosis is the most common endemic mycosis in North America, but also occurs throughout Central and South America. In the USA, the disease is most prevalent in states surrounding the Mississippi and Ohio Rivers. Other endemic regions include parts of Africa, Australia, India and Malaysia. *H. capsulatum* var. *duboisii* is restricted to the continent of Africa.

H. capsulatum var. *capsulatum* grows in soil and in culture at 25–30°C as a mould and as an intracellular yeast in animal tissues. The small oval yeast phase cells (2–4 μm diameter) can also be produced in vitro by culture at 37°C on blood agar or other enriched media containing cysteine. In culture the mould colonies are fluffy, white or buff-brown; the mycelium is septate and two types of unicellular asexual spores are usually produced: large round, tuberculate macroconidia (8–15 μm in diameter) are most prominent and are diagnostic, but smaller, broadly elliptical, smooth-walled microconidia (2–4 μm in diameter) are also present in primary isolates (Fig. 60.12). *H. capsulatum* var. *duboisii* is morphologically identical to *H. capsulatum* var. *capsulatum* in its mycelial phase, but the yeast phase has larger cells (10–15 μm in diameter).

Epidemiology

Infection results from the inhalation of spores; the incubation period is 1–3 weeks. The major risk factor is environmental exposure; longer and more intense exposures usually result in more severe pulmonary disease. Most reported outbreaks have been associated with exposures to sites contaminated with *H. capsulatum* or have followed activities that disturbed accumulations of bird or bat guano, such as building demolition, soil excavation and caving.

The most serious disseminated forms of the disease are more common among individuals with underlying cell-mediated immunological deficiencies, such as persons with HIV infection, transplant recipients, and those receiving immunosuppressive treatments.

Clinical features

There is a wide spectrum of disease, ranging from a transient pulmonary infection that subsides without treatment, to chronic pulmonary infection, or to more widespread disseminated disease. Many healthy individuals develop no symptoms when exposed to *H. capsulatum* in an endemic setting. Higher levels of exposure result in an acute symptomatic and often severe flu-like illness, with fever, chills, non-productive cough and fatigue. The symptoms usually disappear within a few weeks, but patients are frequently left with discrete, calcified lesions in the lung.

Disseminated histoplasmosis may range from an acute illness that is fatal within a few weeks if left untreated (often seen in infants, persons with AIDS and solid-organ transplant recipients) to an indolent, chronic illness that can affect a wide range of sites. Hepatic infection is common in non-immunosuppressed individuals and adrenal gland destruction is a frequent problem. Mucosal

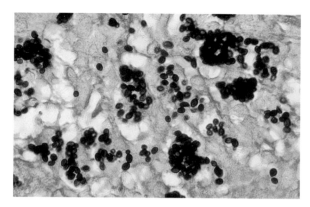

Fig. 60.13 Microscopical appearance of *Histoplasma capsulatum* yeast cells in tissue.

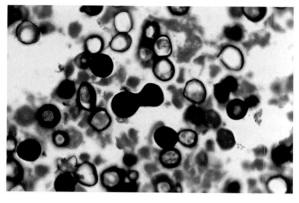

Fig. 60.14 Microscopical appearance of *Blastomyces dermatitidis* yeast cells in tissue, showing broad-based budding.

ulcers are found in more than 60% of these patients; central nervous system disease occurs in 5–20% of patients.

The clinical features of *H. capsulatum* var. *duboisii* infection differ from those of var. *capsulatum* infection. The illness is indolent in onset and the predominant sites affected are the skin and bones. Those with more widespread infection involving the liver, spleen and other organs have a febrile wasting illness that is fatal within weeks or months if left untreated.

Laboratory diagnosis

Microscopy of smears of sputum or pus should be stained by the Wright or Giemsa procedure. Blood smears may be positive for *H. capsulatum*, especially in persons with AIDS. Liver or lung biopsies stained with periodic acid–Schiff or Grocott–Gomori methenamine–silver may provide a rapid diagnosis of disseminated histoplasmosis in some patients. *H. capsulatum* is seen as small, oval yeast cells, often within macrophages or monocytes (Fig. 60.13).

Specimens should be cultured on Sabouraud agar at 25–30°C to obtain the mycelial phase. Mycelial colonies develop within 1–2 weeks but cultures should be retained for 4 weeks before discarding. The fungus is identified by its colonial morphology and the presence of the characteristic macroconidia and microconidia. Culture at 37°C for the yeast phase is not used for primary isolation, although conversion from the mould to yeast phase is useful to confirm the identity of isolates. Mould cultures of *H. capsulatum* are a hazard to laboratory staff and consequently screw-capped slopes rather than Petri dishes should be used for isolation.

Serological tests are useful, but cross-reactions can occur, mainly with *C. immitis*. Antibody tests fail to detect antibodies in up to 50% of immunosuppressed individuals. Tests for antigen detection in the urine by ELISA are useful in disseminated histoplasmosis but are not widely available.

Treatment

Intravenous amphotericin B is recommended for treatment of the most severe forms of disseminated histoplasmosis. Itraconazole is widely used in non-immunocompromised patients with milder forms of disseminated disease and for the continuation of treatment in those who have responded to amphotericin B.

BLASTOMYCOSIS

This disease is caused by *B. dermatitidis*, a dimorphic soil-inhabiting fungus. The largest number of cases of blastomycosis has been reported from North America, but the disease is also endemic in Africa, and parts of Central and South America. In the USA, the organism is most commonly found in states surrounding the Mississippi and Ohio Rivers; in Canada, the disease occurs in the provinces that border the Great Lakes.

In culture at 25–30°C *B. dermatitidis* grows as a mould with a septate mycelium. The colony varies in texture from floccose to smooth and from white to brown in colour. Asexual conidia are produced on lateral hyphal branches of variable length; the oval or pear-shaped conidia are 2–10 μm in diameter. In tissue and in culture at 37°C the fungus grows as a large round yeast (5–15 μm in diameter) that characteristically produces broad-based buds from a single pole on the mother cell (Fig. 60.14).

Epidemiology

Infection results from inhalation of spores; the incubation period is 4–6 weeks. The disease is more commonly seen in adults than in children, and often occurs in individuals with an outdoor occupation or recreational interest.

Clinical features

Acute pulmonary blastomycosis usually presents as a non-specific influenza-like illness, similar to that seen with histoplasmosis or coccidioidomycosis. Most otherwise healthy persons recover after 2–12 weeks of symptoms, but some return months later with infection of other sites. Other patients with acute blastomycosis fail to recover and develop chronic pulmonary disease or disseminated infection.

The skin and bones are the most common sites of disseminated disease. The skin is involved in more than 70% of cases; the characteristic lesions are typically raised, with a well demarcated edge. It is from these skin lesions that the diagnosis is most often made. Osteomyelitis occurs in about 30% of patients, with the spine, ribs and long bones being the commonest sites of infection. Arthritis occurs in about 10% of patients. Meningitis is rare, except in immunocompromised individuals.

Laboratory diagnosis

Direct microscopy of pus, scrapings from skin lesions, or sputum usually shows thick-walled yeast cells 5–15 μm in diameter that characteristically produce buds on a broad base; the buds remain attached until they are almost the size of the parent cell, often forming chains of three or four cells. In biopsy material the yeasts are best seen in stained sections.

B. dermatitidis will grow in culture on Sabouraud agar (or blood agar) without cycloheximide, to which the fungus is sensitive. The mycelial phase develops slowly at 25–30°C and cultures must be retained for 6 weeks before discarding. Test-tube slopes rather than Petri dishes are used for culture. Identification is usually confirmed by subculture at 37°C to convert it to the yeast phase.

The most useful serological test is the precipitin test using the A antigen of *B. dermatitidis*. However, a negative result does not rule out the diagnosis because the sensitivity ranges from 30% for localized infections to 90% for cases of disseminated blastomycosis.

Treatment

Intravenous amphotericin B is used to treat all forms of blastomycosis and is the drug of choice in serious life-threatening infection. Itraconazole follow-on therapy is given once the patient improves. Itraconazole is also the drug of choice in less serious infection that does not involve the central nervous system.

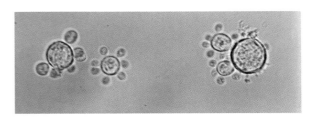

Fig. 60.15 Microscopical appearance of *Paracoccidioides brasiliensis* yeast cells in sputum, showing multipolar budding.

PARACOCCIDIOIDOMYCOSIS

This is a chronic granulomatous infection caused by the dimorphic fungus *P. brasiliensis* that may involve the lungs, mucosa, skin and lymphatic system. The disease may be fatal if untreated. Although *P. brasiliensis* has been isolated from soil, understanding of its precise environmental reservoir remains limited. The endemic region extends from Mexico to Argentina, but the disease is seen most frequently in the subtropical upland rainforests of Brazil and Central America.

P. brasiliensis grows in the mycelial phase in culture at 25–30°C and in the yeast phase in tissue or at 37°C on brain–heart infusion or blood agar. The mould colonies are slow growing with a variable colonial morphology, although most are white and velvety to floccose in texture with a pale brown reverse. Spore production is usually sparse and best seen in 8–10-week cultures. Asexual conidia may be produced but are not characteristic, and identification depends on conversion from the mycelial to the yeast phase. The yeast phase consists of oval or globose cells 2–30 μm in diameter, with small buds attached by a narrow neck encircling the parent cell (Fig. 60.15).

Epidemiology

Infection is usually acquired by inhalation; the incubation period is unknown. More than 90% of cases of symptomatic disease occur in men, most of whom have outdoor occupations; oestrogen-mediated inhibition of the mould to yeast transformation could help to account for this.

Clinical features

The lungs are the usual initial site of *P. brasiliensis* infection, but the organism then spreads through the lymphatics to the regional lymph nodes. In most cases the primary infection is asymptomatic. There is evidence of prolonged latent infection before overt disease develops, and a mild, self-limiting pulmonary form of paracoccidioidomycosis probably exists. Children and adolescents sometimes present with an acute disseminated form of infection in which superficial and/or visceral lymph node enlargement is the major manifestation. This presentation is also seen in immunocompromised patients. It has a poor prognosis.

In adults, paracoccidioidomycosis usually presents as an ulcerative granulomatous infection of the oral and nasal mucosa and adjacent skin. In 80% of cases the disease involves the lungs. In some, the liver and spleen, intestines, adrenals, bones and joints, and central nervous system are also involved. The disease is slowly progressive, and may take months or even years to become established.

Laboratory diagnosis

Microscopy of sputum or pus, crusts and biopsies from granulomatous lesions usually reveals numerous yeast cells showing the characteristic multipolar budding, which is diagnostic. In culture the mycelial and yeast phases both develop slowly and cultures must be retained for 6 weeks before discarding. The mould phase can be isolated on Sabouraud agar supplemented with yeast extract at 25–30°C, but colonies may take 2–4 weeks to appear. Serological tests are useful for diagnosis and for monitoring the response to therapy.

Treatment

The choice of therapy depends on the site of infection and its severity. Oral itraconazole is the drug of choice, although amphotericin B remains useful for severe or refractory infections. Oral ketoconazole is almost as effective, but less well tolerated than itraconazole.

ASPERGILLOSIS

There are more than 200 species of *Aspergillus* but fewer than 20 have been implicated in human disease; the most important are *A. fumigatus*, *A. flavus*, *A. terreus*, *A. niger* and *A. nidulans*. In immunocompromised individuals, inhalation of spores can give rise to life-threatening invasive infection of the lungs or sinuses and dissemination to other organs often follows (*invasive aspergillosis*). In non-immunocompromised persons, these moulds can cause localized infection of the lungs, sinuses and other sites. Human disease can also result from non-infectious mechanisms: inhalation of spores can cause allergic symptoms in both atopic and non-atopic individuals.

Aspergillus species are ubiquitous in the environment, growing in the soil, on plants, and on decomposing organic matter. These moulds are often found in the outdoor and indoor air, in water, on food items, and in dust. All grow in nature and in culture as moulds with septate hyphae and distinctive asexual sporing structures,

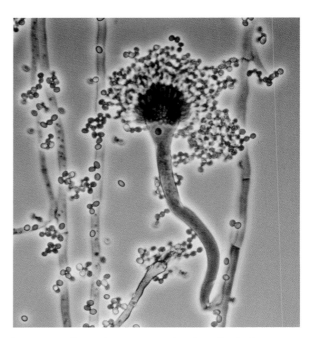

Fig. 60.16 Microscopical appearance of an *Aspergillus fumigatus* conidiophore. (Reproduced with permission from Richardson M D, Warnock D W, Campbell C K 1995 *Slide Atlas of Fungal Infection: Systemic Fungal Infections*. Blackwell Science, Oxford.)

termed conidiophores, that bear long chains of conidia (Fig. 60.16).

Epidemiology

Inhalation of *Aspergillus* conidia is the usual mode of infection; less frequently, infection follows the traumatic implantation of spores as in corneal infection (see p. 605), or inadvertent inoculation as in endocarditis.

Invasive aspergillosis has emerged as a major problem in several groups of immunocompromised patients, including those with acute leukaemia, stem-cell and solid-organ transplant recipients, and children with chronic granulomatous disease. The likelihood of aspergillosis developing in these individuals depends on a number of host factors, the most important of which is the level of immunosuppression. The mortality rate is high, ranging from 50% to 100% in almost all groups of immunocompromised patients.

Clinical features

Invasive aspergillosis

This form occurs in severely immunocompromised individuals who have a serious underlying illness. *A.*

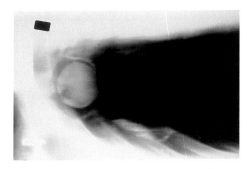

Fig. 60.17 Radiological appearance of an aspergilloma. (Reproduced with permission from Richardson M D, Warnock D W, Campbell C K 1995 *Slide Atlas of Fungal Infection: Systemic Fungal Infections.* Blackwell Science, Oxford.)

fumigatus is the species most frequently involved. The commonest initial presentation in the neutropenic patient is an unremitting fever (>38°C), without any respiratory tract symptoms, that fails to respond to broad-spectrum antibiotics.

The lung is the sole site of infection in 70% of patients, but dissemination of infection to other organs often occurs; the central nervous system is involved in 10–20% of cases. There is widespread destructive growth of *Aspergillus* species in lung tissue and the fungus invades blood vessels, causing thrombosis and infarction; septic emboli may spread the infection to other organs. Invasive aspergillosis has a poor prognosis; early diagnosis is essential for successful management.

Aspergilloma

In this form of aspergillosis, also referred to as *fungus ball*, the fungus colonizes pre-existing (often tuberculous) cavities in the lung and forms a compact ball of mycelium, eventually surrounded by a dense fibrous wall (Fig. 60.17).

Aspergillomas are usually solitary. Patients are either asymptomatic or have only a moderate cough and sputum production. Occasional haemoptysis may occur, especially when the fungus is actively growing, and haemorrhage following invasion of a blood vessel is one of the fatal complications of this condition. Surgical resection is most often used to treat this condition.

Sinusitis

Aspergillus species, particularly *A. flavus* and *A. fumigatus*, may colonize and invade the paranasal sinuses; the infection may spread through the bone to the orbit of the eye and brain. Acute invasive sinusitis is a rapidly progressive disease, most commonly seen in immunocompromised persons. There are also several forms of slowly pro-gressive chronic *Aspergillus* sinusitis that occur in immuno-competent individuals.

Allergic bronchopulmonary aspergillosis

Allergy to *Aspergillus* species is usually seen in atopic individuals with raised immunoglobulin (Ig) E levels; about 10–20% of asthmatics react to *A. fumigatus*. The condition is a form of asthma with pulmonary eosinophilia that manifests as episodes of bronchial obstruction and lung consolidation; the fungus grows in the airways to produce mucous plugs of fungal mycelium that may block off segments of lung tissue and that, when coughed up, are a diagnostic feature.

Laboratory diagnosis

The value of the laboratory in diagnosis varies according to the clinical form of aspergillosis; the diagnosis of invasive disease is particularly difficult.

Direct microscopy

In potassium hydroxide preparations (preferably with Calcofluor to enhance detection) of sputum the fungus appears as non-pigmented septate hyphae, 3–5 μm in diameter, with characteristic dichotomous branching and an irregular outline; rarely the characteristic sporing heads of *Aspergillus* species are present.

- In allergic aspergillosis there is usually abundant fungus in the sputum and mycelial plugs may also be present.
- In aspergilloma, fungus may be difficult to find on microscopy.
- In invasive aspergillosis, microscopy is often negative.

Biopsy may provide a definitive diagnosis, although many clinicians are reluctant to undertake this procedure because of the associated risk in immunosuppressed patients. In tissue sections *Aspergillus* species are best seen after staining with periodic acid–Schiff or methenamine–silver.

Culture

Aspergillus species grow readily at 25–37°C on Sabouraud agar without cycloheximide; colonies appear after 1–2 days. Isolates can be identified by their colonial appearance and micromorphology. The ability of *A. fumigatus* to grow well at 45°C can be used to help identify this species or to isolate it selectively.

As aspergilli are among the commonest laboratory contaminants, quantification of the amount of fungus in sputum helps to confirm the relevance of a positive culture. However, all isolates from immunocompromised patients must be taken seriously and acted upon.

Large quantities of fungus are usually recovered from the sputum of patients with allergic aspergillosis, but cultures from those with aspergilloma or invasive disease are commonly negative or yield only a few colonies. Blood cultures are negative in invasive disease.

Skin tests

Skin tests with *A. fumigatus* antigen are useful for the diagnosis of allergic aspergillosis. All patients give an immediate type I reaction and 70% of those with pulmonary eosinophilia also give a delayed type III Arthus reaction.

Serological tests

Immunodiffusion, CIE and ELISA are widely used for the detection of antibodies in the diagnosis of all forms of aspergillosis, particularly aspergilloma and allergic bronchopulmonary aspergillosis. Tests for *Aspergillus* antibodies are seldom helpful in the diagnosis of invasive infection in immunocompromised patients

Antigen detection has also been used successfully for diagnosis of invasive aspergillosis by techniques such as ELISA and latex agglutination. However, with ELISA sensitivity is currently less than 80% and nucleic acid amplification methods are now increasingly used for diagnosis of invasive aspergillosis.

Treatment

In invasive aspergillosis, the treatment of choice has long been intravenous amphotericin B (conventional or liposomal). Voriconazole proved superior to amphotericin B in a large clinical trial, and is approved for the treatment of this disease.

Aspergilloma is treated by surgical excision because antifungal therapy is of little value, but because of the significant morbidity and mortality with this procedure it is reserved for patients with episodes of life-threatening haemoptysis. Allergic forms of aspergillosis are treated with corticosteroids.

INVASIVE CANDIDOSIS

In addition to causing mucosal and cutaneous infections (see pp. 602–604), *Candida* species can cause acute or chronic invasive infections in immunocompromised or debilitated individuals. This may be confined to one organ or become widespread (disseminated candidosis).

Epidemiology

In most cases invasive candidosis is endogenous in origin, but transmission of organisms from person to person can also occur. Hospital outbreaks of infection have sometimes been linked to contaminated medical devices, such as vascular catheters, and/or parenteral nutrition. There are also reports of cross-infection due to hand carriage by health-care workers.

Invasive candidosis is a significant problem in several distinct groups of hospitalized patients:

- neutropenic cancer patients
- stem cell and liver transplant recipients
- patients receiving intensive care.

Invasive candidosis is now more common in patients in intensive care than among neutropenic individuals. The reduced incidence of the disease among the latter group has been attributed to the widespread use of fluconazole prophylaxis. The predominant pathogen in all groups is still *C. albicans*, but the proportion of serious infections due to less azole-susceptible species, such as *C. glabrata*, has increased.

Many risk factors for invasive candidosis have been identified. These can be divided into host-related and health care-related factors:

- underlying immunosuppression
- low birth-weight
- intravascular catheterization
- broad-spectrum antibiotic use
- total parenteral nutrition
- haemodialysis.

Among adult patients cared for in surgical intensive care units, *Candida* bloodstream infection has a case fatality rate of about 40%.

Clinical features

Infection may be localized, for instance in the urinary tract, liver, heart or meninges, or may be widely disseminated and associated with a septicaemia (*candidaemia*). Invasive candidosis is difficult to diagnose and treat, and for some forms the prognosis is poor.

Disseminated infection is most commonly seen in seriously ill individuals who usually have one or more indwelling vascular catheters, although these are not necessarily the source of the infection. Many cases are thought to arise from translocation of organisms across the wall of the intestinal tract.

Adults with candidaemia usually present with a persistent fever that fails to respond to broad-spectrum antibiotics, but with few other symptoms or clinical signs. One sign of invasive candidosis is the presence of white lesions within the eye (*Candida* endophthalmitis) (Fig. 60.18). These are found in up to 45% of patients in the intensive care unit, but are seldom seen in neutropenic individuals. Other useful signs are the nodular cutaneous

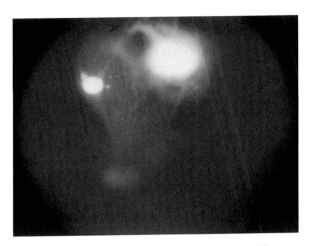

Fig. 60.18 Fundoscopic appearance of *Candida* endophthalmitis. (Reproduced with permission from Richardson M D, Warnock D W, Campbell C K 1995 *Slide Atlas of Fungal Infection: Systemic Fungal Infections*. Blackwell Science, Oxford.)

lesions that occur in about 10% of neutropenic individuals with disseminated *Candida* infection.

Other manifestations include:

- meningitis
- renal abscess
- myocarditis
- osteomyelitis
- arthritis.

Invasive candidosis is a common complication in infants of low birth-weight (<1000 g) requiring prolonged neonatal intensive care. Meningitis occurs more frequently than in older patients and is sometimes associated with arthritis and osteomyelitis.

Laboratory diagnosis

Candida species may be present as commensals in the absence of infection, so that isolation from clinical material, except from sites that are normally sterile, is of little significance. Similarly, antibodies to *Candida* species can be detected in uninfected individuals because of their exposure to commensal yeasts, although a rise in antibody titre or high titres may be of diagnostic significance. In suspected invasive candidosis, samples from as many sources as possible should be examined by direct microscopy and culture. Results should always be interpreted in the light of clinical findings.

Direct microscopy

Appropriate samples are examined microscopically in potassium hydroxide or after Gram staining. In tissue sections, the fungus is seen best in stained preparations. Hyphae are often abundant, but their presence in sputum or urine does not confirm that the yeast is present as a pathogen.

Culture

Candida species grow readily in culture at 37°C on common isolation media, such as Sabouraud dextrose agar. Blood cultures provide the most reliable evidence of invasive infection, although repeated attempts to isolate the organism may be required.

Isolation of the yeast from otherwise sterile sites provides reliable evidence for the diagnosis, but cultures obtained from urine, faeces and sputum are of less value unless done quantitatively over a period of time. Cell counts of the yeast in urine in excess of 10^4 per millilitre are usually taken to indicate urinary tract infection, except in those with an indwelling urinary catheter. As *Candida* species multiply rapidly in clinical material it is important that specimens are processed as soon as possible after collection.

Serological tests

Currently available tests lack specificity and sensitivity, and the results must be interpreted with care. A positive test does not necessarily indicate infection because the antigens used do not differentiate between antibodies formed during mucosal colonization and those produced during deep infection. Similarly, a negative antibody test does not necessarily rule out the possibility of invasive candidosis in immunocompromised patients who are incapable of mounting an adequate antibody response.

Antigen tests, based mainly on ELISA or latex agglutination, that detect cell wall mannan or cytoplasmic components have been developed and are used for diagnosis. Nucleic acid detection methods are increasingly being used to diagnose invasive candidosis, although their place in diagnosis is still being evaluated.

Treatment

The treatments of choice for most forms of invasive candidosis are:

- intravenous caspofungin
- intravenous amphotericin B (conventional or liposomal)
- intravenous or oral fluconazole.

Amphotericin B can be used in combination with flucytosine, but flucytosine is not used alone as resistance may develop during treatment. Voriconazole has also been used successfully. Removal of existing intravenous catheters is desirable if feasible, especially in non-neutropenic patients.

The choice of antifungal treatment depends on both the clinical status of the patient and the species of infecting organism.

- All *Candida* species are susceptible to caspofungin.
- *C. albicans*, *C. parapsilosis* and *C. tropicalis* are susceptible to amphotericin B and fluconazole.
- *C. glabrata* often becomes resistant to fluconazole during therapy.

CRYPTOCOCCOSIS

Cryptococcosis, caused by the encapsulated yeast *Cryptococcus neoformans*, is most frequently recognized as a disease of the central nervous system, although the primary site of infection is the lungs. The disease occurs sporadically throughout the world but it is currently seen most often in persons with AIDS.

There are four serotypes of *C. neoformans* (A–D) that represent two varieties of the organism, namely, *C. neoformans* var. *neoformans* (A & D) and *C. neoformans* var. *gattii* (B & C). Most infections are caused by *C. neoformans* var. *neoformans*, which is commonly found in the droppings of wild and domesticated birds throughout the world. Pigeons carry *C. neoformans* in their crops, but do not appear to become infected, probably because of their high body temperature. *C. neoformans* var. *gattii* has been isolated from decaying wood in the red gum group of Eucalyptus trees. These trees are indigenous to Australia, but have been planted in numerous other countries, including the USA, Africa, India and China. The distribution of human infections largely coincides with the distribution of these trees.

Epidemiology

Infection is acquired by inhalation; the incubation period is unknown. The likelihood that an infection with *C. neoformans* will develop after inhalation depends largely on host factors. Even before the advent of AIDS, infections with var. *neoformans* tended to occur in individuals with abnormalities of T lymphocyte function, such as are found in persons with lymphoma, and those receiving corticosteroid therapy. The major risk factor for development of infection with *C. neoformans* var. *gattii* appears to be environmental exposure, although there is indirect evidence that unidentified host factors contribute to the higher incidence of disease in Australian Aboriginals.

With the advent of the AIDS epidemic, cryptococcosis became the most common cause of meningitis in hospitals in which persons with HIV infection are treated. Although the incidence of the disease has declined in developed countries where combination antiretroviral treatment is available, the incidence is rising in many developing countries afflicted with large epidemics of HIV infection.

Clinical features

A mild self-limiting pulmonary infection is believed to be the commonest form of cryptococcosis. In symptomatic pulmonary infection there are no clear diagnostic features. Lesions may take the form of small discrete nodules, which may heal with a residual scar or may become enlarged, encapsulated and chronic (*cryptococcoma* form). An acute pneumonic type of disease has also been described.

The meningeal form of cryptococcosis can occur in apparently healthy individuals, but occurs most frequently in immunocompromised persons. Chronic meningitis or meningo-encephalitis develops insidiously with headaches and low-grade fever, followed by changes in mental state, visual disturbances and eventually coma. The disease may last from a few months to several years, but the outcome is always fatal unless it is treated. Patients with AIDS and cryptococcosis generally develop a chronic meningeal form with milder symptoms.

Although predominantly a disease of the central nervous system, lesions of the skin, bones and other deep sites may also occur; in its disseminated form, the disease may resemble tuberculosis. Rarely, lesions of skin and bones may occur without any evidence of infection elsewhere.

Laboratory diagnosis

C. neoformans is readily demonstrated in cerebrospinal fluid (CSF) or other material by direct microscopy, culture or serological tests for capsular antigen. The yeast load is generally higher in patients with AIDS. The cellular reaction and chemical changes in CSF usually resemble those seen in tuberculous meningitis. The yeast cells of *C. neoformans* are round, 4–10 μm in diameter and surrounded by a mucopolysaccharide capsule. The width of the capsule varies and is greatest in vivo and on rich media in vitro.

In unstained wet preparations of CSF mixed with a drop of Indian ink or nigrosine, the capsule can be seen as a clear halo around the yeast cells (Fig. 60.19). Capsulate yeasts are seen in the CSF of about 60% of patients with cryptococcosis (higher in AIDS), but the capsule may be difficult to visualize in some cases. Sputum, pus or brain tissue should be examined after digestion in potassium hydroxide and here the capsulate yeasts are often delineated by the cellular debris. For examination of tissue sections it is best to use a specific fungal stain such

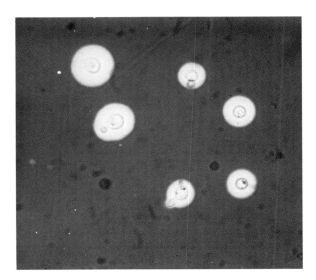

Fig. 60.19 Microscopic appearance of encapsulated *Cryptococcus neoformans* cells in an Indian ink preparation of CSF. (Reproduced with permission from Richardson M D, Warnock D W, Campbell C K 1995 *Slide Atlas of Fungal Infection: Systemic Fungal Infections.* Blackwell Science, Oxford.)

as periodic acid–Schiff; alcian blue and mucicarmine stain the capsular material, enabling the organisms to be differentiated from *H. capsulatum* and *B. dermatitidis*.

The yeast is easily cultured from CSF although large volumes or multiple samples may be required in some cases; in patients with AIDS it is also useful to culture blood. On Sabouraud agar (without cycloheximide) cultured at 25–30°C and 37°C, colonies normally appear within 2–3 days, but cultures should not be discarded for 3 weeks. In culture, *C. neoformans* appears as creamy white to yellow–brown colonies, which are mucoid in strains with well developed capsules and dry in strains that lack prominent capsules. Buds appear at any point on the cell surface but hyphae or pseudohyphae are not normally produced. Preliminary identification depends on demonstration of the capsule but this may be absent or difficult to see. *C. neoformans* can be identified with commercial kits or distinguished from other yeasts by its lack of fermentative ability, its ability to produce urease, to grow at 37°C and to assimilate inositol.

The latex agglutination test for the detection of cryptococcal polysaccharide antigen in CSF or blood is highly sensitive and specific for the diagnosis of cryptococcal meningitis and disseminated forms of the disease, and gives better results than microscopy and culture. In AIDS, the test is positive in well over 90% of infected patients; titres of over 10^6 may be detected.

Treatment

Intravenous amphotericin B in combination with flucytosine is usually the treatment of choice for individuals with meningeal or disseminated cryptococcosis. Oral fluconazole is widely used for the continuation of treatment in those who have responded to amphotericin B. Patients with AIDS commonly relapse after the initial course of therapy and many require lifelong maintenance treatment with fluconazole.

ZYGOMYCOSIS

Zygomycosis, also known as *mucormycosis*, is a relatively rare, opportunistic infection caused by saprophytic mould fungi, notably species of *Rhizopus* and *Absidia*. These moulds are ubiquitous in the soil and on decomposing organic matter. They are characterized by having broad, aseptate hyphae, with large numbers of asexual spores inside a sporangium which develops at the end of an aerial hypha.

Epidemiology

Most infections follow inhalation of spores; less frequently, infection follows traumatic inoculation into the skin and soft tissue. Major risk factors include:

- prolonged or profound neutropenia
- uncontrolled diabetes mellitus
- other forms of metabolic acidosis
- burns.

Certain predisposing conditions seem to be more commonly associated with particular clinical forms of disease; for example, persons with diabetic ketoacidosis often develop rhinocerebral mucormycosis, whereas neutropenic individuals often develop pulmonary or disseminated disease.

Clinical features

The best known form of the disease is rhinocerebral zygomycosis. There is rapid and extensive tissue destruction, most commonly spreading from the nasal mucosa to the turbinate bone, paranasal sinuses, orbit and brain (Fig. 60.20). The condition is fatal if untreated and, although the prognosis has improved over recent years, many diagnoses are still made at necropsy.

Pulmonary and disseminated infections can occur in severely immunocompromised individuals. Primary cutaneous infections have also been reported; these are uncommon but often result in extensive necrotizing

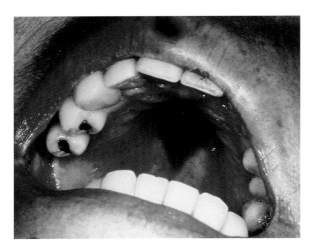

Fig. 60.20 Necrotic palatal lesion in a case of rhinocerebral zygomycosis. (Reproduced with permission from Richardson M D, Warnock D W, Campbell C K 1995 *Slide Atlas of Fungal Infection: Systemic Fungal Infections.* Blackwell Science, Oxford.)

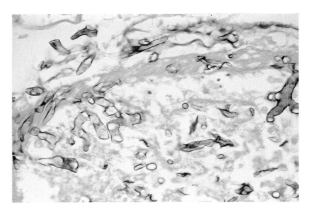

Fig. 60.21 Microscopic appearance of zygomycosis. (Reproduced with permission from Richardson M D, Warnock D W, Campbell C K 1995 *Slide Atlas of Fungal Infection: Systemic Fungal Infections.* Blackwell Science, Oxford.)

fasciitis or disseminated disease. They usually occur in patients with burns or other forms of local trauma.

Laboratory diagnosis

Recognition of the fungus in tissue by microscopy is considerably more reliable than culture, but material such as nasal discharge or sputum seldom contains much fungal material and examination of a biopsy is usually necessary for a firm diagnosis. Direct examination of curetted or biopsy material in potassium hydroxide may reveal the characteristic broad, aseptate, branched and sometimes distorted hyphae. However, they are seen much more clearly when stained with methenamine–silver; the hyphae of these fungi do not stain with periodic acid–Schiff (Fig. 60.21).

The fungi are readily isolated on Sabouraud dextrose agar at 37°C, but isolation is of little diagnostic significance in the absence of strong supporting clinical evidence of infection. There are no established serological tests.

Treatment

Successful treatment depends on early diagnosis of the infection to allow prompt therapy with high doses of intravenous amphotericin B (conventional or liposomal), control of any underlying disorder, such as diabetes, and aggressive surgical intervention.

PNEUMOCYSTOSIS

Pneumocystis is an opportunistic pathogen with some of the features of protozoa, but comparative DNA sequence analysis showed it to be more closely related to the fungi. The organism was originally described as a cause of atypical pneumonia in malnourished infants, but came to prominence in the 1980s as a common cause of pneumonia, which was commonly fatal, in patients with AIDS. With the advent of reliable antiretroviral therapy, the incidence of the disease in HIV-positive individuals has declined.

Based on its morphology, which is similar to that of protozoa, the life cycle of *Pneumocystis* is divided into three stages:

1. the cyst, a spherical or crescent-shaped form (5–7 μm in diameter)
2. the sporozoite, up to eight of which develop within each cyst
3. the trophozoite, found outside the cyst.

When openings appear in the cyst wall, the excysted sporozoites become trophozoites. All of these forms reside within the alveoli of the lungs. As the organism is not cultivatable in vitro, its life cycle has not been fully elucidated.

The finding that *Pneumocystis* organisms from different mammalian hosts are genetically quite dissimilar has led to a name change from *P. carinii* (which infects rats) to *P. jirovecii* for the organisms that infect man.

Epidemiology

Infection is probably acquired by inhalation. Serological and PCR studies indicate that most human beings become subclinically infected with *Pneumocystis* during childhood and that this infection is usually well contained by an intact immune system. The occurrence of clinical disease is related to the extent of immunosuppression, especially impairment of cell-mediated immunity. It may be due to primary infection, re-infection or reactivation. Pneumocystosis has a global distribution.

Clinical features

The clinical presentation of *Pneumocystis* pneumonia is non-specific. Symptoms include:

- fever
- non-productive cough
- dyspnoea
- shortness of breath.

Patients with AIDS have a more indolent course with longer duration of symptoms than patients receiving immunosuppressive drugs. Without treatment the course is progressive and usually ends in death.

In addition to pneumonia, *Pneumocystis* infection may disseminate to the lymph nodes, liver, spleen, bone marrow, adrenal gland, intestines and meninges. Extra-pulmonary disease occurs predominantly in patients with advanced HIV infection and in those not taking co-trimoxazole prophylaxis or receiving aerosolized pentamidine.

Laboratory diagnosis

Diagnosis usually depends on the identification of typical octonucleate cysts or trophozoites in tissues or body fluids. Organisms are detected by immunofluorescent staining of broncho-alveolar lavage fluid or induced sputum smears. Molecular diagnosis by PCR methods is being introduced in some centres.

Treatment

Co-trimoxazole and intravenous pentamidine are the most effective therapies. The former is as potent as the latter, but less toxic. Other regimens include atovaquone, trimetrexate, the combination of trimethoprim and dapsone, and the combination of clindamycin and primaquine.

Co-trimoxazole is the preferred prophylactic agent, but patients with AIDS or those undergoing solid-organ or bone marrow transplantion may suffer unacceptable side effects to the high doses used. Aerosolized pentamidine is also used for prophylaxis.

OTHER OPPORTUNIST FUNGI

Penicillium marneffei causes serious disseminated disease with characteristic papular skin lesions in patients with AIDS in South-East Asia. The fungus is dimorphic, forming yeast-like cells that are often intracellular, resembl-ing histoplasmosis, in infected tissues. Treatment is with amphotericin B, followed by itraconazole to prevent relapse.

Almost any fungus may invade a severely immuno-compromised host, and infections with many common fungi, including *Fusarium* species, *Trichosporon asahii* and *Pseudallescheria boydii*, have been reported. Diagnosis is made by culture of the causative organism from clinical specimens and serological tests play little part. Tissue sections are often not very helpful, either because the causal fungi have no special features to enable identification or because they resemble other fungal pathogens.

Infections are usually treated speculatively, and some-times successfully, with amphotericin B.

KEY POINTS

- Most infections are caused by fungi that grow as saprophytes in the environment. *Superficial*, *subcutaneous* and *systemic* patterns of infection are recognized.
- Fungi take the form of *yeasts*, which grow by *budding*, *moulds*, which grow as filamentous extensions called *hyphae*, forming a *mycelium*, or *dimorphic* fungi, which can grow as yeast or mould forms.
- Pathogenic fungi can establish an infection in all exposed individuals; others are opportunist pathogens that cause disease only in a compromised host.
- Many fungal diseases have a global distribution, but some are endemic to specific geographical regions.
- Subcutaneous fungal infections are acquired by traumatic implantation; systemic infections are usually acquired by inhalation.
- Only dermatophytosis, a common superficial infection of the skin, nails and scalp hair, is truly contagious.
- Some yeasts are human commensals and cause endogenous infection when there is some imbalance in the host.
- The most frequently encountered fungal agents in the UK are *Candida albicans*, dermatophytes, *Aspergillus* spp., *Cryptococcus neoformans* and *Pneumocystis jirovecii*.
- Diagnosis of fungal infections is based on a combination of clinical observation and laboratory investigation, which may include direct microscopy, histology, culture, PCR and serology. Early recognition of systemic infections in immunocompromised persons is a major challenge.

RECOMMENDED READING

Ajello L, Hay R J (eds) 1998 *Microbiology and Microbial Infections*, 9th edn, Vol. 4 *Medical Mycology*. Arnold, London

Anaissie E J, McGinnis M R, Pfaller M A (eds) 2003 *Clinical Mycology*. Churchill Livingstone, Philadelphia

Dismukes W E, Pappas P G, Sobel J D (eds) 2003 *Clinical Mycology*. Oxford University Press, Oxford

Kibbler C C, Mackenzie D W R, Odds F C (eds) 1996 *Principles and Practice of Clinical Mycology*. Wiley, Chichester

Kwon-Chung K J, Bennett J E 1992 *Medical Mycology*. Lea & Febiger, Philadelphia

Richardson M D, Johnson E M 2000 *Pocket Guide to Fungal Infection*. Blackwell Science, Oxford

Richardson M D, Warnock D W 2003 *Fungal Infection: Diagnosis and Management*, 3rd edn. Blackwell Science, Oxford

Internet sites

Canadian National Centre for Mycology. World of Dermatophyte: A Pictorial. http://www.provlab.ab.ca/mycol/tutorials/derm/dermhome.htm

Doctor Fungus. http://www.doctorfungus.org/

Fungal Research Trust. The Aspergillus Website. http://www.aspergillus.org.uk/

University of Adelaide. Mycology Online. http://www.mycology.adelaide.edu.au/

Valley Fever Center for Excellence. Coccidioidomycosis. http://www.vfce.arizona.edu/

Protozoa

Malaria; toxoplasmosis; cryptosporidiosis; amoebiasis; trypanosomiasis; leishmaniasis; giardiasis; trichomoniasis

D. Greenwood

Infection with pathogenic protozoa exacts an enormous toll of human suffering, notably, but not exclusively, in the tropics. Numerically the most important of the life-threatening protozoan diseases is malaria, which is responsible for at least 1 million deaths a year, mostly in young children in Africa.

Pathogenic protozoan parasites are conveniently dealt with by placing them in four groups: *sporozoa, amoebae, flagellates* and a miscellaneous group of other protozoa that may cause human disease (Table 61.1).

SPOROZOA

This group includes the malaria parasites and related coccidia, which exhibit a complex life cycle involving alternating cycles of asexual division (*schizogony*) and sexual development (*sporogony*). In malaria parasites, the sexual cycle takes place in the female anopheline mosquito (Fig. 61.1).

MALARIA PARASITES

Description

Four species are encountered in human disease: *Plasmodium falciparum*, which is responsible for most fatalities; *P. vivax* and *P. ovale*, both of which cause *benign tertian malaria* (febrile episodes typically occurring at 48-h intervals); and *P. malariae*, which causes *quartan malaria* (febrile episodes typically occurring at 72-h intervals). The appearances of trophozoites of the four species as seen in Romanowsky-stained films of peripheral blood are illustrated in Figures 61.2–61.7.

Life cycle

When an infected mosquito bites, *sporozoites* present in the salivary glands enter the bloodstream and are carried to the liver, where they invade liver parenchyma cells. They undergo a process of multiple nuclear division,

followed by cytoplasmic division (*schizogony*); when this is complete, the liver cell ruptures, releasing several thousand individual parasites (*merozoites*) into the bloodstream. The merozoites penetrate red blood cells and adopt a typical 'signet-ring' morphology (see Fig. 61.3).

In the case of *P. vivax* and *P. ovale*, some parasites in the liver remain dormant (*hypnozoites*) and the cycle of

Table 61.1 Principal protozoan pathogens of man

Group	Species	Disease
Sporozoa	*Plasmodium falciparum*	Malignant tertian malaria
	P. vivax	Benign tertian malaria
	P. ovale	Benign tertian malaria
	P. malariae	Quartan malaria
	Toxoplasma gondii	Toxoplasmosis
	Isospora belli	Diarrhoea
	Cryptosporidium parvum	Diarrhoea
	Cyclospora cayetanensis	Diarrhoea
Amoebae	*Entamoeba histolytica*	Amoebic dysentery; liver abscess
	Naegleria fowleri[a]	Meningo-encephalitis
	Acanthamoeba spp.[a]	Keratitis
	Balamuthia mandrillaris[a]	Encephalitis
	Blastocystis hominis[b]	Pathogenicity doubtful
Flagellates	*Giardia lamblia*	Diarrhoea, malabsorption
	Trichomonas vaginalis	Vaginitis, urethritis
	Trypanosoma brucei gambiense	Sleeping sickness
	T. brucei rhodesiense	Sleeping sickness
	T. cruzi	Chagas' disease
	Leishmania spp.	See Table 61.4
Others	*Babesia microti*[a]	Babesiosis
	B. divergens[a]	Babesiosis
	Balantidium coli[a]	Balantidial dysentery
	Encephalitozoon cuniculi[a]	Microsporidiosis
	Enterocytozoon bieneusi[a]	Microsporidiosis
	Nosema connori[a]	Microsporidiosis

[a]Organisms rarely encountered in human disease.
[b]Taxonomic status uncertain.

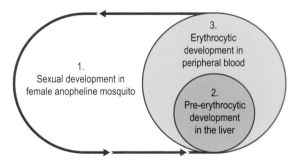

Fig. 61.1 Schematic representation of the life cycle of malaria parasites.

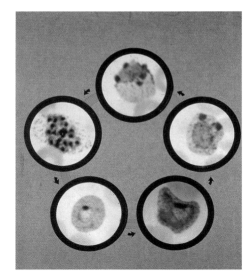

Fig. 61.2 Stages in the erythrocytic cycle of *Plasmodium vivax.*

pre-erythrocytic schizogony is completed only after a long delay. Such parasites are responsible for the relapses of tertian malaria that may occur usually within 2 years of the initial infection.

In the bloodstream, the young ring forms (*trophozoites*) develop and start to undergo nuclear division (*erythrocytic schizogony*). Depending on the species, about 8 to 24 nuclei are produced before cytoplasmic division occurs, and the red cell ruptures to release the individual merozoites, which then infect fresh red blood cells (see Fig. 61.2).

Instead of entering the cycle of erythrocytic schizogony, some merozoites develop within red cells into male or female *gametocytes*. These do not develop further in the human host, but when the insect vector ingests the blood, the nuclear material and cytoplasm of the male *gametocytes* differentiate to produce several individual gametes, which give it the appearance of a flagellate body (*exflagellating male gametocyte*). The gametes become detached and penetrate the female gametocyte, which elongates into a zygotic form, the *ookinete*. This penetrates the mid-gut wall of the mosquito and settles on the body cavity side as an *oocyst*, within which numerous *sporozoites* are formed. When mature, the oocyst ruptures, releasing the sporozoites into the body cavity, from where some find their way to the salivary glands.

P. falciparum differs from the other forms of malaria parasite in that developing erythrocytic schizonts form aggregates in the capillaries of the brain and other internal organs, so that normally only relatively young ring forms or gametocytes (which are typically crescent shaped) are found in peripheral blood.

The cycle of erythrocytic schizogony takes 48 h, except in the case of *P. malariae*, in which the cycle occupies 72 h. Since febrile episodes occur shortly after red cell rupture, this explains the characteristic periodic fevers. However, with *P. falciparum*, the cycles of different broods of parasite do not become synchronized as they do in other forms of malaria, and typical tertian fevers are not usual in falciparum malaria.

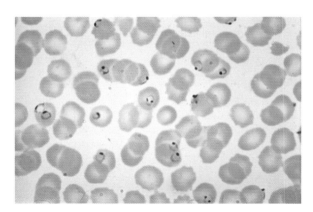

Fig. 61.3 Ring form trophozoites of *P. falciparum.*

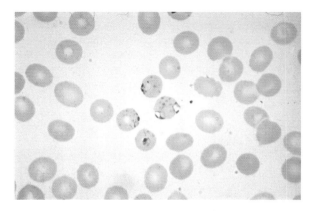

Fig. 61.4 Trophozoites of *P. falciparum.* Note peripheral location of the parasites (*appliqué* or *accolé* forms) and light stippling of the red cells (*Maurer's spots*).

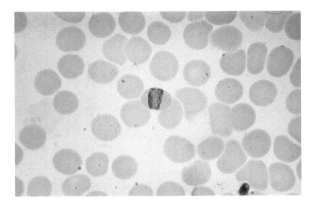

Fig. 61.7 Band-form trophozoite of *P. malariae*.

Laboratory diagnosis

Acute falciparum malaria is a medical emergency that demands immediate diagnosis and treatment. To establish the diagnosis, a drop of peripheral blood is spread on a glass slide. The smear should not be too thick; a useful criterion is that print should be just visible through it. The smear is allowed to dry thoroughly and stained by Field's method. This is an aqueous Romanowsky stain, which stains the parasites very rapidly and haemolyses the red cells, so that the parasites are easy to detect despite the thickness of the film.

With experience, the species of malaria can usually be determined from a thick blood film, but some of the characteristic features used to establish the identity of the parasites (Table 61.2), such as the typical stippling of the red cell that accompanies infection with *P. vivax* or *P. ovale*, are better observed in a conventional thin blood film stained by one of the many modifications of Giemsa's or Leishman's stain. The water used to dilute the stain should be at pH 7.2 (not 6.8 as used for haematological purposes).

Various other diagnostic tests have been devised for the rapid diagnosis of malaria including simple 'dipstick' tests, which are sufficiently reliable to be used by inexperienced staff (perhaps even by patients themselves) or in field conditions where microscopy is not available. Tests that detect *P. falciparum* alone or discriminate between *P. falciparum* and *P. vivax* are available. They are not foolproof, and may remain positive for some time after successful treatment. Visualization of the parasites in Romanowsky-stained peripheral blood films remains the most reliable method.

Pathogenesis

Malaria is characterized by severe chills, high fever and sweating, often accompanied by headache, muscle pains

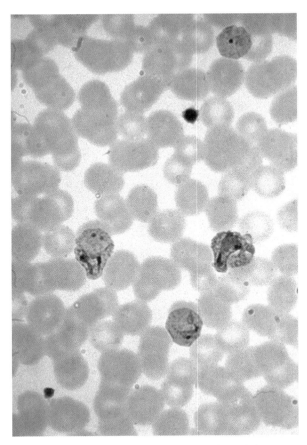

Fig. 61.5 Amoeboid trophozoites of *P. vivax*. Note the marked enlargement of the parasitized red cells and the intense stippling (*Schüffner's dots*).

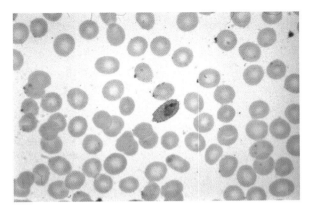

Fig. 61.6 Trophozoite of *P. ovale*. Note the fimbriate, oval-shaped red cell and marked stippling (*James' stippling*).

Table 61.2 Differential characteristics of human malaria parasites as seen in Romanowsky-stained thin films of peripheral blood (see also Figs 60.2–60.7)

Species	Morphology of trophozoite	Morphology of red cell	Stippling of red cell	Morphology of gametocyte	No. of merozoites in mature schizont
P. falciparum	Ring forms only	Normal	Maurer's spots[a]	Crescentic	(16–24)[b]
P. vivax	Rings, becoming amoeboid during development	Enlarged	Schüffner's dots	Large, round	16–24
P. ovale	Rings, becoming compact during development	Slightly enlarged; sometimes oval with fimbriate edge	James's stippling[c]	Round	8–12
P. malariae	Rings, becoming compact, stretched across red cell	Normal or slightly shrunken	(Ziemann's dots)[d]	Small, round	8–12

[a]Maurer's spots (or clefts) are relatively scanty and accompany more mature ring forms.
[b]Mature schizonts of P. falciparum are rarely seen in peripheral blood.
[c]James's stippling is similar to the intense stippling of Schüffner's dots.
[d]Ziemann's dots are rarely seen, except in intensely stained preparations.

and vomiting. Falciparum malaria, unlike the other forms, may progress (especially in primary infections) to coma, convulsions and death. This condition, *cerebral malaria*, is associated with the adherence of parasitized red blood cells to the endothelium of brain capillaries. However, the precise mechanism for the pathological sequelae that follow, including the marked increase in the production of tumour necrosis factor that occurs, is not known with certainty.

Falciparum malaria is notoriously varied in its presentation, so that a definitive laboratory diagnosis is most important. Various systems may be affected. Severe anaemia and renal failure are common complications, and hypoglycaemia, pulmonary oedema and gastro-enteritis may be present.

Individuals that are homozygous or heterozygous for the sickle cell gene have a much reduced susceptibility to infection with *P. falciparum*, and this has provided a selective advantage for the maintenance of sickle cell disease in holo-endemic areas. Similarly, individuals whose red cells lack the antigen known as the *Duffy factor* are protected from infection with *P. vivax*. In parts of tropical Africa, where most of the population are Duffy factor negative, *P. vivax* is rare, although the related *P. ovale* is found, especially in West Africa.

Treatment

For many years the standard treatment for acute malaria was chloroquine. However, resistance to that drug in *P. falciparum* (and, less commonly, in *P. vivax*) is now widespread and other agents have to be used. Quinine (or quinidine), the traditional remedy that has been available for centuries in the form of cinchona bark, is widely used for the therapy of acute falciparum malaria. Some

antibiotics, including tetracyclines and clindamycin, exhibit anti-malarial activity and are used as an adjunct to quinine therapy. Mefloquine and halofantrine are active against chloroquine-resistant strains, but resistance to them is increasing in prevalence and both are associated with occasional problems of toxicity. Derivatives of artemisinin (a natural product from the plant *Artemisia annua*), including artemether and sodium artesunate, are quick acting and effective. Combination treatments that target different functions in the parasite are increasingly in vogue. Various combinations of artemisinin derivatives with other antimalarial agents are successfully used in endemic areas, as is the combination of atovaquone and proguanil.

Treatment of acute malaria with chloroquine, quinine or other antimalarials will not eliminate parasites in the liver. For this purpose the 8-aminoquinoline drug primaquine must be used. This agent carries the risk of precipitating haemolysis in individuals who are deficient in the enzyme glucose-6-phosphate dehydrogenase.

Prophylaxis

Antimalarial prophylaxis is essential for non-immune travellers to malarious areas. Chloroquine and the antifolate drugs pyrimethamine (often combined with sulfadoxine or dapsone) and proguanil were formerly used widely, but the widespread occurrence of resistance to these agents has made it difficult to offer definitive advice to travellers, particularly those going to regions in which *P. falciparum* is prevalent. Common recommendations include:

- mefloquine, once a week
- the combination of daily proguanil and weekly chloroquine
- the combination of atovaquone and proguanil daily.

Others recommend daily doxycycline or even prima-quine. Whatever prophylactic guidance is given, it should be combined with advice to bring any fever to medical attention; to avoid exposure to mosquito bites by wearing long clothing in the evening when the insects are most active; to use insect repellents; and to sleep under mosquito netting impregnated with insecticide.

Because parasites in the pre-erythrocytic stage of development escape the action of most prophylactic drugs, prophylaxis should continue for at least 4 weeks after leaving a malarious area. This will effectively prevent the development of falciparum malaria, although relapses of other types may occur up to 2 years after exposure.

An effective vaccine has long been sought, but is still awaited.

COCCIDIA

The coccidia are related to malaria parasites and share alternating sexual and asexual phases of development. They are not, however, transmitted by insects and infection is usually acquired by ingesting mature oocysts.

Toxoplasma gondii

This is a coccidian parasite of the intestinal tract of the cat that is transmissible to many other mammals. It occurs worldwide. Serological evidence suggests that human infection is common, presumably as a transient febrile illness or a subclinical attack. Occasionally, more severe infection occurs.

- Intra-uterine toxoplasmosis is an important cause of stillbirth and congenital abnormality.
- Ocular disease is a rare but serious condition.
- Cerebral toxoplasmosis sometimes occurs in immunocompromised patients, probably through reactivation of latent infection.

Mature oocysts excreted by infected cats contain two sporocysts, within which *tachyzoites* develop. On ingestion the tachyzoites pass to the bloodstream and lymphatics to invade macrophages, in which they multiply (Fig. 61.8). As the immune response develops, other cells are infected and tissue cysts containing slowly metabolizing *bradyzoites* are formed. Infection is acquired by ingestion of oocysts, or of tissue cysts in undercooked meat. Intra-uterine infection is acquired transplacentally.

It is difficult to find toxoplasmas in clinical material. Diagnosis of acute infection is made by demonstration of a rising titre of serum antibodies to *Tox. gondii*. The *Sabin–Feldman dye exclusion test* recognizes the ability of serum antibody to kill viable toxoplasmas. Other serological tests, including an enzyme-linked immuno-

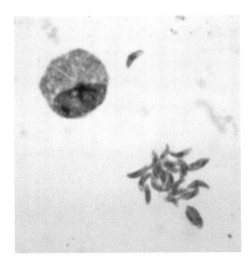

Fig. 61.8 Tachyzoites of *Toxoplasma gondii* in a macrophage (top left) and lying free (bottom right.)

sorbent assay (ELISA), are available and have the advantage that they avoid the use of live toxoplasmas. The polymerase chain reaction is useful in the diagnosis of intra-uterine and cerebral infections.

The combination of pyrimethamine and a sulphonamide is effective against active tachyzoites. Spiramycin is also effective and may be preferred during pregnancy. Clindamycin, azithromycin and atovaquone, usually in combination with pyrimethamine, offer alternatives in patients with cerebral toxoplasmosis.

Isospora belli

This coccidian parasite usually causes mild self-limiting diarrhoea, but is occasionally associated with more severe infection, particularly in patients with acquired immune deficiency syndrome (AIDS). It is more common in areas of poor sanitation. The characteristic oocysts can be seen in faecal 'wet' mounts, but are poorly refractile and easily missed. Co-trimoxazole is usually effective if antimicrobial treatment is necessary.

Cryptosporidium parvum

Cryptosporidia are common animal parasites. One species, *Cryptosporidium parvum*, causes diarrhoea in man. Infection is usually water-borne or acquired from animals. Large numbers of oocysts are often present in faeces; they are partially acid-fast and can be demonstrated by modifications of the Ziehl–Neelsen method (see p. 211).

The infection usually responds to symptomatic treatment, with fluid replacement if necessary. In severely immunocompromised patients, cryptosporidia may cause severe life-threatening diarrhoea for which nitazoxanide or macrolides such as azithromycin may offer some benefit.

Cyclospora cayetanensis

Unlike cryptosporidia, cyclospora develop intracellularly in the gut mucosa. The immature oocysts are excreted in the faeces as round bodies about 10 μm in diameter, with a characteristic 'mulberry' appearance. They are more variably acid-fast than are cryptosporidia.

Cyclospora cayetanensis causes diarrhoea and is associated with poor sanitation. As with other coccidian parasites, infection is more severe in the immunocompromised. Mild infection is treated symptomatically, with rehydration if necessary. Co-trimoxazole appears to be effective in serious infection.

Sarcocystis species

The animal parasites Sarcocystis bovihominis and S. suihominis occasionally invade the human intestinal tract or muscle. Infection is usually subclinical and discovered accidentally.

AMOEBAE

ENTAMOEBA HISTOLYTICA

This is the most important amoebic parasite of man. The amoebae invade the colonic mucosa, producing characteristic ulcerative lesions and a profuse bloody diarrhoea (amoebic dysentery). Systemic infection may arise, leading to abscess formation in internal organs, notably the liver. Such disease may arise in the absence of frank dysentery.

Laboratory diagnosis

In acute amoebiasis, blood-stained mucus or colonic scrapings from ulcerated areas are examined by direct microscopy. The material should be examined within 2 h of collection. Entamoeba histolytica may be recognized by its active movement, pushing out finger-like pseudopodia and, sometimes, progressing across the microscope field. If mucosal invasion has occurred, the amoebae usually contain ingested red blood cells, but these may be absent if infection is confined to the gut lumen. The nucleus is not usually visible in unstained 'wet' preparations, but in fixed smears stained with haematoxylin it is seen as a delicate ring of chromatin with a central karyosome (Fig. 61.9a).

Typical amoebic trophozoites may also be seen in aspirates of liver abscess. The pus often has a distinctive red–brown 'anchovy sauce' appearance. As the amoebae actively multiply in the walls of the abscess, they are most likely to be found in the last few drops of pus drained from the lesion.

In the intestinal carrier state, active amoebae are usually absent, but the encysted form, by which infection is spread, may be found. They are spherical, about 10–15 μ in diameter, and contain one to four of the nuclei typical of Entamoeba species – a circular ring with a central dot. Young uninucleate cysts may also contain a large glycogen vacuole and, in fresh specimens, cysts of all stages of development may exhibit one or more thick, blunt-ended chromatoidal bars (Fig. 61.9b).

Although demonstration of active amoebae or cysts is the best way to make a definitive diagnosis, serology is also sometimes helpful, particularly in systemic disease. Various immunodiagnostic tests have been described, but they are usually performed only in reference centres. Culture of E. histolytica is unhelpful as a diagnostic procedure.

Treatment

Not all strains of E. histolytica are invasive and some (now classified as E. dispar) never cause disease. Nevertheless, it is not possible readily to distinguish pathogenic from non-pathogenic strains in asymptomatic cyst excreters and, at least in areas of the world in which amoebiasis is uncommon, it is prudent to treat all excreters, particularly if they are food handlers. For this purpose, diloxanide furoate is often used.

Acute amoebiasis is usually effectively treated with metronidazole or tinidazole. Chloroquine is also useful in amoebic liver abscess. Older drugs, including emetine and dehydroemetine, are effective, but more toxic.

NON–PATHOGENIC INTESTINAL AMOEBAE

Although there are occasional reports of diarrhoea associated with other intestinal amoebae, notably Dientamoeba fragilis (an amoeba flagellate), most occur as commensals and are important only because of potential confusion with E. histolytica. The differential characteristics of non-pathogenic intestinal amoebae are compared with those of E. histolytica in Table 61.3. The greatest opportunity for confusion arises with the other intestinal Entamoeba species, E. hartmanni and E. coli. E. hartmanni is morphologically identical to E. histolytica, but is smaller and the trophozoite never contains ingested red blood cells. E. coli is somewhat larger than E. histolytica, particularly in the cyst form. The trophozoites are more sluggish than those of E. histolytica, and mature cysts contain up to eight

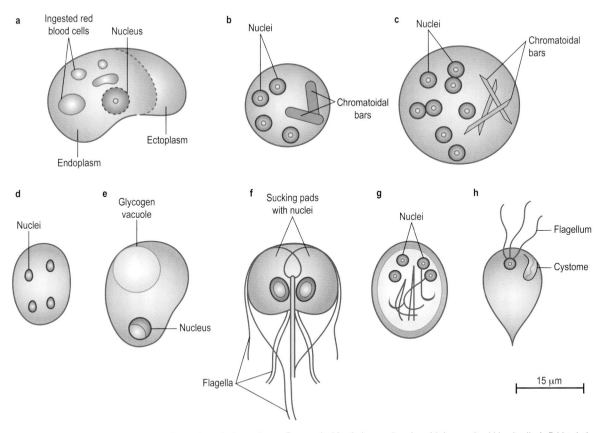

Fig. 61.9 Diagrammatic representation of some intestinal parasites: **a** *Entamoeba histolytica*, trophozoite with ingested red blood cells; **b** *E. histolytica*, mature cyst; **c** *E. coli*, mature cyst; **d** *Endolimax nana*, mature cyst; **e** *Iodamoeba bütschlii*, mature cyst; **f** *Giardia lamblia*, trophozoite; **g** *G. lamblia*, mature cyst; **h** *Chilomastix mesnili*, trophozoite.

Table 61.3 Differential characteristics of intestinal amoebae

| Species | Trophozoites | | Cysts | | | |
	Size (μm)	Ingested red blood cells	Size (μm)	No. of nuclei[a]	Chromatoidal bars	Nuclear morphology
Entamoeba histolytica[b]	10–40	+	10–15	4	Solid, blunt-ended	Fine ring of chromatin with central karyosome
E. hartmanni	4–10	−	6–10	4	As above	As above
E. coli	10–40	−	15–25	8	Slender, pointed	As above, but karyosome may be eccentric
E. gingivalis[c]	10–25	−	No cyst stage			
Iodamoeba bütschlii	10–20	−	10–15	1	None	Chromatin massed at one end of ring
Endolimax nana	5–12	−	5–8	4	None	Small shadowy masses of chromatin
Dientamoeba fragilis	5–10	−	No cyst stage			Ring containing several chromatin granules

[a]Refers to mature cyst.
[b]Including *E. dispar*.
[c]*E. gingivalis* is a commensal of the mouth.

nuclei; chromatoidal bars, if present, are fine and pointed, rather like slivers of broken glass (Fig. 61.9c).

Blastocystis hominis, an organism commonly found in faeces and formerly thought to be a yeast, is probably an amoeba. Any pathogenic role is the subject of dispute, but it has been associated with diarrhoea in the absence of other known pathogens. Metronidazole is said to be useful if true infection is suspected.

FREE-LIVING AMOEBAE

Environmental amoebae belonging to the genus *Naegleria* (usually *N. fowleri*) have occasionally been implicated in meningo-encephalitis. Rare cases of granulomatous encephalitis caused by *Balamuthia mandrillaris* and *Acanthamoeba* species have also been described, often, but not exclusively, in immunocompromised patients. The outcome in all kinds of amoebic encephalitis is ordinarily fatal, although amphotericin B has been successfully used in infections with *N. fowleri*.

Acanthamoeba spp. more commonly cause keratitis, sometimes following use of contaminated cleaning fluids for soft contact lenses. Optimal antimicrobial chemotherapy remains to be defined, but topical propamidine in combination with neomycin or other agents has been used.

FLAGELLATES

GIARDIA LAMBLIA (SYN. G. INTESTINALIS, G. DUODENALIS)

This intestinal parasite lives attached to the mucosal surface of the upper small intestine. Vast numbers may be present, and their presence may lead to malabsorption of fat and chronic diarrhoea. Young infants may be particularly severely affected. Infection is usually water-borne.

The trophozoite is kite-shaped, with two nucleated sucking pads and four pairs of flagella (Fig. 61.9f). Trophozoites may be found in duodenal aspirate, but examination of faeces usually reveals the cyst form by which the disease is transmitted. This is oval, about $10 \times 8\ \mu m$, and contains up to four nuclei as well as the remains of the skeletal structure of the trophozoite (Fig. 61.9g).

Cysts of other, non-pathogenic, intestinal protozoa, including *Chilomastix mesnili*, *Enteromonas hominis* and *Retortamonas intestinalis*, may be mistaken for *G. lamblia*, but they are usually smaller and lack the regular oval shape and characteristic internal morphology. These non-pathogenic protozoa may also be found as trophozoites during microscopy of diarrhoeic faeces, but the most common intestinal flagellate is *Trichomonas hominis*,

which is recognizable by its undulating membrane. There is no cyst form.

Giardiasis can be treated with 5-nitroimidazoles such as metronidazole or, on the rare occasions when this fails (and re-infection is excluded), with mepacrine (quinacrine). Surprisingly, the anthelminthic, albendazole, also appears to be effective.

TRICHOMONAS VAGINALIS

T. vaginalis is a flagellate protozoon with four anterior flagella and one lateral flagellum which is attached to the surface of the parasite to form an undulating membrane. There is no cyst form; the parasite is transmitted by sexual intercourse.

As the name suggests, *T. vaginalis* is predominantly a vaginal parasite, although urethritis may occur in the male consorts of infected women. The organism is responsible for a mild vaginitis, with discharge, which ordinarily responds to treatment with metronidazole or tinidazole.

T. vaginalis is readily identified by its characteristic motility in untreated 'wet' films of vaginal discharge and can be cultivated in appropriate culture media.

TRYPANOSOMES

In contrast to the flagellates already described, trypanosomes have a complex life cycle involving an insect vector. The diseases that are caused in man, African trypanosomiasis (*sleeping sickness*) and South American trypanosomiasis (*Chagas' disease*), are restricted in distribution according to the habitat of the insect host.

African trypanosomiasis

African sleeping sickness is caused by trypanosomes that are subspecies of *Trypanosoma brucei*, an important aetiological agent of *nagana* in cattle in tropical Africa. Tsetse flies (see pp. 650–651) act as the insect vector. The human parasites are *T. brucei gambiense*, which occurs in riverine areas of west and central Africa, and *T. brucei rhodesiense*, a parasite of the savannah plains of east Africa, where cattle and wild antelope act as reservoirs of infection.

Pathogenesis

Following the bite of an infected tsetse fly, a localized *trypanosomal chancre* may appear transiently, but invasion of the bloodstream rapidly occurs. The parasites multiply in blood, and at this stage there may be non-specific symptoms with occasional febrile episodes and some

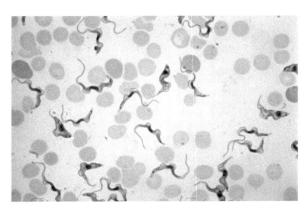

Fig. 61.10 Trypomastigotes of *Trypanosoma brucei rhodesiense* in mouse blood.

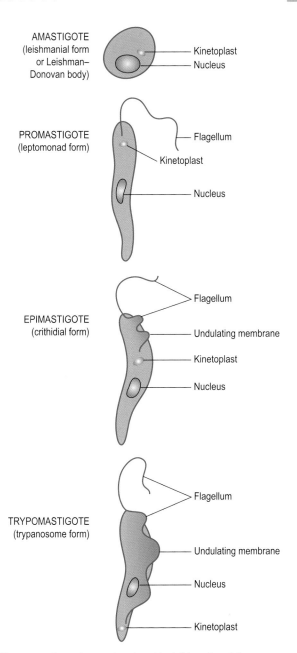

Fig. 61.11 Forms that may be adopted by *Leishmania* and *Trypanosoma* spp. In parentheses are given the terms for the forms under the old nomenclature now superseded.

lymphadenitis. Swollen lymph glands in the posterior triangle of the neck (*Winterbottom's sign*) are often present in *T. brucei gambiense* infection. If untreated, the disease progresses inexorably to involve the central nervous system with the classical signs of sleeping sickness and, ultimately, death.

Infection with *T. brucei rhodesiense* tends to follow a more acute, fulminating course over a period of a few months, whereas *T. brucei gambiense* infection usually progresses slowly, sometimes over several years.

Laboratory diagnosis

During the parasitaemic stage, sparse trypanosomes may be found in peripheral blood in unstained 'wet' mounts or in smears stained by the Giemsa or Leishman methods. Examination of lymph node exudate may be helpful. Once the disease has progressed to involve the central nervous system, examination of cerebrospinal fluid reveals a lymphocytic exudate, often with *morula cells* (plasma cells) and sparse motile trypanosomes.

The parasites have a characteristic morphology: they are elongated, about 20–30 μm in length, with a single anterior flagellum arising via an undulating membrane from a basal body situated near a posteriorly placed kinetoplast (Figs 61.10 & 61.11).

In-vitro cultivation is unreliable, but animal inoculation is sometimes useful, particularly with *T. brucei rhodesiense*, which infects laboratory mice more readily than *T. brucei gambiense*. Various immunodiagnostic tests have been described, but they are not as reliable as direct microscopy in establishing a definitive diagnosis.

Treatment

In the early parasitaemic stage, the infection is amenable to treatment with suramin or pentamidine, but once the disease has progressed to sleeping sickness the trivalent arsenicals, melarsoprol or tryparsamide, are used. Less toxic alternatives are clearly required; eflornithine is effective in *T. brucei gambiense* infections, but not in disease caused by *T. brucei rhodesiense*.

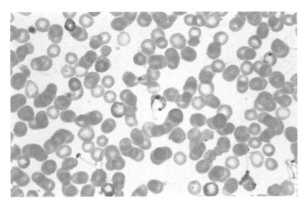

Fig. 61.12 Trypomastigote of *T. cruzi* in blood. Note the prominent kinetoplast and the 'C' shape adopted by the parasite.

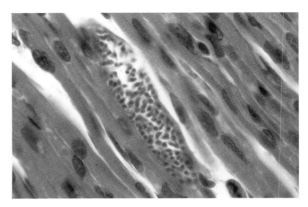

Fig. 61.13 Amastigotes of *T. cruzi* in heart muscle.

South American trypanosomiasis

Chagas' disease, caused by *T. cruzi*, is quite different from African trypanosomiasis. The insect vectors are various species of reduviid bugs (see pp. 648–649). The trypanosomes are not transmitted by the bite, but are present in the bug's faeces, which the unwitting sleeper rubs into the bite wound. The trypanosomes enter the bloodstream, but do not multiply there; instead, they invade cells of the reticulo-endothelial system and muscle, where they lose their flagellum and associated undulating membrane and adopt a more rounded shape (Fig. 61.11). This morphological form is called an *amastigote*, and suggests a phylogenetic relationship with *Leishmania* species (see below). The amastigotes multiply in muscle and are liberated from ruptured cells as trypanosomal forms (*trypomastigotes*), which disseminate the infection and provide the parasitaemia needed to infect fresh reduviid bugs when they next feed.

Pathogenesis

Chagas' disease is a chronic condition characterized by extensive cardiomyopathy, sometimes with gross distension of other organs (e.g. mega-oesophagus and megacolon). Death is usually from heart failure.

Laboratory diagnosis

Trypomastigotes may be seen in peripheral blood, but are often extremely sparse. They are shorter than those of the *T. brucei* group and have a characteristically large kinetoplast (Fig. 61.12). Unlike the African trypanosomes, *T. cruzi* can be grown in vitro in the rich blood agar medium also used to isolate leishmania (see below). Biopsy of skeletal muscle may be performed but is usually of little value. The appearance of amastigotes in heart muscle is shown in Figure 61.13.

T. cruzi is infective to laboratory mice. Alternatively, a procedure known as *xenodiagnosis* may be used, in which uninfected reduviid bugs are allowed to feed on the patient and after about 3–4 weeks the gut contents of the bug are examined for trypanosomes.

Immunofluorescence, ELISA and other tests may be used for presumptive serological diagnosis, but false-positive results are common. Assays based on the polymerase chain reaction have been developed and are more specific.

Treatment

There is no reliable chemotherapy for Chagas' disease. The nitrofuran derivative nifurtimox and the imidazole compound benznidazole have been used with modest success.

Leishmania species

Leishmania species are intracellular parasites of the reticulo-endothelial system. They are related to trypanosomes, but exist in only two morphological forms: *amastigotes* (non-flagellate forms), which occur in the infected lesion, and *promastigotes* (flagellate forms that lack an undulating membrane), which occur in the insect vector or in laboratory culture (Figs 61.11, 61.14 & 61.15).

The parasites are transmitted by sandflies (see p. 650) in various parts of the world, including the Middle East, India, South America, the Mediterranean littoral and parts of Africa.

Pathogenesis

Several distinct types of disease are recognized (Table 61.4), although they are caused by morphologically identical parasites. The taxonomic relationships between the various forms have still not been entirely clarified. *Cutaneous leishmaniasis (oriental sore)* is the least troublesome, causing a boil-like swelling on the face or other exposed part of the body. The central part of the lesion may become secondarily infected with bacteria, but the leishmania organisms reside in the raised, indurated edge of the lesion. The sore usually heals spontaneously, leaving a scar, but with some species a more severe *disseminated cutaneous leishmaniasis* may occur. Parasites of the *Leishmania mexicana* complex may cause a destructive lesion of the outer ear (*Chiclero's ulcer*).

In *mucocutaneous leishmaniasis (espundia)*, which is associated with the *L. braziliensis* complex, disfiguring lesions of the mouth and nose may be caused. However, the most serious form of leishmaniasis is *visceral leishmaniasis (kala azar)*, which is a life-threatening disease involving the whole of the reticulo-endothelial system. A late complication of kala azar, *post-kala azar dermal leishmaniasis*, may be confused with leprosy or other skin conditions.

Laboratory diagnosis

In the cutaneous or mucocutaneous form of the disease, typical intracellular amastigotes may be recognized in Giemsa-stained smears of material obtained from tissues at the margin of the lesion (see Fig. 61.14). Free amastigotes are commonly seen because of rupture of the macrophage host cell. Material should also be cultured in Novy, MacNeal and Nicolle's (NNN) medium or a modification thereof. This is a rabbit blood agar containing antibiotic to prevent bacterial contamination and a buffered salt overlay solution in which the parasites grow as promastigotes (see Fig. 61.15). Incubation is maintained for up to 3 weeks at room temperature (*not* 37°C).

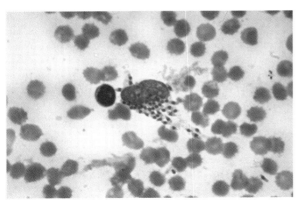

Fig. 61.14 Amastigotes of *Leishmania tropica* in a ruptured macrophage from a cutaneous lesion (*oriental sore*).

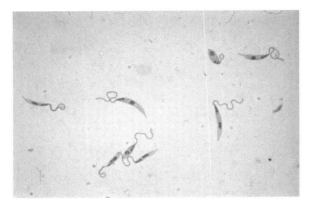

Fig. 61.15 Promastigotes of *L. tropica* from laboratory culture in Novy, MacNeal and Nicolle's (NNN) medium.

Table 61.4 Leishmania species involved in human disease

Species	Form of disease	Common names	Main geographical distribution
Leishmania tropica	Cutaneous	Oriental sore, Baghdad boil, Delhi boil, etc.	Middle East, central Asia
L. major	Cutaneous		Africa, Indian subcontinent, central Asia, Ethiopia, Kenya
L. aethiopica	Cutaneous, DCL		Middle East, Africa, Indian subcontinent
L. donovani	Visceral	Kala azar, Dum-dum fever	Mediterranean coast, Middle East, China
L. infantum[a]	Visceral		Tropical South America
L. chagasi	Visceral		
L. mexicana complex	Cutaneous, DCL	Chiclero's ulcer	Central America, Amazon basin
L. braziliensis complex	Mucocutaneous	Espundia	Tropical South America
L. peruviana	Cutaneous	Uta	Western Peru

DCL, disseminated cutaneous leishmaniasis.
[a]*L. infantum* may be a subspecies of *L. donovani*.

In kala azar, spleen puncture is the most reliable method of diagnosis, but sternal marrow aspirate is safer and is usually preferred. Smears and cultures are made and examined as for cutaneous leishmaniasis. Various serological tests have been designed, but demonstration of the parasite by microscopy or culture is preferable whenever possible.

Treatment

The pentavalent antimony compounds, sodium stibo-gluconate and meglumine antimoniate, have been used traditionally in all forms of leishmaniasis, but they are toxic and therapy often fails. Amphotericin B is effective, but poorly tolerated, and the less toxic lipid-based formulations are preferred. Antifungal azoles and paromomycin (aminosidine) have been used with some success in cutaneous forms of disease. A phosphocholine derivative, miltefosine, offers considerable promise in kala azar.

OTHER PATHOGENIC PROTOZOA

BABESIA SPECIES

Babesiae are predominantly animal parasites related to the piroplasmas that cause theileriasis in wild and domestic animals in many parts of the world. They are intracellular parasites living within red blood cells and are transmitted by ixodid ticks (see p. 652). In stained blood films they superficially resemble young ring forms of plasmodia.

Human infection is uncommon. European cases have mostly been in patients whose resistance was impaired by lack of a functioning spleen; the causative parasite was usually *Babesia divergens*. In contrast, babesiosis caused by *B. microti* has been reported in otherwise healthy persons in parts of the USA.

In immunocompetent individuals the disease is usually self-limiting, so that specific treatment is not required. Optimal treatment for more serious cases has not been properly defined, but the combination of quinine with clindamycin has been used successfully.

BALANTIDIUM COLI

Balantidium coli is the only ciliate protozoon that is pathogenic to man. It is a common parasite of the pig,

and human infections have usually been traced to contact with these animals. The infective form is a large (about 50 µm in diameter) thick-walled cyst. The trophozoite inhabits the lumen of the gut and may attack the colonic mucosa in much the same way as *E. histolytica*, to cause balantidial dysentery. Many highly motile ciliate trophozoites are readily seen in untreated 'wet' films of diarrhoeic faeces. Tetracyclines and metronidazole are said to offer effective treatment.

MICROSPORIDIA

These animal and insect parasites are represented by several genera, including *Encephalitozoon*, *Enterocytozoon* and *Nosema*. They are, on rare occasions, implicated in opportunistic infections of immunocompromised patients, especially those with AIDS. Infections of the eye, meninges and other organs have been reported. Albendazole may be useful in treatment, but is not always curative.

KEY POINTS

- Protozoa are unicellular eukaryotic organisms.
- Most protozoal infections require laboratory confirmation for diagnosis; toxoplasmosis is diagnosed serologically.
- Malaria kills more than 1 million people each year.
- Acute malaria is a medical emergency demanding immediate diagnosis and treatment; quinine or artemisinin derivatives, usually in combination with other antimalarial drugs, are effective.
- African trypanosomiasis (sleeping sickness) is treated with toxic arsenical drugs; the West African form responds to eflornithine.
- South American trypanosomiasis (Chagas' disease) is a chronic condition that is difficult to treat.
- Leishmaniasis takes various forms; the most dangerous is visceral leishmaniasis (kala azar). Antimonial compounds or antifungal drugs are used for treatment.
- Amoebiasis, giardiasis and trichomoniasis occur worldwide; they usually respond to 5-nitroimidazole drugs such as metronidazole.

RECOMMENDED READING

Chiodini P L, Moody A, Manser D W 2001 *Atlas of Medical Helminthology and Protozoology*, 4th edn. Churchill Livingstone, Edinburgh

Cook G C (ed.) 2002 *Manson's Tropical Diseases*, 21st edn. W B Saunders, London

Croft S L, Coombs G H 2003 Leishmaniasis – current chemotherapy and recent advances in the search for novel drugs. *Trends in Parasitology* 19: 502–508

Goodgame R W 1996 Understanding intestinal spore-forming protozoa: cryptosporidia, microsporidia, isospora, and cyclospora. *Annals of Internal Medicine* 124: 429–441

Greenwood B M, Bojang K, Whitty C J M, Targett G A T 2005 Malaria. *Lancet* 365: 1487–1498

Guerrant R L, Walker D H, Weller P F 2006 *Tropical Infectious Diseases*, 2nd edn. Churchill Livingstone, Philadelphia

Peters W, Pasvol G 2001 *Tropical Medicine and Parasitology*, 5th edn. Mosby, London

Stanley S L 2003 Amoebiasis. *Lancet* 361: 1025–1034

Warrell D, Gilles H M 2002 *Essential Malariology*, 4th edn. Hodder Arnold, London

World Health Organization 1991 *Basic Laboratory Methods in Medical Parasitology*. WHO, Geneva

CD-ROMs

Wellcome Trust 2006 *Malaria*, 3rd edn (Topics in International Health series). Available from Wellcome Trust, London (www.wellcome.ac.uk/tih) and (for developing countries) from TALC (Teaching-aids at Low Cost; www.talcuk.org)

Wellcome Trust 2000 *Leishmaniasis* (Topics in International Health series). Available from Wellcome Trust, London (www.wellcome.ac.uk/tih) and (for developing countries) from TALC (Teaching-aids at Low Cost; www.talcuk.org)

Internet sites

Malaria Centre, London School of Hygiene and Tropical Medicine. http://www.lshtm.ac.uk/malaria/index.htm

National Travel Health Network and Centre. http://www.nathnac.org.uk

Division of Parasitic Diseases, US Centers for Disease Control and Prevention. Professional Information. http://www.cdc.gov/ncidod/dpd/professional/default.htm

World Bank. Rolling Back Malaria: the World Bank Global Strategy & Booster Program. http://siteresources.worldbank.org/INTMALARIA/Resources/377501-1114188195065/WB-Malaria-Strategy&BoosterProgram-Lite.pdf

62 Helminths

Intestinal worm infections; filariasis; schistosomiasis; hydatid disease

D. Greenwood

Medical helminthology is concerned with the study of parasitic worms. These creatures are responsible for an enormous burden of infection throughout the world and, although few helminthic infections are life threatening, their impact on human health is incalculable. Most helminths have no independent existence outside the host and are therefore truly parasitic. As they rely on the host for sustenance, it is not in their interest to cause the host harm; consequently they do not usually exhibit great virulence and are characterized more by the novel methods that they have evolved to prevent rejection by the host defences. The pathogenic manifestations of helminthic disease, which can none the less be considerable, are ordinarily due to physical factors related to the location of the worms, their lifestyle or their size (see Ch. 11).

There are two major groups of helminth: *nematodes*, or roundworms, and *platyhelminths*, or flatworms. Flatworms are, in their turn, represented by two classes: *trematodes* (flukes) and *cestodes* (tapeworms).

NEMATODES

The principal nematode parasites of man are conveniently considered under two headings: intestinal nematodes and tissue nematodes.

INTESTINAL NEMATODES

Infection with intestinal roundworms (Table 62.1) is generally associated with conditions of poor hygiene. Such infections are extremely common, particularly throughout the tropics and subtropics, although several are also found in temperate regions. Low worm burdens are generally asymptomatic, but heavy infections may cause problems, especially in young children in whom they have been associated with impaired development.

Ascaris lumbricoides

This is the common roundworm, which infects more than a billion people in the world. The adults are large and fleshy and, as with many nematodes (other than the hookworm group), the smaller male can be recognized by his characteristically crooked tail. The eggs (ova) are produced in huge numbers; they are thick-walled, bile-stained and typically exhibit a corrugated albuminous coat (Fig. 62.1a). In the absence of a male worm, the female produces infertile eggs, which are more elongated and irregular than the fertile variety (Fig. 62.1b).

Table 62.1 Principal intestinal nematodes of man

Species	Common name	Relevant examination
Ancylostoma duodenale	Hookworm	Stool concentration (ova)
Ascaris lumbricoides	Common roundworm	Stool concentration (ova)
Enterobius vermicularis	Threadworm	Peri-anal swab (ova), adult worms on stool
Necator americanus	Hookworm	Stool concentration (ova)
Strongyloides stercoralis	–	Stool concentration (larvae) or culture
Toxocara canis[a]	Dog roundworm	Serology
Trichostrongylus spp.	Hookworm	Stool concentration (ova)
Trichuris trichiura	Whipworm	Stool concentration (ova)

[a]Not an intestinal parasite of man; causes visceral larva migrans (see text).

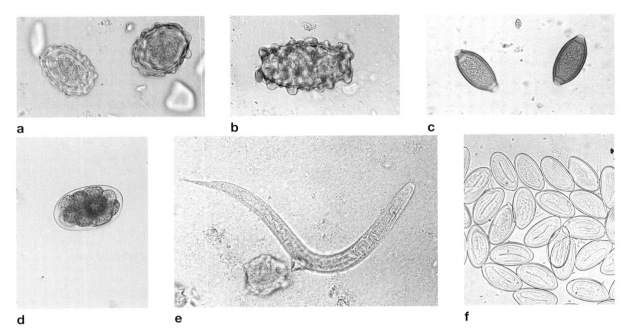

Fig. 62.1 Eggs of intestinal helminths: a *Ascaris lumbricoides* (fertile eggs); b *A. lumbricoides* (infertile egg); c *Trichuris trichiura*; d hookworm; e *Strongyloides stercoralis* (larva); f *Enterobius vermicularis*.

In warm, moist conditions, infective larvae develop within fertile eggs, but do not hatch. Such eggs can survive for long periods in soil. If ingested, the eggs hatch in the duodenum and the larvae penetrate the gut mucosa to reach the bloodstream. They are carried to the pulmonary circulation, where they gain access to the lung and undergo two moults before migrating via the trachea to the intestinal tract. Having completed their round trip, they mature in the gut lumen and live for several years.

Ascaris lumbricoides is a well adapted parasite that is usually not pathogenic in the ordinary sense. However, pneumonic symptoms may accompany the migratory phase, and the adult worms may invade the biliary and pancreatic ducts. Moreover, heavy infection with these large worms can cause intestinal obstruction. Allergy is also sometimes a problem.

The dog ascarid, *Toxocara canis*, may accidentally infect man. Larvae hatch in the small intestine and penetrate the gut wall, but they are unable to complete their migratory phase. Instead, they find their way to remote parts of the body, a condition known as *visceral larva migrans*. Occasionally the larvae reach the eye and cause serious retinal lesions. Larvae of several other roundworms, including *Angiostrongylus*, *Gnathostoma* and *Anisakis* species, are occasionally implicated in visceral larva migrans in some parts of the world.

Trichuris trichiura

This is the common whipworm, often found together with ascaris. The adults live with the head (the 'whip' end of the worm) embedded in the colonic mucosa. Each female lays thousands of characteristic 'tea-tray' eggs (Fig. 62.1c) every day. Like those of ascaris, they develop infective larvae in warm, moist conditions, but the ova do not hatch outside the body. However, after ingestion and hatching, there is no migratory phase and adult worms develop directly in the large intestine.

Infection is usually trivial, although massive infections can cause rectal prolapse in young children, and a form of dysentery is described.

Hookworm

The two human hookworms, *Ancylostoma duodenale* and *Necator americanus*, are widely distributed throughout the tropics and subtropics. The two species produce indistinguishable thin-walled eggs (Fig. 62.1d) that hatch in soil. The larvae undergo several moults before infective larvae are produced. These are capable of penetrating unbroken skin, and in this way they gain access to the bloodstream to begin a migratory phase similar to that of ascaris. When they reach the gut they attach by their mouthparts to the mucosa of the small intestine.

Hookworms ingest blood and, moreover, move from site to site in the gut mucosa, leaving behind small bleeding lesions. These two facts are responsible for the chief pathological manifestation of heavy infection with hookworms: iron deficiency anaemia.

Larvae of animal hookworms, notably the dog hookworm, *A. caninum*, may penetrate human skin, but do not migrate further. They do, however, cause irritation by wandering locally through subcutaneous tissue to cause *cutaneous larva migrans*.

Trichostrongylus species

Various species of *Trichostrongylus* have been associated with human disease, particularly in the Middle East. The eggs are similar to those of hookworm, but more elongated.

Strongyloides stercoralis

This parasite is also related to the hookworms but differs in several important respects. There is a distinct free-living phase in the life cycle, during which males and females reproduce. Human infections arise after penetration of infective larvae through skin and there is a migratory phase involving the lungs. However, human infection appears to be restricted to female worms, which attach to the gut mucosa and produce eggs that contain fully developed larvae; these hatch within the intestinal lumen so that larvae, not eggs, are found in faecal samples (Fig. 62.1e). Infection can persist for many years, probably because some larvae can develop sufficiently within the body to initiate a fresh cycle of development and cause auto-infection.

Symptoms are usually benign, but in debilitated individuals larvae may be activated to penetrate the gut wall and invade other organs, a serious condition known as *hyperinfection*.

Enterobius vermicularis

This is the common threadworm, which infects children throughout the world. It has the simplest life cycle of all intestinal worms. Adults live in the large intestine and are occasionally found in the appendix. Mature, gravid females crawl through the anus at night and lay their eggs in the peri-anal area. The eggs are characteristically flattened on one side (Fig. 62.1f) and usually contain fully developed larvae. Ingestion of these eggs initiates a fresh infection. Symptoms are restricted to itching (*pruritus ani*) associated with the deposition of eggs.

As eggs are not discharged by the worm into faeces, faecal examination is not appropriate in the laboratory diagnosis of threadworm infection. The diagnosis is established by finding the characteristic 'threads' on the surface of formed stools, or by examination of swabs (or Sellotape impressions) of unwashed peri-anal skin for the characteristic ova.

Treatment of intestinal nematode infections

In endemic areas, the need to treat intestinal worm infections has to be balanced against the severity of symptoms (if any), the inevitability of re-infection, and the use of scarce medical resources. In the industrially developed world, where the infections are less common and mostly imported, a more liberal approach to treatment can be adopted.

The range of options is listed in Table 62.2. Most effective (and expensive) are the benzimidazole derivatives, especially albendazole and mebendazole.

TISSUE NEMATODES

This group includes the filarial worms, the guinea-worm (*Dracunculus medinensis*) and *Trichinella spiralis* (Table 62.3).

Filarial worms have a complex life cycle involving developmental stages in an insect vector. They vary considerably in their pathogenic effects, but some are responsible for disabling diseases that have a major impact on communities living in endemic areas.

Wuchereria bancrofti

This filarial worm is transmitted by the bite of various species of mosquito throughout the tropical belt of the world. It is believed that more than 100 million people are infected. The larvae invade the lymphatics, usually of the lower limbs, where they develop into adult worms. The presence of adult worms causes lymphatic blockage and gross lymphoedema, which sometimes leads to the bizarre deformities associated with bancroftian filariasis, *elephantiasis*.

Embryonic forms (*microfilariae*) are liberated into the bloodstream. They retain the elastic egg membrane as a sheath, which covers the whole larva (Fig. 62.2a). Microfilariae remain in the pulmonary circulation during the day, emerging into the peripheral circulation only at night, to coincide with the biting habits of the insect vector. The physiological basis of this nocturnal periodicity is not understood, but it can be reversed by altered sleep patterns in, for example, night-shift workers. Moreover, strains of *Wuchereria bancrofti* encountered in some Pacific islands do not exhibit a nocturnal periodicity. Aside from these exceptions, blood for examination for *W. bancrofti* must

Table 62.2 Spectrum of activity of drugs used in the treatment of intestinal nematode infections

Drug	Ancylostoma duodenale	Necator americanus	Ascaris lumbricoides	Strongyloides stercoralis	Trichuris trichiura	Enterobius vermicularis
Piperazine	–	–	+++	–	–	+++
Levamisole	++	++	+++	–	–	–
Pyrantel pamoate	++	++	+++	+	++	+++
Tiabendazole	++	++	++	++	–	++
Mebendazole	++	++	+++	+	++	+++
Albendazole	+++	++	+++	++	++	+++

From Greenwood D, Finch R G, Davey P G, Wilcox M H 2007 *Antimicrobial Chemotherapy*, 5th edn. Oxford University Press, Oxford.
+++, Highly effective; +, poorly effective; –, no useful activity.

Table 62.3 Principal tissue nematodes of man

Species	Intermediate host	Geographical distribution	Relevant examination
Wuchereria bancrofti	Mosquitoes	Tropical belt	Night blood
Loa loa	*Chrysops* spp.	West and Central Africa	Day blood
Brugia malayi	Mosquitoes	South-East Asia	Night blood
Mansonella perstans	*Culicoides* spp.	Tropical Africa, South America	Blood
M. ozzardi	*Culicoides* spp.	West Indies, South America	Blood
Onchocerca volvulus	*Simulium* spp.	Tropical Africa, Central America	Skin shavings
M. streptocerca	*Culicoides* spp.	West and Central Africa	Skin shavings
Dracunculus medinensis	Water fleas	Africa, Indian subcontinent	Adult worm when mature
Trichinella spiralis	None[a]	Worldwide	Muscle biopsy, serology

[a]Pork forms the chief reservoir.

be taken during the night, optimally between midnight and 2 a.m.

Loa loa

This worm is restricted in distribution to central and western parts of tropical Africa, where it is transmitted by biting flies (*Chrysops* species; see p. 651). The adult worms live in subcutaneous tissue and wander round the body, provoking localized reactions known as *Calabar swellings* and sometimes migrating across the front of the eye.

The sheathed microfilariae of *Loa loa* (Fig. 62.2b) exhibit diurnal periodicity, so that, unlike those of *W. bancrofti*, they appear in peripheral blood only during the day.

Brugia malayi

This parasite is probably related to *W. bancrofti*. It is transmitted by mosquitoes in parts of India, the Far East and South-East Asia. Adult worms inhabit the lymphatics and, like *W. bancrofti*, can cause elephantiasis. Microfilaraemia usually shows a nocturnal periodicity.

Onchocerca volvulus

This filarial worm is common in parts of tropical Africa and Central America. It is transmitted by *Simulium damnosum* and related species of black-fly (see p. 651). Adult worms develop in subcutaneous and connective tissue, and often become encapsulated in nodules, which form on bony parts of the body such as the hip, elbow and (particularly in Central America) the head. The microfilariae are not found in blood, but live in the superficial layers of the skin causing itching and, in heavy chronic infections, gross thickening of the skin. The eye is commonly invaded by microfilariae, which may cause corneal and retinal lesions that lead to blindness. Because the vector breeds by rivers, the condition is known as *river blindness*.

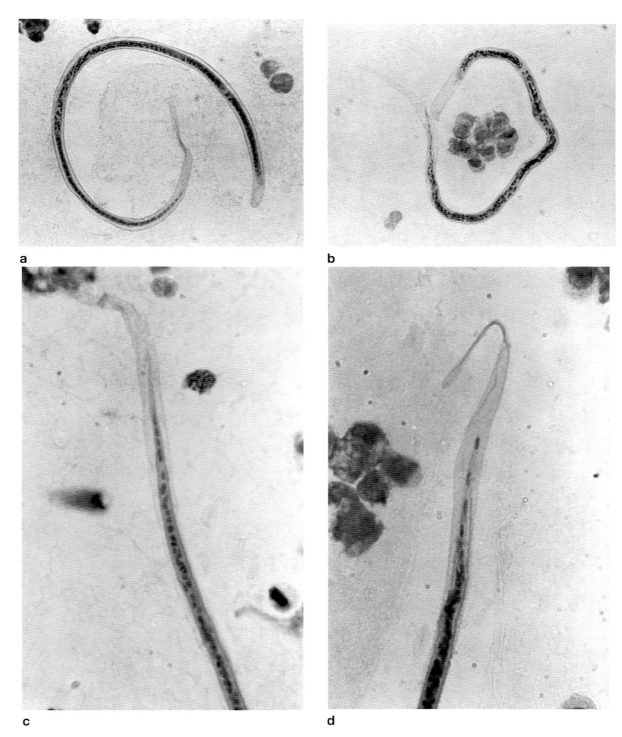

Fig. 62.2 Sheathed microfilariae of **a** *Wuchereria bancrofti* and **b** *Loa loa*; **c** tail of microfilaria of *W. bancrofti* showing tip devoid of somatic nuclei; **d** tail of microfilaria of *L. loa* showing nuclei extending to tip of tail.

Table 62.4 Differential features of human microfilariae

Species	Site in human host	Periodicity	Sheath	Nuclei in tip of tail	Length (µm)
Wuchereria bancrofti	Blood	Nocturnal[a]	Present	Absent	250–300
Loa loa	Blood	Diurnal	Present	Present	250–300
Brugia malayi	Blood	(Nocturnal)[b]	Present	Present[c]	200–250
Mansonella perstans	Blood	Non-periodic	Absent	Present	150–200
M. ozzardi	Blood	Non-periodic	Absent	Absent	180–220
Onchocerca volvulus	Skin	Non-periodic	Absent	Absent	250–300
M. streptocerca	Skin	Non-periodic	Absent	Present	180–240

[a]Subperiodic forms occur in the Pacific Islands.
[b]Partial nocturnal periodicity.
[c]Two small, well separated nuclei in the tip of the tail.

If nodules are present, diagnosis can be made by finding macroscopic worms within an excised nodule. Otherwise, superficial slivers of skin, taken from calves, buttocks and shoulders, are suspended in a drop of saline and examined microscopically for motile microfilariae.

Mansonella species

Mansonella perstans is widespread throughout tropical Africa and parts of South America; the related *M. ozzardi* is restricted to parts of the West Indies and South America. They are transmitted by biting midges (*Culicoides* species; see p. 650). The unsheathed microfilariae appear in the bloodstream and exhibit no periodicity. They are generally regarded as non-pathogenic.

M. streptocerca causes skin infections similar to those of *O. volvulus*, although the symptoms are usually milder. It is restricted to parts of western and central Africa.

Differential characteristics of microfilariae

The microfilariae of filarial worms can be differentiated in stained preparations of clinical material by various criteria, the most useful of which are the presence or absence of a sheath and the disposition of the somatic nuclei in the tip of the tail (Table 62.4 & Fig. 62.2c,d). Giemsa stain is suitable for the demonstration of somatic nuclei, but hot (60°C) haematoxylin is necessary to stain the sheath.

Treatment of filariasis

Diethylcarbamazine (DEC) has been used for many years for the treatment of all forms of filariasis. It effectively kills microfilariae, but is not reliably lethal to adult worms. It is relatively non-toxic, but death of the microfilariae is often accompanied by a severe allergic reaction (*Mazzotti reaction*), especially in onchocerciasis. Suramin kills the adult worms, but is much more toxic than DEC.

The treatment of onchocerciasis has been revolutionized by use of the veterinary anthelminthic ivermectin. This drug is effective in a single oral dose and is less likely than DEC to elicit a severe reaction. Periodic administration of ivermectin, together with vector control measures, has had an important impact on onchocerciasis in endemic areas.

Ivermectin and albendazole are effective in other forms of filariasis, although neither drug kills adult worms. As they exhibit activity against intestinal nematodes, including *A. lumbricoides*, these may be incidentally expelled during treatment. Albendazole is being used, alone or in combination with ivermectin, in campaigns aimed at eradicating lymphatic filariasis.

Surprisingly, tetracyclines also have an effect in filariasis, apparently by inhibiting endosymbiotic bacteria (*Wohlbachia* species) that are essential for the fertility of the worms.

Dracunculus medinensis

This is the *guinea-worm*. The infective larvae develop within water fleas of the genus *Cyclops*, and human infection is normally acquired through infected drinking water. The larvae penetrate the gut mucosa and grow to maturity in connective tissue, usually of the lower limbs. The male is small and insignificant, but the female may reach a length of 1 m. After fertilization, the female worm incubates the larvae to maturity and, when ready to give birth, emerges to the skin surface to provoke an intensely irritating blister. When the sufferer immerses the blister in water, the uterus of the female worm bursts, liberating up to 1 million larvae, which are ingested by water fleas to continue the cycle.

Attempts can be made to wind out the dead worm over several days, but breakage of the worm often occurs, and pyogenic cocci may be carried into the tissues to cause a cellulitis. Chemotherapy is not usually helpful, other than to treat secondary bacterial infection. Prevention is the best approach; health education campaigns and the provision of safe water have much reduced the prevalence of this disease, and complete eradication is now possible.

Trichinella spiralis

Unlike most parasitic worms, *Trichinella spiralis* has an extremely wide host range. Human infections are usually acquired by eating undercooked pork products, although other meat, including bear and walrus meat, has been incriminated. The infected larvae lie dormant in skeletal muscle (Fig. 62.3) and are released when the meat is digested. Male and female worms develop to maturity attached to the mucosa of the small intestine. The female is viviparous, producing numerous larvae during a lifespan of only a few weeks. The larvae penetrate the gut wall and migrate to skeletal muscle, where they enter the quiescent phase. Most of the symptoms of trichinosis, which can be severe – even life threatening – are associated with the migration of larvae.

Mebendazole is said to be effective against the adult worms and the larvae, but treatment is unsatisfactory. Symptoms usually develop only during the invasive stage, and measures to control the sequelae of invasion are more important than anthelminthic therapy.

Fig. 62.3 Larvae of *Trichinella spiralis* in muscle.

TREMATODES

The flukes (Table 62.5) are a diverse group of worms that share a similar life cycle involving a snail host and, often, a second intermediate host that provides the vehicle for the transmission of infection. Most flukes have a restricted geographical distribution that reflects the habitat of the appropriate type of snail.

Table 62.5 Principal trematode parasites of man

Species	Common name	Intermediate host		Geographical distribution	Relevant examination
		First[a]	Second		
Clonorchis sinensis	Chinese liver fluke	*Bithynia* sp.	Freshwater fish	Far East	Stool concentration
Fasciola hepatica	Sheep liver fluke	*Lymnaea* sp.	Vegetation	Worldwide	Stool concentration, serology
Fasciolopsis buski	Giant intestinal fluke	*Segmentina* sp. etc.	Water chestnut	Far East	Stool concentration
Paragonimus westermani	Lung fluke	*Semisulcospira* sp.	Crabs and crayfish	Chiefly Far East	Sputum
Schistosoma mansoni	Bilharzia	*Biomphalaria* sp.	None (water)	Africa, West Indies, South America	Stool concentration, rectal biopsy
S. haematobium	Bilharzia	*Bulinus* sp.	None (water)	Africa	Terminal urine (midday)
S. japonicum	Bilharzia	*Oncomelania* sp.	None (water)	Far East	Stool concentration, rectal biopsy

[a]The first intermediate host is a snail in each case.

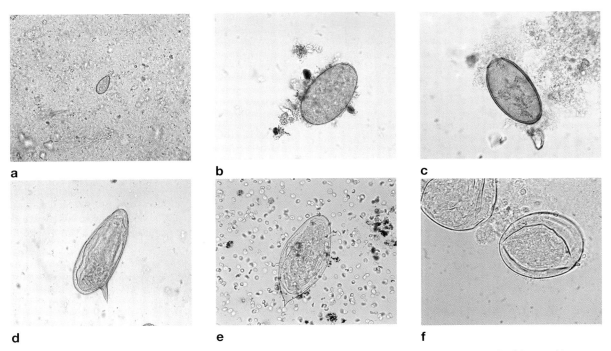

Fig. 62.4 Eggs of trematodes: **a** *Clonorchis sinensis*; **b** *Fasciola hepatica*; **c** *Paragonimus westermani*; **d** *Schistosoma mansoni*; **e** *S. haematobium*; **f** *S. japonicum*.

Most trematodes are hermaphrodite, but the most important human flukes, the schistosomes, are differentiated into separate sexes.

Life cycle

When excreted, trematode eggs often contain a fully developed ciliated organism called a miracidium, although in some species immature eggs are produced that require a period of development before the miracidium is formed. In water, the miracidium escapes, either through a lid-like *operculum* in the egg shell or (in the case of the schistosomes) by osmotic rupture of the egg. The miracidium penetrates the appropriate species of snail and undergoes several stages of asexual reproduction before emerging as a free-swimming body called a *cercaria*. The cercariae encyst in the muscle of fish (*Clonorchis sinensis*), crabs and crayfish (*Paragonimus westermani*), water chestnuts (*Fasciolopsis buski*) or vegetation (*Fasciola hepatica*), and man becomes infected by ingesting the encysted metacercariae. In the case of *Schistosoma* species, the cercariae remain in water and penetrate unbroken skin to gain access to the body.

Clonorchis sinensis (syn. *Opisthorchis sinensis*)

This is the Chinese liver fluke. Infection is acquired from uncooked freshwater fish, notably carp. The meta-

cercariae excyst in the small intestine and pass into the bile ducts, where they mature. Typical small, flask-shaped eggs with a prominent operculum (Fig. 62.4a) are excreted in large numbers into the faeces.

Infection is commonly asymptomatic, but fibrosis of the bile ducts with impairment of liver function may occur in heavy, chronic infections. As with most fluke infections, praziquantel is emerging as the drug of choice for treatment.

A closely related fluke, *Opisthorchis felineus*, which is a parasite of the cat, has been associated with human disease in parts of eastern Europe.

Fasciola hepatica

This is the cosmopolitan liver fluke of sheep. Human infections have usually been associated with eating wild watercress from infected sheep pastures. The adult worm is larger than *C. sinensis*, and lighter infections can cause biliary fibrosis and obstructive jaundice. The large, immature eggs with an indistinct operculum (Fig. 62.4b) may be found in faeces, but are usually sparse.

Unlike other trematode infections, fascioliasis does not reliably respond to praziquantel, and treatment with the veterinary anthelminthic triclabendazole or the more toxic chlorophenol derivative bithionol may be required.

Paragonimus westermani

This is the lung fluke, which is found in parts of the Far East. Closely related species have occasionally been implicated in human disease in parts of Africa and South America. Human infection follows ingestion of raw, infected muscle of freshwater crabs and crayfish. The metacercariae penetrate through the gut wall and diaphragm to reach the lung, where they develop to maturity. Occasionally the larvae find their way to the brain. Pulmonary infection usually provokes the production of sputum, in which the characteristic large eggs (Fig. 62.4c) can be found, often associated with flecks of altered blood. Praziquantel is used for treatment.

Intestinal flukes

Several genera of intestinal flukes cause human infection, particularly in the Far East. *Fasciolopsis buski* is found in restricted foci in China and South-East Asia. Infection is often acquired by the habit of opening water chestnuts with the teeth. The adult flukes live attached to the wall of the small intestine and produce a large number of eggs that resemble those of *F. hepatica*.

Other intestinal flukes include *Gastrodiscoides hominis*, *Heterophyes heterophyes*, *Metagonimus yokogawai* and various species of *Echinostoma*. Infection is usually asymptomatic unless the worm burden is large. Such evidence as exists suggests that praziquantel is effective in these cases.

Schistosoma species

The schistosomes, or *blood flukes*, also known as *bilharzia* after the discoverer, Theodor Bilharz, are the most important of the pathogenic trematodes. At least 200 million people are infected, principally in Africa, where *Schistosoma mansoni* and *S. haematobium* are widespread, and *S. intercalatum* is encountered in some areas. *S. mansoni* is also found in parts of the West Indies and South America; *S. japonicum* and the related *S. mekongi* are restricted to the Far East.

Human infection follows exposure to cercariae in water harbouring infected snails. The cercariae penetrate the skin, often causing a transient dermatitis, called *swimmer's itch*. Once in the bloodstream, the schistosomula migrate to the liver, where they develop into mature male and female worms. The integument of the mature male worm is adapted in the form of two long flaps, the *gynaecophoral canal*, in which the female is held. The mature worms migrate to the small veins of the rectum (*S. mansoni, S. intercalatum, S. japonicum* and *S. mekongi*) or the bladder (*S. haematobium*). Eggs, which contain a fully developed miracidium, are passed through the rectal mucosa on to the surface of colonic faeces or through the bladder wall into the urine.

The ova of *S. mansoni* (Fig. 62.4d) are large (about 140 μm long) and possess a characteristic lateral spine, whereas those of *S. haematobium* (Fig. 62.4e) and *S. intercalatum* have a terminal spine. The smaller, more rounded, eggs of *S. japonicum* and *S. mekongi* do not have a prominent spine, but may exhibit a rudimentary nipple-like appendage (Fig. 62.4f).

Pathogenesis

The adult worms adopt the subterfuge of coating themselves with host antigens to evade attack by host defences; in themselves, the adults are innocuous. Most of the serious manifestations of schistosomiasis are associated with the deposition of eggs, with the formation of granulomata and fibrotic lesions of the liver, bladder or other organs. Such effects may herald malignant changes. Heavy infection with *S. mansoni* may give rise to *schistosomal dysentery*, whereas *S. haematobium* infections are commonly accompanied by a marked haematuria (visible in Fig. 62.4e).

Laboratory diagnosis

Ova of rectal schistosomes can be sought on the surface of formed faeces. Blood-stained mucus should be examined, if present. Alternatively, microscopical examination of snips of rectal mucosa teased out in a drop of saline on a microscope slide may reveal viable or calcified eggs.

For the diagnosis of infection with *S. haematobium*, the last few drops of urine at the end of micturition (*terminal urine*) are most likely to be rich in ova. Excretion is said to be maximal around midday.

To test for viability of the eggs after treatment, the ova can be hatched in water (the motile miracidia can be seen with the help of a hand-lens) or the eggs can be examined microscopically for the characteristic flickering movement of excretory 'flame cells'.

Various serodiagnostic tests, including enzyme-linked immunosorbent assay (ELISA), are available, but are no substitute for demonstration of the ova. Antigen detection methods have also been developed.

Treatment

Praziquantel is effective against all the human schistosomes and is the drug of choice. Because of its lack of toxicity and simplicity in administration. praziquantel has been used, together with molluscicides and water purification, in control programmes.

Other compounds are more selective in their action: metrifonate (an organophosphate compound) is active

Table 62.6 Principal cestode parasites of man

Species	Common name	Intermediate host	Relevant examination
Taenia saginata	Beef tapeworm	Cattle	Mature segments on stool
T. solium	Pork tapeworm	Pig	
Diphyllobothrium latum	Fish tapeworm	*Cyclops* spp./fish	Ova in stool
Hymenolepis nana	Dwarf tapeworm	None	Ova in stool
Echinococcus granulosus[a]	Hydatid worm	Sheep, man	Radiology; serology

[a]Dog tapeworm; man is one of the intermediate hosts.

against *S. haematobium*; *oxamniquine* is effective in *S. mansoni* infection.

CESTODES

The species of tapeworm most commonly involved in human infection are listed in Table 62.6.

Taenia species

Taenia saginata, the *beef tapeworm*, is much more prevalent than the related *T. solium*, the *pork tapeworm*. Both have a relatively simple life cycle, alternating between man and the intermediate host. Human infection is acquired by eating raw or undercooked beef or pork containing the encysted larval stage, the *cysticercus*. The larvae hatch in the small intestine and attach to the mucosal surface by four suckers on the head (*scolex*) of the worm. The scolex of *T. solium* additionally carries a crown of hooklets. The worm grows backwards from the head, first producing immature segments (*proglottids*), which continue to develop as they become more distant from the head. When sexually mature, the proglottids, which exhibit both male and female characteristics, cross-fertilize one another, and eggs start to be produced in the uterine canal. This becomes grossly distended as more eggs are produced, so that the fully gravid segments at the end of the worm become nothing more than bags full of eggs. The complete chain of segments is known as a *strobila*, and may measure 10 m or more.

Eggs are not laid. They are retained within the proglottids, which become detached from the end of the worm and are passed with the faeces. Animals become infected by ingesting the eggs from pastures that are contaminated with inadequately treated sewage or by birds that scavenge in untreated sewage.

Considering the size of the worm, infection is usually remarkably asymptomatic. However, in the case of *T. solium*, eggs may hatch in the human host and form cysticerci. When these lodge in the brain, they may cause a serious epileptiform disease, *cerebral cysticercosis*.

Laboratory diagnosis

Taenia infection is usually diagnosed by finding the typical segments in faeces. Since eggs are not laid, faecal examination for ova is inappropriate. *T. saginata* can usually be differentiated from *T. solium* if the segment is pressed between two microscope slides and examined macroscopically. In the case of *T. saginata*, numerous branchings of the central uterine canal are evident, whereas there are usually far fewer branchings with *T. solium* (Fig. 62.5). The eggs (Fig. 62.6a) are thick walled and contain an *oncosphere* with six hooklets. The eggs of *T. saginata* and *T. solium* are indistinguishable.

Treatment

A single dose of praziquantel is usually successful. Niclosamide is also used, but this drug causes the worm to disintegrate, with the consequent theoretical (but unproven) risk of auto-infection in the case of *T. solium* through the intraluminal release of eggs. Treatment of cerebral cysticercosis is problematical, but albendazole and praziquantel have been used successfully.

Diphyllobothrium latum

This is the *fish tapeworm*, which is prevalent in lakeland areas where freshwater fish is eaten raw. The life cycle is reminiscent of that of the trematodes. The mature adult worm, which may attain a length of 10 m, lays numerous operculate eggs within which a ciliated body called a *coracidium* develops. This hatches in water and is ingested by the water flea (*Cyclops* species). After a period of development, the larva awaits ingestion by a freshwater fish in which it invades the muscle as an infective *plerocercoid larva* or *sparganum*.

Human infection is usually asymptomatic, although a form of pernicious anaemia caused by competition for dietary vitamin B_{12} has been described.

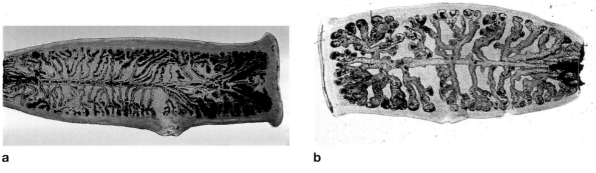

Fig. 62.5 Segments of **a** *Taenia saginata* and **b** *T. solium*. The uterine canal has been injected with Indian ink to show the branchings.

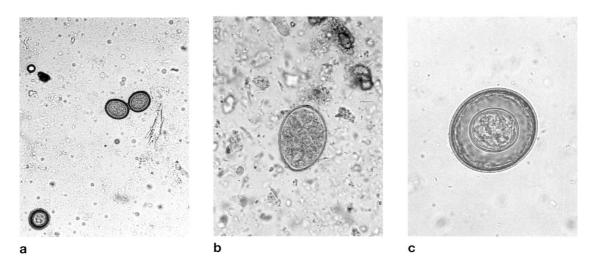

Fig. 62.6 Eggs of cestodes: **a** *Taenia* species; **b** *Diphyllobothrium latum*; **c** *Hymenolepis nana*.

The characteristic immature eggs have an indistinct operculum (Fig. 62.6b) and are usually present in large numbers in faeces. Occasionally, a length of the worm may break off and be passed in the stool.

Niclosamide or praziquantel is used for treatment.

Hymenolepis nana

In contrast to the enormous length of *Taenia* species and *D. latum*, *Hymenolepis nana* is only 2–4 cm long, and is consequently known as the *dwarf tapeworm*. It has a very simple life cycle with no known intermediate host. The characteristic 'poached egg' ova (Fig. 62.6c) are directly infective, and it is surprising that infection is not more common.

Infection is usually asymptomatic; heavy infections can be treated with praziquantel or niclosamide.

A slightly larger species, *H. diminuta*, is occasionally found in man. This is a parasite of small rodents and is transmitted by their fleas.

Echinococcus granulosus

This is the tapeworm of the dog and other canine species, and, unusually, man is an intermediate host. It is a small worm, consisting usually of just four segments and measuring only a few millimetres in length. Sheep are the usual intermediate hosts. Other animals, including man, may become infected, especially in sheep-farming areas, where the cycle of transmission is maintained between sheep and dogs.

After ingestion of the eggs, which resemble those of *Taenia* species, larvae hatch in the small intestine, penetrate the gut mucosa and are carried by the bloodstream to various

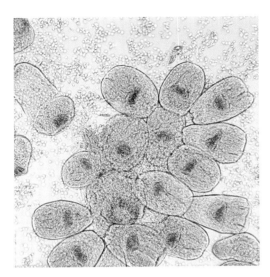

Fig. 62.7 Scolices of *Echinococcus granulosus* from hydatid cyst.

organs (commonly the liver), where they are filtered out. The larva starts to grow, eventually forming a cystic cavity, the *hydatid cyst*. The inner wall of the cyst contains the *germinal layer*, from which develop *brood capsules* that bud off and fall into the cyst cavity. Within these brood capsules new scolices develop, and some of these may initiate the formation of daughter cysts within the main cavity. The young cyst may die and calcify, but it often continues to grow inexorably, eventually seriously compromising the function of the organ in which it is situated.

In certain parts of the world, notably the arctic regions of North America and Siberia, infection with a related canine tapeworm, *E. multilocularis*, is encountered. The hydatid cyst infiltrates the surrounding tissue, making it difficult to remove surgically.

Laboratory diagnosis

The diagnosis is usually made on clinical and radiological evidence. Examination of the cyst fluid (*hydatid sand*) reveals the typical invaginated scolices (Fig. 62.7), but diagnostic puncture of cysts is not recommended because of the risk of spillage (see below).

Imaging techniques supported by serological tests offer the best means of diagnosis. ELISA is the preferred laboratory method, but other serodiagnostic tests are also available. A skin test with antigen derived from hydatid fluid (*Casoni test*) was formerly used, but is unreliable.

Treatment

Cysts of *E. granulosus* can often be removed surgically, but accidental spillage of viable scolices into body cavities may cause an anaphylactic reaction and, moreover, is likely to lead to the development of fresh cysts. For this reason the hydatid cyst is first injected with a scolicidal agent, such as hypertonic saline or ethanol.

The relative impermeability of the cyst militates against successful chemotherapy, but some success has been obtained with benzimidazole derivatives, notably albendazole, and with praziquantel.

KEY POINTS

- Helminths are multicellular parasitic worms that often have complex life cycles.
- Intestinal nematodes (roundworms) are extremely common, especially in conditions of poor hygiene.
- Heavy infections with hookworm can give rise to severe anaemia; benzimidazoles offer effective therapy.
- Serious infections caused by tissue nematodes include bancroftian filariasis (elephantiasis) and onchocerciasis (river blindness); they are treated with diethylcarbamazine or ivermectin.
- Trematodes (flukes) cause schistosomiasis (bilharzia) and other diseases such as Chinese liver fluke infection.
- Cestodes (tapeworms) are often several metres long, but usually cause little pathology.
- Most trematode and cestode infections respond to praziquantel.
- Hydatid disease is caused by a dog tapeworm. Treatment is by surgery.

RECOMMENDED READING

Cook G C, Zumla A I (eds) 2002 *Manson's Tropical Diseases*, 21st edn. W B Saunders, London

Crompton D W T 1999 How much helminthiasis is there in the world? *Journal of Parasitology* 85: 397–403

Fleck S L, Moody A H 1988 *Diagnostic Techniques in Medical Parasitology*. Wright, London

Francis J, Barrett S P, Chiodini S L 2003 Best practice guidelines for the examination of specimens for the diagnosis of parasitic infections in routine diagnostic laboratories. *Journal of Clinical Pathology* 56: 888–891

Guerrant R L, Walker D H, Weller P F 2006 *Tropical Infectious Diseases*, 2nd edn. Churchill Livingstone, Philadelphia

Hoerauf, A, Büttner DW, Adjei O, Pearlman E 2003 Onchocerciasis. *British Medical Journal* 326: 207–210

Muller R 2001 *Worms and Disease*, 2nd edn. CABI Publishing, Wallingford

Peters W, Pasvol G 2001 *Tropical Medicine and Parasitology*, 5th edn. Mosby, London

Stoll N R 1947 This wormy world. *Journal of Parasitology* 33: 1–18 (reprinted *Journal of Parasitology* 1999; 85: 392–396)

World Health Organization 1991 *Basic Laboratory Methods in Medical Parasitology*. WHO, Geneva

CD–ROM

Wellcome Trust 1999 *Schistosomiasis* (Topics in International Health series). Available from Wellcome Trust, London (www.wellcome. ac.uk/tih) and (for developing countries) from TALC (Teaching-aids at Low Cost; www.talcuk.org)

Internet sites

Schistosomiasis Research Group, University of Cambridge. Helminth Infections of Man. http://www.path.cam.ac.uk/~schisto/General_ Parasitology/Hm.helminths.html

Division of Parasitic Diseases, Centers for Disease Control and Prevention. Professional Information. http://www.cdc.gov/ncidod/dpd/professional/ default.htm

63

Arthropods

Arthropod-borne diseases; ectoparasitic infections; allergy

D. Greenwood

Arthropods are animals with jointed legs, segmented bodies and chitinous exoskeletons. They are hugely diverse and incredibly numerous – more than 850 000 species have been described (probably only one-tenth of the true number) and a swarm of locusts alone may contain 10^9 individuals. Strictly speaking, the term 'arthropod' includes lobsters, crabs, millipedes and centipedes, but these seldom cause much serious mischief and most medical interest centres on insects and arachnids. A simplified classification scheme is shown in Table 63.1.

MEDICAL IMPORTANCE OF ARTHROPODS

Insects and other arthropods are mainly of importance in human disease in three ways:

1. as vectors of the agents of bacterial, viral or parasitic infection
2. as parasites in their own right, spending part or all of their lifespan on man
3. as instigators of allergic responses that vary in severity.

In addition, many arthropods have a considerable nuisance effect because of their biting or stinging habits, and these occasionally give rise to serious, even life-threatening, reactions.

The larvae (maggots) of some common flies that feed on decomposing matter have been used for centuries to treat infected lesions. There is renewed interest in this phenomenon, as the maggots not only scavenge dead tissue, but also appear to secrete factors conducive to wound healing.

Abnormal fear of insects or other arthropods (e.g. arachnophobia; excessive fear of spiders) is well recognized, as are delusions of infestation with these creatures. The mere mention of head lice can make people scratch their scalps. Distinguishing between the real and the imaginary can test the diagnostic acumen of the attending physician. Persistent cases may need psychiatric referral.

Arthropods as disease vectors

Mechanical transmission

Insects such as flies, ants and cockroaches that are attracted by food can transmit pathogenic micro-organisms passively. The common house fly, *Musca domestica*, is a particular nuisance because of its predilection for decaying matter, its mobility and its habit, shared by a number of other flies, of regurgitating food material and defecating on food. Flies undoubtedly play a role in the transmission of diseases such as shigellosis and trachoma.

Intermediate hosts

A wide variety of arthropods act as obligatory hosts in the transmission of viral, bacterial, protozoal and helminthic disease. Their role in individual diseases is dealt with in appropriate chapters elsewhere in the book.

Arthropods as ectoparasites

Many insects, ticks and mites pester man to obtain a blood meal or spend part or all of their life in association with human beings. Several fleas, lice and mites are among those that are adapted in various ways for life on man. Occasionally, the human acts as the host of the larval stages of certain insects, causing a condition known as *myiasis*.

Arthropods as allergens

Biting or stinging insects and other arthropods can give rise to severe reactions in hypersensitive individuals. Anaphylactic reactions and multiple stings need immediate treatment with adrenaline (epinephrine). The hairs of several caterpillar species found in various parts of the world are irritant and may give rise to urticaria if brushed against. Contrary to popular belief, the bites of mosquitoes (gnats), fleas and some other insects are painless. The intense irritation that may develop is an

Table 63.1 Simplified classification of arthropods of medical importance

Class	Members of class	Medical importance
Insects	Ants, bees, wasps (Hymenoptera)	Venomous bites and stings
	Beetles (Coleoptera)	Some secrete fluids causing blisters
	Bugs (Hemiptera)	Bites; vectors of Chagas' disease
	Butterflies and moths (Lepidoptera)	Urticaria (caterpillars)
	Cockroaches (Dictyoptera)	Mechanical vectors of disease
	Fleas (Siphonaptera)	Ectoparasites; vector of plague
	Flies and gnats (Diptera)	Vectors of many viral and parasitic diseases; myiasis
	Lice (Phthiraptera)	Ectoparasites; vectors of typhus, trench fever, relapsing fever
Arachnids	Spiders and scorpions	Venomous bites and stings
	Ticks	Vectors of rickettsiae, borreliae
	Mites	Scabies; allergy; vectors of scrub typhus, rickettsial pox
Pentastomes	Tongue worms	Animal parasites; human infestations rare
Myriapods	Centipedes and millipedes	Some cause painful bites or secrete fluids causing blisters
Crustacea	Crabs, crayfish	Intermediate host of lung fluke
	Copepods	Intermediate host of fish tapeworm and guinea-worm

immunological reaction elicited in most, but not all, subjects. Persons travelling to malarious areas should beware of claiming that they never get bitten and hence do not need protection.

House dust mites (usually *Dermatophagoides pteronyssinus* in Europe; more commonly *D. farinae* in Japan and America) flourish in centrally heated homes with wall-to-wall carpeting. They feed on flakes of skin and have been incriminated as a cause of asthma in atopic subjects. The allergens, which include a cysteine protease, are secreted in the mites' faeces.

INSECTS

Insects are the most numerous and familiar form of arthropod life. They are characterized by having six legs and segmented bodies. The legs (and wings, if present) are carried on the thoracic segment between the head, which bears sucking or biting mouthparts, and the abdomen. Most insects, apart from lice and bugs, undergo complete metamorphosis, developing from eggs to adults through morphologically distinct larval and pupal (chrysalis) stages.

Ants, bees and wasps (Hymenoptera)

These insects have little medical relevance apart from their propensity to retaliate to disturbance with defensive bites or stings. These can be serious if the subject is hypersensitive, but are usually trivial. Particularly notorious are the African honey bees, *Apis mellifera scutellata*, and the South American fire ants, *Solenopsis* spp., both of which have been introduced into the USA. Ants also forage wherever food is to be found and frequently infest kitchens and food stores.

Beetles (Coleoptera)

Some beetles act as intermediate host of the dwarf tapeworm *Hymenolepis diminuta*, an uncommon human parasite of minor importance. Otherwise, their only real medical significance lies in the fact that the body fluids of certain beetles can cause blistering of the skin. One such beetle, 'Spanish fly', is the source of cantharidin, a vesiculating agent.

Bugs (Hemiptera)

The bugs of importance in human medicine are blood-sucking species. Most familiar throughout the world is the bed bug, *Cimex lectularius*, together with its tropical cousin, *C. hemipterus*. They have a body that is flattened dorsoventrally and from a distance they resemble brown lentils. Bed bugs live in cracks and crevices of walls, floorboards and furniture, from where they emerge to take a periodic blood meal whenever it is on offer. The adults are long-lived and can survive up to a year without a blood meal. They are usually spread between premises in infested furniture. There is no evidence that they transmit disease.

Reduviid bugs (colloquially known as *kissing bugs* or *assassin bugs*) transmit Chagas' disease in South America

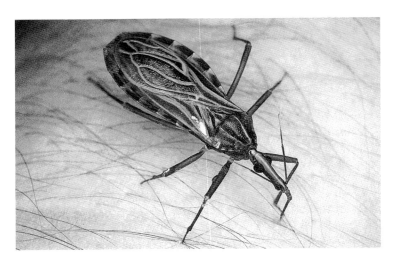

Fig. 63.1 A reduviid bug feeding on human skin. These insects transmit Chagas' disease in South America. (Photograph courtesy of Dr H.-J. Grundmann, University of Nottingham Medical School.)

(see p. 630). Various species of *Triatoma*, *Rhodnius* and *Panstrongylus* are implicated. They are about 2.5 cm in length – much larger than bed bugs – and, unlike them, they have wings (Fig. 63.1). They are usually active at night, settling on the face of an unsuspecting sleeper to take a blood meal and to defecate. The infective trypanosomes are in the hindgut and the bitten person becomes infected by rubbing the bug's faeces into the irritating bite wound.

Butterflies and moths (Lepidoptera)

Adult butterflies and moths are of no medical significance, although some tropical moths may feed on the discharge from the eyes of mammals, including man. The hairs of certain caterpillars can cause skin rashes, as noted above.

Cockroaches (Dictyoptera)

Cockroaches are of little medical significance, although they can harbour pathogens and infest food stores. Laboratory workers in the tropics are familiar with their habit of removing the blood from blood slides left exposed on the bench.

Fleas (Siphonaptera)

Fleas are small blood-sucking parasites. Their laterally flattened bodies and lack of wings enable them to negotiate the hairs and feathers of their animal hosts. Well developed hind legs enable them to jump from host to host (Fig. 63.2). Many fleas will feed on man if given the opportunity, as those who have been attacked by the common cat flea *Ctenocephalides felis* bear witness. However, the species that is adapted for life on man

Fig. 63.2 The rat flea *Xenopsylla cheopsis*, vector of plague.

is the human flea, *Pulex irritans*, which is still common throughout the world. Female fleas of another species, *Tunga penetrans*, attack man once they have been fertilized. The fleas burrow into the skin, or under the toenails, of the human host and are known as *jiggers*. The abdomen of the gravid female becomes grossly distended with eggs, causing pain, irritation and, sometimes, secondary infection. Jigger fleas are common in dry, sandy soil, mainly in Africa and parts of Central and South America.

Human fleas are seldom implicated in the transmission of disease, but some other species are important disease vectors. Most notorious is the rat flea, *Xenopsylla cheopsis*, which is the most important, but not the sole, vector of plague (see p. 343). Some forms of typhus are also transmitted by *X. cheopsis* and other fleas (see Table 40.1, p. 386).

Table 63.2 Principal infections associated with biting flies, mosquitoes and midges

Type of insect	Diseases transmitted
Biting midges (*Culicoides* spp.)	Filariasis (*Mansonella* spp.)
Black-flies (*Simulium* spp.)	Onchocerciasis
Deer flies (*Chrysops* spp.)	Loiasis
Mosquitoes:	
Anopheline (*Anopheles* spp.)	Malaria; bancroftian filariasis
Culicine (*Culex*, *Aedes*, *Mansonia* spp.)	Bancroftian and brugian filariasis; yellow fever, dengue and other arbovirus infections
Sandflies (*Phlebotomus*, *Lutzomyia* spp.)	Leishmaniasis; Oroya fever; sandfly fever
Tsetse flies (*Glossina* spp.)	African trypanosomiasis

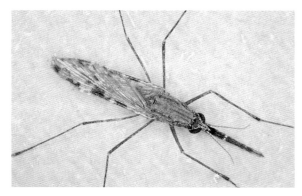

Fig. 63.3 A female anopheline mosquito, the vector of human malaria. (Photograph courtesy of Dr H.-J. Grundmann, University of Nottingham Medical School.)

Flies and gnats (Diptera)

Dipterous insects are strong and active flyers, with a pair of wings used for flying and an additional vestigial pair, known as *halteres*, which are used as organs of balance. The developmental cycle from egg to adult fly involves complete metamorphosis. As well as being among the most annoying of insect pests, many biting flies have an obligate role in the transmission of a wide variety of important infections (Table 63.2). The larvae of certain flies may also infest wounds; others are able to penetrate skin to cause myiasis (see below).

Mosquitoes

Mosquitoes are readily recognized by a long needle-like proboscis (Fig. 63.3). Adult males and females both feed on plant juices, but the female needs blood for the development of her eggs, and is also a voracious predator on a wide variety of vertebrate animals throughout the world. Mosquitoes of importance in human medicine are divided into two broad types: *anopheline* mosquitoes, numerous species of which transmit malaria (see pp. 621–625), and *culicine* mosquitoes, which are the vectors of many so-called *arbovirus* infections (Ch. 51). Both anopheline and culicine mosquitoes also act as the intermediate hosts of certain filarial worms (pp. 636–639).

Female mosquitoes lay their eggs on water; larvae and pupae are both aquatic. Most anopheline mosquitoes prefer relatively large expanses of water that do not dry up, but many culicine mosquitoes, particularly *Aedes* spp., will breed in small pockets of water, such as tree-holes, water butts, etc. The adults have a wide flight range and may be found several kilometres from their breeding ground. Mosquitoes capable of transmitting malaria are found throughout the world, even within the Arctic Circle, but indigenous disease is nowadays restricted to the tropical and subtropical belt.

Midges

Biting midges are tiny flies that are able to cause a nuisance out of all proportion to their size. The females attack in swarms, usually in the evening, and may give rise to painful reactions. Like mosquitoes, they are mostly aquatic. One genus, *Culicoides* spp., transmits filarial worms of *Mansonella* spp. (see p. 639).

Sandflies

Sandflies are small enough to penetrate most mosquito netting. They are more demanding in their habitat than mosquitoes and have a restricted flight range, so that the diseases they transmit – notably kala azar and other forms of leishmaniasis (see p. 630), bartonellosis (p. 339) and sandfly fever (p. 521) – tend to be localized in distribution. Female flies suck blood, usually at night, and breed in dark, moist areas, often in or around human dwellings. Species associated with disease transmission in Africa, the Middle East, Asia and the Mediterranean littoral belong to the genus *Phlebotomus*. In Central and South America, *Lutzomyia* spp. act as vectors of leishmaniasis and Oroya fever, a form of bartonellosis.

Other biting flies

Although flies that are capable of inflicting a painful bite, such as the stable fly *Stomoxys calcitrans*, are found throughout the world, species that are important vectors of human disease are restricted in distribution to areas of the tropics. The tsetse flies, *Glossina* spp., are found in the so-called 'fly-belts' of sub-Saharan Africa, where they are responsible for trypanosomiasis in man as well as in cattle and other animals (see p. 628). Unusually, both males and females feed on blood. The vectors of human trypanosomiasis ('sleeping sickness') in areas of West Africa are riverine species such as *Glossina palpalis*, whereas *G. morsitans*, which prefers the savannah plains and woodlands, is the principal vector in the eastern part of the continent.

Other biting flies responsible for transmission of disease in Africa include species of *Chrysops (deer flies* or *mango flies)* and *Simulium (black-flies)*. *Chrysops* spp. belong to the tabanid group and are related to horse flies, which themselves can evoke a severe reaction in man although they are not known to transmit disease. The species that act as vectors of the filarial worm *Loa loa* (see p. 637), *Chrysops dimidiata* and *C. silacea*, breed in marshy areas of the rainforests of tropical West and Central Africa. The females are said to be attracted by the smoke of fires and inflict a painful bite.

The breeding grounds of black-flies are the banks of rivers, where females attach their eggs to vegetation. They are small hump-backed flies that often attack in swarms. Only the female feeds on blood. *Simulium damnosum* and *S. naevi* are important vectors of onchocerciasis (*river blindness*; see pp. 637 and 639) in parts of tropical Africa. Related species transmit the disease in Central and South America.

Myiasis

Given the chance, many common flies will lay their eggs on the exposed tissues of ulcers and sores, which consequently become the breeding ground for maggots. The condition is known as *semi-specific myiasis* to distinguish it from *specific myiasis*, in which man and other animals act as obligate hosts for the larval stage of development. *Accidental myiasis* is said to occur when larvae are ingested with food, or invade orifices such as the urogenital tract.

Myiasis is of great economic importance in animal husbandry throughout the world, but human infestation is largely a problem of the tropics. Specific myiasis in Africa is most commonly caused by *Cordylobia anthropophaga* (the *tumbu fly*). In Central and South America *Dermatobia hominis* (the *human bot fly*) is the usual culprit. *C. anthropophaga* lays its eggs in soil or dust, but *D. hominis* attaches its eggs to other insects, including mosquitoes, that visit man. When the larvae hatch they burrow into the skin of individuals with whom they are in contact. The larvae remain there, breathing through spiracles at the posterior end until they are ready to pupate and emerge. Larvae of another African fly, *Auchmeromyia luteola* (the *Congo floor maggot*), also parasitize man, but do not penetrate the skin, preferring instead to suck the blood of unsuspecting sleepers.

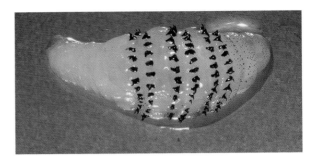

Fig. 63.4 Larva of *Dermatobia hominis*, from a case of human myiasis.

Cutaneous myiasis is a transient condition causing boil-like lesions in the skin. The body of the larvae is furnished with spines (Fig. 63.4), which make them difficult to remove, but covering the posterior spiracles with paraffin oil prevents them from breathing and encourages them to emerge with the help of digital pressure. Care should be taken not to allow the lesion to become secondarily infected with bacteria.

Lice (Phthiraptera)

Lice are wingless insects that undergo incomplete metamorphosis during their development. The ones that parasitize man are blood-sucking species with flattened bodies and short legs that are adapted to cling to hairs. Body lice and head lice are variants of the same species, *Pediculus humanus*, although the body louse, *P. humanus corporis*, is somewhat larger than the head louse, *P. humanus capitis*, and there are other minor differences. A third species, *Phthirus pubis*, is quite distinct morphologically, living up to its common description as the 'crab' louse (Fig. 63.5). Head lice are usually confined to the hairs of the scalp, but body lice live in clothing covering the body, rather than on the skin itself. *Ph. pubis*, as the name suggests, is usually found on pubic hairs, but may also infest other hairy parts, including eyelashes. All types attach their characteristic eggs (*nits*) to body hairs, and effective treatment involves removal of the nits as well as dealing with the adults.

Crab lice are not known to be involved in disease transmission, but body and head lice are the classic vectors of epidemic typhus (see p. 389) and relapsing fever (p. 365). Lice also cause irritation, and continuous scratching may lead to various forms of infective dermatitis.

Lice are very common throughout the world and head lice, in particular, often spread quickly between school children, even in affluent areas. Treatment with insecticides such as permethrin, malathion and carbaryl may be effective, but resistance occurs. There are also fears about the possible neurotoxicity of malathion and the mutagenic potential

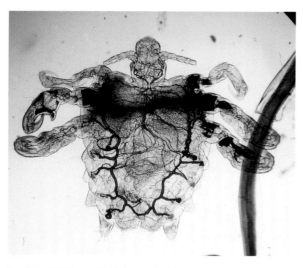

Fig. 63.5 *Phthirus pubis*, the human 'crab' louse.

of carbaryl. Repeated 'wet-combing' with a fine-tooth comb after shampooing the hair may eventually succeed in eliminating the infestation, although re-infection from untreated family members or school contacts is common. Aqueous solutions of malathion or carbaryl may also be used to treat crab lice, but the whole body should be treated. As the infestation is transmitted by intimate contact, sexual partners should also be treated.

ARACHNIDS

This group includes spiders, scorpions, ticks and mites. Unlike insects, the adult forms have eight legs and are invariably wingless. They have two main body regions – cephalothorax and abdomen – which in mites and ticks are fused to give the appearance of a single segment. They develop by incomplete metamorphosis, and immature forms resemble small versions of the adults. However, ticks and mites acquire the full complement of eight legs only as they mature from first-stage larvae to nymphs during their progression to adulthood.

Mites

Despite their name, mites are variable in size, although many, including those commonly implicated in human disease, are so small as to be almost invisible to the naked eye. Parasitic varieties include *Sarcoptes scabiei* (the *itch mite*) and *Demodex folliculorum* (the *blackhead mite*). Other species occupy a wide variety of natural habitats. Human contact with certain species may lead to an intense pruritus or dermatitis. Some mites transmit scrub

typhus and rickettsialpox (see p. 390). House dust mites have attracted considerable attention as a precipitating cause of atopic disease (see p. 648).

The human itch mite, *S. scabiei*, is related to mites causing mange in various animals. It is the cause of scabies, an infestation of the skin that is still very prevalent in many countries. After fertilization on the surface of the skin, the gravid female mite burrows into the epidermis, eventually leaving behind a trail of about 40 eggs. The larvae usually hatch in 3–4 days, leave the burrow and pass through nymphal stages to reach adulthood in hair follicles. Burrowing females cause intense itching, often in folds of skin, especially between the fingers, and there may be secondary bacterial infection. Elderly and immuno-compromised patients may develop a severe crusting infestation known as *Norwegian scabies*, which causes outbreaks in institutions and may be misdiagnosed as psoriasis.

Application of an aqueous solution of malathion or permethrin is often successful therapy, but household contacts should also be treated. Norwegian scabies can be treated systemically with the anthelminthic agent ivermectin, which is widely used in animal husbandry for the control of ectoparasites.

Demodex folliculorum, the blackhead mite, has an elongated body adapted for its life in sebaceous follicles, commonly around the nose or eyelids. They seldom cause much pathology, but can be treated with permethrin application or sulphur preparations.

Ticks

Ticks are essentially large mites, but they are much more important as vectors of human disease. They are conveniently classified into two main families: *hard (ixodid) ticks*, which have a chitinous shield (*scutum*) on the back (Fig. 63.6), and *soft (argasid) ticks*, which lack this feature. In addition, the head parts of soft ticks are hidden on the ventral surface and are not visible from above. Ticks are obligate blood-feeders. They parasitize a very wide variety of animals in nature and many will attack man, given the opportunity. The initial bite is usually painless, but it can give rise to a serious reaction. In some countries, notably Canada, the USA and Australia, *tick paralysis*, a poorly understood condition associated with the bite of ixodid ticks, is occasionally reported. It is likely that a neurotoxin is responsible.

Ixodid ticks, which transmit many rickettsiae of the spotted fever group (see p. 390), as well as the agents of Q fever (pp. 392–393), Lyme disease (p. 366), tularaemia (p. 349), babesiosis (p. 632) and some arboviruses (Table 51.2, p. 508), are very tenacious. Adults, especially the females, gorge on blood for long periods, and efforts to remove them manually often leave the head embedded in the

Fig. 63.6 *Dermacentor andersoni*, the vector of Rocky Mountain spotted fever and a cause of tick paralysis. These are known as 'hard ticks' because of the presence of the dorsal scutum, prominent in the male (right), but much reduced in the female (left).

skin. However, the various developmental stages – larvae, nymphs and adults – may not remain on the same host and some of the most important species involved in human disease transmission, such as *Dermacentor andersoni* and *Ixodes ricinus*, are known as *three host ticks*. Larvae that acquire micro-organisms remain infected through the nymphal and adult stages (*trans-stadial transmission*); the adult can, in turn, pass on infection through the egg (*transovarial transmission*). Thus, spread of infection to new hosts may be very efficient.

Argasid ticks prefer to attack the host at night and do not remain attached to the host after feeding. They are relatively long-lived and inhabit dry, dusty environments, mainly in hot countries. The most important species from a medical point of view is *Ornithodoros moubata*, the main vector of tick-borne borreliosis (relapsing fever) in tropical Africa (see p. 365). Other species of *Ornithodoros* transmit American forms of relapsing fever and some are notorious for their voracious feeding habits and extremely painful bites.

Spiders and scorpions

Although all spiders kill their prey by injecting venom, the toxin is usually innocuous to man, and the mouthparts (*chelicerae*) are seldom robust enough to allow penetration of human skin. The sting of scorpions, which is carried at the end of an elongated extension of the abdomen, can penetrate skin, but the venom is usually of low potency (although sufficient to make children ill).

Some spiders and scorpions have painful bites or stings and a few can cause serious, occasionally fatal, illness

by virtue of powerful neurotoxins. The most dangerous spider is the Australian funnel web spider, *Atrax robustus*, but species of *Latrodectus* (including types of *black widow spider*, found in many areas of the world, and the Australian *redback spider*) can also inflict a serious bite. The venom of some *Lycosa* and *Loxosceles* spiders encountered in the USA and South America may cause tissue necrosis.

The most dangerous scorpions belong to the large Buthidae family. Buthid scorpions with dangerous stings are most commonly found in parts of Africa, the Middle East, the southern states of the USA and Central America. They are nocturnal creatures and sting as a defensive reaction. Most human cases of scorpion bite occur when the arthropod seeks shelter in shoes or other clothing.

Spider and scorpion wounds seldom need more than supportive treatment. Antivenoms are sometimes available in areas where dangerous species are prevalent. Intravenous calcium gluconate has been used successfully to relieve the painful muscle spasm associated with bites of black widow spiders.

Other arthropods

Pentastomes (*tongueworms*) are important parasites of animals, but human infestation is very rare. A few species of centipede can inflict a painful bite and some millipedes can secrete a fluid capable of raising blisters. Crustaceans (crabs and crayfish) are of interest in human medicine mainly as intermediate hosts of *Paragonimus westermani*, the lung fluke (see p. 642). Copepods (*water fleas*) are similarly important only as hosts of the guinea-worm, *Dracunculus medinensis* (p. 639) and the fish tapeworm, *Diphyllobothrium latum* (p. 643).

KEY POINTS

- Medically important arthropods include insects (bugs, flies, gnats, lice and fleas) and arachnids (spiders, ticks and mites).
- Many arthropods act as accidental or obligatory vectors of bacteria, viruses, protozoa or helminths.
- Certain fleas, lice and mites are human parasites.
- Some arthropods cause allergies (e.g. house dust mites); others may cause painful bites or stings.
- Myisais is a condition caused by invasion of the body by insect larvae.

RECOMMENDED READING

Burgess N R H, Cowan G O 1993 *A Colour Atlas of Medical Entomology.* Chapman & Hall, London

Goddard J 2002 *Physician's Guide to Arthropods of Medical Importance*, 4th edn. CRC Press, Boca Raton

Kettle D S 1995 *Medical and Veterinary Entomology*, 2nd edn. CABI publications, Wallingford

Service M W 2004 *Medical Entomology for Students*, 3rd edn. Cambridge University Press, Cambridge

Zinsser H 1935 *Rats, Lice and History.* Routledge, London

Internet site

Iowa State University Entomology Index of Internet Resources. http://www.ent.iastate.edu/list/directory/114/vid/5 (an extensive listing of medical entomology sites).

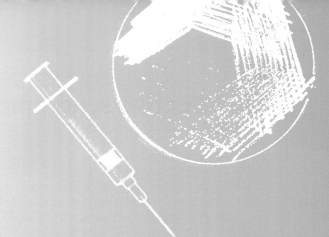

PART 6

DIAGNOSIS, TREATMENT AND CONTROL OF INFECTION

64 Infective syndromes

R. C. B. Slack

Throughout this book, infections have been dealt with as appropriate according to the micro-organisms involved. In this chapter, the study of infection is considered by syndromes associated with the major organs in order to emphasize the variety of microbes that may affect different body systems. The subject is presented in broad outline, and the reader should refer back to earlier chapters for a more extensive account of specific themes.

In infectious diseases, as in other branches of clinical practice, a diagnosis may be obvious and require little investigation, for example a case of chickenpox during an epidemic, or may be established only after laboratory and radiological examination, as with a patient with pyrexia of unknown origin. Figure 64.1 shows a flow chart of a rational approach to diagnosis and management of a patient with infection. In practice, treatment is often started before isolating and identifying the pathogen, and in many viral conditions the exact cause can be determined only during convalescence, either by serological tests or after prolonged growth in tissue culture. The availability of rapid methods of diagnosis and new chemotherapeutic agents has made the study of infectious disease even more important.

SPECIFIC SYNDROMES

UPPER RESPIRATORY TRACT

The upper respiratory tract is frequently the site of general and localized infections. Indeed, this group of ailments is among the most common presenting to domiciliary practice. It is the primary site of infection for most viral diseases, which are spread by sneezing, coughing or direct contact with materials contaminated by respiratory secretions. Although the majority of such symptoms are viral in origin, secondary bacterial infection may often follow, particularly in the very young and malnourished. Resident bacteria in the upper respiratory tract such as *Haemophilus influenzae*, *Streptococcus pyogenes* and *Str. pneumoniae* are the most common causes. Figure 64.2

shows the anatomical sites of respiratory infection and the appropriate specimens to be taken for microbiology.

Sore throat

Bacterial acute tonsillitis or pharyngitis is commonly due to *Str. pyogenes* (group A), and a few are caused by groups C and G. Viruses are even more common causes of sore throats, especially the milder, non-exudative forms. However, it is important to make a definitive diagnosis of streptococcal pharyngitis for two reasons:

1. *Str. pyogenes* remains sensitive to penicillin, which should be used for treatment.
2. Group A β-haemolytic streptococci, if untreated, may give rise to septic complications such as a peritonsillar abscess, or to immune complex disease (glomerulonephritis, rheumatic fever).

However, since *Str. pyogenes* may account for only 20% of patients presenting with typical symptoms, many courses of antibiotics would be prescribed unnecessarily if all suspected cases were treated. Even in experienced hands it is difficult to predict on clinical grounds alone which cases are streptococcal. This has led to three approaches:

1. Treat all children with a penicillin, or erythromycin if allergic. Adults are treated if the throat looks very inflamed or there is pus on the tonsils.
2. Swabs are taken for culture of streptococci, and antibiotics are started as above but stopped if *Str. pyogenes* is not found.
3. Swabs are taken for rapid diagnosis (by antigen detection) and penicillin commenced only if the throat swab is positive.

There are pros and cons to each method. One drawback to the early use of antibiotics is that the patient may react to the drug. If ampicillin is used, there is a strong chance of a skin reaction if the sore throat is the harbinger of glandular fever. In defence of the antibiotic lobby, there is no doubt that complications of streptococcal disease are seen far less commonly where there is access to medical

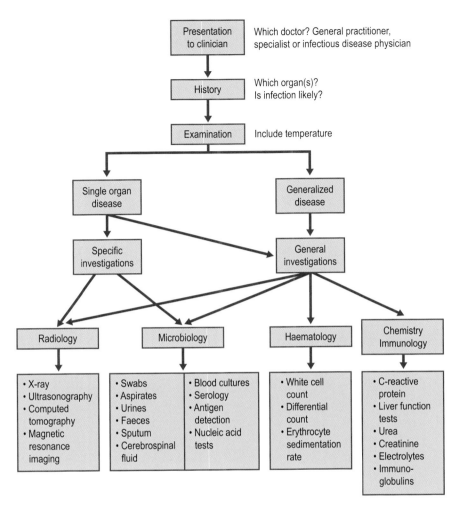

Fig. 64.1 Flow diagram for the diagnosis of infection.

service and pharmacies. However, these improvements have occurred in concert with better housing and social conditions.

There is less need to make a virological diagnosis as specific therapy is limited. However, epidemiological studies of patients with throat symptoms have revealed how common and varied are the viruses in the respiratory tract. With the advent of antiviral agents active against influenza viruses, interest has increased in developing rapid tests for influenza A.

In the severely ill child with toxaemia and a membrane, diphtheria must be considered and treatment should not await laboratory confirmation. Moreover, the laboratory needs to do special tests to isolate and identify *Corynebacterium diphtheriae*, and communication (by telephone, if possible) between the clinician and laboratory is essential. Other corynebacteria such as *C. ulcerans* and *Arcanobacterium haemolyticum* may rarely cause ulcerated sore throats. In the sexually active, gonococcal pharyngitis should not be missed and again the laboratory needs to be told as they will use special selective media for *Neisseria gonorrhoeae*.

Common cold (coryza)

This common complaint, characterized by a nasal discharge (acute rhinitis) that is usually watery with scanty cells, afflicts people of all ages who congregate together. Since there are many types of rhinovirus, coronavirus, adenovirus, etc., and since immunity may be short lived, individuals in a crowded environment, such as at school and university or travelling on public transport, may suffer three or four clinical infections a year. Most people do not seek medical help, knowing that there is little to offer. This is a condition for which many 'alternative' remedies are tried, from garlic to peppermint and vitamin supplements.

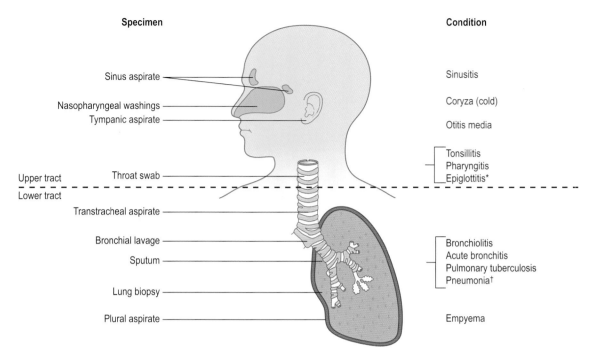

Fig. 64.2 Microbial infections of the respiratory tract and the appropriate specimens for laboratory investigation.

Bacterial superinfection with pneumococci and *H. influenzae* can occur in the nasopharynx but is symptomatic only when the sinuses or middle ear are involved. Pharyngitis and, occasionally, tracheobronchitis may occur with a cold or develop in more susceptible individuals. Respiratory syncytial virus (RSV) may cause upper respiratory symptoms in children and adults, but in those contracting the virus for the first time (usually infants under 1 year of age) acute bronchiolitis is common.

Sinusitis and otitis media

Direct extension of a viral or bacterial infection from the nasopharynx into frontal and maxillary sinuses in adults and into the middle ear in children is not uncommon. Obtaining adequate material for microbiology is difficult and requires the expertise of ear, nose and throat specialists. In most cases of acute sinusitis or otitis media the microbial cause is not found. It is assumed that severe pain and discharge of pus from the nose or ear is suggestive of bacterial infection and antibiotics are usually given to cover streptococci (mainly *Str. pneumoniae*) and *H. influenzae*. In adults with recurrent or chronic sinusitis, anaerobes (peptostreptococci or bacteroides) are often found in sinus washings.

LOWER RESPIRATORY TRACT

Epiglottitis

Although situated in the upper part of the respiratory tract, *epiglottitis* behaves like a serious systemic infection that requires urgent admission to hospital and treatment. The diagnosis should be made clinically in a toxic child (usually under 5 years old) with respiratory obstruction and stridor. The most useful investigation is blood culture, which invariably grows *H. influenzae* type b unless antibiotics have been given or the child has been immunized. The epiglottis, which is swollen and cherry-red in appearance, should be examined only by skilled paediatricians prepared for a respiratory arrest. A lateral radiograph shows a soft tissue swelling in the throat.

Laryngotracheobronchitis

In adults, some viral infections cause acute laryngitis with voice loss or tracheitis with a dry cough. Respiratory tract infection in children is usually generalized and presents as *croup*. This may lead to respiratory obstruction and, as with haemophilus epiglottitis, urgent admission to hospital is necessary. The condition may be caused by a variety of respiratory viruses, with para-influenza viruses, adenoviruses and enteroviruses the most common. RSV can also cause croup, but more commonly this virus attacks infants in the first few months of life. In the UK there are winter epidemics of *acute bronchiolitis* due to RSV. This clinical syndrome starts as a cold which is followed by wheezing. In older children, asthmatic attacks are often precipitated by viral respiratory infections.

Whooping cough

Pertussis may be confused with croup, and, as special specimens need to be taken, the laboratory must be informed. A pernasal swab in special transport medium is required, and *Bordetella pertussis* will grow only on enriched culture media in conditions of high humidity.

Acute bronchitis

Acute exacerbation of chronic obstructive pulmonary disease (COPD) is the commonest adult lower respiratory infection. It is invariably due to pneumococci, non-encapsulated haemophili, or both. Sometimes, acute bronchitis follows a viral infection. Although often examined, expectorated sputum from ambulatory patients with COPD is almost always a waste of effort.

Patients with cystic fibrosis have frequent exacerbations and, in these situations, sputum examination is valuable because the causative bacteria (*Staphylococcus aureus*, *H. influenzae*, *Pseudomonas aeruginosa* or *Burkholderia cepacia*) may show variable antimicrobial resistance and appropriate therapy is essential.

Acute pneumonias

It is sometimes possible on clinical and radiological grounds to distinguish between *lobar pneumonia* (pneumococcal), *bronchopneumonia* (staphylococcal and klebsiellal) and '*atypical' pneumonias* (mycoplasmal and chlamydial). This division is of importance in guiding primary treatment, but should not give the clinician a blinkered view in investigation as some cases will invariably not follow a textbook! Expectorated sputum is often poorly collected and may not yield the pathogen. Blood cultures should always be taken from patients with pneumonia, but may not yield an organism, especially if antibiotics have been started. *Antigen detection* in urine or sputum by fluorescent antibodies, immuno-electrophoresis, latex agglutination or enzyme-linked immunosorbent assay (ELISA) is a useful method for rapid diagnosis but may lack sensitivity. New amplified DNA detection methods are likely to improve diagnostic accuracy in respiratory infections. At present most cases of 'atypical' pneumonia are diagnosed by obtaining acute and convalescent sera

and finding a significant titre or a rising titre of antibodies. *Legionnaires' disease* may be diagnosed by culture of *Legionella pneumophila* from sputum or a biopsy, but this has too low a sensitivity. Urine antigen detection by ELISA and DNA amplification of respiratory secretions have become more widely available methods of diagnosis. However, as with most pneumonias, treatment must often be given on clinical suspicion.

Chronic chest disease

Tuberculosis must always be considered in any patient with fever, prolonged cough and weight loss. Again, a request for examination for mycobacteria must be included on the request card.

In the immunocompromised, various opportunist pathogens may give rise to respiratory infection. *Pneumocystis carinii* is demonstrated by fluorescent antibody staining of bronchial lavage. Fungal infections are diagnosed by growth of the pathogen or by serology. Special investigations are required for all these situations and, unless the diagnosis is considered, the laboratory cannot assist the clinician.

GASTRO–INTESTINAL INFECTIONS

Acute diarrhoea with or without vomiting is a common complaint. Microbial causes, either by multiplication in the intestine or from the effects of preformed toxin, are the most important reasons for acute gastro-intestinal upset in an otherwise healthy individual. The aetiology of inflammatory bowel disease such as ulcerative colitis has not been established, although a patient may present with symptoms similar to those of an infectious diarrhoea; however, the natural history and chronicity of the condition usually make the distinction obvious.

Although many new causes of bowel infection have been discovered in the past 25 years, the majority of food-related and short-lived episodes do not yield a microbial cause. In part this is due to the wide range of viruses, bacteria and protozoa that may be sought (Table 64.1). A search for all causes involves extensive and expensive laboratory effort, and this is often considered unnecessary for a condition that is usually self-limiting and relatively harmless.

Toxin-mediated disease of microbial origin ranges in severity from relatively trivial episodes of food poisoning caused by:

- enterotoxin-producing strains of *Staph. aureus*
- *Clostridium perfringens*
- *Bacillus cereus*;

to the life-threatening systemic diseases:

Table 64.1 Common microbial causes of infectious intestinal disease (see individual chapters for details)

Viruses	Norovirus (SRSV)
	Rotavirus
	Astrovirus
	Calicivirus (SRSV)
Bacteria	*Salmonella enterica* serotypes
	Campylobacter spp.
	Clostridium difficile
	Shigella spp.
	Escherichia coli (ETEC, VTEC)
	Vibrio cholerae
	V. parahaemolyticus
	Yersinia enterocolitica
Protozoa	*Cryptosporidium parvum*
	Entamoeba histolytica
	Giardia lamblia

SRSV, small round structured viruses.
ETEC, enterotoxigenic *Esch. coli*; VTEC, Verotoxigenic *Esch. coli*.

Table 64.2 Common causes of urinary tract infection (approximate percentages)

Organism	Domiciliary	Hospital
Escherichia coli	70–80	50
Proteus mirabilis	10	1–5
Klebsiella spp.	1–5	5–10
Staphylococcus saprophyticus	10–15	0
Staph. epidermidis	1–5	10–20
Enterococci	1–5	10–20
Other coliforms	<1	5–10
Pseudomonas aeruginosa	1–2	5–10

- botulism caused by *Cl. botulinum*
- severe *pseudomembranous colitis* caused by *Cl. difficile* (often antibiotic-associated).

There are, in addition, many non-microbial causes of food poisoning, such as that due to the ingestion of certain toadstools, undercooked red kidney beans or various types of fish (*ciguatera toxin, scombrotoxin*); most notorious is the puffer fish which, during part of its reproductive cycle, produces a neurotoxin that is responsible for more than 100 deaths a year in Japan, where the delicacy *fugu* is enjoyed.

It may be possible on clinical grounds to distinguish between patients with dysentery, when bloody diarrhoea and mucus are found, and those with watery diarrhoea owing to the toxic effects of the pathogen on the small intestinal mucosa, leading to the accumulation of fluid in the bowel. Some conditions, such as staphylococcal food poisoning and some viral illnesses, present largely with vomiting. A specific cause is also suspected if there is a history of foreign travel, if the case forms part of an outbreak associated with food or water, or if the individual has a relevant food history.

Travellers' diarrhoea encompasses many clinical and microbial causes, but the commonest organisms implicated are enterotoxigenic strains of *Escherichia coli* (ETEC). However, a host of microbes must be considered. Some, such as salmonellae, are found worldwide; others, such as vibrios, have a more limited distribution.

More chronic intestinal infections contracted in warmer climates usually do not present with diarrhoea but with vague abdominal symptoms. Many helminths and protozoa are found on screening faeces for other pathogens.

URINARY TRACT INFECTIONS

The diagnosis of urinary tract infection cannot be made without bacteriological examination of the urine because many patients with frequency dysuria syndrome have sterile urine and, conversely, asymptomatic bacteriuria is a common condition. Infection is most commonly caused by members of the Enterobacteriaceae (Table 64.2), but there are great variations in antimicrobial susceptibility and control of chemotherapy requires laboratory examination. Occasionally, *Mycobacterium tuberculosis* invades the kidney; appropriate tests must be carried out if this is suspected.

Infection of the urinary tract may be confined to particular anatomical sites, for example urethritis or renal abscess. Alternatively, the urine may become infected and, in cases of obstruction or reflux, bacteria may ascend from the bladder to give rise to kidney infections. *Urethritis* is really a genital infection and is most commonly due to sexually transmitted organisms such as chlamydia or *N. gonorrhoeae*. Confirmation depends on obtaining, by swabbing or scraping, a sample of urethral discharge for microscopy and culture. *Metastatic abscesses* in the kidney and *perinephric abscesses* cannot usually be diagnosed by urine examination, although pyuria may be present. Specific radiological or surgical exploration is necessary, as with any localized infection.

Throughout most of their lifespan women suffer far more attacks of urinary tract infection than men. In young adult women who become sexually active, frequency and dysuria are common reasons for seeking medical attention. Only about one-half of these yield an organism in 'significant' numbers, usually defined as more than 10^7 organisms per litre. The commonest cause by far is *Esch. coli* (see Table 64.2), some strains of which possess specific uropathogenic determinants. It has been suggested that many of the culture-negative infections are due to coliforms, which are present in small numbers in urine. Infection of the urine is common and may be asymptomatic in many elderly patients.

INFECTIONS OF THE CENTRAL NERVOUS SYSTEM

Meningitis

Meningeal irritation may occur in association with other acute infections (*meningism*) or with non-infective conditions such as subarachnoid haemorrhage. In the early stages of meningococcal disease, signs of meningitis may be absent, yet examination of the cerebrospinal fluid (CSF) yields *N. meningitidis*. Thus, CSF and blood cultures should be examined in all suspected cases. There are contra-indications to performing a lumbar puncture in any patient with raised intracranial pressure because of the danger of herniation through the foramen magnum (*coning*). Thus, lumbar puncture should be carried out only in hospital. In fulminating disease, especially if it is meningococcal, it is prudent to give penicillin as soon as the diagnosis is suspected and before admission. This has led to a greatly reduced yield by culture of suspected cases. Antigen and DNA detection have become increasingly important in confirming the diagnosis.

It is often possible to consider the likely pathogen on clinical and epidemiological grounds. *H. influenzae* type b occurs almost always in infants aged 6–24 months, although use of the Hib vaccine in many countries has almost made this a disease of the past. *Str. pneumoniae* is seen generally in the very young and the elderly. *Meningococcal meningitis* is characteristically a disease of children and young adults. In the neonate, coliforms (mainly *Esch. coli* K1), *Listeria monocytogenes*, group B streptococci and pneumococci may be found. In infants a few months old, salmonella meningitis is an important condition in some countries with a warm climate.

If no readily cultivated organism is found, but the CSF shows an increased number of cells, the syndrome of *aseptic meningitis* is present. Table 64.3 shows some of the causes of this condition. If the symptoms are short in duration, viruses are most likely. If the child has been unwell for more than 1 week, *tuberculous meningitis* must be considered. This is one of the most difficult and important microbiological diagnoses to make. Interferon assays of activated lymphocytes from peripheral blood of patients with tuberculosis are a useful adjunct to the clinical diagnosis in such situations.

In the immunocompromised, listeria meningitis may be seen in the adult and *Cryptococcus neoformans* in all age groups, but particularly in those with human immunodeficiency virus infection. Central nervous system disease in patients with acquired immune deficiency syndrome is a very complicated differential diagnosis (Table 64.4).

Table 64.3 Causes of aseptic meningitis

Viruses	Enteroviruses (echoviruses, polioviruses, coxsackieviruses) Mumps (including post-immunization) Herpes (herpes simplex and varicella-zoster) Arboviruses
Spiral bacteria	Syphilis (*Treponema pallidum*) Leptospira (*Leptospira canicola*)
Other bacteria	Partially treated with antibiotics Tuberculous (*Mycobacterium tuberculosis*) Brain abscess
Fungi	*Cryptococcus neoformans*
Protozoa	*Acanthamoeba, Naegleria, Toxoplasma gondii*
Non-infective	Lymphomas, leukaemias Metastatic and primary neoplasms Collagen-vascular diseases

Table 64.4 Causes of neurological damage in HIV-infected patients

Direct HIV infection	Subacute encephalomyelitis (AIDS–dementia complex)
Opportunist infections Viruses	Cytomegalovirus Herpes simplex Varicella-zoster Papovavirus
Bacteria	*Treponema pallidum* (syphilis)
Fungi	*Cryptococcus neoformans*
Protozoa	*Toxoplasma gondii*

Malignancy
  Primary
  Secondary
	Brain lymphoma
	Kaposi's sarcoma
	Systemic lymphoma

Cerebral infections

Encephalitis may extend into the meninges with signs and CSF findings of an aseptic meningitis. Many viral infections may, however, infect only the brain cortex and clinical symptoms may be vague: loss of consciousness, fits, localized paralysis. This can also occur in toxaemia, cerebral malaria, electrolyte disturbances or vascular accidents.

In western Europe, herpes simplex or varicella-zoster viruses are the most common causes of encephalitis, but in many parts of the world arboviruses such as Japanese B encephalitis virus are important (see Ch. 51).

Abscesses in the brain or subdural space may arise from haematogenous spread during bacteraemia or by direct extension, either through the cribriform plate from the nasopharynx or from sinuses or the middle ear. They may be clinically silent or present as a space-occupying lesion accompanied by fever and systemic upset. Scanning by computed tomography or magnetic resonance imaging and early neurosurgical intervention will reduce complications, and with appropriate systemic antibiotics given for a prolonged period the success rate nowadays is good.

SKIN AND SOFT TISSUE INFECTIONS

Human skin acts as an excellent barrier to infection. Some parasites, such as hookworm larvae and schistosome cercariae, can penetrate skin to initiate infection. This may also be true of some bacteria, notably *Treponema pallidum*, although in primary syphilis the spirochaete probably enters through minute abrasions, which are present even in healthy skin. Primary skin infection such as *impetigo* is due to *Staph. aureus* or *Str. pyogenes*, or both, gaining access to abrasions, usually in children. This may occur in association with ectoparasite infestation, in particular scabies. Dermatophyte fungi are specialized to grow well in keratinized tissue.

Skin lesions are a feature of some virus infections, such as warts, herpes simplex and molluscum contagiosum. In other virus diseases, including rubella, measles, chickenpox (and, before its eradication, smallpox), a characteristic rash follows the viraemic phase of the illness. *Wound infections* may be accidental or postoperative and many organisms can cause sepsis. Even after surgery many wounds are infected with the endogenous flora of the patient. Swabs, or preferably pus, obtained directly from the wound or abscess, are adequate to find the causative organisms. *Anaerobic* sepsis occurs most commonly following amputations or in contaminated traumatic wounds, particularly if the blood supply has been compromised or the bowel perforated. In classical *gas gangrene*, bubbles of gas may be felt in the wound and surrounding tissues, and the muscle and fascia have a black necrotic appearance. Less florid examples of anaerobes causing extensive cellulitis are more commonly seen in situations where deep wounds are contaminated with endogenous flora.

GENITAL TRACT INFECTIONS

In the male, acute urethritis is a common condition that is usually due to a sexually transmitted microbe such as *N. gonorrhoeae*, *Chlamydia trachomatis* or *Ureaplasma urealyticum*. If untreated, these organisms can cause prostatitis or epididymitis, and gonorrhoea may produce unpleasant consequences such as urethral stricture or sterility. Genital ulcers in both sexes may be due to herpes simplex virus, syphilis or chancroid (*H. ducreyi*).

The more complicated female reproductive organs are subjected to many more infections with a greater scope for sequelae. Vaginitis may present as vaginal discharge or irritation, and often these symptoms are due to infections that are not always acquired exogenously. *Trichomonas vaginalis* is the most common sexually acquired microbe, although both *N. gonorrhoeae* and chlamydia may also present as discharge. Thrush due to *Candida* species is especially common in pregnancy and in diabetics. It is usually an endogenous condition due to disturbances in the normal commensal flora. Another cause of vaginal discharge, but usually without inflammatory cells and irritation, is associated with an alteration of local pH with proliferation of *Gardnerella vaginalis* and anaerobic spiral bacteria, now termed *Mobiluncus* species. The alkaline conditions and characteristic amines found in *bacterial vaginosis* allow a diagnosis to be made easily on examination of the patient. This condition, which used to be called non-specific vaginitis, is a common condition in sexually active women, although probably not a venereal disease in the usual sense.

The endocervical canal is the site of infection with *N. gonorrhoeae* and *C. trachomatis* in the sexually mature woman. During parturition both organisms may be passed to the baby's eyes to give rise to *ophthalmia neonatorum*. The cervix may also be infected with human papillomavirus, the cause of genital warts. Some types are associated with a high risk of cervical cancer (see Ch. 45).

Ascending genital infection may present as *acute salpingitis* with fever and pelvic pain. On vaginal examination, there is referred lower abdominal pain on moving the cervix (*cervical excitation*) and tenderness in the iliac fossa on abdominal palpation. Signs and symptoms in cases due to *C. trachomatis* are much less pronounced than in those due to gonococci. Some women may develop chronic *pelvic inflammatory disease* (PID) without having suffered a recognizable acute episode. PID, although most often initiated by these two common sexually transmitted pathogens, is usually a polymicrobial infection, in which endogenous commensals, particularly anaerobes, play an important role.

EYE INFECTIONS

Various microbes may cause acute conjunctivitis. During birth *N. gonorrhoeae* and *C. trachomatis* may be passed to the baby's eyes from the maternal genital tract to give rise to ophthalmia neonatorum. In the newborn, *Staph. aureus* is commonly found in 'sticky eyes', either as a primary cause of conjunctivitis or after infection with another

Table 64.5 Causes of choroidoretinitis

Viruses	Cytomegalovirus, rubella
Bacteria	*Treponema pallidum*
Protozoa	*Toxoplasma gondii*
Helminths	*Toxocara canis, Onchocerca volvulus*

pathogen. In older infants and children, *H. influenzae* and *Str. pneumoniae* are common. Chlamydiae give rise to *trachoma*, the commonest cause of blindness in the world, and to a milder form of inclusion blennorrhoea in sexually active individuals.

Primary viral conjunctivitis often occurs in epidemics, when certain types of adenovirus are implicated. This is usually a mild condition with few sequelae, compared with the keratitis due to herpes simplex virus or in shingles when the ophthalmic division of the trigeminal nerve is infected with varicella-zoster virus.

Corneal damage due to fungi as well as herpesviruses is seen in immunosuppressed patients, and keratitis caused by free-living amoebae (*Acanthamoeba* species), although rare, is becoming more common, particularly in wearers of contact lenses.

Penetrating injuries of the eye and ophthalmic surgery may introduce a wide range of bacteria and fungi into the chambers of the eye that may give rise to *hypopyon* (pus in the eye). This condition requires prompt surgical drainage and instillation of appropriate antibiotics such as gentamicin. *Ps. aeruginosa* and *Proteus* species are among the more common organisms isolated.

Infections of the back of the eye (choroidoretinitis) are seen in many diverse infectious diseases (Table 64.5).

SYSTEMIC AND GENERAL SYNDROMES

Pyrexia of unknown origin

Pyrexia of unknown origin may be defined as a significant fever (greater than 38°C) for a few days without an obvious cause, that is, no apparent infection of an organ or system. In the classical studies of pyrexia of unknown origin, only patients with persistent fever for at least 3 weeks were included. These chronic cases are often due to non-infective causes such as malignancy (especially lymphomata) or auto-immune and connective tissue diseases (such as systemic lupus erythematosus).

In determining an infective aetiology, some of the most important questions to be asked of the patient are:

1. Have you been abroad recently?
2. What is your occupation (especially whether contact with animals is involved)?
3. What immunizations have you had – in particular, have you had BCG?
4. Have you or your family ever had tuberculosis?
5. Are you taking, or have you recently taken, any drugs (especially antibiotics)?

Character of fever

The individual with suspected pyrexia of unknown origin should be admitted to hospital so that measurements can be made regularly by skilled staff. Rarely, malingerers may be found out and drug reactions discovered by controlling intake. Rhythmical fevers such as the quartan fever (every 72 h) of *Plasmodium malariae* or undulant fever of *Brucella melitensis* may rarely be found, pointing to the aetiology. More commonly, fevers are intermittent with rises at the end of the day and falls after rigors or extensive sweating.

The degree of temperature depends also on the host response as well as the pyrogens produced by microbes. Generally, the older the patient the less able they are to mount a pyrexia. Many elderly patients with septicaemia may have normal or subnormal temperatures, whereas infants can have fevers of 40°C and febrile convulsions with otherwise mild respiratory viral infections.

Endocarditis

Infections of the tissue of the heart usually involve damaged valves, either after rheumatic fever or with atheroma. Another important group of patients are those who have had heart surgery, in particular prosthetic valve replacement. In addition, injecting drug users or patients who have had indwelling vascular devices are liable to bacteraemia; occasionally, endocarditis may follow. The most common causative organisms are listed in Table 64.6.

Septicaemia

It is not clinically useful to distinguish between *bacteraemia*, organisms isolated from the bloodstream, *septicaemia*, which is a clinical syndrome, and *endotoxaemia*, which is circulating bacterial endotoxin. The spectrum of clinical disease ranges from hypotensive shock and disseminated intravascular coagulation with a high mortality rate, to transient bacteraemia, which may occur in healthy individuals during dental manipulations.

The vascular compartment is sterile and usually intact. Microbes gain entry from breakages of blood vessels adjacent to skin or mucosal surfaces, or by phagocytic cells carrying organisms into capillaries or the lymphatic system. Active multiplication within the bloodstream probably occurs only terminally, but in many cases of septicaemia high numbers of bacteria are recovered from

Table 64.6 Some causes of infective endocarditis (approximate percentages)

Organism	Non-operative	IDU or surgery
Viridans group of streptococci	70	35
Enterococci	5	3
Other streptococci (group G, F)	10	<1
Staphylococcus epidermidis	10	25
Staph. aureus	5	25
Gram-positive rods (diphtheroids)	<1	5
Haemophilus spp. and other fastidious Gram-negative organisms[a]	<1[b]	<1
Gram-negative bacilli (coliforms, *Pseudomonas* spp.)	0	5
Coxiella burnetii (Q fever)	<1[b]	0
Chlamydia psittaci	<1	0
Fungi (*Candida* spp.)	<1	2

IDU, injecting drug user.

[a]*Neisseria, Brucella, Cardiobacterium, Streptobacillus* spp.

[b]These are rough UK figures; in some parts of the Middle East, brucellae and Q fever cause significant numbers of infective endocarditides.

combined lead to disseminated intravascular coagulation and hypotension. This process becomes irreversible and results in failure of all major organs. Patients die from a variety of terminal events that make up the syndrome of *septic shock*. The main microbial causes are listed with approximate frequency in Table 64.7.

Clinical features may occasionally suggest the aetiological agent, for example the characteristic purpuric rash of meningococcal disease and the black lesions (*ecthyma gangrenosum*) seen on the skin of compromised patients with pseudomonas septicaemia, but in the majority of bacteraemias the agent can be determined only after blood culture. Sometimes, prior antibiotic therapy may render cultures negative and new methods of antigen detection or gene probes may be useful. Non-specific investigations, such as those shown in Figure 64.1, may offer some indication that the cause of the illness is infective. C-reactive protein, an acute-phase protein whose serum concentration is often greatly increased during bacterial infections, may be the most useful of these but, as with peripheral leucocyte counts, there are a significant number of errant results.

blood cultures, which often only sample 10 ml at a time. This occurs from a heavily contaminated site such as an indwelling urinary catheter which releases bacteria into veins on movement. Septic shock may be due to Gram-negative lipids (endotoxins) or Gram-positive toxins (e.g. staphylococcal enterotoxin), which are usually proteins. The end result of both is to initiate a cascade of events involving cytokines, especially tumour necrosis factor and interleukin-2, vascular mediators and platelets, which

Imported infections

An important group of patients with fever are those who have recently returned from abroad. In whichever country a doctor may practise, travellers with unfamiliar disease will be encountered. A knowledge of medical geography is useful but conditions vary greatly within one country and with time. Up-to-date information is held, often on computer, by communicable disease centres and tropical disease hospitals and schools. The World Health

Table 64.7 Major causes of septicaemia (approximate percentages)

Organism	Community acquired	Hospital acquired	Sources and comments
Escherichia coli	35	30	UTI (catheters), biliary tract
Other enterobacteria	5	10	UTI, chest
Pseudomonas spp.	5	5	UTI, immunocompromised
Other Gram-negative rods	2	5	Ventilator pneumonia
Neisseria meningitidis	3	0	Characteristic skin rash
Staphylococcus aureus	30	25	Vascular, postoperative
Staph. epidermidis	<1	20	Vascular devices
Streptococcus pneumoniae	10	0	Pneumonia
Str. pyogenes	5	2	Skin, soft tissue
Other Gram-positive cocci	2	0	Skin
Listeria monocytogenes	<1	0	Bowel, foods
Clostridium spp.	<1	0	Bowel, gangrenous wounds
Bacteroides spp.	3	1	Bowel, pelvis, wounds
Mixed infection	10	7	Bowel, intensive care

Data from Dr P. Ispahani, Nottingham Public Health Laboratory, UK.
UTI, urinary tract infection.

Table 64.8 Some important infective conditions imported to temperate regions from the tropics

Causative organism	Tourists[a]	Expatriates[a]	Immigrants[a]
Viruses	Hepatitis A Influenza	HIV Yellow fever (other arboviruses)	Hepatitis B Haemorrhagic fever
Rickettsiae and chlamydiae	Tick typhus	Q fever	Trachoma
Bacteria	Legionnaires' disease Toxigenic *Escherichia coli*	Brucellosis Shigellosis	Tuberculosis, enteric fever Cholera
Protozoa	Cryptosporidiosis Falciparum malaria Cutaneous leishmaniasis	Giardiasis Malaria (all) Schistosomiasis	Amoebiasis Vivax malaria Kala azar (visceral leishmaniasis)
Helminths	–	Tapeworm Filariasis Stronglyloidiasis	Roundworm Hookworm
Ectoparasites (ticks, mites and insects)	Ticks	Myiasis Jigger flea	Scabies
Fungi	–	Dermatophytosis Histoplasmosis	Mycetoma

[a]These categories are not mutually exclusive.

Organization and the Centers for Disease Control, Atlanta, USA, publish international notification data and maps (see Ch. 69, p. 711 for websites). Undoubtedly the most important condition to diagnose is malaria due to *Plasmodium falciparum*, which may be rapidly fatal without appropriate treatment in the non-immune subject. The wide distribution of drug resistance in *P. falciparum* has led to difficulties in giving adequate prophylaxis and in treating an acute attack. Other common febrile illnesses that are imported into northern countries are typhoid and paratyphoid. These do not usually present with diarrhoea, so the possibility of an enteric fever may not be considered. It is also obvious, but sometimes overlooked, that the fever may not be related to the travel history and that the cause is a microbe that could have been caught at home.

Table 64.8 lists some of the more common causes of infectious diseases imported from tropical countries into temperate regions where the diseases are not normally transmitted. There are three groups of patients that are considered separately, although the separation of diseases and microbes is not exclusive:

1.   Short-term travellers or *tourists* who usually visit major cities or special holiday areas, staying in good accommodation and having minimal contact with the indigenous population.

2.   Long-term visitors who may be engaged in lengthy overland trips or be working abroad as *expatriates*.

Table 64.9 Some common sites of infection in pyrexia of unknown origin

Abdomen	Subphrenic abscess Appendiceal abscess Ileal tuberculosis Pelvic abscess
Liver and biliary tract	Intrahepatic abscess Empyema of gallbladder Ascending cholangitis Cholecystitis Viral hepatitis
Kidney and urinary tract	Perinephric abscess Renal tuberculosis Pyelonephritis (especially children)
Bones	Vertebral osteomyelitis Tuberculosis Prosthetic infections
Cardiovascular	Endocarditis Graft infections
Respiratory	Tuberculosis Empyema and lung abscess
Nervous system	Cryptococcal or tuberculous meningitis Brain or spinal abscess

3. *Immigrants* who were brought up abroad and visit or have residence in the host country; also settled immigrants who pay short-term visits to their country of origin.

Individuals who travel abroad vary in their risk behaviour and their exposure to potential pathogens. Generally, advice given by travel operators, tourist offices, embassies and medical sources has improved greatly in the past few years. Companies sending out expatriate workers tend to look after their staff well. Nevertheless, many tourists (up to 50% in some studies) have episodes of travellers' diarrhoea, which in some cases results in admission to hospital. The major groups who are missed in preventive programmes are overland travellers and immigrants returning to their homeland, often with young families who have never been exposed to the infectious risks of their parents' home. Immigrants returning for visits to malarious areas seldom take prophylactic advice, believing themselves to be immune, but protective immunity wanes with prolonged absence.

Cryptogenic infections

Some of the commonly encountered sites of infection that may give rise to fever, and must be considered in the differential diagnosis of pyrexia of unknown origin, are listed in Table 64.9.

KEY POINTS

- Clinically distinct infective syndromes can generally be caused by several different organisms. Sometimes the specific microbial aetiology may be apparent on clinical grounds alone (e.g. several viral exanthems). More usually a systematic and hierarchical approach is necessary.
- Whether a local or generalized (systemic) process is involved, the medical history, clinical signs (especially temperature) and non-microbiological investigations (e.g. white cell count, inflammatory markers and radiological findings) are often used to determine whether infection should be considered in the differential diagnosis.
- In further establishing the differential diagnosis, the time course of symptoms (acute, subacute, chronic) and potential exposure of the patient to endogenous (e.g. surgery) or exogenous (e.g. travel) sources of infection are critical factors.
- Key localized infective syndromes in which a microbiological diagnosis is commonly attempted include: sore throat; lower respiratory tract infections; intestinal infections; urinary and genital tract infections; meningitis and other central nervous system infections; eye, skin, soft tissue, bone and joint infections.
- Prominent systemic and general syndromes that may demand extensive microbiological investigation include pyrexia of unknown origin, endocarditis and septicaemia.

RECOMMENDED READING

Armstrong D, Cohen J (eds) 1999 *Infectious Diseases*. Mosby, St Louis

Christie A B 1987 *Infectious Diseases: Epidemiology and Clinical Practice*, 4th edn, Vols 1 and 2. Churchill Livingstone, Edinburgh

Conlon C, Snydman D 2000 *Colour Atlas and Text of Infectious Diseases*. Mosby, St Louis

Long S S, Pickering L K, Prober C G (eds) 1997 *Principles and Practice of Paediatric Infectious Diseases*. Churchill Livingstone, New York

Mandell G L, Bennett J E, Dolin R 2004 *Principles and Practice of Infectious Diseases*, 6th edn. Churchill Livingstone, New York

Shanson D C 1998 *Microbiology in Clinical Practice*, 4th edn. Arnold, London

Warrell D A, Firth J D, Benz E J, Weatherall D J 2005 *Oxford Textbook of Medicine*, 4th edn. Oxford University Press, Oxford

65 Diagnostic procedures

R. C. B. Slack

The role of the laboratory in assisting clinicians in the diagnosis of infection is illustrated in the specimen flow diagram shown in Figure 65.1. The choice of specimen depends on following the principles outlined in Chapter 64. The microbiology laboratory requires sufficient information on the request card accompanying the specimen to use the optimal methods necessary for identification of potential pathogens in particular infective syndromes. At the most basic level, it is obvious that, when a swab is received in the laboratory, it is necessary to know whether it comes from the throat or the vagina! But additional information is also essential: is the patient being investigated for pharyngitis or diphtheria; for vaginal discharge or septic abortion? Furthermore, the specimen must be obtained with care and transported to the laboratory without delay in an appropriate manner. The value of the result is in direct proportion to the attention given to these details, as well as to the skill and efficiency of the laboratory. For further details of laboratory methods, including specimen containers and culture media used, the reader is encouraged to consult the recommended reading section.

COLLECTION OF SPECIMENS

Samples for microbiological examination need to be collected carefully, if possible without contamination with commensals or from external sources. Some points to remember with specimens from individual sources are shown in Table 65.1. It is essential to use sterile containers that are leak-proof and able to withstand transportation through the post if necessary. It is more convenient for both the clinician and the microbiologist if the laboratory provides request cards, containers and an efficient transport system. There is a need for staff to be aware of safety regulations and for all parties to understand who has responsibility for each step of the process and how to minimize handling by untrained people. Special precautions required for 'high risk' specimens need to be defined by the laboratory and hospital management.

Storage of clinical material must be separate from food and drugs, and this may necessitate provision of additional refrigerator space and transport facilities.

Food, water and environment

The examination of non-clinical specimens is beyond the scope of this book and readers are referred to appropriate reading. However, an outbreak of gastro-intestinal disease inevitably leads to the question of identifying the source and comparing strains isolated. When disease is due to preformed toxins, as with staphylococcal food poisoning, faecal examination is unhelpful, and the diagnosis can be made only by testing the food. Frequently, the offending item has been discarded and the examination of food and water related to specific patients is often unrewarding. Routine sampling of water sources and potentially contaminated food such as poultry at various critical points of production is essential in maintaining good public health. As part of the investigation of an outbreak it may be necessary to collect samples under controlled conditions that can be used as evidence in any prosecution. Laboratories need to be able to receive and process such specimens in an approved way.

TRANSPORT

Many microbes may perish on transit from the host's body to a laboratory incubator. Some contaminants, especially coliforms, may overgrow the pathogen and so mask its presence. These two constraints make it essential that any material for cultivation of microbes is transported as quickly as possible to the laboratory in a manner expected to protect the viability of any pathogens. Such problems may be minimized by the use of antigen or gene probe detection because of the relative stability of the chemical structures identified.

The ideal situation is to bring the patient to the laboratory for specimen collection, or to take the laboratory to the clinic. Both approaches are used for special purposes

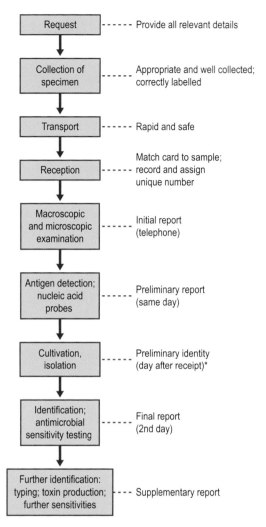

Request	----- Provide all relevant details
Collection of specimen	----- Appropriate and well collected; correctly labelled
Transport	----- Rapid and safe
Reception	----- Match card to sample; record and assign unique number
Macroscopic and microscopic examination	----- Initial report (telephone)
Antigen detection; nucleic acid probes	----- Preliminary report (same day)
Cultivation, isolation	----- Preliminary identity (day after receipt)*
Identification; antimicrobial sensitivity testing	----- Final report (2nd day)
Further identification: typing; toxin production; further sensitivities	--- Supplementary report

*Viruses, fungi and some bacteria may take longer

Fig. 65.1 Steps in the isolation and identification of pathogens from an infected patient.

but are obviously inconvenient for many patients, and inappropriate and costly for complicated techniques such as virus isolation which need specialized (and safe) facilities.

To overcome any drawbacks due to delay in reaching the microbiology department the following methods may be used:

1. *Transport media* (Table 65.2).
2. *Boric acid.* The addition of boric acid to urine at a concentration of 1.8% (v/w) will stop bacterial multiplication, but lower concentrations are ineffective and higher ones may kill the pathogen.

3. *Dip slides*. These provide a convenient way of inoculating urines at the clinic. They comprise small plastic spoons or strips holding a thin layer of agar which is dipped into the urine and then put in a screw-topped bottle for transport. The agar adsorbs a fixed volume of urine and, after incubation, colony counts of bacteria give a semi-quantitative estimation of numbers.
4. *Refrigeration*. Storage at 4°C before processing will prevent multiplication of most bacteria. However, delicate microbes such as neisseriae may not survive whereas certain organisms, notably listeriae, flourish at low temperatures.
5. *Freezing*. Temperatures of −70°C or below, which can be achieved in liquid nitrogen or special deep freezes, will preserve many microbes, provided they are protected by a stabilizing fluid such as serum or glycerol.

RECEPTION

The importance of good documentation cannot be over-stressed. No matter how well the specimen was taken, transported and processed in the laboratory, the end result depends on communication between people. The clinician making the request must give complete details on the request card and specimen to reduce errors. Staff receiving specimens in the laboratory must match them with the cards and record them into a book or computer. This is usually done by assigning a unique number to each specimen and labelling both the specimen and the request. When parts of the specimen are separated from the original bottle (e.g. after centrifugation of serum), the laboratory number becomes the only recognizable identification. Transcription errors are far more common than is supposed and are especially important for requests for human immunodeficiency virus (HIV) or syphilis serology, which may have disastrous consequences and medicolegal implications if wrong results are given.

EXAMINATION

Looking at clinical material with the naked eye or hand-lens should be part of the examination of a patient at the bedside. Sadly, many doctors delegate this to the laboratory staff. Many unnecessary laboratory tests could be avoided if unsuitable specimens were rejected on the ward or in the general practitioner's surgery. These would include: crystal-clear urine from patients with 'cystitis', well formed stools from patients with 'diarrhoea', and mouth washings or saliva from patients with respiratory symptoms.

Table 65.1 Some important points to remember in the collection of specimens for microbiological examination

Respiratory secretions

Nasal swab (anterior)	Only for carriage of staphylococci and streptococci
Nasopharyngeal swab	For pertussis and meningococci
External ear swab	Wide range of microbes, including fungi
Myringotomy and sinus samples	As for abscesses – including anaerobes
Throat (pharyngeal) swab	Specify if only for streptococci; mention if diphtheria possible; use special transport media for virology
Saliva	Used for antibody detection (gingival swab); otherwise discard
Laryngeal swab	Specify if for mycobacteria
Expectorated sputum	Often poorly collected; specify mycobacteria, legionellae, pneumocystis
Transtracheal aspirate, bronchoscopy specimens, lung biopsy	Specify likely diagnosis; ask for specific tests
Pleural fluid	Treat as pus; always look for mycobacteria

Gastro-intestinal specimens

Vomitus	Only for virology
Gastric washings	For mycobacteria (particularly in children)
Gastric biopsy	For *Helicobacter pylori*
Duodenal or jejunal aspirates	Protozoa (*Giardia lamblia*, microsporidia, etc.)
Liver aspirates	As for pus (anaerobes); consider amoebae
Spleen puncture	For *Leishmania* spp.
Rectal biopsy	Schistosomiasis
Rectal swab	Only for gonococci and chlamydia
Colonic biopsy	Histopathological diagnosis of amoebiasis, pseudomembranous colitis (*Clostridium difficile*)
Colonic scrapings	Protozoa; amoebic trophozoites (deliver to laboratory immediately)
Faeces	Specify possible diagnosis; ask for clostridial toxins, parasite examination if suspected
Peri-anal swab	For eggs of threadworm

Urine

Mid-stream (MSU)	Suitable for most patients
Clean catch	Infants and elderly – increased contamination
Suprapubic aspirate	Infants and neonates
Ureteric/bladder washout	To localize infection
Prostatic massage	Collect samples before, during and at end of micturition
Terminal urine	Schistosome ova, chlamydial DNA amplification
Complete early morning or 24-h urine	Mycobacteria (tubercle)

Central nervous system

Cerebrospinal fluid by spinal tap	For meningitis collect sample for protein and glucose – test blood sugar simultaneously, specify virology, fungi or syphilis serology
Ventricular tap	Specify if through an indwelling shunt or catheter
Brain abscess	As for pus (include anaerobes)

Skin and soft tissue

Skin scraping or nail clipping	Dermatophyte fungi
Skin swab	Rarely valuable without pus
Skin snips	Onchocerciasis – seek advice
Vesicle fluid	Suitable for electron microscopy for viruses
Wound swab	Obtain pus if possible; record site
Pus, tissues, aspirates	Describe site and any relevant operative details

Genital

Urethral swab	Pus for gonococci, scrape for chlamydia
Vaginal swab (adult)	Candida, trichomonas, bacterial vaginosis or chlamydia DNA
Vaginal swab (prepubertal)	State age; caution required if abuse possible
Cervical swab	Separate media for chlamydia
Ulcer scrape	Immediate dark-ground microscopy; separate media for virology or chancroid
Uterine secretions	Specify puerperium or post-abortion
Pelvic aspirates	As for pus
Laparoscopy specimens	Include chlamydia specimen

Eye

Conjunctival swab	Separate virology; scrape for chlamydia
Aspirates	As for pus

Table 65.1 Some important points to remember in the collection of specimens for microbiological examination (*cont'd*)

Blood

Culture	Strict aseptic technique; take large sample in special media before antibiotics
Bone marrow	Valuable for leishmania, mycobacteria, brucella
Film	Malaria (thick and thin), filaria, borrelia, trypanosomes
Whole blood (in anticoagulant)	Filaria (day or night samples as appropriate)
Serum antigen	Rapid diagnosis of many microbial diseases (e.g. hepatitis B)
Serum antibody	Retrospective diagnosis of common viral diseases, syphilis and other selected infections; need rising titre or specific IgM

Table 65.2 Types of transport medium

Type of organism	Medium	Comments
Bacteria	Stuart's semi-solid agar	Contains charcoal to inactivate toxic material
Anaerobic bacteria	Various systems, including gassed-out tubes and anaerobic bags	Not widely used, but essential for some strict anaerobes
Viruses	Buffered salts solution containing serum	Contains antibotics to control bacteria and fungi
Chlamydiae	Similar to viral transport medium, but without agents that inhibit chlamydiae	Chlamydial antigen media contain detergent to lyse infected cells
Protozoa, helminths	Merthiolate–iodine–formalin	Kills active protozoa, but preserves cysts and ova in a form suitable for concentration and microscopy

It was once common practice to carry out certain basic investigations in ward side-rooms, and kits offering 'near-patient testing' are becoming available. However, there are cogent arguments against bedside pathology, including issues of safety, time, competence and quality control.

Microscopy

Microscopy is an important part of the examination of many specimens. For bacteriology, the Gram and acid-fast (Ziehl–Neelsen or auramine) stains are usually sufficient, but for the demonstration of fungi or parasites special stains or concentration techniques may be required. 'Wet' mounts (i.e. unstained preparations of fluid material) are used widely when looking at cells in urine, cerebrospinal fluid (CSF), faeces and vaginal secretions. None of these procedures takes more than 5 or 10 min and all are inexpensive in terms of reagent costs and capital equipment. They are therefore ideal rapid methods and new diagnostic techniques have to be judged against microscopy. An initial report, such as 'Gram-negative diplococci and pus cells seen' from meningitic CSF, can be issued within minutes of receiving the specimen and will aid the clinician in confirming the diagnosis and starting appropriate antibiotics.

Similarly, a rapid diagnosis of falciparum malaria can be life-saving. Indeed, suspected pyogenic meningitis or falciparum malaria are among the few conditions for which it is clearly justifiable to call upon emergency laboratory services outside normal working hours.

Electron microscopy requires more elaborate preparation than does light microscopy and is therefore much slower. It is, however, valuable in the relatively rapid diagnosis of certain viral infections, including viral diarrhoea. The non-specific nature of electron microscopy gives it an advantage over virus-specific techniques in that any type of viral agent, if present in sufficient numbers, may be recognized.

Non-culture methods

There are many situations where isolation of microbes in vitro or in tissue culture is impossible or insufficiently sensitive to make a microbiological diagnosis. Some microbes are so fastidious or slow-growing that useful information cannot be given to clinicians. The isolation of many viruses requires laborious tissue culture methods that are too slow to influence patient management. Microscopy is usually of low sensitivity; for example, the threshold of detection of acid-fast bacilli in sputum is about 10^5 microbes per millilitre. Microscopy also has low specificity; for example, Gram staining of faeces would yield millions of Gram-negative rods, but it is not possible to recognize those that are pathogenic by this means.

To overcome these deficiencies probes have been developed that combine a part that reacts with a specific microbial structure and a part that will produce a signal (colour, fluorescence or radio-activity) after the reaction. Many microbes are detectable in this way. Antibodies labelled with fluorescent molecules are widely used in diagnostic virology (e.g. for respiratory secretions to find respiratory syncytial virus). Enzyme-linked antibodies are available as enzyme-linked immunosorbent assay (ELISA) kits for chlamydia detection, and many other *immunoprobes* are commercially produced or under investigation. The explosion in molecular biology has led to the widespread availability of DNA and RNA probes, which have evolved from complicated research techniques requiring radio-isotopes to relatively simple methods that can be carried out with minimal expertise. The use of amplified nucleic acid tests, such as the *polymerase chain reaction* (PCR), has greatly increased the sensitivity of probes. Provided a unique sequence of nucleotides is used, the method is highly specific. None the less, as the new technology becomes more commonplace traditional methods will not be discarded. It is more likely that it will complement them.

One of the chief advantages of probes – their specificity – is also one of their major disadvantages. For example, in the investigation of diarrhoea it would be impossibly laborious to have to use a separate probe for each of the possible microbial causes. Furthermore, use of probes leaves little scope for detection of the unexpected and hampers the discovery of previously unsuspected aetiological agents of disease. The advent of 'chip' technology (*micro-arrays*) may overcome this deficiency.

Serology

In situations in which microbial isolation is impossible and probes are not available, evidence of infection may be obtained by finding a rise in antibody titre or the presence of specific immunoglobulin (Ig) M. Serology is still the most common method of diagnosing the causes of 'atypical' pneumonia (mycoplasmal pneumonia, psittacosis, Legionnaires' disease), syphilis, brucellosis and many viral infections, including HIV. It is preferable to take a blood sample early in the illness (the *acute serum*) and another 10–14 days after the onset (the *convalescent serum*); a four-fold or greater rise in antibody titre in the second specimen is diagnostic of acute infection. With some infections, such as HIV and hepatitis B and C, much longer times must elapse.

Isolation of micro-organisms

The basis of the study of medical microbiology was laid over a century ago by the isolation of microbes in pure culture outside the host animal. The methods used by the fathers of bacteriology have been adapted, simplified and, in some cases, automated for the modern diagnostic laboratory. The principles remain the same: use sterile equipment and media (with cell lines if necessary) and add clinical material. After incubation at 37°C for a variable time, from a few hours for enterobacteria to weeks for mycobacteria and some viruses and fungi, a visible effect will be produced. This might be colonies growing on agar or a cytopathic effect in tissue culture. The skill of microbiology is in identifying the microbes responsible for the effect.

There are limits to the methods that can be used in a routine hospital laboratory to isolate microbes. The choice of media used is dictated by the specimen and by the clinical condition of the patient. For example, in some areas of the world there is little point in looking for *Corynebacterium diphtheriae* in every throat swab, so that the specific media needed are not used routinely. It is therefore incumbent on the doctor to make sure that the laboratory is alerted to look for diphtheria bacilli in any suspicious case. Similarly, clinicians need to know which microbes are routinely sought from particular specimens so that they can make a special request if they suspect the unusual.

'New' causes of illness are often found and new methods of investigating old diseases regularly appear on the market. It is a difficult decision for the laboratory manager to assess at what point a 'new' pathogen becomes sufficiently important to be sought routinely, or to balance the advantages of new (and expensive) technology against cheap and well tried techniques.

Identification

The full identification of each microbe isolated in a clinical laboratory is both uneconomic and unnecessary. Short-cuts must be made to satisfy the clinical demand of a final report that the doctor can understand and that is available in time to influence management of the patient. In practice, most laboratories use simple and incomplete methods of identification, depending on the level of useful information required. Typing of isolates is for epidemiological or other special reasons, and this is usually done in national reference centres using standardized methods.

Examples of the extent of identification are shown in Table 65.3. This shows that the same organism may not be identified even to the genus level, or it may have extensive genetic investigation depending on the reason for the request. In a cost-conscious climate you get what you pay for and what the service thinks you need!

Table 65.3 Examples of bacterial identification in medical laboratories

Reason for request	Extent of identification	
	Example 1	Example 2
Test of sterility	Bacteria present	
Initial blood culture report	Gram-negative rod	Gram-positive cocci
Urine examination	'Coliforms'	Staphylococci
Wound swab	*Escherichia coli*	*Staphylococcus aureus*
Outbreak epidemiology	*Esch. coli* O157	*Staph. aureus* phage type 80/81
Pathogenicity tests	*Esch. coli* O157 Vero toxin-producing	*Staph. aureus* enterotoxin A-positive

Antimicrobial sensitivity testing

One advantage of good bacterial identification is that it helps greatly in choosing antibiotics, and in some situations in-vitro tests of susceptibility may be unnecessary. However, the widespread occurrence of bacterial resistance, even in genera such as *Neisseria* and *Haemophilus* in which sensitivity to β-lactam antibiotics was previously assumed, has meant that tests are usually performed on all significant clinical isolates of bacteria. The relevance of information generated in the extremely artificial conditions of the laboratory is discussed in Chapter 66.

As there are technical problems in carrying out sensitivity tests at the same time as the primary culture, the report will be delayed for at least 24 h after isolation of the pathogen. Tests of slow-growing bacteria such as mycobacteria take much longer and tests involving fungi, viruses and protozoa are not ordinarily available. In general, all patients with acute symptoms will receive treatment before the report returns to the doctor so that the result will merely confirm that correct treatment had been chosen. Sometimes, the laboratory report will enable speculative antimicrobial therapy to be modified, for instance by allowing one or more drugs in a precautionary mixture to be discontinued.

A frequent difficulty facing microbiologists is which, out of a rapidly increasing number of agents, should be tested and, of those tested, which should be reported. Most laboratories test a few representative compounds from among the many penicillins, cephalosporins, aminoglycosides, tetracyclines, quinolones, etc., and restrict those reported to the clinician to two or three agents selected for their appropriateness in the particular infection. In this way, institutional antibiotic policies are reinforced and the impact of the promotional activities of pharmaceutical companies is lessened.

REPORTS OF RESULTS

Just as it is important to obtain as accurate a result as possible, so it is vital to transfer the information rapidly to the user of the laboratory in a form that is easily intelligible. Laboratory workers need to tailor their reports to suit their customers and to recognize their needs. Trained microbiologists are used to the vagaries of microbial taxonomy and the profusion of antibiotics with similar names, but physicians and surgeons are easily confused by laboratory jargon and changes in nomenclature. On the other hand, clinicians rightly expect laboratory reports to be explicit and helpful. As in so many spheres of health care, a spirit of mutual respect and co-operation is in the best interests of the patients, and this is most likely to happen if microbiologists regularly visit the wards and clinicians are encouraged to discuss problems with laboratory staff.

In these days of computers and facsimile machines it is attractive to use the latest technology to transfer results, but the personal touch is essential, particularly for important results such as positive blood or CSF cultures. Clinicians dealing with difficult problems enjoy the reassurance that the laboratory has found the cause of a patient's illness, and advice on management is generally well received. Seeing the problem for oneself and discussing it with the responsible doctors is the ideal. The telephone is a satisfactory substitute in many cases and is far preferable to a report form printed by a computer and arriving after the crisis is over. It behoves us all to communicate better and faster.

NOTIFICATION OF INFECTIOUS DISEASES

In addition to making a clinical diagnosis and treating the individual there is a need with some infectious diseases to determine the source and prevent further spread. In some countries there are public health laws specifying the method of reporting these. The list of *notifiable diseases* in England and Wales is shown in Box 65.1. This reporting system requires a clinical diagnosis, although laboratory confirmation may be sought separately. In England and Wales each health district has a consultant responsible for communicable disease control, who is usually the 'proper officer' for the local government authority. Notifications are sent by this officer to the Health Protection Agency (HPA),

BOX 65.1 List of notifiable diseases in England and Wales

Acute encephalitis	Paratyphoid fever
Acute poliomyelitis	Plague
Anthrax	Rabies
Cholera	Relapsing fever
Diphtheria	Rubella
Dysentery (amoebic or bacillary)	Scarlet fever
Food poisoning	Smallpox
Leprosy	Tetanus
Leptospirosis	Tuberculosis
Malaria	Typhoid fever
Measles	Typhus
Meningitis	Viral haemorrhagic
Meningococcal septicaemia	fever
(without meningitis)	Viral hepatitis
Mumps	Whooping cough
Ophthalmia neonatorum	Yellow fever

formerly the Communicable Disease Surveillance Centre (CDSC).

There is also an obligation, which in some circumstances may be statutory, for laboratories to report isolates of certain pathogens to central authorities for surveillance purposes. For HIV/AIDS the system is voluntary and confidential, yet more than 90% of cases are reported to the HPA. In an outbreak, or where the pathogen is highly infectious, it is essential for the head of the laboratory to inform both the clinician looking after the patient and local communicable disease control staff. In this way a co-ordinated approach can be made to prevent further spread of infection.

KEY POINTS

- The microbiology laboratory will analyse clinical specimens in a manner directed to supporting the appropriate management of infections from the clinical and the public health perspective. Effort is directed towards timely reporting relevant to the clinical situation but analytical approaches are constrained by cost.
- Selection of the appropriate specimen may require specialist advice from the laboratory. A direct sample from the infected area or secretions from it are preferred; swabs, aspirates, blood samples and scrapings are commonly analysed.
- Careful attention to providing the laboratory with the information required to select the appropriate analysis, labelling the sample and transporting it to the laboratory in a timely fashion or preserving its content will minimize wasted effort on behalf of both the clinician and the laboratory.
- Direct, non-culture-dependent analyses, including microscopy, can be rapid and can support management in urgent situations. Microscopy following Gram or acid-fast stains is universally available and may be highly informative. Availability of specific antigen and nucleic acid detection will depend on local facilities. Both may provide rapid results.
- Detection of the immune response to specific infections is generally achieved by detecting antibodies (serology). Sampling, analysis and interpretation are critically dependent on the time course of infection.
- Isolation of micro-organisms is often attempted and provides a highly sensitive and specific means of determining the cause of many infective syndromes. Although isolation is time consuming, it facilitates determination of antimicrobial susceptibilities and typing of isolates.
- Because of their public health significance, the reporting of certain *notifiable diseases* to the appropriate authorities is a legal requirement.

RECOMMENDED READING

American Society for Microbiology 2003 *Manual of Clinical Microbiology*, 8th edn. American Society for Microbiology, Washington, DC
Collee J G, Fraser A G, Marmion B P, Simmons A 1996 *Practical Medical Microbiology*, 14th edn. Churchill Livingstone, Edinburgh

Forbes B A, Sahm D F, Weissfeld A S 2002 *Bailey and Scott's Diagnostic Microbiology*, 10th edn. Mosby, St Louis
Johnson F B 1990 Transport of viral specimens. *Clinical Microbiology Reviews* 3: 120–131

66 Strategy of antimicrobial chemotherapy

R. C. B. Slack

In the previous chapter the work of the laboratory in assisting doctors in making a definitive microbiological diagnosis was described and summarized in Figure 65.1. Ideally, management of an infection should involve these stages plus the choice of an appropriate antimicrobial agent (if necessary) and follow-up tests of cure. Chapter 5 listed the different classes of antimicrobial agents from which the doctor may prescribe. The objectives of an antibiotic strategy are to implement clinical guidelines that cover the treatment of an individual patient and policies based on these which will have a maximum public health effect. These include cost-effectiveness and limiting the spread of antibiotic resistance.

CHOICE OF TREATMENT

Infections are so common in general practice that it is not practical for each patient to be fully investigated and a rational choice of antimicrobial agent made on the results. Because many have a viral aetiology and most respond rapidly to simple symptomatic relief, the choice is not which agent to use but whether to prescribe at all. Many patients who make the effort to see a doctor expect drug treatment. Although the time taken to educate the public and convince the individual is often longer than that needed to write a prescription, it is important that prescribers do not yield to pressure when antibiotics are not indicated. Figure 66.1 shows a flow diagram of choice of antibiotics that is applicable to many common conditions. There are three main points in the diagram:

1. Patients may be categorized on clinical grounds into those with mild, moderate or severe infection, and also into those who were previously healthy or have underlying disease that may affect their response to therapy, such as impaired immunity or the presence of a foreign body. In the latter situation it is more important to send specimens to the laboratory and start empirical therapy to cover likely pathogens.

2. The organisms that cause many common clinical syndromes have predictable antibiotic sensitivities, allowing a rational choice to be made.
3. The results of laboratory tests may lead to modification or withdrawal of chemotherapy.

Although 'best guess' or empirical therapy is commenced before the results of laboratory tests are available, it is wrong to argue that this approach is incorrect as early administration of an antimicrobial agent is sometimes life-saving.

LABORATORY MONITORING

Susceptibility tests

As well as providing routine tests of sensitivity on individual isolates, the laboratory should provide a general picture of the prevalence of different pathogens occurring in the local population and their pattern of resistance. Different lists should be prepared and circulated regularly to general practitioners, hospital consultants and specialized units. The methods used and limitations of susceptibility testing are outlined in Chapter 5. Sometimes individual patients need to be monitored for the efficacy of treatment and to determine whether failure has occurred owing to acquired resistance or replacement with a more resistant organism. This is especially important in the assessment of new agents in clinical trials.

Antibiotic assay

Rapid methods are now available to determine the amount of antibiotic in serum and other body fluids during treatment. These assays can be used to monitor the adequacy of dosage or to avoid possible toxic effects of overdosage.

In practice, it is necessary to carry out such assays only in a small minority of patients in hospital: patients who are seriously ill with infections such as endocarditis, or those receiving potentially toxic agents such as aminoglycosides and vancomycin. Urine samples may be

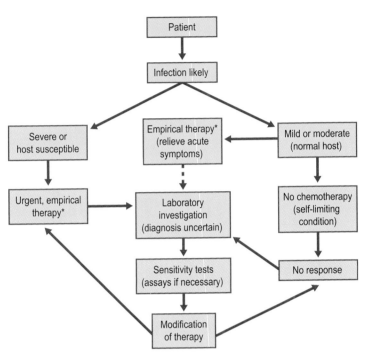

Fig. 66.1 General strategy of antimicrobial chemotherapy.

*Choice based on predictable sensitivity or local resistance patterns.
Broken line indicates a course of action that will depend on circumstances

screened for antibacterial activity. This can be of value in determining patient compliance especially for tuberculosis therapy.

The interpretation of the results of antibiotic assays may be difficult if the patient is receiving multiple antibiotics. In such cases, the power of the body fluid to inhibit or kill the infecting organism in vitro may be a more useful measurement of adequate dosage; for example, in bacterial endocarditis, the patient's serum should kill the causative organism at a dilution of 1 in 4 or more.

It is important to know the relationship of the sample being assayed to the time of dosage: maximum or *peak* levels are usually measured in serum 0.5–1 h after a dose; minimum or *trough* levels are from a sample taken shortly before the next dose.

Assays should always be carried out in close collaboration with the microbiologist to ensure the correct timing of samples and to avoid errors of interpretation.

Clinical correlation of laboratory tests

The epithet *sensitive* or *resistant* is a clinical description; in laboratory tests it is usual to select a concentration of the agent that is known to be easily attained in serum,

other body fluids or tissue after normal recommended dosage. Organisms susceptible to this, or lower concentrations, are regarded as sensitive; those able to grow are resistant. These terms relate to the *minimum inhibitory concentration* (MIC) of the microbe (see p. 65). The correlation between this interpretation and clinical results is generally good, but not perfect. Discrepancies may be due to a failure to attain adequate concentrations at the site of infection, resulting in clinical failure. However, there are many factors on which the outcome of the host–parasite relationship in infection depends, and it would be surprising if a complete correlation between in-vitro tests and the results of chemotherapy was observed. The best correlation occurs with penicillin treatment of acute gonorrhoea in males, where outcome is easy to assess and single-dose therapy was standard.

PREDICTABLE SENSITIVITY

Once the identity of an infecting organism is known, the sensitivity pattern is often predictable with a fair degree of accuracy. Not surprisingly, however, the susceptibility of common pathogens may change with time. Thus, pre-

Table 66.1 Examples of organisms and antimicrobial agents for which susceptibility is usually predictable

Organism	Antimicrobial agents normally active
Streptococci	Penicillin, erythromycin, vancomycin
Enterococci	Ampicillin, vancomycin
Anaerobic cocci	Penicillin, erythromycin, metronidazole
Staphylococci	Flucloxacillin, clindamycin, fusidic acid, gentamicin, vancomycin
Haemophilus influenzae	Co-amoxiclav, tetracycline, cefotaxime, erythromycin
Escherichia coli *Proteus mirabilis*	Gentamicin, cefotaxime, co-amoxiclav, ciprofloxacin
Pseudomonas aeruginosa	Gentamicin, ceftazidime, azlocillin, ciprofloxacin, imipenem
Bacteroides fragilis	Metronidazole, cefoxitin, co-amoxiclav
Rickettsiae Chlamydiae	Tetracyclines, macrolides, rifampicin
Mycoplasmas	Tetracyclines, erythromycin
Candida albicans	Nystatin, amphotericin B, azoles
Herpes simplex virus	Aciclovir

Table 66.2 Examples of organisms and antimicrobial agents for which resistance is usually predictable

Organism	Antimicrobial agents **NOT** normally active
Streptococci Enterococci	Aminoglycosides, nalidixic acid, aztreonam
Staphylococci	Most penicillins[a], nalidixic acid, aztreonam
Escherichia coli	Penicillin, vancomycin, erythromycin, metronidazole
Klebsiella spp.	Most penicillins, erythromycin, metronidazole
Pseudomonas aeruginosa	Most penicillins and cephalosporins, erythromycin, metronidazole
Anaerobes	Aminoglycosides, aztreonam, nalidixic acid, fluoroquinolones

[a]Except methicillin and isoxazolylpenicillins.

dictable sensitivity of micro-organisms, although useful, is a changing concept and shows local variations, so that the advice of local microbiologists should be sought.

Organisms with predictable sensitivity patterns

Streptococcus pyogenes is a good example of an organism that has retained a sensitivity pattern that has changed little in 60 years. Thus, penicillin is always the drug of choice for the treatment of haemolytic streptococcal infections. Other β-lactam antibiotics are also active against *Str. pyogenes*. Erythromycin is an alternative if the patient cannot be given penicillin because of allergy, but an increasing number of strains in some areas show some resistance to this antibiotic in vitro.

Until recently, most strains of *Str. pneumoniae* were highly sensitive to penicillin. An increasing number of penicillin-resistant pneumococci are now found in cases of invasive disease in some institutions, and are posing a major problem worldwide.

Reproducible patterns of sensitivity to antimicrobial drugs can be shown for many other groups of bacteria and offer a useful guide to the choice of chemotherapy (Table 66.1). However, exceptions to these patterns occur and resistant variants may become prevalent in some localities, particularly under the pressure of intensive antibiotic usage.

In the same way that organisms may exhibit predictable sensitivity to some agents, they may be predictably resistant. Some examples are shown in Table 66.2.

ORGANISMS OF VARIABLE SENSITIVITY

Many groups of pathogenic bacteria, notably staphylococci and enterobacteria, vary unpredictably in their sensitivity to antimicrobial drugs. With these organisms there is a greater need to carry out sensitivity tests on individual isolates.

Staphylococci

Staphylococci, including coagulase-negative strains, show great variability in susceptibility, and this can severely limit choice of antibiotics. Four main kinds of staphylococci may be encountered:

1. *Penicillin-sensitive strains.* Before the widespread use of penicillin in the 1940s, most *Staphylococcus aureus* isolates were sensitive. Usually these strains are susceptible to other antistaphylococcal agents (macrolides, aminoglycosides and fusidic acid). Penicillin-sensitive strains now account for about 10% of isolates.

2. β-*Lactamase-producing strains.* Staphylococcal penicillinase confers resistance to all penicillins except methicillin, the isoxazolyl group (cloxacillin, etc.) and nafcillin. These strains usually retain susceptibility to cephalosporins. Most *Staph. aureus* and *Staph. epidermidis* strains found in clinical practice belong to this group.

3. *Methicillin-resistant strains* (MRSA). Methicillin, the first penicillinase-stable penicillin, is used as the test compound to identify these strains in the laboratory. Strains of *Staph. aureus* resistant to methicillin are also resistant to virtually all β-lactam agents. MRSA have epidemiological significance as a group and have been isolated from hospital outbreaks in different countries. Many of these epidemic strains are multiresistant,

exhibiting resistance to aminoglycosides, macrolides and other antistaphylococcal agents, including the topical compound, mupirocin, which has been used to eradicate the organism from carriers of MRSA. Vancomycin, teicoplanin or linezolid may be the only agents available for treatment.

4. *Other antibiotic-resistant strains.* Multiple resistance to antistaphylococcal antibiotics may be seen in strains that are methicillin sensitive. They are found in similar situations to MRSA, namely hospital units with severely debilitated patients and problems with cross-infection. Such resistance patterns are more common in coagulase-negative staphylococci.

Enterobacteria

Escherichia coli includes many strains of variable sensitivity, but among the general population in the UK half of the strains isolated from urinary tract infections are sensitive to ampicillin and amoxicillin, and an even higher proportion are sensitive to cephalosporins, co-amoxiclav (the combination of amoxicillin and clavulanic acid) and trimethoprim.

Variable sensitivity is also a feature of other enterobacteria. There has been an increase in *Esch. coli* resistant to expanded-spectrum β-lactams in many countries. Where these organisms are also quinolone and trimethoprim resistant, treatment with oral agents is limited. Multiresistant typhoid is a particular problem in many countries where the disease is endemic. If there is doubt about the clinical response, or the sensitivity of the strain, laboratory tests should be carried out.

Other organisms

Several pathogens that were formerly reliably sensitive to standard agents, including neisseriae, *Haemophilus influenzae*, pneumococci, enterococci, mycobacteria and *Plasmodium falciparum*, are now commonly resistant to first-line therapy. In some cases, options for treatment of infection with these organisms have become severely constrained.

ANTIBIOTIC POLICIES AND THE CONTROL OF RESISTANCE

The benefits of the use of antibiotics, not only in medicine but also in animal husbandry, must be taken into account in considering the merits of restrictive policies undertaken to control antibiotic resistance.

There is, without doubt, a problem of resistance among many commonly occurring micro-organisms. However, most of the problems of resistance occur among patients within closed communities such as hospitals. This is most marked in areas of intensive care, where large amounts of antibiotics are used in highly susceptible patients with low immunity to infection. In the general community, multiple resistance to antibiotics is presently not a major problem in the UK, but the spread of MRSA out of hospitals into nursing and residential homes may herald a change in this situation.

Accordingly, attention should be directed to those hospital units in which the problem of resistance is significant and which may be the most important source of spread to the general community. As resistance in bacteria has many of the features of an epidemic disease, the application of classical rules and precautions for the prevention of infection may halt this spread.

Restrictive policies

The use of antibiotics by general practitioners and clinicians is influenced by the prescribing habits of other medical colleagues and by the promotional efforts of the pharmaceutical companies. Thus, an ordered and systematic use of antimicrobial drugs might be of benefit in the general strategy of chemotherapy. As the use of antibiotics acts as a powerful selective factor in the emergence and spread of resistant micro-organisms, restriction of use should have the opposite effect. The acceptance of this thesis has encouraged restrictive policies involving temporary bans on the use of certain antibiotics. Such policies may also lead to a reduced chance of prescription error and some cost benefits.

Rotational policies

Periodic changes of antibiotics used in treatment might also help to avoid the emergence of resistant strains by altering the selective pressures and discouraging opportunistic pathogens such as *Pseudomonas* and *Acinetobacter* species. Such a rotational policy might help to retain the therapeutic value of antibiotics over a longer period. The availability of a large range of clinically useful antibacterial substances (Table 66.3) makes this kind of policy more practicable.

PROPHYLACTIC USE OF ANTIBIOTICS

The unnecessary prophylactic use of antibiotics should be discouraged as this may result in increased selection of resistant variants or superinfection with resistant flora. However, there are several circumstances in which chemoprophylaxis is clearly beneficial.

A widely accepted use of antimicrobial agents for the prevention of infection is in patients who have suffered from

Table 66.3   Main groups of antibiotics available for clinical use

Antibacterial agent	Route of administration	Antibacterial spectrum[a]	Some indications for therapy
Phenoxymethylpenicillin	O	+	Streptococcal infections
Benzylpenicillin	P	+	Community-acquired pneumonias or meningitis
Ampicillin Amoxicillin	O, P	+ and –	Respiratory, hepatic, biliary infections
Flucloxacillin	O, P	+	Staphylococcal infections
Co-amoxiclav	O, P	+ and –	Infections due to β-lactamase-producing organisms
Imipenem	P	+ and –	Serious infections with mixed organisms
Cephalosporins	O, P	+ and –	Respiratory and urinary infections; staphylococcal infections
Aminoglycosides	P	–	Gram-negative coliform infections; endocarditis (with penicillin); pseudomonas infections
Macrolides	O, P	+	Streptococcal and staphylococcal infections and those due to legionellae, campylobacters, chlamydiae and coxiellae
Tetracyclines	O, P	+ and –	Respiratory infections and those due to chlamydiae, mycoplasmas and rickettsiae
Chloramphenicol	O, P	+ and –	Serious infections due to *Haemophilus influenzae*, salmonellae and rickettsiae
Fusidic acid	O, T	+	Staphylococcal infections
Glycopeptides	O, P	+	Resistant staphylococcal infections; *Clostridium difficile* enteritis
Trimethoprim Co-trimoxazole	O, P	+ and –	Urinary and respiratory infections; salmonellosis
Imidazoles	O, P	+ and –	Anaerobic infections
Sulphonamides	O, T	+ and –	Intestinal and eye infections
Fluoroquinolones	O, P	+ and –	Urinary infections; pseudomonal infections; salmonellosis

O, oral; P, parenteral; T, topical; +, active mainly against Gram-positive bacteria; –, active mainly against Gram-negative bacteria.
[a]See also Table 5.1, p. 53.

rheumatic fever, or are thought otherwise to be at risk of rheumatic carditis, to prevent further streptococcal infection. Among other indications for chemoprophylaxis are the use of peri-operative antibiotics in patients undergoing joint replacement to reduce the chance of potentially disastrous infection. Similarly, patients undergoing lower bowel resection receive peri-operative treatment with combinations of agents intended to suppress the lower bowel flora. The clinical and laboratory evidence for the benefit of these kinds of prophylaxis is now well established.

Chemoprophylaxis is also justified in preventing pneumococcal infection in splenectomized patients and pneumocystis pneumonia in people with human immunodeficiency virus (HIV) infection. In healthy individuals who are close contacts of people with infectious diseases such as meningococcal meningitis, short courses of antibiotics are justified to prevent acquisition of the organism or to eradicate carrier status. Travellers to regions in which malaria is endemic should take antimalarial prophylaxis.

HOST FACTORS INFLUENCING RESPONSE

In most individuals with normal immune systems an antimicrobial drug assists in a more rapid recovery from infection, but the choice of agent is often not critical providing it has some effect on the causative organisms. In very serious infections or when the patient is at a disadvantage because of impaired or absent immunity, the role of the antibiotic becomes more important and more care should be taken in its selection, dosage and administration.

Problems of toxicity

The final choice of the most appropriate antimicrobial drug may be influenced by the history and response of the patient. Known hypersensitivity to a drug such as penicillin means that neither this antibiotic nor other penicillins can be given safely, and alternatives must be used. Side effects such as nausea, vomiting, diarrhoea or pruritus may be severe enough to warrant a change in treatment. Some antibiotics have potentially serious side effects, such as the ototoxicity of the aminoglycosides, and care must be taken to avoid these by laboratory monitoring of drug levels, as such toxicity is usually dose related. Intercurrent disease may require modification of the dosage of certain antibiotics because of specific organ deficiency, such as liver failure or impaired renal function. Ototoxicity of the aminoglycosides must always be borne in mind in patients who have poor renal function.

Failure to reach the site of infection

The absorption of oral antibiotics varies widely from patient to patient and, if poor, may be a cause of treatment failure. Alternatively, infection may be localized within a large collection of pus or at an anatomical site that is penetrated by antibiotics with difficulty. Obstruction to the flow of body fluids may likewise militate against success; this is important in infections of the urinary tract, biliary system and central nervous system. In some of these cases more radical interference may be indicated to relieve the obstruction.

Alteration of normal flora

Antibiotic therapy may upset the patient's microflora and result not only in the selection of resistant strains of commensals, such as staphylococci on the skin or *Esch. coli* in the intestinal tract, but also colonization with species not normally present. This can lead to antibiotic-induced infection, such as candidosis. Diarrhoea is frequently associated with antibiotic therapy and reflects disturbance of the normal bowel flora. In a few patients the clinical condition can be severe and proceed to pseudomembranous colitis associated with the toxins of *Clostridium difficile* (see p. 249).

Intravenous administration

When antibiotics are given intravenously, they should normally be administered directly into the vein and not added to other infusion fluids; otherwise adequate blood levels of the drug may not be attained owing to excretion outpacing administration. There is also the possibility of incompatibility between the antibiotic and the contents of the fluid. In all cases, the manufacturer's instructions and recommendations about the administration of the drug should be followed closely.

COMBINATIONS OF ANTIBIOTICS

The clinical benefits of the use of combinations of antibacterial agents tend to be exaggerated. The combination of two or more agents has been long accepted in the treatment of tuberculosis, so limiting the selection of mutants resistant to the individual components. The use of β-lactam antibiotics with an aminoglycoside in the treatment of streptococcal endocarditis is also accepted, as the mixture is more bactericidal than the individual components.

A combination of a β-lactam antibiotic with a β-lactamase inhibitor may prevent destruction of the antibiotic. Thus, the enzyme inhibitor clavulanic acid in combination with amoxicillin (co-amoxiclav) restores the activity of the antibiotic against many β-lactamase-producing bacteria.

Potentiation of the antibacterial effect by combinations is referred to as *synergy*; some combinations exhibit a lesser effect than the individual components, and this is called *antagonism*. These interactions are generally displayed in vitro, and it is difficult to establish evidence of advantages or disadvantages in the patient. Thus, the combination of trimethoprim and sulfamethoxazole (co-trimoxazole) can be shown to be synergistic in the test tube, but it has been difficult to demonstrate any clinical benefit, and trimethoprim is now often used on its own to avoid the chance of toxic reactions to sulphonamides.

ANTIVIRAL THERAPY

An increasing number of agents are being used in the prophylaxis or treatment of viral infections (see pp. 61–64).

Aciclovir is the most widely used agent at present, and is prescribed for the treatment of herpes simplex and herpes zoster. This drug can be given orally, but in severely ill patients it is administered intravenously, for example in treating herpes encephalitis. Aciclovir is also used in the prophylaxis and treatment of varicella, particularly in immunocompromised patients. The related compound ganciclovir is available for the treatment of cytomegalovirus infection, but is associated with considerable toxicity.

Amantadine is active against influenza A virus, but not against influenza B virus. This drug, and the related rimantadine, have been used chiefly for the prophylaxis of influenza. The new agents, zanamivir and oseltamivir, are active against both A and B viruses and have been used to modify infection and prevent spread of influenza.

Ribavirin is useful in the treatment of respiratory syncytial virus infections in children, hepatitis B and C (in combination with interferon) and some haemorrhagic fever viruses, such as Lassa fever.

Zidovudine (azidothymidine) is an inhibitor of HIV and was the first drug used in the management of patients with acquired immune deficiency syndrome. There are now a number of other antiretroviral agents used in combination (see Ch. 5).

ANTIFUNGAL THERAPY

Superficial fungal infections are very common. Systematic fungal disease is relatively rare in the UK, except in patients who are immunosuppressed or otherwise compromised. There are few effective antifungal agents that can be used systemically (see Table 5.4, p. 61 and Ch. 60) and some have considerable toxicity.

For many years the chemotherapy of superficial fungal infections depended on preparations of benzoic, salicylic or undecanoic acid. Griseofulvin, given orally, sometimes over long periods, is more effective for the treatment of dermatophytoses; the allylamine, terbinafine, is at least as effective in these conditions. Nystatin and other polyenes are used for the topical treatment of superficial candidosis (thrush). Imidazoles such as clotrimazole are also used in the treatment of vaginal yeast infections.

Amphotericin B is used parenterally for the treatment of systemic candidosis, cryptococccal infections and aspergillosis. Flucytosine is also active against yeasts and has been used in combination with amphotericin B. Amphotericin B is toxic and treatment has to be monitored carefully to obtain the most satisfactory clinical results. Liposomal formulations of the drug appear to be safer and more effective.

Ketoconazole, and triazoles such as itraconazole and fluconazole, also have a fairly wide range of antifungal activity and are being used more widely in the treatment of systemic fungal disease because of their relative lack of toxicity, although the development of resistance may be a problem.

CHEMOTHERAPY OF SYSTEMIC INFECTIONS

Serious generalized infection

Patients with serious and often overwhelming generalized infection include those with *bacterial shock syndrome* and others with the symptomatology of acute infection without localizing signs. These syndromes may be related to postoperative infection, to instrumentation, or to the aggravation of a previously mild infection (e.g. extension of a middle ear infection to the brain). Generalized infection may also follow a reduction in the natural resistance of the patient. Such a situation may arise in renal failure, immune deficiency states, blood dyscrasias and neoplastic disease. Specific measures used in the treatment of these diseases may further reduce resistance to infection and the clinical features characteristic of infection may be muted or absent.

Appropriate specimens should be submitted to the laboratory before treatment is begun, in an effort to isolate the causative organisms. However, these patients may require immediate treatment and often there is little indication of the nature of the infecting organism; more than one species may be involved, especially when the source of the infection is within the abdomen. Empirical chemotherapy must therefore cover Gram-positive cocci, Gram-negative bacilli and anaerobes such as bacteroides. A combination of amoxicillin, an aminoglycoside and metronidazole is suitable. Co-amoxiclav improves the spectrum of amoxicillin and allows omission of the metronidazole. Alternatively, an expanded-spectrum cephalosporin such as cefotaxime can be used. Therapy may be subsequently changed according to the results of laboratory tests. If the patient fails to respond after treatment for 3 days, the use of alternative agents should be considered.

In some patients in intensive care and in the immunocompromised, the possibility of systematic fungal infection or activation of dormant viruses, such as cytomegalovirus, must be considered.

It must always be remembered that bacterial toxins are unaffected by antibacterial therapy. Supportive treatment will include correction of fluid and electrolyte imbalance and appropriate treatment of any co-existing organ failure. In future, immunotherapy such as the use of anti-endotoxins and drugs that alter the cytokine and clotting cascades may find a place in treatment.

Infective endocarditis

Whenever possible, treatment of endocarditis should be related to the results of sensitivity tests made on the organism isolated from the blood. Formerly, viridans streptococci were the most common isolates and penicillin was the drug of choice. Viridans streptococci are now isolated from only about one-third of patients and the organisms from the remainder are usually relatively resistant to penicillin. Bactericidal therapy with a penicillin in combination with an aminoglycoside is usually indicated. It is sometimes helpful to carry out bactericidal tests against the causative organism and it may also be useful to monitor the progress of the patient by assays of the bactericidal activity of the serum against the causative organism.

After operations for the replacement of heart valves it may be difficult to isolate an organism and there may be doubt as to whether or not infection has become superimposed. The risk of withholding treatment, however, is so great that empirical treatment may be indicated, and as staphylococcal infection may occur in such circumstances it is best to include an antistaphylococcal antibiotic. Injecting drug users are also prone to staphylococcal endocarditis.

For the prevention of bacterial endocarditis in patients at risk after dental extraction, 3 g amoxicillin, given orally 1 h before surgery, is recommended.

Rare causes of infective endocarditis in which blood cultures are negative include infections with *Coxiella burnetti* or certain chlamydiae. The diagnosis of these conditions depends largely on serological evidence as it is difficult to isolate the organisms. An aetiological diagnosis is important, as treatment with an appropriate antibiotic, usually tetracycline, may be life-saving.

Urinary tract infections

Uncomplicated cystitis

Uncomplicated urinary infections and asymptomatic bacteriuria are common in adolescent and adult women seen in general practice and maternity clinics. Most infections are due to *Esch. coli*, and most are sensitive to a wide range of drugs. In most cases eradication of the organism is achieved after a short course of therapy with an oral agent such as trimethoprim. The remainder will often respond to a second course of treatment, but if the bacteriuria persists fuller urological investigation is required as failure of treatment is more usually related to an abnormality of the urinary system than to resistance of the organism.

Infections following catheterization

Infection of the urinary tract sometimes occurs after instrumentation such as catheterization or cystoscopy; it is almost unavoidable if indwelling catheters are used. Often the strains causing these infections are derived from the hospital environment, and include resistant strains of Gram-negative bacilli, sometimes in mixed culture. Frequently, the patient is not greatly inconvenienced by such infection, and chemotherapy is not indicated in most areas. If therapy is required, elimination of the organisms can be difficult, particularly if there is residual pathology or a degree of urinary obstruction. Sensitivity tests must be carried out to assist in the choice of therapy. The most useful agents are cephalosporins, trimethoprim and ciprofloxacin; for more serious or refractory cases, in which there is the hazard of systemic spread of the infection, parenteral treatment with an aminoglycoside or an expanded-spectrum cephalosporin must be considered.

Recurrent infection

In chronic pyelonephritis and recurrent bacteriuria it may be important to control infection by continuous treatment with antibiotics such as trimethoprim. Unfortunately, some patients may become re-infected with resistant species so that alternative drugs must be used. Thus, long-term therapy may require periodic bacteriological re-assessment and sensitivity tests on the flora isolated to indicate when changes in therapy are required.

Respiratory infections

Respiratory tract infections are extremely common, and, although many are primary virus infections, those that are more severe and prolonged usually indicate secondary bacterial invasion. Laboratory diagnosis of these conditions is important because effective treatment depends on use of an antibiotic specifically active against the causative organisms.

Upper respiratory tract infections

A number of bacteria may be associated with sore throat, pharyngitis and sinusitis, including *Str. pyogenes, Str. pneumoniae, H. influenzae* and Vincent's organisms. The most important infections from the point of view of the development of sequelae are those due to *Str. pyogenes*, for which the treatment of choice is penicillin given for at least 7 days to ensure eradication of the organism. All of the other important bacterial pathogens also respond to this treatment, with the exception of *H. influenzae*, for which treatment with amoxicillin, tetracycline, co-trimoxazole or erythromycin is appropriate.

If for some reason a penicillin cannot be prescribed, the antibiotic of choice is erythromycin or co-trimoxazole. Ampicillin and amoxicillin should be avoided if there is any likelihood of glandular fever because these antibiotics tend to cause a rash and prolong the disease. Tetracyclines should be avoided in young children because of the risk of discoloration of developing teeth.

Lower respiratory tract infections

Lobar pneumonia is most frequently caused by pneumococci, and penicillin is the drug of choice; erthyromycin or cefotaxime is a good alternative, as are some of the newer fluoroquinolones such as levofloxacin or moxifloxacin. Infections due to klebsiellae or other organisms require treatment with antibiotics according to the results of sensitivity tests; cefotaxime or cefuroxime are usually effective if laboratory confirmation is not available. Other coliform bacilli are rarely involved in pulmonary infections, although they often colonize the upper respiratory tract; they seldom require specific treatment.

Among atypical pneumonias, mycoplasma infections respond best to a tetracycline or erythromycin; legionella infections respond to erythromycin, alone or combined with rifampicin.

Bronchopneumonia is most frequently associated with pneumococci or haemophilus, so that amoxicillin, tetracycline or trimethoprim should be effective. More rarely, bronchopneumonia is caused by *Staph. aureus*, and flucloxacillin, fusidic acid or clindamycin is urgently required for treatment.

Acute exacerbations of chronic bronchitis are almost invariably associated with either *Str. pneumoniae* or *H. influenzae*. Amoxicillin, tetracycline or trimethoprim is usually effective.

In bronchiectasis, lung abscess or cystic fibrosis, antimicrobial treatment should be prescribed according to laboratory culture and sensitivity test results. Combinations of antibiotics may be effective in some of these patients. In pulmonary tuberculosis, combination treatment, usually with rifampicin, isoniazid and pyrazinamide, is mandatory.

Meningitis

A Gram film and cell count of cerebrospinal fluid (CSF) can usually differentiate viral from bacterial meningitis. If bacteria cannot be seen, initial treatment is with cefotaxime or ceftriaxone. This choice ensures adequate treatment of infections with Gram-negative bacilli (most common in neonates) and of infections with *Neisseria meningitidis, H. influenzae* and *Str. pneumoniae*, which are the most common causes of meningitis in young children. Care must be taken to differentiate *Listeria monocytogenes* infection, as this organism is less sensitive to penicillin and cefotaxime; a combination of amoxicillin and gentamicin is usually used.

When the causative organism has been isolated, therapy may be altered to penicillin for meningococcal or pneumococcal infections. *Pseudomonas aeruginosa* meningitis may occur occasionally as a nosocomial infection and should be treated with full doses of gentamicin plus a β-lactam agent such as azlocillin. It is rarely necessary to give intrathecal antibiotics.

Where there is increased CSF pressure, as in spina bifida, shunts are inserted to facilitate the circulation of fluid. These often become contaminated, usually with staphylococci, probably derived from the skin. Anti-staphylococcal antibiotics are used to control the growth of the infecting organism in anticipation of replacement of the prosthesis.

Intestinal infections

Mild bacillary dysentery, salmonella food poisoning and other forms of bacterial diarrhoea do not ordinarily require chemotherapy. Indeed, such therapy may make intestinal carriage more likely and increase the risk of the selection of antibiotic-resistant strains. The most important treatment is correction of fluid balance.

Patients with invasive infection (e.g. typhoid and para-typhoid fever) and those with severe bacillary dysentery or cholera may warrant treatment (see appropriate chapters). Even potentially fatal intestinal infections with *Esch. coli* O157 fare worse with antimicrobials than fluids alone. This may be due to excess toxin release.

Many intestinal infections are due to viruses, for which there is presently no specific chemotherapy. Crypto-sporidiosis is now being diagnosed more frequently, but is usually self-limiting except in the immunocompromised. Infection with other protozoa, such as *Giardia lamblia* or *Entamoeba histolytica*, requires treatment with metronidazole.

Spread of infection from the bowel may give rise to serious infections and associated toxaemia. If subphrenic, retrocolic or pelvic abscesses form, drainage is the most important aspect of treatment, but antibiotics can assist recovery of the patient. Cephalosporins or an aminoglyoside plus ampicillin are useful in empirical treatment, and metronidazole should be added if bacteroides is likely to be present. Acute peritonitis after non-specific inflammation of the bowel, such as appendicitis, is usually treated effectively with amoxicillin (with or without clavulanate) or cefotaxime combined with metronidazole.

Liver infections

Bacterial infections of the liver and portal pyaemia are best treated with large doses of cefuroxime, cefotaxime or an antibiotic chosen on the basis of laboratory findings (e.g. ampicillin against enterococci). Liver abscesses occur most frequently by spread via the portal tract of intestinal Gram-negative bacilli, or by retrograde spread in the biliary passages of streptococci (especially *Str. milleri*) and anaerobes from the gallbladder. Ampicillin is concentrated selectively in the bile, and is often useful in treatment of cholecystitis. Amoebic abscess requires prompt and specific treatment with metronidazole, and its possibility should always be kept in mind, particularly in a patient who has been abroad.

Infection caused by the hepatitis viruses B and C has been treated by α-interferon with ribavirin and lamivudine, with variable success.

Bone and joint infections

Most infections of bones and joints are due to *Staph. aureus*. Large doses of flucloxacillin should be given in combination with either fusidic acid or clindamycin unless antibiotic-resistant strains are involved, which may limit the choice to a glycopeptide or linezolid.

Penicillin in prolonged high dosage is the drug of choice for streptococcal arthritis. Where other organisms, such as *H. influenzae*, neisseriae or Gram-negative rods, are involved, specifically directed therapy is required.

Genital tract infections

Venereal infections such as syphilis and gonorrhoea are traditionally treated with penicillin, but a proportion of isolates of gonococci are now resistant to this agent, in which case ceftriaxone, co-amoxiclav or ciprofloxacin should be used.

Non-specific infections of the genital tract are common in women; aerobic and anaerobic bacteria, *Candida* species or *Trichomonas vaginalis*, may be involved. Frequently, an abnormal flora is isolated in the absence of an inflammatory exudate, and it may be difficult to decide on the necessity for chemotherapy. Bacterial vaginosis is such a clinical syndrome associated with *Mobiluncus* spp., which respond to metronidazole or amoxicillin. Vaginal candidiasis is usually treated topically with nystatin or an antifungal imidazole. In trichomoniasis, metronidazole is indicated.

More serious pelvic infection in women is often associated with chlamydiae, and tetracycline or erythromycin is the drug of choice. Azithromycin is a useful single-dose agent for uncomplicated genital infection due to chlamydiae. Infection of the fallopian tubes usually involves anaerobic bacteria, and metronidazole or co-amoxiclav may be prescribed.

Surgical wound infections

Most surgical infections are caused by the patient's own organisms, but some arise exogenously, often by cross-infection. In orthopaedic units most exogenous infections are due to staphylococci, but in gastro-intestinal units Gram-negative bacilli and anaerobes are more common. The situation may change from time to time as a result of ecological movements in the microbial flora within the unit, associated with selective pressures of antibiotic use.

Specimens from infected lesions should always be sent to the laboratory, because the identity of the isolate may have epidemiological significance as well as being of importance in the management of the patient. The microbiologist should always be informed of any therapy as this may affect the interpretation of bacteriological tests.

Superficial infections

Skin and soft tissue infections are common in general practice. Most are of bacterial origin, although some have a fungal or viral aetiology.

Common lesions such as boils, carbuncles, impetigo and infected wounds are associated with *Staph. aureus* and *Str. pyogenes*. Systemic treatment with appropriate antibiotics may be indicated in some patients, but topical treatment is often effective. The use of antibiotics commonly prescribed for systemic infections should be avoided in favour of topical antiseptics or topical antibiotics such as mupirocin.

KEY POINTS

- A central aim in choosing antimicrobial therapy is to match the most appropriate agent to the specific microbial aetiology. In practice, this is constrained by urgency of the situation, the diagnostic information available and policy decisions about antimicrobial use that are made on a local basis consistent with national policies.
- The initial decision on whether or not to give an antimicrobial and on the agent chosen is generally made before microbiological information is available but, where possible, after microbiological samples have been taken.
- In the UK, local and specific guidance is generally available for chemotherapy of common infective syndromes. The strategy reflects the likely causal agents and their local susceptibility patterns, the need to control antimicrobial resistance, and cost. This guidance is reflected in the local *antibiotic policy*.
- The laboratory will assist in modifying or initiating therapy by determining in vitro susceptibilities and by monitoring antimicrobial drug concentrations in specific circumstances. These analyses may lead to modification of therapy.
- Certain organisms such as streptococci and anaerobes have predictable susceptibilities, whereas others, such as enterobacteria, must be monitored continuously.
- Prophylactic antimicrobials are recommended in highly defined settings and following precise regimens. Ad hoc usage is to be avoided.
- Antimicrobial combinations may be given to increase efficacy, extend the spectrum of cover or prevent the development of resistance during therapy.

RECOMMENDED READING

Finch R G, Greenwood D, Norrby S R, Whitley R J 2002 *Antibiotic and Chemotherapy*, 8th edn. Churchill Livingstone, Edinburgh

Greenwood D, Finch R G, Davey P G, Wilcox M H 2007 *Antimicrobial Chemotherapy*, 5th edn. Oxford University Press, Oxford

Kucers A, Crowe S M, Grayson M L, Hoy J F 1997 *The Use of Antibiotics*, 5th edn. Butterworth-Heinemann, Oxford

Mandell G L, Douglas R G, Dolin R (eds) 2004 *Principles and Practice of Infectious Disease*, 6th edn. Churchill Livingstone, New York

Internet sites

World Health Organization. Essential Drugs Monitor. http://mednet2.who.int/edmonitor/

British Society for Antimicrobial Chemotherapy. http://www.bsac.org.uk/

67 Epidemiology and control of community infections

D. Reid and D. Goldberg

When you can measure what you are speaking about and express it in numbers, you know something about it; when you cannot express it in numbers, your knowledge is of a meagre and unsatisfactory kind.

> William Thomson, Lord Kelvin
> (Popular Lectures and Addresses, 1891)

Good surveillance does not necessarily mean the making of right decisions but it reduces the chances of wrong ones.

> Alexander Langmuir, 1963

Once is happenstance, twice is coincidence, the third time it's enemy action.

> Ian Fleming
> (Goldfinger, Jonathan Cape, 1959)

Attempts to observe and record diseases in order to devise means of determining their cause and control have a long history. Hippocrates (460–361 BC), 'the father of medical science', and Herodotus (484–425 BC), 'the father of history', both related environmental factors to health. Hippocrates, when writing of the occurrence of diseases, distinguished between the 'steady state', the 'endemic state' and the abrupt change in incidence, the 'epidemic'.

Probably the first public health measures based on case reports of infectious diseases taken by a European government occurred in 1348 when the Republic of Venice excluded ships with affected people on board in order to control outbreaks of pneumonic plague (the *black death*). Fifty years later, again in Venice, the concept of *quarantine* was introduced when ships from plague-stricken areas had to stay outside the harbours for 40 days (*quaranta giorni*).

In addition, because of the fear of a plague epidemic, the first of the *Bills of Mortality*, in which causes of death were recorded, was published in London in 1592. In 1662 John Graunt (1620–1674), in his book *Natural and Political Observations Made Upon the Bills of Mortality*, was the first to count the number of persons dying in London from specific illnesses and to advocate the value of obtaining numerical data on a population in order to study the causes of disease (Table 67.1). In 1837 the office of the Registrar General was established

to develop the work started by John Graunt; the English physician William Farr (1807–1883) added reports to those of the Registrar General that dealt with infectious diseases, occupational diseases, accidents or hazardous work conditions.

The importance of keen observation of disease in order to deduce the likely cause has been demonstrated on many occasions. In 1849, 34 years before the identification of *Vibrio cholerae* by Robert Koch (1843–1910), John Snow (1813–58), a London physician, proved by epidemiological observation that cholera was mainly spread by drinking infected water and not through the air in the form of miasmas, as was commonly thought at the time. Similarly, William Budd (1811–1880), a general practitioner from Devon, showed in 1873 how typhoid was caused, even

Table 67.1 Selection of causes of death in London taken from the Bills of Mortality, 1632

Causes of death	No. of deaths
Chrisomes[a] and infancy	2268
Consumption[b]	1797
Fever	1108
Aged	628
Smallpox	531
Teeth	470
Abortive and stillborn	445
Bloody flux[c], scouring[d] and flux	348
Dropsy[e] and swelling	267
Convulsions	241
Childbed	171
Measles	80
Ague[f]	43
King's evil[g]	38

[a]A child who died during the first month of life or a child who died unbaptized.
[b]Usually pulmonary tuberculosis.
[c]Dysentery.
[d]Diarrhoea.
[e]Oedema.
[f]Malaria.
[g]Tuberculosis of the skin.

though it was not until 1885 that the typhoid bacillus was first isolated in the laboratory. More recently, William Pickles (1885–1969), a general practitioner in Wensleydale, Yorkshire, was able to elucidate many of the epidemiological characteristics of hepatitis and other infections well before microbiological advances were to confirm his observations.

From these beginnings the surveillance of infection has assumed national and international proportions. In the UK, information on microbial disease is collated for England by the Health Protection Agency, for Scotland by Health Protection Scotland, for Wales by the Infection and Communicable Disease Services and for Northern Ireland by the Communicable Disease Surveillance Centre. The Royal College of General Practitioners also undertakes regular recording of disease reported voluntarily by various general practices.

Elsewhere, national surveillance is carried out in different countries, for example at the Centers for Disease Control and Prevention (CDC) in Atlanta, USA. In May 2005 the European Centre for Disease Prevention and Control in Stockholm was established and, on a worldwide basis, the World Health Organization (WHO) provides important liaison and support. This international co-operation is vital as 'germs do not recognize boundaries'.

Probably the most outstanding achievement of international surveillance was the development of a programme for smallpox eradication. The multidisciplinary approach adopted by the WHO, in which programmes were community based with measurable goals and constant monitoring, resulted in the last endemic case being recorded in October 1977; smallpox was officially declared eradicated in December 1979.

EPIDEMIOLOGY: DEFINITIONS AND PRINCIPLES

Epidemiology is usually defined as *the study of the nature, distribution, causation, mode of transfer, prevention and control of disease*. It has also been regarded as 'the natural history of disease' or as 'the human face of ecology'. Closely linked with the study of epidemiology is the concept of surveillance, which is probably the most effective infection control technique available. Surveillance is defined as: *the epidemiological study of a disease as a dynamic process involving the ecology of the infectious agent, the host, the reservoirs, the vectors as well as the complex mechanisms concerned in the spread of infection and the extent to which this spread will occur*. There are three main elements of surveillance of infection:

1. the systematic collection of pertinent data
2. the orderly consolidation and evaluation of the data

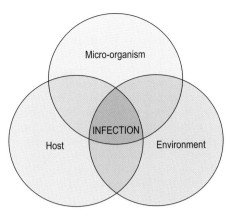

Fig. 67.1 The three main factors involved in the infectious process.

3. the prompt dissemination of the findings, especially to those who can take appropriate action.

Surveillance provides for the recognition of acute problems requiring immediate local, national or international action, and for further assessment by revealing trends or facilitating forecasts. It also provides a rational basis for planning and implementing efficient control measures and for their evaluation and continuing assessment. Although particularly appropriate to the study of infectious diseases, epidemiological principles are also used to elucidate the causes of non-communicable diseases.

The infectious process is a dynamic state involving three main factors: the micro-organism, the host and the environment (Fig. 67.1).

The micro-organism

Since few micro-organisms are harmful to man, the concept of *virulence* (the degree of pathogenicity of an infectious agent indicated by fatality rates and/or its ability to invade and damage the tissues of the host) must be recognized. The degree of virulence depends on *invasiveness* (the capacity of the organism to spread widely through the body) and *toxigenicity* (the toxin-producing property of the organism) (see Ch. 13). A further variable is the *dosage* of the organism, and this is closely related to the virulence. A small number of organisms of high virulence is usually sufficient to cause disease in a susceptible person, whereas if the organism is of low virulence it often fails to cause disease. Another variable is the *portal of entry*. Many organisms have a predilection for a particular tissue or organ. For example, the causal organism of typhoid fever, *Salmonella enterica* serotype Typhi. usually causes typhoid only when it enters the human body through the mouth in food or water.

The host

The reaction of the host to a micro-organism will depend on the ability to resist infection. The individual may not possess sufficient resistance against a particular pathogenic agent to prevent contraction of infection when exposed to the organism. Alternatively, the individual may possess specific protective antibodies or cellular immunity as a result of previous infection or immunization. However, immunity is relative and may be overwhelmed by an excessive dose of the infectious agent or if the person is infected via an unusual portal of entry; it may also be impaired by immunosuppressive drug therapy, concurrent disease or the ageing process.

Herd immunity

Herd immunity is an important element in the balance between the host population and the micro-organism, and represents the degree to which the community is susceptible or not to an infectious disease as a result of members of the population having acquired active immunity from either previous infection or prophylactic immunization (see pp. 705–706).

Herd immunity can be measured:

1. *Indirectly* from the age distribution and incidence pattern of the disease if the disease is clinically distinct and reasonably common. This is an insensitive and inadequate method for infections that manifest subclinically.
2. *Directly* from assessments of immunity in defined population groups by antibody surveys (seroepidemiology) or skin tests; these may show 'immunity gaps' and provide an early warning of susceptibility in the population. Although it may be difficult to interpret the data in absolute terms of immunity and susceptibility, the observations can be standardized and reveal trends and differences between various defined population groups in place and time.

The decision whether to introduce herd immunity artificially by immunization against a particular disease will depend on several epidemiological principles.

- The disease must carry a substantial risk.
- The risk of contracting the disease must be considerable.
- The vaccine must be effective.
- The vaccine must be safe.

The effectiveness and safety of immunization programmes are monitored by observing the expected and actual effects of such programmes on disease transmission patterns in the community by appropriate epidemiological techniques.

The environment

The environment plays a major role in the causation, spread and control of infection. In the UK, the virtual disappearance of relapsing fever, plague and cholera, the rarity of indigenous typhoid fever and the relative infrequency of bacillary dysentery are all indications of the improvements that have taken place in environmental conditions.

The decrease in overcrowding and infestation, together with the demand for cleaner water supplies and better sanitation, have been of paramount importance in producing these dramatic advances. This is well illustrated in the case of tuberculosis, which was declining before the availability of chemotherapy and mass bacille Calmette–Guérin (BCG) vaccination in countries where socio-economic conditions were improving (Fig. 67.2). Paradoxically, better living conditions may unexpectedly create new problems; for example, poliovirus infection, previously experienced mainly in early childhood, is usually postponed in more favoured communities to older ages when paralysis is a more likely complication unless there is an adequate immunization programme.

In recent years, areas of urban deprivation, particularly in western countries, have been blighted by the culture of illicit drug injection and its associated infections, such as human immunodeficiency virus (HIV) and hepatitis C, which are acquired through the sharing of injecting equipment.

THE SPREAD OF INFECTION

Infection spreads in well defined epidemiological patterns. A knowledge of these will lead to an understanding of the best methods of control, or even eradication, and enables an estimate to be made of the likelihood of this happening.

Infection spread directly from one person to another

Among this group can be included such highly infectious diseases as measles. Infection is passed directly from a person with the disease to a susceptible contact. Diseases in this category are usually clinically apparent and healthy carriers are not a feature. When it is possible to diminish the number of susceptibles in the target population then eradication becomes feasible, as has happened with smallpox.

Infection in which healthy carriers are involved

Because apparently healthy individuals may harbour the bacilli responsible for such diseases as typhoid, para-

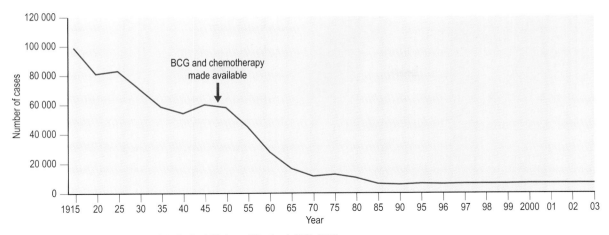

Fig. 67.2   Notifications of tuberculosis in England, Wales and Scotland, 1915–2003.

typhoid and diphtheria, often for long periods after having acquired the infection, it is possible for such infections to be transmitted to others and the source remains undetected. For certain infections, such as hepatitis C virus infection, the healthy carrier state may last several decades.

Infection in which persons harbour the organism before the onset of clinical illness

Organisms such as *Streptococcus pneumoniae* may not cause the person any harm until an event such as a skull fracture allows for the transfer of the bacterium from the middle ear to the cerebrospinal space where it can cause a potentially fatal meningitis.

Infection derived from animal sources

Diseases derived from animals, such as leptospirosis, Q fever, anthrax, rabies and brucellosis, are known as *zoonoses*. These diseases are spread by direct contact with the animal concerned or indirectly by such means as the ingestion of infected milk, contact with infected bone products, etc.

Infections derived from environmental sources

The spread of legionellae from cooling towers and air conditioning units to cause legionnaires' disease is an example of illness derived from an infected environment.

By dealing appropriately with the infected source by use of biocides in the water, the population can be protected.

OUTBREAKS OF INFECTION

The crowding together of human beings (or for that matter animals, fish or birds) provides the necessary conditions to allow micro-organisms to multiply and spread. When human beings led nomadic lives there was less opportunity for outbreaks to occur; the main opportunities came when large numbers gathered for a pilgrimage or had other reasons for a meeting. These clusterings facilitated the spread of infection, resulting in outbreaks; the subsequent dispersal of the group enabled the causative organism to be carried elsewhere.

The threat of outbreaks in overcrowded and difficult conditions is particularly well illustrated in military history; on many occasions the germ has been as important in determining the outcome of a campaign as the sword or gun. The typhoid bacillus caused severe effects during both the American Civil War (1861–1865) and the Boer War (1899–1902). The use of typhoid vaccine in the latter years of the First World War meant that the main impact of typhoid in this war subsided after 1916. Similarly, typhus was rife in the Civil War in Britain (1642–1649), when both the Parliamentary and the Royalist armies were affected. The pandemic of influenza in 1918–1919, in which about 700 million people were affected with approximately 22 million deaths, was a scourge of military camps and affected many servicemen returning home from the First World War.

Nomenclature of outbreaks

The term *outbreak* is often confused with other epidemiological terms used to enumerate infection:

1. *Sporadic case*: a person whose illness is not apparently connected with similar illnesses in another person.
2. *Outbreak*: the occurrence of cases of a disease associated in time or location among a group of persons. A *household outbreak* involves two or more persons

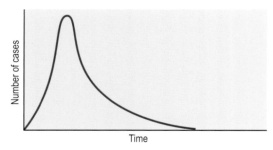

Fig. 67.3 Epidemic curve apparent when there is an explosive (common or point source) outbreak.

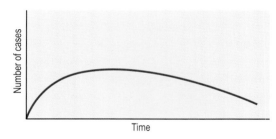

Fig. 67.4 Epidemic curve apparent when there is person-to-person spread of infection.

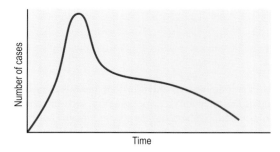

Fig. 67.5 Epidemic curve apparent when there is person-to-person spread subsequent to a common source outbreak.

resident in the same private household and not apparently connected with any other case or outbreak. A *general outbreak* involves two or more persons who are not confined to one private household.

3. *Epidemic*: the large-scale temporary increase in the occurrence of a disease in a community or region that is clearly in excess of normal expectancy.

4. *Pandemic*: the occurrence of a disease that is clearly in excess of normal expectancy and is spread over a whole geographical area, usually crossing national boundaries.

Types of outbreak

There are three main patterns of outbreak that may be revealed by the construction of graphs of occurrence of cases over time:

1. *The explosive outbreak*. This is characterized by the occurrence of a large proportion of cases in a relatively short period of time (Fig. 67.3); there is a sharp rise and fall in the number of infected persons, because the usual cause of such an event is a common source that infects the people concerned. This type of outbreak is also frequently termed a *common source outbreak* or a *point source outbreak*. This pattern of infection is often discovered when water or food becomes contaminated, although other vehicles of infection can also be responsible for this type of outbreak.

2. *Person-to-person spread*. Outbreaks caused by infections that are spread from person to person usually have a more protracted course, taking longer than explosive outbreaks to build up and to subside. An infective agent may be passed from person to person by a variety of routes. Diseases such as dysentery, hepatitis type A and gastro-enteritis, which are usually spread by the faecal–oral route, often follow this pattern of spread (Fig. 67.4).

3. *Explosive outbreaks with subsequent person-to-person spread*. This pattern is often apparent when there is contamination of a common water or food source and the initial cases subsequently infect their contacts. Thus,

the pattern of the outbreak is a combination of that seen with an explosive outbreak, but followed by a slower decline (Fig. 67.5).

Analysis of outbreaks

The investigation of an outbreak should be approached in a logical and methodical way. The cause may be elucidated by determining details of the *persons* involved, the *place* where they had been and the *time* when they became ill.

The fundamental pieces of information that should be sought whenever an outbreak occurs are as follows:

1. *WHO gets infected?* What is their age? For example, if a possible food-borne outbreak affects mainly children, could the source be milk or ice-cream?

2. *WHERE were those who became infected?* Where have they recently been? For example, in a hospital outbreak were they all in the same surgical ward? Could a member of the operating staff be a carrier of a pathogen? In a community outbreak of legionnaires' disease were those affected living downwind from a contaminated source of infection? (Fig. 67.6).

3. *WHEN did the infection occur?* By knowing the incubation period of the infection it may be possible to trace back to an event that was attended by all those affected.

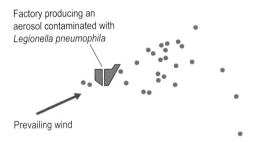

Fig. 67.6 Occurrence of legionnaires' disease in persons living downwind from a factory with an evaporative condenser contaminated with *Legionella pneumophila*.

4. *WHAT was the common factor?* For example, in a food poisoning episode, the ingestion of an article of food by most of those affected but not by those unaffected may be a vital piece of evidence.

5. *HOW did those involved become infected?* For example, abscess formation among recently immunized persons might be due to contaminated vaccine.

6. *WHY did the infection occur?* For example, the reheating of meat may be the cause of a *Clostridium perfringens* food-poisoning outbreak.

Investigation of outbreaks

In the investigation of outbreaks it is important to have a standardized approach to the various steps involved. Such an approach might have the following as a basis:

1. *Verify the diagnosis.* It is always prudent to confirm that the clinical history is compatible with the diagnosis. Occasional 'pseudo-epidemics' can occur, sometimes resulting from contamination of specimens.

2. *Establish the existence of an outbreak.* The increased interest of an investigator or a change in the mechanism of reporting can sometimes result in an increase in the number of reports of illness. It is important to check the previous level of investigation of a clinical entity or alteration in laboratory methods.

3. *Establish the extent of an outbreak.* Often the number of cases notified is only a proportion of the total number of those affected. It is necessary to seek out the additional cases or vital information may be lost.

4. *Identify common characteristics or experiences of the affected persons.* An individual history from each confirmed or suspected case is required to detect any common factor among those affected (e.g. eating the same item of food).

5. *Investigate the source and vehicle of infection.* In addition to ascertaining the general characteristics of the material suspected as being the source or vehicle of infection, appropriate laboratory investigation will often have to be done. Good co-operation with laboratory staff

is vitally important. Increasingly, putative links between individuals who present with the same infection or infectious disease are confirmed or dispelled through the typing or sequencing of the genes of the organism involved.

6. *Analyse the findings.* The data should be analysed by various epidemiological criteria, especially persons, time and place. Denominators should be obtained to calculate attack rates.

7. *Construct a hypothesis.* On the basis of the evidence a hypothesis should be constructed concerning the origin of the outbreak. This may be confirmed by laboratory findings but action to control the outbreak may be needed in advance of such findings.

Control of outbreaks

The investigation of an outbreak should be carried out as swiftly as possible so that adequate control measures can be started without delay. Knowledge of the *source of infection*, the *route of transmission* and the *person at risk* should allow appropriate action to be taken to achieve success.

Sources of infection

These may be:

- human cases or carriers
- animal cases or carriers
- the environment.

If the initial cases have readily identifiable clinical features (e.g. measles) then control is often easier as it is much more likely that the index case will be located.

On the other hand, it is more difficult to control diseases in which apparently healthy carriers are responsible, as it is necessary to search for an infected person who may be asymptomatic.

It may be important to isolate the case or carrier, and possibly to institute appropriate treatment, until the patient is no longer infectious. The degree of isolation will depend on the type of disease, as not all infections require strict isolation. For example, a patient with a highly infectious disease may require very strict isolation whereas the salmonella excreter will usually need only to cease food handling activities and observe a high standard of personal hygiene until free from infection. In contrast to 'isolation', the term 'quarantine' applies to restrictions on the healthy contacts of an infectious disease.

If an animal reservoir is responsible, action has to be directed at ensuring that the source of infection is eradicated, withdrawn from consumption or rendered harmless, for example by the pasteurization of milk or the adequate cooking of meat.

When the environment is the source of an outbreak the control measures required will depend on the nature of

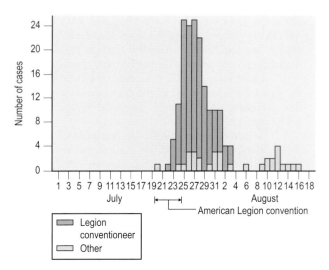

Fig. 67.7 Legionnaires' disease among those associated with a convention in a hotel in Philadelphia (July to August 1976); an example of an explosive outbreak.

infection and the mode of spread. In recent years, water-borne spread of legionnaires' disease (e.g. from shower-heads, air-conditioning systems or droplet spread from cooling towers) has become increasingly recognized (Fig. 67.7). Environmental measures, such as the use of biocides, can destroy the causative legionellae at their source and so prevent further cases.

The hospital setting is particularly dangerous as the presence of compromised patients can result in tragic consequences. Moreover, the increasing use of invasive techniques and the appearance of antibiotic-resistant strains of micro-organisms further compound the problem. In a survey of hospital patients in England and Wales, 9% of all patients acquired infection while in hospital. The early detection of infection by effective surveillance, the emphasis on the cleanest possible environment and awareness among the staff of potential problems are among the measures that need to be stressed to control infection among hospital patients (see Ch. 68).

Route of transmission

Infection may be spread by:

- direct or indirect contact
- air-borne transmission
- food- and water-borne transmission
- insect-borne transmission
- percutaneous transmission
- transplacental transmission.

There are various ways in which the routes of transmission occurring during an outbreak may be blocked by measures appropriate to the route involved: effective handwashing; disinfection or disposal, if necessary, of the patient's belongings; and strict adherence to high standards of

personal hygiene by the *contacts* of a case are important measures.

When the disease is air-borne, overcrowding should be avoided and, where appropriate, dormitory or ward beds should be well spaced out. In certain instances major isolation measures, such as those adopted to control the outbreak of severe acute respiratory syndrome (SARS) in 2003, need to be implemented. Indeed, the spread of the coronavirus, the air-borne infection responsible for SARS, throughout regions of the Far East (China, Vietnam and Singapore) and North America (Toronto), before it was brought under control, demonstrates how effectively international travel and overcrowding in major cities can facilitate the transmission of serious air-borne infections.

Food and *water* should be as free as possible from infection; otherwise their consumption can cause major outbreaks. To prevent *insect-borne transmission*, insect eradication policies and repellents should be considered. HIV is an infection that, commonly, is transmitted through the percutaneous, sexual and transplacental routes; the elimination of infected blood donations by blood donor testing, the provision of sterile needles and syringes for infected drug users, the use of condoms and the treatment of HIV-infected pregnant women with antiretroviral therapy are measures that prevent the spread of this infection through these routes.

Persons at risk

Where indicated and feasible, susceptible persons at risk should be protected as soon as possible. Among the measures available, the following may need to be considered:

- immunization
- chemoprophylaxis for close contacts.

Immunity against several infectious diseases can be obtained by either active or passive immunization (see Ch. 69). Examples of rapid effective protection of communities by active immunization are the 'ring vaccination' policy used to protect close contacts of poliomyelitis and so stop more widespread dissemination of poliovirus, and the early (within 72 h of exposure) administration of measles vaccine to close contacts in an institution. Passive immunization, usually by means of human specific immunoglobulin, gives rapid protection to contacts of certain infections, such as hepatitis B, although protection is short lived. Chemoprophylaxis is effective in the protection of close contacts of meningococcal infections and diphtheria.

MATHEMATICAL MODELS

Mathematical modelling techniques attempt to define, by use of relatively simple estimates and assumptions, the dynamic conditions governing transmission of communicable agents.

With the advent of powerful computers it is possible to calculate many values of likely variables and to present them graphically. Details of the multiplication and growth rates of micro-organisms and the spread of infection under natural or experimental conditions need to be known. Measurable factors include:

- the number of infective persons or sources of introduction of infection
- the proportion of susceptible persons in a community at risk
- the duration of immunity
- the introduction of new susceptibles
- the removal rate of infective persons (by isolation, immunity or death)
- the response to vaccines and chemotherapeutic agents.

Mathematical models can be used to predict institutional outbreaks and epidemics.

This approach has been used to forecast the number of cases of acquired immune deficiency syndrome, variant Creutzfeldt–Jakob disease and bovine spongiform encephalopathy that are likely to occur, the effect of control measures such as immunization, and the number of cases over time.

ASSOCIATION AND CAUSATION OF INFECTION

A problem commonly encountered by microbiologists and epidemiologists is the attribution of an infectious disease to a particular micro-organism. How do we deter-

Table 67.2 Deaths in London during the cholera epidemic of 1854 according to source of water supply

Water source	No. of houses supplied	Deaths
Southwark and Vauxhall Company (polluted)	10 000	71
Lambeth Company (non-polluted)	10 000	5

mine whether the relationship is one of causation or merely a chance association?

Koch addressed this when he formulated his 'postulates' in 1891. These state that:

1. The organism must always be found in the given disease.
2. The organism must be isolated in pure culture.
3. The organism must reproduce the given disease after inoculation of a pure culture into a susceptible animal.
4. The organism must be recoverable from the animal so inoculated.

For many organisms pathogenic to man it is not possible to fulfil all of Koch's postulates; moreover, they are not applicable to the study of the transmission of infection within a population. For this purpose it is more appropriate to consider the following factors, suggested by the medical statistician Bradford Hill, to establish whether a disease is caused by a particular infectious agent:

1. *Strength.* What is the strength of the association? During the cholera epidemic in London in 1854, John Snow compared the death rates among persons drinking the sewage-polluted drinking water of the Southwark and Vauxhall Company and those receiving the purer water of the Lambeth Company; he discovered that the rate in the former was 14 times greater (Table 67.2). This strength of association allowed Snow to consider that polluted water was a cause of cholera, although at that time the causative organism itself had not been identified.

2. *Consistency.* Similar observations, made by different people at different times in different places, add confidence to a conclusion that causation is likely.

3. *Specificity.* If the association is limited to a specific group of persons, with a specific type of illness, who have all been subjected to the same specific infection, then a cause-and-effect relationship can be more strongly suspected.

4. *Temporality.* This can be of especial importance when persons in particular occupations become infected (e.g. leptospirosis in fishworkers). The history of working

in a particular environment *before* infection rather than vice versa is particularly relevant.

5. *Biological gradient.* If a dose–response curve is apparent then the evidence for causation is much stronger.

6. *Plausibility.* Is the possibility biologically plausible? The likelihood of veterinary surgeons becoming ill with a zoonotic infection from affected cattle which they have recently been treating seems biologically plausible.

7. *Coherence.* If all the evidence is coherent (e.g. if the same micro-organism is isolated from the index case, the vehicle of transmission and from the victims), this is strong support for causation.

8. *Experiment.* Is the frequency of infection reduced if certain preventive measures are taken? The beneficial effects of the pasteurization of milk to diminish the number of cases of milk-borne salmonellosis is presumptive evidence of a zoonotic relationship.

9. *Analogy.* Has there been similar evidence in the past? The known capacity of the rubella virus to cause congenital abnormalities in the infants of infected mothers makes it easier to accept the possibility of other viruses causing similar problems if maternal infection occurs.

CONCLUSION

Because of the multifactorial causation of infection it is usually necessary to study the epidemiology of infection in a *multidisciplinary* manner. The microbiologist, the clinician, the epidemiologist, the infection control nurse, the veterinarian, the environmental health officer, and other appropriate personnel, must all be involved; the extent of the involvement will depend on the nature of the infection. Success will depend on the expertise and co-operation of these members of the team.

KEY POINTS

- The surveillance of infection provides the basis for appropriate investigation and preventive action.
- The infectious process involves three main factors: the micro-organism, the host and the environment.
- Herd immunity indicates the degree to which a community is susceptible to infection.
- Infection is spread in five ways: from person to person, from healthy carriers, from animal sources, from environmental sources and as a result of an organism situated in an area of the body where it is harmless gaining access to a more dangerous site.
- Infection may manifest itself as: a sporadic case, an outbreak, an epidemic or a pandemic.
- Outbreaks have three main patterns of spread: from a single source, from one person to another, or from a single source with subsequent person-to-person spread.
- The epidemiological investigation of an outbreak involves an analysis concerning: the persons involved, the place it occurred and the time those infected became ill.
- Vigilance and high-quality surveillance are necessary in order to have an early warning of emerging infections.

RECOMMENDED READING

Begg N, Blair I, Reintjes R, Hawker J, Weinberg J 2005 *Communicable Disease Control Handbook.* Blackwell Science, Oxford

Detels R, McEwen J, Beaglehole R, Tanaka H 2002 *Oxford Textbook of Public Health.* Oxford University Press

Giesecke J 2001 *Modern Infectious Disease Epidemiology*, 2nd edn. Edward Arnold, London

Heymann D L (ed.) 2004 *Control of Communicable Diseases Manual*, 18th edn. American Public Health Association, Washington, DC

Ksiazek T G, Erdman D, Goldsmith C S, Zaki S R, Peret T, Emery S et al 2003 A novel coronavirus associated with severe acute respiratory syndrome. *New England Journal of Medicine* 348: 1953–1966

Last J M, Abrahamson J H (eds) 2001 *A Dictionary of Epidemiology.* Oxford University Press, Oxford

Mandell G, Bennet J, Dolin R 2004 *Principles and Practice of Infectious Diseases*, 6th edn. Elsevier Churchill Livingstone, Philadelphia

Mims C A, Nash A, Stephen J 2000 *Mims' Pathogenesis of Infectious Disease*, 5th edn. Academic Press, London

Pickles W N 1939 *Epidemiology in Country Practice.* Wright, Bristol (re-issued 1972 by the Devonshire Press, Torquay)

Internet sites

Centers for Disease Control and Prevention. http://www.cdc.gov/

European Centre for Disease Prevention and Control. http://www.ecdc.eu.int/

Health Protection Agency. http://www.hpa.org.uk/

Health Protection Scotland. http://www.hps.scot.nhs.uk/

World Health Organization. www.who.int/en/

All have links to many other websites worldwide.

68 Hospital infection

R. C. B. Slack

The battle between man and microbe is at its most obvious in institutions where vulnerable people are crowded together. Historically, hospitals have a notorious reputation for infection. The hazards of puerperal sepsis and the horrors of septic infection in the pre-listerian era have been well documented; admission to hospital in the mid-nineteenth century was associated with the fear of gangrene and death.

Since then, surgical and medical techniques have developed dramatically, basic standards of building and hygiene have greatly improved, and the identification and treatment of most infecting micro-organisms have become possible. Despite such changes, infection acquired in hospitals remains a major cause of morbidity and mortality, leading directly or indirectly to an enormous increase in the cost of hospital care and to the emergence of new health hazards for the community. In the past two decades great advances in biomedical technology and therapeutics have produced increased numbers of highly susceptible patients requiring treatment in hospitals, and this is aggravated by the occurrence of transferable resistance to antibiotics in pathogenic bacteria and the emergence of new pathogens transmitted by a variety of routes. In spite of these advances in medical care, in many countries pressures on health-care facilities and shortages of trained staff make it difficult to practise adequate infection control. There has also been a mistaken view among many health professionals that the advent of the antibiotic era made such precautions unnecessary, and many studies have shown poor compliance with simple hygiene.

CLASSIFICATION

To measure the extent of hospital and health care-associated infection and conduct surveillance, the following definitions should be considered:

1. Community-acquired infections – either those contracted and developing outside hospitals that require admission of the patient (e.g. pneumococcal pneumonia) or those that become clinically apparent within 48 h of admission when the patient has been admitted to hospital for other reasons (e.g. chickenpox or zoster).

2. Infections contracted and developing within hospital (e.g. device-associated bacteraemias).

3. Infections contracted in hospital but not becoming clinically apparent until after the patient has been discharged (e.g. many postoperative wound infections).

4. Infections contracted by hospital staff as a consequence of their work, whether or not this involves direct contact with patients (e.g. hepatitis B).

On average, around 9% of all hospital patients develop an infection as a result of their stay in hospital. Urinary, respiratory and wound infections are the most common.

FACTORS THAT INFLUENCE INFECTION

Hospital infection, also known as *nosocomial infection*, may be exogenous or endogenous in origin. The exogenous source may be another person in the hospital (*cross-infection*) or a contaminated item of equipment or building service (*environmental infection*). A high proportion of clinically apparent hospital infections are endogenous (*self-infection*), the infecting organism being derived from the patient's own skin, gastro-intestinal or upper respiratory flora.

Most infections acquired in hospital are caused by micro-organisms that are commonly present as commensals in the general population. Thus, contact with micro-organisms is seldom the sole or main event predisposing to infection. Various risk factors, alone or in combination, influence the frequency and nature of hospital infection. In comparing rates of hospital-associated infection it is important to be aware of the frequencies of risk factors, such as age, drug treatment or pre-existing diseases in the population surveyed, as well as the medical or surgical procedures used.

CONTACT WITH OTHER PATIENTS AND STAFF

In common with any large institution or workplace, the patients and staff of a hospital share many facilities in close or crowded conditions. Outbreaks of diarrhoeal and food-borne disease may be traced to a common source via the hospital water or food supplies. The specific role of hospitals in admitting infected patients or carriers for treatment clearly serves as a potential source of infection for others. Patients with comparable susceptibility to infection tend to be concentrated in the same area, for instance in neonatal units, burns units or urological wards, where infected and non-infected patients may be cared for by the same staff, thus creating numerous opportunities for the spread of micro-organisms by direct contact. The more susceptible patients usually require the most intensive care with far more daily contacts with staff, who act as vectors in the transmission of microbes like insects spreading parasites.

Inanimate reservoirs of infection

Equipment and materials in use in hospitals often become contaminated with micro-organisms, which may subsequently be transferred to susceptible body sites on patients. Gram-positive cocci, derived from skin scales of the hospital population, are found in the air, dust and on surfaces where they may survive along with fungal and bacterial spores of environmental origin. Gram-negative aerobic bacilli are common in moist situations and in fluids, where they often survive for long periods, and may even multiply in the presence of minimal nutrients. An important example of this is legionellae in hospital domestic water supplies. Awareness of the common reservoirs of environmental and contaminating hospital micro-organisms provides the basis for maintaining standards of hygiene (cleaning, disinfection, sterilization) throughout the hospital, as well as good engineering and building.

Role of antibiotic treatment

At least 30% of hospital patients receive antibiotics, and this exerts strong selective pressures on the microbial flora, especially of the gastro-intestinal tract, leading to the development of antibiotic-associated diarrhoea due to *Clostridium difficile*, one of the commonest causes of outbreaks of hospital infection. Sensitive species or strains of micro-organisms that normally maintain a protective function on the skin and other mucosal surfaces tend to be eliminated, whereas those that are more resistant survive and become endemic in the hospital population. This may

Table 68.1 Commonly occurring micro-organisms in hospital infection

Urinary tract infections	*Escherichia coli*
	Klebsiella, Serratia, Proteus spp.
	Pseudomonas aeruginosa
	Enterococcus spp.
	Candida albicans
Respiratory infections	*Haemophilus influenzae*
	Streptococcus pneumoniae
	Staphylococcus aureus
	Enterobacteriaceae
	Respiratory viruses
	Fungi (*Candida* spp., *Aspergillus*)
Wounds and skin sepsis	*Staph. aureus*
	Str. pyogenes
	Esch. coli
	Proteus spp.
	Anaerobes
	Enterococcus spp.
	Coagulase-negative staphylococci
Gastro-intestinal infections	*Salmonella* serotypes
	Clostridium difficile
	Viruses (Norwalk-like)

restrict the range of agents available for treatment and lead to the transmission of plasmid-mediated antibiotic resistance into strains that show increased virulence, survival and spread within the hospital.

MICRO-ORGANISMS CAUSING HOSPITAL INFECTION

The most important micro-organisms responsible for hospital infection are listed in Table 68.1.

Outbreaks of *Staphylococcus aureus* infection were often seen in surgical wards and maternity units before the advent of penicillinase-stable penicillins such as methicillin in the early 1960s. Some strains, such as the notorious phage type 80/81, demonstrated particular virulence and colonizing capabilities. Subsequently, epidemic or pandemic strains characterized by resistance to methicillin (MRSA) have been found in many hospitals worldwide, presenting a daunting challenge. Some strains are better able to colonize patients or staff than to produce systemic disease. This illustrates the adaptation and evolutionary changes possible within common hospital bacteria, revealed by careful epidemiological typing and observation. The changing patterns may be related to particular events and infection control measures, so that lessons can be learnt for future control and prevention.

Table 68.2 Hospital infection: sources and spread

Route	Source	Examples of disease
Aerial (from persons) Droplets Skin scales	Mouth Nose Skin exudate, infected lesion	Measles, tuberculosis, pneumonia Staphylococcal sepsis Staphylococcal and streptococcal sepsis
Aerial (from inanimate sources) Particles	Respiratory equipment Air-conditioning plant	Gram-negative respiratory infection Legionnaires' disease, fungal infections
Contact (from persons) Direct spread Indirect via equipment	Respiratory secretions Faeces, urine, skin and wound exudate	Staphylococcal and streptococcal sepsis Enterobacterial and viral diarrhoea, *Pseudomonas aeruginosa* sepsis
Contact (environmental source)	Equipment, food, medicaments, fluids	Enterobacterial sepsis (*Klebsiella*/*Serratia*/*Enterobacter* spp.) *Ps. aeruginosa* and other pseudomonads
Inoculation	Sharp injury, blood products	Hepatitis B, HIV, malaria

With the advent of more elaborate surgery and intensive care, combined with the use of broad-spectrum antibiotics and immunosuppressive drugs, Gram-negative bacteria increased in importance. Many, such as *Pseudomonas aeruginosa*, are *opportunists* capable of causing infection in compromised patients. Such organisms may be found in the patient's own flora, or in damp environmental sites, including patient equipment and medicaments. They may exhibit natural resistance to many antibiotics and antiseptics, and have the ability to colonize traumatized skin such as burns and areas with poor tissue viability such as decubitus ulcers and bed sores.

In recent years, groups of micro-organisms that formerly played no recognized part in hospital infection have emerged. These include the coagulase-negative staphylococci and *Acinetobacter baumanii* present in normal skin flora. Viral or fungal infection, particularly of the immunocompromised patient, has become more important. *Legionella pneumophila*, disseminated from environmental sources such as cooling towers and hot water systems, causes sporadic cases or outbreaks of respiratory infection in hospitals. Awareness of the risks of blood-borne viruses, including hepatitis B and C, human immunodeficiency virus (HIV) and of many agents such as cytomegalovirus, which can be transmitted by organ or cellular transplants, has increased in patients and in staff. The possible risk of iatrogenic spread of the prion causing Creutzfeldt–Jakob disease has persuaded the UK government to take stringent measures on decontamination of surgical instruments, and the US blood transfusion service to ban British donors.

ROUTES OF TRANSMISSION

The hospital offers many opportunities for the exchange of microbes, many of which are harmless and a normal part of the balance between human beings and their environment. For there to be a significant risk of infection, several factors, including a susceptible host and the appropriate inoculum of infecting micro-organism, must be linked via an appropriate route of transmission. Understanding of the sources and transmission routes of hospital infection enables efforts to be concentrated in more effective preventive measures.

Common routes of transmission for different micro-organisms are shown in Table 68.2.

Air-borne transmission

Infections may be spread:

- by air-borne transmission from the respiratory tract (talking, coughing, sneezing)
- from the skin by natural shedding of skin scales during wound dressing or bed-making
- by aerosols from equipment such as respiratory apparatus and air-conditioning plants.

Infectious agents may be dispersed as small particles or droplets over long distances (Fig. 68.1). Staphylococci survive well on mucosal secretions, skin scales and dried pus, and may be redistributed in the air after initial settlement during periods of increased activity (Fig. 68.2). Gram-negative bacilli do not generally survive desiccation

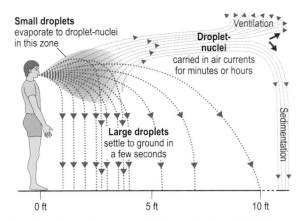

Fig. 68.1 Spread of respiratory infections by droplets and droplet nuclei.

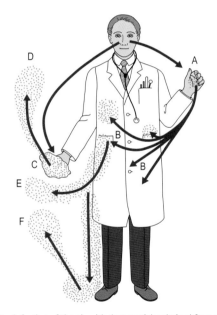

Fig. 68.2 Infection of the air with dust particles derived from nasal and oral secretions contaminating hands, handkerchief, clothing and surrounding surfaces: A, hand soiled with secretions from lips or nose-picking; B, clothing contaminated by hand; C, soiled handkerchief; D, infected dust from handkerchief; E, dust from clothing (e.g. from near handkerchief pocket); F, infected dust raised after settling on floor.

in air, and this route of transmission is therefore limited to conditions of high humidity such as ventilatory equipment, showers or other fine water aerosols.

Contact spread

The most common routes of transmission for hospital infection are:

- by *direct* contact spread from person to person
- by *indirect* contact spread via contaminated hands or equipment.

Human secretions as well as contaminated dust particles or fluids may be carried on thermometers, bed-pans, bed-linen, cutlery or other shared items. Hands and, to a lesser extent, clothing of hospital staff serve as vectors of Gram-negative and Gram-positive infection. Procedures involving contact with mucosal surfaces (e.g. insertion of a urinary catheter) may introduce micro-organisms from the contaminated hands of the operator or from the patient's own urethral flora into the normally sterile bladder.

Food-borne spread

Infection may originate in the hospital kitchen, or in special diets, infant feeds, kitchen or commercial supplies. Deteriorating hygiene standards may support the proliferation of flies, cockroaches and other insects or rodents that damage stored products and act as carriers of microbes.

Blood-borne spread

The accidental transmission of infections such as HIV, hepatitis B or C by needlestick or contaminated 'sharp' injuries has been well documented. In areas with a high prevalence of malaria, syphilis or these viruses, stringent precautions should be taken to minimize transmission between patients by the strict use of single-use items, and from health-care workers to patients and vice versa by minimizing needlestick episodes.

Self-infection and cross-infection

The interaction between different sources of infection may be illustrated by the example of a patient undergoing lower bowel surgery. *Self-infection* may occur due to transfer into the wound of staphylococci (or occasionally streptococci) carried in the patient's nose and distributed over the skin, or of coliform bacilli and anaerobes released from the bowel during surgery. Alternatively, *cross-infection* may result from staphylococci or coliform bacilli derived from other patients or healthy staff carriers. The organisms may be transferred into the wound *during operation* through the surgeon's punctured gloves or moistened gown, on imperfectly sterilized surgical instruments and materials, or by air-borne theatre dust. *After surgery*, organisms may be transferred in the ward from contaminated bed-linen, by air-borne ward dust or in consequence of a faulty wound dressing technique.

Of all the possible routes, by far the most likely in this example is self-infection from the patient's own bowel flora and it is therefore against this route that most specific preventive measures in colorectal surgery are directed. Understanding of possible sources of infection and the methods available to block transmission to susceptible sites forms the basis of hospital infection control.

Cross-infection is caused more often by 'hospital' strains selected for characteristics of antimicrobial resistance and virulence. An important example at present is MRSA, which can be identified easily by the microbiology laboratory, and which should have a system for alerting the infection control team who need to collect the following basic epidemiological data:

- patient details
- site and extent of infection
- dates of admission, operative procedures and first recognition of infection
- specimens and laboratory isolates and typing results
- ward and staff details.

The clustering of cases according to a common surgical team or location in the ward may suggest a common source and may be the first firm indication of an outbreak of hospital infection.

Soon after admission to hospital, individuals commonly become contaminated with the 'hospital flora'. This has been shown with *Staph. aureus* in studies of patients before and during hospital treatment. Patients who need to stay longer in hospital, such as those requiring intensive care or the elderly, are less able to withstand infection, and the risks of hospital infection are greater in these patients.

PREVENTION AND CONTROL

The infection control policy

The establishment of an effective infection control organization is the responsibility of good management of any hospital. There will normally be two parts:

1. An *infection control committee*, meeting regularly to formulate and update policies for the whole hospital on matters having implications for infection control, and to manage outbreaks of nosocomial infection.

2. An *infection control team* of workers, headed by the *infection control doctor* (usually the microbiologist), to take day-to-day responsibility for this policy.

The functions of this team include surveillance and control of infection and monitoring of hygiene practices, advising the infection control committee on matters

of policy relating to the prevention of infection and the education of all staff in the microbiologically safe performance of procedures. The *infection control nurse* is a key member of this team. Close working links between the microbiology laboratory, infection control nurse, and the different clinical specialties and support services (including sterile services, laundry, pharmacy and engineering) are important to establish and maintain the infection control policy, and to ensure that it is rationally based and that the recommended procedures are practicable. It is important for all members of the committee to ensure that everyone in the organization makes infection control and hospital hygiene their business. Campaigns should be launched locally to raise the profile of proven methods of control such as hand-washing. Some of the control measures in which the infection control team should be involved are shown in Figure 68.3.

Decontamination and sterilization

The provision of sterile instruments, dressings and fluids is of fundamental importance in hospital practice. Items that are re-used for procedures in normally sterile sites, such as surgical equipment, must be decontaminated by thorough washing before sterilization by heat. High-vacuum autoclaves have become accepted hospital practice. The development of these sterilizers for processing wrapped goods facilitated the provision of a centralized service of sterile supply to wards, complementing the existing theatre service. The availability of a wide range of pre-packed single-use items (syringes, needles, catheters and drainage bags) sterilized commercially by γ-irradiation or ethylene oxide has further improved aseptic procedures and removed the need for reprocessing items that are difficult to clean and therefore impossible to sterilize. The scientific basis of sterilization is mentioned in more detail in Chapter 4.

Most fluids for topical use or intravenous administration are now prepared commercially or in regional units where standards of quality control and efficiency for bulk processes are more readily achieved than in individual hospital pharmacies.

Aseptic techniques

The provision of sterile equipment will not prevent the spread of infection if there is carelessness in its use. Wherever possible, *no-touch* techniques must be used, coupled with strict personal hygiene on the part of the operator. These routines are rigidly laid down in operating theatre practice and may be modified as required for other procedures such as wound dressing and insertion of intravenous catheters.

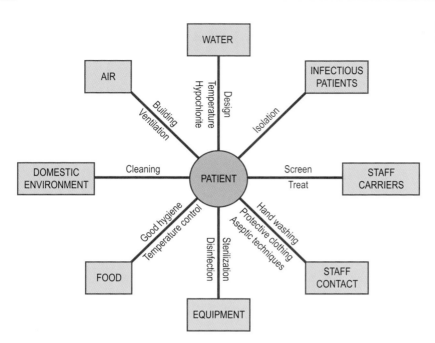

Fig. 68.3 Control measures to reduce exogenous hospital-acquired infection.

Cleaning and disinfection

The general hospital environment can be kept in good order by attention to basic cleaning, waste disposal and laundry. The use of chemical disinfectants for walls, floors and furniture is necessary only in special instances, such as spillages of body fluids from patients with blood-borne virus infections. Ward equipment such as bed-pan washer–disinfectors and dishwashers should be monitored to ensure reliable performance, and cleaning materials such as mopheads and cloths should be heat-disinfected and stored dry after use. Pre-cleaning of contaminated instruments and equipment, preferably by means of an automatic washing process with an ultrasonicator, is an essential step before disinfection or sterilization.

Skin disinfection and antiseptics

The ease of acquisition and transfer of transient hospital contaminants, particularly Gram-negative bacilli on the hands of staff, is an important factor in the spread of hospital infection. Thorough hand-washing after any procedure involving nursing care or close contact with the patient is essential. Alcohol-based hand antiseptics or 'rubs' have been introduced in wards where routine hand-washing with water and detergent is not practicable. Gloves may be worn for many dirty contact procedures, such as emptying a urinary drainage bag or bed-pan, although it should not be forgotten that the gloved hand may also become colonized by transient hospital flora.

Procedures for pre-operative disinfection of the patient's skin and for surgical scrubs are mandatory within the operating theatre. Dilute 'in use' solutions of antiseptics may readily become colonized with Gram-negative bacteria and should be replaced regularly. Ideally, *single-use* preparations should be used. Restriction should be placed on the indiscriminate use of antiseptics and disinfectants by means of a *disinfectant policy* agreed by pharmacists, microbiologists and key users, such as theatre staff.

Decontamination and disinfectant policy

All hospitals should agree a policy to ensure that all clinical staff are familiar with the agents used and procedures involved in decontamination. The policy should consider the following:

1. The sources (equipment, skin and environment) for which the choice of process is required.
2. The processes and products available for sterilization and disinfection. One objective of the policy is to include a limited range of options and chemicals to be used.
3. The category of process required for each item in a simple table: sterilization for surgical instruments, heat disinfection for laundry and crockery, cleaning for floors and furniture.
4. The specific products and method to be used for each item and staff responsible.

Effective implementation of the policy requires liaison and training of staff and regular updating as new methods and agents become available. Safety considerations for staff and patients require that a risk assessment is made of procedures and chemicals to minimize hazards and comply with local health and safety regulations. This includes the use of fume cupboards, air-scavenging equipment and sealed washer–disinfectors. These requirements take decontamination of equipment out of small clinics into centralized sterile supply departments and increase the use of disposables in medical practice.

Prophylactic antibiotics

Widespread and haphazard use of antibiotics hastens the emergence of antibiotic-resistant bacteria and increases both the incidence of toxic side effects and the cost of treatment. However, rational antibiotic prophylaxis plays an important role in infection control. Specific indications include peri-operative prophylaxis in gastro-intestinal and gynaecological surgery directed predominantly against anaerobic infection and for patients known to have bacteriuria at the time of urological surgery or instrumentation, directed against the urine isolate. An *antibiotic policy* that limits the choice of broad-spectrum agents is important for both prophylaxis and treatment (see Ch. 66).

Protective clothing

Different activities within the hospital require different degrees of protection to staff and patients. In operating theatres the wearing of sterile gowns, gloves, head-gear and face-masks minimizes the shedding of micro-organisms. The properties of fabrics available for theatre use have improved, and now include close-weave ventile fabrics that are comfortable to wear and allow evaporation of moisture. 'Total protection' of the operating site may be considered for certain high-risk clean surgery such as hip replacements, during which the surgical team may wear exhaust-ventilated suits and operate under conditions of ultra-clean laminar air-flow.

For many ward procedures in which there may be soiling, or for simple *barrier nursing* of patients with communicable diseases, plastic aprons and gloves are used. Gloves, face-masks and goggles are also indicated for specific procedures in which contact with blood is likely through splashes or aerosols, such as dental procedures. These should conform to international standards, and staff should be trained in their use and disposal.

Isolation

The isolation policy should list facilities and procedures needed to prevent the spread of specific infections to other patients (*source isolation*) and to protect susceptible or immunocompromised patients (*protective isolation*). Effective isolation demands a highly disciplined approach by all staff to ensure that none of the barriers to transmission (air-borne, direct and indirect contact) is breached. Multi-bedded rooms may be used, and even wards converted during hospital outbreaks, but the simplest solution wherever possible is to use single rooms.

'Cubicle' isolation, by which the patient is nursed alone in a room separated by a door and corridor from other patients, confers a substantial measure of protection. Preferably, each isolation room has its own toilet and washing facility. Clean, filtered air is supplied to the room, which should be at negative pressure (*exhaust ventilated*) to the corridor for source isolation or at positive pressure (*pressure ventilated*) to the corridor for protective isolation. If, however, there is a small airlock vestibule separating the room from the outer corridor, then exhaust ventilation of the airlock will give effective isolation for either situation. The vestibule or lobby should contain a wash-basin and include space for gowning and equipment.

In some critical situations such as bone marrow transplant units, where air-borne contamination with environmental fungal spores is a problem, the efficiency of air filtration may be increased and laminar airflow maintained as a barrier around the patient. Stringent isolation, such as a plastic tent or 'Trexler' isolator, is required only for patients with highly contagious infections.

Hospital building and design

The routine maintenance of the hospital building is important, ensuring that surfaces wherever possible are smooth, impervious and easy to clean. Major rebuilding works on or near the hospital site may generate dust containing fungal and bacterial spores, with implications for specialized units serving immunocompromised patients. Close communications with the works department and hospital administration are necessary to co-ordinate any protective action. When a new hospital or modification of existing building is planned, the infection control team should be closely involved in discussing the plans. In many countries, guidance on new building design exists to minimize potential hospital-acquired infection. Areas requiring special attention include operating theatres, kitchens, acute wards, laboratories and air-conditioning systems. The risk of legionnaires' disease is reduced by regular flushing all outlets and installing water supplies that circulate below 20°C for the cold and above 60°C for the hot circuit.

Equipment

Any object or item of equipment for clinical use should be assessed to determine the appropriate method, frequency and site of decontamination. Wherever possible, heat processes are preferred, although this may be precluded for certain thermolabile items such as fibre-optic endoscopes. Dedicated washer–disinfectors in sealed units are necessary for these items.

Personnel

An occupational health service in hospitals should screen staff before employment and offer appropriate immunization. Hepatitis B vaccine should be given to all health-care workers. Those at special risk performing exposure-prone procedures, such as invasive surgery, should be screened for blood-borne viruses. All staff (including medical students) should receive occupational health advice and protection. Staff who have contracted infections such as impetigo or diarrhoea should report and be screened if necessary. Those who sustain needlestick injuries from potentially contaminated sources should have access to advice and post-exposure prophylaxis with antiviral agents or immunization.

Monitoring

Routine microbiological monitoring of the environment is of little benefit, although monitoring of the physical performance of air-conditioning plants and machinery used for disinfection and sterilization is essential. In the event of an outbreak of hospital infection such as legionnaires' disease, more specific monitoring targeted at the known or likely causative micro-organism should be considered.

Microbiological screening of staff or patients is not undertaken routinely, but it may be needed for specific purposes: to detect carriers of MRSA and hepatitis viruses in those performing some types of surgery or where transmission to patients has occurred.

Surveillance and the role of the laboratory

The detection and characterization of hospital infection incidents or outbreaks rely on laboratory data that alert the infection control team to unusual clusters of infection, or to the sporadic appearance of organisms that may present a particular infection risk or management problem. This is sometimes referred to as the 'alert organism' system. Bacterial typing schemes and antibiograms (see Ch. 3) are very important in this regard. In some situations mandatory surveillance of cases of infection with particular pathogens such as MRSA or C. difficile may be useful, to compare hospitals or units and reduce hospital infection by regular, timely feedback to clinicians and managers. Regular visits to the wards are also important to record data on infected patients from whom no specimens

have been received and to respond to problems as they occur. Such visits also serve to provide opportunities for practical teaching, which is another important element of the infection control team's responsibility.

EFFICACY OF INFECTION CONTROL

The evidence base in the literature for acceptable proof of efficacy for infection control measures is limited. These include sterilization, hand-washing, closed-drainage systems for urinary catheters, intravenous catheter care, peri-operative antibiotic prophylaxis for contaminated wounds and techniques for the care of equipment used in respiratory therapy. Isolation techniques are assumed to be reasonable as suggested by experience or inference. Measures that are now considered to be ineffective include the chemical disinfection of floors, walls and sinks, and routine environmental monitoring.

Effective surveillance and action by the infection control team have been shown to reduce infection rates. One important role of the team is to monitor compliance

---

KEY POINTS

- Hospitals constitute a special environment where the epidemiology of infection is distinct. The chief contributing factors are the accumulation of patients with particular features, the special activities undertaken in hospitals, and the special environment created by the patients and the activities.
- Patients may constitute a special hazard because they are infectious, or they may be unusually susceptible to infection because they have particular conditions or are receiving immunosuppressive treatments.
- Special activities include surgery and extensive use of intravascular devices. Needlestick injuries are a constant hazard.
- The extensive use of antibiotics and disinfectants, and the need to reuse equipment and areas that may become contaminated, contribute to the special environment.
- Certain organisms such as methicillin-resistant *Staphylococcus aureus* (MRSA), *Pseudomonas aeruginosa* and *Clostridium difficile* are important agents of hospital-acquired infection owing to the factors outlined above.
- Careful and detailed attention must be paid to controlling the routes of transmission of infection, through the establishment and maintenance of an *infection control policy*.

with practices known to be effective and to eliminate the many rituals or less effective practices that may even increase the incidence or cost of cross-infection. As further advances occur in medical care, and limited health-care resources are spread across hospital and community needs, innovations in infection control will need to be evaluated for efficacy and cost-effectiveness. With this understanding it is possible that hospital infection can be controlled and largely prevented. The dictum of Florence Nightingale, made over a century ago, that 'the very first requirement in a hospital is that it should do the sick no harm', remains the goal.

RECOMMENDED READING

Ayliffe G A J, Taylor L J, Babb J 1999 *Hospital Acquired Infection*, 3rd edn. Arnold, London

Ayliffe G A J, Fraise A, Mitchell K, Geddes A M 2000 *Control of Hospital Infection*, 4th edn. Arnold, London

Department of Health and Public Health Laboratory Service 1995 *Hospital Infection Control. Guidance on the Control of Infection in Hospitals.* Department of Health and Public Health Laboratory Service, London

Russell A D, Hugo W B, Ayliffe G A J, Lambert P, Maillard J-Y, Fraise A P 2004 Principles and *Practice of Disinfection, Preservation and Sterilization*, 4th edn. Blackwell, Oxford

Wenzel R P (ed.) 1997 *Prevention and Control of Nosocomial Infection*, 3rd edn. Lippincott, Williams and Wilkins, Baltimore

Internet sites

Centers for Disease Control and Prevention. Infection Control in Healthcare Settings. http://www.cdc.gov/ncidod/dhqp/index.html

Department of Health. Healthcare associated infection. http://www.dh.gov.uk/PolicyAndGuidance/HealthAndSocialCareTopics/HealthcareAcquiredInfection/fs/en

Hospital Infection Society. http://www.his.org.uk/

Infection Control Nurses Association. http://www.icna.co.uk/

Medicines and Healthcare Products Regulatory Agency. http://www.mhra.gov.uk/

Thames Valley University. Evidence based practice in infection control (epic). http://www.epic.tvu.ac.uk/

69 Immunization

R. C. B. Slack

Immunization is only one of the measures that may be used for the control of infectious diseases (see Ch. 67). It has dramatically and successfully rid the world of the scourge of smallpox and is well on the way to eradicating poliomyelitis. Its cost and efficacy must be assessed against other forms of defence (Box. 69.1), such as:

- environmental sanitation
- safe sewage disposal
- a secure water supply
- food hygiene
- clean air and adequate ventilation
- good animal husbandry with effective quarantine arrangements where necessary
- insect vector control
- improved nutrition.

Much of the morbidity and mortality of infection is still either not assuredly preventable by immunization or presents great difficulties. For example, many of the diarrhoeal diseases and respiratory infections that take a heavy toll of life and health among young children in poor and overcrowded communities are not amenable to control by specific vaccines, and neither are the common bacterial infections associated with haemolytic streptococci, staphylococci and coliform bacilli, or many virus infections. However, some spectacular successes can be claimed, and encouraging efforts are being made to extend the range of success.

RATIONALE OF IMMUNIZATION

The objective is to produce, without harm to the recipient, a degree of resistance sufficient to prevent a clinical attack of the natural infection and to prevent the spread of infection to susceptibles in the community. There is therefore both a personal gain from being immunized and a public health benefit to the population. The degree of resistance conferred may not protect against an overwhelming challenge, but exposure may help to boost immunity.

PASSIVE IMMUNIZATION

Artificial passive immunization is used in clinical practice when it is considered necessary to protect a patient at short notice and for a limited period. Antibodies, which may be antitoxic, antibacterial or antiviral, in preparations of human or animal serum are injected to give temporary protection. Human preparations are referred to as *homologous*, and are much less likely to give rise to the adverse reactions occasionally associated with the injection of animal (*heterologous*) sera. An additional advantage of homologous antisera is that, although they do not confer durable protection, their effect may persist for 3–6 months, whereas the protection afforded by a heterologous serum is likely to last for only a few weeks.

Antiserum raised in the horse against diphtheria toxin (*equine diphtheria antitoxin*) is available for the prophylaxis and treatment of diphtheria. A similar heterologous antiserum is available for emergency use in cases of suspected botulism and to protect those thought to be at risk. Equine tetanus antitoxin is still used in some countries, but it should be abandoned in favour of human tetanus

---

BOX 69.1 Approaches to the prevention of infectious disease: an indication of priorities

- Preventive engineering: water, sewage, ventilation, food production and food processing
- Surveillance and diagnostic awareness
- Prompt management: recognition, treatment, isolation and contact tracing where necessary
- Improvement and maintenance of good socio-economic conditions: housing, nutrition, education, medical and social care
- Increased resistance to infection by appropriate active immunization policies and in some special circumstances by passive immunization

immunoglobulin (see Ch. 22, p. 247). It is most important to give an intended recipient of equine serum a prior test dose to exclude hypersensitive subjects who may have been sensitized by a previous dose of equine serum and may suffer from 'serum sickness'.

Pooled immunoglobulins

Protective levels of antibody to a range of diseases are present in pooled normal human serum. *Human normal immunoglobulin* (HNIG) is available for the short-term prophylaxis of hepatitis A in contacts or travellers who intend to visit countries where hepatitis A is common. However, active immunization is a much better alternative against hepatitis A for frequent travellers to areas where hepatitis A is common. HNIG also protects those with agammaglobulinaemia.

Specific immunoglobulins

Preparations of specific immunoglobulins are available for passive immunization against the following:

- tetanus (human tetanus immunoglobulin; HTIG)
- hepatitis B (HBIG)
- rabies (HRIG)
- varicella–zoster (ZIG).

ACTIVE IMMUNIZATION

Types of vaccine

Toxoids

If the signs and symptoms of a disease can be attributed essentially to the effects of a single toxin, a modified form of the toxin that preserves its antigenicity but has lost its toxicity (a *toxoid*) provides the key to successful active immunization against the disease. This has been spectacularly successful with tetanus and diphtheria.

Inactivated vaccines

If the disease is not mediated by a single toxin, it may be possible to stimulate the production of protective antibodies by using the killed (inactivated) organisms. This is done as a routine with vaccines against pertussis (whooping cough), influenza and the inactivated polio (Salk) vaccine.

Attenuated live vaccines

In some cases, the inactivation procedure to make a killed vaccine destroys or modifies the protective antigenicity (*immunogenicity*) of the organisms. Hence, another approach is to use suspensions of living organisms that are reduced in their virulence (*attenuated*) but still immunogenic. This strategy has yielded mumps, measles and rubella vaccines (now combined), the live-virus polio (Sabin) vaccine, and yellow fever vaccine.

Sometimes it is possible to use a related organism with shared antigens. Thus, the vaccinia virus vaccine was used to eradicate smallpox, and a bovine tubercle bacillus was modified by Calmette and Guérin to make bacille Calmette–Guérin (BCG), which protects man against tuberculosis and leprosy (see pp. 214 & 219).

Special procedures

Some vaccines, such as influenza (virus) vaccine, can be refined by a process that removes unwanted protein and other reactive material but retains the important protective antigens. Some others must be conjugated to proteins to render them immunogenic. Some vaccines, such as hepatitis B vaccine, can be bio-engineered. There is much current interest in the possibility of developing *subunit vaccines* consisting of purified fragments of the major immunogenic components of micro-organisms, particularly viruses.

Immune response

Antibodies against the agents of some bacterial and viral infections may be present in the mother's blood and be passively acquired by the baby (see Part 2 of this book). This gives some protection to the infant at a time when it is poorly equipped to produce specific antibodies, but may interfere to a varying extent with the infant's capacity to respond to the stimulus of injected or ingested vaccines in the very early months of life. Although the capacity of the infant to produce specific antibody to injected antigens is poorly developed in the first few months of life, this problem can often be resolved by the use of *adjuvants*. Thus, effective responses are produced to powerful antigens such as alum-adsorbed toxoids. Pertussis whole-cell vaccines have an adjuvant effect of their own, so the combination of adsorbed toxoids and whole-cell pertussis vaccine (the triple diphtheria, tetanus and pertussis vaccine; DTP) is an effective immunizing complex that can be given at 2, 3 and 4 months of age to cover the period when the harmful potential of pertussis is greatest (see Ch. 32). The tissues of the newborn respond effectively to BCG vaccine because the protection here is cell mediated. The use of a high-potency measles vaccine at 6 months of age is likely to be exploited in countries where measles kills or severely injures the very young.

When a good specific antibody response is being sought to a toxoid or killed antigen, the usual procedure is to give three doses of the antigen at spaced intervals. The first or 'priming' dose evokes a low level of antibody after a latent period of about 2 weeks, but the second dose elicits a much greater (secondary) antibody response, and this is further boosted by the third dose. The efficacy of injected antigen preparations can be enhanced by slow-release agents such as mineral carriers that have adjuvant effects. With most antigens, the response is better if the first two doses are separated by an interval of a month or two. A third dose is generally recommended at some time thereafter, and further booster doses may be given to maintain immunity.

Duration of immunity

After an effective course of active immunization, a protective amount of antibody may persist in the blood for some years and a subsequent booster injection may maintain protection for a further decade. Much depends upon circumstances, which vary for different vaccines and different groups of people. The duration of active protection cannot be equated absolutely with the presence of demonstrable antibody because factors such as the sensitivity of the test and the actual protective role of the antibody detected have to be taken into consideration.

Age of commencement of active immunization

This must take account of the immaturity of the antibody-forming system in the very early months (see above) and the infectious challenges that a child may encounter early in life and in later years. The start of any immunization programme must be adjusted to the known epidemiology of the diseases that are prevalent in the country in which it is to be instituted, and it must also be related to any serious infective challenges that may be imported from time to time. Poliomyelitis is a good example of a disease that has been largely eradicated from many communities; however, the disease is quick to strike back if the immunization shield is lowered.

CONTROLLED STUDIES OF VACCINES

Combined field and laboratory studies aim to provide confidence in the efficacy of vaccines. A field trial can show only whether or not the actual preparation of vaccine used was successful under the circumstances prevailing at the time. Accordingly, trials require very sophisticated design and much care in their execution. They are very expensive. In order to satisfy the requirement for reproducibility, the method of preparation of the vaccine must be meticulously described and controlled. Much

work continues to define and to refine laboratory tests for the protectiveness of a particular vaccine that might be correlated with its efficacy in field trials. The laboratory test that gives results most closely corresponding to the protective value in the field trials may then be adopted as the test for standardizing future batches of vaccine.

Manufacturers of vaccines for commercial use are required in the UK and in many other countries to satisfy certain standards relating to the purity, safety, potency and stability of their products. In addition, the World Health Organization (WHO) has established internationally agreed requirements. At the national level, a system for continuing surveillance of the efficacy of prophylactic vaccines with continuing notification of any suspected adverse reactions is essential.

CONTRA-INDICATIONS TO THE USE OF VACCINES

In the delivery of an important immunization programme there is a balance between the rights of the individual (and often this is the carer, not the person being immunized) and the benefit to the public. It is necessary not to take too legalistic a view of theoretical hazards and thus err on the side of opting out. The risk of opting out of an immunization schedule should be clearly appreciated and shown to be greater than the adverse effects.

There are some useful general principles about contra-indications.

- Do not give a vaccine to a patient with an acute illness.
- Do be sure that the postponed immunization is subsequently given.
- Do not give a live vaccine to a pregnant woman, unless there is a clear balance of risk in favour of vaccination.
- Avoid giving any vaccine in the first trimester of pregnancy.
- Do not give live vaccines to patients receiving immunosuppressive drugs or irradiation, or to patients suffering from malignant conditions of the reticulo-endothelial system – delay until after successful therapy when they are in remission.

Experience with human immunodeficiency virus (HIV) antibody-positive patients with or without the signs and symptoms of the acquired immune deficiency syndrome (AIDS) indicates so far that measles, mumps, rubella and polio (live virus) vaccines can be given, but that BCG vaccine should not be given. Inactivated vaccines (e.g. pneumococcal) are not contra-indicated and are of great benefit.

Hazards of immunization

Possible adverse reactions to immunization are:

- mild or moderate pain at the site of injection
- fever and malaise for a day or two afterwards
- anaphylactic reactions are very rare but, as they may be fatal, doctors and nurses should be aware of the possibility and should be prepared for such an emergency by carrying drugs and equipment for resuscitation (a 'shock box') to all immunization sessions.

During the first few years of life when many vaccines are given, children tend to have various health problems that include occasional febrile convulsions and may, sadly, include an unexplained cot death or other tragedy. It is inevitable that some of these events will coincide with the period shortly after a vaccine was given to a child and the possibility of a causal relationship will be entertained. The probability of such a link certainly deserves to be considered, but it is most important to bear in mind that the issue is emotive and that ill-balanced or ill-informed adverse publicity can do irreparable harm to an immunization programme. For example, annual notifications of whooping cough in England and Wales dropped from more than 100 000 in the early 1950s to some tens of thousands in the late 1950s as the vaccine gained ground. By the early 1970s (and after modification of the vaccine to take account of a possible deficiency in its protective cover), whooping cough was largely controlled in the UK, with about 75% of children immunized. In the late 1970s, after some ill-informed adverse publicity, acceptance rates for the vaccine fell steeply to 30% or lower, and the uptake of other vaccines was also affected. Epidemics of whooping cough followed in the UK in 1978 and 1982, with tens of thousands of children affected and a number of deaths. As a result of efforts to restore confidence in pertussis vaccine, uptake figures increased in the 1980s and the disease again began to be brought under control. Preliminary results of tests with new (acellular) pertussis vaccines indicate that these are safe and probably more effective.

Over the past few years there has been considerable publicity about claims that autism and inflammatory bowel disease are associated with MMR vaccine. Although several well conducted studies have failed to confirm the original study that first drew attention to this hazard, the media in the UK have supported the hypothesis and cover of the vaccine has fallen to the point that these three childhood infectious diseases may re-emerge. Mathematical models of measles show that if the number of susceptibles in the population exceeds 5% there is a possibility of transmission.

Children who are most vulnerable are likely to be least protected. Pertussis, polio and measles spread most effectively when living conditions are overcrowded and unhygienic. To some extent, these diseases depend upon population density for their spread. Some underprivileged groups in a community are in very real danger when the average rate of vaccine uptake falls, because they often represent the extreme end of the fall in vaccine cover.

In view of the stringency of the regulations that control the quality and efficacy of vaccines, the probability of error lies more with the vaccinator than with the producer. There is a special obligation to ensure that a vaccine is properly stored, properly reconstituted (if relevant) and properly administered (Box 69.2). Many preparations rapidly lose their potency if frozen and thawed or exposed to temperature variation. This means that it is vital to maintain the 'cold chain' in which the vaccines are held at 2–8°C. The appropriate instructions and local protocols should be followed in detail. *An injectable vaccine must be given with a sterile syringe and needle; a separate sterile syringe and needle must be used for each injection; and the equipment must be disposed of properly.* The dangers of contamination with blood-borne viruses such as HIV have led to the search for alternative methods of delivery by air jets or via mucosal surfaces.

> **BOX 69.2** Safety considerations
>
> - Use a separate sterile syringe and needle
> - Avoid errors: check the vial personally
> - Check 'cold chain'
> - Consider the patient's history: note pregnancy and various contra-indications
> - Keep careful records, including batch number

Site of injection

The site of injection will vary with the vaccine to be given. Specific instructions should be followed. In general, and with the exception of BCG, injectable vaccines are given by intramuscular or deep subcutaneous injection. The anterolateral aspect of the thigh or upper arm is the preferred site for infants. Some doctors and nurses use the upper outer quadrant of the buttock, but fat in this area may interfere with the efficacy of the vaccine, especially hepatitis A and B. In future, the technological advances in delivery methods mentioned above will revolutionize the immunization clinics feared by countless school children.

HERD IMMUNITY

When most of the people in a community are immune to a particular infection that is spread from person to person,

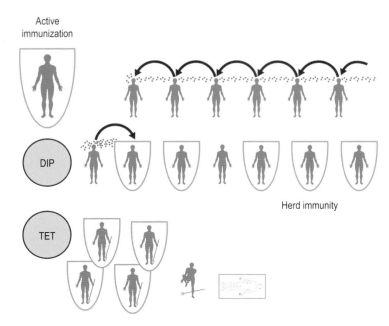

Active immunization

DIP

TET

Herd immunity

Fig. 69.1 If almost all of the members of a community receive active immunization against a disease that is normally transmitted from person to person, the resulting herd immunity confers some advantage even upon an occasional non-immune member because rapid transmission of the disease through the community is prevented. This is true for diphtheria, but not for tetanus, which is not dependent on person-to-person spread.

the natural transmission of the infection is inhibited effectively. Thus, if almost all children in a residential school have been immunized against measles, the school is most unlikely to have an outbreak of measles; even the few children who have not been immunized will enjoy a measure of protection afforded by the general herd immunity in that they will not be challenged within the school. This will apply only so long as the school population is composed largely of immune pupils and a non-immune pupil does not encounter a visitor who is infected with measles. If there is an influx of susceptibles, the level of herd immunity will fall and the general protection will be lost. When the pupils go into other communities at holiday times, the non-immune individuals are liable to get measles at their first contact with an infective case.

For herd immunity to operate well in a community or a country, vaccine uptake rates must exceed 90%. For some highly transmissible infections, such as measles, uptake rates above 95% are the target. Bear in mind that herd immunity operates only for infections transmitted from person to person. Tetanus is not transmitted in this way; a non-immune person is fully vulnerable to tetanus even if he or she is surrounded by fully immunized colleagues in a closed community (Fig. 69.1).

IMMUNIZATION PROGRAMMES

An immunization campaign carried out without provision for its continuation as a routine procedure will not give

satisfactory results unless complete eradication of the disease is achieved. Thus, in planning immunization schedules, consideration must be given to ensuring that the general public is receptive and understands the policy. It is essential to secure the trust and co-operation of parents who have to bring their children to the doctor or clinic for a series of inoculations and who will undoubtedly seek reassurance that the benefits are considerable and the risks negligible. Parental consent must be obtained for each immunization.

In the planning and execution of a programme, immunological points that merit special attention are:

- the use of combined antigens and the simultaneous administration of killed and live vaccines
- the incorporation of adjuvants in killed vaccines and toxoids
- the age of commencement
- the dosage and spacing of antigens.

IMMUNIZATION SCHEDULES

The provision of a programme of active immunization in a community should be governed by considerations of need, efficacy, safety and ease of administration. An overriding consideration is the cost and the availability of skilled staff and the safe delivery and supply of vaccines. Circumstances vary widely in different countries, and priorities vary. The WHO Expanded Programme on Immunization has been adopted by most countries of the world as a minimum

Table 69.1   Schedule of immunization for children in the UK until September 2006

Age	Vaccine	Notes
During year 1	Diphtheria Tetanus Pertussis	DTP vaccine: start at 2 months; second dose at 3 months and third dose at 4 months, intramuscularly
	Hib vaccine MenC Polio vaccine	Give at the same times as the DTP injections
During year 2	Measles Mumps Rubella	MMR vaccine: give one injection of combined live vaccine at age 12–15 months
At 4–5 years	Diphtheria Tetanus Pertussis MMR Polio vaccine	Pre-school booster
At 13–18 years	Tetanus/low-dose diphtheria (Td) Polio	Booster Booster

schedule to protect children in parts of the world where transmission rates are particularly high. Single measles is given at about 9 months rather than the combined MMR given at 12–15 months in industrialized countries. This is a balance between mean age of infection and response to the vaccine. In addition, the WHO recommends hepatitis B immunization from birth (see below).

In the UK, it is generally agreed that protection of the susceptible population against diphtheria, whooping cough and tetanus, poliomyelitis, *Haemophilus influenzae* type b (with an inactivated conjugated Hib vaccine), *Neisseria meningitidis* group C (with protein conjugated to group polysaccharide as menC), and mumps, measles and rubella (with a combined live vaccine) merit priority in the very early years of life. BCG vaccine is offered to some high-risk populations neonatally. In the UK it is now recommended that the DTP, Hib and polio vaccine are started early as a single injection given at the same time as menC, at 2 months, with further doses at 3 and 4 months (Table 69.1). This secures earlier protection against pertussis in the months when the disease is most dangerous to the young child. This schedule changed when conjugate pneumococcal vaccine was added in September 2006.

NOTES ON SOME VACCINES IN COMMON USE

Adsorbed tetanus toxoid

The preparation in routine use in the UK is adsorbed on to an aluminium salt; it is more effective and less reactive than simple toxoid. It is a component of the triple DTP vaccine. A course of three injections given at intervals of about 4 weeks and boosters before, and on leaving, school (i.e. a total of five doses) should protect for life, although studies show that the elderly have low levels of protective antibody. Prevention of tetanus should be considered in relation to the management of wounds (see Ch. 22, pp. 246–247).

Diphtheria toxoid

The adsorbed preparation used in the UK and incorporated into the triple DTP vaccine contains aluminium phosphate and affords good protection, which is boosted at school entry age. This preparation is not used for adults; a dilute adsorbed diphtheria vaccine preparation is used for people who are more than 10 years old and is now incorporated with tetanus toxoid for school leavers as Td. Active immunization of adults against diphtheria is not practised in the UK as a routine, unless there is a potentially high occupational risk of encountering the organism in laboratory or clinical work, or unless a person has been in contact with a case of diphtheria or is travelling to an endemic area and protection is deemed necessary. A course of erythromycin gives further protection to a close contact, and this has replaced the use of diphtheria antitoxin in contacts. The falling level of immunity to diphtheria among adults in communities that regard themselves as having good preventive medical services gives some cause for concern and calls for continuing vigilance, particularly because of large outbreaks in eastern Europe.

Pertussis vaccine

The vaccines in general use are either whole-cell preparations of killed *Bordetella pertussis* or acellular vaccines containing antigenic components of the organism. The protection afforded by the whole-cell vaccines is generally acknowledged to be considerable, but the acellular vaccine seems to be more effective and appears to have fewer adverse reactions so it is now included in the UK schedule. Adverse reactions that may be associated with pertussis vaccine include soreness at the site of injection, irritability and pyrexia. Reactions such as persistent screaming, shock, vomiting and convulsions have been reported, but the association with pertussis vaccine is not invariably clear.

Haemophilus influenzae type b (Hib) vaccine

Invasive encapsulated strains of *H. influenzae*, almost always of serotype b, are associated with meningitis, bacteraemia, epiglottitis and other serious infections particularly affecting young children. The risks of severe morbidity and mortality are highest in children in the age range of 3–18 months. Hib conjugate vaccines containing the capsular polysaccharide linked to a protein are immunogenic and reduce pharyngeal carriage of the organism. The Hib vaccines used in the UK are given combined as part of the primary immunization. Children in the age range of 1–4 years who have not received Hib vaccine can be protected by a single dose of Hib vaccine. Routine immunization of older children or adults with Hib vaccine is not generally recommended unless they are immunocompromised.

Meningococcal group C (MenC) vaccine

In 1999, following an increase in fatal cases of invasive meningococcal disease, especially that due to group C, the UK government was the first in the world to introduce MenC vaccine to the routine programme. Although it is now given as part of primary immunization at 2, 3 and 4 months, when the campaign started it was necessary to protect the whole population up to 18 years of age, and the timing of the dose in the UK schedule changed recently to include a booster at 12 months. The design of the antigen is similar to that of Hib, in that the relatively poorly immunogenic polysaccharide is covalently linked to a protein carrier that produces a prolonged immune response in naive infants. As with Hib there are few adverse reactions and the introduction universally to children has led to a great reduction in cases of meningitis and carriage in the nasopharynx of group C meningococci.

Poliomyelitis vaccines

The control of poliomyelitis is one of the great success stories of active immunization, and the worldwide eradication of poliomyelitis is a goal of the WHO that has nearly been achieved. The Salk vaccine came first; this is a killed mixture of the three types of poliovirus (1, 2 and 3). A course of three injections is given at appropriate intervals, and in the UK it is combined with the other infant vaccines.

The oral polio vaccine was favoured in the UK and many other countries. It is the live-attenuated form developed by Sabin. This is a mixture of the three types; it is given orally on three occasions, usually at the same time as the infant vaccinations in the early months. As it is a live preparation, the vaccine must be stored and held according to the manufacturer's instructions. Oral polio vaccine colonizes the gut and gives rise to local and humoral antibodies. The faeces contain live virus for some time, and poliomyelitis caused by a vaccine strain is a rare, but recognized, hazard, which is why at the end stage of eradication of the disease it is recommended for countries to swap to the injectable form.

Pneumococcal vaccines

The 23-valent polysaccharide vaccine has been available for more than two decades, but its place in universal protection of vulnerable patients has not been established. Splenectomized and immunocompromised people should be immunized routinely, preferably before surgery or therapy. A conjugate vaccine produced against seven types of pneumococcus is commercially available in many countries and has been used either for universal vaccination or to protect infants at highest risk. There is evidence from North America that the universal infant programme has produced a significant decrease in pneumococcal disease in the elderly.

Measles, mumps and rubella (MMR) vaccine

This is a mixture of live-attenuated strains of these three viruses in freeze-dried form; it has to be stored at 2–8°C (not frozen), reconstituted according to the manufacturer's instructions and used promptly. One injection of the mixture is given between 12 and 15 months in the UK schedule, and the second dose in the fifth year, before entry to school.

The measles component may give rise to fever or malaise and sometimes a rash about a week after inoculation. The mumps component may cause some parotid swelling, seen about 3 weeks after inoculation. Very occasional cases of mumps vaccine-associated meningo-encephalitis occur, generally at 3 weeks after use of the Urabe strain. The condition is usually benign, but there may be confusion with other more serious syndromes in this age group. The rubella component is not usually associated with reactions in this age group, although temporary malaise, mild fever and arthralgia occurring about the ninth day after vaccination have been noted in some older recipients of rubella vaccine.

It is hoped that high vaccine coverage with two doses of MMR vaccine in the UK schedule will eradicate measles, mumps and rubella in this country, as it nearly has done in North America. Much depends upon public faith in the immunization programme, which has been undermined by media-led campaigns about the safety of MMR.

All seronegative women of child-bearing age and seronegative professional attendants who might come into contact with pregnant women should be offered rubella vaccine after a test for immunity. Although it is not thought to be teratogenic, any woman of child-bearing age who is given the vaccine should avoid pregnancy for a month.

BCG

The attenuated strain of bovine tubercle bacillus known as bacille Calmette–Guérin (BCG) produces cross-immunity to human tuberculosis and has contributed significantly to the control of that disease and to the other important mycobacterial disease, leprosy, in many countries. An intradermal injection of the live-attenuated vaccine is given on the lateral aspect of the arm at the level of the deltoid insertion, but not higher, or on the upper lateral surface of the thigh. Instruction on the reconstitution of the freeze-dried vaccine, the dosage and the detailed technique of giving a truly intradermal injection, should be observed most carefully. With the exception of children under 5 years, any recipient of BCG vaccination should have been tested for hypersensitivity to tuberculin and found to be negative. Tuberculin tests include the *Mantoux test*, in which diluted tuberculin is injected intradermally, and the *Heaf test*, which is done with a multiple puncture apparatus (see p. 210).

Hepatitis B vaccine

For the protection of groups of people considered to be at special risk of acquiring hepatitis B, a bio-engineered vaccine is now in common use. A course of three intramuscular injections is given (not into the buttock) at intervals of 1 and 5 months; thus, it takes 6 months to complete the course, and this is of practical importance. Following needlestick injuries or known contact with the virus it is possible to give an accelerated course, although efficacy may not be as good. Protective antibody responses are generally achieved in about 90% of those given the vaccine, with a range of responses from weak to strong, but there is a worrying minority of non-responders (10–15% in those over 40 years of age). Antibody responses should be checked at least 6 weeks after completion of a course, and it may be prudent to give poor responders or non-responders a booster dose or a repeat course. Non-

---

BOX 69.3   Vaccines for active immunization of people at special risk

Anthrax	Pneumococcal infection[a]
Cholera	Q fever
Hepatitis A[a]	Rabies
Hepatitis B[a]	Tick-borne encephalitis
Influenza[a]	Typhoid[a]
Japanese B encephalitis	Typhus
Meningococcal infection[a]	Varicella–zoster
Plague	Yellow fever

[a]Indicates some associated problems (see text).

---

response rates are particularly worrying among patients on maintenance haemodialysis, for whom a higher dose preparation is available. In the general population, the protection afforded to responders is thought to last for about 5 years, when a booster dose of vaccine may be given. Established policy must await further experience with hepatitis vaccines, which in many countries are being given universally as part of the routine schedule. Great success has been claimed in Asia in reducing prevalence and preventing cancer.

Other vaccines

Various vaccines are available for the protection of special groups of people or for individuals in special circumstances (Box 69.3). The reader is referred to the appropriate chapters for more detailed consideration of such topics as:

- prevention of tetanus in wounded patients
- indications for pneumococcal vaccine
- management of rabies
- control of Q fever
- protection of those at special risk from hepatitis B and varicella–zoster
- protection of the immunocompromised, such as patients with HIV/AIDS
- prevention of cervical cancer by human papillomavirus vaccines.

PROTECTING THE TRAVELLER

An intending traveller to another country should seek advice in advance about the prevailing diseases and the precautions that should be taken. Advice on immunization is unlikely to be of much help if the traveller unwisely runs the risk of drinking raw water or eating uncooked salads and vegetables in a country where sanitation is inadequate and water supplies are insecure.

In advising travellers about exotic diseases, do not forget that diseases now uncommon in countries with developed medical services may still be common in countries that lack such services. Health-care workers need to include the risk of diphtheria and check immunity to tuberculosis, hepatitis B, measles and poliomyelitis. If the traveller is going to Central Africa or Central America, bear in mind the need for protection against yellow fever. Other vaccines that may be indicated include those that will afford some protection against typhoid. Some of the special vaccines listed in Box 69.3 may also merit inclusion, especially if the traveller's activities when abroad are likely to expose him or her to the relevant diseases. Japanese B encephalitis vaccine may provoke more adverse events than the risk entailed. Travellers to rural areas in the tropics must always put insecticides and antimalarials at the top of their list of necessities because, although there is some progress in the development of an effective malaria vaccine, none is available commercially. Simple practical advice is often of more value than a typhoid or hepatitis immunization.

UNRESOLVED PROBLEMS

Nothing is perfect (Box 69.4). Some vaccines are more imperfect than others, and some circumstances pose special problems. The influenza virus has shifts and drifts in its antigenic pattern, so vaccines must be updated frequently. Then decisions have to be made on the patients who merit this special protection. As these include old people and many other patients with chronic medical conditions, the logistics are quite difficult. Most countries with a temperate climate immunize the elderly and vulnerable before winter, and this has been shown in the USA to reduce mortality and hospital admissions.

Cholera vaccines have had a chequered record, with very little evidence of real efficacy for the traveller, who is much better protected by a basic knowledge of hygiene. New developments with oral vaccines are more promising and may protect against enterotoxigenic *Esch. coli.*

Acute purulent (bacterial) meningitis is a dramatic clinical problem. The MenC vaccine has afforded protection against group C meningitis, but a group B vaccine is not available at the time of writing, and group B strains are more common in many areas.

Several vaccines have been developed against various herpesviruses, including herpes simplex, varicella–zoster and Epstein–Barr viruses. There have been many difficulties, but a live-attenuated varicella vaccine has been developed that is useful for susceptible staff who work in paediatric or maternity wards, and has been given universally to all infants in some countries.

A vaccine effective against HIV is being sought urgently in the fight against AIDS. There have been many disappointments, and there is now a glimmer of hope. Delivery of vaccines without using traditional needles and syringes is another target that is getting nearer and will save infections caused by the re-use of unsterile equipment.

KEY POINTS

• Immunization is a major means by which infections may be controlled and, in some cases, eradicated.
• Passive immunization involves the administration of immunoglobulin-containing preparations that provide protection against one or more infections. The immunity is short-lived. Preparations providing protection against tetanus, hepatitis B, rabies, varicella–zoster and vaccinia are widely available.
• Active immunization produces more enduring immunity. Live-attenuated, killed (inactivated) and subunit (e.g. toxoid or capsular antigen) vaccines are used widely.
• The use of particular vaccines is contra-indicated in certain individuals, such as the use of live vaccines in immunocompromised individuals.
• Where immunization prevents transmission, the objective of immunization programmes is to achieve *herd immunity* such that, once a threshold level of immunization has been achieved, no further transmission can take place.
• The standard immunization schedule in the UK provides for active immunization in childhood against diphtheria, tetanus, pertussis, seven serotypes of pneumococci, *Haemophilus influenzae* capsule type B (Hib), *Neisseria meningitidis* capsule type C, polio, and measles, mumps and rubella.
• Additional active immunization may be considered against hepatitis B, varicella, tuberculosis and influenza for specific groups, and travellers are routinely offered additional vaccines specific to the locations visited.

RECOMMENDED READING

British Medical Association and Royal Pharmaceutical Society of Great Britain 2006 Immunological products and vaccines. In *British National Formulary*. British Medical Association and Royal Pharmaceutical Society of Great Britain, London (revised at intervals of about 6 months). Online. Available: http://www.bnf.org/

Centers for Disease Control and Prevention 2005 *Health Information for International Travel 2005–6*. US Department of Health and Human Services, Public Health Service, Atlanta

Department of Health, Welsh Office, Scottish Home and Health Department 2001 *Health Information for Overseas Travel*. HMSO, London

Department of Health, Welsh Office, Scottish Home and Health Department, DHSS (Northern Ireland) 1996 *Immunisation Against Infectious Disease*. HMSO, London. Updated Green Book available online at http://www.dh.gov.uk/PolicyAndGuidance/HealthAndSocialCareTopics/GreenBook/fs/en

Kassianos G C 2001 *Immunization: Childhood and Traveller's Health*, 4th edn. Blackwell, Oxford

Mackett M, Williamson J D 1995 *Human Vaccines and Vaccination*. Bios Scientific Publications, Oxford

Nicholl A, Rudd P (eds) 1989 *British Paediatric Association Manual on Infections and Immunizations in Children*. Oxford University Press, Oxford

Plotkin S, Mortimer E, Orenstein W (eds) 2003 *Vaccines*, 4th edn. WB Saunders, Philadelphia

World Health Organization 2000 *International Travel and Health. Vaccination Requirements and Travel Advice*. WHO, Geneva

Internet sites

Centers for Disease Control and Prevention. Travelers' Health http://www.cdc.gov/travel/

Department of Health (England). http://www.dh.gov.uk/

Health Protection Agency. http:www.hpa.org.uk/

Health Protection Scotland. http://www.hps.scot.nhs.uk/

National Health Service (Scotland). Fit for Travel. http://www.fitfortravel.scot.nhs.uk/

National Health Service (UK). Immunisation Information http://www.immunisation.org.uk/

National Travel Health Network and Centre. Protecting the Health of British Travellers/ http://www.nathnac.org/

World Health Organization. Immunization, Vaccines and Biologicals. http://www.who.int/immunization/en

World Health Organization. International Travel and Health. http://www.who.int/ith/

INDEX